f'_c = specified 28-day compressive strength of concrete

f_{un} = factored normal (compression or tension) stress

f_{ut} = nominal tension stress at factored load, R_{ut}/A_b

f_{uv} = nominal shear stress at factored load, R_{uv}/A_b

f_v = service load shear stress, R/A_b or V/A_w (bolts)

f_{vs} = service load shear flow (kips/in.)(Sec. 11.11)

F_a = allowable service load axial stress in ASD, Eqs. 6.7.9 or 6.7.10

F_b, F_{bx}, F_{by} = allowable bending stress M/S in ASD at service load, bending about the x- or y-axis according to the second subscript

F_{cr} = critical stress in compression; buckling stress

F'_e = Eq. 12.14.11

F_{EXX} = weld metal tensile strength

F_r = residual stress

FS = factor of safety

F_u = tensile strength of structural steel; tensile strength of base metal (Chap. 5)

F_u^b = tensile strength of bolt material

F'_{ut} = design *stress* limit on R_{ut}/A_b in tension on bolt subject to combined shear and tension, values in Table 4.14.1

F_v = allowable shear stress (for ASD)

F'_v = allowable shear stress in the presence of tension (slip-critical connections)

F_t = allowable tension stress in bolt

F_y, F_{yf}, F_{yw} = yield stress; for beam flange, F_{yb}; for column web, F_{yc}; for flange, F_{yf}; for longitudinal reinforcing bars, F_{yr}; for stiffeners, F_{yst}; for tension flange, F_{yt}; for web or weld metal, F_{yw}

g = gage distance for fastener holes measured transverse to direction of load

G = shear modulus of elasticity, $E/[2(1 + \mu)]$

h = unsupported web height (For unsymmetrical sections it relates to the compression side of the neutral axis.) (See LRFD-B5.1); overall depth (Fig. 6.13.1); distance between centers of flanges (Chap. 8); depth of web plate (Chap. 11); story height (Chaps. 14 and 15)

h_c = twice the distance from the neutral axis to the inside face of the compression flange less the fillet or corner radius

H_u = factored lateral force causing sway deflection

I, I_x, I_y = moment of inertia, about the x- or y-axis, respectively

I_f = moment of inertia of one flange about y-axis (Chap. 8)

I_g = girder moment of inertia (Chaps. 14 and 15)

I_p = polar moment of inertia about shear center

I_{st} = moment of inertia of the cross-sectional area of a transverse stiffener as defined following Eq. 11.11.11

I_{tr} = transformed cracked section moment of inertia of composite section (Chap. 16)

I_{ry} = product of inertia, $\int xy\,dA$ (Sec. 7.10)

J = torsion constant

k = $\sqrt{P/EI}$ (Secs. 6.2 and 12.2); distance from outer face of flange to web toe of fillet (Sec. 7.8); plate buckling coefficient, Eq. 6.14.28; spring constant (Sec. 9.13); spring constant = $\Sigma P_n/h$, Eq. 14.5.9

k_c = coefficient in λ_r for welded I-shape, $4/\sqrt{h/t_w}$ (Tables 7.4.1 and 9.6.2)

k_1, k_2 = coefficients relating to fillet weld strength, Eqs. 5.17.4 and 5.17.7

K, K_x, K_y, K_z = effective length factor (Sec. 6.9), with respect to the x-, y-, or z-axes, respectively

K_n = modified effective length factor used in design of column supporting a "leaner" column, Eq. 14.5.11

L = length; span

L_b = laterally unbraced length

L_c = ASD: maximum laterally unbraced length for using $F_b = 0.66F_y$

L_e = end distance measured in direction of line of force (Chap. 4)

L_p = LRFD: maximum laterally unbraced length for using $M_n = M_p$, Eq. 9.6.3

L_{pd} = LRFD: maximum laterally unbraced length for using plastic analysis, Eq. 9.6.2

L_r = LRFD: maximum laterally unbraced length for using $M_n = M_r$

L_u	=	ASD: maximum laterally unbraced length for using allowable stress $F_b = 0.60F_y$ when $C_b = 1$
L_w	=	length of fillet weld
m	=	number of shear planes (Chap. 4); uniformly distributed service load torsional moment (Chap. 8)
m_u	=	factored uniformly distributed torsional moment (Chap. 8)
M	=	bending moment; service load moment (unfactored) for ASD
M_1, M_2	=	smaller moment M_1 and larger moment M_2 at the ends of a laterally unbraced segment
M_A, M_B, M_C	=	moment (absolute value) at $1/4, 1/2,$ and $3/4$ points, respectively, of laterally unbraced segment, Eq. 9.6.11
M_{cr}	=	elastic lateral-torsional buckling moment strength
M_E	=	equivalent constant moment, $C_m M_2$ for beam-column in braced frame subject to end moments M_1 and M_2 only (Sec. 12.4)
M_f	=	lateral bending moment on one flange, $(EC_w/h)(d^2\phi/dz^2)$
M_i	=	primary bending moment from first-order analysis
$M_{\ell t}$	=	first-order moment in sway analysis under H_u force
$M_{\max}$	=	maximum moment in the unbraced segment (Sec. 9.6)
M_n	=	nominal moment strength
M'_n	=	nominal moment strength in the presence of shear
M_{nt}	=	primary factored moment for the no translation case of beam-column
M_{nx}, M_{ny}	=	nominal moment strengths about x- and y-axes, respectively (Secs. 7.11 & 9.14)
M_p, M_{px}, M_{py}	=	plastic moment strength, ZF_y, with respect to the x- and y-axes, according to second subscript
M_r	=	moment strength when extreme fiber reaches $(F_y - F_r)$
M_s	=	pure torsional moment (St. Venant torsion) (Chap. 8)
M_u, M_{ux}, M_{uy}	=	factored service load moment, about the x- or y-axes, according to second subscript
M_{uEQ}	=	equivalent factored bending moment on beam-column
M_{uf}	=	factored moment on one flange
M_w	=	warping torsional moment (lateral bending effect) (Chap. 8)
M_x	=	moment about x-axis; bending moment from stress σ_x (Sec. 6.14)
M_{xy}	=	torsional moment from shear stress τ_{xy} (Sec. 6.14)
M_y	=	bending moment from stress σ_y (Sec. 6.14); moment about y-axis when biaxial bending is being considered (Chapters 8 and 9; Sec. 12.9); nominal moment strength M_n when extreme fiber reaches F_y, $F_y S_x$ (except for biaxial bending)
M_z	=	bending moment or torsional moment at a location z along axis of member; for torsion, $M_z = M_s + M_w$
n	=	modulus of elasticity ratio, E_s/E_c; number of fasteners (Chap. 4)
N	=	bearing length (Sec. 7.8)
p	=	pitch (spacing) of bolts; connector spacing (Chap. 16)
P	=	service axial load; factored load per bolt (Chap. 4)
P_{bf}	=	design strength of column web to resist a concentrated factored load
P_{cr}	=	critical buckling load; compression force at buckling
P_e, P_{ex}, P_{ey}	=	Euler load $= \pi^2 EA_g/(KL/r)^2$ for axis of bending (using two subscripts for biaxial bending)
P_{e1}, P_{e2}	=	Euler load $= \pi^2 EA_g/(KL/r)^2$ for axis of bending, for use with magnification factors B_1 and B_2 according to the second subscript
P_{EQ}	=	ASD: equivalent column load for beam-column
P_n	=	nominal strength of an axially loaded compression member, $F_{cr}A_g$; nominal strength of weld configuration (Fig. 5.17.2)
P_u	=	factored axial load (Sec. 1.9); factored reaction or load
P_{uEQ}	=	LRFD: equivalent factored column load for beam-column
P_y	=	yield load, $F_y A_g$ (Chap. 12)
q	=	distance from shear center to mid-thickness of web on channel (Chap. 8)
Q	=	form factor, $Q_a Q_s$ (Sec. 6.18); first moment of area (i.e., statistical moment $\int y\, dA$) about the neutral axis from extreme fiber to section at which elastic shear stress is computed, (see Sec. 7.7)
Q_a	=	shape factor for stiffened compression element (Sec. 6.18)
Q_f	=	moment of area of one-half flange about y-axis (Sec. 8.5)

STEEL STRUCTURES

Design and Behavior

Fourth Edition

Emphasizing Load and Resistance Factor Design

Charles G. Salmon
University of Wisconsin-Madison

John E. Johnson
University of Wisconsin-Madison

HarperCollins*CollegePublishers*

Sponsoring Editor: T. Michael Slaughter
Project Management: Interactive Composition Corporation
Design Administrator: Jess Schaal
Cover Design: Kay D. Fulton
Cover Photo: Stuart McCall/Tony Stone Images, Inc.
Glen Canyon Bridge, spanning the Colorado River (AZ)
Production Administrator: Randee Wire
Compositor: Interactive Composition Corporation
Printer and Binder: R. R. Donnelley & Sons Company
Cover Printer: Phoenix Color Corporation

Steel Structures: Design and Behavior, Emphasizing Load and Resistance Factor Design, Fourth Edition

Library of Congress Cataloging-in-Publication Data

Salmon, Charles G.
    Steel structures : design and behavior / Charles G. Salmon, John
E. Johnson.—4th ed.
        p.   cm.
    Includes bibliographical references and index.
    ISBN 0-673-99786-3 (hc)
    1. Building, Iron and steel. 2. Steel, Structural. I. Johnson,
John Edwin, 1931– . II. Title.
TA684.S24   1995
624.1′821—dc20   95-37679
                                                                CIP

4 5 6 7 8-DOC-9998

Contents

CHAPTER 5
Welding   188

CHAPTER 6
Compression Members   276

CHAPTER 7
Beams: Laterally Supported   370

CHAPTER 8
Torsion   424

CHAPTER 9
Lateral-Torsional Buckling of Beams   479

CHAPTER 10
Continuous Beams   571

CHAPTER 11
Plate Girders   610

CHAPTER 12
Combined Bending and Axial Load   702

CHAPTER 13
Connections   788

Preface

The publication of this fourth edition reflects the continuing changes occurring in design requirements for structural steel, particularly the first significant updating by the American Institute of Steel Construction (AISC) of the *Load and Resistance Factor Design (LRFD) Specification for Structural Steel Buildings.*

Design of structural steel members has developed over the past 95 years from a simple approach involving a few basic properties of steel and elementary mathematics, to a sophisticated treatment demanding a thorough knowledge of structural and material behavior. Present design practice utilizes knowledge of mechanics of materials, structural analysis, and particularly, structural stability, in combination with nationally recognized design rules for safety. The most widely used design rules are those of the American Institute of Steel Construction (AISC), given in *Load and Resistance Factor Design Specification for Structural Steel Buildings* and *Specification for Structural Steel Buildings—Allowable Stress Design and Plastic Design,* referred to hereafter as LRFD Specification and ASD Specification, respectively.

The specific occurrence dictating this fourth edition is the publication of the 1993 LRFD Specification (effective December 1, 1993) with Commentary, along with the corresponding LRFD handbook, *Manual of Steel Construction—Load and Resistance Factor Design,* 2nd Edition, 1994, Volume I (*Structural Members, Specifications, & Codes*) and Volume II (*Connections*). References are continued to the 1989 ASD Specification (effective June 1, 1989) with Commentary; and the corresponding ASD handbook, *Manual of Steel Construction—Allowable Stress Design and Plastic Design,* 9th Edition, 1989. Steel members and components are selected from these handbooks, referred to hereafter as *LRFD Manual* and *ASD Manual.*

The fourth edition follows the same philosophical approach that has gained wide acceptance of users since the first edition was published in 1971. This edition continues to strive to present in a logical manner the theoretical background needed for developing and explaining design requirements, particularly those of the 1993 LRFD Specification. Beginning with coverage of background material, including references to pertinent research, the development of specific formulas used in the AISC Specifications is followed by a generous number of design examples explaining in detail the process of selecting minimum weight members to satisfy given conditions.

Emphasis throughout this fourth edition is on the 1993 LRFD Specification. That specification is based on statistical studies of loads and the resistances of steel structures subject to various types of load effects, such as bending moment, shear, axial force, and torsional moment. The rational treatment of both loads (including load effects) and resistances results in steel structures having more uniform safety throughout. This modern philosophy of design, discussed only briefly in one section of the second edition, is moving toward being the predominant approach to design.

Considerable emphasis has been placed on presenting for the beginner, as well as the advanced student, the necessary elastic and inelastic stability concepts, the understanding of which is essential to properly apply steel design rules. The same concepts are applicable whether design is according to the LRFD Specification or the ASD Specification. The explanation of stability concepts is incorporated into the chapters in such a way that the reader may either study in detail the stability concepts in logical sequence, or omit or postpone study of sections containing detailed development, merely accepting qualitative explanation and proceeding directly to design.

As this fourth edition is prepared, a majority of design is still done according to the traditional Allowable Stress Design (i.e., Working Stress Design). That method focuses on service (working) loads and elastically computed stresses, comparing those stresses with allowable limiting values. However, the logical trend over the next several years should be toward the more rational Load and Resistance Factor Design Specification. Strength design philosophy (reflected in the 1986 and 1993 LRFD Specifications) uses factored service loads and compares the strength provided with such factored loads (or load effects). The strength in any given case depends on the "limit state", or mode of failure, such as yielding, fracture, or buckling. The traditional "plastic design", included as Chapter N of the 1989 ASD Specification, is an option integrally included as part of the LRFD Specification.

Throughout the text, the theory and background material, being common to both the LRFD and ASD philosophies of design, have been integrated. The specific design provisions and illustrative examples are, however, treated generally in separate sections within the chapters so that the user may study either the Allowable Stress Design or the Load and Resistance Factor Design portions separately.

The fourth edition continues the use of SI units as an addition to the primary use of Inch-Pound units. Although neither the LRFD nor the ASD Specification contains SI units, some use of SI units appears in the text. LRFD and ASD formulas have their SI equivalent (conversions are the practical ones made by the authors) given as a footnote on the text page containing the AISC-specified Inch-Pound units. Tables and diagrams generally contain both Inch-Pound and SI units.

Depending on the proficiency required of the student, this textbook provides material for two courses of three or four semester-credit hours each. It is suggested that the beginning course in steel structures for undergraduate students might contain the material of Chapters 1 through 7, 9, 10, 12 and 16, except Sections 6.4, 6.6, 6.12 to 6.19, 7.9 to 7.11, 9.3 to 9.5, 9.8, 9.12 to 9.14, and 12.6 to 12.7. The second course would review some of the same topics of the first course, but more rapidly, emphasizing items omitted in the first course. In addition, the remaining chapters—namely Chapter 8 on torsion, Chapter 11 on plate girders, Chapter 13 on connections, Chapter 14 on frames, and Chapter 15 on frame design—are suggested for inclusion. The primary philosophy emphasized in both courses should be Load and Resistance Factor Design.

The reader will need ready access to the *LRFD Manual** throughout the study of the text, particularly when working with the examples. However, it is not the objective of this text that the reader become proficient in the routine use of tables; the tables serve only as a guide to obtaining experience with variation of design parameters and as an aid in arriving at good design. The LRFD Specification and Commentary are contained in the *LRFD Manual* and are therefore not included in this book, except for various individual provisions quoted where they are explained.

The direct use of the computer is not specifically employed anywhere in the text. The authors believe the study of basic principles in the classroom is of the highest priority. However, the reader may find that acquiring the data base of standard section properties, available for purchase from AISC, will be helpful. The authors recommend the use of a spreadsheet software, such as Lotus 1-2-3† or Microsoft†† Excel, along with the use of the database properties.

Features of this fourth edition are: (1) detailed presentation of strength-related background and design rules for Load and Resistance Factor Design; (2) an integrated treatment of both the 1993 Load and Resistance Factor Design Specification along with the 1989 Allowable Stress Design Specification.

Other special features of this text are (3) comprehensive treatment of design of I-shaped members subject to torsion (Chapter 8), including a simplified practical method; (4) detailed treatment of plate girder theory as it relates to Load and Resistance Factor Design (Chapter 11) and a comprehensive design example of a two-span continuous girder using two different grades of steel; (5) extensive treatment of connections (Chapter 13), including significant discussion and illustration of the design of components. Chapters which were extensively rewritten for the third edition to reflect LRFD approaches of the 1986 specification, have been updated for the 1993 LRFD Specification and improved for better readability and understanding as a result of suggestions from users.

The authors are indebted to students, colleagues, and other users of the first three editions who have suggested improvements of wording, identified errors, and recommended items for inclusion or deletion. The suggestions have been carefully considered, resulting in this complete revision. The continued cooperation and help of AISC through Nestor Iwankiw, Director, Research and Codes, is also appreciated.

The authors are greatly indebted to Dr. Robert E. Abendroth of Iowa State University for his extensive and detailed suggestions to correct errors and improve readability regarding all chapters. The detailed suggestions from Dr. Patrick D. Zuraski of Louisiana State University regarding Chapters 1–4 are sincerely appreciated.

Special thanks are due Dr. David C. Salmon of the faculty of the University of Nebraska-Lincoln for his many significant suggestions and advice regarding the entire manuscript.

**Manual of Steel Construction, Load and Resistance Factor Design,* 2nd edition, 1994. Since nearly continuous reference is made to the *LRFD Manual* (which also contains the 1993 LRFD Specification and Commentary), the reader will find it desirable to secure a copy of the two-volume document from the American Institute of Steel Construction, Inc., One East Wacker Drive, Suite 3100, Chicago, IL 60601-2001. The LRFD Specification and Commentary are also available in a separate paper cover document. The *ASD Manual* and the paper cover ASD Specification and Commentary are also available from AISC.

†Lotus 1-2-3 is a registered trademark of Lotus Development Co.

††Microsoft is a registered trademark of the Microsoft Corporation.

The authors also acknowledge with thanks the comments and suggestions to eliminate errors, improve clarity, and generally improve usability, by Dr. Thomas M. Murray of Virginia Polytechnic Institute, Dr. Gary L. Kraus of the University of Nebraska-Lincoln, Dr. Abdul-Hamid Zureick of Georgia Institute of Technology, and Dr. Gregory G. Dierlein of Cornell University.

Users of this fourth edition are urged to communicate with the authors regarding all aspects of this book, particularly on identification of errors and suggestions for improvement.

The senior author affectionately dedicates this book in memory of his late wife, Bette Salmon, for her patience and encouragement through the first three editions.

Charles G. Salmon
John E. Johnson
February 1995

Conversion Factors

Some Conversion Factors, between Inch-Pound Units and SI Units, Useful in
Structural Steel Design

	To Convert	To	Multiply by
Forces	kip force	kN	4.448
	lb	N	4.448
	kN	kip	0.2248
Stresses	ksi	MPa (i.e., N/mm²)	6.895
	psi	MPa	0.006895
	MPa	ksi	0.1450
	MPa	psi	145.0
Moments	ft-kip	kN·m	1.356
	kN·m	ft-kip	0.7376
Uniform Loading	kip/ft	kN/m	14.59
	kN/m	kip/ft	0.06852
	kip/ft²	kN/m²	47.88
	psf	N/m²	47.88
	kN/m²	kip/ft²	0.02089

For proper use of SI, see *Standard for Use of the International System of Units (SI)* (*the Modernized Metric System*) (ASTM E380-93), American Society for Testing and Materials, Philadelphia, 1993. Also see *Standard Practice for the Use of Metric (SI) Units in Buildings Design and Construction* (*Committee E-6 Supplement to E380*) (ANSI/ASTM E621-78), American Society for Testing and Materials, Philadelphia, 1978.

Basis of Conversions (ASTM E380): 1 in. = 25.4 mm; 1 lb force = 4.448 221 615 260 5 newtons.

Basic SI units relating to structural steel design:

Quantity	Unit	Symbol	
length	metre	m	
mass	kilogram	kg	
time	second	s	

Derived SI units relating to structural design:

Quantity	Unit	Symbol	Formula
force	newton	N	$kg \cdot m/s^2$
pressure, stress	pascal	Pa	N/m^2
energy, or work	joule	J	$N \cdot m$

Chapter 1

Introduction

1.1 STRUCTURAL DESIGN

Structural design may be defined as a mixture of art and science, combining the experienced engineer's intuitive feeling for the behavior of a structure with a sound knowledge of the principles of statics, dynamics, mechanics of materials, and structural analysis, to produce a safe economical structure that will serve its intended purpose.

Until about 1850, structural design was largely an art relying on intuition to determine the size and arrangement of the structural elements. Early man-made structures essentially conformed to those which could also be observed in nature, such as beams and arches. As the principles governing the behavior of structures and structural materials have become better understood, design procedures have become more scientific.

Computations involving scientific principles should serve as a *guide* to decision making and not be followed blindly. The art or intuitive ability of the experienced engineer is utilized to make the decisions, guided by the computational results.

1.2 PRINCIPLES OF DESIGN

Design is a process by which an optimum solution is obtained. In this text the concern is with the design of structures—in particular, *steel* structures. In any design, certain criteria must be established to evaluate whether or not an optimum has been achieved. For a structure, typical criteria may be (a) minimum cost; (b) minimum weight; (c) minimum construction time; (d) minimum labor; (e) minimum cost of manufacture of owner's products; and (f) maximum efficiency of operation to owner. Usually several criteria are involved, each of which may require weighting. Observing the above possible criteria, it may be apparent that setting clearly measurable criteria (such as weight and cost) for establishing an optimum frequently will be difficult, and perhaps impossible. In most practical situations, the evaluation must be qualitative.

If a specific objective criterion can be expressed mathematically, then optimization techniques may be employed to obtain a maximum or minimum for the objective

1350 ft World Trade Center, New York.
(Photo by C. G. Salmon)

function. Optimization procedures and techniques comprise an entire subject that is outside the scope of this text. The criterion of minimum weight is emphasized throughout, under the general assumption that minimum material represents minimum cost. Other subjective criteria must be kept in mind, even though the integration of behavioral principles with design of structural steel elements in this text utilizes only simple objective criteria, such as weight or cost.

Design Procedure

The design procedure may be considered to be composed of two parts—functional design and structural framework design. Functional design ensures that intended results are achieved, such as (a) adequate working areas and clearances; (b) proper ventilation and/or air conditioning; (c) adequate transportation facilities, such as elevators, stairways, and cranes or materials handling equipment; (d) adequate lighting; and (e) aesthetics.

The structural framework design is the selection of the arrangement and sizes of structural elements so that service loads may be safely carried, and displacements are within acceptable limits.

The iterative design procedure may be outlined as follows:

1. *Planning.* Establishment of the functions for which the structure must serve. Set criteria against which to measure the resulting design for being an optimum.
2. *Preliminary structural configuration.* Arrangement of the elements to serve the functions in step 1.
3. *Establishment of the loads* to be carried.
4. *Preliminary member selection.* Based on the decisions of steps 1, 2, and 3 selection of the member sizes to satisfy an objective criterion, such as least weight or cost.
5. *Analysis.* Structural analysis involving modeling the loads and the structural framework to obtain internal forces and any desired deflections.
6. *Evaluation.* Are all strength and serviceability requirements satisfied and is the result optimum? Compare the result with predetermined criteria.
7. *Redesign.* Repetition of any part of the sequence 1 through 6 found necessary or desirable as a result of evaluation. Steps 1 through 6 represent an iterative process. Usually in this text only steps 3 through 6 will be subject to this iteration since the structural configuration and external loading will be prescribed.
8. *Final decision.* The determination of whether or not an optimum design has been achieved.

1.3 HISTORICAL BACKGROUND OF STEEL STRUCTURES

Metal as a structural material began with cast iron, used on a 100-ft (30-m) arch span which was built in England in 1777–1779 [1.1].* A number of cast-iron bridges were built during the period 1780–1820, mostly arch-shaped with main girders consisting of individual cast-iron pieces forming bars or trusses. Cast iron was also used for chain links on suspension bridges until about 1840.

Wrought iron began replacing cast iron soon after 1840, the earliest important example being the Brittania Bridge over Menai Straits in Wales, which was built in 1846–1850. This was a tubular girder bridge having spans 230–460–460–230 ft (70–140– 140–70 m), which was made from wrought-iron plates and angles.

The process of rolling various shapes was developing as cast iron and wrought iron received wider usage. Bars were rolled on an industrial scale beginning about 1780. The rolling of rails began about 1820 and was extended to I-shapes by the 1870s.

The development of the Bessemer process (1855), the introduction of a basic liner in the Bessemer converter (1870), and the open-hearth furnace brought wide-

---

* Numbers in brackets refer to the Selected References at the end of the chapter.

spread use of iron ore products in building materials. Since 1890, steel has replaced wrought iron as the principal metallic building material. Currently (1995), steels having yield stresses varying from 24,000 to 100,000 pounds per square inch, psi (165 to 690 megapascals,* MPa), are available for structural uses. The various steels, their uses and their properties are discussed in Chapter 2.

1.4 LOADS

The determination of the loads to which a structure or structural element will be subjected is, at best an estimate. Even if the loads are well known at one location in a structure, the distribution of load from element-to-element throughout the structure usually requires assumptions and approximations. Some of the most common kinds of loads are discussed in the following sections.

Dead Load

Dead load is a fixed-position gravity service load, so called because it acts continuously toward the earth when the structure is in service. The weight of the structure is considered dead load, as well as attachments to the structure such as pipes, electrical conduit, air-conditioning and heating ducts, lighting fixtures, floor covering, roof covering, and suspended ceilings; that is, all items that remain throughout the life of the structure.

Dead loads are usually known accurately but not until the design has been completed. Under steps 3 through 6 of the design procedure discussed in Sec. 1.2, the weight of the structure or structural element must be estimated, preliminary section selected, weight recomputed, and member selection revised if necessary. The dead load of attachments is usually known with reasonable accuracy prior to the design.

Live Load

Gravity loads acting when the structure is in service, but varying in magnitude and location, are termed *live loads*. Examples of live loads are human occupants, furniture, movable equipment, vehicles, and stored goods. Some live loads may be practically permanent, others may be highly transient. Because of the unknown nature of the magnitude, location, and density of live load items, realistic magnitudes and the positions of such loads are very difficult to determine.

Because of the public concern for adequate safety, live loads to be taken as service loads in design are usually prescribed by state and local building codes. These loads are generally empirical and conservative, based on experience and accepted practice rather than accurately computed values. Wherever local codes do not apply, or do not exist, the provisions from one of several regional and national building codes may be used. One such widely recognized code, *Minimum Design Loads for Buildings and Other Structures* ASCE 7 (formerly ANSI A58.1 published by the American

---

* MPa, megapascals, are equivalent to Newtons per square millimeter, N/mm², in SI units.

TABLE 1.4.1 TYPICAL MINIMUM UNIFORMLY DISTRIBUTED LIVE LOADS
(Adapted from Ref. 1.2)

Occupancy or use	Live load	
	psf	Pa*
1. Hotel guest rooms School classrooms Private apartments Hospital private rooms	40	1900
2. Offices	50	2400
3. Assembly halls, fixed seat Library reading rooms	60	2900
4. Corridors, above first floor in shcools, libraries, and hospitals	80	3800
5. Assembly areas; theater lobbies Dining rooms and restaurants Office building lobbies Main floor, retails stores Assembly hall, movable seats	100	4800
6. Wholesale stores, all floors Manufacturing, light Storage warehouses, light	125	6000
7. Armories and drill halls Stage floors Library stack rooms	150	7200
8. Manufacturing, heavy Sidewalks and driveways subject to trucking Storage warehouses, heavy	250	12000

*SI values are approximate conversions 1 psf (lb/sq ft) = 47.9 Pa.

National Standards Institute), for the past few years has been developed under the jurisdiction of the American Society of Civil Engineers. This code will henceforth be referred to as the ASCE 7 Standard. This Standard is updated from time to time, most recently in 1993, making ASCE 7-93 [1.2] the current specific reference. Some typical live loads from ASCE 7 are presented in Table 1.4.1.

Live load when applied to a structure should be positioned to give the maximum effect, including partial loading, alternate span loading, or full span loading as may be necessary. The simplified assumption of full uniform loading everywhere should be used only when it agrees with reality or is an appropriate approximation. The probability of having the prescribed loading applied uniformly over an entire floor, or over all floors of a building simultaneously, is almost nonexistent. Most codes recognize this by allowing for some percentage reduction from full loading. For instance, for live loads of 100 psf or more ASCE 7 [1.2] allows members having an influence area of 400 sq ft or more to be designed for a reduced live load according to Eq. 1.4.1, as follows:

$$L = L_0 \left[0.25 + \frac{15}{\sqrt{A_I}} \right] \tag{1.4.1}$$

where L = reduced live load per sq ft of area supported by the member
L_0 = unreduced live load per sq ft of area supported by the member (from Table 1.4.1)
A_I = influence area, sq ft

The *influence area* A_I is four times the *tributary area* (the area which distributes load to member being considered) for a column, two times the tributary area for a beam, and is equal to the panel area for a two-way slab. The reduced live load L shall not be less than 50% of the live load L_0 for members supporting one floor, nor less than 40% of the live load L_0 otherwise.

Highway Live Loads

Highway vehicle loading in the United States has been standardized by the American Association of State Highway and Transportation Officials (AASHTO) [1.3] into standard truck loads and lane loads that approximate a series of trucks. There are two systems, designated H and HS, that are identified by the number of axles per truck. The H system has two axles, whereas the HS system has three axles per truck. There are several classes of loading; however, the usual ones are known as H20 and HS20, shown in Fig. 1.4.1.

In designing a given bridge, either one truck loading is applied to the entire structure, or the lane loading is applied. When the lane loading is used, the uniform portion is distributed over as much of the span or spans as will cause the maximum effect. In addition, the *one* concentrated load (for maximum negative moment on continuous spans, a second concentrated load is also used) is positioned for the most severe loading effect. The load distribution across the width of a bridge to its various

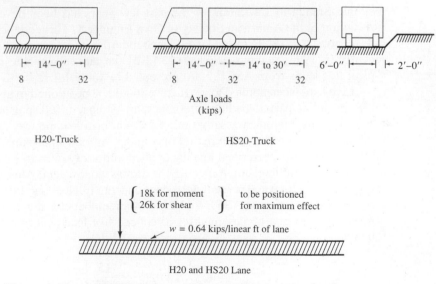

Figure 1.4.1   AASHTO highway H20 and HS20 loadings [1.3].
(1 kip = 4.45 kN)

supporting members is taken in accordance with semiempirical rules that depend on the type of bridge deck and supporting structure.

The single truck loading provides the effect of a heavy concentrated load and usually governs on relatively short spans. The uniform lane load is to simulate a line of traffic, and the added concentrated load is to account for the possibility of one extra heavy vehicle in the line of traffic. These loads have been used with no apparent difficulty since 1944, before which time a line of trucks was actually used for the loading. On the interstate system of highways, a military loading is also used that consists of two 24-kip (107-kN) axle loads spaced 4 ft (1.2 m) apart.

Railroad bridges are designed to carry a similar semiempirical loading known as the Cooper E72 train, consisting of a series of concentrated loads a fixed distance apart followed by uniform loading. This loading is prescribed by the American Railway Engineering Association (AREA) [1.4].

Impact

The term *impact* as ordinarily used in structural design refers to the dynamic effect of a suddenly applied load. In the building of a structure, the materials are added slowly; people entering a building are also considered a gradual loading. Dead loads are static loads; i.e., they have no effect other than weight. Live loads may be either static or they may have a dynamic effect. Persons and furniture would be treated as static live load, but cranes and various types of machinery also have dynamic effects.

Consider the spring-mass system of Fig. 1.4.2a, where the spring may be thought of as analogous to an elastic beam. When load is gradually applied (i.e., static loading) the mass (weight) deflects an amount x and the load on the spring (beam) is equal to the weight W. In Fig. 1.4.2b the load is suddenly applied (dynamic loading), and the maximum deflection is $2x$; i.e., the maximum load on the spring (beam) is $2W$. In this case the mass vibrates in simple harmonic motion with its neutral position equal to its static deflected position. In real structures, the harmonic (vibratory) motion is damped out (reduced to zero) very rapidly. Once the motion has stopped, the force remaining in the spring is the weight W. To account for the increased force during the time the

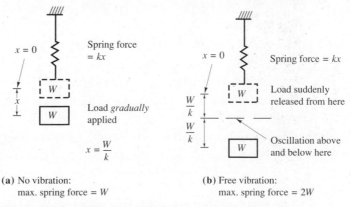

(a) No vibration:
max. spring force = W

(b) Free vibration:
max. spring force = $2W$

Figure 1.4.2 Comparison of static and dynamic loading.

member is in motion, a load equal to twice the static load W should be used—add 100% of the static load to represent the dynamic effect. This is called a 100% impact factor.

Any live load that can have a dynamic effect should be increased by an impact factor. While a dynamic analysis of a structure could be made, such a procedure is unnecessary in ordinary design. Thus, empirical formulas and impact factors are usually used. In cases where the dynamic effect is small (say where impact would be less than about 20%), it is ordinarily accounted for by using a conservative (higher) value for the specified live load. The dynamic effects of persons in buildings and of slow-moving vehicles in parking garages are examples where ordinary design live load is conservative, and usually no explicit impact factor is added.

For highway bridge design, however, impact is always to be considered. AASHTO [1.3] prescribes empirically that the impact factor expressed as a portion of live load is

$$I = \frac{50}{L + 125} \leq 0.30 \tag{1.4.2}$$

In Eq. 1.4.2, L (expressed in feet) is the length of the portion of the span that is loaded to give the maximum effect on the member. Since vehicles travel directly on the superstructure, all parts of it are subjected to vibration and must be designed to include impact. The substructure, including all portions not rigidly attached to the superstructure such as abutments, retaining walls, and piers, are assumed to have adequate damping or be sufficiently remote from the application point of the dynamic load so that impact might not be considered. Again, conservative static loads may account for the smaller dynamic effects.

In buildings, impact is explicitly provided for primarily in the design of supports for cranes and heavy machinery. The American Institute of Steel Construction (AISC) Specifications* [1.5, 1.16] provided for an increase in the live load to account for impact. The *Load and Resistance Factor Design Specification* (LRFD-A4.2) and the *Allowable Stress Design Specification* (ASD-A4.2) state that if not otherwise specified, the live load increase shall be:

For supports of elevators and elevator machinery	100%
For supports of light machinery, shaft or motor driven, not less than	20%
For supports of reciprocating machinery or power driven units, not less than	50%
For hangers supporting floors and balconies	33%
For cab-operated traveling crane support girders and their connections	25%
For pendant-operated traveling crane support girders and their connections	10%

---

* The reader will require continued reference to the *AISC Specifications* (ASD and LRFD) which are contained, respectively in the AISC *ASD Manual* [1.7] and AISC *LRFD Manual* [1.18]. These two books may be purchased from AISC, One East Wacker Drive, Suite 3100, Chicago, IL 60601-2001.

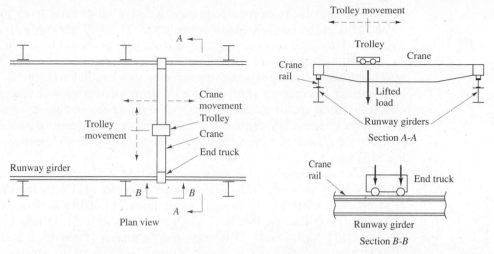

Figure 1.4.3 Crane arrangement, showing movements that contribute impact loading.

In the design of crane runway girders (see Fig. 1.4.3) and their connections, the horizontal forces caused by moving crane trolleys must be considered. Both LFRD and ASD-A4.3 prescribe using a minimum of "20% of the sum of weights of the lifted load and of the crane trolley, but exclusive of other parts of the crane. The force shall be assumed to be applied at the top of the rails, acting in either direction normal to the runway rails, and shall be distributed with due regard for lateral stiffness of the structure supporting the rails."

In addition, due to acceleration and deceleration of the entire crane, a longitudinal tractive force is transmitted to the runway girder through friction of the end truck wheels with the crane rail. LRFD and ASD-A4.3 require this force, unless otherwise specified, to be a minimum of 10% of the maximum wheel loads of the crane applied at the top of the rail.

Snow Load

The live loading for which roofs are designed is either totally or primarily a snow load. Since snow has a variable specific gravity, even if one knows the depth of snow for which design is to be made, the load per unit area of roof is at best only a guess.

The best procedure for establishing snow load for design is to follow the ASCE 7 Standard [1.2]. This standard uses a map of the United States giving isolines of ground snow corresponding to a 50-year mean recurrence interval for use in designing most permanent structures. The ground snow is then multiplied by a coefficient that includes the effect of roof slope, wind exposure, nonuniform accumulation on pitched or curved roofs, multiple series roofs, and multilevel roofs and roof areas adjacent to projections on a roof level.

It is apparent that the steeper the roof the less snow can accumulate. Also, partial snow loading must be considered in addition to full loading, if such loading can occur and would cause maximum effects. Wind may also act on a structure that is carrying snow load. It is unlikely, however, that maximum snow and wind loads would act simultaneously.

In general, the basic snow load used in design varies from 30 to 40 psf (1400 to 1900 MPa) in the northern and eastern states to 20 psf (960 MPa) or less in the southern states. Flat roofs in normally warm climates should be designed for 20 psf (960 MPa) even when such accumulation of snow may seem doubtful. This loading may be thought of as due to people gathered on such a roof. Furthermore, though wind is frequently ignored as a vertical force on a roof, nevertheless it may cause such an effect. For these reasons, a 20 psf (960 MPa) minimum loading, even though it may not always be snow, is reasonable. Local codes, actual weather conditions, ASCE 7 [1.2], or the *Canadian Structural Design Manual* [1.9], should be used when designing for snow.

Other snow load information has been provided by Lew, Simiu, and Ellingwood in the *Building Structural Design Handbook* (Chapter 2) [1.10], and in the works of O'Rourke and Stiefel [1.39], Templin and Schriever [1.40], O'Rourke, Tobiasson, and Wood [1.41], O'Rourke, Redfield, and von Bradsky [1.42], O'Rourke, Speck, and Stiefel [1.43], Sack [1.44], O'Rourke and Galanakis [1.45], and Sack and Giever [1.46].

Wind Load

All structures are subject to wind load, but it is usually only those more than three or four stories high, as well as long bridges, for which special consideration of wind is required.

On any typical building of rectangular plan and elevation, wind exerts pressure on the windward side and suction on the leeward side, as well as either uplift or downward pressure on the roof. For most ordinary situations, vertical roof loading from wind is neglected on the assumption that snow loading will require a greater strength than wind loading. This assumption is not true for southern climates where the vertical loading due to wind must be included. Furthermore, the total lateral wind load, windward and leeward effect, is commonly assumed to be applied to the windward face of the building.

In accordance with Bernoulli's theorem for an ideal fluid striking an object, the increase in static pressure equals the decrease in dynamic pressure, or

$$q = \frac{1}{2} \rho V^2 \tag{1.4.3}$$

where q is the dynamic pressure on the object, ρ is the mass density of air (specific weight $w = 0.07651$ pcf at sea level and 15°C), and V is the wind velocity. In terms of velocity V in miles per hour, the dynamic pressure q (psf) would be

$$q = \frac{1}{2}\left(\frac{0.07651}{32.2}\right)\left(\frac{5280V}{3600}\right)^2 = 0.0026V^2 \tag{1.4.4*}$$

In design of usual types of buildings, the dynamic pressure q is commonly converted into equivalent static pressure p, which may be expressed [1.9]

$$p = qC_e C_g C_p \tag{1.4.5}$$

---

*In SI units, $q = 0.63V^2$, for q in MPa and V in m/sec. (1.4.4) where C_e is an exposure factor that varies from 1.0 (for 0–40-ft height) to 2.0 (for 740–1200-ft height); C_g is a gust factor, such as 2.0 for structural members and 2.5 for small elements including cladding; and C_p is a shape factor for the building as a whole. Excellent details of application of wind loading to structures are available in the ASCE 7 Standard [1.2] and in the *National Building Code of Canada* [1.9].

The commonly used static wind pressure of 20 psf, as specified by many building codes, corresponds to a velocity of 88 miles per hour (mph) from Eq. 1.4.4. An exposure factor C_e of 1.0, a gust factor C_g of 2.0, and a shape factor C_p of 1.3 for an airtight building, along with a 20 psf equivalent static pressure p, will give from Eq. 1.4.5 a dynamic pressure q of 7.7 psf, which corresponds, using Eq. 1.4.4, to a wind velocity of 55 mph. For all buildings having nonplanar surfaces, plane surfaces inclined to the wind direction, or surfaces having significant openings, special determination of the wind forces should be made using such sources as the ASCE 7 Standard [1.2], or the *National Building Code of Canada* [1.9]. For more extensive treatment of wind loads, the reader is referred to the Task Committee on Wind Forces [1.36], Lew, Simiu, and Ellingwood in the *Building Structural Design Handbook* [1.10], Mehta [1.37], and Stathopoulos, Surry, and Davenport [1.38].

Earthquake Load

An earthquake consists of horizontal and vertical ground motions, with the vertical motion usually having the much smaller magnitude. Because the horizontal motion of the ground causes the most significant effect, it is that effect which is often thought of as earthquake load. When the ground under an object (structure) having mass suddenly moves, the inertia of the mass tends to resist the movement, as shown in Fig. 1.4.4. A shear force is developed between the ground and the mass.

Most building codes having earthquake provisions require that the designer either (1) use a dynamic analysis of the structure, or (2) for usual generally rectangular medium height buildings, use an empirical lateral base shear force CW. The dynamics of earthquake action on structures is outside the scope of this text, and the reader is referred to Chopra [1.49] and Clough and Penzien [1.50].

The equivalent lateral base shear force procedure for designing to resist earthquakes has traditionally been used by most building codes to simplify the design process. For many years, a widely used source has been the Structural Engineers

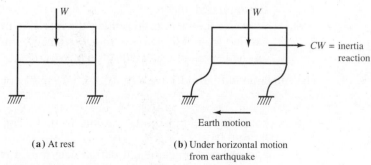

(a) At rest   (b) Under horizontal motion from earthquake

Figure 1.4.4 Force developed by earthquake.

Association of California (SEAOC) recommendations [1.47], the latest version of which is 1990. ASCE 7 Standard [1.2] contains an equivalent lateral force procedure for "Buildings designated as regular up to 240 feet", wherein the seismic base shear V is given by the following:

$$V = C_s W \qquad (1.4.6)$$

where C_s = seismic design coefficient, varying from around 0.005 for low velocity-related acceleration coefficient (say, 0.05) on good soil (say, rock) with a good seismic-resisting structural system (say, moment-resisting frame) for the maximum 240 ft high "regular" building, to around 0.35 for high velocity-related acceleration coefficient (say, 0.20) on poor soil (say, soft clay or silt) with poor moment-resisting system (say, unreinforced masonry) for a 120 ft high "regular" building.

W = total dead load of the building, including partitions, and portions of other loads as defined in ASCE 7–93[1.2].

The seismic coefficient C_s is given by

$$C_s = \frac{1.2 A_v S}{R T^{2/3}} \qquad (1.4.7)$$

but need not be greater than the following:

$$C_s = \frac{2.5 A_a}{R} \qquad (1.4.8)$$

where A_v = coefficient representing effective peak *velocity-related acceleration,* varying from 0.05 to 0.40, in accordance with a contour map based on the degree of expected seismicity

S = coefficient relating to the soil characteristics at the site, varying from 1.0 to 2.0 as the soil differs from rock to soft clay or silt.

R = response modification factor relating to the seismic force-resisting structural system, varying from 1.25 in an unreinforced masonry bearing shear wall system to 8 for a moment-resisting steel frame incorporating special framing requirements of ASCE 7-93 [1.2].

T = fundamental period of the building which, according to ASCE 7-93, "shall be established using the structural properties and deformational characteristics of the resisting elements in a properly substantiated analysis."

A_a = seismic coefficient representing the effective peak *acceleration,* varying from 0.05 to 0.40, in accordance with a contour map based on the degree of expected seismicity

The fundamental period T used in Eq. (1.4.7) shall not exceed

$$T_{max} = C_a T_a \qquad (1.4.9)$$

where C_a = coefficient relating to A_v, varying from 1.2 for $A_v = 0.4$ to 1.7 for $A_v = 0.05$.

T_a = approximate fundamental period determined from

$$T_a = C_T h_n^{3/4} \qquad (1.4.10)$$

where C_T = coefficient relating to the seismic force resisting structural system, varying from 0.035 for the most effective to 0.020 for the least effective system

h_n = height in feet above the base to the highest level of the building.

The above is intended to show generally the variables involved in earthquake forces. ASCE 7-93[1.2] contains an entirely new (as compared to ASCE 7-88 and other prior earthquake codes) procedure for earthquake design, containing many details which are outside the scope of this text.

Note that the base shear force method is for "regular" buildings. Irregular buildings are those which contain (1) *plan structural irregularities,* such as torsional irregularity, re-entrant corners, diaphragm discontinuity, out-of-plane offsets, and nonparallel systems, and/or (2) *vertical structural irregularities,* such as stiffness irregularity (soft story), mass irregularity, geometric irregularity, in-plane discontinuity in vertical lateral force-resisting elements, and discontinuity in lateral strength (weak story). The base shear force method is also limited to buildings not exceeding 250 ft in height.

After the base shear force V is determined, the vertical distribution of seismic forces must be determined. The seismic design story shear must include direct shear as well as torsion. The building must be designed to resist overturning effects caused by seismic forces. Also, story drifts, and where required, member forces and moments due to P-delta effects must be determined.

ASCE 7-93 is based on the *NEHRP Recommended Provisions for the Development of Seismic Regulations for New Buildings* [1.52], which is the definitive source for seismic design. Various traditional building codes for earthquake-resistant design are compared by Chopra and Cruz [1.48]. Many states have adopted the *Uniform Building Code* (UBC) [1.51], which contains provisions for seismic design generally based on the ASCE 7 Standard [1.2]. For steel design, the *Seismic Provisions for Structural Steel Buildings* [1.53] is available. Other information on steel-related earthquake codes is provided by Popov [1.54] and Marsh [1.55].

1.5 TYPES OF STRUCTURAL STEEL MEMBERS

As discussed in Sec 1.2, the function of a structure is the principal factor determining the structural configuration. Using the structural configuration along with the design loads, individual components are selected to properly support and transmit loads throughout the structure. Steel members are selected from among the standard rolled shapes adopted by the American Institute of Steel Construction (AISC) (also given by American Society for Testing and Materials [ASTM] A6 Specification). Of course, welding permits combining plates and/or other rolled shapes to obtain any shape the designer may require.

Typical rolled shapes, the dimensions for which are found in the AISC Manual* [1.7, 1.18], are shown in Fig. 1.5.1. The most commonly used section is the wide-

---

* When reference is made to *AISC Manual,* the material is available in both Refs. 1.7 and 1.18.

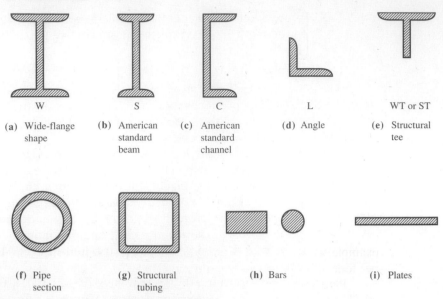

Figure 1.5.1   Standard rolled shapes.

flange shape (Fig. 1.5.1a) which is formed by hot rolling in the steel mill. The wide-flange shape is designated by the nominal depth and the weight per foot, such as a W18×97 which is nominally 18 in. deep (actual depth = 18.59 in. according to *AISC Manual*) and weighs 97 pounds per foot. (In SI units the W18×97 section could be designated W460×144, meaning nominally 460 mm deep and having a mass of 144 kg/m.) Two sets of dimensions are found in the *AISC Manual:* one set stated in decimals for the designer to use in computations, and another set expressed in fractions ($\frac{1}{16}$ in. as the smallest increment) for the detailer to use on plans and shop drawings. Rolled W shapes are also designated by ASTM A6/A6M* [1.8] in accordance with web thickness as Groups 1 through 5, with the thinnest web sections in Group 1.

The American Standard beam (Fig. 1.5.1b), commonly called the I-beam, has relatively narrow and sloping flanges and a thick web compared to the wide-flange shape. Use of I-beams has become uncommon because of excessive material in the web and relative lack of lateral stiffness resulting from the narrow flanges.

The channel (Fig. 1.5.1c) and angle (Fig. 1.5.1d) are commonly used either alone or in combination with other sections. The channel is designated, for example, as C12×20.7, a nominal 12-in. deep channel having a weight of 20.7 pounds per foot. Angles are designated by their leg length (long leg first) and thickness, such as, L6×4×$\frac{3}{8}$.

The structural tee (Fig. 1.5.1e) is made by cutting wide-flange or I-beams in half and is commonly used for chord members in trusses. The tee is designated, for

*The M refers to the SI version of the ASTM Standard.

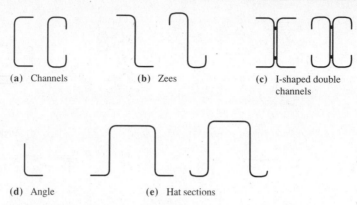

(a) Channels   (b) Zees   (c) I-shaped double
                                           channels

(d) Angle   (e) Hat sections

Figure 1.5.2   Some cold-formed shapes.

example, as WT5×44, where the 5 is the nominal depth and 44 is the weight in pounds per foot; this tee being cut from a W10×88.

Pipe sections (Fig. 1.5.1f) are designated "standard," "extra strong," and "double-extra strong" in accordance with the thickness and are also nominally prescribed by diameter; thus 10-in.-diam double-extra strong is an example of a particular pipe size.

Structural tubing (Fig. 1.5.1g) is used where pleasing architectural appearance is desired with exposed steel. Tubing is designated by outside dimensions and thickness, such as structural tubing, $8×6×\frac{1}{4}$.

The sections shown in Fig. 1.5.1 are all hot-rolled; that is, they are formed from hot billet steel (blocks of steel) by passing through rolls numerous times to obtain the final shapes. Many other shapes are cold-formed from plate material having a thickness not exceeding 1 in., as shown in Fig. 1.5.2.

Regarding size and designation of cold-formed steel members, there are no truly standard shapes even though the properties of many common shapes are given in the *Cold-Formed Steel Design Manual* [1.12]. Various manufacturers produce many proprietary shapes.

Tension Members

The tension member occurs commonly as a chord member in a truss, as diagonal bracing in many types of structures, as direct support for balconies, as cables in suspended roof systems, and as suspension bridge main cables and suspenders that support the roadway. Typical cross-sections of tension members are shown in Fig. 1.5.3, and their design (except for special factors relating to suspension-type cable supported structures) is treated in Chapter 3.

Compression Members

Because compression member strength is a function of the cross-sectional shape (radius of gyration), the area is generally spread out as much as is practical. Chord members in trusses, and many interior columns in buildings, are examples of members

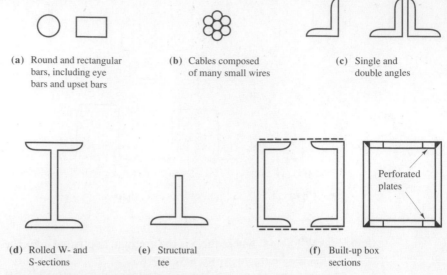

(a) Round and rectangular
      bars, including eye
      bars and upset bars

(b) Cables composed
      of many small wires

(c) Single and
      double angles

(d) Rolled W- and
      S-sections

(e) Structural
      tee

(f) Built-up box
      sections

Perforated
plates

Figure 1.5.3   Typical tension members.

subject to axial compression. Even under the most ideal condition, pure axial compression is not attainable; so, design for "axial" loading assumes the effect of any small simultaneous bending may be neglected. Typical cross-sections of compression members are shown in Fig. 1.5.4, and their behavior and design are treated in Chapter 6.

Beams

Beams are members subjected to transverse loading and are most efficient when their area is distributed so as to be located at the greatest practical distance from the neutral axis. The most common beam sections are the wide-flange (W) and I-beams (S) (Fig. 1.5.5a), as well as smaller rolled I-shaped sections designated as "miscellaneous shapes" (M).

For deeper and thinner-webbed sections than can economically be rolled, welded I-shaped sections (Fig. 1.5.5b) are used, including stiffened plate girders.

For moderate spans carrying light loads, open-web "joists" are often used (Fig. 1.5.5c). These are parallel chord truss-type members used for the support of floors and roofs. The steel may be hot-rolled or cold-formed. Such joists are designated "K-Series," "LH-Series," and "DLH-Series." The K-Series is suitable for members having the direct support of floors and roof decks in buildings. The LH-Series and DLH-Series are known as Longspan and Deep Longspan, respectively. Longspan Steel Joists are shop-fabricated trusses used "for the direct support of floor or roof slabs or decks between walls, beams, and main structural members" [1.13]. Deep Longspan Joists are used "for the direct support of roof slabs or decks between wall, beams and main structural members" [1.13]. The design of the chords for K-Series

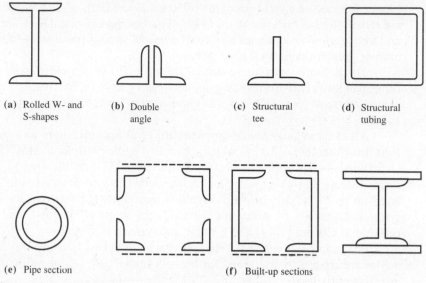

(a) Rolled W- and   (b) Double   (c) Structural   (d) Structural
    S-shapes   angle   tee   tubing

(e) Pipe section   (f) Built-up sections

Figure 1.5.4   Typical compression members.

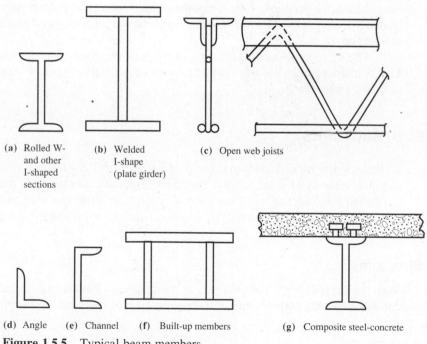

(a) Rolled W-   (b) Welded   (c) Open web joists
    and other   I-shape
    I-shaped   (plate girder)
    sections

(d) Angle   (e) Channel   (f) Built-up members   (g) Composite steel-concrete

Figure 1.5.5   Typical beam members.

trusses is based on a yield strength* of 50 ksi (345 MPa), while the web sections may use either 36 (248 MPa) or 50 ksi (345 MPa). For the LH- and DLH-Series, the chord and web sections design must be based on a yield strength of at least 36 ksi (248 MPa) but not greater than 50 ksi (345 MPa).

The K-Series joists have depths from 8 to 30 in. for clear spans to 60 ft. The Longspan joists (LH-Series) have depths from 18 to 48 in. for clear spans to 96 ft. The Deep Longspan joists (DLH-Series) have depths from 52 to 72 in. for clear spans to 144 ft.

All of these joists are designed according to Specifications adopted by the Steel Joist Institute (SJI) [1.13], which generally agree with the AISC Specifications [1.5, 1.16] for hot-rolled steels, and with the AISI Specification [1.11] for cold-formed steels. *Designing with Steel Joists, Joist Girders, Steel Deck* by Fisher, West, and Van de Pas [1.14] provides excellent treatment of joists and joist-related floor systems. Dynamics of structures is outside the scope of this text, and the reader is referred to Chopra [1.49] and Clough and Penzien [1.50].

For beams (known as lintels) carrying loads across window and door openings, angles are frequently used; and for beams (known as girts) in wall panels, channels are frequently used.

Bending and Axial Load

When simultaneous action of tension or compression along with bending occurs, a combined load problem arises and the type of member used will be dependent on the type of load that predominates. A member subjected to axial compression and bending is usually referred to as a *beam-column,* the behavior and design of which is dealt with in Chapter 12.

The aforementioned illustration of types of members to resist various kinds of load is intended only to show common and representative types of members and not to be all inclusive.

1.6 STEEL STRUCTURES

Structures may be divided into three general categories: (a) framed structures, where elements may consist of tension members, columns, beams, and members under combined bending and axial load; (b) shell-type structures, where axial forces predominate; and (c) suspension-type structures, where axial tension predominates the principal support system.

Framed Structures

Most typical building construction is in this category. The multistory building usually consists of beams and columns, either rigidly connected or having simple end connec-

---

* Refer to Sec. 2.4 for definition.

tions along with diagonal bracing to provide stability. Even though a multistory building is three-dimensional, it usually is designed to be much stiffer in one direction than the other; thus it may reasonably be treated as a series of plane frames. However, if the framing is such that the behavior of the members in one plane substantially influences the behavior in another plane, the frame must be treated as a three-dimensional space frame.

Industrial buildings and special one-story buildings such as churches, schools, and arenas, generally are either wholly or partly framed structures. Particularly the roof system may be a series of plane trusses (see Fig. 1.6.1), a space truss (see Fig. 1.6.2), a dome (see Fig. 1.6.3), or it may be part of a flat or gabled one-story rigid frame.

Bridges are mostly framed structures, such as beams and plate girders (see Fig. 1.6.4), or trusses, usually continuous (see Fig. 1.6.5).

This text is devoted to behavior and design of elements in framed structures.

Figure 1.6.1 Floor joists (plane trusses) and steel decking. (Photo by C. G. Salmon)

Figure 1.6.2   Space truss roof erected in sections; also shows plate girder at lower level containing vertical stiffener plates and special stiffening around rectangular holes through girder web. Upjohn Office Building, Kalamazoo, Michigan. (Photo courtesy Whitehead and Kales Company)

Shell-Type Structures

In this type of structure, the shell serves a use function in addition to participation in carrying loads. One common type where the main stress is tension is the containment vessel used to store liquids (for both high and low temperatures), of which the elevated water tank is a notable example. Storage bins, tanks, and the hulls of ships are other examples. On many shell-type structures, a framed structure may be used in conjunction with the shell.

On walls and flat roofs, the "skin" elements may be in compression while they act together with a framework. The aircraft body is another such example.

Shell-type structures are usually designed by a specialist and are not within the scope of this text.

Suspension-Type Structures

In the suspension-type structure, tension cables are major supporting elements. A roof may be cable-supported, as shown in Fig. 1.6.6. Probably the most common structure of this type is the suspension bridge, as shown in Fig. 1.6.7. Usually a subsystem of

Figure 1.6.3   Dome roof, Brown University auditorium. (Photo courtesy Bethlehem Steel Corporation)

Figure 1.6.4   Continuous orthotropic plate girder across Mississippi River at St. Louis, Missouri. (Photo by C. G. Salmon)

Figure 1.6.5   Continuous truss bridge. Outerbridge Crossing, Staten Island, New York. (Photo by C. G. Salmon)

Figure 1.6.6   Cable-suspended roof for Madison Square Garden Sports and Entertainment Center, New York. (Photo courtesy Bethlehem Steel Corporation)

Figure 1.6.7   Suspension bridge. Golden Gate Bridge, San Francisco, California. (Photo by C. G. Salmon)

the structure consists of a framed structure, as in the stiffening truss for the suspension bridge. Since the tension element is the most efficient way of carrying load, structures utilizing this concept are increasingly being used.

Many unusual structures utilizing various combinations of framed, shell-type, and suspension-type structures have been built. However, the typical designer must principally understand the design and behavior of framed structures.

1.7 SPECIFICATIONS AND BUILDING CODES

Structural steel design of buildings in the United States is principally based on the specifications of the American Institute of Steel Construction (AISC) [1.5, 1.16]. AISC is comprised of steel fabricator and manufacturing companies, as well as individuals interested in steel design and research. The AISC Specifications [1.5, 1.16] are the result of the combined judgment of researchers and practicing engineers. The research efforts have been synthesized into practical design procedures to provide a safe, economical structure. The advent of the digital computer in design practice has made feasible more elaborate design rules. The current specifications which are referred to throughout this book are the 1989 *Specification for Structural Steel Buildings—Allowable Stress Design and Plastic Design* [1.5] and the 1993 *Load and Resistance Factor Design Specification for Structural Steel Buildings* [1.16].

A specification containing a set of rules is intended to insure safety; however, the designer must understand the behavior for which the rule applies, otherwise an absurd, a grossly conservative, and sometimes unsafe design may result. The authors contend that it is virtually impossible to write rules that fully apply to every situation. Behavioral understanding must come first; application of rules then follows. No matter what set of rules is applicable, *the designer has the ultimate responsibility for a safe structure.*

A specification when adopted by AISC is actually a set of recommendations put forth by a highly respected group of experts in the field of steel research and design. Only when governmental bodies, such as city, state, and federal agencies, who have legal responsibility for public safety, adopt or incorporate a specification such as the 1993 Load and Resistance Factor Design Specification [1.16] into their building codes does it become legally official.

The design of steel bridges is generally in accordance with specifications of the American Association of State Highway and Transportation Officials (AASHTO) [1.3]. This is a legal set of rules because it has been adopted by the states (usually the state highway departments have this responsibility).

Railroad bridges are designed in accordance with the specifications adopted by the American Railway Engineering Association (AREA) [1.4]. In this case the railroads have the responsibility for safety and through their own organization adopt the rules to insure safe designs.

The term *building code* is sometimes used synonymously with specifications. More correctly a building code is a broadly based document, either a legal document such as a state or local building code, or a document widely recognized even though not legal which covers the same wide range of topics as the state or local building code. Building codes generally treat all facets relating to safety, such as structural design, architectural details, fire protection, heating and air conditioning, plumbing and sanitation, and lighting. On the other hand, specifications frequently refer to rules set forth by the architect or engineer that pertain to only one particular building while under construction. Building codes also ordinarily prescribe standard loads for which the structure is to be designed, as discussed in Sec. 1.4.

The reader should not be disturbed by the interchangeable use of building code and specification, but should clearly understand that which is legally required for design and that which could be thought of as recommended practice.

1.8 PHILOSOPHIES OF DESIGN

Two philosophies of design are in current use: the *working stress design* (referred to by AISC as *Allowable Stress Design*) and *limit states design* (referred to by AISC as *Load and Resistance Factor Design*). Working stress design has been the principal philosophy used during the past 100 years. During the past 20 years or so, structural design has been moving toward a more rational and probability-based design procedure referred to as "limit states" design. Haaijer [1.27] and Kennedy [1.28, 1.29] have presented the current status of the limit states concept and its use in design. Limit states design includes the methods commonly referred to as "ultimate strength design," "strength design," "plastic design," "load factor design," "limit design," and the recent "load and resistance factor design (LRFD)."

Structures and structural members must have adequate strength, as well as adequate stiffness and toughness to permit proper functioning during the service life of the structure. The design must provide some reserve strength above that which is needed to carry the service loads; that is, the structure must provide for the possibility of overload. Overloads can arise from changing the use for which a particular structure was designed, from underestimation of the effects of loads by oversimplifications in structural analysis, and from variations in construction procedures. In addition, there must be a provision for the possibility of understrength. Deviations in the dimensions of members, even though within accepted tolerances, can result in a member having less than its computed strength. The materials (steel for members, bolts, and welds) may have less strength than used in the design calculations. A steel section may occasionally have a yield stress below the minimum specified value, but still within the statistically acceptable limits.

Structural design must provide for adequate safety no matter what philosophy of design is used. Provision must be made for both *overload* and *understrength*. The study of what constitutes the proper formulation of structural safety has been continuing during the past 30 years [1.20, 1.21]. The main thrust has been to examine by various probabilistic methods the chances of "failure" occurring in a member, connection, or system.

Rather than refer to "failure" the term "limit state" is preferred. *Limit states* means "those conditions of a structure at which it ceases to fulfill its intended function" [1.28]. Limit states are generally divided into two categories: *strength* and *serviceability*. Strength (i.e., safety) limit states are such behavioral phenomena as achieving ductile maximum strength (i.e., plastic strength), buckling, fatigue, fracture, overturning, and sliding. Serviceability limit states are those concerned with occupancy of a building, such as deflection, vibration, permanent deformation, and cracking.

Both the loads acting on the structure and its resistance (strength) to loads are variables that must be considered. In general, a thorough analysis of all uncertainties that might influence achieving a "limit state" is not practical, or perhaps even possible. The current approach to a simplified method for obtaining a probability-based assessment of structural safety uses *first-order second-moment reliability methods* [1.32]. Such methods assume the load (or load effect) Q and the resistance R are random variables. Typical frequency distributions of these random variables are shown in Fig. 1.8.1. When the resistance R exceeds the load (or load effect) Q there will be a margin of safety. Unless R exceeds Q by a large amount, there will be some probability that R may be less than Q.

Structural "failure" (achievement of a limit state) may then be examined by comparing R with Q, or in logarithmic form observing $\ln(R/Q)$, as shown in Fig. 1.8.2. "Failure" is represented by the cross-hatched region. The distance between the failure line and the mean value of the function $[\ln(R/Q)]$ is defined as a multiple β of the standard deviation σ of the function. The multiplier β is called the *reliability index*. The larger is β the greater is the margin of safety.

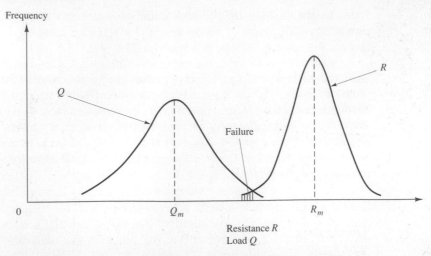

Figure 1.8.1   Frequency distributions of load Q and resistance R.

As summarized by Pinkham [1.21], the reliability index β is useful in several ways:

1. It can give an indication of the consistency of safety for various components and systems using traditional design methods.
2. It can be used to establish new methods which will have consistent margins of safety.
3. It can be used to vary in a rational manner the margins of safety for those components and systems having a greater or lesser need for safety than that required in ordinary situations.

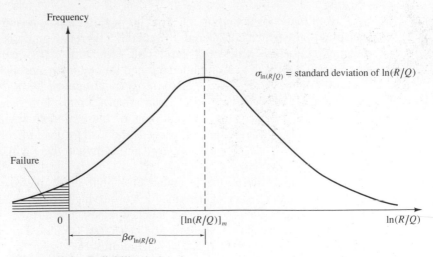

Figure 1.8.2   Reliability index β.

In general, the expression for the structural safety requirement may be written as

$$\phi R_n \geq \sum \gamma_i Q_i \qquad (1.8.1)$$

where the left side of Eq. 1.8.1 represents the *resistance,* or strength, of the component or system, and the right side represents the *load* expected to be carried. On the strength side, the nominal resistance R_n is multiplied by a resistance (strength reduction) factor ϕ to obtain the *design strength* (also called *usable strength* or *usable resistance*). On the load side of the equation, the various load effects Q_i (such as dead load, live load, and snow load) are multiplied by overload factors γ_i to obtain the sum $\sum \gamma_i Q_i$ of *factored loads.*

AISC—Load and Resistance Factor Design (LRFD)

During the past 15 years, the general "limit states design" approach has continued to gain acceptance, particularly for steel design in the United States with the adoption in 1986 of a Load and Resistance Factor Design Specification by AISC. The latest version of that Specification is 1993 [1.16], referred to throughout this book as the AISC LRFD Specification. In Canada, limit states design for steel has been used since 1974, and since 1978 has been the only method used; the latest edition is 1989 [1.19].

The AISC LRFD Specification was developed under the leadership of T. V. Galambos [1.22–1.26]. The adaptation of probabilistic methods to steel design and the development of that Specification are explained by Galambos [1.23, 1.25] and by Galambos and Ravindra [1.24, 1.26].

The safety requirement of the LRFD Specification is given by Eq. 1.8.1. This means the design strength ϕR_n provided by the resulting design must at least equal the sum $\sum \gamma_i Q_i$ of the applied factored service loads. The subscript i indicates that there are terms for each type of load Q_i acting, such as dead load D, live load L, wind load W, snow load S, and earthquake load E. The γ_i may be different for each load type.

The AISC LRFD Specification is based on the following:

1. A probability-based model [1.24, 1.26, 1.32–1.34].
2. Calibration with the 1978 AISC Allowable Stress Design (ASD) Specification.
3. Evaluation using judgment and past experience, along with studies of representative structures conducted by design offices.

The development of probability-based load criteria by Galambos, Ellingwood, MacGregor, and Cornell [1.32–1.34] led to the factored load combinations of the 1982 ANSI Standard, which has become ASCE 7 [1.2]. The ANSI Standard was developed for use in design with *all* structural materials. It is reasonable that the probability of various loads acting in combination, as well as the probability of overload with certain types of loads, should be unrelated to the material of which a structure is built. With this concept in mind, the AISC LRFD Specification adopted the ANSI factored load combinations. LRFD-A4.1 specifies that the following combinations shall be investigated:

$$1.4D \tag{1.8.2}$$

$$1.2D + 1.6L + 0.5(L_r \text{ or } S \text{ or } R) \tag{1.8.3}$$

$$1.2D + 1.6(L_r \text{ or } S \text{ or } R) + (0.5L \text{ or } 0.8W) \tag{1.8.4}$$

$$1.2D + 1.3W + 0.5L + 0.5(L_r \text{ or } S \text{ or } R) \tag{1.8.5}$$

$$1.2D \pm 1.0E + 0.5L + 0.2S \tag{1.8.6}$$

$$0.9D \pm (1.3W \text{ or } 1.0E) \tag{1.8.7}$$

where the nominal *service loads* indicated by Eqs. 1.8.2 through 1.8.7 are

D = dead load (gravity load from the weight of structural elements and permanent attachments)

L = live load (gravity occupancy and movable equipment load)

L_r = roof live load

W = wind load

S = snow load

E = earthquake load

R = rainwater or ice load

Note that D, L, W, S, etc. are loads in a general sense, which includes bending moment, shear, axial force, and torsional moment. Sometimes these internal forces are called *load effects*. Thus, the symbol D means dead load, dead load moment, dead load shear, dead load axial force, etc. An explanation of the statistics relating to snow and wind load factors is given by Ravindra, Cornell, and Galambos [1.35]. The factors for earthquake E are reduced from 1.5 in the 1986 LRFD Specification to 1.0 in the 1993 Specification. This reflects the new earthquake loads of National Earthquake Hazards Reduction Program (NEHRP) [1.52] and the *AISC Seismic Provisions* [1.53], which specify larger earthquake forces than traditionally used.

AISC—Allowable Stress Design (ASD)

The traditional method of the AISC Specification has been *allowable stress design* (also called *working stress design*). In ASD the focus is on service load conditions (i.e., unit stresses assuming an elastic structure) when satisfying the safety requirement (adequate strength) for the structure. The AISC 1989 Specification for Allowable Stress Design is referred to in this book as the ASD Specification.

For Allowable Stress Design, Eq. 1.8.1 may be reformulated as follows:

$$\frac{\phi R_n}{\gamma} \geq \sum Q_i \tag{1.8.8}$$

In this philosophy all loads are assumed to have the same average variability. The entire variability of the loads and the strengths is placed on the strength side of the equation. To examine the equation in terms of Allowable Stress Design for beams, the left side would represent nominal beam strength M_n divided by a factor of safety FS (equal to γ/ϕ), and the right side would represent the service load bending moment M resulting from all types of loads. Thus, Eq. 1.8.8 would become

$$\frac{M_n}{\text{FS}} \geq M \tag{1.8.9}$$

The term *Allowable Stress Design* implies an elastic stress calculation. Equation 1.8.9 may be divided by I/c (i.e., the moment of inertia I divided by the distance c from the neutral axis to the extreme fiber) to obtain stress units. Thus, if one assumes the nominal strength M_n is reached when the extreme fiber stress is the yield stress F_y (i.e., $M_n = F_y I/c$), Eq. 1.8.9 will become

$$\frac{F_y}{\text{FS}} \frac{I/c}{I/c} \geq \frac{M}{I/c} \tag{1.8.10}$$

or

$$\frac{F_y}{\text{FS}} \geq \left[f_b = \frac{Mc}{I} \right] \tag{1.8.11}$$

In ASD the F_y/FS would be the allowable stress F_b and f_b would be the computed elastic stress under full service load. If the nominal strength M_n had been based on achievement of a stress F_{cr} less than F_y because of, say, buckling, then the allowable stress F_b would be F_{cr}/FS. Thus, the safety criterion in ASD may be written

$$f_b \leq \left[F_b = \frac{F_y}{\text{FS}} \quad \text{or} \quad F_b = \frac{F_{cr}}{\text{FS}} \right] \tag{1.8.12}$$

The allowable stresses of the ASD Specification are derived from the strength capable of being achieved if the structure is overloaded. When the section is ductile and buckling does not occur, strains greater than the "first yielding" strain $\epsilon_y = F_y/E_s$ can exist on the section (E_s is the modulus of elasticity). Such ductile inelastic behavior *may permit* higher loads to be carried than possible if the structure had remained entirely elastic. In such cases the allowable stress is adjusted upward. When the strength is limited by buckling or some other behavior such that the stress does not reach yield stress, the allowable stress is adjusted downward.

Serviceability requirements such as deflection limits are always investigated at service load conditions, whether the LRFD or the ASD design procedure has been used to satisfy safety requirements.

AISC—Plastic Design

Traditionally, Part 2 of the AISC Specification was called *Plastic Design*. The 1989 *Specification for Structural Steel Buildings* [1.5] contains Plastic Design in Chapter N. Plastic Design is a special case of limit states design, wherein the limit state for strength is the achievement of *plastic moment strength* M_p. Plastic moment strength is the moment strength when all fibers of the cross-section are at the yield stress F_y (one side of the neutral axis in tension and the other side in compression). Plastic design does not permit using other limit states, such as instability, fatigue, or brittle fracture. The design philosophy as used by AISC applies to flexural members, including beam-columns, and for such members may be expressed by Eq. 1.8.1. Then letting $R_n = M_p$ and $\gamma_i/\phi = 1.7$, Eq. 1.8.1 becomes

$$M_p \geq 1.7 \sum Q_i \tag{1.8.13}$$

The provisions for overload and for understrength are combined into a single factor 1.7 used for all gravity loads. The nominal strength *must* be the plastic moment strength M_p. Since plastic design is a special case of limit states design and is covered more rationally in the AISC LRFD Specification, it is no longer treated as a special topic as in previous editions of this book. Plastic design becomes a component of LRFD.

The limit states design philosophy as codified in LRFD provides the designer with a more rational approach to design than has been available in ASD or Plastic Design, whose philosophies are outlined in Sec. 1.9. Beedle [1.30] provides an excellent summary of the advantages of using LRFD.

1.9 FACTORS FOR SAFETY—ASD AND LRFD COMPARED

Allowable Stress Design (ASD)

The "safety factor" FS used in Eqs. 1.8.9 through 1.8.12 was not determined consciously by using probabilistic methods. The values used in the AISC ASD Specification have been in use for many years and are the result of experience and judgment. It is clear that the safety required must be a combination of economics and statistics. Obviously, it is not economically feasible to design a structure so that the probability of failure is zero. Prior to the development of the AISC LRFD Specification [1.16], the AISC Specifications from 1924 through 1978 did not give a rationale for the allowable stresses prescribed.

One may state that the minimum resistance must exceed the maximum applied load by some prescribed amount. Suppose the actual load exceeds the service load by an amount ΔQ, and the actual resistance is less than the computed resistance by an amount ΔR. A structure that is just adequate would have

$$R_n - \Delta R_n = Q + \Delta Q$$
$$R_n(1 - \Delta R_n/R_n) = Q(1 + \Delta Q/Q) \tag{1.9.1}$$

The margin of safety, or "safety factor," would be the ratio of the nominal strength R_n to nominal service load Q; or

$$\text{FS} = \frac{R_n}{Q} = \frac{1 + \Delta Q/Q}{1 - \Delta R_n/R_n} \tag{1.9.2}$$

Equation 1.9.2 illustrates the effect of overload ($\Delta Q/Q$) and understrength ($\Delta R_n/R_n$); however it does not identify the factors contributing to either. If one assumes that occasional overload ($\Delta Q/Q$) may be 40% greater than its nominal value, and that an occasional understrength ($\Delta R_n/R_n$) may be 15% less than its nominal value, then

$$\text{FS} = \frac{1 + 0.4}{1 - 0.15} = \frac{1.4}{0.85} = 1.65$$

The above is an oversimplification but it shows a possible scenario for obtaining the traditional AISC value of FS = 1.67 used as the basic value in Allowable Stress Design. Dividing by 1.67 as indicated in Eq. 1.8.12 gives a multiplier of 0.60 on F_y or F_{cr}.

The basic value of 1.67 is used for tension members and beams, and it is the lower bound for zero-length columns. A value of 1.92 is used for long columns, and values from 2.5 to 3 are used for connections. However, it must be noted that using these values for γ/ϕ in Eq. 1.8.8 still leaves the "real" safety against "failure" *unknown*!

Load and Resistance Factor Design (LRFD)

As discussed in Sec. 1.8, the factors for overload are variable depending upon the type of load, and the factored load combinations that must be considered are those given by the ASCE 7 Standard [1.2] and LRFD-A4.1, and presented as Eqs. 1.8.2 through 1.8.7. The other part of the safety-related provisions is the ϕ factor, known as the *resistance factor*. The resistance factor ϕ varies with the type of member and with the limit state being considered. Some representative resistance factors ϕ are as follows.

Tension Members (LRFD-D1)

$$\phi_t = 0.90 \quad \text{for } yielding \text{ limit state}$$
$$\phi_t = 0.75 \quad \text{for } fracture \text{ limit state}$$

Compression Members (LRFD-E2)

$$\phi_c = 0.85$$

Beams (LRFD-F1.1, F1.2, and F2.2)

$$\phi_b = 0.90 \quad \text{for flexure}$$
$$\phi_v = 0.90 \quad \text{for shear}$$

Welds (LRFD-Table J2.5)

$$\phi = \text{same as for type of action; i.e., tension, shear etc.}$$

Fasteners (LRFD-Table J3.2 and J3.7)

$$\phi = 0.75$$

In order to establish adequate safety using probabilistic methods, the natural logarithm of the resistance R divided by the load Q, that is, $\ln(R/Q)$ as shown in Fig. 1.8.2, may be treated as a random variable. It is simpler than working with two groups (R and Q) of random variables as in Fig. 1.8.1. When $\ln(R/Q) < 0$, the limit state has been reached, and the shaded area in Fig. 1.8.2 is the probability of this event. The method used to develop LRFD uses the *mean values* R_m and Q_m and the *standard deviations* σ_R and σ_Q of the resistance and load, respectively. Frequently, the mean values and standard deviations can be estimated even though the actual distributions cannot be obtained. From these estimated quantities, the standard deviation σ of the $\ln(R/Q)$ may be approximated as

$$\sigma_{\ln(R/Q)} \approx \sqrt{V_R^2 + V_Q^2} \tag{1.9.3}$$

where $V_R = \sigma_R/R_m$
$V_Q = \sigma_Q/Q_m$

The margin of safety is the distance from the origin to the mean, as shown in Fig. 1.8.2, and is expressed as a multiple β of $\sigma_{\ln(R/Q)}$. The distance representing the margin of safety may be approximated [1.17] as

$$\beta\sigma_{\ln(R/Q)} \approx \beta\sqrt{V_R^2 + V_Q^2} = \ln(R_m/Q_m) \tag{1.9.4}$$

Thus, the larger the distance the smaller the probability of reaching the limit state. The multiplier β is called the *reliability index*. The expression for β from Eq. 1.9.4 becomes

$$\beta = \frac{\ln(R_m/Q_m)}{\sqrt{V_R^2 + V_Q^2}} \tag{1.9.5}$$

More discussion of the development of Eq. 1.9.5 is given in the LRFD Commentary [1.17], by Ravindra and Galambos [1.26], and in NBS Special Publication 577 [1.32]. The treatment of the theory of probability is outside the scope of this book.

Using the factored load combinations given by the ASCE 7 Standard [1.2], the AISC Task Force and Specification Committee calibrated the LRFD Specification [1.16] to generally agree with past experience. Thus, the resistance factors ϕ were set in LRFD with the objective of obtaining the following values of β:

Load combinations	Objective reliability index β	
Dead load + live load (or snow load)	3.0	for members
	4.5	for connections
Dead load + live load + wind load	2.5	for members
Dead load + live load + earthquake load	1.75	for members

Because of a lower probability of wind or earthquake occurring with full gravity load, the reliability index β was made lower for those cases. The β values for connections were made higher than for members in keeping with tradition to make connections stronger than members.

The LRFD Specification (LRFD-A4.) uses six factored load combinations, given as Eqs. 1.8.2 through 1.8.7. This was necessary to account for each of the separate loads (dead, live, roof, wind additive to gravity, wind opposite to gravity, and earthquake) acting at its maximum lifetime value. Loads other than dead load and the load being maximized will act at an "arbitrary point-in-time" value. The "arbitrary point-in-time" value is that value which can be expected to be on the structure at any time. The arbitrary point-in-time value of live load (L) might be as low as one-quarter of its mean maximum lifetime load but its distribution widely varies. The arbitrary point-in-time wind (W) is the maximum daily wind. The "lifetime maximum" is taken as the 50-year recurrence value.

Thus, each factored load combination and its corresponding load occurring at its 50-year maximum are as follows:

LRFD Eq.	Load combination	Load at its lifetime (50 year) maximum
(A4-1)	$1.4D$	Dead load D during construction; other loads not present.
(A4-2)	$1.2D + 1.6L + 0.5S$	Live load L
(A4-3)	$1.2D + 1.6S + (0.8W$ or $0.5L)$	Roof load; i.e., snow load S or rain R other than ponding effect.
(A4-4)	$1.2D + 1.3W + 0.5L + 0.5S$	Wind load W *additive* to dead load.
(A4-5)*	$1.2D + 1.0E + 0.5L + 0.2S$	Earthquake load E *additive* to dead load.
(A4-6)*	$0.9D - (1.3W$ or $1.0E)$	Wind load W or earthquake load E *opposite* to dead load.

* The sign following $1.2D$ or $0.9D$ is to be taken $+$ or $-$ such as to provide for the more severe effect.

Note: Where snow S is used in the above equations, except in Formula (A4-5), the meaning is snow S OR roof live load L_r OR rain R other than ponding.

Comparison of LRFD with ASD for Tension Members

The comparison of safety obtained for tension members designed by the two AISC methods is indicative of the general result expected. Direct comparisons are more difficult in design of other types of members because the nominal strengths R_n are not necessarily the same in the two methods.

For tension members acted upon by gravity dead and live loads, the resistance factor $\phi = 0.90$, and using Eq. 1.8.3 gives for LRFD

$$1.2D + 1.6L = 0.90R_n \qquad\qquad [1.8.3]$$
$$1.33D + 1.78L = R_n \qquad \text{LRFD}$$

In ASD the factor of safety FS = 1.67 for axial tension, which gives from Eq. 1.8.8 where $(\gamma/\phi$ is the factor of safety)

$$R_n/1.67 = \sum Q = D + L \qquad\qquad [1.8.8]$$

or

$$1.67D + 1.67L = R_n \qquad \text{ASD}$$

Next, dividing Eq. 1.8.3 by Eq. 1.8.8 gives

$$\frac{\text{LRFD}}{\text{ASD}} = \frac{1.33D + 1.78L}{1.67D + 1.67L} = \frac{0.8 + 1.07(L/D)}{1 + (L/D)} \qquad (1.9.6)$$

Since this is a gravity load comparison, LRFD formula (A4-1) [Eq. 1.8.2] must also be used as L/D approaches zero. Thus, Eq. 1.8.2 gives

$$1.4D = 0.90R_n \qquad\qquad [1.8.2]$$
$$1.56D = R_n \qquad \text{LRFD}$$

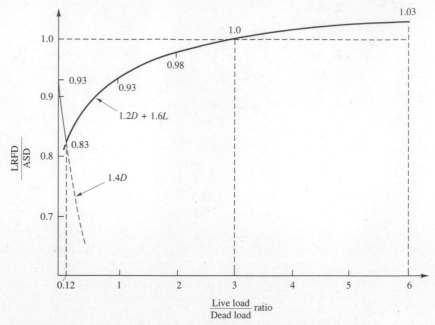

Figure 1.9.1   Comparison of load and resistance factor design with allowable stress design for tension members.

Dividing LRFD by ASD gives

$$\frac{\text{LRFD}}{\text{ASD}} = \frac{1.56D}{1.67D + 1.67L} = \frac{0.93}{1 + (L/D)} \tag{1.9.7}$$

Equations 1.9.6 and 1.9.7 are shown plotted in Fig. 1.9.1. The design of tension members will be about the same in both LRFD and ASD when the live load to dead load ratio (L/D) is about 3. As the L/D ratio becomes lower (that is, dead load becomes more predominant) there will be economy in using LRFD. With L/D ratio larger than 3, ASD will be slightly more economical, but rarely by more than about 3%.

1.10 WHY SHOULD LRFD BE USED?

The many advantages of LRFD are well-expressed by Beedle [1.30], whose listing is the basis of the following:

1. LRFD is another "tool" for structural engineers to use in steel design. Why not have the same tools (variable overload factors and resistance factors) available for steel design as are available for concrete design?

2. Adoption of LRFD is not mandatory but provides a flexibility of options to the designer. The marketplace will dictate whether or not LRFD will become the sole method.
3. ASD is an approximate way to account for what LRFD does in a more rational way. The use of plastic design concepts in ASD has made ASD such that it no longer may be called an "elastic design" method.
4. The rationality of LRFD has always been attractive, and becomes an incentive permitting the better and more economical use of material for some load combinations and structural configurations. It will also likely lead to having safer structures in view of the arbitrary practice under ASD of combining dead and live loads and treating them the same.
5. Using multiple load factor combinations should lead to economy.
6. LRFD will facilitate the input of new information on loads and load variations as such information becomes available. Considerable knowledge of the resistance of steel structures is available. On the other hand, our knowledge of loads and their variation is much less. Separating the loading from the resistance allows one to be changed without the other if that should be desired.
7. Changes in overload factors and resistance factors ϕ are much easier to make than to change the allowable stress in ASD.
8. LRFD makes design in all materials more compatible. The variability of loads is actually unrelated to the material used in the design. Future specifications not in the limit states format for any material will put that material at a disadvantage in design.
9. LRFD provides the framework to handle unusual loads that may not be covered by the Specification. The design may have uncertainty relating to the resistance of the structure, in which case the resistance factors may be modified. On the other hand, the uncertainty may relate to the loads and different overload factors may be used.
10. Future adjustments in the calibration of the method can be made without much complication. Calibration for LRFD was done for an *average* situation but might be adjusted in the future.
11. Economy is likely to result for low live load to dead load ratios. For high live load to dead load ratios there will be slightly greater costs.
12. Safer structures may result under LRFD because the method should lead to a better awareness of structural behavior.
13. Design practice is still at the beginning with regard to serviceability limit states; however, at least LRFD provides the approach.

The rationality of LRFD and its many advantages over ASD (as outlined by Beedle [1.30]) are indicative that the design philosophy will relegate ASD to the background in the next few years. It is important, however, that the designer understand both philosophies of design because many structures will continue to be designed using ASD and the designer may frequently need to evaluate structures of the past. Heger [1.31] provides some interesting thoughts on the difficulties of bridging the gap between the theory of statistics and probability and the real world of actual structures.

1.11 ANALYSIS OF THE STRUCTURE

In general, the structural analysis to obtain the service loads (or load *effects* bending moment, shear, axial force, and torsional moment) on the members is performed the same for LRFD as for ASD. Elastic methods of structural analysis are used except when the limit state is the *plastic collapse mechanism* as described in Chapter 10. A *first-order analysis* [1.15] is sufficient in usual framed structures that are braced against sway. In a first-order analysis, equilibrium equations are based on the original geometry of the structure. This means that the designer is assuming the internal forces (moments, shears, etc.) are not sufficiently affected by the change in shape of the structure to justify more complicated analysis. When the elastic displacements are small compared to the dimensions, this approximation will be satisfactory.

The most common situation where a *second-order* effect must be considered is in a multistory structure that must rely on the stiffnesses of the interacting beams and columns to resist sway under lateral loading (wind and/or earthquake). This is the so-called *unbraced frame*. In this case the lateral displacement Δ (also called *sway* or *drift*) causes additional bending moments due to the gravity loads (ΣP) acting at positions that have been displaced by an amount Δ. The analysis must include this $P\Delta$ secondary effect. There are varying degrees of sophistication that are used in analysis to include the $P\Delta$ effects. In both ASD and LRFD the second-order effects may be computed as a part of the analysis or they may be accounted for using formulas in the Specifications [1.5, 1.16] or Commentaries [1.6, 1.17].

The emphasis in this book is on designing the members to have adequate strength and proper serviceability, rather than on structural analysis. The reader is referred to Wang and Salmon [1.15] for basic structural analysis topics. Other references specifically related to analysis for unbraced frames are given in Chapter 14.

Most examples in this book use given service loads, or service load effects, acting on the structural member to be designed. These values are to be assumed the result of structural analysis. The service loads are to be factored using Eqs. 1.8.2 through 1.8.7 for design in LRFD, but used as given in ASD.

SELECTED REFERENCES

1.1. Hans Straub. *A History of Civil Engineering.* Cambridge, MA: M.I.T. Press, 1964 (pp. 173–180).

1.2. ASCE. *Minimum Design Loads for Buildings and Other Structures,* ASCE Standard 7-93. New York: American Society of Civil Engineers (345 E. 47th Street, New York, NY 10017), 1993.

1.3. AASHTO. *Standard Specifications for Highway Bridges,* 15th ed. Washington, DC: The American Association of State Highway and Transportation Officials (444 North Capitol Street, N.W., Suite 249, Washington, DC 20001), 1992.

1.4. AREA. *Manual for Railway Engineering,* Chapter 15 (Specifications for Steel Railway Bridges). Washington, DC: American Railway Engineering Association, 1990.

1.5.   AISC. *Specification for Structural Steel Buildings—Allowable Stress Design and Plastic Design* (June 1, 1989). Chicago: American Institute of Steel Construction (One East Wacker Drive, Suite 3100, Chicago, IL 60601-2001), 1989.

1.6.   AISC. *Commentary on the Specification for Structural Steel Buildings— Allowable Stress Design and Plastic Design* (June 1, 1989). Chicago: American Institute of Steel Construction, 1989.

1.7.   AISC. *Manual of Steel Construction, Allowable Stress Design,* 9th ed. Chicago. American Institute of Steel Construction, 1989.

1.8.   ASTM. *Standard Specification for General Requirements for Rolled Steel Plates, Shapes, Sheet Piling, and Bars for Structural Use,* A6/A6M-94a. Philadelphia, PA: American Society for Testing and Materials, 1994.

1.9.   National Research Council of Canada. *Canadian Structural Design Manual,* Supplement No. 4 to the National Building Code of Canada, Ottawa, Canada, 1985.

1.10. H. S. Lew, Emil Simiu, and Bruce Ellingwood. "Loads," Chapter 2, *Building Structural Design Handbook,* Richard N. White and Charles G. Salmon, Ed. New York: John Wiley & Sons, 1987 (pp. 9–43).

1.11. AISI. *Load and Resistance Factor Design Specification for Cold-Formed Steel Structural Members.* Washington, DC: American Iron and Steel Institute, March 16, 1991.

1.12. AISI. *Cold-Formed Steel Design Manual,* Part I, Specification; Part II, Commentary; Part III, Supplementary Information; Part IV, Illustrative Examples; and Part V, Charts and Tables. New York: American Iron and Steel Institute, 1982.

1.13. SJI. *Standard Specifications Load Tables & Weight Tables for Steel Joists & Joist Girders.* Myrtle Beach, SC: Steel Joist Institute, 1990. (Suite A, 1205 48th Avenue North, Myrtle Beach, SC 29577)

1.14. James M. Fisher, Michael A. West, and Julius P. Van de Pas. *Designing With Steel Joists, Joist Girders, Steel Deck.* Charlotte, NC: Nucor Corporation, 1991.

1.15. Chu-Kia Wang and Charles G. Salmon. *Introductory Structural Analysis.* Englewood Cliffs, New Jersey: Prentice-Hall, Inc., 1984.

Load and Resistance Factor Design

1.16. AISC. *Load and Resistance Factor Design Specification for Structural Steel Buildings.* Chicago: American Institute of Steel Construction, December 1, 1993.

1.17. AISC. *Commentary on the Load and Resistance Factor Design Specification for Structural Steel Buildings* (December 1, 1993). Chicago: American Institute of Steel Construction, 1993.

1.18. AISC. *Manual of Steel Construction, Load & Resistance Factor Design,* 2nd ed. Volumes 1 and 2. Chicago: American Institute of Steel Construction, 1994.

1.19. CSA. *Limit States Design of Steel Structures,* Appendices G, H, and I, CSA S16.1-M89. Rexdale, Ontario, Canada: Canadian Standards Association, 1989.

1.20. ASCE Task Committee on Structural Safety of the Administrative Committee on Analysis and Design of the Structural Division. "Structural Safety- A Literature Review," *Journal of the Structural Division,* ASCE, **98,** ST4 (April 1972), 845–884.

1.21. Clarkson W. Pinkham. "Design Philosophies," Chapter 3, *Building Structural Design Handbook,* Richard N. White and Charles G. Salmon, Ed. New York: John Wiley & Sons, 1987 (pp. 44–54).

1.22. C. W. Pinkham and W. C. Hansell. "An Introduction to Load and Resistance Factor Design for Steel Buildings," *Engineering Journal,* AISC, **15,** 1 (First Quarter 1978), 2–7.

1.23. T. V. Galambos. "Load Factor Design of Steel Buildings," *Engineering Journal,* AISC, **9,** 3 (July, 1972), 108–113.

1.24. Theodore V. Galambos and M. K. Ravindra. "Proposed Criteria for Load and Resistance Factor Design," *Engineering Journal,* AISC, **15,** 1 (First Quarter 1978), 8–17.

1.25. Theodore V. Galambos. "Load and Resistance Factor Design," *Engineering Journal,* AISC, **18,** 3 (Third Quarter 1981), 74–82.

1.26. Mayasandra K. Ravindra and Theodore V. Galambos. "Load and Resistance Factor Design for Steel," *Journal of the Structural Division,* ASCE, **104,** ST9 (September 1978), 1337–1353.

1.27. Geerhard Haaijer. "Limit States Design- A Tool for Reducing the Complexity of Steel Structures," paper presented at AISC National Engineering Conference, March 4, 1983.

1.28. D. J. Laurie Kennedy. "Limit States Design of Steel Structures in Canada," *Journal of Structural Engineering,* ASCE, **110,** 2 (February 1984), 275–290.

1.29. D. J. Laurie Kennedy. "North American Limit States Design," *Proceedings, The 1985 International Engineering Symposium on Structural Steel.* Chicago: American Institute of Steel Construction, May 22–24, 1985.

1.30. Lynn S. Beedle. "Why LRFD?", *Modern Steel Construction,* AISC, **26,** 4(4th Quarter 1986), 30–31.

1.31. Frank J. Heger. "Proposed AISC LRFD Design Criteria," *Journal of the Structural Division,* ASCE, **106,** ST3 (March 1980), 729–734. (Disc. **106,** ST12 (December 1980), 2576–2577)

1.32. Bruce Ellingwood, Theodore V. Galambos, James G. MacGregor, and C. Allin Cornell. *Development of a Probability Based Load Criterion for American National Standard A58,* NBS Special Publication 577. Washington, DC: US Department of Commerce, National Bureau of Standards, June 1980.

1.33. Theodore V. Galambos, Bruce Ellingwood, James G. MacGregor, and C. Allin Cornell. "Probability Based Load Criteria: Assessment of Current Design Practice," *Journal of the Structural Division,* ASCE, **108,** ST5 (May 1982), 959–977.

1.34. Bruce Ellingwood, James G. MacGregor, Theodore V. Galambos, and C. Allin Cornell. "Probability Based Load Criteria: Load Factors and Load Combinations," *Journal of the Structural Division,* ASCE, **108,** ST5 (May 1982), 978–997.

1.35. Mayasandra K. Ravindra, C. Allin Cornell, and Theodore V. Galambos. "Wind and Snow Load Factors for Use in LRFD," *Journal of the Structural Division,* ASCE, **104,** ST9 (September 1978), 1443–1457.

Wind Loads

1.36. ASCE. Task Committee on Wind Forces, Committee on Loads and Stresses, Structural Division. "Wind Forces on Structures," *Journal of the Structural Division,* ASCE, **84,** ST4 (July 1958) (Preliminary Report); and Final Report, *Transactions,* ASCE, **126,** pt. II (1961), 1124–1198.

1.37. Kishor C. Mehta. "Wind Load Provisions ANSI #A58.1–1982," *Journal of Structural Engineering,* **110,** 4 (April 1984), 769–784.

1.38. Theodore Stathopoulos, David Surry, and Alan G. Davenport. "Effective Wind Loads on Flat Roofs," *Journal of the Structural Division,* ASCE, **107,** ST2 (February 1981), 281–298.

Snow Loads

1.39. Michael J. O'Rourke and Ulrich Stiefel. "Roof Snow Loads for Structural Design," *Journal of Structural Engineering,* ASCE, **109,** 7 (July 1983), 1527–1537.

1.40. J. T. Templin and W. R. Schriever. "Loads due to Drifted Snow," *Journal of the Structural Division,* ASCE, **108,** ST8 (August 1982), 1916–1925.

1.41. Michael O'Rourke, Wayne Tobiasson, and Evelyn Wood. "Proposed Code Provisions for Drifted Snow Loads," *Journal of Structural Engineering,* **112,** 9 (September 1986), 2080–2092.

1.42. Michael J. O'Rourke, Robert Redfield, and Peter von Bradsky. "Uniform Snow Loads on Structures," *Journal of the Structural Division,* ASCE, **108,** ST12 (December 1982), 2781–2798.

1.43. Michael J. O'Rourke, Robert S. Speck, Jr., and Ulrich Stiefel. "Drift Snow Loads on Multilevel Roofs," *Journal of Structural Engineering,* ASCE, **111,** 2 (February 1985), 290–306.

1.44. R. L. Sack. "Snow Loads on Sloped Roofs," *Journal of Structural Engineering,* ASCE, **114,** 3 (March 1988), 501–517.

1.45. Michael O'Rourke and Ioannis Galanakis. "Roof Snowdrifts due to Blizzards," *Journal of Structural Engineering,* ASCE, **116,** 3 (March 1990), 641–658.

1.46. Ronald L. Sack and Paul M. Giever. "Predicting Roof Snow Loads on Gabled Structures," *Journal of Structural Engineering,* ASCE, **116,** 10 (October 1990), 2763–2779. Errata, **116,** 11 (November 1990), 3249–3250.

Earthquake Load

1.47. SEAOC. *Recommended Lateral Force Requirements and Commentary.* San Francisco, CA: Seismology Committee, Structural Engineers Association of California, Los Angeles, CA, 1990.

1.48. Anil K. Chopra and Ernesto F. Cruz. "Evaluation of Building Code Formulas for Earthquake Forces," *Journal of Structural Engineering,* ASCE, **112,** 8 (August 1986), 1881–1899.

1.49. Anil K. Chopra. *Dynamics of Structures, Theory and Applications to Earthquake Engineering.* Englewood Cliffs, NJ: Prentice-Hall, Inc., 1995.

1.50. R. W. Clough and Joseph Penzien. *Dynamics of Structures*. New York: McGraw-Hill, 1975.

1.51. UBC. *Uniform Building Code*. Whittier, CA: International Conference of Building Officials, 1991.

1.52. BSSC. *NEHRP* (National Earthquake Hazards Reduction Program) *Recommended Provisions for the Development of Seismic Regulations for New Buildings*. Washington, DC: Building Seismic Safety Council, Federal Emergency Management Agency, 1991.

1.53. AISC. *Seismic Provisions for Structural Steel Buildings*. Chicago, IL: American Institute of Steel Construction, 1992.

1.54. Egor P. Popov. "U.S. Seismic Steel Codes," *Engineering Journal,* AISC, **28,** 3 (3rd Quarter 1991), 119–128.

1.55. James W. Marsh. "Earthquakes: Steel Structures Performance and Design Code Developments," *Engineering Journal,* AISC, **30,** 2 (2nd Quarter 1993), 56–65.

Chapter 2

Steels and Properties

2.1 STRUCTURAL STEELS

During most of the period from the introduction of structural steel as a major building material until about 1960, the steel used was classified as a carbon steel with the ASTM (American Society for Testing and Materials) designation A7, and had a minimum specified yield stress of 33 ksi. Most designers merely referred to "steel" without further identification, and the AISC Specification prescribed allowable stresses and procedures only for A7 steel. Other structural steels, such as a special corrosion resistant low alloy steel (A242) and a more readily weldable steel (A373), were available but they were rarely used in buildings. Bridge design made occasional use of these other steels.

Today (1995) the many steels available to the designer permit use of increased strength material in highly stressed regions rather than increase the size of members. The designer can decide whether maximum rigidity or least weight is the more desirable attribute. Corrosion resistance, hence elimination of frequent painting, may be a highly important factor. Some steels oxidize to form a dense protective coating that prevents further oxidation (corrosion), acquiring a pleasing even-textured dark red-brown appearance. Since painting is not required, it may be economical to use these "weathering steels" even though the initial cost is somewhat higher than traditional carbon steels.

Certain steels provide better weldability than others; some are more suitable than others for pressure vessels, either at temperatures well above or well below room temperatures.

Structural steels are referred to by ASTM designations, and also by many proprietary names. For design purposes the yield stress in tension is the material property that specifications, such as AISC, use to establish strength or allowable stress. The term *yield stress* is used to include either "yield point," the well-defined deviation from perfect elasticity exhibited by most of the common structural steels; or "yield strength", the unit stress at a certain offset strain for steels having no well-defined yield point. Today (1995) steels are readily available having yield stresses from 24 to 100 ksi (170 to 690 MPa).

John Hancock Center, Chicago, showing exterior diagonal
bracing. (Photo by C. G. Salmon)

Steels for structural use in hot-rolled applications may be classified as *carbon
steels, high-strength low-alloy steels,* and *alloy steels.* The general requirements for
such steels are covered under ASTM A6/A6M Specification [1.8]. Table 2.1.1 lists the
common steels, their minimum yield stresses, and tensile strengths. Their common
uses are given in Table 2.1.2.

Carbon Steels

Carbon steels are divided into four categories based on the percentage of carbon: low
carbon (less than 0.15%); mild carbon (0.15–0.29%); medium carbon (0.30–0.59%);
and high carbon (0.60–1.70%). Structural carbon steels are in the mild carbon category; a steel such as A36 has maximum carbon varying from 0.25 to 0.29% depending
on thickness. Structural carbon steels exhibit definite yield points as shown on curve
(a) of Fig. 2.1.1. Increased carbon percent raises the yield stress but reduces ductility,
making welding more difficult.

The carbon steels given in Table 2.1.1 are A36 [2.1], A53 [2.2], A500 [2.9], A501 [2.10], A529 [2.12], A570 [2.13], A611 [2.18], and A709 [2.20], Grade 36.

High-Strength Low-Alloy Steels

This category includes steels having yield stresses from 40 to 70 ksi (275 to 480 MPa), exhibiting the well-defined yield point shown in curve (b) of Fig. 2.1.1, the same as shown by carbon steels. The addition to carbon steels of small amounts of alloy elements such as chromium, columbium, copper, manganese, molybdenum, nickel, phosphorus, vanadium, or zirconium, improves some of the mechanical properties. Whereas carbon steels gain their strength by increasing carbon content, the alloy elements create increased strength from a fine rather than coarse microstructure obtained during cooling of the steel. High-strength low-alloy steels are used in the as-rolled or normalized condition; i.e., no heat treatment is used.

The high-strength low-alloy steels of Table 2.1.1 are A242 [2.3], A441 [2.6], A572 [2.14], A588 [2.15], A606 [2.16], A607 [2.17], A618 [2.19], and A709 [2.20], Grades 50 and 50 W.

Alloy Steels

Low-alloy steels may be quenched and tempered to obtain yield strengths of 80 to 110 ksi (550 to 760 MPa). Yield strength is usually defined as the stress at 0.2% offset strain, since these steels do not exhibit a well-defined yield point. A typical stress-strain curve is shown in Fig. 2.1.1, curve (c). These steels are weldable with proper procedures, and ordinarily require no additional heat treatment after they have been welded. For special uses, stress relieving may occasionally be required. Some carbon steels, such as certain pressure vessel steels, may be quenched and tempered to give yield strengths in the 80 ksi (550 MPa) range, but most steels of this strength are low-alloy steels. These low-alloy steels generally have a maximum carbon content of about 0.20% to limit the hardness of any coarse-grain microstructure (martensite) that may form during heat treating or welding, thus reducing the danger of cracking.

The heat treatment consists of quenching [rapid cooling with water or oil from at least 1650°F (900°C) to about 300–400°F]; then tempering by reheating to at least 1150°F (620°C) and allowing to cool. Tempering, even though reducing the strength and hardness somewhat from the quenched material, greatly improves the toughness and ductility. Reduction in strength and hardness with increasing temperature is somewhat counteracted by the occurrence of a secondary hardening, resulting from precipitation of fine columbium, titanium, or vanadium carbides. This precipitation begins at about 950°F (510°C) and accelerates up to about 1250°F (680°C). Tempering at or near 1250°F to get maximum benefit from precipitating carbides may result in entering the transformation zone, thus producing the weaker microstructure that would have been obtained without quenching and tempering.

In summary, the quenching produces martensite, a very hard, strong, and brittle microstructure; reheating reduces the strength and hardness somewhat while increasing the toughness and ductility. For more detailed information concerning the metallurgy of the quenching and tempering process, the reader is referred to the *Welding Handbook* [2.23]. The quenched and tempered alloy steels of Table 2.1.1 are A514 [2.11], A709 [2.20], Grades 100 and 100 W, A852 [2.22], and A913 [2.46].

TABLE 2.1.1 PROPERTIES OF STEELS USED FOR BUILDINGS AND BRIDGES

ASTM[†] designation	F_y Minimum yield stress ksi (MPa)[‡]	F_u Tensile strength ksi (MPa)[‡]	Maximum thickness for plates in. (mm)	ASTM A6 groups* for shapes
A36	32 (220)	58–80 (400–550)	Over 8 (200)	—
	36 (250)	58–80 (400–550)	To 8 (200)	All
A53 Grade B	35 (240)	60 (415)		
A242	42 (290)	63 (435)	Over $1\frac{1}{2}$ to 4 (40 to 100)	4, 5
	46 (315)	67 (460)	Over $\frac{3}{4}$ to $1\frac{1}{2}$ (20 to 40)	3
	50 (345)	70 (480)	To $\frac{3}{4}$ (20)	1, 2
A441 Discontinued 1989; replaced by A572				
A500 Grade A	33 (228)	45 (310)	Round	
Grade B	42 (290)	58 (400)	Round	
Grade C	46 (317)	62 (427)	Round	
Grade A	39 (269)	45 (310)	Shaped	
Grade B	46 (317)	58 (400)	Shaped	
Grade C	50 (345)	62 (427)	Shaped	
A510	36 (250)	58 (400)		
A514	90 (620)	100–130 (690–895)	Over $2\frac{1}{2}$ to 6 (65 to 150)	
	100 (690)	110–130 (760–895)	To $2\frac{1}{2}$ (65)	
A529 Grade 42	42 (290)	60–85 (415–585)	To $\frac{1}{2}$ (13)	1
Grade 50	50 (345)	70–100 (485–690)	To 1 (25)	1, 2
A570 Grade 40	40 (275)	55 (380)		
Grade 45	45 (310)	60 (415)		
Grade 50	50 (345)	65 (450)		
A572 Grade 42	42 (290)	60 (415)	To 6 (150)	All
Grade 50	50 (345)	65 (450)	To 4 (100)	All
Grade 60	60 (415)	75 (520)	To $1\frac{1}{4}$ (32)	1, 2, 3
Grade 65	65 (450)	80 (550)	To $1\frac{1}{4}$ (32)	1, 2, 3

TABLE 2.1.1 (*Continued*)

ASTM[†] designation	F_y Minimum yield stress ksi (Mpa)[‡]	F_u Tensile strength ksi (MPa)[‡]	Maximum thickness for plates in. (mm)	ASTM A6 groups* for shapes
A588	42 (290)	63 (435)	Over 5 to 8 (125 to 200)	—
	46 (315)	67 (460)	Over 4 to 5 (100 to 125)	—
	50 (345)	70 (485)	To 4 (100)	All
A606	45 (310)	65 (450)		
	50 (345)	70 (480)		
A607 Grade 45	45 (310)	60 (410)		
Grade 50	50 (340)	65 (450)		
Grade 55	55 (380)	70 (480)		
Grade 60	60 (410)	75 (520)		
Grade 65	65 (450)	80 (550)		
Grade 70	70 (480)	85 (590)		
A611 Grade C	33 (230)	48 (330)		
Grade D	40 (275)	52 (360)		
Grade E	80 (550)	82 (565)		
A618 Grades I&II	50 (345)	70 (485)	To $\frac{3}{4}$ in. walls	
Grade III	50 (345)	65 (450)		
A709 Grade 36	36 (250)	58–80 (400–550)	To 4 (100)	All
Grade 50	50 (345)	65 (450)	To 4 (100)	All
Grade 50W	50 (345)	70 (485)	To 4 (100)	All
Grade 70W	70 (485)	90–110 (620–760)	To 4 (100)	
Grade 100 & 100W	90 (620)	100–130 (690–895)	Over $2\frac{1}{2}$ to 4 (64–102)	
Grade 100 & 100W	100 (690)	110–130 (760–895)	To $2\frac{1}{2}$ (64)	
A852	70 (485)	90–110 (620–760)	To 4 (100)	
A913 Grade 60	60 (415)	75 (520)		All
Grade 65	65 (450)	80 (550)		
Grade 70	70 (485)	90 (620)		

[†] All steels listed are approved under the AISC Specification [1.5, 1.16] except A611 and A709.
*ASTM A6/A6M [1.8] places structural rolled shapes (W, M, S, HP, C, MC, and L) in Groups 1 through 5 according to size for tensile property classification. All rolled flanged sections having at least one cross-section dimension 3 in. (75 mm) or greater are included. The size basis for groups is approximately the web thickness corresponding to the maximum thickness for plates, with the thinnest web sections in Group 1. The specific sections included in each group are given in ASTM A6/A6M [1.8] and in the AISC Manuals [1.7, 1.18].
[‡] All SI values are those given in the particular ASTM Specification.

TABLE 2.1.2   USES OF VARIOUS STEELS

ASTM* designation	Common usage
A36 Carbon steel	General structural purposes; bolted and welded, mainly for buildings
A53 Carbon steel	Welded and seamless pipe
A242 High-strength low-alloy steel	Welded and bolted bridge construction where corrosion resistance is desired; essentially superseded by A709, Grade 50W
A500 Carbon steel	Cold-formed welded and seamless round, square, rectangular, or special shape structural tubing for bolted and welded general structural purposes
A501 Carbon steel	Hot-formed welded and seamless square, rectangular, round, or special shape structural tubing for bolted and welded general structural purposes
A514 Alloy steel, quenched and tempered	Plates in thicknesses of 6 in. (150 mm) and under, primarily for welded bridges; largely superseded by A709 for bridges
A529 Carbon steel	Plates and bars $\frac{1}{2}$ in. (13 mm) and less in thickness or diameter and Group 1 shapes [1.8] for use in bolted and welded metal building system frames and trusses
A570 Carbon steel	Hot-rolled sheet and strip cut in lengths or coils; for cold-formed sections [maximum thickness 0.229 in. (6 mm)]
A572 High-strength, low-alloy, columbium or vanadium steel	Structural shapes, plates, sheet piling, and bars for bolted and welded buildings; welded bridges in Grades 42 and 50 only; essentially superseded by A709, Grade 50
A588 High-strength low-alloy steel	Stuctural shapes, plates, and bars for welded buildings and bridges where weight savings or added durability are needed; atmospheric corrosion resistance is about four times that of A36 steel; essentially superseded by A709, Grade 50 for bridges
A606 High-strength low-alloy steel	Hot- and cold-rolled sheet and strip in lengths or coils; for cold-formed sections, where enhanced durability is desired; atmospheric corrosion resistance for Type 2 at least twice that of carbon steel; and for Type 4 at least four times that of carbon steel
A607 High-strength low-alloy columbium or vanadium steel	Hot- and cold-rolled sheet and strip in lengths or coils; for cold-formed sections, where greater strength and weight savings are important; atmospheric corrosion resistance (without copper) is the same as carbon steel; with copper, corrosion resistance is twice that of carbon steel
A611 Carbon steel	Cold-rolled sheet in cut lengths or coils for making cold-formed sections
A618 High-strength low-alloy steel	Hot-formed welded and seamless square, rectangular, round, or special shape structural tubing for bolted and welded general structural purposes; Grade II has corrosion resistance about twice that of carbon steel; Grade I has corrosion resistance about four times that of carbon steel; Grade III for enhanced corrosion resistance may have copper specified
A709 Carbon; high-strength low-alloy; and quenched and tempered alloy	Structural shapes, plates, and bars in Grades 36, 50, and 50W for use in bridges; plates in Grades 100 and 100W for use in bridges; when supplementary requirements are used, requirements of A36, A572, A588, and A514 are exceeded; Grades 50W and 100W are weathering steels
A852 High-strength, low-alloy; and quenched and tempered alloy	Plates to 4 in. thick for welded and bolted construction where atmospheric corrosion resistance is desired
A913 High-strength low-alloy steel	Structural shapes for bolted and welded construction

*All steels listed are approved under the AISC Specifications [1.5, 1.16] except A611 and A709.

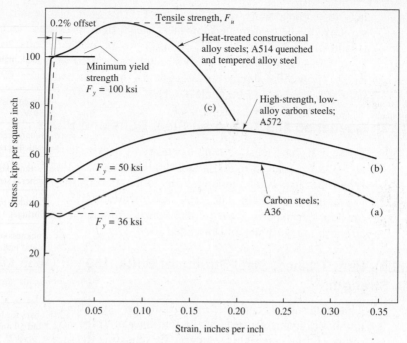

Figure 2.1.1 Typical stress-strain curves.

2.2 FASTENER STEELS

The detailed treatment of the design of threaded fasteners appears in Chapter 4. A brief description of the materials used for bolts appears in the following paragraphs. The headings are the ASTM specification exact titles.

A307 [2.4], Carbon Steel Bolts and Studs, 60,000 psi Tensile Strength

This material is used for what are commonly referred to as "machine bolts." These are usually used only for temporary installations. Included are Grade A bolts for general applications, which have a *minimum* tensile strength of 60 ksi (415 MPa); and Grade B bolts for flanged joints in piping systems where one or both flanges are cast iron. The Grade B bolts have a *maximum* tensile strength limitation of 100 ksi (700 MPa). No well-defined yield point is exhibited by these bolts, and no minimum yield strength (for instance, 0.2% offset strength) is specified.

A325 [2.5], High-Strength Bolts for Structural Steel Joints

This quenched and tempered medium carbon steel is used for bolts commonly known as "high-strength structural bolts," or high-strength bolts. This material has maximum carbon of 0.30%. It is heat-treated by quenching and then by reheating (tempering) to a temperature of at least 800°F. This steel behaves in a tension test more similarly to the heat-treated low-alloy steels than to carbon steel. It has an ultimate tensile strength of 105 ksi (733 MPa) ($1\frac{1}{8}$ to $1\frac{1}{2}$-in.-diam bolts) to 120 ksi (838 MPa) ($\frac{1}{2}$ to 1-in.-diam bolts). Its yield strength, measured at 0.2% offset, is prescribed at 81 ksi (566 MPa) minimum for $1\frac{1}{8}$ to $1\frac{1}{2}$-in.-diam bolts, and 92 ksi (643 MPa) for bolts $\frac{1}{2}$ to 1 in. diam (see Table 4.1.1).

A449 [2.7], Quenched and Tempered Steel Bolts and Studs

These bolts have tensile strengths and yield stresses (strength at 0.2% offset) the same as A325 for bolts $1\frac{1}{2}$ in. diam and smaller; however, they have the regular (instead of heavy) hexagon head and longer thread length of A307 bolts. They are also available in diameters up to 3 in. The AISC Specifications [1.5, 1.16] permit use of A449 bolts only for certain structural joints requiring diameters exceeding $1\frac{1}{2}$ in. and for high-strength anchor bolts and threaded rods.

A490 [2.8], Heat Treated, Steel Structural Bolts, 150 ksi (1035 MPa) Tensile Strength

This material has carbon content that may range up to 0.53% for $1\frac{1}{2}$ in.-diam bolts, and has alloying elements in amounts similar to the A514 [2.11] steels. After quenching in oil, the material is tempered by reheating to at least 900°F. The minimum yield strength, obtained by 0.2% offset, ranges from 115 ksi (803 MPa) (over $2\frac{1}{2}$ in. to 4 in. diam) to 130 ksi (908 MPa) (for $2\frac{1}{2}$ in. diam and under).

Galvanized High-Strength Bolts

In order to provide corrosion protection, A325 bolts may be galvanized. Hot-dip galvanizing requires the molten zinc temperature to be in the range of the heat treatment temperature; thus, the mechanical properties obtained by heat treatment may be diminished. Whenever galvanized bolts are used, the nuts must be "oversized." If the nuts are also galvanized, they must be "double oversized."

Steels having tensile strength in the range of 200 ksi or higher are subject to hydrogen embrittlement when hydrogen is permitted to remain in the steel and high tensile stress is applied. The introduction of hydrogen occurs during the pickling operation of the galvanizing process and the subsequent "sealing-in" of the hydrogen and zinc coating [2.24]. The minimum tensile strength of A325 bolts is well below the critical 200 ksi range. On the other hand, A490 bolts have a maximum tensile strength of 170 ksi, a value considered too close to the critical range. Thus, *galvanizing of A490 bolts is not permitted*.

2.3 WELD ELECTRODE AND FILLER MATERIAL

The detailed treatment of welding and welded connections appears in Chapter 5. The electrodes used in shielded metal arc welding (SMAW) (see Sec. 5.2) also serve as the filler material and are covered by American Welding Society (AWS) A5.1 and A5.5 Specifications [2.25]. Such consumable electrodes are classified E60XX,

TABLE 2.3.1   ELECTRODES USED FOR WELDING*

	Process						
Shielded metal arc welding (SMAW) AWS A5.1 or A5.5	Submerged arc welding (SAW) AWS A5.17 or A5.23	Gas metal arc welding (GMAW) AWS A5.18 or A5.28	Flux cored arc welding (FCAW) AWS A5.20 or A5.29	Minimum yield stress		Minimum tensile strength	
				(ksi)	(MPa)	(ksi)	(MPa)
E60XX			E6XT-X	50	345	62 min	425
	F6XX-EXXX			50	345	62–80	425–550
E70XX		ER70S-X	E7XT-X	60	415	72 min	495
	F7XX-EXXX			60	415	70–90	485–655
E80XX				67	460	80 min	550
	F8XX-EXXX		E8XT	68	470	80–100	550–690
		ER80S		65	450	80 min	550
E100XX				87	600	100 min	690
	F10XX-EXXX			88	605	100–130	690–895
		ER100S		90	620	100 min	690
			E10XT	88	605	100–120	690–830
E110X				97	670	110 min	760
	F11XX-EXXX			98	675	110–130	760–895
		ER110S		98	675	110 min	760
			E11XT	98	675	110–125	760–860

* Filler metal requirement given by AWS D1.1 [2.25], Table 4.1.1 to match the various structural steels.

E70XX, E80XX, E90XX, E100XX, and E110XX. The "E" denotes electrode. The first two digits indicate the tensile strength in ksi; thus the tensile strength ranges from 60 to 110 ksi (414 to 760 MPa). The "X's" represent numbers indicating the usage of the electrode.

For submerged arc welding (SAW) (see Sec. 5.2), the electrodes which also serve as filler material are specified under AWS A5.17 and A5.23. The weld-electrode combinations are designated F6XX-EXXX, F7XX-EXXX, etc. up to F11XX-EXXX. The "F" designates a granular flux material that shields the weld as it is made. The first one or two of the three digits following the "F" indicate the tensile strength (6 means 60 ksi, 11 means 110 ksi). The "E" stands for electrode and the other X's represent numbers relating to use. The yield stresses and tensile strengths of commonly used electrodes are given in Table 2.3.1.

2.4   STRESS-STRAIN BEHAVIOR (TENSION TEST) AT ATMOSPHERIC TEMPERATURES

Typical stress-strain curves for tension are shown in Fig 2.1.1 for the three categories of steel already discussed: carbon, high-strength low-alloy, and heat-treated high-strength low-alloy. The same behavior occurs in compression when support is provided so as to preclude buckling. The portion of each of the stress-strain curves of Fig. 2.1.1 that is utilized in ordinary design is shown enlarged in Fig. 2.4.1.

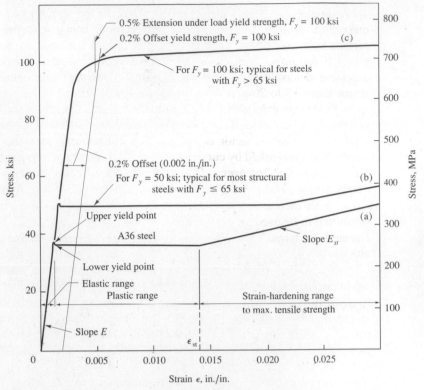

0.5% Extension under load yield strength, $F_y = 100$ ksi
0.2% Offset yield strength, $F_y = 100$ ksi
(c)
For $F_y = 100$ ksi; typical for steels with $F_y > 65$ ksi
0.2% Offset (0.002 in./in.)
For $F_y = 50$ ksi; typical for most structural steels with $F_y \leq 65$ ksi
(b)
(a)
Upper yield point
A36 steel
Slope E_{st}
Lower yield point
Elastic range
Plastic range
Strain-hardening range to max. tensile strength
Slope E
ϵ_{st}

Stress, ksi
Stress, MPa
Strain ϵ, in./in.

Figure 2.4.1   Enlarged typical stress-strain curves for different yield stresses.

The stress-strain curves of Fig. 2.1.1 are determined using a unit stress obtained by dividing the load by the original cross-sectional area of the specimen, and the strain (inches per inch) is obtained as the elongation divided by the original length. Such curves are known as *engineering stress-strain* curves and rise to a maximum stress level (known as the tensile strength) and then fall off with increasing strain until they terminate as the specimen breaks. Insofar as the material itself is concerned, the unit stress continues to rise until failure occurs. The so-called *true-stress/true-strain* curve is obtained by using the actual cross-section even after necking down begins and using the instantaneous incremental strain.

Stress-strain curves (as per Fig. 2.4.1) show a straight line relationship up to a point known as the *proportional limit,* which essentially coincides with the yield point for most structural steels with yield points not exceeding 65 ksi (450 MPa). For the quenched and tempered low-alloy steels, the deviation from a straight line occurs gradually, as in curve (c), Fig. 2.4.1. Since the term yield point is not appropriate to curve (c), *yield strength* is used for the stress at an offset strain of 0.2%; or alternatively, a 0.5% extension under load, as shown in Fig 2.4.1. *Yield stress* is the general term to include the unit stress at a yield point, when one exists, or the yield strength.

The ratio of stress to strain in the initial straight line region is known as the modulus of elasticity, or Young's modulus, E, which for structural steels may be taken approximately as 29,000 ksi (200,000 MPa). In the straight-line region, loading and unloading results in no permanent deformation; hence it is the *elastic range*. The service load unit stress in steel design is always intended to be safely below the proportional limit, even though in order to ascertain safety factors against failure or excessive deformation, knowledge is required of the stress-strain behavior up to a strain about 15 to 20 times the maximum elastic strain.

For steels exhibiting yield points, as curves (a) and (b) of Fig. 2.4.1, the long plateau for which essentially constant stress exists is known as the *plastic range*. The load and resistance factor design method consciously uses this range. The higher strength steels typified by curve (c) of Fig. 2.4.1 also have a region that might be called the plastic range; however, in this zone the stress is continuously increasing (instead of remaining constant) as strain increases. For lack of having a region of constant stress with increasing strain, the steels whose yield stress exceeds 65 ksi are not permitted to be used for plastic analysis (LRFD-A5.1 and ASD-N2). Plastic analysis (treated in Chaps. 7 and 10) relies on the ability of steel to deform (strain) at constant stress.

For strains greater than 15 to 20 times the maximum elastic strain, the stress again increases but with a much flatter slope than the original elastic slope. This increase in strength is called *strain hardening,* which continues up to the tensile strength. The slope of the stress-strain curve is known as the strain-hardening modulus, E_{st}. Average values for this modulus and the strain ϵ_{st} at which it begins have been determined [2.26] for two steels: A36 steel, $E_{st} = 900$ ksi (6200 MPa) at $\epsilon_{st} = 0.014$ in. per in.; and for A441, $E_{st} = 700$ ksi (4800 MPa) at $\epsilon_{st} = 0.021$ in. per in. The strain-hardening range is not consciously used in design, but certain of the buckling limitations are conservatively derived to preclude buckling even at strains well beyond onset of strain hardening.

The stress-strain curve also indicates the *ductility*. Ductility is defined as the amount of permanent strain (i.e., strain exceeding proportional limit) up to the point of fracture. Measurement of ductility is obtained from the tension test by determining the percent elongation (comparing final and original lengths over a specified gage distance) of the specimen. Ductility is important because it permits yielding locally due to high stresses and thus allows the stress distribution to change. Design procedures based on inelastic behavior require large inherent ductility, particularly for treatment of stresses near holes or abrupt change in member shape, as well as for design of connections.

2.5 MATERIAL TOUGHNESS

The use of steels having higher strength than A36 without heat treatment has resulted in problems relating to lack of ductility and material fracture [2.28]; at least the use of such steels requires the structural engineer to be more conscious of material behavior.

In structural steel design, toughness is a measure of the ability of steel to resist fracture; i.e., to absorb energy. According to Rolfe [2.27], material toughness is defined as "the resistance to unstable crack propagation in the presence of a notch."

Unstable crack propagation produces brittle fracture, as opposed to stable crack growth of a subcritical crack from fatigue.

For uniaxial tension, toughness can be expressed as the total area under the stress-strain curve out to the fracture point where the diagram terminates. Since uniaxial tension rarely exists in real structures, a more useful index of toughness is based on the more complex stress condition at the root of a notch.

Notch toughness is the measure of the resistance of a metal to the start and propagation of a crack at the base of a standard notch, commonly using the Charpy V-notch test. This test uses a small rectangular simply supported beam having a V-notch at midlength. The bar is fractured by a blow from a swinging pendulum. The amount of energy absorbed is calculated from the height the pendulum raises after breaking the specimen. The amount of energy absorbed will increase with increasing temperature at which the test is conducted.

Though the Charpy V-notch test has been a common means of determining notch toughness, other fracture criteria and more recently fracture mechanics have been used [2.27]. Barsom and Rolfe [2.29] and Barsom [2.30, 2.31] have excellently presented the important factors relating to fracture of steel.

Figure 2.5.1 shows the typical relationship between temperature and toughness, and also shows the transition from ductile to brittle behavior, such as one may obtain from the Charpy V-notch test. The temperature at the point where the slope is steepest (point A of Fig. 2.5.1) is the transition temperature. Since brittleness and ductility are qualitative terms, the various structural steels have different requirements for ductility at various temperatures depending on their service environment (loading, temperature, stress and strain levels, loading rate, and number of load repetitions).

For example, a moderate amount of ductility may be required for ordinary structures where very low temperatures are not expected; in such cases, 15 ft-lb has

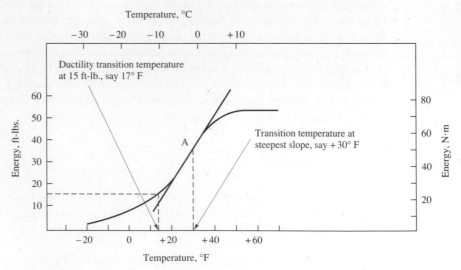

Figure 2.5.1   Transition temperature curve for carbon steel obtained from Charpy V-notch impact tests. (Adapted from Ref. 2.26)

commonly been the energy absorption required. The corresponding temperature obtained from the test results shown in Fig. 2.5.1 would be about 17°F. The temperature at which marked decrease in slope begins to occur is known as the *ductility transition temperature.* This would indicate that the material may be expected to be brittle when service temperatures are below 17°F.

2.6 YIELD STRENGTH FOR MULTIAXIAL STATES OF STRESS

Only when the load-carrying member is subject to uniaxial tensile stress can the properties from the tension test be expected to be identical with those of the structural member. It is easy to forget that yielding in a real structure is usually *not* the well defined behavior observed in the tension test. Yielding is commonly assumed to be achieved when any one component of stress reaches the uniaxial value F_y.

For all states of stress other than uniaxial, a definition of yielding is needed. These definitions, and there are frequently several for a given state of stress, are called *yield conditions* (or *theories of failure*) and are equations of interaction between the stresses acting.

Energy-of-Distortion (Huber–von Mises–Hencky) Yield Criterion

This most commonly accepted theory gives the uniaxial yield stress in terms of the three principal stresses. The yield criterion* may be stated

$$\sigma_y^2 = \tfrac{1}{2}[(\sigma_1 - \sigma_2)^2 + (\sigma_2 - \sigma_3)^2 + (\sigma_3 - \sigma_1)^2] \tag{2.6.1}$$

where σ_1, σ_2, σ_3 are the tensile or compressive stresses that act in the three principal directions; i.e., the stresses that act in the three mutually perpendicular planes of zero shear, and σ_y is the "yield stress" that may be compared with the uniaxial value F_y.

For most structural design situations, one of the principal stresses is either zero or small enough to be neglected; hence Eq. 2.6.1 reduces to the following for the case of the plane stress (all stresses considered are acting in a plane)

$$\sigma_y^2 = \sigma_1^2 + \sigma_2^2 - \sigma_1\sigma_2 \tag{2.6.2}$$

When stresses on thin plates are involved, the principal stress acting transverse to the plane of the plate is usually zero (at least to first-order approximation). Flexural stresses on beams assume zero principal stress perpendicular to the plane of bending. Furthermore, structural shapes (Fig. 1.5.1) are comprised of thin plate elements, so that each is subject to Eq. 2.6.2. The plane stress yield criterion, Eq. 2.6.2, is the one used throughout the remaining chapters where needed, and is illustrated in Fig. 2.6.1.

---

*See Arthur P. Boresi, Richard J. Schmidt, and Omar M. Sidebottom, *Advanced Mechanics of Materials,* 5th ed. New York: John Wiley & Sons, Inc., 1993, pp. 133–134.

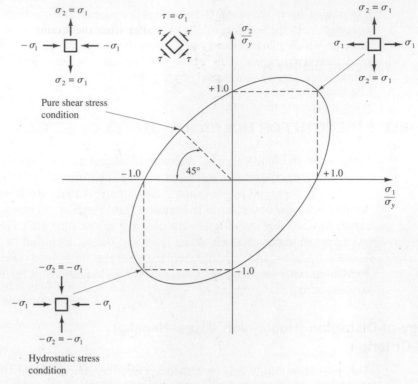

Figure 2.6.1 Energy-of-distortion yield criterion for plane stress.

Shear Yield Stress

The yield point for pure shear can be determined from a stress-strain curve with shear loading, or if the multiaxial yield criterion is known, that relationship can be used. Pure shear occurs on 45° planes to the principal planes when $\sigma_2 = -\sigma_1$, and the shear stress $\tau = \sigma_1$. Substitution of $\sigma_2 = -\sigma_1$ into Eq. 2.6.2 gives

$$\sigma_y^2 = \sigma_1^2 + \sigma_1^2 - \sigma_1(-\sigma_1) = 3\sigma_1^2 \tag{2.6.3}$$

$$\sigma_1 = \tau = \sigma_y/\sqrt{3} = \text{shear yield} \tag{2.6.4}$$

which indicates that the yield condition for shear stress acting alone is

$$\tau_y = \sigma_y/\sqrt{3} = 0.58\sigma_y \tag{2.6.5}$$

Poisson's Ratio, μ

When stress is applied in one direction, strains are induced not only in the direction of applied stress but also in the other two mutually perpendicular directions. The usual value of μ used is that obtained from the uniaxial stress condition, where it is the ratio of the transverse strain to longitudinal strain under load. For structural steels, Poisson's ratio is approximately 0.3 in the elastic range where the material is compressible and approaches 0.5 when in the plastic range where the material is essentially incompressible (i.e., constant resistance no matter what the strain).

Shear Modulus of Elasticity

Loading in pure shear produces a stress-strain curve with a straight line portion whose slope represents the shear modulus of elasticity. If Poisson's ratio μ and the tension-compression modulus of elasticity E are known, the shear modulus G is defined by the theory of elasticity as

$$G = \frac{E}{2(1 + \mu)} \tag{2.6.6}$$

which for structural steel is just over 11,000 ksi (75,800 MPa).

2.7 HIGH TEMPERATURE BEHAVIOR

The design of structures to serve under atmospheric temperature rarely involves concern about high temperature behavior. Knowledge of such behavior is desirable when specifying welding procedures, and is necessary when concerned with the effects of fire.

When temperatures exceed about 200°F (93°C), the stress-strain curve begins to become nonlinear, gradually eliminating the well-defined yield point. The modulus of elasticity, yield strength, and tensile strength all reduce as temperature increases. The range from 800 to 1000°F (430 to 540°C) is where the rate of decrease is maximum. While each steel, because of its different chemistry and microstructure, behaves somewhat differently, the general relationships are shown in Fig. 2.7.1. Steels having relatively high percentages of carbon, such as A36, exhibit "strain aging" in the range 300 to 700°F (150 to 370°C). This is evidenced by a relative rise in yield strength and tensile strength is that range. Tensile strength may rise to about 10% above that at room temperature and yield strength may recover to about its room temperature value when the temperature reaches 500 to 600°F (260 to 320°C). Strain aging results in decreased ductility.

The modulus of elasticity decrease is moderate up to 1000°F (540°C); thereafter it decreases rapidly. More importantly, at temperatures above about 500 to 600°F (260 to 320°C), steels exhibit deformation which increases with increasing time under load, a phenomenon known as *creep*. Creep is well known in concrete structures; and its effect in steel, which does not occur at atmospheric temperatures, increases with increasing temperature.

Other high temperature effects are (a) improved notch impact resistance up to about 150 to 200°F (65 to 95°C), as discussed in Sec. 2.5; (b) increased brittleness due to metallurgical changes, such as carbide precipitation discussed in Sec. 2.1, begins to occur at about 950°F (510°C); and (c) corrosion resistance of structural steels increases for temperatures up to about 1000°F (540°C). Most steels are used in applications below 1000°F, and some heat treated steels should be kept below about 800°F (430°C).

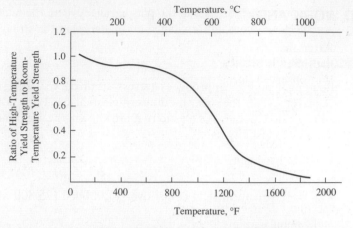

(a) Average Effect of Temperature on Yield Strength

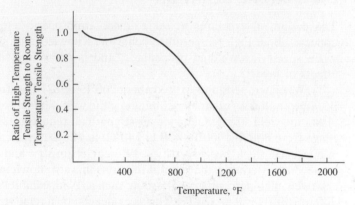

(b) Average Effect of Temperature on Tensile Strength

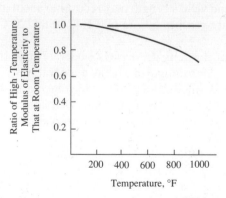

(c) Typical Effect of Temperature on Modulus of Elasticity

Figure 2.7.1   Typical effects of high temperature on stress-strain curve properties of structural steels. (Adapted from Ref. 2.26)

2.8 COLD WORK AND STRAIN HARDENING

After the strain $\epsilon_y = F_y/E_s$ at first yield has been exceeded appreciably, and the specimen is unloaded, reloading may give a stress-strain relationship differing from that observed during the initial loading. Elastic loading and unloading results in no residual strain; however, initial loading beyond the yield point, such as to point *A* of Fig. 2.8.1, results in unloading to a strain at point *B*. A permanent set *OB* has occurred. The ductility capacity has been reduced from a strain *OF* to the strain *BF*. Reloading exhibits behavior as if the stress-strain origin were at point *B*; the plastic zone prior to strain hardening is also reduced.

   When loading has occurred until point *C* is reached, unloading follows the dashed line to point *D*; i.e., the origin for a new loading is now point *D*. The length of the line *CD* is greater, indicating that the yield point has increased. The increased yield point is referred to as a strain hardening effect; the ductility remaining when loading from point *D* is severely reduced from its original value prior to the initial loading. The process of loading beyond the elastic range to cause a change in available ductility, when done at atmospheric temperature, is known as *cold work*. Since real structures are not loaded in uniaxial tension-compression, the cold work effect is much more complex, and any theoretical study of it is outside the scope of the text.

   When structural shapes are made by cold-forming from plates at atmospheric temperature, inelastic deformations occur at the bends. Cold working into the strain hardening range at the bend locations increases the yield strength, which design

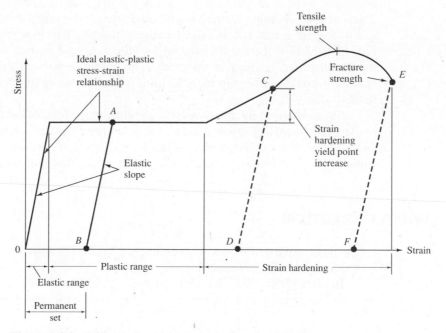

Figure 2.8.1   Effects of straining beyond the elastic range.

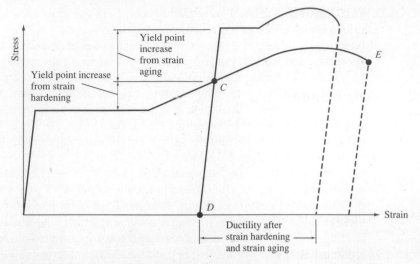

Figure 2.8.2   Effect of strain aging after straining into strain-hardening range and unloading.

specifications may permit taking into account. The *Specification for the Design of Cold-Formed Steel Structural Members** [1.11] has such provisions.

Upon unloading and after a period of time, the steel will have acquired different properties from those represented by points *D, C,* and *E* of Fig. 2.8.1 by a phenomenon known as *strain aging*. Strain aging, as shown in Fig. 2.8.2, produces an additional increase in yield point, restores a plastic zone of constant stress, and gives a new strain hardening zone at an elevated stress. The original shape of the stress-strain diagram is restored, but the ductility is reduced. The new stress-strain diagram may be used as if it were the original for analyzing cold-formed sections, as long as the ductility that remains is sufficient. The corner regions of cold-formed shapes generally would not require high ductility for rotational strain about the axis of the bend.

Stress relieving by annealing will eliminate the effects of cold work should that be desired. Annealing involves heating to a temperature above transformation range and allowing slow cooling; a recrystallization occurs to restore the original properties. Bittence [2.32] provides an excellent summary of the basics of heat treating.

2.9 BRITTLE FRACTURE

As has been discussed in several sections, steel that is ordinarily ductile can become brittle under various conditions.

Barsom [2.30, 2.31], Barsom and Rolfe [2.29], and Rolfe [2.27] have provided an excellent summary of fracture and fatigue control for structural engineers. Rolfe

---

*Referred to henceforth as the AISI Specification.

[2.27] defines *brittle fracture* as "a type of catastrophic failure that occurs without prior plastic deformation and at extremely high speeds." Fracture behavior is affected by temperature, loading rate, stress level, flaw size, plate thickness or constraint, joint geometry, and workmanship.

Effect of Temperature

Notch toughness, as determined by the Charpy-impact energy vs temperature curves (see Sec. 2.5), is an indication of the susceptibility to brittle fracture. Temperature is a vital factor in several ways: (a) the value below which notch toughness is inadequate; (b) the 600 to 800°F (320 to 430°C) range causes formation of brittle microstructure; and (c) over 1000°F (540°C) causes precipitation of carbides of alloying elements to give more brittle microstructure. The other temperature factors have already been discussed in earlier sections.

Effect of Multiaxial Stress

The complex stress condition found in usual structures, particularly at joints, is another major factor affecting brittleness. The *Primer on Brittle Fracture* [2.33] has provided an excellent rational presentation of this and forms the basis for what follows. The engineering stress-strain curve is for uniaxial stress; prior to fracture a necking down occurs, as shown in Fig. 2.9.1a. If biaxial lateral loading as shown in Fig. 2.9.1b could be applied, plastic necking down could be suppressed to the point where the bar would break in a brittle manner without elongation and without reduction in area. The fracture stress based on the unreduced cross-sectional area would be the same high value as that based on the necked-down cross section in the uniaxial tension case. The unit stress would be far above the nominal maximum tensile strength

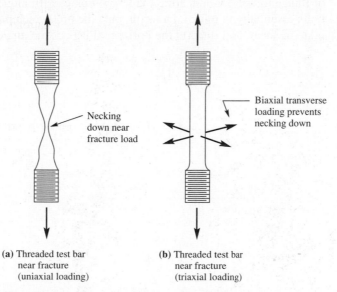

(a) Threaded test bar
near fracture
(uniaxial loading)

(b) Threaded test bar
near fracture
(triaxial loading)

Figure 2.9.1   Uniaxial and triaxial loading.

of the engineering stress-strain curve, which is always computed on the basis of original cross section.

Also the effects of notches have been alluded to in the discussion of notch toughness in Sec. 2.5. The notch serves somewhat the same purpose as the theoretical triaxial loading of Fig. 2.9.1b, in that it restrains plastic flow which otherwise would occur and thus at some higher stress may likely fail in a brittle manner. Figure 2.9.2 shows the effect of a notch in a tensile test specimen. The cross-sectional area at the base of the notch corresponds to the area of the original specimen of Fig. 2.9.1b. The reduced section tries to become narrower as the axial tension increases, but is resisted by the diagonal pull that develops in the corners, as shown in Fig. 2.9.2. The test bar will fail at high stress by brittle fracture.

Notches can occur in real structures by use of unfilleted corners in design or from improperly made welds that may crack. Such occurrences can lead to brittleness. Notches and cracked welds can, however, be minimized by good design and welding procedures.

Unusual configurations and changes in section should be made gradually so the stress flow lines are not required to make abrupt changes. Whenever the complexity is such as to give rise to three-dimensional stresses, the tendency for brittleness increases. Castings, for instance, have the reputation for brittleness. Primarily this is because of the built-in three-dimensional continuity.

Multiaxial Stress Induced by Welding

In general, welding creates a built-in restraint that gives rise to biaxial and triaxial stress and strain conditions, which result in brittle behavior. To illustrate, consider the loaded simply supported beam of Fig. 2.9.3, which in turn supports a plate in tension. Due to flexure, the bottom flange of the beam is in tension; therefore, the stress at point *A* is uniaxial tension (neglecting the small effects of beam width and attachment of flange to web). Connecting the tension plate with angles and bolts puts the flange bolts essentially in uniaxial tension, puts the bolt which passes through the suspender plate in shear, and distorts the horizontal legs of the angles in bending; so that there

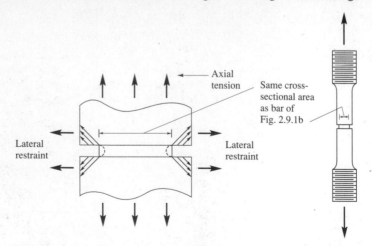

Figure 2.9.2   Effect of notch on uniaxial tension test.

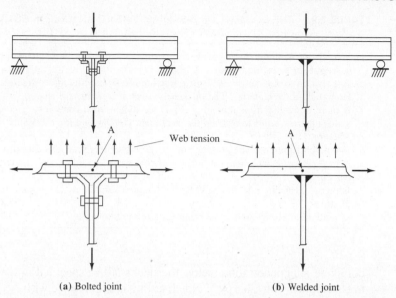

(a) Bolted joint   (b) Welded joint

Figure 2.9.3   Comparison of stress conditions in bolted and welded joints.

is no appreciable effect on the stress at point *A*. In other words, the stress conditions in the connection of Fig. 2.9.3a are approximately uniaxial in nature.

Next, consider the tensile suspender plate welded to the tension flange of the beam, as in Fig. 2.9.3b. The stress at point *A* is now biaxial because of the direct attachment to the flange at that point. The weld region, therefore, is subject to triaxial stress; biaxial from the directly applied loads, plus the resistance to deformation along the axis of the welds resulting from continuous attachment (Poisson's ratio effect). The design of welded joints should consider the possibilities of brittleness due to three-dimensional stressing. The subject of *lamellar tearing* is treated in Sec. 2.10.

Effect of Thickness

As discussed in Sec. 2.6, thin plates may usually be assumed to be in a state of plane stress in which the three dimensional stress effects may be ignored. This is not generally the case for thick plate elements for which three-dimensional stress contributes to brittleness. Brittleness in thick plates also increases due to the manufacturing process. The slower cooling rate produces a coarser microstructure, and a higher carbon content is required to achieve the strength otherwise obtained by hot working in thin plates.

The very thick rolled W-shapes (ASTM A6/A6M [1.8], Groups 4 and 5); i.e., the so-called "jumbo shapes," exhibit low fracture toughness at the core of the thick flange to web junction and the center of the web adjacent to it [2.34]. This low fracture toughness may cause brittle failures when these heavy W-shapes are used as tension members. For this reason their use is intended only for compression members [2.35].

When ASTM A6/A6M, Groups 4 and 5 rolled shapes are to be "used as members subject to primary tensile stresses due to tension or flexure, toughness need not be specified if splices are made by bolting. If such members are spliced using

TABLE 2.9.1   THE ELEMENT OF RISK: FACTORS TO ANALYZE IN ESTIMATING
SERIOUSNESS OF BRITTLE FRACTURE (ADAPTED FROM REF. 2.33)

1. What is the minimum anticipated service temperature? The lower the temperature, the greater the susceptibility to brittle fracture.
2. Are tension stresses involved? Brittle fracture can occur only under condition of tensile stress.
3. How thick is the material? The thicker the steel, the greater the susceptibility to brittle fracture.
4. Is there three-dimensional continuity? Three-dimensional continuity, giving rise to multiaxial states of stress, tends to restrain the steel from yielding and increases susceptibility to brittle fracture.
5. Are notches present? The presence of sharp notches increases susceptibility to brittle fracture.
6. Is loading applied at a high rate? The higher the rate of loading, the greater susceptibility to brittle fracture.
7. Is there a changing rate of stress? Brittle fracture occurs only under conditions of increasing rate of stress.
8. Is welding involved? Weld cracks can act as severe notches.

complete joint penetration welds, the steel shall be specified in the contract documents to be supplied with Charpy V-Notch testing in accordance with ASTM A6, Supplementary Requirement S5." (LRFD-A3.1c) The Bethlehem Steel Technical Bulletin, *Use of Heavy Structural Shapes in Tension Applications* [2.36] provides additional guidance. Even though jumbo sections were not originally intended for tension applications, designers use them in such situations.

Effect of Dynamic Loading

The stress-strain properties referred to so far have been for static loading slowly applied. More rapid loading, such as that of forge drop-hammers, earthquake, or nuclear blast changes the stress-strain properties. Ordinarily, the increased strain rate from dynamic loading increases the yield point, tensile strength, and ductility. At about 600°F (320°C) there will be a small decrease in strength. Some increased brittleness has been noted with high strain rate, but it seems principally associated with other factors already discussed, such as notches where stress concentrations exist and the temperature effect on toughness. The more important factor relating to dynamic load application is not that a rapidly increasing strain rate occurs, but that it is combined with a rapidly *decreasing* strain rate. The effect of stress *variation* is discussed in the section on fatigue.

Table 2.9.1, from Ref. 2.32, provides a list of factors "to help determine whether or not the risk of brittle fracture is serious and requires special design considerations."

2.10 LAMELLAR TEARING

Lamellar tearing is a form of brittle fracture occurring "in planes essentially parallel to the rolled surface of a plate under high through thickness loading." [2.37]. Because strains resulting from service loads are well below ϵ_y, normal loads are not believed to initiate or propagate lamellar tears, and something else must be responsible. In a highly restrained welded joint "thru-thickness" strains ϵ are induced by weld metal shrinkage. The localized strains resulting from weld metal shrinkage, which can be several times larger than yield strain ϵ_y, are the source of the problem.

The subject of lamellar tearing has received considerable attention since the early 1970s, resulting in a tendency for structural engineers to blame lamellar tearing for many brittle fractures. The AISC has provided an excellent summary of the

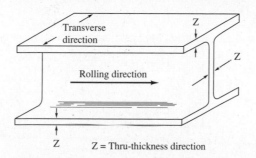

Figure 2.10.1   Definition of direction terminology. (From Ref. 2.38)

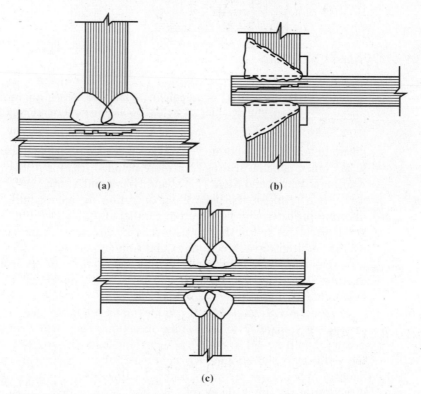

Figure 2.10.2   Joints showing typical lamellar tears resulting from shrinkage of large welds in thick material under high restraint. (From Ref. 2.38)

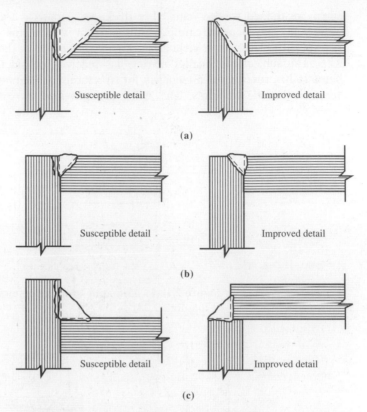

Figure 2.10.3   Susceptibility to lamellar tearing can be reduced by careful detailing of welded connections. (From Ref. 2.38)

phenomenon [2.38]. Thornton [2.39] has provided design and supervision procedures to minimize lamellar tearing. For more detailed treatment, the reader is referred to Kaufman, Pense, and Stout [2.37] and Holby and Smith [2.40].

As a result of the hot rolling operation in manufacture, steel sections have different properties in the direction parallel to rolling (see Fig. 2.10.1), in the transverse direction, and in the "thru-thickness" direction. In the elastic range, both the rolling and transverse directions exhibit similar behavior, with the elastic limit for the transverse direction being only slightly below that for the rolling direction. The ductility (strain capability), however, in the "thru-thickness" direction may be well below that for the rolling direction.

Generally, I-shaped steel sections are adequately ductile when loaded either parallel or transverse to the rolling direction. They will deform locally to strains greater than the yield strain (F_y/E_s), carrying load with some of the material acting at the yield stress and bringing adjacent material into participation if added strength is needed. When, however, the strain is localized for instance in the "thru-thickness" direction at one thick flange of a section, a restrained situation exists because the strain cannot redistribute from the flange through the web to the opposite flange. The large localized "thru-thickness" strain may exceed the yield point strain, causing decohesion and leading to a lamellar tear.

Figure 2.10.2 shows conditions that promote lamellar tearing in welded joints. Internal joint restraint that inhibits large strains ϵ resulting from weld shrinkage can potentially cause lamellar tearing. Figure 2.10.3 indicates weld shrinkage in the "thru-thickness" direction, increasing susceptibility to lamellar tearing. The weld detail should be made such that weld shrinkage occurs in the rolling direction so that the shrinkage pulls on the fibers longitudinally in their strongest orientation. References 2.38 and 2.41 suggest ways of avoiding the problem.

2.11 FATIGUE STRENGTH

Repeated loading and unloading, primarily in *tension,* may eventually result in failure even if the yield stress is never exceeded. The term *fatigue* means failure under cyclic loading. It is a progressive failure, the final stage of which is unstable crack propagation. The fatigue strength is governed by three variables: (1) the number of cycles of loading, (2) the *range* of service load stress (the difference between the maximum and minimum stress), and (3) the initial size of a flaw. A flaw is a discontinuity, such as an extremely small crack.

In welded assemblies, a flaw could be the "notch" intersection of two elements or a "discontinuity" such as a bolt hole. Flaws may be the result of poorly made welds, rough edges resulting from shearing, punching, or flame cutting, or small holes. Such flaws may be of no concern; however, under many cycles of loading the flaw (notch effect) may give rise to a crack that increases in length with each cycle of load and reduces the section carrying the load, consequently increasing the stress intensity on the uncracked part. The fatigue strength is more dependent on the localized state of stress than is the static strength. Fatigue is always a service load consideration; the actual service load state of stress is what determines crack propagation.

The grade of steel has no apparent affect on the number of cycles to failure, and the effect of minimum stress (attributable to dead load) is considered to be negligible for design purposes. On the other hand, the specimen geometry, including the surface condition and internal soundness of the weld, have a significant effect. These factors are reflected in the *Structural Welding Code* [2.25] rules for welded structure design.

Work by Zuraski and Johnson [2.42] evaluating the remaining life in steel bridges has shown that under certain conditions repeated stressing in steel sections can actually increase their fatigue life. The phenomenon, known as *coaxing,* was first studied by Sinclair [2.43] and results from repeatedly stressing near, but below, the fatigue limit and gradually increasing the stress.

The AISC Specifications [1.5, 1.16] in LRFD-App. K3 and ASD-App. K4 prescribe no fatigue effect for fewer than 20,000 cycles, which is approximately two applications a day for 25 years. Since most loadings in buildings are in that category, fatigue is generally not considered. The exceptions are crane-runway girders and structures supporting machinery. Fatigue is always considered in the design of highway bridges, which are expected to have in excess of 100,000 cycles of loading.

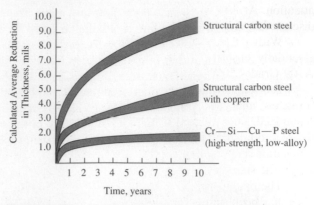

Figure 2.12.1 Comparative corrosion of steels in an industrial atmosphere. Shaded areas indicate range for individual specimens. (Adapted from Ref. 2.45)

Volume 1 of the *Welding Handbook* [2.23, p. 402] shows several good examples of the fatigue relationships for welded plate girders and cover-plated beams. Extended discussion of fatigue is given by Barsom and Rolfe [2.29].

2.12 CORROSION RESISTANCE AND WEATHERING STEELS

Since the earliest uses of steel, one of the important drawbacks was that painting was required to prevent the deterioration of the metal by corrosion (rusting). The lower-strength carbon steels were inexpensive but very vulnerable to corrosion. Corrosion resistance may be improved by the addition of copper as an alloy element. However, copper-bearing carbon steel is too expensive for general use.

High-strength low-alloy steels have several times [2.45] the corrosion resistance of structural carbon steel, with or without the addition of copper, as shown in Fig. 2.12.1. The high-strength low-alloy steels do not pit as severely as carbon steels and the rust that forms becomes a protective coating to prevent further deterioration. With certain alloy elements the high-strength low-alloy steel will develop an oxide protective coating that is pleasing in appearance and is described as follows*: "It is a very dense corrosion—actually a deeply colored brown, red, purple. . . . It has a texture and color which cannot be reproduced artificially—a character only nature can give, as with stone, marble, and granite." When steels are to be unpainted and left exposed they are called weathering steels.

As might be expected, the corrosion properties of any steel, including the weathering steels, are dependent on the chemical composition, the degree of pollution in the atmosphere, and the frequency of wetting and drying of the steel.

Since its first major use in 1958, for the Administrative Center for Deere & Company in Moline, Illinois, the use of weathering steel has received considerable

---

Architectural Record, August 1962.

attention. At first such steels were specified under ASTM A242, which as previously discussed is very general, allowing a wide variation in chemistry.

With the adoption of A588 steel in 1969, and A709 in 1975, A242 is now essentially obsolete. A588 is generally used for weathering steel in buildings and A709 Grades 50W and 100W for weathering steel in bridges (see Table 2.1.2).

Fabrication and erection of weathering steel requires care. Unsightly gouges, scratches, and dents should be avoided. Painting, even for identification, should be minimized, since all marks must be removed after the erection is completed. Scale and discoloration from welding also must be removed. The extra expense resulting from fabrication and erection is offset by the elimination of painting at intervals during the life of the structure.

The practice of using weathering steels, including the results of 30 years experience, has been summarized by Coburn [2.44], who presents the following "rules":

1. For optimum performance in the unpainted condition, the structure should be boldly exposed to the elements.
2. The development of the protective oxide film is best achieved under normal exposure, wherein the surfaces are wet at night by dew formation and dry during daylight hours.
3. Because this wet-dry cycle cannot occur when the steel, regardless of its grade, is buried in the soil or immersed in water, the protective oxide will not form and the performance will resemble that of mild carbon steel exposed to the same conditions.

SELECTED REFERENCES

2.1. ASTM. *Standard Specification for Structural Steel* (A36/A36M-94). Philadelphia, PA: American Society for Testing and Materials, 1994.

2.2. ASTM. *Standard Specification for Pipe, Steel, Black and Hot-Dipped, Zinc-Coated Welded and Seamless* (A53-93a). Philadelphia, PA: American Society for Testing and Materials, 1993.

2.3. ASTM. *Standard Specification for High-Strength Low-Alloy Structural Steel* (A242/A242M-93). Philadelphia, PA: American Society for Testing and Materials, 1993.

2.4. ASTM. *Standard Specification for Carbon Steel Bolts and Studs, 60 000 psi Tensile Strength* (A307-92a). Philadelphia, PA: American Society for Testing and Materials, 1992.

2.5. ASTM. *Standard Specification for Structural Bolts, Steel, Heat Treated, 120/105 ksi Minimum Tensile Strength* (A325/A325M-92a). Philadelphia, PA: American Society for Testing and Materials, 1992.

2.6. ASTM. *Standard Specification for High-Strength Low-Alloy Structural Manganese Vanadium Steel* (A441-85). Philadelphia, PA: American Society for Testing and Materials, 1985.

2.7. ASTM. *Standard Specification for Quenched and Tempered Steel Bolts and Studs* (A449-92b). Philadelphia, PA: American Society for Testing and Materials, 1992.

2.8.   ASTM. *Standard Specification for Heat-Treated Steel Structural Bolts, 150 ksi Minimum Tensile Strength* (A490-92): also *Specification for High-Strength Steel Bolts, Classes 10.9 and 10.9.3, for Structural Steel Joints [Metric]* (A490M-92). Philadelphia, PA: American Society for Testing and Materials, 1992.

2.9.   ASTM. *Standard Specification for Cold-Formed Welded and Seamless Carbon Steel Structural Tubing in Rounds and Shapes* (A500-93). Philadelphia, PA: American Society for Testing and Materials, 1993.

2.10. ASTM. *Standard Specification for Hot-Formed Welded and Seamless Carbon Steel Structural Tubing* (A501-93). Philadelphia, PA: American Society for Testing and Materials, 1993.

2.11. ASTM. *Standard Specification for High-Yield Strength, Quenched and Tempered Alloy Steel Plate, Suitable for Welding* (A514/A514M-93a). Philadelphia, PA: American Society for Testing and Materials, 1993.

2.12. ASTM. *Standard Specification for High-Strength Carbon-Manganese Steel of Structural Quality* (A529/A529M-94). Philadelphia, PA: American Society for Testing and Materials, 1994.

2.13. ASTM. *Standard Specification for Steel, Sheet and Strip, Carbon, Hot-Rolled, Structural Quality* (A570-88). Philadelphia, PA: American Society for Testing and Materials, 1988.

2.14. ASTM. *Standard Specification for High-Strength Low-Alloy Columbium-Vanadium Steels of Structural Quality* (A572/A572M-94b). Philadelphia, PA: American Society for Testing and Materials, 1994.

2.15. ASTM. *Standard Specification for High-Strength Low-Alloy Structural Steel with 50 ksi [345 MPa] Minimum Yield Point to 4 in. [100 mm] Thick* (A588/A588M-94). Philadelphia, PA: American Society for Testing and Materials, 1994.

2.16. ASTM. *Standard Specification for Steel, Sheet and Strip, High-Strength, Low-Alloy, Hot-Rolled and Cold-Rolled, with Improved Atmospheric Corrosion Resistance* (A606-91a). Philadelphia, PA: American Society for Testing and Materials, 1991.

2.17. ASTM. *Standard Specification for Steel, Sheet and Strip, High-Strength, Low-Alloy, Columbium or Vanadium, or Both, Hot-Rolled and Cold-Rolled* (A607-92a). Philadelphia, PA: American Society for Testing and Materials, 1992.

2.18. ASTM. *Specification for Steel, Sheet, Carbon, Cold-Rolled, Structural Quality* (A611-94). Philadelphia, PA: American Society for Testing and Materials, 1994.

2.19. ASTM. *Standard Specification for Hot-Formed Welded and Seamless High-Strength Low-Alloy Structural Tubing* (A618-92). Philadelphia, PA: American Society for Testing and Materials, 1992.

2.20. ASTM. *Standard Specification for Structural Steel for Bridges* (A709/A709M-94a). Philadelphia, PA: American Society for Testing and Materials, 1994.

2.21. ASTM. *Standard Specification for Steel Sheet and Strip, High-Strength, Low-Alloy, Hot-Rolled, and Steel Sheet, Cold-Rolled, High-Strength, Low-Alloy, with Improved Formability* (A715-91). Philadelphia, PA: American Society for Testing and Materials, 1991.

2.22. ASTM. *Standard Specification for Quenched and Tempered Low-Alloy Structural Steel Plate with 70 ksi [485 MPa] Minimum Yield Strength to 4 in. [100 mm] Thick* (A852/A852M-93). Philadelphia, PA: American Society for Testing and Materials, 1993.

2.23. AWS. *Welding Handbook,* 8th ed., Vol. 1, *Welding Technology,* 1987. Vol. 2, *Welding Processes,* 1991. Miami, FL: American Welding Society, 1987, 1991. (550 N. W. LeJeune Road, P.O. Box 351040, Miami, FL 33135).

2.24. Research Council on Structural Connections. *Commentary on Specifications for Structural Joints Using ASTM A325 or A490 Bolts.* Chicago: American Institute of Steel Construction, June 8, 1988.

2.25. AWS. *Structural Welding Code—Steel,* Thirteenth Edition, Effective December 30, 1993 (ANSI/AWS D1.1-94). Miami, FL: American Welding Society, 1994.

2.26. R. L. Brockenbrough and B. G. Johnston. *Steel Design Manual.* Pittsburgh, PA: United States Steel Corporation, 1968. (Chap. 1).

2.27. S. T. Rolfe. "Fracture and Fatigue Control in Steel Structures," *Engineering Journal,* AISC, **14,** 1 (1st Quarter 1977), 2–15.

2.28. K. A. Godfrey, Jr. "High Strength Steel: Crisis or No?", *Civil Engineering,* May 1985, 50–53.

2.29. John M. Barsom and Stanley T. Rolfe. *Fracture and Fatigue Control in Structures Applications of Fracture Mechanics,* 2nd ed. Englewood Cliffs, New Jersey: Prentice-Hall, Inc., 1987.

2.30. John M. Barsom. "Material Considerations in Structural Steel Design," *Proceedings, National Engineering Conference & Conference of Operating Personnel.* Chicago: American Institute of Steel Construction, April 29–May 2, 1987, 1-1 through 1-15.

2.31. J. M. Barsom. "Material Considerations in Structural Steel Design," *Engineering Journal,* AISC, **24,** 3 (3rd Quarter 1987), 127–139.

2.32. John C. Bittence. "The Basics of Heat Treating- What it Does- How it Works- Where to Specify it," *Machine Design,* January 24, 1974, 106–111; February 7, 1974, 117–121.

2.33. *A Primer on Brittle Fracture,* Booklet 1960-A, Steel Design File, Bethlehem Steel Corporation, Bethlehem, PA.

2.34. John W. Fishcr and Alan W. Pense, "Experience with Use of Heavy W Shapes in Tension," *Engineering Journal,* AISC, **24,** 2 (2nd Quarter 1987), 63–77.

2.35. "The Use of Jumbo Shapes in Non-column Applications," *Engineering Journal,* AISC, **23,** 3 (3rd Quarter 1986), 96.

2.36. "Use of Heavy Structural Shapes in Tension Applications" *Construction Marketing Technical Bulletin TB-312,* Bethlehem Steel Corporation, June 1992.

2.37. E. J. Kaufman, A. W. Pense, and R. D. Stout. "An Evaluation of Factors Significant to Lamellar Tearing, *Welding Journal,* **60,** March 1981, Research Supplement, 43s–49s.

2.38. "Commentary on Highly Restrained Welded Connections," *Engineering Journal,* AISC, **10,** 3 (3rd Quarter 1973), 61–73.

2.39. Charles H. Thornton. "Quality Control in Design and Supervision Can Eliminate Lamellar Tearing," *Engineering Journal,* AISC, **10,** 4 (4th Quarter 1973), 112–116.

2.40. E. Holby and J. F. Smith. "Lamellar Tearing- The Problem Nobody Seems to Want to Talk About," *Welding Journal,* **59,** February 1980, 37–44.

2.41. "Causes and Prevention of Lamellar Tearing," *Civil Engineering,* April, 1982, 74–75.

2.42. P. D. Zuraski and J. E. Johnson, "Research on the Remaining Life in Steel Bridges," *Proceedings, ASCE Specialty Conference on Probabilistic Mechanics and Structural Reliability.* Berkeley, CA: January 11–13, 1984, 414–418.

2.43. G. M. Sinclair. "An Investigation of the Coaxing Effect in Fatigue of Metals," *Proceedings,* ASTM, **52** (1952), 743–758.

2.44. Seymour Coburn. "Theory and Practice in Use of Weathering Steels." *Proceedings, National Engineering Conference & Conference of Operating Personnel.* Chicago: American Institute of Steel Construction, April 29- May 2, 1987, 14-1 through 14-25.

2.45. C. P. Larrabee. "Corrosion Resistance of High-Strength Low-Alloy Steels as Influenced by Composition and Environment," *Corrosion,* **9,** August 1953, 259–271.

2.46. ASTM. *Standard Specification for High-Strength Low-Alloy Steel Shapes of Structural Quality, Produced by Quenching and Self-Tempering Process (QST)* (A913/A913M-93). Philadelphia, PA: American Society for Testing and Materials, 1993.

Chapter 3

Tension Members

3.1 INTRODUCTION

Tension members are encountered in most steel structures. They occur as principal structural members in bridge and roof trusses, in truss structures such as transmission towers and wind bracing systems in multistoried buildings. They frequently appear as tie rods to stiffen a trussed floor system or to provide intermediate support for a wall girt system. Tension members may consist of a single structural shape or they may be built up from a number of structural shapes. The cross-sections of some typical tension members are shown in Fig. 3.1.1.

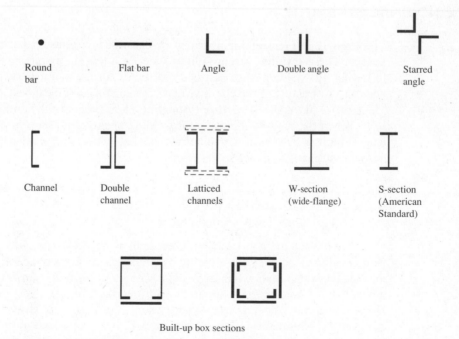

Figure 3.1.1 Cross-section of typical tension members.

Structural steel framework at intermediate floor level. This level suspended from roof space truss with tension rods. (Photo by C. G. Salmon)

3.2 NOMINAL STRENGTH

The strength of a tension member may be described in terms of the "limit states" that govern. The controlling strength limit state for a tension member will be one of the following: (a) yielding of the gross cross-section of the member away from the connection, (b) fracture of the effective net area (i.e., through the holes) at the connection [3.1], or (c) block shear fracture through the bolt holes at the connection.

When the limit state is general yielding of the gross section over the member length, typified by a tension member without holes (i.e., with welded connections), the nominal strength T_n may be expressed

$$T_n = F_y A_g \tag{3.2.1}$$

where F_y = yield stress
       A_g = gross cross-sectional area

For tension members having holes, such as for bolts, the reduced cross-section is referred to as the *net area*. Holes in a member cause stress concentrations at service load as, for example, shown in Fig. 3.2.1a. Theory of elasticity shows that tensile stress adjacent to a hole will be about three times the average stress on the net area. However, as each fiber reaches yield strain $\epsilon_y = F_y/E_s$, its stress then becomes a constant F_y, with deformation continuing with increasing load until finally all fibers have achieved or exceeded the strain ϵ_y (Fig. 3.2.1b).

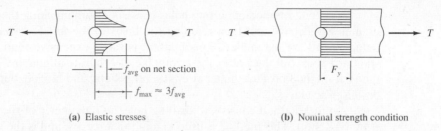

(a) Elastic stresses

(b) Nominal strength condition

Figure 3.2.1   Stress distribution with holes present.

When the limit state is a localized yielding resulting in a fracture through the *effective* net area of a tension member having holes, the nominal strength T_n may be expressed

$$T_n = F_u A_e \qquad\qquad (3.2.2)$$

where F_u = specified minimum tensile strength (see Fig. 2.1.1)
  A_e = effective net area = $U A_n$ (see Secs. 3.4 and 3.5)
  A_n = net area
  U = reduction coefficient (an efficiency factor)

Because of *strain hardening,* that is, the rise in resistance when the tensile strain becomes large (Sec. 2.4), the actual strength of a ductile member may exceed that indicated by Eq. 3.2.1 [3.1]. However, the large elongations resulting from yielding of the member along its entire length may cause the member ends to move unacceptably far apart and distress the structure; thus, the member no longer serves its intended purpose. Either unrestrained yielding or fracture through reduced section at holes may limit the structural usefulness of the member. Traditionally, a higher margin of safety has been used in design when considering the fracture limit state than for the yielding limit state.

3.3 NET AREA

Whenever a tension member is to be fastened by means of bolts or rivets, holes must be provided at the connection. As a result, the member cross-sectional area at the connection is reduced and the strength of the member *may* also be reduced depending on the size and location of the holes.

Several methods are used to cut holes. The most common and least expensive method is to punch *standard* holes $\frac{1}{16}$ in. (1.6 mm) larger than the diameter of the rivet or bolt. In general, the plate thickness is less than the punch diameter. During the punching operation, the metal at the edge of the hole is damaged. This is accounted for in design by assuming that the extent of the damage is limited to a radial distance of $\frac{1}{32}$ in. (0.8 mm) around the hole. Therefore the total width to be deducted (LRFD and ASD-B2) is to be taken as the nominal dimension of the *hole* normal to the direction of applied load plus $\frac{1}{16}$ in. (1.6 mm). For fasteners in standard holes, the total deduction is equal to the fastener diameter plus $\frac{1}{8}$ in. (3.2 mm).

A second method of cutting holes consists of subpunching them $\frac{3}{16}$ in. (4.8 mm) diameter undersize and then reaming the holes to the finished size after the pieces being joined are assembled. This method is more expensive than that of punching standard holes but does offer the advantage of accurate alignment. This method and the next method produce better strength, both static and fatigue, but this is ignored in design procedures.

A third method consists of drilling holes to a diameter of the bolt or rivet plus $\frac{1}{32}$ in. (0.8 mm). This method is used to join thick pieces, and is the most expensive of the common methods.

When greater latitude is needed in meeting dimensional tolerances during erection, larger than *standard* holes can be used with high strength bolts $\frac{1}{2}$ in. diameter and larger without adversely affecting the performance. The maximum sizes for *oversized, short-slotted,* and *long-slotted* holes are specified in LRFD and ASD-J3.2.

EXAMPLE 3.3.1 _____

What is the net area A_n for the tension member shown in Fig. 3.3.1?

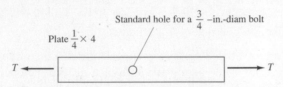

Figure 3.3.1   Tension member for Example 3.3.1.

Solution.

$$A_g = 4(0.25) = 1.0 \text{ sq in.}$$

$$\text{Width to be deducted for hole} = \tfrac{3}{4} + \tfrac{1}{8} = \tfrac{7}{8} \text{ in.}$$

$$A_n = A_g - (\text{width for hole})(\text{thickness of plate})$$
$$= 1.0 - 0.875 \, (0.25) = 0.78 \text{ sq in.} \quad \blacksquare\blacksquare$$

3.4 EFFECT OF STAGGERED HOLES ON NET AREA

Whenever there is more than one hole and the holes are *not* lined up transverse to the loading direction, more than one potential failure line may exist. The controlling failure line is that which gives the largest stress on an effective net area. In many cases, the critical failure path is also the path that has the minimum net area.

In Fig. 3.4.1a the failure line is along the section $A{-}B$. In Fig. 3.4.1b showing two lines of staggered holes, the failure line might be through one hole (section $A{-}B$) or it might be along a diagonal path, $A{-}C$. At first glance one might think section $A{-}B$ is critical since the path $A{-}B$ is obviously shorter than path $A{-}C$. However, from path $A{-}B$, only one hole would be deducted while two holes would have to be deducted from path $A{-}C$. In order to determine the controlling section, both paths $A{-}B$ and $A{-}C$ must be investigated. Accurate checking of strength along path $A{-}C$ is complex. However, a simplified empirical relationship proposed by Cochrane [3.2] has been

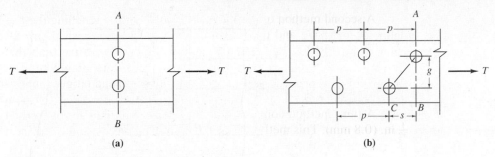

Figure 3.4.1 Paths of failure on net section.

adopted by LRFD and ASD-B2 to account for the difference between the path $A-C$ and the path $A-B$, expressed as a length correction,

$$\frac{s^2}{4g}$$

where s is the stagger, or spacing of adjacent holes parallel to the loading direction (see Fig. 3.4.1), and g is the gage distance transverse to the loading direction. Thus the net lengths of paths $A-B$ and $A-C$ would be

Net length of $A-B$ = length of $(A-B)$ − (width of hole + $\frac{1}{16}$ in.)

Net length of $A-C$ = length of $(A-B)$ − 2(width of hole + $\frac{1}{16}$ in.) + $\frac{s^2}{4g}$.

The minimum net area would then be determined from the minimum net length multiplied by the thickness of the plate.

In the years since Cochrane proposed the simple $s^2/4g$ expression, many investigators have proposed other rules [3.3–3.6] but none of them gives significantly better results, and all are more complicated.

Consistent with the general trend toward using strength-related design approaches, the work of Bijlaard [3.7] and others [3.8–3.10] has provided limit analysis theories to obtain net area in tension. These theories do not deviate from the $s^2/4g$ method by more than 10 to 15 percent.

The reader is referred to McGuire [3.11] for a more complete coverage of this subject of net section through staggered lines of fasteners.

EXAMPLE 3.4.1 _____

Determine the minimum net area of the plate shown in Fig. 3.4.2, assuming $\frac{15}{16}$-in.-diam holes are located as shown.

Solution. According to LRFD and ASD-B2, the width used in deducting for holes is the hole diameter plus $\frac{1}{16}$ in., and the staggered length correction is $s^2/4g$.

Path AD (two holes) :

$$\left[12 - 2\left(\frac{15}{16} + \frac{1}{16}\right) \right]0.25 = 2.50 \text{ sq in.}$$

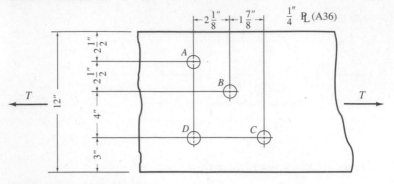

Figure 3.4.2   Example 3.4.1.

Path *ABD* (three holes; two staggers) :

$$\left[12 - 3\left(\frac{15}{16} + \frac{1}{16}\right) + \frac{(2.125)^2}{4(2.5)} + \frac{(2.125)^2}{4(4)}\right]0.25 = 2.43 \text{ sq in.}$$

Path *ABC* (three holes; two staggers) :

$$\left[12 - 3\left(\frac{15}{16} + \frac{1}{16}\right) + \frac{(2.125)^2}{4(2.5)} + \frac{(1.875)^2}{4(4)}\right]0.25 = 2.42 \text{ sq in.}$$
(controls) ■■

Angles

When holes are staggered on two legs of an angle, the gage length g for use in the $s^2/4g$ expression is obtained by using a length between the centers of the holes measured along the centerline of the angle thickness, i.e., the distance *A–B* in Fig. 3.4.3. Thus the gage distance g is

$$g = g_a - \frac{t}{2} + g_b - \frac{t}{2} = g_a + g_b - t \qquad (3.4.1)$$

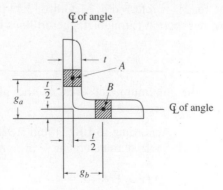

Figure 3.4.3   Gage dimensions for an angle.

TABLE 3.4.1 USUAL GAGES* FOR ANGLES, INCHES (FROM AISC MANUAL[†])

Leg	8	7	6	5	4	$3\frac{1}{2}$	3	$2\frac{1}{2}$	2	$1\frac{3}{4}$	$1\frac{1}{2}$	$1\frac{3}{8}$	$1\frac{1}{4}$	1
g_1	$4\frac{1}{2}$	4	$3\frac{1}{2}$	3	$2\frac{1}{2}$	2	$1\frac{3}{4}$	$1\frac{3}{8}$	$1\frac{1}{8}$	1	$\frac{7}{8}$	$\frac{7}{8}$	$\frac{3}{4}$	$\frac{5}{8}$
g_2	3	$2\frac{1}{2}$	$2\frac{1}{4}$	2										
g_3	3	3	$2\frac{1}{2}$	$1\frac{3}{4}$										

*Other gages are permitted to suit specific requirements subject to clearances and edge distance limitations.
[†]*LRFD Manual* [1.18], p. 9-13

Every rolled angle has a standard value for the location of holes (i.e., gage distances g_a and g_b), depending on the length of the leg and the number of lines of holes. Table 3.4.1 shows *usual gages* for angles as listed in the AISC Manual*.

EXAMPLE 3.4.2

Determine the net area A_n for the angle given in Fig. 3.4.4 if $\frac{15}{16}$-in.-diam holes are used.

Solution. For net area calculation the angle may be visualized as being flattened into a plate as shown in Fig. 3.4.5:

$$A_n = A_g - \sum Dt + \sum \frac{s^2}{4g}t$$

where D is the width to be deducted for the hole.
Path AC:

$$4.75 - 2\left(\frac{15}{16} + \frac{1}{16}\right)0.5 = 3.75 \text{ sq in.}$$

Path ABC:

$$4.75 - 3\left(\frac{15}{16} + \frac{1}{16}\right)0.5 + \left[\frac{(3)^2}{4(2.5)} + \frac{(3)^2}{4(4.25)}\right]0.5 = 3.96 \text{ sq in.}$$

Since the smallest A_n is 3.75 sq in., that value governs.

LRFD Manual [1.18]. p. 9–13; *ASD Manual* [1.7]. p. 1–52.

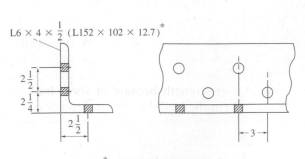

$L6 \times 4 \times \frac{1}{2}$ (L152 × 102 × 12.7)*

*legs and thickness in mm

Figure 3.4.4   Example 3.4.2.

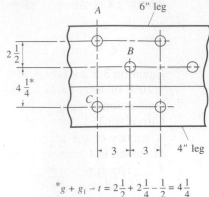

$^*g + g_1 - t = 2\frac{1}{2} + 2\frac{1}{4} - \frac{1}{2} = 4\frac{1}{4}$

Figure 3.4.5   Angle for Example 3.4.2 with legs shown "flattened" into one plane.   ■ ■

3.5 EFFECTIVE NET AREA

The net area as computed in Secs. 3.3 and 3.4 gives the reduced section that resists tension but still may not correctly reflect the strength. That is particularly true when the tension member has a profile consisting of elements not in a common plane and where the tensile load is transmitted at the end of the member by connection to some but not all of the elements. An angle section having connection to one leg only is an example of such a situation. For such cases the tensile force is not uniformly distributed over the net area.

LRFD and ASD-B3 provide that the *effective net area* A_e be computed as

$$A_e = UA_n \tag{3.5.1}$$

where U = reduction coefficient
A_n = net area.

The above equation logically applies for *both* fastener connections having holes and for welded connections. For welded connections, the net area equals the gross area A_g since there are no holes.

When a tensile load is applied eccentrically to a wide plate, the stress distribution across the width of the plate is nonuniform. The mechanism by which stress gets transmitted from the location of the applied load to sections distant from the load is by shear stresses acting in the plane of the plate. The fact that the stress is lower the farther the location is from the applied load means that the shear transfer "lags" or is inefficient. Thus, the nonuniformity of stress in wide plates or plate elements of rolled sections when a tensile load is applied nonuniformly, is referred to as "shear lag".

For *bolted* or *riveted connections,* the reduction coefficient U relates to the eccentricity $\bar{x}$ of loading in the connection. Whenever tension is transmitted through some but not all of the cross-sectional elements, LRFD-B3.2. indicates the following shall be used,

$$U = 1 - \frac{\bar{x}}{L} \le 0.9 \tag{3.5.2}$$

where $\bar{x}$ = distance from centroid of element being connected eccentrically to plane of load transfer (see Fig. 3.5.1 and also LRFD Commentary Figs. C-B3.1 and C-B3.2 for added guidance to determine $\bar{x}$.)

L = length of connection in the direction of loading.

Equation 3.5.2 is based on the work of Munse and Chesson [3.9] and correlates within ±10% of tests [3.1]. A review of shear lag research for both bolted and welded connections, along with design recommendations, is provided by Easterling and Giroux [3.15]. In addition to the reduction in strength because of shear lag, the efficiency of the fasteners is reduced in long (in the load direction) connections, as discussed in Chapter 4, where detailed treatment of bolted tension members occurs.

The approximate values of U (used prior to the 1993 LRFD Specification) for bolted and riveted connections to I-shaped sections, channels, and angles, are still considered acceptable, according to LRFD Commentary-B3.

For short tension members (connecting elements), such as splice and gusset plates, where the elements of the cross-section lie essentially in a common plane, the effective net area is taken equal to A_n, but may not exceed 85% of the gross area A_g

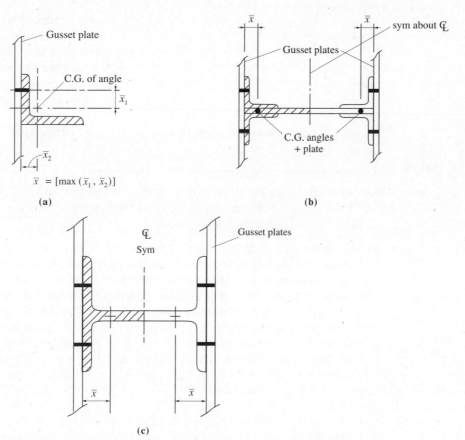

Figure 3.5.1   Eccentricity in joints; determination of $\bar{x}$ for computing U, Eq. 3.5.2.

(LRFD-J5.2.b and ASD-B3). Tests [3.1] have shown that when any holes are present in such short elements where general yielding on the gross section cannot occur, there will be at least 15% reduction in strength from that obtained based on yielding of the gross section.

For *welded tension connections,* there are three categories:

(1) Load transmitted to a member other than a plate by longitudinal welds, or by longitudinal welds in combination with transverse welds

$$A_e = UA_n = UA_g \tag{3.5.3}$$

Although not specifically stated in the *LRFD Specification,* the intention is for Eq. 3.5.2 to apply also for welded connections when shear lag is present. This intention is indicated both in the LRFD-Commentary B3, and *LRFD Manual,* p. 2-11.

(2) Load transmitted only by transverse welds

$$A_e = UA_n = A_{con} \tag{3.5.4}$$

where $A_{con} =$ area of directly connected elements. In this case, the shear lag effect is approximated indirectly by using the reduced area A_{con}.

(3) Load transmitted to a plate by longitudinal welds along both sides of the plate spaced apart such that $l \geq w$

$$A_e = UA_g \tag{3.5.5}$$

where $l =$ length of weld along one side of plate
         $w =$ distance between longitudinal welds (i.e., plate width)
         $U = 1.00$ For $l \geq 2w$
            $= 0.87$ For $2w > l \geq 1.5w$
            $= 0.75$ For $1.5w > l \geq w$

Welded connections for tension members are treated in Chapter 5 on welding.

EXAMPLE 3.5.1 ──

Determine the reduction factor U to be applied in computing the effective net area for a W14×82 section connected by plates at its two flanges, as shown in Fig. 3.5.2. There are three bolts along each connection line.

Solution.   In this case, two elements (the flanges) of the cross-section are connected but one (the web) is not connected. In accordance with Eq. 3.5.1, there is reduced efficiency of carrying load. The reduction factor U must be computed using Eq. 3.5.2. Because each flange connection can be thought of as a load on the tributary portion of the W shape, the section may be treated as two structural tees, as in Fig. 3.5.1c. The half W shape corresponds to a structural tee WT7×41, whose centroidal distance $\bar{x}$ is given by the AISC Manual as 1.39 in. The length L of the connection is 6 in. Thus, Eq. 3.5.2 gives

$$U = 1 - \frac{1.39}{6.0} = 0.77 \leq 0.9 \quad \text{OK}$$

LRFD Commentary-B3 indicates the 1986 LRFD Specification values may still be used. Thus, for W, M, or S shapes having flange widths not less than

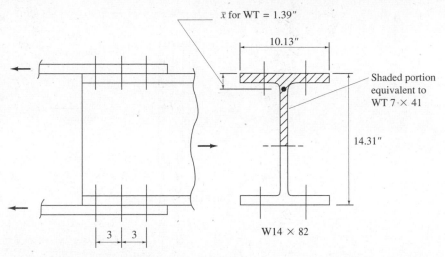

Figure 3.5.2 Example 3.5.1.

two-thirds of the depth, and structural tees cut from these shapes, $U = 0.90$ when there are at least three fasteners per line in the direction of stress. For this example,

$$\frac{b_f}{d} = \frac{\text{flange width}}{\text{section depth}} = \frac{10.13}{14.31} = 0.71 > 0.67 \quad \text{OK}$$

Thus, the reduction factor U could be taken as 0.90. However, the Specification approach is the more correct one and seems mandatory (using the word "shall"). Thus, the 0.77 value should be used. ■■

3.6 TEARING FAILURE AT BOLT HOLES

When thin plates are attached by bolts, a tearing limit state, known as *block shear,* may control the strength of a tension member, or the tension region at the end connection of a beam (see Chapter 13). Referring to Fig. 3.6.1a, the angle tension member attached to a gusset plate may have a tearing failure along the bolt holes, section a–b–c. The tearing (shear rupture) strength on section a–b plus the tensile yielding strength on section b–c will give the total resistance to a block shear failure. Tests [3.12, 3.13] have shown it to be reasonable to add the strength in tension yielding in one plane to the shear rupture strength of the perpendicular plane.

The four holes in the plate of Fig. 3.6.1b and c will contribute to a tear-out failure if the sum of the shear strengths along a–b and c–d plus the tensile strength along b–c is less than either of the strengths in general yielding of the member (Eq. 3.2.1) or fracture along e–b–c–f (Eq. 3.2.2).

Combination shear and tension tearing failures are uncommon in tension members; however, this combination mode frequently controls the design of bolted end connections to the thin webs of beams.

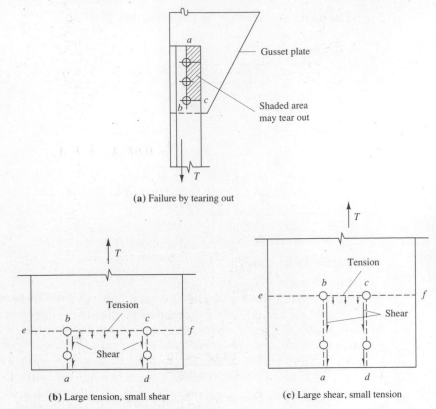

(a) Failure by tearing out

(b) Large tension, small shear

(c) Large shear, small tension

Figure 3.6.1 Tearing failure limit state.

Tests on block shear failure in angle members [3.14] have shown that block shear controls for short connections (i.e., two or fewer connectors per gage line). These tests also indicate that shear lag is a factor in block shear failure of tension members, which is not currently accounted for by the LRFD Specification [1.18]. Furthermore, adding $s^2/4g$ to the net tensile area for block shear calculations on staggered bolts may be unconservative, depending on the direction of the stagger with respect to the failure path.

LRFD-J4.3. provides for considering two block shear failure modes: (1) fracture along the tensile plane ($b-c$ in Fig. 3.6.1b) followed by yielding along the shear planes ($a-b$ and $c-d$ in Fig. 3.6.1b), and (2) fracture along the shear planes ($a-b$ and $c-d$ in Fig. 3.6.1c) followed by yielding along the tensile plane ($b-c$ in Fig. 3.6.1c). Block shear is a fracture limit state and the failure mode which governs in that which has the greater fracture strength.

Consistent with the limit states discussed in Sec. 3.2, the gross area is used for the yielding limit state and the net area is used for the fracture limit state. In addition, following the energy-of-distortion theory (Eq. 2.6.5), the shear yield stress τ_y is taken as $0.6F_y$. Similarly, the shear strength τ_u is taken as $0.6F_u$.

The nominal strength T_n in tension is thus given by:

1. Shear yielding—tension fracture ($F_u A_{nt} \geq 0.6 F_u A_{nv}$)

$$T_n = 0.6 F_y A_{gv} + F_u A_{nt} \qquad (3.6.1)$$

or

2. Shear fracture—tension yielding ($F_u A_{nt} < 0.6 F_u A_{nv}$)

$$T_n = 0.6 F_u A_{nv} + F_y A_{gt} \qquad (3.6.2)$$

where A_{gv} = gross area acted upon by shear
A_{gt} = gross area acted upon by tension
A_{nv} = net area acted upon by shear
A_{nt} = net area acted upon by tension
F_u = specified (ASTM) minimum tensile strength
F_y = specified (ASTM) minimum yield stress.

Equation 3.6.1 indicates fracture ($F_u A_{nt}$) on the net tensile area followed by yielding ($0.6 F_y A_{gv}$) along the shear planes. Equation 3.6.2 indicates fracture ($0.6 F_u A_{nv}$) on the net shear area followed by yielding ($F_y A_{gt}$) on the gross tension area. Because yielding cannot occur until after fracture has taken place, the appropriate equation to use is the one having the greater ratio of fracture strength to yield strength.

The comparison criterion that provides the basis for using either Eq. 3.6.1 or Eq. 3.6.2 is slightly different from the aforementioned rationale. Instead of comparing the fracture and yield components of the strength within a single equation, the fracture components of Eqs. 3.6.1 and 3.6.2 are compared. The correlation between using the equation having the larger proportion of fracture to yield strength, and the comparison of fracture strengths used for Eqs. 3.6.1 and 3.6.2 is as follows.

Requiring a larger proportion of fracture strength to yield strength in Eq. 3.6.1 than in Eq. 3.6.2 means that

$$\left(\frac{F_u}{F_y} \frac{A_{nt}}{0.6 A_{gv}} \right) \geq \left(\frac{F_u}{F_y} \frac{0.6 A_{nv}}{A_{gt}} \right) \qquad (3.6.3)$$

Canceling the ratio (F_u/F_y) from both sides and multiplying through by the denominators gives

$$(A_{nt})^2 \left(\frac{A_{gt}}{A_{nt}} \right) \geq (0.6 A_{nv})^2 \left(\frac{A_{gv}}{A_{nv}} \right) \qquad (3.6.4)$$

Taking the square root of both sides and multiplying each side by F_u produces an expression similar to the criterion of Eqs. 3.6.1 and 3.6.2. Thus,

$$F_u A_{nt} \geq 0.6 F_u A_{nv} \sqrt{\frac{A_{gv}/A_{nv}}{A_{gt}/A_{nt}}} \qquad (3.6.5)$$

Therefore, the criterion used for selecting either Eq. 3.6.1 or Eq. 3.6.2 is the same as choosing the equation with the greater ratio of fracture strength to yield strength, assuming the ratio of gross area to net area in tension is equal to the ratio of gross area to net area in shear. This is approximately true when the hole spacings in the perpendicular directions are uniform and nearly equal.

3.7 STIFFNESS AS A DESIGN CRITERION

Even though stability is not a criterion in the design of tension members, it is still necessary to limit their length to prevent a member from becoming too flexible both during erection and final use of the structure. Tension members that are too long may sag excessively because of their own weight. In addition, they may also vibrate when subjected to wind forces as in an open truss or when supporting vibrating equipment such as fans or compressors.

To reduce the problems associated with excessive deflections and vibrations a stiffness criterion was established. This criterion is based on the slenderness ratio, L/r, of a member where L is the length and r the least radius of gyration ($r = \sqrt{I/A}$). The *preferable* maximum slenderness ratio is 300 for members whose design is based on tensile force (LRFD and ASD-B7). This limitation does not apply to rods in tension.

In applying the stiffness criterion to tension members, the highest slenderness ratio must be used. A symmetrical member may have two different radii of gyration, and for nonsymmetrical members one must consider the weakest principal axis. When a tension member is built up from a number of sections, the radius of gyration must be computed using the moment of inertia I and the cross-sectional area A. The value for r will be with respect to the same axis as that used to calculate the moment of inertia.

3.8 LOAD TRANSFER AT CONNECTIONS

Normally the holes in tension members are those for rivets or bolts to transfer load from one tension member into another.

Although the detailed treatment of fasteners and their behavior is in Chapter 4, the basic assumption is that each equal size fastener transfers an equal share of the load whenever the fasteners are arranged symmetrically with respect to the centroidal axis of a tension member. The following example is to illustrate the idea and its relationship to net area calculations.

EXAMPLE 3.8.1 _____

Calculate the governing net area for plate A of the single lap joint in Fig. 3.8.1 and show free-body diagrams of portions of plate A with sections taken through each line of holes. Assume that plate B has adequate net area and does not control the strength T.

Solution.   The full tensile force T in plate A acts on section 1–1 of Fig. 3.8.1. Examination of other sections in plate A to the left of section 1–1 will involve *less than* 100% of T acting, since part of that force will have already been

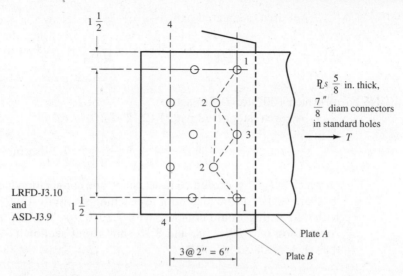

Figure 3.8.1   Single lap connection for Example 3.8.1.

transferred from plate A to plate B. At section 4–4, 100% of T must now be acting in plate B while only 20% of T acts in plate A. Since there must be zero force acting on the end of plate A a short distance to the left of section 4–4, the force T must have been entirely transferred to plate B over the distance from sections 1–1 to 4–4. The free bodies of the various segments are shown in Fig. 3.8.2.

Deduction in width for 1 hole = Diam of hole $+ \frac{1}{16}$ in.

$\qquad\qquad\qquad\qquad\qquad$ = Diam of fastener $+ \frac{1}{8}$ in.   for *standard* hole

$\qquad\qquad\qquad\qquad\qquad$ = $\frac{7}{8} + \frac{1}{8} = 1$ in.

$\qquad$ Net area (section 1–1) = $\frac{5}{8}(15 - 3) = 7.50$ sq in.

on which 100% of T acts (Fig. 3.8.2d).

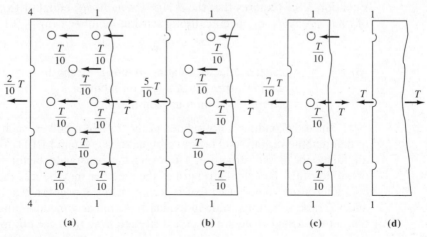

Figure 3.8.2   Load distribution in plate A.

Net area (staggered path 1–2–3–2–1);

$$= \frac{5}{8}\left[15 - 5(1) + 4\frac{(2)^2}{4(3)}\right] = 7.08 \text{ sq in.}$$

$$\uparrow\!\!\rule{2cm}{0.4pt}\quad s^2/4g$$

on which 100% of T also acts.

Net area (staggered path 1–2–2–1):

$$= \frac{5}{8}\left[15 - 4 + 2\frac{(2)^2}{4(3)}\right] = 7.29 \text{ sq in.}$$

on which 0.9 of T is presumed to act, since one connector has already transferred its share (0.10) of the load prior to reaching section 1–2–2–1. The 7.29 sq in. with $0.9T$ acting would compare with $7.29/0.9 = 8.10$ sq in. with T acting. A comparison of 7.50, 7.08, and 8.10 shows that section 1–2–3–2–1 governs; then $A_n = 7.08$ sq in. ■■

3.9 LOAD AND RESISTANCE FACTOR DESIGN—TENSION MEMBERS

The general philosophy of Load and Resistance Factor Design (LRFD) was described in Secs. 1.8 and 1.9. Equation 1.8.1 gives the structural safety requirement, as follows:

$$\phi R_n \geq \sum \gamma_i Q_i \qquad [1.8.1]$$

where ϕ = resistance factor (strength reduction factor)
R_n = nominal resistance (strength)
γ_i = overload factors (LRFD-A4.1)
Q_i = loads (such as dead load, live load, wind load, earthquake load) or load effects (such as bending moment, shear, axial force, and torsional moment resulting from the various loads)

Equation 1.8.1 requires that the *design strength* ϕR_n equal or exceed the summation of *factored loads,* or specifically for tension members, Eq. 1.8.1 becomes

$$\phi_t T_n \geq T_u \qquad (3.9.1)$$

where ϕ_t = resistance factor relating to tensile strength
T_n = nominal strength of a tension member
T_u = factored load on a tension member

Since the loading is not related to the type of member, such as tension member or column, the various load factor combinations given in LRFD-A4.1 are the same for all members in the structure. However, the resistance factor ϕ accounts for the possibility that the actual strength of the member may be less than the theoretically computed strength because of variations in material properties and dimensional tolerances. These variations while individually within accepted tolerance limits may combine in the actual structure to give a strength less than the computed value. Neither the ϕ factor nor the overload factors γ are intended to account for careless errors in design or construction. The reliability of designs of bolted steel tension members using LRFD has been reported by Bennett and Najem-Clarke [3.16].

The design strength $\phi_t T_n$ according to LRFD-D1 is the smaller of that based on *yielding in the gross section,*

$$\phi_t T_n = \phi_t F_y A_g = 0.90 F_y A_g \tag{3.9.2}$$

or *fracture in the net section,*

$$\phi_t T_n = \phi_t F_u A_e = 0.75 F_u A_e \tag{3.9.3}$$

Note that the resistance factor ϕ_t is 0.90 for the yielding limit state and is 0.75 for the fracture limit state.

In addition, the designer must take into account rupture strength (tension, shear, or a combination of both) along a potential tear-out path (see Sec. 3.6). The design strength requirements of LRFD-J4 are:

1. Shear rupture design strength ϕV_n:

$$\phi V_n = \phi(0.6 F_u) A_{nv} \tag{3.9.4}$$

2. Tension rupture design strength ϕT_n:

$$\phi T_n = \phi F_u A_{nt} \tag{3.9.5}$$

3. Shear—tension combination design strength ϕR_{bs}:
 a. When $F_u A_{nt} \geq 0.6 F_u A_{nv}$,

$$\phi R_{bs} = \phi(0.6 F_y A_{gv} + F_u A_{nt}) \tag{3.9.6}$$

 which means shear yielding—tension fracture controls.
 b. When $0.6 F_u A_{nv} > F_u A_{nt}$,

$$\phi R_{bs} = \phi(0.6 F_u A_{nv} + F_y A_{gt}) \tag{3.9.7}$$

 which means shear fracture—tension yielding controls.

where A_{gv} = gross area acted upon by shear
 A_{gt} = gross area acted upon by tension
 A_{nv} = net area acted upon by shear
 A_{nt} = net area acted upon by tension
 ϕ = 0.75 for the fracture limit state (Eqs. 3.9.4 through 3.9.7)
 F_u = specified (ASTM) minimum tensile strength
 F_y = specified (ASTM) minimum yield stress

In the 1993 *LRFD Manual* (Tables 8–46 and 8–47), the deduction for holes is computed in accordance with LRFD-B2, par. 2, the same as for the tension fracture limit state for the member, as discussed in Section 3.3. In the 1986 *LRFD Manual,* the implied "recommendation" was to deduct the *actual hole;* i.e., the bolt diameter plus 1/16 in. for standard holes when checking block shear.

The Specification [1.16] sections used in Load and Resistance Factor Design of tension members are summarized in Table 3.9.1.

TABLE 3.9.1   TENSION MEMBERS—
AISC SPECIFICATIONS REFERENCES

	Specification section	
Topic	LRFD [1.16]	ASD [1.5]
Net area	B2	B2
Effective net area	B3	B3
Limiting slenderness ratio	B7	B7
Tensile strength	D1	D1
Built-up members	D2	D2
Pin-connected member; eyebars	D3	D3
Threaded rods	J3	J3
Block shear	J4	J4
Connecting elements	J5	J5
Fatigue	K3	K4

EXAMPLE 3.9.1

Determine the service load capacity in tension for an L6×4×$\frac{1}{2}$ of A572 Grade 50 steel connected with $\frac{7}{8}$-in.-diam bolts in standard holes as shown in Fig. 3.9.1. Use AISC Load and Resistance Factor Design, and assume the live load to dead load ratio is 3.0.

Solution.   The angle tension member is connected to a *gusset plate,* typical of truss joints. The gusset plate is the plate at the intersection of members to which they are connected.

The maximum strength will be based on either section 1–1 with one hole deducted, or on the staggered section 1–2 through two holes. The governing section will have 100% of load T acting on it.

For section 1–1,

$$A_n = A_g - 1 \text{ hole} = 4.75 - \left(\frac{7}{8} + \frac{1}{8}\right)0.50 = 4.25 \text{ sq in.}$$

For section 1–2,

$$A_n = A_g - 2 \text{ holes} + (s^2/4g)t$$

$$= 4.75 - 2\left(\frac{7}{8} + \frac{1}{8}\right)0.50 + \frac{(2)^2}{4(2.5)}(0.50) = 3.95 \text{ sq in.}$$

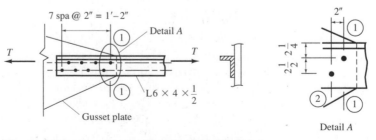

Figure 3.9.1   Tension member for Example 3.9.1.

The two design strengths to be used in accordance with LRFD-D1 are given by Eq. 3.9.2 based on *general yielding* on the gross section of the member,

$$\phi_t T_n = \phi_t F_y A_g$$

$$= 0.90(50)(4.75) = 214 \text{ kips}$$

[3.9.2]

and by Eq. 3.9.3 based on *fracture* at the connection,

$$\phi_t T_n = \phi_t F_u A_e$$

[3.9.3]

The angle does not have both legs connected to transfer the tensile force, so the effective net area is less than the computed net area, accounting for the eccentricity at the connection.

The effective net area A_e is

$$A_e = U A_n$$

where, according to LRFD-B3.2 the reduction coefficient U is to be computed from Eq. 3.5.2:

$$U = 1 - \frac{\bar{x}}{L}$$

[3.5.2]

where the distance $\bar{x}$ from centroid of element being connected eccentrically to plane of load transfer (see Fig. 3.5.1) is for the case of the angle 0.987 in. The length L of the joint is 14 in. Equation 3.5.2 then gives

$$U = 1 - \frac{\bar{x}}{L} = 1 - \frac{0.987}{14} = 0.93 > 0.9 \text{ max;}$$

Use U = 0.9.

$$A_e = U A_n = 0.90(3.95) = 3.56 \text{ sq in.}$$

Thus, Eq. 3.9.3 gives

$$\phi_t T_n = \phi_t F_u A_e = 0.75(65)(3.56) = 173 \text{ kips}$$

Thus, the controlling $\phi_t T_n$ is the smaller of the values from Eqs. 3.9.2 (214 kips) and 3.9.3 (173 kips),

$$\phi_t T_n = 173 \text{ kips}$$

The overload factors relate the design strength to the service loads or load effects. Using the gravity load combination Eq. 1.8.3 [LRFD-Eq (A4-2)], the factored load T_u is

$$T_u = \sum \gamma_i Q_i = 1.2D + 1.6L + 0.5(L_r \text{ or } S \text{ or } R)$$

[1.8.3]

where in this example the roof loading L_r, S (snow), and R (rain) are not involved. The live load (L) is given as three times dead load (D). Thus, applying the safety requirement, Eq. 3.9.1; that is, letting $\phi_t T_n$ equal factored load T_u gives

$$\phi_t T_n = 1.2D + 1.6L = 1.2D + 1.6(3D) = 6.0D$$
$$173 = 6.0D$$
$$D = 28.8 \text{ kips}$$
$$L = 3D = 3(28.8) = 86.4 \text{ kips}$$

The total safe service load T is

$$T = D + L = 28.8 + 86.4 = 115 \text{ kips.}$$

If this angle connection consisted of a very few large fasteners or if either the gusset plate or the angle were thin elements, the shear rupture limit state of LRFD-J4, represented by Eq. 3.9.4 through 3.9.7, might give a lower strength than the lesser of Eqs. 3.9.2 and 3.9.3. ■■

EXAMPLE 3.9.2 _____

Investigate the shear rupture failure mode on the angle L4×4×$\frac{1}{4}$ attached with three $\frac{7}{8}$-in.-diam bolts to a $\frac{3}{8}$-in. gusset plate, as shown in Fig. 3.9.2. The material is A36 steel.

Solution. The usual general yielding and fracture limit states governed by the lesser of Eqs. 3.9.2. and 3.9.3 give

$$\phi_t T_n = \phi_t F_y A_g = 0.90(36)(1.94) = 62.9 \text{ kips}$$
$$\phi_t T_n = \phi_t F_u A_e = \phi_t F_u U A_n$$

where

$$U = 1 - \frac{\bar{x}}{L} = 1 - \left(\frac{1.09}{6}\right) = 0.82 < 0.90 \text{ max; Use } 0.82$$

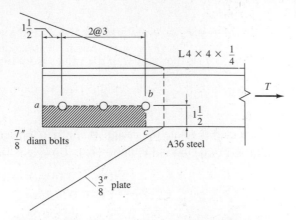

Figure 3.9.2　Tension member of Example 3.9.2.

Thus,

$$\phi_t T_n = \phi_t F_u U A_n$$
$$= 0.75(58)(0.82)(1.94 - 0.25) = 60.3 \text{ kips}$$

The block shear potential failure along path a–b–c of Fig. 3.9.2 must be investigated according to LRFD-J4. Calculating the net areas A_{nv} and A_{nt},

$$A_{nv} = (\text{length } a\text{–}b \text{ less 2.5 holes}) \times \text{thickness}$$
$$= \left[7.5 - 2.5 \left(0.875 + \frac{1}{8} \right) \right] 0.25 = 1.25 \text{ sq in.}$$

$$A_{nt} = (\text{length } b\text{–}c \text{ less 0.5 holes}) \times \text{thickness}$$
$$= \left[1.5 - 0.5 \left(0.875 + \frac{1}{8} \right) \right] 0.25 = 0.25 \text{ sq in.}$$

Compare $0.6 F_u A_{nv}$ with $F_u A_{nt}$,

$$[0.6 F_u A_{nv} = 0.6(58)1.25 = 43.5] > [F_u A_{nt} = 58(0.25) = 14.5]$$

which means Eq. 3.9.7 [shear fracture, tension yielding] controls,

$$\phi R_{bs} = \phi(0.6 F_u A_{nv} + F_y A_{gt}) \qquad [3.9.7]$$
$$\phi R_{bs} = 0.75(43.5 + F_y A_{gt})$$
$$= 0.75[43.5 + 36(1.5)0.25] = 0.75(57.0) = 42.8 \text{ kips}$$

Thus, block shear [42.8 kips] controls over tension yielding on gross section [62.9 kips] or fracture on effective net section [60.3 kips]. ■■

EXAMPLE 3.9.3 ———

Select a tension diagonal member for a roof truss of A572 Grade 50 steel using AISC Load and Resistance Factor Design. The axial tension is 60 kips dead load and 6 kips live load and the member is 12 ft long. Assume $\frac{7}{8}$-in.-diam bolts are located on a single gage line in standard holes. Assume the preferable limit on slenderness ratio L/r is 240 (*not* an LRFD limit).

a. Select the lightest single angle member.
b. Select the lightest double angle member having legs separated by $\frac{1}{4}$ in. back-to-back.

Solution. For tension members, the design strength requirement is

$$\phi_t T_n \geq T_u$$

where the factored load T_u may be governed by Eq. 1.8.3 in general, or Eq. 1.8.2 when the proportion of dead load is large, as in this case; thus,

$$T_u = 1.2D + 1.6L = 1.2(60) + 1.6(6) = 82 \text{ kips}$$

or

$$T_u = 1.4D = 1.4(60) = 84 \text{ kips (controls)}$$

In this case, the factored load to be designed for is 84 kips.

The strength of the members may be controlled by either

$$\phi_t T_n = \phi_t F_y A_g = 0.90(50)A_g$$

or

$$\phi_t T_n = \phi_t F_u A_e = 0.75(65)A_e$$

The design strength ϕT_n must equal the factored load T_u. The area requirements become

$$\text{Required } A_g = \frac{84}{0.90(50)} = 1.87 \text{ sq in.}$$

$$\text{Required } A_e = \frac{84}{0.75(65)} = 1.72 \text{ sq in.}$$

Estimating U for either a single or double angle section to be 0.85, in accordance with LRFD Commentary-B3,

$$\text{Required } A_n = \frac{A_e}{U} = \frac{1.72}{0.85} = 2.03 \text{ sq in.}$$

The net area requirement obviously controls since it exceeds the gross area requirement.

Also, the minimum r to satisfy the given limitation of $L/r = 240$ is

$$\min r = \frac{L}{240} = \frac{12(12)}{240} = 0.6 \text{ in.}$$

(a) Select single angle member. The required gross area in each case depends on the area deducted for one hole, which in turn depends on the thickness. The following tabular procedure may be found useful in making the selection:

Standard thickness t	Deduction for one hole	Required gross area	Choices from AISC manual single angle properties	
$\frac{5}{16}$	0.313*	2.34†	L4×4×$\frac{5}{16}$,	$A = 2.40$, $r = 0.79$‡
$\frac{3}{8}$	0.375	2.40	L3$\frac{1}{2}$×3$\frac{1}{2}$×$\frac{3}{8}$,	$A = 2.48$, $r = 0.69$
			L4×3×$\frac{3}{8}$,	$A = 2.48$, $r = 0.64$
$\frac{7}{16}$	0.438	2.47	L3×3×$\frac{7}{16}$,	$A = 2.43$, $r = 0.59$
$\frac{1}{2}$	0.500	2.53		

* $(\frac{7}{8} + \frac{1}{8})0.3125 = 0.3125$ sq in.
† Required $A_g = 2.03 + 0.31 = 2.34$ sq in.
‡ *Note:* Min $r = r_z$ for single angles.

Use L4×4×$\frac{5}{16}$ single angle member (least area, therefore lightest)

(b) Select double angle member. For this type of section two holes must be deducted. Selection should be made from the double angle properties in AISC Manual.

Standard thickness t	Deduction for two holes	Required gross area	Choices from AISC manual double angle properties	
$\frac{1}{4}$	0.500	2.53	L3×3×$\frac{1}{4}$,	$A = 2.88, r = 0.93$
			L3×2$\frac{1}{2}$×$\frac{1}{4}$,	$A = 2.63, r = 0.95$
$\frac{5}{16}$	0.625	2.65	L2$\frac{1}{2}$×2×$\frac{5}{16}$,	$A = 2.62, r = 0.78$
$\frac{3}{8}$	0.750	2.78		

Use 2—L3×2$\frac{1}{2}$×$\frac{1}{4}$ with long legs back-to-back.

The L2$\frac{1}{2}$×2×$\frac{5}{16}$ is slightly understrength. Block shear and the reduction factor U can be calculated only after the number of fasteners has been determined. ■■

3.10 TENSION RODS

A common and simple tension member is the threaded rod. Such rods are usually secondary members where the required strength is small, such as (a) sag rods to help support purlins in industrial buildings (Fig. 3.10.1a); (b) vertical ties to help support girts in industrial building walls (Fig. 3.10.1b); (c) hangers, such as tie rods supporting a balcony (Fig. 3.10.1c); and (d) tie rods to resist the thrust of an arch.

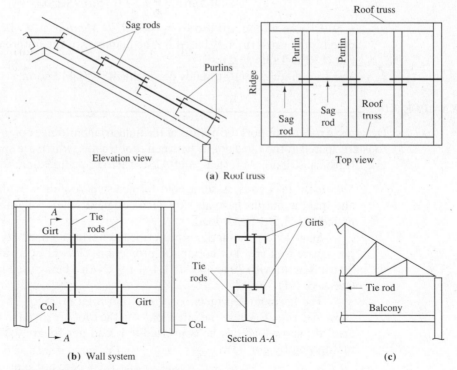

Figure 3.10.1 Uses of tension rods.

Tie rods are frequently used with an initial tension as diagonal wind bracing in walls, roofs, and towers. The initial tension effectively adds to the stiffness and reduces deflection and vibrational motion, which tends to cause fatigue failures in the connections. Such initial tension can be obtained by designing the member something on the order of $\frac{1}{16}$ in. short for a 20-ft length or by the use of turnbuckles which can be tightened after construction.

EXAMPLE 3.10.1

Select the diameter for a threaded round rod of A36 steel to carry an axial tension of 6 kips live load and 3 kips dead load. Use AISC Load and Resistance Factor Design.

Solution. The design strength of a threaded rod is given by LRFD-J3.6 (Table J3.2) as

$$\phi_t T_n = 0.75 A_b F_n = 0.75 A_b (0.75 F_u) \qquad (a)$$

The factored load T_u to be carried is

$$T_u = 1.2D + 1.6L = 1.2(3) + 1.6(6) = 13.2 \text{ kips}$$

Using the minimum tensile strength F_u for A36 steel as 58 ksi from Table 2.1.1 (or LRFD "NUMERICAL VALUES" TABLE 2), and equating the factored load T_u to the design strength $\phi_t T_n$, gives the required gross area A_b from Eq. (a) as

$$\text{Required } A_b = \frac{\text{Required } \phi_t T_n}{0.75(0.75 F_u)} = \frac{13.2}{0.75(0.75)(58)} = 0.40 \text{ sq in.}$$

Choose a threaded rod from the *LRFD Manual,* p. 8–17. The area computed is the gross area A_b based on the diameter of the unthreaded body of the rod (LRFD-Table J3.2).
Use $\frac{3}{4}$-in.-diam rod (10 threads per inch)($A_b = 0.442$ sq in.).   ■■

EXAMPLE 3.10.2

Design sag rods to support the purlins of the industrial building roof of Fig. 3.10.2. Sag rods are spaced at the third points between roof trusses, which are spaced 24 ft apart. Use 20 psf snow load, A36 steel, and AISC LRFD Specification.

Solution. (a) Loads. Assume cold-formed steel roofing is used, weighing 3 psf, and that the purlins have already been designed. Their weight may be approximated as a 3.5 psf roof load.

Snow load customarily is prescribed as having an intensity given in pounds per square foot (psf) of horizontal projection. Generally, a value not less than 20 psf is used, with 30 to 40 psf (1.4 to 1.9 kN/m²) being used in northern areas (see Sec. 1.4).

The horizontal projection of the roof area is 25cos 25° over which the snow load acts. Because the other loads on the roof are given in terms of the roof area, the snow load can be converted to a load per square foot of roof area by multiplying by cos 25°,

$$20(\cos 25°) = 18.1 \text{ psf of roof}$$

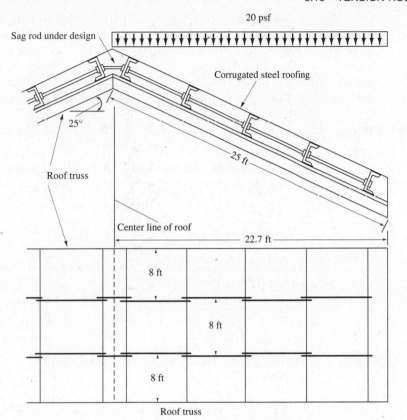

Figure 3.10.2   Roof and sag rods for Example 3.10.2.

(b) Compute factored load on roof. The structure weight is dead load. Because there are no live loads other than snow, Eq. 1.8.3 will control,

$$1.2D + 1.6(L_r \text{ or } S \text{ or } R) \qquad [1.8.3]$$
$$1.2(3 + 3.5) + 1.6(18.1) = 36.8 \text{ psf}$$

(c) Compute factored load to be carried by a single rod. Because the rods are spaced 8 ft apart, the load carried by each rod is that load over a 4 ft strip of roof on either side of the rod. The sag rod indicated in Fig. 3.10.2 must carry the load for the entire length (i.e., 25 ft) of roof. The vertical factored load associated with a line of sag rods is

$$36.8 \text{ times tributary area (in sq ft)}$$
$$36.8(25)(8)/1000 = 7.4 \text{ kips}$$

The top rod parallel to the roof carries only that component of the load parallel to the roof, equal to

$$7.4(\sin 25°) = 7.4(0.423) = 3.1 \text{ kips}$$

The component of the vertical load perpendicular to the roof is carried by beam action in the purlins. The horizontal rod at the roof peak, indicated in Fig. 3.10.2, must carry $3.1/(\cos 25°)$,

$$T_u = 3.1/(\cos 25°) = 3.1/(0.906) = 3.4 \text{ kips}$$

(d) Select the diameter or rod. The strength requirement of LRFD-J3.6 is

$$\phi_t T_n \geq T_u$$
$$\phi_t T_n = 0.75(0.75F_u)A_b$$

Obtaining F_u from Table 2.1.1 (or LRFD "NUMERICAL VALUES" TABLE 2), and equating the factored load T_u to the design strength $\phi_t T_n$, gives the required area A_b as

$$\text{Required } A_b = \frac{T_u}{0.75(0.75F_u)} = \frac{3.4}{0.75(0.75)(58)} = 0.11 \text{ sq in.}$$

From *LRFD Manual* [1.18], Table 8–7, p. 8–17, select a $\frac{3}{8}$-in. diam threaded rod having $A_b = 0.11$ sq in.
Use $\frac{3}{8}$-in.-diam rod ($A_b = 0.110$ sq in.).   ■■

3.11 ALLOWABLE STRESS DESIGN—TENSION MEMBERS

The general philosophy of Allowable Stress Design (ASD) was described in Secs. 1.8 and 1.9. Equation 1.8.8 gives the structural safety requirement, as follows:

$$\frac{\phi R_n}{\gamma} \geq \sum Q_i \qquad [1.8.8]$$

which expresses that the design strength ϕR_n divided by a factor γ for overload must exceed the sum of the service loads. In the allowable stress design method, the safety provision is γ/ϕ.

For tension members, taking the factor of safety FS $= \gamma/\phi$, the nominal strength R_n as T_n for tension members, and $\sum Q_i$ equal to the service load T in tension, Eq. 1.8.8 becomes

$$\frac{T_n}{\text{FS}} \geq T \qquad (3.11.1)$$

The nominal strength T_n for tension members may be controlled by either of Eqs. 3.2.1 or 3.2.2,

$$T_n = F_y A_g \qquad [3.2.1]$$
$$T_n = F_u A_e \qquad [3.2.2]$$

where A_g = gross cross-sectional area
A_e = effective net area = UA_n (see Sec. 3.5)
A_n = net area (see Sec. 3.4)
U = reduction coefficient (efficiency factor) (ASD-B3)

Using a factor of safety FS of 1.67, substituting Eq. 3.2.1 into Eq. 3.11.1, and dividing by A_g to obtain a *stress format* gives

$$\frac{F_y A_g}{1.67 A_g} \geq \frac{T}{A_g} \tag{3.11.2}$$

Using a larger factor of safety FS = 2.0 as typically used for connections, substituting Eq. 3.2.2 into Eq. 3.11.1, and dividing by A_e to obtain a *stress format* gives

$$\frac{F_u A_e}{2.00 A_e} \geq \frac{T}{A_e} \tag{3.11.3}$$

Then, defining f_a as computed service load stress, T/A_g or T/A_e, Eqs. 3.11.2 and 3.11.3 become the requirements of ASD-D1,

$$\left[f_a = \frac{T}{A_g} \right] \leq 0.60 F_y \tag{3.11.4}$$

$$\left[f_a = \frac{T}{A_e} \right] \leq 0.50 F_u \tag{3.11.5}$$

The computation of net area A_n is discussed in Sec. 3.5 and the effective net area is treated in Sec. 3.6

The Specification [1.5] sections used in Allowable Stress Design of tension members are summarized in Table 3.9.1.

EXAMPLE 3.11.1

Investigate the angle $4 \times 4 \times \frac{5}{16}$ selected in Example 3.9.3 to serve as a single angle tension member of A572 Grade 50 steel to carry 60 kips dead load and 6 kips live load. Assume $\frac{7}{8}$-in-diam bolts will be located on a single gage line in standard holes to attach the member to a gusset plate.

Solution.

(a) Compute the effective net area of the member. The net area is

$$A_n - A_g - 1 \text{ hole} = 2.40 - (0.875 + 0.125)0.3125 = 2.09 \text{ sq in.}$$

The member is attached to the gusset plate along one leg (i.e., one of its two elements); therefore, the nonuniform stress across the section reduces the efficiency of the member to carry load. According to ASD-B3, the effective net area is 0.85 of the actual net area.

Thus, the effective net area A_e is

$$A_e = U A_n = 0.85(2.09) = 1.78 \text{ sq in.}$$

(b) Check the service load stresses. For A572 Grade 50 steel, $F_y = 50$ ksi and $F_u = 65$ ksi,

$$f_a = \frac{T}{A_g} = \frac{66}{2.40} = 27.5 \text{ ksi} < [F_a = 0.60 F_y = 30 \text{ ksi}] \quad \text{OK}$$

$$f_a = \frac{T}{A_e} = \frac{66}{1.78} = 37.1 \text{ ksi} > [F_a = 0.50 F_u = 32.5 \text{ ksi}] \quad \text{NG}$$

Thus, the section is not adequate by Allowable Stress Design and a larger section would be required. ■■

SELECTED REFERENCES

3.1. Geoffrey L. Kulak, John W. Fisher, and John H. A. Struik. *Guide to Design Criteria for Bolted and Riveted Joints,* 2nd ed. New York: John Wiley & Sons, 1987.

3.2. V. H. Cochrane. "Rules for Rivet Hole Deductions in Tension Members," *Engineering News-Record,* **89,** November 16, 1922, 847–848.

3.3. W. M. Wilson. Discussion of "Tension Tests of Large Rivet Joints," *Transactions,* ASCE, **105** (1942), 1268.

3.4. W. M. Wilson, W. H. Munse, and M. A. Cayci. "A Study of the Practical Efficiency under Static Loading of Riveted Joints Connecting Plates." *Bulletin 402,* U. of Illinois Engineering Experiment Station, Urbana, IL, 1952.

3.5. F. W. Schutz. "Effective Net Section of Riveted Joints," *Proceedings,* Second Illinois Structural Engineering Conference, November, 1952.

3.6. "Here's a Better Way to Design Splices," *Engineering News-Record,* **150,** Part I, January 8, 1953, 41.

3.7. P. P. Bijlaard. Discussion of "Investigation and Limit Analysis of Net Area in Tension," *Transactions,* ASCE, **120** (1955), 1156–1163.

3.8. G.W. Brady and D. C. Drucker. "Investigation and Limit Analysis of Net Area in Tension," *Transactions,* ASCE, **120** (1955), 1133–1154.

3.9. W. H. Munse and E. Chesson, Jr. "Riveted and Bolted Joints: Net Section Design," *Journal of the Structural Division,* ASCE, **89,** ST2 (February 1963), 107–126.

3.10. E. Chesson and W. H. Munse. "Behavior of Riveted Connections in Truss-Type Members," *Journal of the Structural Division,* ASCE, **83,** STI (January 1957), Paper 1150, 1–61; also *Transactions,* ASCE, **123** (1958), 1087–1128.

3.11. William McGuire. *Steel Structures.* Englewood Cliffs, NJ: Prentice-Hall, Inc., 1968 (pp. 310–328).

3.12. James M. Ricles and Joseph A. Yura. "Strength of Double-Row Bolted-Web Connections," *Journal of Structural Engineering,* ASCE, **109,** 1 (January 1983), 126–142.

3.13. Steve G. Hardash and Reidar Bjorhovde. "New Design Criteria for Gusset Plates in Tension," *Engineering Journal,* AISC, **22,** 2 (2nd Quarter 1985), 77–94.

3.14. Howard I. Epstein. "An Experimental Study of Block Shear Failure of Angles in Tension," *Engineering Journal,* AISC, **29,** 2 (2nd Quarter 1992), 75–84.

3.15. W. Samuel Easterling and Lisa Gonzales Giroux. "Shear Lag Effects in Steel Tension Members," *Engineering Journal,* AISC, **30,** 3 (3rd Quarter), 1993, 77–89.

3.16. Richard M. Bennett and F. Shima Najem-Clarke. "Reliability of Bolted Steel Tension Members," *Journal of Structural Engineering,* ASCE, **113,** 8 (August 1987), 1865–1872.

PROBLEMS

All problems are to be done according to the AISC Load and Resistance Factor Design or Allowable Stress Design, as indicated by the instructor. Assume fastener strength is adequate and does not control. All holes are *standard* holes. Values of yield stress F_y and tensile strength F_u are available in Table 2.1.1. For all problems where the total number of holes is not known, assume rupture strength of LRFD-J4 does not control. Where needed, assume distances from center of hole to end of piece are $1\frac{1}{2}$ in.

3.1. Compute the maximum acceptable tensile service load that may act on a single angle L6×4×$\frac{3}{4}$ that is welded along only one leg to a gusset plate; thus, there are no holes. The service live load is three times the dead load. Solve for (a) A36 steel and (b) A572 Grade 50 steel.

3.2. Compute the maximum acceptable tensile service load the angle in Prob. 3.1 may carry when connected on both legs. The 4-in. leg contains a single gage line of $\frac{7}{8}$-in.-diam bolts, and the 6-in. leg contains a double gage line of $\frac{7}{8}$-in.-diam bolts. Assume no stagger of bolts, and that all bolts participate in carrying load.

3.3. Compute the maximum acceptable service load on an A36 steel plate tension member $\frac{1}{4}$ in. × 12 in. having a single line of holes parallel to the direction of loading. The load is 25% dead load and 75% live load, and $\frac{7}{8}$-in.-diam bolts are used.

3.4. Compute the net area A_n for the plate (a connecting element according to LRFD and ASD-J5) shown in the accompanying figure. Then compute the maximum value for service load T when A36 steel is used, the live load is four times the dead load, and the holes are $\frac{13}{16}$ in. diameter.

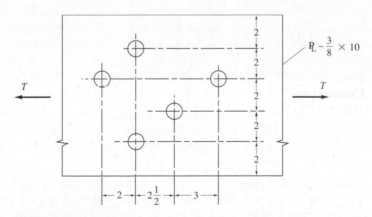

Problems 3.4 and 3.5

3.5. Repeat Prob. 3.4 using A572 Grade 60 steel and $\frac{15}{16}$-in.-diam holes.

3.6. Select a pair of angles to support a tensile live load (LL) and dead load (DL) for the case assigned by the instructor. Assume the angles are separated back-to-back $\frac{3}{8}$ in. by a connected gusset plate, and that the connection is welded. Assume the slenderness ratio is desired to not exceed 300.

Case	DL (kips)	LL (kips)	Steel	Length (ft)
1	70	20	A36	20
2	65	22	A36	30
3	70	20	A572 Gr 60	20
4	48	30	A36	22
5	50	30	A572 Gr 50	22
6	80	30	A36	20
7	80	100	A572 Gr 50	28

3.7. Select a single angle (for the case assigned by the instructor) to support a tensile load. A single gage line of at least three bolts is to be used.

Case	DL (kips)	LL (kips)	Steel	Bolt diameter (in.)	Length (ft)
1	15	40	A36	3/4	15
2	15	40	A572 Gr 50	3/4	15
3	15	40	A36	3/4	20
4	15	40	A572 Gr 60	7/8	25
5	10	30	A36	7/8	20
6	10	30	A572 Gr 50	7/8	20
7	12	35	A36	7/8	18
8	12	35	A572 Gr 60	7/8	18

3.8. Select a standard threaded rod to carry a tensile force T of 4 kips dead load and 6 kips live load. Use A572 Grade 50 steel.

3.9. Select a standard threaded rod to carry a tensile force T of 2 kips dead load and 4 kips live load. Use A36 steel.

3.10. Design sag rods to support the purlins of an industrial building roof whose span and slope are shown in the accompanying figure. Sag rods are placed at $\frac{1}{3}$ points between roof trusses, which are spaced 30 ft apart. Assume roofing and purlin weight is 9 psf of roof surface. Use standard threaded rods and A36 steel. The snow load to be carried is 20, 30 or 40 psf of horizontal projection, whichever is appropriate for your locale.

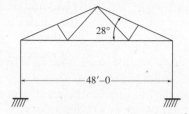

Problem 3.10

3.11. Determine the maximum allowable tensile load (20% dead load, 80% live load) for a single C15×33.9 fastened to a $\frac{1}{2}$-in. gusset plate as in the accompanying figure. Use A36 steel and assume holes are for $\frac{3}{4}$-in.-diam bolts. Base answer on tension strength of the channel, and include shear rupture strength.

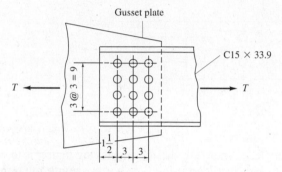

Problem 3.11

3.12. Repeat Prob. 3.11 using a C10×25 attached to a $\frac{5}{8}$-in. gusset plate. Assume the 12 bolts are in 3 lines parallel to the direction of loading, with the same 3 in. spacing.

3.13. Determine the tensile load (85% live load; 15% dead load) permitted by AISC for a pair of angles L6×4×$\frac{3}{8}$ attached to a gusset plate as shown in the accompanying figure. Use A36 steel and $\frac{3}{4}$-in.-diam bolts on standard gage lines whose distances are given in Table 3.4.1. The force T is transmitted to the gusset plate by fasteners on lines A and B; assume only open holes in the 4-in. (outstanding) legs.

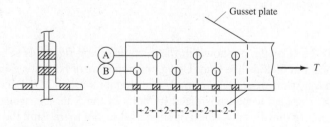

Problem 3.13

3.14. Repeat Prob. 3.13 using a pair of L8×6×$\frac{3}{4}$ angles with staggered $\frac{7}{8}$-in.-diam bolts in the 8-in. leg.

3.15. Given the splice shown in the accompanying figure:
  a. Determine the maximum capacity T (25% dead load, 75% live load) based on the A36 steel plates having holes arranged as shown.
  b. What value of s should be specified to provide the maximum capacity T as computed in part (a), if the final design is to have $s_1 = s_2 = s$?

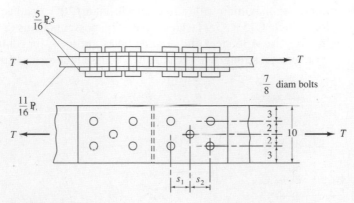

Problem 3.15

3.16. An L5×3$\frac{1}{2}$×$\frac{1}{2}$ angle, as shown in the accompanying figure, is to carry 20 kips dead load and 70 kips live load using the shortest length of connection using two gage lines of bolts in the 5-in. leg. What is the minimum acceptable stagger, theoretical and specified ($\frac{1}{2}$-in. multiples), using A572 Grade 50 steel?

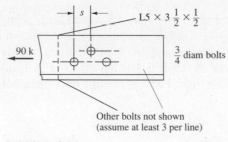

Other bolts not shown
(assume at least 3 per line)

Problem 3.16

3.17. An L5×3$\frac{1}{2}$×$\frac{1}{2}$ angle, as shown in the accompanying figure, is to carry 20 kips dead load and 60 kips live load. Using one gage line of holes for $\frac{7}{8}$-in.-diam bolts in each leg, what would be the minimum stagger s required to accomplish this? Consider the load to be transferred by bolts in the 5-in. leg, while the holes in the 3$\frac{1}{2}$-in. leg may be considered open ones (i.e., not to transmit the tensile load). Use A36 steel.

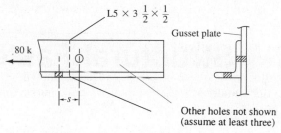

Problem 3.17

3.18. Compute the minimum value of s that could theoretically be used on the angle of the accompanying figure such that the maximum factored tensile force T_u may be carried. Assume m is large enough so that a failure pattern through the open hole will not govern. Include consideration of shear rupture strength.

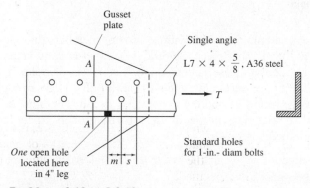

Problems 3.18 and 3.19

3.19. Assuming that s for the angle of Prob. 3.18 is made as large as required in Prob. 3.18, compute the minimum distance m required so that the open hole in the 4-in. leg will not reduce the strength below its maximum possible value. If Prob. 3.18 is not solved, assume s is 3.75 in.

3.20. Design an eyebar to carry 24 kips dead load and 76 kips live load, using flame-cut A572 Grade 50 steel plate. (Refer to LRFD or ASD-D3.)

Chapter 4

Structural Fasteners

4.1 TYPES OF FASTENERS

Every structure is an assemblage of individual parts or members that must be fastened together, usually at the member ends. Welding is one method and is treated in Chapter 5. The other method is to use fasteners, such as rivets or bolts. This chapter is concerned with bolting; in particular, high-strength bolts. High-strength bolts have replaced rivets as the means of making nonwelded structural connections. However, for completeness, a brief description of the other fasteners, including rivets and unfinished machine bolts, is given.

High-Strength Bolts

The two basic types of high-strength bolts are designated as ASTM A325 [2.5] and A490 [2.8], the material properties of which are discussed in Sec. 2.2 and summarized in Table 4.1.1. These bolts are heavy hexagon-head bolts, used with heavy semifinished hexagon nuts, as shown in Fig. 4.1.1b. The threaded portion is shorter than for bolts in nonstructural applications, and may be cut or rolled. A325 bolts are of heat-treated *medium carbon* steel having an approximate yield strength of 81 to 92 ksi (560 to 630 MPa) depending on diameter. A490 bolts are also heat-treated but are of *alloy* steel having an approximate yield strength of 115 to 130 ksi (790 to 900 MPa) depending on diameter. A449 bolts are occasionally used when diameters over $1\frac{1}{2}$ in. up to 3 in. are needed, and also for anchor bolts and threaded rods.

High-strength bolts range in diameter from $\frac{1}{2}$ to $1\frac{1}{2}$ in. (3 in. for A449). The most common diameters used in building construction are $\frac{3}{4}$ in. and $\frac{7}{8}$ in., whereas the most common sizes in bridge design are $\frac{7}{8}$ in. and 1 in.

High-strength bolts are usually tightened to develop a specified tensile stress in them, which results in a predictable clamping force on the joint. The actual transfer of service loads through a joint is, therefore, due to the friction developed in the pieces being joined. Joints containing high-strength bolts are designed either as *slip-critical* (formerly called *friction-type*), where high slip resistance at service load is desired; or as *bearing-type,* where high slip resistance at service load is unnecessary.

TABLE 4.1.1 PROPERTIES OF BOLTS

ASTM designation	Bolt diameter in. (mm)	Proof load,[a] length measurement[b] method, ksi (MPa)	Proof load,[a] yield strength[c] method, ksi (MPa)	Minimum tensile strength, ksi (MPa)
A307 [2.4], low-carbon steel				
Grades A and B	$\frac{1}{4}$ to 4	—	—	60
	(6.35 to 104)			
A325 [2.5], high-strength steel				
Types 1, 2, and 3	$\frac{1}{2}$ to 1	85	92	120
	(12.7 to 25.4)	(585)	(635)	(825)
Types 1, 2, and 3	$1\frac{1}{8}$ to $1\frac{1}{2}$	74	81	105
	(28.6 to 38.1)	(510)	(560)	(725)
A449 [2.7], quenched and	$\frac{1}{4}$ to 1	85	92	120
tempered steel	(6.35 to 25.4)	(585)	(635)	(825)
(*Note:* AISC[d] permits	$1\frac{1}{8}$ to $1\frac{1}{2}$	74	81	105
use only for bolts	(28.6 to 38.1)	(510)	(560)	(725)
larger than $1\frac{1}{2}$ in.	$1\frac{3}{4}$ to 3	55	58	90
and for threaded rods	(6.35 to 76.2)	(380)	(400)	(620)
and anchor bolts)				
A490 [2.8], quenched and	$\frac{1}{2}$ to $1\frac{1}{2}$	120	130	150
tempered alloy steel	(12.7 to 38.1)	(825)	(895)	(1035)

[a] Actual proof load and tensile load obtained by multiplying given stress value by the tensile stress area A_s; $A_s = 0.785\,[d_b - (0.9743/n)]^2$, where A_s = stress area in square inches, d_b = nominal diameter of bolt in inches, and n = number of threads per inch.

[b] 0.5% extension under load.

[c] 0.2% offset value.

[d] LRFD and ASD-A3.3 and J3.

Figure 4.1.1 Types of fasteners.

Rivets

For many years rivets were the accepted means of connecting members but today (1995) they are virtually obsolete in the United States. Undriven rivets are formed from bar steel, a cylindrical shaft with a head formed on one end, as shown in Fig. 4.1.1a. Rivet steel is a mild carbon steel designated by ASTM as A502 Grade 1 (F_y = 28 ksi) (190 MPa) and Grade 2 (F_y = 38 ksi) (260 MPa), with the minimum specified yield strengths based on bar stock as rolled. The forming of undriven rivets and the driving of rivets cause changes in the mechanical properties.

Installation requires heating the rivet to a light cherry-red color, inserting it into a hole and then applying pressure to the preformed head while at the same time squeezing the plain end of the rivet to form a rounded head. During this process the shank of the rivet completely or nearly fills the hole into which it had been inserted. Upon cooling, the rivet shrinks, thereby providing a clamping force. However, the amount of clamping produced by the cooling of the rivet varies from rivet to rivet and therefore cannot be counted on in design calculations.

Unfinished Bolts

Bolts of low-carbon steel designated ASTM A307 [2.4] are the least expensive bolt. They may not, however, produce the least expensive connection since more are required in a particular connection. Their primary use is in light structures, secondary or bracing members, platforms, catwalks, purlins, girts, small trusses, and similar applications in which the loads are primarily small and static in nature. Such bolts are also used as temporary fitting-up fasteners in cases where high-strength bolts, rivets, or welding may be the permanent means of connection. Unfinished bolts are sometimes called common, machine, or rough bolts and may come with square heads and square nuts.

Ribbed Bolts

Bolts of ordinary rivet steel having a rounded head and raised ribs parallel to the shank were used for many years as an alternative to rivets. The actual diameter of a ribbed bolt is slightly larger than the hole into which it is driven. In driving, the bolt actually cuts into the edges around the hole, producing a relatively tight fit. The ribbed bolt was particularly useful in bearing-type connections and in connections that had stress reversals.

A modern variation of the ribbed bolt is the *interference-body bolt* shown in Fig. 4.1.1c, which is of A325 bolt steel and, instead of longitudinal ribs, has serrations around the shank as well as parallel to the shank. Because of the serrations around the shank through the ribs, this bolt is often called an *interrupted-rib* bolt. Ribbed bolts are difficult to drive when several layers of plates are to be connected. The A325 interference-body bolt may also be more difficult to insert through several plates; however, it is used when tight fit of the bolt in the hole is desired, and it permits tightening the nut without simultaneously holding the bolt head as may be required with smooth loose-fitting, ordinary A325 bolts. These bolts are, however, rarely used in ordinary steel structures.

4.2 HISTORICAL BACKGROUND OF HIGH-STRENGTH BOLTS

The first experiments indicating the possibility of using high-strength bolts in steel-framed construction were reported by Batho and Bateman [4.1] in 1934. They concluded that bolts with a minimum yield strength of 54 ksi (370 MPa) could be relied on to prevent slip between the connected parts. Follow-up tests by Wilson and Thomas [4.2] substantiated the earlier work by reporting that high-strength bolts smaller in diameter than the holes in which they were inserted had fatigue strengths equal to that of well-driven rivets, provided that the bolts were sufficiently pretensioned.

The next major step occurred in 1947 with the formation of the Research Council on Riveted and Bolted Structural Joints. This organization began by using and extrapolating information from studies of riveted joints–in particular, the extensive annotated 1945 bibliography by De Jonge [4.3].

By 1950 the concept of using high-strength bolts and a summary of research and behavior was presented [4.4] to practicing engineers and the steel-fabrication industry. The first Specification in 1951 permitted replacement of rivets with bolts on a one-to-one basis. It was conservatively assumed that friction transfer of the load was necessary in all joints under service load conditions. The factor of safety against slip was established at a high enough level so that good fatigue resistance (i.e., no slip under varying stress or stress reversal occurring during many load cycles) was provided in every joint, similar to or better than that shown by riveted joints.

In 1954 a revision was made in the Specification to include the use of flat washers on 1:20 sloping surfaces and to allow the use of impact wrenches for installing bolts. Also, the 1954 revision permitted the surfaces in contact to be painted when the bolts were used in a *bearing-type* connection; i.e., when the strength of the connection was to be based on the bolt in bearing against the side of the hole.

In 1956 Munse [4.5] summarized bolt behavior and concluded that bolts must have as high an initial tension as practicable if high-strength bolts were to be efficient and economical. By 1960 the minimum bolt tension was increased. The *bearing-type* connection was recognized as an acceptable substitute for a riveted connection, and the connection designed on the basis of slip resistance, known then as a *friction-type* connection, would probably only be necessary when direct tension acts on the bolts or when stress reversals occur.

Also, in 1960 a simple installation procedure, known as the *turn-of-the-nut method,* was introduced as an alternative to the torque wrench method previously required. The economics of high-strength bolting further improved when only one washer [located under the element (head or nut) being turned], instead of the previously required two, could be used if the turn-of-the-nut method was used.

In 1962 the requirement for washers was eliminated except in special situations [4.6]. In 1964 the higher strength A490 bolt was introduced.

The 1985 and 1988 Specifications of the Research Council on Structural Connections (RCSC) [4.7, 4.8] changed the philosophy of design for bearing-type connections and for friction-type connections, now called *slip-critical* connections. A detailed review of the historical background is provided in the RCSC Commentary [4.9]. The *Guide to Design Criteria for Bolted and Riveted Joints* by Kulak, Fisher, and Struik [3.1] (hereinafter referred to as the *Guide*) summarizes research and makes recommendations which generally form the basis for current design of bolted connections.

4.3 CAUSES OF RIVET OBSOLESCENCE

Riveting is a method of connecting members at a joint by inserting ductile metal pins into holes in the pieces being joined and forming a head at each end to prevent the joint from coming apart. Typical types of rivets are shown in Fig. 4.3.1 (see also Fig. 4.1.1a).

Riveting required a crew of four or five experienced persons. On the other hand, the crews required for high-strength bolt installation do not need to be highly skilled. Inspection was difficult, and cutting out and replacing bad rivets was an expensive procedure. Even the preheating immediately prior to driving is critical in developing the necessary tightness after cooling.

The principal factor that delayed immediate acceptance of high-strength bolts was the high cost of the material including two hardened washers. In the early 1950s the reduced labor cost for installing bolts did not offset the higher bolt material cost. After the washers could be reduced to one or eliminated and the greater strength of a bolt over that of a rivet could be utilized in design, high-strength bolts became economical. Now (1995) with even higher labor cost and connection design generally requiring fewer bolts than would be required for riverts, the economy is clearly with high-strength bolts.

Welding, as treated in Chapter 5, has played an important role in reducing the use of all fasteners, both rivets and bolts.

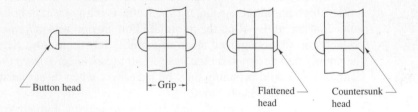

Button head   ← Grip →   Flattened   Countersunk
                                    head   head

Figure 4.3.1   Types of rivets.

4.4 DETAILS OF HIGH-STRENGTH BOLTS

Both the most commonly used A325 bolt and the occasionally used A490 bolt are heavy hexagon (hex) head bolts with heavy hexagon nuts, identified by the ASTM designation on the top of the head as shown in Fig. 4.4.1.

Heavy hex bolts have shorter threaded portions than other bolts; this reduces the probability of having threads occur where maximum strength is required across the shank of the bolt. The heavy hex bolt and nut are shown in Fig. 4.4.1. Requirements for marking bolts and nuts, including manufacturer's identification, are given in Fig. 4.4.2.

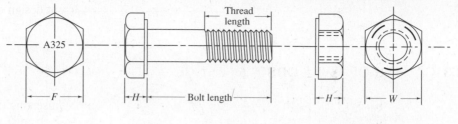

Nut may be chamfered
on both faces

Figure 4.4.1   Heavy hex structural bolt and heavy hex nut. (From Ref. 4.9)

The A325 and A490 bolts are available as Types 1, 2, or 3. The A325 Type 1 is the medium-carbon steel bolt that has been available for many years and is the one that would be supplied if not otherwise specified. Similarly for A490, Type 1 is the usual alloy steel bolt. Type 2 for both A325 and A490 is a low-carbon martensite steel alternative to Type 1 for atmospheric temperature applications. For elevated temperature applications, Type 1 must be used. Type 3 for both A325 and A490 is a weathering steel bolt having corrosion resistance comparable to A588 weathering steel (see Table 2.1.1).

Occasionally, ASTM A449 bolts are used where diameters larger than $1\frac{1}{2}$ in. are required.

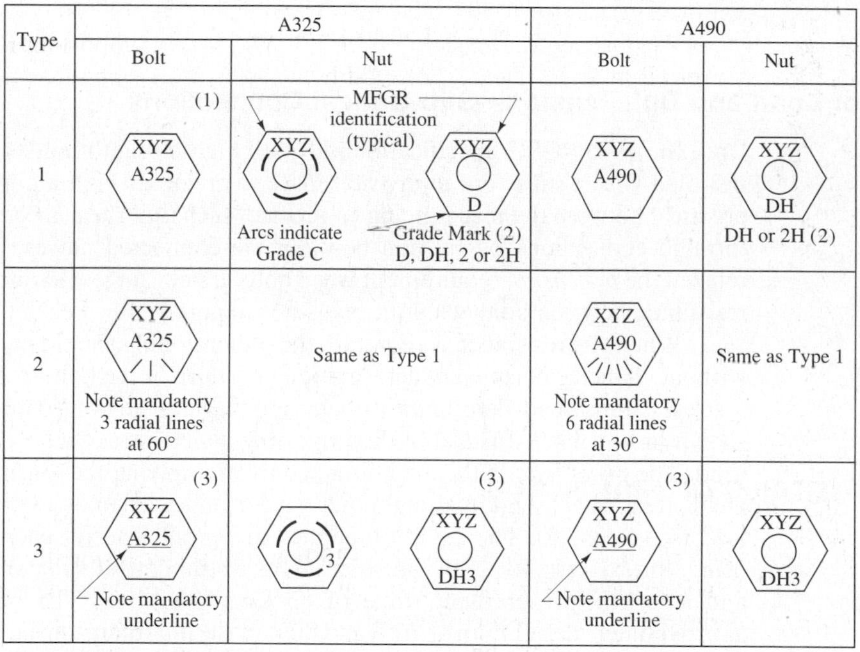

| Type | A325 | | A490 | |
	Bolt	Nut	Bolt	Nut
1	(1) XYZ A325	MFGR identification (typical) XYZ / Arcs indicate Grade C XYZ D / Grade Mark (2) D, DH, 2 or 2H	XYZ A490	XYZ DH DH or 2H (2)
2	XYZ A325 / \ Note mandatory 3 radial lines at 60°	Same as Type 1	XYZ A490 // \\ Note mandatory 6 radial lines at 30°	Same as Type 1
3	(3) XYZ A325 Note mandatory underline	(3) XYZ 3 XYZ DH3	(3) XYZ A490 Note mandatory underline	(3) XYZ DH3

(1) Additional optional 3 radial lines at 120° may be added.
(2) Type 3 also acceptable.
(3) Additional optional mark indicating weathering grade may be added.

Figure 4.4.2   Required markings for acceptable bolt and nut assemblies. (From Ref. 4.9)

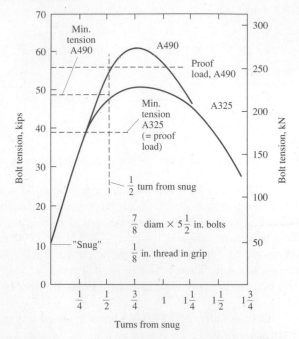

Figure 4.4.3   Typical load vs nut rotation relationships for A325 and A490 bolts. (From Kulak,
Fisher, and Struik [3.1])

Proof Load and Bolt Tension—Slip-Critical Connections

Until the 1985 RCSC Specification [4.8], all high-strength bolts were required to be installed with a sufficient *pretension* force to create as high a compression force as practical between the pieces being connected, such that shear forces were transmitted through connections by friction between the connected pieces. That Specification relaxed the *pretension* requirement when bolts are not subject to direct tension and slip resistance between connected pieces is not required [4.8, Sec. 5].

When slip resistance is required, the pretensioning should be as high as possible without chancing permanent deformation or failure of the bolt. Bolt material exhibits a stress-strain (load-deformation) behavior that has no well-defined yield point, as shown in Fig. 4.4.3. Instead of directly using a yield stress, a so-called *proof load* is used. The *proof load* is the load obtained by multiplying the tensile stress area* by a yield stress established by using either a 0.2% offset strain or a 0.5% extension under load (see Sec. 2.4). The ASTM tabulates this proof load for each diameter fastener using, for example, for $\frac{1}{2}$- to 1-in.-diam bolts a strain offset value of 92 ksi (630 MPa) and a length measurement value of 85 ksi (590 MPa). The proof load stress is approximately a minimum of 70% and 80% of the minimum tensile strengths for A325 and A490 bolts, respectively.

---

* Tensile stress area = $0.785 + \left(d_b - \dfrac{0.9743}{n}\right)^2$ where n = number of threads per inch

TABLE 4.4.1 MINIMUM BOLT TENSION* FOR
FULLY-TIGHTENED BOLTS
(LRFD-TABLE J3.1 AND
ASD-TABLE J3.7)

Bolt size		A325 bolts		A490 bolts	
(in.)	(mm)	(kips)	(kN)	(kips)	(kN)
1/2	12.7	12	53	15	67
5/8	15.9	19	85	24	107
3/4	19.1	28	125	35	156
7/8	22.2	39	173	49	218
1	25.4	51	227	64	285
1 1/8	28.6	56	249	80	356
1 1/4	31.8	71	316	102	454
1 3/8	34.9	85	378	121	538
1 1/2	38.1	103	458	148	658

*Equal to 70% of minimum tensile strength of bolts, rounded off to the nearest kip, as specified in ASTM Specifications for A325 and A490 bolts for UNC (unified standard *coarse* screw threads under ANSI B1.1; see *LRFD Manual* [1.18], p. 8–17).

Since the early 1950s, the minimum required pretension equals the proof load for A325 bolts. Using the turn-of-the-nut installation method (discussed in Sec. 4.5) no difficulty is encountered in obtaining proof load for these bolts with $\frac{1}{2}$ turn of the nut from snug position, as shown in Fig. 4.4.3. With the A490 bolt, however, the $\frac{1}{2}$ turn from snug may not achieve the proof load. Also for long bolts, more than $\frac{1}{2}$ turn from snug will be required to achieve the same tension as for short bolts.

AISC (LRFD-Table J3.1; ASD-Table J3.7) requires slip-critical connections to be pretensioned to 70% of the minimum tensile strength, as given in Table 4.4.1. This equals the proof load for A325 bolts and about 85 to 90% of proof load for A490 bolts.

The magnitude of pretension that is desirable and necessary has been the subject of considerable study by researchers [3.1].

4.5 INSTALLATION PROCEDURES

Connections Not Requiring Full Pretensioning

When slip-resistant connections are not required [4.7, Sec. 8(c)] and when the bolts are not subject to direct tension, bolts are permitted to be tightened "snug tight". This is a tightness that exists "when all plies in a joint are in firm contact," which is further defined as the result of "a few impacts of an impact wrench or the full effort of a man using an ordinary spud wrench" [4.7].

Connections Requiring Full Precompression

There are four general methods for installing high-strength bolts to obtain the pretension indicated in Table 4.4.1. These are the *turn-of-the-nut tightening, calibrated wrench tightening, installation of alternative design bolts,* and *direct tension indicator tightening* [4.7].

The *turn-of-the-nut method* is the simplest. Developed in the 1950s and 1960s, this method obtains the specified pretension by a *specified rotation* of the nut from the "snug tight" condition, which causes a specified strain in the bolt. Although snugness or initial tightness can vary due to the surface condition of the pieces being tightened, this variation does not significantly affect the clamping force, as may be seen from Fig. 4.5.1. The clamping force of 48.6 kips (220 kN) corresponding to a $\frac{1}{2}$ turn-of-the-nut occurs at a sufficiently large bolt elongation (i.e., along the horizontal portion of the curve) that any variation in the initial snug tightness has an insignificant effect on the clamping force.

Calibrated wrench tightening uses manual torque wrenches and power wrenches adjusted to stall at a specified torque. For a given torque, variations in bolt tension have been found [3.1, p. 52] to range as high as ±30% with an average variation of ±10%. The RCSC Specification [4.7] therefore requires that calibrated wrenches be set to produce a bolt tension at least 5% in excess of the values specified in Table 4.4.2. Furthermore, calibrated wrenches must be calibrated at least *daily* and a hardened washer must be used under the element (head or nut) being tightened.

One may wonder whether there is danger of having inadequate reserve strength if the pretension exceeds the proof load; i.e., when it approaches 90% of tensile strength. Figure 4.5.2 shows the effect of various turns of the nut with the margin of safety indicated. If the calibrated wrench method is used, *strength* is the critical factor, with the typical safety margin shown in Fig. 4.5.2. The possibility of overtorquing the bolts with power wrenches is not considered a problem since such overtorquing usually fractures the bolts and they are replaced during installation. In the turn-of-the-nut method, *deformation* is the critical factor with the typical safety margin shown in Fig. 4.5.2. For either installation process, one can expect a minimum of $2\frac{1}{4}$ turns from snug to fracture. When the turn-of-the-nut method is used and bolts are tensioned using $\frac{1}{8}$ turn increments, frequently as many as four turns may be obtained from snug to fracture. The turn-of-the-nut method is the cheapest, is more reliable,

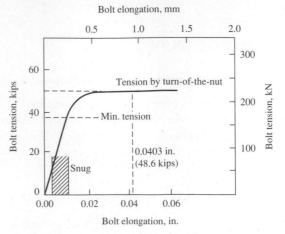

Figure 4.5.1   Bolt elongations in a typical test joint. (From Fisher, Ramseier, and Beedle [4.11])

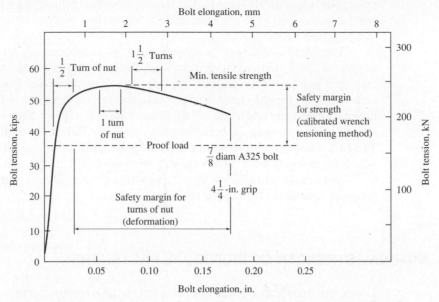

Figure 4.5.2 A325 bolt behavior. (Adapted from Rumpf and Fisher [4.12])

and is generally preferred over the calibrated wrench method. The approved [4.7] nut rotations are indicated in Table 4.5.1.

The third general category of installation technique, the *installation of alternative design bolts,* uses proprietary fasteners designed to indirectly indicate the bolt

TABLE 4.5.1. NUT ROTATION* FROM SNUG TIGHT CONDITION (FROM REF. 4.7)

Bolt length (underside of head to end of bolt)	Disposition of outer faces of bolted parts		
	Both faces normal to bolt axis	One face normal to bolt axis and other face sloped not more than 1:20 (beveled washer not used)	Both faces sloped not more than 1:20 from normal to bolt axis (beveled washer not used)
Up to and including 4 diameters	$\frac{1}{3}$ turn	$\frac{1}{2}$ turn	$\frac{2}{3}$ turn
Over 4 diameters but not exceeding 8 diameters	$\frac{1}{2}$ turn	$\frac{2}{3}$ turn	$\frac{5}{6}$ turn
Over 8 diameters but not exceeding 12 diameters†	$\frac{2}{3}$ turn	$\frac{5}{6}$ turn	1 turn

* Nut rotation is relative to bolt regardless of the element (nut or bolt) being turned. For bolts installed by $\frac{1}{2}$ turn and less, the tolerance should be plus or minus 30°; for bolts installed by $\frac{2}{3}$ turn and more, the tolerance should be plus or minus 45°.

† No research has been performed by the Council to establish the turn-of-the-nut procedure for bolt lengths exceeding 12 diameters. Therefore, the required rotation must be determined by actual test in a suitable tension measuring device which simulates conditions of solidly fitted steel.

tension or automatically provide the required tension. A procedure for qualifying such bolts is prescribed [4.7]. Sometimes the alternative design feature is a twist-off or yielding-type element.

The fourth category is the *direct tension indicator tightening*. Again, a procedure is specified [4.7] to qualify such devices. Typically, a hardened washer is used containing a number of protrusions on one face. The washer is inserted between the element being turned (head or nut) and the gripped material with the protrusions bearing against the underside of the element leaving a gap maintained by the protrusions. Upon tightening the bolt, the protrusions are flattened and the gap is reduced. The bolt tension is determined by measuring (using a feeler gage) the remaining gap, which for properly tensioned bolts will be about 0.015 in. (0.38 mm) or less [4.13].

In all cases, installation must begin at the most rigid part of the connection and progress systematically toward the least rigid areas.

4.6 NOMINAL STRENGTH OF INDIVIDUAL FASTENERS

Loads are transferred from one member to another by means of the connection between them. A few typical connections are shown in Fig. 4.6.1.

The simplest device for transferring load from one steel piece to another is with a pin (a cylindrical piece of steel) inserted in holes that are aligned in the two pieces as shown in Fig. 4.6.2. The cotter pins shown would prevent the pin from sliding out. Load would be transferred by bearing of the shank of the pin against the side of the hole. From the free bodies of the pin it may be noted the transfer between plate *A* and *B* is actually made via shear on the pin (the slight rotation of the pin due to unbalanced moment would be relatively negligible). There would be negligible friction between plates *A* and *B*. The earliest steel structures, particularly trusses, were actually connected by pins.

When a high-strength bolt is installed to have a specified initial tension, there will be an initial precompression between the pieces being joined, as shown in Fig. 4.6.3. A transfer of plate tension loads *P* as shown in Fig. 4.6.3 may then occur entirely via friction at service-load levels, and there may be no bearing of the bolt shank against the side of the hole. Until the friction force μT is overcome, the shear strength of the bolt and the bearing strength of the plate will not affect the ability to transfer load across the shear plane between plates.

The AISC Specifications [1.5, 1.16] recognize two general categories of performance requirement for high-strength bolted connections, for many years known as *bearing-type* and *friction-type*. The 1986 LRFD and 1989 ASD Specifications renamed the friction-type connection as a *slip-critical* connection.

The strength of all high-strength bolted connections in transmitting shear forces across a shear plane between steel elements is the same whether the connection is a *bearing-type* or a *slip-critical* connection. The slip-critical connection has *in addition* the serviceability requirement that slip must not occur at service load.

The possible "limit states", or failure modes, that may control the strength of a bolted connection are shown in Fig. 4.6.4.

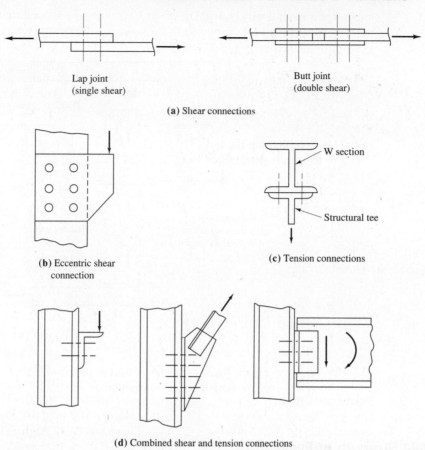

Lap joint
(single shear)

Butt joint
(double shear)

(**a**) Shear connections

(**b**) Eccentric shear
connection

(**c**) Tension connections

(**d**) Combined shear and tension connections

Figure 4.6.1   Typical bolted connections.

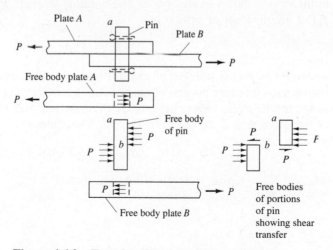

Figure 4.6.2   Transfer of load in pin connections.

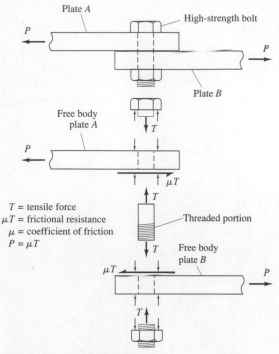

Figure 4.6.3 Transfer of load in pretensioned high-strength bolted connection.

Tensile Strength of Fasteners

In accordance with the fracture limit state in tension discussed in Sec. 3.2 and the failure mode shown in Fig. 4.6.4e, the nominal strength R_n (here using R_n instead of T_n) of one fastener in tension is

$$R_n = F_u^b A_n \qquad [3.2.2]$$

where F_u^b = tensile strength of the bolt material. The net area A_n should be the area through the threaded portion of the bolt, known as the "tensile stress area"*. The ratio of the tensile stress area to the gross area A_g ranges from 0.75 to 0.79. Thus, in terms of the gross area A_b of one bolt, Eq. 3.2.2 becomes

$$R_n = F_u^b(0.75A_b) \qquad (4.6.1)$$

where $\quad F_u^b$ = tensile strength of the bolt material
$\qquad A_b$ = gross cross-sectional area of one bolt

---

* Tensile stress area $= 0.785 + \left(d_b - \dfrac{0.9743}{n}\right)^2$ where n = number of threads per inch.

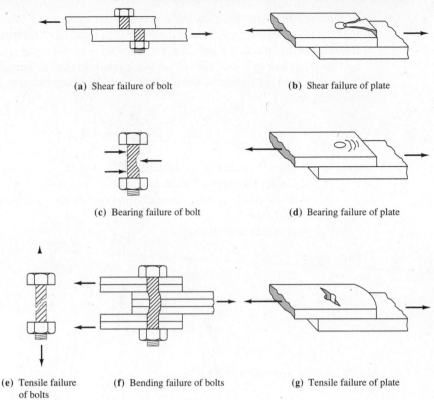

(a) Shear failure of bolt

(b) Shear failure of plate

(c) Bearing failure of bolt

(d) Bearing failure of plate

(e) Tensile failure
of bolts

(f) Bending failure of bolts

(g) Tensile failure of plate

Figure 4.6.4   Possible modes of failure of bolted connections.

Shear Strength of Fasteners

In accordance with the fracture limit state as the basis for fastener strength and the failure mode shown in Fig. 4.6.4a, the nominal strength R_n for one fastener will be the ultimate shear stress τ_u across the gross area A_b of the bolt times the number m of shear planes; thus,

$$R_n = mA_b\tau_u = mA_b(0.62F_u^b) \tag{4.6.2}$$

Note that the ultimate shear strength was found experimentally [3.1] to be about 62% of ultimate tensile strength; about the same ratio as for the yield strengths between shear and tension (see Sec. 2.6).

Bearing Strength

The bearing limit state relates to deformation around a bolt hole, as shown in Fig. 4.6.4d. A shear tear-out failure as shown in Fig. 4.6.4b is closely related to a bearing failure.

The bearing strength R_n is the force applied against the side of the hole to split or tear the plate. The larger the end distance L_e, measured from the center of the hole to the edge of the plate, the smaller the possibility of having a splitting failure.

Referring to Fig. 4.6.5, the actual tearing would occur along lines 1−1 and 2−2. As a lower bound for strength, the angle α could be taken as zero, giving the nominal strength R_n as

$$R_n = 2t\left[L_e - \frac{d}{2}\right]\tau_u^p \qquad (4.6.3)$$

where τ_u^p = shear strength of plate material $\approx 0.62F_u$
F_u = tensile strength of plate material
L_e = distance along line of force from edge of the connected part to the center of a hole
d = nominal bolt diameter.

Thus,

$$R_n = 2t\left[L_e - \frac{d}{2}\right](0.62F_u) \qquad (4.6.4)$$

$$R_n = 1.24F_u dt\left[\frac{L_e}{d} - \frac{1}{2}\right] \qquad (4.6.5)$$

which may be approximated as

$$R_n = F_u dt\left[\frac{L_e}{d}\right] = F_u t L_e \qquad (4.6.6)$$

which applies for a single bolt hole or for the bolt hole nearest the edge when there are two or more bolt holes in the line of force.

Experience and tradition has recommended that the center-to-center spacing of bolts be a minimum of $2\frac{2}{3}$ diameters [LRFD-J3.3; ASD-J3.8]. When $L_e = 2.67d$ is used in Eq. 4.6.5, the nominal bearing strength becomes

$$R_n = 3.0F_u dt \qquad (4.6.7)$$

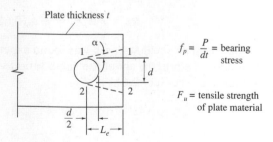

Figure 4.6.5 Bearing strength related to end distance.

which is the basic expression when tear-out is prevented. When a bearing strength represented by Eq. 4.6.7 is achieved and no rupture occurs, the elongation of the hole may be excessive. Thus, Eq. 4.6.7 should be used only when deformation around the hole is *not* of concern. Preferably lower strengths should be utilized.

4.7 LOAD AND RESISTANCE FACTOR DESIGN—FASTENERS

The general philosophy of Load and Resistance Factor Design (LRFD) was described in Secs. 1.8. and 1.9. Equation 1.8.1 gives the structural safety requirement, as follows:

$$\phi R_n \geq \sum \gamma_i Q_i \qquad [1.8.1]$$

where ϕ = resistance factor (strength reduction factor)
   R_n = nominal resistance (strength)
   γ_i = overload factors (LRFD-A4.1)
   Q_i = loads (such as dead load, live load, wind load, earthquake load) or load effects (such as bending moment, shear, axial force, and torsional moment resulting from the various loads)

Equation 1.8.1 requires the *design strength* ϕR_n to equal or exceed the summation of *factored loads,* or specifically for fasteners, Eq. 1.8.1 becomes

$$\phi R_n \geq P_u \qquad (4.7.1)$$

where ϕ = resistance factor, 0.75 for fracture in tension, shear on high-strength bolts, and bearing of bolt against side of hole
   R_n = nominal strength of one fastener
   P_u = factored load on one fastener

Since the loading is not related to the type of member, such as tension member or column, the various load factor combinations given in LRFD-A4.1 are the same for all members in the structure. However, the resistance factor ϕ accounts for the possibility that the actual strength of the fastener (or member) may be less than the theoretically computed strength because of variations in material properties and dimensional tolerances. These variations, while individually within accepted tolerance limits, may combine in the actual structure to give a strength less than the computed value. Neither the ϕ factor nor the overload factors γ are intended to account for careless errors in design or construction.

The strength of a fastener may be based on (1) shear, (2) bearing, (3) tension, or (4) combined shear and tension. The nominal strengths in tension, shear, and bearing were presented in Sec. 4.6.

Design Shear Strength—No Threads in Shear Planes

The nominal shear strength R_n for a single fastener, given by Eq. 4.6.2, was used by the 1986 LRFD Specification along with a ϕ factor of 0.65 to obtain the design shear strength for connection design. The 0.65 is lower than the ϕ factor normally used for shear-related behavior because of the calibration of the method to experimental results and past design practice. Particularly, the strength of long connections, up to 50 in., is lower than indicated by the sum of the strengths of individual fasteners. LRFD indicates no explicit adjustment for long connections unless they exceed 50 in.

The 1993 LRFD Specification uses $\phi = 0.75$, the commonly used ϕ factor for shear. This change requires the nominal shear strength per fastener be reduced to account for the approximately 20% reduction in strength of connections up to 50 in. long (see Ref. 3.1, p.100).

Thus, the nominal strength in shear is taken at 0.8 of that given by Eq. 4.6.2,

$$R_n = [\text{reduction for connection length}](0.62F_u^b)mA_b$$
$$= (0.8)(0.62F_u^b)mA_b = (0.50F_u^b)mA_b \tag{4.7.2}$$

and the design strength is

$$\phi R_n = 0.75(0.50F_u^b)mA_b \tag{4.7.3}$$

where $\phi = 0.75$, the standard ϕ value for shear
F_u^b = tensile strength of the bolt material (120 ksi for A325 bolts, 150 ksi for A490 bolts)
m = the number of shear planes participating [usually one (*single shear*) or two (*double shear*) as in Fig. 4.6.1a]
A_b = gross cross-sectional area across the unthreaded shank of the bolt

Design Shear Strength—Threads in Shear Planes

When threads are in the shear plane, the area at the root of the threads should be used in place of A_b. The thread root area is approximately 70% to 75% of the gross (shank) area. The nominal shear strength then becomes

$$R_n = [\text{reduction for connection length}](0.62F_u^b)m(0.75A_b)$$
$$= (0.8)(0.62F_u^b)m(0.75A_b) = (0.37F_u^b)mA_b \tag{4.7.4}$$

Because both the 0.8 connection length effect and the 0.75 for area through the threads are approximate, the 1993 LRFD Specification uses

$$R_n = (0.40F_u^b)mA_b \tag{4.7.5}$$

and the design strength is

$$\phi R_n = 0.75(0.40F_u^b)mA_b \tag{4.7.6}$$

Design Tension Strength

The design strength ϕR_n *based on the tension strength of the fastener,* according to LRFD-J3.6 and developed as Eq. 4.6.1, is

$$\phi R_n = \phi F_u^b(0.75A_b) = 0.75F_u^b(0.75A_b) \tag{4.7.7}$$

or

$$\phi R_n = 0.75(0.75F_u^b)A_b \tag{4.7.8}$$

where $\phi = 0.75$, a value for the tensile fracture mode as discussed in Chap. 3

F_u^b = tensile strength of the bolt material (120 ksi for A325 bolts; 150 ksi for A490 bolts)

A_b = gross cross-sectional area across the unthreaded shank of the bolt

Note that the $0.75A_b$ represents the area through the threaded portion of the bolt. LRFD-Table J3.2 indicates that the "Tensile Strength" is 90 ksi and 113 ksi, for A325 and A490 bolts, respectively; that is, $0.75F_u^b$. As is apparent from Eq. 4.7.7, multiplying the gross area A_b by $0.75F_u^b$ gives the correct value for R_n. Using the reduced stress on the gross area gives the same result as using the correct stress on the reduced area.

The design strengths for tension and shear on A325 and A490 bolts are summarized in Table 4.7.1.

Design Bearing Strength

Although the nominal bearing strength R_n is given by Eq. 4.6.7, excessive deformation around the hole may result. To prevent bolt elongations exceeding 0.25 in., LRFD-J3.10 uses reduced nominal strengths based on a deformation limit state for *usual situations*. The several categories in LRFD-J3.10 are as follows:

1. *Usual conditions* based on the deformation limit state, according to LRFD-Formula (J3-1a). This applies for all holes except long-slotted holes perpendicular to the line of force, where end distance L_e is at least 1.5 times the bolt diameter d, the center-to-center spacing s is at least $3d$, and there are two or more bolts in the line of force.

$$\phi R_n = \phi(2.4dtF_u) \tag{4.7.9}$$

TABLE 4.7.1 DESIGN STRENGTH* OF A325 AND A490 HIGH-STRENGTH BOLTS

Fastener	F_u^b (ksi)	Tension strength (ksi) $\phi = 0.75$	Shear strength (ksi) $\phi = 0.75$
A325 bolts, when threads are *not* excluded from shear planes	120	$\phi(0.75\,F_u^b)$ $0.75(90.0) = 67.5$	$\phi(0.40\,F_u^b)$ $0.75(48) = 36.0$
A325 bolts, when threads *are* excluded from shear planes	120	$\phi(0.75\,F_u^b)$ $0.75(90.0) = 67.5$	$\phi(0.50\,F_u^b)$ $0.75(60) = 45.0$
A490 bolts, when threads are *not* excluded from shear planes	150	$\phi(0.75\,F_u^b)$ $0.75(113) = 84.8$	$\phi(0.40\,F_u^b)$ $0.75(60) = 45.0$
A490 bolts, when threads *are* excluded from shear planes	150	$\phi(0.75\,F_u^b)$ $0.75(113) = 84.8$	$\phi(0.50\,F_u^b)$ $0.75(75) = 56.3$

* Design strength ϕR_n equals stress in table times gross bolt cross-sectional area A_b.

where $\phi = 0.75$

d = nominal diameter of bolt at unthreaded area

t = thickness of part against which bolt bears

F_u = tensile strength of connected part against which bolt bears

L_e = distance along line of force from the edge of the connected part to the center of a standard hole or the center of a short- and long-slotted hole perpendicular to the line of force.

2. Deformation limit state for long-slotted holes perpendicular to the line of force, where end distance L_e is at least 1.5 times the bolt diameter d, the center-to-center spacing s is at least $3d$, and there are two or more bolts in the line of force, according to LRFD-Formula (J3-1d),

$$\phi R_n = \phi(2.0dtF_u) \tag{4.7.10}$$

where $\phi = 0.75$

3. Strength limit state for the bolt nearest the edge, according to LRFD-Formulas (J3-1b), (J3-2a), and (J3-2c)

$$\phi R_n = \phi L_e t F_u \tag{4.7.11}$$

where $\phi = 0.75$. The upper limit to Eq. 4.7.11 is the appropriate value from Eqs. 4.7.9, 4.7.10, or 4.7.12.

4. Strength limit state when hole elongation exceeding 0.25 in. and hole "ovalization" can be tolerated, LRFD-Formulas (J3-1b) and (J3-1c) give,

$$\phi R_n = \phi(3.0dtF_u) \tag{4.7.12}$$

where $\phi = 0.75$.

Minimum Spacing of Bolts in Line of Transmitted Force

When the spacing of bolts in the direction of the transmitted forces is at least 3 diameters, Eqs. 4.7.9 or 4.7.10 are used. Under all other conditions, Eq. 4.7.11 applies. Solving Eq. 4.7.11 for the minimum distance L_e from the center of one fastener to the edge of the adjacent hole,

$$L_e \geq \frac{R_n}{F_u t} \tag{4.7.13}$$

Then, adding the radius $d/2$ of the bolt to Eq. 4.7.13 gives the minimum center-to-center spacing,

$$\text{Spacing} \geq \frac{R_n}{F_u t} + \frac{d}{2} \tag{4.7.14}$$

Because R_n in Eq. 4.7.14 is the *required* nominal strength, which equals the factored load P acting on one bolt divided by the resistance factor ϕ, Eq. 4.7.14 (LRFD-J3.10) becomes

$$\text{Spacing} \geq \frac{P}{\phi F_u t} + \frac{d}{2} \tag{4.7.15}$$

where $\phi = 0.75$

  $P =$ factored load acting on one bolt

  $F_u =$ tensile strength of *plate* material

  $t =$ thickness of plate material

  $d =$ diameter of the bolt

Equation 4.7.15 is the same as LRFD Formula (J3-1c) after substituting P/ϕ for R_n and solving for s. The minimum spacing of bolts in a line is *preferably* [LRFD-J3.3] 3 bolt diameters and shall not be less than $2\frac{2}{3}$ diameters.

Minimum End Distance in Direction of Transmitted Force

Equation 4.7.13 gives the requirement for end distance in the direction of the force on the bolt to prevent a rupture. When the usually accepted strengths given by Eqs. 4.7.9 and 4.7.10 are used, the minimum end distances must be at least $1\frac{1}{2}$ diameters. When higher bearing strengths are used, Eq. 4.7.13 gives the minimum end distance, as follows:

$$L_e \geq \frac{R_n}{F_u t} \qquad [4.7.13]$$

where the *required* nominal strength is the factored load divided by the resistance factor ϕ. Thus, Eq. 4.7.13 becomes

$$L_e \geq \frac{P}{\phi F_u t} \qquad (4.7.16)$$

where $\phi = 0.75$

  $P =$ factored load per bolt

  $F_u =$ tensile strength of the plate material

  $t =$ thickness of plate material

The end distance actually used must be the larger of that computed from a strength requirement and the minimum prescribed by Table 4.7.2 [LRFD-Table J3.4].

Equations 4.7.13 through 4.7.16 are *design* equations giving required spacing and end distance; LRFD-J3.10 gives those equations as R_n equations if the spacing and end distance is known. The designer will usually want to obtain the spacing and end distance necessary to carry the given factored loads P.

Maximum Edge Distance

In accordance with LRFD-J3.5, the maximum distance from the center of a bolt to the nearest edge is $12t$, where t is the thickness of the connected part, and this edge distance may not exceed 6 in. The purpose of this requirement is to prevent corrosion resulting from moisture entering the joint. The two contact surfaces of a joint may not be perfectly flat, and the clamping action will be lower when the bolts are far apart (or far from an edge).

Maximum Spacing of Connectors

The maximum longitudinal spacing of connectors between elements in continuous contact when the elements consist of a plate and a shape, or two plates, is given by LRFD-J3.5.

TABLE 4.7.2   MINIMUM EDGE DISTANCES* (CENTER OF
STANDARD HOLE** TO EDGE OF CONNECTED
PART) (ADAPTED FROM LRFD-TABLE J3.4;
ASD-TABLE J3.5)

| Nominal rivet or bolt diameter | | Minimum edge distance | | | |
| | | At sheared edges | | At rolled edges of plates, shapes or bars, or gas cut edges† | |
(in.)	(mm)	(in.)	(mm)	(in.)	(mm)
1/2	12.7	7/8	22.2	3/4	19.1
5/8	15.9	1 1/8	28.6	7/8	22.2
3/4	19.1	1 1/4	31.8	1	25.4
7/8	22.2	1 1/2‡	38.1	1 1/8	28.6
1	25.4	1 3/4‡	44.4	1 1/4	31.8
1 1/8	28.6	2	50.8	1 1/2	38.1
1 1/4	31.8	2 1/4	57.2	1 5/8	41.3
Over 1 1/4	Over 31.8	1 3/4 × diam		1 1/4 × diam	

*Lesser edge distances are permitted to be used provided Formulas from LRFD J3.10, as appropriate, are satisfied. (No applicable for ASD)

** When oversize or slotted holes are used, see LRFD-J3.10; ASD-J3.9.

† All edge distances in this column may be reduced $\frac{1}{8}$ in. (3.2 mm) when the hole is at a point where factored load does not exceed 25% of the maximum design strength in the element.

‡ These are permitted to be $1\frac{1}{4}$ in. at the ends of beam connection angles and shear end plates.

(a) For painted members or unpainted members not subject to corrosion,

$$s \leq 24t \leq 12 \text{ in.} \tag{4.7.17}$$

(b) For unpainted members of weathering steel subject to atmospheric corrosion,

$$s \leq 14t \leq 7 \text{ in.} \tag{4.7.18}$$

where t is the thickness of the thinner element.

4.8 EXAMPLES—TENSION MEMBER BEARING-TYPE CONNECTIONS—LRFD

The following examples illustrate the AISC Load and Resistance Design method for connections of tension members. The principles related to member strength are discussed in Sec. 3.9 and those that are fastener related are in Sec. 4.7.

EXAMPLE 4.8.1

Compute the tensile service load capacity for the bearing-type connection of two members in Fig. 4.8.1 if (a) the bolt threads are excluded from the shear plane and (b) the bolt threads are included in the shear plane. Use AISC LRFD Specification with $\frac{7}{8}$-in.-diam A325 bolts in standard holes and A572 Grade 50 steel plates. The service live load is three times the service dead load.

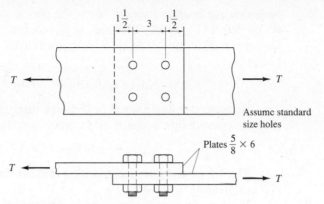

Figure 4.8.1 Example 4.8.1.

Solution

(a) Threads excluded from shear plane. First compute the strength of the plates as tension members (Chapter 3):

$$A_g = 6(0.625) = 3.75 \text{ sq in.}$$
$$A_n = [6 - 2(\tfrac{7}{8} + \tfrac{1}{8})]0.625 = 2.50 \text{ sq in.}$$
$$A_e = A_n = 2.50 \text{ sq in.}$$

Then, using Eqs. 3.9.2 and 3.9.3 gives

$$\phi T_n = \phi F_y A_g = 0.90(50)(3.75) = 169 \text{ kips}$$
$$\phi T_n = \phi F_u A_e = 0.75(65)(2.50) = 122 \text{ kips (controls)}$$

The action of the entire connection is a shear transfer of load between the two plates. The plane of contact may be thought of as the shear plane. When there is a single plane of contact involved in the load transfer, it is referred to as "single shear."

The design strength of the bolts in shear and bearing must be computed using Eqs. 4.7.3 and 4.7.9. The design strength ϕR_n per bolt in single shear for A325-X* is

$$\phi R_n = \phi(0.50F_u^b)mA_b \qquad [4.7.3]$$
$$= 0.75(0.50)(120)(1)(0.6013) = 27.1 \text{ kips/bolt}$$

Since $L_e \geq 1.5d$ and center-to-center spacing $s > 3d$, the design bearing strength for each bolt is

$$\phi R_n = \phi(2.4F_u dt)$$
$$= 0.75(2.4)(65)(0.875)(0.625) = 64 \text{ kips/bolt}$$

---

*Standard designation for an A325 bolt in a bearing-type connection with threads *eXcluded* from the shear planes.

Note that for bearing strength, F_u is the tensile strength of the plate material; 65 ksi for A572 Grade 50. Thus, shear controls and the design strength per fastener is 27.1 kips. The total design strength based on the fasteners then becomes

$$\phi T_n = (\text{No. of fasteners})\, R_n = 4(27.1) = 108 \text{ kips}$$

Because the single shear bolt design strength ($\phi T_n = 108$ kips) is lower than the plate design strength as a tension member ($\phi T_n = 122$ kips),

$$\phi T_n = 108 \text{ kips (for A325-X).}$$

The 1.5 in. end distance provided satisfies the minimum for sheared edges ($1\frac{1}{2}$ in.) given in Table 4.7.2.

Rupture shear strength, known as *block shear,* is not checked here (see Sec. 3.6); it is not likely to control when $\frac{5}{8}$-in. thickness of pieces is used.

To relate the design strength to the service load, Eq. 3.9.1 must be used,

$$\phi T_n \geq T_u \qquad [3.9.1]$$

Setting the factored service load T_u equal to the design strength ϕT_n will give the maximum acceptable service load. Using the gravity dead and live load factored load combination, Eq. 1.8.3 [LRFD-Formula(A4-2)],

$$T_u = 1.2D + 1.6L \qquad [1.8.3]$$

Substituting into Eq. 1.8.3 the live load L as three times the dead load D, and using ϕT_n as 108 kips in Eq. 3.9.1 gives

$$\phi T_n = T_u$$
$$108 = 1.2D + 1.6(3D) = 6.0D$$
$$D = 18 \text{ kips}$$
$$L = 3D = 54 \text{ kips}$$

Thus, the maximum total service load T permitted is

$$T = D + L = 18 + 54 = 72 \text{ kips (for A325-X)}$$

(b) Threads included in shear plane. The fastener design strength in single shear is less than computed in part (a). The design strength ϕR_n per bolt in single shear for A325-N* is

$$\phi R_n = \phi(0.40 F_u^b)mA_b \qquad [4.7.6]$$
$$= 0.75(0.40)(120)(1)(0.6013) = 21.6 \text{ kips/bolt}$$

Strength in single shear again governs, since there are no changes in the net section tension member strength, or in bearing strength. Since ϕR_n is 0.8 of that

---

*Standard designation for an A325 bolt in a bearing-type connection with threads *iNcluded* in the shear planes.

in part (a), the service load capacity T is

$$T = 0.8(72) = 57.6 \text{ kips (for A325-N)} \quad \blacksquare\blacksquare$$

EXAMPLE 4.8.2 _____

Determine the number of $\frac{3}{4}$-in.-diam A325 bolts required to develop the full strength of the A572 Grade 65 steel plates in Fig. 4.8.2 (a portion of a double lap splice connection) for a bearing-type connection with threads excluded from the shear planes. Use LRFD Specification and assume a double row of bolts with standard size holes.

Solution. It is determined by inspection that the cross-sectional area of the center plate is less than the sum of the areas of the two outer plates. Therefore in this case one need check only the strength of the center plate:

$$A_n = [6 - 2(\tfrac{3}{4} + \tfrac{1}{8})]0.375 = 1.59 \text{ sq in.}$$

$$\text{Max } A_n = 0.85A_{\text{gross}} = 0.85(6)0.375 = 1.91 \text{ sq in.}$$

$$A_e = A_n = 1.59 \text{ sq in.}$$

Then, using Eqs. 3.9.2 and 3.9.3 gives

$$\phi T_n = \phi F_y A_g = 0.90(65)(6)(0.375) = 132 \text{ kips}$$

$$\phi T_n = \phi F_u A_e = 0.75(80)(1.59) = 96 \text{ kips (controls)}$$

Thus, the design strength of the plates in tension is 96 kips.

In this case the action of the entire connection may be thought of as a shear transfer between plates occurring at the *two* planes of contact. When connectors are positioned so that they cross two shear planes of contact, it is called "double shear." *Double shear* is a symmetrically loaded situation insofar as the shear planes and directions of shear transfer are concerned, whereas the single shear case is unsymmetrical (Fig. 4.8.1).

The design strength in double shear (per bolt) is

$$\phi R_n = \phi(0.50F_u^b)mA_b = 0.75(0.50)(120)(2)(0.4418) = 39.8 \text{ kips/bolt}$$

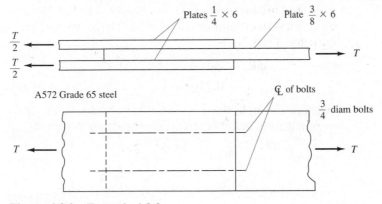

Figure 4.8.2   Example 4.8.2.

The design strength in bearing on the $\frac{3}{8}$-in. plate is

$$\phi R_n = \phi(2.4 F_u dt) = 0.75(2.4)(80)(0.75)(0.375) = 40.5 \text{ kips/bolt}$$

$$\text{Number of bolts} = \frac{96}{39.8} = 2.4$$

The end distance must be at least 1.25 in. in accordance with Table 4.7.2. <u>Use 4—$\frac{3}{4}$-in.-diam A325 bolts (A325-X).</u>

The shear rupture limit state (block shear) should also be checked; however, since bearing does not control in other regards, the shear rupture mode is unlikely to control. ■■

EXAMPLE 4.8.3 _____

Determine the number of $\frac{3}{4}$-in.-diam A325 bolts in standard size holes required to carry 7 kips dead load and 43 kips live load on the plates in Fig. 4.8.3 if A36 steel is used. Assume the portion of the double lap splice is a bearing-type connection with threads excluded from the shear plane and a double row of bolts. Use Load and Resistance Factor Design.

Solution.   The factored load T_u that must be carried is

$$T_u = 1.2D + 1.6L = 1.2(7) + 1.6(43) = 77 \text{ kips}$$

In this example, the center plate will control the tension strength:

$$A_g = 10(0.25) = 2.50 \text{ sq in.}$$
$$A_n = [10 - 2(\tfrac{3}{4} + \tfrac{1}{8})]0.25 = 2.06 \text{ sq in.}$$
$$\text{Max } A_n = 0.85 A_g = 0.85(2.50) = 2.13 \text{ sq in.}$$
$$\therefore A_e = A_n = 2.06 \text{ sq in.}$$
$$\phi T_n = \phi F_y A_g = 0.90(36)(2.50) = 81 \text{ kips (controls)}$$
$$\phi T_n = \phi F_u A_e = 0.75(58)(2.06) = 90 \text{ kips}$$

and this exceeds the factored load on the members (i.e., $\phi T_n > T_u$).

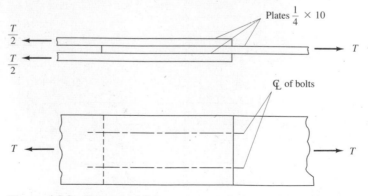

Figure 4.8.3   Example 4.8.3.

For shear, using A325-X,

$$\phi R_n = \phi(0.50F_u^b)mA_b = 0.75(0.50)(120)(2)(0.4418) = 39.8 \text{ kips/bolt}$$

For bearing on the $\frac{1}{4}$-in. plate, when $L_e \geq 1.5d$ and $s \geq 3d$,

$$\phi R_n = \phi(2.4F_u dt) = 0.75(2.4)(58)(0.75)(0.25) = 19.6 \text{ kips/bolt}$$

$$\text{Number of bolts} = \frac{77}{19.6} = 3.9 \quad \underline{\textit{Use } 4 \text{ bolts}}$$

The shear rupture limit state (block shear) should also be checked; particularly here, since the plate ($\frac{1}{4}$-in.) is thin and heavily loaded. Referring to Fig. 4.8.4, the preliminary arrangement for four bolts is to use 2 in. edge distances at the sides and end, and 3 in. longitudinal spacing.

Investigate the block shear strength of the resulting arrangement using LRFD-J4.3.

The block shear potential failure along path $a-b-c$ and $d-e-f$ of Fig. 4.8.4 must be investigated according to LRFD-J4. Calculating the net areas A_{nv} and A_{nt},

$$A_{nv} = (\text{length } b-c \text{ and } e-f \text{ less } 1.5 \text{ holes}) \text{ thickness}$$

$$= 2\left[5 - 1.5\left(0.75 + \frac{1}{8}\right)\right]0.25 = 1.84 \text{ sq in.}$$

$$A_{nt} = (\text{length } a-b \text{ and } d-e \text{ less } 0.5 \text{ holes}) \text{ thickness}$$

$$= 2\left[2 - 0.5\left(0.75 + \frac{1}{8}\right)\right]0.25 = 0.78 \text{ sq in.}$$

Compare $0.6F_u A_{nv}$ with $F_u A_{nt}$,

$$[0.6F_u A_{nv} = 0.6(58)1.84 = 64.0] > [F_u A_{nt} = 58(0.78) = 45.2],$$

which means Eq. 3.9.7 [shear fracture, tension yielding] controls,

$$\phi R_{bs} = \phi(0.6F_u A_{nv} + F_y A_{gt}) \qquad\qquad [3.9.7]$$

$$\phi R_{bs} = 0.75(64.0 + F_y A_{gt})$$

$$= 0.75[64.0 + 36(4.0)0.25] = 0.75(100.0) = 75 \text{ kips}.$$

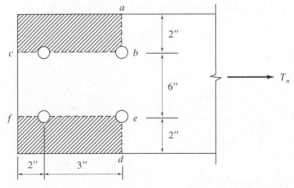

Figure 4.8.4 Block shear failure on plate of Example 4.8.3.

Thus, block shear [75 kips] controls over tension yielding on gross section [81 kips] or fracture on effective net section [90 kips]. The design strength is slightly below the factored load [77 kips] that must be carried; an unsatisfactory situation. Increase end distance to 2.5. in.

Use 2 in. edge distances along the sides, 3 in. longitudinal spacing, and $2\frac{1}{2}$ in. end distance. ■■

4.9 SLIP-CRITICAL JOINTS

When slip resistance at service load is desired, the joint is referred to as a *slip-critical joint*. In the 1978 AISC Specification, such connections were termed *friction-type connections*. All tensioned high-strength bolted connections actually resist load by friction. Referring to Fig. 4.6.3, the pretension force T in the bolt equals the clamping force between the pieces being joined. The resistance to shear is a frictional force μT, where μ is the coefficient of friction.

The coefficient of friction, or more properly the *slip coefficient,* depends on the surface condition, with such items as mill scale, oil, paint, or special surface treatments determining the value of μ.

Slip is defined as occurring when "the friction bond is definitely broken and the two surfaces slip with respect to one another by a relatively large amount" [4.14]. The range of μ varies from 0.2 to 0.6 depending on the surface condition [3.1].

To avoid directly using the coefficient of friction and to permit design of slip-critical joints using the same general approach as for bearing-type connections, the friction force μT is divided by the bolt gross cross-sectional area A_b to obtain a so-called "shear stress" on the bolt.

EXAMPLE 4.9.1 _____

Determine the amount of force P required to cause slip of a $\frac{7}{8}$-in.-diam A325 bolt loaded as in Fig. 4.9.1, if the slip coefficient μ is 0.33 (a typical value for the usual "clean mill scale" surface condition). Using the service load force P, compute the "shear stress", $f_v = P/A_b$.

Solution.   Using the pretension load from Table 4.4.2,

$$T_b = 39 \text{ kips}$$
$$P = \mu T_b = 0.33(39) = 12.9 \text{ kips}$$

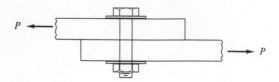

Figure 4.9.1   Example 4.9.1.

Since the overall action of the connection is a shearing effect, the "shear stress" f_v in the $\frac{7}{8}$-in.-diam bolt at the load causing slip to begin is

$$f_v = \frac{P}{A_b} = \frac{12.9}{0.6013} = 21.4 \text{ ksi} \quad \blacksquare\blacksquare$$

The AISC Specifications [1.5, 1.16] use the "shear stress" approach to provide adequate slip resistance in joints where slip at *service load* cannot be tolerated. The limit state of slip in the joint is a serviceability requirement. The actual failure of fasteners in a joint will be as discussed in Sec. 4.6; that is, a shear failure of the bolt, tension failure of the bolt, or bearing failure in the connected material.

Because resistance to slip in "slip-critical" connections is a serviceability (rather than strength) limit state that must be investigated at service load, the limiting "shear stresses" to be used are *in concept* the same for LRFD as were used in the 1978 ASD Specification. The LRFD maximum acceptable service load "shear stress" F_v for Class A surface condition (see Ref. 4.7, Sec. 5(b), for data on other surface conditions) appears as Table 4.9.1.

As may be noted, for the situation of Example 4.9.1 the LRFD-J3.8a allowable "shear stress" F_v of 17.0 ksi for A325 bolts is used for slip-critical connections containing standard holes. The margin of safety against slip provided would be $21.4/17.0 = 1.26$. Since slip is a serviceability requirement, rather than a strength requirement, a margin of safety lower than that used for strength is reasonable. The AISC has used for this situation a safety factor against slip about 70% of that used for strength.

Overcoming slip does *not* imply that a failure mode has been reached. However, when connections are subject to stress reversal there is greater concern regarding any slip at service load. Repeated loading may introduce fatigue concerns if slip is occurring, particularly when oversized or slotted holes are used.

The factor of safety against slip computed above is typical; for other bolt sizes and for both A325 and A490 bolts the variation may be observed from Table 4.9.2. The clamping force is the initial tension from Table 4.4.2.

TABLE 4.9.1   MAXIMUM ACCEPTABLE SERVICE LOAD "SHEAR STRESS" F_v[†]
FOR HIGH-STRENGTH BOLTS IN SLIP-CRITICAL CONNECTIONS[‡]
(FROM LRFD SPECIFICATION, TABLE J3.6)

Type of bolt	Nominal Resistance* to shear (ksi)		
	Standard size holes	Oversized and short-slotted holes	Long-slotted holes
A325	17	15	12
A490	21	18	15

[†] F_v is the unfactored service load shear divided by the nominal bolt area A_b.
[‡] For Class A surface condition.
* For each shear plane.

TABLE 4.9.2 FACTOR OF SAFETY AGAINST SLIP, SLIP-CRITICAL JOINTS (COEFFICIENT OF SLIP = 0.33)

Bolt diameter (in.)	Nominal area (sq in.)	Nominal area (mm²)	Factor of safety A325	Factor of safety A490
5/8	0.3068	197.9	1.20	1.23
3/4	0.4418	285.0	1.23	1.24
7/8	0.6013	387.9	1.26	1.28
1	0.7854	506.7	1.26	1.28
1 1/8	0.9940	641.3	1.09	1.26
1 1/4	1.2272	791.7	1.12	1.31

The reliability index β (see Sec. 1.8) is higher for slip-resistance when bolts are installed by the turn-of-the-nut method than when they are installed by calibrated wrench [4.16]. The AISC Specifications [1.5, 1.16] do not recognize the difference; the values assume a 10% probability of slip with bolts installed by the calibrated wrench method. The slip-critical nominal resistances F_v in LRFD-J3.8a and ASD-J3.6 are rounded lower and slightly reduced from those used in 1978 ASD-1.5.2.2.

Because slip-resistance certainly must relate to the contact area between pieces, the F_v values used for standard holes are reduced when larger holes reduce the contact area. Dimensions for oversized and slotted holes are given in LRFD-Table J3.3. The F_v values for oversized and short-slotted holes are 85% and those for long-slotted holes are approximately 70% of the values for standard holes.

Design of Slip-Critical Connections—LRFD Specification

The design of slip-critical connections requires full consideration of the strength limit states. The strength of the fasteners in shear, bearing, and direct tension, as discussed in Secs. 4.6 and 4.7 must be investigated. Sufficient strength must be provided to resist *factored loads. In addition,* the service load that must be transferred by friction without slip must not exceed that indicated in LRFD-J3.8a, presented here in Table 4.9.1. Galambos, Reinhold, and Ellingwood [4.16] have presented an excellent statistical study of connection slip as a serviceability limit state.

Though it is illogical to calculate for a service load design criterion (slip resistance) using factored loads, the 1993 LRFD Specification has added this option. LRFD-J3.9b directs the designer to LRFD-Appendix J3.9b, where the design slip resistance ϕR_{str} per bolt at factored service loads is given as

$$\phi R_{str} = \phi 1.13 \mu T_i m \qquad (4.9.1)$$

where R_{str} = nominal slip resistance per bolt at factored loads

        m = number of slip (shear) planes

        T_i = minimum fastener initial tension given in Table 4.4.2 (LRFD-Table J3.1)

        μ = mean slip coefficient, as applicable, or as established by tests

          = 0.33 for Class A surface condition

          = 0.50 for Class B surface condition

          = 0.40 for Class C surface condition

ϕ = 1.0 for standard holes

= 0.85 for oversize and short-slotted holes

= 0.70 for long-slotted holes transverse to load

= 0.60 for long-slotted holes parallel to load

EXAMPLE 4.9.2

Determine the tensile capacity of the connection previously investigated in Example 4.8.1 as a bearing-type connection (Fig. 4.8.1); however, consider it as a slip-critical connection with the usual clean mill scale (Class A) surface condition. Use $\frac{7}{8}$-in.-diam A325 bolts in standard holes with A572 Grade 50 plate material. Use Load and Resistance Factor Design.

Solution. The tension member (plates) design strength ϕT_n was determined in Example 4.8.1 to be

$$\phi T_n = 122 \text{ kips (based on } \phi F_u A_e)$$

From Table 4.9.1, the service load capacity per bolt in single shear is

$$R = mF_v A_b = 1(17)(0.6013) = 10.2 \text{ kips/bolt}$$
$$T = 4(10.2) = 40.9 \text{ kips}$$

This service load capacity T based on the slip-resistance is lower than the bolt-related strength determined service load $T = 72$ kips computed in Example 4.8.1 for A325-X; thus, $T = 40.9$ kips. ■■

The requirements for spacing, end, and edge distances are all identical whether the joint is a bearing-type connection or a slip-critical connection. Since the factored load per bolt will be lower when slip-resistance governs the number of fasteners used, the bearing-related equations for spacing of fasteners and end distance will permit smaller spacing and end distances. Of course, the minimums are then more likely to control. Generally, slip-resistance controls in slip-critical connections, rather than strength in shear or bearing.

EXAMPLE 4.9.3

Redesign the connection for Fig. 4.8.2 as a slip-critical connection using *oversized* holes and a Class A surface condition. Use $\frac{3}{4}$-in.-diam A325 bolts in a double row and AISC LRFD Specification. The live load is four times the dead load.

Solution. The design strength ϕT_n based on plate strength in tension was computed in Example 4.8.2 to be 96 kips. Using the relationship between factored load and service load, Eq. 1.8.3,

$$T_u = 1.2D + 1.6L = 96 \text{ kips}$$
$$= 1.2D + 1.6(4D) = 7.6D$$
$$D = \frac{96}{7.6} = 12.6 \text{ kips}$$
$$T = D + L = 12.6 + 4(12.6) = 63 \text{ kips}$$

The service load capacity R per bolt based on shear (double shear) in a slip-critical connection is, from Table 4.9.1,

$$R = F_v m A_b = 15.0(2)0.4418 = 13.3 \text{ kips/bolt}$$

$$\text{Number of bolts} = \frac{T}{R} = \frac{63}{13.3} = 4.7$$

Use 6—$\frac{3}{4}$-in.-diam bolts (A325-SC).

Note that bolts in a slip-critical connection are designated with an SC following the ASTM Specification number, as indicated above. This connection only required four bolts as a bearing-type connection in Example 4.8.2. ■■

4.10 ALLOWABLE STRESS DESIGN—FASTENERS

The general philosophy of Allowable Stress Design (ASD) was described in Secs. 1.8 and 1.9. Equation 1.8.8 gives the structural safety requirement, as follows:

$$\frac{\phi R_n}{\gamma} \geq \sum Q_i \qquad [1.8.8]$$

which expresses that the design strength ϕR_n divided by a factor of safety γ must exceed the sum of the service loads. In the allowable stress design method, the safety provision is γ/ϕ.

In terms of fasteners, the right hand side of Eq. 1.8.8 becomes the total service load R *per fastener* in shear, tension, or bearing, as the case may be. Then dividing both sides by the safety provision γ/ϕ gives

$$\left[\frac{R_n}{\gamma/\phi} = \frac{R_n}{\text{FS}} \right] \geq R \qquad (4.10.1)$$

Since the underlying philosophy of the allowable stress method is to compute "stresses" at service load, Eq. 4.10.1 may be divided by the bolt cross-sectional area, and the sides of the equation may be interchanged (reversing the inequality) giving

$$\left[f = \frac{R}{A_b} \right] \leq \left[F = \frac{R_n}{(\text{FS})A_b} \right] \qquad (4.10.2)$$

Thus, the "nominal" stress f has the units of force per unit area, and the right side of the equation becomes an *allowable stress F*. Note that only when the bolt is installed without initial tension will the actual stress be R/A_b; initial tension in the bolt will cause part or all of the service load to be transferred by friction, giving rise to the fictitious "nominal" stress f acting on the bolt cross-sectional area. In either case, Eq. 4.10.2 properly expresses the safety requirement.

Shear Strength of Fasteners

The strength in shear of a high-strength bolt is given by Eq. 4.6.2 when threads are excluded from the shear planes,

$$R_n = m A_b (0.62 F_u^b) \qquad [4.6.2]$$

TABLE 4.10.1 ALLOWABLE SHEAR STRESS F_v FOR
BEARING-TYPE* CONNECTIONS
(FROM ASD-TABLE J3.2)

	F_v	
Connection	(ksi)	(MPa)
A325-N, threads *iNcluded* in shear planes	21	140
A325-X, threads *eXcluded* from shear planes	30	210
A490-N, threads *iNcluded* in shear planes	28	190
A490-X, threads *eXcluded* from shear planes	40	280

* Bolts in slip-critical joints (i.e., friction-type connections) must also satisfy these allowable stresses, *in addition to* satisfying the allowable stresses relating to slip resistance as discussed in Sec. 4.9.

In terms of the working stress (or allowable stress) design method, the strength is divided by a safety factor γ/ϕ. Thus, using the traditional factor of safety of about 2.5 for medium length joints (typically 2.0 for short joints to 3.0 for long joints), Eq. 4.6.2 substituted into Eq. 4.10.2 gives the allowable shear stress F_v,

$$F_v = \frac{0.62F_u^b}{2.5} = 0.25\,F_u^b \qquad (4.10.3)$$

which for A325 ($F_u^b = 120$ ksi) becomes 30 ksi, and for A490 ($F_u^b = 150$ ksi) becomes 37.5 ksi. ASD-J3.4 uses $F_v = 30$ ksi for A325-X (threads eXcluded from shear planes) and $F_v = 40$ ksi for A490-X. The adjustment from 37.5 (i.e., $0.25F_u^b$) to 40 reflects approximation in the 2.5 factor of safety, and what constitutes the connection length to be used as *standard*.

When threads are possible within the shear planes, the cross-sectional area should be reduced to that measured through the threads. Either a smaller area should be used in computing f or the allowable stress F_v must be reduced. AISC has chosen to have the nominal stress f continue to be computed using the gross area A_b; thus, the allowable stress must be reduced. The ratio of tensile stress area* to the gross area is approximately 0.75 to 0.79; therefore, the allowable stress F_v is conservatively taken as 70% of the A325-X or A490-X values to give $F_v = 21$ ksi for A325-N (threads iNcluded within shear planes) and $F_v = 28$ ksi for A490-N.

The allowable shear stresses F_v for A325 and A490 bolts in bearing-type connections are summarized in Table 4.10.1.

Bearing Strength of Fasteners

The nominal strength in bearing, from Eq. 4.7.9, is for usual conditions in standard and short-slotted holes,

$$R_n = 2.4F_u dt$$

---

* Tensile stress area $= 0.7.85 + \left(d_b - \dfrac{0.9743}{n}\right)^2$ where n = number of threads per inch.

Following the general working stress format of Eq. 4.10.1, the strength in bearing is divided by the factor of safety FS and both sides are divided by the nominal bearing area dt to obtain

$$\left[f_p = \frac{R}{dt} \right] \le \left[F_p = \frac{R_n}{(FS)dt} \right]$$

(4.10.4)

where f_p = nominal bearing stress under service load (the subscript p is used for *pressure*, rather than b which is used later for *bending*)

d = nominal diameter of bolt

t = thickness of plate against which bolt bears

F_p = allowable bearing stress

When the factor of safety of 2 traditionally used for bearing is applied, the allowable bearing stress F_p becomes

$$F_p = \frac{2.4F_u dt}{2.0dt} = 1.2F_u$$

(4.10.5)

which is the value of ASD-J3.7 for standard holes. When deformation around the hole is not a consideration, the following higher value is permitted:

$$F_p = 1.5F_u$$

(4.10.6)

The 25% increase in F_p is consistent with the increase from 2.4 to 3.0 associated with the coefficients in Eqs. 4.7.9 and 4.7.12, respectively.

Slip-Critical (i.e., Friction-Type) Connections

The limitations on shear stress that are imposed by Allowable Stress Design to prevent slip are identical to those imposed by Load and Resistance Factor Design. This is to be expected because resistance to slip at *service load* is the performance requirement being enforced. Consequently, the "shear stress" limits in Table 4.9.1 are also the ASD allowable shear stresses F_v for slip-critical connections that appear in ASD-Table J3.2.

Of course, slip-critical connections must also be designed so that they have adequate *strength*. This is accomplished by satisfying the allowable stress limitations for shear (Table 4.10.1) and bearing ($f_p \le F_p$) in bearing-type connections. For Class A surface conditions, the strength-based limitations on shear stress in Table 4.10.1 will be satisfied automatically because the allowable stresses associated with the prevention of slip are all smaller than the allowable shear stresses for bearing-type connections. Thus, the usual procedure for slip-critical connection design in ASD will be to determine the number of fasteners based upon the allowable stresses in Table 4.9.1 and then to check for adequate strength in bearing, comparing to the allowable stress in either Eq. 4.10.5 or 4.10.6. Only for relatively thin plates would the number of fasteners have to be increased.

Minimum Spacing and Minimum End Distance of Bolts in Line of Transmitted Force

These requirements are essentially the same for ASD as for LRFD, as described in Sec. 4.7.

4.11 EXAMPLES—TENSION MEMBERS USING ALLOWABLE STRESS DESIGN

EXAMPLE 4.11.1

Investigate the tension member connection of Example 4.8.1 (Fig. 4.8.1) to carry a total service load of 75 kips. The connection is a bearing-type connection, with threads excluded from the shear planes, using $\frac{7}{8}$-in.-diam A325 bolts in standard holes. The plates are A572 Grade 50 steel. Use the AISC ASD Specification.

Solution

(a) Tension member capacity. The areas as computed in Example 4.8.1 are

$$A_e = A_n = 2.50 \text{ sq in.}$$
$$A_g = 3.75 \text{ sq in.}$$

Using Eqs. 3.11.4 and 3.11.5,

$$f_a = \frac{T}{A_g} = \frac{75}{3.75} = 20 \text{ ksi} < [0.60F_y = 30 \text{ ksi}] \quad \text{OK}$$

$$f_a = \frac{T}{A_e} = \frac{75}{2.50} = 30 \text{ ksi} < [0.50F_u = 32.5 \text{ ksi}] \quad \text{OK}$$

(b) Fasteners. The allowable capacity R in single shear is, from Table 4.10.1,

$$R\text{(single shear)} = mA_bF_v = 1(0.6013)30 = 18.0 \text{ kips/bolt}$$

The allowable capacity in bearing when deformation around the hole *is* of concern (ASD-J3.7.1), and $L_e \geq 1.5d$ and $s \geq 3d$, is

$$R\text{(bearing)} = 1.2F_u dt = 1.2(65)(\tfrac{7}{8})(\tfrac{5}{8}) = 42.7 \text{ kips/bolt}$$

The shear value is the smaller of the two (18.0 and 42.7) and therefore controls:

$$R = 18.0 \text{ kips/bolt}$$
$$T = \text{(Number of bolts)}R = 4(18.0) = 72 \text{ kips}$$

Thus, a service load of 75 kips will make the nominal shear stress exceed the allowable value of 30 ksi by about 4%, generally too high to accept. Note that this connection investigated by Load and Resistance Factor Design in Example 4.8.1 also obtained a service load capacity of 72 kips, assuming that live load represents 75% of the total service load. ■■

EXAMPLE 4.11.2

Investigate the acceptability of the connection of Fig. 4.8.1 to serve as a slip-critical connection carrying a service load tension of 41 kips. The connection consists of four $\frac{7}{8}$-in.-diam A325 bolts in standard holes connecting A572 Grade 50 steel plates having Class A surface condition. Use the AISC ASD Specification.

Solution

(a) Tension member capacity. The areas as computed in Example 4.8.1 are

$$A_e = A_n = 2.50 \text{ sq in.}$$
$$A_g = 3.75 \text{ sq in.}$$

Using Eqs. 3.11.4 and 3.11.5,

$$f_a = \frac{T}{A_g} = \frac{41}{3.75} = 10.9 \text{ ksi} < [0.60F_y = 30 \text{ ksi}] \quad \text{OK}$$

$$f_a = \frac{T}{A_e} = \frac{41}{2.50} = 16.4 \text{ ksi} < [0.50F_u = 32.5 \text{ ksi}] \quad \text{OK}$$

(b) Fasteners—slip resistance. The allowable capacity R in single shear is, from Table 4.9.1.

$$R(\text{single shear}) = mA_b F_v = 1(0.6013)17 = 10.2 \text{ kips/bolt}$$

(c) Fasteners—strength (i.e., check as if a bearing-type connection). The allowable capacity R in single shear is, from Table 4.10.1, for threads possibly within the shear planes,

$$R(\text{single shear}) = mA_b F_v = 1(0.6013)21 = 12.6 \text{ kips/bolt}$$

The allowable capacity in bearing (for $L_e \geq 1.5d$ and $s \geq d$) is

$$R(\text{bearing}) = 1.2F_u dt = 1.2(65)(\tfrac{7}{8})(\tfrac{5}{8}) = 42.7 \text{ kips}$$

(d) Allowable service load capacity. The result here is the usual situation; i.e., the slip-critical shear capacity governs. Since the serviceability-related value of 10.2 kips is less than strength-related values of 12.6 and 42.7 kips, the allowable load is 10.2 kips/bolt. Thus,

$$T = (\text{Number of bolts})R = 4(10.2) = 40.8 \text{ kips}$$

The applied load of 41 kips is essentially the allowable service load permitted on the member with its connection. ■■

4.12 ECCENTRIC SHEAR

When the load P is applied on a line of action that does not pass through the center of gravity of a bolt group, there will be an eccentric loading effect, such as in Fig. 4.12.1. A load P at an eccentricity e, as shown in Fig. 4.12.2, is statically equivalent to a moment P times e plus a concentric load P both acting on the connection. Since both the moment and the concentric load contribute shear effects on the bolt group, the situation is referred to as *eccentric shear*.

Only in the past few years has significant experimental work [3.1, 4.17, 4.18] been done to evaluate the strength of such joints. Essentially, two approaches are available to the designer: (1) the traditional elastic (vector) analysis assuming no friction with the plates rigid and the fasteners elastic; and (2) a strength analysis

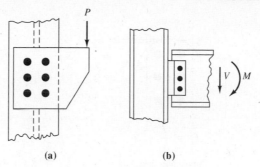

Figure 4.12.1   Typical eccentric shear connections.

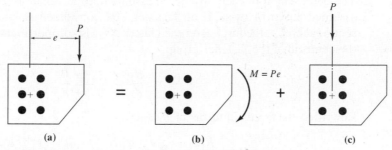

Figure 4.12.2   Combined moment and direct shear.

(plastic analysis) wherein it is assumed the eccentrically loaded fastener group rotates about an instantaneous center of rotation, and the deformation at each fastener is proportional to its distance from the center of rotation.

Since the AISC Specifications [1.5, 1.16] prescribe the capacity (allowable in ASD and design strength in LRFD) for a fastener but do not specify the means of analysis, any rational approach may be used by the designer. Several common approaches are treated in the remainder of this section.

Traditional Elastic (Vector) Analysis

For many years eccentrically loaded fastener groups have been analyzed by considering the fastener group areas as an elastic cross-section subjected to direct shear and torsion. The stresses resulting are nominal in the sense that they have stress units (say, psi) and provide a guide to safety but are not real stresses because the service loads are actually carried by friction. This elastic analysis method has been used because it makes use of simple mechanics of materials concepts and has been found to be a conservative procedure.

To develop the equations for use in this procedure, consider first the connection acted upon by the moment M, as shown in Fig. 4.12.3. Neglecting friction between the plates, the moment equals the sum of the forces shown in Fig. 4.12.3b times their distances to the centroid (CG) of the fastener areas:

$$M = R_1 d_1 + R_2 d_2 + \cdots + R_6 d_6 = \sum R d \qquad (4.12.1)$$

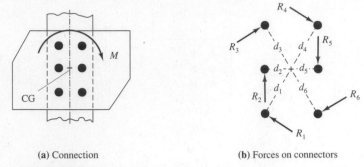

(a) Connection   **(b)** Forces on connectors

Figure 4.12.3   Pure moment connection.

The deformation in each fastener is assumed proportional to its distance d from the assumed center of twist. If all fasteners are considered elastic and of equal cross-sectional area A, the force R on each fastener is also proportional to its distance d from the centroid of the fastener group,

$$\frac{R_1}{d_1} = \frac{R_2}{d_2} = \cdots = \frac{R_6}{d_6} \qquad (4.12.2)$$

Rewriting the forces in terms of R_1 and d_1,

$$R_1 = \frac{R_1 d_1}{d_1}; \quad R_2 = \frac{R_1 d_2}{d_1}; \quad \ldots; \quad R_6 = \frac{R_1 d_6}{d_1} \qquad (4.12.3)$$

Substituting Eqs. 4.12.3 into Eq. 4.12.1 gives

$$M = \frac{R_1 d_1^2}{d_1} + \frac{R_1 d_2^2}{d_1} + \cdots + \frac{R_1 d_6^2}{d_1}$$

$$= \frac{R_1}{d_1}[d_1^2 + d_2^2 + d_3^2 + \cdots + d_6^2] \qquad (4.12.4a)$$

$$= \frac{R_1}{d_1} \sum d^2 \qquad (4.12.4b)$$

The force in fastener 1 is therefore

$$R_1 = \frac{M d_1}{\sum d^2} \qquad (4.12.5a)$$

and by similar reasoning, the forces on the other fasteners are

$$R_2 = \frac{M d_2}{\sum d^2}; \quad R_3 = \frac{M d_3}{\sum d^2}; \quad \ldots; \quad R_6 = \frac{M d_6}{\sum d^2} \qquad (4.12.5b)$$

or in general,

$$R = \frac{M d}{\sum d^2} \qquad (4.12.6)$$

which gives the force R on the fastener at the distance d from the center of rotation.

Note that if stress is desired, Eq. 4.12.6 may be divided by the bolt cross-sectional area A_b to give stress $f = R/A_b$. The denominator then would become $\Sigma\,Ad^2$, which is the polar moment of intertia J about the center of rotation for the series of bolt cross-sectional areas. Equation 4.12.6 is essentially the familiar mechanics of materials formula for torsion on a circular shaft, Tr/J, which is discussed in Sec. 8.2. The torsional moment T is M; the radius r from the center of rotation to the point at which the stress is computed would be the distance d.

It is usually convenient to work with the horizontal and vertical components of R, R_x and R_y, respectively, obtained when the horizontal and vertical components of d, x and y, respectively, are used in Eq. 4.12.6. From Fig. 4.12.4,

$$R_x = \frac{y}{d}R \quad \text{and} \quad R_y = \frac{x}{d}R \tag{4.12.7}$$

Substituting Eq. 4.12.7 into Eq. 4.12.6 gives

$$R_x = \frac{My}{\Sigma d^2} \quad \text{and} \quad R_y = \frac{Mx}{\Sigma d^2} \tag{4.12.8}$$

Noting that $d^2 = x^2 + y^2$, Eqs. 4.12.8 may be written

$$R_x = \frac{My}{\Sigma x^2 + \Sigma y^2} \tag{4.12.9a}$$

$$R_y = \frac{Mx}{\Sigma x^2 + \Sigma y^2} \tag{4.12.9b}$$

By taking the vector sum of R_x and R_y, the total force R on the fastener becomes

$$R = \sqrt{R_x^2 + R_y^2} \tag{4.12.10}$$

To compute the total force on a fastener in an eccentric shear connection such as shown in Fig. 4.12.2a, the direct shear force R_v is

$$R_v = \frac{P}{\Sigma N} \tag{4.12.11}$$

where N is the number of fasteners in the group. The total resultant force R then becomes

$$R = \sqrt{[R_y + R_v]^2 + R_x^2} \tag{4.12.12}$$

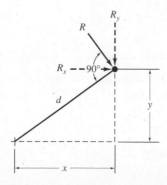

Figure 4.12.4 Horizontal and vertical components of force R.

EXAMPLE 4.12.1 ⎯⎯⎯⎯⎯⎯⎯⎯⎯⎯⎯⎯⎯⎯⎯⎯⎯⎯⎯⎯⎯⎯⎯⎯⎯⎯⎯⎯⎯⎯⎯⎯⎯⎯⎯⎯⎯⎯

Use the elastic (vector) method to compute the maximum force R on any bolt in the eccentrically loaded bolt group of Fig. 4.12.5a. The bolts are all the same size.

Solution. From Fig. 4.12.5 it can be seen that the upper and lower right fasteners are the most highly stressed. Since these two fasteners are equally stressed, only one need be investigated; check upper-right fastener. The eccentricity e, as measured from the centroid (assumed center of rotation), is

$$e = 3 + 2 = 5 \text{ in.}$$

$$M = 24(5) = 120 \text{ in.-kips}$$

$$\Sigma x^2 + \Sigma y^2 = 6(2)^2 + 4(3)^2 = 60 \text{ in.}^2$$

$$R_x = \frac{My}{\Sigma x^2 + \Sigma y^2} = \frac{120(3)}{60} = 6.0 \text{ kips} \rightarrow$$

$$R_y = \frac{Mx}{\Sigma x^2 + \Sigma y^2} = \frac{120(2)}{60} = 4.0 \text{ kips} \downarrow$$

$$R_v = \frac{P}{\Sigma N} = \frac{24}{6} = 4.0 \text{ kips} \downarrow$$

$$R = \sqrt{[R_y + R_v]^2 + R_x^2}$$

$$R = \sqrt{(4.0 + 4.0)^2 + (6.0)^2} = 10.0 \text{ kips}$$

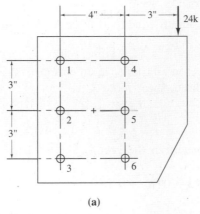

(a)

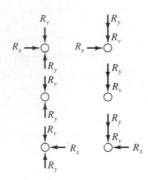

(b) Forces acting on fasteners

Figure 4.12.5   Example 4.12.1. ■■

EXAMPLE 4.12.2 ⎯⎯⎯⎯⎯⎯⎯⎯⎯⎯⎯⎯⎯⎯⎯⎯⎯⎯⎯⎯⎯⎯⎯⎯⎯⎯⎯⎯⎯⎯⎯⎯⎯⎯⎯⎯⎯⎯

Use the elastic (vector) method to compute the force R on the top right bolt in the eccentrically loaded bolt group of Fig. 4.12.6a. The bolts are all the same size.

Solution

$$e = 6.6 \text{ in.}$$

$$M = 10(6.60) = 66.0 \text{ in.-kips}$$

$$\Sigma d^2 = 4(3.60)^2 = 51.8 \text{ in.}^2$$

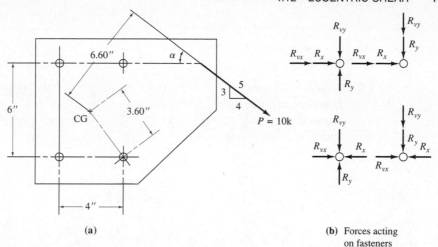

(a)

(b) Forces acting
on fasteners

Figure 4.12.6 Example 4.12.2.

$$R_x = \frac{My}{\Sigma d^2} = \frac{66.0(3)}{51.8} = 3.82 \text{ kips} \rightarrow$$

$$R_y = \frac{Mx}{\Sigma d^2} = \frac{66.0(2)}{51.8} = 2.55 \text{ kips} \downarrow$$

$$R_{vx} = \frac{P \cos \alpha}{\Sigma N} = \frac{10(0.8)}{4} = 2.00 \text{ kips} \rightarrow$$

$$R_{vy} = \frac{P \sin \alpha}{\Sigma N} = \frac{10(0.6)}{4} = 1.50 \text{ kips} \downarrow$$

$$R = \sqrt{[R_y + R_{vy}]^2 + [R_x + R_{vx}]^2}$$

$$R = \sqrt{(2.55 + 1.50)^2 + (3.82 + 2.00)^2} = 7.09 \text{ kips} \quad \blacksquare\blacksquare$$

Ultimate Strength Analysis

This method, also called plastic analysis, currently is recognized by the *Guide* [3.1]
as the most rational one.

The application of the load P causes both a translation and a rotation of the
fastener group. The translation and rotation can be reduced to a pure rotation about
a point defined as the *instantaneous center of rotation* (see Fig. 4.12.7).

The requirements for equilibrium are as follows:

$$\Sigma F_H = 0; \qquad \sum_{i=1}^{n} R_i \sin \theta_i - P \sin \delta = 0 \tag{4.12.13}$$

$$\Sigma F_V = 0; \qquad \sum_{i=1}^{n} R_i \cos \theta_i - P \cos \delta = 0 \tag{4.12.14}$$

$$\Sigma M = 0; \qquad \sum_{i=1}^{n} R_i d_i - P(e + x_0 \cos \delta + y_0 \sin \delta) = 0 \tag{4.12.15}$$

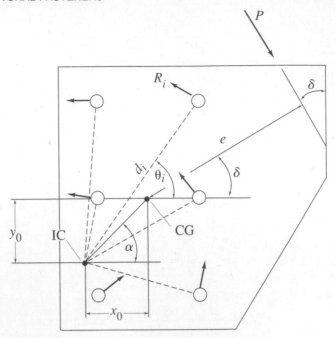

Figure 4.12.7 Instantaneous center of rotation.

These three equations (Eqs. 4.12.13 to 4.12.15) contain three unknowns (P, x_0, and y_0) and thus exactly determine the location (x_0, y_0) of the instantaneous center and the magnitude of the applied force P. When either the resistance R_i is proportional to the deformation, or when the angle δ is equal to zero or ninety degrees, the angle α is equal to the angle δ and Eq. 4.12.15 reduces to

$$\sum M = 0; \qquad \sum_{i=1}^{n} R_i d_i - P(e + r_0) = 0 \qquad (4.12.16)$$

where r_0 is the distance between the instantaneous center and the centroid (CG) of the connection (r_0 shown in Fig. 4.12.8).

Actually the concept of instantaneous center is identical to the elastic (vector) method when the resistance R_i is assumed proportional to the deformation (i.e., stress is proportional to strain). For either the elastic or the strength method, the deformation is proportional to the distance d_i from the instantaneous center of rotation.

For the strength analysis, two approaches have been used [3.1]. For a *bearing-type* connection, slip is neglected so that the *deformation* of each fastener is proportional to its distance from the instantaneous center. The resistance of each fastener is related to its deformation according to its load-deformation relationship. An expression proposed by Fisher [4.21] and used by Crawford and Kulak [4.17] for this load R vs deformation Δ response is

$$R_i = R_{ult}(1 - e^{-10\Delta})^{0.55} \qquad (4.12.17)$$

where $R_{ult} = \tau_u A_b$. The coefficients 10 and 0.55 were experimentally determined and the maximum Δ at failure was about 0.34 in. [3.1]. Note that e in Eq. 4.12.17 is the

Naperian base (2.718) and not the eccentricity. For A325 bolts, the ultimate shear strength τ_u is approximately 70% of the tensile strength (120 ksi minimum). Actually, the experimental work directly obtained $\tau_u A_b$ to be 74 kips for $\frac{3}{4}$-in.-diam A325 bolts in double shear, making τ_u equal to 0.7 of 120 ksi.

The experimental work relating to Eq. 4.12.17 used bolts loaded symmetrically (that is, in double shear); however, the general strength method could use *any* appropriate load R vs deformation Δ relationship, not necessarily Eq. 4.12.17.

EXAMPLE 4.12.3

Illustrate the general strength method by determining the nominal strength load P_n that may be applied to the fastener group of Fig. 4.12.5. Use Eq. 4.12.17 as the load-deformation expression, and assume that the maximum deformation $\Delta_{\max}$ at failure is 0.34 in. Assume full shank cross-sections of the bolts resist shear.

Solution. For $\frac{7}{8}$-in.-diam A325 bolts, Eq. 4.12.17 becomes

$$R_i = 0.7(120)(0.6013)(1 - e^{-10\Delta})^{0.55}$$
$$= 50.5(1 - e^{-10\Delta})^{0.55}$$

The load is applied in the y-direction; therefore $\delta = 0$ (Fig. 4.12.7). Using y_i/d_i for $\sin \theta_i$ and x_i/d_i for $\cos \theta_i$, Eqs. 4.12.13 through 4.12.15 become

$$\sum R_i \frac{y_i}{d_i} = 0 \qquad\qquad (4.12.18)$$

$$\sum R_i \frac{x_i}{d_i} = P_n \qquad\qquad (4.12.19)$$

$$\sum R_i d_i = P_n(e + r_0) \qquad\qquad (4.12.20)$$

Also, a basic deformation assumption is

$$\Delta_i = \frac{d_i}{d_{\max}} \Delta_{\max} = \frac{d_i}{d_{\max}} (0.34)$$

(a) Since an iterative process will be required to solve Eqs. (4.12.18 through 4.12.20), let the first trial $r_0 = 3$ in. (see Fig. 4.12.8).

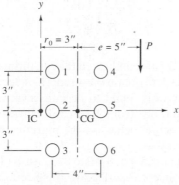

Trial $r_0 = 3$ in.

Figure 4.12.8 Strength method for Example 4.12.3.

Fasteners	x_i	y_i	d_i	Δ_i	R_i	$\dfrac{R_i x_i}{d_i}$	$R_i d_i$
1	1	3	3.162	0.184	45.9	14.53	145.3
2	1	0	1.0	0.058	32.2	32.23	32.2
3	1	−3	3.162	0.184	45.9	14.53	145.3
4	5	3	5.831	0.340	49.6	42.51	289.1
5	5	0	5.0	0.292	49.0	48.99	244.9
6	5	−3	5.831	0.340	49.6	42.51	289.1
						195.30	1145.9

$$\text{Eq. 4.12.19:} \qquad P_n = 195 \text{ kips}$$

$$\text{Eq. 4.12.20:} \qquad P_n = \frac{1145.9}{8} = 143 \text{ kips}$$

Since the values are not identical, further trials are required. One will generally find that P_n from Eq. 4.12.20 is relatively close to the correct value even though Eqs. 4.12.19 and 4.12.20 give values that are not very close. A trial value of r_0 that will give P_n between the two values but close to the value from Eq. 4.12.20 will make the calculation converge rather rapidly. The correct value is $r_0 = 2.06$ in., as shown in the following table:

Fasteners	x_i	y_i	d_i	Δ_i	R_i	$\dfrac{R_i x_i}{d_i}$	$R_i d_i$
1	0.06	3	3.001	0.202	46.7	0.93	140.2
2	0.06	0	0.060	0.004	8.6	8.55	0.5
3	0.06	−3	3.001	0.202	46.7	0.93	140.2
4	4.06	3	5.048	0.340	49.6	39.87	250.3
5	4.06	0	4.060	0.273	48.7	48.68	197.6
6	4.06	−3	5.048	0.340	49.6	39.87	250.3
						138.85	979.0

$$\text{Eq. 4.12.19:} \qquad P_n = 139 \text{ kips}$$

$$\text{Eq. 4.12.20} \qquad P_n = \frac{979.0}{7.06} = 139 \text{ kips}$$

Thus, $P_n = 139$ kips. ■■

Studies [4.17, 4.18, 4.20, 4.22] have indicated that an ultimate strength (plastic) analysis is the most rational approach to obtain the strength of eccentric shear connections. The elastic (vector) analysis was found to be conservative, making the ratio between strength and service load range from 2.5 to 3.0. However, since that elastic method does not properly reflect actual behavior, the margins of safety are variable from case to case, though conservative.

LRFD Manual Tables 8-18 to 8-25, pp. 8-40 to 8-87, provide for eccentric shear based on the ultimate strength concept described in this section. Brandt [4.19] has provided a practical procedure for making a general solution to these problems.

Slip-Critical Connections

The same strength procedure is recommended [3.1] for slip-critical connections as for bearing-type connections. Since slip-resistance is a serviceability requirement, not a strength requirement, it is logical to investigate the *strength* of either type connection by the ultimate strength approach described above.

A preferred procedure for analysis of slip-critical connections at service load is to consider the resistance R_i as constant for all fasteners, say, at whatever is the maximum acceptable "shear stress" F_v (see Table 4.9.1 from RCSC [4.7]). Since the bolts in a slip-critical joint must be installed with initial tension, there will be a fairly uniform clamping action between the pieces being joined.

EXAMPLE 4.12.4

Repeat Example 4.12.3 (Fig. 4.12.5) using $R_i = R_s$ as for a slip-critical connection using the instantaneous center approach, similar to the ultimate strength method.

Solution. For $R_i = R_s$ and $\delta = 0$, Eqs. 4.12.13 through 4.12.15 become

$$R_s \sum \frac{y_i}{d_i} = 0 \qquad (4.12.21)$$

$$R_s \sum \frac{x_i}{d_i} - P = 0 \qquad (4.12.22)$$

$$R_s \sum d_i - P(e + r_0) = 0 \qquad (4.12.23)$$

Try $r_0 = 2$ in. Referring to Fig. 4.12.8,

Fastener	x_i	y_i	d_i	$\dfrac{x_i}{d_i}$
1	0	3	3.0	0
2	0	0	0	0
3	0	−3	3.0	0
4	4	3	5.0	0.8
5	4	0	4.0	1.0
6	4	−3	5.0	0.8
			$\overline{20.0}$	$\overline{2.6}$

From Eq. 4.12.22,

$$P = R_s \sum \frac{x_i}{d_i} = 2.6 R_s$$

From Eq. 4.12.23,

$$P = \frac{R_s \sum d_i}{e + r_0} = \frac{R_s(20)}{7} = 2.86 R_s$$

For this assumption, fastener No. 2 is at the center of rotation and therefore is not involved in Eq. 4.12.23. Also, fastener No. 2 is assumed to have *no* contribution to Eq. 4.12.22. When r_0 is assumed slightly larger than 2.0, Eq. 4.12.22 gives $P = 3.6R_s$ because $x_i/d_i = 1.0$ for fastener No. 2. When r_0 is assumed slightly smaller than 2.0, Eq. 4.12.22 gives $P = 1.6R_s$ because $x_i/d_i = -1$ for the same fastener. Thus the value $P = 2.86R_s$ from Eq. 4.12.23 is accepted as the answer (i.e., $r_0 = 2.0$ in.). If a factor of safety is applied, R_s may be interpreted as the allowable value for a slip-critical connection and P the safe applied load. For $\frac{7}{8}$-in.-diam fasteners, using $F_v = 17$ ksi (see Table 4.9.1), gives for single shear,

$$R_{ss} = 17(0.6013) = 10.2 \text{ kips}$$

and the service load capacity would be

$$P = 2.86(10.2) = 29.2 \text{ kips} \quad \blacksquare\blacksquare$$

Load and Resistance Factor Design

EXAMPLE 4.12.5

Compare the service load capacities P of the eccentric shear connection of Fig. 4.12.5 when investigated by various methods. Solve assuming the connection is (a) a bearing-type connection with threads excluded from the shear plane, and (b) a slip-critical (friction-type) connection. Assume the live load is 80% and dead load is 20% of the total. Assume the plates are thick enough that bearing on the plates does not control. Use $\frac{7}{8}$-in.-diam A325 bolts and AISC LRFD Specification.

Solution

(a) Elastic analysis—bearing-type connection.

$$R = 10 \text{ kips (fasteners 4 and 6 in Example 4.12.1)}$$

The design strength ϕR_n in shear on the bolt for a bearing-type connection (A325-X) is

$$\phi R_n = \phi(0.50F_u^b)mA_b \qquad [4.7.3]$$
$$= 0.75(0.50)(120)(1)0.6013 = 27.1 \text{ kips/bolt}$$

The factored service load R_u is

$$R_u = 1.2D + 1.6L = 1.2(2) + 1.6(8) = 15.2 \text{ kips}$$

Since the factored load (15.2 kips) on the critical fastener is well below the design strength of 27.1 kips, the load P on the eccentric connection may be increased proportionally. Thus,

$$P = 24(27.1/15.2) = 42.8 \text{ kips}$$

(b) Elastic analysis—slip-critical connection. The design strength ϕR_n is still 27.1 kips/bolt as computed in (a). The "shear stress" F_v from Table 4.9.1 is 17 ksi for A325 bolts in standard holes. Thus, the service load bolt capacity for slip resistance is

$$\text{Allowable } R = F_v mA_b = 17(1)0.6013 = 10.2 \text{ kips/bolt}$$

The computed maximum load per bolt is 10 kips, very close to the limit for the slip-critical connection. Strength in shear does not govern. Thus,

$$P = 24(10.2/10) = 24.5 \text{ kips}$$

(c) Ultimate strength—bearing-type connection. The design strength ϕR_n is still 27.1 kips/bolt as computed in (a), which means the maximum nominal resistance R_n per bolt is $27.1/\phi = 36.1$ kips. Thus, P_n from Example 4.12.3 must be reduced in the ratio of 36.1 to maximum R_i (i.e., 49.6 kips). Thus,

$$\phi P_n = 0.75(139)(36.1/49.6) = 75.9 \text{ kips}$$

The factored load on the connection is

$$P_u = 1.2(0.2P) + 1.6(0.8P) = 1.52P$$

Equating P_u to ϕP_n gives

$$P = 75.9/1.52 = 49.9 \text{ kips}$$

(d) Instantaneous center—slip-critical connection. The service load capacity from Example 4.12.4 is

$$P = 29.2 \text{ kips}$$

(e) *LRFD Manual* [1.18] Table 8-20 (p. 8-52) for $\theta = 0°$—bearing-type connection. Since the horizontal spacing of fasteners is 4 in., the LRFD tables are not directly applicable. One *could* interpolate between Table 8-19 for 3 in. and Table 8-20 for $5\frac{1}{2}$ in.

	Table 8-19	Table 8-20
$e = 5$ in.	(p. 8-46)	(p. 8-52)
$n = 3$ bolts	Angle = 0°	Angle = 0°
Coefficient, $C =$	2.59	2.96

$$\text{Interpolated, } C = 2.59 + (2.96 - 2.59)(1/2.5) = 2.74$$
$$\phi P_n = C\phi r_v$$

where $\phi r_v = \phi R_n = $ design strength of one fastener. Thus,

$$\phi P_n = 2.74(27.1) = 74.3 \text{ kips}$$
$$P = 74.3/1.52 = 48.9 \text{ kips}$$

In this case, the interpolated design strength is less than the value computed in Example 4.12.3; a conservative result but adequate within the accuracy that behavior can be predicted.

(f) *LRFD Manual* [1.18] tables—slip-critical connection. The interpolated coefficient $C = 2.74$ from (e) is multiplied by the service load capacity (10.2 kips) per bolt from (b). thus,

$$P = 2.74(10.2) = 27.9 \text{ kips}$$

(g) Summary.

Procedure	Bearing-type load P	Slip-critical load P
Elastic (vector) method	42.8 kips	24.5 kips
Strength analysis	49.9 kips	—
Friction analysis	—	29.2 kips
LRFD Manual—interpolated	48.9 kips	27.9 kips

Thus, one may note the elastic (vector) method produces the most conservative result. ■■

Design Formula for Moment on Single Line of Fasteners

The *LRFD Manual* [1.18] provides Tables 8-18 to 8-25 that allow the designer to determine the number of connectors required for a given load and eccentricity. In unusual cases for which the tables do not apply, or when they may not be readily available, it is desirable to have a simple alternative method to use. The following development from Shedd* provides a useful simple formula.

Consider a single line of n equally spaced fasteners subjected to moment alone, as shown in Fig. 4.12.9. Since with uniform spacing the resistance of the fasteners is uniform from top to bottom, and according to Eq. 4.12.6, $R = Md/(\Sigma d^2)$, the force varies linearly as shown in Fig. 4.12.9.

Assuming R is the force in the outermost fastener, and that it represents the accumulation of stress that would occur on a rectangular resisting section over the distance p, one may designate the average load per inch of height at the outermost fastener as R/p.

Using similar triangles the load per inch at the extreme fiber may be determined,

$$\text{Extreme fiber value} = \frac{R}{p}\left(\frac{n}{n-1}\right) \tag{4.12.24}$$

The tensile force is the area of the triangle represented by the force per inch diagram,

$$T = \frac{1}{2}\left(\frac{np}{2}\right)\left(\frac{R}{p}\right)\left(\frac{n}{n-1}\right) = \frac{Rn^2}{4(n-1)} \tag{4.12.25}$$

The internal resisting moment is

$$M = T(\tfrac{2}{3}np) \tag{4.12.26}$$

Substitution of Eq. 4.12.25 into Eq. 4.12.26 gives

$$M = \frac{Rn^2}{4(n-1)}\left(\frac{2}{3}np\right) = \frac{Rn^3p}{6(n-1)} \tag{4.12.27}$$

Solving Eq. 4.12.27 for n^2, one obtains

$$n = \sqrt{\frac{6M}{Rp}\left(\frac{n-1}{n}\right)} \tag{4.12.28}$$

---

*Thomas C. Shedd, *Structural Design in Steel* (John Wiley & Sons, New York 1934), p. 287.

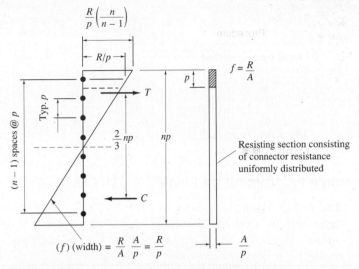

Figure 4.12.9   Moment on a single line of fasteners.

which as a first approximation becomes

$$n = \sqrt{\frac{6M}{Rp}} \qquad (4.12.29)$$

which is suggested for design use.

Since Eq. 4.12.29 is for moment alone acting on a single row of fasteners, the numerical value for R to be used in it should be adjusted to account for direct shear and for more than one row of fasteners. It is suggested to use a reduced effective R for the direct-shear effect and use an increased effective R for the effect of lateral spread. For lateral spread use a multiplier on R of 1.0 for one line up to about 2.0 for a square array of connectors.

More complicated formulas have been developed to compute the maximum stress or force on a fastener, but no direct solution for the number of connectors or the number of rows is possible from such equations.

EXAMPLE 4.12.6 ──

Determine the required number of $\frac{7}{8}$-in.-diam A325 bolts for one vertical line of fasteners Ⓐ — Ⓐ in the bracket shown in Fig. 4.12.10. Assume it to be a bearing-type connection with threads included in the shear planes (A325-N). Use AISC Load and Resistance Factor Design.

Solution

(a) Factored load. Using the gravity load equation, Eq. 1.8.3,

$$P_u = 1.2(7) + 1.6(41) = 74 \text{ kips}$$

(b) Design strength of a fastener in a bearing-type connection.

$$\phi R_n = \phi(0.40F_u^b)mA_b \quad \text{(shear strength—double shear)}$$
$$= 0.75(0.40)(120)(2)0.6013 = 43.3 \text{ kips} \quad \text{(controls)}$$

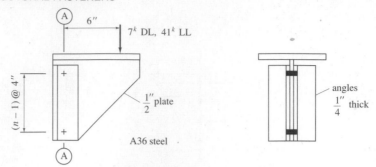

Figure 4.12.10   Example 4.12.6.

$$\phi R_n = \phi(2.4F_u dt) \quad \text{(bearing strength)}$$
$$= 0.75(2.4)(58)(0.875)0.5 = 45.7 \text{ kips}$$

(c) Estimate the number of bolts required, using Eq. 4.12.29.

$$n = \sqrt{\frac{6M}{Rp}} = \sqrt{\frac{6M_u}{\phi R_n p}} = \sqrt{\frac{6(74)(6)}{43.3(4)}} = 3.9$$

The R value (here ϕR_n) has *not* been adjusted for the direct shear effect; try 4 fasteners.

(d) Check the adequacy using the general ultimate strength analysis. Starting with the assumption that the instantaneous center is 3 in. to the left of the vertical line of fasteners, several iterations are required to obtain satisfaction of Eqs. 4.12.18 through 4.12.20. The final results are tabulated below. It is noted that quite a few iterations may be needed to get the values of P_n from Eqs. 4.12.19 and 4.12.20 exactly equal; however, the initial assumption of $r_0 = 3$ in. gave $P_n = 111$ kips from Eq. 4.12.20. The authors have found that even the first trial gives a reasonable approximation of the answer.

Verify that answer is $r_0 = 2.55$ in. to the left of the line of fasteners:

Fastener	x_i	y_i	d_i	Δ_i	R_i	$\dfrac{R_i x_i}{d_i}$	$R_i d_i$
1	2.55	6.00	6.519	0.340	49.6	19.39	323.2
2	2.55	2.00	3.241	0.169	45.2	35.53	146.3
3	2.55	−2.00	3.241	0.169	45.2	35.53	146.3
4	2.55	−6.00	6.519	0.340	49.6	19.39	323.2
						109.84	939.1

Eq. 4.12.19: $P_n = 110$ kips

Eq. 4.12.20: $P_n = 110$ kips   (Equal proves solution)

Note that the solution above is for single shear and has assumed that no threads are in the shear planes (A325-X). Also, the maximum R_i, even though logically computed using Eq. 4.12.17, cannot exceed the LRFD-Table J3.2 specified value (based on $0.50F_u^b = 60$ ksi),

$$R_n = 0.50F_u^b A_b = 60(0.6013) = 36.1 \text{ kips}$$

Then, P_n must be reduced in proportion that 36.1 kips is to the maximum R_i (i.e., 49.6 kips) in the table above,

$$P_n = 110(36.1/49.6) = 80.1 \text{ kips}$$

for A325-X in *single shear*.

For A325-N, this value must be multiplied by 2 for the double shear case of this example, and multiplied by 0.7 because threads are possible in the shear planes (A325-N).

Compare ϕP_n with P_u using the shear-related ϕ value since shear controlled the fastener strenth,

$$[\phi P_n = 0.75(80.1)2(0.7) = 84 \text{ kips}] > [P_u = 74 \text{ kips}] \quad \text{OK}$$

The above shows that 4 fasteners in a line are more than adequate. Investigation (not shown) for 3 fasteners indicates that 3 fasteners are not adequate.

(e) Check the adequacy using the elastic (vector) method. The direct shear component R_{us} from Eq. 4.12.11 is

$$R_{us} = \frac{P_u}{\Sigma N} = \frac{74}{4} = 18.5 \text{ kips} \downarrow$$

The moment component R_{ux} from Eq. 4.12.9a (noting that M equals P_u times e) is

$$R_{ux} = \frac{P_u e y}{\Sigma x^2 + \Sigma y^2} \qquad [4.12.9a]$$

$$\Sigma x^2 + \Sigma y^2 = 2[(2)^2 + (6)^2] = 80 \text{ in.}^2$$

$$R_{ux} = \frac{74(6)6}{80} = 33.3 \text{ kips} \rightarrow$$

Then, using Eq. 4.12.12, the resultant is obtained,

$$R_u = \sqrt{(18.5)^2 + (33.3)^2} = 38.1 \text{ kips} < [\phi R_n = 43.3 \text{ kips}]$$

Thus, the factored load R_u on the most heavily loaded bolt does not exceed the design strengh $\phi R_n = 43.3$ kips for A325-N. Thus, 4 fasteners are acceptable.

It must be noted that the edge distance L_e measured in the direction of the resultant force must satisfy Eq. 4.7.13 (LRFD-J3.10) requiring that $L_e \geq R_u/(\phi F_u t)$. Note that the factored force R_u on the most heavily loaded bolt is used as P in the original equation.

Use 4—$\frac{7}{8}$-in.-diam A325-N bolts @ 4-in. pitch. ■■

EXAMPLE 4.12.7 _____

Determine the required number of $\frac{3}{4}$-in.-diam A325 bolts in standard holes for the bracket plate of Fig. 4.12.11, assuming 4 vertical rows. Use a slip-critical connection

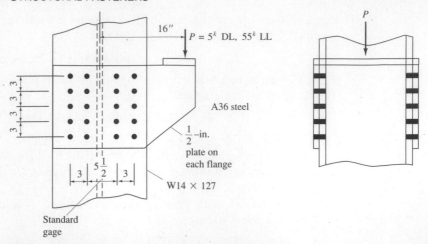

Figure 4.12.11   Eccentric shear connection of Example 4.12.7.

with clean mill scale (Class A) surface condition, and use AISC Load and Resistance Factor Design.

Solution

(a) Factored load. Since one-half the load is carried by each plate in a single shear situation, use one-half the total load in all calculations for one plate. Using the gravity load equation, Eq. 1.8.3,

$$P_u = 1.2(5) + 1.6(55) = 94 \text{ kips} \quad \text{(or 47 kips/plate)}$$

(b) Design strength of a fastener in a slip-critical connection. For strength, values are as for bearing-type connections. Assume no threads are to be in the shear planes (A325-X). For single shear,

$$\phi R_n = \phi(0.50F_u^b)mA_b \quad \text{(shear strength—single shear)}$$
$$= 0.75(0.50)(120)(1)(0.4418) = 19.9 \text{ kips} \quad \text{(controls)}$$
$$\phi R_n = \phi(2.4F_u dt) \quad \text{(bearing strength)}$$
$$= 0.75(2.4)(58)(0.75)0.5 = 39.2 \text{ kips}$$

For slip resistance, the service load capacity is

$$R = mF_v A_b = 1.0(17)(0.4418) = 7.5 \text{ kips}$$

(c) Estimate the number of bolts required, using Eq. 4.12.29. It is assumed that slip resistance at service load is the controlling limit state; strength will later be checked.

$$n = \sqrt{\frac{6M}{Rp}} = \sqrt{\frac{6(30/4)(16)}{7.5(3)}} = 5.7$$

In the above equation, the service load per plate is 30 kips, and the load per line of fasteners is 30/4, which must be used since Eq. 4.12.29 applies to one line of fasteners. No adjustment in R was made either for direct shear or for several lines of fasteners. Try 5 bolts per row.

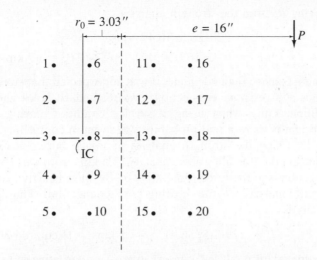

Figure 4.12.12 Fastener numbering for Example 4.12.7.

(d) Check the adequacy using the general ultimate strength analysis. Referring to Fig. 4.12.12, and using the same method illustrated in Example 4.12.6, the value of r_0 is found to be 3.03 in. to the left of the centroid of the fastener group:

Fastener	x_i	y_i	d_i	Δ_i	R_i	$\dfrac{R_i x_i}{d_i}$	$R_i d_i$
1	−2.72	6.00	6.588	0.211	34.6	−14.27	227.6
2	−2.72	3.00	4.049	0.129	31.1	−20.90	126.0
6	0.28	6.00	6.007	0.192	34.0	1.59	204.3
7	0.28	3.00	3.013	0.096	28.5	2.65	85.8
11	5.78	6.00	8.331	0.266	35.7	24.74	297.1
12	5.78	3.00	6.512	0.208	34.5	30.61	224.6
16	8.78	6.00	10.634	0.340	36.4	30.07	387.3
17	8.78	3.00	9.278	0.297	36.0	34.11	334.5
					$\Sigma =$	88.60	1887.2
	Multiply the above times 2 for symmetry $=$					177.20	3774.4
3	−2.72	0.00	2.720	0.087	27.5	−27.53	74.9
8	0.28	0.00	0.280	0.009	9.6	9.60	2.7
13	5.78	0.00	5.780	0.185	33.8	33.77	195.2
18	8.78	0.00	8.780	0.281	35.9	35.86	314.9
					$\Sigma =$	228.90	4362.1

Eq. 4.12.18: $P_n = 228.9$ kips

Eq. 4.12.19: $P_n = 4362.1/(16 + 3.03) = 229.2$ kips

Note that the solution above is for single shear and has assumed that no threads are in the shear planes (A325-X). In addition, consider that LRFD limits the maximum strength of a bolt in shear to

$$R_n = 0.50 F_u^b A_b = 0.50(120)0.4418 = 26.5 \text{ kips}$$

Reducing P_n from the strength analysis.

$$P_n = 229(26.5/36.4) = 167 \text{ kips}$$
$$[\phi P_n = 0.75(167) = 125 \text{ kips}] > [P_u = 47 \text{ kips}] \quad \text{OK}$$

Strength is more than adequate; it was not expected that strength would govern for this slip-resistant connection having standard holes and Class A surface condition. Only when using a surface condition having a very high slip coefficient is there a real probability of strength controlling.

(e) Using the strength analysis approach but reducing the capacity in proportion that the allowable capacity [7.5 kips from part (b)] of a slip-critical connection is to the force (36.4 kips) on the most heavily loaded fastener in the strength analysis of the bearing-type connection. Thus, the service load capacity is

$$P = 229(7.5/36.4) = 47.1 \text{ kips} > [\text{Required 30 kips}]$$

This procedure is always conservative. An alternative is to do an analysis as illustrated in Example 4.12.4 where the resistance of each fastener is the same.

(f) Check using *LRFD Manual* Table 8-25 (p. 8-82) for eccentrically loaded bolt groups. The horizontal spacing of fasteners is not exactly the same as in Table 8-25; however, when the external dimensions of the array are about the same, and the vertical spacing of bolts agrees with the tables, the coefficient will be found to be about the same. If the LRFD table is assumed applicable as the authors believe it is for this case, enter the table with the number ($n = 5$) of fasteners in a vertical line, and the eccentricity ($e = 16$ in.) of the load from the center of gravity of the group of fasteners,

Find coefficient $C = 6.15$
$$\phi P_n = C\phi R_n = 6.15(19.9) = 122 \text{ kips}$$

which compares with $\phi P_n = 125$ kips for a bearing-type connection from the analysis in (d).

For the slip-critical connection of this example,

$$P = CR = 6.15(7.5) = 46 \text{ kips}$$

which compares with $P = 47.1$ kips computed in (e).

(g) Investigate the connection using the elastic (vector) method. Since the slip-critical service load condition has already been shown to govern, the illustration of the use of the elastic (vector) method is performed for the service load (30 kips on one plate).

Check 5 bolts per row by elastic method:

$$R_s = \frac{P}{n} = \frac{30}{20} = 1.50 \text{ kips} \downarrow$$
$$\sum x^2 + \sum y^2 = 10[(2.75)^2 + (5.75)^2] + 8[(3)^2 + (6)^2] = 766 \text{ in.}^2$$
$$R_x = \frac{My}{\sum x^2 + \sum y^2} = \frac{30(16)6}{766} = 3.76 \text{ kips} \rightarrow$$

$$R_y = \frac{Mx}{\Sigma x^2 + \Sigma y^2} = \frac{30(16)5.75}{766} = 3.60 \text{ kips} \downarrow$$

$$\text{Actual } R = \sqrt{(1.50 + 3.60)^2 + (3.76)^2} = 6.34 \text{ kips} < 7.5 \quad \text{OK}$$

Use $5-\frac{3}{4}$-diam A325 bolts per row. The fasteners (3, 8, 13, and 18) located on the *x*-axis could have been omitted; however, the regular pattern of Fig. 4.12.11 is preferred. ■ ■

4.13 FASTENERS ACTING IN AXIAL TENSION

Axial tension occurring without simultaneous shear exists in fasteners for tension members such as hangers (see Fig. 4.6.1c) or other members whose line of action is perpendicular to the member to which it is fastened. When such tension members are not perpendicular to their connecting members, the fasteners are subjected to both axial tension and shear. The latter, more typical case, is discussed in Sec. 4.14.

Section 4.6.1 discusses the nominal strength R_n for fasteners subject to axial tension; Eq. 4.6.1 gives the strength as

$$R_n = F_u^b(0.75A_b) \qquad [4.6.1]$$

When Eq. 4.6.1 is used in Load and Resistance Factor Design (LRFD-Table J3.2), the design strength ϕR_n using the resistance factor ϕ of 0.75 is

$$\phi R_n = 0.75F_u^b(0.75A_b) \qquad (4.13.1)$$

Values of $\phi(0.75F_u^b)$ have been given in Table 4.7.1.

When Eq. 4.6.1 is used in Allowable Stress Design (ASD-Table J3.2), the service load capacity R is the nominal strength R_n divided by the factor of safety of 2. Thus,

$$R = 0.5F_u^b(0.75A_b) = 0.375F_u^b A_b$$

$$R = F_t A_b \qquad (4.13.2)$$

where F_t = allowable service load nominal stress, T/A_b
= $0.375(120) \approx 44$ ksi (ASD-Table J3.2) for A 325 bolts
= $0.375(150) \approx 54$ ksi (ASD-Table J3.2) for A490 bolts
A^b = fastener gross cross-sectional area
F_u^b = tensile strength of fastener material (120 ksi for A325 and 150 ksi for A490 bolts)
T = service tensile load on a fastener

Fasteners subject to axial tension must be pretensioned according to Table 4.4.2 whether the design is for a bearing-type connection or a slip-critical connection [4.7].

Prestress Effect of High-Strength Bolts Under External Tension

To understand the effect of an externally applied load on a pretensioned high-strength bolt, consider a single bolt and the tributary portion of the connected plates as shown in Fig. 4.13.1a. The pieces being joined are of thickness *t* and the area of contact

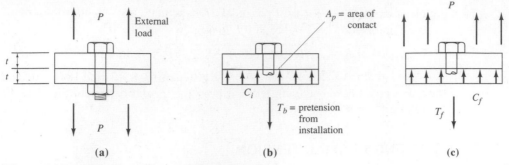

Figure 4.13.1   Prestress effect on bolted joint.

between the pieces is A_p. Prior to applying external load, the situation is as shown in Fig. 4.13.1b, where the bolt has been installed to have a pretension force T_b (values as in Table 4.4.2). The pieces being joined are initially compressed an amount C_i. For equilibrium,

$$C_i = T_b \tag{4.13.3}$$

The external load P is then applied and the forces acting are shown in Fig. 4.13.1c. This time equilibrium requires

$$P + C_f = T_f \tag{4.13.4}$$

where the subscript f refers to final conditions after application of the load P.

The force P acting on the system lengthens the bolt an amount δ_b between the underside of the bolt head and the surface of contact between the two connected plates,

$$\delta_b = \frac{T_f - T_b}{A_b E_b} \tag{4.13.5}$$

where E_b = modulus of elasticity of the bolt
     T_f = final force in the bolt after external load is applied.

At the same time, the compression between the plates decreases and the plate thickness increases an amount δ_p,

$$\delta_p = \frac{C_i - C_f}{A_p E_p} t \tag{4.13.6}$$

where E_p = modulus of elasticity of the plate material
     C_f = final compression force between the pieces being joined, after external load is applied

If contact is maintained, compatibility of deformation requires $\delta_b = \delta_p$; thus, Eq. 4.13.5 equated to Eq. 4.13.6 gives

$$\frac{T_f - T_b}{A_b E_b} = \frac{C_i - C_f}{A_p E_p} \tag{4.13.7}$$

Next, substitution of Eq. 4.13.3 for C_i and Eq. 4.13.4 for C_f into Eq. 4.13.7 gives

$$\frac{T_f - T_b}{A_b E_b} = \frac{T_b - T_f + P}{A_p E_p} \tag{4.13.8}$$

The moduli of elasticity E_b for the bolt and E_p for the plate are essentially the same and may be eliminated. Then solving for T_f gives

$$(T_f - T_b)\frac{A_p}{A_b} = T_b - T_f + P \tag{4.13.9}$$

$$T_f\left(1 + \frac{A_p}{A_b}\right) = T_b\left(1 + \frac{A_p}{A_b}\right) + P$$

$$T_f = T_b + \frac{P}{1 + A_p/A_b} \tag{4.13.10}$$

EXAMPLE 4.13.1

Assume $\frac{7}{8}$-in.-diam A325 bolts are used in a direct tension situation such as in Fig. 4.13.2. With bolts spaced 3 in. apart and having $1\frac{1}{2}$ -in. edge distances, the tributary area of contact may reasonably be about 9 sq in. If the maximum external tensile load permitted by AISC Load and Resistance Factor Design is applied, how much does the bolt tension increase? Assume the service load is 20% dead load and 80% live load.

Solution

(a) Design strength ϕR_n. Using Eq. 4.13.1,

$$\phi R_n = 0.75F_u^b(0.75A_b) = 0.75(120)(0.75)0.6013 = 40.6 \text{ kips}$$

(b) Permissible service load R per bolt. Equating the design strength ϕR_n to the factored load R_u gives

$$R_u = 1.2(0.2R) + 1.6(0.8R) = 1.52R$$

$$1.52R = 40.6 \text{ kips}$$

$$R = 26.7 \text{ kips}$$

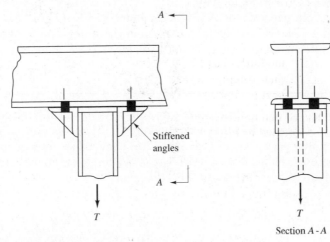

Figure 4.13.2 Example 4.13.1.

(c) Initial tensile force in $\frac{7}{8}$-in.-diam A325 bolt. From Table 4.4.2,

$$T_b = 39 \text{ kips}$$

(d) Determine final tensile force in bolt. The ratio of plate contact area to bolt area is

$$\frac{A_p}{A_b} = \frac{9}{0.6013} = 15$$

This neglects subtracting the bolt area from the total tributary area, but little difference results. Using Eq. 4.13.10 with the load P per fastener equal to its maximum value R gives

$$T_f = 39 + \frac{26.7}{1 + 15} = 39 + 1.7 = 40.7 \text{ kips}$$

The increase in tension is 4.3%. The variation in actual pretension from installation may be expected to exceed this amount, so that this increase is not of concern. Furthermore the tributary area used for the example (9 sq in.) is probably the minimum one might encounter in practice, since less than a 3-in. pitch and gage is rarely used.

The important conclusion from this example is that *no significant increase in bolt tension arises until the external load equals or exceeds the pretension force,* in which case the pieces do not remain in contact and the applied force equals the bolt tension.

If the connection can distort and give rise to "prying forces" these must also be considered. (See LRFD-J3 and ASD-J4 and the treatment in the "Split-Beam Tee Connections" part of Sec. 13.6.) In the situation of Example 4.13.1, the approximate factor of safety against overcoming initial compression between pieces is

$$\text{FS} = \frac{T_b}{P} = \frac{39}{26.7} \approx 1.5$$

In general, for A325 bolts under LRFD-J3 or ASD-J4, the margin of safety against service load exceeding the proof load is approximately 1.5 for diameters up to 1 in. and approximately 1.3 for diameters over 1 in. ■■

EXAMPLE 4.13.2 ───

Determine the required number of $\frac{3}{4}$-in.-diam A490 bolts for the connection shown in Fig. 4.13.3. Assume that the pieces making up the connection are adequate, and very stiff such that prying forces (see Sec. 13.6) can be disregarded, and that the nominal tensile stresses on the bolts govern. Use Load and Resistance Factor Design assuming the load is 10% dead load and 90% live load.

Solution
(a) Design strength ϕR_n per bolt. Using Eq. 4.13.1,

$$\phi R_n = 0.75 F_u^b (0.75 A_b) = 0.75(150)(0.75)0.4418 = 37.3 \text{ kips}$$

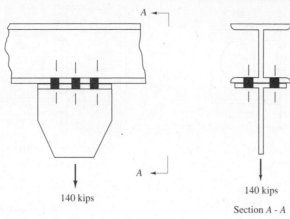

140 kips

140 kips

Section A - A

Figure 4.13.3 Example 4.13.2.

(b) Calculate the factored tension force T_u.

$$T_u = 1.2(0.1)(140) + 1.6(0.9)(140) = 218 \text{ kips}$$

(c) Determine the number n of bolts required.

$$n = \frac{T_u}{\phi R_n} = \frac{218}{37.3} = 5.9, \quad \text{say } 6$$

Use 6—$\frac{3}{4}$-in.-diam A490 bolts. ■■

4.14 COMBINED SHEAR AND TENSION

In a large number of commonly used connections, both shear and tension occur and must be considered in their design. Figure 4.14.1 shows a few typical connections in which the connectors are simultaneously subjected to both shear and tension. The connection shown in Fig. 4.14.1a is a common one where two angles join the beam web to the column flange. From the moment force indicated in the figure, the upper fasteners are subjected to tension in proportion to the magnitude of the applied moment. However, one may recall from structural analysis that only a small amount of end rotation on a beam is necessary to change from a fixed end to a hinged end condition. In addition, the web carries only a small part of the bending moment. Thus, one may intuitively sense that the moment shown will be relieved before a significant tension force can be developed in the fasteners. Such connections are used when little end moment is desired to be transmitted. An exception to this occurs in the case of a very deep beam such as a plate girder.

Referring next to Fig. 4.14.1b in the which the applied moment is transmitted through the flanges of the beam, the situation is different. In this case a large applied moment is intended to be transmitted so the connection is made at the flanges, the elements carrying most of the moment. Chapter 13 treats this type of connection. Figure 4.14.1c and 4.14.1d typify the two types of fastener loading in combined shear and tension which are developed in the following parts of this section.

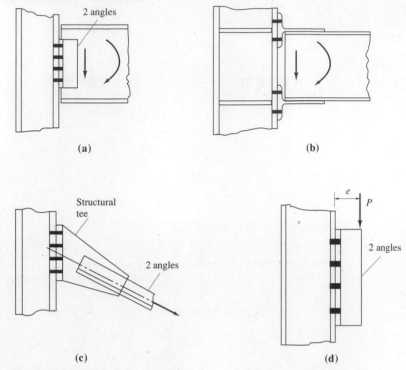

Figure 4.14.1   Typical combined shear and tension connections.

Bearing-Type Connections—LRFD Design

In earlier sections, the nominal strengths of fasteners loaded separately in shear and tension were treated. When the full strength in tension is required, the full strength in shear is not simultaneously available. Based on experimental studies [4.23, 3.1], the strength interaction equation may be represented by a circular relationship

$$\left[\frac{R_{ut}}{\phi_t R_{nt}}\right]^2 + \left[\frac{R_{uv}}{\phi_v R_{nv}}\right]^2 \leq 1.0 \tag{4.14.1}$$

where R_{ut} = factored tension load on bolt
R_{uv} = factored shear load on bolt
$\phi_t R_{nt}$ = design strength of bolt in tension alone (ϕ_t = 0.75)
$\phi_v R_{nv}$ = design strength of bolt in shear alone (ϕ_v = 0.75)

The nominal strength R_{nt} of bolts in tension is given by Eq. 4.6.1,

$$R_{nt} = R_n = F_u^b(0.75A_b) \tag{4.6.1}$$

The nominal strength R_{nv} of bolts in shear is given by Eq. 4.7.2,

$$R_{nv} = R_n = mA_b(0.50F_u^b) \tag{4.7.2}$$

for *no threads in shear planes;* and Eq. 4.7.5

$$R_{nv} = R_n = mA_b(0.40F_u^b) \qquad [4.7.5]$$

for *threads possible within shear planes.*

The AISC Specifications [1.5, 1.16] have simplified the circular interaction relationship of Eq. 4.14.1 into a straight line which requires a reduction only in the most severe loading cases. The straight-line expression is

$$\frac{R_{ut}}{\phi_t R_{nt}} + \frac{R_{uv}}{\phi_v R_{nv}} \le C \qquad (4.14.2)$$

where C is a constant. Comparison of the straight line with the circle appears in Fig. 4.14.2. Multiplying Eq. 4.14.2 by $\phi_t R_{nt}$ and solving for R_{ut} gives

$$R_{ut} \le C\phi_t R_{nt} - \frac{\phi_t R_{nt}}{\phi_v R_{nv}} R_{uv} \qquad (4.14.3)$$

The AISC Specifications [1.5, 1.16] have used the *stress format* by dividing all terms by the gross cross-sectional bolt area A_b. In addition, when the expressions for $\phi_t R_{nt}$ and $\phi_v R_{nv}$ are substituted, Eq. 4.14.3 becomes

$$\frac{R_{ut}}{A_b} \le C \frac{\phi(0.75F_u^b)A_b}{A_b} - \frac{\phi(0.75F_u^b)A_b}{0.75(0.50F_u^b)A_b} \frac{R_{uv}}{A_b} \qquad (4.14.4)$$

$$f_{ut} \le [\phi F'_{ut} = \phi(0.75F_u^b C - 2f_{uv})] \qquad (4.14.5)$$

When threads are in the shear planes, using Eq. 4.7.5 for R_n gives 2.5 instead of 2. The Specifications [1.5, 1.16] have used $C = 1.3$ to obtain the straight line of Fig. 4.14.2. The tension stress limit $\phi F'_{ut}$ then becomes

$$\phi F'_{ut} = \phi[0.75(120)1.3 - 2f_{uv}] \qquad (4.14.6)$$

$$\phi F'_{ut} = \phi[117 - 2f_{uv}] \qquad (4.14.7)$$

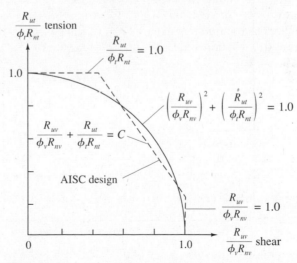

Figure 4.14.2 Nondimensional shear—tension strength interaction curve: bearing-type connections.

LRFD-J3.7 and LRFD-Table J3.5 gives for A325-X,

$$\phi F'_{ut} = \phi[117 - 1.5 f_{uv}] \le \phi 90 \tag{4.14.8}$$

The use of 1.5 instead of 2.0 reflects the fact that in the development above, the reduced "pseudo" shear strength reflecting medium length connection was used, whereas in the combined shear and tension case, it is more rational to use the actual strength $0.62 F_u^b A_b$. The use of the latter would reduce the 2.0 to 1.6. Straightline approximation is always a judgement call regarding the coefficients to use. When threads are in the shear planes, LRFD-J3.7 and LRFD-Table J3.5 give

$$\phi F'_{ut} = \phi[117 - 1.9 f_{uv}] \le \phi 90 \tag{4.14.9}$$

Equations 4.14.8 and 4.14.9 for A325 bolts are shown in Fig. 4.14.3 with the $\phi = 0.75$ included. The equations form $\phi F'_{ut}$ from LRFD-Table J3.5 are summarized in Table 4.14.1.

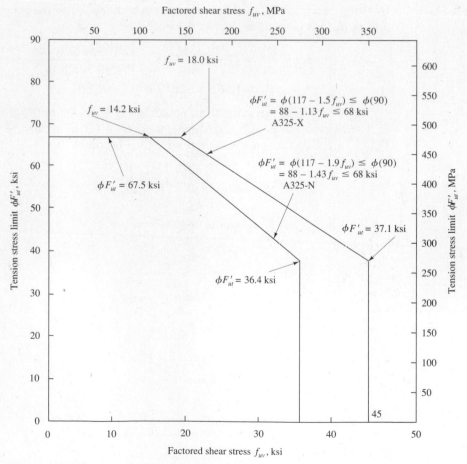

Figure 4.14.3 Interaction relationship for combined shear and tension in A325-X and A325-N bearing-type connections.

TABLE 4.14.1 DESIGN TENSION STRESS‡ LIMIT $\phi F'_{ut}$ IN THE PRESENCE OF FACTORED SHEAR STRESS f_{uv} (FOR BEARING-TYPE CONNECTIONS) (from LRFD-TABLE J3.5)

Fastener	$\phi F'_{ut}$	
	(ksi)	(MPa)
A307 bolts	$\phi(59 - 1.9 f_{uv}) \le \phi(45)$	$\phi(407 - 1.9 f_{uv}) \le \phi(310)$
A325-N bolts (threads *not* excluded)	$\phi(117 - 1.9 f_{uv}) \le \phi(90)$	$\phi(807 - 1.9 f_{uv}) \le \phi(621)$
A325-X bolts (threads excluded)	$\phi(117 - 1.5 f_{uv}) \le \phi(90)$	$\phi(807 - 1.5 f_{uv}) \le \phi(621)$
A490-N bolts (threads *not* excluded)	$\phi(147 - 1.9 f_{uv}) \le \phi(113)$	$\phi(1010 - 1.9 f_{uv}) \le \phi(779)$
A490-X bolts (threads excluded)	$\phi(147 - 1.5 f_{uv}) \le \phi(113)$	$\phi(1010 - 1.5 f_{uv}) \le \phi(779)$

* Note that $\phi = 0.75$
‡ Nominal stress due to factored load acting on gross bolt cross-sectional area, $f_{ut} = R_{ut}/A_b$

Slip-Critical Connections

Again, a straight-line interaction relationship is used; but one that is more conservative than the type of Eq. 4.14.2 for bearing-type connections. The constant C is reduced from 1.3 to 1.0 for slip-critical connections. In addition, since slip resistance is a service load consideration, the numerator terms of Eq. 4.14.2 become service loads T and V (tension and shear per bolt) that may simultaneously act, and the denominator terms become the maximum service load forces permitted in slip-critical connections subject to tension or shear acting alone:

$$\left[\frac{V}{F_v A_b}\right] + \left[\frac{T}{T_b}\right] \le 1.0 \qquad (4.14.10)$$

where F_v = maximum nominal service load shear stress V/A_b permitted in slip-critical connections subject to shear only (LRFD-J3.9)

The use of the initial tension force T_b from installation of the bolts reflects the "prestress" concept discussed in Sec. 4.13. Until the external load on a bolt exceeds the precompression force between the pieces, the tension force in the bolt will not change significantly from its initial tension T_b.

Solving Eq. 4.14.8 for the maximum service load shear stress V/A_b that may exist *in the presence of tension,* gives

$$\left[f_v = \frac{V}{A_b}\right] \le F_v \left[1.0 - \frac{T}{T_b}\right] \qquad (4.14.11)$$

Substituting the values for F_v, the allowable service load shear stress in the absence of tension for bolts in standard holes, one obtains from Eq. 4.14.9 the limitations of LRFD-J3.9a:

$$f_v \le \left[F'_v = 17\left(1 - \frac{T}{T_b}\right)\right] \quad \text{A325 bolts in standard holes} \qquad (4.14.12)$$

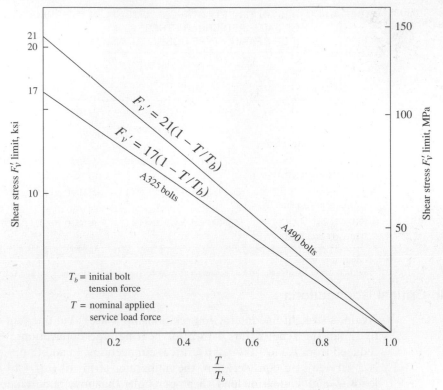

Figure 4.14.4   Combined shear and tension on slip-critical connections with standard holes (LRFD-J8 and J9).

$$ f_v \leq \left[F_v' = 21 \left(1 - \frac{T}{T_b} \right) \right] \quad \text{A490 bolts in standard holes} \qquad (4.14.13) $$

These equations are shown in Fig. 4.14.4.

EXAMPLE 4.14.1 _____

Using Load and Resistance Factor Design, determine the adequacy of the fasteners in Fig. 4.14.5 when $\frac{7}{8}$-in.-diam A325 bolts are used in (a) a bearing-type connection (A325-X) with threads eXcluded from the shear planes, and (b) a slip-critical connection (A325-SC) with Class A surface condition and standard holes. Assume the strength of the column flange and the ST section do not govern the answer. Neglect prying action (see Sec. 13.6). The gravity loading is 10% dead and 90% live load.

   Solution

   (a) Check connection as a bearing-type connection (A325-X). Obtaining the tension and shear components of the factored applied force,

$$ P_u = 1.2(0.1)(75) + 1.6(0.9)(75) = 117 \text{ kips} $$

$$ \text{Tension component} = P_{ux} = 0.8(117) = 93.6 \text{ kips} $$

$$ \text{Shear component} = P_{uy} = 0.6(117) = 70.2 \text{ kips} $$

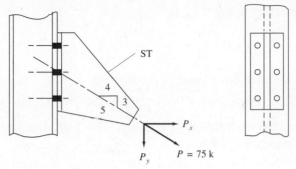

Figure 4.14.5 Example 4.14.1.

The factored loads T_u and V_u per bolt are

$$\text{Tension } T_u = 93.6/6 = 15.6 \text{ kips/bolt}$$
$$\text{Shear } V_u = 70.2/6 = 11.7 \text{ kips/bolt}$$

Using Table 4.14.1,

$$\phi F'_{ut} = \phi[117 - 1.5 f_{uv}] \leq \phi 90$$
$$= 0.75(117) - 0.75(1.5) f_{uv} \leq 0.75(90)$$
$$= 87.8 - 1.13 f_{uv} \leq 67.5 \text{ ksi}$$

Multiplying by A_b for $\frac{7}{8}$-in.-diam bolts, the above becomes

$$\text{Max } T_u = \phi F'_{ut} A_b$$
$$= 87.8(0.6013) - 1.13 V_u \leq 67.5(0.6013)$$
$$= 52.8 - 1.13 V_u \leq 40.6 \text{ kips}$$
$$T_u = 15.6 < [\text{Max } T_u = 52.8 - 1.13(11.7) = 39.6 \text{ kips}]$$

Note that the 39.6 kips limit from the linear interaction equation is applicable since it does not exceed the upper limit on that value of 40.6 kips. In addition, the shear value 11.7 kips/bolt must be checked. The design strength (Eq. 4.7.1) in single shear is

$$\phi R_{nv} = 0.75(0.50 F_u^b) m A_b$$
$$= 0.75(0.50)(120)(1)0.6013 = 27.1 \text{ kips} > 11.7 \text{ kips} \quad \text{OK}$$

The connection is very conservatively designed as a bearing-type connection.

(b) Check connection as a slip-critical connection (A325-SC). For strength, a slip-critical connection must satisfy the same strength-related criteria of a bearing-type connection. The shear strength was investigated in (a); bearing strength was given in the problem statement as not controlling.

For serviceability, the adequacy of slip resistance must be checked. The service load forces per bolt are

$$T = 0.8(75)/6 = 10 \text{ kips/bolt}$$
$$V = 0.6(75)/6 = 7.5 \text{ kips/bolt}$$

From LRFD-J3.9, Eq. 4.14.12,

$$F_v' = 17\left[1 - \frac{T}{T_b}\right] = 17\left[1 - \frac{10}{39}\right] = 12.6 \text{ ksi}$$

The service load shear capacity per bolt is

$$\text{Max } V = F_v'A_b = 12.6(0.6013) = 7.6 \text{ kips/bolt}$$

which exceeds $V = 7.5$ kips; therefore, the bolts are satisfactory as a slip-critical connection. ■■

EXAMPLE 4.14.2

Determine the maximum value of the load P in Example 4.14.1 assuming (a) a slip-critical connection, and (b) a bearing-type connection (A325-N) with threads possible in the shear planes. Use AISC Load and Resistance Factor Design and assume Class A surface condition with standard holes.

Solution

(a) Slip-critical connection. The service load forces are

$$\text{Tension component} = P_x = 0.8P$$
$$\text{Shear component} = P_y = 0.6P$$

The service load per bolt in shear is

$$V = 0.6P/6 = 0.10P \text{ per bolt}$$

The service load per bolt in tension is

$$T = 0.8P/6 = 0.133P \text{ per bolt}$$

$$F_v' = 17\left(1 - \frac{T}{T_b}\right) \qquad\qquad [4.14.10]$$

$$F_v' = 17\left(1 - \frac{0.133P}{39}\right) = 17 - 0.058P$$

The maximum service load shear V per bolt permitted on $\frac{7}{8}$-in.-diam bolts is

$$\text{Max } V = F_v'A_b = (17 - 0.058P)0.6013 = 10.2 - 0.035P$$

Equating V to Max V gives

$$0.10P = 10.2 - 0.035P$$
$$P = 76 \text{ kips}$$

The strength in shear, tension, and bearing must also be checked as if this were a bearing-type connection. For some very slip-resistant surface conditions the, strength as a bearing-type connection may control instead of slip resistance; a relatively unusual situation.

(b) Bearing-type connection (A325-N) with threads possible in the shear planes. The factored loads are

$$P_u = 1.2(0.1)P + 1.6(0.9)P = 1.56P \text{ kips}$$

$$\text{Tension component} = P_{ux} = 0.8(1.56P) = 1.25P \text{ kips}$$
$$\text{Shear component} = P_{uy} = 0.6(1.56P) = 0.94P \text{ kips}$$

The factored loads per bolt are

$$T_u = 1.25P/6 = 0.208P$$
$$V_u = 0.94P/6 = 0.156P$$

From Table 4.14.1 (LRFD-J3.7), using $\phi = 0.75$,

$$\phi F'_{ut} = \phi[117 - 1.9f_{uv}] \leq \phi 90$$
$$= 87.8 - 1.43 f_{uv} \leq 67.5 \text{ ksi}$$

Multiplying by A_b for $\frac{7}{8}$-in.-diam bolts, the above becomes

$$\text{Max } T_u = \phi F'_{ut} A_b$$
$$= 87.8(0.6013) - 1.43V_u \leq 67.5(0.6013)$$
$$= 52.8 - 1.43V_u \leq 40.6 \text{ kips}$$

Equating T_u to Max T_u gives

$$0.208P = 52.8 - 1.43(0.156P) = 52.8 - 0.223$$
$$P = 123 \text{ kips}$$

Check that the Max T_u from linear interaction does not exceed the 40.6 kips upper limit,

$$\text{Max } T_u = 52.8 - 0.223(123) = 25.4 \text{ kips} < 40.6 \text{ kips}\quad \text{OK}$$

In addition, the shear strength must be checked,

$$\phi R_n = 0.75(0.40F_u^b)mA_b$$
$$= 0.75(0.40)(120)(1)(0.6013) = 21.6 \text{ kips}$$
$$V_u = 0.156P = 0.156(123) = 19.2 \text{ kips} < 21.1 \text{ kips}\quad \text{OK}$$

Thus, the maximum value of the service load P is 76 kips as a slip-critical connection and 123 kips as a bearing-type (A325-N) connection. ■■

EXAMPLE 4.14.3

Determine the number of $\frac{3}{4}$-in.-diam A325 bolts required to carry a shear force consisting of 14 kips dead load and 56 kips live load, and a tension force of 24 kips dead load and 96 kips live load. The connection is to be designed such that the resultant force acts through the centroid of the connection. Use Load and Resistance Factor Design: (a) design as a bearing-type (A325-X) connection, and (b) design as a slip-critical (A325-SC) connection having Class A surface condition and standard holes.

Solution.   In this design it seems apparent that since both the tension and the shear forces are of comparable magnitude, it is likely that neither the maximum shear strength nor the maximum tension strength of the fasteners may be used, as may be observed from Fig. 4.14.3. The following approach [4.24] may be used when design aids are not available.

(a) Bearing-type connection. From Table 4.14.1 (LRFD-Table J3.7), and using $\phi = 0.75$, the interaction criterion for maximum nominal stress in tension under factored loads (threads excluded from shear plane) is

$$\phi F'_{ut} = 87.8 - 1.13 f_{uv} \le 67.5 \text{ ksi} \quad \text{(A325-X)}$$

and the design shear strength is

$$\phi R_{nv} = 0.75(0.50 F^b_u) m A_b$$

The linear interaction equations for $\phi F'_{ut}$ may, in general be expressed as

$$\phi F'_{ut} = C_1 - C_2 f_{uv} \le \phi F_{ut} \tag{a}$$

where C_1 and C_2 are constants.

Convert Eq. (a) into a force equation by multiplying by the bolt area A_b;

$$\phi F'_{ut} A_b = C_1 A_b - C_2 f_{uv} A_b \le \phi F_{ut} A_b \tag{b}$$

or, if ΣA_b represents the total area of all bolts, then $\phi F'_{ut} \Sigma A_b = \Sigma T_u$ and $f_{uv} \Sigma A_b = \Sigma V_u$, giving

$$\Sigma T_u = C_1 \Sigma A_b - C_2 \Sigma V_u \le \phi R_{nt} \tag{c}$$

where ΣT_u and ΣV_u are the total factored tension and shear forces, respectively, applied to the connection. Solving Eq. (c) for ΣA_b gives

$$\Sigma A_b = \frac{\Sigma T_u + C_2 \Sigma V_u}{C_1} \tag{d}$$

which is the basic *design equation* for bearing-type connections. The values of C_1 are 87.8 and 110 for A325 and A490, respectively, and the values of C_2 are 1.13 when threads are eXcluded (A325-X) from the shear planes and 1.43 when threads are iNcluded (A325-N) in the shear planes.

For A325-X connection, Eq. (d) becomes

$$\Sigma A_b = \frac{\Sigma T_u + 1.13 \Sigma V_u}{87.8} \tag{e}$$

For this example, the factored service loads are

$$\Sigma T_u = 1.2(24) + 1.6(96) = 182 \text{ kips}$$
$$\Sigma V_u = 1.2(14) + 1.6(56) = 106 \text{ kips}$$

and using Eq. (d) which is expected to govern gives

$$\text{Required } \Sigma A_b = \frac{\Sigma T_u + 1.13 \Sigma V_u}{87.8} = \frac{182 + 1.13(106)}{87.8} = 3.4 \text{ sq in.}$$

The use of Eq. (e) satisfies the tension requirement. The shear strength must also be checked.

$$\phi R_{nv} = 0.75(0.50)(120)(1) A_b = 45.0 A_b \text{ kips/bolt}$$

The total bolt area ΣA_b required to carry ΣV_u would be

$$\text{Required } \Sigma A_b = \frac{106}{45.0} = 2.4 \text{ sq in.} < 2.9 \text{ sq in.}$$

As expected, the tension force limitation based on the linear interaction equation controlled. The number n of bolts required is

$$\text{Required } n = \frac{3.4}{0.4418} = 7.7 \text{ bolts}$$

Use 8—$\frac{3}{4}$-in.-diam A325-X bolts for a bearing-type connection.

(b) Slip-critical connection. All strength requirements as checked in (a) for a bearing-type connection apply to the slip-critical connection. The interaction relation, Eq. 4.14.10 from LRFD-J3.9, is

$$F_c' = 17\left(1 - \frac{T}{T_b}\right) \tag{4.14.12}$$

For $\frac{3}{4}$-in.-diam bolts, $T_b = 28$ kips. Multiplying Eq. 4.14.12 by A_b gives

$$F_v'A_b = 17A_b\left(1 - \frac{T(0.4418)}{28A_b}\right) \tag{f}$$

Note that in Eq. (f) A_b could be the total area ΣA_b of all participating bolts and T could be ΣT. However, T_b is the initial tension in *one* bolt; thus, one area (0.4418) appears in the numerator to give stress units. This will be correct whether the other variables are $\Sigma T/\Sigma A_b$ for all bolts or T/A_b for one bolt. If the area A_b is considered the total area ΣA_b of all participating bolts, then Eq. (f) may be written

$$\Sigma V = 17\Sigma A_b - 0.27\Sigma T \tag{g}$$

Note that $0.27 = 17(0.4418)/28$. Solving for the total bolt area ΣA_b required gives

$$\text{Required } \Sigma A_b = \frac{\Sigma V + 0.27\Sigma T}{17} \tag{h}$$

which may be considered the *design equation* for A325 bolts in slip-critical connections having Class A surface condition and standard holes.

For A490 bolts, the denominator of Eq. (h) becomes 21. Though the $\frac{3}{4}$-in.-diam bolt was used in this development, it is noted that A_b/T_b is sufficiently constant to use Eq. (h) as the design equation for all sizes of bolts.

For this example,

$$\text{Required } \Sigma A_b = \frac{\Sigma V + 0.27\Sigma T}{17} = \frac{70 + 0.27(120)}{17} = 6.0 \text{ sq in.}$$

The use of Eq. (h) satisfies the shear requirement for slip resistance. The maximum tension must also be checked. The design tensile strength ϕR_{nt} per bolt is

$$\phi R_{nt} = 0.75(0.75F_u^b)A_b = 0.75(0.75)(120)A_b = 67.5A_b \text{ kips}$$

Using the factored load $\Sigma T_u = 182$ kips from part (a), the total bolt area required is

$$\text{Required } \Sigma A_b = \frac{182}{67.5} = 2.7 \text{ sq in.} < 6.0 \text{ sq in.}$$

It is clear that slip resistance in shear controls. The number of bolts required is

$$\text{Required } n = \frac{6.0}{0.4418} = 13.6 \text{ bolts}$$

Use 14—$\frac{3}{4}$-in.-diam A325 bolts for this slip-critical connection.   ■■

4.15 SHEAR AND TENSION FROM ECCENTRIC LOADING

In a bracket connection such as in Fig. 4.14.1a and d, the eccentric load produces both shear and tension in the upper fasteners. As in most other connections, the manner in which the pieces behave is complex. However, nominal forces carried by the fasteners are usually determined by using one of two approaches: (1) that of neglecting any initial tension in the fasteners or (2) that of considering the initial pretension forces in the fasteners. When fasteners such as A307 bolts are used, the amount of initial tension present is usually small and of an indeterminable amount. Therefore, in this case, the neglect of any initial tension is reasonable and gives conservative results. On the other hand, when high-strength bolts are used the initial pretension forces exist and should be recognized.

If initial tension does not exist to any appreciable degree, the application of moment Pe (Fig. 4.14.1d) will produce a tension that is maximum at the top bolts. Near the bottom of the connection, compression would exist between the pieces being joined with little effect directly on the bolts. The direct shear would be carried nearly entirely by the bolts since little friction would exist from bolt installation. The use of A307 bolts having little initial tension is rare in important connections having shear in combination with moment-induced tension; thus no further treatment is given to the analysis neglecting initial tension.

Veillette and DeWolf [4.24] have conducted tests on tee connections with bolts loaded in shear and tension.

Tension from Bending Moment Considering Initial Tension

Consider the service load moment M applied to the bracket of Fig. 4.15.1 to cause tension on the upper bolts (maximum in the top one of each row of bolts shown as black dots). High-strength bolts used as such fasteners are required to be installed with a prescribed initial tension in them whether the joint is considered a bearing-type or a slip-critical connection (RCSG [4.7], Sec. 8). This tension in each bolt will precompress the plates or sections being joined. For the situation of Fig. 4.15.1, the neutral axis under the action of moment M will occur at the centroid (CG) of the contact area; that is, at $d/2$ for the rectangular contact area shown.

The initial bearing pressure f_{bi} as shown in Fig. 4.15.1c is assumed to be uniform over the contact area bd and is equal to

$$f_{bi} = \frac{\Sigma T_b}{bd} \tag{4.15.1}$$

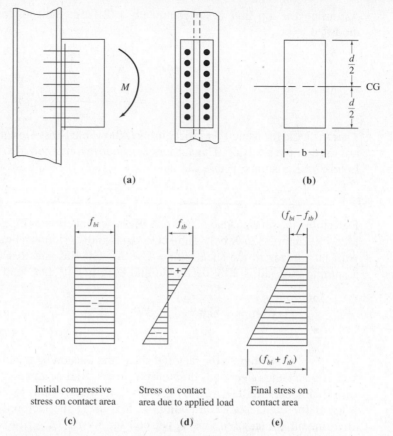

Figure 4.15.1   Stresses on contact surface of moment-resisting connection, considering initial tension in the bolts.

where ΣT_b = the pretension load times the number of bolts. The tensile stress f_{tb} at the top due to the applied moment is

$$f_{tb} = \frac{Md/2}{I} = \frac{6M}{bd^2} \tag{4.15.2}$$

and should not exceed f_{bi} if compression between the pieces is to remain at the top.

The load T on the top bolt is equal to the product of the bolt tributary area (width b times bolt spacing p) times f_{tb}. Thus,

$$T = f_{tb}bp \tag{4.15.3}$$

Substituting Eq. 4.15.2 into Eq. 4.15.3 gives the load on the top bolt as

$$T = \frac{6M}{bd^2}bp = \frac{6Mp}{d^2} \tag{4.15.4}$$

Assuming the top bolt is approximately $p/2$ from the top, the value of T can be modified to be

$$T_{\text{modified}} = T\left(\frac{d - p}{d}\right)$$

$$= \frac{6Mp}{d^2}\left(\frac{d - p}{d}\right) \tag{4.15.5}$$

There is logic for using this procedure in Allowable Sress Design where service loads are used in the analysis. The authors consider this method also acceptable for use in Load and Resistance Factor Design.

EXAMPLE 4.15.1 _____

Determine the service load capacity P for the connection of Fig. 4.15.2 if the fasteners are $\frac{3}{4}$-in.-diam A325-X bolts subject to shear and tension in a bearing-type connection with no threads in the shear plane. Use AISC Load and Resistance Factor Design, assuming the load is 20% dead load and 80% gravity live load.

Solution

(a) Compute factored load P_u.

$$P_u = 1.2(0.2P) + 1.6(0.8P) = 1.52P$$

(b) Compute the factored shear and tension on the bolts using the assumption that applied loads do not overcome initial compression between the pieces being joined. Referring to Fig. 4.15.2, the neutral axis for flexure is at mid-depth of the contact area. Equation 4.15.5, using the factored moment M_u and the maximum factored load T_u per bolt for M and T_{modified}, respectively, and noting the moment M_u equals the load P_u times the eccentricity e of 3 in., gives

$$T_u = \frac{6M_u p}{d^2}\left[\frac{d - p}{d}\right] = \frac{6(1.52P)(3)3}{2(12)^2}\left[\frac{12 - 3}{12}\right] = 0.214P$$

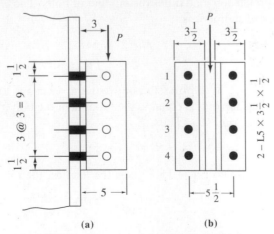

(a)   (b)

Figure 4.15.2 Shear and moment-induced tension connection for Example 4.15.1.

Note that since there are two vertical lines of fasteners, there is a 2 in the denominator of the above equation.

The shear component taken equally by all bolts is

$$V_u = \frac{P_u}{\Sigma n} = \frac{1.52\, P}{8} = 0.190P$$

(c) Design strengths ϕR_n of bolts in shear and tension. In shear the design strength ϕR_{nv} is

$$\phi R_{nv} = 0.75(0.50\, F_u^b)mA_b$$
$$= 0.75(0.50)(120)(1)0.4418 = 19.9 \text{ kips}$$

and in tension the design strength $\phi F'_{ut} A_b$ is reduced from its maximum value by virtue of the simultaneously acting shear; thus, from Table 4.14.1 using $\phi = 0.75$ and multiplying by A_b,

$$\phi F'_{ut} A_b = 87.8 A_b - 1.13 f_{uv} A_b \le 67.5 A_b$$
$$\phi F'_{ut} A_b = 87.8(0.4418) - 1.13 V_u \le 67.5(0.4418)$$
$$\text{Max } T_u = 38.8 - 1.13 V_u \le 29.8 \text{ kips}$$

Assuming the interaction equation controls, solve for P by equating T_u to Max T_u,

$$0.214P = 38.8 - 1.13(0.190P)$$
$$P = 90.5 \text{ kips}$$

Then also check maximum V_u and the upper limit (29.8 kips/bolt) on T_u:

$$V_u = 0.190P = 0.190(90.5) = 17.2 \text{ kips} < 19.9 \text{ kips} \quad \text{OK}$$
$$T_u = 0.214P = 0.214(90.5) = 19.4 \text{ kips} < 29.8 \text{ kips} \quad \text{OK}$$

Therefore, the service load capacity P is 90.5 kips. ■■

Considering Initial Tension—Simplified Procedure

As long as the initial compression between plates resulting from initial tension in installed bolts is not totally counteracted by external load, one may compute tensile stress in a bolt by the flexure formula $f = My/I$ as if only the bolt cross-sectional areas comprise the resisting section:

$$f_t = \frac{My}{I} = \frac{My}{\Sigma A_b y^2} \tag{4.15.6}$$

When all fasteners are the same size (as is usual), A_b may be combined with f_t to obtain the tensile force T in a bolt. Thus,

$$T = A_b f_t = \frac{My}{\Sigma y^2} \tag{4.15.7}$$

To show that Eq. 4.15.7 is identical to Eq. 4.15.5, let the depth d of the contact area equal np, where n is the number of fasteners in one line; Eq. 4.15.5 then becomes

$$T = \frac{6Mp}{n^2 p^2}\left(\frac{np - p}{np}\right) = \frac{12M}{n^3 p^2}\left(\frac{p(n-1)}{2}\right) \tag{4.15.8}$$

Note that $p(n - 1)/2$ is the distance from mid-depth to the outermost fastener and corresponds to y of Eq. 4.15.6. Further, a single line of fasteners spaced at p apart may be treated as a rectangular resisting section of width A/p and depth np. The moment of inertia of such a section would be

$$I = \frac{1}{12}\left(\frac{A}{p}\right)(np)^3 \qquad (4.15.9)$$

which corresponds approximately to the moment of inertia of the bolt areas, $\Sigma A_b y^2$. Thus, Eqs. 4.15.7 and 4.15.8 are essentially the same. Based on this reasoning, design Eq. 4.12.28 may also be used to estimate required number of fasteners when fasteners are subject to moment causing tension or shear and tension.

EXAMPLE 4.15.2

For the connection of the bracket of Fig. 4.15.3 to the column, determine the number of $\frac{7}{8}$-in.-diam A325 bolts required to transmit the shear and tension forces. Use 3-in. vertical pitch. (a) Use bearing-type (A325-X) connection with threads excluded from the shear plane, and (b) use slip-critical (A325-SC) connection. Use Load and Resistance Factor Design assuming the load is 52 kips gravity live load and 8 kips dead load.

Solution

(a) Bearing-type (A325-X) connection. The factored load P_u is

$$P_u = 1.2(8) + 1.6(52) = 92.8 \text{ kips}$$

The design shear strength ϕR_n for fasteners subject to shear alone is

$$\phi R_{nv} = 0.75(0.50 F_u^b) m A_b$$
$$= 0.75(0.50)(120)(1)0.6013 = 27.1 \text{ kips}$$

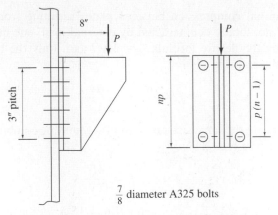

$\frac{7}{8}$ diameter A325 bolts

Figure 4.15.3 Example 4.15.2. Design for shear and tension.

The design tension strength ϕR_n for fasteners subject to tension alone is

$$\phi R_{nt} = 0.75(0.75F_u^b)A_b$$
$$= 0.75(0.75)(120)0.6013 = 40.6 \text{ kips}$$

Noting that M for Eq. 4.12.29 is $92.8(8)/2 = 371$ in.-kips per vertical line of fasteners, the number n of fasteners per line is approximately

$$n = \sqrt{\frac{6M}{Rp}} = \sqrt{\frac{6(371)}{40.6(3)}} = 4.3 \quad \text{required for } M \text{ alone}$$

$$n = \frac{P}{2R} = \frac{92.8}{2(27.1)} = 1.7 \quad \text{required for shear alone}$$

Try 10 bolts (5 per line):

$$\sum y^2 = 4[(3)^2 + (6)^2] = 180 \text{ in.}^2$$

$$R_{ut} = \frac{M_u y}{\sum y^2} = \frac{92.8(8)6}{180} = 24.7 \text{ kips} < 40.6 \text{ kips} \quad \text{OK}$$

$$R_{uv} = \frac{P_u}{\sum n} = \frac{92.8}{10} = 9.3 \text{ kips} < 27.1 \text{ kips} \quad \text{OK}$$

Next check the interaction between shear and tension. From Table 4.14.1,

$$\phi F'_{ut} = 0.75(117 - 1.5f_{uv}) \leq 0.75(90)$$
$$\phi F'_{ut}A_b = 87.8(0.6013) - 1.13R_{uv} \leq 67.5(0.6013)$$
$$= 52.8 - 1.13R_{uv} \leq 40.6 \text{ kips}$$
$$= 52.8 - 1.13(9.3) = 42.3 \text{ kips} > [R_{ut} = 24.7 \text{ kips}] \quad \text{OK}$$

One could check 4 bolts per line,

$$\sum y^2 = 4[(1.5)^2 + (4.5)^2] = 90 \text{ in.}^2$$

$$R_{ut} = \frac{M_u y}{\sum y^2} = \frac{92.8(8)4.5}{90} = 37.1 \text{ kips} < [\phi R_{nt} = 40.6 \text{ kips}] \quad \text{OK}$$

Clearly, 4 bolts per line are acceptable.
Use 8—$\frac{7}{8}$-in.-diam A325 bolts, 4 per row.

(b) Slip-critical connection. The strength requirements examined in (a) are still applicable here; however, usually the service load limitations relating to slip resistance will govern. Equation 4.12.28 may be used with service loads and allowable resistances entered,

$$M = 60(8)/2 = 240 \text{ in.-kips per bolt line}$$
$$R_v = F_v m A_b = 17(1)(0.6013) = 10.2 \text{ kips/bolt}$$

The reduced value for shear in the presence of tension is

$$F'_v = F_v(1 - T/T_b) \quad \text{(LRFD-J3.9)}$$

Since strength must be checked even though the joint is a slip-critical one, the number of fasteners to resist tension is the same calculation as in (a) where

4.3 fasteners per line were indicated. It is more likely here that shear will control; thus, estimating R at somewhat less than 10.2 kips, say 8 kips,

$$n = \frac{P}{2R} = \frac{60}{2(8)} = 3.8 \quad \text{estimate for shear alone}$$

With about 4 bolts required for shear and 4 required for tension, more than 4 should be investigated. Try 5 bolts per line. The maximum service load per bolt in tension and shear is as follows:

$$R_t = \frac{My}{\Sigma y^2} = \frac{60(8)6}{180} = 16.0 \text{ kips}$$

$$R_v = \frac{P}{\Sigma n} = \frac{60}{10} = 6.0 \text{ kips}$$

$$F_v' = F_v(1 - T/T_b)$$

$$= 17(1 - 16.0/39) = 10.0 \text{ ksi}$$

$$\text{Max } R_v = F_v'A_b = 10.0(0.6013)$$

$$= 6.0 \text{ kips} \approx [R_v = 6.0 \text{ kips}] \quad \text{OK}$$

Thus, 5 bolts per line exactly meets the slip resistance requirement and is more than adequate for the strength requirements in shear and tension.

Use 10—$\frac{7}{8}$-in.-diam A325 bolts, 5 per row. ■■

SELECTED REFERENCES

4.1. C. Batho and E. H. Bateman. *Investigations on Bolts and Bolted Joints,* Second Report of the Steel Structures Research Committee. London; His Majesty's Stationery Office, 1934.

4.2. W. M. Wilson and F. P. Thomas. "Fatigue Tests on Riveted Joints," *Bulletin 302.* Urbana, IL: University of Illinois, Engineering Experiment Station, 1938.

4.3. A. E. R. De Jonge. Riveted Joints: *A Critical Review of the Literature Covering Their Development.* New York: American Society of Mechanical Engineers, 1945.

4.4. AISC. "Symposium on High-Strength Bolts," *Proceedings of AISC National Engineering Conference.* New York: American Institute of Steel Construction, 1950, 22–43.

4.5. William H. Munse. "Research on Bolted Connections," *Transactions,* ASCE, **121** (1956), 1255–1266.

4.6. "Rivets and High-Strength Bolts, A Symposium," *Transactions,* ASCE, **126,** Part II (1961), 693–820.

4.7. Research Council on Structural Connections. *Load and Resistance Factor Design Specification for Structural Joints Using ASTM A325 or A490 Bolts.* Chicago, IL: American Institute of Steel Construction, 1988.

4.8. Research Council on Structural Connections. *Allowable Stress Design Specification for Structural Joints Using ASTM A325 or A490 Bolts.* Chicago, IL: American Institute of Steel Construction, November 13, 1985.

4.9. Research Council on Structural Connections. *Commentary on Specifications for Structural Joints Using ASTM A325 or A490 Bolts.* Chicago, IL: American Institute of Steel Construction, 1988.

4.10. John W. Fisher, Theodore V. Galambos, Geoffrey L. Kulak, and Mayasandra K. Ravindra. "Load and Resistance Factor Design Criteria for Connectors," *Journal of the Structural Division,* ASCE, **104**, ST9 (September 1978), 1427–1441.

4.11. J. W. Fisher, P. O. Ramseier, and L. S. Beedle. "Strength of A440 Steel Joints Fastened with A325 Bolts," *Publications,* IABSE, **23** (1963).

4.12. John L. Rumpf and John W. Fisher, "Calibration of A325 Bolts," *Journal of the Structural Division,* ASCE, **89,** ST6 (December 1963), 215–234.

4.13. John H. A. Struik, Abayomi O. Oyeledun, and John W. Fisher. "Bolt Tension Control with a Direct Tension Indicator," *Engineering Journal,* AISC, **10,** 1 (First Quarter 1973), 1–5.

4.14. Desi D. Vasarhelyi and Kah Ching Chiang. "Coefficient of Friction in Joints of Various Steels," *Journal of the Structural Division,* ASCE, **93,** ST4 (August 1967), 227–243.

4.15. Joseph A. Yura and Karl H. Frank. "Testing Method to Determine the Slip Coefficient for Coatings Used in Bolted Joints," *Engineering Journal,* AISC, **22,** 3 (3rd Quarter 1985), 151–155. (also available in AISC *LRFD Manual* [1.18], p. 6-389)

4.16. Theodore V. Galambos, T. A. Reinhold, and Bruce Ellingwood. "Serviceability Limit States: Connection Slip," *Journal of the Structural Division,* ASCE, **108,** ST12 (December 1982), 2668–2680.

4.17. Sherwood F. Crawford and Geoffrey L. Kulak. "Eccentrically Loaded Bolted Connections," *Journal of the Structural Division,* ASCE, **97,** ST3 (March 1971), 765–783.

4.18. Geoffrey L. Kulak. "Eccentrically Loaded Slip-Resistant Connections," *Engineering Journal,* AISC, **12,** 2 (2nd Quarter 1975), 52–55.

4.19. G. Donald Brandt. "Rapid Determination of Ultimate Strength of Eccentrically Loaded Bolt Groups," *Engineering Journal,* AISC, **19,** 2 (2nd Quarter 1982), 94–100. Disc. by Cedric Marsh, *Engineering Journal,* **19,** 4 (4th Quarter 1982), 214–215; Nestor Iwankiw, *Engineering Journal,* **20,** 1 (1st Quarter 1983), 46; 2 (2nd Quarter 1983), 88.

4.20. Avigdor Rutenberg. "Nonlinear Analysis of Eccentric Bolted Connections," *Engineering Journal,* AISC, **21,** 4 (4th Quarter 1984), 227–236.

4.21. J. W. Fisher. "Behavior of Fasteners and Plates with Holes," *Journal of the Structural Division,* ASCE **91,** ST6 (December 1965), 265–286.

4.22. Carl L. Shermer. "Plastic Behavior of Eccentrically-Loaded Connections," *Engineering Journal,* AISC, **8,** 2 (April 1971), 48–51.

4.23. Eugene Chesson, Jr., Norberto L. Faustino, and William H. Munse. "High-Strength Bolts Subjected to Tension and Shear," *Journal of the Structural Division,* ASCE, **91,** ST5 (October 1965), 155–180.

4.24. John R. Veillette and John T. DeWolf. "Eccentrically Loaded High Strength Bolted Connections," *Journal of Structural Engineering,* ASCE, **111,** 5 (May 1985), 1003–1018.

PROBLEMS

All problems are to be done according to the AISC Load and Resistance Factor Design or Allowable Stress Design, as indicated by the instructor. All given loads are service loads unless otherwise indicated. All holes are standard holes and surface condition is clean mill scale (Class A) unless otherwise indicated. When an ultimate strength analyis is requested in an Allowable Stress Design problem, use a factor of safety of 2.5 to obtain the allowable value. Where needed, assume distance from center of hole to nearest edge (edge or end distance) is $1\frac{1}{2}$ in. unless otherwise given.

Values of yield stress F_y and tensile strength F_u for member steels are available in Table 2.1.1. For A325 bolts, $F_u^b = 120$ ksi minimum for bolts $\frac{1}{2}$-to 1-in. diameter, and 105 ksi minimum for $1\frac{1}{8}$-to $1\frac{1}{2}$-in. diameter. For A490 bolts, $F_u^b = 150$ ksi minimum for $\frac{1}{2}$-to $1\frac{1}{2}$-in. diameter.

4.1. Determine the service load tension capacity of the connection of the accompanying figure for the case assigned by the instructor. Investigate as bearing-type connection with threads excluded (X) from the shear planes, bearing-type with threads included (N) in the shear planes, or as a slip-critical (SC) connection, as indicated under the heading "Type Connection." Specify the minimum dimensions A and B appropriate for the connection.

Case	% Dead load	% Live load	Plate steel	Bolt diameter (in.)	Type connection
1	10	90	A36	3/4	A325-X
2	15	85	A36	3/4	A325-SC
3	15	85	A572 Gr 50	3/4	A325-N
4	20	80	A572 Gr 50	3/4	A325-SC
5	20	80	A572 Gr 60	7/8	A325-X
6	40	60	A572 Gr 60	7/8	A325-SC
7	15	85	A572 Gr 65	7/8	A490-X
8	15	85	A572 Gr 65	7/8	A490-SC

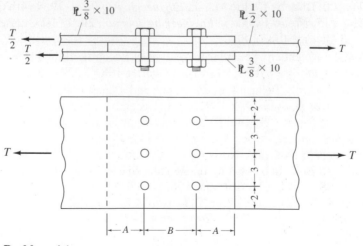

Problem 4.1

4.2. For any of the cases solved in Prob. 4.1, determine the factor of safety against slip at service load. Do you expect slip to occur at service load? If bolt strength had controlled would slip have been expected at service load?

4.3. Determine the service load capacity T for the butt splice of Prob. 3.15 when $s_1 = s_2 = 2$ in. and A325 bolts are used with no threads in the shear planes. Specify the end distances required and evaluate whether or not the given stagger of 2 in. is sufficient. If the 2 in. is not adequate, specify the value to be used. Use (a) a bearing-type connection (A325-X), and (b) a slip-critical (A325-SC) connection.

4.4. For the case assigned by the instructor, determine the number of bolts required to develop the full capacity of the double angle tension member shown in the accompanying figure. Use a double row of bolts without stagger. Detail the connection.

Case	% Dead load	% Live load	Angle steel	Bolt diameter (in.)	Type connection
1	10	90	A572 Gr 50	3/4	A325-X
2	15	85	A572 Gr 50	3/4	A325-SC
3	30	70	A572 Gr 50	7/8	A325-N
4	40	60	A572 Gr 50	7/8	A325-SC

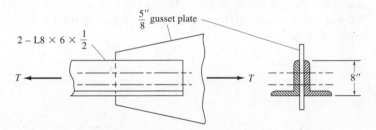

Problem 4.4

4.5. For the single angle tension member of A572 Grade 50 steel in the accompanying figure, how many $\frac{3}{4}$-in.-diam A325 bolts are required for the connection? The load T is 80 kips live load and 20 kips dead load. Assume a slip-critical connection (A325-SC) is to be used. Design the shortest feasible overlap of pieces for the connection, and detail it.

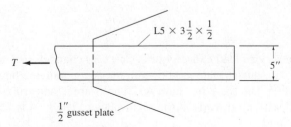

Problems 4.5 and 4.6

4.6. Solve Prob. 4.5 as a bearing-type (A490-X) connection using $\frac{5}{8}$-in.-diam A490 bolts.

4.7. For the single angle tension member of A36 steel in the accompanying figure, determine the number of $\frac{7}{8}$-in.-diam A325 bolts required in a bearing-type connection (A325-N) where threads may exist in the shear plane. The load is 7 kips dead load and 70 kips live load. Assume there are three $\frac{15}{16}$-in.-diam *empty holes* in the outstanding leg that are not a part of the connection. The bolts carrying the load T are to be in a single line with the first bolt located a distance s ahead of the first empty hole. Detail the connection.

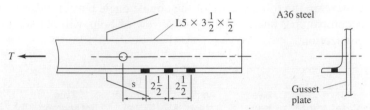

Problem 4.7

4.8. Design and detail the double lap splice shown, to develop maximum tension capacity assuming the load is 20% dead load and 80% live load. The steel is A36 and $\frac{7}{8}$-in.-diam A325 bolts are to be used in a bearing-type connection (A325-X) with no threads in the shear planes. What is the resulting service load capacity of the joint?

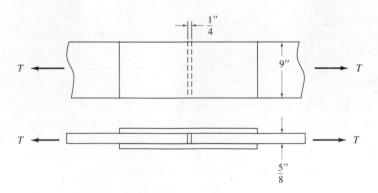

Problem 4.8

4.9. Determine the service load capacity of 2-C10×20 channels as a truss member attached to a $\frac{7}{8}$-in. gusset plate as shown in the accompanying figure. The steel is A572 Grade 65, and the $\frac{7}{8}$-in.-diam A325 bolts are in a bearing-type connection (A325-X) with no threads in the shear planes. The load is 25% dead load and 75% live load.

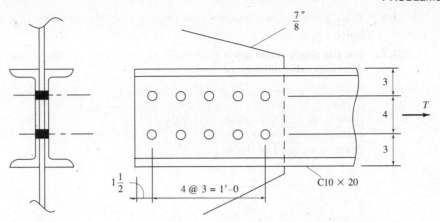

Problem 4.9

4.10. Compute the maximum service load P causing eccentric shear on the connection of the accompanying figure. The loading is 25% dead load and 75% live load. The bolts are $\frac{7}{8}$-in.-diam A325 bolts in a bearing-type connection (A325-X) with threads excluded from the shear plane. Assume the bracket plate has adequate strength.
   a. Use the elastic (vector) method.
   b. Use the strength method with rotation about the instantaneous center.
   c. Use *LRFD Manual* Tables 8-18 to 8-25, with interpolation.

4.11. For the A36 steel bracket plate of the accompanying figure, calculate the maximum service load P (15% dead load and 85% live load) when $\frac{7}{8}$-in.-diam A325 bolts are used in a bearing-type connection (A325-N) with threads included in the shear planes.
   a. Use the elastic (vector) method.
   b. Use the ultimate strength method.
   c. Use *LRFD Manual* Tables 8-18 to 8-25.

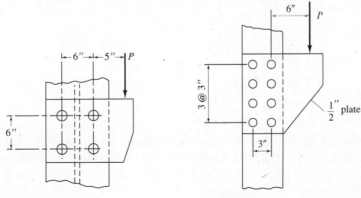

Problem 4.10   **Problem 4.11**

4.12. Repeat Prob. 4.11, except use only 6 bolts instead of 8; that is, 2 at 3 in. vertically instead of 3 at 3 in.

4.13. Repeat Prob. 4.11 as a slip-critical (A325-SC) connection instead of a bearing-type connection.

4.14. Select the proper diameter A490 bolts for a bearing-type connection (A490-X) if the loading is 10% dead load and 90% live load.
   a. Use the elastic (vector) method.
   b. Use ultimate strength analysis.
   c. Use *LRFD Manual* Tables 8–18 to 8–25.

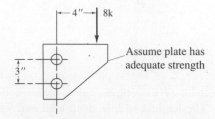

Problem 4.14

4.15. Assuming the fasteners control the capacity, determine the bolt size required for the connection shown when the load is 10% dead load and 90% live load. The connection is a bearing-type (A325-X) containing A325 bolts.
   a. Use the elastic (vector) method.
   b. Use ultimate strength analysis.
   c. Use *LRFD Manual* Tables 8-18 to 8-25.

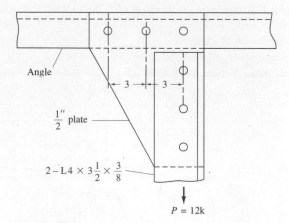

Problem 4.15

4.16. For the connection of the accompanying figure containing $\frac{3}{4}$-in.-diam A325 bolts in a slip-critical connection, determine the service load capacity P (15% dead load and 85% live load) by the following methods, and compare the results:
   a. Use the elastic (vector) method.

b. Use the instantaneous center—constant slip resistance method for slip-critical connections (see Example 4.12.4).

c. Use *LRFD Manual* Tables 8-18 to 8-25.

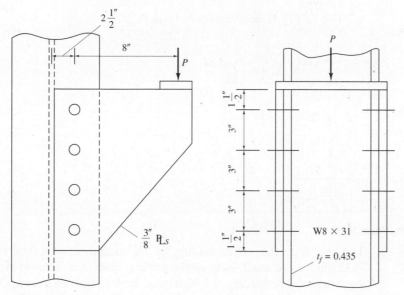

Problem 4.16

4.17. For the eccentric shear loading of the accompanying figure, two vertical lines of $\frac{7}{8}$-in.-diam A325 bolts are used having a 3-in. spacing. Select the proper number of bolts for a bearing-type (A325-X) connection. The load is 40% dead load and 60% live load.

    a. Use the elastic (vector) method.

    b. Use the ultimate strength method.

    c. Use *LRFD Manual* Tables 8-18 to 8-25.

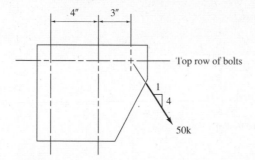

Problem 4.17

4.18. Repeat Prob. 4.17 using $\frac{3}{4}$-in.-diam A490 bolts.

4.19. For the eccentric shear loading of the accompanying figure, $\frac{7}{8}$-in.-diam A325 bolts are used in two vertical lines in a bearing-type (A325-X) connection. The loading is 10 kips dead load and 30 kips live load.

a. Determine the adequacy of the design using basic principles of the elastic (vector) method.
b. Compute the service load capacity of the connection using ultimate strength analysis.
c. Use *LRFD Manual* Tables 8-18 to 8-25.

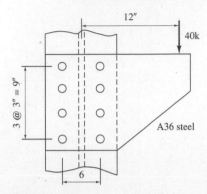

Problem 4.19

4.20. For the eccentric shear loading of the accompanying figure, $\frac{7}{8}$-in.-diam A325 bolts are used in a single vertical line in a bearing-type (A325-X) connection. The loading is 7 kips dead load 33 kips live load. Determine the number of bolts required. What thickness of pieces is required to avoid having bearing control and still use minimum edge distances (see LRFD-Table J3.4 or ASD-Table J3.5).
a. Use the elastic (vector) method.
b. Use the ultimate strength method.
c. Use *LRFD Manual* Tables 8-18 to 8-25.

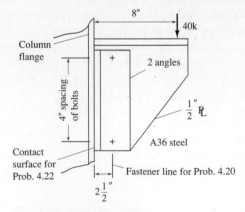

Problems 4.20 and 4.22

4.21. For the connection of the accompanying figure, subject to direct tension and shear, two angles, L4×3$\frac{1}{2}$×$\frac{5}{8}$, are used to carry their maximum capacity as a tension member of A36 steel. Assume the connection of the angles to the structural tee (WT) web will be along a single gage line as shown. Determine the number and positioning of $\frac{7}{8}$-in.-diam A325 bolts to attach the WT to the flange of a W section. The flanges of the WT and the W shape are both $\frac{3}{4}$-in. thick and A36 steel is used.

a. Use a bearing-type (A325-X) connection.

b. Use a slip-critical (A325-SC) connection (20% DL; 80% LL).

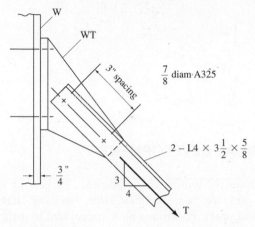

Problem 4.21

4.22. For the eccentric connection of the accompanying figure, causing shear and tension on the fasteners connecting the angles to the column flange; determine the number and spacing of $\frac{7}{8}$-in.-diam bolts required to make the connection to the $\frac{3}{4}$-in. column flange. The load is 5 kips dead load and 35 kips live load.

a. Use a bearing-type (A325-X) connection.

b. Use a slip-critical (A325-SC) connection.

4.23. For the eccentric connection of the accompanying figure, causing shear and tension on the fasteners connecting the structural tee (WT) to the column flange, determine the number of $\frac{7}{8}$-in.-diam bolts spaced vertically at 3 in. required to make the connection to the $\frac{5}{8}$-in. column flange. The load is 10 kips dead load and 60 kips live load.

a. Use a bearing-type (A325-X) connection.

b. Use a slip-critical (A325-SC) connection.

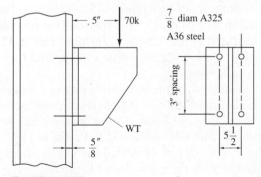

Problem 4.23

Chapter 5

Welding

5.1 INTRODUCTION AND HISTORICAL DEVELOPMENT

The process of welding denotes the joining of metal pieces by heating to a plastic or fluid state, with or without pressure. In its simplest form, "welding" has been known and used for several thousand years. Historians have speculated that the early Egyptians may have first used pressure welding about 5500 B.C. in making copper pipes from sheets by overlapping the edges and hammering. Winterton [5.1] has reported that Egyptian art objects dating about 3000 B.C. have been found on which gold foil has been hammered and fused onto the base copper. This type of welding, called *forge welding*, was man's first process to join pieces of metal together. A well-known early example of forge welding is the Damascus sword which was made by forging layers of iron with different properties. Interestingly, forge welding was sufficiently well developed and important enough to the early Romans that they named one of their gods Vulcan (the god of fire and metalworking) to represent that art. In recent times, the word vulcanizing has been used in reference to treating rubber with sulfur but originally was used to mean "to harden." Today, forge welding is practically a forgotten art in which the village blacksmith was the last major practitioner.

Little progress in welding technology had been made until 1877, prior to which time most of the then known processes such as forge welding and brazing had been used for at least 3000 years. The origin of resistance welding began around 1877 when Professor Elihu Thompson began a set of experiments [5.2] reversing the polarity of transformer coils. He received his first patent [5.3] in 1885 and the first resistance butt welding machine was demonstrated at the American Institute Fair in 1887. In 1889 Charles Coffin [5.2] was issued a patent for flash-butt welding and this became one of the important butt joining processes.

Zerner, in 1885, introduced the carbon arc welding process, making use of two carbon electrodes, and N. G. Slavianoff [5.5] in 1888 in Russia was the first to use the metal arc process using uncoated, bare electrodes. Coffin, working independently also investigated the metal arc process and was issued a U.S. Patent in 1892. In 1889, A. P. Strohmeyer [5.2] introduced the concept of coated metal electrodes to eliminate many of the problems associated with the use of bare electrodes.

Welded truss as part of roof space truss system. (Photo by C. G. Salmon)

Thomas Fletcher [5.1] in 1887 used a blowpipe, burning hydrogen and oxygen and showed that he could successfully cut or melt metal. In 1901–1903 Fouche and Picard developed torches that could be used with acetylene and, thus, the era of oxyacetylene welding and cutting began.

The period between 1903 and 1918 saw the use of welding primarily as a method of repair, the greatest impetus occurring during World War I (1914–1918). Welding techniques proved to be especially adapted to repairing ships that had been damaged.

After World War I (1918), there was continued experimentation with electrodes and various gases to shield the arc and the weld area, resulting in the development of gas tungsten arc welding and gas metal arc welding (see Sec. 5.2). During the period 1930 to 1950 many improvements occurred, including in 1932 [5.5] the introduction of the use of granular flux to protect the weld, which when coupled to the use of a continuously fed electrode resulted in the development of submerged arc welding (see Sec. 5.2), where the arc is buried under the granular flux. This common method of the 1990s was patented in 1935.

In the 1990s automation has become a significant factor in welding technology and extensive use of welding robotics is occurring.

There are many welding processes available to join various metals and their alloys. Those of particular interest in welding structural steel, and of interest to structural engineers in general, are discussed in Sec. 5.2.

5.2 BASIC PROCESSES

Welding is the process of joining materials (usually metals) by heating them to suitable temperatures such that the materials coalesce into one material. There may or may not be pressure, and there may or may not be filler material applied. *Arc welding* is the general term for the many processes that use electrical energy in the form of an electric arc to generate the heat necessary for welding.

This section treats those processes used in arc welding carbon and low-alloy steel for buildings and bridges. For some situations involving light-gage steel, resistance welding may also be important. More extensive descriptions than those that follow are available in the *Welding Handbook* [2.23, 5.10].

Shielded Metal Arc Welding (SMAW)

Shielded metal arc welding is one of the oldest, simplest, and perhaps most versatile types for welding structural steel. The SMAW process is often referred to as the *manual stick electrode process*. Heating is accomplished by means of an electric arc between a coated electrode and the materials being joined. The welding circuit is shown in Fig. 5.2.1a.

The coated electrode is consumed as the metal is transferred from the electrode to the base material during the welding process. The electrode wire becomes filler material and the coating is converted partly into a shielding gas, partly into slag, and some part is absorbed by the weld metal. The coating is a clay-like mixture of silicate binders and powdered materials, such as fluorides, carbonates, oxides, metal alloys, and cellulose. The mixture is extruded and baked to produce a dry, hard, concentric coating.

The transfer of metal from electrode to the work being welded is induced by molecular attraction and surface tension, without application of pressure. The shielding of the arc prevents atmospheric contamination of the molten metal in the arc stream and in the arc pool. It prevents nitrogen and oxygen from being picked up and forming nitrides and oxides which may cause embrittlement.

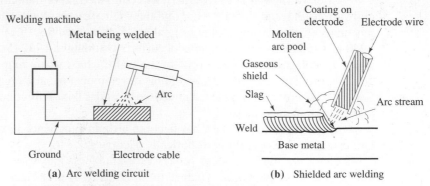

(a) Arc welding circuit   (b) Shielded arc welding

Figure 5.2.1   Shielded metal arc welding (SMAW).

The electrode coating may perform the following functions:

1. Produces a gaseous shield to exclude air and stabilize the arc.
2. Introduces other materials, such as deoxidizers, to refine the grain structure of the weld metal.
3. Produces a blanket of slag over the molten pool and the solidified weld to protect it from oxygen and nitrogen in the air, and also retards cooling.

The electrode material is specified under various American Welding Society specifications that are listed in AWS [2.25] Table 4.1.1, and is summarized in Table 2.3.1. The designations such as E60XX or E70XX indicate 60 ksi and 70 ksi, respectively, for tensile strength. The X's refer to factors such as the suitable welding positions, recommended power supply, type of coating, and type of arc characteristics. Morgan [5.6] has provided an excellent guide to classification and use of mild steel coated electrodes. Table 5.13.1 indicates which coated electrodes should be used with each particular structural steel.

For welding high-carbon or low-alloy steels, low-hydrogen electrodes are required by AWS [2.25] to be used with SMAW for all steels having yield stresses higher than 36 ksi (248 MPa). The low-hydrogen electrode is a rod with a carbonate of soda, or "lime," coating. This electrode requires a different technique from that using the conventional electrode in that a short arc must be made and globular-type, rather than a spray-type, deposition of metal occurs. It is desirable in design because the as-welded mechanical properties have been found to be superior to properties obtained using other types of electrode coatings.

Submerged Arc Welding (SAW)

In the SAW process the arc is not visible because it is covered by a blanket of granular, fusible material, as shown in Fig. 5.2.2. The bare metal electrode is consumable in that it is deposited as filler material. The end of the electrode is kept continuously shielded by the molten flux over which is deposited a layer of unfused flux in its granular condition.

The granular flux, which is a special feature of this method, is usually laid automatically along the seam ahead of the advancing electrode, and provides a cover that allows the weld to be made without spatter, sparks, or smoke. This flux material protects the weld pool against the atmosphere, serves to clean the weld metal, and modifies the chemical composition of the weld metal [2.23, Vol. 2, pp. 207–208].

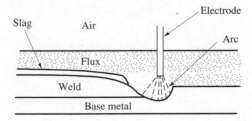

Figure 5.2.2 Submerged arc welding (SAW).

Welds made by the submerged arc process have uniform high quality; exhibiting good ductility, high impact strength, high density, and good corrosion resistance. Mechanical properties of the weld are consistently as good as the base material.

The combinations of bare-rod electrodes and granular flux are classified under AWS A5.17 or A5.23. They are designated FXXX-EXXX where the first X following the F is the first digit of the tensile strength (i.e., 7 for 70 ksi), the second X is a letter indicating the condition of heat treatment (i.e., A for as-welded and P for postweld heat treated), and the third X indicates the lowest temperature at which impact strength of the weld metal meets or exceeds 20 ft-lb (27 J). When the third X is 6, for example, it means the Charpy V-notch impact strength is at least 20 ft-lb (27 J) at -60°F (-51°C). The three Xs following the letter E indicate properties of the electrode. The designations appear in Table 5.13.1 under the SAW process.

The submerged arc method is commonly used in shop-fabricated steel operations using automatic or semiautomatic equipment.

Gas Metal Arc Welding (GMAW)

In the GMAW process the electrode is a continuous wire that is fed from a coil through the electrode holder, a gun-shaped device as shown in Fig. 5.2.3. The shielding is entirely from an externally supplied gas or gas mixture. The distinguishing features of this method are the high rates at which filler metal can be transferred, and the gaseous shield that is uniformly provided around the molten weld. Special uses of this method are described by Craig [5.7], Lyttle [5.8], and Dillenbeck and Castagno [5.9].

Originally, this method was used only with inert gas shielding, hence, the name MIG (metal inert gas) has been used. Reactive gases alone are generally not practical; the exception is CO_2 (carbon dioxide). The use of CO_2 has become extensive for welding of steels, either alone or in a mixture with inert gases.

Argon as a shielding gas works for welding virtually all metals; however, it is not recommended for steels because of its expense and the fact that other shielding gases and gas mixtures are acceptable. For welding carbon steel and some low-alloy steels, recent research [5.9] indicates the best overall performance is obtained using 80% CO_2 and 20% helium. Traditionally, it has been recommended [5.10] to use either a

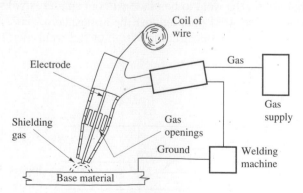

Figure 5.2.3   Gas metal arc welding (GMAW).

mixture of 75% argon and 25% CO_2, or 100% CO_2. For low-alloy steels where toughness is important, the recommendation [5.10] is to use a mixture of 60 to 70% helium, 25 to 30% argon, and 4 to 5% CO_2.

The shielding gas serves the following functions in addition to protecting the molten metal from the atmosphere.

1. Controls the arc and metal-transfer characteristics.
2. Affects penetration, width of fusion, and shape of the weld region.
3. Affects the speed of welding.
4. Controls undercutting.

By mixing an inert gas with a reactive gas the arc may be made more stable and the spatter during metal transfer may be reduced. The use of CO_2 alone for welding steel is the least expensive procedure because of its lower cost for shielding gas, higher welding speed, better joint penetration, and sound deposits with good mechanical properties. The only disadvantage is that it gives harsh and excessive spatter.

The electrode material for welding carbon steels is an uncoated mild steel, deoxidized carbon manganese steel covered under AWS A5.18 and listed in Table 5.13.1 (see also Table 2.3.1). For welding low-alloy steel a deoxidized low-alloy material is necessary.

The GMAW process using CO_2 shielding is good for the lower carbon and low-alloy steels usually used in buildings and bridges.

Flux Cored Arc Welding (FCAW)

The FCAW process, developed in 1958, is similar to GMAW, except that the continuously fed filler metal electrode is tubular and contains the flux material within its core. The core material provides the same functions as does the coating in SMAW or the granular flux in SAW. For a continuously fed wire, an outside coating would not remain bonded to the wire. Gas shielding is provided by the flux core (self-shielded) but additional shielding is frequently provided by CO_2 gas. Flux cored arc welding has become a useful procedure for field welding in severe cold weather conditions [5.11] as well as to speed up high-rise construction [5.12].

The electrodes used for obtaining weld metal having minimum specified yield strengths of 60 ksi (415 MPa) or less are covered under AWS A5.20, and are designated E6XT or E7XT, for tensile strengths of 60 or 70 ksi, respectively. When it is desirable to produce weld metal having yield strength exceeding 60 ksi (415 MPa), the electrodes are covered under AWS A5.29, and are designated E8XT, E10XT, and E11XT, having tensile strengths of 80, 100, and 110 ksi (550, 690, and 760 MPa), respectively.

Electrogas Welding (EGW)

The EGW process is a machine process used primarily for vertical position welding, shown in Fig. 5.2.4. Either flux cored or solid electrodes may be used. This method is used to obtain a single pass weld such as for the splice in a heavy column section. Weld metal is deposited into a cavity created by the separated plate edges on two sides

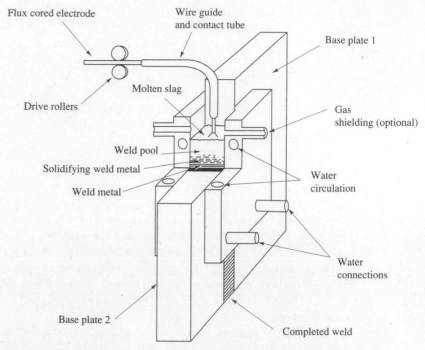

Figure 5.2.4 Electrogas welding with a flux cored electrode. (From *Welding Handbook* [2.23])

and water-cooled "shoes" or guides to keep the molten metal in its proper location on the other two sides. The gas shielding is provided either by the flux cored electrode, by externally supplied gas, or both.

Electroslag Welding (ESW)

The ESW process, shown in Fig. 5.2.5, is similar to electrogas welding, except that the welding is actually done by the heat produced through the resistance of the slag to the flow of current. The molten conductive slag protects the weld and melts the filler metal and the plate edges. Since solid slag is not conductive, an arc is required to start the process by melting the slag and heating the plates. Since resistance heating is used for all but this initial stage, the ESW is really not an arc welding process. The side guides, or "shoes", may be nonconsumable as in Fig. 5.2.5, or they may be consumable. The electroslag process allows welding nearly any thickness of material in one pass; both electrogas and electroslag welding become economical as the thickness of weld required becomes large. Because of the slow weld travel speed used in this process, a weld with relatively coarse grain structure and low notch toughness is the result.

An excellent review of electroslag welding has been provided by Raman [5.14]. Schilling and Klippstein [5.15] have reported research on electroslag welding for bridges, and Pense, Wood, and Fisher [5.16] reported experience with electroslag welding on welded bridges.

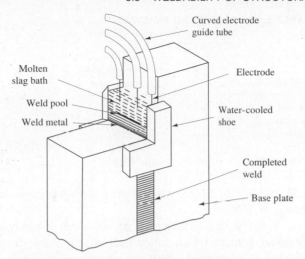

Figure 5.2.5 Nonconsumable guide method of electroslag welding (three electrodes). (From *Welding Handbook* [2.23])

Stud Welding

The most commonly used process of welding a metal stud to a base material is known as arc stud welding, an essentially automatic process but similar in characteristics to the SMAW process. The stud serves as the electrode and an electric arc is created from the end of the stud to the plate. The stud is contained in a gun which controls the timing during the process. Shielding is accomplished by placing a ceramic ferrule around the end of the stud in the gun. The gun is placed in position and the arc is created, during which time the ceramic ferrule contains the molten metal. After a short instant of time, the gun drives the stud into the molten pool and the weld is completed leaving a small fillet around the stud. Full penetration across the shank of the stud is obtained and the weld is completed usually in less than one second.

5.3 WELDABILITY OF STRUCTURAL STEEL

Most of the ASTM-specification construction steels can be welded without special precautions or special procedures. Section 5.13 discusses the need to select the proper electrode to join a particular grade of steel and a summary of the "matching" electrodes and the base steel is given in Table 5.13.1.

The *weldability* of a steel is a measure of the ease of producing a crack-free and sound structural joint. Some of the readily available structural steels are more suited to welding than others, and are discussed in Chapter 2. Welding procedures should be based on a steel's chemistry instead of the published maximum alloy content, since most mill runs are usually below the maximum alloy limits set by its specification. Table 5.3.1 shows the ideal chemical analysis of the carbon steels. Most mild steels fall

TABLE 5.3.1   PREFERRED ANALYSIS OF
CARBON STEEL [5.17] FOR
GOOD WELDABILITY

Element	Normal range (%)	Percent requiring special care
Carbon	0.06–0.25	0.35
Manganese	0.35–0.80	1.40
Silicon	0.10 max	0.30
Sulfur	0.035 max	0.050
Phosphorus	0.030 max	0.040

well within this range, while higher-strength steels may exceed the ideal analysis shown in Table 5.3.1.

When a mill produces a run of steel, it maintains a complete record of its chemical content which follows all shapes made from the particular ingot. If the designer is concerned about the chemistry of a particular grade of steel, he may request a Mill Test Report. Any variation in chemical content above the ideal values may be evaluated, and special welding procedures be set up to insure a properly welded joint.

5.4 TYPES OF JOINTS

The type of joint depends on factors such as the size and shape of the members coming into the joint, the type of loading, the amount of joint area available for welding, and the relative costs for various types of welds. There are five basic types of welded joints, although many variations and combinations are found in practice. The five basic types are the butt, lap, tee, corner, and edge joints, as shown in Fig. 5.4.1.

Butt Joints

The butt joint is used mainly to join the ends of flat plates of the same or nearly the same thicknesses. The principal advantage of this type of joint is to eliminate the f eccentricity developed in single lap joints as shown in Fig. 5.4.1b. When used in

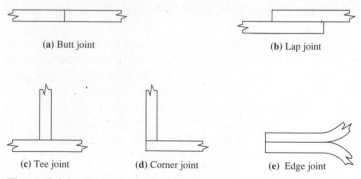

(a) Butt joint   (b) Lap joint

(c) Tee joint   (d) Corner joint   (e) Edge joint

Figure 5.4.1   Basic types of welded joints.

conjunction with full penetration groove welds, butt joints minimize the size of a connection and are usually more esthetically pleasing than built-up joints. Their principal disadvantage lies in the fact that the edges to be connected must usually be specially prepared (beveled, or ground flat) and very carefully aligned prior to welding. Little adjustment is possible and the pieces must be carefully detailed and fabricated. As a result, most butt joints are made in the shop where the welding process can be more accurately controlled.

Lap Joints

The lap joint, shown in Fig. 5.4.2, is the most common type. It has two principal advantages:

1. *Ease of fitting.* Pieces being joined do not require the preciseness in fabricating as do the other types of joints. The pieces can be slightly shifted to accommodate minor errors in fabrication or to make adjustments in length.

2. *Ease of joining.* The edges of the pieces being jointed do not need special preparation and are usually sheared or flame cut. Lap joints utilize fillet welds and are therefore equally well suited to shop or field welding. The pieces being joined are in

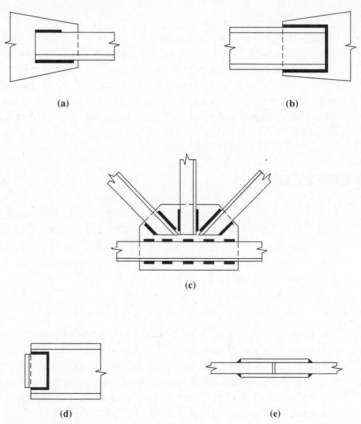

Figure 5.4.2   Examples of lap joints.

most cases simply clamped together without the use of special jigs. Occasionally the pieces are positioned by a small number of erection bolts which may be either left in place or removed after the welding is completed.

A further advantage of the lap joint is the ease in which plates of different thickness can be joined, such as in the double lap joint in Fig. 5.4.2e. The reader should especially note the truss connection shown in Fig. 5.4.2c and consider the difficulty in making such a connection by any other type of joint.

Tee Joints

This type of joint is used to fabricate built-up sections such as tees, I-shapes, plate girders, bearing stiffeners, hangers, brackets, and in general, pieces framing in at right angles as shown in Fig. 5.4.1c. This type of joint is especially useful in that it permits sections to be built up of flat plates that can be joined by either fillet or groove welds.

Corner Joints

Corner joints are used principally to form built-up rectangular box sections such as used for columns and for beams required to resist high torsional forces.

Edge Joints

Edge joints are generally not structural but are most frequently used to keep two or more plates in a given plane or to maintain initial alignment.

As the reader can infer from the previous discussions, the variations and combinations of the five basic types of joints are virtually infinite. Since there is usually more than one way to connect one structural member to another, the designer is left with the decision for selecting the best joint (or combination joints) in each given situation.

5.5 TYPES OF WELDS

The four types of welds are the groove, fillet, slot, and plug welds as shown in Fig. 5.5.1. Each type of weld has specific advantages that determine the extent of its use. Roughly, the four types represent the following percentages of welded construction: groove welds, 15%; fillet welds, 80%; the remaining 5% are made up of the slot, plug, and other special welds.

Groove Welds

The principal use of groove welds is to connect structural members that are aligned in the same plane. Since groove welds are usually intended to transmit the full load of the members they join, the weld should have the same strength as the pieces joined. Such a groove weld is known as a *complete joint penetration groove weld*. When joints are designed so that groove welds do not extend completely through the thickness of the pieces being joined, such welds are referred to as *partial joint penetration groove welds*. For these, special design requirements apply.

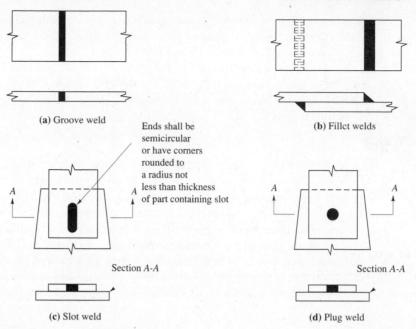

Figure 5.5.1 Types of welds.

There are many variations of groove welds and each is classified according to its particular shape. Most groove welds require a specific edge preparation and are named accordingly. Figure 5.5.2 shows several types of groove welds and indicates the groove preparations required for each. The selection of the proper groove weld is dependent on the welding process used, the cost of edge preparations, and the cost of making the weld. Groove welds may also be used in tee connections as shown in Fig. 5.5.3.

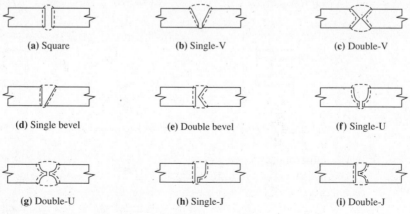

Figure 5.5.2 Types of groove welds.

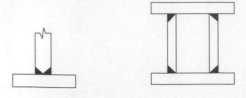

Figure 5.5.3 Use of groove welds in tee joints.

Fillet Welds

Fillet welds owing to their overall economy, ease of fabricating, and adaptability are the most widely used. A few uses of fillet welds are shown in Fig. 5.5.4. They generally require less precision in the "fitting up" because of the overlapping of pieces, whereas the groove weld requires careful alignment with specified gap (root opening) between pieces. The fillet weld is particularly advantageous to welding in the field or in realigning members or connections that were fabricated within accepted tolerances but which may not fit as accurately as desired. In addition, the edges of pieces being

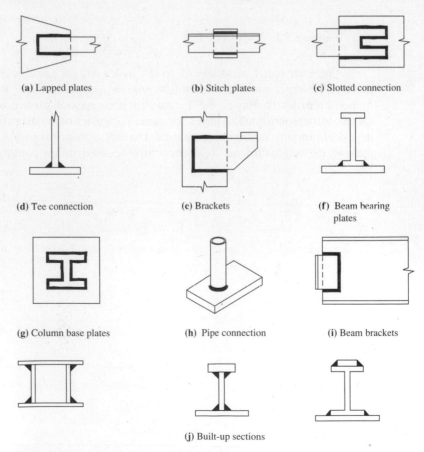

(a) Lapped plates   (b) Stitch plates   (c) Slotted connection

(d) Tee connection   (e) Brackets   (f) Beam bearing plates

(g) Column base plates   (h) Pipe connection   (i) Beam brackets

(j) Built-up sections

Figure 5.5.4   Typical uses of fillet welds.

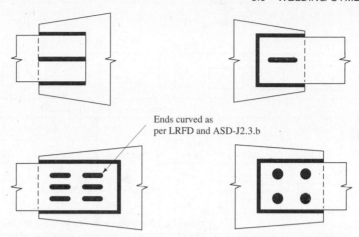

Figure 5.5.5   Slot and plug welds in combination with fillet welds.

joined seldom need special preparation such as beveling or squaring since the edge conditions resulting from flame cutting or from shear cutting procedures are generally adequate.

Slot and Plug Welds

Slot and plug welds may be used exclusively in a connection as shown in Figs. 5.5.1c and d, or they may be used in combination with fillet welds as shown in Fig. 5.5.5. A principal use for plug or slot welds is to transmit shear in a lap joint when the size of the connection limits the length available for fillet or other edge welds. Slot and plug welds are also useful in preventing overlapping parts from buckling.

5.6 WELDING SYMBOLS

Before a connection or joint is welded, the designer must in some way be able to instruct the steel detailer and the fabricator as to the type and size of weld required. The basic types of welds and some of their variations are discussed in Sec. 5.5. If individual and detailed instructions were needed each time a connection was made, the task of providing directions for making the joint would indeed be formidable.

The need for a simple and yet accurate method for communicating between the designer and fabricator gave rise to the use of shorthand symbols that characterize the type and size of weld. As a result, the American Welding Society standard symbols, shown in Fig. 5.6.1, indicate the type, size, length, and location of weld, as well as any special instructions.

Most of the commonly made connections do not require special instructions and are typically specified as shown in Fig. 5.6.2. For a more detailed use of welding symbols the reader is referred to *Standard Symbols for Welding, Brazing and Nondestructive Examination,* A2.4 [5.39].

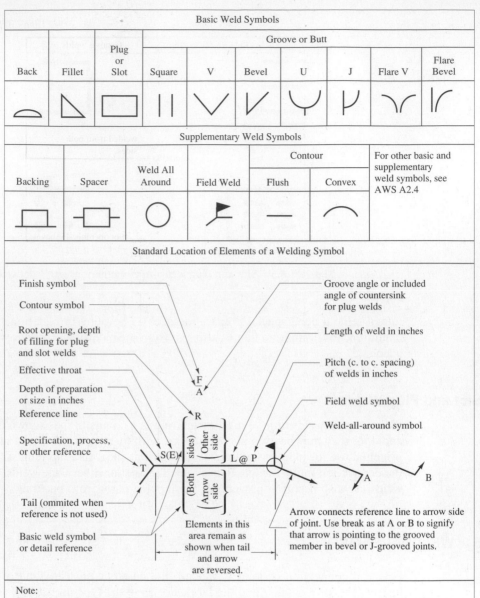

Basic Weld Symbols									
			Groove or Butt						
Back	Fillet	Plug or Slot	Square	V	Bevel	U	J	Flare V	Flare Bevel

Supplementary Weld Symbols						
				Contour		For other basic and supplementary weld symbols, see AWS A2.4
Backing	Spacer	Weld All Around	Field Weld	Flush	Convex	

Standard Location of Elements of a Welding Symbol

Finish symbol

Contour symbol

Root opening, depth of filling for plug and slot welds

Effective throat

Depth of preparation or size in inches

Reference line

Specification, process, or other reference

Tail (ommited when reference is not used)

Basic weld symbol or detail reference

Groove angle or included angle of countersink for plug welds

Length of weld in inches

Pitch (c. to c. spacing) of welds in inches

Field weld symbol

Weld-all-around symbol

Elements in this area remain as shown when tail and arrow are reversed.

Arrow connects reference line to arrow side of joint. Use break as at Λ or B to signify that arrow is pointing to the grooved member in bevel or J-grooved joints.

F
A
R
S(E)
(Other sides) (Other side)
(Both sides) (Arrow side)
T
L @ P
A
B

Note:
Size, weld symbol, length of weld, and spacing must read in that order, from left to right, along the reference line. Neither orientation of reference nor location of the arrow alters this rule.

The perpendicular leg of $\triangle$, V, P, lr, weld symbols must be at left.

Arrow and other side welds are of the same size unless otherwise shown. Dimensions of fillet welds must be shown on both the arrow side and the other side symbol.
The point of the field weld symbol must point toward the tail.

Symbols apply between abrupt changes in direction of welding unless governed by the "all around" symbol or otherwise dimensioned.

These symbols do not explicitly provide for the case that frequently occurs in structural work, where duplicate material (such as stiffeners) occurs on the far side of a web or gusset plate. The fabricating industry has adopted this convention: that when the billing of the detail material discloses the existence of a member on the far side as well as on the near side, the welding shown for the near side shall be duplicated on the far side.

Figure 5.6.1 Standard welding symbols. (From *LRFD Manual* [1.18, p. 8–135]). Additional welding symbol information is available from AWS-A2.4 [5.42].

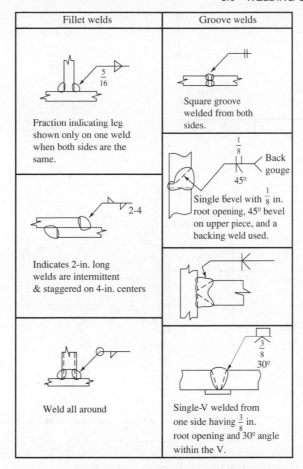

Figure 5.6.2 Common uses of welding symbols.

The reader may feel that the number of symbols is burdensome. However, the system of designating welds is broken down into a few basic types that are built up to give a complete set of instructions. Whenever a particular connection is used in many parts of a structure, it may only be necessary to show a typical detail as shown in Fig. 5.6.3a. Whenever special connections are used, they should be detailed sufficiently to leave no doubt as to the designer's intentions, as shown in Fig. 5.6.3b.

In Fig. 5.6.3b the designer specified that the plug weld be made in the shop and ground flush while the double bevel weld connecting the gusset plate to the column be made in the field. Since the designer did not specify whether the fillet welds attaching the angle to the gusset plate were to be made in the shop or in the field, the steel fabricator would be free to make the decision. However, in this particular detail, it would be better to make the fillet welds in the shop since the plug weld might be overstressed during the field erection process. In general, as many welds as feasible will be made in the shop because of economic considerations. Therefore it is important that the designer specify those welds that are to be *field welded*.

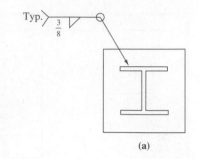

(a)

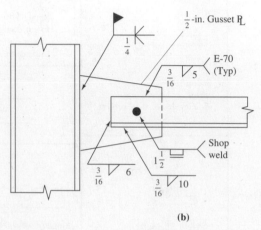

(b)

Figure 5.6.3   Details showing use of welding symbols.

5.7 FACTORS AFFECTING THE QUALITY OF WELDED CONNECTIONS

Obtaining a satisfactory welded connection requires the combination of many individual skills, beginning with the actual design of the weld and ending with the welding operation. The structural engineer needs to be aware of the factors that affect the quality of a weld and design the connections accordingly.

Proper Electrodes, Welding Apparatus, and Procedures

After the proper electrode material is specified to match the strength of the steel in the pieces being joined (see Sec. 5.13), the diameter of the welding electrode must be selected. The particular size of the electrode selected is based on the size of the weld to be made and on the electrical current output of the welding apparatus. Since most welding machines have controls for reducing the current output, electrodes smaller than the maximum capability can easily be accommodated and should be used.

Since the weld metal in arc welding is deposited by the electromagnetic field and not by gravity, the welder is not limited to the flat or horizontal welding positions. The four basic welding positions are shown in Fig. 5.7.1. The designer should avoid whenever possible the overhead position, since it is the most difficult one. Joints

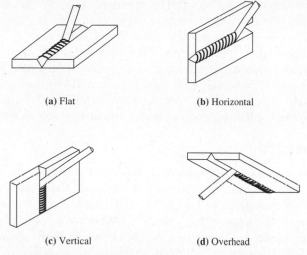

(a) Flat   **(b)** Horizontal

(c) Vertical   **(d)** Overhead

Figure 5.7.1   Welding positions.

welds may require any welding position depending on the orientation of the connection. The welding position for field welds should be carefully considered by the
designer.

Proper Edge Preparation

Typical edge preparations provided for groove welds are shown in Fig. 5.7.2. The root
opening R is the separation of the pieces being joined and is provided for electrode
accessibility to the base of a joint. The smaller the root opening the greater must be

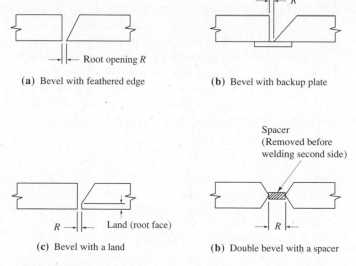

(a) Bevel with feathered edge   **(b)** Bevel with backup plate

(c) Bevel with a land   **(b)** Double bevel with a spacer

Figure 5.7.2   Typical edge preparations for groove welds.

the angle of the bevel. The feathered edge as shown in Fig. 5.7.2a is subject to burn-through unless a backup plate is provided as shown in Fig. 5.7.2b. Backup strips are commonly used when the welding is to be done from one side only. The problem of burn-through is lessened if the bevel is provided a land as shown in Fig. 5.7.2c. The welder should *not* provide a backup plate when a land is provided, since there would be a good possibility that a gas pocket would be formed, preventing a full penetration weld. Occasionally a spacer, as shown in Fig. 5.7.2d, is provided to prevent burn-through but is gouged out before the second side is welded.

Control of Distortion

Another factor affecting weld quality is shrinkage. If a single bead is put down in a continuous manner on a plate, it will cause the plate to distort as shown in Fig. 5.7.3. Such distortions will occur unless care is exercised in both the design of the joint and the welding procedure. Figure 5.7.4 shows the result of using unsymmetrical welds as compared to symmetrical welds. Although there are many techniques available for minimizing distortion, the most common one is that of staggering intermittent welds as shown in Fig. 5.7.5a, and then returning to fill in the spaces as shown in Fig. 5.7.5b, a typical sequence being shown. For many structures, such as plate girders, short segments of weld (though not usually regular intermittent welds) may be used at strategic locations to give enough strength to hold all pieces in place; then the continuous lines of weld are placed.

To minimize shrinkage and to insure adequate ductility, the AWS Code (Table 4.3 of Ref. 5.25) has established minimum preheat and interpass temperatures. For welds requiring more than one progression (pass) of a welding operation along a joint, the interpass temperature is the temperature of the deposited weld when the next pass is begun.

The following summarizes ways of minimizing distortion:

1. Reduce the shrinkage forces by
   a. Using minimum weld metal; for grooves use no greater root opening than necessary; do not overweld.
   b. Using as few passes as possible.
   c. Using proper edge preparation and fit-up.
   d. Using intermittent weld, at least for preliminary connection.
   e. Using backstepping; depositing weld segments toward the previously completed weld; i.e., depositing in the direction opposite to the progress of welding the joint.

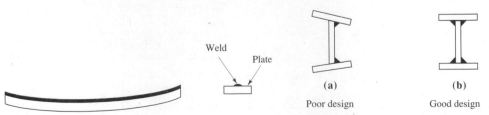

Weld
Plate

(a)
Poor design

(b)
Good design

Figure 5.7.3 Distortion of plate.   **Figure 5.7.4** Effect of weld placement.

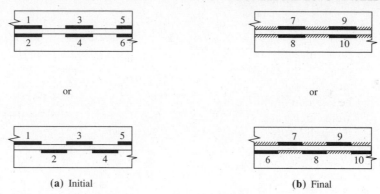

Figure 5.7.5 Sequences for intermittent welds.

2. Allow for the shrinkage to occur by
   a. Tipping the plates so after shrinkage occurs they will be correctly aligned.
   b. Using prebending of pieces.
3. Balance shrinkage forces by
   a. Using symmetry in welding; fillets on each side of a piece contribute counteracting effects.
   b. Using scattered weld segments.
   c. Using peening; stretching the metal by series of blows.
   d. Using clamps, jigs, etc.; this forces weld metal to stretch as it cools.

Blodgett [5.18] has provided a more extensive treatment of ways to minimize distortion.

5.8 POSSIBLE DEFECTS IN WELDS

Unless good welding techniques and procedures are used, a number of possible defects may result relating to discontinuities within the weld. Some of the more common defects are: incomplete fusion, inadequate joint penetration, porosity, undercutting, inclusion of slag, and cracks. Examples of these defects are shown in Fig. 5.8.1.

Incomplete Fusion

Incomplete fusion is the failure of the base metal and the adjacent weld metal to fuse together completely. This may occur if the surfaces to be joined have not been properly cleaned and are coated with mill scale, slag, oxides, or other foreign materials. Another cause of this defect is the use of welding equipment of insufficient current, so that base metal does not reach melting point. Too rapid a rate of welding will also have the same effect.

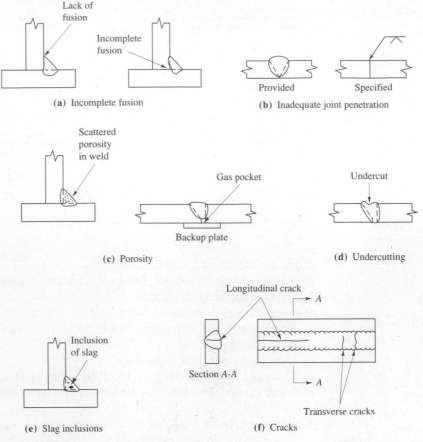

Figure 5.8.1   Possible weld defects.

Inadequate Joint Penetration

Inadequate joint penetration means the weld extends a shallower distance through the depth of the groove than specified, as shown in Fig. 5.8.1, where complete penetration was specified. Partial joint penetration is acceptable only when it is so specified.

This defect, relating primarily to groove welds, occurs from use of an unsuitable groove design for the selected welding process, excessively large electrodes, insufficient welding current, or excessive welding rates. Joint designs prequalified by AWS [5.25, Sec. 2.6 through 2.10] should always be used.

Porosity

Porosity occurs when voids or a number of small gas pockets are trapped during the cooling process. This defect results from using excessively high current or too long an arc length. Porosity may occur uniformly dispersed through the weld, or it may be a large pocket concentrated at the root of a fillet weld or at the root adjacent to a backup plate in a groove weld. The latter is caused by poor welding procedures and careless use of backup plates.

Undercutting

Undercutting means a groove melted into the base material adjacent to the toe of a weld and left unfilled by weld metal. The use of excessive current or an excessively long arc may burn or dig away a portion of the base metal. This defect is easily detected visually and can be corrected by depositing additional weld material.

Slag Inclusion

Slag is formed during the welding process as a result of chemical reactions of the melted electrode coating and consists of metal oxides and other compounds. Having a lower density than the molten weld metal, the slag normally floats to the surface, where upon cooling, it is easily removed by the welder. However, too rapid a cooling of the joint may trap the slag before it can rise to the surface. Overhead welds as shown in Fig. 5.7.1d are especially subject to slag inclusion and must be carefully inspected. When several passes are necessary to obtain the desired weld size, the welder must remove slag between each pass. Failure to properly do so is a common cause of slag inclusion.

Cracks

Cracks are breaks in the weld metal, either longitudinal or transverse to the line of weld, that result from internal stress. Cracks may also extend from the weld metal into the base metal or may be entirely in the base metal in the vicinity of the weld. Cracks are perhaps the most harmful of weld defects; however, tiny cracks called *microfissures* may not have any detrimental effect.

Some cracks form as the weld begins to solidify, generally caused by brittle constituents, either brittle states of iron or alloying elements, forming along the grain boundaries. More uniform heating and slower cooling will prevent the "hot" cracks from forming.

Cracks may also form at room temperature parallel to but under the weld in the base material. These cracks arise in low-alloy steels from the combined effects of hydrogen, a brittle martensite microstructure, and restraint to shrinkage and distortion. Use of low-hydrogen electrodes along with proper preheating and postheating will minimize such "cold" cracking.

5.9 INSPECTION AND CONTROL

The enormous success and growth in recent years in the area of structural welding of buildings and bridges could not have occurred without some means of inspection and control. The welding industry has led in the development of guidelines which, if followed, virtually insure a sound weld. The inspection and control procedure should begin before the first arc is struck, continue throughout the welding procedure, and if necessary, a pretest of the joint should be made to assure its satisfactory performance.

Since such close supervision is not possible on every weld made, the following suggestions will serve as a guideline to achieve good structural welds:

1. Establish good welding procedures.
2. Use only prequalified welders.
3. Use qualified inspectors and have them present.
4. Use special inspection techniques when necessary.

Good welding procedures can be developed from recommendations from the AWS, AISC, and the manufacturers of welding supplies and equipment. The procedure to be followed will depend on the chemical and physical properties of the materials, the types and sizes of weld, and the particular equipment used.

All welders should be required to have passed an American Welding Society Qualification Test before being permitted to make a structural connection. Although this is usually considered adequate, it doesn't prove the ability of the welder to make welds at the actual job site, particularly if the welds are unusual or difficult and were not specified in the Qualification Test. Happily, most welding contractors exercise control over their welders in such situations.

The use of qualified welding inspectors at a job site generally has the effect of causing welders to perform their best work, feeling that the inspector is able to recognize the quality of their welds. The welding inspector should be a competent welder and be able to recognize possible defects. Any poor or suspicious welds should be cut out and replaced.

The simplest and least expensive method of inspection is *visual* but it is dependent on the competence of the observer and weld appearance may be deceiving [5.20]. In some cases a poor weld may be overlooked. A welding gage such as shown in Fig. 5.9.1a offers a rapid means of checking the size of fillet welds.

On more important structures, for welds where fatigue is an important consideration, or welds whose failure could be catastrophic, more rigid inspection techniques [5.20] should be used. Some of the useful ones are the ultrasonic, radiographic, and magnetic particle methods. The *ultrasonic* method [5.21–5.23] passes ultra-high-frequency sound waves through the weldment. Defects in a particular weld will reflect the sound waves while a weld without defects will not impede passage of the waves. *Radiographic* methods include the use of both X-rays and gamma rays. In this method

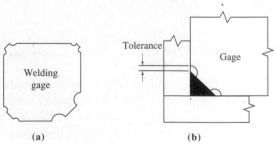

(a)   (b)

Figure 5.9.1   Checking size of fillet welds.

the radiating source is placed on one side of the weld and a photographic plate on the other. This method is expensive and requires special precautions be taken because of the hazards of radiation. However, the method is reliable and furnishes a permanent record. The *magnetic particle* testing method [5.24] uses iron powder which is spread around the welded area and polarized by passing an electric current through the weld. Small local poles will be formed at the edges of any defects, and this may be interpreted by an experienced observer.

The *Welding Handbook*, Vol. 1 [2.23, pp. 468–515] provides an extensive review, of the many nondestructive testing methods.

5.10 ECONOMICS OF WELDED BUILT-UP MEMBERS AND CONNECTIONS

Overall economy in welded connections is difficult to evaluate. Some of the factors to be considered such as the amount of electrode material used can easily be computed while other factors such as the value to be placed on esthetics may be intangible. The actual economy of welded connections must be viewed from a broad aspect and include the overall design of the structural system.

Welded connections are usually neater in appearance, providing a less cluttered effect in contrast to bolted connections. Fig. 5.10.1 shows a comparison between a section of a bolted plate girder and a section of a welded plate girder. Besides the neater appearance of the welded joint, welded connections offer the designer more freedom to be innovative in his entire design concept. The designer is not bound to standard sections but may build up any cross section deemed to be most advantageous. Similarly, the best configuration to transmit the loads from one member to another can be used.

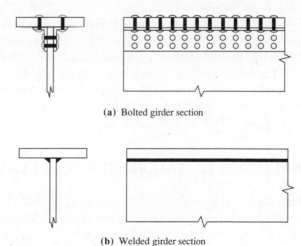

(a) Bolted girder section

(b) Welded girder section

Figure 5.10.1 Comparison between plate girders.

Welded connections generally eliminate the need for holes in members except possibly for erection purposes. Since it is usually the holes at the ends that govern the design of a bolted tension member, a welded connection will generally result in a member with a smaller cross-section.

Welded connections can sometime reduce field construction costs by the fact that members may be shifted slightly to accommodate minor errors in fabrication or erection. Also, members may be shortened by cutting and rejoined by suitable welding, as well as lengthened by splicing a piece of the same cross-section.

In addition, several direct factors influence the cost of welding. Generally, welding performed in the shop is less expensive than field welding. Some reasons for this are availability in the shop of automatic welding machines, a more pleasant and less hostile environment (the weather), and the availability of special jigs for holding the pieces to be welded in a more favorable position. Also, work can be scheduled for a continuous operation, whereas field welding must often wait for cranes and special erection equipment. Other operations such as the proper preheating of pieces to be welded can be difficult if not impossible to perform in the field. Other factors that influence welding costs are:

1. Cost of preparing the edges to be welded.
2. The amount of weld material required.
3. The ratio of the actual arc time to overall welding time.
4. The amount of handling required.
5. General overhead costs.

The factors listed above are generally unknown to the designer because the fabricator is usually not selected until after the design has been completed. However, the designer must still make decisions—should short length large size fillet welds or longer length smaller size welds be used? Should large size fillet welds or groove welds be used? If the decision is to use groove welds then the proper and most economical type must be selected.

In most instances the designer is not as concerned with the specific cost of a weld type as with the *relative* cost. Donnelly [5.26] has developed factors relating the cost of fillet and groove welds of common sizes to the cost of a single-pass $\frac{1}{4}$-in. fillet weld.

Currently (1995), welded connections are used for the vast majority of shop connections and a sizable, through not a majority of of field connections. More extensive treatment of welding cost is available in the *Welding Handbook,* Vol. 1 [2.23, Chap. 8] and from Blodgett [5.27].

5.11 SIZE AND LENGTH LIMITATIONS FOR FILLET WELDS

Since all welding involves the heating of the metal pieces, prevention of too rapid a rate of cooling is of fundamental importance to achieving a good weld. Consider the two extreme thicknesses of plates in Fig. 5.11.1, each of which has received a bead of fillet weld. Most heat energy given off during the welding process is absorbed by plates being joined. The thicker plate dissipates the heat vertically as well as horizontally

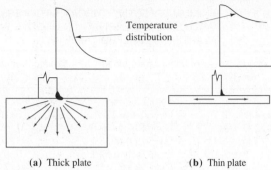

(a) Thick plate   (b) Thin plate

Figure 5.11.1   Effect of thickness on cooling rate.

whereas the thinner plate is essentially limited to a horizontal dissipation. In other words, the thicker the plate, the faster heat is removed from the welding area, thereby lowering the temperature in the region of the weld. Since a minimum temperature is required to cause the base metal to become molten, it is therefore necessary to provide as a minimum, a weld of sufficient size (and heat content) to prevent the plate from removing the heat at a faster rate than it is being supplied. Unless a proper temperature is maintained in the area being welded a lack of fusion will result.

Minimum Weld Size

To help insure fusion and minimize distortion the AWS [5.25] and AISC Specifications [1.5, 1.16] provide for a minimum size weld based on the thicker of the pieces being joined. The requirements for fillet weld based on the leg dimension a of the fillet and for partial-joint-penetration groove weld based on the effective throat (see Sec. 5.12) are given in Tables 5.11.1 and 5.11.2.

Maximum Fillet Weld Size Along Edges

The *maximum* size of fillet weld used *along the edges* of pieces being joined is limited (LRFD and ASD-J2.2b) in order to prevent the melting of the base material at the

TABLE 5.11.1   MINIMUM SIZE* OF FILLET WELDS
(ADAPTED FROM LFRD-TABLE J2.4)

| Material thickness of thicker part joined | | Minimum size** of fillet weld*** | |
(in.)	(mm)	(in.)	(mm)
To 1/4 inclusive	To 6.4 inclusive	1/8	3
Over 1/4 to 1/2	Over 6.4 to 12.7	3/16	5
Over 1/2 to 3/4	Over 12.7 to 19.0	1/4	6
Over 3/4	Over 19.0	5/16	8

*See LRFD-J2.2b for maximum size of fillet welds.
**Leg dimension; single pass weld.
***According to AWS-2.7, weld size need not exceed the thickness of the thinner part joined.

TABLE 5.11.2   MINIMUM EFFECTIVE THROAT THICKNESS OF
PARTIAL-JOINT-PENETRATION GROOVE WELDS
(ADAPTED FROM LFRD-TABLE J2.3)

Material thickness of thicker part joined		Minimum effective throat thickness*	
(in.)	(mm)	(in.)	(mm)
To 1/4 inclusive	To 6.4 inclusive	1/8	3
Over 1/4 to 1/2	Over 6.4 to 12.7	3/16	5
Over 1/2 to 3/4	Over 12.7 to 19.0	1/4	6
Over 3/4 to 1-1/2	Over 19.0 to 38.1	5/16	8
Over 1-1/2 to 2-1/4	Over 38.1 to 57.1	3/8	10
Over 2-1/4 to 6	Over 57.1 to 152	1/2	13
Over 6	Over 152	5/8	16

* See LRFD-J2.1 for effective throat thickness. For J- and U-joints, and Bevel- or V-joints having included angle at root of groove $\geq 60°$, effective throat is depth of chamfer. For Bevel-or V-joints having included angle $< 60°$ but $\geq 45°$, effective throat is depth of chamfer minus 1/8 in. (3 mm).

location where the fillet would meet the corner of the plate if the fillet were made the full plate thickness. The maximum permitted is (see Fig. 5.11.2):

1. Along edges of material *less* than $\frac{1}{4}$-in. (6.4-mm) thick, the maximum size is equal to the thickness of the material.
2. Along edges of material $\frac{1}{4}$ in. (6.4 mm) or more in thickness, the maximum size shall be $\frac{1}{16}$ in. (1.6 mm) less than the thickness of the material, unless the weld is especially designated on the drawings to be built out to obtain full throat thickness.

Minimum Effective Length of Fillet Welds

When placing a fillet weld, the welder builds up the weld to its full dimension as near the beginning of the weld as practicable. However, there is always a slight tapering off in the region where the weld is started and where it ends. Therefore, the minimum effective length of a fillet weld is four times the nominal size (LRFD and ASD-J2.2.b).

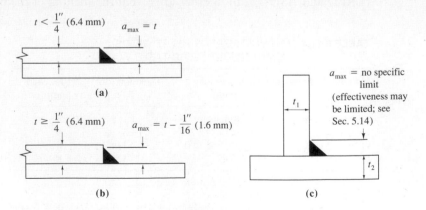

Figure 5.11.2   Maximum weld size.

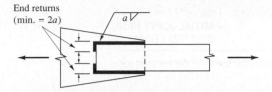

Figure 5.11.3   Use of end returns.

If this requirement is not met, the size of the weld shall be considered to be one-fourth of the effective length.

LRFD and ASD-J2.2.b recommend the use of end returns, whenever practicable, as shown in Fig. 5.11.3. For other limitations the reader is referred to the AISC Specifications [1.5, 1.16].

5.12 EFFECTIVE AREAS OF WELDS

The strengths of the various types of welds discussed in Sec. 5.5 are based on *effective areas*. The effective area of a groove or fillet weld is the product of the *effective throat* dimension t_e times the length of the weld.

The effective throat dimension depends on the nominal size and the shape of the weld, and may be thought of as the minimum width of the expected failure plane.

Groove Welds

The effective throat dimension of a full penetration groove weld is the thickness of the thinner part joined, as shown in Figs. 5.12.1a and b. For a partial joint penetration groove weld, the effective throat may be less than the depth of the chamfer. For example, when bevel- or V-joint grooves have an included angle at the root of the groove less than 60° but not less than 45° when SMAW or SAW processes are used, or when the GMAW or FCAW processes are used in vertical or overhead positions, the effective throat is the depth of the chamfer less $\frac{1}{8}$ in. When the included angle is 60° or more, the effective throat is the full depth of chamfer for all four processes mentioned. Effective throat requirements for the groove situations mentioned, as well as others, are given in LRFD and ASD-J2.1a.

Fillet Welds

The effective throat dimension of a fillet weld is nominally the shortest distance from the root to the face of the weld, as shown in Fig. 5.12.2. Assuming the fillet weld to have equal legs of nominal size a, the effective throat t_e is $0.707a$. If the fillet weld is designed to be unsymmetrical (a rare situation) with unequal legs, as shown in Fig. 5.12.2b, the value of t_e must be computed from the diagrammatic shape of the weld. The effective throat dimensions for fillet welds made by the submerged arc

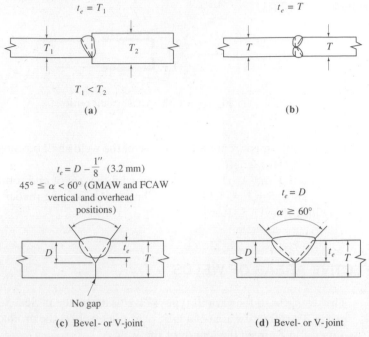

$$t_e = T_1$$

$$T_1$$ $$T_2$$

$$T_1 < T_2$$

(a)

$$t_e = T$$

$$T$$ $$T$$

(b)

$$t_e = D - \frac{1''}{8} \ (3.2 \text{ mm})$$
$45° \leq \alpha < 60°$ (GMAW and FCAW vertical and overhead positions)

$$D$$ $$t_e$$ $$T$$

No gap

(c) Bevel- or V-joint

$$t_e = D$$

$$\alpha \geq 60°$$

$$D$$ $$t_e$$ $$T$$

(d) Bevel- or V-joint

Figure 5.12.1   Effective throat dimensions for groove welds (LRFD and ASD-J21a) made by SMAW, SAW, GMAW, and FCAW.

(SAW) process are modified by LRFD and ASD-J2.2a as follows to account for theinherently superior quality of such welds:

1. For fillet welds the leg size equal to or less than $\frac{3}{8}$ in. (9.5 mm), the effective throat dimension shall be taken as equal to the leg size a.
2. For fillet welds larger than $\frac{3}{8}$ in. (9.5 mm), the effective throat dimension shall be taken as the theoretical throat dimension plus 0.11 in. (2.8 mm) (i.e., $0.707a + 0.11$ for symmetric welds).

Plug and Slot Welds

The effective shearing area of plug or slot welds is their nominal area (sometimes called *faying surface*) in the shearing plane. The resistance of plug or slot welds is the product of the nominal cross-section times the stress on that area.

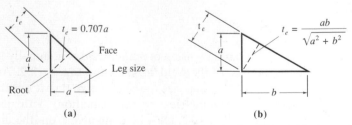

$$t_e = 0.707a$$

$$t_e$$

Face

$$a$$

Leg size

Root $$a$$

(a)

$$t_e = \frac{ab}{\sqrt{a^2 + b^2}}$$

$$t_e$$

$$a$$

$$b$$

(b)

Figure 5.12.2   Effective throat dimensions for fillet welds (except by submerged arc process).

5.13 NOMINAL STRENGTH OF WELDS

Since welds must transmit the entire load from one member to another, welds must be sized accordingly and be formed from the correct electrode material. For design purposes fillet welds are assumed to transmit loads through *shear stress* on the effective area no matter how the fillets are oriented on the structural connection. Groove welds transmit loads exactly as in the pieces they join.

The electrode material used in welds should have properties of the base material. When properties are comparable, the weld metal is referred to as "matching" weld metal. Table 5.13.1 gives "matching" weld metal for many ASTM structural steels used in buildings and bridges.

Strength of Groove Welds

Complete joint penetration groove welds are considered to have the same strength on the *effective area* as the pieces being joined. Welds subject to tension normal to the effective area must be made with "matching" weld metal, where the properties of the weld metal are comparable to those of the base metal. In compression, where stability of the compression member is usually the major factor, the weld metal strength is permitted to be one classification (10 ksi) lower than the "matching" base metal requirement. The authors recommend, however, that matching electrodes be used when the member may possibly be used in tension in the future.

Thus, the nominal strength R_{nw} of weld per inch of length is based on yielding of the base metal, which gives

$$R_{nw} = t_e F_y \qquad \text{tension or compression} \qquad (5.13.1)$$
$$R_{nw} = t_e(0.60F_y) \quad \text{shear} \qquad (5.13.2)$$

where t_e = effective thickness (see Sec. 5.11)
F_y = yield stress of the base metal

Note that as discussed in Sec. 2.6, the shear yield stress is taken as approximately $\frac{2}{3}$ (actually 0.6) of the tension-compression yield stress. When "matching" electrode material is used, the weld material is somewhat stronger than the base material; thus the strength of the welded joint is controlled by the base material properties.

When the groove welded joint is subject to shear, the weld metal is permitted (AWS-Table 8.4.1 [2.25]) to be of lower strength than the base metal; i.e., less than "matching" weld metal, in which case the strength F_{EXX} of the weld material must be used instead of F_y in computing R_{nw}.

Strength of Fillet Welds

The strength R_{nw} of a fillet weld per inch of length is based on assumption that failure of such a weld is by shear on the effective area whether the shear transfer is parallel to or perpendicular to the axis of the line of fillet weld. In fact, the strength is greater for shear transfer perpendicular to the weld axis; however, for simplicity the situations are treated the same. Thus, fillet weld strength may be controlled by the weld electrode strength *or* by the base material shear strength. The nominal strength R_{nw} per

TABLE 5.13.1 MATCHING FILLER METAL REQUIREMENTS (ADAPTED FROM AWS-TABLE 4.1.1 [2.25])

Group	Base metal steel specification*	Welding process‡‡			
		Shielded metal arc welding (SMAW)	Submerged arc welding (SAW)	Gas metal arc welding (GMAW)	Flux cored arc welding (FCAW)
I	ASTM A36, A53 Grade B, A500, A501, A529, A570 Grades 40, 45, and 50 A709 Grade 36	AWS A5.1 or A5.5 E60XX or E70XX	AWS A5.17 or A5.23 F6XX- or F7XX-EXXX	AWS A5.18 ER70S-X	AWS A5.20 E6XT-X and E7XT-X (except -2 -3, -10 -GS)
II	ASTM A242,+ A572 Grades 42 and 50 A588 A709 Grades 50 and 50W	AWS A5.1 or A5.5 E70XX‖	AWS A5.17 or A5.23 F7XX-EXXX	AWS A5.18 ER70S-X	AWS A5.20 E7XT-X (except -2, -3, -10, -GS)
III	ASTM A572 Grades 60 and 65	AWS A5.5 E80XX-X‖	AWS A5.23 F8XX-EXX-XX‖	AWS A5.28 ER80S-X¶	AWS A5.29 E8XTX-X‖
IV	ASTM A514 (over 2½ in. thick), A709 Grades 100 and 100W (2½ to 4 in.)	AWS A5.5 E100XX-X‖	AWS A5.23 F10XX-EXX-XX‖	AWS A5.28 ER100S-X¶	AWS A5.29 E10XTX-X¶
V	ASTM A514 (2½ in. and under), A709 Grades 100 and 100W (2½ in. and under)	AWS A5.5 E110XX-X‖	AWS A5.23 F11XX-EXX-XX‖	AWS A5.28 ER110S-X¶	AWS A5.29 E11XTX-X¶

* In joints involving base metals of different groups, low-hydrogen filler metal requirements applicable to the lower strength group may be used. The low-hydrogen processes shall be subject to the technique requirements applicable to the higher strength group.
† When welds are to be stress relieved, the deposited weld metal shall not exceed 0.05 percent vanadium.
‡ See AWS D1.1–94, Sec. 4.16 for electroslag and electrogas weld metal requirements.
‖ Low hydrogen classifications only.
¶ Deposited weld metal shall have a minimum impact strength of 20 ft-lb (27.1J) at 0°F (−18°C) when Charpy V-notch specimens are required.
+ Special welding materials and procedures may be required to match the notch toughness of base metal or for atmospheric corrosion and weathering characteristics.

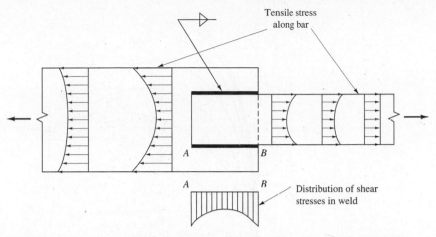

Figure 5.13.1 Typical stress distribution in a lap joint with longitudinal fillet welds.

inch of weld length may be expressed

$$R_{nw} = t_e(0.60F_{\text{EXX}}) \quad \text{weld metal} \tag{5.13.3}$$

or

$$R_{nw} = t(0.60F_u) \quad \text{base metal.} \tag{5.13.4}$$

The stress distribution in welded joints is complex and nonuniform. Figure 5.13.1 shows the typical stress distribution at service load for longitudinal fillet welds in a lap joint. The actual variation of shear stress in the weld from points A to B depends on the length of the weld as well as the ratio of the widths of the plates being joined. Figure 5.13.2 shows the typical shear variation for fillet welds loaded transverse to the weld axis.

The load-deformation relationship of a fillet weld has been studied by Butler, Pal, and Kulak [5.29], Kulak and Timmler [5.30], Swannell [5.31, 5.32], and Neis [5.33], and is shown in Fig. 5.13.3, where the reader may observe that the strength is related to the angle θ to the weld axis at which the weld is loaded. The longitudinal

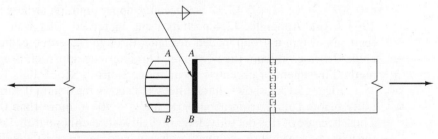

Figure 5.13.2 Typical stress distribution in a lap joint with transverse fillet welds.

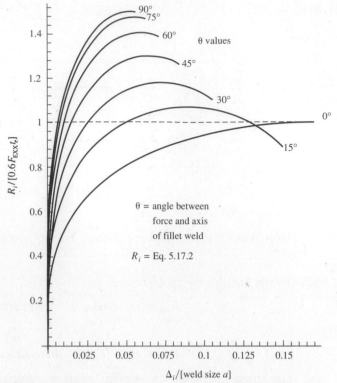

Figure 5.13.3   Load-deformation relationship for fillet welds.

welds in Fig. 5.13.1 are loaded at 0° (parallel to the weld axis); Fig. 5.13.3 shows considerable ductility (large deformation capability) for such welds. Figure 5.13.2 shows transverse welds loaded at 90° (perpendicular to the weld axis); Fig. 5.13.3 shows greater strength but considerably less deformation capability for such welds.

For all AISC Specifications, both ASD and LRFD, prior to the 1993 LRFD Specification, the additional strength of fillet welds loaded at angles θ greater than 0° was not utilized. The shear strength of the weld segment per unit length, taken at $t_e(0.60F_{EXX})$ (i.e., Eq. 5.13.3), has been the upper limit, regardless of the angle θ. The 1993 LRFD-Appendix J2.4 permits using increased fillet weld strength for weld segments within a group loaded in-plane, through the group center of gravity, at an angle θ measured from the weld axis. This alternate approach is used in the strength method treatment of eccentric shear connections in Sec. 5.17.

Figure 5.13.3 shows that a fillet weld shows maximum deformation Δ_i/a capability before failure from about 0.06 for $\theta = 90°$ to more than 0.15 when $\theta = 0°$. Thus, because of this ductility, lines of weld loaded either parallel or perpendicular to the axis of the weld are assumed for design purposes to resist equally anywhere along their length. Even though the *elastic* distribution of load along the length of weld is not uniform, plastic deformation will permit this simplified designed treatment.

5.14 LOAD AND RESISTANCE FACTOR DESIGN—WELDS

The general philosophy of Load and Resistance Factor Design (LRFD) was described in Secs. 1.8 and 1.9. Equation 1.8.1 gives the structural safety requirement, as follows:

$$\phi R_n \geq \sum \gamma_i Q_i \qquad [1.8.1]$$

where ϕ = resistance factor (strength reduction factor), values of which for welds are given in Table 5.14.1
R_n = nominal resistance (strength) = R_{nw} for welds
γ_i = overload factors (LRFD-A4.1)
Q_i = service loads (see Secs. 1.8 and 1.9)

Equation 1.8.1 requires the *design strength* ϕR_n to equal or exceed the summation of *factored loads,* or specifically for welds, Eq. 1.8.1 becomes

$$\phi R_{nw} \geq R_u \qquad (5.14.1)$$

where ϕ = resistance factor (for welds the ϕ factor is the same as for the material it connects; that is, 0.90 for the yielding limit state and 0.75 for the fracture limit state)
R_{nw} = the nominal strength per unit length of weld, but not to exceed the nominal strength per unit length of adjacent base material
R_u = factored load per unit length of weld

Groove Welds

The *design strength* (LRFD-J2.4.) per unit length of complete penetration groove welds depends on the type of stress that is applied:

1. *Tension* and *compression* normal to effective area,* and tension and compression parallel to axis of weld:

$$\phi R_{nw} = 0.90 t_e F_y \quad \text{base material} \qquad (5.14.2)$$
$$\phi R_{nw} = 0.90 t_e F_{yw} \quad \text{weld metal} \qquad (5.14.3)$$

where F_y and F_{yw} are the yield stresses of the base and weld metals, respectively. When "matching" welds as indicated in Table 5.13.1 are used, F_{yw} is taken as the yield stress of the base material.

2. *Shear* on effective area*:

$$\phi R_{nw} = 0.90 t_e \tau_y \qquad \text{base metal}$$
$$= 0.90 t_e (0.60 F_y) \qquad (5.14.4)$$
$$\phi R_{nw} = 0.80 t_e (0.60 F_{EXX}) \quad \text{weld metal} \qquad (5.14.5)$$

---

*See Sec. 5.12 for definition of effective area.

TABLE 5.14.1 DESIGN STRENGTH OF WELDS
(FROM *LRFD SPECIFICATION* [1.16], LRFD-Table J2.5, p. 6–78)

Types of Weld and Stress [a]	Material	Resistance Factor ϕ	Nominal Strength F_{BM} or F_W	Required Weld Strength Level [b, c]
Complete-Joint-Penetration Groove Weld				
Tension normal to effective area	Base	0.90	F_y	Matching weld must be used.
Compression normal to effective area	Base	0.90	F_y	Weld metal with a strength level equal to or less than matching weld metal is permitted to be used.
Tension or compression parallel to axis of weld				
Shear on effective area	Base Weld electrode	0.90 0.80	$0.60F_y$ $0.60F_{EXX}$	
Partial-Joint-Penetration Groove Weld				
Compression normal to effective area	Base	0.90	F_y	Weld metal with a strength level equal to or less than matching weld metal is permitted to be used.
Tension or compression parallel to axis of weld [d]				
Shear parallel to axis of weld	Base Weld electrode	0.75	[e] $0.60F_{EXX}$	
Tension normal to effective area	Base Weld electrode	0.90 0.80	F_y $0.60F_{EXX}$	
Fillet Welds				
Shear on effective area	Base Weld electrode	0.75	[f] $0.60F_{EXX}$	Weld metal with a strength level equal to or less than matching weld metal is permitted to be used.
Tension or compression parallel to axis of weld [d]	Base	0.90	F_y	
Plug or Slot Welds				
Shear parallel to faying surfaces (on effective area)	Base Weld electrode	0.75	[e] $0.60F_{EXX}$	Weld metal with a strength level equal to or less than matching weld metal is permitted to be used.

[a] For definition of effective area, see Section J2.

[b] For matching weld metal, see Table 4.1, AWS D1.1

[c] Weld metal one strength level stronger than matching weld metal is permitted.

[d] Fillet welds and partial-joint-penetration groove welds joining component elements of built-up members, such as flange-to-web connections, are not required to be designed with to the tensile or compressive stress in these elements parallel to the axis of the welds.

[e] The design of connected material is groverned by Sections J4 and J5.

[f] For alternative design strength, see Appendix J2.4.

Note that in the above, the shear yield stress is taken approximately as $0.6F_y$ for the base metal and 0.6 of the electrode *tensile strength* for the weld metal; thus, the lower ϕ factor 0.80 is used for the weld metal.

Partial penetration groove welds are similarly treated and the summary of design strengths is given in Table 5.14.1.

Fillet Welds

The *design strength* (LRFD-J2.4.) per unit length of a fillet weld is based on the shear resistance through the throat of the weld, as follows:

$$\phi R_{nw} = 0.75t_e(0.60F_{EXX}) \quad \text{fillet weld} \tag{5.14.6}$$

but not greater than the shear rupture strength (LRFD-J4.) of the adjacent base metal.

$$\phi R_{nw} = 0.75t(0.60F_u) \quad \text{base metal} \tag{5.14.7}$$

where
t_e = effective throat dimension (see Sec. 5.12)
F_{EXX} = tensile strength of electrode material
t = thickness of base material along which weld is placed
F_u = tensile strength of base metal

Note that the limit state for fillet welds is fracture through the throat of the fillet; thus, the ϕ factor is 0.75. Values of ϕR_{nw} based on Eq. 5.14.6 for various electrode strengths are given in Tables 5.14.2 and 5.14.3 for shielded metal arc welding (SMAW) and in Table 5.14.4 for submerged arc welding (SAW).

When a fillet weld is loaded through its center of gravity but at an angle θ measured from the weld axis, or weld segments within a weld group are loaded

TABLE 5.14.2   DESIGN SHEAR STRENGTH ϕR_{nw} OF FILLET WELD, (kips/in.) (SHIELDED METAL ARC WELDING)

Nominal size (in.)	Effective throat (in.)	Minimum tensile strength of weld (ksi)					
		60	70	80	90	100	110
1/8	0.088[a]	2.38[b]	2.77	3.17	3.56	3.96	4.36
3/16	0.133	3.58	4.18	4.77	5.37	5.97	6.56
1/4	0.177	4.77	5.57	6.36	7.16	7.95	8.75
5/16	0.221	5.97	6.96	7.95	8.95	9.94	10.94
3/8	0.265	7.16	8.35	9.54	10.74	11.93	13.12
7/16	0.309	8.35	9.74	11.14	12.53	13.92	15.31
1/2	0.354	9.54	11.14	12.73	14.32	15.91	17.50
9/16	0.398	10.74	12.53	14.32	16.11	17.90	19.69
5/8	0.442	11.93	13.92	15.91	17.90	19.88	21.87
11/16	0.486	13.12	15.31	17.50	19.69	21.87	24.06
3/4	0.530	14.32	16.70	19.09	21.48	23.86	26.25

[a] t_e = 0.707 times leg size a = 0.707(1/8) = 0.088 in.
[b] $\phi t_e(0.60 F_{EXX})$ = 0.75(0.707a)(0.60 times tensile strength)
      = 0.75(0.707)(1/8)(0.60)(60) = 2.38 kips/in.

TABLE 5.14.3   DESIGN SHEAR STRENGTH OF FILLET WELDS, ϕR_{nw} (N/mm)
(SHIELDED METAL ARC WELDING)

Nominal size (mm)	Effective throat (mm)	Minimum tensile strength of weld (MPa)					
		415	485	550	620	690	760
3	2.12[a]	396[b]	463	525	592	659	725
4	2.83	528	617	700	789	878	967
5	3.54	660	772	875	986	1098	1209
6	4.24	792	926	1050	1184	1317	1451
8	5.66	1056	1234	1400	1578	1756	1934
10	7.07	1320	1543	1750	1973	2195	2418
12	8.48	1584	1852	2100	2367	2634	2902
14	9.90	1848	2160	2450	2762	3073	3385
16	11.31	2113	2469	2800	3156	3512	3869
18	12.73	2377	2777	3150	3551	3951	4352
20	14.14	2641	3086	3500	3945	4390	4836

[a] $t_e = 0.707$ times leg size a $= 0.707(3) = 2.12$ mm.
[b] $\phi t_e(0.60F_{EXX}) = 0.75(0.707a)(0.60$ times tensile strength)
$= 0.75(0.707)(2.12)(0.60)415 = 396$ N/mm.

in-plane, the strength of each weld segment is permitted to be modified from that given by Eq. 5.14.6, in accordance with the provisions of LRFD-Appendix J2.4. This alternate approach is used in the strength method treatment of eccentric shear connections in Sec. 5.17. Kennedy, Miazga, and Lesik [5.37, 5.38, 5.39] have presented the latest information on fillet weld shear strength.

LRFD-Table J2.5 containing the design strengths of welds is presented as Table 5.14.1.

TABLE 5.14.4   DESIGN SHEAR STRENGTH OF FILLET WELD, ϕR_{nw} (kips/in.)
(SUBMERGED ARC WELDING)

Nominal size (in.)	Effective throat (in.)	Minimum tensile strength of weld (ksi)					
		60	70	80	90	100	110
1/8	0.125[a]	3.38[c]	3.94	4.50	5.06	5.63	6.19
3/16	0.188[a]	5.06	5.91	6.75	7.59	8.44	9.28
1/4	0.250[a]	6.75	7.88	9.00	10.13	11.25	12.38
5/16	0.313[a]	8.44	9.84	11.25	12.66	14.06	15.47
3/8	0.375[a]	10.13	11.81	13.50	15.19	16.88	18.56
7/16	0.419[b]	11.32	13.21	15.10	16.98	18.87	20.76
1/2	0.463	12.51	14.60	16.69	18.77	20.86	22.94
9/16	0.508	13.71	15.99	18.28	20.56	22.85	25.13
5/8	0.552	14.90	17.38	19.87	22.35	24.83	27.32
11/16	0.596	16.09	18.78	21.46	24.14	26.82	29.51
3/4	0.640	17.29	20.17	23.05	25.93	28.81	31.69

[a] $t_e = a =$ leg size for sizes $\leq 3/8$ in.
[b] $t_e = 0.707a + 0.11$ for sizes $> 3/8$ in.
[c] $\phi t_e(0.60F_{EXX}) = 0.75(a)(0.60$ times tensile strength)
$= 0.75(1/8)(0.60)(60) = 3.38$ kips/in.

EXAMPLE 5.14.1 _____

Determine the effective throat dimension of $\frac{7}{16}$-in. fillet weld made by (a) shielded metal arc welding (SMAW), and (b) submerged arc welding (SAW), in accordance with the AISC Specifications [1.5, 1.16].

Solution
(a) $t_e = 0.707a = 0.707(0.4375)\ = 0.309$ in.
(b) $t_e = 0.707a + 0.11 = 0.707(0.4375) + 0.11 = 0.419$ in.
These values are in agreement with those given in Tables 5.14.2 (for SMAW) and 5.14.3 (for SAW). ■■

EXAMPLE 5.14.2 _____

Determine the design shear strength ϕR_{nw} of a $\frac{3}{8}$-in. fillet weld produced by (a) shielded metal arc welding, and (b) submerged arc welding. Assume E70 electrodes having minimum tensile strength F_{EXX} of 70 ksi are used, according to AISC LRFD Specification.

Solution
(a) SMAW process. $t_e = 0.707a = 0.707(0.375) = 0.265$ in. According to Eq. 5.14.6,

$$\phi R_{nw} = \phi t_e(0.60 F_{\text{EXX}}) = 0.75(0.265)(0.60)70 = 8.35 \text{ kips/in.}$$

(b) SAW process. $t_e = a = 0.375$ in.

$$\phi R_{nw} = \phi t_e(0.60 F_{\text{EXX}}) = 0.75(0.375)(0.60)70 = 11.81 \text{ kips/in.}$$

These values agree with those in Tables 5.14.2 and 5.14.3. ■■

EXAMPLE 5.14.3 _____

Determine the design shear strength ϕR_n for a $\frac{3}{4}$-in.-diam plug weld using E70 electrode material. Use AISC LRFD Specification.

Solution. Assuming the weld diameter D satisfies the limitations of LRFD-J2.3b relating to the dimension of the piece in which the plug weld is made,

$$\phi R_n = 0.75(\text{area of faying surface}, \pi D^2/4)(0.60 F_{\text{EXX}})$$
$$= 0.75\ (0.4418)(0.60)70 = 13.9 \text{ kips}\ ■■$$

Maximum Effective Fillet Weld Size

In Sec. 5.11 the limitations on maximum and minimum fillet weld size and length relating to practical design considerations were given. Those requirements relate to the size of weld that is actually placed. Regarding strength, however, no welds of whatever size may be designed using a strength greater than available on the adjacent base material.

Consider the two lines of fillet weld transmitting the shear V across section $a-a$ of Fig. 5.14.1a. The design strength ϕR_{nw} of the fillet weld is given by Eq. 5.14.6. The design shear rupture strength $\phi F_n A_{nv}$ of the base material is $0.75(0.6F_u) \times$(net area in base material subject to shear) according to LRFD-J4. Equating the capacity per inch

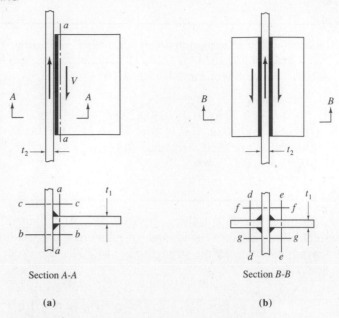

Figure 5.14.1 Critical sections for possible overstressing of base material.

of the weld metal to the shear capacity per inch in the base material gives for shielded metal arc welding

$$\phi R_{nw}(\text{weld}) = \phi R_n \text{ (base metal)}$$

$$2\phi a(0.707)(0.60F_{\text{EXX}}) = \phi(0.6F_u)t_1 \qquad (5.14.8)$$

$$a_{\max\text{eff}} = \frac{0.60F_u t_1}{2(0.707)0.60F_{\text{EXX}}} = 0.707\frac{F_u t_1}{F_{\text{EXX}}} \qquad (5.14.9)$$

where t_1 = thickness of base material
        F_u = tensile strength of base material
    F_{EXX} = tensile strength of electrode material (70 ksi for E70 electrodes)

Sections $b-b$ and $c-c$ will not be critical since two lines of weld transfer load across two sections, as shown in the equation for maximum effective weld size across those sections,

$$\phi a(0.707)(0.60F_{\text{EXX}}) = \phi(0.6F_u)t_2 \qquad (5.14.10)$$

$$a_{\max\text{eff}} = 1.41\frac{F_u t_2}{F_{\text{EXX}}} \qquad (5.14.11)$$

Considering the four fillet welds of Fig. 5.14.1b, sections $d-d$ and $e-e$ are the same as section $a-a$; therefore, Eq. 5.14.9 applies. On sections $f-f$ and $g-g$ four fillet welds transfer load across two sections. Thus,

$$\phi 4a(0.707)(0.60F_{\text{EXX}}) = \phi 2(0.6F_u)t_2 \qquad (5.14.12)$$

and again Eq. 5.14.9 is the result.

Even when fillet welds connect members that are in tension, the transfer of load by means of the weld is a shear transfer to the base pieces when the fillet welds are

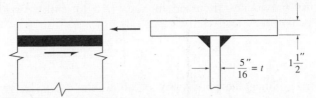

Figure 5.14.2 Example 5.14.4.

parallel to the direction of the load. For such cases, the maximum effective weld size concept still applies.

EXAMPLE 5.14.4

Determine the design shear strength ϕR_{nw} to be used for the flange to web connection in Fig. 5.14.2. The plates are A36 steel and electrodes having $F_{EXX} = 70$ ksi are to be used with (a) shielded metal arc welding (SMAW) and (b) submerged arc welding (SAW).

Solution.   Minimum weld size $= a_{\min} = \frac{5}{16}$ in. (LRFD-Table J2.4)
(a) SMAW process. Equation 5.14.9 applies,

$$a_{\max\,eff} = 0.707 \frac{F_u t_w}{F_{EXX}} = 0.707 \frac{58 t_w}{70} = 0.59 t_w$$

$$= 0.59(5/16) = 0.183 \text{ in.}$$

$$\phi R_{nw} = 0.75(0.183)(0.707)(0.60)(70) = 4.08 \text{ kips/in.}$$

or, for two fillets, the design strength is $2(4.08) = 9.16$ kips/in. Thus, even though a $\frac{5}{16}$-in. fillet weld must be placed, its strength in design may not exceed the strength assuming $a = 0.183$ in.
(b) SAW process. The effective throat dimension equals the weld size, Equating the weld strength to the plate strength gives

$$2a(0.60)70 = 0.60(58)t_w$$

$$a_{\max\,eff} = 0.414 t_w = 0.414(5/16) = 0.129 \text{ in.}$$

$$\phi R_{nw} = 0.75(0.129)(0.60)70 - 4.08 \text{ kips/in.}$$

or, 9.16 kips/in. for two fillets. Again, the same result is obtained as in (a) because 9.16 kips/in. is the design strength based on the $\frac{5}{16}$-in. web plate, which is well below the strength of the weld metal. ■■

5.15 ALLOWABLE STRESS DESIGN—WELDS

The general philosophy of Allowable Stress Design (ASD) was described in Secs. 1.8 and 1.9. Equation 1.8.8 gives the structural safety requirement, as follows:

$$\frac{\phi R_n}{\gamma} \geq \sum Q_i \qquad\qquad [1.8.8]$$

As previously discussed in Sec. 4.10 for bolts, dividing Eq. 1.8.8 by the safety provision γ/ϕ gives

$$\left[\frac{R_n}{\gamma/\phi} = \frac{R_n}{FS}\right] \geq R \qquad [4.10.1]$$

Since the philosophy of allowable stress design is to compute "stresses" at service load, Eq. 4.10.1 may be divided by the "effective" area of the weld per inch, the sides of the equation interchanged (reversing the inequality) giving

$$\left[f = \frac{R}{t_e}\right] \leq \left[F = \text{Allowable stress} = \frac{R_n}{(FS)t_e}\right] \qquad (5.15.1)$$

or the service load R per inch of weld may not exceed the *allowable* load R_w per inch of weld. This may be expressed

$$R \leq [R_w = \text{Allowable stress times } t_e] \qquad (5.15.2)$$

The allowable stresses for (1) shear on the effective area of all welds and (2) the tensile stress normal to the axis on the effective area of a partial joint penetration groove weld are equal to 0.30 times the electrode tensile strength. However, the stress in the adjacent base metal may not exceed $0.60F_y$ for tension or $0.40F_y$ for shear. The summary of allowable stresses is given in Table 5.15.1, and R_w values of allowable shear resistance per inch for various electrode strengths are given for SMAW and SAW, respectively, in Tables 5.15.2 and 5.15.3.

The limitations on maximum and minimum size welds were discussed in Sec. 5.11, and the effective areas were discussed in Sec. 5.12.

The general summary of AISC Allowable Stress Design provisions relating to welds is in Table 5.14.4.

EXAMPLE 5.15.1 _____

Determine the allowable shear resistance of a $\frac{3}{8}$-in. fillet weld produced by (a) shielded metal arc process and (b) submerged arc process. Assume use of E70 electrodes having minimum tensile strength F_{EXX} of 70 ksi, and use Allowable Stress Design.

Solution

(a) SMAW process. $t_e = 0.707a = 0.707(0.375) = 0.265$ in.

$$R_w = t_e(0.30F_{EXX}) = 0.265(0.3)70 = 0.265(21) = 5.57 \text{ kips/in.}$$

(b) SAW process. $t_e = a = 0.375$ in.

$$R_w = t_e(0.30F_{EXX}) = 0.375(21.0) = 7.87 \text{ kips/in.}$$

These values may be checked by referring to Tables 5.15.2 and 5.15.3. ■■

One may determine the nominal safety provision (γ/ϕ) by comparing the design strength ϕR_{nw} in Load and Resistance Factor Design (Sec. 5.14) with the allowable load R_w. Using the values computed in Examples 5.14.2 and 5.15.1 for $\frac{3}{8}$-in. fillet weld with the SMAW process and E70 electrodes,

$$\phi R_{nw} = 8.35 \text{ kips/in.} \qquad \text{LRFD Design Strength}$$
$$R_w = 5.57 \text{ kips/in.} \qquad \text{ASD Allowable Load}$$

TABLE 5.15.1 ALLOWABLE STRESSES ON EFFECTIVE AREA OF WELDS[f]
(ASD [1.5])–TABLE J2.5 (USED BY PERMISSION OF AISC)

Type of Weld and Stress[a]	Allowable Stress	Required Weld Strength Level[b, c]
Complete-Penetration Groove Welds		
Tension normal to effective area	Same as base metal	"Matching" weld metal shall be used.
Compression normal to effective area	Same as base metal	Weld metal with a strength level equal to or less than "matching" weld metal is permitted.
Tension or compression parallel to axis of weld	Same as base metal	
Shear on effective area	0.30 × nominal tensile strength of weld metal (ksi)	
Partial-Penetration Groove Welds[d]		
Compression normal to effective area	Same as base metal	Weld metal with a strength level equal to or less than "matching" weld metal is permitted.
Tension or compression parallel to axis of weld[e]	Same as base metal	
Shear parallel to axis of weld	0.30 × nominal tensile strength of weld metal (ksi)	
Tension normal to effective area	0.30 × nominal tensile strength of weld metal (ksi), except tensile stress on base metal shall not exceed 0.60 × yield stress of base metal	
Fillet Welds		
Shear on effective area	0.30 × nominal tensile strength of weld metal (ksi)	Weld metal with a strength level equal to or less than "matching" weld metal is permitted.
Tension or compression Parallel to axis of weld[e]	Same as base metal	
Plug and Slot Welds		
Shear parallel to laying surfaces (on effective area)	0.30 × nominal tensile strength of weld metal (ksi)	Weld metal with a strength level equal to or less than "matching" weld metal is permitted.

[a] For definition of effective area. See ASD-J2.

[b] For "matching" weld metal. See Table 4.1.1. AWS D1.1 [2.25].

[c] Weld metal one strength level stronger than "matching" weld metal will be permitted.

[d] See ASD-J2.1b for a limitation on use of partial-penetration groove welded joints.

[e] Fillet welds, and partial-penetration groove welds joining the component elements of built-up members, such as flange-to-web connections, may be designed without regard to the tensile or compressive stress in these elements parallel to the axis of the welds.

[f] The design of connected material is governed by ASD-D through ASD-G.

TABLE 5.15.2   ALLOWABLE RESISTANCE R_w OF FILLET WELDS, (kips/in.)
(SHIELDED METAL ARC WELDING)

Nominal size (in.)	Effective throat (ASD-J2.2a) (in.)	Minimum tensile strength of weld (ksi)					
		60	70	80	90	100	110
1/8	0.088	1.59	1.86	2.12	2.39	2.69	2.92
3/16	0.132	2.38	2.78	3.18	3.58	3.97	4.37
1/4	0.177	3.18	3.71	4.24	4.77	5.30	5.83
5/16	0.221	3.98	4.64	5.30	5.96	6.63	7.30
3/8	0.265	4.77	5.57	6.36	7.16	7.95	8.75
7/16	0.309	5.57	6.49	7.42	8.35	9.28	10.21
1/2	0.353	6.36	7.42	8.48	9.54	10.60	11.66
9/16	0.398	7.16	8.35	9.54	10.74	11.93	13.12
5/8	0.442	7.95	9.28	10.61	11.93	13.26	14.58
11/16	0.486	8.75	10.21	11.67	13.12	14.58	16.04
3/4	0.530	9.54	11.13	12.72	14.31	15.91	17.50

From Eq. 5.15.1 and using 0.75 as the ϕ factor,

$$\text{FS} = \frac{R_{nw}}{R_w} = \frac{8.35}{0.75(5.57)} = 2.00$$

Thus, the nominal strength R_{nw} used in LRFD is twice the allowable load R_w used in ASD. Typically, the allowable service load is obtained by dividing the strength by a desired factor; for connections 2.0 has traditionally been used. Once the safe service load has been established, that load may be divided by the area to obtain allowable stress.

TABLE 5.15.3   ALLOWABLE RESISTANCE R_w OF FILLET WELDS, (kips/in.)
(SUBMERGED ARC WELDING)

Nominal size (in.)	Effective throat (ASD-J2.2a) (in.)	Minimum tensile strength of weld (ksi)					
		60	70	80	90	100	110
1/8	0.125	2.25	2.62	3.00	3.37	3.75	4.12
3/16	0.187	3.37	3.94	4.50	5.06	5.62	6.19
1/4	0.250	4.50	5.25	6.00	6.75	7.50	8.25
5/16	0.312	5.62	6.56	7.50	8.44	9.37	10.31
3/8	0.375	6.75	7.87	9.00	10.12	11.25	12.37
7/16	0.419	7.55	8.80	10.06	11.32	12.58	13.84
1/2	0.463	8.34	9.73	11.12	12.51	13.90	15.30
9/16	0.508	9.14	10.66	12.18	13.71	15.23	16.75
5/8	0.552	9.93	11.59	13.25	14.90	16.56	18.21
11/16	0.596	10.73	12.52	14.31	16.09	17.88	19.67
3/4	0.640	11.52	13.44	15.36	17.28	19.21	21.13

5.16 WELDS CONNECTING MEMBERS SUBJECT TO DIRECT AXIAL LOAD

In the design of welds connecting tension or compression members, the welds should be at least as strong as the members they connect and the connection should not introduce significant eccentricity of loading.

Groove Welds

In the case of full joint penetration groove welds as shown in Fig. 5.5.2, the full strength of the cross-section may be developed by selecting the proper electrode material corresponding to the base material as indicated in Table 5.13.1, and specifying an AWS prequalified joint.

EXAMPLE 5.16.1

For the plate tension member (Fig. 5.16.1) carrying axial serivce loads of 60 kips live load and 12 kips dead load, select the required thickness of the plates (A572 Grade 50 steel), the proper electrode material, and specify a proper AWS prequalified groove joint. Use Load and Resistance Factor Design.

Solution

(a) Compute the factored load to be carried.

$$T_u = 1.2(12) + 1.6(60) = 110 \text{ kips}$$

(b) Compute the thickness required for the plates.

$$\phi T_n = \phi F_y A_g = 0.90(50)A_g \qquad \text{(yielding limit state)}$$
$$\phi T_n = \phi F_u A_e = 0.75(65)A_e \qquad \text{(fracture limit state)}$$

Since there are no holes and no eccentricity of loading, the effective net area A_e equals the gross area A_g. Thus, from the above two equations it is noted that $0.90(50) < 0.75(65)$; therefore,

$$\text{Required } A_g = \frac{T_u}{0.90(50)} = \frac{110}{45} = 2.45 \text{ sq in.}$$

$$\text{Required } t = \frac{2.45}{6} = 0.41 \text{ in.} \quad \underline{\textit{Use } \tfrac{7}{16} \times 6 \text{ plates.}}$$

(c) Select electrode and an AWS prequalified joint. From Table 5.13.1, use F7XX-EXXX ($F_{EXX} = 70$ ksi) flux electrode combination. Referring to *LRFD*

60k LL   6"   60k LL
12k DL   12k DL

Figure 5.16.1   Example 5.16.1.

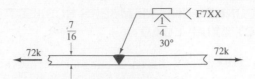

Figure 5.16.2 Design sketch for Example 5.16.1.

Manual [1.18] section "Prequalified Welded Joints" (Table 8-36) or AWS [2.25] Figs. 2.4 and 2.5, select a prequalified single-V-groove weld designated B-L2a-S. The designation B refers to a butt joint, L refers to limited thickness of material for this weld (in this case 2 in. maximum), and S refers to submerged arc welding. The weld requires that a backup plate be used. The details and welding symbol are shown in Fig. 5.16.2. ∎∎

EXAMPLE 5.16.2 ──────────────────────────────────────

Repeat Example 5.16.1, except use A572 Grade 65 plates, a square-groove weld, and submerged arc welding (SAW).

Solution

(a) Determine the required plate thickness.

$$\text{Required } A_g = \frac{T_u}{0.90(65)} = \frac{110}{58.5} = 1.89 \text{ sq in.}$$

$$\text{Required } t = \frac{1.89}{6} = 0.31 \text{ in.} \quad \underline{Use \tfrac{5}{16} \times 6 \text{ plates.}}$$

(b) Select electrode and specify the proper prequalified joint. From Table 5.13.1, use F8XX-EXXX flux electrode combination. From *LRFD Manual* [1.18] or AWS [2.25] Table 2.9.1, select the square-groove weld designated B-L1-S (*LRFD Manual* p. 8–137) as indicated in Fig. 5.16.3. This weld has zero root opening and is prequalified for material no thicker than $\frac{3}{8}$ in. ∎∎

Note: In Examples 5.16.1 and 5.16.2 it was not necessary to include the weld size or the length of the welds since they are to be made full joint penetration and the full width of the plates unless otherwise specified.

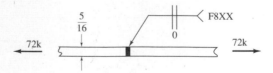

Figure 5.16.3 Design sketch for Example 5.16.2.

EXAMPLE 5.16.3

Determine the service load capacity, assuming the load is 70% live load, of the tee connection shown in Figure 5.16.4 and detail the proper double-bevel-groove weld for the SMAW process. Assume the flange of the tee does not control design. Use Load and Resistance Factor Design.

Solution

(a) Determine the strength of the $\frac{3}{4} \times 8$ plate.

$$\phi T_n = \phi F_y A_g = 0.90(0.75)8 = 194 \text{ kips}$$
$$T_u = 1.2T_D + 1.6T_L = 1.2(0.2T) + 1.6(0.8T) = 1.52T$$
$$\phi T_n = T_u$$
$$T = \frac{194}{1.52} = 128 \text{ kips}$$

The maximum service load tension force permitted by LRFD is 128 kips.

(b) Select the electrode material and select a proper prequalified AWS joint. From Table 5.13.1, use E60 electrodes. From the *LRFD Manual* [1.18, p. 8–143] or AWS [2.25] Fig. 2.4, select the double-bevel-groove joint designated TC-U5a.

Note: On the basis of strength only, a single $\frac{3}{4}$-in. bevel (TC-U4b) could have been used instead of the double-bevel-groove weld specified. However, welding the stem of the tee from one side only may cause excessive warping and introduces eccentricity into the connection.   ■■

Fillet Welds

The design for fillet welds is based on the nominal shear stress on the effective area of the fillet weld as discussed in Sec. 5.12. The selection of the size of the fillet weld is based on the thickness of the pieces being joined and the available length over which the fillet weld can be made. Other factors such as the type of welding equipment used,

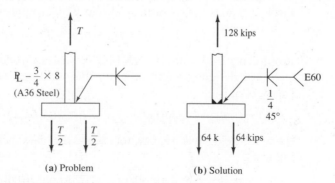

(a) Problem   (b) Solution

Figure 5.16.4   Example 5.16.3.

whether the welds are to be made in the field or in the shop, and the size of other welds being made will also influence the size of fillet specified. Large fillet welds require larger diameter electrodes, which in turn require larger and bulkier welding equipment, not necessarily convenient for field use. The most economical size of fillet weld is usually the one that can be made in one pass; about $\frac{5}{16}$ in. for SMAW and $\frac{1}{2}$ in. for SAW. Also, if a certain size of fillet weld is used in adjacent areas to the particular joint in question, it is advisable to use the same size, since then the same electrodes and welding equipment could be used, and the welder would not have to alter his procedure to accommodate a larger or smaller weld. In addition, inspection of the welds is further simplified.

EXAMPLE 5.16.4 _____

Determine the size and length of the fillet weld for the lap joint shown in Fig. 5.16.5 using the submerged arc (SAW) process if the plates are A36 steel. Use AISC Load and Resistance Factor Design.

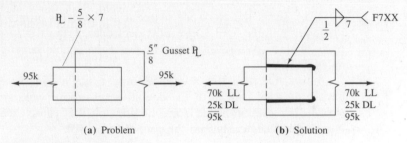

Figure 5.16.5   Example 5.16.4.

Solution.   Referring to Sec. 5.11, LRFD-J2.2b gives the following limits:

$$\text{Maximum size} = \tfrac{5}{8} - \tfrac{1}{16} = \tfrac{9}{16} \text{ in.}$$
$$\text{Minimum size} = \tfrac{1}{4} \text{ in.}$$

Use $\frac{1}{2}$-in. fillet weld, since that is about the maximum size that can be made in one pass by the SAW process.

Since the fillet weld exceeds $\frac{3}{8}$ in., LRFD-J2.2a states that the effective throat t_e equals the theoretical throat $(0.707a)$ plus 0.11 in.

$$t_e = 0.707(0.50) + 0.11 = 0.464 \text{ in.}$$

From Table 5.13.1, use F7XX-EXXX flux electrode combination.

The design strength of $\frac{1}{2}$-in. fillet weld per inch of length, according to Eq. 5.14.6, is

$$\phi R_{nw} = \phi t_e(0.60 F_{\text{EXX}}) = 0.75(0.464)42 = 14.6 \text{ kips/in.}$$

Since the weld capacity may not exceed the plate shear rupture strength according to Eq. 5.14.7,

$$\text{Max } \phi R_{nw} = \phi t(0.60 F_u) = 0.75(0.625)(0.60)58 = 16.3 \text{ kips/in.}$$

The weld strength controls.

The factored tensile load to be carried is

$$T_u = 1.2(25) + 1.6(70) = 142 \text{ kips}$$

The total length L_w of fillet weld required is

$$L_w = \frac{142}{14.6} = 9.7 \text{ in.}$$

To satisfy the strength requirement, only 5 in. of weld are needed on each side. However, to avoid excessive nonuniformity of loading (i.e., shear lag, Sec. 3.5) in the plates, LRFD-J2.2b requires "If longitudinal fillet welds are used alone in end connections of flat-bar tension members, the length of each fillet weld shall be not less than the perpendicular distance between them." In fact, the transverse distance between longitudinal fillet welds may not exceed one-half the length of the weld without reducing the strength of the connection in accordance with LRFD-B3.

In this example, use $\frac{1}{2}$-in. fillet weld, 7 in. on each side as shown in Fig. 5.16.5b. Note that returns are shown in the figure; however, no dimension is specified and they are *not* included in the strength computation. Returns of this type should be made in order to ensure the full weld strength at the ends of lines of weld. ■■

EXAMPLE 5.16.5 _____

Rework Example 5.16.4 using $\frac{1}{4}$-in. fillet welds.

Solution.   Since the nominal size is less than $\frac{3}{8}$ in., the effective throat t_e is the full leg dimension when using submerged arc welding.

$$\phi R_{nw} = \phi t_e(0.60F_{\text{EXX}}) = 0.75(0.25)42 = 7.87 \text{ kips/in.}$$

which is well below the plate shear rupture strength of 16.3 kips/in. The total length L_w of fillet weld required is

$$L_w = \frac{142}{7.87} = 18.0 \text{ in.}$$

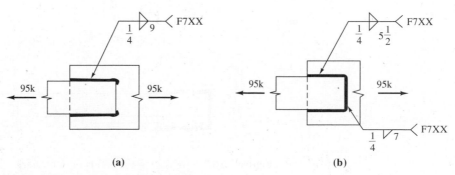

(a)   (b)

Figure 5.16.6   Solutions to Example 5.16.5.

Two possible solutions are shown in Fig. 5.16.6, both of which provide 19 in. of $\frac{1}{4}$-in. fillet welds. The solution in Fig. 5.16.6b is preferred since it is more compact and reduces the overall length of the connection, giving better stress distribution. ■■

Balanced Connection

In a number of cases, members subjected to direct axial stress are themselves unsymmetrical and cause eccentricities in welded connections. This discussion relates to largely in-plane forces such as those within a truss. Consider the angle tension member shown in Fig. 5.16.7 welded as indicated. The force T applied at some distance from the connection will act along the centroid of the member as shown. The force T will be resisted by the forces F_1, F_2, and F_3 developed by the weld lines. The forces F_1 and F_3 are assumed to act at the edges of the angle rather than more correctly at the center of the effective throat. The force F_2 will act at the centroid of the weld length which is located at $d/2$. Taking moments about point A located on the bottom edge of the member and considering clockwise moments positive,

$$\sum M_A = -F_1 d - F_2 d/2 + Ty = 0 \tag{5.16.1}$$

or

$$F_1 = \frac{Ty}{d} - \frac{F_2}{2} \tag{5.16.2}$$

The force F_2 is equal to the resistance R_w of the weld per inch times the length L_w of the weld:

$$F_2 = R_w L_{w2} \tag{5.16.3}$$

Horizontal force equilibrium gives

$$\sum F_H = T - F_1 - F_2 - F_3 = 0 \tag{5.16.4}$$

Solving Eqs. 5.16.1 and 5.16.4 simultaneously gives

$$F_3 = T\left(1 - \frac{y}{d}\right) - \frac{F_2}{2} \tag{5.16.5}$$

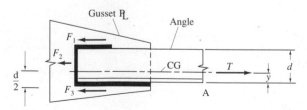

Figure 5.16.7   Balancing the welds on a tension member connection.

Designing the connection shown in Fig. 5.16.7 to eliminate eccentricity caused by the unsymmetrical weld is called *balancing the welds*. The procedure for balancing the welds may be summarized as follows:

1. After selecting the proper weld size and electrode, compute the force resisted by the end weld F_2 (if any) using Eq. 5.16.3.
2. Compute F_1 using Eq. 5.16.2.
3. Compute F_3 using Eq. 5.16.5, or

$$F_3 = T - F_1 - F_2 \qquad (5.16.6)$$

4. Compute the lengths, L_{w1} and L_{w3}, on the basis of

$$L_{w1} = \frac{F_1}{R_w} \qquad (5.16.7a)$$

and

$$L_{w3} = \frac{F_3}{R_w} \qquad (5.16.7b)$$

An alternative to the above is to compute the total length L_w of weld required to carry the load, subtract the length on the end, and then allocate the remaining required length to F_1 and F_3 in inverse proportion to the distances from the center of gravity.

Note that approximately balanced welds are "desirable"; however, LRFD-J1.8. does *not* require it for ". . . end connections of statically-loaded single angle, double angle and similar members." Temple and Sakla [5.43] have discussed the effects of balanced connections on angle compression members.

The foregoing discussion of balanced welds is valid for both Load and Resistance Factor Design and Allowable Stress Design. The service load T used in ASD would become factored load T_u used in LRFD; allowable resistance R_w in ASD would become design strength ϕR_{nw} in LRFD.

EXAMPLE 5.16.6

Design the fillet welds to develop the full strength of the angle shown in Fig. 5.16.8 minimizing the effect of eccentricity. Assume the gusset plate does not govern and the SMAW process is used. Use AISC Load and Resistance Factor Design.

Solution

(a) Compute the strength of the member. Even though the connection is to be made to only one leg of the angle, giving rise to shear lag in the outstanding leg, it is not clear that the LRFD-B3 intends U to be taken less than 1.0. The authors believe it is the intention to account for shear lag in both bolted and welded connections. Thus, following LRFD-Commentary-B3(b), estimate $U = 0.85$. A more accurate value of U can be computed according to Eq. 3.5.2 after the weld length has been determined.

The design strength of the angle member is the smaller of the following:

$$\phi T_n = 0.90 F_y A_g = 0.90(50)3.61 = 162 \text{ kips}$$
$$\phi T_n = 0.75 F_u A_e = 0.75(65)[0.85(3.61)] = 150 \text{ kips (controls)}$$

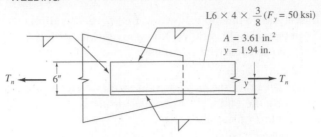

Figure 5.16.8   Example 5.16.6.

    (b) Select weld size and compute strength.

$$\text{Min size fillet weld} = \tfrac{3}{16} \text{ in. (Table 5.11.1)}$$

$$\text{Max size fillet weld} = \tfrac{3}{8} - \tfrac{1}{16} = \tfrac{5}{16} \text{ in. (Fig. 5.11.2)}$$

Use $\tfrac{3}{16}$-in. fillet weld with E70 electrodes. The design strength ϕR_{nw} is

$$\phi R_{nw} = \phi t_e(0.60F_{\text{EXX}}) = 0.75(\tfrac{3}{16})(0.707)42 = 4.18 \text{ kips/in.}$$

which cannot exceed the shear rupture strength of the base metal,

$$\text{Max } \phi R_{nw} = \phi t(0.60F_u) = 0.75(0.375)(0.60)65 = 11.0 \text{ kips/in.}$$

The weld controls.

    (c) Determine the lengths of weld to be used for the connection. Referring to Fig. 5.16.9,

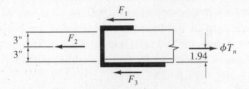

Figure 5.16.9   Balancing the welds for Example 5.16.6.

$$F_2 = \phi R_{nw} L_w = 4.18(6) = 25.1 \text{ kips}$$

From moment equilibrium about the back of the angle (at F_3),

$$F_1 = \frac{150(1.94) - 25.1(3)}{6} = 36.0 \text{ kips}$$

Summation of forces gives

$$F_3 = T_u - F_1 - F_2 = 150 - 36.0 - 25.1 = 88.9 \text{ kips}$$

$$L_{w1} = \frac{F_1}{\phi R_{nw}} = \frac{36.0}{4.18} = 8.6 \text{ in.} \quad \underline{Use \ 9 \ in.}$$

$$L_{w3} = \frac{F_3}{\phi R_{nw}} = \frac{88.9}{4.18} = 21.2 \text{ in.} \quad \underline{Use \ 22 \ in.}$$

Use welds as shown in Fig. 5.16.10, though for better economy the largest welds that can be placed in one pass are preferred and this would also reduce the connection length.   ■ ■

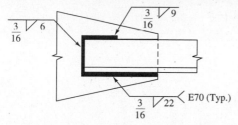

Figure 5.16.10   Solution for Example 5.16.6.

EXAMPLE 5.16.7

Rework Example 5.16.6 if the weld at the end of the angle is omitted, and the SAW process is used instead of SMAW.

Solution.   This time try $\frac{5}{16}$ in. as more economical and which can still be placed in one pass. Using the forces in Fig. 5.16.11.

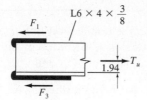

Figure 5.16.11   Forces acting for Example 5.16.7.

$$F_1 = \frac{150(1.94)}{6} = 48.5 \text{ kips}$$

$$F_3 = T_u - F_1 - F_2 = 150 - 48.5 = 101.5 \text{ kips}$$

$$\phi R_{nw} = \phi t_e(0.60 F_{EXX}) = 0.75(\tfrac{5}{16})42 = 9.84 \text{ kips/in.}$$

which is less than the shear rupture strength (11.0 kips/in.) of the angle; therefore, the weld controls.

$$L_{w1} = \frac{F_1}{\phi R_{nw}} = \frac{48.5}{9.84} = 4.9 \text{ in.} \qquad \underline{\textit{Use 5 in.}}$$

$$L_{w3} = \frac{F_3}{\phi R_{nw}} = \frac{101.5}{9.84} = 10.3 \text{ in.} \quad \underline{\textit{Use 11 in.}}$$

Use welds as summarized in Fig. 5.16.12. Note that small returns at the ends of weld lines are made to ensure full strength of the specified weld lengths. Unless the return length is specified, the returns would not be included in the strength computation.   ■■

Slot and Plug Welds

Slot and plug welds have their strength based on the area in the shearing plane between the plates being joined. As indicated in Sec. 5.5, their principal use is in lap joints. Plug welds are also occasionally used to fill up holes in connections, such as beam-to-

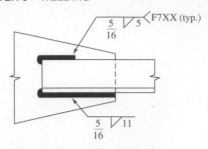

Figure 5.16.12   Solution for Example 5.16.7.

column angles where temporary erection bolts had been placed to align the members prior to welding. The strength of such welds may or may not be included in the design strength of a joint. As a rule, plug and slot welds are designed to work together with other welds, usually fillet welds, in lap joints as shown in Fig. 5.5.5.

EXAMPLE 5.16.8

Determine the service load T permitted on the connection in Fig. 5.16.13 if the load is 80% live load and 20% dead load. The steel is A36 and Load and Resistance Factor Design is to be used. See comment regarding returns at end of Ex. 5.16.7.

Solution.   The design strength ϕR_{nw} per inch supplied by the $\frac{1}{2}$-in. fillet welds is

$$\phi R_{nw} = \phi t_e(0.60F_{EXX})$$
$$= 0.75[0.5(0.707) + 0.11)]42 = 14.6 \text{ kips/in.}$$

but not to exceed the shear rupture strength of the plate,

$$\text{Max } \phi R_{nw} = \phi t(0.60F_u) = 0.75(0.75)(0.60)58 = 19.6 \text{ kips/in.}$$

The strength provided by the fillet welds is

$$T_1 = L_w(\phi R_{nw}) = 10(14.6) = 146 \text{ kips}$$

The design strength ϕR_n provided by the $1\frac{1}{2}$-in. diam plug weld is

$$T_2 = \phi R_n = 0.75\frac{\pi(1.5)^2}{4}(0.60)70 = 74 \text{ kips}$$

The design strength based on the weld is

$$\phi T_n = T_1 + T_2 = 146 + 74 = 220 \text{ kips}$$

Check the tensile capacity of the plate:

$$\phi T_n = 0.90F_y A_g = 0.90(36)(9)0.75 = 219 \text{ kips (controls)}$$
$$\phi T_n = 0.75F_u A_e = 0.75F_u U A_g = 0.75(58)(1.0)(9)0.75 = 294 \text{ kips}$$

which makes the service load capacity T

$$219 = 1.2(0.2T) + 1.6(0.8T) = 1.52T$$
$$T = 144 \text{ kips} \quad \blacksquare\blacksquare$$

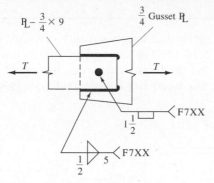

Figure 5.16.13 Example 5.16.8.

EXAMPLE 5.16.9

Compute the service load capacity of the connection shown in Fig. 5.16.14 when A573 Grade 50 steel and welding by the SMAW process are used. Assume the service load is 83% live load and 17% dead load. Use AISC Load and Resistance Factor Design. See comment regarding weld returns at end of Ex. 5.16.7.

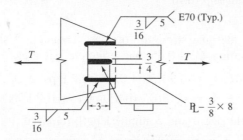

Figure 5.16.14 Example 5.16.9.

Solution.   From Table 5.14.2, a $\frac{3}{16}$-in. E70 fillet weld provides $\phi R_{nw} =$ 4.18 kips/in. The resistance T_1 provided by the fillet welds is

$$T_1 = L_w(\phi R_{nw}) = 2(5)(4.18) = 41.8 \text{ kips}$$

The resistance T_2 provided by the $\frac{3}{4}$-in. wide slot weld is

$$\text{Faying area} = \frac{\pi(3/4)^2}{8} + 2.25(\tfrac{3}{4}) = 1.91 \text{ sq in.}$$

$$T_2 = \phi R_n = 0.75(1.91)(0.60)70 = 60.1 \text{ kips}$$

$$\phi T_n = T_1 + T_2 = 41.8 + 60.1 = 101.9 \text{ kips (controls)}$$

Note that slot has a semicircular end (LRFD-J2.3b, par. 4).

Check the tensile capacity of the plate:

$$\phi T_n = 0.90 F_y A_g = 0.90(50)(8)0.375 = 135 \text{ kips}$$
$$\phi T_n = 0.75 F_u A_e = 0.75(65)(8)0.375 = 146 \text{ kips}$$

Thus, the service load capacity T is

$$101.9 = 1.2(0.17T) + 1.6(0.83T) = 1.53T$$
$$T = 67 \text{ kips} \quad \blacksquare\blacksquare$$

EXAMPLE 5.16.10 _____

Design an end connection to develop the full tensile strength of a C8×13.75 in a lap length of 5 in. as shown in Fig. 5.16.15. The channel of A572 Grade 50 steel is connected to a $\frac{3}{8}$-in. gusset plate, and the fillet welds are to be made by the SMAW process and may not exceed $\frac{3}{8}$-in. Use Load and Resistance Factor Design.

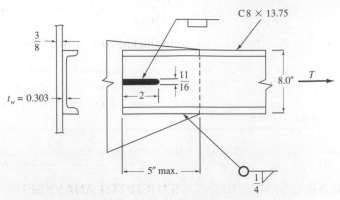

Figure 5.16.15   Solution for Example 5.16.10.

Solution

(a) Compute the design strength of the channel. Using LRFD-Commentary B3, estimating $U = 0.85$, the effective net area A_e is $0.85A_g$.

$$\phi T_n = 0.90 F_y A_g = 0.90(50)4.04 = 182 \text{ kips}$$
$$\phi T_n = 0.75 F_u A_e = 0.75(65)(0.85)4.04 = 167 \text{ kips}$$

(b) Select fillet weld size a and compute length required.

Min $a = \frac{3}{16}$ in.   (Table 5.11.1)

Max $a = 0.303 - \frac{1}{16} = 0.24$ in.,   say $\frac{1}{4}$ in. (Fig. 5.11.2)

While $\frac{1}{4}$-in. weld must be used on one end along the channel web, $\frac{3}{8}$-in. weld could be used along the flanges. It is better not to mix the fillet sizes, so try $\frac{1}{4}$ in. all around.

$$\phi R_{nw} = \phi t_e (0.60 F_{EXX}) = 0.75(\tfrac{1}{4})(0.707)42 = 5.57 \text{ kips/in.}$$

which cannot exceed the shear rupture strength of the base metal,

$$\text{Max } \phi R_{nw} = \phi t(0.60 F_u) = 0.75(0.375)(0.60)65 = 11.0 \text{ kips/in.}$$

The weld controls.

$$\text{Required } L_w = \frac{T_u}{\phi R_{nw}} = \frac{167}{5.57} = 30 \text{ in.}$$

Since the length all around is only 26 in., additional capacity from fillet weld in a slot, slot welds, or plug welds, is necessary.

(c) Slot weld. Try a slot weld in accordance with LRFD-J2.3b.

Min width of slot $= (t + \frac{5}{16})$ (rounded to next odd $\frac{1}{16}$ in.)

$$= 0.303 + 0.3125 = 0.6155, \quad \text{say } \tfrac{11}{16} \text{ in.}$$

Max width of slot $= 2\frac{1}{4}$ (weld thickness) $= 2\frac{1}{4}(0.303) = 0.68$ in.

Load T_u to be carried by slot weld,

$$\text{Required } T_u = 167 - (26 - 0.68)5.57 = 26 \text{ kips}$$

Try $\frac{11}{16}$-in. width of slot and *estimate* the slot area as rectangular even though the end must be rounded:

$$\text{Length required} = \frac{26}{0.75(11/16)42} = 1.2 \text{ in.}$$

$$\text{Max length of slot} = 10 \text{ (weld thickness)} = 10(0.303) = 3.03 \text{ in.}$$

Use a slot weld $\frac{11}{16} \times 2$. The final design is shown in Fig. 5.16.15. Note that the interior end of the slot must be semicircular or have corners rounded to a radius not less than the thickness of the part containing slot (LRFD or ASD-J2.3b). ■■

5.17 ECCENTRIC SHEAR CONNECTIONS—STRENGTH ANALYSIS

There are many situations where the loading of fillet welds is neither parallel to nor transverse to the axis of the fillet welds, as shown in Fig. 5.17.1. Analysis of such eccentric loading cases is complicated by the fact that, as shown in Fig. 5.13.4, the load-deformation behavior is a function of the angle θ between the direction of the resistance and the axis of the fillet weld.

In a manner similar to that used for eccentric loading on bolted connections (see Sec. 4.12), the strength of an eccentrically loaded fillet weld configuration can be

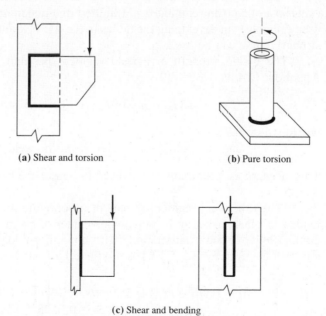

(a) Shear and torsion   (b) Pure torsion

(c) Shear and bending

Figure 5.17.1   Types of eccentric loading.

determined by locating the instantaneous center of rotation, using the load-deformation relationship of a fillet weld. The resistance R_i of a weld segment at any distance from the instantaneous center is proportional to such distance and acts in a direction perpendicular to the radial distance to the segment. Unlike bolts, however, where the load-deformation relationship is independent of the direction the shear force acts on the circular bolt cross-section, the fillet weld strength depends on the angle between the applied force and the axis of the weld resisting it.

Until the availability of calculators that could readily obtain the values from complicated formulas, a trial and error procedure to determine the strength of eccentrically loaded connections, either bolted or welded, was not feasible. Thus, traditionally the elastic vector analysis similar to that discussed for bolts was used for eccentrically loaded welds, as is explained in the next section (Sec. 5.18).

Though the AISC Specifications [1.5, 1.16] have in the past not described the method of analysis to be used for fillet welds eccentrically loaded in shear, the most rational procedure is one using strength analysis. The work of Butler, Pal, and Kulak [5.28, 5.29] formed the basis for the 1978 *ASD Manual* tables. Tide [5.34, 5.35] and Brandt [5.36] have presented detailed treatment of the *ASD Manual* approach. Brandt [5.36] has also developed a computer program. The 1986 *LRFD Manual* was based on later work by Kulak and Timmler [5.30]. The equations have been revised in 1993 [LRFD-Appendix J2.4] to reflect the work of Miazga, Lesik, and Kennedy [5.37, 5.38, 5.39].

A more complicated situation of eccentric shear parallel to one of the axes of the weld configuration combined with a force acting at 90° to the shear force has been treated by Loomis, Thornton, and Kane [5.40]. Their design aids are primarily for beam connection angle welds, including C-shaped, L-shaped, and lines, where the beam is subject to axial tension or compression.

The general strength analysis procedure is unchanged; however, the strength (kips/in.) of a given segment is not limited to a maximum value of $0.6F_{EXX}t_e$ as was used previously in developing the tables in the AISC Manuals. Instead, the theoretical strength curve can be used.

The design strength of a weld segment per unit length is given by LRFD-Appendix J2.4 as

$$\phi R_{nw} = \phi 0.60 F_{EXX}(1.0 + 0.50\sin^{1.5}\theta)t_e \qquad (5.17.1)$$

where $\phi = 0.75$

θ = angle of loading measured from the weld longitudinal axis, degrees

This strength as a function of θ is used in lieu of the constant strength discussed in Sec. 5.14.

When the weld segment is part of a configuration subject to eccentric shear, loaded in-plane, using an instantaneous center of rotation procedure which satisfies compatibility of deformation along with nonuniform load-deformation behavior, the strength as given by Eq. 5.17.1 is modified by LRFD-Appendix J2.4 to become

$$R_i = 0.60 F_{EXX}t_e(1.0 + 0.50\sin^{1.5}\theta)\left[\frac{\Delta_i}{\Delta_m}\left(1.9 - 0.9\frac{\Delta_i}{\Delta_m}\right)\right]^{0.3} \qquad (5.17.2)$$

where R_i = nominal strength of weld segment i, kips/in.

θ = angle of loading measured from the weld longitudinal axis, degrees

Δ_i = deformation of element $i = r_i \dfrac{\Delta_u}{r_{crit}}$

r_{crit} = distance from instantaneous center of rotation to weld element having minimum ratio Δ_u/r_i

$\Delta_m = 0.209(\theta + 2)^{-0.32}a$
= deformation of element at maximum strength, in.

$\Delta_u = 1.087(\theta + 6)^{-0.65}a \leq 0.17a$
= deformation of element when fracture is imminent, usually in element farthest from instantaneous center of rotation

a − leg size of fillet weld, in.

The procedure is as follows:

1. Divide the weld configuration into segments, say 1-in.-long segments.

2. Select a trial location for the instantaneous center of rotation (see Fig. 5.17.2).

3. Assume the resisting force R_i or R_j at any weld segment acts in a direction perpendicular to the radial line from the instantaneous center to the centroid of the weld segment.

4. Compute the angle θ (in degrees) between the direction of the resisting force R_i or R_j and the axis of the weld.

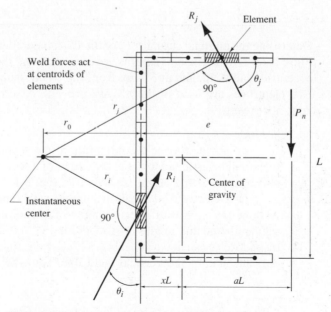

Figure 5.17.2   Resistance R of a fillet weld segment.

5. Compute the deformations Δ_m and Δ_u which can occur at that particular θ of the weld segment.

6. Deformations on weld segments are assumed to vary linearly with the distance from the instantaneous center to the centroid of the weld segment. Thus, the critical segment is the one where the ratio of its Δ_u to its radial distance r_i is the *smallest*.

7. Compatible deformations Δ_i are then computed at each weld segment.

8. Compute the nominal strength R_i of each weld segment using Eq. 5.17.2.

9. Using statics, compute the load P_n that represents the nominal strength of the connection when the load is applied at the given eccentricity e. For P_n applied in the y-direction as shown in Fig. 5.17.2, the statics equations are

$$\sum M = 0; \qquad P_n(e + r_0) = \sum R_i r_i + \sum R_j r_j \qquad (5.17.3)$$

$$P_n = \frac{\sum R_i r_i + \sum R_j r_j}{e + r_0} \qquad (5.17.4)$$

$$\sum F_y = 0; \qquad P_n = \sum (R_i)_y + \sum (R_j)_y \qquad (5.17.5)$$

$$P_n = \sum R_i \cos \theta_i + \sum R_j \sin \theta_j \qquad (5.17.6)$$

10. Compare the values of P_n from Eqs. 5.17.4 and 5.17.6. If they are equal the solution is correct. If they are not equal, revise the trial value of r_0 and repeat the process until Eqs. 5.17.4 and 5.17.6 give the same result.

For cases where P_n is not applied perpedicular to an axis of symmetry of the weld configuration, the three equations of equilibrium must be solved simultaneously for the x and y location of the instantaneous center and the load P_n.

EXAMPLE 5.17.1

Determine the nominal strength P_n of the C-shaped fillet weld configuration shown in Fig. 5.17.3. The horizontal lengths are 3 in. each and the vertical length is 14 in. The eccentric load is applied at 3.5 in. from the 14-in. length. The weld size is $\frac{1}{4}$ in. and E70 electrodes are used in the SMAW process. Assume the base material strength does not govern.

Solution.

   (a) Divide the weld configuration into segments 1-in. long. The resisting force will be assumed to act at the center of a segment. The instantaneous center will be used as the origin, as shown in Fig. 5.17.3.

   (b) Select a trial location for the instantaneous center (IC) of rotation. Assume say $r_0 = 6$ in. Then compute the coordinates of the centroids of the segments and the angles θ_i and θ_j for the vertical and horizontal segments, respectively, as shown in Table 5.17.1. Note that IC will generally be on the opposite side of the vertical weld line from the point of action of the load.*

---

* The use of a spreadsheet program will allow a rapid solution of this problem by trial and error.

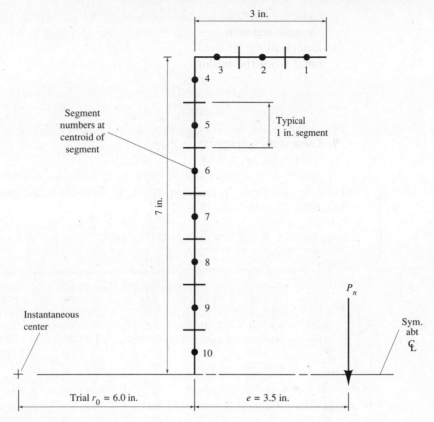

Figure 5.17.3 Weld segments for Example 5.17.1.

TABLE 5.17.1 GEOMETRY FOR TRIAL $1(r_0 = 6$ in.) OF EXAMPLE 5.17.1

Horizontal segments	Length (in.)	x (in.)	y (in.)	r_i (in.)	θ_i (radians)	θ_i (degrees)
1	1	8.50	7.00	11.011	0.882	50.5
2	1	7.50	7.00	10.259	0.820	47.0
3	1	6.50	7.00	9.552	0.748	42.9

Vertical segments	Length (in.)	x (in.)	y (in.)	r_i (in.)	θ_i (radians)	θ_i (degrees)
4	1	6.00	6.50	8.846	0.825	47.3
5	1	6.00	5.50	8.139	0.742	42.5
6	1	6.00	4.50	7.500	0.644	36.9
7	1	6.00	3.50	6.946	0.528	30.3
8	1	6.00	2.50	6.500	0.395	22.6
9	1	6.00	1.50	6.185	0.245	14.0
10	1	6.00	0.50	6.021	0.083	4.8

(c) Compute the maximum deformations Δ_m and Δ_u which can occur at each θ value for the segments. Illustrate for segment 1, where $\theta = 50.5°$ and a = weld leg = 0.25 in.,

$$\Delta_m = 0.209(\theta + 2)^{-0.32}a$$
$$= 0.209(50.5 + 2)^{-0.32}(0.25) = 0.0155$$
$$\Delta_u = 1.087(\theta + 6)^{-0.65}a \leq 0.17a$$
$$= 1.087(50.5 + 6)^{-0.65}(0.25) = 0.0217$$
$$< [0.17(0.25) = 0.0425] \quad \text{OK}$$

(d) Compute the ratio Δ_u/r_i for each segment, determine the minimum value for that ratio. In this case, segment 1 has the minimum value (0.00179). The results are in Table 5.17.2.

TABLE 5.17.2   RESISTANCE AT EACH SEGMENT-TRIAL (r_0 = 6.0 in.)

Segment	Δ_m	Δ_u	Δ_u/r_i	Δ_i	Δ_i/Δ_m	R_i
1	0.01471	0.01973	0.00179	0.01973	1.34165	9.72
2	0.01504	0.02058	0.00201	0.01839	1.22229	9.68
3	0.01547	0.02169	0.00227	0.01712	1.10673	9.51
4	0.01501	0.02050	0.00232	0.01585	1.05609	9.77
5	0.01551	0.02180	0.00268	0.01459	0.94053	9.46
6	0.01620	0.02362	0.00315	0.01344	0.82987	9.03
7	0.01719	0.02634	0.00379	0.01245	0.72407	8.49
8	0.01874	0.03071	0.00473	0.01165	0.62144	7.87
9	0.02150	0.03872	0.00626	0.01108	0.51549	7.19
10	0.02834	0.04250	0.00706	0.01079	0.38070	6.42

(e) Compute Δ_i. Take the distance r_i to segment 1 as r_{crit} because segment 1 has the minimum Δ_u/r_i of 0.00179. Compute Δ_i as follows:

$$\Delta_i = \text{deformation of element } i = r_i\frac{\Delta_u}{r_{crit}} = r_i(0.00179)$$

Results are in Table 5.17.2.

(f) Compute Δ_i/Δ_m and R_i using Eq. 5.17.2,

$$R_i = 0.60F_{EXX}t_e(1.0 + 0.50\sin^{1.5}\theta)\left[\frac{\Delta_i}{\Delta_m}\left(1.9 - 0.9\frac{\Delta_i}{\Delta_m}\right)\right]^{0.3}$$

In this example, E70 electrodes and 1/4-in. fillet weld are used, making F_{EXX} = 70 ksi and $t_e = 0.25(0.707)$ in.

(g) Compute check of statics to determine correctness of trial value of r_0. Using Eqs. 5.17.4 and 5.17.6, the results are in Table 5.17.3.

From Eq. 5.17.4, rotational equilibrium gives

$$P_n = 2(721.94)/(6.0 + 3.5) = 152.0 \text{ kips}$$

From Eq. 5.17.6, force equilibrium in the y-direction gives

$$P_n = 2(69.78) = 139.6 \text{ kips}$$

TABLE 5.17.3 CHECK OF STATICS FOR TRIAL ($r_0 = 6.0$ in.)

Segment	R_i	r_i	x	y	$(R_i)_x$	$(R_i)_y$	$R_i r_i$
1	9.72	11.01	8.5	7	6.18	7.51	107.07
2	9.68	10.26	7.5	7	6.60	7.08	99.29
3	9.51	9.55	6.5	7	6.97	6.47	90.83
4	9.77	8.85	6	6.5	7.18	6.63	86.42
5	9.46	8.14	6	5.5	6.39	6.97	76.99
6	9.03	7.50	6	4.5	5.42	7.22	67.71
7	8.49	6.95	6	3.5	4.28	7.33	58.97
8	7.87	6.50	6	2.5	3.03	7.26	51.13
9	7.19	6.18	6	1.5	1.74	6.97	44.46
10	6.42	6.08	6	1	1.06	6.33	39.06
					48.84	69.78	721.94

Since the values of P_n from the statics equations are not equal, the assumed location of the instantaneous center is not correct. Try a new value of r_0 and respeat the analysis until the values from Eqs. 5.17.4 and 5.17.6 and identical. Since the applied load P_n is in the y-direction for this example, the summation of forces in the x-direction is automatically satisfied; 48.84 kips acting in one direction from the segments on one side of the axis of symmetry, and the same force acting at 180° from the segments on the other side of the axis of symmetry.

(h) After several trials the solution for r_0 is determined to be 8.34 in. Table 5.17.4 shows the final solution.

TABLE 5.17.4 SOLUTION FOR EXAMPLE 5.17.1 ($r_0 = 8.34$ in.)

Segment	θ_i	Δ_u	Δ_u/r_i	Δ_i	R_i	$(R_i)_y$	$R_i r_i$
1	56.3	0.0185	0.00142	0.0184	10.13	8.51	130.66
2	53.6	0.0191	0.00156	0.0172	10.11	8.24	122.06
3	50.5	0.0197	0.00173	0.0160	10.00	7.84	112.80
4	39.1	0.0228	0.00220	0.0150	9.19	7.24	97.12
5	34.5	0.0245	0.00250	0.0142	8.85	7.38	88.37
6	29.4	0.0268	0.00288	0.0135	8.45	7.44	80.07
7	23.7	0.0300	0.00338	0.0129	8.01	7.38	72.42
8	17.4	0.0350	0.00410	0.0124	7.53	7.22	65.59
9	10.6	0.0425	0.00502	0.0121	7.03	6.92	59.58
10	3.6	0.0425	0.00509	0.0119	6.43	6.39	54.03
						74.56	882.70

From Eq. 5.17.4, rotational equilibrium gives

$$P_n = 2(882.70)/(8.34 + 3.5) = 149.11 \text{ kips}$$

From Eq. 5.17.6, force equilibrium in the y-direction gives

$$P_n = 2(74.56) = 149.11 \text{ kips}$$

Thus, the nominal strength P_n is 149 kips. ■■

EXAMPLE 5.17.2 _____

Determine the maximum service load P that the eccentric shear connection of Fig. 5.17.4 may be permitted to carry using strength analysis and the AISC Load and Resistance Factor Design Specification. The live load is three times the dead load. The weld is $\frac{1}{4}$ in. and E70 electrodes are used in the SMAW process. Assume the base material strength does not govern.

Solution

(a) Strength analysis using basic concepts. The nominal strength of this connection as determined in Example 5.17.1 is

$$P_n = 149 \text{ kips}$$

(b) Use *LRFD Manual* [1.18] tables, p. 8-187. For $\frac{1}{4}$-in. weld using E70 electrodes,

$$a = (e - xL)/L = (3.5 - 0.45)/14 = 0.218$$
$$k = kL/L \qquad = 3.0/14 = 0.214$$

	$k - 0.2$	0.214	0.3	
$a = 0.2$	1.98		2.33	
0.218	1.94	1.99	2.99	$C = 1.99$
0.25	1.88		2.22	

Table value $= \phi P_n = CC_1DL = 1.99(1.0)(4)14 = 111$ kips

where $C_1 =$ coefficient for electrode $=$ (Electrode used)/70
$D =$ number of $\frac{1}{16}$s of an inch in weld size
$L =$ length of vertical weld (in.)

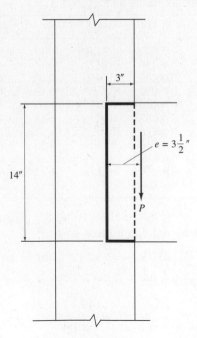

Figure 5.17.4   Loading for Example 5.17.2.

(c) Summary. Compare the values of the design strength ϕP_n.

1. Strength analysis P_n after completing analysis:
$$\phi P_n = 0.75(149) = 112 \text{ kips}$$

2. *LRFD Manual* tables:
$$\phi P_n = 111 \text{ kips}$$

(d) Compute the safe service load P using the factored gravity load combination, LRFD-A4.1, Formula (A4-2). For the given 75% live load and 25% dead load,

$$P_u = 1.2D + 1.6L$$
$$\phi P_n = P_u = 1.2(0.25P) + 1.6(0.75P) = 1.5P$$
$$P = \frac{\phi P_n}{1.5} = \frac{112}{1.5} = 75 \text{ kips} \quad \blacksquare\blacksquare$$

5.18 ECCENTRIC SHEAR CONNECTIONS— ELASTIC (VECTOR) ANALYSIS

The traditional elastic vector analysis is easier than the strength method to carry out when the computer is not available, or when the AISC Manual tables are not available. The elastic vector method is conservative, sometimes excessively so.

The elastic method has the following assumptions:

1. Each segment of weld, if of the same size, resists a concentrically applied load with an equal force. This concept was used for welds on tension members in Sec. 5.16.
2. The rotation caused by torsional moment is assumed to occur about the centroid of the weld configuration.
3. The load on a weld segment caused by the torsional moment is assumed to be proportional to the distance from the centroid of the weld configuration.
4. The direction of the force on a weld segment caused by torsion is assumed to be perpendicular to the radial distance from the centroid of the weld configuration.
5. The components of the forces caused by direct load and by torsion are combined vectorially to obtain a resultant force.

It will be convenient to think of this analysis using the principles of mechanics on a homogeneous material, combining direct shear with torsion. Beginning with the stresses on a homogeneous section,

$$f' = \frac{P}{A} = \text{stress due to direct shear} \tag{5.18.1}$$

$$f'' = \frac{Tr}{I_p} = \text{stress due to torsional moment} \tag{5.18.2}$$

where r = radial distance from the centroid to point of stress
         I_p = polar moment of inertia

For computing nominal stresses or forces on weld segments the *locations* of the lines of weld are defined by edges along which the fillets are placed, rather than to the center of the effective throat. This makes little difference, since the throat dimension is usually small.

For the general case shown in Fig. 5.18.1, the components of stress caused by direct shear are

$$f'_x = \frac{P_x}{A} \tag{5.18.3a}$$

$$f'_y = \frac{P_y}{A} \tag{5.18.3b}$$

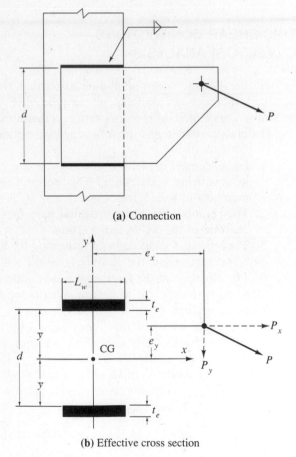

(a) Connection

(b) Effective cross section

Figure 5.18.1   Eccentric bracket connection.

The x- and y-components of f'' resulting from torsion are

$$f''_x = \frac{Ty}{I_p} = \frac{(P_x e_y + P_y e_x)y}{I_p}$$ (5.18.4a)

$$f''_y = \frac{Tx}{I_p} = \frac{(P_x e_y + P_y e_x)x}{I_p}$$ (5.18.4b)

where

$$I_p = I_x + I_y = \sum I_{xx} + \sum A\bar{y}^2 + \sum I_{yy} + \sum A\bar{x}^2$$ (5.18.5)

In Eq. 5.18.5, $\bar{x}$ and $\bar{y}$ refer to distances from the center of gravity of the weld group to the center of gravity of the individual weld segments. I_{xx} and I_{yy} refer to the moments of inertia of the individual segments with respect to their own centroidal axes.

Thus, for the situation of Fig. 5.18.2, Eq. 5.18.5 becomes

$$I_p = 2\left[\frac{L_w(t_e)^3}{12}\right] + 2[L_w(t_e)(\bar{y})^2] + 2\left[\frac{t_e(L_w)^3}{12}\right]$$

$$= \frac{t_e}{6}[L_w(t_e)^2 + 12L_w(\bar{y})^2 = L_w^3]$$ (5.18.6)

For practical situations, the first term of Eq. 5.18.6 is neglected because, with t_e small, the term is not significant compared to the other terms. Hence

$$I_p \approx \frac{t_e}{6}[12L_w(\bar{y})^2 + L_w^3]$$ (5.18.7)

Note that I_p equals the throat thickness t_e times the property of *lines;* i.e., an element having length but having a width of unity. Actually, the area A in Eqs. 5.18.3 equals the thickness t_e times the total length of the weld configuration; and in Eq. 5.18.5 the polar moment of inertia equals the thickness t_e times the polar moment of inertia of the configuration as *lines.* When the stress f is multiplied by t_e, it becomes a force R per unit length, say, kips/in.

Treating the welds making up the effective cross-section in Fig. 5.18.2 as line welds (i.e., as in deriving Eq. 5.18.7 with $t_e = 1$) and using the general terms b and

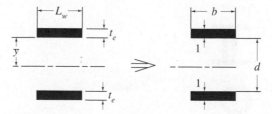

Figure 5.18.2 Treatment of weld configuration as lines having unit thickness.

TABLE 5.18.1 PROPERTIES OF WELDS TREATED AS LINES

Section b = width; d = depth	Section modulus $I_x/\bar{y}$	Polar moment of inertia, I_p about center of gravity
1.	$S = \dfrac{d^2}{6}$	$I_p = \dfrac{d^3}{12}$
2.	$S = \dfrac{d^2}{3}$	$I_p = \dfrac{d(3b^2 + d^2)}{6}$
3.	$S = bd$	$I_p = \dfrac{b(3d^2 + b^2)}{6}$
4. $\bar{y} = \dfrac{d^2}{2(b + d)}$ $\bar{x} = \dfrac{b^2}{2(b + d)}$	$S = \dfrac{4bd + d^2}{6}$	$I_p = \dfrac{(b + d)^4 - 6b^2d^2}{12(b + d)}$
5. $\bar{x} = \dfrac{b^2}{2b + d}$	$S = bd + \dfrac{d^2}{6}$	$I_p = \dfrac{8b^3 + 6bd^2 + d^3}{12} - \dfrac{b^4}{2b + d}$
6. $\bar{y} = \dfrac{d^2}{b + 2b}$	$S = \dfrac{2bd + d^2}{3}$	$I_p = \dfrac{b^3 + 6b^2d + 8d^3}{12} - \dfrac{d^4}{2d + b}$
7.	$S = bd + \dfrac{d^2}{3}$	$I_p = \dfrac{(b + d)^3}{6}$
8. $\bar{y} = \dfrac{d^2}{b + 2d}$	$S = \dfrac{2bd + d^2}{3}$	$I_p = \dfrac{b^3 + 8d^3}{12} - \dfrac{d^4}{b + 2d}$
9.	$S = bd + \dfrac{d^2}{3}$	$I_p = \dfrac{b^3 + 3bd^2 + d^3}{6}$
10.	$S = \pi r^2$	$I_p = 2\pi r^3$

d, as shown in Fig. 5.18.2, Eq. 5.18.7 becomes

$$I_p \approx \frac{1}{6}\left[12b\left(\frac{d}{2}\right)^2 + b^3\right] = \frac{b}{6}[3d^2 + b^2] \tag{5.18.8}$$

Table 5.18.1 gives I_p values treated as properties of lines for other common weld configurations.

EXAMPLE 5.18.1

Compute the maximum load (kips/in.) on the weld configuration shown for the bracket in Fig. 5.18.3 using the elastic (vector) method. Assume the plate thickness does not affect the result.

Solution

(a) Locate the centroid of the configuration. The maximum force R will occur at points A and B. The properties of lines will be used. Taking moments about the vertical weld,

$$\bar{x} = \frac{2(6)3}{2(6) + 8} = 1.8 \text{ in.}$$

(b) Compute the area (length) and the polar moment of inertia about the centroid of the configuration.

$$L = 2(6) + 8 = 20 \text{ in.}$$

$$I_p = \underbrace{\frac{(8)^3}{12} + 2[6(4)^2]}_{I_x} + \underbrace{2\left[\frac{(6)^3}{12}\right] + 2[6(1.2)^2] + 8(1.8)^2}_{I_y} = 314 \text{ in.}^3$$

(c) Compute the components of force on the weld at points A and B. From the direct shear,

$$R_v = \frac{P}{L} = \frac{15}{20} = 0.75 \text{ kips/in.} \downarrow$$

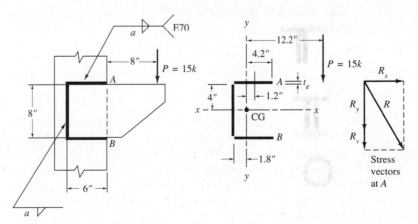

Figure 5.18.3   Example 5.18.1.

From the torsion T about the centroid of the configuration,

$$R_x = \frac{Ty}{I_p} = \frac{15(12.2)4}{314} = 2.33 \text{ kips/in. } \rightarrow$$

$$R_y = \frac{Tx}{I_p} = \frac{15(12.2)4.2}{314} = 2.45 \text{ kips/in. } \downarrow$$

The vector sum gives the resultant force R,

$$R = \sqrt{(2.33)^2 + (2.45 + 0.75)^2} = 3.96 \text{ kips/in. } \blacksquare\blacksquare$$

EXAMPLE 5.18.2 _____

Determine the weld size required for the bracket of Fig. 5.18.3 when the service load P is 15 kips (80% live load and 20% dead load). Compare the results using (a) elastic (vector) analysis from Example 5.18.1 and (b) strength analysis as described in Sec. 5.17, both with AISC Load and Resistance Factor Design. Assume the plate thickness does not affect the result (stiffened seats and bracket plates are treated in Secs. 13.4 and 13.5.)

Solution

(a) Elastic (vector) method. The gravity factored load according to LRFD is computed,

$$P_u = 1.2D + 1.6L = 1.2(0.2)15 + 1.6(0.8)15 = 22.8 \text{ kips}$$

The maximum load (kips/in.) on the weld due to the factored P will be (using the result from Example 5.18.1 for $P = 15$ kips),

$$R_u = 3.96(22.8/15) = 6.02 \text{ kips/in.}$$

$$\text{Weld resistance } \phi R_{nw} = 0.75a(0.707)(0.60\ F_{EXX})$$
$$= 0.75a(0.707)(0.60)70 = 22.3a$$

The weld size required is obtained by equating the design strength and the factored load.

$$\phi R_{nw} = R_u$$
$$22.3a = 6.02; \qquad a = 0.27, \quad \text{say } \tfrac{5}{16} \text{ in.}$$

Use $\tfrac{5}{16}$-in. E70 fillet welds.

(b) Strength analysis. Divide the horizontal 6-in. weld into six parts and the vertical 8-in. length into 16 segments of 1/2-in. each, though it is considered adequate to use 1-in. segments. Using Eq. 5.17.2 and solving by trial in the manner used for Example 5.17.1, the instantaneous center is found to be located -0.1134 in. from the vertical weld line. Negative means the center is on the same side of the vertical weld line as the applied load P. The solution details are in Tables 5.18.2 and 5.18.3.

TABLE 5.18.2 GEOMETRY AND PARTIAL RESULTS, EXAMPLE 5.18.2

Seg. No.	Length (in.)	x (in.)	y (in.)	r_i (in.)	θ (deg)	Δ_u (in.)	Δ_u/r_i
1	1.0	0.387	4.00	4.02	5.5	0.0425	0.01058
2	1.0	1.387	4.00	4.23	19.1	0.0334	0.00790
3	1.0	2.387	4.00	4.66	30.8	0.0261	0.00560
4	1.0	3.387	4.00	5.24	40.3	0.0225	0.00429
5	1.0	4.387	4.00	5.94	47.6	0.0204	0.00344
6	1.0	5.387	4.00	6.71	53.4	0.0191	0.00285
7	0.5	−0.113	3.75	3.75	88.3	0.0142	0.00377
8	0.5	−0.113	3.25	3.25	88.0	0.0142	0.00436
9	0.5	−0.113	2.75	2.75	87.6	0.0142	0.00516
10	0.5	−0.113	2.25	2.25	87.1	0.0143	0.00633
11	0.5	−0.113	1.75	1.75	86.3	0.0143	0.00818
12	0.5	−0.113	1.25	1.26	84.8	0.0145	0.01155
13	0.5	−0.113	0.75	0.76	81.4	0.0149	0.01960
14	0.5	−0.113	0.25	0.27	65.6	0.0169	0.06165

TABLE 5.18.3 SOLUTION FOR EXAMPLE 5.18.2 ($r_0 = -0.1134$ in.)

Segment	Δ_m	Δ_i	R_i	$(R_i)_y$	$R_i r_i$
1	0.0274	0.0114	6.58	0.63	26.44
2	0.0197	0.0121	7.67	2.51	32.46
3	0.0171	0.0133	8.60	4.41	40.07
4	0.0158	0.0149	9.33	6.03	48.89
5	0.0150	0.0169	9.78	7.22	58.03
6	0.0145	0.0191	9.91	7.95	66.46
7	0.0124	0.0107	5.52	−0.17	20.69
8	0.0124	0.0093	5.42	−0.19	17.64
9	0.0124	0.0078	5.28	−0.22	14.54
10	0.0124	0.0064	5.09	−0.26	11.46
11	0.0125	0.0050	4.81	−0.31	8.44
12	0.0125	0.0036	4.43	−0.40	5.56
13	0.0127	0.0022	3.85	−0.58	2.92
14	0.0136	0.0008	2.72	−1.12	0.75
				25.52	354.35

From Eq. 5.17.4, and multiplying by 2 because of symmetry, gives

$$P_n = \frac{\Sigma R_i r_i}{e + r_0} = \frac{2(354.35)}{14.0 - 0.1134} = 51.03 \text{ kips}$$

$$P_n = \sum (R_i)_y = 2(25.52) = 51.04 \text{ kips}$$

The nominal strength $P_n = 51$ kips. This is the strength using $\frac{1}{4}$-in. weld with E70 electrodes and the SMAW process. The design strength ϕP_n is $0.75(51) =$

38.3 kips. The weld size required is

$$P_u = 22.8 \text{ kips} = \phi P_n = 38.3 \frac{a}{0.25}$$

$$\text{Required weld size } a = \frac{22.8(0.25)}{38.3} = 0.15 \text{ in.}$$

(c) Use *LRFD Manual* [1.18] tables, p. 8–187. For $\frac{1}{4}$-in. weld using E70 electrodes,

$$a = (e - xL)/L = (14.0 - 1.80)/8 = 1.525$$
$$k = kL/L \quad\quad = 6.0/8 = 0.75$$

$k = 0.7$	0.75	0.8
$a = 1.4$	1.21		1.36
1.525	1.12	1.20	1.27	$C = 1.20$
1.6	1.07		1.21

$$\text{Table value} = \phi P_n = C C_1 DL = 1.20(1.0)(4)8 = 38.4 \text{ kips}$$

As may be noted, the LRFD tables give essentially the same result as obtained in part (b).

(d) Summary. The weld size a required using E70 electrodes and the SMAW process is

$$\text{Elastic (vector) method, required } a = \tfrac{5}{16} \text{ in.}$$
$$\text{Strength analysis,} \quad\quad \text{required } a = \tfrac{3}{16} \text{ in.}$$

As will always be the case, the elastic vector method is conservative. ■■

5.19 LOADS APPLIED ECCENTRIC TO THE PLANE OF WELDS

When an applied load is eccentric to the *plane* of the weld configuration, as in Fig. 5.19.1, the strength method of analysis may still be used as long as the plane of the welds is rigid. The weld plane is rigid in Fig. 5.19.1 because the welds are on each

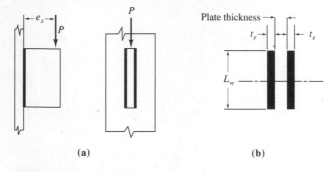

(a)   (b)

Figure 5.19.1   Welds in shear and bending.

side of a plate; i.e., there is sufficient rigidity between the two lines of weld such that there will be no bending *of the material being welded* in the plane of the welds.

As discussed in Sec. 5.17, the strength of a segment of weld depends on the angle θ_i of the resisting force R_i to the axis of the weld. It makes no difference whether R_i acts at an angle to the plane of the welds (Fig. 5.19.1) or whether it acts in the plane of the welds (eccentric shear as in Secs. 5.17 and 5.18).

The situation in Fig. 5.19.1 is commonly referred to as shear and bending, which is actually the stress condition on the bracket plate supporting the load P. The welds must carry the loads in the same manner that the members being connected carries them. The stresses are shown in Fig. 5.19.2.

For loading of the weld configuration in shear and tension, one must realize that the weld segments subject to compression are not free to rotate; thus, if a strength analysis is made, the compression region should be assumed to have a compressive stress distribution between the pieces being welded. Dawe and Kulak [5.41] reported that relatively good agreement with tests was obtained using any of a triangular, parabolic, or rectangular stress distribution on the compression side of the neutral axis.

The *LRFD Manual* [1.18] does not contain tables for the case of shear and bending. The tables for eccentric shear are suggested [1.18] to be used for all cases where "the connection material between the welds is solid and does not bend in the plane of the welds."

Alternatively, the elastic (vector) method is conservative and relatively easy to use for loading in shear and tension.

Thus, there are four methods (Nos. 2 and 3 are really identical) suggested for use:

1. Strength analysis dividing the weld on the tension side of the neutral axis into segments the resistance of each depending on the angle the resistance makes with the weld axis using the formulas in Sec. 5.17. The weld on the compression side of the neutral axis is assumed to have only a resistance parallel to the weld axis. The compression force from bending is assumed

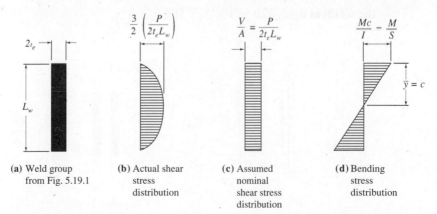

(a) Weld group
  from Fig. 5.19.1

(b) Actual shear
  stress
  distribution

(c) Assumed
  nominal
  shear stress
  distribution

(d) Bending
  stress
  distribution

Figure 5.19.2   Stresses on vertical lines of weld acting in shear and bending.

to be carried by direct compression of the pieces being welded using a triangular distribution with the yield stress at the extreme compression fiber. The instantaneous center is then located by trial in a manner similar to that illustrated for eccentric shear. Details of the procedure are described by Dawe and Kulak [5.41].

2. Strength analysis exactly as for eccentric shear. In this procedure all weld segments, both on the tension and compression sides of the neutral axis, are assumed to have resistance. This procedure will give the values in the *LRFD Manual* tables.

3. *LRFD Manual* tables.

4. Elastic (vector) analysis.

EXAMPLE 5.19.1 _____

Compute the size of E70 fillet weld required for the shear and tension connection in Fig. 5.19.3a using the SMAW process. Assume the column and the bracket plate do not control (stiffened seats and bracket plates are treated in Secs. 13.4 and 13.5.). The load is 80% live load and 20% dead load, and Load and Resistance Factor Design is to be used.

Solution

(a) Compute the factored load P_u.

$$P_u = 1.2(0.2)10 + 1.6(0.8)10 = 15.2 \text{ kips}$$

(b) Use the traditional elastic (vector) method. The weld segments are treated as lines having a thickness of unity (1.0). The direct shear component is assumed to be carried equally by each segment of weld,

$$(R_n)_v = \frac{P}{A} = \frac{P}{2(1)L_w} = \frac{15.2}{2(1)10} = 0.76 \text{ kips/in.}$$

The tension component (horizontal) due to the moment Pe is

$$(R_n)_t = \frac{Mc}{I} = \frac{15.2(6)5}{166.7} = 2.74 \text{ kips/in.}$$

where $I = \frac{1}{12}[2(1)(10)^3] = 166.7 \text{ in.}^4$

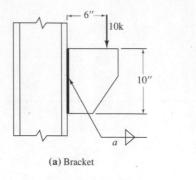
(a) Bracket

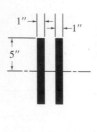
(b) Assumed weld cross section

Figure 5.19.3   Example 5.19.1.

The resultant force is
$$\text{Required } \phi R_n = \sqrt{(0.76)^2 + (2.74)^2} = 2.84 \text{ kips/in.}$$

The design strength of E70 electrode fillet weld is
$$\phi R_{nw} = \phi t_e(0.60 F_{\text{EXX}}) = 0.75(0.707)a(0.60)70 = 22.3a$$

and the fillet weld size a required is
$$\text{Required } a = \frac{2.84}{22.3} = 0.13 \text{ in.}, \quad \text{say } \tfrac{3}{16} \text{ in.}$$

(c) Use *LRFD Manual* [1.18] tables, p. 8–163. For weld using E70 electrodes.

$$e/L \text{ (Table symbol } a) = 6/10 = 0.60$$
$$k = 0$$
$$\text{Find } C = 1.50$$

$$\text{Table value} = \phi P_n = CC_1 DL = 1.50(1.0)(D)10 = 15.0D \text{ kips}$$
$$\text{Required } D = 15.2/15.0 = 1.0$$
$$\text{Required weld size } a = 1.0/16 = 0.06 \text{ in.}, \text{ say } \tfrac{1}{8} \text{ in.}$$

The elastic (vector) method is as expected more conservative than the strength method represented by the by the *LRFD Manual* tables. The minimum desirable size to be used for this situation is probably $\tfrac{3}{16}$ in. *Use $\tfrac{3}{16}$-in. E70 fillet welds.* ■■

Design for Lines of Weld Subject to Bending Moment

Even when there are moderate returns at the top of lines of fillet weld, an estimate of the length required may be obtained by using the same approach as used to determine the number of bolts in a line in Sec. 4.12. In Fig. 4.12.9, R/p has units kips/in. which becomes ϕR_{nw}, the design strength at the top of the lines of weld.

For moment alone on one line of weld,

$$R = \frac{M}{S} = \frac{M}{(\frac{1}{6}L_w^2)} \text{ kips/in.} \tag{5.19.1}$$

Since the maximum value of R is ϕR_{nw}

$$\phi R_{nw} = \frac{6M}{L_w^2}$$

$$\text{Required } L_w = \sqrt{\frac{6M}{\phi R_{nw}}} \tag{5.19.2}$$

Equation 5.19.2 for welds corresponds to Eq. 4.12.29 for bolts. Since it is correct only for moment alone, R_{nw} should be entered as a reduced value to account for direct shear.

EXAMPLE 5.19.2 _____

Determine the length L required to carry the load indicated in Fig. 5.19.4, when 75% of the load is live load and 25% is dead load. The weld to be used is $\tfrac{5}{16}$-in. E70 fillet weld. Use AISC Load and Resistance Factor Design.

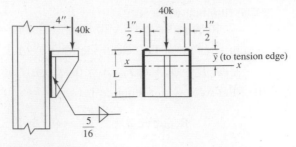

Figure 5.19.4 Example 5.19.2.

Solution

(a) Compute factored load P_u:

$$P_u = 1.2(0.25)40 + 1.6(0.75)40 = 60 \text{ kips}$$

(b) Estimate length of weld L required by using Eq. 5.19.2:

$$\phi R_{nw} = 0.75(a)(0.707)(0.60)70 = 22.3a \text{ kips/in.}$$
$$= 22.3(5/16) = 6.96 \text{ kips/in.}$$
$$M_u = 60(4) = 240 \text{ in.-kips per 2 lines of weld}$$

$$\text{Required } L \approx \sqrt{\frac{6M_u}{\phi R_{nw}}} = \sqrt{\frac{6(240/2)}{6(\text{est.})}} = 11 \text{ in.}$$

A reduced value of ϕR_{nw} has been used to account for the direct shear effect. Since the $\frac{1}{2}$-in. returns at the top add something; try $L = 10$ in.

When the returns have a specified dimension as is the case here, the weld is to be placed to provide full strength over the specified dimension. The returns resist the moment $P_n e$ but probably not much to resist P_n as a shear force. For these small returns, it is optional with the designer whether or not to include them in the strength computation.

(c) If the returns are neglected, the *LRFD Manual* [1.18] tables, p. 8–163, can be used to obtain an approximate result, as follws:

$$e/L \text{ (Table symbol } a) = 4/10 = 0.40$$
$$k = 0$$
$$\text{Find } C = 2.00$$

Table value $= \phi P_n = CC_1 DL = 2.00(1.0)(D)10 = 20.0 \, D$ kips

Required $D = 60/20 = 3.0$

Required weld size $a = \dfrac{3.0}{16} = \frac{3}{16}$ in.

For $L = 8$ in.,

$$\text{Find } C = 1.72, \quad \text{required } a = 0.22 \text{ in.}, \quad \text{say } \tfrac{1}{4} \text{ in.}$$

(d) Elastic (vector) method. The actual weld configuration has the $\frac{1}{2}$-in. returns which make the center of gravity of the weld configuration lie closer to the top than the mid-depth assumed in part (c). LRFD Tables 8–38, "Coefficients C for Eccentrically Loaded Weld Groups," are indicated to be used also when load is not in the plane of weld group. This is acceptable for two vertical lines. However, for other weld configurations having the load applied out of plane, it is prudent to use a conservative elastic (vector) analysis.

Locate the center of gravity of the configuration,

$$\bar{y} = \frac{2(10)5}{2(10 + 0.5)} = \frac{100}{21} = 4.76 \text{ in.}$$

The direct shear component $(R_n)_v$ is computed assuming that none of the shear is carried by the returns,

$$(R_n)_v = \frac{P}{2L} = \frac{60}{2(10)} = 3.00 \text{ kips/in.}$$

The tension component $(R_n)_t$ due to the moment Pe is

$$I_x = \frac{2L^3}{12} + 2L(5 - 4.67)^2 + 2(0.5)(4.76)^2$$

$$= \frac{(10)^3}{6} + 20(0.24)^2 + (4.76)^2 = 190.5 \text{ in.}^3$$

$$(R_n)_t = \frac{60(4)4.76}{190.5} = 6.00 \text{ kips/in.}$$

The resultant force is

$$\text{Required } \phi R_n = \sqrt{(3.00)^2 + (6.00)^2} = 6.71 \text{ kips/in.}$$

The design strength of E70 electrode $\frac{5}{16}$-in. fillet weld is

$$\phi R_{nw} = \phi t_e (0.60 F_{\text{EXX}})$$
$$= 0.75(0.707)(\tfrac{5}{16})(0.60)70 = 6.97 \text{ kips/in.}$$

The design strength ϕR_{nw} exceeds the 6.71 kips/in. required; thus, $L = 10$ in. is adequate.
Use $L = 10$ in. ■■

Additional treatment of eccentric load on welds is to be found in Chapter 13 on connections.

SELECTED REFERENCES

5.1. K. Winterton. "A Brief History of Welding Technology," *Welding and Metal Fabrication,* November 1962; December 1962.
5.2. "100 Years of Metalworking—Welding, Brazing and Joining," *The Iron Age,* June 1955.
5.3. H. Carpmael. *Electric Welding and Welding Appliances.* London: D. Van Nostrand Company, 1920.

5.4. Preston M. Hall. "77 Years of Resistance Welding," *The Welding Engineer,* February 1954, 54–55; March 1954, 36–37; April 1954, 62–63.

5.5. W. L. Miskoe. "The Centenary of Modern Welding, 1885–1985—A Commemoration." *Welding Journal,* **65,** April 1986, 19–24.

5.6. D. W. Morgan. "Classification and Use of Mild Steel Covered Electrodes." *Welding Journal,* **55,** December 1976, 1035–1038.

5.7. E. Craig. "A Unique Mode of GMAW Transfer." *Welding Journal,* **66,** September 1987, 51–55.

5.8. K. A. Lyttle. "GMAW—A Versatile Process on the Move," *Welding Journal,* **62,** March 1983, 15–23.

5.9. V. R. Dillenbeck and L. Castagno. "The Effects of Various Shielding Gases and Associated Mixtures in GMA Welding of Mild Steel." *Welding Journal,* **66,** September 1987, 45–49. (See also letter from Nils Larson, Chairman, Committees C50 and A55, American Welding Society, and author's reply, **67,** March 1988, 6.)

5.10. AWS. *Welding Handbook,* 7th ed., Vols. 2 and 3. Miami, FL: American Welding Society, 1978, 1980.

5.11. _____. "Office Building Columns Field Spliced with Self-Shielded Welding Wire," *Welding Journal,* **65,** October 1986, 53–54.

5.12. _____. "Self-Shielded FCAW Speeds High-Rise Construction," *Welding Journal,* **63,** April 1984, 47–49.

5.13. _____. "Self-Shielded FCA Welding is a Breeze in the Windy City," *Welding Journal,* **67,** March 1988, 47–48.

5.14. A. Raman. "Electroslag Welds: Problems and Cures," *Welding Journal,* **60,** December 1981, 17–21.

5.15. C. G. Schilling and K. H. Klippstein. "Tests of Electroslag-Welded Bridge Girders," *Welding Journal,* **60,** December 1981, 23–30.

5.16. A. W. Pense, J. D. Wood, and J. W. Fisher. "Recent Experiences with Electroslag Welded Bridges," *Welding Journal,* **60,** December 1981, 33–42.

5.17. Omer W. Blodgett, "Distortion . . . How to Minimize it with Sound Design Practices and Controlled Welding Procedures Plus Proven Methods for Straightening Distorted Members," *Bulletin G261.* Cleveland, OH: The Lincoln Electric Company. (No date)

5.18. Omer W. Blodgett. "Shrinkage Control in Welding," *Civil Engineering,* November 1960, 56–61.

5.19. E. R. Holby. "Weld Appearances May Be Deceiving," *Welding Journal,* **63,** May 1984, 33–36.

5.20. J. E. Jones. "Inspecting for Fatigue," *Welding Journal,* **62,** May 1983, 21–24.

5.21. R. Fenn. "Ultrasonic Monitoring and Control During Arc Welding," *Welding Journal,* **64,** September 1985, 18–22.

5.22. Paul D. Watson. "Design for Welding Examination," *Welding Journal,* **61,** February 1982, 32–35.

5.23. C. M. Fortunko and R. E. Schramm. "Ultrasonic Nondestructive Evaluations of Butt Welds Using Electromagnetic-Acoustic Transducers," *Welding Journal,* **61,** February 1982, 39–46.

5.24. Ronald Selner. "Dye Penetrant and Magnetic Particle Inspection," *Welding Journal,* **61,** February 1982, 28–31.

5.25. AWS. *Structural Welding Code—Steel* (ANSI/AWS D1.1-94), 13th edition. Miami, FL: American Welding Society, 1994 [same as 2.24].

5.26. J. A. Donnelly. "Determining the Cost of Welded Joints," *Engineering Journal,* AISC, **5,** 4 (October 1968), 146–147.

5.27. Omer W. Blodgett. "How to Determine the Cost of Welding," *Bulletin G610.* Cleveland, OH: The Lincoln Electric Company. (No date)

5.28. L. J. Butler and G. L. Kulak. "Strength of Fillet Welds as a Function of Direction of Load," *Welding Journal* (Welding Research Supplement), **36,** May 1971, 231s–234s.

5.29. Lorne J. Butler, Shubendu Pal, and Geoffrey L. Kulak. "Eccentrically Loaded Welded Connections," *Journal of the Structural Division,* ASCE, **98,** ST5 (May 1972), 989–1005.

5.30. G. L. Kulak and P. A. Timmler. "Tests on Eccentrically Loaded Fillet Welds," Structural Engineering Report No. 124, Department of Civil Engineering, University of Alberta, Edmonton, Alberta, December, 1984 (23 pp).

5.31. Peter Swannel. "Rational Design of Fillet Weld Groups," *Journal of the Structural Division.* ASCE, **107,** ST5 (May 1981), 789–802. Disc. **108,** ST5 (May 1982), 1197–1198.

5.32. Peter Swannel. "Weld Group Behavior," *Journal of the Structural Division,* ASCE, **107,** ST5 (May 1981), 803–815.

5.33. Vernon V. Neis. "New Constitutive Law for Equal Leg Fillet Welds," *Journal of Structural Engineering,* **111,** 8, August 1985, 1747–1759.

5.34. Raymond H. R. Tide. "Eccentrically Loaded Weld Groups—AISC Design Tables," *Engineering Journal,* AISC, **17,** 4 (4th Quarter 1980), 90–95.

5.35. Raymond H. R. Tide. Disc. of "Rational Design of Fillet Weld Groups." *Journal of the Structural Division,* ASCE, **108,** ST5 (May 1982), 1197–1198.

5.36. G. Donald Brandt. "A General Solution for Eccentric Loads on Weld Groups." *Engineering Journal,* AISC, **19,** 3 (3rd Quarter 1982), 150–159.

5.37. G. S. Miazga and D. J. L. Kennedy. "Behavior of Fillet Welds as a Function of the Angle of Loading," *Canadian Journal of Civil Engineering,* **16,** 1989, 583–599.

5.38. D. F. Lesik and D. J. L. Kennedy. "Ultimate Strength of Fillet Welded Connections Loaded in Plane," *Canadian Journal of Civil Engineering,* **17,** 1990, 55–67.

5.39. D. J. L. Kennedy, G. S. Miazga, and D. F. Lesik. "Discussion of Fillet Weld Shear Strength," *Welding Journal,* **55,** May 1990, 44–46.

5.40. Kenneth M. Loomis, William A. Thornton, and Thomas Kane. "A Design Aid for Connection Angle Welds Subjected to Combined Shear and Axial Loads," *Engineering Journal,* AISC, **22,** 4 (4th Quarter 1985), 178–196.

5.41. John L. Dawe and Geoffrey L. Kulak. "Welded Connections under Combined Shear and Moment," *Journal of the Structural Division,* ASCE, **100,** ST4 (April 1974), 727–741.

5.42. AWS. *Symbols for Welding, Brazing and Nondestructive Examination* (A2.4–93). Miami, FL: American Welding Society, 1993.

5.43. Murray E. Temple and Sherief S. S. Sakla. "Balanced and Unbalanced Welds for Angle Compression Members," *Canadian Journal of Civil Engineering,* **21,** 1994, 396–403.

PROBLEMS

All problems are to be done according to the AISC Load and Resistance Factor Design or Allowable Stress Design, as indicated by the instructor. All given loads are service loads unless otherwise indicated. Whenever possible, show all answers on a design sketch (drawn to scale) using appropriate welding symbols.

5.1. Specifically identify the AWS D1.1 Joint Designation for each of the following "prequalified" butt joints made by the submerged arc process (SAW). Specify the proper thickness for each of the plates, as well as the groove angle, root opening, and other requirements for the welds. Draw the cross-section for each weld. The given loads are 85% live load and 15% dead load. Refer to AWS D1.1 [2.24], *ASD Manual* [1.7] section "WELDED JOINTS," or *LRFD Manual* [1.18] Table 8-36 "Prequalified Welded Joints". The given joints are (a) Square-groove weld, complete penetration; (b) Single-V-groove weld, complete penetration; (c) Single-V-groove weld, complete penetration; and (d) Double-V-groove weld, complete penetration.

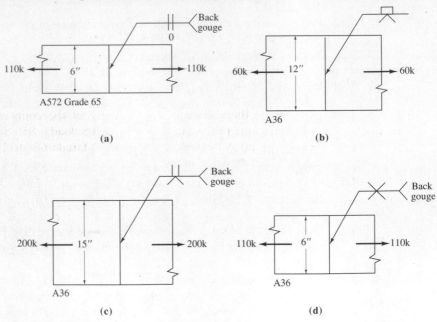

Problem 5.1

5.2. Specifically identify the AWS D1.1 Joint Designation for each of the following "prequalified" butt joints made by the shielded metal arc process (SMAW). Specify the proper thickness for each of the plates, or determine the service load capacity assuming the load is 85% live load and 15% dead load. Indicate any thickness-related limitations and draw the cross-section for each weld. Refer to AWS D1.1 [2.24], *ASD Manual* [1.7] section "WELDED JOINTS," or *LRFD Manual* [1.18] Table 8-36 "Prequalified Welded Joints." The given joints are (a) Square-groove weld, complete penetration; (b) Double-bevel-groove weld, complete penetration; (c) Double-bevel-groove weld, partial penetration; and (d) Single-V-groove weld, partial penetration.

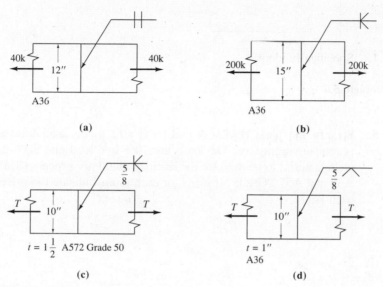

(a)   (b)

(c)   (d)

Problem 5.2

5.3. Determine the service load capacity T of the connection shown when the submerged arc process (SAW) is used. The load is 85% live load and 15% dead load. Use (a) A36 steel, and (b) A572 Grade 65 steel. Assume appropriate electrode material is used.

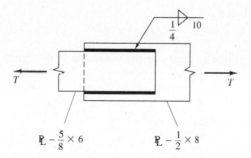

Problem 5.3

5.4. Specify the fillet welds required to develop the strength of the connection shown. State the maximum service load T permitted to be carried. The load is 90% live load and 10% dead load. Specify the proper flux-electrode material using the submerged arc process.

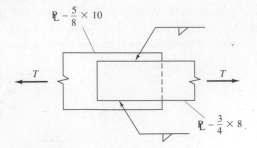

$\mathbb{R} - \dfrac{5}{8} \times 10$

T T

$\mathbb{R} - \dfrac{3}{4} \times 8$

Problem 5.4

5.5. Specify the plate thickness and weld size to be used for the joints in the accompanying figure. The loads are 70% live load and 30% dead load. State weld material to be used for the shielded metal arc process (SMAW). Compare A36 and A572 Grade 50 steels for each joint. Indicate the preferred design for each joint.

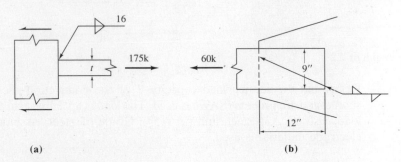

16

t 175k   60k   9″

12″

(a)   (b)

Problem 5.5

5.6. For the joints in the accompanying figure, satisfy the requirements stated in Prob. 5.5.

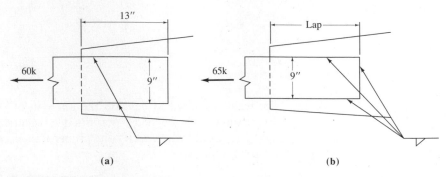

13″   Lap

60k   9″   65k   9″

(a)   (b)

Problem 5.6

5.7. Design the reinforced lap joint shown in the accompanying figure. The plates are 7 in. wide of A36 steel and the SMAW process is used. Refer to AWS Joint Designation BTC-P4 (*LRFD Manual* p. 8–150). The given load is 25% dead load and 75% live load.

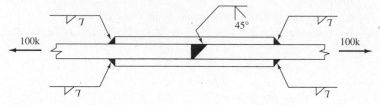

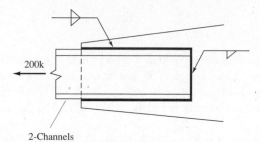

Problem 5.7

5.8. Select a pair of channels and design the fillet welds using the SMAW process. The loading is 85% live load and 15% dead load. Compare for (a) A36 steel and (b) A572 Grade 60 steel.

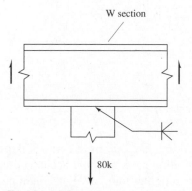

2-Channels

Problem 5.8

5.9. Design the tension plate attached to the wide-flange (W) section as well as the welds, assuming the SMAW process is used. The load is 70% live load and 30% dead load.
   a. Use A572 Grade 42 steel.
   b. Use A572 Grade 65 steel.
   c. Use A572 Grade 42 steel, with fillet welds instead of groove weld.
   d. Use A572 Grade 65 steel, with fillet welds instead of groove weld.

W section

80k

Problem 5.9

5.10. A $5 \times 3\frac{1}{2} \times \frac{3}{8}$ angle of A572 Grade 50 steel is connected by its long leg to a $\frac{5}{16}$-in. gusset plate. Develop the maximum service load capacity (25% dead load; 75% live load) of the angle and use a balanced fillet welded connection with the SMAW process. State the service load capacity. Use the following arrangements:

  a. $\frac{5}{16}$-in. weld on toe and back, with none on end.

  b. $\frac{1}{4}$-in. weld on toe and $\frac{3}{8}$-in. weld on back, and none on end.

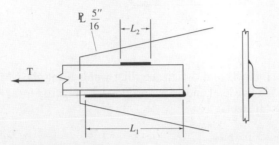

Problem 5.10

5.11. Design a balanced connection for two $7 \times 4 \times \frac{1}{2}$ angles connected by their long legs to a $\frac{3}{8}$-in. gusset plate. Develop the maximum service load capacity (20% dead load; 80% live load) and state its value. Use A572 Grade 60 steel and the SMAW process. Detail the joint to balance the loads using the shortest possible overlap.

5.12. Design the welds indicated to develop the full strength of the angles and minimize eccentricity. Assume service load is 20% dead load and 80% live load. Use the SAW welding process.

  a. Use A36 steel.

  b. Use A572 Grade 50 steel.

  c. Use A572 Grade 65 steel.

  d. Use A36 steel, but omit weld on the end of angles.

  e. Use A572 Grade 65 steel, but omit weld on the ends of angles.

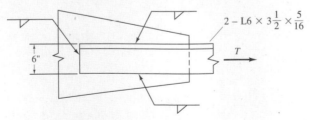

Problem 5.12

5.13. Assume a 9-in.-wide plate used in a lap joint must carry 30 kips dead load and 115 kips live load, and a possibility exists of some accidental eccentricity that cannot be computed. To insure a tighter joint, a $1\frac{1}{4}$-in.-diam plug weld is to be used. Determine the thickness of the plate, the amount of lap, and the weld size for the best joint. Assume the gusset plate to which the 9-in. plate is welded does not control any of the design. Use A572 Grade 50 steel and the SMAW process.

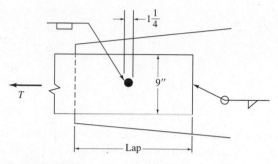

Problem 5.13

5.14. Determine the minimum length of slot in order to develop the full strength of a C12×20.7 welded to a $\frac{3}{8}$-in plate. Use the same size fillet weld over the entire length, and assume it is to be placed by the SMAW process. Assume service load is 35% dead load and 65% live load.

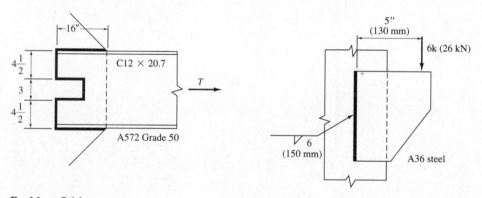

Problem 5.14   **Problem 5.15**

5.15. For the connection shown in the accompanying figure, what is the maximum required design strength ϕR_{nw}? The load is 90% live load and 10% dead load. What weld size is indicated if E70 electrodes are used with the SMAW process?

a. Use strength analysis (i.e., locate the instantaneous center by trial).

b. Use *ASD Manual* tables, "Eccentric Loads on Weld Groups," or *LRFD Manual* Table 8–38, "Coefficients C for Eccentrically Loaded Weld Groups."

c. Use elastic (vector) method.

5.16. For the connection in the accompanying figure, satisfy the requirements of Prob. 5.15.

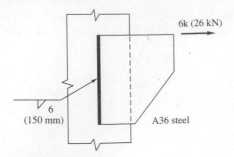

Problem 5.16

5.17. For the bracket shown in the accompanying figure, calculate the service load capacity *P* (90% live load and 10% dead load) based on the weld. Neglect the returns at ends and assume the SMAW process is to be used.

    a. Use strength analysis (i.e., locate the instantaneous center by trial).

    b. Use *ASD Manual* table, "Eccentric Loads on Weld Groups," or *LRFD Manual* Table 8–38.

    c. Use elastic (vector) method.

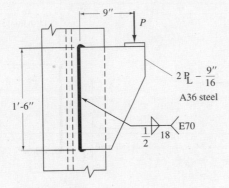

Problem 5.17

5.18. Compute the maximum acceptable service load *P* for the connection of the accompanying figure. The load is 85% live load and 15% dead load. *Ignore the effect of returns at the lower end of the connection.*

    a. Use strength analysis (i.e., locate the instantaneous center by trial).

    b. Use *ASD Manual* table, "Eccentric Loads on Weld Groups," or *LRFD Manual* Table 8–42.

    c. Use elastic (vector) method.

5.19. Repeat Prob. 5.18, except use $\frac{3}{8}$-in. fillet weld on the side and $\frac{1}{4}$-in. on the end. The steel is A572 Grade 50.

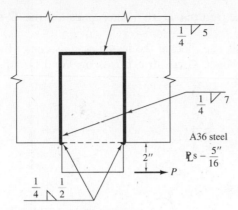

Problems 5.18 and 5.19

5.20. Compute the service load capacity P for the welded bracket of the accompanying figure. The load is 70% live load and 30% dead load. Neglect the returns at the outer ends of the C-shaped weld configuration. The weld size is $\frac{3}{8}$ in. and E70 electrodes are used in the shielded metal arc process.

    a. Use strength analysis (i.e., locate the instantaneous center by trial).

    b. Use *ASD Manual* tables, "Eccentric Loads on Weld Groups," or *LRFD Manual* Table 8–42.

    c. Use elastic (vector) method.

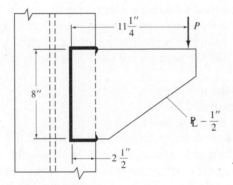

Problem 5.20

5.21. Repeat Prob. 5.20 if the vertical dimension of the weld configuration is 12 in. instead of 8 in.

5.22. Repeat Prob. 5.20 if the vertical dimension of the weld configuration is 4 in. instead of 8 in.

5.23. Compute the theoretical weld size required for the bracket of the accompanying figure when the SMAW process is used. The load is 60% live load and 40%

dead load. Neglect the returns at the outer ends of the C-shaped weld configuration.

a. Use strength analysis (i.e., locate the instantaneous center by trial).

b. Use *ASD Manual* tables, "Eccentric Loads on Weld Groups," or *LRFD Manual* Table 8–38.

c. Use elastic (vector) method.

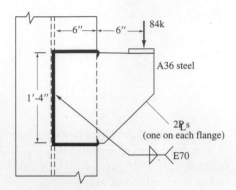

Problem 5.23

5.24. Repeat Prob. 5.23 if the vertical dimension is 12 in. instead of 16 in.

5.25. Repeat Prob. 5.23 if the vertical dimension is 8 in. instead of 16 in.

5.26. Repeat Prob. 5.23 if the horizontal dimension is 4 in. instead of 16 in.

5.27. Use the elastic (vector) method to derive a general expression for the required weld size on the seat angle of the accompanying figure in terms of the factored load P_u, the leg length L, and the eccentricity e of the applied load. Assume E70 electrodes with the SMAW process are used. Use the following assumptions:

a. Ignore the returns at the top;

b. Use an average return of $L/12$; and

c. Use a return equal to twice the weld size.

    If $e = 2\frac{3}{4}$ in. and $L = 6$ in., determine the weld size needed to carry $P = 38$ kips (80% live load and 20% dead load). For the weld size selected, check the capacity P using all three assumptions, and also compare with the result using the *ASD Manual*, "Seated Beam Connections," or *LRFD Manual* Table 9–7, "All-Welded Unstiffened Seated Connections."

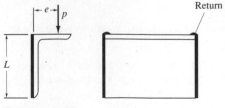

Problem 5.27

5.28. For the bracket of the accompanying figure, determine the length L required when using $\frac{5}{16}$-in. fillet weld with the SMAW process. The load is 70% live load and 30% dead load. Verify your result using the following procedures:

    a. Strength analysis (i.e., locate the instantaneous center by trial).

    b. *ASD Manual* tables, "Eccentric Loads on Weld Groups," or *LRFD Manual* Table 8–38.

    c. Elastic (vector) method.

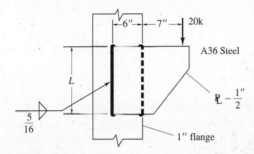

Problem 5.28

5.29. Repeat Prob. 5.28 if the service load is 40 kips.

5.30. For the bracket of the accompanying figure, satisfy the requirements of Prob. 5.28. Note that A572 Grade 50 steel is used instead of A36.

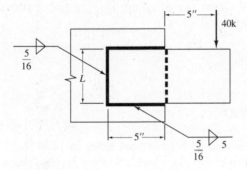

Problem 5.30

Chapter 6

Compression Members

Part I: Columns

6.1 GENERAL

In this chapter, members subjected to axial compression forces are to be treated. Referred to by various terms, such as *column, stanchion, post,* and *strut,* these members are rarely if ever actually carrying only axial compression. However, whenever the loading is so arranged that either the end rotational restraint is negligible or the loading is symmetrically applied from members framing in at the column ends, and bending may be considered negligible compared to the direct compression, the member can safely be designed as a concentrically loaded column.

It is well known from basic mechanics of materials that only very short columns can be loaded to their yield stress; the usual situation is that buckling, or sudden bending as a result of instability, occurs prior to developing the full material strength of the member. Thus, a sound knowledge of compression member stability is necessary for those designing in structural steel.

6.2 EULER ELASTIC BUCKLING
AND HISTORICAL BACKGROUND

Column buckling theory originated with Leonhard Euler in 1744 [6.1]. An initially straight concentrically loaded member, in which all fibers remain elastic until buckling occurs, is slightly bent as shown in Fig. 6.2.1. Although Euler dealt with a member built-in at one end and simply supported at the other, the same logic is applied here to the pin-end column, which having zero end rotational restraint is the member having least buckling strength.

At any location z, the bending moment M_z on the member bent slightly about the x principal axis is

$$M_z = Py \tag{6.2.1}$$

and since

$$\frac{d^2y}{dz^2} = -\frac{M_z}{EI} \tag{6.2.2}$$

Steel framework, showing particularly the exterior columns, Southeast Recreation Facility, University of Wisconsin-Madison Campus. (Photo by C. G. Salmon)

the differential equation becomes

$$\frac{d^2y}{dz^2} + \frac{P}{EI}y = 0 \tag{6.2.3}$$

After letting $k^2 = P/EI$, the solution of this second-order linear differential equation may be expressed

$$y = A\sin kz + B\cos kz \tag{6.2.4}$$

Applying boundary conditions, (a) $y = 0$ at $z = 0$; and (b) $y = 0$ at $z = L$, one obtains for condition (a), $B = 0$; and for condition (b),

$$0 = A\sin kL \tag{6.2.5}$$

Satisfaction of Eq. 6.2.5 may be accomplished in three possible ways; (a) constant $A = 0$, i.e., no deflection; (b) $kL = 0$, i.e., no applied load; and (c) $kL = N\pi$, the

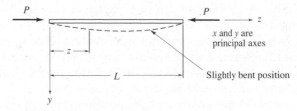

Figure 6.2.1 Euler column.

requirement for buckling to occur. Thus

$$\left(\frac{N\pi}{L}\right)^2 = \frac{P}{EI}$$

$$P = \frac{N^2\pi^2 EI}{L^2} \tag{6.2.6}$$

The fundamental buckling mode, a single-curvature deflection ($y = A \sin \pi z/L$ from Eq. 6.2.4.), will occur when $N = 1$; thus the Euler critical load for a column with both ends pinned is

$$P_{cr} = \frac{\pi^2 EI}{L^2} \tag{6.2.7}$$

or in terms of average compressive stress, using $I = A_g r^2$

$$F_{cr} = \frac{P_{cr}}{A_g} = \frac{\pi^2 E}{(L/r)^2} \tag{6.2.8}$$

Euler's approach was generally ignored for design because test results did not agree with it; columns of ordinary length used in design were not as strong as Eq. 6.2.7 would indicate.

Considère and Engesser [6.2, 6.3] in 1889 independently realized that portions of usual length columns become inelastic prior to buckling and that a value of E should be used that could account for some of the compressed fibers being strained beyond the proportional limit. It was thus consciously recognized that in fact *ordinary length* columns fail by inelastic buckling rather than by elastic buckling.

Complete understanding of the behavior of concentrically loaded columns, however, was not achieved until 1946 when Shanley [6.4, 6.5] offered the explanation that now seems obvious. He reasoned that it was actually possible for a column to bend and still have increasing axial compression, but that it *begins* to bend upon reaching what is commonly referred to as the *buckling load,* which includes inelastic effects on some or all fibers of the cross-section. These inelastic effects are discussed in detail in Sec. 6.4

An extensive historical review of the development of column theory is given by B. G. Johnston [6.6].

6.3 BASIC COLUMN STRENGTH

To determine a basic column strength, certain conditions may be assumed for the ideal column [6.7]. With regard to material, it may be assumed (1) there are the same compressive stress-strain properties throughout the section; (2) no initial internal stresses exist such as those due to cooling after rolling and those due to welding. Regarding shape and end conditions, it may be assumed (3) the column is perfectly straight and prismatic; (4) the load resultant acts through the centroidal axis of the member until the member begins to bend; (5) the end conditions must be determinate so that a definite equivalent pinned length may be established. Further assumptions regarding buckling may be made, as (6) the small deflection theory of ordinary bending is applicable and shear may be neglected; and (7) twisting or distortion of the cross-section does not occur during bending.

Once the foregoing assumptions have been made, it is now agreed [6.8] that the strength of a column may be expressed by

$$P_{cr} = \frac{\pi^2 E_t}{(KL/r)^2} A_g = F_{cr} A_g \tag{6.3.1}$$

where
E_t = tangent modulus of elasticity at stress P_{cr}/A_g
A_g = gross cross-sectional area of member
KL/r = effective (or equivalent pinned-end) slenderness ratio
K = effective length factor (treated in Sec. 6.9)
L = length of member
r = radius of gyration = $\sqrt{I/A_g}$
I = moment of inertia

It is well known that long compression members fail by elastic buckling and that short stubby compression members may be loaded until the material yields or perhaps even into the strain-hardening range. However, in the vast majority of usual situations, failure occurs by buckling after a portion of the cross-section has yielded. This is known as *inelastic buckling*.

Actually, buckling under axial load occurs only when the aforementioned assumptions (1) through (7) apply. Columns are usually an integral part of a structure and as such *cannot* behave entirely independently. The practical use of the term *buckling* is that it is the boundary between stable and unstable deflections of a compression member, rather than the instantaneous condition that occurs in the isolated slender elastic rod.

As previously mentioned, for many years theoretical determinations of column strength did not agree with test results. Test results included effects of initial crookedness of the member, accidental eccentricity of load, end restraint, local or lateral buckling, and residual stress. A typical curve of observed strengths was as shown in Fig. 6.3.1. Design formulas, therefore, were based on such empirical results. Various

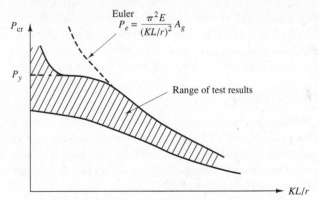

Figure 6.3.1 Typical range of column strength vs slenderness ratio.

straight-line and parabolic formulas have been used, as well as other more complex expressions, in order to fit the curve of test results in a reasonably accurate, yet practical manner.

In summary, Euler elastic buckling governs the strength for large slenderness ratios, yield strength $P_y = F_y A_g$ controls for short columns, and a transition curve must be used for inelastic buckling.

6.4* INELASTIC BUCKLING

Since ordinary length columns buckle when some of their fibers are inelastic, having a modulus of elasticity less than their initial elastic value, the logic of Engesser, Considère, and Shanley is explained in this section, generally following Bleich [6.9, pp. 8–20].

Basic Tangent Modulus Theory

Euler's theory pertained only to situations where compressive stress below the elastic limit acts uniformly over the cross-section when unstable equilibrium occurs. Engesser [6.3] and Considère [6.2] were the first to utilize the possibility of a variable modulus of elasticity. In Engesser's tangent modulus theory the column remains straight up to the instant of failure and the modulus of elasticity at failure is the tangent to the stress-strain curve. The relationships are shown in Fig. 6.4.1. The theory prescribed that at a certain stress, $F_{cr} = P_{cr}/A_g$, the member could acquire an unstable deflected shape and that the deformation at F_{cr} is governed by $E_t = df/d\epsilon$. Thus Engesser modified Euler's equation to become

$$F_{cr} = \frac{P_t}{A_g} = \frac{\pi^2 E_t}{(KL/r)^2} \qquad (6.4.1)$$

where P_t is the tangent modulus load, and E_t is the tangent modulus of elasticity at stress F_{cr}.

This theory, however, still did not agree with test results, giving computed loads lower than measured ultimate capacities. The principal assumption that caused this tangent modulus theory to be considered erroneous is that as the member changes from a straight to bent form, no strain reversal takes place. In 1895 Engesser changed his theory, reasoning that during bending some fibers undergo increased strain (lowered tangent modulus) and some fibers are unloaded (higher modulus at the reduced strain); therefore a combined value should be used for the modulus.

Double Modulus Theory

To examine the process of column bending at stresses beyond the elastic limit, consider the section of Fig. 6.4.2 from which Engesser's double modulus, or "reduced" modulus, is developed. This concept had logic to it which was generally accepted but gave computed strengths higher than test values. Not until Shanley's explanation was the inconsistency resolved.

---

*Sections so marked may be omitted without loss of continuity.

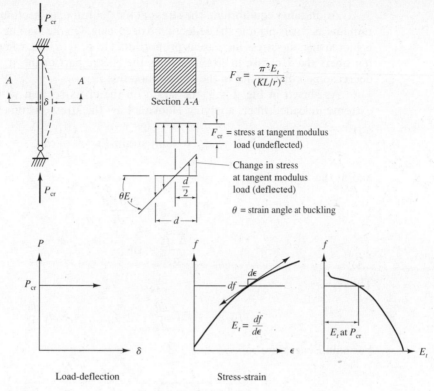

Figure 6.4.1   Engesser original tangent modulus theory, 1889.

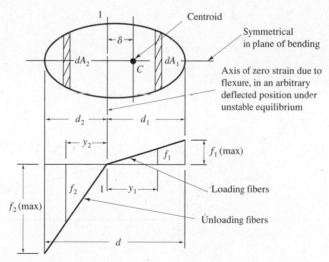

Figure 6.4.2   Stress distribution in condition of unstable equilibrium (double modulus theory).

At unstable equilibrium, the stress at the neutral axis (section 1–1 of Fig. 6.4.2) remains as it was prior to the deflection δ occurring. On the loading fibers where strain is increasing, the stress increase is proportional to $E_t = df/d\epsilon$, whereas on the unloading fibers the decrease in strain relieves the elastic part of the strain; thus the stress decrease is proportional to the elastic modulus E.

As shown in Fig. 6.4.3, the strain on the cross-section will be linear. At the extreme unloaded fiber, applying Hooke's Law the stress becomes

$$f_{2(\text{max})} = (\text{unit strain})E = \frac{\Delta\, dz}{dz}E \tag{6.4.2}$$

and at the loaded fiber,

$$f_{1(\text{max})} = \frac{\Delta\, dz\, d_1}{d_2}\frac{E_t}{dz} \tag{6.4.3}$$

$$\frac{\Delta\, dz}{d_2} = d\theta \tag{6.4.4}$$

Thus

$$f_{2(\text{max})} = Ed_2\frac{d\theta}{dz}; \qquad f_{1(\text{max})} = E_t d_1\frac{d\theta}{dz} \tag{6.4.5}$$

For small curvature,

$$\frac{1}{\text{radius of curvature}} = \frac{M}{E_r I} = \frac{d\theta}{dz} = \frac{d^2 y}{dz^2} \tag{6.4.6}$$

where E_r = Engesser's double modulus.

The internal resisting moment for the stress condition of Fig. 6.4.2 gives

$$M = -Py = \int_0^{d_1} f_1(y_1 - \delta)\, dA_1 + \int_0^{d_2} f_2(y_2 + \delta)\, dA_2 \tag{6.4.7}$$

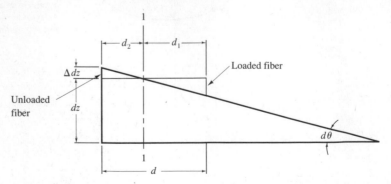

Figure 6.4.3 An element dz along the axis of the column in the unstable equilibrium position.

and from linear stress distribution and Eq. 6.4.5,

$$f_1 = f_{1(\max)}\frac{y_1}{d_1} = E_t d_1 \frac{d^2 y}{dz^2}\frac{y_1}{d_1}$$

$$f_2 = f_{2(\max)}\frac{y_2}{d_2} = E d_2 \frac{d^2 y}{dz^2}\frac{y_2}{d_2}$$

(6.4.8)

Thus Eq. 6.4.7 becomes

$$-Py = E_t\frac{d^2 y}{dz^2}\int_0^{d_1} y_1(y_1 - \delta)\,dA_1 + E\frac{d^2 y}{dz^2}\int_0^{d_2} y_2(y_2 + \delta)\,dA_2 \qquad (6.4.9)$$

Force equilibrium requires

$$\int_0^{d_1} f_1\,dA_1 = \int_0^{d_2} f_2\,dA_2 \qquad (6.4.10)$$

which, using Eq. 6.4.8, gives

$$E_t\frac{d^2 y}{dz^2}\int_0^{d_1} y_1\,dA_1 = E\frac{d^2 y}{dz^2}\int_0^{d_2} y_2\,dA_2 \qquad (6.4.11)$$

Using Eq. 6.4.11, it is seen the terms involving δ cancel each other in Eq. 6.4.9, thus giving

$$-Py = E_t\frac{d^2 y}{dz^2}\int_0^{d_1} y_1^2\,dA_1 + E\frac{d^2 y}{dz^2}\int_0^{d_2} y_2^2\,dA_2$$

$$\therefore \frac{d^2 y}{dz^2}\left[E_t\int_0^{d_1} y_1^2\,dA_1 + E\int_0^{d_2} y_2^2\,dA_2\right] + Py = 0 \qquad (6.4.12)$$

Equation 6.4.12 is obviously of the same form as the elastic buckling equation, Eq. 6.2.3. Thus for the double modulus theory,

$$P_{\mathrm{cr}} = \frac{\pi^2}{L^2}\left[E_t\int_0^{d_1} y_1^2\,d\Lambda_1 + E\int_0^{d_2} y_2^2\,dA_2\right] \qquad (6.4.13)$$

Shanley Concept—True Column Behavior

To understand the actual behavior of a column as explained by Shanley [6.4] in 1946, consider the rectangular section of Fig. 6.4.4 subjected to axial compression. For loads below the tangent modulus load P_t, the ideal column remains perfectly straight with zero deflection (point A of Fig. 6.4.4a). The load P_t at point A may most correctly be defined as follows [6.10]: "The tangent modulus load is the smallest value of axial load at which bifurcation of the equilibrium positions can occur regardless of whether or not the transition to the bent position requires an increase of axial load." Consider that at onset of bending (infinitesimal curvature) there will be an infinitesimal increase in axial strain and stress Δf_1. By the time the curvature becomes finite, i.e., the point N moves to N_1, some strain reversal must occur if the column cross-section is to develop a resisting moment to maintain equilibrium with the moment due to the external load $P\delta$. For small but finite values of curvature, the increment of load represented by

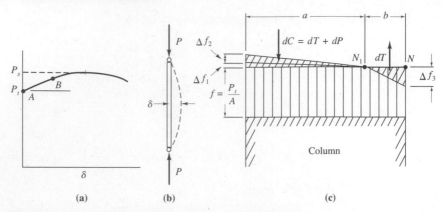

Figure 6.4.4 Shanley concept—true column behavior.

stress on the area of increasing strain exceeds the increment of load represented by stress on the area of decreasing strain; thus P is increased by an amount dP (point B of Fig. 6.4.4a). As each increment of curvature takes place P will further increase as long as $dC > dT$. The increased compressive force dC is computed using the tangent modulus E_t while in the region of strain reversal the elastic modulus E is used to compute dT. The double modulus theory, which similarly treated loading and unloading fibers, did not accept $dC > dT$, but rather only considered equilibrium positions near the perfectly straight one.

For practical purposes, the increase of capacity from P_t to P_s (Fig. 6.4.4a) can be neglected for design use. Therefore, the tangent modulus load may be treated as the critical load, i.e., the load at which bending begins.

6.5 RESIDUAL STRESS

Residual stresses are stresses that remain in a member after it has been formed into a finished product. Such stresses result from plastic deformations, which in structural steel may result from several sources: (1) uneven cooling which occurs after hot rolling of structural shapes; (2) cold bending or cambering during fabrication; (3) punching of holes and cutting operations during fabrication; and (4) welding. Under ordinary conditions those residual stresses resulting from uneven cooling and welding are the most important. Actually the important residual stresses due to welding are really the result of uneven cooling.

The mechanism of residual stress due to uneven cooling is treated in the *Welding Handbook* [2.23, Volume 1, Chap. 7] and the effect of residual stress on compression structural members appears in the *Guide to Stability Design Criteria for Metal Structures* [6.8, pp. 33–45], prepared by the Structural Stability Research Council. This latter publication will be extensively referred to as the *SSRC Guide*.

In wide-flange or H-shaped sections, after hot rolling, the flanges, being the thicker parts, cool more slowly than the web region. Furthermore, the flange tips having greater exposure to the air cool more rapidly than the region at the junction of

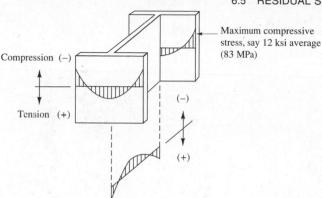

Figure 6.5.1 Typical residual stress pattern on rolled shapes.

flange to web. Consequently, compressive residual stress exists at flange tips and at mid-depth of the web (the regions that cool fastest), while tensile residual stress exists in the flange and the web at the regions where they join. Figure 6.5.1 shows typical residual stress distribution on rolled beams. Considerable variation can be expected as the true pattern will be a function of the dimensions of the section.

At this point one might wonder whether the general column strength equation (Eq. 6.3.1) discussed in the preceding section still is applicable. The theory is applicable, but all fibers in the cross-section cannot be considered as stressed to the same level under the action of the compressive service load. The tangent modulus E_t on one fiber is not the same as that on an adjacent fiber.

In a rolled steel shape the influence of residual stress on the stress-strain curve is shown in Fig. 6.5.2. using average stress on the gross area as the ordinate. It is noted that residual stress in an elastic-plastic material such as steel gives the same effect as that obtained for a material such as aluminum, which is not linearly elastic when it contains no residual stress. Thus, assuming the tangent modulus concept applies, column strength may be said to be based on inelastic buckling because the average stress-strain curve is nonlinear when maximum column strength is reached.

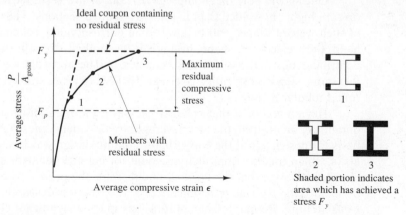

Figure 6.5.2 Influence of residual stress on average stress-strain curve.

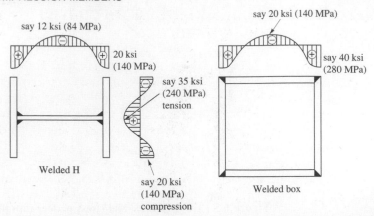

Figure 6.5.3 Typical residual stress distribution in welded shapes.

Whereas it was once believed the nonlinear portion of the average stress-strain curve for axially loaded compression members was due entirely to initial curvature and accidental eccentricity, Huber and Beedle [6.11] have verified that residual stress is the primary cause, and the other factors have a relatively minor effect. Residual stresses at flange tips of rolled shapes have been measured as high as 20 ksi (138 MPa), a high percentage of the minimum specified yield stress for steels such as A36. Residual stresses are essentially independent of yield stress, depending instead on cross-sectional dimensions and configuration since those factors govern cooling rates [6.12].

Welding of built-up shapes is an even greater contributor to residual stress than cooling of hot-rolled H-shapes [6.13]. The plates themselves generally have little residual stress initially because of relatively uniform cooling after rolling. However, after the heat is applied to make the welds, the subsequent nonuniform cooling and restraint against distortion cause high residual stresses. Figure 6.5.3 shows typical residual stress patterns on welded H and box built-up shapes.

One should note that compressive residual stresses typically occurring at flange tips are higher in welded than in rolled H-shaped sections. Thus the column strength of such welded shapes will be lower than corresponding rolled shapes. On the other hand, the welded-box shape, having tensile residual stress in the corner regions that contribute most to the stiffness as a column, will be stronger than a rolled shape having the same slenderness ratio. Sherman [6.14.6.15] has studied residual stresses on rolled tubular members.

Having accepted that residual stresses exist, such information must be used to obtain a column strength curve (average stress vs slenderness ratio) that can form the basis for design. Until the early 1950s, column design was based on many formulas, all of which tried to empirically account for column behavior exhibited by tests. By clearly indicating that the tangent modulus was the proper criterion for strength and by identifying the role of residual stress, the Column Research Council (now Structural Stability Research Council [6.8] has made a significant contribution.

6.6* DEVELOPMENT OF COLUMN STRENGTH CURVES INCLUDING RESIDUAL STRESS

The following analytical approach, patterned after Huber and Beedle [6.11] and Beedle and Tall [6.16], is intended to show the logic to obtain a column strength equation. Column strength can be obtained by two general methods. One method is to use the residual stress distribution, either the actual variation from measurements or a mathematical model, along with the stress-strain relationship for the material (say a small test specimen of the steel).

The other method is to determine experimentally an average stress-strain relationship by testing short lengths of rolled shapes containing residual stress. Column strength can then be determined from the test results using the tangent modulus of the experimental curve in combination with the appropriate slenderness ratio. Knowledge of the residual stress pattern is not used in this second approach.

Yu and Tall [6.17] have discussed these approaches in detail. Johnston [6.18] and Batterman and Johnston [6.19] have treated the tangent modulus application to in-elastic buckling of columns.

The following development is made with the objective of obtaining a relationship between average externally applied stress and the slenderness ratio. Thus, the capacity of a member can be obtained by a simple multiplication of safe stress times gross area, without regard to what the actual stress is at any point in the cross-section or what is the true residual stress pattern.

As a starting point consider steel which as a material is perfectly elastic until a certain strain ϵ_y is achieved and then is plastic (i.e., constant stress with increasing strain). A coupon cut from the web of a rolled shape exhibits such behavior, as shown by the dotted lines in Fig. 6.6.1. The solid lines indicate the behavior of an H-shaped rolled section including residual stress.

To account for the effects of early yielding because of residual stress, consider one fiber at a distance x from the axis of zero strain caused by bending (Fig. 6.6.2). The bending is taken as an infinitesimal amount consistent with equilibrium at the

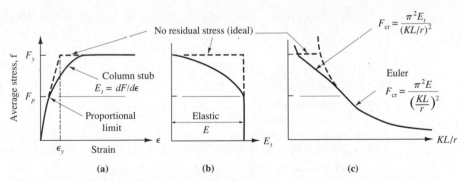

Figure 6.6.1 Comparison of coupon with H-shaped rolled section containing residual stress.

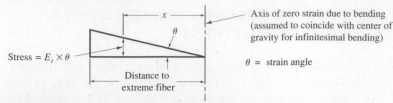

Figure 6.6.2   Stress on fiber at distance x from axis of zero strain due to bending.

tangent modulus load. The bending moment contribution from stress on the one fiber is

$$dM = \text{(stress)(area)(moment arm)} = (\theta E_t x)(dA)(x) \qquad (6.6.1)$$

which for the entire cross-section becomes

$$M = \int_A \theta E_t x^2 \, dA = \theta \int_A E_t x^2 \, dA \qquad (6.6.2)$$

From elementary bending theory, the radius of curvature is

$$R = \frac{1}{\theta}$$

$$\theta = \frac{1}{R} = \frac{M}{\text{equivalent } EI} = \frac{M}{E'I} \qquad (6.6.3)$$

Thus

$$E'I = \frac{M}{\theta} = \int_A E_t x^2 \, dA$$

$$E' = \frac{1}{I} \int_A E_t x^2 \, dA \qquad (6.6.4)$$

which may be called the *effective modulus* and used in Fg. 6.4.1 as an equivalent value for E_t.

If the idealized elastic-plastic f-ϵ curve of Fig. 6.6.1a (dotted) is used (for $f < F_y$, $E_t = E$ and for $f = F_y$, $E_t = 0$) the bending stiffness of yielded parts becomes zero; however, the buckling strength will be the same as a column whose moment of inertia I_e is the moment of inertia of the portion remaining elastic. Equation 6.6.4 then becomes

$$E' = \frac{E}{I} \int_{A \text{ (elastic part only)}} x^2 \, dA = \qquad E \frac{I_e}{I} \qquad (6.6.5)$$

The stress at which the column may begin to bend, from Eq. 6.3.1, is

$$P_{cr} = \left(\frac{\pi^2 E}{(KL/r)^2} \frac{\int_A x^2 \, dA}{I} \right) A_g = F_{cr} A_g \qquad (6.6.6)$$

$$P_{cr} = \left(\frac{\pi^2 E (I_e/I)}{(KL/r)^2} \right) A_g = F_{cr} A_g \qquad (6.6.7)$$

In order for Eq. 6.6.7 to be useful, the relationship between F_{cr} and I_e must be established.

Case A. Buckling about Weak Axis

A reasonable assumption will be that the flanges become fully plastic before the web yields (see Fig. 6.6.3).

Let k = proportion of the flange remaining elastic = $2x_0/b = A_e/A_f$. Then Eq. 6.6.5 becomes

$$E\frac{I_e}{I} = E\frac{t_f(2x_0)^3}{12}\left(\frac{12}{t_f b^3}\right) = Ek^3 \tag{6.6.8}$$

if the web is neglected in computing I. Applying the tangent modulus definition,

$$E_t = \frac{\text{nominal incremental stress}}{\text{incremental elastic strain}} = \frac{dP/A}{\dfrac{dP/A_e}{E}} = \frac{A_e E}{A} \tag{6.6.9}$$

$$E_t A = A_e E = (A_w + 2kA_f)E \tag{6.6.10}$$

where A_w = web area
A_f = gross area of one flange
A = total gross area of section

Solving Eq. 6.6.10 for k and using Eq. 6.6.8 in Eq. 6.6.7 gives

$$k = \frac{E_t A}{2EA_f} - \frac{A_w}{2A_f} \tag{6.6.11}$$

$$F_{cr} = \frac{\pi^2 Ek^3}{(KL/r)^2} = \frac{\pi^2 E}{(KL/r)^2}\left[\frac{AE_t}{2A_f E} - \frac{A_w}{2A_f}\right]^3 \tag{6.6.12}$$

which includes the elastic web effect, for buckling with respect to the weak axis (y–y).

Case B. Buckling about Strong Axis

Again, assuming the web is elastic, but neglecting its contribution toward the moment of inertia gives approximately

$$E\frac{I_e}{I} \approx E\frac{2A_e(d/2)^2}{2A_f(d/2)^2} = Ek \tag{6.6.13}$$

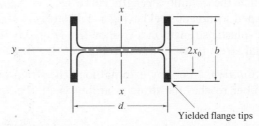

Figure 6.6.3   Portion of section that has yielded.

If the elastic web is included,

$$E\frac{I_e}{I} = E\left[\frac{2kA_f(d^2/4) + t_w d^3/12}{2A_f(d^2/4) + t_w d^3/12}\right] \qquad (6.6.14)$$

$$= E\left[\frac{2kA_f + A_w/3}{2A_f + A_w/3}\right] \qquad (6.6.15)$$

Using tangent modulus definition and Eq. 6.6.10,

$$2kA_f = \frac{E_t A}{E} - A_w$$

which, upon eliminating the $2kA_f$ term from Eq. 6.6.15, gives

$$E\frac{I_e}{I} = \left[\frac{E_t A/E - 2A_w/3}{2A_f + A_w/3}\right]E \qquad (6.6.16)$$

Thus

$$F_{cr} = \frac{\pi^2 Ek}{(KL/r)^2} \qquad (6.6.17)$$

is the approximate equation using Eq. 6.6.11 for k, or more exactly using Eq. 6.6.16 in Eq. 6.6.7 gives

$$F_{cr} = \frac{\pi^2 E}{(KL/r)^2}\left[\frac{E_t A/E - 2A_w/3}{2A_f + A_w/3}\right] \qquad (6.6.18)$$

for buckling with respect to the strong axis (x–x).

From the foregoing development it is apparent that two equations are necessary to properly determine column strength of I-shaped sections: one for strong-axis buckling and one for weak-axis buckling. Although the value I_e/I is not itself a function of the residual stress distribution provided that it satisfies the general geometric requirement as shown in Fig. 6.6.3; nevertheless, the critical stress F_{cr}, computed as the buckling load divided by the gross area, has a relationship with KL/r that does depend on residual stress.

Note that if the stress-strain curve for the material is not elastic-plastic, i.e., if E_t is neither E nor zero, then the more general equation, Eq. 6.6.4, must be used instead of Eq. 6.6.5.

EXAMPLE 6.6.1 _____

Establish the column strength curve (F_{cr} vs KL/r) for weak axis buckling of an H-shaped section of steel having a yield stress of 100 ksi (690 MPa) exhibiting perfect elastic-plastic strength in a coupon test (Fig. 6.6.4b), and having the very simplified residual stress pattern shown in Fig. 6.6.4a. Neglect the contribution of the web.

Solution.   For any external load the strain on every fiber is the same. Until a fiber reaches the strain ϵ_y at first yield, the applied load is

$$P = \int_A f\, dA = fA$$

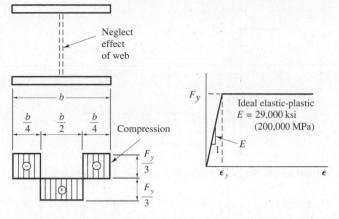

(a) Residual-stress pattern   (b) Coupon stress-strain diagram

Figure 6.6.4   Data for Example 6.6.1.

After a portion has become plastic, the applied load is

$$P = (A - A_e)F_y + \int_{A_e} f \, dA$$

In this problem for $F_{cr} = P/A \le 2F_y/3$ the entire section remains elastic, $E_t = E$, in which case E' is EI_e/I and $I_e = I$; thus,

$$F_{cr} = \frac{2F_y}{3} = \frac{\pi^2 E}{(KL/r)^2}$$

$$\frac{KL}{r} = \sqrt{\frac{\pi^2(29,000)}{(2/3)(100)}} = 65.4 \quad \text{(point 1, Fig. 6.6.5)}$$

When $F_{cr} = P/A > 2F_y/3$, the flange tips have yielded, making I_e less than I; thus

$$\frac{I_e}{I} = \frac{(b/2)^3}{b^3} = \frac{1}{8}$$

$$F_{cr} = \frac{2F_y}{3} = \frac{\pi^2 E(I_e/I)}{(KL/r)^2} = \frac{\pi^2 E}{8(KL/r)^2}$$

$$\frac{KL}{r} = 23.2 \quad \text{(point 2, Fig. 6.6.5)}$$

for average stress infinitesimally greater than $2F_y/3$. When $F_{cr} = P/A = F_y$,

$$F_{cr} = F_y = \frac{\pi^2 E}{8(KL/r)^2}$$

$$\frac{KL}{r} = 18.9 \quad \text{(point 3, Fig. 6.6.5)}$$

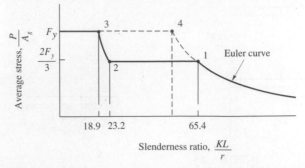

Figure 6.6.5 Column strength curve for Example 6.6.1.

when the total load $P = F_y A$. The results are shown in Fig. 6.6.5. If there had been no residual stress at $F_{cr} = F_y$,

$$\frac{KL}{r} = 53.5 \quad \text{(point 4, Fig. 6.6.5)} \quad \blacksquare\blacksquare$$

EXAMPLE 6.6.2 _____

Establish the column strength curve for the more realistic linear distribution of residual stress shown in Fig. 6.6.6. Consider weak-axis buckling of an H-shaped section of steel for both (a) $F_y = 36$ ksi (250 MPa) and (b) $F_y = 100$ ksi (690 MPa). Neglect the effect of the web,

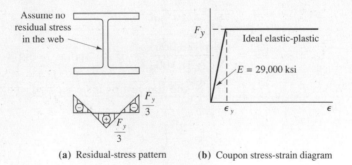

(a) Residual-stress pattern   (b) Coupon stress-strain diagram

Figure 6.6.6 Data for Example 6.6.2.

Solution. For an average superimposed stress $f = P/A \leq 2F_y/3$, the entire section remains elastic (Fig. 6.6.7a); therefore $E_t = E$, and

$$F_{cr} = \frac{2F_y}{3} = \frac{\pi^2 E}{(KL/r)^2}$$

For an average stress due to applied load greater than $2F_y/3$, part of the cross-section is plastic and part elastic, as in Fig. 6.6.7b. During this stage, the *change in stress is not the same on all fibers,* because the modulus of elasticity is not the

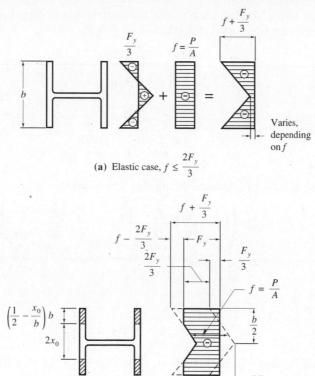

(a) Elastic case, $f \leq \dfrac{2F_y}{3}$

(b) Elasto-plastic case, $f > \dfrac{2F_y}{3}$

f = superimposed stress on elastic fibers

Figure 6.6.7 Stress distribution with linear residual stress.

same on all fibers.

$$F_{cr} = \frac{\pi^2 E I_e / I}{(KL/r)^2}$$

$$\frac{I_e}{I} = \frac{2(1/12)(2x_0)^3 t}{2(1/12)b^3 t} = \frac{8x_0^3}{b^3}$$

neglecting the effect of the web,

$$F_{cr} = \frac{8\pi^2 E (x_0/b)^3}{(KL/r)^2} \tag{a}$$

which gives F_{cr} as a function of two variables, x_0/b and KL/r. An additional relationship is required. The total load during the elasto-plastic stage can be expressed

$$P_{cr} = 2\left[fbt - 2\left(\frac{1}{2}\right)\left(f - \frac{2F_y}{3}\right)\left(\frac{1}{2} - \frac{x_0}{b}\right)bt \right] \tag{b}$$

which is the shaded area of the stress diagram in Fig. 6.6.7b. From similar triangles on the dotted triangle of Fig. 6.6.7b,

$$\frac{f - 2F_y/3}{\left(\dfrac{1}{2} - \dfrac{x_0}{b}\right)b} = \frac{2F_y/3}{b/2}$$

Solving for f,

$$f = \left[1 - \frac{x_0}{b}\right]\frac{4F_y}{3} \tag{c}$$

Using Eq. (c) to eliminate f from Eq. (b) gives

$$P_{cr} = 2bt\left\{\left(1 - \frac{x_0}{b}\right)\frac{4F_y}{3} - \left[\left(1 - \frac{x_0}{b}\right)\frac{4F_y}{3} - \frac{2F_y}{3}\right]\left(\frac{1}{2} - \frac{x_0}{b}\right)\right\}$$

$$= A_g F_y\left[1 - \frac{4}{3}\left(\frac{x_0}{b}\right)^2\right] \tag{d}$$

Thus

$$F_{cr} = \frac{P}{A_g} = F_y\left[1 - \frac{4}{3}\left(\frac{x_0}{b}\right)^2\right] \tag{e}$$

which is used in combination with Eq. (a). The results are presented in Fig. 6.6.8.

$\dfrac{x_0}{b}$	F_{cr}	F_{cr} for $F_y = 36$ ksi (ksi)	$\dfrac{KL}{r}$	F_{cr} for $F_y = 100$ ksi (ksi)	$\dfrac{KL}{r}$
0.50	$0.667F_y$	24.0	109.2	66.7	65.4
0.45	0.730	26.3	89.0	73.0	53.4
0.40	0.787	28.3	72.0	78.7	43.1
0.35	0.837	30.2	57.0	83.7	34.2
0.30	0.880	31.7	44.1	88.0	26.5
0.25	0.917	33.0	32.9	91.7	19.7
0.20	0.947	34.1	23.2	94.7	13.9
0.10	0.987	35.5	8.0	98.7	4.8

If the web of the section were to be included, I_e/I could easily include the web terms. Furthermore, Eq. (b) could also have included the web terms. Such inclusion of the effect of the web brings in the variable A_w/A_f and in most cases the effect is small.

Finally, curves similar to those of Fig. 6.6.8 can be obtained by using an average stress-strain curve for a short length of rolled shape as referred to earlier in this section, in which case Eqs. 6.6.12 and 6.6.18 can be used with the E_t obtained from the "cross-section" stress-strain curve. ∎∎

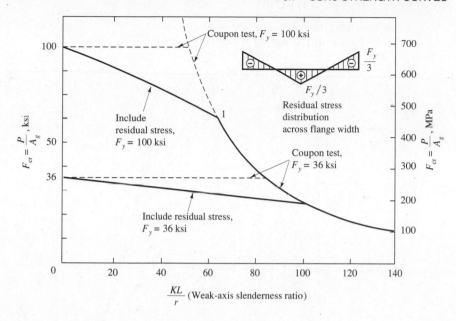

Figure 6.6.8 Column strength curves showing effect of residual stress ($E = 29{,}000$ ksi). Solution for Example 6.6.2.

6.7 STRUCTURAL STABILITY RESEARCH COUNCIL (SSRC) STRENGTH CURVES

Parabolic Equation—Basis for Allowable Stress Design

Based upon the methods discussed in Sec. 6.6, column strength curves can be obtained for weak- or strong-axis buckling with various distributions of residual stress. For most practical situations it has been reported that an assumed linear distribution of residual stress in the flanges results in a reasonable average column curve [6.20]. Furthermore, the development in the previous section (Eqs. 6.6.12 and 6.6.17) shows that for the *same* slenderness ratio, H-shaped column sections allowed to bend in the weak direction can carry less load than columns permitted to bend only in the strong direction. Compressive residual stress which is greatest at the flange tips accounts for this strength difference.

Typical column strength curves for parabolic and linear distribution of residual stress across the flange are shown in Fig. 6.7.1. For structural carbon steels, the average value of the maximum compressive residual stress is 12 to 13 ksi (83 to 90 MPa), corresponding roughly to $0.3F_y$. For the high-strength steels residual stress will generally be a lower fraction of the yield stress.

Since 1960, structural steel design according to the AISC Allowable Stress Design Specifications has used the SSRC parabolic equation based on the one proposed by Bleich [6.9]. The SSRC parabolic curve is

$$F_{cr} = F_y\left[1 - \frac{F_y}{4\pi^2 E}\left(\frac{KL}{r}\right)^2\right] \tag{6.7.1}$$

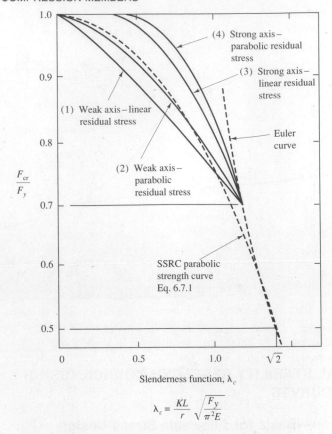

Figure 6.7.1 Column strength curves for H-shaped sections having compressive residual stress at flange tips. (Adapted from Ref. 6.20, p. 39)

Equation 6.7.1 is compared in Fig. 6.7.1 with other curves that distinguish between residual stress patterns and axes of bending. The SSRC curve gives fairly good agreement with the weak axis curve for H-shaped sections, particularly when it is noted that parabolic residual stress is more representative of the actual situation than linear residual stress.

From Fig. 6.7.1 one may note that to provide the same degree of safety for all columns, different strength curves would be required depending on the expected residual stress distribution, the shape of the section, and the axis of bending when the column buckles.

Note is made that Fig. 6.7.1 introduces the slenderness function λ_c, which has been adopted as the slenderness parameter (instead of KL/r) by the AISC Load and Resistance Factor Design Specification. The slenderness parameter λ_c is defined as

$$\lambda_c^2 = \frac{F_y}{F_{cr}(\text{Euler})} = \frac{F_y}{\dfrac{\pi^2 E}{(KL/r)^2}} \tag{6.7.2}$$

$$\lambda_c = \frac{KL}{r}\sqrt{\frac{F_y}{\pi^2 E}} \tag{6.7.3}$$

In terms of the slenderness parameter λ_c, the SSRC parabola becomes

$$\frac{F_{cr}}{F_y} = 1 - \frac{\lambda_c^2}{4} \qquad \text{for } \lambda_c \leq \sqrt{2} \tag{6.7.4}$$

Note that $\lambda_c = \sqrt{2}$ when the parabola and the Euler hyperbola become tangent to each other. Thus Eq. 6.7.4 applies for $\lambda_c \leq \sqrt{2}$; for greater values of λ_c, the Euler equation applies,

$$\frac{F_{cr}}{F_y} = \frac{1}{\lambda_c^2} \qquad \text{for } \lambda \geq \sqrt{2} \tag{6.7.5}$$

Strength Equation—Basis for Load and Resistance Factor Design

The parabolic equation, Eq. 6.7.4, provides a reasonable approximation for a column strength curve that provides a transition between elastic buckling and yielding, reflecting essentially the effect of residual stress. Traditionally, accidental eccentricity and initial crookedness were accounted for by using an increased safety factor as the slenderness increased. Load and resistance factor design philosophy provides for a constant margin of safety for *all* columns. If the strength truly varies with slenderness, then the nominal strength P_n should account for it.

Bjorhovde, as reported in the *SSRC Guide* [6.8], showed that three column strength curves would be sufficient to approximate the strength for all practical shapes. These are referred to as SSRC Curves 1, 2, and 3 and the details are to be found in the *SSRC Guide* [6.8, pp. 59–62].

In the development of the load and resistance factor design specification, the AISC Specification Committee decided to continue using only one column strength curve for steel design. An equation was established to fit closely the SSRC Curve 2 modified to reflect an initial out-of-straightness of about 1/1500.

The nominal strength P_n of rolled shape compression members (LRFD-E2) is given by

$$P_n = A_g F_{cr} \tag{6.7.6}$$

1. For $\lambda_c \leq 1.5$,

$$F_{cr} = (0.658^{\lambda_c^2})F_y \tag{6.7.7}$$

2. For $\lambda_c > 1.5$,

$$F_{cr} = \left[\frac{0.877}{\lambda_c^2}\right]F_y \tag{6.7.8}$$

The reader may observe in Fig. 6.7.2 the comparison between the new curves, Eqs. 6.7.7 and 6.7.8, and the strength F_{cr} obtained by increasing the traditional allowable stress equation by a factor of 1.67. In allowable stress design, ASD-E2 gives

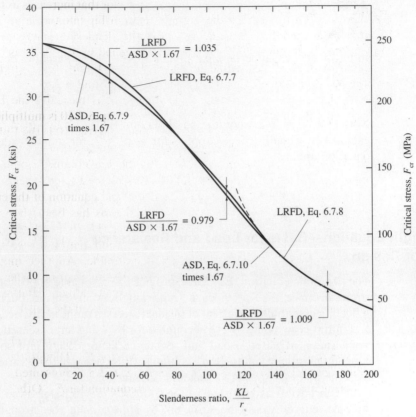

Figure 6.7.2 Comparison of column critical stress F_{cr} for using LRFD Eqs. 6.7.7 and 6.7.8 with 1.67 times ASD Eqs. 6.7.9 and 6.7.10, for $F_y = 36$ ksi.

the allowable axial compression stress F_a as

1. For $\dfrac{KL}{r} < C_c$,

$$F_a = \frac{\left[1 - \dfrac{(KL/r)^2}{2C_c^2}\right]F_y}{\dfrac{5}{3} + \dfrac{3(KL/r)}{8C_c} - \dfrac{(KL/r)^3}{8C_c^3}} \qquad (6.7.9)$$

2. For $\dfrac{KL}{r} > C_c$,

$$F_a = \frac{12\pi^2 E}{23(KL/r)^2} \qquad (6.7.10)$$

where

$$C_c = \sqrt{\frac{2\pi^2 E}{F_y}} \qquad (6.7.11)$$

The denominator of Eq. 6.7.9 is the safety factor that increases as the slenderness ratio increases. Putting part of the strength relationship into the safety factor hides the fact that the nominal strength decreases as the slenderness increases because of initial out-of-straightness. The actual strength implied by Eq. 6.7.9 is that equation multiplied by 1.67 (the nominal factor of safety). Figure 6.7.2 compares $1.67F_a$ using Eqs. 6.7.9 and 6.7.10 with F_{cr} using Eqs. 6.7.7 and 6.7.8.

The long column equation, Eq. 6.7.10, is basically the Euler equation. For consistency with traditional design, the F_a of Eq. 6.7.10 is multiplied by 1.67 to obtain an expression for F_{cr} for long columns. Since Eq. 6.7.10 is the Euler equation divided by 1.92, multiplying by 1.67 gives the Euler equation times (1.67/1.92), or 0.87 times the Euler equation. This gives essentially the long column nominal buckling stress F_{cr} equation, Eq. 6.7.8, used in Load and Resistance Factor Design.

The establishment of an acceptable *single* equation of the critical stress F_{cr} has been the subject of some controversy since, as has been shown, the shape of the cross-section and the method of manufacture (i.e., hot-rolling or welding) influence the strength. Furthermore, residual stress and out-of-straightness are significant influencing parameters but are not quantifiable. Another major factor affecting column strength is end restraint, particularly in situations where the joints are *not* rigid; Bjorhovde [6.21] has provided extensive treatment relating to practical design situations. Bjorhovde [6.22, 6.23] also has reviewed the entire subject in the context of load and resistance factor design.

More recently, Hall [6.24] has compared data from physical tests and has presented statistically developed expressions for F_{cr}. Though the actual AISC LRFD equations are not referred to by Hall, Fig. 6.7.3 is presented here to compare the experimental data [6.24] with the LRFD equation for F_{cr}. Other recent proposals for

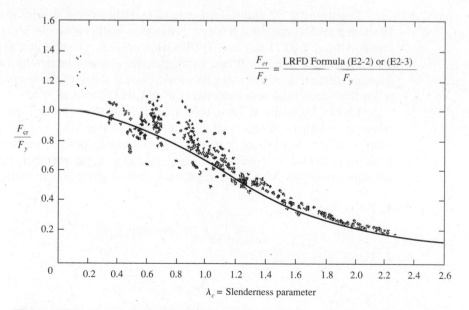

Figure 6.7.3 Comparison of LRFD equations for F_{cr} for columns with data from physical tests. (Test data from Hall [6.24])

column strength curves have been given by Rondal and Maquoi [6.26] and Rotter [6.25], as well as the three SSRC Curves 1, 2, and 3 mentioned previously [6.8]. Lui and Chen [6.27] have discussed the design of columns with imperfections using a beam-column (see Chap. 12) approach.

Columns Having Varying Axial Load, Stepped Columns, and Tapered Columns

Special treatment is required when the axial load varies along the length and/or the member is not prismatic over the entire column length. The reader is referred to Anderson and Woodward [6.28] and Castiglioni [6.29] for stepped columns, Shrivastava [6.30] for columns having varying axial load, Sandhu [6.31] for columns with an intermediate axial load, and Ermopoulos [6.32] for tapered bars under stepped axial loads. Tapered-web members are covered by LRFD-F3.

6.8 LOAD AND RESISTANCE FACTOR DESIGN

The strength requirement in load and resistance factor design according to LRFD-E2 may be stated

$$\phi_c P_n \geq P_u \tag{6.8.1}$$

where $\phi_c = 0.85$
P_n = nominal strength = $A_g F_{cr}$
F_{cr} = Eqs. 6.7.7 and 6.7.8
P_u = factored service load (see Sec. 1.9)

Equations 6.7.7 and 6.7.8 for F_{cr} are applicable in design of ordinary rolled H-shaped section columns; however, when thin-walled plate elements are used in the cross-section, LRFD-E2 and LRFD-Appendix B provide for using a reduced efficiency of the section. When a thin element exhibits instability (local buckling), such an element does not carry its proportionate share of the load. This may occur when the width/thickness ratio limits λ_r of LRFD-Table B5.1 are exceeded.

LRFD Appendix B introduces a reduction factor Q into Eqs. 6.7.7 and 6.7.8 when the width/thickness limitations of LRFD-Table B5.1 are not satisfied. Thus, when local buckling of one or more plate components of the cross-section may occur prior to achieving the overall buckling strength of the member, the F_y in the F_{cr} equations becomes $\sqrt{Q} F_y$, giving for the critical stress F_{cr}, the following:

1. For $\lambda_c \sqrt{Q} \leq 1.5$,

$$F_{cr} = (0.658^{Q\lambda_c^2})QF_y \tag{6.8.2}$$

2. For $\lambda_c \sqrt{Q} > 1.5$,

$$F_{cr} = \left(\frac{0.877}{\lambda_c^2}\right)F_y \tag{6.8.3}$$

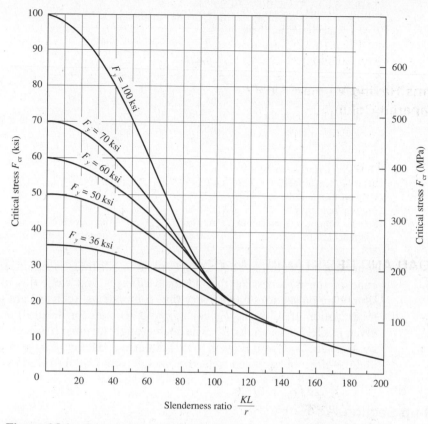

Figure 6.8.1 Critical column stress F_{cr} vs KL/r according to Load and Resistance Factor Design, for various yield stresses.

Since $Q = 1$ for all rolled H-shaped sections (standard W, S, and M shapes) listed in the *LRFD Manual* Column Load Tables for $F_y = 36$ and 50 ksi, the development of the logic behind the use of the Q factor is reserved for Sec. 6.18 in Part II on plate buckling.

The critical stress F_{cr} equations, Eqs. 6.7.7 and 6.7.8, are shown in Fig. 6.8.1 for $F_y = 36, 50, 60, 70$, and 100 ksi. LRFD Specification section references for axial compression are given in Table 6.8.1.

Singly Symmetric Double Angles and Tees

Hot-rolled double angles and tee sections usually buckle in the flexural mode based on KL/r with respect to the x- or y-axis. However, since the shear center (see Sec. 8.4) does not coincide with the centroid of the section, a torsional effect is possible. Though the subject of torsion is not treated until Chapter 8, the reader should note that the *LRFD Manual* tables "COLUMNS" giving design axial strength $\phi_c P_n$ for double angles and structural tees include the effect of the flexural-torsional buckling limit

TABLE 6.8.1   AXIAL COMPRESSION
—AISC SPECIFICATION REFERENCES

Topic	Specification sections	
	LRFD [1.16]	ASD [1.5]
Local buckling limits for "noncompact" sections	B5.1	B5.1
Local buckling limits for "compact" sections	B5.1	B5.1
Slenderness limits	B7	B7
Braced frame, definition	C2.1	C2.1
Unbraced frame, definition	C2.2	C2.2
Effective length factors	C2	C2
Column formulas, basic	E2	E2
Built-up members	E4	E4
Slender compression elements	Appendix B5.3	Appendix B5.2
Alignment chart	Commentary C2	Commentary C2

state in the calculation of F_{cr}. Since torsional stiffness is significantly related to the cube of the thickness of the elements (flanges and web) this effect will only be significant on sections having very thin components, such as light-gage sections. The modification of λ_c in the F_{cr} equations to account for flexural-torsional buckling is given in LRFD-Appendix E3. The double-angle member has been specifically treated by Astaneh, Goel, and Hanson [6.40], Zahn and Haaijer [6.41], and Galambos [6.87], and the effect of balanced welded connections has been treated by Temple and Sakla [5.43].

Built-up Sections

The compression member strength of built-up sections is affected by the shear transfer strength of the fasteners attaching the elements together, and the slip-resistance of these connectors. A double angle compression chord member of a truss can behave as a single element compression member only when the two angles are adequately attached to each other so that when the flexural buckling limit state is reached there will be no relative axial movement (shear deformation) of one angle relative to another. The strength of built-up sections as affected by fastener strength, spacing, and installation (slip-resistance) of bolts has been studied by Libove [6.39] and Aslani and Goel [6.89, 6.95], and design rules are given by Duan and Chen [6.82].

Column Strength as Affected by Connector Spacing

When the controlling buckling mode "involves relative deformation that produces shear forces in the connectors between individual shapes," a modified slenderness ratio $(KL/r)_m$ is used. For the double angle section, this will be the case when the controlling flexural buckling occurs in the plane parallel to the outstanding legs (y-axis for *LRFD Manual* properties). When the controlling flexural buckling limit state is based on slenderness KL/r in the plane parallel to the back-to-back legs (x-axis for *LRFD Manual* properties), the angles will move parallel to each other and the connectors will not affect the compression strength.

The modified slenderness ratio $(KL/r)_m$ is also dependent on the slip-resistance of the connections. LRFD-E4 provides:

1. For *snug-tight bolted connections:*

$$\left(\frac{KL}{r}\right)_m = \sqrt{\left(\frac{KL}{r}\right)_0^2 + \left(\frac{a}{r_i}\right)^2} \tag{6.8.4}$$

2. For *welded connectors* and for *fully-tightened bolted connections:*

$$\left(\frac{KL}{r}\right)_m = \sqrt{\left(\frac{KL}{r}\right)_0^2 + 0.82\frac{\alpha^2}{(1+\alpha^2)}\left(\frac{a}{r_{ib}}\right)^2} \tag{6.8.5}$$

where

$\left(\dfrac{KL}{r}\right)_0$ = slenderness ratio of the built-up member acting as a unit (for the y-axis in double angle members)

$\dfrac{a}{r_i}$ = largest slenderness ratio of an individual component (for the z-axis of a single angle for the double angle member)

$\left(\dfrac{KL}{r}\right)_m$ = modified (increased) slenderness ratio based on the connectors (for double angle members this replaces KL/r based on y-axis)

a = distance between connectors measured along member length

r_i = minimum radius of gyration of individual component (the z-axis for an angle)

r_{ib} = radius of gyration of individual component relative to its centroidal axis parallel to member axis of buckling (for double angle members this is r_y)

$\alpha = h/(2r_{ib})$ = separation ratio

h = distance between centroids of individual components perpendicular to the member axis of buckling

$\dfrac{a}{r_{ib}}$ = column slenderness ratio of individual component relative to its centroidal axis parallel to member axis of buckling

The application of the modified slenderness ratio to the double angle member applies only when $(KL/r)_y$ exceeds $(KL/r)_x$; however, the modified ratio should also be checked when $(KL/r)_x$ is the larger by only a small amount. In effect, the modified value *always* replaces the $(KL/r)_y$ for the double angle member.

Tubular Sections

The formulas for F_{cr} are applicable for tubular sections. Additional information on tubular compression members is available in the work of Sherman [6.14, 6.15, 6.33], as well as Snyder and Lee [6.34, 6.35], Chen and Ross [6.36], and Ross, Chen, and Tall [6.37]. For round columns, see Galambos [6.38].

Single Angle Sections

The strength of single-angle struts (i.e. compression members) has received considerable attention in recent years. For many years AISC recommended against using single-angle compression members because of their having the torsional (see Chap. 8) or flexural-torsional (see Chap. 9) limit states for strength. The full treatment of single angles is outside the scope of this book.

For the design of single-angle members subject to compression, bending, and beam-column action, the reader is referred to *Specification for Load and Resistance Factor Design of Single-Angle Members* [6.83] published by AISC.

Many researchers contributed to development of that specification, including Kennedy and Murty [6.42, 6.43], Woolcock and Kitipornchai [6.44], El-Tayem and Goel [6.45, 6.46], Chuenmei [6.47], Galambos [6.87], Elgaaly, Dagher, and Davids [6.90, 6.91], Adluri and Madugula [6.92], Zureick [6.96], and Bathon, Mueller, and Kempner [6.97].

Other Non-Symmetrical Sections

Sections such as cruciform (see Smith [6.48]) and Z-sections require special consideration using the provisions of LRFD-Appendix E. The strength of built-up sections connected intermittently by fasteners is affected by such fastener spacing as discussed above.

6.9 EFFECTIVE LENGTH

Discussion of column strength to this point has assumed hinged ends where no moment rotational restraint exists. Zero moment restraint at the ends constitutes the weakest situation for compression members having no transverse movement of one end relative to the other. For such pinned-end columns the equivalent pinned-end length KL is the actual length L; thus $K = 1.0$ as shown in Fig. 6.9.1a. The equivalent pinned-end length is referred to as the *effective length*.

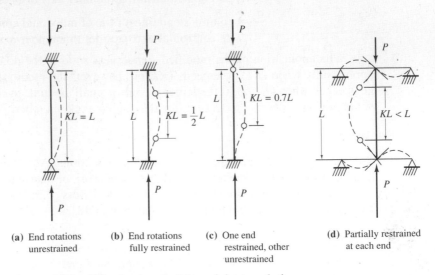

(a) End rotations unrestrained

(b) End rotations fully restrained

(c) One end restrained, other unrestrained

(d) Partially restrained at each end

Figure 6.9.1 Effective length KL; no joint translation.

For most real situations moment restraint at the ends does exist causing the points of zero moment (inflection points) to move away from the restrained ends as shown in Fig. 6.9.1b, c, and d where the effective lengths *KL* then are reduced.

In many situations it is difficult, or perhaps impossible, to adequately evaluate the degree of moment restraint contributed by adjacent members framing into a column, by a footing and soil under it, and indeed the full interaction of all members of a steel frame.

Whether or not the degree of end restraint can be ascertained accurately, the designer must understand the concepts of *braced frame* and *unbraced frame*. A more extensive treatment of frames is given in Chap. 14.

Braced Frame

A *braced frame,* according to LRFD-C2.1, is one in which "lateral stability is provided by diagonal bracing, shear walls or equivalent means." The vertical bracing system must be "adequate" as determined by structural analysis ". . . to prevent buckling of the structure and to maintain the lateral stability of the structure, including the overturning effects of drift, under the factored loads . . . ." Note that a vertical column in a braced frame would have no sideways movement of its top relative to its bottom.

Figure 6.9.1 illustrates effective lengths for columns in a braced frame. Once it has been determined that a frame is braced, the bracing is presumed to provide any needed lateral restraint, as in Fig. 6.9.2a and c; therefore, the joints are assumed not to move laterally (at least in first order structural analysis) and an individual column may be designed as if isolated once the effective length factor *K* has been determined.

From Figs. 6.9.1 and 6.9.2a and c, one may observe that end restraint in braced frames always reduces the distance between inflection points; that is, reduces the effective length *KL* from the pinned-end condition. The effective length factor *K* will always be less than unity.

Unbraced Frame

An *unbraced frame,* according to LRFD-C2.2, is one in which "lateral stability depends upon the bending stiffness of rigidly connected beams and columns." The buckling of an unbraced frame is one of sidesway where, for example, the top of a column moves to the side relative to the bottom. In Fig.6.9.2b and d an unbraced frame is shown having sidesway buckling. The buckled shape and therefore the effective length of the columns will depend on the stiffnesses of the participating members in flexure. The effective length *KL* may be obtained by matching the buckled shape of a column with a portion of the pinned-end column buckled shape. As shown in Fig. 6.9.2, *KL* will always exceed *L*.

To understand why the *minimum* value of *K* in an unbraced frame is theoretically 1.0, examine the rectangular frame of Fig. 6.9.2d. The stiffest situation would be when the beam is infinitely stiff, that is, it cannot bend. The inflection point would then be at mid-height and the buckled shape would be as in Fig.6.9.3a.

The practical situation in an unbraced frame is that *K* is always *greater than unity*. Furthermore, there is no simple way of obtaining a value other than evaluating

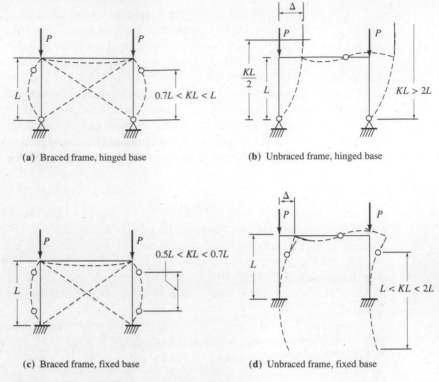

(a) Braced frame, hinged base

(b) Unbraced frame, hinged base

(c) Braced frame, fixed base

(d) Unbraced frame, fixed base

Figure 6.9.2 Effective length KL for frames.

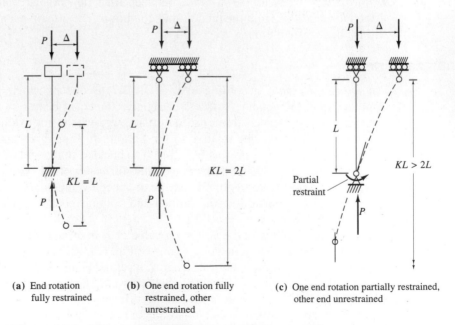

(a) End rotation fully restrained

(b) One end rotation fully restrained, other unrestrained

(c) One end rotation partially restrained, other end unrestrained

Figure 6.9.3 Effective length KL; joint translation possible.

the end restraint. LRFD-C2.2 requires that K "shall be determined by structural analysis."

Alignment Charts for Evaluating Effective Length Factor K

For ordinary design, it is entirely impractical to analyze an entire frame to determine its buckling strength and the effective lengths for the members.

Various investigators have provided charts to permit easy determination of frame buckling loads and effective lengths for commonly encountered situations. Effective length factors K are given by Anderson and Woodward [6.28] for stepped columns, Sandhu [6.31] for columns having an intermediate axial load, Lu [6.49] for gabled frames, Fraser [6.84] for pin-based crane columns, Stoman [6.85] for cross bracings, Rutenberg and Scarlat [6.86] for columns in one-story buildings, and Hassan [6.50] for one-story, one-bay frames, having vertical loads applied to the columns at an intermediate point in addition to the load at the top. Galambos [6.51] has presented them for one- and two-story, one-bay-wide frames, and Gurfinkel and Robinson [6.52] have given K values for the general case of an elastic rotationally restrained column (both with and without sidesway elastic restraint). Switzky and Wang [6.53] have summarized buckling load data from which effective lengths can be obtained for frames one-bay wide and up to five stories high.

The most commonly used procedure for obtaining effective length is to use the alignment charts originally developed by O. J. Julian and L. S. Lawrence, and presented in detail by T. C. Kavanagh [6.54].* The alignment chart method using Fig. 6.9.4 is also suggested by the LRFD Commentary as satisfying the "analysis" requirement of LRFD-C2.2 to get "adequate" K values. Dumonteil [6.93] has presented equations equivalent to the alignment charts, and Lui [6.94] has given a simplified formula alternative.

For simple situations, one may use Fig. 6.9.5 from the *SSRC Guide* [6.8]. For braced frames it is always conservative to take the K factor as unity, and some interpolation is possible from Fig. 6.9.5. For unbraced frames, except perhaps for the flagpole-type situation of Fig. 6.9.5, case (e), an arbitrary selection of K is not satisfactory for design.

Adjustment of Alignment Chart K Factors for Inelastic Column Behavior

The design of columns in ordinary construction involves slenderness ratios KL/r in the inelastic buckling range (i.e., $\lambda_c < 1.5$ or $F_{cr} >$ about $0.5F_y$). The inelastic buckling strength is given by Eq. 6.3.1,

$$F_{cr} = \frac{\pi^2 E_t}{(KL/r)^2} \qquad [6.3.1]$$

where E_t is the tangent modulus of elasticity at stress P_{cr}/A_g. When the alignment chart is used to evaluate K there is an implicit assumption that elastic buckling controls.

---

* The derivation was presented in detail in the 2nd edition (1980) of this text (pp. 843–851).

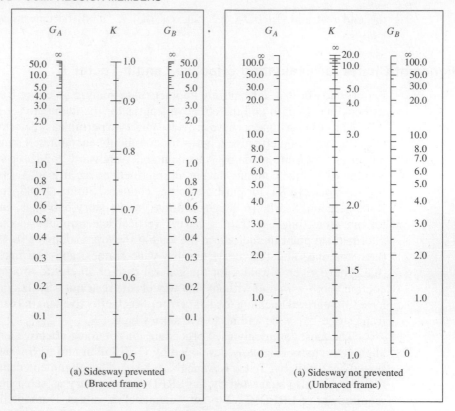

Figure 6.9.4 Alignment Charts for effective length of columns in continuous frames [6.46], where $G = \dfrac{\Sigma I/L,\ \text{columns}}{\Sigma I/L,\ \text{girders}}$

When the column is inelastic and the beam is elastic, an adjustment may be made in the restraint factor G used in the alignment chart. The G factor would then become

$$G_{\text{inelastic}} = \frac{\Sigma(E_t I/L)_{\text{col}}}{\Sigma(EI/L)_{\text{beam}}} = G_{\text{elastic}}\,[\beta_s] \tag{6.9.1}$$

where

$$\beta_s = \frac{E_t}{E} = \frac{F_{\text{cr, inelastic}}}{F_{\text{cr, elastic}}} = \frac{\text{Eq. 6.8.2}}{\text{Eq. 6.8.3}} = \left(\frac{\lambda_c^2}{0.877}\right)\!\left(0.658^{Q\lambda_c^2}\right) \tag{6.9.2}$$

The procedure of using Eq. 6.9.2 and various simplifications for practical use have been discussed by Yura [6.55]. This provoked considerable discussion by Adams [6.56], Johnston [6.57], Disque [6.58], Smith [6.59], Matz [6.60], and Stockwell [6.61]. The LRFD-Commentary C2 endorses Yura's procedure as modified by Disque [6.58]. The Yura-Disque procedure was presented in the context of Allowable Stress

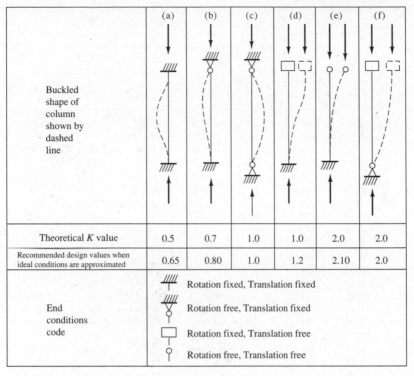

	(a)	(b)	(c)	(d)	(e)	(f)
Theoretical K value	0.5	0.7	1.0	1.0	2.0	2.0
Recommended design values when ideal conditions are approximated	0.65	0.80	1.0	1.2	2.10	2.0

Figure 6.9.5 Effective length factors for centrally loaded columns having various idealized end conditions. (Adapted from Ref. 6.8, p. 52)

Design where E_t/E could be approximated as f_a/F'_e. For LRFD the numerator of Eq. 6.9.2 may be taken as P_u/A_g since if properly designed the strength ϕP_n will approximately equal P_u. Note that Eq. 6.9.2 becomes

$$\beta_s = \frac{E_t}{E} \approx \frac{\phi[F_{\text{cr, short}}]A_g}{\phi[F_{\text{cr, long}}]A_g} \approx \frac{\phi P_n}{\phi[F_{\text{cr, long}}]A_g} = \frac{P_u/A_g}{\phi[\text{Eq. 6.8.3}]} \qquad (6.9.3)$$

Thus, it seems reasonable to use the nominal stress P_u/A_g under factored load as the numerator of Eq. 6.9.2 and ϕ times Eq. 6.8.3 as the denominator.

Values of β_s (Eq. 6.9.2) have been tabulated for A36 steel in Table 6.9.1. The *LRFD Manual* [1.18] calls these "Stiffness Reduction Factors" and also includes values for $F_y = 50$ ksi. Harichandran [6.88] has summarized the procedure, and noted how easily mistakes can be made.

For truss compression members, end restraint may be present and joint translation is prevented so that K might logically be less than 1.0. Under static loading, stresses in all the members remain in the same proportion to one another for various loads. If all members are designed for minimum weight, they will achieve ultimate

TABLE 6.9.1 ADJUSTMENT β_s OF RESTRAINT FACTOR G^* TO ACCOUNT FOR INELASTIC BUCKLING—FOR A36 STEEL

$$\beta_s = \frac{E_t}{E} = \frac{F_{cr,\,inelastic}}{F_{cr,\,elastic}} = \frac{\text{LRFD Formula (E2-2)}}{\text{LRFD Formula (E2-3)}} = \frac{\text{Eq. 6.8.2}}{\text{Eq. 6.8.3}}$$

λ_c	$\dfrac{KL}{r}$	F_{cr}	β_s	λ_c	$\dfrac{KL}{r}$	F_{cr}	β_s
0.20	17.8	35.4	0.04	0.82	73.1	27.2	0.58
0.22	19.6	35.3	0.05	0.84	74.9	26.8	0.60
0.24	21.4	35.1	0.06	0.86	76.7	26.4	0.62
0.26	23.2	35.0	0.07	0.88	78.5	26.0	0.64
0.28	25.0	34.8	0.09	0.90	80.2	25.6	0.66
0.30	26.7	34.7	0.10	0.92	82.0	25.3	0.68
0.32	28.5	34.5	0.11	0.94	83.8	24.9	0.70
0.34	30.3	34.3	0.13	0.96	85.6	24.5	0.71
0.36	32.1	34.1	0.14	0.98	87.4	24.1	0.73
0.38	33.9	33.9	0.15	1.00	89.2	23.7	0.75
0.40	35.7	33.7	0.17	1.02	90.9	23.3	0.77
0.42	37.4	33.4	0.19	1.04	92.7	22.9	0.78
0.44	39.2	33.2	0.20	1.06	94.5	22.5	0.80
0.46	41.0	32.9	0.22	1.08	96.3	22.1	0.82
0.48	42.8	32.7	0.24	1.10	98.1	21.7	0.83
0.50	44.6	32.4	0.26	1.12	99.9	21.3	0.85
0.52	46.4	32.1	0.28	1.14	101.6	20.9	0.86
0.54	48.1	31.9	0.29	1.16	103.4	20.5	0.87
0.56	49.9	31.6	0.31	1.18	105.2	20.1	0.89
0.58	51.7	31.3	0.33	1.20	107.0	19.7	0.90
0.60	53.5	31.0	0.35	1.22	108.8	19.3	0.91
0.62	55.3	30.6	0.37	1.24	110.6	18.9	0.92
0.64	57.1	30.3	0.39	1.26	112.3	18.5	0.93
0.66	58.8	30.0	0.41	1.28	114.1	18.1	0.94
0.68	60.6	29.7	0.43	1.30	115.9	17.7	0.95
0.70	62.4	29.3	0.46	1.32	117.7	17.4	0.96
0.72	64.2	29.0	0.48	1.34	119.5	17.0	0.97
0.74	66.0	28.6	0.50	1.36	121.3	16.6	0.97
0.76	67.8	28.3	0.52	1.38	123.0	16.2	0.98
0.78	69.5	27.9	0.54	1.40	124.8	15.8	0.98
0.80	71.3	27.5	0.56	1.42	126.6	15.5	0.99

*G defined in Fig. 6.9.4 and used with Alignment Charts (Fig. 6.9.4).

capacity simultaneously under live load. Thus restraint offered by members framing at a joint disappears or at least is greatly reduced. The SSRC, therefore, recommends using $K = 1.0$ for members of a truss designed for fixed-position loading. When designing for moving load systems on trusses, K can be reduced to 0.85 because conditions causing maximum stress in the member under consideration will not cause maximum stress in the members framing in to provide restraint [6.8].

When the adjacent members framing at the ends of a column are heavily loaded compression members, they may have a destabilizing effect instead of stabilizing the member being considered: in effect a *negative* G-factor. Bridge and Fraser [6.81] have presented a procedure to account for such negative G-factors.

The application of the *K* factor approach to the unbraced frame containing a "leaner" column is presented in Sec. 14.5.

6.10 LOAD AND RESISTANCE FACTOR DESIGN OF ROLLED SHAPES (W, S, AND M) SUBJECT TO AXIAL COMPRESSION

In this section reference will be made to the *LRFD Manual,* Part 1 which gives properties of the rolled shapes and Part 2 which contains Column Load Tables. All examples of this section use W shapes that satisfy the $b_f/2t_f$ and d/t_w limits to preclude local buckling; thus, $Q = 1.0$. Section 6.18 treats the subject when $Q < 1.0$.

General Procedure

Whether Load and Resistance Factor Design or Allowable Stress Design is used, the strength of a compression member is based on its gross are A_g. The strength is always a function of the effective slenderness ratio KL/r, and for short columns the yield stress F_y of the steel. Since the radius of gyration r depends on the section selected, the design of compression members is an indirect process unless column load tables are available. The general procedure to satisfy Eq. 6.8.1 is:

1. Compute the factored service load P_u using all appropriate load combinations, as discussed in Sec. 1.8.
2. Assume a critical stress F_{cr} based on an assumed KL/r.
3. Compute the gross area A_g required from $P_u/(\phi_c F_{cr})$.
4. Select a section. Note that the width/thickness λ_r limitations of LRFD-Table B5.1 to prevent local buckling must be satisfied. This is discussed in Part II of this chapter, particularly Sec. 6.16.
5. Based on the larger of $(KL/r)_x$ or $(KL/r)_y$ for the section selected, compute the critical stress F_{cr}.
6. Compute the design strength $\phi_c P_n = \phi_c F_{cr} A_g$ for the section.
7. Compare $\phi_c P_n$ with P_u. When the strength provided does not exceed the strength required by more than a few percent the design would be acceptable, otherwise repeat Steps 2 through 7.

EXAMPLE 6.10.1 ────────────────────────────────────

Select the lightest W section of A36 steel to serve as a pinned-end main member column 16 ft long to carry an axial compression load of 95 kips dead load and 100 kips live load in a braced structure, as shown in Fig. 6.10.1. Use Load and Resistance Factor Design and indicate the first three choices.

Solution

(a) Obtain factored loads.

$$P_u = 1.2D + 1.6L = 1.2(95) + 1.6(100) = 274 \text{ kips}$$

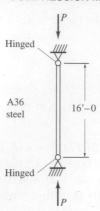

Figure 6.10.1   Example 6.10.1.

(b) Estimate slenderness ratio and obtain estimated F_{cr}. Since the assumption of hinged ends is made, the effective length KL equals the actual length L, i.e., $K = 1.0$. Considering $KL = 16$ ft as a moderately long length, the slenderness ratio might be estimated at about 70 to 80. For rolled W shapes contained in the *LRFD Manual* Column Load Tables for $F_y = 36$ and 50 ksi, $Q = 1.0$, thus the special provisions of LRFD-Appendix B are not involved. From Fig. 6.8.1 one might estimate $F_{cr} \approx 27$ ksi ($KL/r \approx 75$). Alternatively, LRFD Formula (E2-2) or LRFD "NUMERICAL VALUES" TABLE 3-36 could be used. The required area may then be computed.

$$\text{Required } A_g = \frac{P_u}{\phi_c F_{cr}} = \frac{274}{0.85(27)} = 11.9 \text{ sq in.}$$

(c) Select a section. Since buckling in the weak direction [based on $(KL/r)_y$] will control the strength for W shapes when KL is the same with respect to both x- and y-axes, the lightest sections for this design will be those having the least r_x/r_y. A high r_x indicates excessive strength with respect to the strong axis, which may be utilized only by providing additional bracing in the weak direction.

Using the Column Load Tables in the *LRFD Manual* [1.18] Part 2, one might select a W8×48, the lightest W8 that has at least the required area. Furthermore, the W8 sections have the lowest r_x/r_y for a given area. Try W8×48 section.

(d) Check the W8×48 section. $A = 14.1$ sq in.

$$\frac{KL}{r_y} = 92; \qquad F_{cr} = 23.0 \text{ ksi}$$

$$\phi_c P_n = \phi_c F_{cr} A_g = 0.85(23.0)14.1 = 19.5(14.1) = 276 \text{ kips}$$

$$[\phi_c P_n = 276 \text{ kips}] \geq [P_u = 274 \text{ kips}] \qquad \text{OK}$$

Note that *LRFD Manual* "NUMERICAL VALUES" TABLE 3-36 provides a tabulation of $\phi_c F_{cr}$ for values of KL/r.

Since the design strength ϕP_n exceeds the factored load P_u, and since no other section having this area has a lower r_x/r_y, the W8×48 is the lightest section available. Deeper sections will be heavier, as follows:

Section	Area (sq in.)	KL/r_y	F_{cr} (ksi)	$\phi_c P_n$ (kips)	
W8×48	14.1	92	23.0	276	1st choice
W10×49	14.4	76	26.6	326	2nd choice
W12×50	14.7	98	21.7	271	NG 1% short
W14×53	15.6	100	21.3	282	3rd choice

EXAMPLE 6.10.2 _____

Select the lightest W section of A36 steel to serve as a main member 30 ft long to carry an axial compression load of 50 kips dead load and 110 kips live load in a braced structure, as shown in Fig. 6.10.2. The member is assumed pinned at top and bottom and in addition has weak direction support at mid-height. The mid-height brace also prevents twisting of the column at the brace location. Use AISC Load and Resistance Factor Design.

Solution

(a) Obtain factored loads.

$$P_u = 1.2D + 1.6L = 1.2(50) + 1.6(110) = 236 \text{ kips}$$

(b) Select from LRFD Column Load Tables. The effective length factors K for buckling in either the strong or the weak direction equal unity; i.e.,

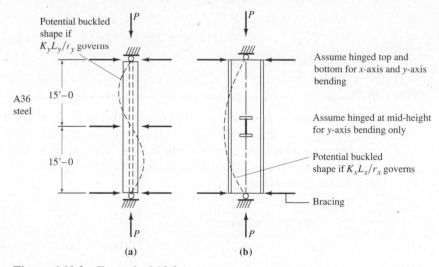

Figure 6.10.2 Example 6.10.2.

$K_x = K_y = 1.0$. Since the Column Load Tables are computed assuming $(KL/r)_y$ controls, enter these tables with the effective length $(KL)_y$. Thus enter with

$$P_u = 236 \text{ kips}; \qquad (KL)_y = 15 \text{ ft}$$

Starting with the shallow W8 sections and working toward the deeper sections, find

W8×40	$\phi_c P_n = 238$ kips	$r_x/r_y = 1.73$
W10×45	$\phi_c P_n = 267$ kips	$r_x/r_y = 2.15$
W12×45	$\phi_c P_n = 257$ kips	$r_x/r_y = 2.65$

Since the actual support conditions are such that $(KL)_x = 2(KL)_y$, if $r_x/r_y \geq 2$, weak axis controls and tabular loads give the correct answer. Thus, W10×45 and W12×45 are obviously acceptable.

Since r_x/r_y for W8×40 is less than 2, strong axis bending controls. The strength may be obtained from the tables by entering with the equivalent $(KL)_y$ that corresponds to $(KL)_x = 30$ ft. For equal strength with respect to the x- and y-axes,

$$\frac{(KL)_x}{r_x} = \frac{(KL)_y}{r_y}$$

$$\text{Equivalent } (KL)_y = \frac{(KL)_x}{r_x/r_y} = \frac{30}{1.73} = 17.3 \text{ ft}$$

For $(KL)_y = 17.3$ ft, the W8×40 has a design strength $\phi_c P_n$ of only 208 kips; therefore it is not acceptable.

(c) Check of section. It is *always* advisable to make a final check of the apparently acceptable section:

$$W10\times45, \qquad (KL/r)_y = 15(12)/2.01 = 89.6, \qquad \phi_c F_{cr} = 20.1 \text{ ksi}$$

$$\phi_c P_n = \phi_c F_{cr} A_g$$
$$= 20.1(13.3) = 267 \text{ kips} > [P_u = 236 \text{ kips}] \qquad \text{OK}$$

Use W10×45.   ∎∎

EXAMPLE 6.10.3 ——————————————————————————————

Select the lightest W section of A572 Grade 60 steel to serve as a main member 22 ft long to carry an axial compression load of 100 kips dead load and 200 kips live load in a braced structure, as shown in Fig. 6.10.3. Assume the member hinged at the top and fixed at the bottom for buckling in either principal direction. Use AISC Load and Resistance Factor Design.

Solution

(a) Obtain factored loads.

$$P_u = 1.2D + 1.6L = 1.2(100) + 1.6(200) = 400 \text{ kips}$$

(b) Select the section. For this problem there are no AISC Column Load Tables available for the direct selection of a section. To get an estimate of the required section, use the Column Load Tables for $F_y = 50$ ksi. Since the mem-

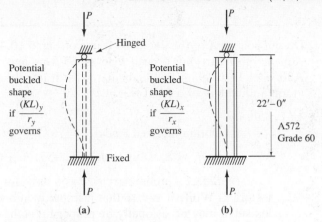

Figure 6.10.3   Example 6.10.3.

ber is fixed at one end, in accordance with Fig. 6.10.3, assume $K = 0.8$ Select for $(KL)_y = 0.8(22) = 17.6$ ft and

$$\text{Required } \phi_c P_n \approx P_u \left[\frac{F_y = 50}{F_y = 60} \right] = 440 \left[\frac{50}{60} \right] = 367 \text{ kips}$$

Try
$$\text{W10×49: } \quad \phi_c P_n = 369 \text{ kips for } F_y = 50 \text{ ksi}$$
$$\text{W12×53: } \quad \phi_c P_n = 390 \text{ kips for } F_y = 50 \text{ ksi}$$

There can be no assurance that sections selected using the ratio 50/60 will satisfy design requirements; however, such a procedure will serve as a first trial. Check W10×49 for $P_u = 400$ kips with $F_y = 60$ ksi. Using Eq. 6.7.3 to obtain λ_c,

$$\lambda_c = \frac{KL}{r_y} \sqrt{\frac{F_y}{\pi^2 E}} = \frac{17.6(12)}{2.54} \sqrt{\frac{60}{\pi^2 29,000}} = 1.20$$

Then using AISC "NUMERICAL VALUES" TABLE 4, obtain $\phi_c F_{cr}/F_y$,

$$\phi_c F_{cr}/F_y = 0.465$$
$$\phi_c F_{cr} = 0.465(60) = 27.9 \text{ ksi}$$
$$\phi_c P_n = \phi_c F_{cr} A_g = 27.9(14.4) = 400 \text{ kips}$$

Thus, the W10×49 is not acceptable.
   Check W10×54,

$$\frac{KL}{r_y} = \frac{0.8(22)12}{2.56} = 82.5; \qquad \lambda_c = 1.19$$

$$\phi_c F_{cr}/F_y = 0.468$$
$$\phi_c F_{cr} = 0.468(60) = 28.1 \text{ ksi}$$
$$\phi_c P_n = \phi_c F_{cr} A_g = 28.1(15.8) = 443 \text{ kips} > [P_u = 440 \text{ kips}] \quad \text{OK}$$

Use W10×54, A572 Grade 60. Though the calculation is not shown here, $Q = 1.0$ for this section when $F_y = 60$ ksi. The limits for $b_f/2t_f$ and d/t_w are discussed in Sec. 6.16, and the limits appear in Table 6.16.2. ■■

EXAMPLE 6.10.4

Design column *A* of the unbraced frame of Fig. 6.10.4 as an axially loaded compression member carrying a dead load of 55 kips and a live load of 220 kips using A36 steel. In the plane perpendicular to the frame the system is braced, with supports at top and bottom of a 21-ft height.

Solution

(a) Obtain factored loads.

$$P_u = 1.2D + 1.6L = 1.2(55) + 1.6(220) = 418 \text{ kips}$$

(b) Select a preliminary size as a basis for evaluating the effective length factors *K*. While it is rare that a frame member would be designed as axially loaded, it may occasionally be proper for some interior members having symmetrical loading. Note is also made that the axially loaded member is one boundary for the more typical beam-column interaction formula discussed in Chapter 12.

It is given that $(KL)_y = 21$ ft for the plane perpendicular to the frame: thus, a preliminary member might be determined from interpolation in the *LRFD Manual* Column Load Tables.

$$(KL)_y = 21 \text{ ft} \qquad \text{Find W12×72,} \qquad \phi_c P_n = 449 \text{ kips}$$

(c) Evaluate the effective length factor *K*. Using $I = 597$ in.⁴ for W12×72, compute G_{top} in accordance with Fig. 6.9.4 and as discussed in LRFD Commentary C2.

$$G_{\text{top}} = \frac{\Sigma(EI/L)_{\text{columns}}}{\Sigma(EI/L)_{\text{beams}}} = \frac{597/21}{2(882)/40} = 0.64$$

$$G_{\text{bottom}} = 10 \quad (\text{according to LRFD Fig. C-C2.2})$$

Using Fig. 6.9.4,

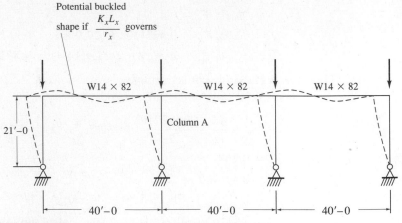

Figure 6.10.4 Unbraced frame for Example 6.10.4.

Find $K_x = 1.8$

$(KL)_x = 1.8(21) = 37.8$ ft

For equal strength about each axis,

$$\frac{r_x}{r_y} = \frac{37.8}{21} = 1.8 \quad \text{(coincidentally the same as } K_x\text{)}$$

(d) Select sections from LRFD Column Load Tables.

W12×72: $r_x/r_y < 1.8$ Equivalent $(KL)_y = 37.8/1.75 = 21.6$ ft

$\phi_c P_n = 440$ kips OK

W14×74: $r_x/r_y > 1.8$ $(KL)_y = 21$ ft $\phi_c P_n = 387$ kips NG

(e) Make final check on W12×72, $I = 597$ in⁴. At this point it may be noted that the Yura–Disque modification on G for inelastic buckling could have been made. The lower the slenderness ratio the greater will be the reduction on G. The following will illustrate the procedure. Compute the slenderness ratios,

$$\frac{(KL)_y}{r_y} = \frac{1.8(21)12}{5.31} = 85.4; \quad \frac{(KL)_x}{r_x} = \frac{1.0(21)12}{3.04} = 82.9$$

From Table 6.9.1 for the slenderness ratio 85.4, find $\beta_s = 0.71$. Now the elastic G value must be reduced; thus,

$$G_{\text{inelastic}} = G_{\text{elastic}}(\beta_s) = 0.64(0.71) = 0.45$$

From the Alignment Chart, Fig. 6.9.4, find $K_x = 1.75$. This will make no significant difference to ϕP_n for this problem. KL/r becomes 83.1,

$\phi_c F_{\text{cr}} = 21.3$ ksi from LRFD "NUMERICAL VALUES" TABLE 3-36

$\phi_c P_n = \phi_c F_{\text{cr}} A_g = 21.3(21.1) = 449$ kips $> [P_u = 418$ kips] OK

Use W12×72, A36. ■■

6.11 ALLOWABLE STRESS DESIGN

The safety requirement for axially loaded columns in Allowable Stress Design (ASD) according to ASD-E2 may be stated

$$f_a \leq F_a \tag{6.11.1}$$

where f_a = service load compression stress = P/A_g
P = service load axial compression force
A_g = gross cross-sectional area of column
F_a = allowable stress at service load = Eqs. 6.7.9 or 6.7.10

Equations 6.7.9 and 6.7.10 are used for typical rolled W sections satisfying the local buckling limitations on width/thickness ratios for plate compression elements given in ASD-B5.1. Table 6.8.1 gives ASD Specification section references for members subject to axial compression.

EXAMPLE 6.11.1

Check the adequacy of the W10×45 section for the condtions of Fig. 6.10.2 if the Allowable Stress Design were used. This is the section selected by LRFD in Example 6.10.2. A36 steel is used and the service loads are 50 kips dead load and 110 kips live load.

Solution

(a) Compute slenderness ratios.

$$\frac{(KL)_x}{r_x} = \frac{1.0(30)12}{4.32} = 83.3 \qquad \frac{(KL)_y}{r_y} = \frac{1.0(15)12}{2.01} = 89.6$$

(b) Compute allowable stress F_a. Compare the KL/r with C_c (Eq. 6.7.11) to determine whether the short or long column formula applies:

$$C_c = \sqrt{\frac{2\pi^2 E}{F_y}} = \sqrt{\frac{2\pi^2 29,000}{36}} = 126 \qquad [6.7.11]$$

Since the controlling KL/r of 89.6 is less than C_c, the allowable stress is based on the parabolic equation for inelastic equation for inelastic buckling, Eq. 6.7.9. Thus, by calculation or from ASD "NUMERICAL VALUES" TABLE 3,

$$F_a = \frac{\left[1 - \frac{(KL/r)^2}{2C_c^2}\right]F_y}{\frac{5}{3} + \frac{3(KL/r)}{8C_c} - \frac{(KL/r)^3}{8C_c^3}} = 14.3 \text{ ksi} \qquad [6.7.9]$$

(c) Comparison of stresses.

$$f_a = \frac{P}{A_g} = \frac{160}{13.3} = 12.0 \text{ ksi} < [F_a = 14.3 \text{ ksi}]$$

The W10×45 certainly is not overstressed; however, the service load stress f_a is low. If the next lighter W10 is checked it will be found satisfactory. Check W10×39:

$$\frac{(KL)_y}{r_y} = \frac{1.0(15)12}{1.98} = 90.9 \quad \text{giving} \quad F_a = 14.1 \text{ ksi}$$

$$f_a = \frac{P}{A_g} = \frac{160}{11.5} = 13.9 \text{ ksi} < [F_a = 14.1 \text{ ksi}] \quad \text{OK}$$

In this case Allowable Stress Design is less conservative than Load and Resistance Factor Design.
Use W10×39. ■■

6.12* SHEAR EFFECT

When built-up members are connected together by means of lacing bars, the objective is to make all of the components act as a unit. As a compression member bends, a shearing component of the axial load arises. The magnitude of the shear effect in reducing column strength is proportional to the amount of deformation that can be attributed to shear.

According to the *SSRC Guide* [6.8], shear in columns is caused by:

1. Lateral load, resulting from wind, earthquake gravity, or other cause.
2. Slope, with respect to the line of thrust, due both to unintentional initial curvature and added curvature developed during the buckling process.
3. End eccentricity of load, introduced either by the end connections or fabrication imperfections.

Certainly the shear from lateral load must always be considered in design. Items 2 and 3 should at the least be estimated. AISC Specifications [1.5, 1.16] require providing an arbitrary shear resistance (in addition to any computable shear) of 2% of the compressive strength of the member for lacing bars in latticed columns (see Fig. 6.12.1).

Solid-webbed sections, such as W shapes, have less shear deformation than do latticed columns using lacing bars and/or batten plates.

Furthermore, as shown later, shear has an insignificant effect on reducing column strength for solid-webbed shapes and may safely be neglected. The shear effect should not, however, be neglected for latticed columns.

To include the effect of shear, the curvature resulting from shear should be added to the buckling curvature to obtain the total curvature. It is well known that

$$V = \frac{dM_z}{dz} = P\frac{dy}{dz} \qquad (6.12.1)$$

after recognizing that $M_z = Py$ from Eq. 6.2.1 (see Fig. 6.2.1).

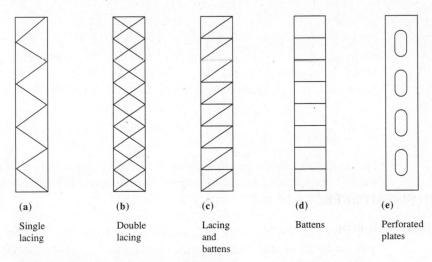

(a)	(b)	(c)	(d)	(e)
Single lacing	Double lacing	Lacing and battens	Battens	Perforated plates

Figure 6.12.1   Types of latticed columns.

The slope θ due to shear deformation is

$$\theta = \frac{\text{shear stress}}{\text{shear modulus}} = \frac{\beta_v V}{AG} \qquad (6.12.2)$$

where β_v is a factor to correct for nonuniform stress across various cross-sectional shapes. The shear contribution to curvature becomes

$$\frac{d\theta}{dz} = \frac{\beta_v}{AG}\frac{dV}{dz} = \frac{P\beta_v}{AG}\frac{d^2y}{dz^2} \qquad (6.12.3)$$

The total curvature is the sum of Eqs. 6.2.3 and 6.12.3,

$$\frac{d^2y}{dz^2} = -\frac{Py}{EI} + \frac{P\beta_v}{AG}\frac{d^2y}{dz^2}$$

which gives

$$\frac{d^2y}{dz^2} + \frac{P}{EI}\left[\frac{1}{1 - P\beta_v/AG}\right]y = 0 \qquad (6.12.4)$$

which is of the same form as Eq. 6.2.3; therefore the modified form of the Euler critical load is

$$P_{cr} = \frac{\pi^2 EI}{L^2}\underbrace{\left[\frac{1}{1 + \dfrac{\beta_v}{AG}\dfrac{\pi^2 EI}{L^2}}\right]}_{\substack{\text{modification for} \\ \text{shear effect}}} \qquad (6.12.5)$$

In accordnace with the previous discussion on basic column strength, G and E can be replaced by the tangent modulus values, G_t and E_t, and $E_t/G_t = 2(1 + \mu)$, and L can be replaced by the effective length KL. Further, combining the shear effect with KL gives

$$F_{cr} = \frac{P_{cr}}{A} = \frac{\pi^2 E_t}{(\alpha_v KL/r)^2} \qquad (6.12.6)$$

where $\alpha_v = \sqrt{1 + 2(1 + \mu)\pi^2\beta_v/(KL/r)^2}$. Thus the shear effect may be accounted for by an adjustment to the effective length. For W shapes when bending about the weak axis, β_v averages about 2. Using $\mu = 0.3$ for steel, typical values for α_v are

$$
\begin{aligned}
KL/r &= 50 & \alpha_v &= 1.01 \\
&= 70 & &= 1.005 \\
&= 100 & &= 1.003
\end{aligned}
$$

For slenderness ratios less than about 50, yielding controls, so that the shear effect on solid H-shaped columns is equivalent to an increase in effective length of less than 1%, which can be safely neglected.

Latticed Columns

The lacing or batten plates used to tie together the main longitudinal compression elements are themselves subject to axial deformation. For instance, from Fig. 6.12.2a,

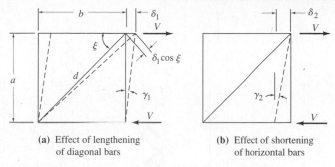

(a) Effect of lengthening
of diagonal bars

(b) Effect of shortening
of horizontal bars

Figure 6.12.2 Shear deformation in laced column.

the lengthening of the diagonal gives a slope γ_1 over the panel length a and from Fig. 6.12.2b the shortening of the horizontal bars gives a slope γ_2 over panel length a. Because the lacing elements are relatively small in cross-section, the stiffness of such members to resist the transverse shear is considerably less than for solid-webbed members. Detailed treatment of the columns with lacing, battens, or perforated plates is available in the *SSRC Guide* [6.8] and elsewhere [6.62–6.64].

The Structural Stability Research Council [6.8] reports the suggestion of Bleich [6.9] that "a conservative estimate of the influence of 60° or 45° lacing, as generally specified in bridge design practice, can be made by modifying the effective length factor" K to a new factor $\alpha_v K$, as follows:

$$\text{For } \frac{KL}{r} > 40, \qquad \alpha_v = \sqrt{1 + 300/(KL/r)^2} \qquad (6.12.7)$$

$$\text{For } \frac{KL}{r} \le 40, \qquad \alpha_v = 1.1 \qquad (6.12.8)$$

Such effective length modification will rarely affect the design of short columns in braced systems.

6.13* DESIGN OF LATTICED MEMBERS

Under most specifications, latticed members are designed according to detailed empirical rules, most of which are related to local buckling requirements. Two examples follow that illustrate some of the provisions of LRFD-E4 (similar to ASD-E4) as well as general procedures for built-up sections. The reader is referred to Blodgett [6.65] who has summarized the AISC provisions along with other information concerning built-up section design.

EXAMPLE 6.13.1 _____

Design a laced column as shown in Fig. 6.13.1, consisting of four angles to carry 100 kips dead load and 475 kips live load axial compression, with an effective length KL of 30 ft. Assume all connections will be welded and use A572 Grade 50 steel.

Solution

(a) Establish the depth h of section. The radius of gyration of a four angle column depends only on h and essentially is independent of thickness. Thus

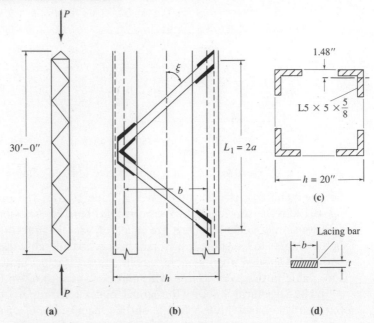

Figure 6.13.1   Details of Example 6.13.1.

selecting h establishes the slenderness ratio, and vice versa. Appendix Table A1 shows the relationships between the radius of gyration and the geometry of the cross-section. Thus from text Appendix Table A1,

$$r \approx 0.42h$$

$$\text{Approximate } \frac{KL}{r} = \frac{360}{0.42h} = \frac{857}{h}$$

$$P_u = 1.2(100) + 1.6(475) = 880 \text{ kips}$$

h (in.)	Approx. KL/r	F_{cr} (ksi)	$\phi_c F_{cr}$[†] (ksi)	Required area* A_g (sq in.)	Angles
24	35.7	45.5	38.7	22.7	L6×6×$\frac{1}{2}$ $A = 23.0$
					L5×5×$\frac{5}{8}$ $A = 23.44$
22	39.0	44.7	38.0	23.1	Same as above
21	40.8	44.3	37.6	23.4	L5×5×$\frac{5}{8}$
20	42.9	43.7	37.2	23.7	L5×5×$\frac{5}{8}$

[†]Also from LRFD "NUMERICAL VALUES" TABLE 3-50.

* Required $A_g = P_u/(\phi_c F_{cr})$.

The 20-in. section is preferred, since the floor area occupied is likely more important than the additional 0.44 sq in. of cross-section. Investigate $4-L5\times5\times\frac{5}{8}$, assuming $Q = 1.0$,

$$I_x = I_y = 4[13.6 + 5.86(10.0 - 1.48)^2] = 1756 \text{ in.}^4$$

$$r_x = r_y = \sqrt{\frac{I}{A}} = \sqrt{\frac{1756}{23.44}} = 8.66 \text{ in.} \quad \text{(Approx } 0.42(20) = 8.40 \text{ in.)}$$

$$\frac{KL}{r} = \frac{1.0(30)12}{8.66} = 41.6; \qquad \phi_c F_{cr} = 37.5 \text{ ksi (TABLE 3-50)}$$

$$\phi_c P_n = \phi_c F_{cr} A_g = 37.5(23.44) = 878 \text{ kips} \approx (P_u = 880 \text{ kips}) \quad \text{OK}$$

Use $4-L5\times5\times\frac{5}{8}$ with $h = 20$ in.

(b) Local buckling control. For $Q = 1.0$ to apply, angles must satisfy width/thickness limits of LRFD-Table B5.1 to prevent local buckling. In this case,

$$\left[\frac{b}{t} = \frac{5}{0.625} = 8\right] < \left[\lambda_r = \frac{76}{\sqrt{50}} = 10.7\right] \quad \text{OK}$$

Development of AISC limits appears in Chapter 6, Part II on plate strength.

(c) Design single lacing. According to LRFD-E4, "The inclination of lacing bars to the axis of the member shall preferably be not less than 60° for single lacing . . . ."

For an inclination of 60° ($\xi = 30°$) (Fig. 6.13.1b), $b = 20 - 2(3) = 14$ in., assuming use of distance between standard gage lines for bolts. This would be approximately center-to-center of welded connection. Thus

$$L_1 \approx 2b \tan 30° = 2(14)0.577 = 16.2 \text{ in.} \quad \text{Use 16 in.}$$

One might prefer to use 15 in. since LRFD-E4 states, "When the distance between the lines of welds . . . is more than 15 in., the lacing shall preferably be double or be made of angles."

For a single angle,

$$\frac{L_1}{r_z} = \frac{16}{0.978} = 16.4 < 41.6 \begin{bmatrix} \text{slenderness ratio for} \\ \text{overall member} \end{bmatrix} \quad \text{OK}$$

According to LRFD-E4, "Lacing shall be proportioned to provide a shearing strength normal to the axis of the member equal to 2% of the compressive design strength of the member."

$$V_u = 0.02(880) = 17.6 \text{ kips} \quad \text{(8.8 kips per side)}$$

The force in one bar is

$$P_u = V_u/\cos \xi \approx V_u/\cos 30° = 8.8/0.866 = 10.2 \text{ kips}$$

$$\frac{L}{r} \leq 140 \quad \text{for single lacing}$$

$$r = \sqrt{\frac{I}{A}} = \sqrt{\frac{\frac{1}{12}bt^3}{bt}} = 0.288t \quad \text{(Fig. 6.13.1d)}$$

$$t_{min} = \frac{16.1}{0.288(140)} = 0.397 \text{ in.} \quad \text{Use } \tfrac{7}{16} \text{ in.}$$

For $t = \tfrac{7}{16}$ in., $\dfrac{L}{r} = \dfrac{16.1}{0.288(0.4375)} = 128$

From LRFD "NUMERICAL VALUES" TABLE 3-36, $\phi_c F_{cr} = 13.0$ ksi,

$$\text{Required } A = \frac{10.2}{13.0} = 0.78 \text{ sq in.}$$

$$\text{Width } b = \frac{0.78}{0.4375} = 1.78 \text{ in.}$$

Since no holes are required for connectors, tension on net section need not be investigated for this design.

Use bars $\tfrac{7}{16} \times 1\tfrac{3}{4}$.

(d) Design the plates at ends (LRFD-E4, par. 6). The tie plates should extend along the length of the member a distance equal to the distance b (Fig. 6.13.1) from the end of the member. Use a length of 14 in.

$$t \geq \frac{b}{50} = \frac{14}{50} = 0.28 \text{ in.}$$

Use tie plates $\tfrac{5}{16} \times 14 \times 1'\text{-}8''$.

(e) Examine the effect of shear on the effective length, Eqs. 6.12.7 or 6.12.8.

$$\frac{KL}{r} = 41.6 > 40, \quad \text{Use Eq. 6.12.7}$$

$$\alpha_v = \sqrt{1 + 300/(KL/r)^2}$$
$$= \sqrt{1 + 300/(41.6)^2} = 1.08$$

Thus the effective length should have been increased 8% due to shear. The neglect of end restraint probably is, in most cases, equal to about the same increase in effective length. ■■

EXAMPLE 6.13.2

Redesign the column of Example 6.13.1 using a welded perforated box shape (Fig. 6.13.2). The factored load P_u is 880 kips as computed in Example 6.13.1.

Solution. One important advantage of a welded shape is the number of individual components in the shape is minimized. Four plates can be used for this welded shape, whereas a bolted or riveted section requires four angles in addition to plates.

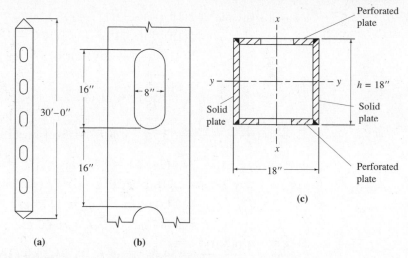

Figure 6.13.2   Details of Example 6.13.2.

(a) Select trial section. Using text Appendix Table A1 for a box shape,

$$r \approx 0.40h$$

$$\frac{KL}{r} = \frac{360}{0.4h} = \frac{900}{h}$$

$$\text{Area} \approx 4ht$$

h (in.)	Approx. KL/r	F_{cr} (ksi)	$\phi_c F_{cr}^\dagger$ (ksi)	Required area* A_g (sq in.)	Plate thickness (in.)
21	42.9	43.7	37.2	23.7	0.28

$$\frac{b}{t}\text{(solid plate)} = \frac{21}{5/16} = 67 > \left(\frac{238}{\sqrt{F_y} - 16.5} = 41.1\right) \quad \text{NG}$$

18	50	41.6	35.4	24.9	0.35
16	56.3	39.7	33.7	26.1	0.41
15	60	38.4	32.7	26.9	0.45

† Also from LRFD "NUMERICAL VALUES" TABLE 3-50.
* Required $A_g = P_u/(\phi_c F_{cr})$.

Without performations, 2PLs—$\frac{7}{16}\times16$ and 2PLs—$\frac{7}{16}\times15$ are probably accept-able, satisfying b/t ratios in accordance with LRFD-B5.1.

Usually, however, such shapes are used on bridges where access is required for maintenance so that perforations may be desirable.

Assume perforations to be 8 in. wide (frequently one-half or less of total width) and $h \approx 18$ in.

Net area available $- 2(18)(\frac{7}{16}) + 2(17 - 8)(\frac{1}{2}) = 24.75$ sq in.

$$I_x = 2(18)(\tfrac{7}{16})(8.719)^2 + \tfrac{2}{12}[(17)^3 - (8)^3]\tfrac{1}{2} = 1564 \text{ in.}^4$$

$$I_y = \tfrac{2}{12}(18)^3(\tfrac{7}{16}) + 2(9)(\tfrac{1}{2})(8.75)^2 = 1114 \text{ in.}^4$$

$$r_y = \sqrt{\frac{1114}{24.75}} = 6.71 \text{ in.}$$

$$\frac{KL}{r} = \frac{1.0(30)12}{6.71} = 53.7; \qquad \phi_c F_{cr} = 34.4 \text{ ksi (TABLE 3-50)}$$

$$\phi_c P_n = \phi_c F_{cr} A_g = 34.4(24.75) = 852 \text{ kips} < [P_u = 880 \text{ kips}] \quad \text{NG}$$

Try 2PLs—$\frac{1}{2}\times18$ and 2PLs—$\frac{1}{2}\times17$ (perforated):

$$A_g = 27.0 \text{ sq in.} \qquad I_y = 1175 \text{ in.}^4$$

$$I_x = 1744 \text{ in.}^4 \qquad r_y = 6.60 \text{ in.}$$

$$\frac{KL}{r} = \frac{1.0(30)12}{6.60} = 54.6; \qquad \phi_c F_{cr} = 34.2 \text{ ksi (TABLE 3-50)}$$

$$\phi_c P_n = \phi_c F_{cr} A_g = 34.2(27.0) = 923 \text{ kips} > [P_u = 880 \text{ kips}] \quad \text{OK}$$

(b) Check proportioning of plates (LRFD-E4).

1. Check b/t ratio for entire perforated plate (LRFD-Table B5.1):

$$\frac{b}{t} = \frac{17}{0.5} = 34 < \left[\lambda_r = \frac{317}{\sqrt{F_y}} = 44.8\right] \quad \text{OK}$$

2. Check b/t ratio for unstiffened portion within hole (LRFD-Table B5.1):

$$\frac{b}{t} = \frac{4.5}{0.5} = 9 < \left[\lambda_r = \frac{76}{\sqrt{F_y}} = 10.7\right] \quad \text{OK}$$

3. Check proportions of access holes (LRFD-E4):

$$\frac{\text{Length of hole}}{\text{Width of hole}} \leq 2 \qquad \text{Length of hole} = 2(8) = 16 \text{ in.-max}$$

$$\begin{pmatrix} \text{Clear distance} \\ \text{between holes} \end{pmatrix} \geq \begin{pmatrix} \text{Transverse distance} \\ \text{between nearest lines} \\ \text{of connecting welds} \end{pmatrix}$$

Clear spacing between holes $= 17$ in. min.

Use 2PLs—$\frac{1}{2}\times18$ (solid) and 2PLs—$\frac{1}{2}\times17$ (perforated) with holes 16″ by 8″ spaced $2'-8''$ center-to-center. ■■

Part II: Plates

6.14* INTRODUCTION TO STABILITY OF PLATES

All column sections, whether rolled shapes or built-up sections, are composed of plate elements. Up to this point in the chapter, consideration has been given only to the possibility of buckling of the member based on the slenderness ratio for the entire cross-section. It may be, however, that a local buckling will occur first in one of the plate elements that make up the cross-section. Such local buckling means that the buckled element will no longer take its proportionate share of any *additional* load the column is to carry; in other words, the efficiency of the cross-section is reduced.

The theory of bending of plates and elastic stability of plates are subjects that should be studied in depth by the advanced student in structural engineering. The brief treatment that follows is intended to give the reader the general idea of plate buckling necessary to properly use and understand current steel specifications. The general approach and terminology follow that of Timoshenko [6.66, 6.67].

Before one can treat the stability problem, the differential equation for bending of plates is required, just as the differential equation for the bending of a beam, Eq. 6.22, was used in the slender column stability treatment in Sec. 6.2.

Differential Equation for Bending of Homogeneous Plates

First, the strains will be obtained in terms of displacements. Let h = plate thickness and u, v, and w equal the displacements in the x, y, and z directions, respectively. Referring to Fig. 6.14.1, consider an element of a plate $dx\,dy$, and assume no stretching of the neutral plane at $z = 0$. Examining a slice $dx\,dy\,dz$ of the plate element located

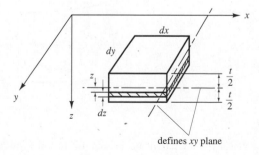

Figure 6.14.1 Plate element, coordinate system definition.

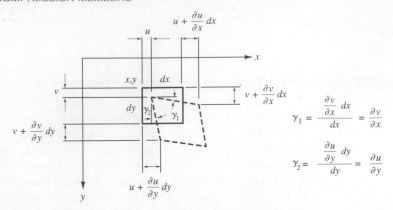

Figure 6.14.2 Deformations of plate element in xy plane.

at a distance z from the neutral plane shows, in Fig. 6.14.2, the coordinate unit strains ϵ_x, ϵ_y, and shearing strain γ_{xy}. Thus

$$\epsilon_x = \frac{u + \dfrac{\partial u}{\partial x}\,dx - u}{dx} = \frac{\partial u}{\partial x} \tag{6.14.1a}$$

$$\epsilon_x = \frac{\partial v}{\partial y} \tag{6.14.1b}$$

$$\gamma_{xy} = \gamma_1 + \gamma_2 = \frac{\partial v}{\partial x} + \frac{\partial u}{\partial y} \tag{6.14.1c}$$

Expressing the displacements in the plane of the plate in terms of the lateral deflection w, as shown in Fig. 6.14.3, and recognizing that positive slope gives a negative displacement u or v, one establishes

$$-u = z\,\frac{\partial w}{\partial x}; \qquad -v = z\,\frac{\partial w}{\partial y} \tag{6.14.2}$$

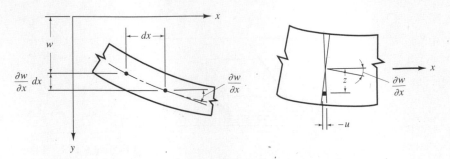

Figure 6.14.3 Deformation of plate element perpendicular to the xy plane.

Substitution of Eqs. 6.14.2 into Eqs. 6.14.1 gives strains in terms of curvatures for x-direction bending, y-direction bending, and twisting

$$\epsilon_x = \frac{\partial u}{\partial x} = -z\frac{\partial^2 w}{\partial x^2} \tag{6.14.3a}$$

$$\epsilon_y = \frac{\partial v}{\partial y} - -z\frac{\partial^2 w}{\partial y^2} \tag{6.14.3b}$$

$$\gamma_{xy} = \frac{\partial v}{\partial x} + \frac{\partial u}{\partial y} = -z\left(\frac{\partial^2 w}{\partial x\,\partial y} + \frac{\partial^2 w}{\partial x\,\partial y}\right) = -2z\frac{\partial^2 w}{\partial x\,\partial y} \tag{6.14.3c}$$

Next, making use of Hooke's Law expressing strains in terms of the stress σ_x, σ_y, normal stresses in the x- and y-directions, and τ_{xy}, the shear stress,

$$\epsilon_x = \frac{1}{E}[\sigma_x - \mu\sigma_y] \tag{6.14.4a}$$

$$\epsilon_y = \frac{1}{E}[-\mu\sigma_x + \sigma_y] \tag{6.14.4b}$$

$$\gamma_{xy} = \tau_{xy}/G \tag{6.14.4c}$$

where μ = Poisson's ratio (see Sec. 2.6) and G = shear modulus of elasticity.

For any stress condition, such as $\sigma_y = -\sigma_x$ that gives pure shear on an element rotated 45° to the x-axis, the work done by the equivalent systems of Fig. 6.14.4 must be a constant:

$$\tfrac{1}{2}\sigma_x\epsilon_x - \tfrac{1}{2}\sigma_x\epsilon_y = \tfrac{1}{2}\tau\gamma \tag{6.14.5}$$

Substituting Eqs. 6.14.4 into Eq. 6.14.5 gives

$$\frac{\sigma_x}{E}(\sigma_x - \mu\sigma_y + \mu\sigma_x - \sigma_y) = \frac{\tau^2}{G}$$

If $\sigma_y = -\sigma_x$, maximum $\tau = \sigma_x$; thus

$$\frac{1}{E}(1 + \mu + \mu + 1) = \frac{1}{G}$$

$$G = \frac{E}{2(1 + \mu)} \tag{6.14.6}$$

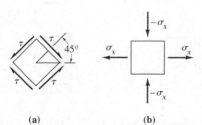

(a)　　　　　(b)　　　　　**Figure 6.14.4** Equivalent systems.

Solving Eqs. 6.14.4 for stresses in terms of strains and substituting Eqs. 6.14.3 for the strains give stresses in terms of curvatures,

$$\sigma_x = \frac{-zE}{1 - \mu^2}\left(\frac{\partial^2 w}{\partial x^2} + \mu\frac{\partial^2 w}{\partial y^2}\right) \tag{6.14.7a}$$

$$\sigma_y = \frac{-zE}{1 - \mu^2}\left(\mu\frac{\partial^2 w}{\partial x^2} + \frac{\partial^2 w}{\partial y^2}\right) \tag{6.14.7b}$$

$$\tau_{xy} = -2zG\frac{\partial^2 w}{\partial x\,\partial y} \tag{6.14.7c}$$

Next it is necessary to relate curvatures and bending moments. Referring to Fig. 6.14.5, and using the right-hand rule for positive twist, it is seen that the moments per unit width are

$$M_x = \int_{-t/2}^{t/2} z\sigma_x\,dz = \frac{-Et^3}{12(1 - \mu^2)}\left(\frac{\partial^2 w}{\partial x^2} + \mu\frac{\partial^2 w}{\partial y^2}\right) \tag{6.14.8a}$$

$$M_y = \int_{-t/2}^{t/2} z\sigma_y\,dz = \frac{-Et^3}{12(1 - \mu^2)}\left(\mu\frac{\partial^2 w}{\partial x^2} + \frac{\partial^2 w}{\partial y^2}\right) \tag{6.14.8b}$$

$$M_{xy} = -\int_{-t/2}^{t/2} \tau_{xy}z\,dz = +2G\left(\frac{t^3}{12}\right)\frac{\partial^2 w}{\partial x\,\partial y} \tag{6.14.8c}$$

Note that plate bending involves double curvature (a dish-shaped deflection surface for the plate). The narrower and longer in plan a plate is, the more the bending causes curvature to be one directional. A beam is a special case of a plate since it has a narrow width and long span. For beams, the Poisson's ratio (μ) effect is neglected. For instance, if the member is narrow in the y-direction and long in the x-direction, Eq. 6.14.8a for the plate would become

$$M_x = \frac{-Et^3}{12}\frac{d^2 w}{dx^2} \tag{6.14.9}$$

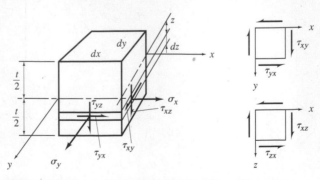

Figure 6.14.5   A plate element in bending. (Note that forces on faces at $x = 0$ and $y = 0$ not shown.)

where the partial derivatives disappear because w is no longer a function of y. If Eq. 6.14.9 is multiplied by the width b to change from moment per unit width in the y-direction to total moment, Eq. 6.14.9 would be the differential equation for beams,

$$M_x = -EI \frac{d^2w}{dx^2} \tag{6.14.10}$$

where $I = t^3 b/12$.

In theory of plates, the sign convention is that M_x is the bending moment caused by *stresses* σ_x acting in the x-direction. In beam theory, the usual practice is to have the subscript on M refer to the axis of zero stress in bending; that is, the neutral axis. Thus, for the span in the x-direction and the width in the y-direction as assumed for Eq. 6.14.10, M_y (instead of M_x) would have been used in beam theory, since the y-axis is the neutral axis about which bending occurs.

To continue the derivation of the plate bending differential equation, consider the equilibrium of all forces and moments acting on the plate element. Moment summation about the x- and y-axes and force summation in the z-direction gives three equations. Figure 6.14.6 is a free body of the plate element showing only the forces involved with moments about the y-axis.

Taking moments about the y-axis gives

$$\cancel{M_x \, dy} + \frac{\partial M_x}{\partial x} dx \, dy - \cancel{M_x \, dy} + \cancel{M_{yx} \, dx} + \frac{\partial M_{yx}}{\partial y} dy \, dx - \cancel{M_{yx} \, dx}$$

$$- \left(Q_x \, dy + \cancel{\frac{\partial Q_x}{\partial x} dx \, dy} \right) dx - \cancel{q \, dx \, dy \frac{dx}{2}} = 0$$

neglect   mom.   neglect
arm

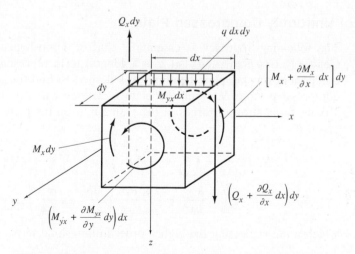

Figure 6.14.6   Free-body forces involved in rotation about y-axis. (Forces involved in rotation about the x-axis not shown.)

Neglecting infinitesimals of higher order and dividing by $dx\,dy$ gives

$$\frac{\partial M_x}{\partial x} + \frac{\partial M_{yx}}{\partial y} - Q_x = 0 \tag{6.14.11}$$

Similarly for moments about the x-axis,

$$\frac{\partial M_y}{\partial y} + \frac{\partial M_{xy}}{\partial x} - Q_y = 0 \tag{6.14.12}$$

Force equilibrium in the z-direction gives

$$\frac{\partial Q_x}{\partial x} + \frac{\partial Q_y}{\partial y} + q = 0 \tag{6.14.13}$$

Using Eqs. 6.14.11 and 6.14.12 for Q_x and Q_y, and substitution into Eq. 6.14.13 gives

$$\frac{\partial^2 M_x}{\partial x^2} + \frac{\partial^2 M_y}{\partial y^2} + 2\frac{\partial^2 M_{xy}}{\partial x\,\partial y} = -q \tag{6.14.14}$$

Defining $D = Et^3/[12(1 - \mu^2)]$ and substituting Eq. 6.14.8 into Eq. 6.14.14 gives the different equation for bending of homogeneous plates,

$$D\left(\frac{\partial^4 w}{\partial x^4} + 2\frac{\partial^4 w}{\partial x^2\,\partial y^2} + \frac{\partial^4 w}{\partial y^4}\right) = q \tag{6.14.15}$$

Equation 6.14.15, if written for a beam of width b, is the differential equation for load,

$$EI\frac{d^4 w}{dx^4} = qb \tag{6.14.16}$$

where qb is the load per unit length along the span of the beam.

Buckling of Uniformly Compressed Plate

The following approach is essentially that of Timoshenko [6.67] as modified by Gerstle [6.68]. Realizing that q is a general term representing the transverse load component causing plate bending, it is desired to find the transverse component of compressive force N_x when the plate is deflected into a slightly buckled position. Taking summation of forces in the z-direction on the plate element of Fig. 6.14.7b gives

$$N_x\,dy\,\frac{\partial w}{\partial x} - \left(N_x + \frac{\partial N_x}{\partial x}\,dx\right)dy\left(\frac{\partial w}{\partial x} + \frac{\partial^2 w}{\partial x^2}\,dx\right) = q\,dx\,dy$$

$$-\left(N_x\frac{\partial^2 w}{\partial x^2} + \frac{\partial N_x}{\partial x}\frac{\partial w}{\partial x} + \frac{\partial N_x}{\partial x}\,dx\,\frac{\partial^2 w}{\partial x^2}\right)dy\,dx = q\,dx\,dy \tag{6.14.17}$$

which upon neglecting the higher-order infinitesimal terms gives

$$q = -N_x\frac{\partial^2 w}{\partial x^2} \tag{6.14.18}$$

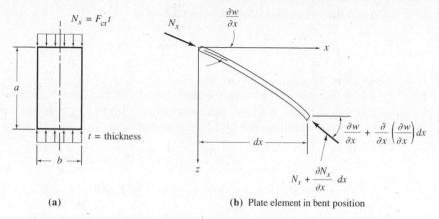

(a)   **(b)** Plate element in bent position

Figure 6.14.7   Uniformly compressed plate.

The differential equation, Eq. 6.14.15, then becomes

$$\frac{\partial^4 w}{\partial x^4} + 2\frac{\partial^4 w}{\partial x^2\, \partial y^2} + \frac{\partial^4 w}{\partial y^4} = -\frac{N_x}{D}\frac{\partial^2 w}{\partial x^2} \qquad (6.14.19)$$

which is a partial differential equation where w is a function of both x and y. The deflection w can be expressed as the product of an x function (X) and a y function (Y). Further, buckling may be assumed to give sinusoidal variation in the x-direction. Thus

$$w = X(x)Y(y) \qquad (6.14.20)$$

Letting

$$X(x) = \sin\frac{m\pi x}{a}$$

where the X function satisfies the zero deflection and zero moment conditions of simple support at $x = 0$ and $x = a$. Substitution of Eq. 6.14.20 into Eq. 6.14.19 gives, after cancelling the common term $\sin m\pi x/a$,

$$\left(\frac{m\pi}{a}\right)^4 Y - 2\left(\frac{m\pi}{a}\right)^2\frac{d^2 Y}{dy^2} + \frac{d^4 Y}{dy^4} = +\frac{N_x}{D}\left(\frac{m\pi}{a}\right)^2 Y$$

$$\frac{d^4 Y}{dy^4} - 2\left(\frac{m\pi}{a}\right)^2\frac{d^2 Y}{dy^2} + \left[\left(\frac{m\pi}{a}\right)^4 - \frac{N_x}{D}\left(\frac{m\pi}{a}\right)^2\right]Y = 0 \qquad (6.14.21)$$

an ordinary fourth-order homogeneous differential equation.

The solution may be expressed in the form

$$Y = C_1 \sinh \alpha y + C_2 \cosh \alpha y + C_3 \sin \beta y + C_4 \cos \beta y \qquad (6.14.22)$$

where

$$\alpha = \sqrt{\left(\frac{m\pi}{a}\right)^2 + \sqrt{\frac{N_x}{D}\left(\frac{m\pi}{a}\right)^2}} \quad \text{and} \quad \beta = \sqrt{-\left(\frac{m\pi}{a}\right)^2 + \sqrt{\frac{N_x}{D}\left(\frac{m\pi}{a}\right)^2}}$$

Thus the entire plate deflection equation is

$$w = \left(\sin \frac{m\pi x}{a}\right)(C_1 \sinh \alpha y + C_2 \cosh \alpha y + C_3 \sin \beta y + C_4 \cos \beta y)$$

$$(6.14.23)$$

which must satisfy boundary conditions. Assuming the x-axis as an axis of symmetry through the plate, i.e., identical support conditions along the two edges parallel to the direction of loading, the odd function coefficients C_1 and C_3 must be zero. Thus

$$w = (C_2 \cosh \alpha y + C_4 \cos \beta y) \sin \frac{m\pi x}{a} \qquad (6.14.24)$$

Using simple support conditions at $y = b/2$ and $y = -b/2$, requires that at $y = \pm b/2$,

$$w = 0 = \left(C_2 \cosh \alpha \frac{b}{2} + C_4 \cos \beta \frac{b}{2}\right) \sin \frac{m\pi x}{a}$$

$$\frac{\partial^2 w}{\partial y^2} = 0 = \left(C_2 \alpha^2 \cosh \alpha \frac{b}{2} - C_4 \beta^2 \cos \beta \frac{b}{2}\right) \sin \frac{m\pi x}{a}$$

$$(6.14.25)$$

For a solution other than $C_2 = C_4 = 0$ it is necessary for the determinant of the coefficients to be zero. Thus

$$(\alpha^2 + \beta^2) \cosh \alpha \frac{b}{2} \cos \beta \frac{b}{2} = 0 \qquad (6.14.26)$$

Since $\alpha^2 \neq -\beta^2$ unless $N_x = 0$ (a trivial solution), and since $\cosh \alpha(b/2) > 1$, the only way Eq. 6.14.26 can be satisfied in the real problem is for

$$\cos \beta \frac{b}{2} = 0$$

Therefore

$$\beta \frac{b}{2} = \frac{\pi}{2}, \frac{3\pi}{2}, \frac{5\pi}{2}, \quad \text{etc.}$$

Using the lowest value of $\beta(b/2)$ and substituting into β as defined below Eq. 6.14.22 gives

$$\frac{b}{2}\sqrt{-\left(\frac{m\pi}{a}\right)^2 + \sqrt{\frac{N_x}{D}\left(\frac{m\pi}{a}\right)^2}} = \frac{\pi}{2}$$

$$\frac{N_x}{D}\left(\frac{m\pi}{a}\right)^2 = \left[\frac{\pi^2}{b^2} + \left(\frac{m\pi}{a}\right)^2\right]^2$$

$$N_x = D\left[\frac{\pi^2 a}{b^2 m\pi} + \frac{m\pi}{a}\right]^2$$

$$N_x = \frac{D\pi^2}{b^2}\left[\frac{1}{m}\frac{a}{b} + m\frac{b}{a}\right]^2 \qquad (6.14.27)$$

Since $N_x = F_{cr}t$ and $D = Et^3/[12(1 - \mu^2)]$, the elastic buckling unit stress may be expressed as

$$F_{cr} = k\frac{\pi^2 E}{12(1 - \mu^2)(b/t)^2} \qquad (6.14.28)$$

where for the specific case treated here, the elastic plate buckling coefficient k is

$$k = \left[\frac{1}{m}\frac{a}{b} + m\frac{b}{a}\right]^2 \qquad (6.14.29)$$

The buckling coefficient k is a function of the type of stress (in this case uniform compression on two opposite edges) and the edge support conditions (in this case simple support on four edges), in addition to the aspect ratio a/b which appears directly in the equation.

The equation for plate buckling, Eq. 6.14.28, is entirely general in terms of k and the development leading up to it for this one case may be considered illustrative of the procedure. The integer m indicates the number of half-waves that occur in the x direction at buckling. Figure 6.14.8 shows that there is a minimum value of k for any given number of half-waves, i.e., the weakest condition. It is noted that this weakest situation occurs when the length is an even multiple of width, and that multiple equals the number of half-waves.

Thus, setting $a/b = m$ gives $k = 4$. Further, as m becomes larger the k equation becomes flatter and approaches a constant value of 4 for large a/b ratio. This gives for the elastic buckling stress equation of plate elements under uniform compression along two edges and simply supported along the two edges parallel to the load,

$$F_{cr} = \frac{4\pi^2 E}{12(1 - \mu^2)(b/t)^2} \qquad (6.14.30)$$

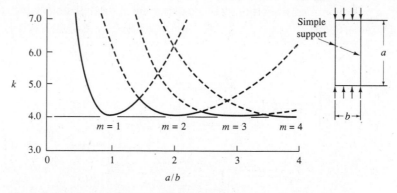

Figure 6.14.8   Buckling coefficient for uniformly compressed plate— simple support longitudinal edges (Eq. 6.14.29).

6.15* STRENGTH OF PLATES UNDER UNIFORM EDGE COMPRESSION

Since rolled shapes, as well as built-up shapes, are composed of plate elements, the column strength of the section based on its overall slenderness ratio can only be achieved if the plate elements do not buckle locally. Local buckling of plate elements can cause premature failure of the entire section, or at the least it will cause stresses to become nonuniform and reduce the overall strength.

In Sec. 6.14 the basic approach to elastic stability of plates was developed. The theoretical elastic buckling stress for a plate was shown to be expressible as

$$F_{cr} = k \frac{\pi^2 E}{12(1 - \mu^2)(b/t)^2} \qquad [6.14.28]$$

where k is a constant depending on type of stress, edge support conditions, and length to width ratio (aspect ratio) of the plate, E the modulus of elasticity, μ Poisson's ratio, and b/t the width/thickness ratio.

In general, plate compression elements can be separated into two categories: (1) stiffened elements; those supported along two edges parallel to the direction of compressive stress; and (2) unstiffened elements; those supported along one edge and free on the other edge parallel to direction of compressive stress. Refer to Fig. 6.15.1 for typical examples of these two situations.

For the elements shown in Fig. 6.15.1 various degrees of edge rotation restraint are present. Figure 6.15.2 shows the variation in k with aspect ratio a/b for most of the idealized edge conditions, i.e., clamped (fixed), simply supported, and free.

Actual plate strength in compression is dependent on many of the same factors that affect overall column strength, particularly residual stress. Figure 6.15.3 shows typical behavior of a compressed plate loaded to its ultimate load. Assuming ideal elastic-plastic material containing no residual stress the stress distribution remains uniform until the elastic buckling stress F_{cr} is reached. Further increase in load can be achieved but the portion of the plate farthest from its side supports will deflect out of its original plane. This out-of-plane deflection causes the stress distribution to be nonuniform even though the load is applied through ends which are rigid and perfectly straight.

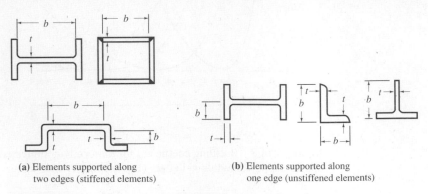

(a) Elements supported along two edges (stiffened elements)

(b) Elements supported along one edge (unstiffened elements)

Figure 6.15.1   Stiffened and unstiffened compression elements.

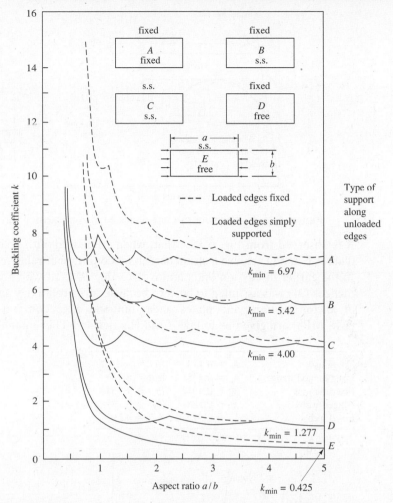

Figure 6.15.2 Elastic buckling coefficients for compression in flat rectangular plates. (Adapted from Gerard and Becker [6.69])

Figure 6.15.3 shows that plate strength under edge compression consists of the sum of two components; (1) elastic or inelastic buckling stress represented by Eq. 6.14.28, and (2) post-buckling strength. Also one should note the higher post-buckling strength as the width-to-thickness ratio b/t becomes larger. For low values of b/t, not only will post-buckling strength vanish, but the entire plate may have yielded and reached the strain-hardening condition, so that F_{cr}/F_y may become greater than unity. For plates *without* residual stress (referring to Fig. 6.15.4) three regions must be considered for establishing strength; elastic buckling (Euler hyperbola), yielding (segments AB, $A'B$, and $A''B$), and strain hardening.

If F_{cr}/F_y is defined as $1/\lambda_c^2$, Eq. 6.14.28 for plates then becomes

$$\lambda_c = \frac{b}{t}\sqrt{\frac{F_y(12)(1-\mu^2)}{\pi^2 E k}} \qquad (6.15.1)$$

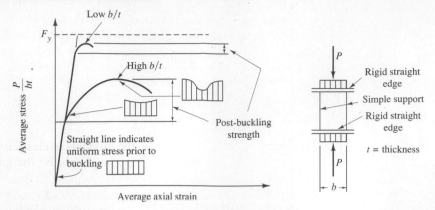

Figure 6.15.3   Behavior of plate under edge compression.

It is observed from Fig. 6.15.4 that, when compared with columns (curve a), plates (curves b and c) achieve a strain hardening condition at relatively higher values of λ_c. In the earlier discussion on columns the value of λ_c at which strain hardening commences (λ_0) was assumed to be zero because of its relatively small value. The values of λ_0 for columns and plates under uniform edge compression for $F_y = 36$ ksi (248 MPa) are given as follows from Haaijer and Thürlimann [6.70]:

Columns	$\lambda_0 = 0.173$	($KL/r = 15.7$)
Long hinged flanges	$\lambda_0 = 0.455$	($b/t = 8.15$)
Fixed flanges	$\lambda_0 = 0.461$	($b/t = 14.3$)
Hinged webs	$\lambda_0 = 0.588$	($b/t = 32.3$)
Fixed webs	$\lambda_0 = 0.579$	($b/t = 42.0$)

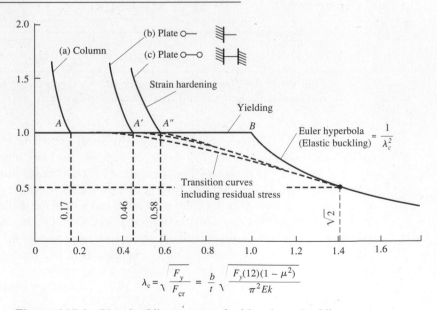

Figure 6.15.4   Plate buckling compared with column buckling. (Adapted from Ref. 6.70)

From the above, the important factor determining λ_0 is whether the plate element is supported along one or both edges parallel to loading, while the degree of rotational restraint along the loaded edge (simply supported or fixed) has essentially no effect. Thus curves b and c of Fig. 6.15.4 each can represent two cases, where point A' has been taken at $\lambda_c = 0.46$ and point A'' at $\lambda_c = 0.58$.

Since plates as well as rolled shapes contain residual stress the true strength should be represented by a transition curve, Fig. 6.15.4, between the Euler curve and the point at which strain hardening commences.

When considering inelastic behavior, the modulus of elasticity used for calculating strain in the direction of maximum stress σ_x should be the tangent modulus E_t. Examination of Eq. 6.14.4a shows that for inelastic strains in the x-direction but elastic strain in the y-direction, E cannot be factored out. Bleich [6.9] has shown the solution for this case of using different E values, and suggests arbitrarily using $\sqrt{E_t/E}$ as a multiplier for Eq. 6.14.28.

In summary, the strength of plates under edge compression may be governed by (1) strain hardening, low values of λ_c; (2) yielding, at $\lambda_c =$ say 0.5 to 0.6; (3) inelastic buckling, represented by the transition curve (some fibers elastic and some yielded); (4) elastic buckling represented by the Euler hyperbola, at λ_c about 1.4; and (5) post-buckling strength with stress redistribution and large deformation, say for λ_c greater than 1.5.

For design purposes, performance criteria must be established to decide what range of λ_c values may be acceptable in design and how conservative (and simple) or liberal (and relatively complicated) should be the specification expressions for plate strength.

6.16 AISC WIDTH/THICKNESS LIMITS λ_r TO ACHIEVE YIELD STRESS WITHOUT LOCAL PLATE BUCKLING

For a better understanding of the background for these requirements the reader is invited to delve into the subject of plate stability and strength as introduced in Secs. 6.14 and 6.15. However, it may be sufficient for many purposes merely to understand that components such as flanges, webs, angles, and cover plates, which are combined to form a column section may themselves buckle locally prior to the entire section achieving its maximum capacity. Typical elements are shown in Fig. 6.15.1. The buckled deflection of uniformly compressed plates is shown in Fig. 6.16.1 where two categories are apparent: (1) "unstiffened" plate elements having one free edge parallel to loading; and (2) "stiffened" plate elements supported along both edges parallel to loading.

Plates in compression behave essentially the same as columns and the basic elastic buckling expression corresponding to the Euler equation for columns has been derived as Eq. 6.14.28,

$$F_{\text{cr}} = k \frac{\pi^2 E}{12(1 - \mu^2)(b/t)^2} \qquad [6.14.28]$$

where k is a constant depending on type of stress, edge conditions, and length to width ratio; μ is Poisson's ratio, and b/t is the width/thickness ratio (see Fig. 6.16.1). Typical k values are given in Fig. 6.15.2.

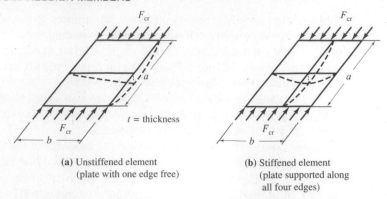

(a) Unstiffened element
(plate with one edge free)

(b) Stiffened element
(plate supported along
all four edges)

Figure 6.16.1   Buckled deflection of uniformly compressed plates.

It is known that for low b/t values, strain hardening is achieved without buckling occurring, for medium values of b/t residual stress and imperfections give rise to inelastic buckling represented by a transition curve, and for large b/t buckling occurs in accordance with Eq. 6.14.28. Actual strength for plates with large b/t ratio exceeds buckling strength, i.e., they exhibit post-buckling strength. Thus strength for plates may be shown in a dimensionless fashion as in Fig. 6.16.2.

To establish design requirements, the desired performance must be ascertained. The local buckling of a column component may logically be prevented prior to achieving full strength of the column based on its overall slenderness ratio KL/r. The performance requirement would then be

$$\underset{\substack{\text{component} \\ \text{element,} \\ \text{i.e., plate}}}{F_{cr}} \geq \underset{\substack{\text{overall} \\ \text{column}}}{F_{cr}} \tag{6.16.1}$$

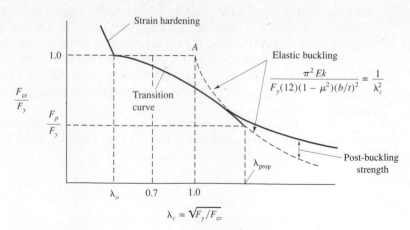

Figure 6.16.2   Dimensionless representation of plate strength in edge compression.

meaning that acceptable b/t ratios would vary depending on the overall slenderness ratio of the column. If post-buckling strength were considered, the relationship would be additionally complicated.

For many years the AISI Specification [1.11] for cold-formed steel has used the approach of Eq. 6.16.1, and also included treatment of post-buckling strength. AISC (LRFD and ASD-Appendix B) also includes similar provisions to consider post-buckling strength. Once buckling of a compression element has occurred, the efficiency of the element to carry load is reduced.

Design limits are generally simplified to assure the compression element will reach yield stress F_y without local buckling occurring, even though the slenderness ratio of a column may prevent the element from reaching yield stress. The width/thickness (b/t) ratios λ to prevent local buckling until the yield stress is reached are the λ_r values of LRFD and the noncompact limits in ASD-B5.

The requirement to achieve yield stress without local buckling is

$$F_{cr} = \frac{k\pi^2 E}{12(1 - \mu^2)(b/t)^2} \geq F_y \tag{6.16.2}$$

Using $\mu = 0.3$ for steel, and $E = 29,000,000$ psi and F_y in psi,

$$\frac{b}{t} \leq 5120 \sqrt{\frac{k}{F_y, \text{psi}}} \tag{6.16.3}*$$

which is represented by point A at $\lambda_c = 1.0$ on Fig. 6.16.2, a point lying above the transition curve. Thus a reduced value of λ_c should be used to minimize the deviation between F_y and the transition curve which accounts for residual stress and imperfections. Thus $\lambda_c = 0.7$ is taken as a rational value, which gives for b/t

$$\frac{b}{t} \leq 5120\lambda_c \sqrt{\frac{k}{F_y}} = 3580 \sqrt{\frac{k}{F_y}} \tag{6.16.4}$$

where F_y is in psi. Table 6.16.1 shows width/thickness ratios for various situations of uniform compression. The coefficients used by AISC since 1969 tend to imply greater accuracy in criteria than is justified. The original coefficients were established using F_y in psi; after some rounding they formed the basis of the 1963 AISC Specification. The present values are obtained by dividing by $\sqrt{1000}$ so that F_y can be used in ksi. Table 6.16.2 gives evaluated limits λ_r from LRFD and ASD-Table B5.1** for several different yield stresses.

For rolled and welded box shapes, *tensile* residual stresses are induced at the corners as discussed in Sec. 6.5. When compressive external loads are applied, the tensile residual stress F_r must first be reduced to zero, afterwhich additional compressive load causes compressive stresses at the corners. Thus, when the nominal stress P/A_g is F_y on the cross-section, it is actually $(F_y - F_r)$ at the important regions near

---

*For SI, $\qquad\qquad\qquad \dfrac{b}{t} \leq 425 \sqrt{\dfrac{k}{F_y}}$ with F_y in MPa $\qquad\qquad\qquad$ (6.16.3)

**ASD-B5.1 does not use the symbols λ for width/thickness ratio or λ_r for "noncompact" limit.

TABLE 6.16.1 AISC WIDTH/THICKNESS RATIO λ_r LIMITS FOR PLATE ELEMENTS SUBJECT TO UNIFORM COMPRESSION[a]

Structural elements (1)	Buckling coefficients k (Fig. 6.15.2) (2)	b/t Eq. 6.16.4 (3)	F_y(psi) (4)	LRFD and ASD-B5 F_y(ksi) (5)	F_y(MPa) (6)
Unstiffened:					
(a) Single angles	0.425	$2340/\sqrt{F_y}$	$2400/\sqrt{F_y}$	$76/\sqrt{F_y}$	$200/\sqrt{F_y}$
(b) Flanges[f]	0.70[b]	$3000/\sqrt{F_y}$	$3000/\sqrt{F_y}$	$95/\sqrt{F_y}$	$250/\sqrt{F_y}$
(c) Stems of tees	1.277	$4050/\sqrt{F_y}$	$4000/\sqrt{F_y}$	$127/\sqrt{F_y}$	$333/\sqrt{F_y}$
Stiffened:					
(a) Uniform thickness flanges, such as tubular sections			$7500/\sqrt{F_y}$[e]	$238/\sqrt{F_y}$	$625/\sqrt{F_y}$
(b) Perforated cover plates	6.97[d]	$9460/\sqrt{F_y}$	$10{,}000/\sqrt{F_y}$	$317/\sqrt{F_y}$	$832/\sqrt{F_y}$
(c) Others	5.0[c]	$8010/\sqrt{F_y}$	$8000/\sqrt{F_y}$	$253/\sqrt{F_y}$	$664/\sqrt{F_y}$

[a] ASD-Table B5.1 does not use symbol λ_r for "noncompact" limit.

[b] Arbitrarily selected to be about one-third of the way between simply supported and fixed along the supported edge.

[c] Edge restraint estimated at about $\frac{1}{3}$ fixed ($k = 4.0$ for simple support, and $k = 6.97$ for fixed—see Fig. 6.15.2).

[d] Consider full fixity—use of net plate width will provide adequate reserve.

[e] Hollow sections generally receive negligble torsional restraint by the thin supporting edges; thus a coefficient somewhat less than 8000 is used.

[f] LRFD-B5.1 in 1993 no longer explicitly mentions *flanges* of rolled I-shaped sections in uniform compression; included are outstanding legs of pairs of angles in contact, flanges of channels in axial compression, and angles and plates projecting from beams or compression members. All of these are unstiffened elements having substantial rotational restraint along their supported edge.

the corners of the box shape. Since the superimposed stress level is actually less than F_y, the limit λ_r could be permitted higher than if the residual stress were compressive.

For stiffened elements, such as uniform thickness plates and perforated plates, the 1986 LRFD Specification used $\sqrt{(F_y - F_r)}$ in place of $\sqrt{F_y}$ in the denominator of the width/thickness ratio limits. The 1989 ASD Specification retained the more conservative equations (without the subtracted F_r) used for the past several decades. Now, the 1993 LRFD Specification has reverted to those limit expressions historically used.

For unstiffened elements, the stress at the free edge is predominant and residual stress is generally low or compressive; thus no reduction F_r is (or was) appropriate.

6.17 AISC WIDTH/THICKNESS LIMIT λ_p TO ACHIEVE SIGNIFICANT PLASTIC DEFORMATION

Sometimes plate elements of the cross-section must not buckle until they have undergone significant compressive strain exceeding the strain ϵ_y at first yield, that is, strain into the plastic region, as shown in Fig. 6.17.1. The lower the width/thickness ratio, the greater the compressive strain ϵ that can be absorbed without buckling. In axially loaded columns there would be no need for the ability to undergo plastic strain because

TABLE 6.16.2 WIDTH/THICKNESS RATIO λ_r LIMITING VALUES[†]
FOR PLATE ELEMENTS* TO REACH YIELD STRESS
IN AXIAL COMPRESSION

	LRFD-B5.1 and ASD-B5.1 F_y(ksi)					
Structural elements	36	42	50	60	65	100
Unstiffened:						
(a) Single angles	12.7	11.7	10.7	9.8	9.4	7.6
(b) Flanges	15.8	14.7	13.4	12.3	11.8	9.5
(c) Stems of tees	21.2	19.6	18.0	16.4	15.8	12.7
Stiffened:						
(a) Uniform thickness flanges, as for tubular sections						
	39.7	36.7	33.7	30.7	29.5	23.8
(b) Perforated plates	52.8	48.9	44.8	40.9	39.3	31.7
(c) Others	42.2	39.0	35.8	32.7	31.4	25.3
	F_y(MPa)					
Structural elements	250	300	350	400	450	700
Unstiffened:						
(a) Single angles	12.6	11.5	10.7	10.0	9.4	7.5
(b) Flanges	15.8	14.4	13.4	12.5	11.8	9.4
(b) Stems of tees	21.1	19.2	17.8	16.7	15.7	12.6
Stiffened:						
(a) Uniform thickness flanges, as for tubular sections						
	39.5	36.1	33.4	31.3	29.5	23.6
(b) Perforated plates	52.6	48.0	44.5	41.6	39.2	31.4
(c) Others	42.0	38.3	35.5	33.2	31.3	25.1

*ASD-Table B5.1 does not use the symbol λ_r for "noncompact" limit.

[†] Values in tables use equations from Table 6.16.1; column (5) for Inch-Pound units, and column (6) for SI units.

the overall strength of the column based on its *KL/r* would not require plastic deformation. However in beams, as discussed in Chapter 7, the flanges might be required to undergo significant plastic strain without having local buckling occur.

Referring to Fig. 6.16.2, λ_c must be restricted not to exceed λ_0 if strain hardening is to be reached without plate buckling. From Fig. 6.15.4, λ_c should not exceed about 0.46 for unstiffened compression elements and 0.58 for stiffened compression elements.

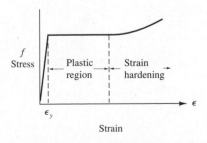

Figure 6.17.1 Plastic and strain-hardening regions of stress-strain relationship for steel.

For *unstiffened elements*, Eq. 6.16.4 with $\lambda_c = 0.46$ gives

$$\frac{b}{t} \leq 2350 \sqrt{\frac{k}{F_y, \text{psi}}} \quad \text{or} \quad 74.3 \sqrt{\frac{k}{F_y, \text{ksi}}} \tag{6.17.1}$$

When $k = 0.425$ (its least value), Eq. 6.17.1 gives

$$\frac{b}{t} \leq \frac{48.5}{\sqrt{F_y, \text{ksi}}} \tag{6.17.2}$$

Since residual stress effects disappear in the plastic range and material imperfections have less effect, Eq. 6.17.2 is an overly severe limitation. Furthermore, the strain at onset of strain hardening is 15 to 20 times ϵ_y and that amount of plastic strain is not necessary even for achieving the plastic moment strength discussed in Chapter 7. The LRFD-B5.1 λ_p limits and ASD-B5.1 compact section limits are intended to achieve compression plastic strain about 7 to 9 times ϵ_y, about one-half the strain necessary to reach strain hardening. Thus, the unstiffened compression element limit λ_p is

$$\frac{b}{t} \leq \frac{65}{\sqrt{F_y, \text{ksi}}} \tag{6.17.3}*$$

For *stiffened elements*, Eq. 6.16.4 with $\lambda_c = 0.58$ gives

$$\frac{b}{t} \leq 2965 \sqrt{\frac{k}{F_y, \text{psi}}} \quad \text{or} \quad 93.7 \sqrt{\frac{k}{F_y, \text{ksi}}} \tag{6.17.4}$$

when $k = 4$, the minimum value assuming edge rotational restraint as the hinged condition (actually the k lies somewhere between values for Cases A and C of Fig. 6.15.2), Eq. 6.17.4 gives

$$\frac{b}{t} \leq \frac{187}{\sqrt{F_y, \text{ksi}}} \tag{6.17.5}$$

LRFD and ASD-B5.1 prescribe the limit for the stiffened compression element as

$$\frac{b}{t} \leq \frac{190}{\sqrt{F_y, \text{ksi}}} \tag{6.17.6}*$$

A tabulation of the limits, Eq. 6.17.3 for unstiffened elements and Eq. 6.17.6 for stiffened elements, appears in Table 6.17.1. Additional discussions relative to local buckling limitations to develop plastic strength are given by Lay [6.71] and McDermott [6.72].

---

*For SI, with F_y in MPa,

$$\frac{b}{t} \leq \frac{171}{\sqrt{F_y}} \tag{6.17.3}$$

$$\frac{b}{t} \leq \frac{500}{\sqrt{F_y}} \tag{6.17.6}$$

TABLE 6.17.1   WIDTH/THICKNESS RATIO λ_p LIMITS
FOR PLATE ELEMENTS TO ACCOMMODATE
PLASTIC STRAIN IN AXIAL COMPRESSION

F_y (ksi)	F_y (MPa)	Unstiffened elements LRFD and ASD-B5.1 (Eq. 6.17.3)	Stiffened elements LRFD and ASD-B5.1 (Eq. 6.17.6)
36	250	10.8	31.7
42	290	10.0	29.3
45	310	9.7	28.3
50	340	9.2	26.9
55	380	8.8	25.6
60	410	8.4	24.5
65	450	8.1	23.6

6.18* AISC PROVISIONS TO ACCOUNT FOR THE BUCKLING AND POST-BUCKLING STRENGTHS OF PLATE ELEMENTS

As discussed in Secs. 6.15 and 6.16, plate elements in compression, either "stiffened" or "unstiffened" (see Fig. 6.16.1), have strength after buckling has occurred, i.e., post-buckling strength. Stiffened elements have a large post-buckling strength while unstiffened elements have only a little. However, since the strength of such elements can be evaluated, there is good reason to provide for its use as has been done in the *Specification for the Design of Cold-Formed Steel Structural Members* [1.11], first introduced in 1946.

From Fig. 6.18.1a it is apparent that the nominal strength P_n of a stiffened element might be expressed as

$$P_n = t \int_0^b f(x)\, dx \tag{6.18.1}$$

involving an integration of a nonuniform stress distribution; or alternatively, an "effective width" concept (Fig. 6.18.1b) may be used:

$$P_n = t b_E f_{max} = A_{eff} f_{max} \quad \text{(stiffened element)} \tag{6.18.2}$$

where b_E = effective width over which the maximum stress may be considered uniform and give the correct total capacity.

Figure 6.18.1c shows that Eq. 6.18.1 is equally valid for the unstiffened element except that the stress distribution is not symmetrical about the center of the element. If a reduced stress $f_{avg} < f_{max}$ is used, the unstiffened element capacity could be written (Fig. 6.18.1d),

$$P_n = t b f_{avg} = A_{gross} f_{avg} \quad \text{(unstiffened element)} \tag{6.18.3}$$

The AISC and the AISI [1.11] have chosen to treat thin elements according to Eqs. 6.18.2 and 6.18.3, although actually either equation could have been used for

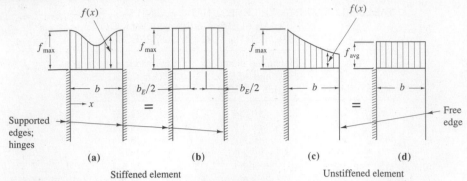

Figure 6.18.1 Plate elements under axial compression, showing actual stress distribution and an equivalent system.

either type of element. Because of the large post-buckling strength of the stiffened element, one can imagine that it *has* buckled and part of the element is no longer active. On the other hand, the unstiffened element, with relatively little post-buckling strength may be thought of as not buckling because of the use of a reduced stress.

Effect on Overall Column Strength

For design, it is desired to use gross section properties; thus for stiffened elements

$$P_n = \frac{A_{\text{eff}}}{A_{\text{gross}}} f_{\max} A_{\text{gross}} = Q_a f_{\max} A_{\text{gross}} \tag{6.18.4}$$

and for unstiffened elements,

$$P_n = \frac{f_{\text{avg}}}{f_{\max}} f_{\max} A_{\text{gross}} = Q_s f_{\max} A_{\text{gross}} \tag{6.18.5}$$

where Q_a and Q_s may be thought of as shape factors, or form factors.

A compression system composed of both stiffened and unstiffened elements would be treated as unstiffened for establishing the stress f_{avg}; then the effective width for the stiffened elements is determined using $f_{\max} = f_{\text{avg}}$. Thus the total capacity would be

$$P_n = f_{\text{avg}} A_{\text{eff}} \tag{6.18.6}$$

which gives

$$P_n = \frac{f_{\text{avg}}}{f_{\max}} (f_{\max}) \frac{A_{\text{eff}}}{A_{\text{gross}}} (A_{\text{gross}}) = Q_s Q_a f_{\max} A_{\text{gross}} \tag{6.18.7}$$

From Eqs. 6.18.4, 6.18.5, and 6.18.7, it is clear that the effect of premature local buckling before the strength of the overall column has been achieved is to multiply the maximum achievable stress by the form factors Q. Neglecting the possibility of strain hardening, the maximum stress is the yield stress, which is therefore to be multiplied by Q.

Load and Resistance Factor Design. For the equations used in LRFD-Appendix E3, λ_c will become $\lambda_c\sqrt{Q}$ and the short column equation becomes Eq. 6.8.2, as follows:

For $\lambda_c\sqrt{Q} \leq 1.5$;

$$F_{cr} = (0.658^{Q\lambda_c^2})QF_y \qquad [6.8.2]$$

Allowable Stress Design. For the basic SSRC parabola used in ASD, Eq. 6.7.1 for short columns becomes:

For $\dfrac{KL}{r} \leq C_c$;

$$F_{cr} = QF_y\left[1 - \frac{QF_y}{4\pi^2E}\left(\frac{KL}{r}\right)^2\right] \qquad (6.18.8)$$

where $Q = Q_sQ_a$. From Eq. 6.18.8, when $F_{cr} = QF_y/2$ the slenderness ratio is

$$C_c = \sqrt{\frac{2\pi^2E}{QF_y}} \qquad (6.18.9)$$

Whenever $\lambda_c\sqrt{Q} > 1.5$ in LRFD, or $KL/r > C_c$ in ASD, the effect of local buckling on overall column strength is negligible; thus, for slender columns the Euler equation is the basis of strength.

Since the form factor Q is used to account for the reduction in efficiency of the cross-section instead of modifying the cross-sectional properties, whether or not Q is less than unity the nominal strength P_n is computed as the product of F_{cr} times A_g, and the radius of gyration r is that of the *gross* section in the computation of KL/r.

Form Factor Q_s for Unstiffened Elements

Referring back to Fig. 6.16.2, one may note that when Q is less than 1.0 it means that $\lambda_c > \lambda_0$. A transition parabola could have been used to compute the reduced stress. For simplification a straight line has been used for single angles as shown by Curve A in Fig. 6.18.2. The straight line uses $\lambda_c = 0.7$ as the maximum for which $F_{cr} = F_y$, and takes the proportional limit λ_{prop} at $\sqrt{2}$, approximately the same as for overall column buckling. However, because of some post-buckling strength the theoretical Euler-type curve (Curve C) has been raised to give the AISC curve (Curve B). Many different expressions could have been used with the same logical results.

LRFD-Appendix B5.3.a and ASD-Appendix B5.2.a give similar stress reduction equations for unstiffened flanges and the stems of tees. These other equations are approximately proportional to $\sqrt{k}$, as may be noted by reference to Eq. 6.16.4 and Table 6.16.1. The table contains the k values used for the other unstiffened cases.

The limiting proportions for tees given in LRFD-Commentary E3 and ASD-Appendix B, Table A-B5.1 are to preclude torsional buckling as a failure mode. This concept is discussed in Sec. 8.11.

Although Q_s has been defined to this point as F_{cr}/F_y, when overall buckling of a column occurs (based on its KL/r) the average stress P_u/A_g is always *less* than F_y. This means local buckling of an unstiffened element will reduce the efficiency of the cross-section *only* when $F_{cr,\,plate}$ for the plate element is less than $F_{cr,\,column}$.

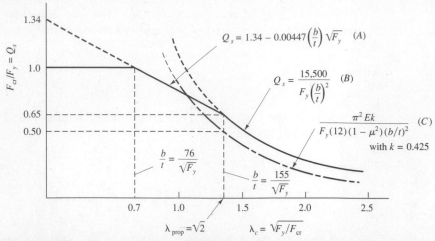

Figure 6.18.2 Plate strength for unstiffened compression element (single angle) having one edge hinged and the other free. (LRFD-Appendix B5.3a and ASD-Appendix B5.2a)

Thus, in general, for columns

$$Q_s = \frac{F_{cr,\,plate}}{F_{cr,\,column}} \geq \frac{F_{cr,\,plate}}{F_y} \qquad (6.18.10)$$

and for compression flanges of beams,

$$Q_s = \frac{F_{cr,\,plate}}{F_{cr,\,beam\,flange}} \geq \frac{F_{cr,\,plate}}{F_y} \qquad (6.18.11)$$

Form Factor Q_a for Stiffened Elements

The concept of using an effective width over which stress may be considered uniform, even though it is actually nonuniform, was developed by von Kármán [6.73] and later modified by Winter [6.74]. Winter's equation, the format of which has been used by the AISI Specification [1.11] since 1946, is

$$\frac{b_E}{t} = 1.9\sqrt{\frac{E}{f}}\left[1.0 - \frac{0.475}{(b/t)}\sqrt{\frac{E}{f}}\right] \qquad (6.18.12)$$

where $f = P_u/A_g$, stress acting on the element (f_{max} of Fig. 6.18.1)
  b/t = actual width/thickness ratio

The form and constants of the equation were essentially determined to agree with experimental results.

Substituting $E = 29,000$ ksi gives

$$\frac{b_E}{t} = \frac{324}{\sqrt{f}}\left[1.0 - \frac{81}{(b/t)\sqrt{f}}\right] \qquad (6.18.13)$$

Experience with light gage steel showed that the coefficient of the second term in Eq. 6.18.13 could be reduced. Since 1962 the equation used by both the AISI Specification [1.11] and the AISC Specification has been essentially unchanged. In LRFD-Appendix B5.3b are given the following:

1. For flanges of square and rectangular sections of uniform thickness:

$$\frac{b_E}{t} = \frac{326}{\sqrt{f}}\left[1.0 - \frac{64.9}{(b/t)\sqrt{f}}\right] \qquad (6.18.14)^*$$

2. For other uniformly compressed elements:

$$\frac{b_E}{t} = \frac{326}{\sqrt{f}}\left[1.0 - \frac{57.2}{(b/t)\sqrt{f}}\right] \qquad (6.18.15)^*$$

The difference between Eqs. 6.18.14 and 6.18.15 can be explained partly by the difference in the assumed rotational restraint (moment along the supported edges, Fig. 6.18.1b).

Additional discussion of the effective width for stiffened elements in compression is available in the work of Korol and Sherbourne [6.75, 6.76], Dawson and Walker [6.77], and Abdel-Sayed [6.78]. Sharp [6.79] has considered stiffened elements having one edge stiffened by a lip. Kalyanaraman, Pekoz, and Winter [6.80] have proposed an effective width expression for unstiffened elements.

Since a column cross-section may include unstiffened elements, which under present design procedures utilize reduced average stress rather than an effective width, the controlling stress on unstiffened elements is used as the applicable maximum stress acting on the stiffened elements. Thus the stress f is

$$f = \frac{P_u}{A_g} \qquad (6.18.16)$$

and the design requirement is that

$$P_u = \phi_c(F_{cr, \text{plate}})A_g \qquad (6.18.17)$$

Substituting Eq. 6.18.17 into Eq. 6.18.16 shows that

$$f = \phi_c F_{cr, \text{plate}} \qquad (6.18.18)$$

and from Eq. 6.18.11,

$$F_{cr, \text{plate}} = Q_s F_{cr, \text{column}} \qquad (6.18.19)$$

Thus,

$$f = \phi_c Q_s F_{cr, \text{column}} \qquad (6.18.20)$$

---

* For SI with f in MPa,

$$\frac{b_E}{t} = \frac{856}{\sqrt{f}}\left[1 - \frac{170}{(b/t)\sqrt{f}}\right] \qquad (6.18.14)$$

$$\frac{b_E}{t} = \frac{856}{\sqrt{f}}\left[1 - \frac{150}{(b/t)\sqrt{f}}\right] \qquad (6.18.15)$$

Note that $Q_s F_{cr, column}$ is F_{cr} in LRFD-Appendix B5.3c. In other words, when $Q < 1$, F_{cr} is redefined as QF_{cr} in the LRFD Specification.

Finally, Q_a as defined by Eq. 6.18.4 is

$$Q_a = \frac{A_{eff}}{A_{gross}} = \frac{b_E t}{bt} \qquad [6.18.4]$$

where $A_{eff} = A_{gross} - \Sigma(b - b_E)t$.

Design Properties

In computing the nominal strength, the following rules apply in accordance with LRFD-Appendix B5.3c.

For axial compression:

1. Use gross area A_g for $P_n = F_{cr}A_g$.
2. Use gross area to compute radius of gyration r for KL/r.

For flexure:

1. Use reduced section properties for beams with flanges containing stiffened elements.

Since the strengths of beams do not include Q factors relating to thin compression elements, it is appropriate to use section properties based on effective area.

For beam-columns:

1. Use gross area for P_n.
2. Use reduced section properties for flexure involving stiffened compression elements for M_{nx} and M_{ny}.
3. Use Q_a and Q_s for determining P_n.
4. For F_{cr} based on lateral-torsional buckling of beams as discussed in Chapter 9, the maximum value of F_{cr} is $Q_s F_{cr}$ when unstiffened compression elements are involved.

Effective Width in Allowable Stress Design

In Allowable Stress Design, the nominal strength P_n would have to be divided by a factor of safety to give the allowable load P. Dividing the nominal strength given by Eq. 6.18.2 by the factor of safety FS, gives

$$P = \frac{P_n}{FS} = b_E t \frac{f_{max}}{FS} \qquad (6.18.21)$$

The term (f_{max}/FS) may be thought of as the allowable stress F_a. The important conclusion here is that the effective width b_E to be used in computing the allowable capacity at service load *is the same b_E used in computing strength P_n*. If Eqs. 6.18.14 or 6.18.15 were used directly, the service load stress f would have to first be multiplied by the factor FS. Instead, ASD-Appendix B5.2b changes the formula by replacing f

with $1.65f$, thus giving for Eq. 6.18.14,

$$\frac{b_E}{t} = \frac{253}{\sqrt{f}}\left[1.0 - \frac{50.3}{(b/t)\sqrt{f}}\right] \qquad (6.18.22)$$

which is ASD-Appendix B Formula (A-B5-7).

6.19 DESIGN OF COMPRESSION MEMBERS AS AFFECTED BY LOCAL BUCKLING PROVISIONS

Design of single and double angle struts, structural tees, welded built-up I-shapes, and most other built-up sections, including box-type sections, involves the close attention to width/thickness limitations to prevent local buckling. Most rolled W, S, and M shapes have proportions such that local buckling will not occur (that is, $Q = 1.0$) prior to achieving the strength of the section based on the overall slenderness ratio KL/r.

The following examples illustrate situations where $Q < 1.0$.

EXAMPLE 6.19.1 _____

A double angle compression chord member for the truss of Fig. 6.19.1 consists of 2—L8×4×$\frac{1}{2}$, having short legs back-to-back. The 28-ft-long member is braced in the plane of the truss every 7 ft, but only at the ends in the transverse direction. Assume the two angles are attached together with fully-tightened bolts and the spacing of connectors is close enough that the double angle member reaches its maximum axial load strength.* Neglect any contribution to lateral support from the roofing. Compute the maximum axial service compression load this member can be permitted to carry. The service load is 30% dead load and 70% gravity live load. Use A572 Grade 50 steel and Load and Resistance Factor Design.

Solution

(a) Check local buckling. The 8-in. legs of this double angle member are unstiffened compression elements. Check whether or not width/thickness ratio λ exceeds λ_r of LRFD-B5,

$$\left(\lambda = \frac{b}{t} = \frac{8}{0.50} = 16.0\right) > \left(\lambda_r = \frac{76}{\sqrt{F_y}} = \frac{76}{\sqrt{50}} = 10.7\right)$$

Since $\lambda > \lambda_r$, local buckling will reduce the section efficiency. Using LRFD-Appendix B5.3.a, the reduction factor Q_s is

$$\left(\frac{b}{t} = 16.0\right) < \left(\frac{155}{\sqrt{F_y}} = \frac{155}{\sqrt{50}} = 21.9\right)$$

---

*The possible reduction in strength relating to the connection between the two angles was discussed in Sec. 6.8 and is covered in LRFD-E4.

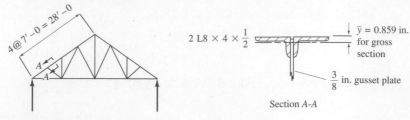

Figure 6.19.1   Example 6.19.1.

$$Q_s = 1.340 - 0.00447\left(\frac{b}{t}\right)\sqrt{F_y}$$

$$= 1.340 - 0.00447(16)\sqrt{50} = 0.834$$

(b) Compute design strength $\phi_c P_n$. For axial compression, properties of the gross section are used. From the AISC Manual properties of single angles and properties of double-angle struts with short legs back-to-back,

$$A_g = 11.5 \text{ sq in.}$$

$$r_x = 1.08 \text{ in.} \qquad r_y = 4.00 \text{ in.} \quad \text{for } \tfrac{3}{8}\text{-in. gusset plate}$$

Assuming $K = 1.0$ for truss members as discussed in Sec. 6.9,

$$\frac{(KL)_x}{r_x} = \frac{1.0(7)12}{1.08} = 78 \qquad \frac{(KL)_y}{r_y} = \frac{1.0(28)12}{4.00} = 84$$

Using LRFD-Formula (A-B5-15), Eq. 6.8.2, with λ_c given by Eq. 6.7.3,

$$\lambda_c = \frac{KL}{r}\sqrt{\frac{F_y}{\pi^2 E}} = 84\sqrt{\frac{50}{\pi^2 29,000}} = 1.11 \qquad [6.7.3]$$

which gives

$$\lambda_c\sqrt{Q} = 1.11\sqrt{0.834} = 1.014 < 1.5$$

For $\lambda_c\sqrt{Q} \le 1.5$;

$$F_{cr} = (0.658^{Q\lambda_c^2})QF_y \qquad [6.8.2]$$

$$F_{cr} = (0.658^{0.834(1.11)^2})(0.834)50 = 27.1 \text{ ksi}$$

$$\phi_c F_{cr} = 0.85(27.1) = 23.1 \text{ ksi}$$

$$\phi_c P_n = \phi_c F_{cr} A_g = 23.1(11.5) = 265 \text{ kips*}$$

$$P_u = 1.2P_D + 1.6P_L = 1.2(0.3P) + 1.6(0.7P) = 1.48P$$

$$\phi_c P_n = P_u = 1.48P$$

$$P = 265/1.48 = 179 \text{ kips (maximum service load capacity)} \quad \blacksquare\blacksquare$$

---

*This load does not agree with that given in the LRFD Manual "COLUMNS" tables for $\phi_c P_n$. Those tabular loads are for a particular number of bolts attaching the two angles together (LRFD-E4) and also include *flexural-torsional buckling* in accordance with LRFD-Appendix E. The maximum $\phi_c P_n$ including flexural-torsional buckling is 259 kips: about 2% lower than computed above. For hot-rolled sections, the authors do not believe it necessary to include this effect. When very thin sections, such as cold-formed light-gage sections, are used flexural-torsional buckling *must* be evaluated.

EXAMPLE 6.19.2

Design a double angle compression member for use as a spreader strut for hoisting large loads, as shown in Fig. 6.19.2. The lifted load is 60 tons, of which 55 tons is live load including impact. The remainder is dead load. Use $F_y = 60$ ksi and Load and Resistance Factor Design.

Solution

(a) Compute factored load P_u. At 4 to 1 slope of cable the compressive load in the strut is 120 tons (240 kips).

$$P_u = 1.2\left(\frac{5}{60}\right)(240) + 1.6\left(\frac{55}{60}\right)(240) = 376 \text{ kips}$$

(b) Estimate the slenderness ratio. Assume $K = 1.0$ and referring to Fig. 6.19.2b, use text Appendix Table A1 to estimate r,

$$r_x \approx 0.29h$$
$$r_y \approx 0.24b$$

If 8-in. legs are used, $r_x \approx 0.29(8) = 2.32$ in.

$$\text{Estimate } \frac{KL}{r} \approx \frac{1.0(20)12}{2.3} = 104$$

(c) Select section. From LRFD "NUMERICAL VALUES" TABLE 3-50, estimate $\phi_c F_{cr} \approx 19.3 \, (60/50) = 23$ ksi.

$$\text{Required } A_g = \frac{P_u}{\phi_c F_{cr}} = \frac{376}{23} = 16.3 \text{ sq in.}$$

Try L8×8×$\frac{5}{8}$, $r_{\min} = 2.49$ in., $A_g = 19.2$ sq in., $KL/r = 96.4$. This is a larger area than seemingly needed; however, the next lighter angle ($A_g = 15.5$ sq in.) will not be adequate. The estimated $\phi_c F_{cr}$ obtained by increasing the $F_y = 50$ ksi value in the ratio of the yield stresses is probably high for KL/r over 100. Applying LRFD-Appendix B3a.(a) for single angles,

$$\left(\lambda = \frac{b}{t} = \frac{8}{0.625} = 12.8\right) < \left(\lambda_r = \frac{76}{\sqrt{F_y}} = \frac{76}{\sqrt{60}} = 9.8\right)$$

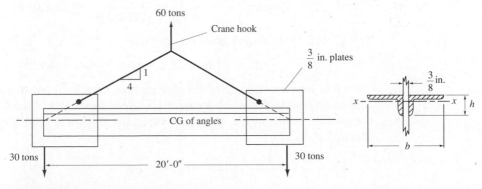

Figure 6.19.2 Example 6.19.2.

$$\left(\frac{b}{t} = 12.8\right) < \left(\frac{155}{\sqrt{F_y}} = \frac{155}{\sqrt{60}} = 20.0\right)$$

$$Q_s = 1.340 - 0.00447(12.8)\sqrt{60} = 0.897$$

$$\lambda_c = 1.40; \qquad \lambda_c\sqrt{Q} = 1.32 < 1.5$$

Thus the short column strength formula, Eq. 6.8.2, controls. Evaluating,

$$F_{cr} = 25.9 \text{ ksi}; \qquad \phi_c F_{cr} = 22.0 \text{ ksi}$$

$$\phi_c P_n = \phi_c F_{cr} A_g = 22.0(19.2) = 423 \text{ kips} > P_u \qquad \text{OK}$$

Use 2–L8×8×$\frac{5}{8}$. Note that the strength is excessive; however, the L8×8×$\frac{1}{2}$ is not adequate. ■■

EXAMPLE 6.19.3

Select the thinnest 12×12 structural tube to carry an axial compression of 60 kips dead load and 250 kips live load. The effective length KL is 18 ft and the member is part of a braced system. Use $F_y = 65$ ksi and Load and Resistance Factor Design. Note that while using $F_y = 65$ ksi illustrates procedure, Grade 50 is the maximum for structural tubes under ASTM A500.

Solution

(a) Compute factored load P_u.

$$P_u = 1.2(60) + 1.6(250) = 472 \text{ kips}$$

(b) Obtain estimate of required area. Use text Appendix Table A1 to estimate radius of gyration,

$$r \approx 0.4h = 0.4(12) = 4.8 \text{ in.}$$

Using Eq. 6.8.2 for F_{cr} with first trial $Q = 1$,

$$\frac{KL}{r} = \frac{18(12)}{4.8} = 45; \qquad F_{cr} = 53.6 \text{ ksi}$$

$$\text{Required } A_g = \frac{P_u}{\phi_c F_{cr}} = \frac{472}{0.85(53.6)} = 10.4 \text{ sq in.}$$

Try 12×12×$\frac{1}{4}$ structural tube (Fig. 6.19.3):

$$A_g = 11.6 \text{ sq in.} \qquad r_x = r_y = 4.78 \text{ in.}$$

(c) Compute strength of selected section. Check b/t on stiffened elements. The width b is the flat width measured between the "roots of the flanges" where the outside radius at the corners is twice the thickness (See footnote, *LRFD Manual*, p. 1-123.)

$$\left(\lambda = \frac{b}{t} = \frac{12 - 4(0.25)}{0.25} = 44\right) > \left(\lambda_r = \frac{238}{\sqrt{F_y}} = 29.5\right)$$

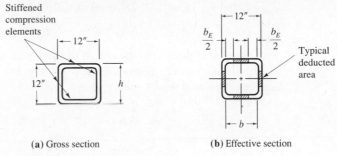

(a) Gross section   (b) Effective section

Figure 6.19.3   Example 6.19.3.

Local buckling may reduce the efficiency of this section. Compute the reduction factor Q_a in accordance with LRFD-Appendix B5.3b. Because the stress f on the stiffened element of a column section will be less than $\phi_c F_y$, the λ_r limitation will actually be higher than 29.5.

For $\dfrac{b}{t} \geq \dfrac{238}{\sqrt{f}}$,

$$\frac{b_E}{t} = \frac{326}{\sqrt{f}}\left[1.0 - \frac{64.9}{(b/t)\sqrt{f}}\right] \qquad [6.18.14]$$

where $f = \phi_c(F_{cr}$ which is a function of Q). For this section having no unstiffened elements $Q_s = 1.0$; thus $Q = Q_a$. Try $f = 0.85(53.6) = 45.6$ ksi as obtained from the first estimate of KL/r.

$$\left(\frac{b}{t} = 44\right) \geq \left(\frac{238}{\sqrt{f}} = \frac{238}{\sqrt{45.6}} = 35.2\right)$$

$$\frac{b_E}{t} = \frac{326}{\sqrt{45.6}}\left[1.0 - \frac{64.9}{(44.0)\sqrt{45.6}}\right] = 37.7$$

Thus local buckling will control the section efficiency for all four sides of the tube. The effective area is

$$A_{\text{eff}} = 11.6 - 4\left(\frac{b}{t} - \frac{b_E}{t}\right)t^2$$

$$= 11.6 - 4(44.0 - 37.7)(0.25)^2 = 10.0 \text{ sq in.}$$

$$Q_a = \frac{A_{\text{eff}}}{A_{\text{gross}}} = \frac{10.0}{11.6} = 0.865$$

Recompute the KL/r value for the trial section,

$$\frac{KL}{r} = \frac{18(12)}{4.78} = 45.2; \quad \lambda_c\sqrt{Q} = 0.633$$

Using F_{cr} from LRFD-Appendix B5.3d,

$$F_{cr} = 47.5 \text{ ksi}; \qquad \phi_c F_{cr} = 40.4 \text{ ksi}$$

Since this does not agree with the assumed value of 45.6 ksi, the effective width is not correct. Repeat using $f = 40.4$ ksi (or something close to 40.4 between 40.4 and 45.6), which gives

$$\frac{b_E}{t} = 39.4; \qquad A_{eff} = 10.45 \text{ sq in.}$$

$$Q_a = 0.901; \qquad \lambda_c \sqrt{Q} = 0.647$$

$$F_{cr} = 49.2 \text{ ksi}; \qquad \phi_c F_{cr} = 41.8 \text{ ksi}$$

This time the computed 41.8 ksi is for practical purposes close enough to the 40.4 ksi assumed.

$$(\phi_c P_n = \phi_c F_{cr} A_g = 41.8(11.6) = 485 \text{ kips}) > (P_u = 472 \text{ kips}) \qquad \text{OK}$$

Use $12 \times 12 \times \frac{1}{4}$ structural tube, $F_y = 65$ ksi. ■ ■

For design, in addition to the properties of rolled structural tubing available in the AISC Manual, there is available the *Manual of Cold Formed Welded Structural Steel Tubing,** containing column and beam load tables.

EXAMPLE 6.19.4 _____

Determine the nominal axial compression strength P_n for the nonstandard shape of Fig. 6.19.4 for an effective length KL equal to 8 ft. Use $F_y = 100$ ksi and Load and Resistance Factor Design.

Solution.   In axial compression this section contains unstiffened compression elements (flanges) and a stiffened compression element (the web). Unstiffened elements must be treated first so that the effective stress level may be determined.

(a) Properties of the gross section.

$I_y = $ negligible (web) $+ 2(\frac{1}{12})(0.5)(10)^3$ (flanges) $= 83.4$ in.^{4}

$A = 11(0.25) + 2(5.0) = 12.75$ sq in.

$r_y = \sqrt{I/A} = \sqrt{83.4/12.75} = 2.56$ in.

(b) Unstiffened elements, LRFD-B5.1,

$$\left(\lambda = \frac{b}{t} = \frac{5.0}{0.5} = 10.0 \right) > \left(\lambda_r = \frac{95}{\sqrt{F_y}} = \frac{95}{\sqrt{100}} = 9.5 \right)$$

---

* First Edition, 1974. Published by the Welded Steel Tube Institute, Inc., Structural Tube Division, 522 Westgate Tower, Cleveland, Ohio 44116.

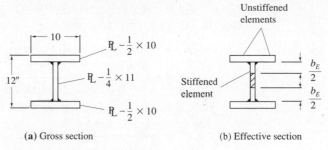

(a) Gross section   (b) Effective section

Figure 6.19.4   Example 6.19.4.

Thus, since $\lambda > \lambda_r$ the local buckling limit state causes reduced efficiency; $Q_s < 1.0$.

However, the web may be overly flexible in its restraint to rotation of the flanges. Thus, LRFD-Appendix B5.3a(c) makes the Q_s factor depend on k_c which depends on h/t_w of the stiffened element (i.e., web) when k_c is between 0.35 and 0.763. Compute h/t_w and k_c,

$$\frac{h}{t_w} = \frac{11}{0.25} = 44$$

$$k_c = \frac{4}{\sqrt{h/t_w}} = \frac{4}{\sqrt{44}} = 0.603 < 0.763$$

When k_c lies between 0.35 and 0.763, as it does in this case, the formulation of Q_s depends on k_c. For

$$\left(\frac{109}{\sqrt{F_y/k_c}} = \frac{109}{\sqrt{100/0.603}} = 8.5\right) < \left(\frac{b}{t} = 10.0\right) < \left(\frac{200}{\sqrt{F_y/k_c}} = \frac{200}{\sqrt{100/0.603}} = 15.5\right)$$

$$Q_s = 1.415 - 0.0038\left(\frac{b}{t}\right)\sqrt{\frac{F_y}{k_c}} \qquad \text{LRFD Formula (A-B5-7)}$$

$$= 1.415 - 0.0038(10.0)\sqrt{\frac{100}{0.603}} = 0.926$$

(c) Stiffened element, LRFD-B5.1:

$$\left(\lambda = \frac{b}{t} = \frac{11}{0.25} = 44\right) > \left(\lambda_r = \frac{253}{\sqrt{100}} = 25.3\right)$$

Thus $Q_a < 1.0$. The stress f that acts on the stiffened element is $\phi_c F_{cr}$ where F_{cr} includes Q, that is, Eq. 6.8.2. A first trial may be made assuming $Q = Q_s$, that is, $Q_a = 1.0$.

$$\frac{KL}{r} = \frac{8(12)}{2.56} = 37.5; \qquad \lambda_c\sqrt{Q} = 0.701\sqrt{0.926} = 0.674$$

From *LRFD Manual* "NUMERICAL VALUES" TABLE 4, find coefficient 0.702 for $\lambda_c = 0.674$; thus,

$$\phi_c F_{cr} = 0.702 Q F_y = 0.702(0.926)100 = 65.0 \text{ ksi}$$

$$\frac{b_E}{t} = \frac{326}{\sqrt{65.0}}\left[1.0 - \frac{57.2}{(44.0)\sqrt{65.0}}\right] = 33.9$$

$$A_{eff} = 12.75 - (44.0 - 33.9)(0.25)^2 = 12.12 \text{ sq in.}$$

$$Q_a = \frac{A_{eff}}{A_{gross}} = \frac{12.12}{12.75} = 0.950$$

The new value of $\phi_c F_{cr}$ for $Q = Q_s Q_a = 0.926(0.950) = 0.880$ is

$$\lambda_c \sqrt{Q} = 0.701\sqrt{0.880} = 0.658$$

$$\phi_c F_{cr} = 0.709 Q F_y = 0.709(0.880)100 = 62.4 \text{ ksi}$$

Another iteration could be made since 62.4 is not identical to 65.0 upon which the effective width b_E was based. The true answer will be slightly above 62.4 and Q will be slightly higher than 0.880. For practical purposes, use 62.4 ksi. Thus,

$$\phi_c P_n = \phi_c F_{cr} A_g = 62.4(12.75) = 796 \text{ kips}$$

$$P_n = 796/\phi_c = 796/0.85 = 936 \text{ kips}$$

SELECTED REFERENCES

6.1. L. Euler. *De Curvis Elasticis, Additamentum I, Methodus Inveniendi Lineas Curvas Maximi Minimive Proprietate Gaudentes.* Lausanne and Geneva, 1744 (pp. 267–268); and "Sur le Forces des Colonnes," *Memoires de l' Academie Royale des Sciences et Belles Lettres,* Vol. 13, Berlin, 1759; English translation of the letter by J. A. Van den Broek, "Euler's Classic Paper 'On the Strength of Columns'," *American Journal of Physics,* **15** (January–February 1947), 309–318.

6.2. A. Considère. "Resistance des pièces comprimèes," *Congrès International des Procèdés de Construction,* Paris, 1891, 3, 371.

6.3. F. Engesser. "Ueber die Knickfestigkeit gerader Stabe," *Zeitschrift des Architekten-und Ingenieur—Vereins zu Hannover,* **35** (1889), 455 and 462; also "Die Knickfestigkeit gerader Stabe," *Zentralblatt der Bauverwaltung,* Berlin (December 5, 1891), 483.

6.4. F. R. Shanley. "The Column Paradox," *Journal of the Aeronautical Sciences,* **13,** 12 (December 1946), 678.

6.5. F. R. Shanley. "Inelastic Column Theory," *Journal of the Aeronautical Sciences,* **14,** 5 (May 1947), 261–264.

6.6. Bruce G. Johnston. "Column Buckling Theory: Historic Highlights," *Journal of Structural Engineering,* **109,** 9 (September 1983), 2086–2096; Disc. by T. H. Lin, Zia Razzaq, and B. G. Johnston, **110,** 8 (August 1984), 1930–1933.

6.7. Bruce G. Johnston. "A Survey of Progress, 1944–51," Bulletin No. 1, Column Research Council, January 1952.

6.8. Theodore V. Galambos, ed. *Guide to Stability Design Criteria for Metal Structures,* 4th ed. New York: John Wiley & Sons, 1988.

6.9. Friedrich Bleich. *Buckling Strength of Metal Structures.* New York: McGraw-Hill Book Company, Inc., 1952.

6.10. Bruce G. Johnston. "Buckling Behavior Above the Tangent Modulus Load," *Journal of the Engineering Mechanics Division,* ASCE, **87,** EM6 (December 1961), 79–98.

6.11. A. W. Huber and L. S. Beedle. "Residual Stress and the Compressive Strength of Steel," *Welding Journal,* December 1954, 589s–614s.

6.12. C. H. Yang, L. S. Beedle, and B. G. Johnston. "Residual Stress and the Yield Strength of Steel Beams," *Welding Journal,* April 1952, 205s–229s.

6.13. N. R. Nagaraja Rao, F. R. Estuar, and L. Tall. "Residual Stresses in Welded Shapes," *Welding Journal,* July 1964, 295s–306s.

6.14. Donald R. Sherman. "Residual Stress Measurement in Tubular Members," *Journal of the Structural Division,* ASCE, **95,** ST4 (April 1969), 635–647.

6.15. Donald R. Sherman. "Residual Stresses and Tubular Compression Members," *Journal of the Structural Division,* ASCE, **97,** ST3 (March 1971), 891–904.

6.16. Lynn S. Beedle and Lambert Tall. "Basic Column Strength," *Journal of the Structural Division,* ASCE **86,** ST7 (July 1960), 139–173. Also *Transactions,* ASCE, **127** (1962), part II, 138–179.

6.17. Ching K. Yu and Lambert Tall. "Signigicance and Application of Stub Column Test Results," *Journal of the Structural Division,* ASCE, **97,** ST7 (July 1971), 1841–1861.

6.18. Bruce G. Johnston. "Inelastic Buckling Gradient," *Journal of the Engineering Mechanics Division,* ASCE, **90,** EM6 (December 1964), 31–47.

6.19. Richard H. Batterman and Bruce G. Johnston. "Behavior and Maximum Strength of Metal Columns," *Journal of the Structural Division,* ASCE, **93,** ST2 (April 1967), 205–230.

6.20. Bruce G. Johnston, ed. *Structural Stability Research Council, Guide to Stability Design Criteria for Metal Structures,* 3rd ed. New York: John Wiley & Sons, Inc., 1976.

6.21. Reidar Bjorhovde. "Effect of End Restraint on Column Strength—Practical Applications," *Engineering Journal,* AISC, **22,** 1 (First Quarter 1984), 1–13.

6.22. Reider Bjorhovde. "The Safety of Steel Columns," *Journal of the Structural Division,* ASCE, **104,** ST3 (March 1978), 463–477.

6.23. Reidar Bjorhovde. "Columns: From Theory to Practice," *Engineering Journal,* AISC, **26,** 1 (First Quarter 1988), 21–34.

6.24. Dann H. Hall. "Proposed Steel Column Strength Criteria," *Journal of the Structural Division,* ASCE, **107,** ST4 (April 1981), 649–670. Disc. by Bruce G. Johnston, **108,** ST4 (April 1982), 956–957; by Zu-Yan Shen and Le-Wu Lu, **108,** ST7 (July 1982), 1680–1681; by author, **108,** ST12 (December 1982), 2853–2855.

6.25. J. Michael Rotter. "Multiple Column Curves by Modifying Factors," *Journal of the Structural Division,* ASCE, **108,** ST7 (July 1982), 1665–1669.

6.26. Jacque Rondal and René Maquoi. "Single Equation for SSRC Column-Strength Curves," *Journal of the Structural Division,* ASCE, **105,** ST1 (January 1979), 247–250.

6.27. E. M. Lui and W. F. Chen. "Simplified Approach to the Analysis and Design of Columns with Imperfections," *Engineering Journal,* AISC, **22,** 2 (Second Quarter 1984), 99–117.

6.28. John P. Anderson and James H. Woodward. "Calculation of Effective Lengths and Effective Slenderness Ratios of Stepped Columns," *Engineering Journal,* AISC, **9,** 3 (October 1972), 157–166.

6.29. Carlo A. Castiglioni. "Stepped Columns: A Simplified Design Method," *Engineering Journal,* AISC, **24,** 1 (First Quarter 1986), 1–8.

6.30. Suresh C. Shrivastava. "Elastic Buckling of a Column Under Varying Axial Force," *Engineering Journal,* AISC, **18,** 1 (First Quarter 1980), 19–21.

6.31. Balbir S. Sandhu. "Effective Length of Columns with Intermediate Axial Load," *Engineering Journal,* AISC, **9,** 3 (October 1972), 154–156.

6.32. John C. Ermopoulos. "Buckling of Tapered Bars Under Stepped Axial Loads," *Journal of Structural Engineering,* ASCE, **112,** 6 (June 1986), 1346–1354.

6.33. D. R. Sherman. *Tentative Criteria for Structural Applications of Steel Tubing and Pipe,* Washington, D.C.: American Iron and Steel Institute, 1976.

6.34. Julian Snyder and Seng-Lip Lee. "Buckling of Elastic-Plastic Tubular Columns," *Journal of the Structural Division,* ASCE, **94,** ST1 (January 1968), 153–173.

6.35. Seng-Lip Lee and Julian Snyder. "Stability of Strain-Hardening Tubular Columns," *Journal of the Structural Division,* ASCE, **94,** ST3 (March 1968), 683–707.

6.36. Wai F. Chen and David A. Ross. "Tests of Fabricated Tubular Columns," *Journal of the Structural Division,* ASCE, **103,** ST3 (March 1977), 619–634.

6.37. David A. Ross. Wai Fah Chen. and Lambert Tall. "Fabricated Tubular Steel Columns," *Journal of the Structural Division,* ASCE, **106,** ST1 (January 1980), 265–282.

6.38. Theodore V. Galambos. "Strength of Round Steel Columns," *Journal of the Structural Division,* ASCE, **91,** ST1 (February 1965), 121–140.

6.39. Charles Libove, "Sparsely Connected Built-Up Columns," *Journal of Structural Engineering,* ASCE, **111,** 3 (March 1985), 609–627.

6.40. Abolhassan Astaneh-Asl, Subhash C. Goel, and Robert D. Hanson. "Cyclic Out-of-Plane Buckling of Double-Angle Bracing," *Journal of Structural Engineering,* ASCE, **111,** 5 (May 1985), 1135–1153.

6.41. Cynthia J. Zahn and Geerhard Haaijer. "Effect of Connector Spacing on Double Angle Compressive Strength," *Materials and Member Behavior,* Proceedings of Structures Congress '87, Orlando, FL, August 17–20, 1987, pp. 199–212.

6.42. John B. Kennedy and Madugula K. S. Murty. "Buckling of Steel Angle and Tee Struts," *Journal of the Structural Division,* ASCE, **98,** ST11 (November 1972), 2507–2522.

6.43. John B. Kennedy and Madugula K. S. Murty. "Buckling of Angles: State of the Art," *Journal of the Structural Division,* ASCE, **108,** ST9 (September 1982),

1967–1980. Disc., *Journal of Structural Engineering,* ASCE, **109,** 8 (August 1983), 2025–2029.

6.44. Scott T. Woolcock and Sritawat Kitipornchai. "Design of Single Angle Web Struts in Trusses," *Journal of Structural Engineering,* ASCE, **112,** 6 (June 1986), 1327–1345. Disc. **113,** 9 (September 1987), 2102–2107.

6.45. Adel A. El-Tayem and Subhash C. Goel. "Effective Length Factor for Design of X-bracing Systems," *Engineering Journal,* AISC, **24,** 4 (First Quarter 1986), 41–45.

6.46. Subhash C. Goel and Adel A. El-Tayem. "Cyclic Load Behavior of Angle X-Bracing," *Journal of Structural Engineering,* ASCE, **112,** 11 (November 1986), 2528–2539.

6.47. Guo Chuenmei. "Elastoplastic Buckling of Single Angle Columns," *Journal of Structural Engineering,* ASCE, **110,** 6 (June 1984), 1391–1395.

6.48. Erling A. Smith. "Buckling of Four Equal-Leg Angle Cruciform Columns," *Journal of Structural Engineering,* ASCE, **109,** 2 (February 1983), 439–450.

6.49. Le-Wu Lu. "Effective Length of Columns in Gabel Frames," *Engineering Journal,* AISC, **2,** 1 (January 1965), 6–7.

6.50. Kamal Hassan. "On the Determination of Buckling Length of Frame Columns," *Publications,* International Association for Bridge and Structural Engineering, **28**-II, 1968, 91–101 (in German).

6.51. Theodore V. Galambos. "Influence of Partial Base Fixity of Frame Stability," *Journal of The Structural Division,* ASCE, **86,** ST5 (May 1960), 85–108.

6.52. German Gurfinkel and Arthur R. Robinson. "Buckling of Elastically Restrained Columns," *Journal of the Structural Division,* ASCE, **91,** ST6 (December 1965), 159–183.

6.53. Harold Switzky and Ping Chun Wang. "Design and Analysis of Frames for Stability," *Journal of the Structural Division,* ASCE, **95,** ST4 (April 1969), 695–713.

6.54. Thomas C. Kavanagh. "Effective Length of Framed Columns," *Transactions,* ASCE, **127** (1962), Part II, 81–101.

6.55. Joseph A. Yura. "The Effective Length of Columns in Unbraced Frames," *Engineering Journal,* AISC, **8,** 2 (April 1971), 37–42; Disc., **9,** 3 (October 1972), 167–168.

6.56. Peter F. Adams. Discussion of "The Effective Length of Columns in Unbraced Frames," by Joseph A. Yura, *Engineering Journal,* AISC, **9,** 1 (January 1972), 40–41.

6.57. Bruce G. Johnston. Discussion of "The Effective Length of Columns in Unbraced Frames," by Joseph A. Yura, *Engineering Journal,* AISC, **9,** 1 (January 1972), 46.

6.58. Robert O. Disque. "Inelastic K-factor for Column Design," *Engineering Journal,* AISC, **10,** 2 (Second Quarter 1973), 33–35.

6.59. C. V. Smith, Jr. "On Inelastic Column Buckling," *Engineering Journal,* AISC, **13,** 3 (Third Quarter 1976), 86–88; Disc., **14,** 1 (First Quarter 1977), 47–48.

6.60. Charles A. Matz. Discussion of "On Inelastic Column Buckling," by C. V. Smith, Jr., *Engineering Journal,* AISC, **14,** 1 (First Quarter 1977), 47–48.

6.61. Frank W. Stockwell, Jr. "Girder Stiffness Distribution for Unbraced Columns," *Engineering Journal,* AISC, **13,** 3 (Third Quarter 1976), 82–85.

6.62. Cyrus Omid'varan. "Discrete Analysis of Latticed Columns," *Journal of the Structural Division,* ASCE, **94,** ST1 (January 1968), 119–132.

6.63. Fung J. Lin, Ernst C. Glauser, and Bruce G. Johnston. "Behavior of Laced and Battened Structural Members," *Journal of the Structural Division,* ASCE, **96,** ST7 (July 1970), 1377–1401.

6.64. Bruce G. Johnston. "Spaced Steel Columns," *Journal of the Structural Division,* ASCE, **97,** ST5 (May 1971), 1465–1479.

6.65. Omer W. Blodgett. *Design of Welded Structures.* Cleveland, Ohio: James F. Lincoln Arc Welding Foundation, 1966.

6.66. S. Timoshenko and S. Woinowsky-Krieger. *Theory of Plates and Shells,* 2nd ed. New York: McGraw-Hill Book Company, Inc., 1959 (pp.79–82).

6.67. Stephen P. Timoshenko and James M. Gere. *Theory of Elastic Stability,* 2nd ed. New York: McGraw-Hill Book Company, Inc., 1961 (pp. 319–328, 351–356).

6.68. Kurt H. Gerstle. *Basic Structural Design.* New York: McGraw-Hill Book Company, Inc., 1967 (pp. 88–90).

6.69. George Gerard and Herbert Becker. *Handbook of Structural Stability,* Part I—*Buckling of Flat Plates,* Tech. Note 3871, National Advisory Committee for Aeronautics, Washington, D.C., July 1957.

6.70. Geerhard Haaijer and Bruno Thürlimann. "On Inelastic Buckling in Steel," *Transactions,* ASCE, **125** (1960), 308–344.

6.71. Maxwell G. Lay. "Flange Local Buckling in Wide-Flange Shapes," *Journal of the Structural Division,* ASCE, **91,** ST6 (December 1965), 95–116.

6.72. John F. McDermott. "Local Plastic Buckling of A514 Steel Members," *Journal of the Structural Division,* ASCE, **95,** ST9 (September 1969), 1837–1850.

6.73. Theodore von Kármán, E. E. Sechler, and L. H. Donnell. "The Strength of Thin Plates in Compression," *Transactions,* ASME, **54,** APM-54-5 (1932), 53.

6.74. G. Winter. "Strength of Thin Compression Flanges," *Transactions,* ASCE, **112** (1947), 527–576.

6.75. Robert M. Korol and Archibald N. Sherbourne. "Strength Predictions of Plates in Uniaxial Compression." *Journal of the Structural Division,* ASCE, **98,** ST9 (September 1972), 1965–1986.

6.76. Archibald N. Sherbourne and Robert M. Korol. "Post-Buckling of Axially Compressed Plates," *Journal of the Structural Division,* ASCE, **98,** ST10 (October 1972), 2223–2234.

6.77. Ralph G. Dawson and Alastair C. Walker. "Post-Buckling of Geometrically Imperfect Plates," *Journal of the Structural Division,* ASCE, **98,** ST1 (January 1972), 75–94.

6.78. George Abdel-Sayed. "Effective Width of Thin Plates in Compression," *Journal of the Structural Division,* ASCE, **95,** ST10 (October 1969), 2183–2203.

6.79. Maurice L. Sharp. "Longitudinal Stiffeners for Compression Members," *Journal of the Structural Division,* ASCE, **92,** ST5 (October 1966), 187–211.

6.80. V. Kalyanaraman, Teoman Pekoz, and George Winter. "Unstiffened Compression Elements," *Journal of the Structural Division,* ASCE, **103,** ST9 (September 1977), 1833–1848.

6.81. Russell Q. Bridge and Donald J. Fraser. "Improved *G*-Factor Method for Evaluating Effective Lengths of Columns," *Journal of Structural Engineering,* **113,** 6 (June 1987), 1341–1356.

6.82. Lian Duan and Wai-Fah Chen. "Design Rules of Built-Up Members in Load and Resistance Factor Design," *Journal of Structural Engineering,* **114,** 11 (November 1988), 2544–2554.

6.83. AISC. *Specification for Load and Resistance Factor Design of Single-Angle Members,* effective December 1, 1993. Chicago: American Institute of Steel Construction, 1993.

6.84. Donald J. Fraser. "Uniform Pin-based Crane Columns, Effective Lengths," *Engineering Journal,* AISC, **26,** 2 (2nd Quarter 1989), 61–65.

6.85. Sayed H. Stoman. "Effective Length Spectra for Cross Bracings," *Journal of Structural Engineering,* ASCE, **115,** 12 (December 1989), 3112–3122.

6.86. A. Rutenberg and A. Scarlat. "Roof Bracing and Effective Length of Columns in One-Story Industrial Buildings," *Journal of Structural Engineering,* ASCE, **116,** 10 (October 1990), 2551–2566.

6.87. T. V. Galambos. "Design of Axially Loaded Compressed Angles," *Structural Stability Research Council Annual Technical Session Proceedings,* 1991.

6.88. Ronald S. Harichandran. "Stiffness Reduction Factor for LRFD of Columns," *Engineering Journal,* AISC, **28,** 3 (3rd Quarter 1991), 129–130.

6.89. Farhang Aslani and Subhash C. Goel. "An Analytical Criterion for Buckling Strength of Built-up Compression Members," *Engineering Journal,* AISC, **28,** 4 (4th Quarter 1991), 159–168.

6.90. M. Elgaaly, H. Dagher, and W. Davids. "Behavior of Single-Angle-Compression Members," *Journal of Structural Engineering,* ASCE, **117,** 12 (December 1991), 3720–3741.

6.91. M. Elgaaly, W. Davids, and H. Dagher. "Non-Slender Single Angle Struts," *Engineering Journal,* AISC, **29,** 2 (2nd Quarter 1992), 49–58.

6.92. Seshu Madhava Rao Adluri and Murty K. S. Madugula. "Eccentrically Loaded Steel Single Angle Struts," *Engineering Journal,* AISC, **29,** 2 (2nd Quarter 1992), 59–66.

6.93. Pierre Dumonteil. "Simple Equations for Effective Length Factors," *Engineering Journal,* AISC, **29,** 3 (3rd Quarter 1992), 111–115. Disc. **30,** 1 (1st Quarter 1993), 37; Errata **30,** 1 (1st Quarter 1993), 38.

6.94. Eric M. Lui. "A Novel Approach for *K* Factor Determination," *Engineering Journal,* AISC, **29,** 3 (3rd Quarter 1992), 150–159.

6.95. Farhang Aslani and Subhash C. Goel. "Analytical Criteria for Stitch Strength of Built-up Compression Members," *Engineering Journal,* AISC, **29,** 4 (4th Quarter 1992), 102–110.

6.96. A. Zureick. "Design Strength of Concentrically Loaded Single Angle Struts," *Engineering Journal,* AISC, **30,** 1 (1st Quarter 1993), 17–30.

6.97. Leander Bathon, Wendelin H. Mueller III, and Leon Kempner, Jr. "Ultimate Load Capacity of Single Steel Angles," *Journal of Structural Engineering,* ASCE, **119,** 1 (January 1993), 279–300.

PROBLEMS

All problems are to be done according to the AISC Load and Resistance Factor Design or Allowable Stress Design, as indicated by the instructor. All given loads are service loads unless otherwise indicated. For each problem, draw the potential buckled shape on a figure showing the column and its restraints for both x and y principal directions. A final check of strength (for LRFD) or stress (for ASD) must be shown in all design problems.

6.1. For the case assigned by the instructor, select the lightest W section to carry an axial compression load as indicated. The member is part of a braced frame. Assume the member as pinned at the top and bottom.

Case	P_D Dead load (kips)	P_L Live load (kips)	Member length (ft)	Steel grade
1	20	80	22	A36
2	20	80	22	A572 Grade 50
3	20	80	14	A36
4	20	80	14	A572 Grade 50
5	60	40	22	A36
6	60	40	14	A36
7	20	80	14	A572 Grade 60
8	20	80	14	A572 Grade 65

6.2. Select the lightest W section to carry a compressive load of 200 kips dead load and 625 kips live load. The effective length KL is 25 ft. Use A36 or A572 Grade 50, whichever is more economical, if the Grade 50 costs 7% more per pound.

6.3. Compute the maximum service axial compression load permitted on the built-up cross-section of the accompanying figure. The load is 30% dead load and 70% live load. The steel used is A572 Grade 50, and the effective lengths are $(KL)_y = 14$ ft and $(KL)_x = 42$ ft.

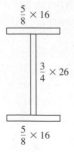

$\frac{5}{8} \times 16$

$\frac{3}{4} \times 26$

$\frac{5}{8} \times 16$

Problem 6.3

6.4. For the data of Prob. 6.1, consider the member to be fixed at the bottom and hinged at the top and part of a braced system. Select the lightest W section for the case assigned.

6.5. Select the lightest W section to serve as an axially loaded column 28 ft long, in a braced frame, with additional lateral support in the weak direction at mid-height. The load to be carried is 65 kips dead load and 150 kips live load. Assume the top and bottom of the column are hinged.
a. Use A36 and indicate first and second choices.
b. Use A572 Grade 60, indicating first and second choices.

6.6. Select the most economical W section to carry an axial compression load of 50 kips dead load and 100 kips live load. The member has $(KL)_x = (KL)_y = 18$ ft and is part of a braced system. Assume that relative costs of various steels are as follows: A36, 1.0; A572 Grade 50, 1.14; A572 Grade 60, 1.20.

6.7. Redesign the column of Prob. 6.6 assuming additional weak direction support at mid-height.

6.8. Select the lightest W section to carry an axial compression load of 60 kips dead load and 250 kips live load. The member is part of a braced frame and is assumed to be pinned at the top and bottom of its 30 ft length, and in addition has lateral support in the weak direction at 14 ft from the bottom. Use (a) A36 steel; (b) $F_y = 50$ ksi; (c) $F_y = 65$ ksi.

6.9. Select the lightest W section to carry an axial compression of 90 kips dead load and 320 kips live load. The member is part of a braced frame. The idealized support conditions are that the member is hinged in both principal directions at the top of a 30 ft height; supported in the weak direction at 14 and 22 ft from the bottom; and fixed in both directions at the bottom. Use (a) A36 steel; (b) $F_y = 50$ ksi; (c) $F_y = 60$ ksi.

6.10. Select the lightest W section for the column shown in the accompanying figure. The loading is 30 kips dead load and 120 kips gravity live load. The member is built into a wall so that it may be considered as continuously braced in the weak direction. *Note:* Not all of the available W sections are included in the AISC Manual Column Load Tables. Use (a) A36 steel; (b) $F_y = 50$ ksi; (c) $F_y = 60$ ksi.

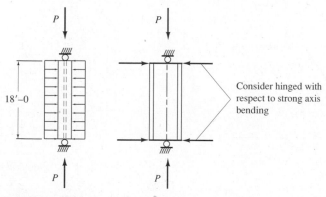

Problem 6.10

6.11. Redesign the column of Prob. 6.2 assuming there is no residual stress or accidental eccentricity such that the column buckling strength may be represented by Euler's equation, Eq. 6.2.8, using KL for L. If designing using LRFD philosophy, use $\phi_c = 0.85$ and the load factor combinations in LRFD-A4.1. If using ASD philosophy, use the denominator of ASD-Formula (E2-1) as the Factor of Safety. Use $F_y = 50$ ksi and take note that F_{cr} cannot exceed F_y.

6.12. Use the tangent modulus theory to compute the column strength relationship (average unit stress F_{cr} on gross area vs slenderness ratio KL/r). Draw the diagram to scale and obtain any F_{cr} values by scaling from your diagram. The steel has $F_y = 50$ ksi but the stress-strain curve for the material is as shown in the accompanying figure. Assume no residual stress. Using your F_{cr} curve, select the lightest W section for the loading and support conditions of Prob. 6.2. If LRFD philosophy is used take $\phi_c = 0.85$ and use the factored load combination of LRFD-A4.1. If ASD philosophy is used, take denominator of ASD-Formula (E2-1) as the Factor of Safety.

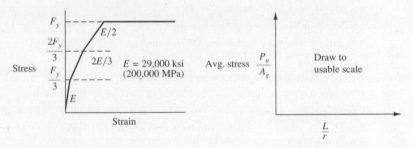

Problem 6.12

6.13. Using the tangent modulus theory: (a) Construct a column strength ($F_{cr} = P_n/A_g$ vs KL/r) for an H-shaped section. Assume weak axis bending $(KL/r)_y$ controls and neglect the effect of the web. Assume the idealized stress-strain relationship shown in the accompanying figure is to be used for each fiber of the cross-section, and the residual stress distribution in the flange

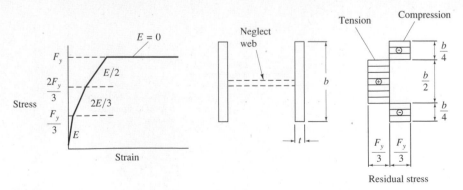

Problem 6.13

is as shown. (b) Select the lightest W section to carry a dead load of 100 kips and a live load of 200 kips with an effective length KL of 30 ft. Use your constructed curve as the relationship between F_{cr} and KL/r. If the LRFD philosophy is used take $\phi_c = 0.85$ and use the factored load combination in LRFD-A4.1. If ASD philosophy is used, take the denominator of ASD-Formula (E2-1) as the Factor of Safety. Use $F_y = 50$ ksi. (c) Solve using the AISC Specification and compare with tangent modulus theory result.

6.14. Follow the same requirements as for Prob. 6.13 to construct a column strength curve. This time the residual stress distribution is linearly varying as shown in the accompanying figure. For (b) and (c) use the loading and support conditions of Prob. 6.2.

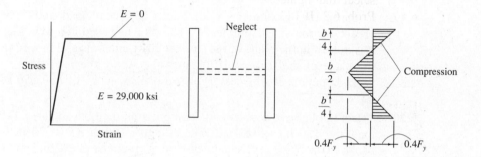

Problem 6.14

6.15. Repeat Prob. 6.14, except for (b) and (c) using the loading and support conditions of Prob. 6.13.

6.16. Design an interior column (use W shape) for a multistory rigid frame. No bracing is provided in the plane of the frame. In the plane perpendicular to the frame, bracing is provided at top, bottom, and mid-height of columns and simple flexible beam-to-column connections are used. The axial compressive load is 400 kips dead load and 1100 kips live load, and bending moments are neglected. Use A572 Grade 50 Steel.

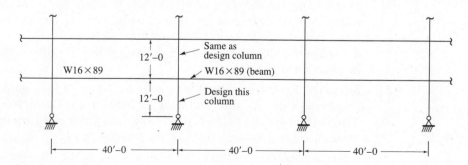

Problem 6.16

6.17. An axial compression load consisting of 100 kips dead load and 400 kips live load is to be carried by a column having an effective length $KL = 28$ ft. Use either A36 or A572 Grade 50 steel, whichever is more economical, if the A572 steel costs 9% more per pound of fabricated steel than A36. Satisfy the width/thickness limits λ_r of LRFD of ASD-B5.1.

a. Design a rolled W section.

b. Design the lightest welded I-shaped section using three plates.

c. Design a welded laced (single lacing) column consisting of four angles.

d. Design a welded box with solid plates.

e. Design a welded box having two perforated plates.

6.18. Design as in Prob. 6.17, except the axial load is 190 kips dead load and 210 kips live load and $KL = 32$ ft.

6.19. Compute the maximum service load (25% dead load and 75% live load) acceptable for a structural tee WT12×38 when used in a truss location where it is braced in the plane of the truss at 20-ft intervals and braced transverse to the plane of the truss at 10-ft intervals. Apply the provisions of LRFD or ASD-Appendix B if necessary. Use (a) A36 steel; (b) $F_y = 50$ ksi; (c) $F_y = 65$ ksi.

6.20. Select the lightest double angle compression member to carry 110 kips dead load and 130 kips live load. The effective length $KL = 20$ ft. Assume the backs of the angles are separated by a $\frac{3}{8}$-in. gusset plate. Indicate the orientation of the angle legs (i.e., short or long legs back-to-back). If angles are selected for all three steels, indicate the economical choice if relative costs are A36 (1.0), $F_y = 50$ (1.07), and $F_y = 60$ (1.10). Use (a) A36 steel; (b) $F_y = 50$ ksi; (c) $F_y = 60$ ksi.

6.21. Design as in Prob. 6.20, except the member must carry 40 kips dead load and 140 kips live load, and $KL = 16$ ft.

6.22. Design as in Prob. 6.20, except the member must carry 30 kips dead load and 50 kips live load, and $KL = 12$ ft.

6.23. Design a top chord member for a roof truss to carry 40 kips dead load, 80 kips live load, and 40 kips wind load acting simultaneously. Assume this loading combination governs the design. The member is braced in the plane of the truss by adjoining web members connecting in at 5-ft intervals. The chord is braced transverse to the plane of the truss at 10-ft intervals. Neglect bending due to roof loads. (*Note:* Refer to LRFD-A4.1 or ASD-A5.2.)

a. Design a double angle member connected to $\frac{1}{2}$-in. gusset plates.

b. Design a structural tee.

6.24. Select the lightest structural tee (WT) for use as a top chord compression member to carry 35 kips dead load and 100 kips live load. Neglect bending. The member has 9-ft effective length for buckling in either the x–x or y–y plane. Use (a) A36 steel; (b) $F_y = 50$ ksi steel; (c) $F_y = 65$ ksi steel.

6.25. Select the lightest structural tee (WT) to serve as the compression chord of a truss to carry 40 kips dead load and 45 kips live load. In the plane of the truss the chord is braced by adjoining web members that frame in at 5-ft intervals.

Perpendicular to the plane of the truss, the chord is braced at 10 ft by a system of lateral purlin supports. Use the most economical of A36 or A572 Grade 65 steels if Grade 65 costs 12% more than A36.

6.26. Compute the service axial compressive load permitted on a $10 \times 10 \times \frac{1}{4}$ structural tube having an effective length $KL = 8$ ft. The load is 60% live load. Use (a) A36 steel; (b) $F_y = 60$ ksi; (c) $F_y = 100$ ksi.

6.27. Compute the service axial compressive load permitted on a $12 \times 8 \times \frac{1}{4}$ structural tube having an effective length $(KL)_y = 7$ ft for weak axis bending, and $(KL)_x = 10$ ft for strong axis bending. The load is 35% dead load and 65% live load. Use (a) A36 steel; (b) $F_y = 65$ ksi; (c) $F_y = 90$ ksi.

6.28. Redesign the column of Prob. 6.1, selecting a structural tube instead of a W section.

6.29. Redesign the column of Prob. 6.6, selecting a structural tube instead of a W section.

6.30. Redesign the column of Prob. 6.7, selecting a structural tube instead of a W section.

6.31. Compute the service axial compressive load permitted on the nonstandard I-shaped section shown in the accompanying figure if the load is 30% dead load and 70% live load. The effective length $(KL)_y = 12$ ft and $(KL)_x = 6$ ft. Use (a) A36 steel; (b) $F_y = 60$ ksi; (c) $F_y = 100$ ksi.

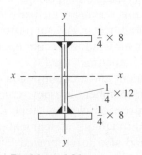

Problem 6.31

6.32. Repeat Prob. 6.31 except use the nonstandard tee section of the accompanying figure.

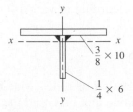

Problem 6.32

Chapter 7

Beams: Laterally Supported

7.1 INTRODUCTION

A beam is defined as a member acted upon primarily by transverse loading, often gravity dead and live load effects. The term transverse loading is taken to include end moments. Thus, beams in a structure may also be referred to as *girders* (usually the most important beams which are frequently at wide spacing); *joists* (usually less important beams which are closely spaced, frequently with truss-type webs); *purlins* (roof beams spanning between trusses); *stringers* (longitudinal bridge beams spanning between floor beams); *girts* (horizontal wall beams serving principally to resist bending due to wind on the side of an industrial building; frequently supporting corrugated siding); and *lintels* (members supporting a wall over window or door openings). Other terms, such as header, trimmer, and rafter, are sometimes used.

A beam is a combination of a tension element and a compression element. The concepts of tension members and compression members are now combined in the treatment as a beam. In this chapter, the compression element, (one flange) that is integrally braced perpendicular to its plane through its attachment to the stable tension flange by means of the web, is assumed also to be braced laterally in the direction perpendicular to the plane of the web. Thus, overall buckling of the compression flange as a column cannot occur prior to its full participation to develop the moment strength of the section. While it is likely true that most beams used in practical situations are adequately braced laterally so that such stability need not be considered, the percentage of stable situations is probably not as high as assumed. The important treatment of lateral stability is found in Chapter 9. Galambos [7.1] has provided an interesting history of beam design according to various editions of the AISC Specification.

7.2 SIMPLE BENDING OF SYMMETRICAL SHAPES

The most common design situations involve selection of rolled wide-flange shapes from the AISC tables, which often becomes routine and may lead the designer into overconfidence in treatment of beams. It is well known that the flexure formula

Beams, including open-web joists, channels, and W (wide-flange) shapes, along with tubular columns. (Photo by C. G. Salmon)

($f = Mc/I$) is applicable to ordinary situations. The stresses on the common sections of Fig. 7.2.1 may be computed by the simple flexure formula when loads are acting in one of the principal directions. When any section with at least one axis of symmetry and loaded through the shear center is subjected to a bending moment in an arbitrary direction, the components M_{xx} and M_{yy}, in the principal directions, can be obtained and the stress computed as

$$f = \frac{M_{xx}}{S_x} + \frac{M_{yy}}{S_y} \tag{7.2.1}$$

where S is the *section modulus*, defined as the moment of inertia I divided by the distance c from the center of gravity to the extreme fiber. The subscripts x and y indicate the axis about which the moment of inertia is computed and from which the distance c is measured (see Fig. 7.2.1). For members without at least one axis of symmetry the reader is referred to Sec. 7.10.

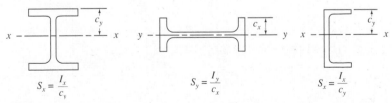

Figure 7.2.1 Elastic section modulus expressions for symmetrical shapes.

7.3 BEHAVIOR OF LATERALLY STABLE BEAMS

When beams have adequate lateral stability of the compression flange, the only stability limit state that might prevent achieving maximum moment strength is local buckling in compression of the flange and/or web plate elements comprising the cross-section.

The stress distribution on a typical wide-flange shape subjected to increasing bending moment is shown in Fig. 7.3.1. In the service load range the section is elastic as in Fig. 7.3.1a, and the elastic condition exists until the stress at the extreme fiber reaches the yield stress F_y (Fig. 7.3.1b). Once the strain ϵ reaches ϵ_y (Fig. 7.3.2), increasing strain induces no increase in stress. This elastic-plastic stress-strain behavior is the accepted idealization for structural steels having yield stresses up to about $F_y = 65$ ksi (448 MPa).

When the yield stress is reached at the extreme fiber (Fig. 7.3.1b), the nominal moment strength M_n is referred to as the *yield moment* M_y and is computed as

$$M_n = M_y = S_x F_y \tag{7.3.1}$$

When the condition of Fig. 7.3.1d is reached, every fiber has a strain equal to or greater than $\epsilon_y = F_y/E_s$, i.e., it is in the plastic range. The nominal moment strength M_n is therefore referred to as the *plastic moment* M_p, and is computed

$$M_p = F_y \int_A y \, dA = F_y Z \tag{7.3.2}$$

where $Z = \int y \, dA$ is called the *plastic modulus*.

Note the ratio M_p/M_y is a property of the cross-sectional shape and is independent of the material properties. This ratio is known as the *shape factor* ξ,

$$\xi = \frac{M_p}{M_y} = \frac{Z}{S} \tag{7.3.3}$$

For wide-flange (W) shapes in flexure about the strong axis $(x-x)$ the shape factor ranges from about 1.09 to about 1.18, with the usual value being about 1.12. One may conservatively say the plastic moment strength M_p of W sections bent about their strong axis is at least 10% greater than the strength M_y when the extreme fiber just reaches the yield stress F_y.

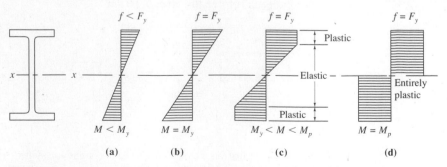

Figure 7.3.1   Stress distribution at different stages of loading.

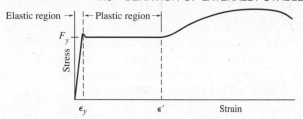

Figure 7.3.2 Stress-strain diagram for most structural steels.

Design procedures have long recognized that beams do exhibit the behavior discussed above. Extensive testing has adequately verified that plastification of the entire cross-section does occur [7.2] (assuming lateral-torsional buckling as treated in Chapter 9 and local buckling as treated in Chapter 6, Part II, do not occur).

EXAMPLE 7.3.1

Determine the shape factor for a rectangular beam of width b and depth d.

Solution.   Referring to Fig. 7.3.3a, the moment M_y at first yield is

$$M_y = \int_A f y \, dA$$

and

$$f = F_y \frac{y}{d/2} = F_y \frac{2y}{d}$$

$$M_y = 2 \int_0^{d/2} \frac{2F_y}{d} y^2 b \, dy = F_y \frac{bd^2}{6} = F_y S$$

From Fig. 7.3.3b,

$$M_p = \int_A f y \, dA = 2 \int_0^{d/2} F_y b \, y \, dy = F_y \frac{bd^2}{4} = F_y Z$$

The shape factor is then

$$\xi = \frac{M_P}{M_y} = \frac{Z}{S} = 1.5$$

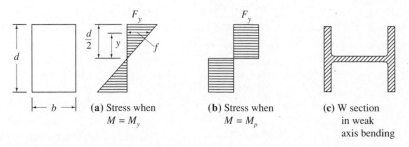

(a) Stress when $M = M_y$

(b) Stress when $M = M_p$

(c) W section in weak axis bending

Figure 7.3.3 A rectangular section and a W section in weak-axis bending.

which illustrates that there is a greater reserve beyond first yield in the bending of a rectangular section than in an I-shaped section bending about its strong axis. The reader is alerted to the fact that the W shape bent about its *weak* axis (y–y) is essentially a rectangular section (two rectangles separated by a distance) (Fig. 7.3.3c). ■■

Once the plastic moment strength M_p has been reached, the section can offer no additional resistance to rotation, behaving as a hinge but with constant resistance M_p, a condition known as a *plastic hinge.* In a statically determinate beam, such as a simply supported one, having one plastic hinge form will make the structure unstable; one real hinge at each end and a plastic hinge in the midspan region will create an unstable situation, known as a *collapse mechanism.* In general, any combination of three hinges, real or plastic, in a span will result in a collapse mechanism.

Referring to Fig. 7.3.4, one may note that the angle of rotation θ (radians/inch) is elastic from service load M until the extreme fiber reaches F_y at M_y, then becomes partially inelastic until the plastic moment M_p is reached. Once the plastic hinge has occurred and the M–θ curve has become horizontal, deflection of the beam (rotation of the plastified cross-section) increases without restraint. At the collapse condition the elastic deformation due to bending on the segments between the ends and midspan is negligible compared to the rotation θ_u occurring at the plastic hinge. Thus, the analysis may treat the collapse situation as two rigid bodies having an angular discontinuity θ_u at midspan. As will be shown later in Sec. 10.2, it is only for statically determinate situations that one can expect every point along the factored moment diagram to be proportional to the elastic moment diagram.

Redistribution of the moments occurs during loading beyond the elastic range in usual statically indeterminate situations; that is, the bending moment diagram after a plastic hinge has occurred will no longer be proportional to the elastic bending moment diagram.

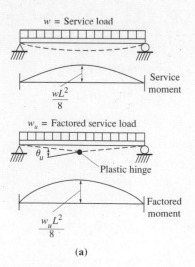

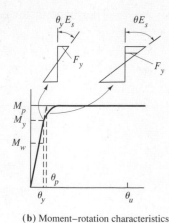

(a)

(b) Moment–rotation characteristics

Figure 7.3.4 Plastic behavior.

As discussed by Yura, Galambos, and Ravindra [7.3], even if the beam has adequate resistance with regard to lateral-torsional buckling (Chapter 9) and to local buckling (Chapter 6, Part II) to achieve the plastic moment strength, the actual limit state will still be failure by lateral-torsional buckling, compression flange local buckling, or web local buckling, but in the plastic range instead of the elastic range. Thus, prevention of failure by any of these instability modes until adequate rotation θ_u has occurred requires limits on the distance between points of lateral support, on the width/thickness ratio of the compression flange, and on the depth/thickness ratio of the web.

7.4 LATERALLY SUPPORTED BEAMS— LOAD AND RESISTANCE FACTOR DESIGN

The strength requirement for beams in load and resistance factor design according to LRFD-F2 may be stated

$$\phi_b M_n \geq M_u \tag{7.4.1}$$

where ϕ_b = resistance (i.e., strength reduction) factor for flexure = 0.90
   M_n = nominal moment strength
   M_u = factored service load moment (see Sec. 1.9)

Compact Sections

The nominal strength M_n for laterally stable "compact sections" according to LRFD-Appendix F1 may be stated

$$M_n = M_p \tag{7.4.2}$$

where M_p = plastic moment strength = ZF_y
   Z = plastic modulus, Eq. 7.3.2
   F_y = specified minimum yield stress

Design must account for the fact that local buckling of the compression flange or local buckling of the web may occur prior to achieving the high compressive strain necessary to develop M_p. When the width/thickness limitations λ_r of LRFD or ASD-B5 are satisfied, achievement of only M_y is assured (that is, local buckling at a stress below F_y is prevented as discussed in Sec. 6.16). The limits λ_r for preventing local buckling in beams are given in Table 7.4.1. The extreme fiber strain is assured only of reaching $\epsilon_y = F_y/E_s$. To achieve greater strain, the values of b/t (referred to generally as λ by LRFD) must be further restricted. To undergo large plastic strain the more severe width/thickness limitations λ_p discussed in Sec. 6.17 and prescribed for "compact sections" must be satisfied, as given in Table 7.4.2.

For the welded (flange and web continuously attached) I-shape, the $b_f/2t_f$ limit in 1993 is a function of h/t_w, instead of independent of that ratio as in earlier Specifications. Rolled I-shapes typically have h/t_w less than 40 and nearly all have that ratio 55 or less; welded I-shapes of similar proportions can be expected to have similar h/t_w ratios. The 1993 limits for welded I-shapes are higher for proportions similar to

TABLE 7.4.1 WIDTH/THICKNESS LIMITS λ_r FOR "NONCOMPACT SECTION" BEAMS TO ACHIEVE F_y AT EXTREME FIBER (LRFD-B5.1)

F_y (ksi)	Flange (Unstiffened) Rolled I-shape $\dfrac{b_f}{2t_f} = \dfrac{141}{\sqrt{F_y - 10}}$	$\dfrac{h}{t_w}$	$k_c{}^*$	Flange (Unstiffened) Welded I-shape $\dfrac{b_f}{2t_f} = \dfrac{162}{\sqrt{(F_{yf} - 16.5)/k_c}}$	Flange (Stiffened) Rolled tube Welded box $\dfrac{b_f}{2t_f} = \dfrac{238}{\sqrt{F_y}}$	Web $\dfrac{h}{t_w} = \dfrac{970}{\sqrt{F_y}}$
36	27.7	161.7	0.35	21.7	39.7	161.7
		100	0.40	23.2		
		40	0.63	29.2		
42	24.9	149.7	0.35	19.0	36.7	149.7
		100	0.40	20.3		
		40	0.63	25.5		
45	23.8	144.6	0.35	18.0	35.5	144.6
		100	0.40	19.2		
		40	0.63	24.1		
50	22.3	137.2	0.35	16.6	33.7	137.2
		100	0.40	17.7		
		40	0.63	22.3		
55	21.0	130.8	0.35	15.4	32.1	130.8
		100	0.40	16.5		
		40	0.63	20.8		
60	19.9	125.2	0.36	14.7	30.7	125.2
		100	0.40	15.5		
		40	0.63	19.5		
65	19.0	120.3	0.36	14.0	29.5	120.3
		100	0.40	14.7		
		40	0.63	18.5		
90	15.8	102.2	0.40	11.9	25.1	102.2
		100	0.40	12.0		
		40	0.63	15.0		
100	14.9	97.0	0.41	11.3	23.8	97.0
		100	0.40	11.2		
		40	0.63	14.1		

$^*k_c = \dfrac{4}{\sqrt{h/t_w}}$, where $0.35 \leq k_c \leq 0.763$

rolled I-shapes, and lower for thin web plate girders. The explanation is that the thinner the web the less rotational restraint it offers to prevent flange local buckling. The reduced limits for $b_f/2t_f$ at the web λ_r limit compared with 1986 LRFD are, for example, 21.7 instead of 24.0 for A36 steel, and 16.6 instead of 18.3 for Grade 50 steel. Since h/t_w is low for rolled I-shapes, the limit independent of h/t_w was retained for simplicity.

Noncompact Sections

The nominal strength M_n for laterally stable "noncompact sections" whose width/thickness ratios λ *exactly equal* the limits λ_r of LRFD-Table B5.1 is the moment strength available when the extreme fiber is at the yield stress F_y. Because of

TABLE 7.4.2 WIDTH/THICKNESS LIMITS λ_p FOR "COMPACT SECTION"* BEAMS TO ACHIEVE SIGNIFICANT PLASTIC STRAIN (LRFD AND ASD-B5)

F_y (ksi)	Unstiffened elements (uniform compression) $\dfrac{b_t}{2t_r} \leq \dfrac{65}{\sqrt{F_y}}$	Stiffened elements (uniform compression) $\dfrac{b}{t_f} \leq \dfrac{190}{\sqrt{F_y}}$	Stiffened elements (bending) $\dfrac{h}{t_w} \leq \dfrac{640}{\sqrt{F_y}}$
36	10.8	31.7	107
42	10.0	29.3	98.8
45	9.7	28.3	95.4
50	9.2	26.9	90.5
55	8.8	25.6	86.3
60	8.4	24.5	82.6
65	8.1	23.6	79.4

* Plastic analysis is restricted by LRFD-A5 and ASD-F1.1 to steels having $F_y \leq 65$ ksi (450 MPa).

residual stress the strength is expressed as

$$M_n = M_r = (F_y - F_r)S \tag{7.4.3}$$

where M_r is the "residual moment" that will cause the extreme fiber stress to rise from its residual stress F_r value when there is no applied load acting to the yield stress F_y. The elastic section modulus S equals the moment of inertia I divided by the distance c from the neutral axis to the extreme fiber.

For *hybrid beams* (see Sec. 11.7), where the flange yield stress F_{yf} is typically higher than the web yield stress F_{yw}, the nominal moment strength M_r must be based on the smaller of $(F_{yf} - F_r)$ or F_{yw}.

Partially Compact Sections

The nominal strength M_n for laterally stable "noncompact sections" whose width/thickness ratios λ are less than λ_r but not as low as λ_p must be linearly interpolated between M_p and M_r, as follows according to LRFD Appendix F1.7:

$$M_n = M_p - (M_p - M_r)\left(\frac{\lambda - \lambda_p}{\lambda_r - \lambda_p}\right) \leq M_p \tag{7.4.4}$$

where $\lambda = b_f/2t_f$ for I-shaped member flanges
         $= h/t_w$ for beam webs
    b_f = flange width
    t_f = flange thickness
    $h = d - 2k$ plus allowance for undersize inside fillet at compression flange (about $\frac{1}{4}$ in.) for rolled I-shaped sections. For practical purposes on rolled sections the tabulated values of h/t_w must be used since this minimum fillet size is not readily available.
    t_w = web thickness
    λ_r = Table 7.4.1 (LRFD and ASD-Table B5.1)
    λ_p = Table 7.4.2 (LRFD and ASD-Table B5.1)

The reader may note that the λ_r expressions for unstiffened flanges given in Table 7.4.1 differ from the derived $95/\sqrt{F_y}$ given for uniform compression in Sec. 6.16. In addition to the introduction of h/t_w as discussed above, the difference arises because the flange is not actually uniformly compressed and in the rolled and welded I-shaped sections residual stress (assumed as 10 ksi for rolled sections and 16.5 ksi for welded sections) is present. The limit on h/t_w for the web arises from the buckling of a stiffened plate under linearly varying stress (bending moment stress), a case which was not included in Sec. 6.16.

Slender Sections

When the width/thickness ratios λ exceed the limits λ_r of LRFD-B5.3, the sections are referred to as "slender" and must be treated in accordance with LRFD-Appendix B.

EXAMPLE 7.4.1 ⎯⎯⎯⎯⎯⎯⎯⎯⎯⎯⎯⎯⎯⎯⎯⎯⎯⎯⎯⎯⎯⎯⎯⎯⎯⎯⎯⎯⎯⎯⎯⎯⎯⎯⎯

Select the lightest W or M section to carry a uniformly distributed dead load of 0.2 kip/ft superimposed (i.e., in addition to the beam weight) and 0.8 kip/ft live load. The simply supported span (Fig. 7.4.1) is 20 ft. The compression flange of the beam is fully supported against lateral movement. Use Load and Resistance Factor Design, and select for the following steels: A36; A572 Grade 50; and A572 Grade 65.

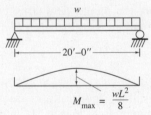

Figure 7.4.1   Example 7.4.1.

Solution

(a) Compute the factored load M_u.

$$w_u = 1.2w_D + 1.6w_L = 1.2(0.2) + 1.6(0.8) = 1.52 \text{ kips/ft}$$
$$M_u = w_u L^2/8 = 1.52(20)^2/8 = 76 \text{ ft-kips (not including beam weight)}$$

(b) A36 steel. Assume "compact section" since the vast majority of rolled sections satisfy $\lambda \leq \lambda_p$ for both the flange and the web. The design strength $\phi_b M_n$ is

$$\phi_b M_n = \phi_b M_p = \phi_b Z_x F_y$$

The design requirement is

$$\text{Required } \phi_b M_n = M_u$$
$$\text{Required } Z_x = \frac{M_u}{\phi_b F_y} = \frac{76(12)}{0.90(36)} = 28.1 \text{ in.}^3$$

Select from *LRFD Manual* table "LOAD FACTOR DESIGN SELECTION TABLE" pp. 4-15 to 4-21:

$$\text{Try W12}\times22: \qquad Z_x = 29.3 \text{ in.}^3$$

Check "compact section" limits λ_p (Table 7.4.2):

$$\left(\lambda = \frac{b_f}{2t_f} = \frac{4.03}{2(0.425)} = 4.7\right) < \left(\lambda_p = \frac{65}{\sqrt{F_y}} = 10.8\right) \quad \text{OK}$$

$$\left(\lambda = \frac{h}{t_w} = 41.8\right) < \left(\lambda_p = \frac{640}{\sqrt{F_y}} = 107\right) \quad \text{OK}$$

Note that the value h/t_w used is tabulated in the *LRFD Manual*. If computed as $(d - 2k + \text{radius of } 0.25)/t_w$, $h/t_w = 41.6$. Use of the *LRFD Manual* values of the property h/t_w is therefore recommended.

Check the strength: Correct the factored moment M_u to include the beam weight.

$$w_u = 1.2(0.222) + 1.6(0.8) = 1.55 \text{ kips/ft}$$
$$M_u = 76(1.55/1.52) = 77.3 \text{ ft-kips}$$
$$M_n = M_p = Z_x F_y = 29.3(36)/12 = 87.9 \text{ ft-kips}$$
$$\phi_b M_n = \phi_b(87.9) = 0.90(87.9) = 79.1 \text{ ft-kips} > M_u \quad \text{OK}$$

Use W12×22, $F_y = 36$ ksi.

(c) A572 Grade 50 steel.

$$\text{Required } Z_x = \frac{M_u}{\phi_b F_y} = \frac{76(12)}{0.90(50)} = 20.3 \text{ in.}^3$$

Select from *LRFD Manual* table "LOAD FACTOR DESIGN SELECTION TABLE Z_x":

$$\text{Try W10}\times19: \qquad Z_x = 21.6 \text{ in.}^3$$

Check "compact section" limits λ_p (Table 7.4.2):

$$\lambda_{\text{flange}} = \frac{b_f}{2t_f} = \frac{4.020}{2(0.395)} = 5.1 < 9.2 \quad \text{OK}$$

$$\lambda_{\text{web}} = \frac{h}{t_w} = 35.4 < 90.5 \quad \text{OK}$$

Check the strength:

$$M_u = 76(1.51/1.52) = 75.5 \text{ ft-kips}$$
$$M_n = M_p = Z_x F_y = 21.6(50)/12 = 90 \text{ ft-kips}$$
$$\phi_b M_n = \phi_b(104) = 0.90(90) = 81 \text{ ft-kips} > M_u \quad \text{OK}$$

Use W10×19, $F_y = 50$ ksi.

(d) A572 Grade 65 steel.

$$\text{Required } Z_x = \frac{M_u}{\phi_b F_y} = \frac{76(12)}{0.90(65)} = 15.6 \text{ in.}^3$$

Select from *LRFD Manual* table "LOAD FACTOR DESIGN SELECTION TABLE":

$$\text{Try W12×14:} \qquad Z_x = 17.4 \text{ in.}^3$$

Check "compact section" limits λ_p (Table 7.4.2):

$$\lambda_{\text{flange}} = \frac{b_f}{2t_f} = \frac{3.97}{2(0.225)} = 8.8 > 8.1 \quad \text{NG}$$

$$\lambda_{\text{web}} = \frac{h}{t_w} = 54.3 > 79.4 \quad \text{OK}$$

In this case the controlling limit state is local buckling of the flange. When $\lambda_r > \lambda > \lambda_p$, as above, the section is classified as "noncompact".

Check the strength. The strength is obtained by interpolation between M_p and M_r using Eq. 7.4.4. First λ_r for the flange must be obtained (from Table 7.4.1) as 19.0, which exceeds $b_f/2t_f$ of 8.8 and the section is "noncompact":

$$\lambda_r = \left(\frac{141}{\sqrt{F_y - 10}} = 19.0 \right)$$

$$M_n = M_p - (M_p - M_r) \left(\frac{\lambda - \lambda_p}{\lambda_r - \lambda_p} \right) \le M_p \qquad [7.4.4]$$

Next M_p and M_r are needed:

$$M_p = Z_x F_y = 17.4(65)/12 = 94.2 \text{ ft-kips}$$
$$M_r = S_x(F_y - F_r) = 14.9(65 - 10)/12 = 68.3 \text{ ft-kips}$$

Then from Eq. 7.4.4,

$$M_n = 94.2 - (94.2 - 68.3)\left(\frac{8.8 - 8.1}{19.0 - 8.1} \right) = 92.5 \text{ ft-kips}$$

which is $0.98M_p$. If in addition the web had $\lambda_r > \lambda > \lambda_p$, then M_n would also have to be computed from Eq. 7.4.4 using λ, λ_r, and λ_p values for the web. The lower of the strengths relating to flange local buckling and web local buckling would be the correct value of M_n. Continuing the check,

$$M_u = 76(1.54/1.52) = 77.1 \text{ ft-kips}$$
$$M_n = 92.5 \text{ ft-kips}$$
$$\phi_b M_n = \phi_b(92.5) = 0.90(92.5) = 83.3 \text{ ft-kips} > M_u \quad \text{OK}$$

Use W12×14, $F_y = 65$ ksi.   ■■

7.5 LATERALLY SUPPORTED BEAMS
—ALLOWABLE STRESS DESIGN

In accordance with the philosophy of Allowable Stress Design described in Secs. 1.8 and 1.9, Eq. 1.8.8 gives the structural safety requirement, as follows:

$$\frac{\phi M_n}{\gamma} \geq \sum Q_i \qquad\qquad [1.8.8]$$

which can be expressed

$$\left(\frac{M_n}{\gamma/\phi} = \frac{M_n}{\text{FS}}\right) \geq M \qquad\qquad (7.5.1)$$

where M_n = nominal moment strength
γ/ϕ = overload factors divided by resistance factor
FS = 1.67 = nominal safety factor in beam design
M = service load bending moment

To obtain Eq. 7.5.1 in "stress" format, divide both sides by the section modulus S; thus,

$$\left(f_b = \frac{M}{S}\right) \leq \left(F_b = \frac{M_n}{(\text{FS})S}\right) \qquad\qquad (7.5.2)$$

Compact Sections

As discussed in Secs. 7.3 and 7.4, the nominal strength of such sections is the plastic moment strength M_p. Equations 7.3.2 and 7.3.3 show that M_p may be expressed in terms of the shape factor ξ, as follows:

$$M_p = ZF_y = \xi SF_y \qquad\qquad (7.5.3)$$

In order to compute and compare stresses, a value of the shape factor ξ must be chosen when in fact each rolled section has a different value. For strong axis bending of I-shaped sections a value of 1.10 was selected. Substituting Eq. 7.5.3 with $\xi = 1.10$ into F_b on the right side of Eq. 7.5.2 and using FS = 1.67 gives the allowable stress F_b.

1. *For I-shaped sections bending about the x–x axis*:

$$F_b = \frac{\xi SF_y}{(\text{FS})S} = \frac{1.10 S_x F_y}{(1.67)S_x} = 0.66F_y \qquad\qquad (7.5.4)$$

In order for sections to qualify as "compact" the width/thickness ratios for the flange and web must not exceed the "compact" limits of ASD-Table B5.1 (the λ_p values of Table 7.4.2).

2. *For I-shaped sections bending about the y–y axis*:

$$F_b = 0.75F_y \qquad\qquad (7.5.5)$$

For bending about the weak axis the shape factor ξ is 1.5 for rectangular sections as shown in Example 7.3.1. In order not to have service load stresses approach too close to the yield stress, the allowable stress F_b for such sections is permitted higher than $0.66F_y$ but not fully in proportion to the shape factor. Only the flange need satisfy "compact" limits of Table 7.4.2.

Noncompact Sections

When the width/thickness ratios exceed the "compact" limits (referred to in LRFD as λ_p limits) of ASD-Table B5.1 but do not exceed the "noncompact" limits of that table, the nominal strength will not reach M_p. As discussed in Sec. 6.16, the "noncompact" limits of ASD-Table 5.1 are intended to assure the section of reaching the yield stress F_y at the extreme fiber. Thus, if the "noncompact" limits of ASD-Table B5.1 (referred to in Sec. 7.4 as λ_r limits) *are exactly met*, the nominal strength M_n is the yield moment M_y. The allowable stress F_b from Eq. 7.5.2 then becomes

$$F_b = \frac{SF_y}{(FS)S} = \frac{F_y}{1.67} = 0.60F_y \tag{7.5.6}$$

Partially Compact Sections

When the width/thickness ratios exceed the "compact" limits but are less than the "noncompact" limits of ASD-Table B5.1 the strength lies between M_y and M_p; therefore, the designer may linearly interpolate between those limits. Note that the "noncompact" limits of ASD-Table 5.1 (see Table 6.16.1) are somewhat more conservative (i.e., lower) than the λ_r limits of LRFD-Table B5.1, though the purpose is the same. The allowable stress relationship is shown in Fig. 7.5.1.

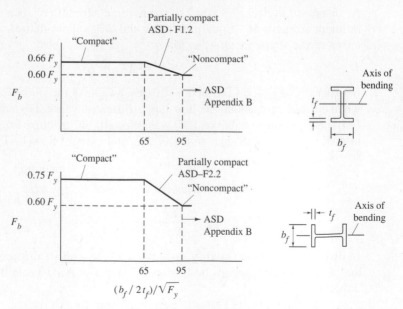

Figure 7.5.1 Allowable stress on H- or I-shaped sections qualifying as "compact sections" except for an excessive width-to-thickness ratio at the unstiffened compression flange element.

Slender Sections

When the width/thickness ratios exceed the "noncompact" limits of ASD-Table B5.1, the sections are referred to as "slender" and must be treated in accordance with ASD-Appendix B.

EXAMPLE 7.5.1

Redesign the beam of Example 7.4.1 using Allowable Stress Design with A36 steel.

Solution

(a) A36 steel. Assume "compact section" since nearly all sections satisfy the width/thickness limits λ_p; thus, the allowable stress F_b would be

$$F_b = 0.66F_y$$

Note that rounded values (i.e., 0.66 times 36 = 23.8 ksi; use 24 ksi) are accepted values in accordance with ASD-"NUMERICAL VALUES" TABLE 1.

The superimposed service load (1 kip/ft) bending moment is

$$M = wL^2/8 = 1.0(20)^2/8 = 50 \text{ ft-kips}$$

$$\text{Required } S_x = \frac{M}{F_b} = \frac{50(12)}{24} = 25 \text{ in.}^3$$

Select from *ASD Manual* "ALLOWABLE STRESS DESIGN SELECTION TABLE," the lightest section having at least $S_x = 25$ in.^{3}:

$$\text{Try W12}\times22: \qquad S_x = 25.4 \text{ in.}^3$$

Check "compact" limits (λ_p) of ASD-Table B5.1:

$$\frac{b_f}{2t_f} = \frac{4.03}{2(0.425)} = 4.7 < 10.8 \quad \text{(Table 7.4.2)} \quad \text{OK}$$

$$\frac{d}{t_w} = \frac{12.31}{0.260} = 47.3 < 107 \quad \text{(Table 7.4.2)} \quad \text{OK}$$

Note that ASD uses overall depth d whereas LRFD uses the unsupported height h of the web even though the limit is the same.

Check the flexural stress:

$$M = 1.022(20)^2/8 = 51.1 \text{ ft-kips} \quad \text{(including beam weight)}$$

$$f_b = \frac{M}{S_x} = \frac{51.1(12)}{25.4} = 24.1 \text{ ksi} \approx [F_b = 24 \text{ ksi}] \quad \text{say OK}$$

Use W12$\times$22, $F_y = 36$ ksi. ∎∎

7.6 SERVICEABILITY OF BEAMS

Serviceability, instead of *strength,* may and often does control the design of beams. Excessive deflection may cause damage to supported nonstructural elements such as partitions, may impair the usefulness of the structure by, for instance, distorting door jambs so that doors will not open or close, or may cause "bouncy" floors. These are serviceability problems, often unrelated to the strength of the floor system.

An excellent appraisal of serviceability of structures is provided by an ASCE Ad Hoc Committee [7.4], Galambos and Ellingwood [7.5] have discussed general serviceability limit states, and Ellingwood [7.6] has provided guidelines for steel structures. Excessive deflection is often indicative of excessive vibration and noise transmission, both serviceability problems.

A recent overall treatment of building floor vibrations, including recommended criteria, is provided by Murray [7.7]. Wright and Walker [7.8], Murray [7.9, 7.10], Ellingwood and Tallin [7.11], Tolaymat [7.12], and Allen [7.13] have treated floor vibrations and the related human response. Hatfield [7.14] has provided a design chart for floor vibration. Allen and Murray [7.54] have given design criteria for vibrations due to walking.

On roofs a major deflection-related concern is ponding of water; this is specifically treated later in this section.

Deflection

Numerous structural analysis methods are available for computing deflections on uniform and variable moment of inertia sections. In general, the maximum deflection in an elastic member may be expressed as

$$\Delta_{max} = \beta_1 \frac{WL^3}{EI} \tag{7.6.1}$$

where W = total *service* load on the span
   L = span length
   E = modulus of elasticity (29,000 ksi or 200,000 MPa for steel)
   I = moment of inertia
   β_1 = coefficient which depends upon the degree of fixity at supports, the variation in moment of inertia along the span, and the distribution of loading. (For a simply supported beam, $\beta_1 = 5/384$; other values are available in the LRFD or ASD Manual section, "BEAM DIAGRAMS AND FORMULAS.")

For continuous beams, the midspan deflection in the common situation of a uniform loading on a prismatic beam with unequal end moments (see Fig. 7.6.1) may be expressed* as

$$\Delta_{midspan} = \frac{5L^2}{48EI} [M_s - 0.1(M_a + M_b)] \tag{7.6.2}$$

Equation 7.6.2 will give satisfactory results when considered to be the maximum deflection for nearly all practical loadings for beams having uniform moment of inertia. Equation 7.6.2 may be verified by the use of a method such as conjugate beam.

---

*See Chu-Kia Wang and Charles G. Salmon, *Reinforced Concrete Design,* 5th ed. (HarperCollins Publishers, New York, 1992), p. 545.

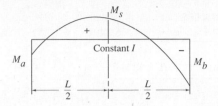

Figure 7.6.1 Typical bending moment diagram for interior span of continuous uniformly loaded beam.

For uniformly loaded simply supported beams, Eq. 7.6.1 becomes

$$\Delta_{\max} = \frac{5wL^4}{384EI} \tag{7.6.3}$$

which upon substitution of $M = wL^2/8$, $f = Mc/I$, and $c = d/2$, gives

$$\Delta_{\max} = \frac{10fL^2}{48Ed} \tag{7.6.4}$$

Equation 7.6.4 can be used as a good approximation for any simply supported beam as long as the maximum stress occurs near midspan. Refer to Table 7.6.1 for typical values.

ASD-L3.1 states, "Beams and girders supporting floors and roofs shall be proportioned with due regard to the deflection produced by design loads." In addition, live load deflection where plastered ceilings are supported is limited to $L/360$.

For the $L/360$ limitation, Eq. 7.6.4, using $E = 29{,}000$ ksi becomes

$$\frac{L}{d} \le \frac{48(29{,}000)}{(10)360f} = \frac{387}{f} \tag{7.6.5}*$$

where f is in ksi.

On the other hand LRFD-L3 contains the more general statement, "Deformations in structural members and structural systems due to service loads shall not impair the serviceability of the structure." Further, LRFD-L says limiting values to ensure serviceability "shall be chosen with due regard to the intended function of the structure."

When considering deflections, it should be remembered that the dead load deflections usually can be accounted for during construction by either cambering (negative bending) or thickening the slab or floor topping. It is only the deflection that occurs due to loads applied after construction is completed that may crack ceilings, partitions, or walls. Discussion of cambering is available from Bethlehem Steel [7.15] and a paper by Ricker [7.16].

Specification requirements for limiting deflections are meager because there is no single or standard value for the tolerable deflection. The acceptable amount must of necessity depend on the type and arrangement of materials being supported.

---

* For SI,
$$\frac{L}{d} = \frac{2668}{f} \text{ for } f \text{ in MPa.} \tag{7.6.5}$$

TABLE 7.6.1   DEFLECTION RELATIONSHIPS
ACCORDING TO EQ. 7.6.4

Δ_{max}	L/d	L/d (f = 22 ksi)	L/d (f = 30 ksi)
$L/360$	$387/f$	17.6	12.9
$L/300$	$464/f$	21.1	15.5
$L/240$	$580/f$	26.3	19.3
$L/200$	$695/f$	31.6	23.2

As a guide only. ASD-Commentary L3.1 suggests the following limitations:
Floor beams and girders, fully stressed:

$$\frac{L}{d} \leq \frac{800}{F_y, \text{ ksi}} \tag{7.6.6}*$$

Floor beams and girders, subject to shock or vibratory loads, supporting large open areas free of partitions or other sources of damping:

$$\frac{L}{d} \leq 20 \tag{7.6.7}$$

Roof purlins, fully stressed, except flat roofs:

$$\frac{L}{d} \leq \frac{1000}{F_y, \text{ ksi}} \tag{7.6.8}*$$

Assuming the service load stress f equals the ASD allowable stress $0.66F_y$ for "compact sections", the ASD-Commentary suggestions of coefficients 800 and 1000 correspond to L/d values of $528/f$ and $660/f$, respectively. Using these L/d values in Eq. 7.6.4 would give simply supported beam limits of about $L/260$ and $L/210$.

On continuous spans it is, of course, the actual deflection that is of importance, not the L/d ratio. For continuous beam deflection, a comparison of Eqs. 7.6.2 and 7.6.4 shows that Eq. 7.6.4 can also be used for continuous beams if the stress f is computed using the equivalent bending moment

$$M_e = M_s - 0.1(M_a + M_b) \tag{7.6.9}$$

EXAMPLE 7.6.1

Select the lightest W section to carry a uniform dead load of 0.5 kip/ft and a live load of 1.0 kip/ft on a simply supported span of 42 ft. Adequate lateral support is provided. The live load deflection is limited to $L/360$. Use A572 Grade 50 and Load and Resistance Factor Design.

---

* For SI, with F_y in MPa,

$$\frac{L}{d} \leq \frac{5500}{F_y} \tag{7.6.6}$$

$$\frac{L}{d} \leq \frac{6900}{F_y} \tag{7.6.8}$$

Solution

(a) Compute the factored moment M_u. Estimating the beam weight at 70 lb/ft:

$$w_u = 1.2w_D + 1.6w_L$$
$$= 1.2(0.5 + 0.07 \text{ est}) + 1.6(1.0) = 2.28 \text{ kips/ft}$$
$$M_u = w_u L^2/8 = 2.28(42)^2/8 = 504 \text{ ft-kips}$$

(b) Compute required plastic modulus Z_x to satisfy strength requirement. Assuming compact section, the design strength $\phi_b M_n$ is

$$\phi_b M_n = \phi_b M_p = \phi_b Z_x F_y$$

$$\text{Required } Z_x = \frac{M_u}{\phi_b F_y} = \frac{504(12)}{0.90(50)} = 134 \text{ in.}^3$$

Select from *LRFD Manual* table "LOAD FACTOR DESIGN SELECTION TABLE":

$$\text{Try W24}\times55: \quad Z_x = 134 \text{ in.}^4$$

The section is compact for $F_y = 50$ ksi.

(c) Compute required moment of inertia I_x to satisfy the deflection limit. The service load moment instead of the factored moment must be used since deflection is of concern when the structure is being used, not when failure is imminent. The service live load moment is

$$M = wL^2/8 = 1.0(4.2)^2/8 = 221 \text{ ft-kips}$$

$$\Delta = \frac{5wL^4}{384EI} = \frac{5ML^2}{48EI}$$

$$\text{Required } I = \frac{5ML^2}{48E\Delta} = \frac{5(221)(42)^2(144)}{48(29,000)(42/360)} = 1724 \text{ in.}^4$$

Select from *LRFD Manual* table "MOMENT OF INERTIA SELECTION TABLE":

$$\text{Try W24}\times68: \quad I_x = 1830 \text{ in.}^4$$

Note that the section required to control deflection is larger than the section required for strength; i.e., deflection controls.

(d) Check the W24×68 section:

$$w_u = 1.2(0.5 + 0.068) + 1.6(1.0) = 2.28 \text{ kips/ft}$$
$$M_u = 503 \text{ ft-kips}$$
$$M_n = M_p = Z_x F_y = 177(50)/12 = 738 \text{ ft-kips}$$
$$\phi_b M_n = \phi_b(738) = 0.90(738) = 664 \text{ ft-kips} > M_u \quad \text{OK}$$

As expected, the strength considerably exceeds the required strength. Check deflection:

$$M(\text{service live load}) = 221 \text{ ft-kips}$$

$$\Delta = \frac{5ML^2}{48EI} = \frac{5(221)(42)^2 1728}{48(29,000)1830} = 1.32 \text{ in.} < \left(\frac{L}{360} = 1.40 \text{ in.}\right) \quad \text{OK}$$

Note that the live load deflection and the limit are quite close (within 6%) but the strength considerably exceeds the requirement (32% greater); that is, *deflection controls*.
Use W24×68, F_y = 50 ksi.   ■■

Ponding of Water on Flat Roofs

When members of a flat roof system deflect, a bowl-shaped volume is created which is capable of retaining water. As water begins to accumulate, deflection increases to provide an increased volumetric capacity. This cyclical process continues until either (1) the succeeding deflection increments become smaller and equilibrium is reached; or (2) succeeding deflection increments are increasing, the system becomes unstable, and collapse occurs. This retention of water which results solely from the deflection of flat roof framing is what is referred to as *ponding*. From a serviceability standpoint, this ponding of water is a major reason for splitting of roof membranes, resulting in costly replacement of both the membrane and the insulation.

To prevent ponding of water accumulated on flat roofs, the 1963 AISC Specification required supporting members to satisfy the limitation

$$\frac{L}{d} \leq \frac{600}{f_b} \tag{7.6.10}$$

where f_b is the computed service load bending stress in ksi. Using Eq. 7.6.4, this would correspond roughly to a deflection limitation $L/240$ on a simply supported span.

Avoidance of ponding is much more complex than indicated by the above limitation. Marino [7.17] has provided an extensive treatment that forms the basis for the AISC provisions of LRFD and ASD-K2. The flat roof is treated as a two-way system of secondary members (say, purlins) elastically supported by primary members (say, girders) which are rigidly supported by walls or columns, as shown in Fig. 7.6.2.

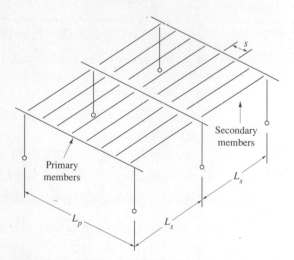

Figure 7.6.2   Flat roof arrangement for ponding analysis. (From Marino [7.17])

LRFD and ASD-K2 give a simple but conservative criterion

$$C_p + 0.9C_s \leq 0.25 \qquad (7.6.11)^*$$

and

$$I_d \geq 25S^4(10^{-6}) \qquad (7.6.12)^*$$

where

$$C_p = \frac{32L_s L_p^4}{10^7 I_p} \qquad (7.6.13)^*$$

$$C_s = \frac{32SL_s^4}{10^7 I_s} \qquad (7.6.14)^*$$

L_p = length of primary member, ft
L_s = length of secondary member, ft
S = spacing of secondary member, ft
I_p = moment of inertia of primary member, in.^{4}
I_s = moment of inertia of secondary member, in.^{4}
I_d = moment of inertia of steel deck supported on secondary members, in.^{4} per ft

The criterion $C_p + 0.9C_s \leq 0.25$ assumes the supporting members to be loaded to full strength before onset of ponding. The terms C_p and C_s indicate the relative stiffnesses of the primary and secondary support systems, respectively. The right hand side of Eq. 7.6.11 is a Stress Index U (See LRFD-Appendix K2) representing the ratio of the increase in stress as a result of ponded water to the service load allowable stress. For instance, assuming that the service load stress in a member supporting ponded water could be permitted to increase to $0.8F_y$ from $0.6F_y$ or $0.66F_y$, the Stress Indexes U would be

$$U = \frac{0.8F_y - 0.66F_y}{0.66F_y} = 0.212$$

$$U = \frac{0.8F_y - 0.6F_y}{0.6F_y} = 0.33$$

A Stress Index U of 0.25 is a reasonable lower limit for that quantity. Design aids appear in LRFD-Appendix K2 as Figs. A-K2.1 and A-K2.2. These diagrams show Eq. 7.6.11 to be conservative, the more so when stresses at onset of ponding are low. Burgett [7.18] has provided graphs for a fast check using Eqs. 7.6.11 and 7.6.12. More recently, Ruddy [7.19] has illustrated the procedure for a concrete floor over metal decking and supported by steel beams and girders.

---

* For SI, with L and S in metres, and I in mm^4, and I_d in mm^4/m,

$$C_p + 0.9C_s \leq 0.25 \qquad (7.6.11)$$

$$I_d \geq 400S^4 \qquad (7.6.12)$$

$$C_p = \frac{5L_s L_p^4}{10^{13} I_p} \qquad (7.6.13)$$

$$C_s = \frac{5SL_s^4}{10^{13} I_s} \qquad (7.6.14)$$

Equation 7.6.12 pertains to roof decking that is supported on secondary members. Since this contributes little to ponding, it may be treated as a one-way system in the manner presented by Chinn [7.20]. When roof decking *is* the secondary member, then it should be treated according to Eq. 7.6.11.

More elaborate mathematical treatment of ponding has been given by Salama and Moody [7.21], Sawyer [7.22, 7.23], Chinn, Mansouri, and Adams [7.24], Avent and Stewart [7.25], and Avent [7.26].

7.7 SHEAR ON ROLLED BEAMS

Whereas long beams may be governed by deflection and medium length beams are usually controlled by flexural strength, short-span beams may be governed by shear.

To review the development of the shear stress equation for symmetrical sections, consider the slice dz of the beam of Fig. 7.7.1a, shown as a free body in Fig. 7.71b. If the unit shear stress v at a section y_1 from the neutral axis is desired, it is observed from Fig. 7.7.1c that

$$dC' = vt\,dz \tag{7.7.1}$$

The horizontal forces arising from bending moment are

$$C' = \int_{y_1}^{y_2} f\,dA$$

$$C' + dC' = \int_{y_1}^{y_2} (f + df)dA$$

Subtracting,

$$dC' = \int_{y_1}^{y_2} df\,dA \tag{7.7.2}$$

$$df = \frac{dMy}{I} \tag{7.7.3}$$

$$dC' = \int_{y_1}^{y_2} \frac{dMy}{I}\,dA = \frac{dM}{I}\int_{y_1}^{y_2} y\,dA \tag{7.7.4}$$

Substituting Eq. 7.7.4 into Eq. 7.7.1 and solving for the shear stress v gives

$$v = \frac{dM}{dz}\left(\frac{1}{tI}\right)\int_{y_1}^{y_2} y\,dA \tag{7.7.5}$$

and upon recognizing that $V = dM/dz$, and letting

$$Q = \int_{y_1}^{y_2} y\,dA$$

the familiar equation

$$v = \frac{VQ}{It} \tag{7.7.6}$$

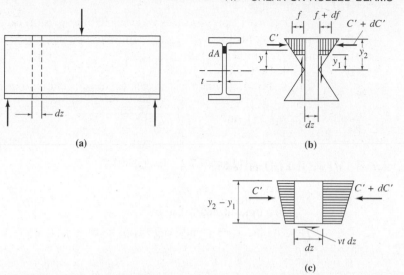

Figure 7.7.1   Flexural stresses involved in derivation of shear stress equation.

is obtained where Q is the first moment of area about the x-axis of the cross-sectional area between the extreme fiber at y_2 (Fig. 7.7.1b) and the particular location at y_1 at which the shear stress is to be determined.

Under usual procedures of steel design, the shear stress is computed as the average value over the gross area of the web neglecting the effect of any fastener holes; thus

$$f_v = \frac{V}{A_w} = \frac{V}{dt_w} \tag{7.7.7}$$

Note that large holes cut in a beam web to permit passage of pipes and ducts require special consideration and their effect may *not* be neglected.

The following example illustrates that in an I-shaped beam most of the shear is carried by the web.

EXAMPLE 7.7.1 _____

Determine the elastic shear stress distribution on a W24×94 beam subjected to a service load shear force of 200 kips. Also compute the portion of the shear carried by the flange and that carried by the web. (See Fig. 7.7.2).

   Solution
   (a) Stress at junction of flange and web.

$$V = 200 \text{ kips}$$
$$Q = 9.065(0.875)(12.155 - 0.4375) = 92.9 \text{ in.}^3$$
$$v = \frac{200(92.9)}{2700(0.515)} = 13.4 \text{ ksi (web)}, \quad v = 0.76 \text{ ksi (flange)}$$

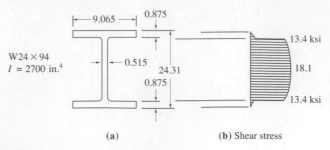

Figure 7.7.2 Example 7.7.1.

(b) Stress at neutral axis.

$$Q = 92.9 + (12.155 - 0.875)^2(0.515)(0.5) = 92.9 + 32.8 = 125.7 \text{ in.}^3$$

$$v = \frac{200(125.7)}{2700(0.515)} = 18.1 \text{ ksi}$$

(c) Shear carried by flanges and web. Using an approximate linear variation,

$$V(\text{flanges}) = 2(\tfrac{1}{2})(0.76)(0.875)(9.065) = 6 \text{ kips}$$
$$V(\text{web}) = 200 - 6 = 194 \text{ kips}$$

In this case, 97% of the shear is carried by the web.

(d) Average shear stress f_v on web.

$$f_v = \frac{V}{dt_w} = \frac{200}{24.31(0.515)} = 16.0 \text{ ksi}$$

which is 11.6% below the maximum value. ■■

Nominal Shear Strength V_n in Rolled Beams

As shown by Example 7.7.1, the web is the element that primarily carries the shear in I-shaped sections. This is true also for the web (or webs) of "singly or doubly symmetric beams . . . and channels subject to shear in the plane of the web." (LRFD-F2.)

As long as the web is stable, that is, instability resulting from shear stress or a combination of shear and bending stress cannot occur, the shear strength V_n of the section is based on overall shear yielding of the web. Thus,

$$V_n = \tau_y\, A_w \tag{7.7.8}$$

where τ_y = shear yield stress of the web steel
A_w = area of the web = dt_w for rolled beams
d = overall depth for rolled beams
t_w = web thickness

According to the "energy of distortion" theory (see Sec. 2.6), the shear yield stress τ_y equals the tension-compression yield stress F_y divided by $\sqrt{3}$ when shear stress acts alone, giving from Eq. 2.6.5,

$$\tau_y = 0.58F_y \qquad\qquad [2.6.5]$$

Thus, it is logical for LRFD-F2.2 to use $\tau_y = 0.6F_y$. Equation 7.7.8 then becomes

$$V_n = 0.6F_{yw}A_w \qquad\qquad (7.7.9)$$

where F_{yw} = yield stress of the web.

Equation 7.7.9 implies that h/t_w ratios do not exceed

$$\frac{h}{t_w} = \frac{418}{\sqrt{F_{yw}}} \qquad\qquad (7.7.10)*$$

The development of Eq. 7.7.10 appears in Chapter 11 (Sec. 11.8) on plate girders. However, typical of plates as developed in Chapter 6 the buckling strength depends on a slenderness ratio; in this case h/t_w as shown in Fig. 7.7.3. From that

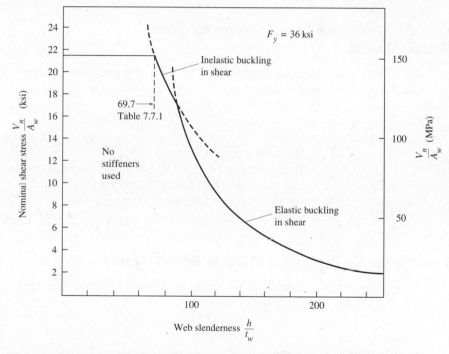

Figure 7.7.3 Nominal shear stress V_n/A_w vs web slenderness h/t_w for A36 steel beams *without* transverse stiffeners.

---

*For SI, with F_y in MPa, $\qquad \dfrac{h}{t_w} = \dfrac{1100}{\sqrt{F_{yw}}}$ $\qquad\qquad (7.7.10)$

TABLE 7.7.1   MAXIMUM h/t_w LIMITS WHEN STIFFENERS ARE NOT USED
(BASED ON LRFD-F2.2 AND ASD-F4)

F_y (ksi)	F_y (MPa)	h/t_w* LRFD-F2.2	h/t_w† ASD-F4
36	248	69.7	63.3
42	290	64.5	58.6
45	310	62.3	56.6
50	345	59.1	53.7
60	414	54.0	49.1
65	448	51.9	47.1
100	689	41.8	38.0

$h = T$

h, for plate girders

For rolled sections

* Equation 7.7.10

† h/t_w limit $= 380/\sqrt{F_y}$, ksi; see Chap. 11 (Sec. 11.8) for derivation.

figure one may note that maximum shear strength is available when h/t_w does not exceed 69.7 (for A36 steel). The h/t_w limits below which Eq. 7.7.9 is applicable for rolled beams without stiffeners are given in Table 7.7.1.

Load and Resistance Factor Design for Shear in Rolled Beams

The shear strength requirement in load and resistance factor design according to LRFD-F2 may be stated

$$\phi_v V_n \geq V_u \tag{7.7.11}$$

where $\phi_v = 0.90$
$V_n =$ nominal strength in shear
$\quad = 0.6F_{yw}A_w$ (i.e., Eq. 7.7.9) for beams without transverse stiffeners and not exceeding h/t_w limits given in Table 7.7.1.
$V_u =$ factored service load shear (see Sec. 1.9)

Allowable Stress Design for Shear in Rolled Beams

The safety requirement for shear in Allowable Stress Design (ASD) according to ASD-F4 may be stated

$$f_v \leq F_v \tag{7.7.12}$$

where $f_v =$ service load average shear stress $= V/A_w$
$V =$ service load shear force
$A_w =$ area of the web $= dt_w$ for rolled beams
$F_v =$ allowable shear stress at service load $= 0.40F_y$ for beams without transverse stiffeners and not exceeding h/t_w limits given in Table 7.7.1

Equation 7.7.12 may be obtained from the strength relationship reformulated as Eq. 1.8.8,

$$\frac{\phi R_n}{\gamma} \geq \sum Q_i \qquad [1.8.8]$$

If γ/ϕ the nominal total factor of safety is taken as the traditional ASD value of 1.67, R_n is V_n for shear, and the right hand side is the total service load shear V, then Eq. 1.8.8 becomes

$$\frac{V_n}{1.67} \geq V \qquad (7.7.13)$$

Dividing both sides by the web area A_w, after noting that $V_n = 0.6F_y A_w$, and reversing the sides of the equation gives

$$\left(f_v = \frac{V}{A_w} \right) \leq \left(F_v = \frac{0.6F_y}{1.67} = 0.36F_y \right) \qquad (7.7.14)$$

Traditional allowable stress design has used two-thirds of the yield stress as the allowable value (2/3 of $0.6F_y$); thus, the allowable shear stress F_v is taken as $0.40F_y$.

The reader may note that the LRFD h/t_w limits for use of maximum shear strength are somewhat higher than the ASD limits for use of maximum allowable stress. This results from an approximation in k used for LRFD; philosophically there is no difference. Rolled beams will generally satisfy the lower limit.

EXAMPLE 7.7.2 _____

Select the lightest W section of A36 steel to carry a live load of 19 kips/ft and dead load of 1 kip/ft (in addition to the weight of the beam). The simply supported span is 5 ft. Lateral bracing is adequate for lateral stability. Use Load and Resistance Factor Design.

Solution. Since the loading is heavy and the span is short, the designer should investigate shear as well as flexure.

(a) Compute factored loads M_u and V_u.

$$w_u = 1.2w_D + 1.6w_L$$
$$= 1.2(1) + 1.6(19) = 32.0 \text{ kips/ft}$$
$$M_u = wL^2/8 = 32.0(5)^2/8 = 100 \text{ ft-kips}$$
$$V_u = wL/2 = 32.0(5)/2 = 80 \text{ kips}$$

With these heavy superimposed loads the beam weight will have little effect.

(b) Select a section for flexure. Assume "compact section"; then,

$$\phi_b M_n = \phi_b M_p = \phi_b Z_x F_y$$

The design requirement is that $\phi_b M_n = M_u$; thus,

$$\text{Required } Z_x = \frac{M_u}{\phi_b F_y} = \frac{100(12)}{0.90(36)} = 37.0 \text{ in.}^3$$

Try W12×26 from *LRFD Manual* "LOAD FACTOR DESIGN SELECTION TABLE" as the lightest beam having $Z_x \geq 37.0$ in.[3] All sections included in that table are compact.

(c) Check shear.

$$\phi_v V_n = \phi_v (0.6 F_y) A_w$$
$$= 0.90(0.6)(36)(12.22)(0.23) = 54.6 \text{ kips}$$

Since $V_u = 80$ kips exceeds the shear strength provided by W12×26, the section is not adequate. The required web area A_w for shear is

$$\text{Required } A_w = \frac{V_u}{\phi_v(0.6 F_y)} = \frac{80}{0.90(21.6)} = 4.12 \text{ sq in.}$$

From the bending moment requirement, the next heavier sections are deeper, such as W14 or W16. If a W16 is selected, its web thickness required will be $4.12/16 = 0.26$ in. Try W16×31, $Z_x = 54.0$ in.[3] For shear,

$$\phi_v V_n = 0.90(0.6)(36)(15.88)(0.275) = 84.9 \text{ kips}$$

which exceeds the factored shear $V_u = 80.1$ kips (including beam weight) and is acceptable.

Note that the h/t_w ratio cannot exceed the value in Table 7.7.1.

$$\frac{h}{t_w} = \frac{\text{unsupported height, } T}{\text{web thickness, } t_w} = \frac{13.625}{0.275} = 49.5$$

This is less than the limit value of 69.7 and confirms the use of Eq. 7.7.9 for the nominal strength V_n in shear. The detailed discussion of the use of intermediate stiffeners when h/t_w exceeds the value from Table 7.7.1 appears in Chapter 11 on plate girders. Note that the value for h/t_w tabulated in the *LRFD Manual* for the W16×31 is 51.6 based on an underestimate of the radius at the junction of flange to web; when the larger tabulated value is available it should preferably be used.

Use W16×31, $F_y = 36$ ksi. ■■

7.8 CONCENTRATED LOADS APPLIED TO ROLLED BEAMS

When concentrated loads are applied to beams, beam bearing at supports, and reactions of beam flanges at connections to columns, a localized yielding from high compressive stress followed by inelastic buckling in the web region adjacent to the toe of a fillet occurs in the vicinity of concentrated loads. This entire behavior was formerly combined under the category "web crippling."

Typical of compression-related situations, there are two possible behaviors; yielding and instability. The recent AISC Specifications consider three categories: (a) local web yielding, (b) web crippling, and (c) sidesway web buckling.

The transmission of concentrated loads in beam-to-column connections is treated in Chapter 13 and concentrated loads on plate girders and related design of bearing stiffeners is treated in Chapter 11.

Load and Resistance Factor Design

The requirement of LRFD-K1.3 may be stated

$$\phi R_n \geq R_u \qquad (7.8.1)$$

where ϕ = resistance (strength reduction) factor
R_n = nominal reaction strength
R_u = factored reaction

Load and Resistance Factor Design—Local Web Yielding

Referring to Fig. 7.8.1, the concentrated reaction R acting on a beam is assumed critical at the toe of the fillet (a distance k from the face of the beam). The load is assumed to distribute along the web at a slope of 2.5 to 1. Prior to 1985 LRFD Specification, the distribution was conservatively taken as 45°. The 1978 ASD and earlier Specifications were based on the work of Lyse and Godfrey [7.27]. Investigators [7.28, 7.29] have shown the 45° slope to be overly conservative; the load actually spreads over a distance $(N + 5k)$ to $(N + 7k)$ for an interior load rather than the distance $(N + 2k)$ formerly used.

The nominal reaction strength R_n based on the yield strength at the toe of the fillet on a rolled I-shaped section is as follows:

1. For *interior loads* where the concentrated load is applied at a distance from the end of the member that is greater than the depth of the member,

$$R_n = (5k + N)F_{yw}t_w \qquad (7.8.2)$$

2. For *end reactions*,

$$R_n = (2.5k + N)F_{yw}t_w \qquad (7.8.3)$$

where k = distance from outer face of flange to web toe of fillet
N = length of bearing $\geq k$ for end beam reaction
F_{yw} = specified minimum yield stress of the web
t_w = web thickness

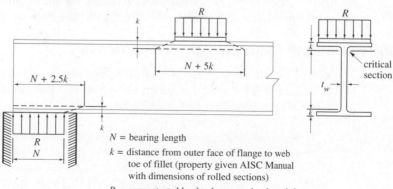

N = bearing length
k = distance from outer face of flange to web toe of fillet (property given AISC Manual with dimensions of rolled sections)
R = concentrated load to be transmitted to girder

Figure 7.8.1 Local web yielding considerations for establishing bearing length.

The resistance (strength reduction) factor ϕ for local web yielding is 1.0; this reflects the traditional lower total safety factor used to control local web yielding than used for overall strength of a member.

Allowable Stress Design—Local Web Yielding

Reformulating the strength relationship, Eq. 7.8.1, as Eq. 1.8.8,

$$\frac{\phi R_n}{\gamma} \geq \left(\sum Q_i = \frac{R_u}{\gamma} = R \right) \tag{7.8.4}$$

where Q_i = service loads
   γ = average overload factor
   R = total service load

Letting the combined safety factor γ/ϕ = FS and dividing Eq. 7.8.4 by the critical area A_c gives

$$\frac{R_n}{(\text{FS})A_c} \geq \left(\frac{R}{A_c} = f_c \right) \tag{7.8.5}$$

Substituting $(5k + N)t_w$ and $(2.5k + N)t_w$ for A_c, for the interior and end situations, respectively, and taking FS = 1.50 give the ASD-K1.3 limitations:

1. For *interior loads,*

$$f_c = \frac{R}{t_w(N + 5k)} \leq 0.66 F_y \tag{7.8.6}$$

2. For *end reactions,*

$$f_c = \frac{R}{t_w(N + 2.5k)} \leq 0.66 F_y \tag{7.8.7}$$

The length N of bearing is not to be taken less than k for end reactions. The 1989 ASD-K1.3 allowable stress for this limit state has been reduced to $0.66F_y$ from the traditional value of $0.75F_y$; somewhat offsetting the increase in bearing length from the traditional $(N + 2k)$ to $(N + 5k)$ for an interior bearing location.

Load and Resistance Factor Design—Web Crippling

To control the stability of the web at concentrated loads, in 1986 a provision was added in LRFD-K1.4 which is based on the work of Roberts [7.30]. The nominal reaction strength based on this stability criterion is:

1. For *interior loads* (i.e., concentrated load acts at $d/2$ or more from member end),

$$R_n = 135t_w^2 \left[1 + 3\left(\frac{N}{d}\right)\left(\frac{t_w}{t_f}\right)^{1.5} \right] \sqrt{\frac{F_{yw}t_f}{t_w}} \tag{7.8.8}$$

2. For *end reactions* (i.e., concentrated load acts less than $d/2$ from member end),

$$R_n = 68t_w^2\left[1 + 3\left(\frac{N}{d}\right)\left(\frac{t_w}{t_f}\right)^{1.5}\right]\sqrt{\frac{F_{yw}t_f}{t_w}} \tag{7.8.9}$$

for $N/d \le 0.2$. A new 1993 equation for the rare case of long bearing length N on a relatively shallow girder is,

$$R_n = 68t_w^2\left[1 + \left(\frac{4N}{d} - 0.2\right)\left(\frac{t_w}{t_f}\right)^{1.5}\right]\sqrt{\frac{F_{yw}t_f}{t_w}} \tag{7.8.10}$$

for $N/d > 0.2$.

where t_f = flange thickness through which the concentrated load is transmitted
          d = overall depth of beam

The resistance (strength reduction) factor ϕ is 0.75.

Allowable Stress Design—Web Crippling

For allowable stress design, ASD-K1.4 gives the allowable service load reaction R as 0.5 of R_n from Eqs. 7.8.8 through 7.8.10.

Load and Resistance Factor Design—Sidesway Web Buckling

The nominal reaction strength R_n based on sidesway web buckling as given by LRFD-K1.5 controls for compressive individual concentrated forces applied to members when relative lateral movement between the loaded compression flange and the tension flange is possible. This phenomenon has been studied by Roberts [7.30], Elgaaly [7.31], and Roberts and Chong [7.32]. Rarely will sidesway web buckling control on a rolled I-shaped section; however, it can influence design of a plate girder.

The nominal reaction strength R_n based on sidesway web buckling in accordance with LRFD-K1.5 is as follows:

1. When the compression flange *is restrained* against rotation:
for $(h/t_w)/(L_b/b_f) \le 2.3$,

$$R_n = \frac{C_r t_w^3 t_f}{h^2}\left[1 + 0.4\left(\frac{h/t_w}{L_b/b_f}\right)^3\right] \tag{7.8.11}$$

for $(h/t_w)/(L_b/b_f) > 2.3$,

$$R_n = \text{no limit}$$

2. When the compression flange *is <u>not</u> restrained* against rotation:
for $(h/t_w)/(L_b/b_f) \le 1.7$,

$$R_n = \frac{C_r t_w^3 t_f}{h^2}\left[0.4\left(\frac{h/t_w}{L_b/b_f}\right)^3\right] \tag{7.8.12}$$

for $(h/t_w)/(L_b/b_f) > 1.7$,

$$R_n = \text{no limit}$$

where L_b = largest laterally unbraced length along either flange at the point of load

b_f = flange width

t_w = web thickness

h = clear distance between flanges less the fillet or corner radius for rolled shapes; distance between adjacent lines of fasteners or the clear distance between flanges when welds are used for built-up shapes

C_r = 960,000 when $M_u < M_y$ at the location of the force, ksi

480,000 when $M_u \geq M_y$ at the location of the force, ksi

For the situation where the flange *is* restrained against rotation, when the factored forced exceeds the design strength ϕR_n, (where ϕ = 0.85), either more local lateral bracing must be used or stiffeners provided. When the flange *is not* restrained against rotation, and the strength is not adequate, local lateral bracing must be provided at both flanges at the point of application of the concentrated forces.

Allowable Stress Design—Sidesway Web Buckling

For allowable stress design, ASD-K1.5 gives equations of a form similar to the LRFD equations for the allowable service load.

EXAMPLE 7.8.1 ———

Determine the size of bearing plate required for an end reaction of 10 kips dead load and 20 kips live load on a W10×26 beam of A36 steel. The beam rests on a concrete wall having a 28-day compressive strength f'_c = 3000 psi. Use Load and Resistance Factor Design.

Solution

(a) Compute factored reaction R_u and required R_n.

$$R_u = 1.2(10) + 1.6(20) = 44 \text{ kips}$$

$$\text{Required } R_n = R_u/\phi = 44/1.0 = 44 \text{ kips}$$

(b) Determine plan dimensions for bearing plate. The bearing length must satisfy the more severe requirement of Eq. 7.8.3 (local web yielding) or Eqs. 7.8.9 or 7.8.10 (web crippling). Solving the simpler Eq. 7.8.3 for the required bearing length N gives

$$N = \frac{R_n}{F_{yw}t_w} - 2.5k = \frac{44}{36(0.260)} - 2.5(0.875) = 2.5 \text{ in.}$$

Try a 3-in. bearing plate. As a practical matter, 3 in. should be considered as minimum bearing length unless clearances require a lesser length.

Before checking Eqs. 7.8.9 or 7.8.10, whichever applies, for web crippling, investigate the requirement for bearing on the concrete. The nominal bearing strength P_p of concrete is given by LRFD-J9 as

$$P_p = 0.85 f_c' A_1$$

where A_1 is the area of steel concentrically bearing on a concrete support. The requirement is that $\phi P_p \geq R_u$ (i.e., the factored bearing load). The strength reduction factor ϕ for bearing is 0.60. Thus, solving for the required bearing area A_1 gives

$$\text{Required } A_1 = \frac{\text{Required } P_p}{0.85 f_c'} = \frac{R_u/\phi}{0.85 f_c'} = \frac{44/0.60}{0.85(3)} = 28.8 \text{ sq in.}$$

For a 3-in. bearing length, the width B (see Fig. 7.8.2) would have to be 9.6 in. Since the beam flange width is only 5.77 in., a width B smaller than 9.6 in. is desirable. A 4-in. bearing length requires a width of 7.2 in. Try a plate 4 in. × 7.5 in. (A_1 = 30 sq. in.).

Now check web crippling on the W10×26:

$$N/d = 4/10.33 > 0.2; \text{ Eq. 7.8.10 applies.}$$

$$R_n = 68(0.260)^2 \left[1 + \left(\frac{4(4)}{10.33} - 0.2 \right) \left(\frac{0.260}{0.440} \right)^{1.5} \right] \sqrt{\frac{36(0.440)}{0.260}} = 57.9 \text{ kips}$$

Then the design strength $\phi R_n = 0.75(57.9) = 43.4$ kips. Since this is less than the factored load $R_u = 44$ kips, the bearing length N must be increased. A 5-in. bearing length N gives $\phi R_n = 64.2$ kips which exceeds R_u and is acceptable. *Use* bearing plate, $N = 5$ in. × $B = 6$ in.

(c) Determine the bearing plate thickness. The uniform (assumed) bearing pressure under factored load is

$$\text{Uniform bearing pressure } p = \frac{44}{5(6)} = 1.47 \text{ ksi}$$

The critical section for bending is taken at the toe of the flange-to-web fillet, a distance k from the mid-thickness of the web, and the beam flange is

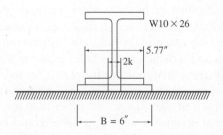

Figure 7.8.2   Example 7.8.1.

assumed not to participate. The bending moment is treated as that acting on a cantilever beam. The factored moment M_u is

$$M_u = \frac{p(B/2 - k)^2 N}{2} = \frac{1.47(3 - 0.875)^2 N}{2} = 3.32N \qquad (a)$$

For a rectangular section, the bending strength requirement is

$$\phi M_n \geq M_u \qquad (b)$$

where $\phi = 0.90$ and $M_n = M_p = ZF_y$. For a rectangular section of width N and thickness t (see Example 7.3.1),

$$Z = Nt^2/4 \qquad (c)$$

To satisfy Eq. (b) above, $\phi ZF_y = 3.32N$; thus,

$$\frac{0.90Nt^2 F_y}{4} = 3.32N$$

and for $F_y = 36$ ksi, the required thickness becomes

$$\text{Required } t = \sqrt{\frac{3.32(4)}{0.90(36)}} = 0.64 \text{ in.} \quad \text{Use } t = \tfrac{3}{4} \text{ in.}$$

Use bearing plate, $\tfrac{3}{4} \times 5 \times 0'-6''$. ■■

Solving for the bearing plate thickness in general, equating ϕM_n to M_u,

$$\phi \frac{Nt^2}{4} F_y = \frac{p(B/2 - k)^2 N}{2}$$

$$\text{Required } t = \sqrt{\frac{2p(B/2 - k)^2}{\phi F_y}} \qquad (7.8.13)$$

7.9 HOLES IN BEAMS

Flange Holes

For tension members the effect of fastener holes has been discussed in Chapter 3, where holes are deducted and net section is used. For compression members, since the fasteners occupy most of the space in the hole, the fasteners are assumed in design to completely fill the holes and a deduction for holes is not made.

When the nominal strength M_n reaches the plastic moment M_p, certainly tension flange holes reduce that strength; however, there will be a shift in neutral axis associated with a loss of strength at one flange, an effect that somewhat counteracts the effect of holes. Traditional design has neglected the effect of holes when holes do not represent a significant proportion of the flange area. When the nominal moment strength M_n is limited to less than M_p by some type of instability the effect of holes is reduced.

Traditionally, AISC Specifications have required deduction for flange holes only when the area of holes exceeds 15% of the gross flange, and then only the area in excess of 15% is deducted.

New in 1993 (1989 for ASD), LRFD-B10 requires use of gross section provided that

$$0.75F_u A_{fn} \geq 0.9F_y A_{fg} \tag{7.9.1}$$

where A_{fn} = net flange area
   A_{fg} = gross flange area

Equation 7.9.1 means that the net flange area must exceed 0.74 and 0.92 of the gross flange area for A36 and A572 Grade 50, respectively, in order to use gross area. That is allowing holes up to 25.5% instead of 15% to be ignored for A36; however, it allows only 8% holes to be ignored in A572 Grade 50. When Eq. 7.9.1 is *not* satisfied, the fracture limit state governs; in such a case some deduction for holes seems appropriate. When the fracture limit state governs, ASD-B10 states "flexural properties shall be based on an effective tension flange area A_{fe} as follows:"

$$A_{fe} = \frac{5F_u}{6F_y} A_{fn} \tag{7.9.2}$$

For A36 steel, Eq. 7.9.2 allows use of an effective flange area 34% higher than the computed net area; however, for A572 Grade 50 the effective flange area can be taken only 8% higher than the computed net area. Logically the effect of holes should depend on the likelihood of the tensile fracture through such holes being the controlling limit state.

Examples of procedure for considering fastener holes appear in the Chapter 13 section on beam and girder splices.

Web Holes

The AISC Specifications [1.5, 1.16] permit neglect of fastener holes located in the web, largely for the same reasons fastener holes in the flange may be neglected. Large holes cut into beam webs are entirely another matter. These holes require special analysis and usually will have to be reinforced by attaching extra plate material, often including stiffeners, around the sides of the hole. Design of large web holes in beams is outside the scope of this text. The recent work of the ASCE Task Committee [7.33, 7.34] should serve as a basic guide for design of beams with web openings. The definitive work related specifically to the AISC LRFD approach is that of Darwin [7.55]. The reader may also refer to the work of Bower [7.35-7.38], Frost and Leffler [7.39], Mandel, Brennan, Wasil, and Antoni [7.40], Cooper and Snell [7.41], Chan and Redwood [7.42, 7.43], Wang, Snell, and Cooper [7.44], Larson and Shah [7.45], Cooper, Snell, and Knostman [7.46], Redwood, Baranda, and Daly [7.47], Redwood and Uenoya [7.48], Daugherty [7.49], and Narayanan and Der-Avanessian [7.50]. Design tables for rectangular holes have been given by Redwood [7.51], and a design example is presented by Kussman and Cooper [7.52].

7.10 GENERAL FLEXURAL THEORY

Thus far, consideration has been given only to symmetrical shapes loaded symmetrically for which $f = Mc/I$ is correct for computing elastic flexural stress. The following development treats the general bending of arbitrary prismatic beams, i.e., beams having any cross-sectional shape with no variation along the length. They are also assumed to be free from twisting.

Consider the straight uniform cross-section beam of Fig. 7.10.1 acted upon by moment applied in the plane $ABCD$ which makes the angle γ with the xz plane. The moments are represented by vectors normal to the plane of action (positive moment defined by using right-hand rule for rotation).

Examine next a portion of the beam of length z as shown in Fig. 7.10.2a. To satisfy equilibrium on the free body of Fig. 7.10.2a requires

$$\sum F_z = 0, \qquad \int_A \sigma \, dA = 0 \qquad (7.10.1)$$

$$\sum M_x = 0, \qquad M_x = \int_A y\sigma \, dA \qquad (7.10.2)$$

$$\sum M_y = 0, \qquad M_y = \int_A x\sigma \, dA \qquad (7.10.3)$$

It is to be noted from Fig. 7.10.2b that the moments M_x and M_x are both positive in accordance with the customary convention of calling positive bending that which causes compression in the top portion of the beam. Also, the subscripts for M designate the axis *about which* bending occurs; i.e., the direction of the moment vector.

Bending in the *yz* Plane Only

If bending occurs in the yz plane, the stress σ is then proportional to y. Thus

$$\sigma = k_1 y \qquad (7.10.4)$$

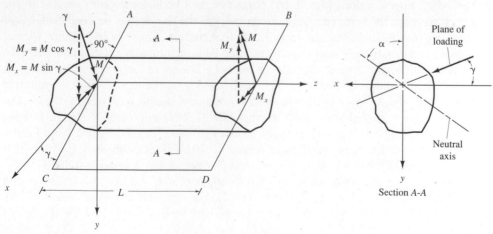

Figure 7.10.1 Prismatic beam under pure bending.

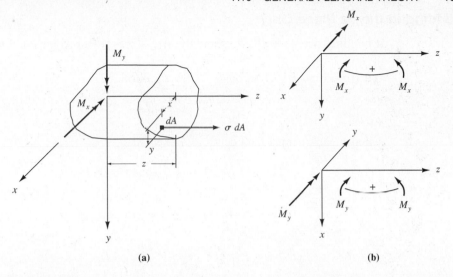

Figure 7.10.2   Free body of a portion of a beam having length z.

Using Eqs. 7.10.1 through 7.10.3 gives

$$k_1 \int_A y\,dA = 0 \tag{7.10.5}$$

$$M_x = k_1 \int_A y^2\,dA = k_1 I_x \tag{7.10.6}$$

$$M_y = k_1 \int_A xy\,dA = k_1 I_{xy} \tag{7.10.7}$$

From the first of the above expressions, $\int_A y\,dA = 0$, meaning x must be a centroidal axis. The stress may then be computed as

$$\sigma = \frac{M_x y}{I_x} \quad \text{or} \quad \frac{M_y y}{I_{xy}} \tag{7.10.8}$$

and the angle γ must be such that

$$\tan \gamma = \frac{M_x}{M_y} = \frac{I_x}{I_{xy}} \tag{7.10.9}$$

As a practical matter it would be unlikely that an unsymmetrical beam section would be located in a plane making the angle γ with the xz plane and have bending occur in the yz plane. Equation 7.10.9 also shows that if a section is used with at least one axis of symmetry (for which $\int_A xy\,dA = I_{xy} = 0$), $\tan \gamma = \infty$, $\gamma = 90°$, meaning the loading and the bending both occur in the yz plane.

An important conclusion here is that only if $I_{xy} = 0$ does the bending occur in the plane of loading. For example, on unsymmetrical shapes such as angles or zees loaded in the xz or yz plane, the plane of loading and the plane of bending will be different.

Bending in the *xz* Plane Only

If bending occurs in the xz plane the stress σ is then proportional to x. Thus

$$\sigma = k_2 x \tag{7.10.10}$$

and from Eqs. 7.10.1 through 7.10.3 are obtained:

$$k_2 \int_A x \, dA = 0 \tag{7.10.11}$$

which means y must be a centroidal axis. Also,

$$M_x = k_2 \int_A xy \, dA = k_2 I_{xy} \tag{7.10.12}$$

$$M_y = k_2 \int_A x^2 \, dA = k_2 I_y \tag{7.10.13}$$

and

$$\tan \gamma = \frac{M_x}{M_y} = \frac{I_{xy}}{I_y} \tag{7.10.14}$$

In the case where $I_{xy} = 0$, $\tan \gamma = 0$, i.e., the loading and bending both occur in the xz plane.

Bending in Neither *xz* Nor *yz* Planes

This is the realistic case when considering unsymmetrical sections in flexure. Since it is assumed all stresses are within the elastic limit, the total stress σ is the sum of the stresses due to bending in each of the xz and yz planes. Thus

$$\sigma = k_1 y + k_2 x \tag{7.10.15}$$

and

$$M_x = k_1 I_x + k_2 I_{xy} \tag{7.10.16}$$

$$M_y = k_1 I_{xy} + k_2 I_y \tag{7.10.17}$$

Solving Eqs. 7.10.16 and 7.10.17 for k_1 and k_2 and substituting into Eq. 7.10.15 gives

$$\sigma = \frac{M_x I_y - M_y I_{xy}}{I_x I_y - I_{xy}^2} y + \frac{M_y I_x - M_x I_{xy}}{I_x I_y - I_{xy}^2} x \tag{7.10.18}$$

which is the general flexure equation. The assumptions inherent in Eq. 7.10.18 are (a) a straight beam; (b) constant cross-section; (c) x- and y-axes are mutually perpendicular centroidal axes; and (d) that stress is proportional to strain and the maximum value is within the proportional limit.

Principal Axes

The principal axes are mutually perpendicular centroidal axes for which the moment of inertia is either a maximum or minimum. Furthermore, these axes are the only mutually perpendicular axes for which the product of inertia I_{xy} is zero. When a section has an axis of symmetry, that axis is a principal axis and Eq. 7.10.18 becomes

$$\sigma = \frac{M_x}{I_x} y + \frac{M_y}{I_y} x \tag{7.10.19}$$

When there is no axis of symmetry, Eq. 7.10.19 can still be used if the principal axes are located and the quantities M_x, M_y, I_x, I_y, x, and y are all corrected so as to refer to the principal axes. Usually such transformations offer no advantage over direct use of Eq. 7.10.18.

Inclination of the Neutral Axis

When the loads acting on a flexural member pass through the centroid of the cross-section but are inclined with respect to either of the principal axes, the stresses may be determined by using Eq. 7.10.18 or Eq. 7.10.19. However, note should be made that the neutral axis is not necessarily perpendicular to the plane of loading. As shown from Eqs. 7.10.9 and 7.10.14 and Fig. 7.10.1,

$$\tan \gamma = \frac{M_x}{M_y} \tag{7.10.20}$$

Since the stress along the neutral axis is equal to zero, σ may be set equal to zero in Eq. 7.10.18. With $\sigma = 0$, solving for $-x/y$ gives

$$-\frac{x}{y} = \left(\frac{M_x I_y - M_y I_{xy}}{I_x I_y - I_{xy}^2} \right) \left(\frac{I_x I_y - I_{xy}^2}{M_y I_x - M_x I_{xy}} \right) \tag{7.10.21}$$

From Fig. 7.10.1, Section A–A, it is seen that at any point on the neutral axis, $\tan \alpha = -x/y$. Dividing both numerator and denominator of the right side of Eq. 7.10.21 by M_y gives

$$\tan \alpha = \frac{\dfrac{M_x}{M_y} I_y - I_{xy}}{I_x - \dfrac{M_x}{M_y} I_{xy}} \tag{7.10.22}$$

Substitution of Eq. 7.10.20 into Eq. 7.10.22 gives

$$\tan \alpha = \frac{I_y \tan \gamma - I_{xy}}{I_x - I_{xy} \tan \gamma} \tag{7.10.23}$$

When investigating a section with one axis of symmetry, $I_{xy} = 0$; Eq. 7.10.23 then becomes

$$\tan \alpha = \frac{I_y}{I_x} \tan \gamma \tag{7.10.24}$$

EXAMPLE 7.10.1 _____

A W18×50 used as a beam is subjected to loads inclined at 5° from the vertical axis as shown in Fig. 7.10.3. Locate the inclination of the neutral axis.

Solution

$$I_x = 800 \text{ in.}^4; \qquad I_y = 40.1 \text{ in.}^4$$

$$\tan 85° = \tan \gamma$$

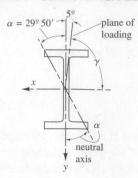

Figure 7.10.3   Biaxially loaded beam of Example 7.10.1.

Using Eq. 7.10.24,

$$\tan \alpha = \frac{I_y}{I_x} \tan \gamma = \frac{40.1}{800} \tan 85° = 0.573$$

$$\alpha = 29.8°(29°50') \quad \blacksquare\blacksquare$$

EXAMPLE 7.10.2 _____

Compute the maximum flexural stress in a $6 \times 4 \times \frac{1}{2}$ angle with the long leg vertically downward when service loaded with 0.5 kip/ft on a simple supported span of 10 ft (see Fig. 7.10.4). Compare the value assuming the angle is completely free to bend in any direction with that obtained assuming bending in only the vertical plane.

Solution

(a) Angle free to bend in any direction. Using Eq. 7.10.18 with AISC properties

$$I_x = 17.4 \text{ in.}^4, \qquad I_y = 6.27 \text{ in.}^4$$

$$I_{xy} = [6(3.00 - 1.99)(-0.987 + 0.25) + 3.5(2.25 - 0.987)(-1.99 + 0.25)]0.50$$
$$= [(-4.47) + (-7.69)]0.50 = -6.08 \text{ in.}^4$$

$$M_x = \tfrac{1}{8}(0.5)(10)^2 = 6.25 \text{ ft-kips} = 75 \text{ in.-kips}$$
$$M_y = 0$$

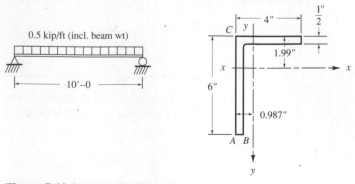

Figure 7.10.4   Data for Example 7.10.2.

Stress at point A:

$$f_A = \frac{M_x(I_y y - I_{xy} x)}{I_x I_y - I_{xy}^2} = \frac{6.25(12)[6.27(+4.01) - (-6.08)(-0.987)]}{17.4(6.27) - (6.08)^2}$$

$$= \frac{75(19.14)}{72.1} = 75(0.265) \doteq +19.9 \text{ ksi (tension)}$$

Stress at point B:

$$f_B = \frac{75[6.27(+4.01) - (-6.08)(-0.487)]}{72.1}$$

$$= 75(0.308) = +23.1 \text{ ksi (tension)}$$

Stress at point C:

$$f_C = \frac{75[6.27(-1.99) - (-6.08)(-0.987)]}{72.1}$$

$$= 75(-0.256) = -19.2 \text{ ksi (compression)}$$

(b) Angle free to bend in any direction. Use alternate method suggested by Gaylord and Gaylord [7.53] (pp. 143–147). Compute the stresses assuming first that the beam bends only in the yz plane. Conveniently let $M_x = 100$ in.-kips. Then, according to Eq. 7.10.9, $M_y = M_x I_{xy}/I_x$ must also be acting (Fig. 7.10.5a) if bending is to occur only in the yz plane.

Since the real loading has only M_x, M_y must be removed by application of an equal and opposite moment, further considering that bending occurs only in the xz plane. This means, according to Eq. 7.10.14, the simultaneous application of $M_x = M_y I_{xy}/I_y$ (see Fig. 7.10.5b).

Bending in vertical plane:

$$M_x = 75.0\left(\frac{100}{66.16}\right) = 113.4 \text{ in.-kips}$$

$$f_{A1} = f_{B1} = \frac{113.4(4.01)}{17.4} = 26.1 \text{ ksi (tension)}$$

$$f_{C1} = \frac{113.4(1.99)}{17.4} = 13.0 \text{ ksi (compression)}$$

Bending in horizontal plane:

$$M_y = -113.4(-6.08)/17.4 = 39.6 \text{ in.-kips}$$

$$f_{A2} = f_{C2} = \frac{39.6(0.987)}{6.27} = 6.2 \text{ ksi (compression)}$$

$$f_{B2} = \frac{39.6(0.487)}{6.27} = 3.1 \text{ ksi (compression)}$$

Total stresses in general bending:

$$f_A = +26.1 - 6.2 = +19.9 \text{ ksi (tension)}$$
$$f_B = +26.1 - 3.1 = +23.0 \text{ ksi (tension)}$$
$$f_C = -13.0 - 6.2 = -19.2 \text{ ksi (compression)}$$

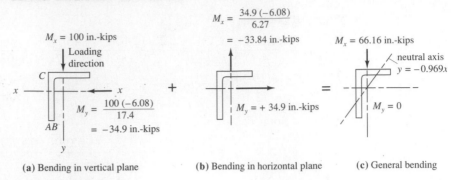

(a) Bending in vertical plane   **(b)** Bending in horizontal plane   **(c)** General bending

Figure 7.10.5   Solution by superposition of bending in the vertical and horizontal planes.

which agree with the values, 19.9, 23.1, and 19.2 ksi, respectively, as computed by the general formula.

The general equation for stress at any point is seen to be

$$f = \frac{113.4y}{17.4} + \frac{39.6x}{6.27}$$

where if $f = 0$, the neutral axis is $y = -0.969x$.

This superimposition method permits the designer to visualize what is taking place. If attached construction constrains an unsymmetrical section to bend in the vertical plane, the restraining moment capacity can be computed by using Eq. 7.10.9.

(c) Angle restrained to bend in the vertical plane:

$$f_A = f_B = \frac{75(4.01)}{17.4} = +17.3 \text{ ksi (tension)}$$

$$f_C = \frac{75(-1.99)}{17.4} = -8.6 \text{ ksi (compression)}$$

Unless the horizontal constraints actually act, the tensile stress at point B is underestimated by 25% and the compressive stress at C is underestimated by 55%.

Frequently designers assume $f = My/I$ is applicable without considering whether or not adequate horizontal restraints are present. Although usually some degree of restraint is present, care should be exercised when investigating unsymmetrical beams. Neglect of the lateral (horizontal) component is always on the unsafe side. ■■

7.11 BIAXIAL BENDING OF SYMMETRIC SECTIONS

Flexural stresses on sections with at least one axis of symmetry and loaded through the centroid may be computed using Eq. 7.10.19, which when modified to give maximum stress σ becomes

$$\sigma = \frac{M_x}{S_x} + \frac{M_y}{S_y} \qquad (7.11.1)$$

where $S_x = I_x/(d/2)$ and $S_y = I_y/(b/2)$ are the section modulus values.

Nominal Strength

The nominal strength of a section subject to biaxial bending is not readily determined. Such strength will certainly depend on the proportions of the section and the relative magnitudes of the applied moments M_x and M_y. The use of an interaction equation such as used for beam-columns is not believed adequately conservative for cases where there is no axial compression or tension. The AISC ASD Specification has traditionally limited the combined stress according to Eq. 7.11.1 to a maximum of $0.60 F_y$. This effectively implies that the nominal strength of the section under combined M_x and M_y is reached when the extreme fiber stress reaches the yield stress F_y. Thus, the entire cross-section will be elastic; no credit is then given to any ability of the cross-section to undergo plastic deformation. Certainly this approach is conservative.

Following the above approach, the elastic stress equation, Eq. 7.11.1, could be used with nominal maximum strength moments, M_{nx} and M_{ny}, and a maximum stress F_y. Thus, the nominal strength could be assumed to be reached when

$$\frac{M_{nx}}{S_x} + \frac{M_{ny}}{S_y} = F_y \qquad (7.11.2)$$

An alternative would require a more complex interaction relationship, such as provided in LRFD-Appendix H3. For biaxial bending of I-shaped members, interaction equations are provided that are influenced by the cross-sectional parameter b_f/d (i.e., flange width to overall depth).

Load and Resistance Factor Design

Method 1.   Many designers will use the beam-column safety criteria of LRFD-H1 for the member loaded in biaxial bending without axial load. In the authors' view that is likely an unconservative approach. The criteria of LRFD-H1 were developed as strength relationships for *compression* members, which simultaneously are acted upon by bending moments; not very close to treating the strength in biaxial bending. A literal reading of the LRFD Specification certainly allows biaxial bending to be covered by LRFD-H1. When one does that, no new principles are encountered in this section of the text.

Method 2.   LRFD-H2 uses the combined stress approach of Eq. 7.11.2 recognizing that the safety provisions include an overload effect and an understrength effect. The *required* M_{nx} equals the factored load M_{ux} divided by the strength reduction factor ϕ. Substitution for M_{nx} and M_{ny} in Eq. 7.11.2 gives

$$\frac{M_{ux}}{\phi_b S_x} + \frac{M_{uy}}{\phi_b S_y} \leq F_y \qquad (7.11.3)$$

where $\phi_b = 0.90$.

For design use, multiply Eq. 7.11.3 by S_x and divide by F_y, giving

$$S_x \geq \frac{M_{ux}}{\phi_b F_y} + \frac{M_{uy}}{\phi_b F_y}\left(\frac{S_x}{S_y}\right) \tag{7.11.4}$$

For selecting standard rolled shapes one finds that for a given depth of section the ratio S_x/S_y is relatively constant. A typical range of such values appears in Table 7.11.1. The lightest sections in any depth will be the narrowest ones and thus have the higher values of S_x/S_y.

An alternate method is suggested by Gaylord and Gaylord [7.53] (p. 168), for use with I-shaped sections (W, M, and S). The quantity S_x/S_y is computed approximately using the beam properties neglecting the web effect. This gives, using $b_f =$ flange width and $d =$ depth,

$$\frac{S_x}{S_y} = \frac{I_x(b_f/2)}{(d/2)I_y} = \frac{2b_f t(d/2)^2(b_f/2)}{(d/2)(2)(1/12)tb_f^3} = \frac{3d}{b_f} \tag{7.11.5}$$

which is suggested to be increased to $3.5d/b_f$ to account for a greater error in the numerator than in the denominator resulting from the neglect of web.

Method 3. The pairs of interaction equations, one a yield criterion and the other a stability criterion, from LRFD-Appendix H3 are given below. Each equation of the pair must be satisfied.

1. For $0.5 \leq b_f/d \leq 1.0$,

$$\left(\frac{M_{ux}}{\phi_b M_{px}}\right)^{1.6} + \left(\frac{M_{uy}}{\phi_b M_{py}}\right)^{1.6} \leq 1.0 \tag{7.11.5}$$

$$\left(\frac{C_{mx}M_{ux}}{\phi_b M_{nx}}\right)^{0.4+b_f/d} + \left(\frac{C_{my}M_{uy}}{\phi_b M_{ny}}\right)^{0.4+b_f/d} \leq 1.0 \tag{7.11.6}$$

where the power $(0.4+b_f/d)$ must equal or exceed 1.0.

TABLE 7.11.1 TYPICAL S_x/S_y VALUES

Shape	Depth d(in.)	S_x/S_y
M	6, 8	5–7
M	10, 12	8–11
M	14, 16	11–12
Light W and M	4–8	3
W	8, 10	3–4
W	12	3–6
W	14 (up to 84 lb/ft)	4–8
W	14 (over 84 lb/ft)	$2\frac{1}{2}$–3
W	16, 18, 21	5–9
W	24, 27	6–10
W	30, 33, 36	7–12
S	6–8	d
S	10–18	$0.75d$
C	up to 7	$1.5d$
C	8–10	$1.25d$
C	12, 15	d

2. For $0.3 \le b_f/d < 0.5$,

$$\frac{M_{ux}}{\phi_b M_{px}} + \frac{M_{uy}}{\phi_b M_{py}} \le 1.0 \qquad (7.11.7)$$

$$\left(\frac{C_{mx} M_{ux}}{\phi_b M_{nx}}\right)^{0.4+b_f/d} + \left(\frac{C_{my} M_{uy}}{\phi_b M_{ny}}\right)^{0.4+b_f/d} \le 1.0 \qquad (7.11.8)$$

where the power $(0.4+b_f/d)$ must equal or exceed 1.0.

3. For $b_f/d < 0.3$,

$$\frac{M_{ux}}{\phi_b M_{px}} + \frac{M_{uy}}{\phi_b M_{py}} \le 1.0 \qquad (7.11.9)$$

$$\frac{C_{mx} M_{ux}}{\phi_b M_{nx}} + \frac{C_{my} M_{uy}}{\phi_b M_{ny}} \le 1.0 \qquad (7.11.10)$$

where M_{ux}, M_{uy} = factored moments about the x- and y-axes, respectively
    M_{px}, M_{py} = plastic moment strength about the x- and y-axes, respectively
    M_{nx}, M_{ny} = nominal moment strength about the x- and y-axes, respectively
        (may be equal to or less than plastic moment strength)
    C_{mx}, C_{my} = moment gradient terms whose meaning is treated in Chapter 12
        on beam-columns. It is always conservative to take these
        equal to 1.0; meaning constant bending moment between
        lateral supports.
    ϕ_b = 0.90 for flexure

For Eqs. 7.11.6, 7.11.8, and 7.11.10, M_{nx} and M_{ny} are to be determined the same as the nominal strength for any uniaxially loaded beam; the maximum values are the plastic moments for all qualifying laterally braced compact sections.

Allowable Stress Design

Dividing the factor of safety $\gamma/\phi = 1.67$ from both sides of Eq. 7.11.3 gives the relationship between the service load combined stress f_b and the allowable stress F_b,

$$\left(f_b = \frac{M_x}{S_x} + \frac{M_y}{S_y}\right) \le \left(F_b = \frac{F_y}{\gamma/\phi} = 0.60F_y\right) \qquad (7.11.8)$$

and the design format, similar to Eq. 7.11.4 becomes

$$S_x \ge \frac{M_x}{F_b} + \frac{M_y}{F_b}\left(\frac{S_x}{S_y}\right) \qquad (7.11.9)$$

EXAMPLE 7.11.1 _____

Select the lightest W or M section to carry service dead load moments $M_x = 15$ ft-kips and $M_y = 5$ ft-kips, and live load moments $M_x = 45$ ft-kips and $M_y = 20$ ft-kips. Consider that adequate lateral bracing is provided to preclude instability. Use steel having $F_y = 50$ ksi. Use Load and Resistance Factor Design.

Solution

(a) Compute factored loads M_{ux} and M_{uy}.

$$M_{ux} = 1.2(15) + 1.6(45) = 90 \text{ ft-kips}$$
$$M_{uy} = 1.2(5) + 1.6(20) = 38 \text{ ft-kips}$$

(b) Using Method 2, determine required section modulus S_x and select section. Equation 7.11.4 gives

$$\text{Required } S_x \geq \frac{90(12)}{0.90(50)} + \frac{38(12)}{0.90(50)}\left(\frac{S_x}{S_y}\right)$$
$$\geq 24 + 10 S_x/S_y$$

From Table 7.11.1 the ratio S_x/S_y can be expected to be on the order of 3 to 4; thus $S_x \approx 54$ to 64 in.^{3}

It will be handy here to use the *ASD Manual* [1.7] table, "ALLOWABLE STRESS DESIGN SELECTION TABLE," to arrive at W16×40, $S_x = 64.7$ in.^{3} For this beam, $S_x/S_y = 64.7/8.25$, and it is apparent that the beam is inadequate. Continuing in the same table, it is noted that the lightest W shapes (boldface type) are inadequate up to about W21×62, which has $S_x/S_y \approx 9$. Thus, 60 lb/ft seems to be the weight range.

Using only the *LRFD Manual* [1.18], one must use the plastic modulus Z_x table, "LOAD FACTOR DESIGN SELECTION TABLE." The section modulus S_x may be approximately converted to Z_x by multiplying by the average shape factor 1.12 (see Sec. 7.3). For this case the initial trial S_x values of 54 to 64 would have become $Z_x \approx 61$ to 72 in.^{3} and the same starting section of W16×40 would have been obtained.

Try W10×60:

$$S_x/S_y = 66.7/23.0 = 2.9$$

Required $S_x = 24 + 10(2.9) = 53$ in.$^3 < 66.7$ in.^{3} OK

For W10×49,

$$S_x/S_y = 54.6/18.7 = 2.92$$

Required $S_x = 24 + 10(2.92) = 53.6$ in.$^3 < 54.6$ in.^{3} OK

Use W10×49, F_y 50 ksi.)

If one had assumed a 10-in. depth as desirable and used Eq. 7.11.4 with a coefficient of 3.5 instead of 3,

$$\text{Required } S_x = 24 + 10(3.5 d/b_f)$$

For W10 sections, d/b_f is either 1 or 1.25.

$$\text{Required } S_x \approx 24 + 35 = 59 \text{ in.}^3$$

which would also have required a check of two sections. The more general approach using typical S_x/S_y values seems most useful unless a specific depth is desired.

(c) Use Method 3 and see if a lighter section will be acceptable. Since there is likely to be some inelastic deformation capability of I-shaped sections, Method 2 is expected to be conservative.

Investigate the W10×49 (which satisfies $\lambda < \lambda_p$ for $F_y = 50$·ksi) selected in part (b),

$$\text{Compute } b_f/d = 10.00/9.98 \approx 1.00$$

Since $0.5 \leq b_f/d \leq 1.0$ is nearly satisfied, Eqs. 7.11.5 and 7.11.6 are applicable. The nominal moment strengths M_{nx} and M_{ny} are the plastic moment strengths for uniaxial bending,

$$M_{nx} = M_{px} = Z_x F_y = 60.4(50)/12 = 252 \text{ ft-kips}$$
$$M_{ny} = M_{py} = Z_y F_y = 28.3(50)/12 = 118 \text{ ft-kips}$$

Because $M_{nx} = M_{px}$, $M_{ny} = M_{py}$, conservatively $C_{mx} = C_{my} = 1.0$, and the power $(0.4 + b_f/d)$ equals 1.4 for Eq. 7.11.6, Eq. 7.11.6 will govern because of the lower power (1.4 vs 1.6) and the same quantities within parenthesis. Substitution into Eq. 7.11.6 using $\phi_b = 0.90$ gives

$$\left(\frac{C_{mx} M_{ux}}{\phi_b M_{nx}}\right)^{1.4} + \left(\frac{C_{my} M_{uy}}{\phi_b M_{ny}}\right)^{1.4} \leq 1.0 \qquad\qquad [7.11.6]$$

$$\left(\frac{1.0(90)}{0.90(252)}\right)^{1.4} + \left(\frac{1.0(38)}{0.90(118)}\right)^{1.4} = (0.397)^{1.4} + (0.358)^{1.4} = 0.51 < 1.0$$

By this criterion the member is considerable overdesigned; select a lighter section. Try W10×39 (satisfies $\lambda < \lambda_p$ for $F_y = 50$ ksi)

$$\left(\frac{1.0(90)}{0.90(195)}\right)^{1.4} + \left(\frac{1.0(38)}{0.90(71.7)}\right)^{1.4} = (0.513)^{1.4} + (0.589)^{1.4} = 0.87 < 1.0$$

Even though the criterion still indicates this section is stronger than necessary, the next lightest (W10×33) gives 1.15 for the criterion.
Use W10×39, $F_y = 50$ ksi. ■■

SELECTED REFERENCES

7.1. T. V. Galambos. "History of Steel Beam Design," *Engineering Journal,* AISC, **14,** 4 (Fourth Quarter 1977), 141–147.

7.2. Joint Committee of Welding Research Council and the American Society of Civil Engineers. *Commentary on Plastic Design in Steel,* 2nd ed., ASCE Manual and Reports on Practice No. 41, New york. 1971.

7.3. Joseph A. Yura, Theodore V. Galambos, and Mayasandra K. Ravindra. "The Bending Resistance of Steel Beams," *Journal of the Structural Division,* ASCE, **104,** ST9 (September 1978), 1355–1370.

7.4. Ad Hoc Committee on Serviceability Research, Committee on Research of the Structural Division. "Structural Serviceability: A Critical Appraisal and Research Needs," *Journal of Structural Engineering,* ASCE, **112,** 12 (December 1986), 2646–2664.

7.5. Theodore V. Galambos and Bruce Ellingwood. "Serviceability Limit States: Deflection," *Journal of Structural Engineering,* ASCE, **112,** 1 (January 1986), 67–84.

7.6. Bruce Ellingwood. "Serviceability Guidelines for Steel Structures," *Engineering Journal,* AISC, **26,** 1 (1st Quarter 1989), 1–8.

7.7. Thomas M. Murray. "Building Floor Vibrations," *Engineering Journal,* AISC, **28,** 3 (3rd Quarter 1991), 102–109. Errata, **28,** 4 (4th Quarter 1991), 176.

7.8. Richard N. Wright and William H. Walker. "Vibration and Deflection of Steel Bridges," *Engineering Journal,* AISC, **9,** 1 (January 1972), 20–31.

7.9. Thomas M. Murray. "Design to Prevent Floor Vibrations," *Engineering Journal,* AISC, **12,** 3 (Third Quarter 1975), 82–87.

7.10. Thomas M. Murray. "Acceptability Criterion for Occupant-Induced Floor Vibrations," *Engineering Journal,* AISC, **18,** 2 (Second Quarter 1981), 62–70.

7.11. Bruce Ellingwood and Andrew Tallin. "Structural Serviceability: Floor Vibrations," *Journal of Structural Engineering,* ASCE, **110,** 2 (February 1984), 401–418. Disc. **111,** 5 (May 1985), 1158–1161.

7.12. Raed A. Tolaymat. "A New Approach to Floor Vibration Analysis," *Engineering Journal.* AISC, **25,** 4 (4th Quarter 1988), 137–143.

7.13. D. E. Allen. "Building Vibrations from Human Activities," *Concrete International,* **12,** 6 (June 1990), 66–73.

7.14. Frank J. Hatfield. "Design Chart for Vibration of Office and Residential Floors," *Engineering Journal,* AISC, **29,** 4 (4th Quarter 1992), 141–144.

7.15. J. W. Larson and R. K. Huzzard. "Economical Use of Cambered Steel Beams," presented at AISC National Steel Construction Conference, March 1990. Available as Bethlehem Steel Corp. *Technical Bulletin TB-309,* May 1990.

7.16. David T. Ricker. "Cambering Steel Beams," *Engineering Journal,* AISC, **26,** 4 (4th Quarter 1989), 136–142.

7.17. Frank J. Marino. "Ponding of Two-Way Roof Systems," *Engineering Journal,* AISC, **3,** 3 (July 1966), 93–100.

7.18. Lewis B. Burgett. "Fast Check for Ponding," *Engineering Journal,* AISC, **10,** 1 (First Quarter 1973), 26–28.

7.19. John L. Ruddy. "Ponding of Concrete Deck Floors," *Engineering Journal,* AISC, **23,** 3 (Third Quarter 1986), 107–115.

7.20. James Chinn. "Failure of Simply-Supported Flat Roofs by Ponding of Rain," *Engineering Journal,* AISC, **2,** 2 (April 1965), 38–41.

7.21. A. E. Salama and M. L. Moody. "Analysis of Beams and Plates for Ponding Loads," *Journal of the Structural Division,* ASCE, **93,** ST1 (February 1967), 109–126.

7.22. D. A. Sawyer. "Ponding of Rainwater on Flexible Roof Systems," *Journal of the Structural Division,* ASCE, **93,** ST1 (February 1967), 122–147.

7.23. D. A. Sawyer. "Roof-Structure Roof-Drainage Interaction," *Journal of the Structural Division,* ASCE, **94,** ST1 (January 1968), 175–198.

7.24. James Chinn, Abdulwahab H. Mansouri, and Staley F. Adams. "Ponding of Liquids on Flat Roofs," *Journal of the Structural Division,* ASCE, **95,** ST5 (May 1969), 797–807.

7.25. R. Richard Avent and William G. Stewart. "Rainwater Ponding on Beam-Girder Roof Systems," *Journal of the Structural Division,* ASCE, **101,** ST9 (September 1975), 1913–1927.

7.26. R. Richard Avent. "Deflection and Ponding of Steel Joists," *Journal of the Structural Division,* ASCE, **102,** ST7 (July 1976), 1399–1410.

7.27. I. Lyse and H. J. Godfrey. "Investigation of Web Buckling in Steel Beams," *Transaction,* ASCE, **100** (1935), 675–706.

7.28. J. D. Graham, A. N. Sherbourne, R.N. Khabbaz, and C. D. Jensen, *Welded Interior Beam-to-Column Connections.* New York: American Institute of Steel Construction, Inc., 1959.

7.29. B. G. Johnston and G. G. Kubo. "Web Crippling at Seat Angle Supports," Fritz Laboratory Report No. 192A2, Lehigh University, Bethlehem, Pa., 1941.

7.30. T. M. Roberts. "Slender Plate Girders Subjected to Edges Loading," *Proceedings of the Institution of Civil Engineers,* Part 2, September 1981, 71.

7.31. M. Elgaaly. "Web Design Under Compressive Edge Loads," *Engineering Journal,* AISC, **20,** 4 (Fourth Quarter 1983), 153–171.

7.32. Terence M. Roberts and Chooi K. Chong. "Collapse of Plate Girders Under Edge Loading," *Journal of the Structural Division,* ASCE, **107,** ST8 (August 1981), 1503–1509.

7.33. ASCE Task Committee on Design Criteria for Composite Structures in Steel and Concrete. "Proposed Specification for Structural Steel Beams with Web Openings," *Journal of Structural Engineering,* ASCE, **118,** 12 (December 1992), 3315–3324.

7.34. ASCE Task Committee on Design Criteria for Composite Structures in Steel and Concrete. "Commentary on Proposed Specification for Structural Steel Beams with Web Openings (with Design Example)," *Journal of Structural Engineering,* ASCE, **118,** 12 (December 1992), 3325–3349.

7.35. John E. Bower. "Elastic Stresses Around Holes in Wide-Flange Beams," *Journal of the Structural Division,* ASCE, **92,** ST2 (April 1966), 85–101.

7.36. John E. Bower. "Experimental Stresses in Wide-Flange Beams with Holes," *Journal of the Structural Division,* ASCE, **92,** ST5 (October 1966), 167–186.

7.37. John E. Bower. "Ultimate Strength of Beams with Rectangular Holes," *Journal of the Structural Division,* ASCE, **94,** ST6 (June 1968), 1315–1337.

7.38. John E. Bower, Chairman, Subcommittee on Beams with Web Openings of the Task Committee on Flexural Members. "Suggested Design Guides for Beams with Web Holes," *Journal of the Structural Division,* ASCE, **97,** ST11 (November 1971), 2707–2728. Disc., **99,** ST6 (June 1973), 1312–1315.

7.39. Ronald W. Frost and Robert E. Leffler. "Fatigue Tests of Beams with Rectangular Web Holes," *Journal of the Structural Division,* ASCE, **97,** ST2 (February 1971), 509–527.

7.40. James A. Mandel, Paul J. Brennan, Benjamin A. Wasil, and Charles M. Antoni. "Stress Distribution in Castellated Beams," *Journal of the Structural Division,* ASCE, **97,** ST7 (July 1971), 1947–1967.

7.41. Peter B. Cooper and Robert R. Snell. "Tests on Beams with Reinforced Web Openings," *Journal of the Structural Division,* ASCE, **98,** ST3 (March 1972), 611–632.

7.42. Peter W. Chan and Richard G. Redwood. "Stresses in Beams with Circular Eccentric Web Holes," *Journal of the Structural Division,* ASCE, **100,** ST1 (January 1974), 231–248.

7.43. Richard G. Redwood and Peter W. Chan. "Design Aids for Beams with Circular Eccentric Web Holes," *Journal of the Structural Division,* ASCE, **100,** ST2 (February 1974), 297–303.

7.44. Tsong-Miin Wang, Robert R. Snell, and Peter B. Cooper. "Strength of Beams with Eccentric Reinforced Holes," *Journal of the Structural Division,* ASCE, **101,** ST9 (September 1975), 1783–1800.

7.45. Marvin A. Larson and Kirit N. Shah. "Plastic Design of Web Openings in Steel Beams," *Journal of the Structural Division,* ASCE, **102,** ST5 (May 1976). 1031–1041.

7.46. Peter B. Cooper, Robert R. Snell, and Harry D. Knostman. "Failure Tests on Beams with Eccentric Web Holes," "*Journal of the Structural Division,*" ASCE, **103,** ST9 (September 1977), 1731–1738.

7.47. Richard G. Redwood, Hernan Baranda, and Michael J. Daly. "Tests of Thin-Webbed Beams with Unreinforced Holes," *Journal of the Structural Division,* ASCE, **104,** ST3 (March 1978), 577–595.

7.48. Richard G. Redwood and Minoru Uenoya. "Critical Loads for Webs with Holes," *Journal of the Structural Division,* ASCE, **105,** ST10 (October 1979), 2053–2067.

7.49. Brian K. Daugherty. "Elastic Deformation of Beams with Web Openings," *Journal of the Structural Division,* ASCE, **106,** ST1 (January 1980), 301–312.

7.50. Rangachari Narayanan and Norire Gara-Verni Der-Avanessian. "Design of Slender Webs Having Rectangular Holes." *Journal of Structural Engineering,* ASCE, **111,** 4 (April 1985), 777–787.

7.51. R. G. Redwood. "Tables for Plastic Design of Beams with Rectangular Holes." *Engineering Journal,* AISC, **9,** 1 (January 1972), 2–19.

7.52. Richard L. Kussman and Peter B. Cooper. "Design Example for Beams with Web Openings," *Engineering Journal,* AISC, **13,** 2 (Second Quarter 1976), 48–56.

7.53. E. H. Gaylord, Jr., and C. N. Gaylord. *Design of Steel Structures.* New York: McGraw-Hill Book Company, Inc., 1957, Chap. 5.

7.54. D. E. Allen and T. M. Murray. "Design Criterion for Vibrations Due to Walking," *Engineering Journal,* AISC, **30,** 4 (4th Quarter 1993), 117–129.

7.55. David Darwin. *Steel and Composite Beams with Web Openings,* Design Guide Series No. 2. Chicago, IL: American Institute of Steel Construction, 1990, 63 pp.

PROBLEMS

All problems are to be done according to the AISC Load and Resistance Factor Design or Allowable Stress Design, as indicated by the instructor. All given loads are service loads unless otherwise indicated. For all problems *assume adequate lateral support of the compression flange such that lateral stability does not control.* Assume all standard sections are equally readily available in the indicated grade of steel (even though actually they are not). A figure showing span and loading is required, and a final check of strength (for LRFD) or stress (for ASD) is required.

7.1. For the case (or cases) assigned by the instructor, select the lightest W section to carry a uniformly distributed dead load w_D in addition to the beam weight, and a uniformly distributed live load w_L as indicated. The member is simply supported and deflection is *not* of concern.

Case	w_D Dead load (kip/ft)	w_L Live load (kip/ft)	Span length (ft)	Steel grade
1	0.2	0.8	35	A36
2	0.2	0.8	35	A572 Grade 50
3	0.2	0.8	35	A572 Grade 65
4	0.2	0.8	35	A514 Grade 100
5	0.2	0.8	55	A36
6	0.2	0.8	55	A572 Grade 50
7	0.2	0.8	55	A572 Grade 65
8	0.2	0.8	55	A514 Grade 100
9	0.4	1.6	30	A36
10	0.4	1.6	30	A572 Grade 65
11	0.8	3.2	30	A36
12	0.8	3.2	30	A572 Grade 65
13	0.4	1.4	60	A36
14	0.4	1.4	60	A572 Grade 50
15	0.4	1.4	60	A572 Grade 65

7.2. A simply supported welded I-section beam carries a concentrated load W at midspan (see accompanying figure). The load is 20% dead load and 80% live load. For the case (or cases) assigned by the instructor, determine the maximum service load W that can be permitted to be carried.

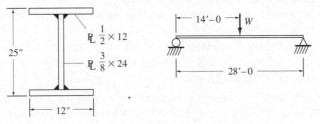

Problem 7.2 Case 1 (for other cases see table)

Case	w_D Flange plates (in.)	w_L Web plate (in.)	Span length (ft)	Steel grade
1	1/2 × 12	3/8 × 24	28	A36
2	1/2 × 14	3/8 × 30	28	A572 Grade 50
3	3/4 × 12	3/8 × 24	28	A607 Grade 70
4	3/4 × 18	1/2 × 30	28	A572 Grade 65

7.3. Repeat Prob. 7.1 (cases 1 to 4 as assigned) additionally assuming the live load deflection may not exceed $L/360$ (*not* an AISC Specification requirement).

7.4. Repeat Prob. 7.1 (cases 5 to 8 as assigned) additionally assuming the live load deflection may not exceed $L/300$ (*not* an AISC Specification requirement).

7.5. For the case (or cases) assigned by the instructor, select the lightest W section to carry a uniformly distributed dead load w_D in addition to the beam weight, and a uniformly distributed live load w_L as indicated. The member is simply supported and deflection is *not* of concern.

Case	w_D Dead load (kip/ft)	w_L Live load (kip/ft)	Span length (ft)	Steel grade
1	2.0	18.0	7	A36
2	2.0	18.0	7	A572 Grade 50
3	2.0	18.0	7	A572 Grade 60
4	2.0	18.0	7	A514 Grade 100
5	8.0	12.0	5	A36
6	8.0	12.0	5	A572 Grade 50
7	8.0	12.0	5	A572 Grade 65

7.6. A W24×94 beam on a 6-ft span (see accompanying figure) underpins a column that brings 80 kips dead load and 200 kips live load to its top flange at a location 2.5 ft from the left support. The column bearing plate is 12 in. measured along the beam, and the bearing plates at the end supports are each 8 in. Investigate this beam of A36 steel for (a) flexure, (b) shear, and (c) satisfactory transmission of the reactions and concentrated load (i.e., local web yielding and web crippling). Specify changes (if any) required to satisfy the AISC Specification.

7.7. A W16×77 section of 36 steel is to serve on a 10-ft simply supported span. The wall bearing length is 10 in. What maximum *slowly* moving concentrated service load (25% dead load; 75% live load) may be carried?

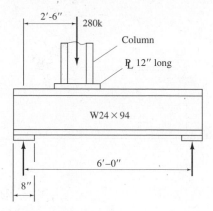

Problem 7.6

7.8. Determine the size bearing plate required for an end reaction of 11 kips dead load and 44 kips live load on a W14×53 beam of A572 Grade 60 steel. The beam rests on a 4-in.-thick concrete wall (f'_c = 3500 psi). Specify thickness in multiples of $\frac{1}{4}$ in. and the length and width to whole inches.

7.9. (LRFD only) For a W12×87 of A36 steel, calculate the design shear strength ϕV_n, and the bearing length N required when the design end reaction ϕR_n equals the design shear strength.

7.10. **through 7.13**.

For the section shown in the accompanying figure, assume uniform loading in the plane of the web (yz plane) for a simply supported span of 12 ft and neglect any torsional effects. Assume the service load acting is 20% dead load and 80% live load, and that the limit state for LRFD occurs when the maximum stress reaches the yield stress F_y at one point. For ASD this means the allowable stress is 0.60 F_y.

a. Determine the maximum uniform service load assuming bending occurs in the plane of loading (yz plane).

b. Use the loading determined in (a) to compute the flexural stress at points designated by letters, assuming the beam is free to bend and not restrained to bend in the yz plane. Use the flexure formula, Eq. 7.10.18.

c. Repeat (b) but use method described in Example 7.10.2(b).

d. If your instructor specifically assigns and discusses this part, locate principal axes, transform the moment into components M'_x and M'_y about the principal axes, compute moments of inertia I'_x and I'_y with respect to these axes, and use $f = M'_x/S'_x + M'_y/S'_y$.

e. State conclusions.

7.14. The given 8×6×$\frac{1}{2}$ angle is positioned with its long leg pointing downward and used as a simply supported beam of 12-ft span. The uniform dead load is 0.1 kip/ft (including angle weight) and gravity live load is 0.5 kip/ft. The horizontal leg is to be restrained by attachments to make the angle bend

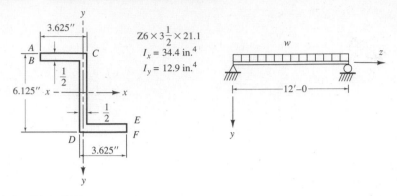

Problem 7.10

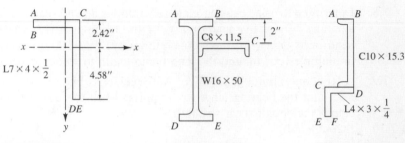

Problem 7.11   **Problem 7.12**   **Problem 7.13**

vertically. Assuming the attachment to the horizontal leg is simply supported, for what service load lateral bending moment must the connection be designed? Consider only the unsymmetrical section effect and neglect any torsion. Assume the controlling limit state is the achievement of yield stress F_y at the extreme fiber.

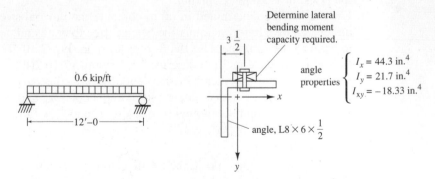

Problem 7.14

7.15.   Select the lightest W8 section of A36 steel to use as a purlin on a roof sloped 30° to the horizontal. The span is 21 ft, the load is uniform 0.13 kip/ft dead load plus the purlin weight and 0.24 kip/ft snow load. Lateral stability is assured by attachment of the roofing to the compression flange. Assume the load acts through the beam centroid, there are no sag rods, and biaxial bending must be assumed. Any torsional effect can be resisted by the roofing and therefore it can be neglected.

7.16. Select the lightest W section to carry moments, $M_x = 145$ ft-kips (15% dead load and 85% live load) and the lateral moment M_f resisted by one flange is 30 ft-kips (20% dead load and 80% live load). To select the beam assume $M_y \approx 2 M_f$ and that torsion is neglected. Use A572 Grade 50 steel and assume lateral stability does not govern.

7.17. Select the lightest W section to carry $M_x = 275$ ft-kips (30% dead load and 70% live load) and lateral moment M_f resisted by one flange is 100 ft-kips (20% dead load and 80% live load). To select the beam assume $M_y \approx 2M_f$. Use A572 Grade 50 steel and assume lateral stability does not control.

7.18. Repeat Prob. 7.17 but select a combination wide flange section and channel as found in the *LRFD Manual* [1.18] under "COMBINATION SECTIONS," pp. 1–106 or 1–107.

Chapter 8

Torsion

8.1 INTRODUCTION

In structural design, torsional moment may, on occasion, be a significant force for which provision must be made. The most efficient shape for carrying a torque is a hollow circular shaft; extensive treatment of torsion and torsion combined with bending and axial force is to be found in most texts on mechanics of materials [8.1].

Frequently torsion is a secondary, though not necessarily a minor effect that must be considered in combination with the action of other forces. The shapes that make good columns and beams, i.e., those that have their material distributed as far from their centroids as practicable, are not equally efficient in resisting torsion. Thin-wall circular and box sections are stronger torsionally than sections with the same area arranged as channel, I, tee, angle, or zee shapes.

When a simple circular solid shaft is twisted, the shearing stress at any point on a transverse cross-section varies directly as the distance from the center of the shaft. Thus, during twisting, the cross-section which is initially planar remains a plane and rotates only about the axis of the shaft.

In 1853 the French engineer Adhémar Jean Barré de Saint-Venant presented to the French Academy of Sciences the classical torsion theory that forms the basis for present-day analysis.* Saint-Venant showed that when a noncircular bar is twisted, a transverse section that was planar prior to twisting does *not* remain plane after twisting. The original cross-section plane surface becomes a warped surface. In torsion situations the out-of-plane, or warping effect, must be considered in addition to the rotation, or pure twisting, effect.[†]

---

*For a summary of Saint-Venant's work, see Isaac Todhunter and Karl Pearson, *A History of the Theory of Elasticity and of the Strength of Materials,* Vol. II, 1893 (reprinted by Dover Publications, Inc., New York, 1960, pp. 17–51).

[†]Throughout Chapters 8 and 9, the symbol ϕ is used for the angle of twist, and should not be confused with the resistance factor ϕ used for Load and Resistance Factor Design. The resistance factor ϕ in these chapters is used subscripted; ϕ_b to indicate bending.

In this chapter primary emphasis is given to the recognition of torsion on the usual structural members, such as I-shaped, channel, angle, and zee sections; how the torsional stresses may be approximated and how such members may be selected to resist torsional effects.

Also included is a brief treatment of torsional stiffness and the computation of torsional stresses on closed thin-wall sections as well as torsional buckling.

8.2 PURE TORSION OF HOMOGENEOUS SECTIONS

A review of shear stress under torsion alone and of torsional stiffness seems a desirable beginning prior to considering structural shapes in locations where the warping of the cross-section is restrained.

Consider a torsional moment T acting on a solid shaft of homogeneous material and uniform cross-section, as shown in Fig. 8.2.1. Assume no out-of-plane warping, or at least that out-of-plane warping has negligible effect on the angle of twist ϕ. This assumption will be nearly correct so long as the cross-section is small compared to the length of the shaft and also that no significant reentrant corners exist. It is further assumed that no distortion of the cross-section occurs during twisting. The rate of twist (twist per unit length) may therefore be expressed as

$$\theta = \text{rate of twist} = \frac{d\phi}{dz} \tag{8.2.1}$$

which can be thought of as torsional curvature (rate of change of angle). Since it is the relative rotation of the cross-sections at z and $z + dz$ that causes strain, the magnitude

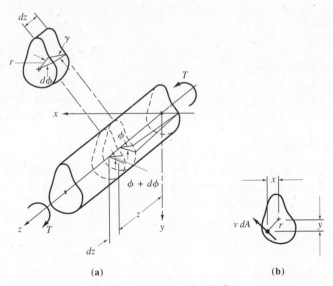

(a)   (b)

Figure 8.2.1 Torsion of a prismatic shaft.

of displacement at a given point is proportional to the distance r from the center of twist. The strain angle γ, or unit shear strain, at any element r from the center is

$$\gamma \, dz = r \, d\phi$$
$$\gamma = r(d\phi/dz) = r\theta \tag{8.2.2}$$

Using the shear modulus G, Hooke's law gives the unit shear stress v as

$$v = \gamma G \tag{8.2.3}$$

Thus as shown in Fig. 8.2.1b, the elemental torque is

$$dT = rv \, dA = r\gamma G \, dA = r^2(d\phi/dz)G \, dA \tag{8.2.4}$$

The total resisting moment for equilibrium is

$$T = \int_A r^2 \frac{d\phi}{dz} G \, dA$$

and since $d\phi/dz$ and G are constants at any section,

$$T = \frac{d\phi}{dz} G \int_A r^2 \, dA = GJ\frac{d\phi}{dz} \tag{8.2.5}$$

where $J = \int_A r^2 dA$. Equation 8.2.5 may be thought of as analogous to flexure, i.e., bending moment M equals rigidity EI times curvature d^2y/dz^2. Here torsional moment T equals torsional rigidity GJ times torsional curvature (rate of change of angle).

Shear stress may then be computed using Eqs. 8.2.2 and 8.2.3,

$$v = \gamma G = r\frac{d\phi}{dz} G \tag{8.2.6}$$

and

$$\frac{d\phi}{dz} = \frac{T}{GJ}$$

which gives

$$v = \frac{Tr}{J} \tag{8.2.7}$$

Thus as long as the assumptions of this development reasonably apply, torsional shear stress is proportional to the radial distance from the center of twist.

Circular Sections

For the specific case of the circular section of diameter t, no warping of the sections occurs (i.e., no assumption is required) and J = polar moment of inertia = $\pi t^4/32$. Thus, for maximum shear stress at $r = t/2$,

$$v_{max} = \frac{16T}{\pi t^3} \tag{8.2.8}$$

Rectangular Sections

The analysis as applied to rectangles becomes complex since the shear stress is affected by warping, though essentially the angle of twist is unaffected.

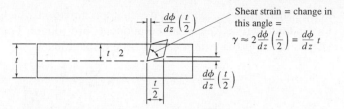

Figure 8.2.2   Torsion of a rectangular section.

As an approximation, consider the element of Fig. 8.2.2 subjected to shear, in which

$$\gamma = t\frac{d\phi}{dz} \tag{8.2.9}$$

For a thin rectangle, neglecting end effects, the shear stress may be expressed as

$$v = \gamma G = tG\frac{d\phi}{dz} \tag{8.2.10}$$

or using Eq. 8.2.5,

$$v = \frac{Tt}{J} \tag{8.2.11}$$

From the theory of elasticity [8.1–8.3], the maximum shear stress v_{max} occurs at the midpoint of the long side of a rectangle and acts parallel to it. The magnitude is a function of the ratio b/t (length/width) and may be expressed as

$$v_{max} = \frac{k_1 T}{bt^2} \tag{8.2.12}$$

and the torsional constant J may be expressed as

$$J = k_2 bt^3 \tag{8.2.13}$$

where the values of k_1 and k_2 may be found in Table 8.2.1.

TABLE 8.2.1   VALUES OF k_1 AND k_2 FOR EQS. 8.2.12 AND 8.2.13

b/t	1.0	1.2	1.5	2.0	2.5	3.0	4.0	5.0	∞
k_1	4.81	4.57	4.33	4.07	3.88	3.75	3.55	3.44	3.00
k_2	0.141	0.166	0.196	0.229	0.249	0.263	0.281	0.291	0.333

I-shaped, Channel, and Tee Sections

As will be observed from a study of Table 8.2.1 the values of k_1 and k_2 become nearly constant for large ratios b/t. Thus the torsional constants for sections composed of thin rectangles may be computed as the sum of the values for the individual components. Such an approach will give an approximation which neglects the contribution in the fillet region where the components are joined. For most common structural shapes this approximation causes little error, thus

$$J \approx \sum \tfrac{1}{3} b t^3 \tag{8.2.14}$$

where b is the long dimension and t the thin dimension of the rectangular elements.

More accurate expressions for various structural shapes have been developed by Lyse and Johnston [8.4], Chang and Johnston [8.5], Kubo, Johnston, and Eney [8.6], and El Darwish and Johnston [8.7].

In addition to the torsional properties in the AISC Manuals, torsional design aid publications are available by AISC [8.8], Hotchkiss [8.9], and Heins and Kuo [8.10].

8.3 SHEAR STRESSES DUE TO BENDING OF THIN-WALL OPEN CROSS-SECTIONS

Before treating the computation of stresses due to torsion of thin-wall open sections restrained from warping, a review of shear stress resulting from general flexure will be developed. Recognition of a torsion situation precedes concern about calculation of resulting stresses. Extensive treatment of thin-wall members of open cross-section is given by Timoshenko [8.11].

Referring to the general thin-wall section of Fig. 8.3.1, where x and y are centroidal axes, consider equilibrium of the element $t\,ds\,dz$ acted upon by flexural stress σ_z and shear stress τ, both of which result from bending moment. The shear

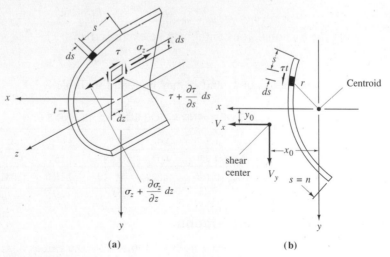

Figure 8.3.1   Stresses on thin-wall open sections in bending.

stress τ multiplied by the thickness t may be termed the *shear flow* τt. Force equilibrium in the z direction requires

$$\frac{\partial (\tau t)}{\partial s} ds\, dz + t\, \frac{\partial \sigma_z}{\partial z} dz\, ds = 0 \tag{8.3.1}$$

or

$$\frac{\partial (\tau t)}{\partial s} = -t\, \frac{\partial \sigma_z}{\partial z} \tag{8.3.2}$$

1. *Assume moment is applied in the yz plane only,* i.e., $M_y = 0$. The flexural stress due to bending, as given by Eq. 7.10.18, is

$$\sigma_z = \frac{M_x}{I_x I_y - I_{xy}^2} (I_y y - I_{xy} x) \tag{7.10.18}$$

$$\frac{\partial \sigma_z}{\partial z} = \frac{\partial M_x / \partial z}{I_x I_y - I_{xy}^2} (I_y y - I_{xy} x) \tag{8.3.3}$$

Recognizing that $V_y = \partial M_x / \partial z$, and substituting Eq. 8.3.3 into Eq. 8.3.2 gives

$$\frac{\partial \tau t}{\partial s} = \frac{-t V_y}{I_x I_y - I_{xy}^2} (I_y y - I_{xy} x) \tag{8.3.4}$$

Integrating to find τt at a distance s from a free edge gives the shear flow τt as

$$\tau t = \frac{-V_y}{I_x I_y - I_{xy}^2} \left[I_y \int_0^s yt\, ds - I_{xy} \int_0^s xt\, ds \right] \tag{8.3.5}$$

2. *Assume moment is applied in the xz plane only,* i.e., $M_x = 0$. The flexural stress due to bending as given by Eq. 7.10.18 is

$$\sigma_z = \frac{M_y}{I_x I_y - I_{xy}^2} (-I_{xy} y + I_x x) \tag{7.10.18}$$

Taking $\partial \sigma_z / \partial z$, recognizing that $V_x = \partial M_y / \partial z$, and integrating to get the shear flow τt, gives in a manner similar to Eq. 8.3.5,

$$\tau t = \frac{+V_x}{I_x I_y - I_{xy}^2} \left[I_{xy} \int_0^s yt\, ds - I_x \int_0^s xt\, ds \right] \tag{8.3.6}$$

3. *Moments applied in both yz and xz planes.* If shear stresses are desired they can be computed by superimposing the results from Eqs. 8.3.5 and 8.3.6.

It is to be observed from Fig. 8.3.1b that equilibrium requires that the shear V_y in the y direction equal the components of τt in the y direction summed over the entire section. Similarly V_x equals the summation of τt components in the x direction. Rotational equilibrium must also be satisfied; the moment about the centroid of the section is (see Fig. 8.31b)

$$\int_0^n (\tau t) r\, ds$$

which will be zero in some cases (such as I-shaped and Z-shaped sections). If such rotational equilibrium is automatically satisfied when the flexural shears act through the centroid, then no torsion will occur simultaneously with bending.

8.4 SHEAR CENTER

The shear center is the location in a cross-section where no torsion occurs when flexural shears act in planes passing through that location. In other words, forces acting through the shear center will cause no torsional stresses to develop, i.e.,

$$\int_0^n (\tau t) r \ ds = 0 \tag{8.4.1}$$

Since the shear center does not necessarily coincide with the centroid of the section, the shear center must be located in order to evaluate the torsional stress. For I-shaped and Z-shaped sections, the shear center coincides with the centroid, but for channels and angles it does not.

Referring to Fig. 8.3.1b, consider the shears V_x and V_y acting at distances from the centroid y_0 and x_0, respectively, such that the torsional moment with respect to the centroid is the same as $\int (\tau t) r \ ds$ integrated from zero to n; thus

$$V_y x_0 - V_x y_0 = \int_0^n (\tau t) r \ ds \tag{8.4.2}$$

In other words, the torsional moment is $(V_y x_0 - V_x y_0)$ when the loads are applied in planes passing through the centroid but is zero if the loads are in planes passing through the shear center, i.e., the point whose coordinates are $x_0 y_0$.

It is observed that the location of shear center is independent of the magnitude or type of loading, but is dependent only on the cross-sectional configuration.

To determine the shear center location, first let one of the shears be zero, say $V_y = 0$; then from Eq. 8.4.2,

$$y_0 = -\frac{1}{V_x} \int_0^n (\tau t) r \ ds \tag{8.4.3}$$

where according to Eq. 8.3.6,

$$\tau t = \frac{V_x}{I_x I_y - I_{xy}^2} \left[I_{xy} \int_0^s yt \ ds - I_x \int_0^s xt \ ds \right]$$

Alternately, letting $V_x = 0$ gives from Eq. 8.4.2,

$$x_0 = \frac{1}{V_y} \int_0^n (\tau t) r \ ds \tag{8.4.4}$$

where according to Eq. 8.3.5,

$$\tau t = \frac{-V_y}{I_x I_y - I_{xy}^2} \left[I_y \int_0^s yt \ ds - I_{xy} \int_0^s xt \ ds \right]$$

EXAMPLE 8.4.1

Locate the shear center for the channel section of Fig. 8.4.1.

Solution. Many practical cases can be solved without using the general formulas, Eqs. 8.4.3 and 8.4.4. Since the shear center location is a problem in equilibrium, moments may most conveniently be taken through a point that eliminates the greatest number of forces. Thus, letting $V_x = 0$ and taking moments about point A of Fig. 8.4.1a, changes the equilibrium equation, Eq. 8.4.2, to

$$V_y q = V_f h = \int_0^b (\tau t) h \, ds \tag{a}$$

where according to Eq. 8.3.5,

$$\tau t = \frac{-V_y}{I_x} \int_0^s yt \, ds = \frac{-V_y}{I_x} yts \tag{b}$$

For these thin-wall sections, the length s along which integration is performed is measured at mid-thickness.

Substituting Eq. (b) into Eq. (a), and using $y = -h/2$ where $t = t_f$ gives

$$V_y q = \int_0^b \frac{-V_y}{I_x} \left(\frac{-h}{2} \right) t_f h s \, ds \tag{c}$$

$$= \frac{V_y t_f h^2}{2I_x} \int_0^b s \, ds = \frac{V_y t_f h^2 b^2}{4I_x}$$

Thus the shear center location along the x-axis is

$$q = \frac{t_f h^2 b^2}{4I_x} \tag{d}$$

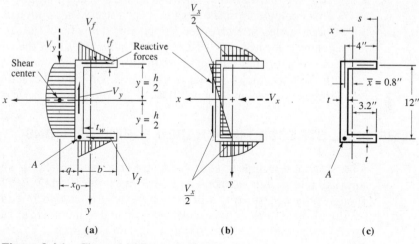

(a)   (b)   (c)

Figure 8.4.1 Channel of Example 8.4.1.

measured in the positive x direction to the left of the channel web.

For the shear center coordinate measured along the y-axis, apply V_x and let $V_y = 0$, and because of symmetry V_x must act at $y = 0$ for equilibrium. To demonstrate, let V_x be applied at the distance y_0 below the x-axis and take moments about point A. Satisfying equilibrium.

$$V_x\left(\frac{h}{2} - y_0\right) = \int_0^b (\tau t)y \, ds \qquad (e)$$

where according to Eq. 8.3.6,

$$\tau t = \frac{-V_s}{I_y} \int_0^s xt \, ds \qquad (f)$$

To illustrate numerically, use $b = 4$ in., $h = 12$ in., and $t = t_w$ in which case the centroid of the channel (refer to Fig. 8.4.1c) is located

$$\bar{x} = \frac{\Sigma Ax}{\Sigma A} = \frac{2b(b/2)t_w}{(h + 2b)t_w} = \frac{2(4)(2)}{12 + 2(4)} = 0.8 \text{ in.}$$

Then, $s = x + 3.2$ in.

$$I_y = [\tfrac{1}{3}(4)^3 2 - 20(0.8)^2]t_w = 29.87t_w$$

$$\tau t = \frac{-V_x t_w}{29.87t_w} \int_0^s (s - 3.2) \, ds = \frac{-V_x}{29.87}\left(\frac{s^2}{2} - 3.2s\right)$$

Substitution of τt into Eq. (e) gives

$$V_x\left(\frac{h}{2} - y_0\right) = \int_0^4 \frac{-V_x}{29.87}\left(\frac{s^2}{2} - 3.2s\right)h \, ds$$

$$= \frac{-V_x h}{29.87}\left(\frac{s^3}{6} - 3.2\frac{s^2}{2}\right)\Big|_0^4 = \frac{+V_x h}{2}$$

Thus it is shown that y_0 is zero. The shear center may also be located as follows. First compute, by integrating over each stress distribution of Fig. 8.4.1, the shear forces acting in each of the component elements of the section. Then the shear center is located such that V_x or V_y counteract all of the shear forces acting on the components to produce equilibrium. In solving for the shear center location, the solution may be made as illustrated, and then checked by verifying that the forces are in equilibrium. ■■

8.5 TORSIONAL STRESSES IN I-SHAPED STEEL SECTIONS

The structural engineer must recognize a torsion situation and be able to apply approximate design methods and perform a stress analysis when necessary, even though only occasionally will torsion be severe enough to control the design of a section. Rolled steel sections under uniform and nonuniform torsion have been studied analytically and experimentally by many investigators.

The development in this section is similar to that of Timoshenko [8.11], Lyse and Johnston [8.4], Kubo, Johnston, and Eney [8.6], Goldberg [8.12], and Chu and Johnson [8.13]. Discussions of some of the practical aspects, along with solutions for various loading and support cases, are given by Hotchkiss [8.9] and Johnston [8.14]; charts for design are available in the handbook from AISC [8.8] and in the paper by Johnston [8.14]; and design tables using the β modified flexure analogy method developed by the authors are presented in Sec. 8.6. Lin [8.15, 8.16] has given additional and expanded β value tables.

Application of load in a plane other than the one through the shear center (see Fig. 8.5.1) will cause the member to twist unless external restraints prevent such twisting. The torsional stress due to twisting consists of both shear and flexural stresses. These stresses must be superimposed on the shear and flexural stresses that exist in the absence of torsion.

Torsion may be categorized into two types: pure torsion, or as it is often called, *Saint-Venant's torsion*, and warping torsion. Pure torsion assumes that a cross-sectional plane prior to application of torsion remains a plane and only element rotation occurs during torsion. A circular shaft subjected to torsion is a situation where pure torsion exists as the only type. Warping torsion is the out-of-plane effect that arises when the flanges are laterally displaced during twisting, analogous to bending from laterally applied loads.

1. *Pure torsion (Saint-Venant's Torsion)*. Just as flexural curvature (change in slope per unit length) can be expressed as $M/EI = d^2y/dz^2$, i.e., moment divided by flexural rigidity equals flexural curvature, in pure torsion the torsional moment M divided by the torsional rigidity GJ equals the torsional curvature (change in angle of twist ϕ per unit length). Recalling previously derived Eq. 8.2.5 for T, which now becomes the component M_s due to pure torsion,

$$M_s = GJ\frac{d\phi}{dz} \tag{8.5.1}$$

where M_s = Pure torsional moment (Saint-Venant torsion)
G = shear modulus of elasticity = $E/[2(1 + \mu)]$, in terms of the tension-compression modulus of elasticity E and Poisson's ratio μ
J = torsional constant (see Sec. 8.2)

In accordance with Eq. 8.2.7, stress due to M_s is proportional to the distance from the center of twist.

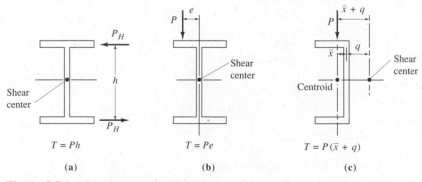

$$T = Ph \qquad\qquad T = Pe \qquad\qquad T = P(\bar{x} + q)$$

(a)   (b)   (c)

Figure 8.5.1 Common torsional loadings.

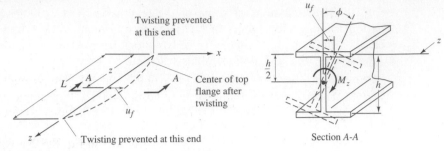

Figure 8.5.2   Torsion of an I-shaped section.

2. *Warping torsion.* A beam subjected to torsion M_z, as in Fig. 8.5.2, will have its compression flange bent in one direction laterally while its tension flange is bent in the other. Whenever the cross-section is such that it would warp (become a nonplanar section) if not restrained, the restrained system has stresses induced. The torsional situation of Fig. 8.5.2 illustrates a beam that is prevented from twisting at each end but the top flange deflects laterally by an amount u_f. This lateral flange bending causes flexural normal stresses (tension and compression) as well as shear stresses across the flange width.

Thus, torsion may be thought of as being composed of two parts: (1) rotation of elements, the pure torsion part, and (2) translation producing lateral bending, the warping part.

3. *Differential equation for torsion on I- and channel-shaped sections.* Consider the deflected position of a flange centerline, as in Fig. 8.5.2, where u_f is the lateral deflection of one of the flanges at a section a distance z from the end of the member; ϕ is the twist angle at the same section, and V_f (Fig. 8.5.3) is the horizontal shear force developed in the flange at the section due to lateral bending. An important assumption is that the web remains a plane during rotation, so that the flanges deflect laterally an equal amount. Thus the web is assumed thick enough compared to the flanges so that is does not bend during twisting as a result of high torsional resistance of the flanges. Except for thin-web plate girders, it has been shown [8.6, 8.17] that assuming no lateral bending in the web, i.e., no effect on the warping torsion component, is sufficiently correct for practical purposes. Since rarely are thin-web plate girders used without stiffeners, and certainly not when torsional stress exists, such cases are not of practical importance.

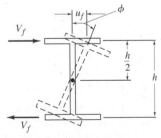

Figure 8.5.3   Warping shear force on I-shaped section.

From geometry,

$$u_f = \phi \frac{h}{2} \qquad (8.5.2)$$

for small values of ϕ. For understanding of torsion on I- and channel-shaped sections, Eq. 8.5.2 is the single most important relationship. The twist angle is directly proportional to the lateral deflection. Torsion boundary conditions are analogous to lateral bending boundary conditions.

Differentiating three times with respect to z in Eq. 8.5.2 gives

$$\frac{d^3 u_f}{dz^3} = \left(\frac{h}{2}\right) \frac{d^3 \phi}{dz^3} \qquad (8.5.3)$$

For one flange the curvature relationship is

$$\frac{d^2 u_f}{dz^2} = \frac{-M_f}{EI_f} \qquad (8.5.4)$$

where M_f is the lateral bending moment on one flange, I_f is the moment of inertia for one flange about the y-axis of the beam, and the minus sign arises from positive bending as shown in Fig. 8.5.2. Also, since $V = dM/dz$,

$$\frac{d^3 u_f}{dz^3} = \frac{-V_f}{EI_f} \qquad (8.5.6)$$

Using Eqs. 8.5.3 and 8.5.6 gives

$$V_f = -EI_f \left(\frac{h}{2}\right) \frac{d^3 \phi}{dz^3} \qquad (8.5.7)$$

Referring to Fig. 8.5.3, the torsional moment component M_w, causing lateral bending of the flanges, equals the flange shear force times the moment arm h. This assumes no shear resistance to warping is contributed by the web,

$$M_w = V_f h = -EI_f \frac{h^2}{2} \frac{d^3 \phi}{dz^3} \qquad (8.5.8)$$

$$= -EC_w \frac{d^3 \phi}{dz^3} \qquad (8.5.9)$$

where $C_w = I_f h^2 / 2$, often referred to as the *warping torsional constant*.

The total torsional moment is composed of the sum of the rotational part M_s and the lateral bending part M_w which from Eqs. 8.5.1 and 8.5.9 give

$$M_z = M_s + M_w = GJ \frac{d\phi}{dz} - EC_w \frac{d^3 \phi}{dz^3} \qquad (8.5.10)$$

the differential equation for torsion. The torsional moment M_z depends on the loading and in usual situations will be a polynomial in z. Expressions for the torsion constant J and warping constant C_w for various shapes are to be found in text, Appendix Table A2.

Rewrite Eq. 8.5.10, dividing by EC_w,

$$\frac{d^3\phi}{dz^3} - \frac{GJ}{EC_w}\frac{d\phi}{dz} = \frac{-M_z}{EC_w} \tag{8.5.11}$$

Letting $\lambda^2 = GJ/EC_w$ ($\lambda = 1/a$ of *Torsion Analysis of Steel Members* [8.8]), and for the homogeneous solution of Eq. 8.5.11 let $\phi_h = Ae^{mz}$,

$$\frac{d^3\phi}{dz^3} - \lambda^2\frac{d\phi}{dz} = 0 \tag{8.5.12}$$

which upon substitution of the homogeneous solution gives

$$Ae^{mz}(m^3 - \lambda^2 m) = 0 \tag{8.5.13}$$

which requires

$$m(m^2 - \lambda^2) = 0; \quad \therefore\ m = 0,\ m = \pm\lambda$$

Thus

$$\phi_h = A_1 e^{\lambda z} + A_2 e^{-\lambda z} + A_3 \tag{8.5.14}$$

which upon using the hyperbolic function identities and regrouping the constants may be expressed as

$$\phi_h = A\sinh \lambda z + B\cosh \lambda z + C \tag{8.5.15}$$

where

$$\lambda = \frac{1}{a} = \sqrt{\frac{GJ}{EC_w}}$$

For the particular solution, since M_z is in general some function of z,

$$M_z = f(z)$$

Let $\phi_p = f_1(z)$, and substitute into Eq. 8.5.11, giving

$$\frac{d^3 f_1(z)}{dz^3} - \lambda^2\frac{d f_1(z)}{dz} = -\frac{1}{EC_w}f(z) \tag{8.5.16}$$

where terms on the left-hand side must be paired with terms on the right side. Rarely will $f_1(z)$ be required to contain higher than second-degree terms.

EXAMPLE 8.5.1 _____

Develop, using the differential equation, the expressions for the twist angle ϕ, as well as the first, second, and third derivatives, for the case of concentrated torsional moment applied at midspan when the ends are torsionally simply supported.

Solution. Referring to Fig. 8.5.4, it is apparent that M_z is constant and equal to $T/2$. Thus let

$$\phi_p = C_1 + C_2 z \quad \text{(any polynomial)} \tag{a}$$

Using Eq. 8.5.11 gives

$$-\lambda^2 C_2 = -\frac{1}{EC_w}\left(\frac{T}{2}\right); \qquad \therefore\ C_2 = \frac{T}{2GJ}$$

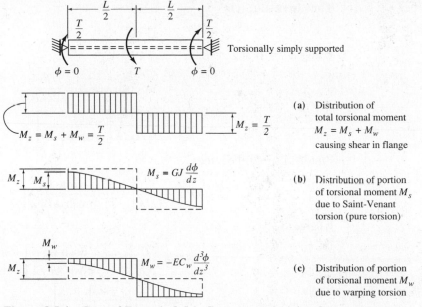

Figure 8.5.4 Case of Example 8.5.1. Concentrated torsional moment at midspan; torsionally simply supported. (Adapted from Hotchkiss [8.9], Fig. 3)

The other constant C_1 may be combined with C of Eq. 8.5.15. The complete solution for this loading is therefore

$$\phi = A\sinh \lambda z + B\cosh \lambda z + C + \frac{T}{2GJ}z \qquad \text{(b)}$$

Consider the boundary conditions for torsional simple support. Thinking of the lateral bending of the flange (since ϕ is proportional to u_f), simple support conditions mean zero moment and deflection at each end, or for torsion.

$$\phi = 0 \qquad \text{at} \quad z = 0 \qquad \text{and} \quad z = L$$

$$\frac{d^2\phi}{dz^2} = \phi'' = 0 \qquad \text{at} \quad z = 0 \qquad \text{and} \quad z = L$$

In this case the differential equation is discontinuous at $L/2$; thus, using zero slope of the flange at $L/2$, i.e., $\phi' = 0$, along with $\phi = 0$ and $\phi'' = 0$ at $z = 0$ will permit solution for the three constants of Eq. (b).

From $\phi = 0$ at $z = 0$,

$$0 = B + C \qquad \text{(c)}$$

Using $\phi'' = 0$ at $z = 0$,

$$\phi'' = A\lambda^2 \sinh \lambda z + B\lambda^2 \cosh \lambda z$$

$$0 = B \qquad \text{(d)}$$

Thus from Eq. (c),

$$C = 0$$

Using $\phi' = 0$ at $z = L/2$,

$$0 = A\lambda \cosh \lambda L/2 + \frac{T}{2GJ} \tag{e}$$

$$A = -\frac{T}{2GJ\lambda}\left[\frac{1}{\cosh \lambda L/2}\right]$$

Finally, Eq. (b) becomes

$$\phi = \frac{T}{2GJ\lambda}\left[\lambda z - \frac{\sinh \lambda z}{\cosh \lambda L/2}\right] \tag{f}$$

Also

$$\phi' = \frac{T}{2GJ}\left[1 - \frac{\cosh \lambda z}{\cosh \lambda L/2}\right] \tag{g}$$

$$\phi'' = \frac{T\lambda}{2GJ}\left[\frac{-\sinh \lambda z}{\cosh \lambda L/2}\right] \tag{h}$$

$$\phi''' = \frac{T\lambda^2}{2GJ}\left[\frac{-\cosh \lambda z}{\cosh \lambda L/2}\right] \tag{i}$$

Thus the solution of the differential equation is illustrated. The stress equations making use of the derivatives are developed in the next section. ■■

4. *Torsional stresses.* The shear stress v_s resulting from the Saint-Venant torsion M_s is computed in accordance with the form of Eq. 8.2.11,

$$v_s = \frac{M_s t}{J} \tag{8.2.11}$$

and using Eq. 8.5.1 gives

$$v_s = Gt\frac{d\phi}{dz} \tag{8.5.17}$$

whose distribution is shown in Fig. 8.5.5a. Though shown uniform fully across the flange, the stress drops sharply to zero at the flange tips.

The shear stress v_w that results from warping varies parabolically across the width of the rectangular flange as shown in Fig. 8.5.5b and may be computed as

$$v_w = \frac{V_f Q_f}{I_f t_f} \tag{8.5.18}$$

where Q_f = statical moment of area about the y-axis.

The negligible shear carried by the web is not considered. For maximum shear stress v_w, which actually acts at the face of web but may be approximated as acting at

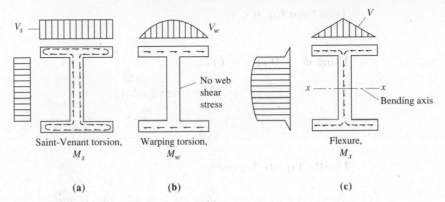

Figure 8.5.5 Direction and distribution of shear stress in I-shaped sections.

the mid-width of the flange, take Q_f (see Fig. 8.5.6) as

$$Q_f = A\bar{x} = \frac{bt_f}{2}\left(\frac{b}{4}\right)$$

Substituting Q_f and V_f from Eq. 8.5.7 into Eq. 8.5.18 gives

$$v_w = E\frac{b^2h}{16}\frac{d^3\phi}{dz^3} \tag{8.5.19}$$

taking the absolute value. The direction of the shear flow has no effect on the combining of shear stresses.

The tension or compression stress due to lateral bending of flanges (i.e., warping of the cross-section as shown in Fig. 8.5.7) may be expressed as

$$f_{bw} = \frac{M_f x}{I_f} \tag{8.5.20}$$

which varies linearly across the flange width as shown in Fig. 8.5.7. The bending moment M_f, the lateral moment acting on one flange, may be obtained by substituting Eq. 8.5.2 into Eq. 8.5.4 and noting that $I_f h^2/2$ is warping torsional constant C_w,

$$M_f = EI_f\left(\frac{h}{2}\right)\frac{d^2\phi}{dz^2} = \frac{EC_w}{h}\frac{d^2\phi}{dz^2} \tag{8.5.21}$$

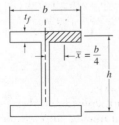

Figure 8.5.6 Dimensions for computation of statical moment of area, Q_f.

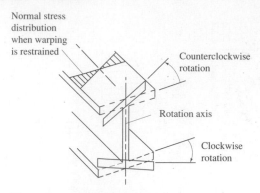

Normal stress
distribution
when warping
is restrained

Counterclockwise
rotation

Rotation axis

Clockwise
rotation

Figure 8.5.7   Warping of cross-section.

The minus sign is dropped since tension occurs on one side while compression occurs on the other.

The maximum stress occurs at $x = b/2$, which when used with Eq. 8.5.21 gives for Eq. 8.5.20,

$$f_{bw} = EI_f\left(\frac{h}{2}\right)\frac{d^2\phi}{dz^2}\left(\frac{b}{2I_f}\right)$$

$$f_{bw} = \frac{Ebh}{4}\frac{d^2\phi}{dz^2} \tag{8.5.22}$$

In a summary, three kinds of stresses arise in any I-shaped or channel section due to torsional loading: (a) shear stresses v_s in web and flanges due to rotation of the elements of the cross-section (Saint-Venant torsional moment, M_s); (b) shear stresses v_w in the flanges due to lateral bending (warping torsional moment, M_w); and (c) normal stresses (tension and compression) f_{bw} due to lateral bending of the flanges (lateral bending moment on flange, M_f).

EXAMPLE 8.5.2 _____

A W18×71 beam on a 24-ft simply supported span is loaded with a concentrated load of 20 kips at midspan. The ends of the member are simply supported with respect to torsional restraint (i.e., $\phi = 0$) and the concentrated load acts with a 2-in. eccentricity from the plane of the web (see Fig. 8.5.8). Compute combined bending and torsional stresses.

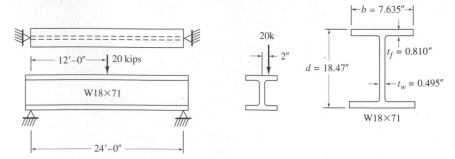

20k

$\leftarrow$ 2″

12′–0″   20 kips

W18×71

24′–0″

$d = 18.47″$

$\leftarrow b = 7.635″ \rightarrow$

$t_f = 0.810″$

$t_w = 0.495″$

W18×71

Figure 8.5.8   Data for Example 8.5.2.

Solution. The differential equation solution for this type of loading and end restraint was obtained in Example 8.5.1. The solution as obtained is

$$\phi = \frac{T}{2GJ\lambda}\left[\lambda z - \frac{\sinh \lambda z}{\cosh \lambda L/2}\right]$$

In accordance with the derivation (see Fig. 8.5.4), T is the applied torsional moment,

$$T = 20(2) = 40 \text{ in.-kips}$$

Recalling from Eq. 8.5.11,

$$\lambda = \sqrt{\frac{GJ}{EC_w}} = \sqrt{\frac{3.39}{2.6(4685)}} = \frac{1}{59.9} = 0.01668$$

where $\dfrac{E}{G} = \dfrac{2E(1 + \mu)}{E} = 2.6$ for $\mu = 0.3$

$$J \approx \sum \frac{bt^3}{3}, \quad \text{Eq. 8.2.14}$$

$$= \tfrac{1}{3}[2(7.635)(0.810)^3 + (18.47 - 1.620)(0.495)^3] = 3.39 \text{ in.}^4$$

$$C_w = \frac{I_f h^2}{2} = \frac{(7.635)^3(0.810)}{12} \frac{(18.47 - 0.810)^2}{2} = 4685 \text{ in.}^6$$

The above values of J and C_w compare with $J = 3.48$ and $C_w = 4700$ given in the AISC Manuals [1.7, 1.18] computed for rectangular flanges using a more exact expression for J including the effect of the fillets at the junction of flange to web. Though different sources give slightly different values for these torsional constants, any of the values are satisfactory for design purposes.

The function values required are

$$\lambda L = 24(12)/59.9 = 4.80$$

z	λz	$\sinh \lambda z$	$\cosh \lambda z$
$0.1L$	0.480	0.499	1.118
$0.2L$	0.960	1.116	1.498
$0.3L$	1.441	1.994	2.231
$0.4L$	1.922	3.343	3.489
$0.5L$	2.402	5.477	5.567

(a) Pure torsion (Saint-Venant torsion). Using Eq. 8.5.17,

$$v_s = Gt \, d\phi/dz$$

$$\frac{d\phi}{dz} = \frac{T}{2GJ}\left(1 - \frac{\cosh \lambda z}{\cosh \lambda L/2}\right)$$

$$v_s = \frac{Tt}{2J}\left(1 - \frac{\cosh \lambda z}{5.567}\right) = \frac{40t}{2(3.39)}\left(1 - \frac{\cosh \lambda z}{5.567}\right)$$

The shear stress v_s is a maximum at $z = 0$ and zero at $z = L/2$:

$$v_s \text{ (flange at } z = 0) = \frac{40(0.810)}{2(3.39)}\left(1 - \frac{1}{5.567}\right) = 3.92 \text{ ksi}$$

$$v_s \text{ (web at } z = 0) = 3.92\frac{0.495}{0.810} = 2.40 \text{ ksi}$$

(b) Lateral bending of flanges (warping torsion). Use Eq. 8.5.19 for shear stress in flanges,

$$v_w = E\frac{b^2 h}{16}\frac{d^3\phi}{dz^3}$$

$$\frac{d^3\phi}{dz^3} = \frac{T\lambda^2}{2GJ}\left(\frac{-\cosh \lambda z}{\cosh \lambda L/2}\right)$$

$$v_w = \frac{T}{2C_w}\frac{b^2 h}{16}\left(\frac{-\cosh \lambda z}{\cosh \lambda L/2}\right)$$

This shear stress acts at mid-width of the flange, and the maximum value occurs at $z = L/2$ while the minimum value is at $z = 0$,

$$v_w \text{ (flange at } z = L/2) = \frac{40}{2(4700)}\left(\frac{(7.635)^2 17.660}{16}\right) = 0.27 \text{ ksi}$$

$$v_w \text{ (flange at } z = 0) = 0.27\frac{1.0}{5.567} = 0.05 \text{ ksi}$$

For normal stress in flanges due to warping, use Eq. 8.5.22:

$$f_{bw} = \frac{Ebh}{4}\frac{d^2\phi}{dz^2}$$

$$\frac{d^2\phi}{dz^2} = \frac{M\lambda}{2GJ}\left[\frac{-\sinh \lambda z}{\cosh \lambda L/2}\right]$$

$$f_{bw} = \frac{M(2.6)\lambda bh}{8J}\left[\frac{\sinh \lambda z}{\cosh \lambda L/2}\right]$$

which is a maximum at $z = L/2$ and zero at $z = 0$. Thus

$$f_{bw} \text{ (flanges at } z = L/2) = \frac{40(2.6)(7.635)(17.660)}{8(3.39)(59.9)}\left[\frac{5.477}{5.567}\right] = 8.49 \text{ ksi}$$

(c) Ordinary flexure. Maximum normal stress is

$$f_b(\text{at } z = L/2) = \frac{PL}{4S_x} = \frac{20(24)(12)}{4(127)} = 11.34 \text{ ksi}$$

The shear stresses due to flexure are constant from $z = 0$ to $L/2$ and are computed by

$$v = \frac{VQ}{It} = \frac{10Q}{1170t} = \frac{Q}{117t}$$

For maximum flange shear stress, taking the more correct value at the face of the web rather than the value at mid-width of the flange,

$$Q = \left(\frac{7.635 - 0.495}{2}\right)(0.810)\left(\frac{17.660}{2}\right) = 25.53 \text{ in.}^3$$

$$v \text{ (flange at } z = 0) = \frac{25.53}{117(0.810)} = 0.27 \text{ ksi}$$

For maximum web shear stress,

$$Q = 7.635(0.810)\left(\frac{17.660}{2}\right) + \frac{16.850}{2}(0.495)\left(\frac{16.850}{4}\right) = 72.18 \text{ in.}^3$$

$$v(\text{web at } z = 0) = \frac{72.18}{117(0.495)} = 1.25 \text{ ksi}$$

A summary of stresses showing combinations is given in Table 8.5.1. ■■

TABLE 8.5.1 SUMMARY OF STRESSES FOR EXAMPLE 8.5.2

Type of Stress	Support ($z = 0$)	Midspan ($z = L/2$)
Compression and tension maximum stresses:		
Vertical bending, f_b	0	11.34
Torsional bending, f_{bw}	0	8.49
		19.83 ksi
Shear stress, web:		
Saint-Venant torsion, v_s	2.40	0
Vertical bending, v	1.25	1.25
	3.65 ksi	
Shear stress, flange:		
Saint-Venant torsion, v_s	3.92	0
Warping torsion, v_w	0.05	0.27
Vertical bending, v	0.27	0.27
	4.24 ksi	

8.6 ANALOGY BETWEEN TORSION AND PLANE BENDING

Because the differential equation solution is time consuming, and really suited only for analysis, design of a beam to include torsion is most conveniently done by making the analogy between torsion and ordinary bending.

Consider that the applied torsional moment T of Fig. 8.6.1 can be converted into a couple P_H times h. The force P_H can then be treated as a lateral load acting on the flange of a beam.

The substitute system will have constant shear over one-half the span, a diagram as given in Fig. 8.5.4a. The true distribution of lateral shear which contributes to

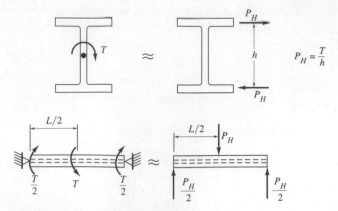

Figure 8.6.1   Analogy between flexure and torsion.

lateral deflection is only that part due to warping as shown in Fig. 8.5.4c. Thus the substitute system overestimates the lateral shear force and consequently overestimates the lateral bending moment M_f which causes tension and compression stresses.

In most practical design situations when it is desirable to include the effect of torsion, the compressive stress due to the warping component is the quantity of most importance. The shear stress contributions are normally not of significance.

EXAMPLE 8.6.1

Compute the stresses on the W18×71 beam of Example 8.5.2 and Fig. 8.5.8 using the flexural analogy rather than the differential equation solution.

Solution.   The substitute system is as shown in Fig. 8.6.2a. The lateral bending moment is then

$$M_f = V_f(L/2) = 1.13(12) = 13.6 \text{ ft-kips}$$

acting on one flange. Twice the moment acting on the entire section gives

$$f_{bw} = \frac{2M_f}{S_y} = \frac{2(13.6)(12)}{15.8} = 20.6 \text{ ksi}$$

For torsional shear stress, since $M_z = T/2 = 20$ in.-kips,

$$v_s = \frac{M_z t}{J} = \frac{20(0.810)}{3.39} = 4.78 \text{ ksi (flange)}$$

$$v_s = 4.78\left(\frac{0.495}{0.810}\right) = 2.92 \text{ ksi (web)}$$

For lateral bending flange shear stress,

$$v_w = \frac{V_f Q_f}{I_f t_f} = \frac{1.13(5.90)}{(30.0)0.810} = 0.27 \text{ ksi}$$

where $Q_f = (7.635/2)(0.810)(7.635/4) = 590 \text{ in.}^3$

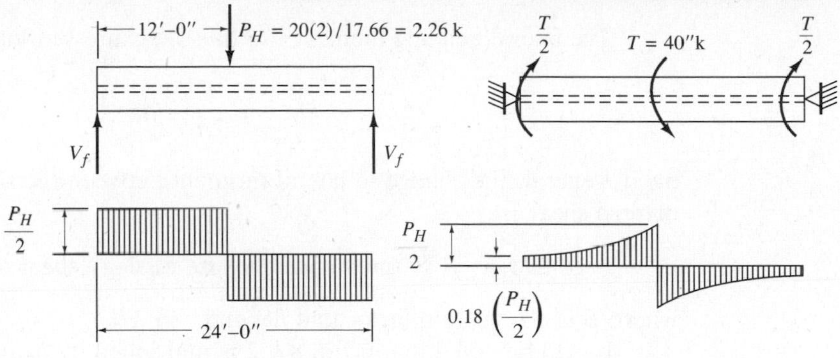

(a) Flexure analogy shear   **(b)** Warping torsion shear

Figure 8.6.2 Comparison of lateral shear on flange due to warping torsion with that from simple lateral flexure analogy.

The results of the two methods are compared as follows:

Type of stress	Flexural analogy	Differential equation
Compression/tension stress = $f_b + f_{bw}$ = 11.3 + 20.6 =	31.9 ksi	19.83 ksi
Web shear stress = $v + v_s$ = 1.25 + 2.92 =	4.17 ksi	3.65 ksi
Flange shear stress = $v + v_s + v_w$ = 0.27 + 4.78 + 0.27 =	5.32 ksi	4.24 ksi

The flexure analogy without modification is obviously a very conservative approach. In some situations it is so excessively conservative as to be practically useless. Furthermore, the most important design item, the lateral bending stress f_{bw} is overestimated by the greatest amount.

The relationship between the flexural analogy and the true torsion problem is best illustrated by referring to Fig. 8.5.4a. Note that the full torsional shear resulting from M_s and M_w is analogous to the lateral flexure problem. Figure 8.5.4b shows the portion of the shear that goes into rotation of elements, while Fig. 8.5.4c shows the portion contributing to lateral flange bending. If one could correctly assess how the shear due to warping torsion compares with the lateral flexure situation, design for torsion could be greatly simplified without being grossly conservative.

Figure 8.6.2b shows the accurate variation of V_f for the problem of Example 8.6.1, computed according to Eq. 8.5.7, whereupon

$$V_f = \frac{T}{2h}\left(\frac{\cosh \lambda z}{\cosh \lambda L/2}\right) \tag{8.6.1}$$

in which T/h can be thought of as the lateral load, which means the shear from the lateral bending analogy is $T/2h$, which is then modified by the hyperbolic function.

The lateral bending moment can thus be expressed for this problem as

$$M_f = \beta \frac{T}{2h}\left(\frac{L}{2}\right) \tag{8.6.2}$$

or, in general, the change in lateral moment between the support and location of zero shear is

$$\Delta M_f = \beta \times (\text{area under flexure analogy shear diagram}) \tag{8.6.3}$$

where β is a reduction factor that depends on λL.

It is to be noted that if Eq. 8.6.2 is multiplied by h, and the concentrated moment T is thought of as a concentrated load, the analogous moment $M_f h$ (sometimes referred to as *bimoment*) equals β times the simple beam moment. Thus the modified flexure analogy gives

$$M_f h = \beta\left(\frac{TL}{4}\right) \tag{8.6.4}$$

for the case of Fig. 8.6.1.   ■■

Tables 8.6.1 through 8.6.5 give "exact" values for β for several common loading and restraint conditions. For other cases Table I of Ref. 8.9 (where M_w equals $M_f h$ above) or the curves of *Torsion Analysis of Steel Members* [8.8] may be used. In Tables 8.6.3 and 8.6.4, m is the applied torsional loading per unit length (say, in.-kips/ft).

EXAMPLE 8.6.2

Recompute the stresses due to torsion on the beam of Example 8.6.1, using the modified flexural analogy method utilizing the β values from Table 8.6.1.

Solution.   The flexure analogy gives

$$M_f = 13.6 \text{ ft-kips}$$

as previously computed.

$$\lambda L = 4.80 \text{ (as computed in Example 8.5.2)}$$

From Table 8.6.1 at $a = 0.5$, $\beta \approx 0.41$, i.e., use about 41 percent of the flexure analogy value. Thus the modified flexure analogy gives

$$M_f = 13.6(0.41) = 5.58 \text{ ft-kips}$$
$$f_{bw} = \frac{2M_f}{S_y} = \frac{2(5.58)12}{15.8} = 8.48 \text{ ksi}$$

which compares favorably with $f_{bw} = 8.49$ ksi as computed by the differential equation solution using $\lambda L = 4.80$. For this case that exactly fits a table case, the β modified flexure analogy is the "exact" value obtained from the differential equation solution value.   ■■

8.7 PRACTICAL SITUATIONS OF TORSIONAL LOADING

There are relatively few occasions in actual practice where the torsional load can cause significant twisting, and frequently these situations arise during construction. In most building construction the members are laterally restrained by attachments along the length of the member and therefore they are not free to twist. Even though torsional loading exists, it may be self-limiting because the rotation cannot exceed the end slope of the transverse attached members.

Torsion exists on spandrel beams, where such loading may be uniformly distributed; torsion also exists where a beam frames into a girder on one side only, or where unequal reactions come to opposite sides of a girder. The design of crane runway girders involves the combination of biaxial bending and torsion, and is illustrated in Sec. 8.8 for laterally stable beams. Any situation where the loading or reaction acts eccentrically to the shear center gives rise to torsion.

Analysis for Torsional Moment

The determination of the torsional moment in a framing system involves an elastic analysis where the joints may be rigid or semi-rigid. While the details of such an analysis are outside the scope of this text, some discussion is necessary so that at least the problem is understood. Goldberg [8.12] has discussed this subject and presented an approximate method suitable for design. Spandrel girders have been treated by Lothers [8.18]. Chen and Jolissaint [8.19] have provided a simple analysis technique for rigid frames.

Consider an example of a floor framing system (similar to Goldberg's [8.12]) as shown in Fig. 8.7.1. Spandrel beam AB is subjected to torsion because of the floor beams framing on only one side. Contrary to some common belief, however, the torsional moment is *not* equal to the beam reaction times its eccentricity from the centerline of the girder web. Moment is transmitted across the joint, and the end moment on the beam must equal the torsional moment on the girder. To attack such a problem one must first determine the relationship between the angle of twist ϕ and the applied torsional moment T.

For example, in Fig. 8.7.1 the loading system causes equal torsional moments at the $\frac{1}{3}$ points on member AB. Assuming the girder torsionally simply supported at ends A and B, using either the differential equation solution formulas or the curves of *Torsion Analysis of Steel Members* [8.8] (Case 3, p. 32), for $\lambda L = 15(12)/51.3 = 3.51$, one finds the angle of twist ϕ at point a,

$$\phi \frac{GJ}{TL} \approx 0.09$$

or

$$\phi_{aa} = 16.2 \frac{T_a}{GJ}$$

for T applied at a. In addition the value of ϕ at point a for T applied at c is

$$\phi_{ac} = 0.07(180) \frac{T_c}{GJ} = 12.6 \frac{T_c}{GJ}$$

TABLE 8.6.1 β VALUES, CONCENTRATED LOAD,
TORSIONAL SIMPLE SUPPORT

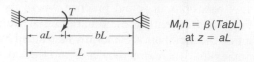

$$M_t h = \beta(TabL)$$
$$\text{at } z = aL$$

λL	β values				
	$a = 0.5$	$a = 0.4$	$a = 0.3$	$a = 0.2$	$a = 0.1$
0.5	0.98	0.98	0.98	0.99	0.99
1.0	0.92	0.93	0.94	0.95	0.97
2.0	0.76	0.77	0.80	0.84	0.91
3.0	0.60	0.62	0.65	0.72	0.83
4.0	0.48	0.50	0.54	0.62	0.76
5.0	0.39	0.41	0.45	0.54	0.70
6.0	0.33	0.34	0.39	0.47	0.65
8.0	0.25	0.26	0.30	0.37	0.55
10.0	0.20	0.21	0.24	0.31	0.48

TABLE 8.6.2 β VALUES, CONCENTRATED LOAD,
TORSIONALLY FIXED SUPPORTS

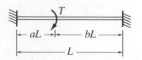

$$M_t h = \beta_1(Tab^2 L)$$
$$\text{at } z = 0$$
$$M_t h = \beta_2(Ta^2 bL)$$
$$\text{at } z = L$$

λL	$a = 0.5$	$a = 0.4$		$a = 0.3$		$a = 0.2$	
	$\beta_1 = \beta_2$	β_1	β_2	β_1	β_2	β_1	β_2
0.5	0.99	1.00	0.99	1.00	0.99	1.00	0.99
1.0	0.98	0.98	0.98	0.98	0.98	0.99	0.98
2.0	0.92	0.93	0.92	0.94	0.92	0.96	0.92
3.0	0.85	0.86	0.84	0.88	0.84	0.91	0.85
4.0	0.76	0.78	0.75	0.81	0.75	0.86	0.77
5.0	0.68	0.70	0.67	0.74	0.67	0.80	0.69
6.0	0.60	0.63	0.59	0.67	0.60	0.75	0.62
8.0	0.48	0.51	0.47	0.56	0.49	0.65	0.52
10.0	0.39	0.42	0.39	0.47	0.41	0.56	0.44

TABLE 8.6.3 β VALUES, UNIFORM LOAD, TORSIONAL SIMPLE SUPPORT

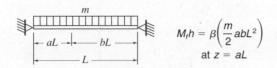

$$M_f h = \beta\left(\frac{m}{2}abL^2\right)$$
at $z = aL$

	β values				
λL	$a = 0.5$	$a = 0.4$	$a = 0.3$	$a = 0.2$	$a = 0.1$
0.5	0.97	0.97	0.98	0.98	0.98
1.0	0.91	0.91	0.91	0.91	0.92
2.0	0.70	0.71	0.71	0.72	0.74
3.0	0.51	0.51	0.52	0.54	0.57
4.0	0.37	0.37	0.38	0.41	0.44
5.0	0.27	0.27	0.29	0.31	0.34
6.0	0.20	0.20	0.22	0.24	0.28
8.0	0.12	0.12	0.13	0.16	0.19
10.0	0.08	0.08	0.09	0.11	0.14

TABLE 8.6.4 β VALUES, UNIFORM LOAD, TORSIONALLY FIXED SUPPORTS

$$M_f h = \beta\left(\frac{m}{12}L^2\right)$$
at $z = 0$ and $z = L$

λL	0.5	1.0	2.0	3.0	4.0	5.0	6.0	8.0
β	0.99	0.98	0.94	0.88	0.81	0.74	0.67	0.56

TABLE 8.6.5 β VALUES, CONCENTRATED LOAD, TORSIONALLY FIXED SUPPORTS

$M_f h = \beta$ (positive moment by flexure theory)
$= \beta[2Ta^2b^2L]$
at $z = aL$

λL	$a = 0.5$	$a = 0.3$	$a = 0.1$
0.5	0.99	1.00	1.00
1.0	0.98	0.99	1.01
2.0	0.92	0.95	1.05
3.0	0.85	0.91	1.10
4.0	0.76	0.85	1.16
5.0	0.68	0.79	1.21
6.0	0.60	0.73	1.25

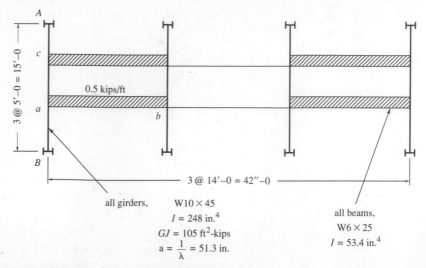

Figure 8.7.1   Plan view of floor framing.

Finally, for $T_a = T_c = T$,

$$\phi_a = (16.2 + 12.6)\frac{T}{GJ} = 28.8\frac{T}{GJ}$$

The twist angle ϕ_a must be compatible with the end slope of the beam; using slope deflection,

$$M_{ab} = M_{Fab} + \frac{2EI}{L_{ab}}\left[-2\phi_a - \phi_b + \frac{3\Delta}{L_{ab}}\right]$$

where M_{Fab} = the fixed-end moment for beam ab at a
ϕ_a = beam slope at a
ϕ_b = beam slope at b
Δ = relative deflection between a and b

After having established the necessary slope deflection equations for moments, joint equilibrium and shear conditions are necessary, after which simultaneous equations must be solved. One such joint equation is that at joint a,

$$T + M_{ab} = 0$$

After solving for the slopes, then the torsional moments can be found; for the torsional moment at a,

$$T = \phi_a GJ/28.8$$

In the Goldberg [8.12] example using members having properties similar to those of Fig. 8.7.1, the value of T obtained was 1.55 in.-kps using an approximate method of satisfying deformation compatibility.

Suppose one had taken the simple beam reaction for member ab, 0.5(7) = 3.5 kips, and assume use of *LRFD Manual* Table 9–2, All-Bolted Double-Angle

Connections (see Fig. 13.21). If the eccentricity had been taken to the bolt line on the outstanding leg ($2\frac{1}{4}$ in.), the torsional moment would have been far too great, while if the eccentricity had been taken as one-half the W10×45 (Fig. 8.7.1) web thickness (0.350/2), the torsional moment would have been far too small.

The proper torsional moment can only be obtained (even approximately) by considering deformation compatibility.

Torsional End Restraint

If a torsional situation is deemed to require analysis, the torsional end restraint must be evaluated. Under LRFD-A2.2 two basic types of construction are permitted: Type FR (fully restrained) which is the traditional "rigid frame" construction; and Type PR (partially restrained) which includes "simple" or "conventional" framing where there is assumed to be negligible flexural restraint at the joint, as well as "semi-rigid" framing where a defined flexural restraint exists that is less than Type FR. In ASD-A2.2 there are three types of construction permitted: Type 1 which is the same as Type FR in LRFD; Type 2 which is "simple" or "conventional" framing; and Type 3 which is "semi-rigid" framing. ASD Types 2 and 3 are included in LRFD Type PR.

The correlation of "simple" and "rigid" framing with torsional restraint is shown in Fig. 8.7.2. Again, the lateral bending analogy will help in visualizing the torsional

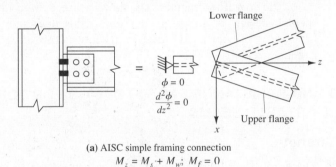

(a) AISC simple framing connection

$$M_z = M_s + M_w; \quad M_f = 0$$

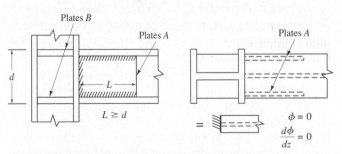

(b) AISC rigid framing connection (LRFD Type FR and ASD Type 1) with additional stiffening plates.

$$M_z = M_w; \quad M_s = 0$$

Figure 8.7.2 Torsional restraint conditions. (Adapted from Hotchkiss [8.9], Fig. 8)

restraint conditions. Figure 8.7.2a shows the analogy situation of zero deflection and zero moment which correspond torsionally to $\phi = 0$ and $d^2\phi/dz^2 = 0$. It is noted that $\phi = 0$ only if the simple connection extends over a significant portion of the beam depth.

Figure 8.7.2b shows the analogy situation of zero deflection and zero slope which corresponds torsionally to $\phi = 0$ and $d\phi/dz = 0$. Hotchkiss [8.9] states the ends of the beam must be boxed in (plates A, Fig. 8.7.2b) so as to assure $d\phi/dz = 0$. Ojalvo [8.20, 8.22] has discussed torsional restraint and indicates that "boxing" is not sufficient to obtain $d\phi/dz = 0$. When "boxing", that is, welding stiffener plates between the toes of flanges and extending them along the beam for a length equal to the beam depth, is used and the beam is then welded to a thick column flange the authors believe it is essentially "torsionally fixed". Furthermore, if the column has flexible flanges, column stiffeners (plates B, Fig. 8.7.2b) should be provided.

The Ojalvo [8.20] suggestion of welding a length of channel, angle, or bent plate between the flanges on one side of the web seems to be a more efficient and economical solution. The ends of the member must be welded against the insides of the flanges, and there must be weld vertically along the edges that bear against the web. Tests of several torsional restraints, including that proposed by Ojalvo [8.20], have been reported by Heins and Potocko [8.21]. Vacharajittiphan and Trahair [8.23] also discuss torsional restraint at I-section joints.

The structural engineer should remember that in practical situations where no special design is made at the ends, the torsional restraint is neither simple $(d^2\phi/dz^2 = 0)$ nor fixed $(d\phi/dz = 0)$ but is, however, usually such that the end twist is nearly zero $(\phi = 0)$.

8.8 LOAD AND RESISTANCE FACTOR DESIGN FOR TORSION— LATERALLY STABLE BEAMS

Nominal Strength

The nominal strength of a section subject to torsion or torsion combined with flexure is not readily determined. Such strength will certainly depend on the proportions of the section and the relative magnitudes of the forces applied. The AISC ASD Specification (such as the 1978 Specification, Sec. 1.5.1.4.4) has traditionally limited the combined stress to a maximum of $0.60F_y$. This implies that the nominal strength of the section under combined bending and torsion is reached when the extreme fiber stress reaches the yield stress F_y. Thus, the entire cross-section will be elastic; no credit is given to ability of the cross-section to undergo plastic deformation. Certainly this approach is conservative. An interesting review of the design of I-shaped beams for combined flexure and torsion is given by Driver and Kennedy [8.35].

LRFD-H2 uses the same limit state by requiring the combined stress computed for factored loads to not exceed $\phi_b F_y$. (Note that in this chapter the strength reduction factor ϕ_b is given the subscript to clearly distinguish it from the angle ϕ of twist.) Thus, the elastic biaxial bending stress equation, Eq. 7.11.1, can be used after converting the torsional moment into a pair of lateral bending moments acting in opposite directions on each flange.

Also, the more detailed criteria of LRFD-Appendix H3, as treated in Sec. 7.11, can be used for design.

In the examples that follow, the beams are assumed to be stable such that the lateral-torsional buckling limit state does not control (see Chapter 9).

EXAMPLE 8.8.1

Select the lightest W section of A36 steel to carry 0.4 kips/ft dead load, in addition to the weight of the beam, and live load of 1.5 kips/ft. The superimposed load is applied eccentrically 7 in. from the center of the web on the simply supported span of 28 ft as shown in Fig. 8.8.1. Assume the ends of the beam have torsional simple support.

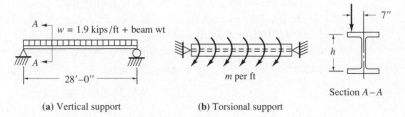

(a) Vertical support   (b) Torsional support

Figure 8.8.1   Conditions for Example 8.8.1.

Solution   (a) Compute factored loads eccentrically applied.

$$w_u = 1.2(0.4) + 1.6(1.5) = 2.88 \text{ kips/ft}$$

(b) Compute factored moment M_{ux}. Estimating the beam weight as 0.13 kips/ft, the moment M_{ux} is

$$M_{ux} = \tfrac{1}{8}wL^2 = \tfrac{1}{8}[2.88 + 1.2(0.13)](28)^2 = 298 \text{ ft-kips}$$

(c) Consider the torsion effect. The factored uniformly distributed torsional moment is

$$m_u = 2.88(7) = 20.2 \text{ in.-kips/ft}$$

Consider m_u/h as the uniformly distributed lateral load acting on *one* flange of the beam. Then using the flexure analogy, the lateral bending moment M_f acting on one flange is

$$M_f = \frac{1}{8}\frac{m_u}{h}L^2 = \frac{1}{8}\frac{20.2}{h}(28)^2 = \frac{1976}{h} \text{ ft-kips}$$

without regard to the modification factor β.

As a first approximation, assume $h = 14$ in. and $\beta = 0.5$ (approximation from Table 8.6.3, for $\lambda L \approx 3$). Thus the modified flexure analogy gives

$$M_f = \beta\frac{1976}{h} = 0.5\left(\frac{1976}{14}\right) = 70.6 \text{ ft-kips}$$

The design acceptability criterion is

$$\frac{M_{ux}}{S_x} + \frac{M_{uy}}{S_y} \leq \phi_b F_y$$

and using the procedure discussed in Sec. 7.11 gives

$$\text{Required } S_x \geq \frac{M_{ux}}{\phi_b F_y} + \frac{M_{uy}}{\phi_b F_y}\left(\frac{S_x}{S_y}\right)$$

$$= \frac{298(12)}{0.90(36)} + \frac{2(70.6)12}{0.90(36)}(3) = 266 \text{ in.}^3$$

in which the ratio S_x/S_y is estimated at 3 (Table 7.11.1) for medium weight W14 sections, and M_f is doubled to give an equivalent moment acting on two flanges.

This would indicate a W14×176 having an $S_x = 281$ in.³ Since the actual S_x/S_y ratio for W14 sections in this weight range is 2.6, the required S_x is then reduced to 245, indicating W14×159.

Using torsional properties in the AISC Manuals [1.7, 1.18] for the W14×159, $\sqrt{GJ/EC_w} = 1/68.3$. Thus,

$$\lambda L = 28(12)/68.3 = 4.9$$

Using Fig. 8.6.3, β is reduced to about 0.3, which further reduces required S_x to about 190 in.³ Try W14×132.

$$\lambda L = 28(12)/73.2 = 4.59$$

$$\beta \approx 0.31 \quad \text{(Table 8.6.3)}$$

$$M_f = \beta \frac{m_u L^2}{8h} = 0.31 \frac{20.2(28)^2}{8(14.66 - 1.030)} = 45.1 \text{ ft-kips}$$

Check design strength criterion under LRFD-H2. Compute the factored bending stress f_{un},

$$f_{un} = \frac{M_{ux}}{S_x} + \frac{M_{uy}}{S_y} = \frac{298(12)}{209} + \frac{45.1(12)}{74.5/2}$$

$$= 17.1 + 14.5 = 31.6 \text{ ksi} < (\phi_b F_y = 32.4 \text{ ksi}) \quad \text{OK}$$

Use W14×132.

Where high torsional strength is required, the wide W14 sections are most suitable. For the *same weight per foot,* deeper sections give a reduced stress from in-plane (of web) flexure but an increased stress from restraint of torsional warping. The W12×132 ($f_{un} = 29.4$ ksi) and the W24×131 ($f_{un} = 30.7$ ksi) give about the same maximum total flexural stress as the above selected beam.

The differential equation solution gives for the factored lateral bending stress f_{un} due to warping torsion 14.3 ksi as compared with 14.5 ksi computed above. The maximum factored flange shear stress f_{uv} is 16.6 ksi, while that in the web is 13.5 ksi, both computed from the differential equation solution. These are acceptable under LRFD-H2,

$$f_{uv} = 16.6 \text{ ksi} < [\phi_b \tau_y = \phi_b(0.6F_y) = 19.4 \text{ ksi}] \quad \text{OK} \quad \blacksquare\blacksquare$$

EXAMPLE 8.8.2

Design a beam having torsionally fixed ends to carry two concentrated loads of 20 kips (5 kips dead load and 15 kips live load) acting eccentric to the plane of the web by 6 in. as shown in Fig. 8.8.2. Assume for conservatism that for in-plane (of web) flexure the beam is simply supported. Use A36 steel. Use Load and Resistance Factor Design.

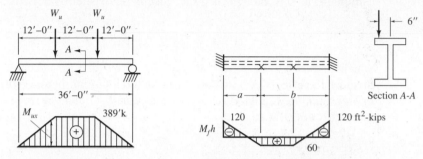

Figure 8.8.2 Loading and factored moments for Example 8.8.2.

Solution

(a) Compute factored loads eccentrically applied.

$$W_u = 1.2(5) + 1.6(15) = 30 \text{ kips}$$

(b) Compute factored moment M_{ux}. Estimating the beam weight as 0.15 kips/ft, the moment M_{ux} is

$$M_{ux} = W_u(12) + \tfrac{1}{8}wL^2 = 30(12) + \tfrac{1}{8}[1.2(0.15)](36)^2 = 389 \text{ ft-kips}$$

(c) Consider the torsion effect. The factored concentrated torsional moment is

$$T_u = 30(6/12) = 15 \text{ ft-kips}$$

Considering T_u/h as the analogous *lateral* concentrated loads acting at *one* flange, the fixed-end moments are computed; thus

$$M_f h \text{ (at ends)} = \frac{T_u ab^2}{L^2} + \frac{T_u a^2 b}{L^2} = \frac{15(12)(24)}{(36)^2}(24 + 12)$$
$$= 80.0 + 40.0 = 120 \text{ ft}^2\text{-kips}$$

and in the positive moment zone (midspan region),

$$M_f h \binom{\text{at concentrated}}{\text{loads}} = \frac{T_u L}{3} - 120.0$$
$$= 15(12) - 120.0 = 60 \text{ ft}^2\text{-kips}$$

The above moments are computed without regard for the β reduction factor; the flexure analogy gives $M_f h$ values as shown in Fig. 8.8.2b. These values are more appropriate than using the expression in Table 8.6.5 because that expression is for *one* concentrated load. The β values from Table 8.6.5 are reasonable, however, since the effect of one load on the torsional stress at the other load is small.

Estimating average λL at about 3, and using $aL = 0.3L$ in Table 8.6.2 for end moments, the modified analogous fixed-end moments become

$$M_f h \text{ (at ends)} = 0.88(80) + 0.84(40) = 70 + 34 = 104 \text{ ft}^2\text{-kips}$$

For positive moment at 12 ft from the support, refer to Table 8.6.5 and estimate β as 0.9, though the exact case being treated is not covered in any of β tables. Thus

$$M_f h \text{ (at } z = 0.3L) \approx 0.9(60) = 54 \text{ ft}^2\text{-kips}$$

which is known to be conservatively high (see Fig. 8.8.3) because the value 54 includes the effects of both concentrated torsional moments. (*Torsion Analysis of Steel Members* [8.8], Case 6, indicates that T_u applied at $z = 0.3L$ has negligible effect at $0.7L$).

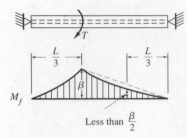

Figure 8.8.3 M_f variation for concentrated torsional moment **T**.

(d) Select the member. Assume $h \approx 14$ in., which makes

$$M_f = 54(12)/14 = 46.3 \text{ ft-kips}$$

$$\text{Required } S_x \geq \frac{M_{ux}}{\phi_b F_y} + \frac{M_{uy}}{\phi_b F_y}\left(\frac{S_x}{S_y}\right)$$

$$= \frac{389(12)}{0.90(36)} + \frac{2(46.3)(12)}{0.90(36)}(2.5)$$

$$= 144 + 86 = 230 \text{ in.}^3$$

A preliminary check of Table 8.6.5 gives about 0.8 for a β estimate. This would reduce the required S_x to 213 in.³ and indicates a W14×132. Try W14×132: $S_x = 209$ in.³:

$$\lambda L = 36(12)/73.2 = 5.90$$

Using Table 8.6.5, find $\beta \approx 0.74$ which gives

$$M_f \text{ (at } z = 0.3L) = \beta\frac{M_f h}{h} = 0.74\frac{60(12)}{14.66 - 1.03} = 39.1 \text{ ft-kips}$$

$$\text{Required } S_x = 144 + \frac{2(39.1)12}{0.90(36)}(2.8) = 225 \text{ in.}^3$$

Thus the next heavier section is indicated; *Use* W14×145.

For a more accurate check of the W14×132 using *Torsion Analysis of Steel Sections* [8.8], Case 6 for $\lambda L = 5.90$ at $z = 0.3L$,

$$M_f h = \left(\frac{\text{Ref. 8.8}}{\text{coeff}}\right)\frac{T_u}{\lambda} = \frac{0.37(15)}{(1/73.2)12} = 33.9 \text{ ft}^2\text{-kips}$$

$$M_f = 33.9(12)/13.63 = 25.8 \text{ ft-kips}$$

$$f_{un} = \frac{389(12)}{209} + \frac{25.8(12)}{74.5/2} = 30.6 \text{ ksi}$$

Since f_{un} does not exceed $\phi_b F_y = 0.90(36) = 32.4$ ksi, the W14×132 would be acceptable by the more exact check.

Also, the stress under factored moment at the supports must be checked. Using Table 8.6.2, find $\beta_1 \approx 0.68$ and $\beta_2 \approx 0.61$ for $\lambda L = 5.87$ (W14×145) and $aL = 0.3L$. Then

$$M_f h = 0.68(80) + 0.61(40) = 54.4 + 24.4 = 78.8 \text{ ft}^2\text{-kips}$$

$$M_f = 78.8(12)/13.69 = 69.1 \text{ ft-kips}$$

Thus the factored moment M_f about the y-axis resisted by one flange gives the factored stress f_{un}

$$f_{un} = \frac{M_f}{S_y/2} = \frac{69.1(12)}{87.3/2} = 19.0 \text{ ksi} < (\phi_b F_y = 32.4 \text{ ksi}) \quad \text{OK} \quad \blacksquare\blacksquare$$

These two examples illustrate that using approximate β values, along with the flexure analogy for lateral bending due to warping torsion, gives sufficiently quick and accurate results for ordinary design. Furthermore, the designer can better visualize what is happening using the flexure analogy rather than working with the hyperbolic functions for ϕ.

The β modified flexure analogy has been expanded by Lin [8.15] where additional β tables are provided. Johnston [8.14] has provided more detailed design aids to compute torsional functions other than the compressive or tensile stress due to restraint of warping; i.e., particularly the shear stress. Johnston also has several excellent detailed design examples. Salmon [8.24] and Lin [8.16] have provided additional insight in their discussions of Johnston's paper. Additional approximate formulas for design are provided by Johnston, Lin, and Galambos [8.25, pp. 330–331].

For additional treatment of combined torsion and flexure, particularly on channel and zee sections, the reader is referred to the work of Lansing [8.26]. For nonprismatic open section members, Evick and Heins [8.27] present solution techniques and give some design information.

Another topic, outside the scope of this text, is the secondary lateral bending moment that arises from the torsional deflection of the compression flange laterally. In the deflected position the compressive force resulting from ordinary flexural moment, M_x, times the lateral flange deflection gives rise to the secondary lateral moment which in turn causes greater lateral deflection. Discussion of this topic appears elsewhere [8.28, 8.14] and is similar to the secondary bending moment that occurs in beam-columns, a subject treated in Chapter 12.

8.9 ALLOWABLE STRESS DESIGN FOR TORSION—LATERALLY STABLE BEAMS

The traditional ASD strength-related requirement (1978 ASD, Sec. 1.5.1.4.6b) is represented by Eq. 7.11.8,

$$\left(f_b = \frac{M_x}{S_x} + \frac{M_y}{S_y} \right) \le \left(F_b = \frac{F_y}{\gamma/\phi_b} = 0.60F_y \right) \qquad [7.11.8]$$

The torsion is converted into equivalent M_y by using the flexure analogy as discussed in Sec. 8.6. The procedure is the same as illustrated for LRFD in Sec. 8.8 except in ASD the service loads are used instead of factored loads to compute the stress f_b. The 1989 ASD Specification [1.5] does not state explicitly how to design for biaxial bending *without* axial load; presumably the formulas for combined bending and axial force (ASD-H2) could be used.

EXAMPLE 8.9.1 ───

Investigate the W14×132 selected in Example 8.8.1 for the loading and conditions of Fig. 8.8.1. Use Allowable Stress Design.

Solution.

(a) Compute the service load moment M_x. Including the 0.132 kip/ft beam weight,

$$M_x = \tfrac{1}{8}wL^2 = \tfrac{1}{8}(1.9 + 0.132)(28)^2 = 199 \text{ ft-kips}$$

(b) Consider the torsion effect. The unformly distributed service load torsional moment m is

$$m = 1.9(7) = 13.3 \text{ in.-kips/ft}$$

Consider m/h as the uniformly distributed lateral load acting on *one* flange of the beam. Then using the β modified flexure analogy, the lateral bending moment M_f acting on one flange is

$$M_f = \beta \frac{1}{8} \frac{m}{h} L^2 = 0.31 \frac{1}{8} \frac{13.3}{13.63} (28)^2 = 29.6 \text{ ft-kips}$$

(c) Check the stress. The design acceptability criterion is

$$f_b = \frac{M_x}{S_x} + \frac{M_f}{S_y/2} \le 0.60F_y$$

$$f_b = \frac{199(12)}{209} + \frac{29.6(12)}{74.5/2} = 20.9 \text{ ksi} < (F_b = 22 \text{ ksi}) \quad \text{OK}$$

Thus the W14×132 is acceptable! ■■

8.10 TORSION IN CLOSED THIN-WALL SECTIONS

In general, where a high torsional stiffness is required, a closed section is preferred over the ordinary open section, such as the I-shape or channel. An excellent general discussion of the torsion phenomena with comparative behavior of open and closed sections is given by Tamberg and Mikluchin [8.29]. Some practical comments relating to closed sections in torsion are given by Siev [8.30]. The high torsional stiffness exhibited by closed sections makes them ideal for aircraft structural components and curved girders in bridges and buildings. This subject is treated in a number of textbooks [8.1–8.3] so that only a brief treatment follows.

In the closed section of Fig. 8.10.1, the walls are assumed thin so that the shearing stress may be assumed uniformly distributed across the thickness t. If the shear stress is τ, then τt is the shear force per unit distance along the wall, usually referred to as *shear flow*. Since only torsional stress is presently being considered, the normal stresses (σ_z of Fig. 8.10.1b) are zero. Since $\sigma_z = 0$, the shear flow τt cannot vary along the wall; i.e., τt is constant.

Referring to Fig. 8.10.1a, the increment of torsional moment contributed by each elements is

$$dT = \tau t \rho \, ds \qquad (8.10.1)$$

Integrating gives the full torsional moment, which is in effect the same as Eq. 8.2.5,

$$T = \tau t \int_s \rho \, ds \qquad (8.10.2)$$

Note that $\frac{1}{2}\rho \, ds$ is the cross-hatched area of the triangular segment in Fig. 8.10.1. Thus the integral

$$\int_s \rho \, ds = 2A \qquad (8.10.3)$$

where $A =$ area enclosed by the walls. Finally,

$$T = 2\tau t A \qquad (8.10.4)$$

If a cut is made in the wall of a closed thin-wall section (Fig. 8.10.2), a relative movement (as in Fig. 8.10.2b) will be produced between the two sides in the axial direction of the member. The unit shear strain along the perimeter is

$$\gamma = \tau/G \qquad (8.10.5)$$

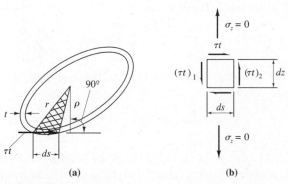

(a)   (b)

Figure 8.10.1 Shear flow in a closed thin-wall section.

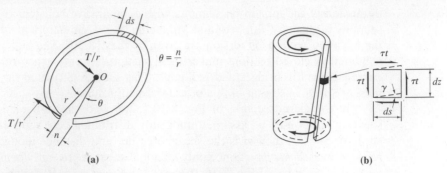

Figure 8.10.2   Forces on a cut thin-wall section.

The internal strain energy for any elemental length ds along the perimeter is

$$dW_i = \tfrac{1}{2}\tau t\gamma \, ds \qquad (8.10.6)$$

$$= \frac{1}{2}\left(\frac{t}{2A}\right)\frac{\tau}{G}\, ds \qquad (8.10.7)$$

The twisting moment T about point 0 can now be replaced by a couple, T/r. The external work done by the couple is

$$dW_e = \frac{1}{2}\left(\frac{T}{r}\right)n = \frac{T\theta}{2} \qquad (8.10.8)$$

Equating internal and external work per unit length gives

$$\frac{T\theta}{2} = \frac{T}{4AG}\int_s \tau \, ds \qquad (8.10.9)$$

$$\theta = \frac{\int_s \tau \, ds}{2AG} = \frac{\tau t \int_s ds/t}{2AG} \qquad (8.10.10)$$

since τt is a constant.

In order to obtain more useful forms of the equations, recall from Sec. 8.2 that

$$T = GJ\theta \qquad [8.2.5]$$

and using Eq. 8.10.10 gives

$$T = GJ\frac{\tau t \int ds/t}{2AG} \qquad (8.10.11)$$

and eliminating T between Eqs. 8.10.4 and 8.10.11 gives, when solving for the torsional constant J,

$$J = \frac{4A^2}{\int_s ds/t} \qquad (8.10.12)$$

The design of closed sections for bending and torsion is outside the scope of this text, and the reader is referred to the work of Felton and Dobbs [8.31]. Shermer [8.32] has shown applications of the closed thin-wall tube concept of torsion applied to trussed structures and structures containing plates and truss forms.

The development of the equations for torsion stiffness and stress in closed thin-wall sections may be developed alternatively using the "membrane analogy" developed by Prandtl [8.1–8.3].

The objective here is to illustrate the high degree of torsional stiffness that closed sections exhibit as compared to open ones.

EXAMPLE 8.10.1 _____

Compare the torsional resisting moment T and the torsional constant J for the sections of Fig. 8.10.3 all having about the same cross-sectional area. The maximum shear stress τ is 14 ksi.

Solution

(a) Circular thin-wall section. Using Eq. 8.10.4,

$$T = 2\tau t A = 2(14)(0.5)[\pi(10)^2/4]\tfrac{1}{12} = 91.6 \text{ ft-kips}$$

$$J = \frac{4A^2}{\displaystyle\int ds/t} = \frac{4(25\pi)^2}{20\pi} = 393 \text{ in.}^4$$

where $\int ds/t = 2\pi(5)/0.5 = 20\pi$.

(b) Rectangular box section.

$$T = 2\tau t A = 2(14)(0.5)(72)\tfrac{1}{12} = 84.0 \text{ ft-kips}$$

$$J = \frac{4A^2}{\displaystyle\int ds/t} = \frac{4(72)^2}{(36/0.5)} = 288 \text{ in.}^4$$

(c) Channel section. Since for this open section,

$$\tau = \frac{Tt}{J}$$

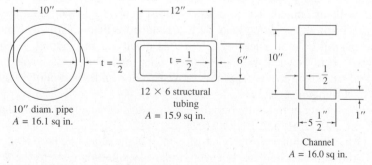

10″ diam. pipe
A = 16.1 sq in.

12 × 6 structural tubing
A = 15.9 sq in.

Channel
A = 16.0 sq in.

Figure 8.10.3   Sections for Example 8.10.1.

the maximum shear stress will be in the flange. Also,

$$J = \sum \tfrac{1}{3} bt^3$$

$$J = \tfrac{1}{3}[10(0.5)^3 + 2(5.5)(1)^3] = 4.1 \text{ in.}^4$$

$$T = \frac{J\tau}{t_f} = \frac{4.1(14)}{(1)(12)} = 4.8 \text{ ft-kips}$$

The circular section is best for torsional capacity, the rectangular box is next; these closed sections have the torsional constant J equal to 96 and 71 times that of the channel, respectively. The resisting moments are 19 and 18, respectively, times that of the channel.   ■■

8.11 TORSION IN SECTIONS WITH OPEN AND CLOSED PARTS

Generally this problem is treated by combining the principles discussed separately for open and closed parts. The procedure to be used for determining resisting moment, stiffness, and shear center location for such sections is presented with examples by Chu and Longinow [8.33]. The following is a summary of pertinent equations:

Total resisting moment is

$$T = \sum_{i=1}^{n} 2\tau_i t_i A_i + GJ\theta \tag{8.11.1}$$

where $J = \sum \tfrac{1}{3} bt^3$ for open parts only.

In addition, each of the closed cells must satisfy Eq. 8.10.14:

$$\int_s \tau_i t_i \frac{ds}{t_i} = 2GA_i\theta \tag{8.11.2}$$

8.12 TORSIONAL BUCKLING

Since the differential equation for torsion was developed earlier in this chapter, and buckling of axially loaded columns has previously been treated, torsional buckling may now be examined. The strength of most centrally loaded columns is reached at the tangent-modulus Euler load with a reduced efficiency if local buckling occurs before overall column buckling occurs, as discussed in Chapter 6. However, some thin-wall sections such as angles, tees, zees, and channels, having relatively low torsional stiffness may, under axial compression, buckle torsionally while the longitudinal axis remains straight.

The subject of torsional buckling is treated extensively by Timoshenko and Gere [6.67, pp. 225–250] and Bleich [6.9].

Using concepts previously developed, it is the objective here to show mathematically that such buckling can occur and identify situations where the designer should be cautious.

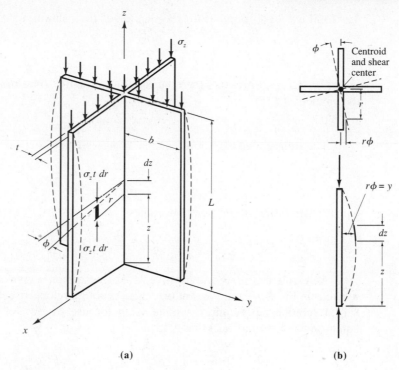

Figure 8.12.1 Torsional buckling.

Consider the doubly symmetrical section in the shape of a cross given in Fig. 8.12.1, whose shear center and centroid coincide. Recalling the Euler equation,

$$EI\frac{d^2y}{dz^2} + Py = 0 \qquad [6.2.3]$$

which differentiated twice becomes

$$EI\frac{d^4y}{dz^4} = -P\frac{d^2y}{dz^2} \qquad (8.12.1)$$

Since $EI\, d^4y/dz^4$ is the loading, the Euler column can be thought of as a beam laterally loaded with the fictitious loading $-P(d^2y/dz^2)$. Thus with the section put in the slightly buckled position (i.e., rotated the angle ϕ at distance z from the end) the compressive force $\sigma_z t\, dr$ on the element $dr\, dz$ is statically equivalent to a lateral load whose intensity per unit length is

$$-(\sigma_z t\, dr)\frac{d^2(r\phi)}{dz^2}$$

The increment of torsional moment about the z-axis tributary to the length dz equals the load times the moment arm r; thus

$$dm_z = -(\sigma_z tr\, dr)\frac{rd^2\phi}{dz^2}dz \qquad (8.12.2)$$

The total torsional moment for the slice dz of the column is

$$m_z = -\sigma_z \frac{d^2\phi}{dz^2} dz \int_A r^2 t \, dr \qquad (8.12.3)$$

Equation 8.12.3 represents the contribution to the torsional moment M_z tributary to the element dz:

$$dM_z = m_z \qquad (8.12.4)$$

The differential equation for torsion on I-shaped sections, Eq. 8.5.10, is

$$M_z = GJ\frac{d\phi}{dz} - EC_w\frac{d^3\phi}{dz^3} \qquad [8.5.10]$$

which when differentiated once becomes

$$\frac{dM_z}{dz} = GJ\frac{d^2\phi}{dz^2} - EC_w\frac{d^4\phi}{dz^4} \qquad (8.12.5)$$

Referring to Fig. 8.5.2, positive M_z at the section z gives a clockwise rotation; whereas in Fig. 8.12.1 at the section z there is counterclockwise rotation. Thus the Eq. 8.12.4 relationship requires a minus sign for use in Eq. 8.12.5, The differential equation for torsional buckling is then

$$\sigma_z \frac{d^2\phi}{dz^2} \int_A r^2 t \, dr = GJ\frac{d^2\phi}{dz^2} - EC_w\frac{d^4\phi}{dz^4}$$

or

$$EC_w\frac{d^4\phi}{dz^4} - \left(GJ - \sigma_z\int_A r^2 t \, dr\right)\frac{d^2\phi}{dz^2} = 0 \qquad (8.12.6)$$

in which $\int_A r^2 t \, dr = I_p$, the polar moment of inertia *about the shear center*. When the centroid coincides with the shear center, Eq. 8.12.6 alone determines the buckling condition.

Note is made that the warping rigidity EC_w is zero (text Appendix Table A2) for shapes consisting of thin rectangular elements intersecting at a common point.

For other cases, Eq. 8.12.6 may be written as

$$\frac{d^4\phi}{dz^4} + p^2\frac{d^2\phi}{dz^2} = 0 \qquad (8.12.7)$$

where

$$p^2 = \frac{\sigma_z I_p - GJ}{EC_w}$$

for which the general solution is

$$\phi = A_1 \sin pz + A_2 \cos pz + A_3 z + A_4 \qquad (8.12.8)$$

Considering the pin-end column, with rotation about z prevented at each end, but with warping not restricted at the ends gives in a manner similar to the Euler column derivation in Chapter 6, that

$$A_2 = A_3 = A_4 = 0$$

and since A_1 cannot also be zero,

$$\sin pL = 0, \quad pL = n\pi$$

The elastic buckling stress $\sigma_{z\,\text{critical}}$ at which torsional buckling occurs is

$$\frac{\pi^2}{L^2} = \frac{\sigma_z I_p - GJ}{EC_w}$$

$$\sigma_{z\,\text{critical}} = \left[\frac{\pi^2 EC_w}{L^2} + GJ\right]\frac{1}{I_p} = F_{ez} \qquad (8.12.9)$$

which is accurate for doubly symmetrical sections whose shear center and centroid coincide, such as I-shaped sections. The symbol F_{ez} is used instead of $\sigma_{z\,\text{critical}}$ in LRFD-Appendix E3. For LRFD use of Eq. 8.12.9, $I_p = I_x + I_y$ and L is the effective length $K_2 L$. For the common single-angle strut, since the distance from centroid to shear center is small, Eq. 8.12.9 will provide a reasonable approximation for the torsional buckling stress. Expressions for the warping constant C_w and the torsion constant J for various shapes are to be found in text Appendix Table A2. Though not in that table, values of the warping constant C_w for a combination W-section with a channel cap (see *LRFD Manual* "COMBINATION SECTIONS" p. 1–105) have been given by Lue and Ellifritt [8.34].

The reader should not lose sight of the fact that the most probable buckling mode is still that occurring at the tangent-modulus Euler load because of lateral bending about the x- or y-axis. Thus the problem involves three critical values of axial load; bending about either principal axis and twisting about the longitudinal axis. On wide-flange sections, torsional buckling may be important on sections with extra wide flanges and short lengths [6.67, pp. 225–250].

In the general case where the shear center does not coincide with the centroid, the buckling failure is actually a combination of torsion and flexure. For this case, the three differential equations, (1) buckling by lateral bending about the x-axis; (2) buckling by lateral bending about the y-axis; and (3) twisting about the shear center, are interdependent. Thus three simultaneous differential equations must be solved to get the buckling loads. The development and solution of the these equations is outside the scope of this text and is adequately treated elsewhere [6.9; 6.67, pp. 225–250].

LRFD Design for Torsional and Flexural-Torsional Buckling

For design, LRFD-E2 implies that the nominal strength of compression members not exceeding the λ_r limits in LRFD-B5.1 (to preclude local buckling prior to reaching the yields stress F_y) can be computed *without considering flexural-torsional buckling*.

When the λ_r limits of LRFD-B5.1 are exceeded, double angle and tee-shaped compression members shall be designed to include the flexural-torsional buckling limit state according to LRFD-E3. Other thin, doubly symmetric sections, such as cruciform and built-up sections, as well as all singly symmetric and unsymmetric columns, must be designed for the limit states of flexural-torsional and torsional buckling in accordance with LRFD-Appendix E3.

For doubly symmetric sections (such as built-up I-sections having thin elements), when the flexural-torsional limit state is evaluated, an equivalent radius of gyration r_E can be compared with r_x and r_y to reduce the computations. The alternative is to compute the elastic flexural-torsional buckling stress F_e from Eq. 8.12.9 [identical LRFD Formula (A-E3–5)] and compare with F_{cr} computed using the larger of $K_x L_x/r_x$ and $K_y L_y/r_y$ in the column formulas of LRFD-E3 (also text Chapter 6). To develop the r_E equation, set Eq. 8.12.9 equal to the Euler equation,

$$\frac{\pi^2 E}{(L/r_E)^2} = \frac{EC_w \pi^2}{I_p L^2} + \frac{GJ}{I_p}$$

$$r_E = \sqrt{\frac{C_w}{I_p} + \frac{GJL^2}{EI_p \pi^2}}$$

which for steel with $E/G = 2.6$ gives

$$r_E = \sqrt{\frac{C_w}{I_p} + 0.04 \frac{JL^2}{I_p}} \qquad (8.12.10)$$

for doubly symmetrical sections. It has been demonstrated that only for short lengths will r_E be lower than r_x and r_y for W shapes [6.9].

Singly symmetric sections may buckle in a combination flexural-torsional mode, which will depend on the Euler column buckling stress F_{ey} for axis of symmetry, and the torsional buckling stress F_{ez} (Eq. 8.12.9). LRFD-Appendix E3 gives the elastic buckling stress F_e for the combined mode as

$$F_e = \frac{F_{ey} + F_{ez}}{2H} \left(1 - \sqrt{1 - \frac{4F_{ey} F_{ez} H}{(F_{ey} + F_{ez})^2}}\right) \qquad (8.12.11)$$

where

$$F_{ey} = \frac{\pi^2 E}{(K_y L/r_y)^2} \qquad (8.12.12)$$

$$F_{ez} = \left(\frac{\pi^2 EC_w}{(K_z L)^2} + GJ\right)\frac{1}{I_p} \qquad (8.12.13)$$

$$H = 1 - \left(\frac{x_0^2 + y_0^2}{\bar{r}_0^2}\right) = \frac{I_x + I_y}{I_p} \qquad (8.12.14)$$

$$I_p = A\bar{r}_0^2 = I_x + I_y + A(x_0^2 + y_0^2) \qquad (8.12.15)$$

E = tension-compression modulus of elasticity, ksi

G = shear modulus of elasticity, ksi

C_w = torsional warping constant, in.^{6}

J = torsion constant, in.^{4}

I_x, I_y = moment of inertia about principal axes, x and y

I_p = polar moment of inertia, Eq. 8.12.15, in.^{4}

K_y, K_z = effective length factors in the y-direction, and for torsional buckling (z-axis)

x_0, y_0 = coordinates of shear center with respect to centroid of section, in.

r_y = radius of gyration about the *axis of symmetry*

r_0 = polar radius of gyration about the shear center

EXAMPLE 8.12.1

For the sections given in Fig. 8.12.2, determine under what conditions, if any, torsional or flexural-torsional buckling is likely to occur under axial compression loading.

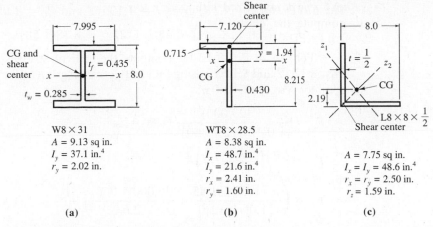

Figure 8.12.2 Sections for Example 8.12.1.

Assume the members pinned at the ends of the unbraced lengths, and free to warp at the ends, fully recognizing that these two assumptions minimize buckling strength.

Solution

(a) W8×31. Since the centroid and shear center coincide, use Eq. 8.12.10 to get the equivalent r_E:

$$J = \tfrac{1}{3}[2(7.995)(0.435)^3 + 7.13(0.285)^3] = 0.494 \text{ in.}^4$$

$$(LRFD \text{ } Manual, J = 0.54 \text{ in.}^4)$$

$$C_w = h^2 I_y/4 = (8.00 - 0.435)^2(37.1)/4 = 531 \text{ in.}^6$$

$$(LRFD \text{ } Manual, C_w = 530 \text{ in.}^6)$$

$$I_p = I_x + I_y = 110 + 37.1 = 147.1 \text{ in.}^4$$

$$r_E = \sqrt{\frac{C_w}{I_p} + 0.04\frac{JL^2}{I_p}} = \sqrt{\frac{531}{147.1} + 0.04\frac{0.494L^2(144)}{147.1}}$$

$$= \sqrt{3.61 + 0.0193L^2}$$

where L is the unsupported length, in feet. Only when L is less than 4.9 ft does r_y exceed r_E; only for a very short column is torsional buckling a possibility, and even when r_E approaches its minimum value (when $L = 0$), which is $r_E(\text{min}) = 1.90$ in., it is only 6% less than r_y.

The result using this section is typical of standard W and S shapes and indicates torsional buckling may generally be neglected for them.

(b) WT8×28.5. The centroid and shear center do not coincide, but the section has one axis of symmetry; use Eq. 8.12.11:

$$J = \tfrac{1}{3}[(7.12)(0.715)^3 + 7.50(0.30)^3] = 1.066 \text{ in.}^3$$

Using the formula from text Appendix Table A2,

$$C_w = \frac{1}{36}\left[\frac{(7.120)^3(0.715)^3}{4} + (7.50)^3(0.430)^3\right] = 1.85 \text{ in.}^6$$

The *LRFD Manual* [1.18] gives $J = 1.10$ in.^{3} and $C_w = 1.99$ in.^{6}

$$I_p = I_x + I_y + Ay_0^2 = 48.7 + 21.6 + 8.38(1.58)^2 = 91.2 \text{ in.}^4$$

For use in Eq. 8.12.11, compute the critical stresses F_{ey} and F_{ez} from Eqs. 8.12.12 and 8.12.13 using for an effective length KL (for y- and z-axes) the common length of 10 ft:

$$F_{ey} = \frac{\pi^2(29,000)}{[1.0(10)(12)/1.60]^2} = 50.9 \text{ ksi}$$

$$F_{ez} = \left(\frac{\pi^2 E C_w}{(K_z L)^2} + GJ\right)\frac{1}{I_p} \qquad\qquad [8.12.13]$$

$$F_{ez} = \left(\frac{\pi^2(29,000)1.85}{[1.0(10)12]^2} + \frac{29,000(1.066)}{2.6}\right)\frac{1}{91.2} = 131 \text{ ksi}$$

The distance y_0 from the centroid of the section to the shear center at the junction of the mid-thicknesses of the flange and the web is

$$y_0 = y - t_f/2 = 1.94 - 0.715/2 = 1.58 \text{ in.}$$
$$x_0 = 0$$
$$I_p = 91.2 \text{ in.}^4 \text{ (computed above)}$$
$$H = \frac{I_x + I_y}{I_p} = \frac{48.7 + 21.6}{91.2} = 0.77$$

Then using Eq. 8.12.11,

$$F_e = \frac{50.9 + 131}{2(0.77)}\left(1 - \sqrt{1 - \frac{4(50.9)(131)0.77}{(50.9 + 131)^2}}\right) = 45.3 \text{ ksi}$$

Equating F_e to Euler's formula, an equivalent radius r_E of gyration for flexural-torsional buckling can be obtained to compare with r_x and r_y,

$$45.3 = \frac{\pi^2 E}{(KL/r_E)^2} = \frac{\pi^2 29,000}{(120/r_E)^2}$$

$$r_E = 1.51 \text{ in.}$$

In this case r_E is less than $r_y = 1.60$ in. and flexural-torsional buckling is critical for ordinary lengths. The same situation will occur with double angle compression members. An equal leg angle (such as in Fig. 8.12.2c) will have z_2 as its axis of symmetry, and the radius of gyration with respect to that axis must be used for r_y in Eq. 8.12.12. ■■

Unsymmetric sections in compression are more complicated and their treatment is outside the scope of this text. Particularly, the single-angle strut is controlled by flexural-torsional buckling; it is for this reason that caution concerning its use is given in the *ASD Manual* [1.7] and the special section is provided in the *LRFD Manual* [1.18, p. 3–104]. AISC has provided a special publication [6.83] relating to the strength of single-angle members. See Sec. 6.8 for comments and references related to single-angle compression members. Note is made that the larger are the width-to-thickness ratios, the greater the possibility of torsional or flexural-torsional buckling being the controlling limit state.

As a conclusion to this treatment, the designer is cautioned about using open sections (torsionally weak sections) in compression having less than two axes of symmetry, particularly when high width/thickness ratios exist for the elements. The width/thickness ratio limits (λ_r in LRFD-B5 and the "noncompact" limits of ASD-B5) for control of local buckling, if not exceeded, provide some control since local buckling of sections such as angles, flanges, and tees is closely related to torsional buckling.

SELECTED REFERENCES

8.1. Arthur P. Boresi, Richard J. Schmidt, and Omar M. Sidebottom. *Advanced Mechanics of Materials,* 5th ed. New York: John Wiley and Sons, Inc., 1993, Chap. 6.

8.2. S. Timoshenko. *Strength of Materials,* Part II, 2nd ed. New York: D. Van Nostrand Company, Inc., 1941 Chap. 6.

8.3. William McGuire. *Steel Structures.* Englewood Cliffs, NJ: Prentice-Hall, Inc., 1968, pp. 346–400.

8.4. Inge Lyse and Bruce G. Johnston. "Structural Beams in Torsion," *Transactions,* ASCE, **101** (1936), 878–926 (includes Discussions).

8.5. F. K. Chang and Bruce G. Johnston. "Torsion of Plate Girders," *Transactions,* ASCE, **118** (1953), 337–396.

8.6. Gerald G. Kubo, Bruce G. Johnston, and William J. Eney. "Nonuniform Torsion of Plate Girders," *Transactions,* ASCE, **121** (1956), 759–785. (Good summary of torsion theory.)

8.7. I. A. El Darwish and Bruce G. Johnston. "Torsion of Structural Shapes," *Journal of the Structural Division,* ASCE, **91,** ST1 (February 1965), 203–227. Errata: **92,** ST1 (February 1966), 471. See also *Transactions,* ASCE, **131** (1966), 428–429 for summary of equations.

8.8. AISC. *Torsion Analysis of Steel Members.* Chicago, IL: American Institute of Steel Construction, 1983.

8.9. John G. Hotchkiss. "Torsion of Rolled Steel Sections in Building Structures," *Engineering Journal,* AISC, **3,** 1 (January 1966), 19–45.

8.10. Conrad P. Heins, Jr. and John T. C. Kuo. "Torsional Properties of Composite-Girders," *Engineering Journal,* AISC, **9,** 2 (April 1972), 79–85.

8.11. S. Timoshenko. "Theory of Bending, Torsion, and Buckling of Thin-Walled Members of Open Cross-Section," *J. Franklin Inst.,* **239,** 3, 4, and 5 (1945), 201–219, 249–268, and 343–361.

8.12. John E. Goldberg. "Torsion of I-Type and H-Type Beams," *Transactions,* ASCE, **118** (1953), 771–793.

8.13. Kuang-Han Chu and Robert B. Johnson. "Torsion in Beams with Open Sections," *Journal of the Structural Division,* ASCE, **100,** ST7 (July 1974), 1397–1419.

8.14. Bruce G. Johnston. "Design of W-Shapes for Combined Bending and Torsion," *Engineering Journal,* AISC, **19,** 2 (Second Quarter 1982), 65–85.

8.15. Philip H. Lin. "Simplified Design for Torsional Loading of Rolled Steel Members," *Engineering Journal,* AISC, **14,** 3 (Third Quarter 1977), 98–107.

8.16. Phil H. Lin. Discussion of "Design of W-Shapes for Combined Bending and Torsion," by Bruce G. Johnston, *Engineering Journal,* AISC, **20,** 2 (Second Quarter 1983), 82–87.

8.17. J. N. Goodier and M. V. Barton. "The Effects of Web Deformation on the Torsion of I-Beams," *J. Appl. Mech.,* March 1944, p. A-35.

8.18. J. E. Lothers. "Torsion in Steel Spandrel Girders," *Transactions,* ASCE, **112** (1947), 345–376.

8.19. Min-Tse Chen and Donald E. Jolissaint, Jr. "Pure and Warping Torsion Analysis of Rigid Frames," *Journal of Structural Engineering,* ASCE, **109,** 8 (August 1983), 1999–2003.

8.20. Morris Ojalvo. Discussion of "Warping and Distortion at I-Section Joints," by P. Vacharajittiphan and N. S. Trahair, *Journal of the Structural Division,* ASCE, **101,** ST1 (January 1975), 343–345.

8.21. Conrad P. Heins and Robert A. Potocko. "Torsional Stiffening of I-Girder Webs," *Journal of the Structural Division,* ASCE, **105,** ST8 (August 1979), 1689–1698.

8.22. Morris Ojalvo. Discussion of "Torsional Stiffening of I-Girder Webs," by C. P. Heins and R. A. Potocko, *Journal of the Structural Division,* ASCE, **106,** ST4 (April 1980), 939.

8.23. Porpan Vacharajittiphan and Nicholas S. Trahair. "Warping and Distortion at I-Section Joints," *Journal of the Structural Division,* ASCE, **100,** ST3(March 1974), 547–564.

8.24. Charles G. Salmon. Discussion of "Design of W-Shapes for Combined Bending and Torsion," by Bruce G. Johnston, *Engineering Journal,* AISC, **19,** 4 (Fourth Quarter 1982), 215–216.

8.25. Bruce G. Johnston, F. J. Lin, and T. V. Galambos. *Basic Steel Design,* 3rd ed. Englewood Cliffs, NJ: Prentice-Hall, Inc., 1986.

8.26. Warner Lansing. "Thin-Walled Members in Combined Torsion and Flexure," *Transactions,* ASCE, **118** (1953), 128–146. (Particular emphasis on channel and zee sections.)

8.27. Donald R. Evick and Conrad P. Heins, Jr. "Torsion of Nonprismatic Beams of Open Section," *Journal of the Structural Division,* ASCE, **98,** ST12 (December 1972), 2769–2784.

8.28. Basil Sourochnikoff. "Strength of I-Beams in Combined Bending and Torsion," *Transactions,* ASCE, **116** (1951), 1319–1342.

8.29. K. G. Tamberg and P. T. Mikluchin, "Torsional Phenomena Analysis and Concrete Structure Design," *Analysis of Structural Systems for Torsion,* SP-35, American Concrete Institute, 1973, 1–102.

8.30. Avinadav Siev. "Torsion in Closed Sections," *Engineering Journal,* AISC, **3,** 1 (January 1966), 46–54.

8.31. Lewis P. Felton and M. W. Dobbs. "Optimum Design of Tubes for Bending and Torsion," *Journal of the Structural Division,* ASCE, **93,** ST4 (August 1967), 185–200.

8.32. Carl L. Shermer. "Torsional Strength and Stiffness of Steel Structures," *Engineering Journal,* AISC, **17,** 2 (Second Quarter 1980), 33–37.

8.33. Kuang-Han Chu and Anatole Longinow. "Torsion in Sections with Open and Closed Parts," *Journal of the Structural Division,* ASCE, **93,** ST6 (December 1967), 213–227.

8.34. Tony Lue and Duane S. Ellifritt. "The Warping Constant for the W-Section with a Channel Cap," *Engineering Journal,* AISC, **30,** 1 (First Quarter 1993), 31–33.

8.35. Robert G. Driver and D. J. Laurie Kennedy. "Combined Flexure and Torsion of I-shaped Steel Beams," *Canadian Journal of Civil Engineering,* **16** (1989), 124–139.

PROBLEMS

All design problems are to be done according to the AISC Load and Resistance Factor Design or Allowable Stress Design, as indicated by the instructor. All given loads are service loads unless othewise indicated. For all problems *assume adequate lateral support of the compression flange such that lateral stability does not control.* Assume all standard sections are equally readily available in the indicated grade of steel (even though actually they are not). A figure showing span and loading is required.

8.1. For the channel shown in the accompanying figure, separately apply V_x and V_y through the centroid of the section. For each shear compute and draw to scale the shear flow τt distribution along each of the elements of the cross-section. On the two separate diagrams (one for V_x and one for V_y) of shear flow distribution compute the total shear force in each element of the cross-section in terms of the applied shear V_x or V_y. Using these computed shear forces calculate the two coordinates of the shear center.

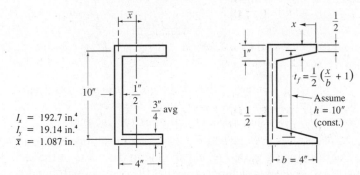

$I_x = 192.7$ in.^{4}
$I_y = 19.14$ in.^{4}
$\bar{x} = 1.087$ in.

Problem 8.1   Problem 8.2

8.2. Repeat the requirements of Prob. 8.1 for the channel with sloping flanges. Comment on the effect of using average thickness instead of the actual sloping flanges for determining shear center on standard rolled channels.

8.3. Repeat the requirements of Prob. 8.1 for the section of the accompanying figure.

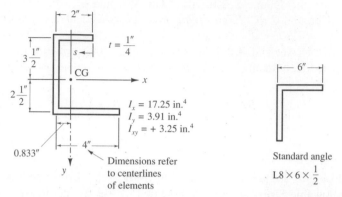

$I_x = 17.25$ in.^{4}
$I_y = 3.91$ in.^{4}
$I_{xy} = + 3.25$ in.^{4}

Dimensions refer to centerlines of elements

Standard angle

$L8 \times 6 \times \dfrac{1}{2}$

Problem 8.3   **Problem 8.4**

8.4. Repeat the requirements of Prob. 8.1 for the section of the accompanying figure.

8.5. Repeat the requirements of Prob. 8.1 for the section of the accompanying figure.

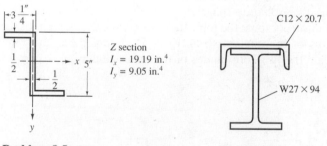

Z section
$I_x = 19.19$ in.^{4}
$I_y = 9.05$ in.^{4}

C12 × 20.7

W27 × 94

Problem 8.5   **Problem 8.6**

8.6. Locate the shear center for the combined W and channel crane girder section. Is there significant error in assuming the shear center lies at the centroid? Use average thickness and constant depth for the channel.

8.7. An MC18×58 is to be used on a 24-ft simply supported span to carry a total load of 0.8 kips/ft, with the load applied in the plane of the web. Suppose the flanges are to have attachments so that the channel will bend vertically about the x-axis. What lateral moment capacity M_y should the attachments be capable of resisting? What percent of M_x does this represent?

8.8. **a.** Develop the torsion differential equation solution for the W section having torsionally fixed ends and an eccentrically applied concentrated load at midspan.

b. Compute the torsion constant J, the warping constant C_w, and λ (i.e., $1/a$); then use them in parts (c) through (f).

c. Compute the combined bending stress, including warping torsion and ordinary flexure components, at $z = 0$, $0.3L$, and $0.5L$ unless otherwise instructed.

d. Compute the maximum shear stress in the web, including Saint-Venant torsional shear and flexural shear, at the same locations indicated in (c).

e. Compute the flange shear force V_f due to warping restraint at the same locations indicated in (c).

f. Compute the maximum shear stress in the flange, including Saint-Venant torsion, warping shear, and vertical flexure flange shear, at the same locations indicated in (c).

g. Give a tabular summary of all stresses.

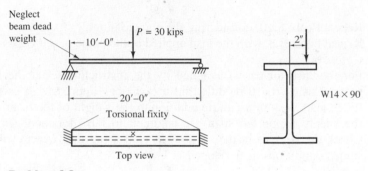

Problem 8.8

8.9. Develop the torsion differential equation solution for the cantilever beam with an eccentric concentrated load at its end, and compute constants and stresses as given in items (b) through (e) of Prob. 8.8. Is there any relationship between this problem and Prob. 8.8?

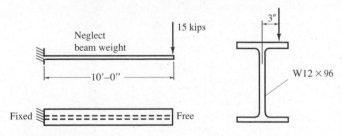

Problem 8.9

8.10. Develop the torsion differential equation solution for the uniformly loaded beam with loading applied eccentrically to the web. Consider the ends torsionally simply supported. Compute constants and stresses as given in items (b) through (e) of Prob. 8.8.

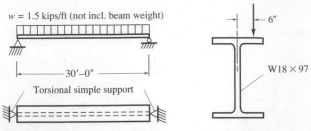

Problem 8.10

8.11. Repeat Prob. 8.10, considering the ends torsionally fixed.

8.12. Repeat Prob. 8.8, with the load applied in the plane of the y-axis of a channel, C15×50.

8.13. For the case (or cases) assigned by the instructor, select the lightest W14 section, using the β modified flexure analogy approach, to carry a concentrated load W at midspan, in addition to the weight of the beam. The ends of the simply supported span are assumed to have torsional simple support. Check the stresses in the section selected using the "exact" solution for assumed conditions as developed in Example 8.5.2.

Case	W_D Dead load (kips)	W_L Live load (kips)	Span length (ft)	F_y Steel yield stress (ksi)	e Eccentricity of loading (in.)
1	5	20	20	36	2
2	5	20	20	36	3
3	5	20	20	36	4
4	5	20	24	36	2
5	5	20	24	36	3
6	5	20	24	36	4
7	7	22	24	50	2
8	7	22	24	50	3
9	7	22	24	50	4
10	10	15	26	50	2
11	10	15	26	50	3
12	10	15	26	50	4
13	10	15	26	50	5
14	10	15	26	50	6
15	10	15	26	50	7

8.14. For the case (or cases) assigned by the instructor, select the lightest W section, using the β modified flexure analogy approach, to carry uniform loading w, in addition to the weight of the beam. The ends of the span are assumed to be *fixed* for both flexure and torsion. If "exact" solution was obtained in Prob. 8.11, check maximum stress using that solution.

Case	w_D Dead load (kips/ft)	w_L Live load (kips/ft)	Span length (ft)	Steel yield stress (ksi)	Eccentricity of loading (in.)
1	0.5	1.5	28	36	7
2	0.5	1.5	28	36	5
3	0.5	1.5	28	36	3
4	0.5	1.5	28	36	2
5	0.35	1.4	26	50	7
6	0.35	1.4	26	50	5
7	0.35	1.4	26	50	3
8	0.35	1.4	26	50	2

8.15. For the case (or cases) listed in Prob. 8.14 and assigned by the instructor, select the lightest W section, using the β modified flexure analogy approach, to carry uniform loading w, in addition to the weight of the beam. The ends of the span are *simply supported* for both flexure and torsion. If the "exact" solution is available (solution to Prob 8.11), check stresses using that solution.

8.16. For the case (or cases) listed in Prob. 8.14 and assigned by the instructor, compare the solution using A36 and A572 Grade 50 steels. Assume the Grade 50 costs 7% more per pound than A36. Indicate the economical solution to the problem. Assume ends simply supported for flexure but fixed for torsion.

8.17. Given the 40-ft simply supported span carrying two symmetrically placed concentrated loads, 20% dead load and 80% live load, as shown in the accompanying figure. If the loads are eccentric to the web by 5 in., and the member is torsionally simply supported, select the lightest W14 section suitable using the β modified flexure analogy method.

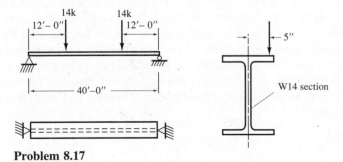

Problem 8.17

8.18. The 30-ft simply supported (for M_x) span is to carry two symmetrically placed concentrated loads of 8 kips dead load and 14 kips live load located 10 ft from the supports. The loads are 6-in. eccentric to the web, and full fixity is assumed for torsional restraint. Select the lightest W section using the β modified flexure analogy method.

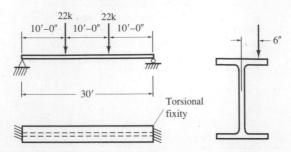

Problem 8.18

8.19. Estimate, using the modified flexure analogy design procedure used for W shapes and crane girders, the uniformly distributed load capacity, 20% dead load and 80% live load, for a C15×50 of A36 steel. The ends of the beam are restrained against rotation ($\phi = 0$), but the ends are free to warp. Check stresses using the differential equation solution.

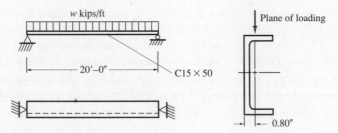

Problem 8.19

8.20. A simply supported beam is to carry 0.2 kip/ft dead load and 0.6 kip/ft live load on a span of 24 ft. A channel section is to be used, and since it is not laterally restrained, torsion is to be considered in the design. Assume the member is torsionally simply supported. After computing the torsional moment, use the generally accepted approximate flexure analogy to make a selection of a channel section from the AISC Manual. Investigate the stress on the selected section using the exact solution based on the stated loading and support condition. Use A36 steel.

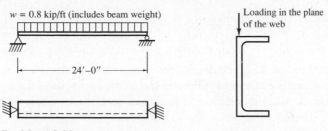

Problem 8.20

8.21. Assume a single W section is to serve as a crane runway girder which carries a vertical loading, as shown. In addition, design must include an axial compressive force of 10 kips and a horizontal force of 3 kips on each wheel applied at $4\frac{1}{4}$ in. above the top of the compression flange. Assume torsional simple support at the ends of the beam. Select the lightest W14 section of A36 steel using the β modified flexure analogy approach. *Note:* All loads except weight of the crane runway girder are live loads.

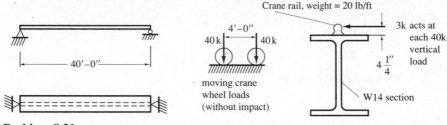

Problem 8.21

8.22. Compare the torsional constants J for each of the sections of the accompanying figure. If the maximum shear stress is 14 ksi, compute the torsional moment capacity of each section. Can this be done for the W30×99? Explain.

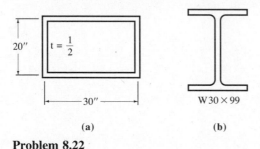

Problem 8.22

8.23. Investigate the possibility of torsional buckling occurring on the following beams: (a) W16×31; (b) W6×16; (c) M8×6.5. For each, plot equivalent slenderness ratio r_E against length of column. Give conclusions.

8.24. Estimate the buckling load, assuming zero residual stress and elastic buckling so that Euler's equation applies, on the following sections for a pin-end length of 10 ft. At what length would torsional buckling be likely to control?
  a. MC10×6.5
  b. L4×4×$\frac{1}{4}$
  c. WT7×15
  d. Zee section shown

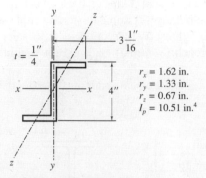

$r_x = 1.62$ in.
$r_y = 1.33$ in.
$r_z = 0.67$ in.
$I_p = 10.51$ in.^{4}

Problem 8.24, Case (d)

Chapter 9

Lateral-Torsional Buckling of Beams

9.1 RATIONAL ANALOGY TO PURE COLUMNS

Emphasis in this chapter is on the lateral stability considerations associated with bending about the strong axis of the section. In beams, as in axially loaded columns, it is not possible to achieve perfect loading, i.e., beams are never perfectly straight, not perfectly homogeneous, and are usually not loaded in exactly the plane that is assumed for design and analysis.

Consider the compression zone of the laterally unsupported beam of Fig. 9.1.1. With the loading in the plane of the web, according to ordinary beam theory, points *A* and *B* are equally stressed. Imperfections in the beam and accidental eccentricity in loading actually result in different stresses at *A* and *B*. Furthermore, residual stresses as discussed in Chapter 6 contribute to unequal stresses across the flange width at any distance from the neutral axis.

In a qualitative way one may look upon the compression flange of a beam as a column, with all the considerations treated in chapter 6. The rectangular flange as a column would ordinarily buckle in its weak direction, by bending about an axis such

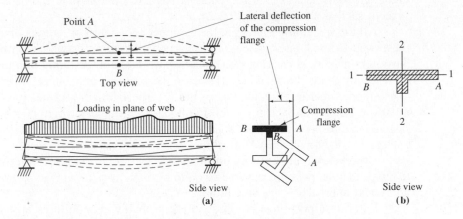

Figure 9.1.1 Beam laterally supported only at its ends.

as 1–1 of Fig. 9.1.1b, but the web provides continuous support to prevent such as buckling. At higher compressive loads the rectangular flange will tend to buckle by bending about axis 2–2 of Fig. 9.1.1b. It is this sudden buckling of the flange about its strong axis in a lateral direction that is commonly referred to as lateral buckling. The analogy between the compression flange of a beam and a column is intended to present only the general behavior for lateral buckling.

In order to evaluate this behavior more precisely, one must realize that the compression flange is not only braced in its weak direction by its attachment via the web to the stable tension flange, but the web also provides continous restraint (rotational and transverse) along the junction of the flange and web (assuming the web and flange plates are continuously attached). Thus the bending stiffness of the web brings the entire section into action when lateral motion commences.

9.2 LATERAL SUPPORT

Rarely does a beam exist with its compression flange entirely free of all restraint. Even when it does not have a positive connection to a floor or roof system, there is still friction between the beam flange and whatever it supports. There are two categories of lateral support that are definite and adequate; these are:

1. Continuous lateral support by embedment of the compression flange in a concrete floor slab (Fig. 9.2.1a and b).
2. Lateral support at intervals (Fig. 9.2.1c through g) provided by cross beams, cross frames, ties, or struts, framing in laterally, where the lateral system is itself adequately stiff and braced.

It is necessary to examine not only the individual beam for adequate bracing, but also the entire system. Figure 9.2.2a shows beam *AB* with a cross beam framing in at midlenght, but buckling of the entire system is still possible unless the system is braced, such as shown in Fig. 9.2.2b.

All too frequently in design, the engineer encounters situations that are none of these well-defined cases. A common unknown situation occurs when heavy beams have light-gage steel decking resistance welded to them; certainly providing a degree of restraint along the member. However the stiffness and lateral strength may be questioned. Other questionable cases are (a) where bracing frames into the beam in question, but it is at or near the tension flange; (b) timber or light-gage decking floor systems that rest on but are not solidly attached to the beams; (c) rigid frames that are enclosed in light-gage metal sheathing.

However, it is better to assume no lateral support in doubtful situations. Alternatively, it may be possible in some cases to evaluate it as an elastic restraint. Such an analytical approach is discussed in Sec. 9.13.

Lateral support must not be ignored; probably most failures in steel structures are the result of inadequate bracing against lateral instability of some type. The engineer is also reminded to consider carefully the construction stage when all of the restraints which may eventually act are not yet in place.

Large girder with stud shear connectors on top flange to be embedded in concrete to provide continuous lateral support (see Fig. 9.2.1b). (Photo by C. G. Salmon)

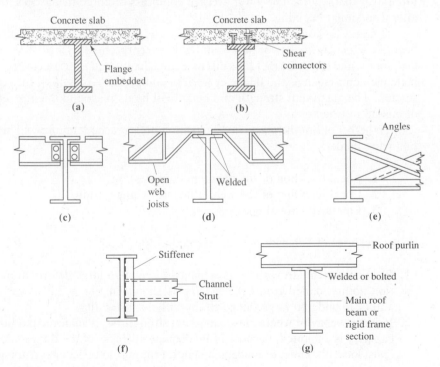

Figure 9.2.1 Types of definite lateral support.

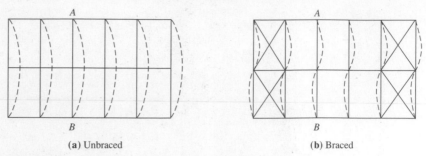

Figure 9.2.2   Lateral buckling of a roof or floor system.

9.3 STRENGTH OF I-SHAPED BEAMS UNDER UNIFORM MOMENT

In the development of design equations. the case of constant (uniform) moment along a laterally unbraced length is usually used as the basic case for lateral-torsional buckling. Using the analogy of the compression flange as a column, the uniform moment causes constant compression in one flange over the entire unbraced length. When there is a moment gradient (variation in moment) the compression force in the flange varies along the unbraced length, resulting in a lower average compression force over that length. The lower average compression force means less likelihood of lateral-torsional buckling.

Lateral-torsional buckling is a limit state that may control the strength of a beam. The general behavior of a beam may be represented by Fig. 9.3.1 from Yura, Galambos, and Ravindra [9.1]. As discussed in Chapter 6, Part II, local buckling of the plate elements (such as the flange or web) in compression may limit the strength of a section. The maximum strength of a beam will be its plastic moment strength M_p as discussed in Chapter 7.

Whether or not the plastic moment strength is reached, failure will be one of the following modes:

1. Local buckling of the flange in compression.
2. Local buckling of the web in flexural compression.
3. Lateral-torsional buckling.

Four categories of behavior are represented in Fig. 9.3.1:

1. Plastic moment strength M_p is achieved *along with* large deformation. Deformation ability, called *rotation capacity* as shown in Fig. 9.3.2, is essentially the ability to undergo large flange strain without instability.
2. Inelastic behavior where plastic moment strength M_p is achieved but little rotation capacity is exhibited, because of inadequate stiffness of the flange and/or web to resist local buckling, or inadequate lateral support to resist lateral-torsional buckling, while the flange is inelastic.

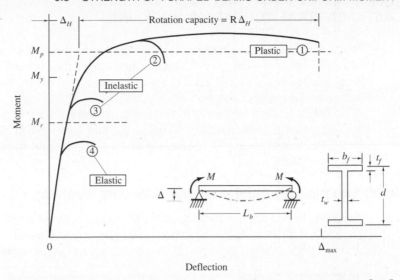

Figure 9.3.1   Beam behavior (From Yura, Galambos, and Ravindra [9.1])

3. Inelastic behavior where the moment strength M_r, the moment above which residual stresses cause inelastic behavior to begin, is reached or exceeded; however, local buckling of the flange or web, or lateral-torsional buckling prevent achieving the plastic moment strength M_p.
4. Elastic behavior where moment strength M_{cr} is controlled by elastic buckling; any or all of local flange buckling, local web buckling, or lateral-torsional buckling.

Most rolled W shapes have low enough slenderness ratios ($b_f/2t_f$ for flange and h/t_w or d/t_w for the web) such that they are categorized as "compact" as discussed in Chapter 7. For such cases, achievement of the plastic moment M_p depends on the laterally unbraced length L_b. A long laterally unbraced length will indicate moment strength M_{cr} controlled by *elastic* lateral-torsional buckling.

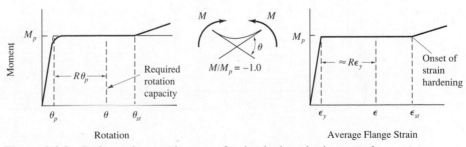

Figure 9.3.2   Deformation requirements for developing plastic strength.

9.4 ELASTIC LATERAL-TORSIONAL BUCKLING

Because structural design requires a thorough understanding of stability, the following development is presented of the basic equation for the elastic lateral-torsional buckling strength M_{cr} of an I-shaped beam under the action of constant (uniform) moment acting in the plane of the web.

More detailed treatment of this uniform moment case as well as other common loading cases is to be found in Timoshenko and Gere [6.67], Bleich [6.9], de Vries [9.2], Hechtman et al. [9.3], Austin et al. [9.4], Clark and Jombock [9.5], Salvadori [9.6], and Galambos [9.7].

Differential Equation for Elastic Lateral-Torsional Buckling

Referring to Fig. 9.4.1, which shows the beam in a buckled position, it is observed that the applied moment M_0 in the yz plane will give rise to moment components $M_{x'}$, $M_{y'}$, and $M_{z'}$, about the x'-, y'-, and z'-axes, respectively. This means there will be bend-

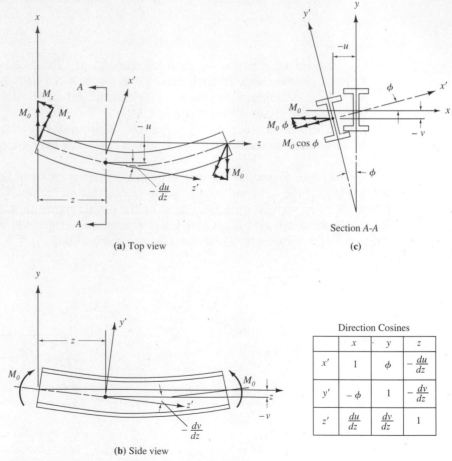

(a) Top view

(c) Section A-A

(b) Side view

Direction Cosines

	x	y	z
x'	1	ϕ	$-\dfrac{du}{dz}$
y'	$-\phi$	1	$-\dfrac{dv}{dz}$
z'	$\dfrac{du}{dz}$	$\dfrac{dv}{dz}$	1

Figure 9.4.1   I-shaped beam in slightly buckled position.

ing curvature in both the $x'z'$ and $y'z'$ planes as well as torsional curvature about the z'-axis. Assuming small deformation, the bending in the $y'z'$ plane (considering the direction cosine is 1 between y'- and y-, and z'- and z-axes) may be written

$$EI_x \frac{d^2v}{dz^2} = M_{x'} = M_0 \tag{9.4.1}$$

where v is the displacement of the centroid in the y direction (see Fig. 9.4.1b).

Also, the curvature in the $x'z'$ plane is

$$EI_y \frac{d^2u}{dz^2} = M_{y'} = M_0\phi \tag{9.4.2}$$

as is seen from Fig. 9.4.1c, where u is the displacement of the centroid in the x direction.

The differential equation for torsion of I-shaped beams was developed in Chapter 8 as Eq. 8.5.10, as follows:

$$M_{z'} = GJ \frac{d\phi}{dz} - EC_w \frac{d^3\phi}{dz^3} \tag{8.5.10}$$

From Fig. 9.4.1 and the direction cosines, the torsional component of M_0 when the beam is slightly buckled is proportional to the slope of the beam in the xz plane:

$$M_{z'} = -\frac{du}{dz} M_0 \tag{9.4.3}$$

which gives for the torsional differential equation

$$-\frac{du}{dz} M_0 = GJ \frac{d\phi}{dz} - EC_w \frac{d^3\phi}{dz^3} \tag{9.4.4}$$

Two assumptions are inherent in Eqs. 9.4.1 and 9.4.2, both of which relate to the assumption of small deformation. It is assumed that properties $I_{x'}$ and $I_{y'}$ equal I_x and I_y, respectively; and also the I_x is large compared to I_y, so that Eq. 9.4.1 is not linked to Eqs. 9.4.2 and 9.4.4. Thus displacement v in the plane of bending does not affect the torsional function ϕ.

Differentiating Eq. 9.4.4 with respect to z gives

$$-\frac{d^2u}{dz^2} M_0 = GJ \frac{d^2\phi}{dz^2} - EC_w \frac{d^4\phi}{dz^4} \tag{9.4.5}$$

From Eq. 9.4.2,

$$\frac{d^2u}{dz^2} = \frac{M_0\phi}{EI_y}$$

which when substituted into Eq. 9.4.5 gives

$$EC_w \frac{d^4\phi}{dz^4} - GJ \frac{d^2\phi}{dz^2} - \frac{M_0^2}{EI_y} \phi = 0 \tag{9.4.6}$$

which is the differential equation for the angle of twist.

To obtain a solution for Eq. 9.4.6, divide by EC_w and let

$$2\alpha = \frac{GJ}{EC_w} \quad \text{and} \quad \beta = \frac{M_0^2}{E^2 C_w I_y} \tag{9.4.7}$$

Equation 9.4.6 then becomes

$$\frac{d^4\phi}{dz^4} - 2\alpha \frac{d^2\phi}{dz^2} - \beta\phi = 0 \tag{9.4.8}$$

Let

$$\left.\begin{array}{c} \phi = Ae^{mz} \\[2mm] \dfrac{d^2\phi}{dz^2} = Am^2 e^{mz} \\[2mm] \dfrac{d^4\phi}{dz^4} = Am^4 e^{mz} \end{array}\right\} \tag{9.4.9}$$

Substitution of Eqs. 9.4.9 into Eq. 9.4.8 gives

$$Ae^{mz}(m^4 - 2\alpha m^2 - \beta) = 0 \tag{9.4.10}$$

Since e^{mz} cannot be zero and A can be zero only if no buckling has occurred (a trivial solution) the bracket expression of Eq. 9.4.10 must be zero:

$$m^4 - 2\alpha m^2 - \beta = 0$$

which gives for the solution

$$m^2 = \alpha \pm \sqrt{\beta + \alpha^2}$$

or

$$m = \pm\sqrt{\alpha \pm \sqrt{\beta + \alpha^2}} \tag{9.4.11}$$

It is apparent from Eq. 9.4.11 that m will consist of two real and two complex roots because

$$\sqrt{\beta + \alpha^2} > \alpha$$

Let

$$n^2 = \alpha + \sqrt{\beta + \alpha^2} \quad \text{(both real roots)} \tag{9.4.12}$$

$$q^2 = -\alpha + \sqrt{\beta + \alpha^2} \quad \text{(real part of complex roots)} \tag{9.4.13}$$

Using the four values for m, the expression for ϕ from Eq. 9.4.9 becomes

$$\phi = A_1 e^{nz} + A_2 e^{-nz} + A_3 e^{iqz} + A_4 e^{-iqz} \tag{9.4.14}$$

The complex exponential functions may be expressed in terms of circular functions,

$$\left.\begin{array}{l} e^{iqz} = \cos qz + i \sin qz \\[2mm] e^{-iqz} = \cos qz - i \sin qz \end{array}\right\} \tag{9.4.15}$$

By using Eqs. 9.4.15 and defining new constants A_3 and A_4 which equal $(A_3 + A_4)$ and $(A_3 i - A_4 i)$, respectively, one obtains

$$\phi = A_1 e^{nz} + A_2 e^{-nz} + A_3 \cos qz + A_4 \sin qz \tag{9.4.16}$$

The constants A_1 and A_4 are determined by the end support conditions. For the case of torsional simple support, i.e., beam ends may not twist, but are free to warp, the conditions are

$$\phi = 0, \quad \frac{d^2\phi}{dz^2} = 0 \quad \text{at } z = 0 \quad \text{and} \quad z = L$$

For $\phi = 0$ at $z = 0$, Eq. 9.4.16 gives

$$0 = A_1 + A_2 + A_3 \tag{9.4.17}$$

For $d^2\phi/dz^2 = 0$ at $z = 0$,

$$0 = A_1 n^2 + A_2 n^2 - A_3 q^2 \tag{9.4.18}$$

Multiplying Eq. 9.4.17 by n^2 and subtracting Eq. 9.4.18 gives

$$0 = A_3(q^2 + n^2), \qquad \therefore \underline{A_3 = 0}$$

Then, from Eq. 9.4.17,

$$A_1 = -A_2 \tag{9.4.19}$$

Thus Eq. 9.4.16 becomes

$$\phi = A_1(e^{nz} - e^{-nz}) + A_4 \sin qz \tag{9.4.20}$$

which may be written

$$\phi = 2A_1 \sinh nz + A_4 \sin qz \tag{9.4.21}$$

At $z = L$, $\phi = 0$; therefore , from Eq. 9.4.21,

$$0 = 2 A_1 \sinh nL + A_4 \sin qL \tag{9.4.22}$$

Also, at $z = L$, $d^2\phi/dz^2 = 0$, which gives

$$0 = 2A_1 n^2 \sinh nL - A_4 q^2 \sin qL \tag{9.4.23}$$

Multiplying Eq. 9.4.22 by q^2 and adding to Eq. 9.4.23 gives

$$2A_1(n^2 + q^2)\sinh nL = 0 \tag{9.4.24}$$

Since $(n^2 + q^2)$ cannot be zero, and $\sinh nL$ can be zero only if $n = 0$, therefore A_1 must be zero:

$$A_1 = -A_2 = 0$$

Finally, from Eq. 9.4.21,

$$\phi = A_4 \sin qL = 0 \tag{9.4.25}$$

If lateral-torsional buckling occurs, A_4 cannot be zero, so that

$$\sin qL = 0 \tag{9.4.26}$$

$$qL = N\pi$$

where N is any integer.

The elastic buckling condition is defined by

$$q = \frac{N\pi}{L} \tag{9.4.27}$$

which for the fundamental buckling mode $N = 1$.

The value of M_0 which satisfies Eq. 9.4.27 is said to be the critical moment:

$$q = \sqrt{-\alpha + \sqrt{\beta + \alpha^2}} = \frac{\pi}{L} \tag{9.4.28}$$

Squaring both sides, and substituting the definitions of α and β, from Eqs. 9.4.7,

$$-\frac{GJ}{2EC_w} + \sqrt{\frac{M_0^2}{E^2 C_w I_y} + \left(\frac{GJ}{2EC_w}\right)^2} = \frac{\pi^2}{L^2} \tag{9.4.29}$$

Solving for $M_0 = M_{cr}$ gives

$$M_{cr}^2 = E^2 C_w I_y \left[\left(\frac{\pi^2}{L^2} + \frac{GJ}{2EC_w}\right)^2 - \left(\frac{GJ}{2EC_w}\right)^2 \right] \tag{9.4.30}$$

$$M_{cr} = \sqrt{\frac{\pi^4 E^2 C_w I_y}{L^4} + \frac{\pi^2 E I_y GJ}{L^2}} \tag{9.4.31}$$

Factoring π/L from inside the root sign gives

$$M_{cr} = \frac{\pi}{L} \sqrt{\left(\frac{\pi E}{L}\right)^2 C_w I_y + E I_y GJ} \tag{9.4.32}$$

Equation 9.4.32 is the elastic lateral-torsional buckling strength for an I-shaped section under the action of constant moment in the plane of the web over the laterally unbraced length L. To adjust for moment gradient, Eq. 9.4.32 may be multiplied by a factor C_b. Thus, in general

$$M_{cr} = C_b \frac{\pi}{L} \sqrt{\left(\frac{\pi E}{L}\right)^2 C_w I_y + E I_y GJ} \tag{9.4.33}$$

9.5 INELASTIC LATERAL-TORSIONAL BUCKLING

When the moment strength is based on some of the fibers of the cross-section reaching a strain ϵ (see Fig. 9.3.2) that is greater than ϵ_y (that is $\epsilon > F_y/E$), buckling is more likely to occur than when the strain $\epsilon < \epsilon_y$. When elements are inelastic the stiffness as related to the modulus of elasticity decreases; therefore, buckling strength decreases. The larger the strain requirement the lower must be the slenderness ratios related to the various types of buckling.

Studies of inelastic lateral-torsional buckling have been made by Galambos [9.8], Lay and Galambos [9.9, 9.10], Massey and Pitman [9.11], Trahair and

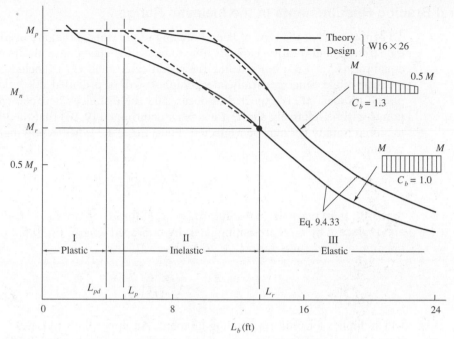

Figure 9.5.1 Beam behavior as related to lateral support. (From Yura, Galambos, and Ravindra [9.1])

Kitipornchai [9.12], Hartmann [9.13], and Ojalvo and Weaver [9.14]. Lateral-torsional buckling as used in Load and Resistance Factor Design is summarized by Yura et al. [9.1].

The strength of I-shaped beams as related to laterally unbraced length L_b as studied by Nethercot and Trahair [9.15] and reported by Yura et al. [9.1] is shwon in Fig. 9.5.1. Though the torsional stiffness is not greatly affected by residual stress, the column-related strength is very much affected, as discussed in Chapter 6. In the presence of residual stress, the maximum *elastic* moment strength M_r is given by

$$M_r = S_x(F_{yf} - F_r) \qquad (9.5.1)*$$

For the same reasons (such as unknown residual stress magnitude and variation, accidental eccentricity, and initial crookedness) as for inelastic column buckling, the range between M_r and M_p is not readily analyzed. Note from Fig. 9.5.1 that when there is uniform moment the decrease in strength due to residual stress is greatest.

When the bending moment acts with a gradient having maximum moment at one end to one-half that amount at the other end the residual stress effect is relatively insignificant. In this case only a small portion of the span near maximum moment will be inelastic; when the applied load stress is lower, the addition of residual stress will be less likely to cause a fiber to reach yield stress.

---

*Note that LRFD-F1.2a. uses F_L in place of $(F_{yf} - F_r)$ to compute M_r. For a hybrid girder (Sec. 11.7) having a web with a much lower yield stress than the flange, M_r must be based on yield in the web; thus, F_L is the lower of $(F_{yf} - F_r)$ or F_{yw}.

Lateral Bracing Requirements In the Inelastic Range

To gain an idea of the type of lateral bracing requirements that are needed to achieve the strength and rotation capacity, one may use the elastic lateral-torsional buckling equation, Eq. 9.4.31, but redefine the rigidities EI_y and GJ to include values in the inelastic range. Since generally lateral support will be provided at locations where the plastic moment M_p is expected to occur, and the distances between lateral support points will be relatively short, it has been determined [9.10] that the term involving torsional rigidity GJ may be neglected. Thus, Eq. 9.4.31, neglecting the second term, becomes

$$M_{cr} = \frac{\pi^2 E}{L^2} \sqrt{C_w I_y} \qquad (9.5.2)$$

Since M_{cr} must reach M_p, substitute $M_p = Z_x F_y$ for M_{cr}. Also, $C_w = I_y h^2/4$, $I_y = A r_y^2$, and replace L by L_b representing laterally unbraced length. Eq. 9.5.2 then gives the maximum slenderness ratio L_b/r_y to achieve M_p,

$$\frac{L_b}{r_y} \leq \sqrt{\frac{\pi^2 E}{2F_y}\left(\frac{hA}{Z_x}\right)} \qquad (9.5.3)$$

which applies for uniform bending moment. An upper limit to Eq. 9.5.3 is obtained if one assumes perfect elasto-plastic steel without residual stress. Taking in Eq. 9.5.3 a conservative (low) value of 1.5 for hA/Z_x along with $E = 29,000$ ksi, gives

$$\frac{L_b}{r_y} \leq \left[\sqrt{\frac{\pi^2\, 29,000(1.5)}{2F_y}} = \frac{460}{\sqrt{F_y}}\right] \qquad (9.5.4)$$

Experimental data [9.16, 9.17] show that a lower limit than Eq. 9.5.4 is necessary to achieve adequate rotation capacity R (Fig. 9.3.2). Thus LRFD-F1.2a has set the limit L_p/r_y as

$$\frac{L_p}{r_y} = \frac{300}{\sqrt{F_{yf}},\ \text{ksi}} \qquad (9.5.5)*$$

where F_{yf} is the flange yield stress. The foregoing is for I-shaped members bent about the x-axis (i.e., strong axis) where the limit state is achievement of plastic moment strength M_p.

When the ability to absorb additional plastic strain is desired, the limit must be lower. In terms of Fig. 9.3.2, L_b/r_y at the Eq. 9.5.5 limit will give rotation capacity somewhat greater than θ_p. When more rotation is required as when plastic analysis, as discussed in Chapter 10, is used the limit must be reduced. The LRFD and ASD Specifications [1.16, 1.5] are based on a rotation capacity factor R (Fig. 9.3.2) approximately 3 when plastic analysis is used.

The elastic modulus of elasticity E in Eq. 9.5.3 should be replaced by the strain-hardening modulus of elasticity E_{st} when the actual strain ϵ approaches the

---

* For SI, with F_{yf} in MPa, $\dfrac{L_p}{r_y} = \dfrac{790}{\sqrt{F_{yf}}}$ $\qquad\qquad$ (9.5.5)

strain hardening strain. Lay and Galambos [9.9] suggested that E_{st} can be approximated as E/F_y. When that is done, evaluating Eq. 9.5.3 using $hA/Z_x = 1.5$ and L_b for L gives,

$$\frac{L_b}{r_y} \leq \frac{2800}{F_y, \text{ ksi}} \tag{9.5.6}$$

which makes the limit vary inversely in a linear manner with F_y instead of inversely as the square root of F_y. Again, there are neglected strength factors such as torsional stiffness and end restraint. Tests by Bansal [9.16] at the University of Texas have established the limit using the same form of Eq. 9.5.6, additionally including provision for moment gradient, as

$$\frac{L_b}{r_y} \leq \frac{3600 + 2200M_1/M_p}{F_y, \text{ ksi}} \tag{9.5.7}^*$$

where $M_1 = $ smaller moment at the ends of the laterally unbraced segment
          (taken positive when moments cause *reverse* curvature)

9.6 LOAD AND RESISTANCE FACTOR DESIGN—I-SHAPED BEAMS SUBJECTED TO STRONG-AXIS BENDING

This section considers the full range of strength from laterally stable beams to situations where lateral-torsional buckling causes considerable strength reduction. Loading in the plane of the web is assumed.

The strength requirement according to LRFD-F1 may be stated

$$\phi_b M_n \geq M_u \tag{9.6.1}$$

where $\phi_b = $ strength reduction factor for flexure = 0.90
          $M_n = $ nominal moment strength
          $M_u = $ factored service load moment (see Sec. 1.9)

Figure 9.6.1 shows the effect of laterally unbraced length L_b on the lateral-torsional buckling strength. Of course, local buckling may result in lower moment strength M_n if the plate element (flange or web) width/thickness ratio are too high.

Case 1: Plastic Moment is Reached ($M_n = M_p$) Along With Large Plastic Rotation Capacity ($R \geq 3$ in Fig. 9.3.2)

The section must be "compact" to prevent local buckling; that is, λ for the flange $(b_f/2t_f)$ and for the web (h/t_w) must not exceed λ_p as discussed in Sec. 6.17 (values in Table 9.6.1) and lateral bracing must be provided such that the laterally unbraced length L_b does not exceed L_{pd}, where from Eq. 9.5.7.

$$L_{pd} = \frac{3600 + 2200M_1/M_p}{F_y, \text{ ksi}} r_y \tag{9.6.2}$$

---

*For SI with F_y in MPa, $\dfrac{L_b}{r_y} \leq \dfrac{24{,}800 + 15{,}200\, M_1/M_p}{F_y}$ (9.5.7)

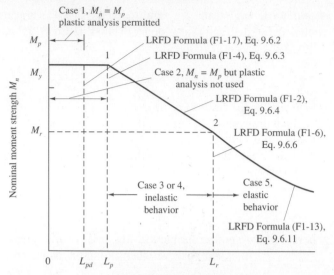

Figure 9.6.1   Nominal strength M_n of "compact" sections as affected by lateral-torsional buckling.

LRFD-F1.2d uses M_2 instead of M_p; however, M_2 will always be M_p under Case 1. In this category, plastic analysis involving redistribution of moments as discussed in Chapter 10 is an option.

Case 2: Plastic Moment is Reached ($M_n = M_p$) But With Relatively Little Rotation Capacity ($R < 3$ in Fig. 9.3.2)

The section must be "compact" to prevent local buckling; that is, λ for the flange ($b_f/2t_f$) and for web (h/t_w) must not exceed λ_p as discussed in Sec. 6.17 and lateral bracing must be provided such that λ (i.e., L_b/r_y) does not exceed λ_p. The three slenderness limits for various yield stresses are given in Table 9.6.1. This latter limit λ_p, from Eq. 9.5.5, means L_b may not exceed L_p when $C_b = 1.0$.

$$L_p = \frac{300}{\sqrt{F_{yf}, \text{ksi}}} r_y \qquad (9.6.3)$$

Case 3: Lateral-Torsional Buckling of "Compact" Sections May Occur in the Inelastic Range ($M_p > M_n \geq M_r$)

This moment strength M_n is approximated as a linear relationship between points 1 (M_p at L_p) and 2 (M_r at L_r) on Fig. 9.6.1. Local buckling must be precluded. Since nearly all of the rolled sections are "compact" that is they have $\lambda \leq \lambda_p$ as discussed in Sec. 6.16, the nominal strength M_n is a linear function of the lateral-torsional buckling strength. Thus,

$$M_n = C_b \left[M_p - (M_p - M_r)\left(\frac{L_b - L_p}{L_r - L_p}\right) \right] \leq M_p \qquad (9.6.4)$$

TABLE 9.6.1   SLENDERNESS RATIO LIMITS λ_p FOR "COMPACT" I-SHAPED BEAMS TO ACHIEVE PLASTIC MOMENT STRENGTH M_p (LRFD-B5.1)[a]

Yield stress F_y	Flange local buckling $\dfrac{b_f}{2t_f} = \dfrac{65}{\sqrt{F_y}}$	Web local buckling $\dfrac{h}{t_w} = \dfrac{640}{\sqrt{F_y}}$	Lateral-torisional buckling $\dfrac{L_b}{r_y} = \dfrac{300}{\sqrt{F_y}}$
36	10.8	107	50.0
42	10.0	98.8	46.3
45	9.7	95.4	44.7
50	9.2	90.5	42.4
55	8.8	86.3	40.5
60	8.4	82.6	38.7
65*	8.1	79.4	37.2

[a] Loading in the plane of the web; $M_p = F_y Z_x$.

* The use of plastic moment strength M_p is restricted by LRFD-A5 and ASD-F1.1 to steels having $F_y \leq 65$ ksi.

where M_r is the moment strength available for service loads when extreme fiber reaches the yield stress F_{yf} (including the residual stress), and may be expressed

$$M_r = (F_{yf} - F_r)S_x \qquad (9.6.5)$$

where F_{yf} = minimum specified yield stress for the flange steel
F_r = compressive residual stress in flange
= 10 ksi for rolled shapes; 16.5 ksi for welded shapes
S_x = elastic section modulus = $I_x/(d/2)$
I_x = moment of inertia about the x-axis
d = overall depth of the section

The length L_r is obtained by equating the maximum elastic moment strength M_r (Eq. 9.6.5) to the elastic lateral-torsional buckling strength M_{cr} (Eq. 9.4.33) and solving for L. Upon performing that operation, one obtains L_r (L in Eq. 9.4.33) as given by LRFD-F1.2a,

$$L_r = \frac{r_y X_1}{(F_{yf} - F_r)}\sqrt{1 + \sqrt{1 + X_2(F_{yf} - F_r)^2}} \qquad (9.6.6)^*$$

where

$$X_1 = \frac{\pi}{S_x}\sqrt{\frac{EGJA}{2}} \qquad (9.6.7)$$

and

$$X_2 = 4\frac{C_w}{I_y}\left(\frac{S_x}{GJ}\right)^2 \qquad (9.6.8)$$

---

* LRFD-F1.2a uses F_L in place of $(F_{yf} - F_r)$, because on hybrid girders (Sec. 11.7) the web often has a lower yield stress than the flange. Since M_r (Eq. 9.6.5) must be based on yield stress in the web, F_L is the lower of $(F_{yf} - F_r)$ or F_{yw}.

Note that X_1 and X_2 are not really physical properties but rather provide a shortened way of writing Eq. 9.6.6; their practical significance is discussed by Hoadley [9.17]. They represent combinations of cross-sectional properties; X_1 and X_2 are given in the *LRFD Manual* [1.18].

Case 4: General Limit State Where Nominal Moment Strength M_n Occurs in the Inelastic Range ($M_p > M_n \geq M_r$)

This condition is relatively uncommon for rolled shapes. When $L_p < L_b < L_r$ for lateral-torsional buckling, *or* $\lambda_p < (\lambda = b_f/2t_f) < \lambda_r$ for flange local buckling, *or* $\lambda_p < (\lambda = h/t_w) < \lambda_r$ for web local buckling, the strength will be in this category. These slenderness limits are given for various yield stresses in Table 9.6.2. LRFD has used the symbol λ to represent the general stability slenderness ratio: $\lambda = L_b/r_y$ when considering lateral-torsional buckling; $\lambda = b_f/2t_f$ when considering flange local buckling; and $\lambda = h/t_w$ when considering web local buckling.

The variation in strength with the generalized slenderness ratio λ is shown in Fig. 9.6.2. When λ lies between λ_p and λ_r for any one or more of the limit states of (1) flange local buckling, (2) web local buckling, or (3) lateral-torsional buckling, the relationship for moment strength is linear. For the limit state of flange or web local buckling, LRFD-Appendix F1 prescribes

$$M_n = M_p - (M_p - M_r)\left(\frac{\lambda - \lambda_p}{\lambda_r - \lambda_p}\right) \qquad (9.6.9)$$

For the limit state of lateral-torsional buckling, the linear relationship of Eq. 9.6.9 is used; however, the result is increased by multiplying by C_b when there is moment gradient. Of course, the maximum M_n is M_p regardless of how steep is the moment gradient. The expression according to LRFD-Appendix F1 is identical to Eq. 9.6.4, except expressed in the form of the ratios λ,

$$M_n = C_b\left[M_p - (M_p - M_r)\left(\frac{\lambda - \lambda_p}{\lambda_r - \lambda_p}\right)\right] \leq M_p \qquad (9.6.10)$$

where

$$C_b = \frac{12.5M_{\max}}{2.5M_{\max} + 3M_A + 4M_B + 3M_C} \qquad (9.6.11)$$

where C_b = a modification factor for non-uniform bending moment variation for a beam segment laterally unbraced excepts at the segment ends, based on *absolute values* of bending moments

$M_{\max}$ = maximum moment in the unbraced segment

M_A = moment at 1/4 point of unbraced segment

M_B = moment at midpoint of unbraced segment

M_C = moment at 3/4 point of unbraced segment

Eq. 9.6.11 (LRFD-Formula C-F1-3) is new in 1993, replacing the following equation which has been used to adjust for non-uniform moment since 1961,

$$C_b = 1.75 + 1.05\left(\frac{M_1}{M_2}\right) + 0.3\left(\frac{M_1}{M_2}\right)^2 \leq 2.3 \qquad (9.6.12)$$

TABLE 9.6.2 SLENDERNESS RATIO LIMITS λ_r FOR "NONCOMPACT SECTION" I-SHAPED BEAMS TO ACHIEVE F_y AT EXTREME FIBER (LRFD-B5.1)

Yield stress F_y (ksi)	Flange local buckling Rolled I-shape $\dfrac{b_f}{2t_f} = \dfrac{141}{\sqrt{F_y - 10}}$	$\dfrac{h}{t_w}$	$k_c{}^*$	Flange local buckling Welded I-shape[†] $\dfrac{b_f}{2t_f} = \dfrac{162}{\sqrt{(F_{yf} - 16.5)/k_c}}$	Web local buckling $\dfrac{h_c}{t_w} = \dfrac{970}{\sqrt{F_y}}$	Lateral torsional buckling $\dfrac{L_b}{r_y}$ — Eq. 9.6.6** (LRFD-F1.2a)
36	27.7	161.7	0.35	21.7	161.7	
		100	0.40	23.2		
		40	0.63	29.2		Values
42	24.9	149.7	0.35	19.0	149.7	must be
		100	0.40	20.3		determined
		40	0.63	25.5		for each
45	23.8	144.6	0.35	18.0	144.6	W shape
		100	0.40	19.2		for a given
		40	0.63	24.1		yield stress
50	22.3	137.2	0.35	16.6	137.2	
		100	0.40	17.7		
		40	0.63	22.3		
55	21.0	130.8	0.35	15.4	130.8	
		100	0.40	16.5		
		40	0.63	20.8		
60	19.9	125.2	0.36	14.7	125.2	
		100	0.40	15.5		
		40	0.63	19.5		
65	19.0	120.3	0.36	14.0	120.3	
		100	0.40	14.7		
		40	0.63	18.5		
90	15.8	102.2	0.40	11.9	102.2	
		100	0.40	12.0		
		40	0.63	15.0		
100	14.9	97.0	0.41	11.3	97.0	
		100	0.40	11.2		
		40	0.63	14.1		

$*k_c = \dfrac{4}{\sqrt{h/t_w}}$, where $0.35 \le k_c \le 0.763$

$**$ Equation 9.6.6, $\dfrac{L_r}{r_y} = \dfrac{X_1}{(F_{yf} - F_r)}\sqrt{1 + \sqrt{1 + X_2(F_{yf} - F_r)^2}}$

[†] F_{yf} = flange yield stress = F_y

The new equation, based on the work of Kirby and Nethercot [9.18], allows a more accurate determination of the modification factor C_b when the unbraced segment has a nonlinear moment variation; however, when the precision with which the loading, the structural model, and the material properties are known, the more detailed equation, Eq. 9.6.11, seems questionable.

The LRFD Commentary-F1.2a says, "It is still satisfactory to use the former C_b factor . . . for straight line moment diagrams within the unbraced length." In Eq. 9.6.12, the ratio M_1/M_2 is negative when the moments cause single curvature; that is, the most severe loading with constant bending moment gives $C_b = 1.0$. A comparison of Eqs. 9.6.11 and 9.6.12 is shown in Table 9.6.3 and Fig. 9.6.3.

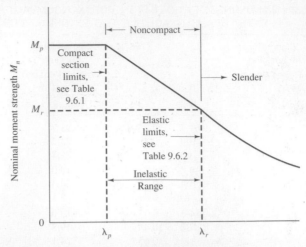

Figure 9.6.2   Nominal strength M_n vs generalized slenderness ratio λ for limit states of flange local buckling, web local buckling, and lateral-torsional buckling.

According to LRFD-F1.2a, C_b is permitted to be taken as 1.0 for all cases; consistent with assuming the most severe loading (constant bending moment). For cantilevers or overhangs where the free end is unbraced, C_b is prescribed to be taken as 1.0.

For laterally unbraced segments having a parabolic moment variation, Eq. 9.6.11 is superior (and more conservative) to using Eq. 9.6.12. Table 9.6.4 gives the 1993 design values from Eq. 9.6.11 for several situations of parabolic unbraced segments.

TABLE 9.6.3   COMPARISON OF C_b FOR LINEAR MOMENT VARIATION

$\dfrac{M_1}{M_2}$	C_b 1993 Eq. 9.6.11	C_b 1986 Eq. 9.6.12
−1.00	1.00	1.00
−0.75	1.11	1.13
−0.50	1.25	1.30
−0.25	1.43	1.51
0.00	1.67	1.75
0.25	2.00	2.03
0.50	2.17	2.30
0.75	2.22	2.30
1.00	2.27	2.30

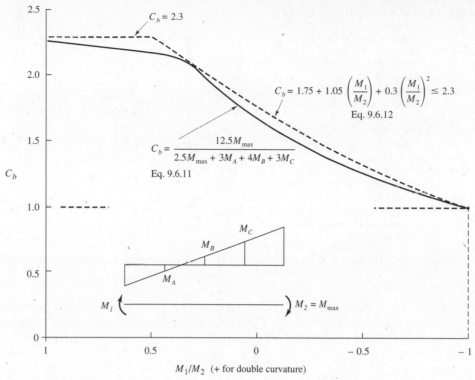

Figure 9.6.3 Comparison of C_b equations for linear variation of moment over laterally unbraced length.

Case 5: General Limit State Where Nominal Moment Strength M_n Equals the Elastic Buckling Strength M_{cr} (i.e., $M_n < M_r$)

There are two subcategories for the case; (1) slenderness ratios λ for flange or web local buckling *do not exceed* λ_r; and (2) slenderness ratios *exceed* λ_r. When λ exceeds λ_r, local buckling will occur prior to the extreme fiber reaching the yield stress and the efficiency of the cross-section is reduced. Such elements are known as "slender" compression elements; a subject treated in Sec. 6.18. All rolled shapes have $\lambda < \lambda_r$ for local buckling; therefore *beams* having $\lambda > \lambda_r$ are not treated in this text. When $\lambda < \lambda_r$ for local buckling (see Table 9.6.2), the nominal moment strength is represented by Eq. 9.4.33, as follows from LRFD-F1.2b:

$$M_n = M_{cr} = C_b \frac{\pi}{L_b} \sqrt{\left(\frac{\pi E}{L_b}\right)^2 C_w I_y + E I_y G J} \qquad [9.4.33]$$

Using the pseudo-properties X_1 and X_2, Eqs. 9.6.7 and 9.6.8, Eq. 9.4.33 becomes

$$M_n = M_{cr} = \frac{C_b S_x X_1 \sqrt{2}}{L_b/r_y} \sqrt{1 + \frac{X_1^2 X_2}{2(L_b/r_y)^2}} \qquad (9.6.13)$$

TABLE 9.6.4 C_b FOR PARABOLIC SEGMENTS
USING LRFD-F1.2a, FORMULA
(C-F1-3), EQ. 9.6.11*

Case 1	Laterally braced at ends; points 1 and 5 only; M_{max} at 3	$C_b = 1.14$
Case 2	Laterally braced at ends and midspan; points 1,3, and 5 only; M_{max} at 3	$C_b = 1.30$
Case 3	Laterally braced at end and 1st quarter point; bracing at points 1 and 2; M_{max} at 2	$C_b = 1.52$
Case 4	Laterally braced at 1st and 2nd quarter points; bracing at points 2 and 3; M_{max} at 3	$C_b = 1.06$
Case 5	Laterally braced at 1st and 3rd quarter points; bracing at points 2 and 4; M_{max} at 3	$C_b = 1.03$

* Values from 1986 LRFD, Eq. 9.6.12 shown in parenthesis.

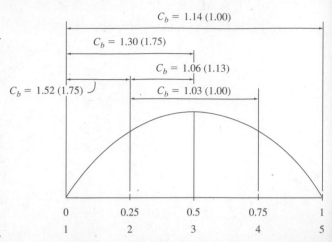

*Values from 1986 LRFD, Eq. 9.6.12, shown in parenthesis.

Note that M_{cr} cannot logically be greater than $C_b M_r$, nor greater than M_p; therefore, Eq. 9.6.13 has those as the meaningfull upper limits. Fig. 9.6.4 shows the nominal strength M_n variation with L_b, emphasizing the effect of C_b.

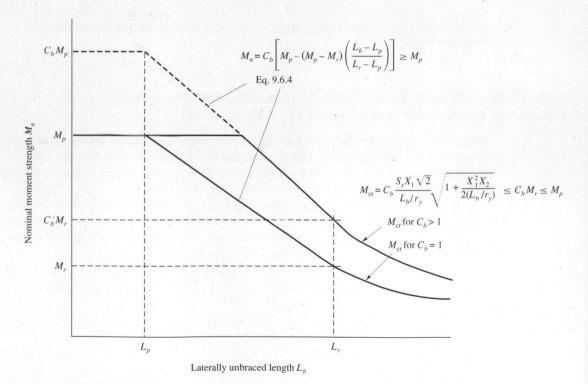

$$M_n = C_b \left[M_p - (M_p - M_r) \left(\frac{L_b - L_p}{L_r - L_p} \right) \right] \geq M_p$$

Eq. 9.6.4

$$M_{cr} = C_b \frac{S_x X_1 \sqrt{2}}{L_b / r_y} \sqrt{1 + \frac{X_1^2 X_2}{2(L_b / r_y)}} \leq C_b M_r \leq M_p$$

M_{cr} for $C_b > 1$

M_{cr} for $C_b = 1$

Figure 9.6.4   Nominal moment strength as affected by C_b.

9.7 ALLOWABLE STRESS DESIGN—I-SHAPED BEAMS SUBJECTED TO STRONG-AXIS BENDING

The general philosophy of Allowable Stress Design (ASD) for beams was presented in Sec. 7.5. The nominal stength expressions for M_n may be divided by an appropriate factor of safety, and divided by the section modulus S_x to convert the expressions into "stress" format.

Case 1: Plastic Moment is Reached ($M_n = M_p$) Along With Large Plastic Rotation Capacity

As shown in Sec. 7.5, achieving the plastic moment strength indicates an allowable stress of $0.66F_y$. The section must be "compact" to prevent local buckling; that is, $b_f/2t_f$ for the flange may not exceed $65/\sqrt{F_y}$ and h/t_w for the web may not exceed $640/\sqrt{F_y}$ in accordance with ASD-B5.1. In addition, according to ASD-F1.1, lateral bracing must be provided such that the laterally unbraced length L_b does not exceed L_c, given as the *smaller* of the following:

$$L_c = \frac{76b_f}{\sqrt{F_y, \text{ksi}}} \qquad (9.7.1)*$$

$$L_c = \frac{20{,}000}{(d/A_f)F_y, \text{ksi}} \qquad (9.7.2)*$$

Equation 9.7.1, which is essentially Eq. 9.5.5 used in LRFD if r_y is approximated as $0.25b_f$, considers lateral stability primarily as column strength of the compression flange. However, since the lateral stability of many rolled sections is governed by the torsional strength, Eq. 9.7.2 is needed for them.

A summary of the "compact section" lateral support requirements is ASD is given in Table 9.7.1.

Case 2: $M_p > M_n \geq (M_y = F_y S_x)$

Beams in this category are those capable of achieving yield strain $\epsilon_y = F_y/E_s$ at the extreme fiber prior to lateral-torsional buckling, but M_p cannot be reached. For case 2 situations, ASD does not permit an interpolated strength between M_y and M_p. Instead the strength is considered to be M_y and the corresponding allowable stress is $0.60F_y$.

In order to achieve a strength $M_n = M_y$ without lateral-torsional buckling, the maximum laterally unbraced length L_u permitted is the length L that makes M_{cr} of Eq. 9.4.31 equal to M_y. Equation 9.4.31 was considered too complex for design use and the use of torsional properties in design was deemed undesirable.

The development of the ASD-F1.3 design equations, ASD Formulas (F1-6 and F1-7) or (F1-8), requires some approximations and transformation of properties in Eq. 9.4.31.

---

* For SI units, with dimensions in mm and F_y in MPa, $\qquad L_c = \dfrac{200b_f}{\sqrt{F_y}} \qquad (9.7.1)$

$$L_c = \frac{138{,}000}{(d/A_f)F_y} \qquad (9.7.2)$$

**TABLE 9.7.1   ASD LATERAL BRACING
REQUIREMENTS FOR
"COMPACT SECTIONS." ASD-F1.1**

F_y (ksi)	Both limits may not be exceeded		F_y (MPa)
	$\dfrac{L_c}{b_f}$	$L_c\left(\dfrac{d}{A_f}\right)$	
36	12.7	556	250
42	11.7	476	290
45	11.3	444	310
50	10.7	400	350
55	10.2	364	380
60	9.8	333	410
65*	9.4	308	450

*I-shaped section loaded in the plane of the web may be assumed to develop M_p only when F_y does not exceed 65 ksi(450 MPa.)

The torsion constant J may be expressed for an I-shaped section by Eq. 8.2.14,

$$J = 2\left(\frac{b_f t_f^3}{3}\right) + \frac{1}{3}d_w t_w^3 = \frac{t_f^2}{3}\left[2b_f t_f + d_w t_w\left(\frac{t_w}{t_f}\right)^2\right] \qquad (9.7.3)$$

where t_f = flange thickness, b_f = flange width, t_w = web thickness, and d_w = web depth. If $t_w/t_f \approx 0.5$, and if the web area is about 20% of the area of the beam ($A_w = 0.2A$), Eq. 9.7.3 becomes

$$J = \frac{t_f^2}{3}[2b_f t_f + d_w t_w - 0.75d_w t_w]$$

$$A = 2b_f t_f + d_w t_w$$

$$A_w = d_w t_w$$

$$0.75A_w = 0.75(0.2A) = 0.15A$$

Then

$$J \approx \frac{t_f^2}{3}[A - 0.15A] \approx 0.28\, At_f^2 \qquad (9.7.4)$$

For other variables,

$$C_w = I_y h^2/4$$

$$G = \text{shear modulus} = \frac{E}{2(1 + \mu)} = \frac{E}{2.6}$$

$$I_y = Ar_y^2$$

$$S_x = 2Ar_x^2/d$$

$$r_x \approx 0.41d \quad \text{(see text Appendix Table A1)}$$

$$h = 0.95d$$

where μ = Poission's ratio (0.3 for steel)
   r_y = radius of gyration with respect to the y-axis
   r_x = radius of gyration with respect to the x-axis
   d = overall depth of beam
   h = distance between centroids of flanges

Substitution into Eq. 9.4.31 and dividing by S_x to obtain stress format gives

$$F_{cr} = \sqrt{\frac{\pi^4 E^2 (I_y h^2/4) I_y}{[2A(0.41d)^2/d]^2 L^4} + \frac{\pi^2 E I_y (E/2.6)(0.28 A t_f^2)}{[2A(0.41d)^2/d]^2 L^2}} \qquad (9.7.5)$$

or

$$F_{cr} = \sqrt{\left(\frac{\pi^2 E I_y h}{4A\,(0.41)^2 dL^2}\right)^2 + \left(\frac{\pi E \sqrt{I_y}\,\sqrt{0.28}\,\sqrt{A}\,t_f}{2A(0.41)^2 dL\,\sqrt{2.6}}\right)^2} \qquad (9.7.6)$$

Then, substituting $I_y = A r_y^2$ and $h = 0.95d$ gives

$$F_{cr} = \sqrt{\left(\frac{\pi^2 (0.95) E}{4(0.41)^2 (L/r_y)^2}\right)^2 + \left(\frac{\pi \sqrt{0.28}\,E}{2(0.41)^2 \sqrt{2.6}\,(Ld/r_y t_f)}\right)^2} \qquad (9.7.7)$$

or

$$F_{cr} = \sqrt{\left(\frac{14E}{(L/r_y)^2}\right)^2 + \left(\frac{3E}{Ld/r_y t_f}\right)^2} \qquad (9.7.8)$$

Equation 9.7.8 is essentially M_{cr}/S_x (Eq. 9.4.31 divided by S_x) where torsional functions have been replaced by more familiar properties. The first term represents column buckling strength and the second term represents torsional strength. The AISC Allowable Stress Design has traditionally used a two-formula procedure for allowable stress. Referring to Fig. 9.7.1, Eq. 9.7.8 may be visualized as the resultant of two vectors at 90° to each other. If the resultant is not used, the next best approach would be to use the larger of the two legs of the right triangle.

   1. *Torsionally Strong Sections.*   For shallow thick flanged sections, torsional strength predominates and the first term of Eq. 9.7.8 may be neglected. Thus,

$$F_{cr} = \frac{3E}{Ld/r_y t_f} \qquad (9.7.9)$$

Since $r_y \approx 0.22 b_f$, $A_f = b_f t_f$, and using L_b instead of L for the laterally unbraced length, Eq. 9.7.9 becomes

$$F_{cr} = \frac{0.66E}{L_b d/A_f} \qquad (9.7.10)$$

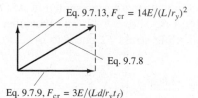

Eq. 9.7.13, $F_{cr} = 14E/(L/r_y)^2$

Eq. 9.7.8

Eq. 9.7.9, $F_{cr} = 3E/(Ld/r_y t_f)$

Figure 9.7.1   Lateral-torsional buckling stress as the resultant of two vectors.

When C_b is included to account for moment gradient, Eq. 9.7.10 becomes

$$F_{cr} = \frac{0.66EC_b}{L_b d/A_f} \tag{9.7.11}$$

Equation 9.4.33 shows that C_b is a direct multiplier on M_{cr}. Next, dividing by the ASD 1.67 basic factor of safety and taking $E = 29{,}000$ ksi gives the allowable stress F_b in ksi,

$$F_b = \frac{12{,}000C_b}{L_b d/A_f} \text{ ksi} \tag{9.7.12}*$$

which is ASD Formula (F1-8).

2. *Torsionally Weak Sections.*   For deep sections having relatively thin flanges and web (such as plate girder sections), column strength of the compression flange predominates and the second term of Eq. 9.7.8 may be neglected. Thus,

$$F_{cr} = \frac{14E}{(L/r_y)^2} \tag{9.7.13}$$

Since only the compression flange is the column, logically the radius of gyration should be that of the compression flange and some portion of the adjacent web. ASD-F1.3 introduces the symbol r_T for this compression portion, defined as "radius of gyration of a section comprising the compression flange plus one-third of the compression web area, taken about as axis in the plane of the web." For doubly symmetrical I-shaped sections r_T may be expressed

$$r_T = \sqrt{\frac{\frac{1}{12} t_f b_f^3}{t_f b_f + \frac{1}{6} t_w (d - 2t_f)}} \tag{9.7.14}$$

The values of r_T are given in the ASD Manual along with other dimensions and properties for rolled shapes. As an approximation for this development of ASD formulas r_T may be taken as $1.2r_y$.

Replacing r_y by $r_T/1.2$ in Eq. 9.7.13 would make that equation very closely the same as the Euler column equation,

$$F_{cr} = \frac{\pi^2 E}{(L/r_T)^2} \tag{9.7.15}$$

This makes it more visible that Eq. 9.7.13 is representative of column strength of the compression flange.

Treatment of column strength as it relates to beams requires application of the principles of column design as discussed in Chapter 6. Recall the use of two design equations, one based on inelastic buckling strength represented in ASD by the SSRC parabola (Eq. 6.7.1) for buckling stress F_{cr} above $F_y/2$, and the other based on the Euler elastic buckling strength for buckling stress F_{cr} equal to or less than $F_y/2$.

---

*For SI units, $$F_b = \frac{83{,}000C_b}{L_b d/A_f} \text{ MPa} \tag{9.7.12}$$

The SSRC parabola with $r = r_T$ and the effective length factor K taken as 1.0 is

$$F_{cr} = F_y \left[1 - \frac{F_y}{4\pi^2 E} \left(\frac{L}{r_T} \right)^2 \right] \tag{9.7.16}$$

Unlike the axial compression loading of Chapter 6 where the magnitude of compression was normally constant over the unbraced length, the compression flange of a beam will have its magnitude varying as the bending moment varies. Thus, for beams the term C_b to account for moment gradient is included in Eq. 9.7.16,

$$F_{cr} = F_y \left[1 - \frac{F_y}{4\pi^2 E C_b} \left(\frac{L}{r_T} \right)^2 \right] \tag{9.7.17}$$

where C_b is given by Eq. 9.6.12 in ASD-F1.3.

To obtain the allowable stress, the strength equation is divided by a factor to get the calculation into the service load range. For small L/r_T the strength should be $M_p = F_y Z$, whereas for large L/r_T the strength will be $M_{cr} = F_{cr} S_x$. As discussed in Chapter 7 when M_p is achievable, the allowable value for elastic service load stresses is $0.66F_y$. By the same reasoning, F_b based on Eq. 9.7.17 should be $\frac{2}{3}$ of F_{cr}, giving

$$F_b = \left[\frac{2}{3} - \frac{(\frac{2}{3})F_y}{4\pi^2 E C_b} \left(\frac{L}{r_T} \right)^2 \right] F_y \tag{9.7.18}$$

For $E = 29{,}000$ ksi and F_y in ksi, Eq. 9.7.18 becomes

$$F_b = \left[\frac{2}{3} - \frac{F_y (L/r_T)^2}{1720(10^3)C_b} \right] F_y \tag{9.7.19}$$

The coefficient in the second term has been adjusted from 1720 to 1530 (i.e., $\frac{3}{4}$ instead of $\frac{2}{3}$ is used on that term) to allow for the fact that as L/r_T gets larger the strength drops below M_y and the extra benefit of the shape factor disappears. When the nominal strength M_n is less than M_y the F_b should be $0.60F_{cr}$ rather than $0.66F_{cr}$. Thus Eq. 9.7.19 becomes

$$F_b = \left[\frac{2}{3} - \frac{F_y (L_b/r_T)^2}{1530(10^3)C_b} \right] F_y \tag{9.7.20}*$$

which is ASD Formula (F1-6). The symbol L_b is used for laterally unbraced length.

For large L_b/r_T when elastic buckling controls, Eq. 9.7.15 would logically be the basis for design. Dividing Eq. 9.7.15 by 1.67 (the basic FS for beams) and inserting C_b for moment gradient gives

$$F_b = \frac{\pi^2 E C_b}{1.67(L_b/r_T)^2} \tag{9.7.21}$$

---

*In SI units, $F_b = \left[\dfrac{2}{3} - \dfrac{F_y (L_b/r_T)^2}{10{,}550(10^3)C_b} \right] F_y$ MPa   (9.7.20) which gives closely the following:

$$F_b = \frac{170,000C_b}{(L_b/r_T)^2} \text{ ksi} \qquad (9.7.22)*$$

which is ASD Formula (F1-7).

The use of 1.67 here for beams rather than the 1.92 used for long axially loaded columns is explainable (a) by realizing that torsional strength is neglected in Eq. 9.7.22, and (b) Eq. 9.7.13 is inherently conservative when lateral-torsional buckling strength is primarily column strength, in that for sections consisting of thin flanges and deep thin web $r_x \approx 0.38d$ would have been more correct than the value $0.41d$ used in the development of Eq. 9.7.13.

Thus lateral-torsional buckling controls the design when the laterally unbraced length L_b exceeds the *larger* value of L_b defined by Eqs. 9.7.12 or 9.7.20 equaling an allowable stress of $0.60F_y$. When the *larger* value of F_b from these two equations exceeds $0.60F_y$ it implies that M_y can be reached prior to lateral-torsional buckling.

The maximum unbraced length L_b for which $F_b = 0.60F_y$ may be obtained by setting Eqs. 9.7.12 and 9.7.20 equal to $F_b = 0.60F_y$. Thus

$$0.60F_y = \frac{12,000C_b}{L_b d/A_f} \qquad [9.7.12]$$

or

$$0.60F_y = \left[\frac{2}{3} - \frac{F_y(L_b/r_T)^2}{1530(10^3)C_b}\right]F_y \qquad [9.7.20]$$

From the first,

$$L_b = \frac{20,000C_b}{(d/A_f)F_y} \qquad (9.7.23)$$

and from the second,

$$L_b = r_T\sqrt{\frac{102,000C_b}{F_y}} \qquad (9.7.24)$$

The *larger* value of L_b from Eqs. 9.7.23 and 9.7.24 is the maximum unbraced length for a Case 2 design situation; i.e, where the nominal strength M_n achievable lies between M_y and M_p.

When $C_b = 1$ (the conservative case of constant bending moment), the larger value of L_b from Eqs. 9.7.23 and 9.7.24 is termed L_u, a property of the section for any

---

*In SI units $F_b = \dfrac{1170(10^3)C_b}{(L_b/r_T)^2}$ MPa   (9.7.22)

specified yield stress F_y,

$$L_u = \frac{20,000}{(d/A_f)(F_y, \text{ksi})}$$

but not less than $\hspace{3cm}$ (9.7.25)*

$$L_u = r_T \sqrt{\frac{102,000}{F_y, \text{ksi}}}$$

The ASD Manual gives values for L_u for $F_y = 36$ ksi and 50 ksi.

Case 3: $M_n < M_y$

For this case, lateral-tosional buckling occurs before M_y can be reached and the allowable stress is less than $0.60F_y$.

Whenever the laterally unsupported length L_b exceeds the *larger* L_u value of Eqs. 9.7.25, the allowable stress F_b is the *larger* stress value of Eqs. 9.7.12 and 9.7.20 for strength controlled by inelastic lateral-torsional buckling, or is the *larger* stress value of Eqs. 9.7.12 and 9.7.22 for strength controlled by elastic lateral-torsional buckling.

The slenderness ratio L_b/r_T at which the criteria for inelastic and elastic lateral-torsional buckling prevention give the same allowable stress is obtained by equating Eqs. 9.7.20 and 9.7.22; thus,

$$\frac{2F_y}{3} - \frac{F_y^2(L_b/r_T)^2}{1530(10^3)C_b} = \frac{170,000C_b}{(L_b/r_T)^2}$$

from which is obtained

$$\frac{L_b}{r_T} = \sqrt{\frac{510,000C_b}{F_y}} \hspace{2cm} (9.7.26)$$

Elastic behavior occurs for slenderness ratio greater than Eq. 9.7.26 and Eq. 9.7.22 controls over Eq. 9.7.20. Table 9.7.2 gives L_b/r_T upper and lower limits and denominator coefficients for the second term in Eq. 9.7.20 [ASD Formula (F1-6)] for various values of F_y.

A summary of the ASD allowable flexural stress vs laterally unbraced length L_b relationships is given in Fig. 9.7.2.

---

*For SI units, with F_y in MPa,

$$L_u = \frac{2900}{(d/A_f)F_y}$$

but not less than $\hspace{3cm}$ (9.7.25)

$$L_u = \sqrt{\frac{14,800}{F_y}}$$

TABLE 9.7.2   LIMITS ON L_b/r_T AND COEFFICIENTS FOR ASD FORMULAS (F1–6 AND 1–7)

F_y (ksi)	$\sqrt{\dfrac{102,000\,C_b}{F_y}}$ Eq. 9.7.24			$\dfrac{1530(10^3)\,C_b}{F_y^2}$, for F_b(F1-6) from Eq. 9.7.20			$\sqrt{\dfrac{510,000\,C_b}{F_y}}$ Eq. 9.7.26		
	C_b			C_b			C_b		
	1.0	1.75	2.3	1.0	1.75	2.3	1.0	1.75	2.3
36	53	70	80	1181	2070	2720	119	158	180*
42	49	65	74	867	1517	1995	110	146	167
50	45	60	68	612	1070	1408	101	134	153
60	41	54	62	425	744	977	92	122	139
65	40	53	61	362	634	832	89	118	135
100	32	42	49	153	268	352	71	94	107

*ASD-B7 limits slenderness ratios for compression members to 200. $L/r_y = 200$ when $L_b/r_T \approx 200/1.2 = 167$.

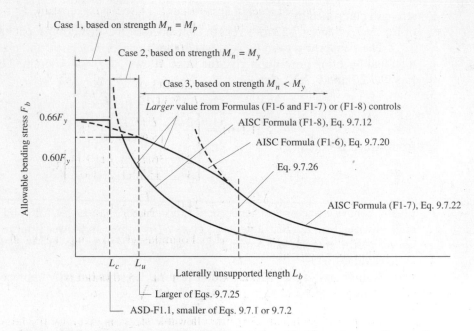

Figure 9.7.2   Summary of ASD criteria for allowable flexural stress for members having an axis of symmetry in, and loaded in, the plane of the web.

EXAMPLE 9.7.1

Calculate and plot the allowable stress relationship for a W14×30 of A36 steel taking $C_b = 1$.

Solution

(a) Calculate L_c. Using the *smaller* value from Eqs. 9.7.1 and 9.7.2 as prescribed in ASD-F1.1,

$$L_c = \frac{76b_f}{\sqrt{F_y}} = \frac{76(6.73)}{\sqrt{36}\,(12)} = 7.1 \text{ ft}$$

or

$$L_c = \frac{20,000}{(d/A_f)F_y} = \frac{20,000}{(5.34)36(12)} = 8.7 \text{ ft}$$

Thus $L_c = 7.1$ ft.

(b) Calculate L_u. Using the *larger* of Eqs. 9.7.25,

$$L_u = \frac{\cdot 20,000}{(d/A_f)F_y} = 8.7 \text{ ft} \quad \text{(from above)}$$

or

$$L_u = r_T\sqrt{\frac{102,000}{F_y}} = \frac{1.74}{12}\sqrt{\frac{102,000}{36}} = 7.7 \text{ ft}$$

Thus $L_u = 8.7$ ft.

(c) Unbraced lengths larger than L_u. Use the *larger* value from Eq. 9.7.12 or 9.7.20, as prescribed in ASD-F1.3, Formulas (F1-6 and F1-7) and (F1-8),

$$F_b(\text{F1-8}) = \frac{12,000C_b}{L_bd/A_f}$$

$$= \frac{12,000(1.0)}{L_b(5.34)12} = \frac{187}{L_b} \text{ ksi}$$

$$F_b(\text{F1-6}) = \left[\frac{2}{3} - \frac{F_y(L_b/r_T)^2}{1530(10^3)C_b}\right]F_y$$

$$= \left[\frac{2}{3} - \frac{36(L_b/1.74)^2 144}{1530(10^3)(1.0)}\right]36$$

$$= 24.0 - \frac{L_b^2}{24.8} \text{ ksi}$$

According to Eq. 9.7.26, ASD Formulas (F1-6) and (F1-7) give the same allowable stress at

$$L_b = r_T\sqrt{\frac{510,000C_b}{F_y}} = \frac{1.74}{12}\sqrt{\frac{510,000(1.0)}{36}} = 17.3 \text{ ft}$$

For L_b larger than 17.3 ft, the allowable stress is given by the larger of

$$F_b(\text{F1-8}) = \frac{187}{L_b} \text{ ksi (computed above)}$$

or

$$F_b(\text{F1-7}) = \frac{17,000C_b}{(L_b/r_T)^2}$$

$$= \frac{170,000(1.0)}{(L_b/1.74)^2 144} = \frac{3570}{L_b^2} \text{ ksi}$$

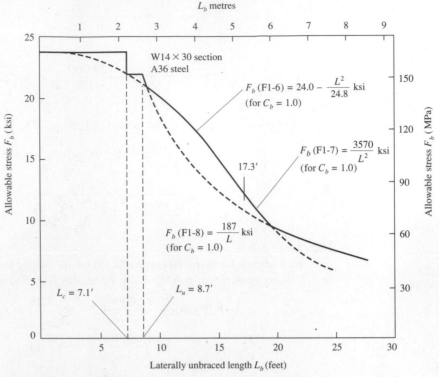

Figure 9.7.3   Allowable bending stress for W14×30 section with $C_b = 1$, as computed in Example 9.7.1.

The complete allowable stress relationship for this W14×30 with $C_b = 1$ is shown in Fig. 9.7.3. ■■

More Precise Analysis

ASD-F1.3 permits the use of "a more precise analysis" which according to the ASD Commentary means the procedure "may be refined to include both St. Venant and warping torsion." This would mean the use of Eq. 9.4.31 in some fashion. If this more precise analysis procedure is desired, the use of Load and Resistance Factor Design as described in Sec. 9.6 will be more practical.

Comparision of ASD and LRFD Strength Relationships

Load and Resistance Factor Design recognizes the moment strength of a rolled I-shaped section in strong axis bending to a greater extent than does Allowable Stress Design. This is particularly true when ASD Formulas (F1-6 and F1-7) or (F1-8) are used, since each reflects only part of the available strength. The comparision for a W14×30 section, the same section whose ASD relationship is shown in detail in Fig. 9.7.3, is given in Fig. 9.7.4. The ASD curve would be higher and much closer to the LRFD relationship if the permission in ASD-F1.3 of using "a more precise analysis" is followed. In that case the two curves for elastic buckling, that is, for L_b

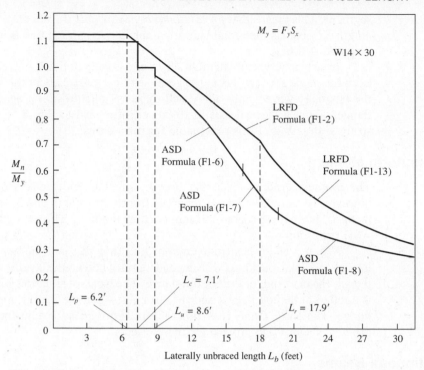

Figure 9.7.4 Comparison of nominal strength M_n used by Load and Resistance Factor Design (LRFD) with that used by Allowable Stress Design (ASD).

greater than L_r, would be identical. However, since usually the "more precise" curve is not used, Fig 9.7.4 shows the practical comparison.

The true design comparison could show a wider difference than the nominal strength used in the two methods may indicate. The strength assumed in ASD is obtained by multiplying the service load capacity $F_b S_x$ by the safety factor 1.67 to obtain the nominal strength M_n. The actual safety factor used in LRFD will be the overload factors γ as discussed in Sec. 1.8 divided by the strength reduction factor ϕ_b for flexure. If the service loading M were 80% live load and 20% dead load, the combined safety factor in LRFD would be

$$M_n = M_u/\phi_b = [1.2(0.2M) + 1.6(0.8M)]/0.90 = 1.69M$$

which would be very close to the 1.67 used in ASD. The range would be from 1.33 for 100% dead load to 1.78 for 100% live load (an impossible situation).

Additionally, the ASD method using $0.66F_y$ as the maximum allowable stress effectively recognizes a shape factor Z_x/S_x for an I-shaped section of 1.1, whereas except in a few cases the actual value exceeds 1.1; for the W14×30 section $Z_x/S_x = 1.13$. In LRFD the actual M_p is used as the upper limit of the strength relationship.

9.8 EFFECTIVE LATERALLY UNBRACED LENGTH

Design equations, such as those of LRFD and ASD-F1, are based on the assumption of torsional simple support (see Figs. 8.5.4 and 8.7.2) at the ends of the unbraced segment. This means that *for torsional behavior only* the effective laterally unbraced length KL_b equals the actual laterally unbraced length L_b; that is, the effective length factor $K = 1.0$.

Nearly any type of lateral brace or end connection will prevent rotation (that is, keep the angle ϕ of twist equal to zero; see Fig. 9.4.1c) about the longitudinal axis of the beam. However, only rarely will there be much restraint against rotation of the flanges about a vertical axis as discussed in regard to Fig. 8.5.7 (i.e., warping restraint); thus $K = 1.0$ is reasonable for most cases.

Nonuniform Moment

The moment gradient does not directly affect laterally unbraced effective length; however, it does affect lateral-torsional buckling strength and is accounted for by the factor C_b when there is non-constant bending moment. This is true for both LRFD and ASD procedures.

When a beam has transverse loading (not just end moments) the strength in lateral-torsional buckling will depend upon whether the load is applied at the top flange, the centroid, or the bottom flange. The weakest situation will be when the load is applied at the top flange, and the strongest when the load is applied at the bottom flange. The *SSRC Guide* [6.8, pp. 157–162] provides for adjustment to the moment gradient factor C_b to provide for these effects.

Continuous Beams

A continuous beam has lateral end restraint moments develop as a result of continuity over several spans. When the adjacent spans are shorter than the span being considered, or at least braced laterally at closer intervals; or the adjacent spans are less severely loaded, some lateral moment restraint may develop.

Typically, however, such end restraint about the $y–y$ axis should not be assumed present since alternate unbraced spans could buckle in opposite directions. For treatment of lateral buckling on continuous beams, the reader is referred to Salvadori [9.6, 9.19], Hartmann [9.20], Trahair [9.21, 9.23], Powell and Klingner [9.22], Roeder and Assadi [9.24], and Fukumoto, Itoh, and Hattori [9.25]. The *SSRC Guide* [6.8, pp. 162–165] presents a method for evaluating the strength of a continuous beam span having intermediate points of lateral support.

In continuous beams, the point of inflection has often been treated as a braced point when design equations did not provide for the effect of moment gradient. In current design using ASD [1.5] or LRFD [1.16], the effect of moment gradient appears in all equations except in the requirements to establish the ability to qualify as a "compact section". The requirements of LRFD-F1.3 where L_p is prescribed and ASD-F1.3 where L_c is prescribed do not include the moment gradient. When applying these provisions one may wish to *consider* the inflection point as a possible braced point.

The authors' present opinion is that whenever moment gradient is explicitly provided for in a design equation (no matter how approximately), the inflection point *should not* be considered a braced point. However, when provision is not made for the effect of moment gradient, generally the inflection may be considered as a braced point. The combination of torsional restraint provided by the floor system attachments and the continuity at the support point of maximum negative moment will normally be adequate to justify treating the inflection point as if it were actually braced *for the purpose of evaluating whether or not L_b exceeds L_p in LRFD or exceeds L_c in ASD*. The important factor is the amount of torsional restraint provided by the floor system at the inflection point.

Regarding endorsement of this suggestion, LRFD-Symbols defines L_p as used in LRFD-F1.2a as "Limiting laterally unbraced length for full plastic bending capactiy, uniform moment case." Thus, there is no implication in either LRFD or ASD for the use of the inflection point as a braced point.

Cantilever Beams Without Lateral Bracing

Unlike the flagpole column where the effective pinned-end length is twice the actual length, the lateral-torsional buckling of a cantilever beam is not even as severe as the unbraced segment under uniform moment. If one considers the analogy to a coumn as discussed in Sec. 9.1, such a result is logical. Since the moment at the free end of the cantilever is zero, the compression force in the flange decreases from a maximum at one end to zero at the free end; thus, the loading is less severe than if the compression force were constant over the entire length.

Clark and Hill [9.26] and *SSRC Guide,* 3rd ed. [6.20] reported that it is conservative to use the full length as the effective laterally unbraced length for lateral-torsional buckling of cantilever beams. However, the conservativeness of using the actual length for a cantilever depends in large part on having *fixed torsional restraint* at the supported end, as well as having the loading applied at the shear center or at the bottom flange. Since such torsional fixity rarely exists, the authors recommend using the actual cantilever length as the effective laterally unbraced length.

Furthermore, the moment gradient factor C_b, Eqs. 9.6.11 and 9.6.12, is *not* applicable for a cantilever; thus C_b should be taken as unity. Obviously, moment gradient has some effect; however, a cantilever inherently has moment that varies from maximum at the support to zero at the free end. The use of the actual length as the effective length already recognizes the moment gradient.

When a more detailed design treatment of lateral-torsional buckling of cantilevers is desired, the *SSRC Guide,* 4th ed. [6.8, pp. 168–169] provides (from a recommendation by Nethercot) effective length factors K in order to use KL instead L in Eq. 9.4.32. Kitipornchai, Dux, and Richter [9.27] have a recent excellent overall review of the subject. Again in that method, $C_b = 1$ would be used. For nonuniform cantilevers, refer to the work of Massey and McGuire [9.28].

9.9   EXAMPLES: LOAD AND RESISTANCE FACTOR DESIGN

Several examples of I-shaped beams are presented to illustrate the Load and Resistance Factor Design method to include the effect of lateral-torsional buckling. Other considerations, such as deflection, shear, and web local yielding and crippling were treated and illustrated in Chapter 7.

EXAMPLE 9.9.1

A simply supported beam is loaded as shown in Fig. 9.9.1. The uniform load is 15% dead load and 85% live load, and the concentrated load is 40% dead load and 60% live load. The beam has transverse lateral support at the ends and every 7'-6" along the span. Select the lightest W section of A36 steel, using Load and Resistance Factor Design.

Solution

(a) Determine the factored moment M_u at midspan and the required nominal strength M_n. Using LRFD-A4.1, Formula (A4-2),

$$w_u = 1.2(0.15)(1.0) + 1.6(0.85)(1.0) = 1.54 \text{ kips/ft}$$

$$W_u = 1.2(0.40)(36) + 1.6(0.60)(36) = 51.8 \text{ kips}$$

$$M_u = \frac{1}{8}(1.54)(30)^2 + \frac{51.8(30)}{4} = 562 \text{ ft-kips}$$

Since the LRFD requirement is

$$\phi_b M_n \geq M_u \qquad [9.6.1]$$

where $\phi_b = 0.90$, the strength reduction factor for flexure, the required nominal strength is

$$\text{Required } M_n = M_u/\phi_b = 562/0.90 = 625 \text{ ft-kips}$$

(b) Estimate whether or not lateral supports are close enough to design the beam using plastic analysis (LRFD-F1.2d) or to use the full plastic moment strength M_p (LRFD-F1.1.1) without plastic analysis. To use either of these provisions the section must be "compact" for local buckling in accordance with LRFD-B5. Since the beam is simply supported, plastic analysis (as described in Chapter 10) does not apply. Assume the beam will be in *Case* 2 (see Sec. 9.6), where $M_n = M_p$. The maximum laterally unbraced length L_b is given by Eq. 9.6.3,

$$L_p = \frac{300}{\sqrt{F_y, \text{ksi}}} r_y = \frac{300}{\sqrt{36}} r_y = 50r_y$$

If one assumes that $r_y \approx 0.22b_f$ (text Appendix Table A1),

$$\text{Min } b_f = \frac{L_p}{50(0.22)} = \frac{7.5(12)}{11} = 8.2 \text{ in.} \quad \text{if } L_p = 7.5 \text{ ft}$$

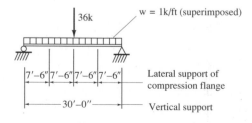

Figure 9.9.1 Examples 9.9.1 and 9.10.1.

Assuming the flange width is at least 8.2 in. the strength can reach the plastic moment M_p.

(c) Select from *LRFD Manual*, "LOAD FACTOR DESIGN SELECTION TABLE." This is an efficient procedure when the designer is certain that the section is "compact" and that $L_b < L_p$. The required plastic modulus Z_x is

$$\text{Required } M_n = \text{Required } M_p = 625 \text{ ft-kips}$$

$$\text{Required } Z_x = \frac{\text{Required } M_p}{F_y} = \frac{625(12)}{36} = 208 \text{ in.}^3$$

The lightest section that has $Z_x \geq 208$ in.³ is W24×84 having $Z_x = 224$ in.³; in addition $b_f > 8.2$ in.

(d) Check the section. The dead weight of the beam must be included; it could have been estimated at the beginning of the design.

$$M_u(\text{dead load}) = 1.2(0.084)(30)^2/8 = 11 \text{ ft-kips}$$
$$M_u = 562 + 11 = 573 \text{ ft-kips}$$
$$M_n = M_p = Z_x F_y = 224(36)/12 = 672 \text{ ft-kips}$$
$$L_p = 50r_y = 50(1.95)/12 = 8.1 \text{ ft} > (L_b = 7.5 \text{ ft}) \quad \text{OK}$$

The other requirements (see Table 9.6.1) for "compact section" are satisfied.

$$[\phi_b M_n = 0.90(672) = 605 \text{ ft-kips}] > (M_u = 573 \text{ ft-kips}) \quad \text{OK}$$

In this case the strength provided is nearly 6% high; however, the W24×84 is the lightest section satisfying the requirements. The flange and web of this section satisfy the "compact section" requirements to prevent local buckling; that is, $\lambda \leq \lambda_p$ as given in Table 9.6.1.
Use W24×84. ■■

EXAMPLE 9.9.2 _____

Select the lightest W section for the simply supported beam of Fig. 9.9.2. The super-imposed load is 0.2 kip/ft dead load and 0.8 kip/ft live load. Lateral support is provided at the ends and at midspan. Assume deflection limitations need not be considered. Use A36 steel and Load and Resistance Factor Design.

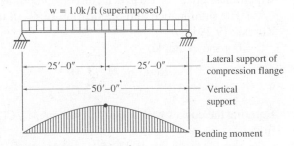

Figure 9.9.2 Data for Examples 9.9.2, 9.9.4, 9.10.2, and 9.10.4.

Solution

(a) Determine the factored moment M_u at midspan and the required design strength $\phi_b M_n$. Estimate the beam weight about 90 lb/ft. Using LRFD-A4.1, Formula (A4-2),

$$w_u = 1.2(0.2 + 0.09) + 1.6(0.8) = 1.63 \text{ kips/ft}$$

$$M_u = 1.63(50)^2/8 = 509 \text{ ft-kips}$$

$$\text{Required } \phi_b M_n = M_u = 509 \text{ ft-kips}$$

(b) Use the LRFD beam curves, "BEAM DESIGN MOMENTS." These curves are plots of the design moment strength $\phi_b M_n$ vs laterally unbraced length L_b for rolled W and M shapes for $F_y = 36$ ksi and $F_y = 50$ ksi. The curves are the nominal strength M_n relationship shown in Fig. 9.6.1 multiplied by the strength reduction factor ϕ_b (= 0.90 for flexure). The moment gradient factor $C_b = 1.0$, the most severe loading case (except for loads applied at the top flange in laterally unbraced locations), is used for the curves.

For this example, C_b will be greater than 1.0 computed in accordance with Eq. 9.6.11 for the unbraced half-span segment,

$$C_b = \frac{12.5 M_{\max}}{2.5 M_{\max} + 3 M_A + 4 M_B + 3 M_c} \qquad [9.6.11]$$

where $M_{\max}$ = 509 ft-kips, maximum at midspan
M_A = 223 ft-kips at 1/4 point of unbraced segment
M_B = 382 ft-kips at midpoint of unbraced segment
M_C = 477 ft-kips at 3/4 point of unbraced segment

$$C_b = \frac{12.5(509)}{2.5(509) + 3(223) + 4(382) + 3(477)} = 1.30$$

The same result can be obtained from Case 3 in Table 9.6.4. This is significantly more conservative (i.e., lower) than using Eq. 9.6.12 from the 1986 LRFD Specification, which gives $C_b = 1.75$.

Referring to Fig. 9.6.4, for situations where $L_b > L_p$, the nominal strength M_n is proportional to C_b up to the maximum strength M_p. To use the LRFD beam curves based on $C_b = 1.0$ for C_b greater than 1.0, divide the M_n equations by C_b. The moment ordinate in the LRFD beam curves is the design strength $\phi_b M_n$ which must equal or exceed factored moment M_u.

The significance and application of C_b in design is discussed in detail by Zuraski [9.29].

For the example,

$$\frac{\text{Reqd } \phi_b M_n}{C_b} = \frac{M_u}{C_b} = \frac{509}{1.30} = 392 \text{ ft-kips}$$

Entering the curves with factored moment $M_u/C_b = 392$ ft-kips and $L_b = 25$ ft, find

$$W\,27\times84, \quad \phi_b M_n \approx 510 \text{ ft-kips}$$

This may well be too conservative; M_n increases linearly with C_b. When $L_b \leq L_r$ and $M_r C_b = M_p$ no further beneficial effect of increasing C_b is possible.

Determine the value of L_r (see Fig. 9.6.1) for the W27×84 using Eq. 9.6.6.

$$L_r = \frac{r_y X_1}{(F_y - F_r)}\sqrt{1 + \sqrt{1 + X_2(F_y - F_r)^2}} \qquad [9.6.6]$$

where $X_1 = 1570$ ksi and $X_2 = 31{,}100 \times 10^{-6}$ in.^{4}/kip^2 as properties given in the *LRFD Manual*, or X_1 and X_2 can be computed using Eqs. 9.6.7 and 9.6.8. Thus,

$$L_r = \frac{2.07(1570)}{(36 - 10)12}\sqrt{1 + \sqrt{1 + \frac{31{,}100(36 - 10)^2}{(10)^6}}} = 24.9 \text{ ft}$$

The value of L_r for a rolled section is also available for $F_y = 36$ and 50 ksi in the LRFD "LOAD FACTOR DESIGN SELECTION TABLE."

Since L_b in this example (25 ft) is very close to L_r, determine the maximum effective C_b. Setting $M_r C_b = M_p$ gives

$$M_r C_b = M_p$$
$$(F_y - F_r)S_x C_b \stackrel{.}{=} Z_x F_y$$

Solving for C_b gives

$$\text{Max effective } C_b = \frac{Z_x F_y}{(F_y - F_r)S_x}$$

which for A36 steel and rolled shapes ($F_r = 10$ ksi) becomes

$$\text{Max effective } C_b = \frac{36}{(36 - 10)}\frac{Z_x}{S_x} = 1.38\frac{Z_x}{S_x}$$

Since $Z_x/S_x \approx 1.12$ as an average, the maximum effective C_b for A36 steel will be about 1.55. Thus, in this example where $C_b = 1.30$, the nominal strength M_n will not likely reach M_p. If it seemed that M_p would be reached, the LRFD "LOAD FACTOR DESIGN SELECTION TABLE" could be used to select the section.

(c) Check the W27×84 section. Including the dead weight of the beam,

$$M_u = 1.62(50)^2/8 = 507 \text{ ft-kips}$$
$$L_r = 24.9 \text{ ft} < (L_b = 25 \text{ ft})$$
$$M_p = Z_x F_y = 244(36)/12 = 732 \text{ ft-kips}$$

Since $L_b > L_r$, $M_n = M_{cr}$ which increases with C_b but cannot exceed $C_b M_r$ nor M_p. Compute M_{cr} using Eq. 9.6.13 which includes C_b,

$$M_n = M_{cr} = \frac{C_b S_x X_1 \sqrt{2}}{L_b/r_y}\sqrt{1 + \frac{X_1^2 X_2}{2(L_b/r_y)^2}} \qquad [9.6.13]$$

$$\frac{L_b}{r_y} = \frac{25(12)}{2.09} = 145$$

$$M_n = M_{cr} = \frac{1.30(213)(1570)\sqrt{2}}{145(12)}\sqrt{1 + \frac{(1570)^2\, 31{,}100}{2(145)^2(10)^6}} = 594 \text{ ft-kips}$$

Since M_{cr} (594 ft-kips) does not exceed the upper limit M_p (732 ft-kips), the nominal strength M_n is

$$M_n = M_{cr} = 594 \text{ ft-kips}$$

$$(\phi_b M_n = 0.90(594) = 535 \text{ ft-kips}) > (M_u = 507 \text{ ft-kips}) \quad \text{OK}$$

The W27×84 is the lightest section satisfying the requirements. In addition to what has been checked, the section must be "compact" to assure that the computed strength is correct. As are nearly all A36 sections, this section is "compact" for local buckling; that is, it satisfies the limits for web and flange given in Table 9.6.1.
Use W27×84. ■■

EXAMPLE 9.9.3 _____

Select an economical W section for the beam of Fig. 9.9.3. Lateral support is provided at the vertical supports, concentrated load points, and at the end of the cantilever. The 26-kip load contains 6 kips dead load and the 11.5-kip load includes 4 kips dead load; the remainder is live load. Use A36 steel and Load and Resistance Factor Design.

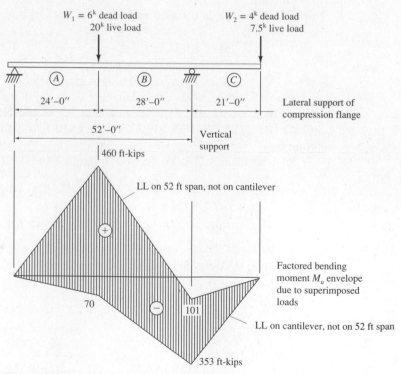

Figure 9.9.3   Data for Example 9.9.3.

Solution

(a) Obtain the bending moment envelope resulting from factored loads. The moment envelope will include two loading cases: (1) dead load plus live load on the 52-ft span and no live load on cantilever; and (2) dead load plus live load on cantilever and no live load on the 52-ft span.

$$W_{u1} = 1.2(6) + 1.6(20) = 39.2 \text{ kips}$$
$$W_{u2} = 1.2(4) + 1.6(7.5) = 16.8 \text{ kips}$$

For loading case 1, the maximum factored moment under load W_1 is

$$M_u \text{ on cantilever} = 1.2(4)21 = 101 \text{ ft-kips}$$
$$M_u \text{ at } W_1 = 39.2(24)(28)/52 - 101(24/52) = 460 \text{ ft-kips}$$

For loading case 2, the maximum factored moment at the cantilever support is

$$M_u = 16.8(21) = 353 \text{ ft-kips}$$

The factored moment envelope is shown in Fig. 9.9.3.

(b) Segment A. Three laterally unbraced segments must be considered since each length is different and is subject to a different maximum factored bending moment M_u. By inspection, segment A controls over segment C; both the moment and the unbraced length are larger on segment A, and the moment gradient is the same; thus, C_b is the same. From the approximate Eq. 9.6.12 of the 1986 *LRFD Specification*, $C_b = 1.75$ by inspection when the moment at one end is zero. Using the more accurate 1993 Eq. 9.6.11, or from Table 9.6.3, $C_b = 1.67$.

$$M_u = 460 \text{ ft-kips}; \qquad C_b = 1.67; \qquad L_b = 24 \text{ ft}$$
$$\text{Required } \phi_b M_n = 460 \text{ ft-kips}$$

Use the *LRFD Manual* charts, "BEAM DESIGN MOMENTS." As in Example 9.9.2, the C_b factor may be combined with the factored moment M_u.

$$\frac{M_u}{C_b} = \frac{460}{1.67} = 275 \text{ ft-kips}$$

Enter the charts with required $\phi_b M_n = M_u/C_b = 275$ ft-kips and $L_b = 24$ ft; find

$$\text{W24} \times 76, \quad \phi_b M_n > 300 \text{ ft-kips}$$

Lighter shallower beams could be selected; however, deflection would then likely control. Determine $L_r = 23.4$ ft for this section from Eq. 9.6.6, or more easily from *LRFD Manual*, "LOAD FACTOR DESIGN SELECTION TABLE."

For segment A, where $L_b = 24$ ft does not significantly exceed L_r, and C_b is more than the maximum effective value 1.55 obtained in Example 9.9.2 for A36 steel, the section will likely have its maximum nominal strength M_p. Use of *LRFD Manual* "LOAD FACTOR DESIGN SELECTION TABLE" is then appropriate. The required plastic modulus is

$$\text{Required } Z_x = \frac{\text{Required } M_p}{F_y} = \frac{(460/0.90)12}{36} = 170 \text{ in.}^3$$

The lightest section having $Z_x \geq 170$ in.[3] is W24×68 which has $Z_x = 177$ in.[3] Because this section has $L_r = 22.4$ ft, lower than the L_r for the W24×76, the strength with $L_b = 24$ ft may not be adequate.

When L_b is near or larger than L_r, the best way of selecting the lightest section will be using the charts, "BEAM DESIGN MOMENTS." To help understand the strength relationships, examine Fig. 9.9.4, where the W24×68 and W24×76 design strengths $\phi_b M_n$ are shown for $C_b = 1.0$ and 1.67.

(c) Segment B. This segment has a steeper moment gradient than segment A, i.e., $C_b > 1.67$; however, the maximum C_b usable is 1.55 for A36 steel. The laterally unbraced length is the longer 28 ft, indicating that segment B will likely control. Using Eq. 9.6.11,

$$C_b = \frac{12.5 M_{\max}}{2.5 M_{\max} + 3 M_A + 4 M_B + 3 M_c} \qquad [9.6.11]$$

or for the ratio $M_1/M_2 = +101/460 = +0.22$ the C_b can be obtained from Fig. 9.6.3 as $C_b \approx 2.0$. Or linearly interpolating the values in Fig. 9.6.3. gives $C_b = 1.96$. Such precision is unjustified. Use $C_b = 2.0$.

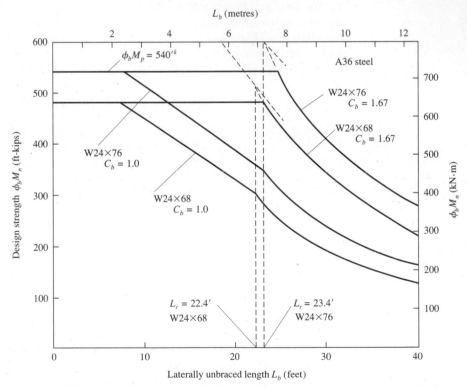

Figure 9.9.4   Comparison of design strengths $\phi_b M_n$ for W24×76 and W24×68 for A36 steel—Example 9.9.3.

An entry to the LRFD "BEAM DESIGN MOMENTS" charts would use $M_u/C_b = 460/2.0 = 230$ ft-kips at $L_b = 28$ ft. Using a W24 section, segment B also requires a W24×76.

(d) Check the W24×76. From the *LRFD Manual*, p. 4-18,

$$L_r = 23.4 \text{ ft}; \quad \phi_b M_p = 540 \text{ ft-kips}; \quad \phi_b M_r = 343 \text{ ft-kips}$$

For all three beams segments, L_b exceeds L_r indicating elastic lateral-torsional buckling is the controlling limit state. Thus, M_{cr} must be computed for both the 24- and 28-ft unbraced lengths (segments A and B). Using Eq. 9.6.13,

$$M_n = M_{cr} = \frac{C_b S_x X_1 \sqrt{2}}{L_b/r_y} \sqrt{1 + \frac{X_1^2 X_2}{2(L_b/r_y)^2}} \qquad [9.6.13]$$

The properties needed are $S_x = 176$ in.^{3}, $X_1 = 1760$ ksi, $X_2 = 18{,}600 \times 10^{-6}$ in.^{4}/kip^2, and $r_y = 1.92$ in.,

$$M_n = M_{cr} = \frac{70{,}090 C_b}{L_b} \sqrt{1 + \frac{738}{L_b^2}}$$

for the two possible controlling segments:

Segment A, $C_b = 1.67$, $L_b = 24$ ft, find $M_n = M_p = 600$ ft-kips

Segment B, $C_b = 2.00$, $L_b = 28$ ft, find $M_n = M_{cr} = 581$ ft-kips

$$[\phi M_n = 0.90(581) = 523 \text{ ft-kips}] > [M_u = 460 + 18(\text{beam wt}) = 478 \text{ ft-kips}] \quad \text{OK}$$

Thus, W24×76 is acceptable. The section satisfies the limits for flange and web given in Table 9.6.1 to preclude local buckling, and therefore is "compact"; i.e., $\lambda \leq \lambda_p$. Therefore, the lateral-torsional buckling limit state controls, as assumed. The W24×68 has $\phi_b M_n$ only 449 ft-kips for segment A and 422 ft-kips for segment B; thus it is not acceptable.
Use W24×76. ∎∎

EXAMPLE 9.9.4

Repeat Example 9.9.2 (See Fig. 9.9.2) except use A572 Grade 60 steel.

Solution

(a) Obtain the factored moment M_u at midspan and the required design strength $\phi_b M_n$. From Example 9.9.2, including the beam weight now estimated as 70 lb/ft (the A36 beam was 76 lb/ft),

$$w_u = 1.2(0.2 + 0.07) + 1.6(0.8) = 1.60 \text{ kips/ft}$$
$$M_u = 1.60(50)^2/8 = 500 \text{ ft-kips}$$
$$\text{Required } \phi_b M_n = M_u = 500 \text{ ft-kips}$$

(b) Select trial section using beam charts, "BEAM DESIGN MOMENTS." Since there are no LRFD tables or charts for $F_y = 60$ ksi, a more general approach must be used. When reaching the plastic moment strength M_p is expected to be the controlling limit state, that strength is proportional to the yield stress. Thus, when $L_b < L_p$, use "LOAD FACTOR DESIGN SELECTION

TABLE." When L_b is expected to exceed L_p, adjust the required design moment strength in the ratio of the yield stress used to $F_y = 50$ ksi for which there are charts. Thus, for this example,

$$\text{Adjusted } M_u \text{ for } F_y = 50 \text{ ksi} \approx 500(50/60) = 417 \text{ ft-kips}$$

The C_b factor may be combined with the factored moment M_u,

$$\frac{\text{Adjusted } M_u \text{ for } F_y = 50 \text{ ksi}}{C_b} = \frac{417}{1.30} = 320 \text{ ft-kips}$$

Enter the charts with required $\phi_b M_n = M_u/C_b = 320$ ft-kips and $L_b = 25$ ft; find

$$W16\times67, \quad \phi_b M_n = 326 \text{ ft-kips}$$
$$W21\times83, \quad \phi_b M_n = 332 \text{ ft-kips}$$
$$W24\times84, \quad \phi_b M_n = 375 \text{ ft-kips}$$

The sections are all located in the charts above and to the right of the entry point. From the *LRFD Manual* "LOAD FACTOR DESIGN SELECTION TABLE," find for $F_y = 50$ ksi:

$$W16\times67, L_r = 23.8 \text{ ft}, \quad \phi_b M_p = 488 \text{ ft-kips}$$
$$\phi_b M_r = 351 \text{ ft-kips}$$
$$W21\times83, L_r = 18.5 \text{ ft}, \quad \phi_b M_p = 735 \text{ ft-kips}$$
$$\phi_b M_r = 513 \text{ ft-kips}$$
$$W24\times84, L_r = 21.5 \text{ ft}, \quad \phi_b M_p = 840 \text{ ft-kips}$$
$$\phi_b M_r = 588 \text{ ft-kips}$$

Try the moderately deep W21 to reduce deflection. Since L_b considerably exceeds the L_r of 18.5 ft for $F_y = 50$ ksi and L_r will be *even lower* for $F_y = 60$ ksi, elastic lateral-torsional buckling strength M_{cr} will likely be the controlling limit state. The variation in L_r with F_y for the W21$\times$83 is shown in Fig. 9.9.5; the two *LRFD Manual* tabulated values are marked as points A and B. Typically, as the yield stress is higher, the L_r is lower, but there is less decrease than if the relationship were linear. For this example, because L_r is low as compared to $L_b = 25$ ft the W21$\times$83 seems unlikely to work. Returning to the charts (*LRFD Manual* p. 4-157), find W18$\times$76 that appears OK and is intermediate between the W16 and W21 tentatively selected above. The W18$\times$76 with $L_r = 24.8$ ft at $F_y = 50$ ksi seems very likely to work.

(c) Check the W18$\times$76. Compute M_p,

$$M_p = Z_x F_y = 163(60)/12 = 815 \text{ ft-kips}$$
$$\phi_b M_p = 0.90(815) = 734 \text{ ft-kips}$$

Compute L_r for $F_y = 60$ ksi from Eq. 9.6.6 below using $r_y = 1.81$ in., $X_1 = 2140$ ksi, and $X_2 = 8380 \times 10^{-6}$ in.^{4}/kips^2 for the W18$\times$76,

$$L_r = \frac{r_y X_1}{(F_y - F_r)} \sqrt{1 + \sqrt{1 + X_2(F_y - F_r)^2}} = 15.4 \text{ ft}$$

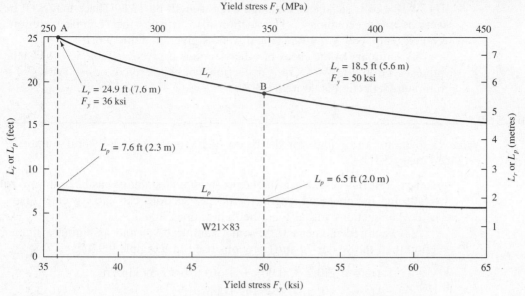

Figure 9.9.5   Variation in L_r and L_p for various yield stresses—W21×83 section.

Since $L_b = 25$ ft for this example exceeds L_r, the elastic lateral-torsional buckling strength M_{cr} must be computed to make the check of this section. Using Eq. 9.6.13 with $C_b = 1.30$ along with the other values used above for L_r gives

$$M_n = M_{cr} = \frac{C_b S_x X_1 \sqrt{2}}{L_b/r_y} \sqrt{1 + \frac{X_1^2 X_2}{2(L_b/r_y)^2}} \qquad [9.6.13]$$

$$M_n = M_{cr} = 625 \text{ ft-kips}; \quad \phi_b M_n = 563 \text{ ft-kips}$$

$$(\phi_b M_n = 563 \text{ ft-kips}) > (M_u = 500 \text{ ft-kips}) \quad \text{OK}$$

TABLE 9.9.1   SUMMARY OF STRENGTH CHECKS FOR EXAMPLE 9.9.4

Section[†]	L_r (ft)	$\phi_b M_{cr}$ (ft-kips)	$\phi_b M_p$ (ft-kips)	$\phi_b M_n$ (ft-kips)	Comment
W24×94	16.0	592	1238	592	18% overstrength
W24×84	15.4	482	1092	482 < 500	NG
W21×93	16.7	527	995	527	OK
W21×83	15.0	432	1092	432 < 500	NG
W18×76	21.5	563	734	563	13% overstrength
W18×71	14.3	313	707	313 < 500	NG
W16×77	21.6	528	675	528	OK
W16×67	19.4	425	634	425 < 500	NG
W14×74	23.5	504	567	504	OK

†All sections in table satisfy $\lambda \le \lambda_p$ for flange and web local buckling (see Table 9.6.1) and are therefore "compact sections" in $F_y = 60$ ksi steel.

The W18×76 is satisfactory though overstrength by 13%. There may well be sections more economical. The section also satisfies the "compact section" requirements $\lambda \leq \lambda_p$ for flange and web as given in Table 9.6.1.

A summary of the check of other sections is given in Table 9.9.1. For the lightest section, *use* W14×74 if deflection can be tolerated; or *use* W24×94 if minimum deflection is desired. ■■

EXAMPLE 9.9.5

What W section can be used for the beam of Example 9.9.4 if lateral support is provided every 5 ft?

Solution.   In this case, since the lateral-torsional stability has been improved by reducing the unbraced length, the deeper sections can carry greater loads. The deeper sections will also be the lighter ones.

(a) Factored moment M_u. From Example 9.9.4 and assuming a lighter section than the 94- or 74-lb/ft sections used in Example 9.9.4,

$$w_u = 1.2(0.2 + 0.06) + 1.6(0.8) = 1.59 \text{ kips/ft}$$
$$M_u = 1.59(50)^2/8 = 497 \text{ ft-kips}$$
$$\text{Required } \phi_b M_n = M_u = 497 \text{ ft-kips}$$

(b) Select a trial section. Assume the section will be adequately braced to achieve its plastic moment strength M_p,

$$\text{Required } Z_x = \frac{\text{Required } M_p}{F_y} = \frac{(497/0.90)12}{60} = 110 \text{ in.}^3$$

From "LOAD FACTOR DESIGN SELECTION TABLE", the lightest section that has $Z_x \geq 110$ in.^{3} is W21×50 having $Z_x = 110$ in.^{3}. Note that $L_p = 4.6$ ft for $F_y = 50$ ksi, and will be lower for $F_y = 60$ ksi. Thus, a section having somewhat larger Z_x must be used. The next lightest section is W24×55 having $Z_x = 134$ in.^{3}.

(d) Check the section. Note that the 5-ft laterally unbraced segment adjacent to midspan will have C_b very close to 1.0 (actually it is 1.02 for $M_1/M_2 = -0.96$); if the check shows the strength to be slightly low, the correct C_b can be obtained from Table 9.6.3 (or it can be computed). Assume $C_b = 1.0$ for the check. The dead weight of the beam must be included,

$$M_u = 1.59(50)^2/8 = 497 \text{ ft-kips}; \qquad \text{Required } M_n = 551 \text{ ft-kips}$$

For the W24×55, using $r_y = 1.34$ in., $X_1 = 1540$ ksi, and $X_2 = 39,600 \times 10^{-6}$ in.^{4}/kips^2,

$$L_p = \frac{300}{\sqrt{F_y, \text{ ksi}}} r_y$$
$$L_p = 4.3 \text{ ft} \quad (\text{for } F_y = 60 \text{ ksi})$$
$$L_r = \frac{r_y X_1}{(F_y - F_r)} \sqrt{1 + \sqrt{1 + X_2(F_y - F_r)^2}} \qquad [9.6.6]$$

$$L_r = 11.4 \text{ ft}$$

$$M_p = Z_x F_y = 134(60)/12 = 670 \text{ ft-kips}$$

$$M_r = (F_y - F_r)S_x = (60 - 10)114/12 = 475 \text{ ft-kips}$$

Since L_b (i.e., 5 ft) exceeds L_p and does not exceed L_r, the strength is determined by linear interpolation according to Eq. 9.6.4,

$$M_n = C_b \left[M_p - (M_p - M_r)\left(\frac{L_b - L_p}{L_r - L_p}\right) \right] \leq M_p \qquad [9.6.4]$$

$$M_n = 661 \text{ ft-kips} \quad (0.968 M_p)$$

$$(\phi_b M_n = 594 \text{ ft-kips}) > (M_u = 497 \text{ ft-kips}) \quad \text{OK}$$

Use W24×55 ($F_y = 60$ ksi), even though it has excess strength. ■■

EXAMPLE 9.9.6

Given the welded I-shaped section of Fig. 9.9.6 used as a 45-ft simply supported beam laterally supported at the one-third points. Determine the service live load the beam may be permitted to carry if the dead load is 0.15 kip/ft including the beam weight. Use Load and Resistance Factor Design. The steel has $F_y = 65$ ksi.

Solution

(a) Compute cross-sectional properties.

$$\text{Area,} \quad A = 28.1 \text{ sq. in.}$$

$$I_x = \frac{1}{12}\left[\left(26 + \frac{5}{8} + \frac{5}{8}\right)^3 (16) - (26)^3\left(16 - \frac{5}{16}\right)\right] = 4003 \text{ in.}^4$$

$$S_x = \frac{I_x}{d/2} = \frac{4003}{13.625} = 294 \text{ in.}^3$$

$$I_y = 2\left(\frac{1}{12}\right)(16)^3(0.625) = 427 \text{ in.}^4$$

$$r_y = \sqrt{\frac{I_y}{A}} = \sqrt{\frac{427}{28.1}} = 3.90 \text{ in.}$$

$$J = \frac{1}{3}[2(16)(0.625)^3 + 26(0.3125)^3] = 2.87 \text{ in.}^4$$

$$C_w = \frac{I_y h^2}{4} = \frac{427(26 + 0.625)^2}{4} = 73,850 \text{ in.}^6$$

$$X_1 = \frac{\pi}{S_x}\sqrt{\frac{EGJA}{2}} = \frac{\pi}{294}\sqrt{\frac{29,000(29,000)(2.87)28.1}{2.6(2)}} = 1220 \text{ ksi}$$

$$X_2 = 4\frac{C_w}{I_y}\left(\frac{S_x}{GJ}\right)^2 = 4\frac{73,850}{427}\left(\frac{294(2.6)}{29,000(2.87)}\right)^2 = 0.0584 \text{ in.}^4/\text{kips}^2$$

(b) Investigate the local flange buckling and local web buckling limit states. Check $\lambda \leq \lambda_p$ according to LRFD-B5.1; the limits are given in Table

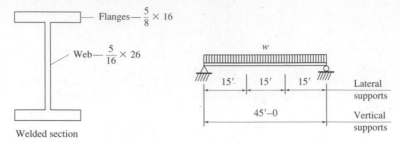

Figure 9.9.6   Data for Example 9.9.6.

9.6.1. For flange local buckling,

$$\left(\lambda = \frac{b_f/2}{t_f} = \frac{16/2}{0.625} = 12.8\right) > \left(\lambda_p = \frac{65}{\sqrt{F_y}} = 8.1\right) \quad \text{NG}$$

For web local buckling,

$$\left(\lambda = \frac{h}{t_w} = \frac{26}{0.3125} = 83.2\right) > \left(\lambda_p = \frac{640}{\sqrt{F_y}} = 79.4\right) \quad \text{NG}$$

Thus, the section is not "compact" with regard to either the flange or the web. Next, check whether the flange and/or the web must be classified as "slender" elements; that is, determine whether λ exceeds λ_r (LRFD-B5.1 and Table 9.6.2). For flange local buckling,

$$\frac{h}{t_w} = 83.2; \qquad k_c = \frac{4}{\sqrt{h/t_w}} = \frac{4}{\sqrt{83.2}} = 0.44 \quad \begin{array}{l} > 0.35 \\ < 0.763 \end{array}$$

$$\lambda_r = \frac{162}{\sqrt{(F_{yf} - 16.5)/k_c}} = \frac{162}{\sqrt{(65 - 16.5)/0.44}} = 15.4$$

$$\left(\lambda = \frac{b_f/2}{t_f} = 12.8\right) < (\lambda_r = 15.4) \quad \text{OK}$$

For web local buckling,

$$\left(\lambda = \frac{h}{t_w} = 83.2\right) < \left(\lambda_r = \frac{970}{\sqrt{F_y}} = 120.3\right) \quad \text{OK}$$

Thus, the slenderness ratios λ for both the flange and the web lie between λ_p and λ_r; the section is "noncompact" with regard to both the flange and the web, and two of the possible controlling limit states for nominal strength will be local buckling of the flange or the web in the inelastic range as shown in Fig. 9.6.2.

(c) Compute the plastic moment strength M_p and the moment strength M_r at the elastic limit (see Fig. 9.6.2).

$$M_p = Z_x F_y = 319(65)/12 = 1728 \text{ ft-kips}$$
$$M_r = (F_y - F_r)S_x = (65 - 16.5)294/12 = 1187 \text{ ft-kips}$$

(d) Compute the nominal strength M_n based on the limit state of *local buckling of the flange*. Using Eq. 9.6.9,

$$M_n = M_p - (M_p - M_r)\left(\frac{\lambda - \lambda_p}{\lambda_r - \lambda_p}\right) \qquad [9.6.9]$$

$$M_n = 1728 - (1728 - 1187)\left(\frac{12.8 - 8.1}{15.4 - 8.1}\right) = 1380 \text{ ft-kips}$$

(e) Compute the nominal strength M_n based on the limit state of *local buckling of the web*. Using Eq. 9.6.9,

$$M_n = 1728 - (1728 - 1187)\left(\frac{83.2 - 79.4}{120.3 - 79.4}\right) = 1678 \text{ ft-kips}$$

(f) Examine the lateral-torsional buckling limit state. Compute L_p using Eq. 9.6.3,

$$L_p = \frac{300}{\sqrt{F_y, \text{ksi}}} r_y = \frac{300(3.90/12)}{\sqrt{65}} = 12.1 \text{ ft}$$

Since $L_b = 15$ ft exceeds L_p, the lateral-torsional buckling limit state must be examined further. Compute L_r using Eq. 9.6.6, with $r_y = 3.90$ in., $X_1 = 1220$ ksi, $X_2 = 0.0584$ in.^{4}/kips^2, $F_r = 16.5$ ksi for a welded section, and $F_y = 65$ ksi.

$$L_r = \frac{r_y X_1}{(F_y - F_r)}\sqrt{1 + \sqrt{1 + X_2(F_y - F_r)^2}} = 29.2 \text{ ft}$$

Thus, the nominal strength M_n based on the limit state of *lateral-torsional buckling* is linearly interpolated between M_p and M_r according to Eq. 9.6.4 using $C_b = 1.0$ for the center 15-ft laterally unbraced segment of the 45-ft span of this example (computed $C_b = 1.01$),

$$M_n = C_b\left[M_p - (M_p - M_r)\left(\frac{L_b - L_p}{L_r - L_p}\right)\right] \le M_p \qquad [9.6.4]$$

$$M_n = 1.0\left[1728 - (1728 - 1187)\left(\frac{15 - 12.1}{29.2 - 12.1}\right)\right] = 1636 \text{ ft-kips}$$

(g) Final evaluation. The nominal strength M_n is the lowest value among the possible controlling limit states: 1380 ft-kips for flange local buckling; 1678 ft-kips for web local buckling: and 1636 ft-kips for lateral-torsional buckling. Thus,

$$M_n = 1380 \text{ ft-kips}$$
$$\phi_b M_n = 1240 \text{ ft-kips}$$
$$\text{Maximum } M_u = \phi_b M_n = 1240 \text{ ft-kips}$$
$$M_u = 1.2M_D + 1.6M_L$$
$$M_D = 0.15(45)^2/8 = 38 \text{ ft-kips}$$

$$M_L = \frac{M_u - 1.2M_D}{1.6} = \frac{1240 - 1.2(38)}{1.6} = 747 \text{ ft-kips}$$

$$w_L = \frac{8M_L}{L^2} = \frac{8(747)}{(45)^2} = 2.95 \text{ kips/ft} \quad \blacksquare\blacksquare$$

9.10 EXAMPLES: ALLOWABLE STRESS DESIGN

Several examples are presented to illustrate allowable stress design to include lateral-torsional buckling as a factor determining the allowable stress. Other considerations, such as deflection, shear, and web crippling were treated and illustrated in Chapter 7. For some additional practical treatment of ASD beam design, see Stockwell [9.30] and Brandt [9.31].

EXAMPLE 9.10.1 ————————————————————————————————————

A simply supported beam is loaded as shown in Fig. 9.9.1. The beam has transverse lateral support at the ends and every 7'-6" along the span. Select the lightest W section of A36 steel, using Allowable stress Design.

Solution

(a) Estimate whether or not lateral supports are close enough to qualify the beam for the maximum allowable bending stress under ASD-F1.1. For "compact section" under Case 1, Eqs. 9.7.1 and 9.7.2,

$$L_c = \text{max unbraced length} = \text{smaller of } \frac{76b_f}{\sqrt{F_y}} \text{ and } \frac{20{,}000}{(d/A_f)F_y}$$

$$= 12.7b_f \quad (\text{for } F_y = 36 \text{ ksi})$$

$$b_{f\min} = \frac{L_c}{12.7} = \frac{7.5(12)}{12.7} = 7.1 \text{ in.} \quad \text{if } L_c = 7.5 \text{ ft}$$

(b) Assume compact section: the allowable bending stress is

$$F_b = 0.66F_y = 0.66(36) = 24 \text{ ksi}$$

$$M = \tfrac{1}{8}(1.0)(30)^2 + \tfrac{1}{4}(36)(30) = 382.5 \text{ ft-kips without beam weight}$$

$$\text{Required } S_x = \frac{M}{F_b} = \frac{382.5(12)}{24} = 191 \text{ in.}^3$$

(c) Select from "ALLOWABLE STRESS DESIGN SELECTION TABLE," *ASD Manual*. This procedure is efficient if the designer is certain "compact section" requirements can be satisfied. Select W24×84, $S_x = 196$ in.^{3}

$$M_{DL} = (0.084)(30)^2/8 = 9.45 \text{ ft-kips}$$

$$f_b = \frac{M}{S_x} = \frac{(382.5 + 9.45)(12)}{196} = 24.0 = F_b \quad \text{OK}$$

$$[L_c = 9.50 \text{ ft}] > [L_b = 7.50 \text{ ft}]$$

Local buckling limits for compact section are also satisfied.
Use W24×84. $\blacksquare\blacksquare$

EXAMPLE 9.10.2

Repeat the selection of the lightest W section for the simply supported beam of Fig. 9.9.2 designed using Load and Resistance Factor Design, except here use Allowable Stress Design. Lateral support is provided at the ends and at midspan. Assume deflection limitations need not be considered. Use A36 steel.

Solution

(a) Use *ASD Manual* beam curves, "ALLOWABLE MOMENTS IN BEAMS." These curves are plots of the allowable moment $F_b S_x$ vs laterally unbraced length L_b for rolled W and M shapes for $F_y = 36$ ksi and $F_y = 50$ ksi. The allowable stress F_b is the larger of the values obtained from ASD Formulas (F1-6 or F1-7) and (F1-8), Eqs, 9.7.20, 9.7.22, and 9.7.12, with the value of C_b taken as 1.0 for the curves.

Assuming the beam weight about 90 lb/ft, midspan bending moment is

$$M = \tfrac{1}{8}(109)(50)^2 = 341 \text{ ft-kips}$$

$$L_b = \text{Laterally unbraced length} = 25 \text{ ft}$$

$$C_b = 1.75 \quad (\text{ASD-F1.3, Eq. 9.6.12})$$

An examination of ASD Formula (F1-6) will show that C_b can be combined with L_b if $L_b/\sqrt{C_b}$ replaces L_b in that equation, Eq. 9.7.20. In ASD Formula (F1-8) C_b can be combined with L_b by replacing L_b by L_b/C_b. Since the curves are for $C_b = 1$, when C_b is not 1.0 the curves may be entered as follows:

1. Use L_b/C_b if Formula (F1-8) controls F_b.
2. Use $L_b/\sqrt{C_b}$ if Formula (F1-6) controls F_b.

Since for W and M rolled shapes torsional strength is usually the predominant factor, most such beams have F_b controlled by ASD Formula (F1-8), it is recommended that the first entry to the curves the with L_b/C_b. As the values are read from the curves, the user can tell from the shape of the curve whether it is the hyperbola of Formula (F1-8) or the parabola of Formula (F1-6), and adjust the entry appropriately.

For this example, in place of L_b enter with

$$\frac{L_b}{C_b} = 14.3 \text{ ft} \quad \text{or} \quad \frac{L_b}{\sqrt{C_b}} = 18.9 \text{ ft}$$

depending on the formula suspected as controlling the allowable stress. In using the curves, the solid line *above* the intersection M and L indicates the lightest section satisfying these requirements.

Entering with $M = 341$ ft-kips and $L_b/C_b = 14.3$ ft, find

$$\text{W27×84,} \quad M > 341 \text{ ft-kips}$$

$$\text{W24×84,} \quad \text{Might work}$$

Note that the W27×84 is controlled by the parabolic curve, F_b (F1-6), and reentering the curves with $L_b/\sqrt{C_b} = 18.9$ ft indicates W27×84 to be inadequate.

(b) Check W24×84, $S_x = 196$ in.[3]

Using ASD Formula (F1-8), Eq. 9.7.12,

$$F_b = \frac{12,000C_b}{L_b d/A_f} = \frac{12,000(1.75)}{25(12)3.47} = 20.2 \text{ ksi} < 0.60F_y \text{ maximum}$$

Though it is not expected to govern based on the *ASD Manual* curves, examine ASD Formula (F1-6), Eq. 9.7.20,

$$\frac{L_b}{r_T} = \frac{25(12)}{2.31} = 129.9; \qquad C_b = 1.75$$

$$\left(\frac{L_b}{r_T}\right)_{\substack{\text{lower} \\ \text{limit}}} = \sqrt{\frac{102,000C_b}{F_y}} = 70$$

$$\left(\frac{L_b}{r_T}\right)_{\substack{\text{upper} \\ \text{limit}}} = \sqrt{\frac{510,000C_b}{F_y}} = 157$$

Since $70 < 129.9 < 157$, parabolic Formula (F1-6) applies,

$$F_b = \left[\frac{2}{3} - \frac{F_y(L_b/r_T)^2}{1530(10^3)C_b}\right]F_y$$

which for A36 steel becomes

$$F_b = 24.0 - \frac{(L_b/r_T)^2}{2070} = 24.0 - \frac{(129.9)^2}{2070} = 15.8 \text{ ksi}$$

which is less than given by Formula (F1-8) and therefore does not control.

Bending stress under applied load, corrected for beam weight, is

$$f_b = \frac{M}{S_x} = \frac{339(12)}{196} = 20.7 \text{ ksi} > [F_b = 20.2 \text{ ksi}]$$

Since the 2.5% overstress may be unacceptable, use the next heavier section. *Use* W24×94. ■ ■

EXAMPLE 9.10.3

Select an economical W section for the beam of Fig. 9.10.1. Lateral support is provided at the vertical supports, concentrated load points, and at the end of the cantilever. Use A36 steel and Allowable Stress Design.

Solution. Three cases must be considered since each of the three laterally unbraced lengths is different and is subject to a different maximum bending moment. Assume the segment containing the largest moment governs.

(a) Segment C:

$$M = 241.5 \text{ ft-kips}, \qquad C_b = 1.75, \qquad L = 21 \text{ ft}$$

Use the *ASD Manual* curves, "ALLOWABLE MOMENTS IN BEAMS," a portion of which is shown in Fig. 9.10.2. As discussed previously, the C_b factor may be combined with L_b, as follows:

$$\frac{L_b}{C_b} = \frac{21}{1.75} = 12 \text{ ft} \qquad \text{if } F_b \text{ (F1-8) governs}$$

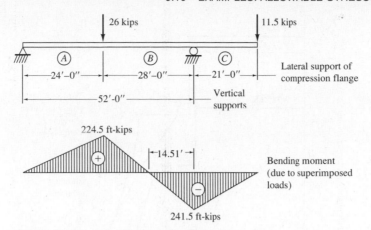

Figure 9.10.1   Data for Example 9.10.3.

or

$$\frac{L_b}{\sqrt{C_b}} = \frac{21}{\sqrt{1.75}} = 15.9 \text{ ft} \quad \text{if } F_b \text{ (F1-6) governs}$$

Try entering curves with $M = 241.5$ ft-kips with $L_b/C_b = 12.0$ ft, select

$$W21 \times 68, \qquad M = 257 \text{ ft-kips}, \qquad L_u = 12.4 \text{ ft}$$

Since $L_b = 21$ ft exceeds L_c (8.7 ft) but $L_b/C_b = 12.0$ ft does not exceed L_u (12.4 ft),

$$F_b = 0.60F_y$$

Note that in the curves, the solid black dot (●) represents L_c and the open circle (○) represents L_u. Also, the values of L_c and L_u for $F_y = 36$ and 50 ksi are given in the *ASD Manual*.

The additional moment due to a beam weight about 76 lb/ft should be included at the this stage.

At the $(-M)$ region,

$$M_{DL} = \tfrac{1}{2}(0.07)(21)^2 = 15.4 \text{ ft kips}$$

At the 26 kips load,

$$M_{DL} = \tfrac{1}{2}(0.07)(24)(28) - 15.4\left(\frac{24}{52}\right) = 16.4 \text{ ft-kips}$$

Reexamine curves with

$$M = 241.5 + 15.4 = 257 \text{ ft-kips}$$

Segment C looks OK.

(b) Segment B: From ASD-F1.3.

$$C_b = 1.75 + 1.05\left(\frac{M_1}{M_2}\right) + 0.3\left(\frac{M_1}{M_2}\right)^2 = 2.99$$

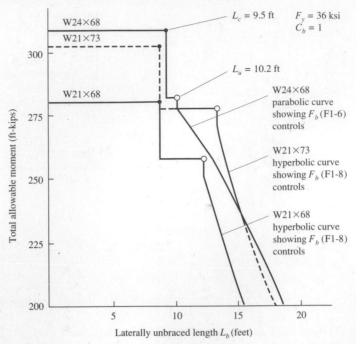

Figure 9.10.2   Portion of *ASD Manual* curves "ALLOWABLE MOMENTS IN BEAMS" used in Example 9.10.3.

Use maximum $C_b = 2.3$.

$$\frac{L_b}{C_b} = \frac{28}{2.3} = 12.2 \text{ ft}; \qquad \frac{L_b}{\sqrt{C_b}} = \frac{28}{\sqrt{2.3}} = 18.5 \text{ ft}$$

Since $L_b/C_b < (L_u = 12.4 \text{ ft})$, this still looks acceptable for $M = 257$ ft-kips.

(c) Segment A: $C_b = 1.75$

$$\frac{L_b}{C_b} = \frac{24}{1.75} = 13.7 \text{ ft}; \qquad \frac{L_b}{\sqrt{C_b}} = \frac{24}{\sqrt{1.75}} = 18.1 \text{ ft}$$

Here L_b/C_b exceeds $L_u = 12.4$ ft for W21×68, meaning F_b will be less than $0.60F_y$.

$$M = 224.5 + 16.4 = 241 \text{ ft-kips}$$

From the curves, the point defined by $M = 241$ and $L_b/C_b = 13.7$ ft lies above the hyperbolic curve for W21×68, indicating that W21×68 is not adequate. A reentry to the curves shows the next solid line curve above the W21×68 is that for W24×68; however, that curve is a parabolic one and for that section

$$M = 241 \text{ ft-kips} \quad \text{at} \quad L_b/\sqrt{C_b} = 18.1 \text{ ft}$$

The W24×68 is not adequate for Segment A. The next section is W21×73 with the hyperbolic curve governing.

$$M > 241 \text{ ft-kips} \quad \text{at} \quad L_b/C_b = 13.7 \text{ ft}$$

(d) Check stresses for W21×73, $S_x = 151$in.^{3}

For Segment A:

$$F_b \text{ (F1-8)} = \frac{12,000C_b}{L_b d/A_f} = \frac{12,000(1.75)}{24(12)3.46} = 21.1 \text{ ksi}$$

or for $L_b/r_T = 24(12)/2.13 = 135.2 < 157$. Thus parabolic Formula (F1-6) is to be used.

$$F_b \text{ (F1-6)} = 24.0 - \frac{(L_b/r_T)^2}{1181C_b} = 24.0 - \frac{(135.2)^2}{1181(1.75)} = 15.2 \text{ ksi}$$

Thus the larger, $F_b = 21.1$ ksi, governs.

$$f_b = \frac{M}{S_x} = \frac{241(12)}{151} = 19.1 \text{ ksi} < F_b \quad \text{OK}$$

For Segment C:

$$F_b \text{ (F1-8)} = \frac{12,000(1.75)}{21(12)3.46} = 24.1 \text{ ksi} > 0.60F_y$$

$$F_b = 0.60F_y = 22 \text{ ksi} \quad \text{governs}$$

(Note that $F_b = 0.60F_y$ may also be determined from $L_b/C_b = 12\text{ft} < L_u$.)

$$f_b = \frac{M}{S_x} = \frac{257(12)}{151} = 20.4 \text{ ksi} < [F_b = 22 \text{ ksi}] \quad \text{OK}$$

For Segment B:

$$M = 257 \text{ ft-kips} \quad \text{same as Segment C}$$
$$L_b/C_b = 12.2 \text{ ft} < L_u; \qquad F_b = 22 \text{ ksi}$$

Stress check identical with Segment C.
Use W21×73. ■■

EXAMPLE 9.10.4 _____

Investigate using Allowable Stress Design the W18×76 section of $F_y = 60$ ksi steel that was acceptable by Load and Resistance Factor Design in Example 9.9.4 (see Fig. 9.9.2). The simply supported span of 50 ft has lateral support at the ends and midspan.

Solution

(a) Determine the service load moment M at midspan.

$$M = wL^2/8 = 1.0(50)^2/8 = 312 \text{ ft-kips}$$

(b) Check compact section requirements for W18×76. Examine L_c, the maximum laterally unbraced length for which $F_b = 0.66F_y$. For $F_y = 50$ ksi, the *ASD Manual* gives $L_c = 9.9$ ft and it will be less than that for $F_y = 60$ ksi. Thus,

$L_b = 25$ ft is well above L_c. To illustrate the L_c computation according to ASD-F1.1.

$$L_c = \frac{20,000}{F_y(d/A_f)} = \frac{20,000}{60(2.43)12} = 11.4 \text{ ft}$$

or

$$L_c = \frac{76b_f}{\sqrt{F_y}} = \frac{76(11.035)}{\sqrt{60}\,(12)} = 9.0 \text{ ft}$$

Thus $L_c = 9.0$ ft, since the *smaller* value governs.

(c) Compute the allowable steress F_b. Since for such a large L_b as 25 ft it is likely that lateral-torsional buckling will control the design, go directly to the ASD Formulas (F1-6 or F1-7) and (F1-8) without computing L_u, the maximum laterally unbraced length at which $F_b = 0.60F_y$ when $C_b = 1.0$. For $F_y = 50$ ksi, $L_u = 13.7$ ft; therefore, for $F_y = 60$ ksi, $L_u < 13.7$ ft.

$$\frac{L_b}{r_T} = \frac{25(12)}{2.95} = 102$$

From Table 9.7.2 with $C_b = 1.75$,

$$54 < 109 < 122 \quad \therefore F_b \text{ (F1-6) applies.}$$

$$F_b \text{ (F1-6)} = \left[\frac{2}{3} - \frac{F_y(L_b/r_T)^2}{1530(10^3)C_b}\right]F_y$$

$$= 40.0 - \frac{(L_b/r_T)^2}{425c_b} \quad \text{for } F_y = 60 \text{ ksi}$$

$$= 40.0 - \frac{(102)^2}{425(1.75)} = 26.0 \text{ ksi}$$

$$F_b \text{ (F1-8)} = \frac{12,000C_b}{L_b d/A_f} = \frac{12,000(1.75)}{25(12)2.43} = 28.8 \text{ ksi}$$

The *larger* value, $F_b = 28.8$ ksi, controls.

(d) Final check and evaluation.

$$M = \tfrac{1}{8}(1.076)(50)^2 = 336 \text{ ft-kips}$$

$$f_b = \frac{M}{S_x} = \frac{336(12)}{146} = 27.6 \text{ ksi} < [F_b = 28.8 \text{ ksi}] \quad \text{OK}$$

The W18×76 is satisfactory. ■■

9.11 WEAK-AXIS BENDING OF I-SHAPED SECTIONS

The treatment thus far in this chapter has dealt with lateral-torsional buckling where instability might occur in a direction perpendicular to the plane of strong-axis bending (that is, the buckling occurs in the weaker direction). When an I-shaped beam is bent about its weak axis (y-axis), that is, bending in a plane *perpendicular* to the plane of the web, making the y-axis the neutral axis, lateral-torsional instability is no longer of concern. The beam will tend to deflect only in the direction of the loading since that is the principal axis orientation offering least resistance. Since lateral instability will not occur on doubly symmetrical I-shaped sections bent about their weak axis, the only factor that might prevent them from achieving the plastic moment condition would be local buckling of the compression portion (unstiffened element) of the flanges.

Load and Resistance Factor Design

According to LRFD-F1 and F1.2, "The lateral-torsional buckling limit state is not applicable to members subject to bending about the minor axis . . ." In addition, the web will not be a compression element when the loading is in a plane parallel to the flanges. Thus, the limit states for I-shaped sections in weak axis bending are (1) development of the plastic moment strength and (2) flange local buckling. In order that the flange local buckling limit state is avoided, the flange λ must not exceed λ_p (LRFD-B5 as given in Table 9.6.1).

Allowable Stress Design

The allowable stress provisions of ASD-F2 are summarized in Section 7.5.

9.12* LATERAL BUCKLING OF CHANNELS, ZEES, MONOSYMMETRIC I-SHAPED SECTIONS, AND TEES

The basic development of lateral buckling strength-related criteria has assumed that loads are applied vertically through the shear center. Furthermore, the resistance to lateral buckling considered that the shear forces which developed in the flanges were equal and the center of twist was located at mid-height.

Channels

Unless loaded through the shear center, a channel is subjected to combined bending and torsion. Since the shear center is not in the plane of the web (see Fig. 8.5.1), usual loadings through the centroid or in the plane of the web give rise to such combined stress. For loads in a plane parallel to the web, lateral buckling must be considered, even if the torsional moment may properly be neglected. The 1976 *SSRC Guide* [6.20] states "if an otherwise laterally unsupported channel has concentrated loads brought in by other members that frame into it, such loads can be considered as being applied at the shear center, *provided that the span of the framing member is measured from the channel shear center and the framing connections are designed for the moment and shear at the connection*."

For design purposes, Hill [9.32] indicates that lateral-torsional buckling equations for symmetrical I-shaped sections may be applied for channels. Such a procedure is stated to err on the unsafe side by about 6% in extreme cases. Both ASD and LRFD adopt this approach. Equation 9.4.33 can be used to represent the elastic lateral-torsional buckling strength of a channel loaded essentially in the plane of its web. The torsion warping constant C_w for a channel has a different expression than the expression for an I-shaped section; the expression is available in text Appendix Table A2. The torsion properties for rolled channels are available in the *LRFD Manual* [1.18] and in the 9th edition of the *ASD Manual*; they were omitted from the 8th edition of the *ASD Manual* (1980).

EXAMPLE 9.12.1

Determine the nominal strength M_n for a channel, C12×20.7 of steel having $F_y =$ 50 ksi, used on a span of 24 ft with concentrated loads at the one-third points as shown in Fig. 9.12.1. It is assumed that the loads act at the shear center of the channel.

Solution

(a) Determine whether or not the section satisfies the "compact section" requirements. For local buckling limit states the flange and web must satisfy the same requirements as for I-shaped sections; i.e., $\lambda \leq \lambda_p$ as given in Table 9.6.1. For the flange,

$$\left(\lambda = \frac{b_f}{t_f} = \frac{2.942}{0.501} = 5.9 \right) < \left(\lambda_p = \frac{65}{\sqrt{F_y}} = 9.2 \right) \quad \text{OK}$$

For the web, λ must not exceed λ_p as given in Table 9.6.1, as follows:

$$\left(\lambda = \frac{h}{t_w} = \frac{\approx T}{t_w} = \frac{9.75}{0.282} = 34.6 \right) < \left(\lambda_p = \frac{640}{\sqrt{F_y}} = 90.5 \right) \quad \text{OK}$$

Note that neither b_f/t_f nor h/t_w is given in the *LRFD Manual* as a property for channel sections. Thus, for h the authors recommend using the dimensional property T to compute the ratio. The webs of channels usually have low ratios h/t_w so that one might almost *assume* $\lambda < \lambda_p$. This channel is a "compact section" for the local buckling limit states.

(b) Examine the lateral-torsional buckling limit state. Plastic moment strength can develop when $L_b \leq L_p$, where according to Eq. 9.6.3,

$$L_p = \frac{300}{\sqrt{F_y}, \text{ksi}} r_y = \frac{300(0.80)/12}{\sqrt{50}} = 2.8 \text{ ft}$$

$$M_p = Z_x F_y = 25.4(50)/12 = 106 \text{ ft-kips}$$

$$M_r = (F_y - F_r)S_x = (50 - 10)21.5/12 = 71.7 \text{ ft-kips}$$

$$L_r = \frac{r_y X_1}{(F_y - F_r)} \sqrt{1 + \sqrt{1 + X_2(F_y - F_r)^2}} \qquad [9.6.6]$$

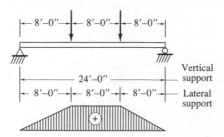

Figure 9.12.1   Example 9.12.1.

where the properties X_1 and X_2 are not tabulated in the *LRFD Manual* for channel sections. Using X_1 according of Eq. 9.6.7,

$$X_1 = \frac{\pi}{S_x}\sqrt{\frac{EGJA}{2}} = \frac{\pi}{21.5}\sqrt{\frac{(29,000)^2(0.37)6.09}{2(2.6)}} = 2790 \text{ ksi}$$

and X_2 according to Eq. 9.6.8,

$$X_2 = 4\frac{C_w}{I_y}\left(\frac{S_x}{GJ}\right)^2 = \frac{4(112)(21.5)^2}{3.88(29,000/2.6)^2(0.37)^2} = 0.00313$$

Then, substituting $X_1 = 2790$ ksi, $X_2 = 0.00313$ in.^{4}/kip^2, $r_y = 0.80$ in, and $F_r = 10$ ksi for a rolled section in Eq. 9.6.6 above, gives

$$L_r = 8.6 \text{ ft}$$

Thus,

$$(L_p = 2.8 \text{ ft}) < (L_b = 8 \text{ ft}) < (L_r = 8.6 \text{ ft})$$

The nominal strength M_n is a linear interpolation between M_p and M_r according to Eq. 9.6.4. The governing laterally unbraced segment in Fig. 9.12.1 is the center one where the maximum moment occurs with a constant moment; therefore, $C_b = 1.0$ for that segment. Evaluating the nominal strength gives

$$M_n = C_b\left[M_p - (M_p - M_r)\left(\frac{L_b - L_p}{L_r - L_p}\right)\right] \le M_p \qquad [9.6.4]$$

$$M_n = 1.0\left[106 - (106 - 71.7)\left(\frac{8.0 - 2.8}{8.6 - 2.8}\right)\right] = 75.4 \text{ ft-kips} \qquad \blacksquare\blacksquare$$

Zees

The zee-section lateral buckling strength is complicated by the fact that loading in the plane of the web causes unsymmetrical bending, resulting because a principal axis does not lie in that plane. The general treatment of buckling under biaxial bending is found in Sec. 9.14. The effect of biaxial bending on zee sections was found by Hill [9.32] to reduce the critical moment M_x to 90–95 percent of the value given by Eq. 9.4.33. In addition, the torsion-bending constant C_w, is different than for channels or I-shaped sections.

For design purposes, in view of the fact that unbraced zees are relatively rare, AISC does not provide for them. The authors have long suggested using one-half the values obtained using M_{cr} from Eq. 9.4.33, which is LRFD Formula (F1-13).

Monosymmetric I-shapes

I-shaped sections symmetrical about the *y*-axis, but unsymmetrical about the *x*-axis, are summarized in the *SSRC Guide* [6.8] and by Clark and Hill [9.26]. The additional variable involved is y_0, the distance from the centroid of the girder cross-section to the shear center (positive if the shear center lies between the centroid and compression flange, otherwise negative).

The 4th edition of the *SSRC Guide* [6.8] provides considerable detail regarding procedure for monosymmetric beams; largely from the work of Galambos [9.7], Anderson and Trahair [9.33], Kitipornchai and Trahair [9.34], and Nethercot [9.35]. More recently, Kitipornchai, Wang, and Trahair [9.36], Wang and Kitipornchai [9.37], and Kitipornchai and Wong-Chung [9.38] have treated the subject.

For practical design purposes, the 1989 ASD and the 1986 LRFD Specifications indicated that it is conservative to use M_{cr} for doubly symmetric I-shapes for a singly symmetric I-shape having a compression flange larger than the tension flange. The 1993 LRFD Specification does *not* make such a statement. Instead the user is directed to LRFD-Appendix F (Table A-F1.1).

LRFD-Appendix F uses an approximate equation for M_{cr} of monosymmetric I-shapes, which derives from theoretical solutions the background of which are discussed by Anderson and Trahair [9.33]. The approximate expression for a simply supported beam under uniform moment is

$$M_{cr} = \frac{\pi \sqrt{EI_y GJ}}{L_b}[B_1 + \sqrt{1 + B_2 + B_1^2}] \tag{9.12.1}$$

where B_1 and B_2 are the terms relating to the non-symmetry of the section (defined following Eq. 9.12.3).

When π, $E = 29{,}000$ ksi, and $G = E/2.6$ are substituted into Eq. 9.12.1, the constant

$$\pi \sqrt{EG} = \pi \sqrt{29{,}000 \frac{29{,}000}{2.6}} = 57{,}000$$

Thus, Eq. 9.12.1 becomes

$$M_{cr} = \frac{57{,}000 C_b \sqrt{I_y J}}{L_b}[B_1 + \sqrt{1 + B_2 + B_1^2}] \tag{9.12.2}$$

When the moment gradient factor C_b is included, the LRFD-Appendix F (Table A-F1.1) equation is

$$M_{cr} = \frac{57{,}000 C_b \sqrt{I_y J}}{L_b}[B_1 + \sqrt{1 + B_2 + B_1^2}] \tag{9.12.3}$$

where
$$B_1 = 2.25\left[2\left(\frac{I_{yc}}{I_y}\right) - 1\right]\left(\frac{h}{L_b}\right)\sqrt{\frac{I_y}{J}}$$

$$B_2 = 25\left(1 - \frac{I_{yc}}{I_y}\right)\left(\frac{I_{yc}}{J}\right)\left(\frac{h}{L_b}\right)^2 \tag{9.12.4}$$

I_{yc} = moment of inertia of compression flange about y axis, or if reverse curvature bending, moment of inertia of smaller flange about y axis.

$C_b = 1.0$ when $I_{yc}/I_y < 0.1$ and $I_{yc}/I_y > 0.9$

otherwise

$$C_b = \frac{12.5 M_{max}}{2.5 M_{max} + 3M_A + 4M_B + 3M_C} \tag{9.6.11}$$

Tee Sections

A T-section *may* be thought of as a monosymmetric I-shaped section that has the moment of inertia I_y of one flange equal to zero. Traditionally, both ASD and LRFD have been vague on how a T-section is to be treated. Rolled structural tees will rarely have strength controlled by the lateral-torsional buckling limit state. Whenever a tee section is loaded in the plane of its web (moment about the x axis) and r_x is less than r_y, there is no limit on laterally unbraced length. LRFD-F1 states, "The lateral-torsional buckling limit state is not applicable to members subject to bending about the minor axis . . ." A significant number of rolled tees are in this category. Ellifrit, Wine, Sputo, and Samuel [9.74] have recently discussed the flexural strength of WT sections.

When the tee having its flange in compression is bent about its major axis, that is, bending in the plane of the stem, and $r_x > r_y$, LRFD-Formula (F1-4), Eq. 9.6.3, applies for L_p,

$$L_p = \frac{300 r_y}{\sqrt{F_{yf}}} \qquad [9.6.3]$$

The indirect LRFD reference to this is in LRFD-Appendix Table A-F1.1, p. 6-115, for singly symmetric I-shaped beams. When L_b does not exceed L_p and λ does not exceed λ_p for the flange local buckling limit state, the yielding limit state controls. One may expect to reach the plastic moment M_p. However, when the shape (see Sec. 7.3) $\xi = Z_x/S_x$ is large, too much plastic deformation may occur at service load. The shape factor ξ for tees may be as high as 2.0. Thus, 1993 LRFD-F1.2c limits the nominal strength M_n to

1. For stem in compression,

$$M_n \leq 1.0 M_y \qquad (9.12.5)$$

2. For stem in tension,

$$M_n \leq 1.5 M_y \qquad (9.12.6)$$

When L_b exceeds L_p but does not exceed L_r, the nominal strength reduces linearly from its maximum defined by Eqs. 9.12.5 and 9.12.6 to M_r defined as the smaller of:

$$\left. \begin{array}{l} M_r = (F_{yf} - F_r) S_{xc} \\ M_r = F_{yf} S_{xt} \end{array} \right\} \qquad (9.12.7)$$

where S_{xc} = section modulus referred to outer fiber of compression flange
S_{xt} = section modulus referred to outer fiber of tension flange

When L_b exceeds L_r, the elastic lateral-torsional buckling limit state controls. Note that L_r is the value of L_b at which $M_{cr} = M_r$ (see also LRFD-Appendix Table A-F1.1, p. 6-115, for singly symmetric members). According to LRFD-F1.2c,

$$M_{cr} = \frac{\pi \sqrt{EI_y GJ}}{L_b} [B + \sqrt{1 + B^2}] \qquad (9.12.8)$$

where

$$B = \pm 2.3 \left(\frac{d}{L_b}\right) \sqrt{\frac{I_y}{J}} \qquad (9.12.9)$$

The plus sign is used for the stem in tension, and the minus sign is for the stem in compression. "If the tip of the stem is in compression anywhere along the unbraced length, use the negative value of B." (LRFD-F1.2c).

EXAMPLE 9.12.2

Investigate the moment strength of a structural tee section WT7×19 both when used with its flange in compression and with the flange in tension. Use Load and Resistance Design and A36 steel. Show how the strength is affected by lateral bracing.

Solution

Case 1: *Flange in Compression.*

(a) Maximum strength. The flange must satisfy that $\lambda \leq \lambda_p$ to preclude the flange local buckling limit state from reducing the strength.

$$\left(\lambda = \frac{b_f}{2t_f} = \frac{6.770}{2(0.515)} = 6.6\right) < \left(\lambda_p = \frac{65}{\sqrt{F_y}} = 10.8\right) \quad \text{OK}$$

Since the stem (web) is primarily in tension, λ for the web does not have to satisfy any limit. Thus, LRFD-F1.2c indicates

$$\text{Max } M_n = 1.5M_y = 1.5S_x F_y = 1.5(4.22)36\frac{1}{12} = 19.0 \text{ ft-kips}$$

Also, the plastic moment strength cannot be exceeded,

$$M_p = Z_x F_y = 7.45(36)\frac{1}{12} = 22.4 \text{ ft-kips}$$

In this case, the shape factor is $22.4(1.5)/19.0 = 1.77$; because of concern about inelastic deformation at service load, LRFD limits the usable increase above first yield to $1.5M_y$ rather than M_p, which is $1.77M_y$.

(b) Examine lateral-torsional buckling limitation on achieving maximum moment strength. There are two cases to consider: (1) When the radius of gyration r_x is less than r_y and the loading is in the plane of the stem, the beam will bend in this weak direction and will have no tendency to buckle laterally. (2) When the axis of symmetry is the weak axis (i.e., $r_x > r_y$) there is some possibility of lateral-torsional buckling.

For the case where $r_x > r_y$, it is conservative to treat the structural tee flange in the same manner as the I-shaped beam flange. Thus, LRFD-Appendix F1 (Table A-F1.1) indicates

$$L_p = \frac{300r_y}{\sqrt{F_{yf}, \text{ksi}}} \qquad [9.6.3]$$

Furthermore, if plastic analysis is considered, LRFD-F1.2d indicates for "singly symmetric I-shaped members with the compression flange larger than the tension flange . . . loaded in the plane of the web,"

$$L_{pd} = \frac{3600 + 2200(M_1/M_2)}{F_y, \text{ksi}} r_y \qquad [9.6.2]$$

For this WT7×19,

$$\frac{r_x}{r_y} = \frac{2.04}{1.55} = 1.3$$

Since the r_x/r_y ratio exceeds unity, lateral-torsional buckling is a possible limit state. Thus, L_{pd} and L_p are computed,

$$L_{pd} = 100r_y = 12.9 \text{ ft} \quad (\text{for } M_1/M_p = 0)$$
$$L_p = 50r_y = 6.5 \text{ ft}$$

It is *unreasonable* to use $L_{pd} > L_p$.

(c) Inelastic lateral-torsional buckling limit state. When the extreme fiber in bending reaches the yield stress F_y, the nominal moment strength M_r is given by Eqs. 9.12.7,

$$\left. \begin{array}{c} M_r = (F_y - F_r)S_{xc} = (36 - 10)\left(\dfrac{23.3}{1.54}\right)\dfrac{1}{12} = 32.8 \text{ ft-kips} \\[2em] M_r = F_y S_{xt} = 36\left(\dfrac{23.3}{7.05 - 1.54}\right)\dfrac{1}{12} = 12.7 \text{ ft-kips} \end{array} \right\}$$

It is clear that the extreme fiber of the stem reaches F_y before the flange extreme fiber reaches $(F_y - F_r)$ because this monosymmetric section has a shorter distance to the compression extreme fiber than to the tension extreme fiber. Either fiber may control; the flange contains residual stress F_r whereas at the extreme tension fiber on the web there will be no significant residual stress.

(d) Calculate the limit L_r at which $M_r = M_{cr}$. When L_b exceeds L_r and λ does not exceed λ_r, LRFD-F1.2c [Formula (F1-15)], Eq. 9.12.8, gives the elastic buckling strength,

$$M_{cr} = \frac{C_b \pi \sqrt{EI_y GJ}}{L_b}[B + \sqrt{1 + B^2}] \leq 1.5M_y \qquad \text{LRFD Formula (F1-15)}$$

where

$$B = \pm 2.3\left(\frac{d}{L_b}\right)\sqrt{\frac{I_y}{J}} \qquad \text{LRFD Formula (F1-16)}$$

The plus sign applies when the stem is in tension as for this case. Evaluating for the WT7×19, taking $C_b = 1.0$ and $E/G = 2.6$ gives,

$$M_{cr} = \frac{1.0\pi E \sqrt{13.3(0.398/2.6)}}{L_b}[B + \sqrt{1 + B^2}]$$

$$M_{cr} = \frac{130,000}{L_b}[B + \sqrt{1 + B^2}]$$

where

$$B = +2.3\left(\frac{7.05}{L_b}\right)\sqrt{\frac{13.3}{0.398}} = +\frac{93.7}{L_b}$$

The nominal strength M_n variation with L_b is shown in Fig. 9.12.2. The value of L_b at which $M_r = M_{cr}$ is defined as L_r. Setting M_{cr} equal to 12.7 ft-kips (i.e, M_r) gives L_r approximately 79 ft. Thus, elastic lateral-torsional buckling will not control in the practical range of laterally unbraced length L_b, which is well below 79 ft. The maximum nominal strength M_n of $1.5M_y$ is applicable for L_b less than about 32 ft. Rarely will L_b exceed 32 ft.

Case 2: Stem in Compression

(e) Plastic moment strength. When the stem is in compression, the indication of LRFD-Table B5.1 is that λ_p has no applicability for "stems of tees". Since

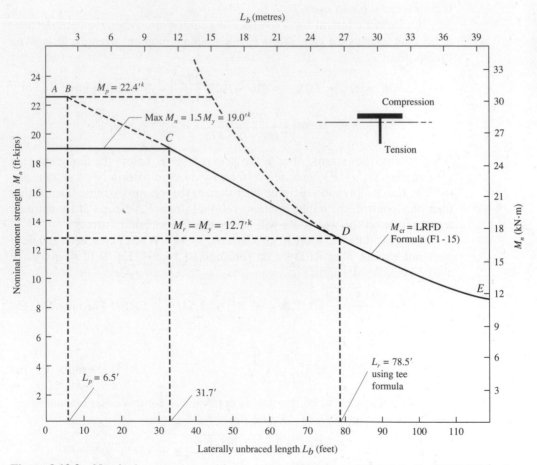

Figure 9.12.2 Nominal moment strength M_n vs laterally unbraced length L_b for WT7×19 having flange in compression.

the reference to tee webs is for a uniform compression situation, reference to λ_p would be inappropriate. However, when the tee web is in compression because it is on the compression side of the neutral axis, the situation is similar to an I-shaped section bent in its weak direction. If the stem of the tee satisfies λ_p for an unstiffened flange, as follows,

$$\lambda_p = \frac{65}{\sqrt{F_y, \text{ksi}}}$$

then it should be acceptable to use the maximum moment strength M_n as high as M_p as long as the extreme fiber in tension does not exceed F_y. However, 1993 LRFD-F1.2c states $M_n \le M_y$, a conservative approach. From a practical point of view, rolled structural tee webs will never satisfy the λ_p limit. However, a tee shape welded of two plates could be made to satisfy the limit.

For this example.

$$\left(\lambda = \frac{d}{t_w} = \frac{7.05}{0.31} = 22.7\right) < \left(\lambda_p = \frac{65}{\sqrt{F_y}} = 10.8\right) \quad \text{NG}$$

Thus the web local buckling limit state precludes developing the plastic moment strength.

(f) Inelastic local buckling of the web. The limit λ_r for the web in compression is, from LRFD-B5.1,

$$\left(\lambda = \frac{d}{t_w} = \frac{7.05}{0.31} = 22.7\right) < \left(\lambda_r = \frac{127}{\sqrt{F_y}} = 21.2\right) \quad \text{NG}$$

Since $\lambda > \lambda_r$, the efficiency of the stem in compression is reduced because local buckling may be expected prior to the yield stress F_y being reached at the extreme fiber. Thus Q_s is less than unity as discussed in Sections 6.18 and 6.19. Using LRFD-Appendix B5.3a, with $d/t_w = 22.7$,

$$\left(\frac{127}{\sqrt{F_y}} = 21.2\right) < 22.7 < \left(\frac{176}{\sqrt{F_y}} = 29.3\right)$$

$$Q_s = 1.908 - 0.00715\left(\frac{d}{t_w}\right)\sqrt{F_y}$$

$$= 1.908 - 0.00715(22.7)\sqrt{36} = 0.93$$

Then M_r, which is controlled by the stem extreme fiber in compression and cannot exceed M_y may be calculated

$$M_r = F_y S_{xt} = F_y I_x/y \le [Q_s F_y S_{xc} = Q_s F_y I_x/(d - y)]$$
$$M_r = Q_s F_y S_{xc} = Q_s F_y I_x/(d - y)$$
$$= 0.93(36)(23.3)/(7.05 - 1.54)/12 = 11.8 \text{ ft-kips}$$

The next step would be to compute L_r, the value of L_b at which $M_{cr} = M_r$.

(g) Elastic lateral-torsional buckling. LRFD Formula (F1-15) for tee sections as used in part (d) above is still valid when the stem is in compression.

However, the constant B must be taken negative $(-)$; thus, the buckling stress will be lower for the tee in this orientation.

$$M_{cr} = \frac{130,000}{L_b}[B + \sqrt{1 + B^2}]$$

$$B = -2.3\left(\frac{7.05}{L_b}\right)\sqrt{\frac{13.3}{0.398}} = -\frac{93.7}{L_b}.$$

Setting M_{cr} equal to $M_r = 11.8$ ft-kips, then solving for L_b gives $L_r = 68$ ft.

The nominal strength relationship for the tee having its stem in compression is not presented in graphical form; however, M_n remains constant at $Q_s F_y S_{xc}$ (that is, 11.8 ft-kips) to L_r, then decreases according to M_{cr}. ∎∎

9.13* LATERAL BRACING DESIGN

The questions of what constitutes bracing and how to design bracing continue to be major concerns of practicing engineers. The subject is included in this chpater because a major item of concern in lateral bracing design is the restraint required to prevent lateral-torsional buckling in beams. The development of this section, however, is applicable to the bracing of columns as well as beams. In Sec. 6.8 the concept of *braced* and *unbraced* systems was briefly discussed in regard to the effective length factor K. In the following discussion, the emphasis in on *braced* systems; that is, the overall structural system is braced by cross bracing or attachment to an adjoining system that is braced. The bracing requirements for frames are treated in Chapter 14. Bracing for individual beams or columns may consist of cross bracing where the axial stiffness of the bracing elements is utilized; it may be provided at discrete locations by flexural members framing in transverse to the member being braced, wherein both axial and flexural stiffnesses of the bracing member are utilized; or such bracing may be provided continously by material such as light gage roof decking or wall panels.

Little is available in specifications but point bracing has been treated by Zuk [9.39], Winter [9.40], Massey [9.41], Pincus [9.42], Galambos [9.43], Urdal [9.44], Lay and Galambos [9.45], Taylor and Ojalvo [9.46], Hartmann [9.47], Mutton and Trahair [9.48], Medland and Segedin [9.49], and Plaut [9.73]. Recent practical treatment has been provided by Yura [9.50], Lutz and Fisher [9.51], Ales and Yura [9.52], Clarke and Bridge [9.53], and Yura [9.72]. What follows is largely a combination of the work of Winter [9.40], Galambos [9.43], and Yura [9.50, 9.72].

Point Bracing for Elastic Columns and Beams

Consider the axially loaded column of Fig. 9.13.1a where the top and bottom of the member are assumed to be supported in such a way that no side movement occurs at one end relative to the other. Such restraint would constitute a *braced* system. The bracing to create such restraint may be considered as a spring at the top that is capable of developing a horizontal reaction equal to the spring constant k times the deflection Δ. When the brace has a large spring constant (that is, the brace is very stiff) the deflection Δ could be close to zero and yet the spring may provide a large enough

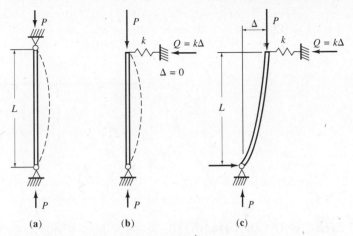

Figure 9.13.1 Bracing for a single-story column.

horizontal force to prevent any side motion (sideway) at the top. This would be the situation in Fig. 9.13.1b. The equilibrium requirement is shown in Fig. 9.13.1c, wherein a sidesway is shown. If one imagines this as a slightly deflected position, then in order to have equilibrium, it is required that

$$P\Delta = QL = (k\Delta)L \qquad (9.13.1)$$

If $(k\Delta)L$ is less than $P\Delta$, sidesway occurs. If $(k\Delta)L$ is greater than $P\Delta$, no sidesway occurs, and the column would be considered braced. The ideal brace, then, would be one that has just enough stiffness k to prevent movement (at the top in this example); that is,

$$k = \frac{P}{L} \qquad (9.13.2)$$

The maximum load for which bracing would be required is the elastic bucking load P_{cr}, or the load causing yielding or inelastic buckling if that is lower than the elastic P_{cr}. Thus the largest required stiffness k_{ideal} is

$$k_{ideal} = \frac{P_{cr}}{L} \qquad (9.13.3)$$

The concept is shown by the plot of P vs kL in Fig. 9.13.2, wherein when k exceeds k_{ideal}, P_{cr} is reached and the column buckles without end translation (sidesway); in other words, it is a *braced* system. When k is less than k_{ideal}, a sidesway deflection will occur such that $P = kL$; in other words, a so-called *unbraced* system. The major treatment of unbraced systems is in Chapter 14, devoted to rigid frames.

Next, extend the concept to a two-story column within a braced system, as shown in Fig. 9.13.3. When no displacement occurs at mid-height, i.e., full bracing is provided, the column will buckle at a load nearly equal to

$$P_{cr} = \frac{\pi^2 EA}{(L/r)^2} \qquad (9.13.4)$$

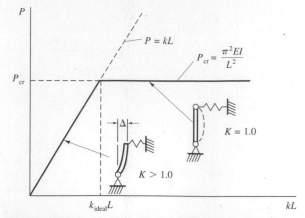

Figure 9.13.2   Brace stiffness relative to concept of "braced" ($K = 1.0$) and "unbraced" ($K > 1.0$) systems for column hinged at top and bottom.

In other words, one may imagine that a hinge exists at mid-height. Buckling occurs when the column snaps into the two half-wave mode of Fig. 9.13.3c.

Taking moments about the imaginary hinge location with the column deflected by an amount Δ, as in Fig. 9.13.3b, gives

$$P_{cr}\Delta = \frac{Q}{2}L \tag{9.13.5}$$

Since $Q = k\Delta$,

$$P_{cr}\Delta = \frac{(k\Delta)L}{2} \tag{9.13.6}$$

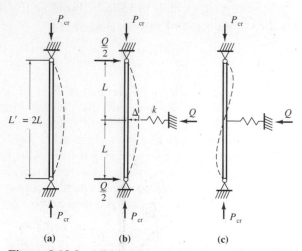

Figure 9.13.3   Mid-height brace for a two-story column.

As in the one-story column, if k_{ideal} is the necessary stiffness to create a nodal point (zero deflection) at mid-height of the two-story column, then

$$k_{ideal} = \frac{2P_{cr}}{L} \tag{9.13.7}$$

For situations with more than two equal spans, the same procedure may be used to obtain k_{ideal}. Examination of Fig. 9.13.4 for three equal spans will show that the spring forces Q can act either in the same or in opposite directions. Assuming they act in the *same* direction (Fig. 9.13.4a), using imaginary hinges at one-third span points, and taking moments when slightly deflected at the brace points, gives

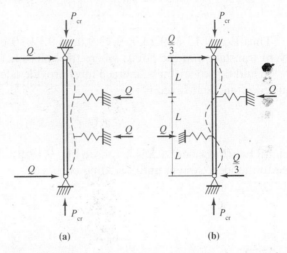

Figure 9.13.4   Column braced to make three equal spans.

$$QL = (k\Delta)L = P_{cr}\Delta; \quad k_{ideal} = \frac{P_{cr}}{L} \tag{9.13.8}$$

Assuming the forces Q acting in *opposite* directions (Fig. 9.13.4b) gives

$$QL/3 = (k\Delta)L = P_{cr}\Delta; \quad k_{ideal} = \frac{3P_{cr}}{L} \tag{9.13.9}$$

The configuration requiring the highest spring constant is the correct one, that which will permit the highest critical load. If a lesser stiffness is used, an alternate buckling mode will occur at a lower load, accompanied by displacement at the springs.

By the same process, k_{ideal} may be determined for any number of equal spans. In general,

$$k_{ideal} = \frac{\beta P_{cr}}{L} \tag{9.13.10}$$

where β varies from 1 for one span to 4 for infinite equal spans. The variation is given in Fig. 9.13.5.

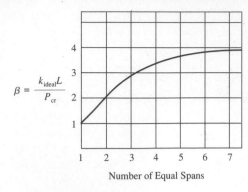

$$\beta = \frac{k_{\text{ideal}}L}{P_{\text{cr}}}$$

Number of Equal Spans

Figure 9.13.5   Variation of required spring constant for column with number of equal unbraced spans.

Thus Eqs. 9.13.3, 9.13.7, 9.13.9, and 9.13.10 give the ideal brace *stiffness* to prevent translation at the points where the braces act.

In addition to stiffness, a brace must provide adequate *strength*. The strength Q required of an ideal brace is

$$Q = k_{\text{ideal}}\Delta \tag{9.13.11}$$

but until buckling occurs, Δ is zero (see Fig. 9.13.6); therefore there will be no brace force in the ideal system until buckling occurs.

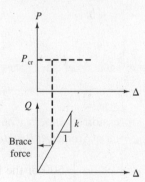

Figure 9.13.6   Brace force relative to column load for ideal system.

Compression members in real structures are not perfectly straight, perfectly aligned vertically, nor perfectly loaded as assumed in calculations; there is always an initial crookednesss. In other words, Δ is not zero even when there is no compressive load P acting. Reexamine the single-story column of Fig. 9.13.1 assuming there is an initial deflection Δ_0 that exists even when P is zero. Then, as shown in Fig. 9.13.7, equilibrium requires

$$(k\Delta)L = P(\Delta + \Delta_0) \tag{9.13.12}$$

for $P = P_{\text{cr}}$,

$$k_{\text{reqd}} = \frac{P_{\text{cr}}}{L}\left(1 + \frac{\Delta_0}{\Delta}\right) \tag{9.13.13}$$

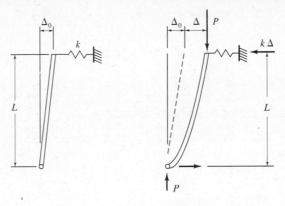

(a) No load applied   **(b)** When load is applied

Figure 9.13.7 Column with initial crookedness Δ_0.

Since $k_{ideal} = P_{cr}/L$, Eq. 9.13.13 then becomes

$$k_{reqd} = k_{ideal}\left(1 + \frac{\Delta_0}{\Delta}\right) \tag{9.13.14}$$

which is the *stiffness* requirement for compression members having initial crookedness Δ_0.

The *strength* requirement is then

$$Q = k_{reqd}\Delta = k_{ideal}\left(1 + \frac{\Delta_0}{\Delta}\right)\Delta$$

$$Q = k_{ideal}(\Delta + \Delta_0) \tag{9.13.15}$$

Normal tolerances on crookedness of compression members would vary from $1/500$ to $1/1000$ of the length [9.41]. The AISC *Code of Standard Practice* * indicates acceptable out-of-plumbness to be $L/500$. Considering accidental eccentricity of loading, Winter [9.41] suggests taking Δ_0 from $1/250$ to $1/500$ of the length.

Load and Resistance Factor Design—Point Bracing

To obtain design equations useful in LRFD, Winter [9.40] has suggests $\Delta = \Delta_0 = L/500$. Substitution of this in Eqs. (9.13.14) and (9.13.15) gives design equations.

1. For *stiffness* k_{reqd},

$$k_{reqd} = 2k_{ideal} \tag{9.13.16}$$

---

* *Code of Standard Practice for Steel Buildings and Bridges,* American Institute of Steel Construction, June 10, 1992 (Section 7.11.3.1).

2. For *nominal strength* Q_n,

$$Q_n = k_{ideal}(2\Delta_0)$$
$$Q_n = k_{ideal}(0.004L) \tag{9.13.17}$$

where $k_{ideal} = \beta P_{cr}/L$.

Allowable Stress Design—-Point Bracing

For ASD a factor of safety FS must be applied so that the service load P may be used instead of P_{cr}. If FS $= 2$, the strength requirement may be expressed in term of service load P; however, the stiffness required will be the same in ASD as in LRFD.

1. *Stiffness requirement,*

$$k_{reqd} = 2k_{ideal} \tag{9.13.18}$$

2. *Strength requirement,*

$$Q = \frac{Q_n}{FS} = \frac{k_{ideal}}{2}(0.004L) \tag{9.13.19}$$

where $k_{ideal} = \beta P_{cr}/L = 2\beta P/L$.

Where the strength of the compression member being braced is controlled by an elastic limit state $F_{cr} \leq F_y$, as would be the case for columns, the foregoing is applicable. When large plastic strain must be accommodated at bracing points, the design suggestions of Lay and Galambos [9.45] for inelastic steel beams, as discussed later in this section, should be applied.

To summarize the procedure:

1. Establish the bracing locations and compute the strength P_{cr} for the compression element (either entire column, or compression flange of beam) being braced.
2. Estimate β from Fig. 9.13.5 based on the number of equal unbraced lengths.
3. Compute $k_{ideal} = \beta P_{cr}/L$.
4. Select bracing area A_{brace} so that a stiffness $2k_{ideal}$ will be obtained. For axial stiffness,

$$k = 2k_{ideal}$$

$$\left(\frac{AE}{L}\right)_{brace} = 2\beta\pi^2\left(\frac{EI}{L}\right)_{\substack{compressed \\ element}}$$

$$\text{Required } A_{brace} = 2\beta\pi^2\left(\frac{E_{comp}}{E_{brace}}\right)\left(\frac{L_{brace}}{L_{comp}}\right)\frac{A_{comp}}{(L_{comp}/r)^2} \tag{9.13.20}$$

If $E_{comp} = E_{brace}$ and $L_{brace} \approx L_{comp}$, then Eq. 9.13.20 becomes

$$\text{Required } A_{brace} \approx 2\beta\pi^2\frac{A_{comp}}{(L_{comp}/r)^2} \tag{9.13.21}$$

5. Verify that the required force can be carried by the brace. Using Eq. 9.13.17,

$$\text{Required } Q_n = k_{\text{ideal}}(0.004L)$$

$$= \frac{\beta P_{\text{cr}}}{L}(0.004L)$$

$$= 0.004\beta P_{\text{cr}} \tag{9.13.22}$$

The nominal strength Q_n of the brace, including consideration of buckling strength, must equal or exceed the value given by Eq. 9.13.22.

EXAMPLE 9.13.1

Design a brace (brace A) to provide lateral support for a W27×84 beam (beam A) positioned as shown in Fig. 9.13.8. Assume the braces are 7.5 ft long, are attached near the compression flange, and are located along the supported beam at the one-third points of a 48-ft span. Use Load and Resistance Factor Design and A36 steel.

Solution. Since the bracing locations are given, the first step is to estimate the force in the compression zone of the beam or beams when the nominal strength M_n of the beams is reached. Since the braces are 16 ft apart, the strength of the beam could be based on F_{cr} less than F_y. For this W27×84 section, $L_p = 8.6$ ft and $L_r = 24.9$ ft. Thus, when braced every 16 ft this beam will have M_n less than M_p. Since stiffness usually controls over strength, it may be practical to assume the entire flange has the stress F_y acting. The stress F_{cr} based on lateral-torsional buckling could be determined if desired; it probably would have been determined when designing the beam. Thus, estimating the total nominal compressive

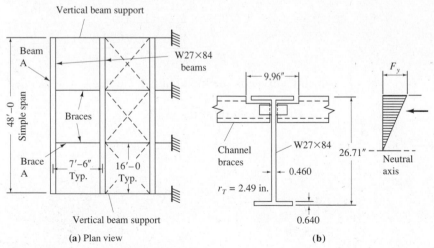

(a) Plan view

(b)

Figure 9.13.8 Data for Example 9.13.1.

load P_{cr} (treat this as the nominal strength P_n of the compression region of the beam) as

$$P_{cr} = F_y A_g/2 = 36(24.8)/2 \approx 450 \text{ kips}$$

The slenderness ratio L_{comp}/r_y of the compression element to be braced, in this case the beam slenderness ratio relating to lateral-torsional buckling, is

$$\frac{L_{comp}}{r_y} = \frac{16(12)}{2.07} = 93$$

For three equal unbraced lengths, estimate $\beta = 3$ from Fig. 9.13.5. Using Eq. 9.13.21.

$$\text{Required } A_{brace} = 2\beta\pi^2 \frac{A_{comp}}{(L_{comp}/r)^2}$$

$$= 2(3)\pi^2 \frac{(24.8/2)}{(93)^2} = 0.085 \text{ sq in.}$$

While any nominal size will satisfy this area requirement, it must be remembered that as a compression element it should preferable not exceed $KL/r = 200$ according to LRFD-B7. When several beams are to be braced, the required A_{brace} for the most heavily loaded brace would be based on the total compression area of all beams to be braced.

$$\text{Min } r_y = \frac{L}{200} = \frac{7.5(12)}{200} = 0.45 \text{ in.}$$

Select C4×5.4 as the lightest section with $r_y \geq 0.45$ in. ($r_y = 0.449$ in.)
    Check stiffness:

$$k_{ideal} = \frac{\beta P_{cr}}{L} = \frac{3(450)}{16(12)} = 7.0 \text{ kips/in.}$$

$$k_{act} = \frac{AE}{L_{brace}} = \frac{1.59(29,000)}{7.5(12)} = 503 \text{ kips/in.}$$

which far exceeds the minimum of 2 times k_{ideal} to account for initial crookedness.
    Check strength of the brace:

$$\frac{L_{brace}}{r_y} = \frac{7.5(12)}{0.45} = 200$$

$$\lambda_c = \frac{KL}{r}\sqrt{\frac{F_y}{\pi^2 E}} = 200\sqrt{\frac{36}{\pi^2 E}} = 2.25 \qquad [6.7.3]$$

For $\lambda_c > 1.5$.

$$F_{cr} = \left(\frac{0.877}{\lambda_c^2}\right)F_y = \left(\frac{0.877}{2.25^2}\right)36 = 6.2 \text{ ksi} \qquad [6.7.8]$$

Actual $Q_n = F_{cr}A_{brace} = 6.2(1.59) = 9.9$ kips

Required $Q_n = 0.004\beta P_{cr}$

$$= 0.004(3)(450) = 5.4 \text{ kips} < 9.9 \text{ kips} \quad \text{OK}$$

Use C4×5.4 for brace A.

For this case the LRFD-B7 recommended maximum slenderness ratio of 200 for compression members assured proper size bracing. ■■

EXAMPLE 9.13.2 _____

For the beam laterally braced by joists as shown in Fig. 9.13.9, determine the amount of weld required so that the joist will adequately brace the beam. The service bending moment is 25 ft-kips dead load and 100 ft-kips live load. The steel is A36. Use Load and Resistance Factor Design.

Solution

(a) Determine stiffness required. Whereas in Example 9.13.1 the compression strength P_{cr} of the flange was used, in this example the strength to be accommodated is computed from the factored load force in the flange. The factored moment M_u is

$$M_u = 1.2(25) + 1.6(100) = 190 \text{ ft-kips}$$

and when the moment is divided by the moment arm distance between flanges, the flange force becomes

$$P_{cr} = \frac{M_u}{\phi_b d} = \frac{190(12)}{0.90(14)} = 181 \text{ kips}$$

For four or more point braces along the compression flange, $\beta = 4$; thus,

$$k_{ideal} = \frac{\beta P_{cr}}{L} = \frac{4(181)}{4} = 181 \text{ kips/ft}$$

$$k_{reqd} = 2k_{ideal} = 362 \text{ kips/ft}$$

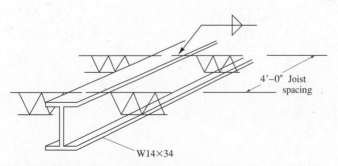

4'–0" Joist
spacing

W14×34

Figure 9.13.9 Beam laterally braced by joists.

Assuming the joists have metal deck adequately attached to them, there will develop in the decking an in-plane shear, known as *diaphragm action,* that will restrain the relative axial movement of two adjacent joists, thereby restraining relative lateral movement between two adjacent laterally braced points on the beam being braced, such as points A and B of Fig. 9.13.10. The adequacy of the attachment of the decking to the joists will determine the degree to which diaphragm action prevents the relative motion of points, such as A and B. If there is zero relative motion, then the requirement for lateral support may be based on $\beta = 1$, i.e., the same as the bracing of a single-story column. In which case, the required stiffness would be

$$k_{\text{reqd}} = 2k_{\text{ideal}} = \frac{2\beta P_{\text{cr}}}{L} = \frac{2(1)(181)}{4} = 90 \text{ kips/ft}$$

Using diaphragm action can greatly reduce the stiffness requirement for point bracing. In this case, if diaphragm action can be developed, the stiffness requirement is one-fourth as much as required without any diaphragm action. Metal deck with $2\frac{1}{2}$-in. concrete fill would provide many times the required stiffness. Even the diaphragm action of the metal deck without the concrete slab is likely to provide more than the 90 kips/ft required. Treatment of diaphragm bracing is generally outside the scope of this text; however, a number of references [9.54–9.65] on diaphragm action are included at the end of this chapter.

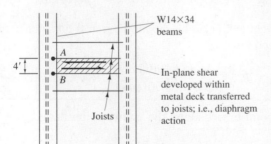

Figure 9.13.10   Diaphragm action of metal deck attached to joists.

(b) Determine weld required to attach joists to top flange of W14×34. The force required is, according to Eq. 9.13.22,

$$\text{Required } Q_n = 0.004\beta P_{\text{cr}} \qquad [9.13.22]$$

Conservatively take $\beta = 4$ as for a series of point braces,

$$\text{Required } Q_n = 0.004(4)181 = 2.9 \text{ kips}$$

Using a $\frac{3}{16}$-in. fillet weld (satisfies LRFD-Table J2.4 for minimum weld size on $\frac{7}{16}$-in. flange thickness) with E70 electrodes and the SMAW process, the nominal strength R_{nw} per inch is

$$R_{nw} = \tfrac{3}{16}(0.707)0.60(70) = 5.57 \text{ kips/in.}$$

$$\text{Required } L_w = \frac{Q_n}{R_{nw}} = \frac{2.9}{5.57} = 0.5 \text{ in.}$$

Use $\frac{3}{16}$-in. weld, E70, $L = 0' - 0\frac{3}{4}''$ on each side of joist bearing seat. This is needed to provide minimum length of 4 times weld size, satistying LRFD-J2.2b.

∎∎

Bracing Requirements for Inelastic Steel Beams

When ability to accommodate large plastic strain is desired at bracing points, such as when plastic strength or plastic analysis is used in design, the procedure in the previous section may not be adequate. Lay and Galambos [9.45] have developed a set of rules for design in cases where such high plastic strain (rotation capacity) is required to be accommodated.

In effect, bracing requirements are based on a rotation capacity R consistent with the beam unbraced length slenderness ratio given by Eq. 9.5.7. It has been found that within the laterally unbraced length "local buckling causes a curtailment of the load capacity of the member and therefore defines the rotation capacity of the beam" [9.45].

The derivation of Lay and Galambos [9.45] has determined the maximum lateral moment that can develop in the compression flange under a *uniform* moment, $M_x = M_p$, by using the strain distributions on the compression flange due to (a) compression due to $M_x = M_p$ and (b) the lateral bending strains when local buckling occurs on the "compression" side.

The design recommendations are:

1. For axial *strength,* the required cross-sectional area is

$$\text{Required } A_{\text{brace}} = \left[\frac{\alpha_{st} - 1}{\alpha_e - \sqrt{\alpha_e}}\right]\left[\frac{2}{3}\right]\frac{A_{\text{comp}}}{(L_{\text{av}}/b)} \qquad (9.13.23)$$

where $L_{\text{av}} = \dfrac{2L_L L_R}{L_L + L_R}$

$\quad L_L$ = unbraced length to left of braced point
$\quad L_R$ = unbraced length to right of braced point
$\quad b$ = width of compression flange
$\quad \alpha_{st}$ = strain at strain hardening divided by yield strain, ϵ_{st}/ϵ_y
        (A value of 12 may be reasonable for steels up to $F_y = 60$ ksi [9.46].)
$\quad \alpha_e$ = elastic modulus divided by strain hardening modulus of elasticity, E_s/E_{st}

2. The axial *stiffness* requirement is satisfied when

$$\frac{L_{\text{brace}}}{L_{\text{av}}} \leq 0.57\left[\frac{\alpha_{st} - 1}{\alpha_e - \sqrt{\alpha_e}}\right]\left[\frac{\text{actual } A_{\text{brace}}}{\text{required } A_{\text{brace}}}\right]\left[\frac{L_a}{b}\right] \qquad (9.13.24)$$

where L_a = longer of the two adjacent unbraced lengths.

In addition to the axial strength and stiffness requirements, Lay and Galambos [9.45] indicated that when only the compression flange is braced there are additional flexural strength and stiffness requirements that must be satisfied. These flexural requirements (not given here) give overly large and deep bracing members. Recent studies [9.47] indicate that flexural requirements are unnecessary for lateral bracing locations away from beam vertical reactions. When the compression flange is braced, point restraint giving the necessary axial strength and stiffness is sufficient. Lateral bracing at vertical supports undoubtedly does need some flexural strength and stiffness to prevent a beam from tipping, but ordinary framing at such locations generally provides adequate flexural strength and stiffness.

EXAMPLE 9.13.3 ⎯⎯⎯⎯⎯⎯⎯⎯⎯⎯⎯⎯⎯⎯⎯⎯⎯⎯⎯⎯⎯⎯⎯⎯⎯⎯⎯⎯⎯⎯⎯⎯⎯⎯⎯⎯⎯⎯⎯

For the beam of Example 9.13.1 (Fig. 9.13.8) determine the brace size required if plastic hinge rotation is required at the bracing points.

Solution

(a) Determine the section required for axial *strength*. Use Eq. 9.13.23,

$$\text{Required } A_{\text{brace}} = \left(\frac{\alpha_{st} - 1}{\alpha_e - \sqrt{\alpha_e}}\right)\left(\frac{2}{3}\right)\frac{A_{\text{comp}}}{(L_{\text{av}}/b)}$$

where $\alpha_{st} = \epsilon_{st}/\epsilon_y$ which may be taken as 12 for $F_y = 36$ ksi and can probably also be used for steels to about $F_y = 60$ ksi. For $\alpha_e = E/E_{st}$, use $E_{st} = 450$ ksi, giving $\alpha_e = 29{,}000/450 = 64$. Certainly, the use of such values is accurate enough for design purposes. For A36 steel, $\alpha_{st} = 12$, $\alpha_e = 64$.

$$\frac{\alpha_{st} - 1}{\alpha_e - \sqrt{\alpha_e}} = \frac{12 - 1}{64 - \sqrt{64}} = 0.2$$

$$L_{\text{av}} = \frac{2L_L L_R}{L_L + L_R} = \frac{2(16)(16)(12)}{16 + 16} = 192 \text{ in.}$$

$$\text{Required } A_{\text{brace}} = 0.2\left(\frac{2}{3}\right)\frac{A_{\text{comp}}}{192/9.96} = 0.007 A_{\text{comp}}$$

$$= 0.007(24.8/2) = 0.09 \text{ sq in.}$$

Practically any rolled shape will satisfy this small requirement.

(b) Examine the requirement for axial *stiffness*. Use Eq. 9.13.24, assuming conservatively the area A_{brace} provided equals exactly the required,

$$\frac{L_{\text{brace}}}{L_{\text{av}}} \leq 0.57\left[\frac{\alpha_{st} - 1}{\alpha_e - \sqrt{\alpha_e}}\right]\left[\frac{\text{actual } A_{\text{brace}}}{\text{required } A_{\text{brace}}}\right]\frac{L_a}{b}$$

$$\leq 0.57(0.2)(1.0)(192/9.96) = 4.06$$

$$L_{\text{brace}} \leq 4.06 L_{\text{av}} = 4.06(16) = 65 > 7.5 \text{ ft for brace} \qquad \text{OK}$$

As in Example 9.13.1, the slenderness ratio recommended limit of 200 for compression members will give a member adequate for bracing purposes. *Use C4×5.4.* ■■

One conclusion regarding lateral bracing for beams and columns is that the requirements for such bracing are easily met. It is more important to provide a brace of *some size* than to be overly concerned about what the size should be.

Empirical procedures have long been used in lieu of a rational investigation of the strength and stiffness requirements for braces. A summary of some of these rules is given by Lay and Galambos [9.45]. The typical rule-of-thumb has been to use a brace having a strength equal to or greater than 2 percent of the compressive strength of the compression element being braced. This seems to be a conservative alternative to an analytical study.

9.14* BIAXIAL BENDING OF DOUBLY SYMMETRIC I-SHAPED SECTIONS

When an I-shaped section is loaded in the plane of its major axis, that is, a moment M_y causes weak axis bending, in combination with strong axis bending M_x from loading in the plane of its web, the strength limit state may be controlled by yielding under combined stress or by lateral-torsional buckling. Two common situations are roof purlins and crane support girders. How does the moment M_y affect lateral stability? When M_y acts alone lateral-torsional buckling is *not* a possible limit state, and most rolled shapes will develop their plastic moment. When M_x acts alone, lateral-torsional buckling may well be the controlling limit state.

This subject of biaxial bending, including torsion, has been treated by Dohrenwend [9.66], Gaylord and Gaylord [9.67], Springfield [9.68], Pastor and DeWolf [9.69], and Razzaq and Galambos [9.70, 9.71]. Many others, including the *SSRC Guide* [6.8, pp. 306–310; 6.20], have treated the subject in the context of beam-columns, which are discussed in Chapter 12.

From Eq. 9.4.32 for pure bending M_x with respect to the strong axis,

$$M_{cr}^2 = M_x^2 = \frac{\pi^2}{L^2}\left[\frac{\pi^2 E^2 C_w I_y}{L^2} + EI_y GJ\right] \tag{9.14.1}$$

or

$$\frac{M_x^2}{EI_y GJ} = \frac{\pi^2}{L^2}\left[1 + \left(\frac{\pi}{\lambda L}\right)^2\right] \tag{9.14.2}$$

where $\lambda = 1/a = \sqrt{GJ/EC_w}$.

When simultaneously a moment M_y is applied, Eq. 9.14.2 becomes [9.65]

$$\frac{M_x^2}{EI_y GJ} + \frac{M_y^2}{EI_x GJ} = \frac{\pi^2}{L^2}\left[1 + \left(\frac{\pi}{\lambda L}\right)^2\right] \tag{9.14.3}$$

which is applicable to I-shaped sections with two axes of symmerty. Furthermore, it is applicable for sections with point symmetry (such as the zee), and is approximately valid for channels when M_y does not exceed $0.25M_x$ [9.67].

Combinations of M_x and M_y which satisfy Eq. 9.14.3 will plot as an ellipse, as shown in Fig. 9.14.1 for a W14×74.

The usual limit state accepted for biaxial bending is the achievement of the yield stress F_y at the extreme fiber under combined stress. Therefore, no matter how stable a beam may be, the combination of moments must satisfy

$$\frac{M_x}{S_x} + \frac{M_y}{S_y} \leq F_y \tag{9.14.4}$$

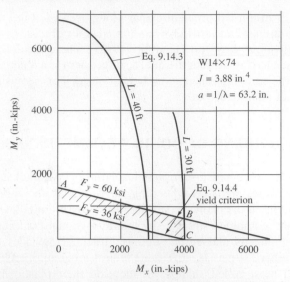

Figure 9.14.1   Lateral buckling strength for biaxial bending of doubly symmetric I-shaped sections.

Since the relationships shown in Fig. 9.14.1 for W14×74 are typical, certain conclusions may be drawn. Consider a laterally unbraced length of 30 ft and steel with $F_y = 60$ ksi. Ideally, the ultimate condition is defined by the lines AB (yielding controls) and BC (buckling controls).

The most important observation is that for ordinary laterally unbraced lengths (say 25 ft or less) the line BC is nearly vertical; therefore *simultaneous application of M_y does not appreciably affect the critical moment M_x.*

Load and Resistance Factor Design

For biaxial bending, LRFD-F1 refers to LRFD-H1, which applies for singly and doubly symmetric members subject to bending and axial force. The interaction equations for combined bending and axial load are treated in Chap. 12. When torsion is included in the loading, LRFD-F1 indicates using LRFD-H2, which applies for "unsymmetric members and members under torsion and combined torsion, flexure, shear, and/or axial force."

Without axial load P_u, LRFD Formula (H1-1b), would be the applicable equation, as follows:

$$\left(\frac{M_{ux}}{\phi_b M_{nx}} + \frac{M_{uy}}{\phi_b M_{ny}} \right) \leq 1.0 \tag{9.14.5}$$

It is logical to use such an interaction equation as Eq. 9.14.5; however, from the discussion at the beginning of this section, such a procedure is conservative. Note that M_{nx} and M_{ny} can be as high as M_{px} and M_{py}, respectively, in accordance with LRFD-F1.

Earlier Specifications, either LRFD or ASD, did not specifically direct the designer to the combined bending and axial load interaction equations for biaxial bending *without* axial load.

The authors, and others [9.67], have long recommended the following procedure be used for biaxial bending of symmetric I-shaped sections subject to biaxial bending without axial compression.

1. For the *yielding limit state* controlling (line *AB* of Fig. 9.14.1):

$$\left(f_{un} = \frac{M_{ux}}{S_x} + \frac{M_{uy}}{S_y}\right) \le \phi_b F_y \tag{9.14.6}$$

2. For the *lateral-torsional buckling limit state* controlling (line *BC* of Fig. 9.14.1):

$$\phi_b M_{nx} \ge M_{ux} \tag{9.14.7}$$

where f_{un} = normal (compressive or tensile) stress under factored loads
   M_{ux} = factored moment about the x-axis (strong axis)
   M_{uy} = factored moment about the y-axis (weak axis)
   ϕ_b = strength reduction factor in bending, 0.90
   M_{nx} = nominal moment strength for a member loaded only in the plane of the web (i.e., strong axis bending) according to LRFD-F1.

LRFD-H2 uses the assumption that the *yielding limit state* controls; thus, the provision requires the combined normal (compressive or tensile) stress f_{un} under factored loads not to exceed $\phi_b F_y$; that is, use Eq. 9.14.6.

A common application of the biaxial bending analysis is when the applied loading includes torsional loading, which contributes to M_{uy}, as discussed in Chap. 8. Equations 9.14.6 and 9.14.7 seem particularly applicable for that situation; though LRFD-H2 indicates to use Eq. 9.14.6 alone. Note that LRFD-H1 (Eq. 9.14.5) is *not* applicable when torsion is included.

In addition, for the *buckling limit state* LRFD-H2 indicates the limit would be $\phi_c F_{cr}$ using the compression member $\phi_c = 0.85$ and the F_{cr} column formulas, Eqs. 6.7.7 and 6.7.8, discussed in Chapter 6.

This latter provision seems not to apply to lateral-torsional buckling under biaxial bending without axial compression.

As an alternative to the combined stress limit state approach, LRFD-H3 provides an alternative interaction relationship that was developed primarily for beam-columns. However, when the axial load P becomes zero, the relationship is suitable for biaxial bending, with or without torsion.

Allowable Stress Design

The same general approach is recommended as described for Load and Resistance Factor Design. This procedure was recommended by Gaylord and Gaylord [9.67] as follows:

1. For yielding controlling (line *AB* of Fig. 9.14.1):

$$\frac{M_x}{S_x} + \frac{M_y}{S_y} \le 0.60 F_y \tag{9.14.8}$$

2. For stability controlling (line BC of Fig. 9.14.1):

$$\frac{M_x}{S_x} \leq \text{ Allowable stresses based on} \atop \text{ASD Formulas (F1-6 and F1-7) or (F1-8)} \qquad (9.14.9)$$

The ASD Specification has no specific provision for biaxial bending without axial compression. One could use the beam-column provisions, or as is commonly and conservatively done, require that the combined stress not exceed the allowable stress based on lateral-torsional buckling for M_x acting alone.

EXAMPLE 9.14.1

Design a W section to serve as a crane support girder to carry a live load moment M_x of 301 ft-kips (without impact) from the crane wheels. In addition, a moment M_f of 30 ft-kips acts in the plane of the top flange as a result of movement of the crane back and forth between the support girders. This moment M_f is based on a lateral force acting on the top flange equal to 10% (the total is 20% with one-half at each end) of the lifted load and crane trolley weight in accordance with LRFD and ASD-A4.3. The moment M_f about the y-axis is assumed to be resisted by one flange; in effect, this is to account for the torsional effect by using the flexure analogy (see Sec. 8.6). The approximation of equivalent systems is shown in Fig. 9.14.2. Assume the simply supported span of 24 ft is laterally braced only at the ends. Use Load and Resistance Factor Design and $F_y = 50$ ksi.

Solution

(a) Compute factored moments. Using LRFD-A4.1 with Formula (A4-2),

Estimated dead load moment = 10 ft-kips

$$M_{ux} = 1.2(10) + 1.6(30)1.25 \quad \text{(using 25\% impact increase)}$$
$$= 614 \text{ ft-kips}$$
$$M_{uf} = 1.6(30) = 48 \text{ ft-kips}$$
$$M_{uy} \approx 2M_{uf} = 96 \text{ ft-kips} \quad \text{(see Fig. 9.14.2b)}$$

(b) Select a section. Because torsion acts LRFD-H2 applies, rather than LRFD-H1, Eq. 9.14.5. Thus, use the approach of Sec. 7.11 with the criterion of Eq. 9.14.6,

$$\text{Required } S_x \geq \frac{M_{ux}}{\phi_b F_y} + \frac{M_{uy}}{\phi_b F_y}\left(\frac{S_x}{S_y}\right)$$

$$\text{Required } S_x \geq \frac{614(12)}{0.90(50)} + \frac{2(48)12}{0.90(50)}\left(\frac{S_x}{S_y}\right)$$

$$\geq 164 + 26(\approx 6) = 320 \text{ in.}^3$$

Try W24×131: $S_x = 329$ in.^{3}

$$\text{Required } S_x \geq \frac{614(12)}{0.90(50)} + \frac{2(48)(12)6.2}{0.90(50)} = 322 \text{ in.}^3$$

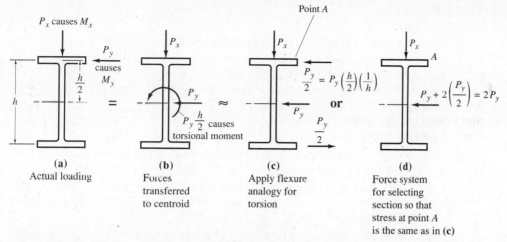

Figure 9.14.2 Approximate equivalent system for biaxial bending and torsion.

(c) Check the strength. The yield limit state appears to be satisfied; however, the stability limit state must be checked and a final check of all criteria should always be made.

Check the *yield limit state* criterion, Eq. 9.14.6, adjusting M_{ux} slightly to reflect the correct beam weight,

$$f_{un} = \frac{M_{ux}}{S_x} + \frac{M_{uy}}{S_y} = \frac{613(12)}{329} + \frac{96(12)}{53.0} = 22.4 + 21.7 = 44.1 \text{ ksi}$$

$$(f_{un} = 44.1 \text{ ksi}) < [\phi_b F_y = 0.90(50) = 45 \text{ ksi}] \quad \text{OK}$$

Check the *lateral-torsional buckling limit state*, Eq. 9.14.7

$$\phi_b M_{nx} \geq M_{ux} \qquad [9.14.6]$$

For the W24×131,

$$M_p = F_y Z_x = 50(370)/12 = 1540 \text{ ft-kips}$$
$$L_p = 300 r_y / \sqrt{F_y} = 300(2.97/12)/\sqrt{50} = 10.5 \text{ ft}$$
$$M_r = (F_y - F_r)S_x = (50 - 10)(329)/12 = 1100 \text{ ft-kips}$$
$$L_r = 29.1 \text{ ft (from LRFD table "LOAD FACTOR DESIGN}$$
$$\text{SELECTION TABLE" or Eq. 9.6.6)}$$

Then using Eq. 9.6.4 with $C_b = 1$,

$$M_n = C_b \left[M_p - (M_p - M_r)\left(\frac{L_b - L_p}{L_r - L_p}\right) \right] \leq M_p \qquad [9.6.4]$$

$$M_n = \left[1542 - (1542 - 1100)\left(\frac{24 - 10.5}{29.1 - 10.5}\right) \right] = 1220 \text{ ft-kips}$$

$$\phi M_n = 0.90(1220) = 1100 \text{ ft-kips} > M_{ux} \quad \text{OK}$$

Use W24×131, $F_y = 50$ ksi. ■■

SELECTED REFERENCES

9.1. Joseph A. Yura, Theodore V. Galambos, and Mayasandra K. Ravindra. "The Bending Resistance of Steel Beams," *Journal of the Structural Division, ASCE,* **104,** ST9 (September 1978), 1355–1370.

9.2. Karl de Vries. "Strength of Beams as Determined by Lateral Buckling," *Transactions, ASCE,* **112** (1947), 1245–1320.

9.3. R. A. Hechtmann, J. S. Hattrup, E. F. Styer, and J. L. Tiedemann. "Lateral Buckling of Rolled Steel Beams," *Transactions, ASCE,* **122** (1957), 823–843.

9.4. W. J. Austin, S. Yegian, and T. P. Tung, "Lateral Buckling of Elastically End-Restrained I Beams," *Transactions, ASCE,* **122** (1957), 374–388.

9.5. J. W. Clark and J. R. Jombock. "Lateral Buckling of I-beams Subjected to Unequal End Moments," *Journal of the Engineering Mechanics Division, ASCE,* **83,** EM3 (July 1957).

9.6. Mario G. Salvadori. "Lateral Buckling of I Beams Under Thrust and Unequal End Moments," *Transactions, ASCE,* **120** (1955), 1165–1182.

9.7. Theodore V. Galambos. *Structural Members and Frames.* Englewood Cliffs, NJ: Prentice-Hall, Inc., 1968.

9.8. Theodore V. Galambos. "Inelastic Lateral Buckling of Beams," *Journal of the Structural Division, ASCE,* **89,** ST5 (October 1963), 217–242.

9.9. Maxwell G. Lay and Theodore V. Galambos. "Inelastic Steel Beams Under Uniform Moment," *Journal of the Structural Division, ASCE,* **91,** ST6 (December 1965), 67–93.

9.10. Maxwell G. Lay and Theodore V. Galambos. "Inelastic Beams Under Moment Gradient," *Journal of the Structural Division, ASCE,* **93,** ST1 (February 1967), 381–399.

9.11. Campbell Massey and F. S. Pitman. "Inelastic Lateral Stability Under a Moment Gradient," *Journal of the Engineering Machanics Division, ASCE,* **92,** EM2 (April 1966), 101–111.

9.12. Nicholas S. Trahair and Sritawat Kitipornchai. "Buckling of Inelastic I-Beams Under Uniform Moment," *Journal of the Structural Division, ASCE,* **98,** ST11 (November 1972), 2551–2566.

9.13. A. J. Hartmann. "Inelastic Flexural-Torsional Buckling," *Journal of the Engineering Mechanics Division, ASCE,* **97,** EM4 (August 1971), 1103–1119.

9.14. Morris Ojalvo and Ronald R. Weaver. "Unbraced Length Requirements for Steel I-Beams," *Journal of the Structural Division, ASCE,* **104,** ST3 (March 1978), 479–490.

9.15. David A. Nethercot and Nicholas S. Trahair. "Inelastic Lateral Buckling of Determinate Beams," *Journal of the Structural Division, ASCE,* **102,** ST4 (April 1976), 701–717.

9.16. J. Bansal. "The Lateral Instability of Continuous Beams," *AISI Report No.* 3, American Iron and Steel Institute, New York, August 1971.

9.17. Peter W. Hoadley. "Practical Significance of LRFD Beam Buckling Factors," *Journal of Structural Engineering, ASCE,* **117,** 3 (March 1991), 988–996.

9.18. P. A. Kirby and D. A. Nethercot. *Design for Structural Stability.* New York: Wiley, 1979.

9.19. Mario G. Salvadori. "Lateral Buckling of Eccentrically Loaded I-Columns," *Transactions,* ASCE, **121,** (1956), 1163–1178.

9.20. A. J. Hartmann. "Elastic Lateral Buckling of Continuous Beams," *Journal of the Structural Division,* ASCE, **93,** ST4 (August 1967), 11–26.

9.21. Nicholas S. Trahair, "Elastic Stability of Continuous Beams," *Journal of the Structural Division,* ASCE, **95,** ST6 (June 1969), 1295–1312.

9.22. Graham Powell and Richard Klingner. "Elastic Lateral Buckling of Steel Beams," *Journal of the Structural Division,* ASCE, **96,** ST9 (September 1970), 1919–1932.

9.23. N. S. Trahair and M. A. Bradford. *The Behavior and Design of Steel Structures* 2nd ed. London: Chapman and Hall, 1988.

9.24. Charles W. Roeder and Mahyar Assadi. "Lateral Stability of I-beams with Partial Support," *Journal of the Structural Division,* ASCE, **108,** ST8 (August 1982), 1768–1779.

9.25. Yushi Fukumoto, Yoshito Itoh, and Ryoji Hattori. "Lateral Buckling on Welded Continuous Beams," *Journal of the Structural Division,* ASCE, **108,** ST10 (October 1982), 2245–2262.

9.26. J. W. Clark and H. N. Hill. "Lateral Buckling of Beams," *Journal of the Structural Division,* Proceedings ASCE, **86,** ST7 (July 1960), 175–196. Also, *Transactions,* ASCE, **127** (1962), Part II, 180–201.

9.27. Sritawat Kitipornchai, Peter F. Dux, and Nevelle J. Richter. "Buckling and Bracing of Cantilevers," *Journal of Structural Engineering,* ASCE, **110,** 9 (September 1984), 2250–2262.

9.28. Campbell Massey and Peter J. McGuire. "Lateral Stability of Nonuniform Cantilevers," *Journal of the Engineering Mechanics Division,* ASCE, **97,** EM3 (June 1971), 673–686.

9.29. Patrick D. Zuraski. "The Significance and Application of C_b in Beam Design," *Engineering Journal,* AISC, **29,** 1 (First Quarter, 1992), 20–25.

9.30. Frank W. Stockwell, Jr. "Simplified Approach to AISC Bending Formulas," *Engineering Journal,* AISC, **11,** 3 (Third Quarter, 1974), 65–66.

9.31. G. Donald Brandt. "Direct Feasible and Optimal Design of Laterally Unsupported Beams," *Engineering Journal,* AISC, **14,** 2 (Second Quarter, 1977), 78–84.

9.32. H. N. Hill. "Lateral Buckling of Channels and Z-Beams," *Transactions,* ASCE, **119** (1954), 829–841.

9.33. John M. Anderson and Nicholas S. Trahair. "Stability of Monosymmetric Beams and Cantilevers," *Journal of the Structural Division,* ASCE, **98,** ST1 (January 1972), 269–286.

9.34. Sritawat Kitipornchai and Nicholas S. Trahair. "Buckling Properties of Monosymmetric I-Beams," *Journal of the Structural Division,* ASCE, **106,** ST5 (May 1980), 941–957.

9.35. D. A. Nethercot. "Elastic Lateral Buckling of Beams," *Beams and Beam Columns—Stability in Strength* (ed. R. Narayanan). Barking, Essex, England: Applied Science Publishers, 1983.

9.36. Sritawat Kitipornchai, Chien Ming Wang, and Nicholas S. Trahair. "Buckling of Monosymmetric I-Breams Under Moment Gradient," *Journal of Structural Engineering,* ASCE, **112,** 4 (April 1986), 781–799.

9.37. Chien Ming Wang and Sritiwat Kitipornchai. "Buckling Capacities of Monosymmetric I-Beams," *Journal of Structural Engineering.* ASCE, **112** (November 1986), 2373–2391.

9.38. Sritawat Kitipornchai and Alain D. Wong-Chung. "Inelastic Buckling of Welded Monosymmetric I-Beams," *Journal of Structural Engineering,* **113,** 4 (April 1987), 740–756.

9.39. William Zuk. "Lateral Bracing Forces on Beams and Columns," *Journal of the Engineering Mechanics Division,* ASCE, **82,** EM3 (July 1956), Proc. Paper No. 1032, 16 pp.

9.40. George Winter. "Lateral Bracing of Columns and Beams," *Transactions,* ASCE, **125** (1960), 807–845.

9.41. Campbell Massey. "Lateral Bracing Force of Steel I-Beams," *Journal of the Engineering Mechanics Division,* ASCE, **88,** EM6 (December 1962), 89–113.

9.42. George Pincus. "On the Lateral Support of Inelastic Columns," *Engineering Journal,* AISC, **1,** 4 (October 1964), 113–115.

9.43. Theodore V. Galambos. "Lateral Support for Tier Building Frames," *Engineering Journal,* AISC, **1,** 1 (January 1964) 16–19; Disc, **1,** 4 (October 1964), 141.

9.44. Tor B. Urdal. "Bracing of Continuous Columns," *Engineering Journal,* AISC, **6,** 3 (July 1969), 80–83.

9.45. Maxwell G. Lay and T. V. Galambos. "Bracing Requirements for Inelastic Steel Beams," *Journal of the Structural Division,* ASCE, **92,** ST2 (April 1966), 207–228.

9.46. Arthur C. Taylor, Jr. and Morris Ojalvo. "Torsional Restraint of Lateral Buckling," *Journal of the Structural Division,* ASCE, **92,** ST2 (April 1966), 115–129.

9.47. A. J. Hartmann. "Experimental Study of Flexural-Torsional Buckling," *Journal of the Structural Division,* ASCE, **96,** ST7 (July 1970), 1481–1493.

9.48. Bruce R. Mutton and Nicholas S. Trahair. "Stiffness Requirements for Lateral Bracing," *Journal of the Structural Division,* ASCE, **99,** ST10 (October 1973), 2167–2182.

9.49. Ian C. Medland and Cecil M. Segedin. "Brace Forces in Interbred Column Structures," *Journal of the Structural Division,* ASCE, **105,** ST7 (July 1979), 1543–1556.

9.50. Joseph A. Yura. "Fundamentals of Beam Bracing," *Is Your Structure Suitably Braced?,* Structural Stability Research Council, 1993 Conference, Milwaukee, Wisconsin, April 6–7, 1993, 20 pp.

9.51. LeRoy A. Lutz and James M. Fisher. "A Unified Approach for Stability Bracing Requirements," *Engineering Journal,* AISC, **22,** 4 (4th Quarter 1985), 163–167.

9.52. Joseph M. Ales, Jr. and Joseph A. Yura. "Bracing Design for Inelastic Structures," *Is Your Structure Suitably Braced?,* SSRC 1993 Conference, Milwaukee, Wisconsin, April 6–7, 1993, 29–37.

9.53. M. J. Clarke and R. Q. Bridge. "Bracing Force and Stiffness Requirements to Develop the Design Strength of Columns," *Is Your Structure Suitably Braced?,* SSRC 1993 Conference, Milwaukee, Wisconsin, April 6–7, 1993, 75–86.

9.54. George Pincus and Gordon P. Fisher, "Behavior of Diaphragm-Braced Columns and Beams," *Journal of the Structural Division,* ASCE, **92,** ST2 (April 1966), 323–350.

9.55. Samuel J. Errera, George Pincus, and Gordon P. Fisher. "Columns and Beams Braced by Diaphragms," *Journal of the Structural Division,* ASCE, **93,** ST1 (February 1967), 295–318.

9.56. Larry D. Luttrell. "Strength and Behavior of Light-Gage Steel Shear Diaphragms," *Cornell Engineering Research Bulletin* No. 67–1, July 1967.

9.57. T. V. S. R. Apparao, Samuel J. Errera, and Gordon P. Fisher. "Columns Braced by Girts and a Diaphragm," *Journal of the Structural Division,* ASCE, **95,** ST5 (May 1969), 965–990.

9.58. Arthur H. Nilson and Albert R. Ammar. "Finite Element Analysis of Metal Deck Shear Diaphragms," *Journal of the Structural Division,* ASCE, **100,** ST4 (April 1974), 711–726.

9.59. David A. Nethercot and Nicholas S. Trahair. "Design of Diaphragm-Braced I-Beams," *Journal of the Structural Division,* ASCE, **101,** ST10 (October 1975), 2045–2061.

9.60. Amir Simaan and Teoman B. Pekoz. "Diaphragm Braced Members and Design of Wall Studs," *Journal of the Structural Division,* ASCE, **102,** ST1 (January 1976), 77–92.

9.61. Samuel J. Errera and Tamirisa V. S. R. Apparao. "Design of I-Shaped Beams with Diaphragm Bracing," *Journal of the Structural Division,* ASCE, **102,** ST4 (April 1976), 769–781.

9.62. J. Michael Davies. "Calculation of Steel Diaphragm Behavior," *Journal of the Structural Division,* ASCE, **102,** ST7 (July 1976), 1411–1430.

9.63. Samuel J. Errera and Tamirisa V. S. R. Apparao. "Design of I-Shaped Columns with Diaphragm Bracing," *Journal of the Structural Division,* ASCE, **102,** ST9 (September 1976), 1685–1701.

9.64. John T. Easley. "Strength and Stiffness of Corrugated Metal Shear Diaphragms," *Journal of the Structural Division,* ASCE, **103,** ST1 (January 1977), 169–180.

9.65. SDI. *Tentative Recommendations for the Design of Steel Deck Diaphragms.* Westchester, IL: Steel Deck Institute, October 1972.

9.66. C. O. Dohrenwend. "Action of Deep Beams Under Combined Vertical, Lateral and Torsional Loads," *Journal of Applied Mechanics,* **8** (1941), A-130.

9.67. E. H. Gaylord, Jr. and C. N. Gaylord. *Design of Steel Structures.* New York: McGraw-Hill Book Company, Inc., 1957 (pp. 169–170).

9.68. John Springfield. "Design of Columns Subject to Biaxial Bending," *Engineering Journal,* AISC, **12,** 3 (Third Quarter 1975), 73–81.

9.69. Thomas P. Pastor and John T. DeWolf. "Beams with Torsional and Flexural Loads," *Journal of the Structural Division,* ASCE, **105,** ST3 (March 1979), 527–538.

9.70. Zia Razzaq and Theodore V. Galambos. "Biaxial Bending of Beams with or without Torsion," *Journal of the Structural Division,* ASCE, **105,** ST11 (November 1979), 2145–2162.

9.71. Zia Razzaq and Theodore V. Galambos. "Biaxial Bending Tests with or without Torsion," *Journal of the Structural Division,* ASCE, **105,** ST11 (November 1979), 2163–2185.

9.72. Joseph A. Yura. "Winter's Bracing Approach Revisited," *Proceedings, 50th Anniversary Conference,* Structural Stability Research Council, Lehigh University, 21–22 June 1994, pp. 375–382.

9.73. R. H. Plaut. "Requirements for Lateral Bracing of Columns with Two Spans," *Journal of Structural Engineering,* ASCE, **119,** 10 (October 1993), 2913–2913–2931.

9.74. Duane S. Ellifritt, Gregory Wine, Thomas Sputo, and Santosh Samuel. "Flexural Strength of WT Sections," *Engineering Journal,* AISC, **29,** 2 (Second Quarter 1992), 67–74.

PROBLEMS

All problems are to be done according to the AISC Load and Resistance Factor Design or Allowable Stress Design, as indicated by the instructor. All given loads are service loads unless otherwise indicated. Assume lateral support consists of translational restraint but not moment (rotational) restraint, unless otherwise indicated. Assume all standard sections are equally readily available in the indicated grade of steel (even though actually they are not). A figure showing span and loading is required, and after making a design selection, a final check of strength (ϕM_n compared with M_u for LRFD) or stress (f_b compared to allowable stress F_b for ASD) is required. *Note:* Live load must always be applied (or not) such that it causes maximum (or minimum) effect.

9.1. For the case (or cases) assigned by the instructor, plot design strength $\phi_b M_n$ (ft-kips) for LRFD or allowable bending stress F_b (ksi) for ASD vs laterally unbraced length L_b (ft). If the ASD problem has been assigned, show both ASD allowable stress relationships: (1) the combination of Formulas (F1-6) and (F1-7), whichever controls; and (2) Formula (F1-8). Show the portions controlling ($\phi_b M_n$ for LRFD or F_b for ASD) with solid lines and the noncontrolling parts with dashed lines. For any case assigned, show relationships for both $C_b = 1.0$ and 2.3.

Case	Section	F_y (ksi)
1	W16×26	36
2	W14×145	36
3	W21×62	36
4	W24×84	36
5	Plate girder: 5/8×12 flanges; 5/16×30 web	36
6	Plate girder: 7/8×16 flanges; 3/4×26 web	36
7	W16×26	50
8	W14×145	50
9	W21×62	50
10	W24×84	50
11	W16×26	65
12	W14×145	65
13	W24×84	65

9.2. For the case (or cases) assigned by the instructor, determine the maximum concentrated service load P that can act at midspan on a simply supported span. Lateral supports exist only at the ends of the span. The service load is 65% live load and 35% dead load.

Case	Section	Span (ft)	F_y (ksi)
1	W21×62	20	36
2	W24×84	24	50
3	W30×99	30	50

9.3. For the case (or cases) assigned by the instructor, select the lightest W section as a beam. Assume only flexure must be considered; i.e., omit treating shear and deflection. The dead load given is *in addition* to the weight of the beam.

Case	w_D Dead load (kip/ft)	w_L Live load (kip/ft)	Span length (ft)	F_y (ksi)	Lateral support
1	0.6	1.4	20	36	Continuous
2	0.6	1.4	20	36	Ends and midspan
3	0.6	1.4	20	36	Ends only
4	0.5	1.0	28	36	Ends and midspan
5	0.5	1.0	28	60	Ends and midspan
6	0.2	0.6	35	36	Continuous
7	0.2	0.6	35	36	Every 7 feet
8	0.2	0.6	35	36	Ends and midspan
9	0.2	0.6	35	65	Continuous
10	0.2	0.6	35	65	Every 7 feet
11	0.2	0.6	35	65	Ends and midspan
12	0.2	0.6	35	100	Continuous
13	0.2	0.6	35	100	Every 7 feet
14	0.2	0.6	35	100	Ends and midspan
15	0	0.65	35	36	Every 5 ft
16	0	0.65	35	36	Ends only
17	0.5	2.0	48	36	Every 16 feet
18	0.5	2.0	48	60	Every 16 feet

9.4. Select the lightest W sections for the situation shown (see figure next page), under the following conditions:
 a. A 36 steel; continuous lateral support
 b. A36 steel; lateral support at ends only
 c. A36 steel; lateral support at ends and at point A
 d. A572 Grade 60 steel; lateral support at ends and point A

9.5. Select the lightest W section for the situation shown in the accompanying figure using (a) A572 Grade 50 steel and (b) steel with $F_y = 65$ ksi, assuming lateral support at the ends and at point A only.

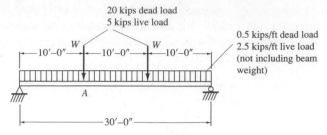

Problem 9.4

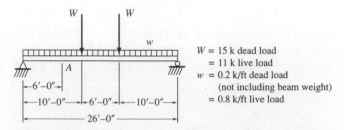

Problem 9.5

9.6. Select the lightest W section for the conditions shown in the accompanying figure. Assume there is no deflection limitation. Use (a) A36 steel and (b) A572 Grade 60 steel.

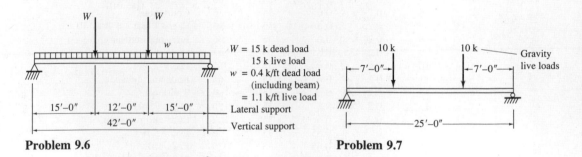

Problem 9.6   **Problem 9.7**

9.7. A floor beam, laterally supported at the ends only and supporting vibration inducing heavy machinery, is subject to the loads shown in the accompanying figure. Select the lightest W section of A36 steel (consider LRFD or ASD-A4.2). Compare the result when there is no deflection limit with that when L/d is limited to a maximum of 20 under full load, a traditional limit to minimize perceptible vibration due to pedestrian traffic.

9.8. For the case (or cases) assigned by the instructor, select the lightest W section to serve as a uniformly loaded library floor beam on a simply supported beam.

Lateral support occurs at the ends and at $L/4$, $L/2$, and $3L/4$. Live load deflection may not exceed $L/300$ (not an LRFD or ASD Specification limit but a design limit for this design). Given dead load moment does *not* include beam weight. Assume $C_b = 1.0$.

Case	M_D Dead load moment (ft-kips)	M_L Live load moment (ft-kips)	Span length (ft)	F_y (ksi)	Deflection limit
1	49	98	28	36	$L/360$
2	49	98	28	60	$L/360$
3	0	240	48	36	$L/300$
4	0	240	48	65	$L/300$
5	50	190	48	36	$L/360$
6	50	190	48	65	$L/360$
7	80	750	60	36	$L/400$
8	80	750	60	50	$L/400$
9	80	750	60	60	$L/400$

9.9. A beam is to serve as a floor beam on a simply supported span of 20 ft. The live load consists of a movable concentrated load (no impact) of 50 kips. Live load deflection may not exceed $L/360$.
   a. Select the lightest W section of A36 steel when continuous lateral support is provided.
   b. Repeat (a) if lateral support is provided only at the ends.

9.10. A W10×33 is to be used as a simply supported beam on a span of 25 ft with lateral support at the ends only. The beam is required to support a plastered ceiling. If the dead load is 0.15 kips/ft (including beam weight), what is the maximum uniform service live load permitted on the beam, using A36 steel? What percentage increase in live load can be gained if the beam is A572 Grade 60 steel? Comment.

9.11. Redesign for loading Cases 1 and/or 2, as assigned, of Prob. 9.8 when lateral support is provided only at the ends and at midspan. In addition to the $L/360$ traditional deflection limit for a plastered ceiling, the architect requires the beam to be no deeper than nominal W12.

9.12. Investigate the beam of the accompanying figure for bending and shear if the section is A572 Grade 50 steel. External lateral support for the beam is

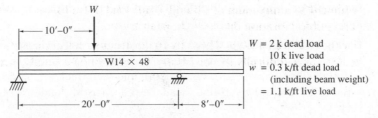

Problem 9.12

provided only at the vertical supports and at the tip of the cantilever. If one additional lateral support were provided at the 12-kip load, how much lighter, if any, could the W14 section be made?

9.13. Select the lightest W section for each of the situations shown in the accompanying figure. The concentrated load W is 5 kips dead load and 15 kips live load. Assume lateral support is provided at the reactions and at the concentrated loads. Use A36 steel.

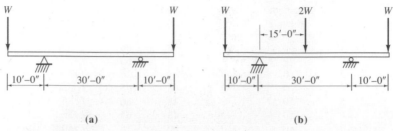

(a)   (b)

Problem 9.13

9.14. Select the lightest W section for the beam of the accompanying figure. Lateral support is provided at concentrated load points, reactions, and at end of cantilever. Use (a) A36 steel and (b) A572 Grade 65 steel.

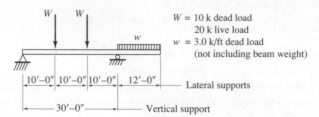

$W = 10$ k dead load
      20 k live load
$w = 3.0$ k/ft dead load
    (not including beam weight)

Problem 9.14

9.15. Determine the nominal strength M_n for LRFD or allowable bending stress F_b for ASD for the channel section and loading in Prob. 8.19 if lateral support exists only at the ends. *Neglecting torsion,* how much larger, if any, section would be required? Use A36 steel.

9.16. Select a channel for the conditions of Prob. 8.20, assuming lateral support at the ends only. Consider torsion in accordance with Chapter 8. Use for safety criterion either of the following, as assigned: (a) for ASD the combined stress due to bending and torsion should not exceed the allowable based on lateral-torsional buckling; or (b) for LRFD the factored load stress f_{un} may not exceed $\phi_b F_y$.

9.17. Design a built-up I-shaped beam with different sized flanges for the conditions of Prob. 9.4, part (b). What percent weight can be saved, if any, by using different sized flanges? Use beam depth and web thickness approximately the same as for the lightest rolled W shape that satisfies loading conditions. A36 steel.

9.18. For the beam selected for a case assigned from Prob. 9.3, estimate the size of bracing (i.e., select a section) required. The bracing frames into both sides and is attached to the compression flange. Length of bracing is 6 ft.

9.19. For the beam designed in the case assigned in Prob. 9.8, estimate the size bracing required. Assume bracing is 12 ft long, frames into both sides of the beam, and is attached to the compression flange. Preferably select channels.

9.20. Determine the adequacy of a W24×84 (with rail, 20 lb/ft) serving as a crane support girder of A36 steel. The simple span is 20 ft with lateral support at the ends only. Use accepted good practice in accounting for the torsional effect of lateral loading. Maximum moments occurring near midspan are
M_x (live load plus impact) = 90 ft-kips
M_x (dead load) = 10 ft-kips
M_f = 10 ft-kips (assumed live load resisted by one flange using top flange lateral loading in accordance with LRFD or ASD-4.3.

9.21. Select the lightest W8 section to be used in an inclined position such that the plane of the web makes an angle of 30 degrees with the plane of loading. The beam is to be of A572 Grade 50 steel, and it has lateral support only at the ends of the 22-ft simple span. The uniform gravity load is 0.3 kips/ft snow load and 0.1 kip/ft dead load, in addition to the beam weight.

9.22. For the case assigned by instructor, design the lightest W section to serve as a crane support girder as shown in the accompanying figure. Assume lateral support at the ends only and that deflection need not be restricted. Assume one-half the lateral force on the crane rail (see LRFD *or* ASD-A4.2 and A4.3 for cab-operated crane) acts at each runway girder. Use A36 steel.

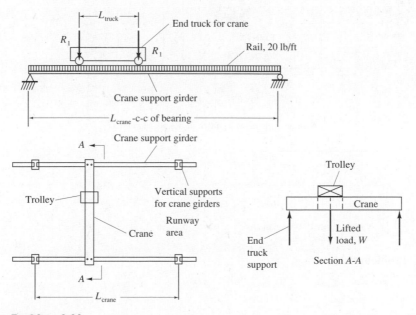

Problem 9.22

Case	Crane capacity W (tons)	Maximum gravity end-truck wheel load R_1 (kips)	Trolley weight (kips)	Span L_{crane} (ft)	Crane end-truck wheel spacing L_{truck} (ft)
1	15	18	6	40	9.5
2	20	27	6	36	8.0
3	30	36	6	32	6.0
4	20	25	5	30	6.0

9.23. Redesign the crane support girder of Prob. 9.22, using a combination channel and W section.

9.24. Design the crane runway girder indicated on the accompanying figure. The two cranes are each 30-ton capacity, with the end-truck wheel spacing as given in the figure. The rails are ASCE 60 lb with clamps (see AISC Manual). The dead load of each crane is 15 kips, equally distributed to its four wheels. Each crane trolley weighs 3 kips. Use A36 steel and (a) select a single W section and (b) select a combination W section and channel, where the channel would have its web flat against the top flange of the W section.

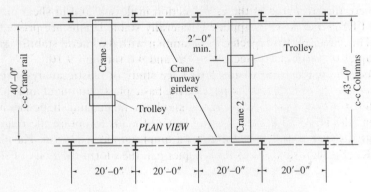

Assume crane runway girders are simply supported

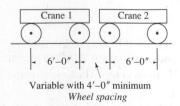

Variable with 4'–0" minimum
Wheel spacing

Problem 9.24

Chapter 10

Continuous Beams

10.1 INTRODUCTION

This chapter brings together theoretical concepts relating to both Load and Resistance Factor Design and Allowable Stress Design that have been presented in Chapters 7 and 9. In addition, plastic analysis is introduced for application to continuous beams; this procedure may be applicable with "compact sections" having adequate lateral bracing. Plastic analysis is permitted under Chapter N of the ASD Specification [1.5] and under LRFD-F1.2d. The reader is assumed to be familiar with elastic methods of statically indeterminate analysis.

Before proceeding into the chapter, the reader should review Secs. 7.3 through 7.5 where the strengths of the cross-section in flexure and in shear are treated, and where LRFD and ASD as applied to laterally stable beams are presented.

The reader is also expected to be familiar with the lateral stability criteria treated in Chapter 9, particularly Secs. 9.1, 9.2, and 9.6 through 9.10.

For the reader who wishes a package study of plastic analysis and the related beam design, Secs. 7.3 and 7.4 provide the basic plastic moment and shear strengths, including the concepts of plastic moment, plastic hinge, and shape factor. Section 7.4 contains the basic approach of using factored loads to obtain the required nominal strength, as well as design examples for a simply supported beam. The plastic analysis and LRFD design sections of this chapter can then form the body of the study.

10.2 PLASTIC STRENGTH OF A STATICALLY INDETERMINATE BEAM

In Sec. 7.3 it was shown that a simply supported beam supporting a concentrated load reached its lmit state (i.e., collapse condition) when the concentrated load is large enough to cause the plastic moment M_p to develop; that is, cause a plastic hinge to develop under the concentrated load.

Referring to Fig. 10.2.1, as the concentrated load increases toward the value W_n that causes the collapse condition, the portion of the beam where the moment does not

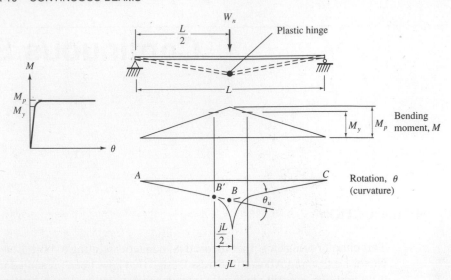

Figure 10.2.1   Moment–curvature relationships for a plastic hinge.

exceed M_y is elastic (meaning straight line θ). In the region near the plastic moment M_p there is an inelastic length jL, and the midspan curvature is large. For practical purposes the beam may be treated as two rigid parts (straight lines) connected by a hinge at B, known as a *plastic hinge*, and having a concentrated angle change, or hinge rotation, θ_u.

The actual length of a plastic hinge is dependent on the shape of the cross-section and can vary from about one-tenth to as much as one-third of the span. As an example, consider the relationship between the bending moments and the beam segments AB and AB' in Fig. 10.2.1:

$$\frac{AB}{M_p} = \frac{AB'}{M_y} \quad \text{or} \quad \frac{L/2}{M_p} = \frac{L/2 - jL/2}{M_y} \tag{10.2.1}$$

Solving for j,

$$j = 1 - \frac{M_y}{M_p} = 1 - \frac{1}{\xi} \tag{10.2.2}$$

where ξ is the shape factor as discussed in Sec. 7.3.

EXAMPLE 10.2.1

Compute the length of the plastic hinge for the beam shown in Fig. 10.2.1 for (a) a W16×40 and (b) a rectangular beam having a width b and a depth d.

Solution
(a) For W16×40,

$$\xi = \frac{M_p}{M_y} = \frac{Z}{S} = \frac{72.9}{64.7} = 1.13$$

From Eq. 10.2.2,

$$jL = L\left(1 - \frac{1}{\xi}\right) = L\left(1 - \frac{1}{1.13}\right) = 0.115L$$

(b) For rectangular section,

$$\xi = \frac{Z}{S} = \frac{\dfrac{bd^2}{4}}{\dfrac{bd^2}{6}} = 1.5$$

$$jL = L\left(1 - \frac{1}{1.5}\right) = 0.333L \quad \blacksquare\blacksquare$$

Even though the distance jL may be as much as one-third of the span, as shown in Example 10.2.1, the simple assumption of a plastic hinge at a point has been amply demonstrated by tests. Beedle [10.1] and Massonnet and Save [10.2] have extensive discussions of theoretical and experimental verification of plastic analysis procedures so individual research references are not included here.

Plastic Limit Load—Equilibrium Method

At the collapse condition with the plastic limit load W_n acting, the requirements of equilibrium are still applicable. Consider first a statically determinate simply supported beam, as shown in Fig. 10.2.2. A collapse condition is achieved when the load W_n is large enough to cause the plastic moment M_p to occur at one location (in this case under the load). When the sufficient number of plastic hinges have been developed to allow instantaneous hinge rotations without developing increased resistance, a mechanism is said to have occurred.

EXAMPLE 10.2.2 _____

Determine the collapse mechanism load W_n for the W21×62 beam of Fig. 10.2.2. Assume $F_y = 36$ ksi.

Solution. Using equilibrium,

$$M_p = \frac{W_n L}{4}$$

$$W_n = \frac{4M_p}{L} = \frac{4(F_y Z)}{L} = \frac{4(36)(144)}{24(12)} = 72 \text{ kips}$$

Using the simplified procedure of considering the behavior as ideally elastic-plastic, the deflection occurring when M_p is reached is based on the elastic equation, which is strictly valid only until M_y (neglecting residual stress) is reached. At a maximum moment of M_y.

$$\Delta_y = \frac{WL^3}{48EI} = \frac{M_y L^2}{12EI} = \frac{F_y IL^2}{c(12)EI} = \frac{F_y L^2}{12cE}$$

$$= \frac{36(24)^2(144)}{12(10.5)(29,000)} = 0.82 \text{ in.}$$

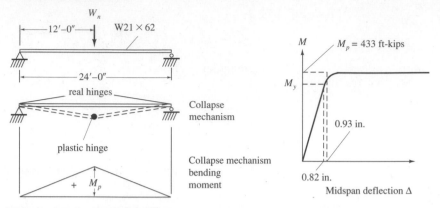

Figure 10.2.2   Example 10.2.2.

Assuming a linear extension until $M = M_p$,

$$\Delta_p = 0.82\frac{Z}{S} = 0.82\frac{144}{127} = 0.82(1.14) = 0.93 \text{ in.}$$

The deflection when M_p is achieved will be higher than this. However, it is the service load deflection that is normally of concern, and this is correctly computed by the usual elastic procedures.

   Note that the collapse mechanism moment diagram for this statically determinate beam is the same shape as that which occurs if the maximum moment is down in the service load range. At whatever load level, from an infinitesimal load to collapse mechanism load, the bending moment at every point remains in a constant proportion to the load.   ■■

EXAMPLE 10.2.3 _____

Determine the collapse mechanism load W_n for the fixed end W16×40 beam of Fig. 10.2.3. Assume $F_y = 26$ ksi.

Solution.   Since the beam is statically indeterminate, three plastic hinges are required to form a mechanism. Using equilibrium,

$$2M_p = \frac{W_n L}{4}$$

$$M_p = F_y Z = 36(72.9)\tfrac{1}{12} = 219 \text{ ft-kips}$$
$$W_n = 8M_p/L = 8(219)/24 = 73 \text{ kips}$$

In order to determine the load-deflection diagram, it is necessary to know the loading history up to the collapse condition. In this case the elastic bending moments, as shown in Fig. 10.2.3, give equal positive and negative bending moment; therefore the three plastic hinges form simultaneously. Moments will increase simultaneously in direct proportion until the collapse condition is reached. Again, as for the statically determinate case, as load is applied the

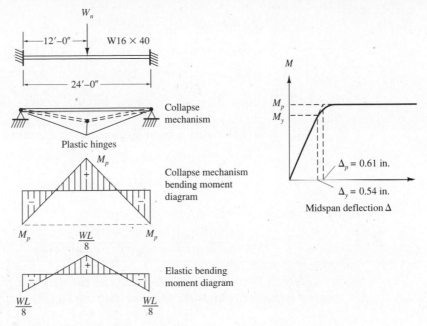

Figure 10.2.3 Example 10.2.3.

bending moment at every point remains in a constant proportion to the load. For this special statically indeterminate case, plastic analysis has no advantage over elastic analysis.

$$\Delta_y = \frac{WL^3}{192EI} = \frac{M_y L^2}{24EI} = \frac{F_y L^2}{24Ec}$$

$$= \frac{36(24)^2(144)}{24(29,000)8} = 0.54 \text{ in.}$$

Assuming a linear extension until $M = M_p$,

$$\Delta_p = 0.54\frac{Z}{S} = 0.54\frac{72.9}{64.7} = 0.61 \text{ in.} \quad \blacksquare\blacksquare$$

EXAMPLE 10.2.4 _____

Determine the load-deflection diagram for the W16×40 beam of Fig. 10.2.4 for loading up to the collapse condition. Assume $F_y = 36$ ksi.

Solution. The collapse mechanism load W_n can be found directly without knowing the sequence of plastic hinge formation. In this case, since the elastic moments are different at the three locations where plastic hinges will form, the plastic hinges will not form simultaneously.

The equilibrium requirement at the collapse mechanism condition gives

$$M_p = \frac{W_n(10)(20)}{30} - M_p$$

$$W_n = 2M_p \frac{3}{20} = 0.3M_p$$

$$M_p = 219 \text{ ft kips} \quad \text{(from Example 10.2.3)}$$

$$W_n = 0.3(219) = 65.7 \text{ kips}$$

The load-deflection diagram, however, requires examination of the loading stages.

Stage 1. From the elastic bending moments, the first plastic hinge will form at point *A*,

$$\frac{40W_1}{9} = M_p = 219 \text{ ft-kips}$$

$$W_1 = 9(219)/40 = 49.3 \text{ kips}$$

$$\Delta_1 = \frac{W_1(10)^3(20)^3}{3(30)^3 EI} = \frac{49.3(8000)(1728)}{3(27)(29,000)(518)} = 0.56 \text{ in.}$$

Stage 2. With $W_1 = 49.3$ kips one can consider that part of the available moment strength at points *B* and *C* has been used up. The strength remaining is

$$M_B = M_p - 80(49.3)/27 = 219 - 146.1 = 72.9 \text{ ft-kips}$$

$$M_C = M_p - 20(49.3)/9 = 219 - 109.6 = 109.4 \text{ ft-kips}$$

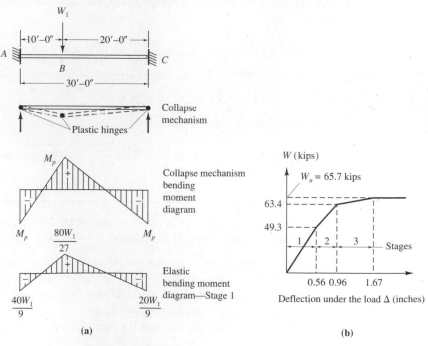

Figure 10.2.4   Example 10.2.4.

As the load is increased above $W_1 = 49.3$ kips, the added load acts on a different elastic system. The moments caused by the additional load are not distributed over the span in the same manner as moments caused by the first 49.3 kips; thus the term "redistribution of moments" is applied. Figure 10.2.5a shows the elastic system and moments for stage 2. It is apparent that the next plastic hinge will form under the load.

$$\frac{140W_2}{27} = 72.9$$

$W_2 = 27(72.9)/140 = 14.1$ kips

$$\Delta_2 = \frac{W_2 a^2 b^3 (3L + a)}{12L^3 EI} = \frac{14.1(10)^2(20)^3(90 + 10)(1728)}{12(30)^3(29,000)(518)} = 0.40 \text{ in.}$$

Stage 3. With a total of $W_1 + W_2 = 49.3 + 14.1 = 63.4$ kips applied, the strength available at C is

$$M_C = M_p - 109.6 \text{ (Stage 1)} - 40(14.1)/9 \text{ (Stage 2)} = 46.7 \text{ ft-kips}$$

$$20W_3 = 46.7; \qquad W_3 = 2.3 \text{ kips}$$

$$\Delta_3 = \frac{W_3 L^3}{3EI} = \frac{2.3(20)^3(1728)}{3(29,000)(518)} = 0.71 \text{ in.}$$

The complete load-deflection diagram appears in Fig. 10.2.4b. The total load to cause the collapse mechanism to form equals

$$W_n = W_1 + W_2 + W_3 = 49.3 + 14.1 + 2.3 = 65.7 \text{ kips}$$

Note that this statically indeterminate case, unlike the previous special case, exhibits the multistage load-deflection diagram typical of statically indeterminate systems. As load is increased moments vary in a different ratio for

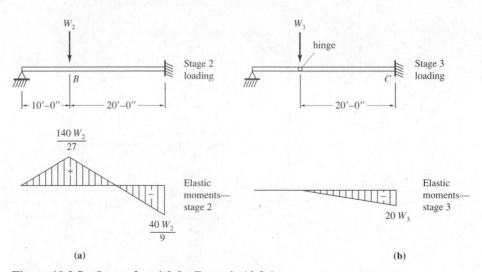

(a)   (b)

Figure 10.2.5 Stages 2 and 3 for Example 10.2.4.

each stage of loading. Thus the collapse mechanism moment diagram is *not* equal to a constant times the elastic moments. Only for statically determinate cases, and a few special statically indeterminate ones can the collapse mechanism moment diagram be obtained by multiplying the elastic moments by a constant. ■■

EXAMPLE 10.2.5 _____

Using the equilibrium method for the continuous beam of Fig. 10.2.6, determine the collapse mechanism load W_n. Assume $F_y = 36$ ksi and that the controlling limit state is the achievement of plastic moment strength.

Solution. Whereas in the previous examples only one mechanism was possible, and therefore obvious, there are many situations where the collapse mechanism is not obvious. Each of the three possible mechanisms will be investigated for this example.

Mechanism 1: Positive moment under the load W_n with a negative moment plastic hinge at the center support,

$$M_p = M_{s1} + \tfrac{1}{2}M_{s2} - \tfrac{1}{3}M_p$$
$$\tfrac{4}{3}M_p = \tfrac{16}{3}W_n + \tfrac{1}{2}(8W_n)$$
$$M_p = 7W_n$$

Mechanism 2: Positive moment plastic hinge at the load $1.5W_n$, with a negative moment plastic hinge at the center support,

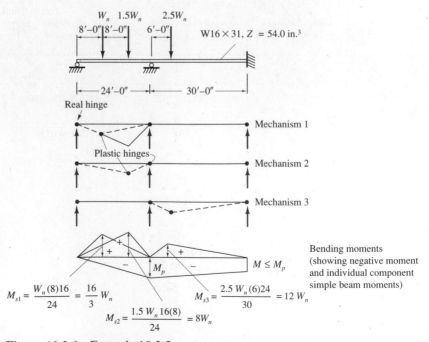

$$M_{s1} = \frac{W_n(8)16}{24} = \frac{16}{3}W_n$$

$$M_{s2} = \frac{1.5\,W_n\,16(8)}{24} = 8W_n$$

$$M_{s3} = \frac{2.5\,W_n(6)24}{30} = 12\,W_n$$

Bending moments (showing negative moment and individual component simple beam moments)

Figure 10.2.6 Example 10.2.5.

$$M_p = \tfrac{1}{2}M_{s1} + M_{s2} - \tfrac{2}{3}M_p$$
$$\tfrac{5}{3}M_p = \tfrac{8}{3}W_n + 8W_n$$
$$M_p = \tfrac{32}{5}W_n = 6.4W_n$$

Mechanism 3: Positive moment plastic hinge at the load $2.5W_n$ with negative moment plastic hinges at the ends of the 30-ft span,

$$M_p = M_{s3} - M_p$$
$$2M_p = 12W_n$$
$$M_p = 6W_n$$

Since the plastic moment capacity required is largest for Mechanism 1, it controls:

$$W_n = \frac{M_p}{7} = \frac{36(54.0)}{7(12)} = 23 \text{ kips}$$

In this case the collapse mechanism occurs in the 24-ft span while the 30-ft span remains stable and elastic. A different section in each span may be a more economical solution. ■■

Plastic Limit Load—Energy Method

The principle of virtual work may also be applied to obtain the plastic limit load W_n in an analysis of a given structure, or to find the required plastic moment M_p in a design problem.

Consider that as the collapse mechanism (i.e., plastic limit) load is reached the beam moves through a virtual displacement δ. For equilibrium, the external work done by the load moving through the virtual displacement must equal the internal strain energy due to the plastic moment rotating through small angles (hinge rotations).

EXAMPLE 10.2.6 _____

Determine the plastic limit load W_n in Example 10.2.2 by the virtual work principle. Referring to Fig. 10.2.7,

$$\text{External work} = \text{Internal work}$$
$$W_n\delta = M_p 2\theta$$
$$W_n\left(\frac{\theta L}{2}\right) = 2M_p\theta$$
$$W_n = 4M_p/L$$

the same as previously obtained.

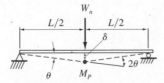

Figure 10.2.7   Example 10.2.6.   ■■

EXAMPLE 10.2.7 _____

Determine the plastic limit load W_n in Example 10.2.3 by the virtual work principle. Referring to Fig. 10.2.8,

$$\text{External work} = \text{Internal work}$$

$$W_n \frac{\theta L}{2} = 2M_p\theta + M_p(2\theta)$$

$$W_n = 8M_p/L$$

which agrees with the previous solution.

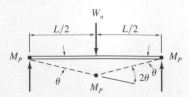

Figure 10.2.8   Example 10.2.7.   ■■

EXAMPLE 10.2.8 _____

Determine the collapse mechanism (i.e., plastic limit) load W_n for Example 10.2.4 by the virtual work method. Referring to Fig. 10.2.9,

$$\text{External work} = \text{Internal work}$$

$$W_n \frac{2\theta L}{3} = M_p(2\theta + \theta + 3\theta)$$

$$W_n = \frac{3}{2L}(6)M_p = 9M_p/L$$

For $L = 30$ ft, $W_n = 0.3M_p$ as before.

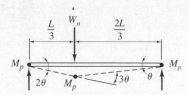

Figure 10.2.9   Example 10.2.8.   ■■

EXAMPLE 10.2.9 _____

Determine the plastic limit load W_n for the continuous beam of Example 10.2.5. Referring to Fig. 10.2.10,

Mechanism 1:

$$W_n(2\theta)(8) + 1.5W_n\theta(8) = M_p(3\theta + \theta)$$

$$28W_n = 4M_p$$

$$M_p = 7W_n$$

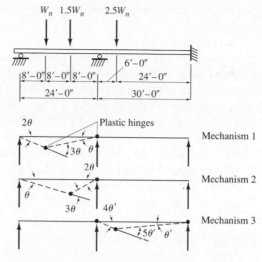

Figure 10.2.10   Example 10.2.9.

Mechanism 2:

$$W_n(\theta)(8) + 1.5W_n 2\theta(8) = M_p(3\theta + 2\theta)$$
$$32W_n = 5M_p$$
$$M_p = 6.4W_n$$

Mechanism 3:

$$2.5W_n(4\theta')(6) = M_p(4\theta' + 5\theta' + \theta')$$
$$60W_n = 10M_p$$
$$M_p = 6W_n \quad \blacksquare\blacksquare$$

The equilibrium and energy procedures are equally applicable for frames where degrees of freedom in sidesway must be considered. In multistory braced frames, the design procedure is to cause the plastic hinges to form in the girders rather than in the columns; thus the girders are treated as continuous beams. Where bracing contributes an axial force to the girder, the beam-column principles of Chapter 12 apply.

10.3 PLASTIC ANALYSIS—
LOAD AND RESISTANCE FACTOR DESIGN EXAMPLES

An elementary example of the design of a simply supported beam that achieves its collapse mechanism was presented in Secs. 7.4 and 7.5. Prior to the adoption of AISC *Load and Resistance Design Specification* in 1986, plastic analysis was used in *plastic design* as discussed in Sec. 1.8. For additional elementary material on the philosophy, procedure, and experimental verification of plastic design, Refs. 10.3 through 10.10 are suggested.

The factored load plastic analysis option within Load and Resistance Factor Design is the most rational approach and is the authors' recommended procedure for continuous beams. Alternatively, factored load *elastic* analysis may be used even when the conditions for plastic analysis are satisfied. The Allowable Stress Design alternative for designing continuous beams is illustrated in Sec. 10.5.

When elastic analysis is used under either LRFD or ASD, adjustments to the elastic analysis bending moment diagram may be made under conditions specified by LRFD-A5.1 or ASD-F1.1. These adjustments attempt to bring the result colse to both the Plastic Design (ASD-N) result and the LRFD plastic analysis result.

The use of plastic analysis to evaluate the moment strength limit state does not preclude other limit states, such as shear, flange and/or web local buckling, lateral-torsional buckling, and particularly the serviceability limit state of deflection, from controlling.

EXAMPLE 10.3.1 ───

Design the lightest W section of A36 steel for use as the two-span continuous beam of Fig. 10.3.1. Lateral support is provided at the ends and midpoint of each span. Use Load and Resistance Factor Design utilizing plastic analysis requirements under LRFD-F1.2d, even though the unbraced lengths L_b are relatively long.

Solution

(a) Compute the factored loads, w_u and W_u. Temporarily neglecting the weight of the beam,

$$w_u = 1.2(0.5) + 1.6(2.5) = 4.60 \text{ kips/ft}$$
$$W_u = 1.2(10) + 1.6(30) = 60 \text{ kips}$$

The required nominal strength is the factored load divided by ϕ_b. In this case,

$$\text{Required } w_n = w_u/\phi_b = 4.60/0.90 = 5.11 \text{ kips/ft}$$
$$\text{Required } W_n = W_u/\phi_b = 60/0.90 = 66.7 \text{ kips}$$

As indicated in the problem statement, plastic analysis is to be used. Evaluate the possible collapse mechanisms; in this case there are two, as shown in Fig. 10.3.1b.

(b) *Mechanism 1:* For a span with uniform load, simply supported at one end, and expected to develop a plastic hinge at the other, the location of maximum positive moment is not self-evident. At an unknown distance x from the discontinuous end, the net positive moment is

$$M_x = \frac{w_n}{2}x(L - x) - \frac{x}{L}M_p \tag{a}$$

Taking $dM_x/dx = 0$ and solving for x, the location of maximum moment is

$$x = \frac{L}{2} - \frac{M_p}{w_n L} \tag{b}$$

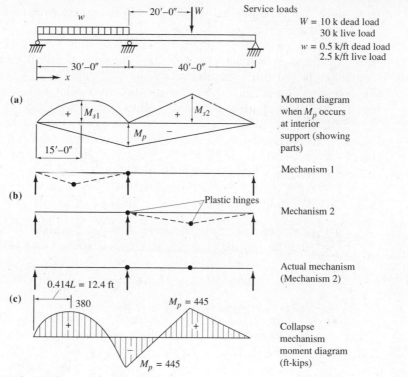

Figure 10.3.1   Example 10.3.1.

Substituting Eq. (b) into Eq. (a) gives

$$M_x(\text{max}) = \frac{w_n}{2}\left(\frac{L}{2} - \frac{M_p}{w_n L}\right)\left(L - \frac{L}{2} + \frac{M_p}{w_n L}\right) - \frac{M_p}{L}\left(\frac{L}{2} - \frac{M_p}{w_n L}\right) \qquad (c)$$

When a plastic hinge develops, $M_x(\text{max}) = M_p$; thus Eq. (c) becomes

$$M_p = \frac{w_n L^2}{8} - \frac{M_p^2}{2w_n L^2} - \frac{M_p}{2} + \frac{M_p^2}{w_n L^2} \qquad (d)$$

Substitute $M_s = w_n L^2/8$ into Eq. (d), giving

$$M_p^2 - 24M_p M_s + 16M_s^2 = 0 \qquad (e)$$

Solving the quadratic for M_p gives

$$M_p = 0.686 M_s \qquad (10.3.1)$$

and the location of the plastic hinge is at

$$x = \frac{L}{2} - \frac{M_p}{w_n L}\left(\frac{L}{L}\right)\frac{8}{8} = \frac{L}{2} - \frac{M_p}{M_s}\left(\frac{L}{8}\right)$$

$$= L\left(\frac{1}{2} - \frac{0.686}{8}\right) = 0.414L$$

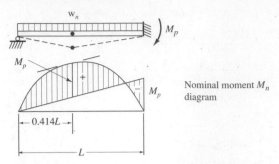

Figure 10.3.2   Collapse condition for uniformly loaded span continuous at one end and simply supported at the other.

The results of this development are shown in Fig. 10.3.2.

For the specific example, Mechanism 1 (Fig. 10.3.1b) requires

$$M_p = 0.686M_{s1}; \qquad M_{s1} = \frac{w_n L^2}{8} = \frac{5.11(30)^2}{8} = 575 \text{ ft-kips}$$

$$M_p = 0.686(575) = 394 \text{ ft-kips}$$

(c) *Mechanism 2:* For the 40-ft span,

$$M_{s2} = \frac{W_n L}{4} = \frac{66.7(40)}{4} = 667 \text{ ft-kips}$$

Equilibrium requires

$$M_p = M_{s2} - \frac{M_p}{2} = 667 - \frac{M_p}{2}$$

$$M_p = 6.67/1.5 = 445 \text{ ft-kips}$$

The largest M_p requirement obtained from the possible mechanisms determines the controlling condition. Thus Mechanism 2 governs. A check by determining the complete collapse mechanism moment diagram should be made to insure that M_p is the maximum moment. If a location is found where the computed nominal moment exceeds M_p, the mechanism used is not correct. In this case, Fig. 10.3.1c shows $M_p = 445$ ft-kips is the maximum, so that Mechanism 2 is the correct one.

The maximum positive moment on the 30-ft span using Eq. (a) is

$$M_u/\phi = \frac{5.11}{2}(12.4)(17.6) - 0.414M_p$$

$$= 558 - 0.4(445) = 380 \text{ ft-kips} < 445 \quad \text{OK}$$

(d) Select section:

$$\text{Required } Z_x = \frac{M_p}{F_y} = \frac{445(12)}{36} = 148 \text{ in.}^3$$

Try W24×62 with $Z_x = 153$ in.^{3} as indicated by the *LRFD Manual*, "LOAD FACTOR DESIGN SELECTION TABLE," for Z_x. Because of the relatively long laterally unbraced lengths, a section having larger r_y may be required.

(e) Check lateral support. When plastic analysis is used, more rotation capacity at the plastic hinges is relied upon to develop the mechanism. LRFD-F1.2d requires that L_b not exceed L_{pd} given by Eq. 9.5.7 [LRFD Formula (F1-17)],

$$L_{pd} = \frac{3600 + 2200(M_1/M_p)}{F_y} r_y \qquad [9.5.7]$$

M_p is used instead of LRFD Formula M_2 because when the formula applies M_2 will always be M_p. In the region of the last hinge to form, and in regions not adjacent to a plastic hinge, Eq. 9.5.7 does not apply.

At unbraced segments adjacent to the last plastic hinge to form, the laterally unbraced length L_b may not exceed L_p if C_b is 1.0; however, when the moment gradient is favorable (i.e., $C_b > 1$), M_p may still be achieved even though $L_b > L_p$. L_p is given by Eq. 9.5.5 [LRFD Formula (F1-4)],

$$L_p = \frac{300}{\sqrt{F_y, \text{ksi}}} r_y \qquad [9.5.5]$$

For this problem, four segments must be considered, as follows:

1. *15 ft at left end:* L_{pd} does not apply for segments containing no plastic hinges. Since the maximum moment of 380 ft-kips is less than M_p, L_b can be greater than L_p. This segment will be checked last since it is not expected to control the design.
2. *15 ft adjacent to interior support:*

$$M_1 = 380 \text{ ft-kips}; \qquad M_p = 445 \text{ ft-kips}$$

$$L_{pd} = \frac{3600 + 2200(380/445)}{36} r_y = 152 r_y$$

$$\text{Minimum } r_y = \frac{L_b}{152} = \frac{15(12)}{152} = 1.18 \text{ in.}$$

3. *20 ft adjacent to interior support:*

$$M_1 = 445 \text{ ft-kips}; \qquad M_p = 445 \text{ ft-kips}$$

$$L_{pd} = \frac{3600 + 2200(445/445)}{36} r_y = 161 r_y$$

$$\text{Minimum } r_y = \frac{L_b}{161} = \frac{20(12)}{161} = 1.49 \text{ in.}$$

4. *20 ft at right end:* Since the maximum positive moment plastic hinge is the last one to form, L_{pd} does not apply. However when L_b exceeds L_p, either the strength decreases due to inelastic lateral-torsional buckling limit state, or the strength still may reach M_p because of a favorable moment gradient.

When L_{pd} does not apply, the section must satisfy LRFD-F1.2a. In this design, the moment gradient gives $C_b = 1.67$ from Table 9.6.3.

Assume the minimum r_y required is 1.49 in. for segment 3; in which case the W24×62 ($r_y = 1.38$ in.) is *not* acceptable for the segments adjacent to the interior support. Increase size to W24×68 ($r_y = 1.87$ in.).

Investigate further for the W24×68,

$$M_p = Z_x F_y = 177(36)/12 = 531 \text{ ft-kips}$$
$$L_p = [300(1.87)/\sqrt{F_y}]/12 = 7.8 \text{ ft}$$
$$M_r = (F_y - F_r)S_x = (36 - 10)154/12 = 334 \text{ ft-kips}$$
$$L_r = \text{Eq. 9.6.6} = 22.4 \text{ ft}$$

(f) Check strength for $L_b = 20$ ft at the right end of the 40-ft span. M_n is controlled by inelastic lateral-torsional buckling since $L_b < L_r$. The strength lies between M_p and M_r when $C_b = 1.0$; however, in this case with $C_b = 1.67$, LRFD-F1.2a. Formula (F1-2) or Eq. 9.6.10. gives

$$M_n = 1.67\left[531 - (531 - 334)\left(\frac{20 - 7.8}{22.4 - 7.8}\right)\right]$$
$$= 1.67(366) = 611 \text{ ft-kips}$$

Since computed M_n exceeds M_p, $M_n = M_p$. Thus, the 20th segment at right end is satisfactory for lateral bracing.

$$(M_n = M_p = 531 \text{ ft-kips}) > (\text{Required } M_p = 445 \text{ ft-kips}) \quad \text{OK}$$

(g) Check strength for $L_b = 15$ ft segment at left end of span 1. Since L_b lies between L_p (7.8 ft) and L_r (22.4 ft), M_n is controlled by inelastic lateral-torsional buckling. For this case, C_b is

$$C_b = \frac{12.5M_{\max}}{2.5M_{\max} + 3M_A + 4M_B + 3M_C} \qquad [9.6.11]$$

The moment gradient is parabolic; thus, from Table 9.6.4, C_b is 1.30. To illustrate the calculation,

$M_{\max}$ = maximum moment in the unbraced segment
M_A = moment at 1/4 point = $0.4375M_{\max}$
M_B = moment at midpoint = $0.75M_{\max}$
M_C = moment at 3/4 point = $0.9375M_{\max}$

$$C_b = \frac{12.5M_{\max}}{2.5M_{\max} + 3(0.4375M_{\max}) + 4(0.75M_{\max}) + 3(0.9375M_{\max})}$$
$$= \frac{12.5M_{\max}}{9.625M_{\max}} = 1.2987 \qquad \therefore C_b = 1.30.$$

Using Eq. 9.6.10,

$$M_n = C_b\left[M_p - (M_p - M_r)\left(\frac{L_b - L_p}{L_r - L_p}\right)\right] \le M_p$$

$$M_n = 1.30\left[531 - (531 - 334)\left(\frac{15 - 7.8}{22.4 - 7.8}\right)\right]$$

$$= 1.30(434) = 564 \text{ ft-kips} > (M_p = 531 \text{ ft-kips})$$

The segment is satisfactory with $M_n = M_p = 531$ ft-kips.

(h) Check moment strength. Note that the factored loads were divided by ϕ_b to obtain the required nominal loads that can be carried when the collapse mechanism (required nominal moment strengths M_n) is reached. Alternatively, the factored moment M_u diagram could have been obtained; that would be appropriate when elastic analysis is used. However, when plastic analysis is used it seems more appropriate to use the M_n diagram involving the collapse mechanism, since only at the collapse mechanism do the plastic hinge and M_p have meaning. In this case, after the beam weight is added, a recalculation of required M_p for Mechanism 2 gives

$$M_{s2} = \frac{W_n L}{4} + \frac{(w_u/\phi)L^2}{8} = 667 + \frac{0.068(1.2/0.90)(40)^2}{8} = 685 \text{ ft-kips}$$

$$M_p = M_{s2} - 0.5M_p = M_{s2}/1.5 = 457 \text{ ft-kips}$$

$$\text{(Provided } M_p = 531 \text{ ft-kips)} > \text{(Required } M_p = 457 \text{ ft-kips)} \quad \text{OK}$$

It is clear that the beam weight has little effect.

(i) Check shear. At the interior end of the 40-ft span the factored load shear V_u is

$$V_u = 33.35 + \frac{445}{40} = 44.5 \text{ kips}$$

$$\frac{h}{t_w} = 50.1 < \left[\frac{418}{\sqrt{F_{yw}}} = 69.7\right]$$

$$V_n = 0.6F_y A_w = 0.6(36)(23.73)0.415 = 213 \text{ kips}$$

$$[\phi_v V_n = 0.90(213) = 191 \text{ kips}] > 44.5 \text{ kips} \quad \text{OK}$$

As expected, except on short spans with heavy loading, shear does not control.

(j) Deflection. If excessive deflection appears to be a problem, *service load* deflection must be computed. Deflection when collapse is imminent is not easily determined, nor is it generally of interest. In this case,

$$\frac{L}{d} = \frac{40(12)}{24} = 20$$

Though L/d is not any guarantee regarding deflection, it does provide some indication of difficulty as discussed in Sec. 7.6. In this case deflection is not likely a problem.
Use W24×68, $F_y = 36$ ksi. ■■

EXAMPLE 10.3.2 _____

Design the lightest constant moment of inertia section required for the continuous beam of Fig. 10.3.3. Lateral supports are provided at the vertical supports and at each

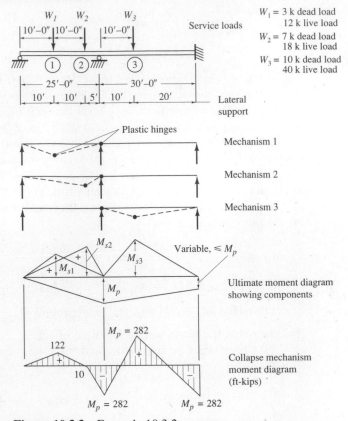

Figure 10.3.3 Example 10.3.2.

concentrated load. The steel has $F_y = 60$ ksi. Use Load and Resistance Factor Design utilizing plastic analysis under LRFD-F1.2d.

Solution

(a) Determine factored loads W_u and associated nominal loads W_n that must be carried when a collapse mechanism is imminent.

$$W_{u1} = 1.2(3) + 1.6(12) = 22.8 \text{ kips}$$
$$W_{u2} = 1.2(7) + 1.6(18) = 37.2 \text{ kips}$$
$$W_{u3} = 1.2(10) + 1.6(40) = 76.0 \text{ kips}$$

Using $\phi_b = 0.90$, the required nominal ultimate loads W_n to be carried are

$$\text{Required } W_{n1} = W_{u1}/\phi_b = 22.8/0.90 = 25.3 \text{ kips}$$
$$\text{Required } W_{n2} = W_{u2}/\phi_b = 37.2/0.90 = 41.3 \text{ kips}$$
$$\text{Required } W_{n3} = W_{u3}/\phi_b = 76.0/0.90 = 84.4 \text{ kips}$$

As indicated in the problem statement, plastic analysis is to be used. Evaluate the possible collapse mechanisms; in this case there are three, as shown in Fig. 10.3.3.

$$M_{s1} = \frac{25.3(10)15}{25} = 152 \text{ ft-kips}$$

$$M_{s2} = \frac{41.3(20)5}{25} = 165 \text{ ft-kips}$$

$$M_{s3} = \frac{84.4(10)(20)}{30} = 563 \text{ ft-kips}$$

(b) *Mechanism 1:* Assume moment under W_1 load equals M_p,

$$M_p = M_{s1} + \tfrac{1}{2}M_{s2} - \tfrac{10}{25}M_p$$
$$\tfrac{35}{25}M_p = 152 + \tfrac{1}{2}(165) = 235$$
$$M_p = \tfrac{25}{35}(235) = 168 \text{ ft-kips}$$

(c) *Mechanism 2:* Assume moment under W_2 load equals M_p,

$$M_p = \tfrac{1}{3}M_{s1} + M_{s2} - \tfrac{20}{25}M_p$$
$$\tfrac{45}{25}M_p = \tfrac{1}{3}(152) + 165 = 216$$
$$M_p = \tfrac{25}{45}(216) = 120 \text{ ft-kips}$$

(d) *Mechanism 3:* Assume moment under W_3 load equals M_p,

$$M_p = M_{s3} - M_p$$
$$2M_p = 563; \qquad M_p = 282 \text{ ft-kips (Controls)}$$

If a section is selected that has $M_p = 282$ ft-kips, the resulting moments under the W_1 and W_2 loads are

$$M_1 = M_{s1} + \tfrac{1}{2}M_{s2} - \tfrac{10}{25}M_p$$
$$= 235 - \tfrac{10}{25}(282) = 122 \text{ ft-kips} < 282 \quad \text{OK}$$
$$M_2 = \tfrac{1}{3}M_{s1} + M_{s2} - \tfrac{20}{25}M_p$$
$$= 216 - \tfrac{20}{25}(282) = -10 \text{ ft-kips} < 282 \quad \text{OK}$$

(e) Select beam.

$$\text{Required } Z_x = \frac{282(12)}{60} = 56.4 \text{ in.}^3$$

Try W18×35 having $Z_x = 66.5$ in.^{3} This section *is compact* for $F_y = 60$ ksi; for flange local buckling,

$$\left(\lambda = \frac{b_f/2}{t_f} = \frac{6.0/2}{0.425} = 7.8 \right) < \left(\lambda_p = \frac{65}{\sqrt{F_y}} = 8.4 \right) \quad \text{OK}$$

For web local buckling,

$$\left(\lambda = \frac{h}{t_w} = 53.5 \right) < \left(\lambda_p = \frac{640}{\sqrt{F_y}} = 82.6 \right) \quad \text{OK}$$

(f) Check lateral support. Although it is likely the plastic hinge at W_3 is the last to form, the designer may not be sure. *The authors suggest that it is prudent to apply the L_{pd} criterion of LRFD-F1.2d to all segments having M_p at either*

end, eliminating the need for knowing the sequence of plastic hinge formation.
The 20-ft segment, being considerably longer than the other unbraced segments, will undoubtedly control lateral-torsional buckling.

$$L_{pd} = \frac{3600 + 2200(M_1/M_p)}{F_y} r_y \qquad [9.5.7]$$

Since the moment gradient in this segment is one of reverse curvature, with $M_1/M_p = +1$,

$$L_{pd} = 5800r_y/60 = 96.7r_y$$

which means for $L_b = 20$ ft, the r_y required would be

$$\text{Required } r_y = 20(12)/96.7 = 2.48 \text{ in.}$$

The W18×35 having r_y only 1.33 in. is not acceptable. None of the W shapes in the range 35 to 45 1b/ft have r_y any higher than 2.0. Since a realistic design having a section with r_y greater than 2.48 in. cannot be achieved, additional bracing must be provided.

If one additional lateral support is provided at 10 ft from the right end of the structure, determine the W shape that will be satisfactory. For the resulting two 10 ft segments that replace the 20-ft segment, $M_1/M_p = 0$; consequently.

$$L_{pd} = 3600r_y/60 = 60r_y$$
$$\text{Required } r_y = 10(12)/60 = 2.00 \text{ in.}$$

This would require W16×67 or W18×76, the same sections needed for maximum strength when L_b was 20 ft. Assume lateral bracing at 5-ft intervals in the 30-ft span. Then at the most critical location near the right support,

$$M_1/M_p = -141/282 = -0.5$$
$$L_{pd} = 2500r_y/60 = 41.7r_y$$
$$\text{Required } r_y = 5(12)/41.7 = 1.44 \text{ in.}$$

The lightest section, W18×35, for the M_p requirement has $r_y = 1.22$ in. and still will not qualify. However, W16×36, the next lightest section has $r_y = 1.52$ in. and is also compact with $F_y = 60$ ksi.
Use 5-ft spacing of lateral bracing on the 30-ft span.

On the 25-ft span, check the laterally unbraced 10-ft segment at the left end. This segment does not have a plastic hinge associated with it. The strength of the W16×36 having $L_b = 10$ ft must be determined.

$$M_p = Z_x F_y = 64(60)/12 = 320 \text{ ft-kips}$$
$$L_p = [300(1.52)/\sqrt{F_y}]/12 = 4.9 \text{ ft}$$
$$M_r = (F_y - F_r)S_x = (60 - 10)56.5/12 = 235 \text{ ft-kips}$$
$$L_r = \text{Eq. } 9.6.6 = 12.4 \text{ ft}$$
$$C_b = 1.67$$

$$M_n = 1.67\left[320 - (320 - 235)\left(\frac{10 - 4.9}{12.4 - 4.9}\right)\right] = 438 \text{ ft-kips}$$

Since computed M_n exceeds M_p, $M_n = M_p$. Since $M_n = 320$ ft-kips exceeds 122 ft-kips under the W_1 load the 10-ft segment is satisfactory.
Use W16×36. $F_y = 60$ ksi. ■■

EXAMPLE 10.3.3

Redesign the beam of Example 10.3.2, using either cover plates or two different sections butt-spliced together. Use plastic analysis under Load and Resistance Factor Design under LRFD-F1.2d. Use $F_y = 60$ ksi.

Solution. Assume that the change of section occurs near the inflection point to the right of the central support, giving rise to different plastic moments M_{p1} and M_{p2} at the two restrained supports (see Fig. 10.3.4.).

(a) Consider first the 25-ft span supporting the W_1 and W_2 loads. Assume the positive moment under the W_1 load reaches M_{p1},

$$M_{p1} = M_{s1} + \tfrac{1}{2}M_{s2} - \tfrac{10}{25}M_{p1}$$
$$= 152 + 82.5 - 0.4M_{p1}$$
$$M_{p1} = 168 \text{ ft-kips}$$

Under the W_2 load the moment is

$$M = \tfrac{1}{3}M_{s1} + M_{s2} - \tfrac{20}{25}M_{p1}$$
$$= 50.7 + 168 - 0.8(168) = 84 \text{ ft kips} < 168 \quad \text{OK}$$

(b) Lateral support requirements. Examine the 10-ft laterally unbraced length between the W_1 and W_2 loads, where the ratio $M_1/M_p = -84/168 = -0.5$. Using Eq. 9.5.7,

$$L_{pd} = \frac{3600 + 2200(-0.5)}{60}r_y = 41.7r_y$$

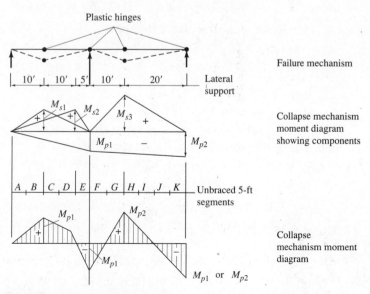

Figure 10.3.4 Example 10.3.3.

For $L_b = 10$ ft, the required r_y is $120/41.7 = 2.88$ in. which cannot be achieved by any realistic design. Increase the lateral support to be every 5 ft. This will be necessary in the 30-ft span also, as shown in Example 10.3.2.

Assuming the plastic hinge $(+M_p)$ between segments B and C (see Fig. 10.3.4) is the last hinge to form, Eq. 9.5.7 is not required by LRFD-F1.2d to be satisfied for those segments. However, in order for the nominal moment strength M_n to reach M_p, L_b may not exceed L_p when $C_b = 1.0$. In this case segment C has the flatter moment gradient, having $M_1/M_2 = -126/168 = -0.75$. Thus, from Table 9.6.3, $C_b = 1.11$; this would mean that L_b could exceed L_p by a small amount such that $1.11M_n$ could still equal M_p, even though M_n might be less than M_p. It will be conservative to require L_b to not exceed L_p for this segment,

$$L_p = \frac{300}{\sqrt{F_y, \text{ksi}}} r_y = \frac{300}{\sqrt{60}} r_y = 38.7 r_y$$

$$\text{Minimum } r_y = \frac{L_b}{38.7} = \frac{5(12)}{38.7} = 1.55 \text{ in.}$$

Referring to text Appendix Table A1, $r_y \approx 0.22 b_f$ to $0.25 b_f$. Thus,

$$\text{Minimum } b_f \text{ required} \approx \frac{1.55}{0.25} \text{ to } \frac{1.55}{0.22} = 6.2 \text{ to } 7 \text{ in.}$$

This requirement applies for segments B and C adjacent to the last plastic hinge to form.

For segment E and F, rotation capacity must be assured by satisfying LRFD-F1.2d; i.e., L_b may not exceed L_{pd}. Segment F has the flatter moment gradient and controls; $M_1/M_p = 0$. Thus,

$$L_{pd} = 3600 r_y/60 = 60 r_y$$
$$\text{Required } r_y = 5(12)/60 = 1.00 \text{ in.}$$

(c) Select section for the 25-ft span.

$$\text{Required } Z_x = \frac{\text{Required } M_p}{F_y} = \frac{168(12)}{60} = 33.6 \text{ in.}^3$$

$$\text{Required } b_f \approx 6.2 \text{ to } 7 \text{ in.}$$

$$\text{Required } r_y = 1.55 \text{ in.}$$

Using the *LRFD Manual,* "LOAD FACTOR DESIGN SELECTION TABLE," find several sections that satisfy the required Z_x:

$$\text{W12}\times26, \quad Z_x = 37.2 \text{ in.}^3$$
$$\text{W14}\times26, \quad Z_x = 40.2 \text{ in.}^3$$
$$\text{W16}\times26, \quad Z_x = 44.2 \text{ in.}^3$$

These sections all have $r_y < 1.55$ in., though W12×26 is close with $r_y = 1.51$ in.

Try W12×26, $r_y = 1.51$ in., and investigate segment C:

$$L_{pd} = \frac{3600 + 2200(-126/168)}{60} r_y = 32.5 r_y = 4.1 \text{ ft}$$

$$L_p = \frac{300}{\sqrt{F_y, \text{ksi}}} r_y = \frac{300(1.51)}{\sqrt{60}(12)} = 4.9 \text{ ft}$$

$C_b = 1.11$ from Table 9.6.3

$M_p = Z_x F_y = 37.2(60)/12 = 186$ ft-kips

$M_r = (F_y - F_r)S_x = (60 - 10)33.4/12 = 139$ ft-kips

$L_r = $ Eq. 9.6.6 $= 12.1$ ft

Since L_b is *between* L_p and L_r, inelastic lateral-torsional buckling controls. The LRFD-Formula (F1-2) linear relationship applies:

$$M_n = 1.11\left[186 - (186 - 139)\left(\frac{5 - 4.9}{12.1 - 4.9}\right)\right]$$
$$= 1.11(185) = 205 \text{ ft-kips} > M_p$$

Since computed M_n exceeds M_p, $M_n = M_p$. Thus segments *B* and *C* are satisfactory for lateral bracing.

Use W12×26, $F_y = 60$ ksi, for the 25-ft span. For the 30-ft span, top and bottom cover plates will be added. An alternative would be to use a heavier section having the same proportions and butt join them by welding or by splice plates.

Since the section used has a greater plastic moment strength than required, the moment diagram under loads W_u/ϕ_b is as shown in Fig. 10.3.5b. The diagram assumes the first plastic hinge is at the support, and that it does form under the application of the required ultimate loads W_u/ϕ_b. The $M_{p1} = 186$ ft-kips as provided is then used in the analysis of the 30-ft span.

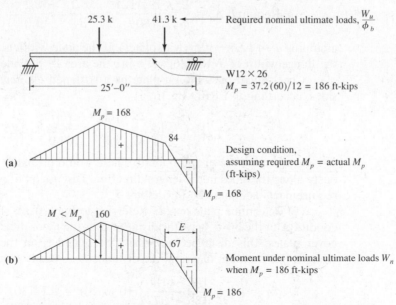

Figure 10.3.5 Comparison of design condition and actual condition for 25-ft span of Example 10.3.3.

In the actual condition, segment E is more severely loaded than in the design condition. However it is still not as severely loaded as segment F.

(d) Determine plastic moment requirement on the 30-ft span. When M_{p2} forms under the W_3 loads, equilibrium requires

$$M_{p2} = M_{s3} - \tfrac{1}{3}(M_{p2} - M_{p1}) - M_{p1} \quad \text{(long cover plates)}$$

if M_{p2} is developed at the fixed end. This would require cover plates to extend to the support. If the cover plates are not extended to the support, equilibrium requires

$$M_{p2} = M_{s3} - M_{p1} \quad \text{(short cover plates)}$$

If long cover plates are used,

$$M_{p2} = 563 - \tfrac{1}{3}M_{p2} + \tfrac{1}{3}(186) - 186; \qquad \tfrac{4}{3}M_{p2} = 439, \qquad M_{p2} = 329 \text{ ft-kips}$$

If shorter cover plates are used,

$$M_{p2} = 563 - 186 = 377 \text{ ft-kips}$$

Since there is little difference in requirements, the cover plates will be made as short as possible. The cover plates, top and bottom, must provide

$$M_p \text{ (cover)} = 377 - 186 = 191 \text{ ft-kips}$$

(e) Select plate size. For an area A_p representing the cover plate at one flange, and a distance d center-to-center of cover plates, the plastic modulus Z becomes

$$Z = 2A_p\left(\frac{d}{2}\right) = A_p d$$

$$\text{Required } A_p = \frac{Z}{d} = \frac{191(12)}{60d} = \frac{38.2}{d} = \frac{38.2}{13.0} = 2.9 \text{ sq in.}$$

assuming $d \approx 12.5 + 0.5$ ($\tfrac{1}{2}$-in. plates). If the plate width is somewhat less than the flange width of W12×26, say 6 in., the area A_p provided will be adequate.

Check local buckling on plate as a stiffened element welded along two sides, according to LRFD-B5.1:

$$\left(\frac{b}{t} = \frac{6}{0.5} = 12\right) < \left(\lambda_p = \frac{190}{\sqrt{F_y}} = \frac{190}{\sqrt{60}} = 24.5\right) \qquad \text{OK}$$

Use $-\tfrac{1}{2} \times 6$ *plates, top and bottom.* These would probably be welded continuously along the sides in the length direction. Discussion of weld sizes and other requirements is contained in Chapter 5.

(f) Determine plate length. Referring to Fig. 10.3.6 showing the collapse condition for the 30-ft span, the distance L_1 is the theoretical length required for cover plates; this distance may easily be scaled from the diagram. For this straight line moment diagram, the distance L_1 is easily computed,

$$L_1 = \left(1 - \frac{186 + 186}{186 + 377}\right)(10 + 20) = 0.34(30) = 10.2 \text{ ft}$$

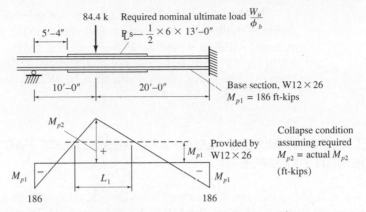

Figure 10.3.6 Collapse condition for 30-ft span of Example 10.3.3.

Extension of the plates in each direction must develop the cover plate's proportion of the strength required and satisfy the requirements of LRFD-B10. The tention or compression strength required from one plate is

$$\text{Required } Z_x = \frac{377(12)}{60} = 75.4 \text{ in.}^3$$

$$\%Z \text{ from plates} = \frac{A_p}{A_p + A_f}(100) = \frac{3.00}{3.00 + 2.47}(100) = 55\%$$

neglecting the web effect. The nominal tensile force T_n required to be developed is

$$T_n \approx \frac{0.55ZF_y}{d} = \frac{0.55(75.4)60}{12.22 + 0.5} = 196 \text{ kips}$$

Assuming no weld across the end of the cover plate and continuous fillet weld along the sides, determine the number of inches of weld necessary to develop the desired force in the plates. As treated in Sec. 5.14, the strength per inch of $\frac{3}{16}$-in. fillet weld made by the submerged arc process using E70 electrodes is

$$\phi_v R_{nw} = 0.75(3/16)(0.60)(70) = 5.91 \text{ kips/in.}$$

Equating the design strength of L_w inches of fillet weld to the design tensile strength $\phi_b T_n$ required to be developed,

$$L_w(5.91) = \phi_b T_n = 0.90(196) = 176 \text{ kips}$$
$$L_w = 176/5.91 = 30 \text{ in.} \quad (2 \text{ lines of 15 in.})$$

The length of plates required is

$$\text{Length} = L_1 + 2(15 \text{ in. on each end}) = 10.2 + 2.5 = 12.7 \text{ ft}$$

Use $-\frac{1}{2} \times 6$ plates, 13 ft long, beginning 5'-4" from support as shown in Fig. 10.3.6.

In this example, the member was not spliced; however, butt splicing the member might be a more economical choice to the use of cover plates. A compromise solution would be to use a section having a Z_x larger than the minimum for the 25-ft span, but not as large as would be required to omit plates entirely. The cover plates could then be thinner and shorter. Splices are discussed in Sec. 10.5. ■■

10.4 ELASTIC ANALYSIS—LOAD AND RESISTANCE FACTOR DESIGN EXAMPLE

Within Load and Resistance Factor Design, either plastic analysis under LRFD-F1.2d or elastic analysis under LRFD-F1.1 through F1.2b may be the appropriate procedure. When sections are "compact" with respect to local buckling and the laterally unbraced length L_b does not exceed L_{pd}, plastic analysis *is permitted* to be used. Plastic analysis is *not required* to be used. Elastic analysis is *always permitted*.

EXAMPLE 10.4.1 _____

Redesign the beam of Example 10.3.1 using Load and Resistance Factor Design but *not utilizing* plastic analysis. Use A36 steel.

Solution

(a) The factored loads W_u and w_u were computed in Example 10.3.1, as follows:

$$w_u = 1.2(0.5) + 1.6(2.5) = 4.60 \text{ kips/ft}$$
$$W_u = 1.2(10) + 1.6(30) = 60 \text{ kips}$$

(b) The elastic analysis bending moment diagram under factored service loads must be computed using a method of statically indeterminate structural analysis. This appears in Fig. 10.4.1.

The four separate laterally unbraced segments and their corresponding C_b values and maximum factored moments M_u are shown in Fig. 10.4.1.

(c) Select a section based on the maximum negative moment. Assuming "compact" section for local buckling, the required plastic modulus Z_x would be

$$M_u = \phi_b M_p = \phi_b Z_x F_y$$

$$\text{Required } Z_x = \frac{M_u}{\phi_b F_y} = \frac{479(12)}{0.90(36)} = 177 \text{ in.}^3$$

Using *LRFD Manual*. "LOAD FACTOR DESIGN SELECTION TABLE." find

$$\text{W24} \times 68, \qquad Z_x = 177 \text{ in.}^3, \qquad r_y = 1.87 \text{ in.}$$

(d) Check lateral support. In order to obtain the plastic moment strength M_p, the laterally unbraced length L_b cannot exceed L_p when $C_b = 1.0$.

$$L_p = \frac{300}{\sqrt{F_y, \text{ksi}}} r_y = \frac{300(1.87)}{\sqrt{36}(12)} = 7.8 \text{ ft}$$

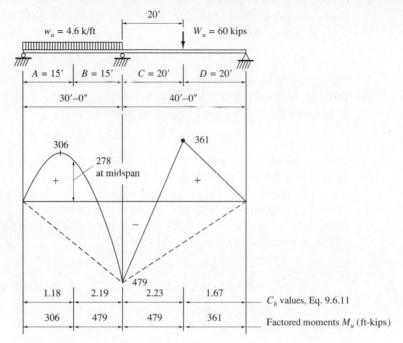

Figure 10.4.1 Example 10.4.1.

At first it may appear that L_p is less than 15 ft and therefore, M_n must be less than M_p. Traditionally since moment gradient does not appear in the L_p equation, the designer would evaluate the rotational restraint at the point of inflection (which in this case is at 6.9 ft from the interior support). If the designer decided that twist of the cross-section was adequately prevented the inflection point could be treated as a braced point. Under LRFD-F1.2a, the full moment gradient across the inflection point to the next actual lateral support is accounted for by C_b. Thus, even when L_b exceeds L_p, the C_b greater than 1.0 may bring the available strength to M_p anyway.

(e) Evaluate the available strength for the W24×68 near the interior support. Using Eq. 9.6.11 for C_b with $M_1/M_2 = +361/479 = +0.79$,

$$C_b \approx 2.23 \text{ from Table 9.6.3}$$
$$M_p = Z_x F_y = 177(36)/12 = 531 \text{ ft-kips}$$
$$L_p = 7.8 \text{ ft} \quad [\text{computed in part (d)}]$$
$$M_r = (F_y - F_r)S_x = (36 - 10)154/12 = 334 \text{ ft-kips}$$
$$L_r = \text{Eq. } 9.6.6 = 22.4 \text{ ft}$$

Since L_b is *between* L_p and L_r, inelastic lateral-torsional buckling controls. The LRFD-Formula (F1-2) linear relationship applies

$$M_n = 2.23\left[531 - (531 - 334)\left(\frac{20 - 7.8}{22.4 - 7.8}\right)\right]$$
$$= 2.23(366) = 816 \text{ ft-kips} > M_p$$

Since computed M_n exceeds M_p, $M_n = M_p$. Thus, segments B and C are satisfactory for lateral bracing.

$$[\phi_b M_p = 0.90(531) = 478 \text{ ft-kips}] \approx (M_u = 479 \text{ ft-kips}) \quad \text{OK}$$

(f) Check segment D. As in segments B and C, L_b is between L_p and L_r; thus, M_n is obtained for this inelastic buckling case by linear interpolation, this time using $C_b = 1.67$ with LRFD-Formula (F1-3),

$$M_n = 1.67(366) = 611 \text{ ft-kips} > M_p$$

Since computed M_n exceeds M_p, $M_n = M_p = 531$ ft-kips. Thus, segment D is satisfactory for lateral bracing.

(g) Check segment A. In this segment, the maximum moment occurs *between* the braced points; thus, C_b could be conservatively taken as 1.0. When computed using Eq. 9.6.11, $C_b = 1.18$. The linear interpolated value using LRFD-Formula (F1-2) is

$$M_n = 1.18(366) = 432 \text{ ft-kips} < M_p$$
$$[\phi_b M_n = 0.90(432) = 389 \text{ ft-kips}] > (M_u = 306 \text{ ft-kips}) \quad \text{OK}$$

When using plastic analysis with these long laterally unbraced lengths, the design was impossible. ■■

10.5 ELASTIC ANALYSIS—ALLOWABLE STRESS DESIGN EXAMPLES

The design of usual continuous beams having relatively close spacing of lateral supports should be done using either (a) plastic analysis under Load and Resistance Factor Design or (b) plastic analysis within Plastic Design covered in Chapter N of the 1989 ASD Specification [1.5]. Allowable Stress Design uses adjustments of the elastic moment diagram to indirectly account for plastic behavior and redistribution of moments. Such adjustments cannot reflect true behavior. Occasionally, for designs involving long laterally unbraced lengths where lateral-torsional buckling is the limit state instead of plastic moment strength, Allowable Stress Design becomes a practical alternative to elastic analysis under Load and Resistance Factor Design.

EXAMPLE 10.5.1 ───

Redesign the beam of Example 10.3.1 (Fig. 10.3.1) using Allowable Stress Design. A36 steel.

Solution.   The elastic bending moment diagram (see Fig. 10.5.1) under service loads must first be computed using a method of statically indeterminate analysis.

The four separate unbraced segments with their C_b values and design moments are shown in Fig. 10.5.1. If lateral bracing is sufficient to permit a plastic moment to form at a given maximum moment location, an allowable stress of $0.66F_y$ is permitted. Further, at the negative moment location if adequate rotation capacity (plastic strain after M_p has been reached) is available,

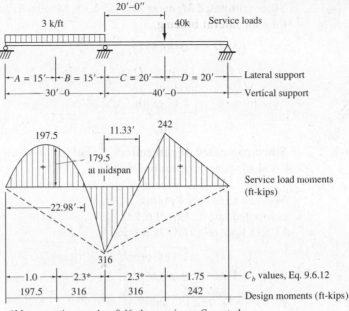

Figure 10.5.1 Example 10.5.1.

redistribution of moments is accounted for by a 10 percent reduction in maximum negative moment, with the corresponding adjustment of positive moments to maintain equilibrium.

(a) Determine required section assuming reduction in $(-M)$ and using $0.66F_y$.

$$\text{Required } S_x = \frac{0.9(316)12}{24} = 142 \text{ in.}^3$$

which is the minimum section modulus required assuming the most favorable conditions. The laterally unbraced length L_b is 15 ft for segment B which is sufficiently large that it is likely to exceed L_c for any practical choice of beam.

(b) Whenever L_b exceeds L_u the suitable approach is to use the *ASD Manual* charts, "ALLOWABLE MOMENTS IN BEAMS." For segment C, enter with

$$M = 316 \text{ ft-kips}$$

$$\frac{L_b}{C_b} = \frac{20}{2.3} = 8.7 \text{ ft} \quad \text{(if } F_b \text{ (F1-6) controls)}$$

$$\frac{L_b}{\sqrt{C_b}} = \frac{20}{\sqrt{2.3}} = 13.2 \text{ ft} \quad \text{(if } F_b \text{ (F1-8) controls)}$$

Noting that 20 ft $> L_c$, try W24×76; $L_u = 11.8$ ft. From the charts, the capacity is

$$M = 322.5 \text{ ft-kips} \quad \text{at } L_b = 8.7 \text{ ft}$$

and $F_b = 0.60F_y$.

(c) Check segment A (Fig. 10.5.1) with W24×76.

$$f_b = \frac{197.5(12)}{176} = 13.5 \text{ ksi}$$

$$\left(\frac{L_b}{C_b} = \frac{15}{1.0} = 15 \text{ ft}\right) > (L_u = 11.8 \text{ ft}); \qquad F_b < 0.60F_y$$

$$F_b(\text{F1-8}) = \frac{12,000C_b}{L_b d/A_f} = \frac{12,000(1.0)}{15(12)3.91} = 17.1 \text{ ksi} > (f_b = 13.5 \text{ ksi}), \quad \text{OK}$$

which indicates segment A is OK and no need to check ASD Formula (F1-6).

(d) Check segment D; W24×76.

$$\left(\frac{L_b}{C_b} = \frac{20}{1.75} = 11.4\right) < (L_u = 11.8 \text{ ft}) \quad \text{OK}$$

$$F_b = 0.60F_y = 22 \text{ ksi}$$

$$f_b = \frac{242(12)}{176} = 16.5 \text{ ksi} < 22 \text{ ksi} \quad \text{OK}$$

(e) Show final check for segments B and C; W24×76;

$$F_b(\text{F1-8}) = \frac{12,000(2.3)}{20(12)(3.91)} = 29.4 \text{ ksi} > 0.60F_y$$

$$F_b = 0.60F_y = 22 \text{ ksi}$$

$$f_b = \frac{316(12)}{176} = 21.5 \text{ ksi} < 22 \text{ ksi} \quad \text{OK}$$

Use W24×76, $F_y = 36$ ksi. ■■

A comparison with Example 10.3.1 where LRFD using plastic analysis was used shows that plastic analysis obtained the lighter section, W24×68, compared with W24×76 for ASD. Example 10.4.1, where LRFD was used with elastic analysis for the same beam, produced the W24×68 section. Primarily the difference results from ASD assuming the limit state to be first yield in the negative moment zone; thus, lateral bracing did not have to ensure large rotation capacity in the plastic range at the interior support. Both LRFD approaches were able to utilize the beam developing its plastic moment strength at the interior support.

EXAMPLE 10.5.2

Redesign the continuous beam of Example 10.3.2 (Fig. 10.3.3) by Allowable Stress Design. $F_y = 60$ ksi.

Solution. The elastic bending moment diagram under service load is shown in Fig. 10.5.2. To get a more appropriate comparison with plastic analysis, lateral support is provided every 5 ft on the 30-ft span and at the load and vertical support points on the 25-ft span.

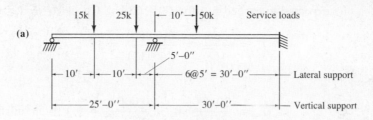

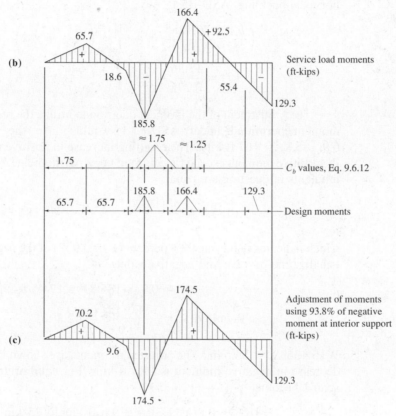

Figure 10.5.2 Example 10.5.2.

(a) Select the section required using 0.9 of maximum negative moment according to ASD-F1.1 with $0.66F_y$. L_c must exceed 5 ft.

$$\text{Required } S_x = \frac{0.9(185.8)(12)}{39.6} = 50.7 \text{ in.}^3$$

Use "ALLOWABLE STRESS DESIGN SELECTION TABLE." Try W16×36; $S_x = 56.5$ in.^{3} This section is "compact" for local buckling limits, since F'_y in that table exceeds $F_y = 60$ ksi. The L_c value for $F_y = 50$ ksi is 6.3 ft which might be expected to exceed 5 ft for $F_y = 60$ ksi. Check to verify acceptability,

$$L_c = \frac{76b_f}{\sqrt{F_y}} = \frac{76(6.985)}{\sqrt{60}\,(12)} = 5.7 \text{ ft}$$

or

$$L_c = \frac{20,000}{(d/A_f)F_y} = \frac{20,000}{5.28(60)12} = 5.3 \text{ ft} \quad \text{(controls)}$$

For local buckling, ASD-B5.1,

$$\left(\frac{b_f}{2t_f} = \frac{6.985}{2(0.430)} = 8.1\right) < \left(\frac{65}{\sqrt{F_y}} = 8.4\right) \quad \text{OK}$$

$$\left(\frac{d}{t_w} = \frac{15.86}{0.295} = 53.8\right) < \left(\frac{640}{\sqrt{F_y}} = 82.6\right) \quad \text{OK}$$

(b) Adjustment of the moment diagram to utilize the reduction in negative moment permitted under ASD-F1.1 would make the negative moment $0.9(185.8) = 167$ ft-kips. The resulting increase in positive moment will make the positive moment exceed the adjusted negative moment. Adjust only until the moments in question are equal:

$$x\,185.8 = \frac{50(10)20}{30} - \frac{129.3}{3} - \frac{2}{3}(185.8x)$$

which indicates that using 93.8 percent ($x = 0.938$) of the negative moment will equalize the positive and negative values at

$$-M = 0.938(185.8) = 174.5 \text{ ft-kips}$$

$$\text{Required } S_x = \frac{174.5(12)}{39.6} = 52.8 \text{ in.}^3$$

W16×36, $S_x = 56.5$ in.^{3} The adjusted moments are shown in Fig. 10.5.2c. The decrease in negative moment is 11.3 ft-kips. For equilibrium, the moments in span 1 become

$$-18.6 - 0.8(-11.3) = -9.6 \text{ ft-kips @ 25-kip load}$$
$$+65.7 - 0.4(-11.3) = +70.2 \text{ ft-kips @ 15-kip load}$$

These are only slightly different than the ASD indication of using the average of the change in negative moments [in this case $0.5(11.3)$] to correct positive moments. Further, the authors believe the intention of the ASD method is followed by adjusting less than the full 10 percent and maintaining equilibrium.

(c) Check 10-ft unbraced lengths on 25-ft span for W16×36:

$$F_b(\text{F1-8}) = \frac{12,000C_b}{L_b d/A_f} = \frac{12,000(1.75)}{10(12)(5.28)} = 33.1 \text{ ksi}$$

$$\frac{L_b}{r_T} = \frac{10(12)}{1.79} = 67.0$$

$$F_b(\text{F1-6}) = 40.0 - \frac{(L_b/r_T)^2}{425C_b}$$

$$= 40.0 - \frac{(67.0)^2}{425(1.75)} = 34.0 \text{ ksi} \quad (\text{controls})$$

$$f_b = \frac{70.2(12)}{56.5} = 14.9 \text{ ksi} < [F_b = 34.0 \text{ ksi}] \quad \text{OK}$$

Use W16×36, $F_y = 60$ ksi. This is the same section obtained using LRFD with plastic analysis. ■■

10.6 SPLICES

While the design of connections is outside the scope of this chapter, the location of and strength requirements for beam splices are appropriately discussed here. It is obvious that if a splice is designed for the moment and shear capacity for the member spliced, full continuity is maintained and no special precautions are necessary. Some designers prefer to use shear splices at points of contraflexure and thus introduce a real hinge at a point of zero moment. There are two reasons why this should be avoided: (1) the point of contraflexure under service load is not at the same location that it occurs under factored loads (i.e., at its mechanism condition); (2) moments obtained assuming continuity are invalid if real hinges are inserted. Hart and Milek [10.11] have provided a good discussion of this problem.

According to LRFD and ASD-J7, connections must be designed for the moments, shears, and axial loads to which they are to be subjected. ASD-J7 refers to service loads, while LRFD-J7 refers to factored loads.

If full continuity is assumed when determining moments, either under service loads or under factored loads, then splices should provide that continuity. *A reduced stiffness at a splice may prevent or reduce transmission of moment across that section and significantly change the resulting moment diagram.*

EXAMPLE 10.6.1

Examine the effect of a shear splice at the point of contraflexure in the 40-ft span of the two-span continuous beam of Example 10.3.1. A36 steel.

Solution

(a) Full dead load plus live load. If full continuity is maintained, partial loading in some spans to account for live load in various locations is unnecessary. Note that the moment $M_p = 445$ ft-kips (see Fig. 10.6.1a) can develop even when the adjacent span has a reduced load. In other words, each span may be treated separately, as long as continuity exists so that the negative moment may be assumed to reach M_p.

(b) Effect of hinge at inflection point. Use of a shear splice in effect creates a hinge for all stages of loading (i.e., transforms the system into a statically determinate one). The maximum negative moment that can develop is limited to the shear at the hinge times the distance to the support. When load on the span with the hinge is reduced, the negative moment that can develop is reduced. As

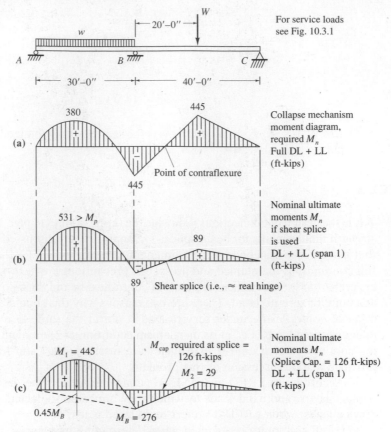

Figure 10.6.1   Example 10.6.1.

shown in Fig. 10.6.1b, when only the dead load is on the 40-ft span, the nominal ultimate load acting is

$$W_n = W_u/\phi_b = 1.2(10)/0.90 = 13.3 \text{ kips}$$

$$\text{Shear at splice} = 13.3(20/30) = 8.9 \text{ kips}$$

$$\text{Moment at support} = 8.9(10) = 89 \text{ ft-kips}$$

With the full factored dead plus live load ($w_u = 4.60$ kips/ft from Example 10.3.1) acting in the adjacent span, the maximum required nominal positive moment strength in the 30-ft span is

$$V_{nAB} = \frac{(4.60/0.90)30}{2} - \frac{89}{30} = 76.7 - 3.0 = 73.7 \text{ kips}$$

$$\text{Required } M_n = \frac{(73.7)^2}{2(4.60/0.90)} = 531 \text{ ft-kips} > M_p \quad \text{NG}$$

Since the original design required $M_p = 445$ ft-kips, the 30-ft span is now inadequate as a result of the inability of the negative moment to exceed 89 ft-kips with the reduced load on the 40-ft span.

(c) Minimum strength required for splice. If the splice design is to provide less than full continuity, Fig. 10.6.1c illustrates the minimum capacity needed. If the section being spliced provides $M_p = 445$ ft-kips, the positive moment in the 30-ft span may not exceed that value. Thus, referring to Fig. 10.6.1c, and assuming the maximum occurs at approximately $0.45L_1$,

$$M_1 = \frac{(4.60/0.90)(0.45)0.55}{2}(30)^2 - 0.45M_B = 445 \text{ ft-kips}$$

$$M_B = \frac{569 - 445}{0.45} = \frac{124}{0.45} = 276 \text{ ft-kips}$$

In order to have this moment develop, a nominal moment strength must be provided at the splice point:

$$M_2 = \frac{(20.0/0.90)30}{4} - \frac{M_B}{2} = 167 - 138 = 29 \text{ ft-kips}$$

$$\text{Required } M_n \text{ at splice} = \frac{276 + 25}{2} - 25 = 126 \text{ ft-kips}$$

If the splice in this LRFD plastic analysis problem is designed for the shear as determined from Fig. 10.6.1a, plus the 126 ft-kip moment, the reduced loading of span BC will not cause unanticipated overload in span AB.

Finally, it is noted that reduced load in span AB has no detrimental effect on span BC.

The authors prefer that continuity be maintained by the design of splices for 100 percent of the moment capacity of members spliced; however, reduced capacities providing more economical designs may be used as long as the resulting designs are checked under possible partial loadings. Other examples and a more detailed procedure are given in Ref. 10.11. ■■

SELECTED REFERENCES

10.1. Lynn S. Beedle. *Plastic Design of Steel Frames*. New York: John Wiley & Sons, Inc., 1958.

10.2. C. E. Massonnet and M. A. Save. *Plastic Analysis and Design*, Vol. I, Beams and Frames. Waltham, Mass: Blaisdell Publishing Company, 1965.

10.3. Bruce G. Johnston. "Strength as a Basis for Structural Design," *Proc. AISC National Engineering Conf.*, 1956, pp. 7–13.

10.4. Bruno Thürlimann. "Simple Plastic Theory," *Proc. AISC National Engineering Conf.*, 1956, pp 13–18.

10.5. Robert L. Ketter. "Analysis and Design Examples," *Proc. AISC National Engineering Conf.*, 1956, pp. 19–35.

10.6. Lynn S. Beedle. "Experimental Verification of Plastic Theory," *Proc. AISC National Engineering Conf.*, 1956, pp 35–49.

10.7. Bruno Thürlimann. "Modifications to 'Simple Plastic Theory'," *Proc. AISC National Engineering Conf.*, 1956, pp. 50–57.

10.8. Frederick S. Merritt. "How to Design Steel by the Plastic Theory." *Engineering News-Record* (Apr. 4, 1957), 38–43.

10.9. Bruce G. Johnston, C. H. Yang, and Lynn S. Beedle. "An Evaluation of Plastic Analysis as Applied to Structural Design," *Welding Journal Research Suppl.* (May 1953), 1–16.

10.10. Joint Committee of Welding Research Council and the American Society of Civil Engineers. *Commentary on Plastic Design in Steel*, 2nd ed., ASCE Manual and Reports on Practice No. 41, New York, 1971.

10.11. Willard H. Hart and William A. Milek. "Splices in Plastically Designed Cotinuous Structures," *Engineering Journal*, AISC **2**, 2 (April 1965), 33–37.

PROBLEMS

All problems are to be done according to the AISC Load and Resistance Factor Design or Allowable Stress Design, as indicated by the instructor. When the LRFD problem is assigned, use *plastic analysis* if possible unless otherwise indicated. All given loads are service loads unless otherwise indicated, and assume the loads are always in the position shown even though uniform live load would rarely exist that way. Assume all standard sections are equally readily available in the indicated grade of steel (even though actually they are not). A figure showing span and loading is required, and after making a design selection a final check of strength (ϕM_n compared with M_u for LRFD) or stress (f_b compared to allowable stress F_b for ASD) is required.

10.1. Determine the maximum value for the service load P of the beam in the accompanying figure. Assume adequate lateral support such that $L_b < L_p$ for LRFD and $L_b < L_c$ for ASD. The load P is 20% dead load and 80% live load.

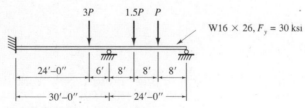

Problem 10.1

10.2. Determine the uniformly distributed service load (40% dead load, 60% live load) a W24×104, $F_y = 50$ ksi, may be permitted to carry as a two-span continuous beam having equal spans of 40 ft. Assume deflection does not control, and that lateral support is provided at the vertical supports and at 15 and 30 ft from the simply supported ends.

10.3. Select the lightest W section for the two-span continuous beam of the accompanying figure. Use $F_y = 50$ ksi. The concentrated loads are 16 kips dead load and 24 kips live load; the uniform load is 0.5 kips/ft dead load and

2.0 kips/ft live load. In the LRFD problem, specify alternative lateral support if necessary in order to use maximum plastic strength in the design.

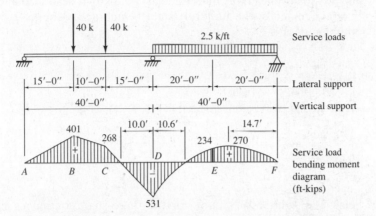

Problem 10.3

10.4. Select the lightest W section for the two-span continuous beam of the accompanying figure. Use $F_y = 60$ ksi. The concentrated loads is 12 kips dead load and 60 kips live load. Lateral support is provided at the vertical supports and under the concentrated load. In the LRFD problem, specify alternative lateral support if necessary in order to use maximum plastic strength in the design.

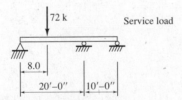

Problem 10.4

10.5. Select the lightest W section for the two-span continuous beam of the accompanying figure. Use $F_y = 50$ ksi. The uniform load is 1 kip/ft dead load and 2 kip/ft live load. Lateral support is provided at the vertical supports and at 10-ft intervals. In the LRFD problem, specify alternative lateral support if necessary in order to use maximum plastic strength in the design.

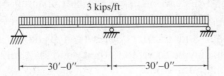

Problem 10.5

10.6.   Select the lightest W section for the two-span continuous beam of the accompanying figure. Use $F_y = 50$ ksi. The concentrated load is 20 kips dead load and 30 kips live load; the uniform load is 0.8 kip/ft dead load and 2.7 kips/ft live load. Specify the uniform laterally unbraced lengths L_b to give the optimum design.

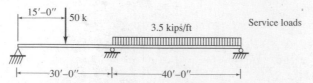

Problem 10.6

10.7.   Select the lightest W section for the three-span continuous beam of the accompanying figure. Use $F_y = 36$ ksi. The 50-kip concentrated loads are 30% dead load and 70% live load, the 20-kip concentrated load is all dead load, and the uniform load is 20% dead load and 80% live load.

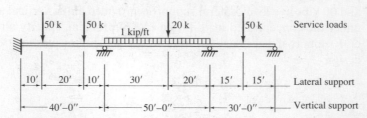

Problem 10.7

10.8.   Select the lightest W section for a three-span (50 ft-65 ft-50 ft) continuous beam to carry a uniform dead load of 2 kips/ft in addition to the beam weight, and a uniform live load of 1.5 kips/ft. In this problem, live load is to be treated in its usual manner; that is, of indefinite length and positioned to give maximum effects. Live load deflection (maximum) may not exceed $L/360$. Lateral support is provided every 5 ft. $F_y = 60$ ksi. After doing the design with $L_b = 5$ ft, give an alternate more economical design prescribing the location of lateral supports.

10.9.   Repeat Prob. 10.8 using steel having $F_y = 50$ ksi and specify lateral bracing for an economical design, instead of using the 5-ft spacing of Prob. 10.8.

10.10.   Assume one splice is required for the beam selected for Prob. 10.3. Specify its location and the shear and moment for which it should be designed. Assume any load or loads may be reduced to their dead load value while other loads remain at their maximum values.

10.11.   Same as Prob. 10.10. but splice the beam of Prob. 10.7.

10.12.   For the beam shown in the accompanying figure, (a) select the lightest W section assuming the same section extending over all three span and (b) redesign, using a smaller base section with welded cover plates where needed. For the cover plate, specify the size, length, and location of the plate. The concentrated loads are 10 kips dead load and 25 kips live load, and the uniform loading is 0.5 kip/ft dead load and 1.5 kip/ft fixed position live load. $F_y = 50$ ksi. Specify the necessary lateral support.

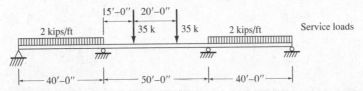

Problem 10.12

Chapter 11

Plate Girders

11.1 INTRODUCTION AND HISTORICAL DEVELOPMENT

A plate girder is a beam built up from plate elements to achieve a more efficient arrangement of material than is possible with rolled beams. Plate girders are economical where spans are long enough to permit saving in cost by proportioning for the particular requirements. Plate girders may be of riveted, bolted, or welded construction. Beginning with early railroad bridges during the period 1870–1900, riveted plate girders (Fig. 11.1.1) composed of angles connected to a web plate, with or without cover plates, were extensively used in the United States on spans from about 50 to 150 ft.

Beginning in the 1950s when welding became more widely used (because of improved quality of welding and shop-fabricating economies resulting from increased

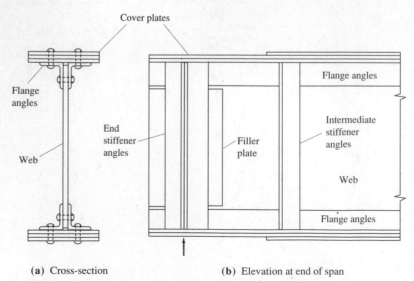

| (a) Cross-section | (b) Elevation at end of span |

Figure 11.1.1 Typical components of riveted plate girder.

Plate girders, showing welded stiffeners in place, rocker bearings for vertical supports at the pier, transverse cross bracing between girders, and hinges to provide a simple support for the spans to the right of the hinges. (Photo by C. G. Salmon)

use of automatic equipment) shop-welded plate girders composed of three plates (Fig. 11.1.2) gradually replaced riveted girders. During this period also, high-tensile-strength bolts were displacing rivets in field construction. Since the 1960s nearly all plate girders are shop welded using two flange plates and one web plate to make an I-shaped cross section.

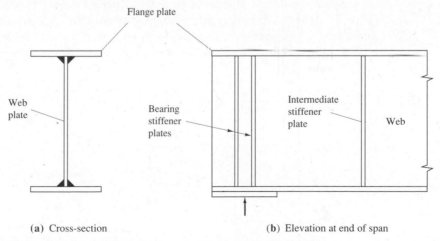

(a) Cross-section (b) Elevation at end of span

Figure 11.1.2 Typical components of a welded plate girder.

Where practically all *riveted* girders were composed of plate and angle components having the same material yield strength, the tendency now with *welded* girders is to combine materials of different strength. By changing materials at various locations *along the span* so that higher strength materials are available at locations of high moment and/or shear, or by using different strength material for flanges than for web (hybrid girders), more efficient and economical girders can be obtained.

Because few railroad bridges are being built today, discussion of economical spans and other dimensioning comments in this chapter is limited to highway bridges, where most are continuous over two or more spans; or to buildings where some spans may be assumed as simply supported but, more frequently, are part of a rigid frame system.

Better understanding of plate girder behavior, higher strength steels, and improved welding techniques have combined to make plate girders economical in many situations formerly thought to be ideal for the truss. Generally, simple spans of 70 to 150 ft (20 to 50 m) have traditionally been the domain for the plate girder. For bridges, continuous spans frequently using haunches (variable depth sections) are now the rule for spans 90 ft or more. There are several three-span continuous plate girders in the United States with center spans exceeding 400 ft, and longer spans are likely to be feasible in the future. The longest plate girder in the world is a three-span continuous structure over the Save River at Belgrade, Yugoslavia, with spans 246–856–246 ft (75–260–75 m). It is a double box girder in cross-section varying in depth from 14 ft 9 in. (4.5 m) at midspan to 31 ft 6 in. (9.6 m) at the pier. The structure replaced a suspension bridge destroyed in World War II.

Three types of plate girders whose design is outside the scope of this chapter are shown in Fig. 11.1.3: (a) the box girder, providing improved torsional stiffness for long-span bridges; (b) the hybrid girder, providing variable material strength in accordance with stresses; and (c) the delta girder, providing improved lateral rigidity for long lengths of lateral unsupport.

Prior to studying the theoretical development in this chapter the reader is advised to review Chapter 6, Part II, where the basic elastic stability of plates is treated.

Since the design of riveted girders has been extensively treated in older texts [11.1, 11.2] and such girders are rarely used at present, emphasis is placed on welded girders. No example of bolted or riveted girder design is given; however, high-strength bolted splices, commonly found in field connections, are treated in Chapter 13.

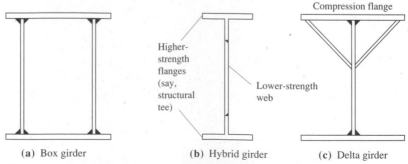

(a) Box girder   (b) Hybrid girder   (c) Delta girder

Figure 11.1.3   Other types of welded plate girders.

11.2 DIFFERENCE BETWEEN BEAM AND PLATE GIRDER

A plate girder is actually a deep beam. The limit states treated in Chapter 9 with regard to beams are still applicable for plate girders. Figure 11.2.1, relating to Load and Resistance Factor Design, shows the nominal strength M_n vs the slenderness ratio λ for the basic flexure limit states, lateral-torsional buckling, flange local buckling, and web local buckling. The relationship for lateral-torsional buckling in Fig. 11.2.1a is valid when a section is "compact" (i.e., $\lambda \le \lambda_p$ in LRFD-B5) with regard to the flange and web local buckling limit states. When the section is "noncompact" (i.e., $\lambda_p < \lambda \le \lambda_r$ in LRFD-B5), the nominal strength M_n must be determined for all three limit states in Fig. 11.2.1; the lowest value controls.

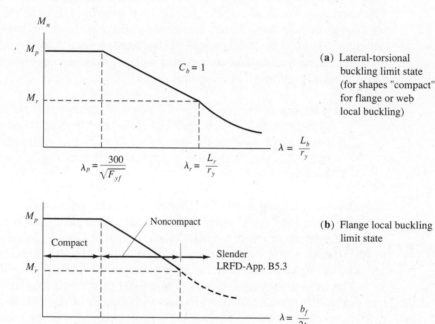

(a) Lateral-torsional buckling limit state (for shapes "compact" for flange or web local buckling)

(b) Flange local buckling limit state

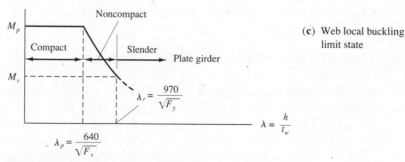

(c) Web local buckling limit state

Figure 11.2.1 Limit states in flexure.

Most rolled shapes are either "compact" or "noncompact"; this entire procedure was discussed and illustrated in Chapters 7 and 9. If the flange is "slender" ($\lambda > \lambda_r$) for flange local buckling, the efficiency is further reduced and $Q < 1$ must be used in accordance with LRFD or ASD-Appendix B5.3. The general treatment of using $Q < 1$ for stiffened and unstiffened elements of a compression member was presented in Chapter 6, Part II.

When the web is "slender" ($\lambda > \lambda_r$), the flexural member must be treated according to the plate girder provisions in LRFD-Appendix G or ASD-G. When λ does not exceed λ_r, the stress on the element can reach the yield stress F_y without elastic buckling occurring.

These concepts are applicable whether LRFD or ASD is used. There are differences in the equations used for lateral-torsional buckling, as discussed in Chapter 9. The specific inclusion of residual stress is not done in ASD, and the symbol λ is not used to refer either to the actual slenderness ratio or its limit. Philosophically λ_p is the "compact" limit in ASD-B5, and λ_r is the "noncompact" limit in ASD-B5.

The flexural and shear strengths of a plate girder are largely related to the web. The "slender" web may cause several problems:

1. Buckling due to bending in the plane of the web will reduce the efficiency of the web to carry its elastic share of the bending moment.
2. Buckling of the compression flange in the vertical direction due to inadequate stiffness of the web to prevent such buckling.
3. Buckling due to shear.

This chapter is largely devoted to treatment of these problems.

Since the 1961 AISC Specification, the design of plate girders has used the same strength approach developed by Basler and others at Lehigh University, even though in ASD the strengths are divided by factors of safety to get the service load limits, and are also divided by the elastic section modulus S to obtain the allowable stresses.

The most distinguishing feature of a plate girder is the use of regularly spaced transverse stiffeners. Stiffeners increase the strength of the web to carry shear. The elastic or inelastic buckling strength of a plate girder web in shear does not represent the maximum strength in shear. There will be significant *post-buckling strength* after buckling (slight out-of-plane deformation) has occurred when properly designed transverse stiffeners are used. The girder will behave like a truss with its web carrying diagonally the tension forces and the stiffeners carrying the compression forces. This truss-like behavior is referred to as *tension-field action*.

11.3 VERTICAL FLANGE BUCKLING LIMIT STATE

The maximum limit on the web slenderness h/t_w is based on the stiffness needed in the plane of the web to prevent the compression flange from buckling vertically (Fig. 11.3.1c). Note that h, the clear unsupported height of the web in a rolled section, is the depth h of the web plate in a welded I-shaped section. Furthermore, some flexural stiffness is needed from the web along the flange-to-web connection to preclude torsional buckling of the flange (Fig. 11.3.1b).

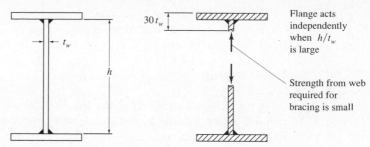

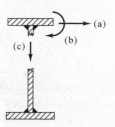

Figure 11.3.2 Vertical buckling of the compression flange.

Figure 11.3.1
(a) Lateral buckling;
(b) torsional buckling;
(c) vertical buckling.

For the purpose of this development, one may imagine that the flange is a compression member independent of the rest of the girder (see Fig. 11.3.2). When the girder is bent, as exaggerated in Fig. 11.3.3, the curvature gives rise to flange force components that cause compression on the edges of the web adjacent to the flanges. When the web remains stable when subject to those compressive flange force components, the flange cannot buckle vertically. In the following derivation the flange itself is assumed to have *zero* stiffness to resist vertical buckling; a conservative procedure.

Referring to Fig. 11.3.3, the deformation $\epsilon_f\,dx$ accumulated over the distance

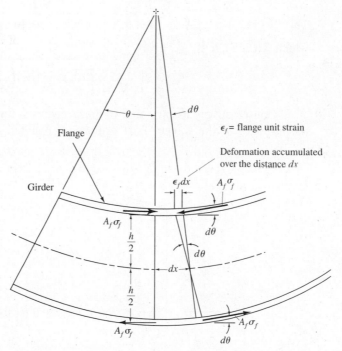

Figure 11.3.3 Flange forces arising from girder curvature.

dx is

$$\epsilon_f \, dx = d\theta \frac{h}{2} \tag{11.3.1}$$

$$d\theta = \frac{2\epsilon_f}{h} dx \tag{11.3.2}$$

As shown in Fig. 11.3.4a, the vertical component causing compression is $\sigma_f A_f \, d\theta$. After dividing by the area $t_w \, dx$ to obtain the compressive stress f_c as shown in Fig. 11.3.4b, one may substitute Eq. 11.3.2 for $d\theta$,

$$f_c = \frac{\sigma_f A_f \, d\theta}{t_w \, dx} = \frac{2\sigma_f A_f \epsilon_f}{t_w h} \tag{11.3.3}$$

Referring again to Eq. 6.14.28, the elastic buckling stress for a plate,

$$F_{cr} = \frac{k\pi^2 E}{12(1 - \mu^2)(b/t)^2} \tag{6.14.28}$$

where $b = h$, $t = t_w$, and $k = 1$ for the case of the Euler plate assumed free along edges parallel to loading, and pinned top and bottom. Thus

$$F_{cr} = \frac{\pi^2 E}{12(1 - \mu^2)(h/t_w)^2} \tag{11.3.4}$$

Equating applied stress, Eq. 11.3.3, to the critical stress, Eq. 11.3.4, gives

$$\frac{2\sigma_f A_f \epsilon_f}{t_w h} = \frac{\pi^2 E}{12(1 - \mu^2)(h/t_w)^2} \tag{11.3.5}$$

which, letting $t_w h = A_w$, gives

$$\frac{h}{t_w} = \sqrt{\frac{\pi^2 E}{24(1 - \mu^2)} \left(\frac{A_w}{A_f}\right) \left(\frac{1}{\sigma_f \epsilon_f}\right)} \tag{11.3.6}$$

Conservatively assume that σ_f must reach the flange yield stress F_{yf} to achieve the strength of the flange. Furthermore, if residual stress F_r exists in the flange

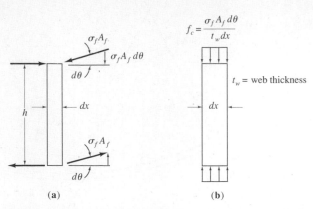

Figure 11.3.4   Effect of flange force component normal to flange plate.

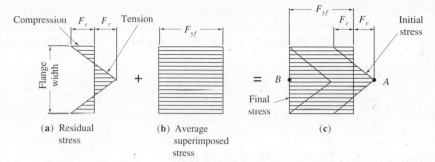

Figure 11.3.5   Effect of residual stress.

distributed as shown in Fig. 11.3.5, then the total flange strain will be that due to the sum of the residual stress plus the yield stress; therefore

$$\epsilon_f = (F_r + F_{yf})/E \tag{11.3.7}$$

It is the strain adjacent to the web that is of concern; in Fig.11.3.5c the change is from F_r in tension (point A) to F_{yf} in compression (point B).

Substitution of $\sigma_f = F_{yf}$, $\epsilon_f =$ Eq. 11.3.7, $E = 29,000$ ksi, and $\mu = 0.3$ into Eq. 11.3.6 gives

$$\frac{h}{t_w} = \frac{19,500\sqrt{A_w/A_f}}{\sqrt{F_{yf}(F_{yf} + F_r)}} \tag{11.3.8}$$

a conservative estimate of the maximum h/t_w to prevent vertical buckling. Basler [11.5] suggests A_w/A_f will rarely be below 0.5 and that $F_r = 16.5$ ksi is realistic. If these substitutions are made,

$$\frac{h}{t_w} = \frac{13,800}{\sqrt{F_{yf}(F_{yf} + 16.5)}} \tag{11.3.9}$$

Equation 11.3.9 has been developed without regard to placement of stiffeners. The effect of stiffeners would certainly be to increase the strength above the elastic buckling strength based on F_{cr} of Eq. 11.3.4. Tests reported by Frost and Schilling [11.6] on hybrid girders having A514 ($F_y = 100$ ksi) flanges indicate that h/t_w can conservatively be accepted in design as 250 if $a/h \leq 1.0$, and 200 for a/h between 1.0 and 1.5. The limitation of 200 may be given for other flange yield stresses as $2000/\sqrt{F_{yf}}$ for $a/h \leq 1.5$.

load and Resistance Factor Design—Maximum H/t_w

Equation 11.3.9 when rounded becomes

$$\frac{h}{t_w} = \frac{14,000}{\sqrt{F_{yf}(F_{yf} + 16.5)}} \tag{11.3.10)*}$$

which is the LRFD-Appendix G1 general limitation (F_{yf} in ksi).

---

*For SI units, with F_{yf} in MPa, $\dfrac{h}{t_w} = \dfrac{96,500}{\sqrt{F_{yf}(F_{yf} + 114)}} \tag{11.3.10}$

In the presence of transverse stiffeners, higher ratios are permitted. Based on the recommendations of the ASCE-AASHO Joint Committee, Subcommittee 1 on Hybrid Girder Design [11.10], LRFD-Appendix G1 gives the maximum h/t_w as

$$\frac{h}{t_w} \leq \frac{2000}{\sqrt{F_{yf}, \text{ksi}}} \tag{11.3.11)*}$$

when stiffener spacing a to web depth h does not exceed 1.5. Values for Eqs. 11.3.10 and 11.3.11 are given in Table 11.3.1.

Allowable Stress Design—Maximum h/t_w

The maximum h/t_w limits in ASD-G1 are identical to those in Load and Resistance Factor Design.

TABLE 11.3.1   MAXIMUM h/t_w LIMITATIONS—
LRFD-APPENDIX G1 AND ASD-G1

F_y (ksi)	h/t_w for Eq. 11.3.10 for $a/h > 1.5$	h/t_w for Eq. 11.3.11 for $a/h \leq 1.5$	F_y (MPa)
36	322	333	248
42	282	309	290
45	266	298	310
50	243	283	345
55	223	270	379
60	207	258	414
65	192	248	448
100	130	200	689

11.4 NOMINAL MOMENT STRENGTH— LOAD AND RESISTANCE FACTOR DESIGN

The concepts relating to the nominal moment strength M_n have been presented in Chapters 7 and 9, and reviewed in Sec. 11.2. Complexity in design arises using M_r and L_r expressed by Eqs. 9.6.5 and 9.6.6, respectively, for plate girders. For rolled beams, not only are the values of M_r and L_r available in the *LRFD Manual* but also all of the properties, including the torsion properties, are readily available. For plate girders all of the properties must be computed for each girder. Thus, while the beam provisions of LRFD-F1 are logically acceptable for the plate girders, simplified rules are given in LRFD-Appendix G2 in order to avoid the use of torsion properties and the accompanying complex expression for L_r.

---

* For SI units, with F_{yf} in MPa,
$$\frac{h}{t_w} \leq \frac{5250}{\sqrt{F_{yf}}} \tag{11.3.11}$$

Since plate girders will usually have "slender" webs; that is, λ will exceed λ_r for web local buckling, the strength cannot exceed that based on reaching yield stress F_y at the extreme fiber. No inelastic behavior is considered possible for design purposes.

The nominal moment strength M_n of slender web plate girders is controlled either by the limit state of yielding at the tension flange or that of buckling at the compression flange, as follows according to LRFD-Appendix G2:

For *yielding of the tension flange*,

$$M_n = F_{yt} S_{xt} R_{PG} \qquad (11.4.1)*$$

For *buckling of the compression flange*,

$$M_n = F_{cr} S_{xc} R_{PG} \qquad (11.4.2)$$

where F_{yt} = yield stress of the tension flange
F_{cr} = buckling stress at the compression flange, controlled by lateral-torsional buckling, flange local buckling, or yielding.
S_{xt} = section modulus referred to the tension flange, I_x/y_t
S_{xc} = section modulus referred to the compression flange, I_x/y_c
I_x = moment of inertia with respect to the x-axis
y_c = distance from the CG of the section to the compression extreme fiber
y_t = distance from the CG of the section to the tension extreme fiber

also

$$R_{PG} = 1 - \frac{a_r}{1200 + 300 a_r} \left(\frac{h_c}{t_w} - \frac{970}{\sqrt{F_{cr}}} \right) \le 1.0 \qquad (11.4.3)$$

where $a_r = A_w/A_f \le 10$
A_f = compression flange area
h_c = twice the distance from the neutral axis to the inside face of the compression flange less the fillet or corner radius
= h, height of web plate for a symmetrical I-shaped plate girder

The factor R_{PG} is to account for the "bend-buckling" effect on the "slender" web and its reduced ability to carry its elastic share of the bending moment. The development is presented in Section 11.6. Equation 11.4.3, from the original work of Basler [11.5], is revised from that used in 1986 LRFD Specification in order to give appropriate R_{PG} values for a_r between 0 and 10; a_r cannot exceed 10 according to LRFD-Appendix G2. The previous equation was specifically applicable for $a_r \le 2.0$.

The determination of the "critical stress" F_{cr} to be used is obtained by dividing the nominal strength M_n by the section modulus S_x. For the "slender" web, the maximum F_{cr} is F_{yf}, the yield stress in the flange.

---

*Note that 1993 LRFD does not use R_{PG} in Eq. 11.4.1 relating to the tension flange. This omission seems inappropriate because the reduction in moment strength R_{PG} resulting from bend-buckling of the web, as discussed in Sec. 11.6, is not related to a specific flange, but rather to the effective moment of inertia, which is reduced making both S_{xc} and S_{xt} reduced.

For the limit state of *lateral-torsional buckling,* LRFD-Appendix G2 gives:

1. For $\lambda \le \lambda_p$,

$$\left(\lambda = \frac{L_b}{r_T}\right) \le \left(\lambda_p = \frac{300}{\sqrt{F_{yf}}}\right) \tag{11.4.4}$$

$$F_{cr} = F_{yf} \tag{11.4.5}$$

2. For $\lambda_p < \lambda \le \lambda_r$,

$$\left(\lambda_p = \frac{300}{\sqrt{F_{yf}}}\right) < \left(\lambda = \frac{L_b}{r_T}\right) \le \left(\lambda_r = \frac{756}{\sqrt{F_{yf}}}\right) \tag{11.4.6}$$

$$F_{cr} = C_b F_{yf} \left[1 - \frac{1}{2}\left(\frac{\lambda - \lambda_p}{\lambda_r - \lambda_p}\right)\right] \le F_{yf} \tag{11.4.7}$$

3. For $\lambda > \lambda_r$,

$$\left(\lambda = \frac{L_b}{r_T}\right) > \left(\lambda_r = \frac{756}{\sqrt{F_{yf}}}\right) \tag{11.4.8}$$

$$F_{cr} = \frac{286,000 C_b}{\left(\dfrac{L_b}{r_T}\right)^2} \tag{11.4.9}$$

For the limit state of *flange local buckling,* LRFD-Appendix G1 gives:

1. For $\lambda \le \lambda_p$,

$$\left(\lambda = \frac{b_f}{2t_f}\right) \le \left(\lambda_p = \frac{65}{\sqrt{F_{yf}}}\right) \tag{11.4.10}$$

$$F_{cr} = F_{yf} \tag{11.4.11}$$

2. For $\lambda_p < \lambda \le \lambda_r$,

$$\left(\lambda_p = \frac{65}{\sqrt{F_{yf}}}\right) < \left(\lambda = \frac{b_f}{2t_f}\right) \le \left(\lambda_r = \frac{230}{\sqrt{F_{yf}/k_c}}\right) \tag{11.4.12}$$

$$F_{cr} = F_{yf} \left[1 - \frac{1}{2}\left(\frac{\lambda - \lambda_p}{\lambda_r - \lambda_p}\right)\right] \le F_{yf} \tag{11.4.13}$$

3. For $\lambda > \lambda_r$,

$$\left(\lambda = \frac{b_f}{2t_f}\right) > \left(\lambda_r = \frac{230}{\sqrt{F_{yf}/k_c}}\right) \tag{11.4.14}$$

$$F_{cr} = \frac{26,200 k_c}{\left(\dfrac{b_f}{2t_f}\right)^2} \tag{11.4.15}$$

where $k_c = 4/\sqrt{h/t_w}$ and $0.35 \le k_c \le 0.763$.

The general relationships of Eqs. 11.4.4 through 11.4.15 are shown in Fig. 11.4.1. Numerical values of λ_p and λ_r for plate girders are given in Table 11.4.1.

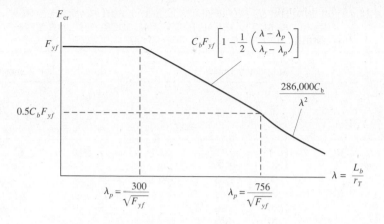

(a) Lateral-torsional buckling limit state

$$F_{cr} = \frac{M_n}{S_x}$$

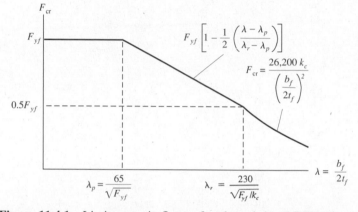

(b) Flange local buckling limit state

$$F_{cr} = \frac{M_n}{S_x}$$

Figure 11.4.1 Limit states in flexure for plate girder (LRFD-Appendix G2), where $k_c = 4/\sqrt{h/t_w}$ and $0.35 \le k_c \le 0.763$.

For the *lateral-torsional buckling limit state,* these plate girder provisions approximate the M_n vs L_b relationship used for rolled beams and shown in Fig. 9.6.1. The plate girder uses $F_{cr} = M_n/S_x$ instead of M_n. Note that λ_p of Eq. 11.4.4 to obtain L_p is identical to Eq. 9.6.3; however, r_T is used for plate girders whereas r_y is used for rolled beams. r_T is the radius of gyration of the compression flange and one-third of the compression portion of the web taken with respect to the y-axis. Equation 11.4.6 for λ_r giving L_r for plate girders avoids the more complplicated Eq. 9.6.6 for rolled beams; and also avoids the necessity of computing torsion properties for plate girders. The plate girder provisions also avoid computing M_r as $S_x(F_y - F_r)$, instead approximating the quantity $(F_y - F_r)$ as $0.5F_y$ for simplicity. The elastic lateral-torsional buckling equation, Eq. 11.4.9, is essentially the Euler column formula recognizing the compression flange (i.e., the flange plate and one-third of the compression portion of the web) as a column by using r_T for the radius of gyration.

Though for the lateral-torsional buckling limit state, the plate girder relationships in LRFD-Appendix G between moment strength and L_b are simplified from those for beams in LRFD-F1, there is no reason why the beam provisions could not also be used for the plate girder. The "slender" web plate would restrict the maximum

TABLE 11.4.1   SLENDERNESS RATIO LIMITS λ_p AND λ_r FOR FLANGE LOCAL BUCKLING AND LATERAL-TORSIONAL BUCKLING LIMIT STATES UNDER LRFD-APPENDIX G2 FOR PLATE GIRDERS

| Yield stress | Flange local buckling | | | | Lateral-torsional buckling | | |
| | λ_p | | | λ_r | λ_p | λ_r |
F_{yf} (ksi)	$\dfrac{b_f}{2t_f} = \dfrac{65}{\sqrt{F_{yf}}}$	$\dfrac{h}{t_w}$	k_c^*	$\dfrac{b_f}{2t_f} = \dfrac{230}{\sqrt{F_{yf}/k_c}}$	$\dfrac{L_b}{r_T} = \dfrac{300}{\sqrt{F_{yf}}}$	$\dfrac{L_b}{r_y} = \dfrac{756}{\sqrt{F_{yf}}}$
36	10.8	161.7†	0.35	22.7	50.0	126
		100	0.40	24.2		
		40	0.63	30.5		
42	10.0	149.7	0.35	21.0	46.3	117
		100	0.40	22.4		
		40	0.63	28.2		
45	9.7	144.6	0.35	20.3	44.7	113
		100	0.40	21.7		
		40	0.63	27.3		
50	9.2	137.2	0.35	19.2	42.4	107
		100	0.40	20.6		
		40	0.63	25.9		
55	8.8	130.8	0.35	18.3	40.5	102
		100	0.40	19.6		
		40	0.63	24.7		
60	8.4	125.2	0.36	17.8	38.7	97.6
		100	0.40	18.8		
		40	0.63	23.6		
65	8.1	120.3	0.36	17.2	37.2	93.8
		100	0.40	18.0		
		40	0.63	22.7		
90	6.9	102.2	0.40	15.2	31.6	79.7
		100	0.40	15.3		
		40	0.63	19.3		
100	6.5	97.0	0.41	14.7	30.0	75.6
		100	0.40	14.5		
		40	0.63	18.3		

$* k_c = \dfrac{4}{\sqrt{h/t_w}}$, where $0.35 \le k_c \le 0.763$

† Maximum h/t_w given for each yield stress $= 970/\sqrt{F_{yf}} = \lambda_r$ (LRFD-TABLE B5.1 for "webs in flexural compression").

moment strength to $F_y S_x$ if the beam provisions of LRFD-F1 were used, rather than M_p when the web plate is "compact" or "noncompact" (that is, when $\lambda \le \lambda_r$ for the web). A comparison in Fig. 11.4.2 of the lateral-torsional buckling provisions of LRFD-Appendix G with LRFD-F1 for a thin web ($\frac{3}{8} \times 100$) plate girder shows good agreement.

For the *flange local buckling limit state*, the plate girder limit λ_p is identical to that used for rolled beams (see Table 9.6.1). The limit λ_r for plate girders approximates the expressions given in Table 9.6.2 for beams. The reduction in strength based on elastic local buckling of the flange, Eq. 11.4.15, is identical to that of LRFD-

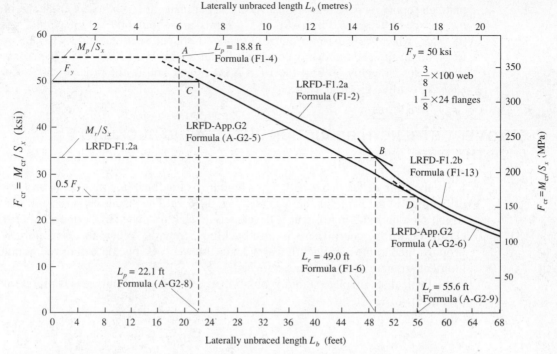

Figure 11.4.2 Comparison of lateral-torsional buckling limit state using plate girder provisions of LRFD-Appendix G with beam provisions of LRFD-F1. Properties of girder: $r_y = 5.32$ in., $r_T = 6.24$ in., $X_1 = 572$ ksi, $X_2 = 1,500,000 \times 10^{-6}$ in.^{4}/kips^2, $M_p = 15,280$ ft-kips, and $M_r = 9,250$ ft-kips (LRFD Formula (F1-7).

Appendix F for welded girders, somewhat lower than used for rolled beams in Appendix F, or for obtaining the efficiency factor Q for unstiffened compression elements in LRFD-Appendix B5.3a.

11.5 MOMENT STRENGTH—ALLOWABLE STRESS DESIGN

The moment strength of plate girders under Allowable Stress Design is treated identically with that presented for beams in Sec. 9.7. Design for the lateral-torsional buckling limit state using the larger of two formulas [ASD-Formula (F1-8), Eq. 9.7.12, or the combination of ASD-Formulas (F1-6) and (F1-7), Eqs. 9.7.20 and 9.7.22], is still applicable; however, generally the deep web plate girder has little torsional strength. Thus, ASD-Formulas (F1-6) and (F1-7), which are based on column strength of the compression flange usually controls. This is particularly true as the laterally unbraced length L_b becomes longer.

The column strength formulas of ASD, Formulas (F1-6) and (F1-7), use r_T as the radius of gyration, exactly as used in LRFD for the plate girder lateral-torsional buckling limit state.

The flange local buckling limit state is considered not to control when $b_f / 2t_f$ does not exceed $95/\sqrt{F_y/k_c}$, where $k_c = 4.05/(h/t_w)^{0.46}$ when $h/t_w > 70$ (as it would for plate girders), in accordance with ASD-B5.2. For plate girders whose web slenderness h/t_w exceeds $970/\sqrt{F_y}$, the ASD $b_f/2t_f$ limits are about 60% of the limits using $95/\sqrt{F_y}$, which ASD uses for flanges of rolled I-shaped sections and LRFD-B5 uses for λ_r on unstiffened uniformly compressed plates. The 1989 ASD limits seem unreasonably low; roughly 60% of the 1978 ASD values, and in the range of 50% of the LRFD λ_r values given in Table 11.4.1.

11.6 MOMENT STRENGTH REDUCTION DUE TO BEND-BUCKLING OF THE WEB

Since the plate girder web usually has a high h/t_w ratio, buckling may occur as a result of bending in the plane of the web (see Fig. 11.6.1). The slenderness ratio λ_r above which such buckling may occur is developed in what follows. Furthermore, after this elastic buckling occurs there is post-buckling strength. When the plate girder is proportioned to most efficiently carry load, the web will buckle before the nominal moment strength of the girder is reached.

Typical of any plate stability situation, the elastic buckling stress is represented by Eq. 6.14.28,

$$F_{cr} = k \frac{\pi^2 E}{12(1 - \mu^2)(b/t)^2} \qquad [6.14.28]$$

where for this case $b = h$.

The theoretical development of the k values for bending in the plane of the plate is given by Timoshenko and Woinowski-Krieger.* For any given type of loading, k varies with the aspect ratio, a/h (see Fig. 11.6.2), and with the support conditions along the edges. If the plate can be considered to have full fixity (full moment resistance against edge rotation) along the edges parallel to the direction of loading (i.e., at the edges joined to the flanges), the minimum value of k is 39.6 for any a/h ratio. If the flanges are assumed to offer no resistance to edge rotation, the minimum k value is 23.9. The variation of k with a/h ratio is given in Fig. 11.6.2.

Thus the critical stress (using $E = 29,000$ ksi and $\mu = 0.3$) may be said to lie between

$$F_{cr} = \frac{627,000}{(h/t)^2} \text{ ksi} \quad \text{for } k = 23.9 \text{ (simple support at flanges)}$$

and

$$F_{cr} = \frac{1038,000}{(h/t)^2} \text{ ksi} \quad \text{for } k = 39.6 \text{ (full fixity at flanges)}$$

While each particular girder will have a different degree of flange restraint, fully welded flange to web connections will surely approach the full fixity case. It will be reasonable then to arbitrarily select a k value closer to 39.6, say 80 percent of the difference toward the higher value. One might say that

---

* Reference 6.66, pp. 373–379.

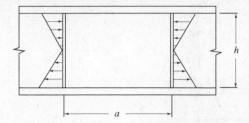

Figure 11.6.1 Web plate under pure moment.

$$F_{cr} = \frac{950,000}{(h/t)^2} \text{ ksi} \qquad (11.6.1)$$

is representative of the stress when elastic buckling is imminent due to bending in the plane of the web. Such "bend buckling" cannot occur if

$$\frac{h}{t_w} \leq \sqrt{\frac{950,000}{F_{cr}, \text{ksi}}} = \frac{970}{\sqrt{F_{cr}}} \qquad (11.6.2)$$

where t_w = web plate thickness.

LRFD-Appendix G2 uses the coefficient 970; however, ASD-G2 uses 982. The latter is obtained by substituting $F_{cr}/1.67$ for the allowable bending stress F_b. The two significant figure 970 seems reasonable. Figure 11.6.3 shows the elastic stability relationship for "bend buckling" of the web.

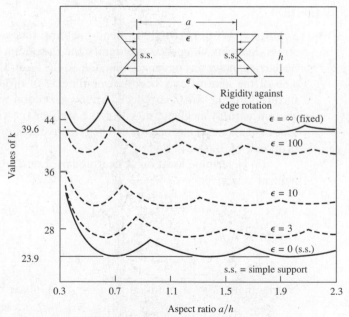

Figure 11.6.2 Buckling coefficients for plates subject to bending in the plane of the plate. (From *Handbook of Structural Stability*, Vol. 1 [6.69], p. 92)

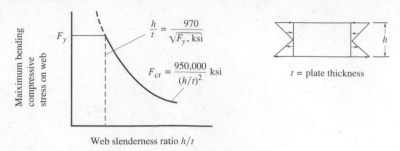

Maximum bending compressive stress on web

$$\frac{h}{t} = \frac{970}{\sqrt{F_y, \text{ksi}}}$$

$$F_{cr} = \frac{950,000}{(h/t)^2} \text{ ksi}$$

t = plate thickness

Web slenderness ratio h/t

Figure 11.6.3 Buckling of plates subject to bending in the plane of the web.

Since the web carries only a small part of the total bending moment to which the girder is subjected, neglect of the transition zone arising from inelastic buckling will not be significant.

Buckling of the web does not end the usefulness of the girder. Fig. 11.6.4 shows the relationship between nominal moment strength M_n vs h/t_w. This figure assumes the lateral-torsional buckling and local flange buckling limit states do not control.

When the post-buckling strength of the girder is considered, the strength is raised from line *BC* of Fig. 11.6.4 to line *BD*. The actual position of line *BD* varies with A_w/A_f, the ratio of the web area to the compression flange area.

EXAMPLE 11.6.1

Using $h/t_w = 320$, determine the expression for M_n/M_y (point *D* of Fig. 11.6.4) in terms of A_w/A_f.

Solution. With $h/t_w = 320$, "bend-buckling" occurs at a low level of flexural stress. Such buckling does not signify the maximum bending moment that can be carried; however, under additional load the flexural stress on the compression side of the neutral axis becomes nonlinear. In order to retain the use of the flexure formula, Mc/I, a reduced effective section must be used. The reduced section entirely neglects much of the web in the region where the buckling (out-of-planeness) has occurred. The effective section shown in Fig. 11.6.5 was proposed by Basler [11.5].

(a) Determine location of neutral axis. Equating static moments about the neutral axis gives

$$A_f(kh) + \frac{t_w(kh)^2}{2} = A_f(1 - k)h + \frac{3}{32}h\left(\frac{61}{64}h - kh\right)t_w$$

Divide by $A_f h$:

$$k + k^2\frac{t_w h}{2A_f} = (1 - k) + \frac{3}{32}\left(\frac{61}{64} - k\right)\frac{t_w h}{A_f}$$

Noting that $t_w h = A_w$ and letting $a_r = A_w/A_f$ gives

$$k^2 + k\left(\frac{4}{a_r} + \frac{3}{16}\right) = \frac{2}{a_r} + \frac{183}{1024}$$

$$k = \sqrt{\frac{192}{1024} + \frac{38}{16a_r} + \frac{4}{a_r^2}} - \left(\frac{3}{32} + \frac{2}{a_r}\right) \tag{a}$$

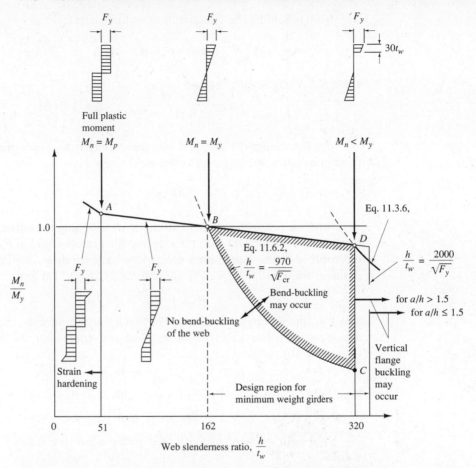

Figure 11.6.4 Nominal moment strength M_n of girders as affected by strength of the web plate resisting bending moment in the plane of the web: A36 steel.

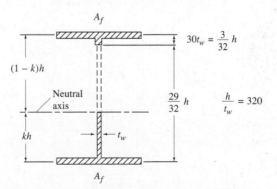

Figure 11.6.5 Effective section in bending when vertical flange buckling is imminent.

(b) Determine the effective moment of inertia.

$$I_e = A_f(kh)^2 + \frac{1}{3}t_w(kh)^3 + A_f(1-k)^2h^2 + \frac{3t_wh}{32}\left(\frac{61}{64}h - kh\right)^2$$

Using $t_wh = A_w$ and $a_r = A_w/A_f$,

$$I_e = A_fh^2\left[\frac{a_r}{3}k^3 + k^2 + (1-k)^2 + \frac{3a_r}{32}\left(\frac{61}{64} - k\right)^2\right] \tag{b}$$

(c) Determine the nominal moment strength M_n. Assuming the extreme fiber in compression stressed to the yield stress F_y,

$$M_n = \frac{F_yI_e}{(1-k)h} \tag{c}$$

(d) Determine the moment strength M_y assuming the entire section elastic (and therefore effective) with the extreme fiber stress equal to F_y. In developing the expression the flange-area concept is used as shown in Fig. 11.6.6. The moment strength of the web is approximately (Fig. 11.6.6a)

$$M_{\text{web}} = fS_x = f(\tfrac{1}{6}t_wh^2) \tag{d}$$

which assumes web depth, distance between flange centroids, and overall depth are the same. The moment strength of the equivalent flange area system (Fig. 11.6.6b) is

$$M_{\text{equiv}} = fA_f'h \tag{e}$$

Equating Eqs. (d) and (e) gives for the equivalent flange area, A_f',

$$A_f' = \tfrac{1}{6}t_wh = \tfrac{1}{6}A_w \tag{f}$$

The total moment strength of a girder where the stress $f = F_y$ then becomes

$$M_y = F_y\left[A_f + \frac{A_w}{6}\right]h \tag{g}$$

$$= F_yA_fh\left(1 + \frac{a_r}{6}\right) \tag{h}$$

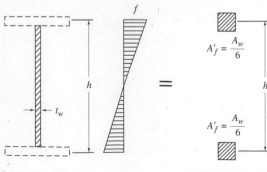

(a) Actual condition   (b) Equivalent condition

Figure 11.6.6   Equivalent flange area to replace web.

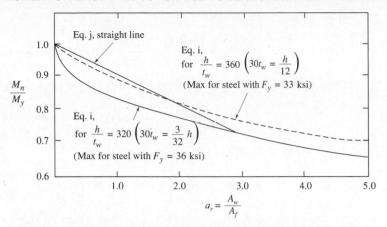

Figure 11.6.7 Reduction in nominal moment strength considering post-buckling strength at maximum h/t_w for A36 steel.

The vertical ordinate of point D in Fig. 11.6.4 is obtained by dividing Eq. (c) by Eq. (h):

$$\frac{M_n}{M_y} = \frac{\frac{a_r}{3}k^3 + k^2 + (1-k)^2 + \frac{3a_r}{32}\left(\frac{61}{64} - k\right)^2}{(1-k)(1 + a_r/6)} \tag{i}$$

which is plotted in Fig. 11.6.7.

From Fig. 11.6.7, the variation in M_n/M_y might be approximated by a straight line for A_w/A_f from zero to three with a slope of $-(1.00 - 0.73)/3.0 = -0.09$.

Thus, at $h/t_w = 320$,

$$\frac{M_n}{M_y} = 1.0 - 0.09\frac{A_w}{A_f} \tag{j}$$

It may be observed that the straight line agrees better for $h/t_w = 360$, the situation for which this linear equation was originally developed [11.5], than it does for $h/t_w = 320$. For higher strength steels, for which the maximum h/t_w to prevent vertical buckling of the flange is less than 360, more of such a stiffer web participates with the compression flange, causing a greater reduction in M_n/M_y.

The linear reduction based on Eq. (j) does not seem conservative, but is within several percent of the more accurate curve using $30t_w$ as the effective depth of web participating with the compression flange.

Tests [11.5] have verified the correctness of this linear reduction method using $h/t_w = 360$ as its basis. ■■

Reduced Nominal Strength M_n when $h/t_w > 970/\sqrt{F_y}$

By reference of Fig. 11.6.4, it may be reasonably assumed that M_n/M_y varies linearly from point B to D. Thus the reduction in M_n/M_y per A_w/A_f per h/t greater than that at point B is

$$\frac{\text{Slope per } A_w/A_f}{320 - 162} = \frac{0.09}{158} = 0.00057 \quad (\text{say } 0.0005)$$

Thus M_n/M_y for the region from point B to D (Fig. 11.6.4) assuming linear variation, is

$$\frac{M_n}{M_y} = 1.0 - 0.0005\frac{A_w}{A_f}\left(\frac{h}{t_w} - \frac{970}{\sqrt{F_y}}\right) \tag{11.6.3}$$

Since $M_y = F_y S_x$, Eq. 11.6.3 may be written

$$M_n = S_x F_y\left[1.0 - 0.0005\frac{A_w}{A_f}\left(\frac{h}{t_w} - \frac{970}{\sqrt{F_y}}\right)\right] \tag{11.6.4}$$

As shown in the development, Eq. 11.6.4 is based on the assumption that the ratio A_w/A_f does not exceed about 3. In the 1993 LRFD Specification, the relationship has been revised to accommodate A_w/A_f up to 10. Thus, in place of the coefficient 0.0005, the coefficient becomes

$$\frac{a_r}{1200 + 300a_r}$$

in accordance with the original development by Basler [11.5]. Thus the more general form of Eq. 11.6.4 is

$$M_n = S_x F_y\left[1 - \frac{a_r}{1200 + 300a_r}\left(\frac{h}{t_w} - \frac{970}{\sqrt{F_y}}\right)\right] \tag{11.6.5}$$

Equation 11.6.5 assumes no influence of the lateral-torsional buckling and flange local buckling limit states. When the controlling limit state prevents the flange stress from reaching F_y, then *that* controlling limit state critical stress F_{cr} should replace F_y in Eq. 11.6.5. Thus, Eq. 11.6.5 becomes

$$M_n = S_x F_{cr}\left[1 - \frac{a_r}{1200 + 300a_r}\left(\frac{h}{t_w} - \frac{970}{\sqrt{F_{cr}}}\right)\right] \tag{11.6.6}$$

$$M_n = S_x F_{cr} R_{PG} \tag{11.6.7}$$

where $R_{PG} = 1 - \dfrac{a_r}{1200 + 300a_r}\left(\dfrac{h}{t_w} - \dfrac{970}{\sqrt{F_{cr}}}\right) \le 1.0$ [11.4.3]

$a_r = A_w/A_f \le 10$
A_f = compression flange area
A_w = web area

This reduction factor R_{PG} is given in LRFD-Appendix G2.

In summary, when h/t_w exceeds $970/\sqrt{F_{cr}}$, one may view Eq. 11.6.7 either as (1) the full section modulus S_x multiplied by a reduced stress $F_{cr}R_{PG}$, or (2) as a reduced section modulus $S_x R_{PG}$ multiplied by the full stress F_{cr}. Philosophically it should be considered the latter; however, in ASD it has traditionally been treated as the former. The idea of using a reduced section when buckling has occurred to cause a non-linear stress distribution, is the same concept used for the stiffened plate element in Chapter 6, Part II.

11.7 NOMINAL MOMENT STRENGTH—HYBRID GIRDERS

As discussed in the last section, a girder having large h/t_w may have its web buckle due to flexural stress, thereby increasing the load-carrying requirement of the compression flange. This extra load-carrying requirement for the flange also may occur when a hybrid girder is used. A hybrid girder is one in which the flanges are of a higher strength steel than the web. The use of a hybrid girder has particular economic advantages in composite construction, as described in Sec. 16.9.

The special behavioral feature of the hybrid girder is the yielding of the lower strength web before the maximum flange strength has been reached. When the moment strength of the hybrid girder is achieved, the web will have participated to a lesser extent than in a girder using only one grade of steel.

Frost and Schilling [11.6], Schilling [11.7], Carskaddan [11.8], and Toprac and Natarajan [11.9] have studied the hybrid girder. The state-of-the-art and design recommendations for hybrid girders have been given by a Joint ASCE-AASHO Joint Committee [11.10].

The principal effect of using a lower yield stress web than is used in the flange is that the onset of yielding in the web will occur prior to yielding in the flanges. For example, when the web is A36 steel and the flanges have $F_y = 100$ ksi, the web may yield at about 40% of the nominal strength M_n based on yielding of the flanges.* This means that the web will yield *even at service load*.

The design of hybrid girders, according to Subcommittee 1 of ASCE-AASHO Joint Committee [11.10], should be based on the moment causing the initiation of flange yielding. This may be accomplished by either of two procedures:

1. The nominal strength M_n may be computed by setting the extreme fiber strain ϵ_{yf} equal to F_{yf}/E_s, where F_{yf} is the yield stress of the flange steel (see Fig. 11.7.1b). At this stage much of the lower yield stress web will have exceeded its yield strain $\epsilon_{yw} = F_{yw}/F_s$, in which case the stress distribution over the depth of the section is nonlinear, as shown in Fig. 11.7.1c.

2. The nominal strength M_n may be computed as a homogeneous elastic section entirely of the flange steel, that is $F_{yf}S_x$, reduced by multiplying by a reduction factor.

Both the AISC Allowable Stress Design (ASD-G2) and Load and Resistance Factor Design (LRFD-Appendix G2) methods use the latter approach.

Thus, the moment strength M_n of an hybrid girder may be expressed as Eq. 11.4.2 (or Eq. 11.6.6) multiplied by a reduction factor R_e to account for yielding in the web prior to reaching the flange yield stress at the extreme fiber,

$$M_n = S_x F_{cr} R_{PG} R_e \tag{11.7.1}$$

where, according, to LRFD-Appendix G2 and ASD-G2,

$$R_e = \frac{12 + a_r(3m - m^3)}{12 + 2a_r} \leq 1.0 \tag{11.7.2}$$

where $a_r = A_w/A_f$ = ratio of the cross-sectional area of the web to the cross-sectional area of one flange

---

*See 2nd edition, pp. 590–592.

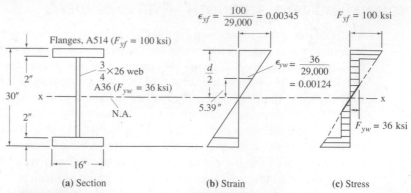

Figure 11.7.1 Hybrid section showing stress and strain when F_{yf} is reached at extreme fiber of section.

$m = F_{yw}/F_{yf} =$ ratio of the yield stress of the web steel to the yield stress of the flange steel

$R_{PG} =$ reduction for web instability when $h/t_w > 970/\sqrt{F_{cr}}$ (see Eq. 11.4.3)

$F_{cr} =$ critical compression flange stress based on lowest value obtained from the lateral-torsional buckling or the flange local buckling limit states.

The reduction factor R_{PG} was recommended by the ASCE-AASHO Joint Committee [11.10], and is somewhat different than the 1986 LRFD Specification expression recommended by Cooper, Galambos, and Ravindra [11.11].

In applying the reduction factor R_{PG} to account for bend-buckling of the web, the derivation of the web slenderness limit of $970/\sqrt{F_y}$ involved the plate yield stress F_y, which means the *web*. On the other hand, examination of Fig. 11.7.1 shows that when the flange reaches F_{yf} the strain on the portions of the web adjacent to the flanges will have strain exceeding the yield strain $\epsilon_{yw} = F_{yw}/E_s$ in the web. Logically the slenderness limit to prevent buckling must then be *lower* than $970/\sqrt{F_{yw}}$. Dawe and Kulak [11.12] have indicated that because of the restraining effect of sturdy flanges the web may be expected to undergo plastic strain without buckling even when the h/t_w ratio is $800/\sqrt{F_{yw}}$. Zahn [11.13] has noted that when $F_{yf} = 50$ ksi (flanges) and $F_{yw} = 36$ ksi (web) the same limiting slenderness ratio is obtained,

$$\left(\frac{800}{\sqrt{F_{yw}}} = 133 \right) \approx \left(\frac{970}{\sqrt{F_{yf}}} = 137 \right)$$

Thus, rather than have a separate limit equation relating to bend-buckling of the web of a hybrid girder using F_{yw}, AISC concluded it would be satisfactory and simpler to use the same limit equation for both homogeneous and hybrid girders. The proper result would be obtained by always using the *flange yield stress F_{yf}* in the h/t_w slenderness limit equation,

$$\left(\frac{h}{t_w} \right)_{\substack{\text{bend-buckling} \\ \text{limit}}} = \frac{970}{\sqrt{F_{yf}}}$$

The special features of hybrid girders relating to composite construction are contained in Chapter 16.

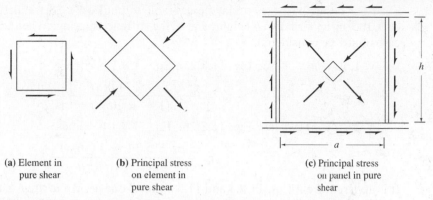

(a) Element in
pure shear (b) Principal stress
on element in
pure shear (c) Principal stress
on panel in pure
shear

Figure 11.8.1 Classical shear theory applied to plate girder web panel.

11.8 NOMINAL SHEAR STRENGTH—ELASTIC AND INELASTIC BUCKLING

Typical of I-shaped sections, the web carries most of the shear. Since the plate girder inherently has a thin web ($h/t_w > 970/\sqrt{F_y}$), stability is of primary concern.

Consider a web plate panel of length a between transverse stiffeners and having a clear height h between longitudinal plate supports (i.e., between flanges, flange and longitudinal stiffener, or between longitudinal stiffeners), as shown in Fig. 11.8.1. In regions of high shear and low bending moment, the buckling strength of the panel may be investigated assuming it is acted upon by shear alone (pure shear), as in Fig. 11.8.1.

Elastic Buckling Under Pure Shear

The elastic buckling stress for any plate is given by Eq. 6.14.28 as

$$F_{cr} = k \frac{\pi^2 E}{12(1 - \mu^2)(b/t)^2} \qquad [6.14.28]$$

where for the case of pure shear (see Fig. 11.8.1), Eq. 6.14.28 may be written (using τ in place of F for shear stress and k_v for k)

$$\tau_{cr} = k_v \frac{\pi^2 E}{12(1 - \mu^2)\left(\dfrac{\text{short dimension}}{t}\right)^2} \qquad (11.8.1)$$

where for the case of edges simply supported (i.e., displacement prevented but rotation about edges unrestrained)

$$k_v = 5.34 + 4.0\left(\frac{\text{short dimension}}{\text{long dimension}}\right)^2 \qquad (11.8.2)$$

the development of which has been given by Timoshenko and Woinowski-Krieger.*

---

* Reference 6.66, pp. 379–385.

For design purposes it may be desirable to put Eqs. 11.8.1 and 11.8.2 in terms of h, the unsupported web height, and a, the stiffener spacing. When this is done two cases must be considered.

1. If $a/h \le 1$ (see Fig. 11.8.2a), Eq. 11.8.1 becomes

$$\tau_{cr} = \frac{\pi^2 E[5.34 + 4.0(a/h)^2]}{12(1 - \mu^2)(a/t)^2} \frac{(h/a)^2}{(h/a)^2} \qquad (11.8.3)$$

2. If $a/h \ge 1$ (see Fig. 11.8.2b), Eq. 11.8.1 becomes

$$\tau_{cr} = \frac{\pi^2 E[5.34 + 4.0(h/a)^2]}{12(1 - \mu^2)(h/t)^2} \qquad (11.8.4)$$

It is apparent from Eqs. 11.8.3 and 11.8.4 that if one desires to use h/t as the stability ratio in the denominator, then two expressions for k_v are necessary. For all ranges of a/h, Eqs. 11.8.3 and 11.8.4 may be written

$$\tau_{cr} = \frac{\pi^2 E k_v}{12(1 - \mu^2)(h/t)^2} \qquad (11.8.5)$$

where

$$k_v = 4.0 + 5.34/(a/h)^2 \quad \text{for } a/h \le 1 \qquad (11.8.6)$$

$$k_v = 4.0/(a/h)^2 + 5.34 \quad \text{for } a/h \ge 1 \qquad (11.8.7)$$

Equations 11.8.6 and 11.8.7 are used in ASD-F4. In LRFD-Appendix G3, those two equations are replaced by the following:

$$k_v = 5 + \frac{5}{(a/h)^2} \qquad (11.8.8)$$

As shown in Table 11.8.1, it is clear that within the accuracy that the theoretical elastic buckling solution agrees with a real plate girder, Eq. 11.8.8 is an acceptable single equation substitute for Eqs. 11.8.6 and 11.8.7.

For use in design equations, Eq. 11.8.5 has been put into nondimensional form, defining C_v as the ratio of shear stress τ_{cr} at buckling to shear yield stress τ_y,

$$C_v = \frac{\tau_{cr}}{\tau_y} = \frac{\pi^2 E k_v}{\tau_y (12)(1 - \mu^2)(h/t)^2} \qquad (11.8.9)$$

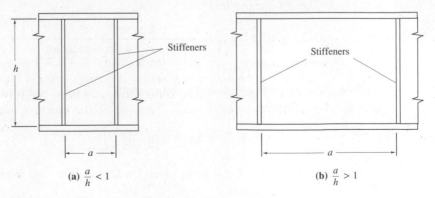

(a) $\dfrac{a}{h} < 1$ **(b)** $\dfrac{a}{h} > 1$

Figure 11.8.2 Two cases of intermediate stiffener spacing.

TABLE 11.8.1   ELASTIC BUCKLING COEFFICIENT k_v
FOR SHEAR STRENGTH

Ratio a/h	ASD Eq. 11.8.6 $a/h \leq 1$	ASD Eq.11.8.7 $a/h > 1$	LRFD Eq.11.8.8	LRFD k / ASD k
0.2	137.5		130.0	0.94
0.4	37.4		36.3	0.97
0.6	18.8		18.9	1.00
0.8	12.3		12.8	1.04
1.0	9.3	9.3	10.0	1.07
1.2		8.1	8.5	1.04
1.4		7.4	7.6	1.02
1.6		6.9	7.0	1.01
1.8		6.6	6.5	1.00
2.0		6.3	6.3	0.99
2.2		6.2	6.0	0.98
2.4		6.0	5.9	0.97
2.6		5.9	5.7	0.97
2.8		5.9	5.6	0.96
3.0		5.8	5.6	0.96

which is C_v for *elastic* stability. Substitution of $E = 29,000$ ksi, $\mu = 0.3$, $\tau_y = 0.6F_{yw}$ (see Eq. 7.7.9), and using the subscript w for both the yield stress F_{yw} and the plate thickness t_w to identify this behavior as occurring in the *web*, gives

$$C_v = \frac{\pi^2(29,000)k_v}{0.6F_{yw}(12)(1 - 0.09)(h/t_w)^2}$$

which gives C_v as given by LRFD-Appendix G3, valid for τ_{cr} below the proportional limit as shown in Fig. 11.8.3,

$$C_v = \frac{44,000k_v}{(h/t_w)^2 F_{yw}} \tag{11.8.10}*$$

ASD-F4 uses the coefficient 45,000 instead of 44,000. The slightly larger value is obtained when τ_y is taken as $F_y/\sqrt{3}$.

Inelastic Buckling under Pure Shear

As in all stability situations, residual stresses and imperfections cause inelastic buckling as critical stresses approach yield stress. A transition curve for inelastic buckling was given by Basler [11.3] based on curve fitting and using test results from Lyse and Godfrey [11.4]. In the transition zone between elastic buckling and yielding,

$$\tau_{cr} = \sqrt{\tau_{\substack{\text{prop.} \\ \text{limit}}} \cdot \tau_{\substack{\text{cr(ideal} \\ \text{elastic)}}}} \tag{11.8.11}$$

---

*For SI units, F_{yw} in MPa, $C_v = \dfrac{303,000k_v}{(h/t_w)^2 F_{yw}}$ (11.8.10)

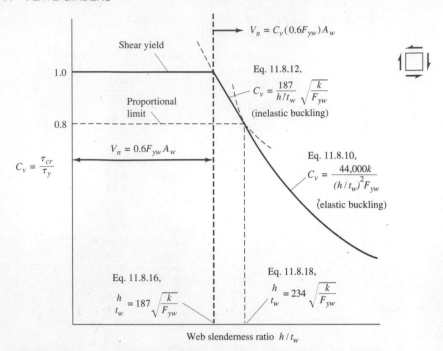

Figure 11.8.3   Buckling of plate girder web resulting from shear alone—LRFD-Appendix G3.

The proportional limit is taken as $0.8\tau_y$, higher than for compression in flanges, because the effect of residual stress is less. Dividing Eq. 11.8.11 by τ_y to obtain C_v and using Eq. 11.8.10 gives

$$C_v = \frac{\tau_{cr}}{\tau_y} = \sqrt{(0.8)\frac{44{,}000k_v}{F_{yw}(h/t_w)^2}}$$

$$= \frac{187}{h/t_w}\sqrt{\frac{k_v}{F_{yw}}} \tag{11.8.12}*$$

which is given by LRFD-Appendix G3 and is shown schematically in Fig. 11.8.3. ASD-F4 uses 190 instead of 187; the two-significant-figure number seems more in keeping with the true accuracy of the relationship. The 187 arises from using 44,000 instead of 45,000 in Eq. 11.8.12.

Nominal Shear Strength—Load and Resistance Factor Design

The nominal shear strength V_n of a girder based on inelastic or elastic buckling of the web may be expressed

$$V_n = \tau_{cr}A_w \tag{11.8.13}$$

---

* For SI units, F_{yw} in MPa, $C_v = \dfrac{491}{h/t_w}\sqrt{\dfrac{k_v}{F_{yw}}}$ (11.8.12)

or using $C_v = \tau_{cr}/\tau_y$,

$$V_n = C_v \tau_y A_w \tag{11.8.14}$$

Approximating τ_y as $0.6F_{yw}$ gives

$$V_n = C_v(0.6F_{yw})A_w \tag{11.8.15}$$

which is LRFD-Formula (A-G3-3). In Eq. 11.8.15, C_v is Eq. 11.8.10 for elastic buckling when $C_v \leq 0.8$, and is Eq. 11.8.12 when $C_v > 0.8$.

Equation 11.8.15 will apply also to rolled beams since rarely would transverse stiffeners be used. If one wishes to have an explicit expression for h/t_w corresponding to $C_v = 1$ (i.e., the web yields in shear and no buckling occurs), Eq. 11.8.12 may be solved for h/t_w when $C_v = 1$,

$$\frac{h}{t_w} = 187\sqrt{\frac{k_v}{F_{yw}}} \tag{11.8.16}$$

When h/t_w does not exceed the value from Eq. 11.8.16, the nominal shear strength is

$$V_n = 0.6F_{yw}A_w \tag{11.8.17}$$

which is LRFD-Formula (A-G3-1) as well as LRFD-Formula (F2-1) for beams. This was first discussed in Chapter 7 (Sec. 7.7).

The h/t_w relationship that divides elastic and inelastic buckling may be obtained by setting C_v equal to 0.8 in Eq. 11.8.10, giving

$$\frac{h}{t_w} = 234\sqrt{\frac{k_v}{F_{yw}}} \tag{11.8.18}$$

The LRFD relationship between buckling strength in shear and web slenderness ratio h/t_w is shown in Fig. 11.8.3.

Shear Strength—Allowable Stress Design

In accordance with Allowable Stress Design philosophy presented in Secs. 1.8 and 1.9, and discussed for shear in Sec. 7.7, the strength relationship may be divided by the safety factor to put the equation in the service load range,

$$\left(\frac{V_n}{\gamma/\phi} = \frac{V_n}{\text{FS}}\right) \geq V \tag{11.8.19}$$

where γ/ϕ = overload factors divided by resistance factor
  FS = 1.67 = nominal safety factor in beam design
  V = service load shear

To obtain Eq. 11.8.19 in "stress" format, divide both sides by the web area A_w; thus,

$$\left(f_v = \frac{V}{A_w}\right) \leq \left(F_v = \frac{V_n}{(\text{FS})A_w}\right) \tag{11.8.20}$$

where the allowable stress F_v may be obtained from putting Eq. 11.8.14 into Eq. 11.8.20,

$$F_v = \frac{C_v \tau_y A_w}{(FS)A_w} = \frac{C_v \tau_y}{FS} \tag{11.8.21}$$

Using $\tau_y = F_y/\sqrt{3}$ instead of $0.6F_y$ as used in LRFD, taking FS = 1.67, and recognizing the traditional upper limit of $0.40F_y$ for F_v,

$$F_v = \frac{F_y C_v}{2.89} \leq 0.40F_y \tag{11.8.22}$$

which is ASD-Formula (F4-2). The expressions for C_v are given by Eq. 11.8.10 for $C_v \leq 0.8$ using 45,000 instead of the LRFD 44,000, and by Eq. 11.8.12 for $C_v > 0.8$ using 190 instead of the LRFD 187,

$$C_v = \frac{45,000k_v}{(F_{yw})(h/t_w)^2} \quad \begin{array}{l} \text{for } C_v \leq 0.8 \\ \text{(elastic buckling)} \end{array} \tag{11.8.23}*$$

$$C_v = \frac{190}{h/t_w}\sqrt{\frac{k_v}{F_{yw}}} \quad \begin{array}{l} \text{for } C_v > 0.8 \\ \text{(inelastic buckling)} \end{array} \tag{11.8.24}*$$

In the C_v expressions, k_v is given by Eqs. 11.8.6 and 11.8.7, as follows:

$$k_v = 4.0 + 5.34/(a/h)^2 \quad \text{for } a/h \leq 1$$
$$k_v = 4.0/(a/h)^2 + 5.34 \quad \text{for } a/h \geq 1$$

Previously, in Table 7.7.1 of Sec. 7.7, maximum h/t_w values were given such that rolled beams without stiffeners could be designed using $F_v = 0.40F_y$. When no stiffeners are used, a/h is large and k_v approaches 5.34. From C_v for inelastic buckling (Eq. 11.8.24),

$$C_v = \frac{190}{h/t_w}\sqrt{\frac{5.34}{F_{yw}}} = \frac{439}{(h/t)\sqrt{F_{yw}}} \tag{11.8.25}$$

Using Eq. 11.8.22,

$$\frac{F_y}{2.89}\frac{439}{(h/t_w)\sqrt{F_{yw}}} = 0.40F_y$$

$$\frac{h}{t_w} = \frac{380}{\sqrt{F_{yw}}} \tag{11.8.26}*$$

Equation 11.8.26 gives the maximum h/t_w values of Table 7.7.1 for Allowable Stress Design.

---

*For SI units, with F_{yw} in MPa,

$$C_v = \frac{310,000k_v}{F_{yw}(h/t_w)^2} \tag{11.8.23}$$

$$C_v = \frac{500}{(h/t)}\sqrt{\frac{k_v}{F_{yw}}} \tag{11.8.24}$$

$$\frac{h}{t_w} = \frac{1000}{\sqrt{F_{yw}}} \tag{11.8.26}$$

11.9 NOMINAL SHEAR STRENGTH—INCLUDING TENSION-FIELD ACTION

The elastic and inelastic buckling strength of the web subject to shear is represented by *ABCD* in Fig. 11.9.1. A plate stiffened by flanges and transverse stiffeners has considerable post-buckling strength.

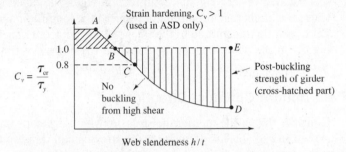

Figure 11.9.1 Shear capacity available, considering post-buckling strength.

According to Basler [11.3], the ability of a plate girder to behave in a manner similar to a truss was recognized as early as 1898. As shown in Fig. 11.9.2, the tension forces are carried by membrane action of the web (referred to as *tension-field action*) while the compression forces are carried by the transverse stiffeners. The work of Basler [11.3] led to a theory that agreed with tests and provides criteria to ensure that truss action can develop. The inclusion of truss action raises the shear strength from that based on buckling (*ABCD* on Fig. 11.9.1) to approach a condition corresponding to shear yield in classical beam theory (*ABE* of Fig. 11.9.1).

The nominal shear strength V_n may be expressed as the sum of the buckling strength V_{cr} and the post-buckled strength V_{tf} from tension-field action,

$$V_n = V_{cr} + V_{tf} \qquad (11.9.1)$$

The nominal buckling strength is given by Eq. 11.8.14 with $V_n = V_{cr}$,

$$V_{cr} = C_v \tau_y A_w \qquad (11.9.2)$$

where $C_v = \tau_{cr}/\tau_y$ and is given by Eqs. 11.8.10 and 11.8.12 for elastic and inelastic buckling, respectively.

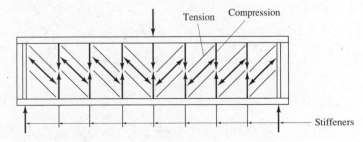

Figure 11.9.2 Tension-field action.

The shear strength V_{tf} arising from the tension-field action in the web develops a band of tensile forces that occur after the web has buckled under diagonal compression (principal stresses in ordinary beam theory). Equilibrium is maintained by the transfer of force to the vertical stiffeners. As the girder load increases, the angle of the tension-field changes to accommodate the greatest carrying capacity. Figure 11.9.3 shows a 50×50-in. (approx. 1.3×1.3 m) panel with a $\frac{1}{4}$-in. (6.4 mm) web which has buckled under diagonal compression when subjected to pure shear. It also illustrates the anchorage requirement wherein the longitudinal component of the tension-field must be transmitted to the flange in the adjacent panel, as shown by the vertical breaks in the whitewash at the flange in the corner of the adjacent panel where the tension-field intersects the stiffener and flange.

Tension-Field Action: Optimum Direction

Consider the tensile membrane stress σ_t which develops in the web at the angle γ, as shown in Fig. 11.9.4. If such tensile stresses can develop over the full height of the web, then the total diagonal tensile force T would be

$$T = \sigma_t t_w h \cos\gamma \qquad (11.9.3)$$

the vertical component of which is the shear force V, given by

$$V = T \sin\gamma = \sigma_t t_w h \cos\gamma \sin\gamma \qquad (11.9.4)$$

Figure 11.9.3 Tension-field in test plate girder. (From Ref. 11.3, Courtesy of Lehigh University)

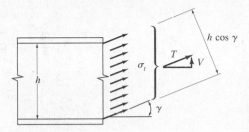

Figure 11.9.4 Membrane stresses in tension-field action.

If such diagonal tensile stresses could develop *along* the flanges, vertical stiffness of the flanges would be required. Since the flanges have little vertical stiffness and are acting to their capacity in resisting flexure on the girder, the tension-field actually can develop only over a band width such that the vertical component can be transferred at the vertical stiffeners. The stiffeners can be designed to carry the necessary compressive force. It will be assumed that the tension-field (or partial tension-field as some may prefer to call it) may develop over the band width s, shown in Fig. 11.9.5a.

The membrane tensile force tributary to one stiffener is $\sigma_t s t_w$, and the partial shear force ΔV_{tf} developed by compression in the stiffener is

$$\Delta V_{tf} = \sigma_t s t_w \sin \gamma \tag{11.9.5}$$

and the angle γ is the angle providing the maximum shear component from the partial tension-field.

From the geometry shown in Fig. 11.9.5b,

$$s = h \cos \gamma - a \sin \gamma \tag{11.9.6}$$

where a = stiffener spacing. Substitution of Eq. 11.9.6 into Eq. 11.9.5 gives

$$\begin{aligned}\Delta V_{tf} &= \sigma_t t_w (h \cos \gamma - a \sin \gamma) \sin \gamma \\ &= \sigma_t t_w \left(\frac{h}{2} \sin 2\gamma - a \sin^2 \gamma\right)\end{aligned} \tag{11.9.7}$$

For maximum ΔV_{tf}, it is required that $d(\Delta V_{tf})/d\gamma = 0$. Thus

$$\frac{d(\Delta V_{tf})}{d\gamma} = \sigma_t t_w \left(\frac{h}{2}(2) \cos 2\gamma - 2a \sin \gamma \cos \gamma\right) = 0 \tag{11.9.8}$$

$$0 = h \cos 2\gamma - a \sin 2\gamma$$

or

$$\tan 2\gamma = \frac{h}{a} = \frac{1}{a/h} \tag{11.9.9}$$

From the trigonometry of Eq. 11.9.9,

$$\sin 2\gamma = \frac{1}{\sqrt{1 + (a/h)^2}} \tag{11.9.10}$$

also

$$\sin^2 \gamma = \frac{1 - \cos 2\gamma}{2} = \frac{1}{2}\left[1 - \frac{a/h}{\sqrt{1 + (a/h)^2}}\right] \tag{11.9.11}$$

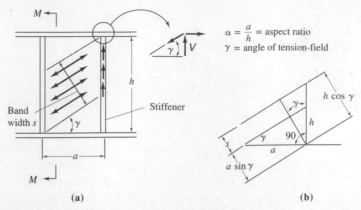

Figure 11.9.5   Forces arising from tension-field.

The maximum contribution ΔV_{tf} from tension-field action is then obtained by substituting Eqs. 11.9.10 and 11.9.11 into Eq. 11.9.7 giving

$$\Delta V_{tf} = \sigma_t \frac{ht_w}{2}[\sqrt{1 + (a/h)^2} - a/h] \qquad (11.9.12)$$

It is not practical to use Eq. 11.9.12 directly, since the shear contribution from the part of the section (such as M–M of Fig. 11.9.5) that cuts through the triangles outside the band s must be added. The state of stress in these triangles is unknown, requiring an alternate approach to finding the total shear V_{tf} when the optimum angle γ is reached.

An alternate way, as used by Basler [11.3], is cut a free body as in Fig. 11.9.6. The section is taken vertically midway between two adjacent stiffeners and horizontally at mid-depth. The mid-depth cut provides access to the tension-field where the state of stress is known, and the shear resultant on each vertical face equals $V_{tf}/2$ from symmetry.

Shear Strength from Tension-Field Action

Using the free body of Fig. 11.9.6, horizontal force equilibrium requires

$$\Delta F_f = (\sigma_t t_w a \sin \gamma)\cos \gamma$$

$$= \sigma_t \frac{t_w a}{2} \sin 2\gamma \qquad (11.9.13)$$

The incremental web force ΔF_w is not used because the web contributes little to the flexural strength of the girder. Rotational equilibrium, taken about point O, requires

$$\Delta F_f \frac{h}{2} - \frac{V_{tf} a}{2} = 0 \qquad (11.9.14)$$

Solving Eq. 11.9.14 for ΔF_f and substituting into Eq. 11.9.13 gives

$$\frac{V_{tf} a}{h} = \sigma_t \frac{t_w a}{2} \sin 2\gamma \qquad (11.9.15)$$

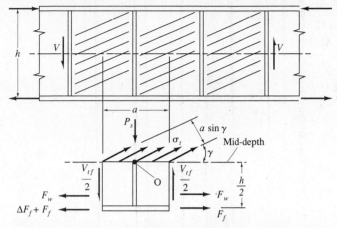

Figure 11.9.6 Force in stiffener resulting from tension-field action.

Solving for V_{tf} and using Eq. 11.9.10 for $\sin 2\gamma$ gives

$$V_{tf} = \sigma_t \frac{ht_w}{2}\left[\frac{1}{\sqrt{1 + (a/h)^2}}\right] \qquad (11.9.16)$$

Failure Condition

The actual state of stress in the web involves both shear stress τ and normal stress σ_t; thus the failure of an element subjected to shear in combination with an inclined tension must be considered, as shown in Fig. 11.9.7. Two basic assumptions are involved: first, τ_{cr} remains at constant value from buckling load to ultimate load and therefore the tension-field stress σ_t acts in addition to the principal stress τ_{cr}; second, the angle γ in Fig. 11.9.7b will be conservatively taken as 45° even though it will always be less than that value.

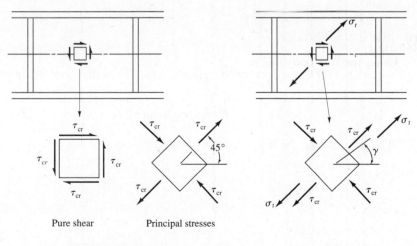

(a) At imminent buckling

(b) At ultimate shear

Figure 11.9.7 State of stress.

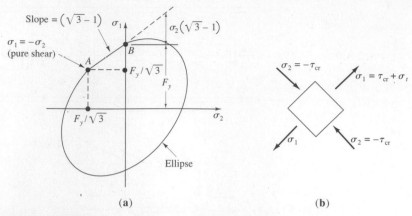

Figure 11.9.8 Energy-of-distortion failure criterion.

The generally accepted relationship for failure in plane stress is the "energy of distortion" theory (discussed in Sec. 2.7) shown as the ellipse in Fig. 11.9.8, which may be written

$$\sigma_1^2 + \sigma_2^2 - \sigma_1\sigma_2 = F_y^2 \tag{11.9.17}$$

where σ_1 and σ_2 are principal stresses. Point A represents the case of shear alone and point B represents tension alone. The actual states of stress in plate girder webs fall on the ellipse between points A and B, and a straight line is a reasonable approximation of the segment AB,

$$\sigma_1 = F_y + \sigma_2(\sqrt{3} - 1) \tag{11.9.18}$$

and for the stress condition that $\sigma_1 = \tau_{cr} + \sigma_t$ and $\sigma_2 = -\tau_{cr}$, Eq. 11.9.18 becomes

$$\frac{\sigma_t}{F_y} = 1 - \frac{\tau_{cr}}{F_y/\sqrt{3}} = 1 - C_v \tag{11.9.19}$$

Force in Stiffener

Using Fig. 11.9.6, vertical force equilibrium requires the force P_s in the stiffener to be

$$P_s = (\sigma_t t_w a \sin \gamma)\sin \gamma \tag{11.9.20}$$

Then, substituting the trigonometric identity,

$$\sin^2 \gamma = \frac{1 - \cos 2\gamma}{2}$$

gives

$$P_s = \sigma_t \left(\frac{at_w}{2}\right)\left[1 - \frac{a/h}{\sqrt{1 + (a/h)^2}}\right] \tag{11.9.21}$$

Substituting Eq. 11.9.19 (using F_{yw} for F_y) into Eq. 11.9.21 gives

$$P_s = \frac{F_{yw}(1 - C_v)at_w}{2}\left[1 - \frac{a/h}{\sqrt{1 + (a/h)^2}}\right]$$ (11.9.22)

which is the force in the stiffener when nominal shear strength V_n is reached, including tension-field action.

More recent work as discussed in the *SSRC Guide* [6.8] has shown that Eq. 11.9.22 may be simplified by using $a/h = 1$; in which case,

$$P_s = 0.5F_{yw}(1 - C_v)at_w\left(1 - \frac{1}{\sqrt{2}}\right)$$ (11.9.23)

$$P_s = 0.15F_{yw}(1 - C_v)at_w$$ (11.9.24)

Nominal Shear Strength, Including Both Buckling and Post-Buckling Strengths

Since thin-web plate girders exhibit some strength in shear before diagonal buckling occurs (V_n from Sec. 11.8) and additional strength in the post-buckling range (V_{tf} from Eq. 11.9.16), their actual strength is the sum of both components. Substituting Eqs. 11.9.2 and 11.9.16 into 11.9.1 gives

$$V_n = ht_w\left[\tau_y C_v + \frac{\sigma_t}{2\sqrt{1 + (a/h)^2}}\right]$$ (11.9.25)

Substituting Eq. 11.9.19 for σ_t and using $\tau_y = F_y/\sqrt{3}$ gives

$$V_n = F_{yw}ht_w\left[\frac{C_v}{\sqrt{3}} + \frac{1 - C_v}{2\sqrt{1 + (a/h)^2}}\right]$$ (11.9.26)

A summary of studies of tension-field mechanisms is presented in the *SSRC Guide* [6.8, p. 194].

Load and Resistance Factor Design

When tension-field action is developed by using appropriately spaced and sized transverse stiffeners, the nominal strength in shear may be expressed by Eq. 11.9.26, Factoring $\sqrt{3}$ from the denominator, and then approximating $F_y/\sqrt{3}$ as $0.6F_y$, gives

$$V_n = 0.6F_{yw}A_w\left(C_v + \frac{1 - C_v}{1.15\sqrt{1 + (a/h)^2}}\right)$$ (11.9.27)

which is LRFD-Formula (A-G3-2) from Appendix G3. Curves for Eq. 11.9.27 are presented in Fig. 11.9.9 for $F_y = 50$ ksi steel.

The force P_s in the intermediate stiffeners when tension-field action is utilized is given by Eq. 11.9.24. The stiffener area A_{st} required is

$$\text{Required } A_{st} = \frac{P_s}{F_{yst}} = \frac{0.15F_{yw}(1 - C_v)at_w}{F_{yst}}$$ (11.9.28)

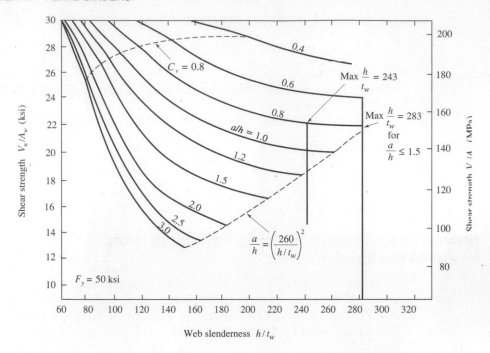

Figure 11.9.9 Nominal shear strength V_n, including tension-field action, divided by web area A_w for plate girders having $F_y = 50$ ksi (345 MPa), according to LRFD-Appendix G3.

When the panel has more strength ϕV_n than is needed to carry the factored shear V_u, the stiffener area A_{st} may be reduced by multiplying by $(V_u/\phi V_n)$. Also, the derivation assumed the stiffener was aligned with the center of the web; when stiffeners are used only on one side or if angle stiffeners are used there is an eccentric effect and the stiffener area must be increased. In addition, the area $(18t_w \times t_w)$ of the web tributary to the stiffener may be subtracted from the required A_{st}. Thus, LRFD-Appendix G4 gives the requirement as

$$\text{Required } A_{st} = \frac{F_{yw}}{F_{yst}}\left(0.15Dht_w(1 - C_v)\frac{V_u}{\phi V_n} - 18t_w^2\right) \qquad (11.9.29)$$

where D = factor to account for eccentric loading on stiffeners
    = 1 for stiffeners in pairs on each side of web
    = 1.8 for single angle stiffeners
    = 2.4 for single plate stiffeners
  F_{yst} = specified yield stress for stiffener material

Note that Eq. 11.9.29 uses h instead of a, a reasonable approximation and a simplification that makes the stiffener area A_{st} required proportional to the web area A_w.

Sometimes stiffeners are alternated on each side of the web to gain better economy or they are placed all on one side to improve esthetics. Referring to Fig. 11.9.10a, the symmetrical pair of stiffeners reaches its yield condition with the force P_s,

$$P_s = 2wtF_{yst} = A_{st}F_{yst} \quad \text{(for concentric load)} \tag{11.9.30}$$

On the other hand, an eccentrically loaded stiffener becomes plastic with a stress distribution as shown in Fig. 11.9.10b. For this case, force equilibrium requires

$$P_s = (w - x)tF_{yst} - xtF_{yst} \tag{a}$$

and moment equilibrium requires

$$xtF_{yst}\left(w - \frac{x}{2}\right) = F_{yst}(w - x)t\left(\frac{w - x}{2}\right) \tag{b}$$

Solving the quadratic gives $x = 0.293w$. Substitution for x in Eq. (a) gives

$$P_s = [(w - 0.293w)t - 0.293wt]F_{yst}$$
$$= 0.414wtF_{yst} = 0.414A'_{st}F_{yst} \quad \text{(for eccentric load)} \tag{11.9.31}$$

If single plate stiffeners are used on one side only, equating Eqs. 11.9.30 and 11.9.31 shows

$$A'_{st} = \frac{A_{st}}{0.414} = 2.4A_{st} \tag{11.9.32}$$

To correct for eccentric loading of stiffeners, the factor D is used in Eq. 11.9.29; 2.4 for single plate stiffeners. For a single angle whose center of gravity is closer to the web, the multiplier D reduces to 1.8.

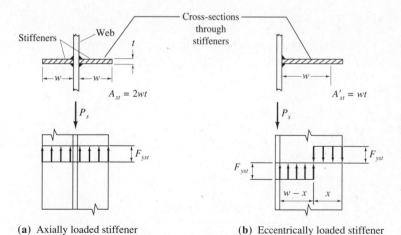

(a) Axially loaded stiffener   (b) Eccentrically loaded stiffener

Figure 11.9.10 Intermediate stiffeners at nominal shear strength V_n, including tension-field action.

Allowable Stress Design

The allowable shear stress when tension-field action is used is based on the strength V_n given by Eq. 11.9.26. When the strength is divided by the factor of safety 1.67 and divided by the web area $A_w = ht_w$, the allowable shear stress F_v is

$$F_v = \frac{F_y}{2.89}\left[C_v + \frac{1 - C_v}{1.15\sqrt{1 + (a/h)^2}} \right] \tag{11.9.33}$$

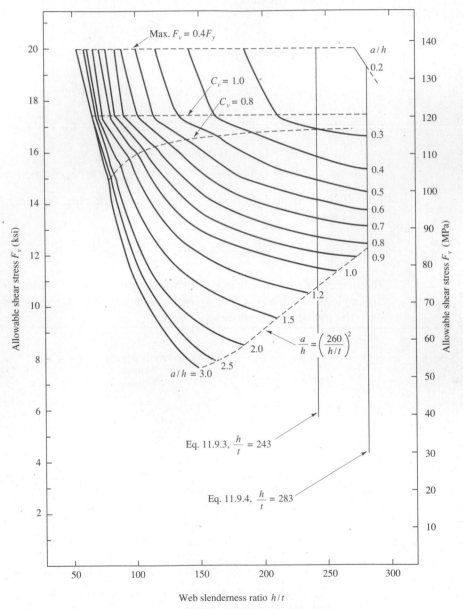

Figure 11.9.11   Allowable shear stress F_v, including tension-field action, for plate girders having $F_y = 50$ ksi (345 MPa), according to ASD-G3.

which is ASD-Formula (G3-1). Equation 11.9.33 is shown graphically in Fig. 11.9.11 for $F_y = 50$ ksi steel.

The maximum force in a stiffener when tension-field action occurs is given by Eq. 11.9.22. This force is reached only when the nominal strength of the girder is reached and the stiffener yields. Thus, the area required for the stiffener is obtained by dividing the force P_s by the yield stress F_{yst} of the stiffener, the same as is done in LRFD. Thus, the stiffener area required is

$$A_{st} = \frac{P_s}{F_{yst}} = \frac{F_{yw}}{F_{yst}} \frac{(1 - C_v)at_w}{2}\left[1 - \frac{a/h}{\sqrt{1 + (a/h)^2}}\right] \qquad (11.9.34)$$

where F_{yw} = yield stress of the web material
        F_{yst} = yield stress of the stiffener material

Equation 11.9.34 may be rewritten, letting $F_{yw}/F_{yst} = Y$, multiplying and dividing by h, and introducing the eccentric stiffener multiplier D when stiffeners are *not* placed in pairs, giving

$$\text{Required } A_{st} = \frac{1 - C_v}{2}\left[\frac{a}{h} - \frac{(\dot{a}/h)^2}{\sqrt{1 + (a/h)^2}}\right]DYht_w \qquad (11.9.35)$$

which is ASD-Formula (G4-2).

11.10 STRENGTH IN COMBINED BENDING AND SHEAR

In the vast majority of cases the nominal strength M_n in bending is not influenced by shear, nor is the nominal shear strength V_n influenced by moment. Particularly, in very slender webs where "bend-buckling" may occur, the bending stress is redistributed as discussed in Sec. 11.6, so that the flanges carry an increased share. The shear strength of the web, however, is not reduced as a result of "bend-buckling" because most of the shear strength is from tension-field action with only a small contribution from the portion of the web adjacent to the flange. In stockier webs no "bend-buckling" may occur, but high web shear in combination with bending may cause yielding of the web adjacent to the flange; again resulting in a transfer of part of the web's share of the bending moment to the flange. The strength of girders subject to combined bending and shear is the subject of the third major paper by Basler [11.17].

Since instability is precluded. a plastic analysis may be used. When subjected to high bending moment, the web yields adjacent to the flange and is, therefore, unable to carry shear. In the mid-depth region of the web the shear causes yielding; thus this part of the web is unable to carry bending moment.

Referring to Fig. 11.10.1, the nominal shear strength V_n' in the presence of bending moment may be expressed as

$$V_n' = \tau_y y_0 t_w \qquad (11.10.1)$$

When *no* bending moment is present. that is $y_0 = h$, the nominal shear strength V_n would be

$$V_n = \tau_y t_w h \qquad (11.10.2)$$

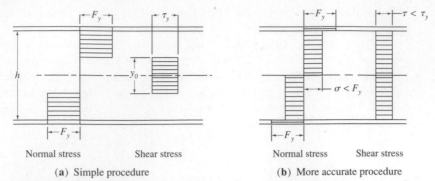

Normal stress   Shear stress   Normal stress   Shear stress

(a) Simple procedure   (b) More accurate procedure

Figure 11.10.1   Shear and moment strengths under combined bending and shear.

Eliminating τ_y from Eqs. 11.10.1 and 11.10.2 gives

$$y_0 = \left(\frac{V'_n}{V_n}\right)h \tag{11.10.3}$$

The nominal moment strength M'_n in the presence of shear from Fig. 11.10.1a is

$$M'_n = A_f F_y h + F_y t_w\left(\frac{h}{2}\right)\left(\frac{h}{2}\right) - F_y t_w\left(\frac{y_0}{2}\right)\left(\frac{y_0}{2}\right) \tag{11.10.4}$$

which upon substitution of Eq. 11.10.3 into Eq. 11.10.4 and letting $A_w = ht_w$ gives

$$M'_n = F_y A_f h\left\{1 + \frac{1}{4}\frac{A_w}{A_f}\left[1 - \left(\frac{V'_n}{V_n}\right)^2\right]\right\} \tag{11.10.5}$$

The nominal strength M_n equals M_y when the extreme fiber reaches the yield stress F_y, and with the web fully participating, is, according to Eq. (g), Example 11.6.1,

$$M_n = M_y = F_y h A_f\left(1 + \frac{1}{6}\frac{A_w}{A_f}\right) \tag{11.10.6}$$

As the percentage of the shear strength V_n that is utilized increases, the nominal moment strength M'_n decreases. In the absence of instability, but in the presence of high shear, the nominal bending moment strength may be expressed as

$$M'_n = M_n\left[\frac{1 + \frac{1}{4}a_r[1 - (V'_n/V_n)^2]}{1 + \frac{1}{6}a_r}\right] \tag{11.10.7}$$

where $a_r = A_w/A_f$.

When $M'_n = M_n$, $V'_n/V_n = 0.577$, or approximately 0.6 When more than 60 percent of the maximum shear strength V_n is used, the available nominal strength M'_n becomes less than M_n. Table 11.10.1 gives some values for M'_n/M_n for various values of a_r in the practical range, with graphical illustration in Fig. 11.10.2.

The relationship in Fig. 11.10.2 may be expressed:
For $V'_n/V_n \leq 0.60$,

$$M'_n = M_n \tag{11.10.8}$$

For $M'_n/M_n \leq 0.75$,

$$V'_n = V_n \tag{11.10.9}$$

TABLE 11.10.1 VALUES OF M'_n/M_n IN ACCORDANCE WITH EQ. 11.8.11 FOR $V'_n/V_n \geq 0.6$

$a_r = \dfrac{A_w}{A_f}$	For $\dfrac{V'_n}{V_n} = 0.8$	For $\dfrac{V'_n}{V_n} = 1.0$
0	1.0	1.0
0.5	0.964	0.923
1.0	0.935	0.856
1.5	0.908	0.800
2.0	0.885	0.750

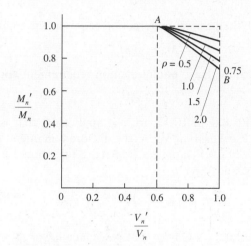

Figure 11.10.2 Moment–shear strength interaction relationship.

When Eqs. 11.10.8 or 11.10.9 are not applicable, the interaction relationship must be used. If one uses a conservative value of $a_r = A_w/A_f = 2.0$ and considers the strength reduction from points A to B of Fig. 11.10.2 as a straight line, the slope of AB would be

$$\text{Slope of } AB = \frac{-0.25}{0.40} = -\frac{5}{8}$$

The reduction equation then becomes

$$\frac{M'_n}{M_n} = 1 - \frac{5}{8}\left(\frac{V'_n}{V_n} - 0.6\right) \leq 1.0 \tag{11.10.10}$$

or

$$\frac{M'_n}{M_n} + \frac{5}{8}\left(\frac{V'_n}{V_n}\right) \leq \left(1.0 + \frac{5}{8}(0.6) = 1.375\right) \tag{11.10.11}$$

where M'_n = nominal flexural strength *in the presence of shear*
   V'_n = nominal shear strength *in the presence of flexure*
   M_n = maximum nominal flexural strength
   V_n = maximum nominal shear strength

Load and Resistance Factor Design

According to the provisions of LRFD-Appendix G5, when plate girder shear strength V_n *depends on inclusion of tension-field action* the web must satisfy the interaction criterion, given by Eqs. 11.10.8, 11.10.9, and 11.10.11, for shear combined with tension from flexure.

Recognizing that in the design of a girder V_n' and M_n' are the *required* strengths in *combination,* these primed terms should be replaced in Eqs. 11.10.8, 11.10.9, and 11.10.11 by V_u/ϕ and M_u/ϕ, respectively. The resistance factor ϕ for this shear-tension relationship is 0.90, the value used for flexure.

Thus, the interaction relationship for LRFD is:

1. For $V_u/\phi V_n \leq 0.60$, the basic noninteraction requirement applies,

$$M_u \leq \phi M_n \qquad (11.10.12)$$

2. For $M_u/\phi M_n \leq 0.75$, the basic noninteraction requirement applies,

$$V_u \leq \phi V_n \qquad (11.10.13)$$

3. When Eqs. 11.10.12 and 11.10.13 are not applicable, that is, either the factored shear V_u exceeds $0.6\phi V_n$ or the factored moment M_u exceeds $0.75\phi M_n$, then Eq. 11.10.11 applies. Replacing the primed terms in Eq. 11.10.11 with M_u/ϕ and V_u/ϕ gives

$$\frac{M_u}{\phi M_n} + 0.625\left(\frac{V_u}{\phi V_n}\right) \leq 1.375 \qquad (11.10.14)$$

which is LRFD-Formula (A-G5-1). The resistance factor $\phi = 0.90$.

Conditions 1 and 2 may be combined into a criterion to determine whether or not Eq. 11.10.14 needs to be investigated. Assuming the inequalities relating to $0.6\phi V_n$ and $0.75\phi M_n$ for Eqs. 11.10.12 and 11.10.13 are *not* satisfied gives a pair of equations below, each followed by the associated basic requirement which must always be satisfied,

$$V_u > \phi 0.6 V_n \qquad (11.10.15a)$$
$$M_u < \phi M_n \qquad (11.10.15b)$$
$$M_u > \phi 0.75 M_n \qquad (11.10.16a)$$
$$V_u < \phi V_n \qquad (11.10.16b)$$

Dividing the first by the second for each of the pairs of Eqs. 11.10.15 and 11.10.16 gives

$$\frac{V_u}{M_u} \geq \frac{0.6 V_n}{M_n} \qquad (11.10.17)$$

$$\frac{M_u}{V_u} \geq \frac{0.75 M_n}{V_n} \qquad (11.10.18)$$

Inverting Eq. 11.10.18, and putting it with Eq. 11.10.17, gives

$$\frac{0.6 V_n}{M_n} \leq \frac{V_u}{M_u} \leq \frac{V_n}{0.75 M_n} \qquad (11.10.19)$$

which is the equivalent of the opening sentence limitations in LRFD-Appendix G5. When the V_u/M_u ratio is within the range of Eq. 11.10.19, Eq. 11.10.14 must be satisfied.

Allowable Stress Design

The same strength interaction equation, Eq. 11.10.11, used for combined shear and tension in LRFD has been used for many years in Allowable Stress Design.

If M_n' equals a service load moment M times a factor for safety FS, the nominal strength M_n equals $F_y S_x$, V_n' equals the service load shear V times FS, and the nominal shear strength V_n equals $\tau_{cr} A_w$, then Eq. 11.10.11 becomes

$$\frac{M(FS)}{F_y S_x} + 0.625\left(\frac{V(FS)}{\tau_{cr} A_w}\right) \leq 1.375 \qquad (11.10.20)$$

Replacing the terms in Eq. 11.10.20 with the service load bending stress f_b for M/S_x, the maximum allowable stress in flexure for a plate girder $0.60F_y$ for F_y/FS, the service load shear stress f_v for V/A_w, and the allowable shear stress F_v for τ_{cr}/FS, Eq. 11.10.20 becomes

$$\frac{f_b}{0.60F_y} + 0.625\left(\frac{f_v}{F_v}\right) \leq 1.375 \qquad (11.10.21)$$

Solving for f_b gives

$$f_b \leq \left(0.825 - 0.375\frac{f_v}{F_v}\right)F_y \leq 0.60F_y \qquad (11.10.22)$$

which is ASD-G5, Formula (G5-1). Note that f_b and f_v are the maximum flexural and shear stresses in the *web*. The adjacent relatively stiff flange prevents stability from influencing the strength of the web under combined stress. When nondimensionalized, Eq. 11.10.22 may be used in the form shown in Fig. 11.10.3 (note comparison with Fig. 11.10.2).

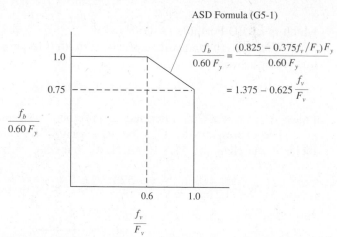

Figure 11.10.3   ASD relationship for combined shear and tension.

11.11 INTERMEDIATE TRANSVERSE STIFFENERS

Plate girders will usually be designed to have intermediate stiffeners as shown in Fig. 11.1.2 and the photo on p. 652. The two stability parameters for the web are h/t_w and a/h as discussed in Secs. 11.8 and 11.9. Buckling resulting from shear can be avoided when these stability parameters are kept low enough; alternatively, the shear stress can be kept below the critical buckling stress τ_{cr}. Since rolled beams have low h/t_w ratios, buckling resulting from shear will not occur. When the spacing a of stiffeners makes a/t_w low enough, and their size is adequate to allow them to act as compression verticals in a truss as discussed in Sec. 11.9, post-buckling strength (tension-field action) is available and may be utilized in design.

Requirements to Omit Intermediate Stiffeners—LRFD

Stiffeners need not be used when the flexural strength of the section can be achieved without diagonal buckling resulting from shear. Stiffeners would not be required, according to LRFD-Appendix G3, and shown in Fig. 11.8.3, when

$$\frac{h}{t_w} \leq 187 \sqrt{\frac{k_v}{F_{yw}}} \qquad [11.8.16]$$

When stiffeners are not used, the buckling coefficient k_v, given by Eq. 11.8.8, is to be taken as 5. This value would be approached when a/h becomes large. When $k_v = 5$, Eq. 11.8.16 becomes

$$\frac{h}{t_w} \leq \frac{418}{\sqrt{F_{yw}}} \qquad (11.11.1)$$

which is the limit given by LRFD-Appendix G4. When the limit of Eq. 11.11.1 is not exceeded, the maximum nominal shear strength V_n is achievable,

$$V_n = 0.6 F_{yw} A_w \qquad [11.8.17]$$

which is LRFD Formula (A-G3-1).

When less than maximum shear strength is required, intermediate stiffeners are not required when

$$V_n \leq C_v (0.6 F_{yw}) A_w \qquad [11.8.15]$$

unless h/t_w exceeds 260. Intermediate stiffeners *are required* when h/t_w exceeds 260.

The expressions for C_v to be used for unstiffened girders are Eqs. 11.8.12 for inelastic buckling and 11.8.10 for elastic buckling with $k_v = 5$, as follows:

1. When $\dfrac{418}{\sqrt{F_{yw}}} \leq \dfrac{h}{t_w} \leq \dfrac{523}{\sqrt{F_{yw}}}$ (i.e., inelastic buckling)

$$C_v = \frac{418}{\dfrac{h}{t_w} \sqrt{F_{yw}}} \qquad (11.11.2)*$$

2. When $\dfrac{h}{t_w} > \dfrac{523}{\sqrt{F_{yw}}}$ (i.e., elastic buckling)

$$C_v = \frac{220,000}{(h/t_w)^2 F_{yw}} \qquad (11.11.3)^*$$

In summary, intermediate stiffeners are *not* required when both of the following requirements are satisfied:

1.
$$\frac{h}{t_w} \le 260 \qquad (11.11.4)$$

2.
$$V_n \le C_v(0.6F_{yw})A_w \qquad (11.11.5)$$

where C_v is given by Eqs. 11.11.2 and 11.11.3. Equation 11.11.4 was recommended by Basler[11.14] as a practical limit. He recommended that fabrication, handling, and erection are facilitated when the smaller panel dimension, a or h, does not exceed $260t_w$. When stiffeners are not used, h is less than a.

Requirements to Omit Intermediate Stiffeners—ASD

Intermediate stiffeners are *not* required under ASD-F4 when both of the following requirements are satisfied:

1.
$$\frac{h}{t_w} \le 260 \qquad (11.11.6)$$

2.
$$f_v \le \frac{F_y C_v}{2.89} \le 0.40F_y \qquad (11.11.7)$$

where $f_v = V/A_w$. The ratio $C_v = \tau_{cr}/\tau_y$ in Eq. 11.11.7 is given by Eqs. 11.8.23 and 11.8.24. When a/h becomes large, k_v approaches 5.34, simplifying C_v to become

$$C_v = \frac{439}{\dfrac{h}{t_w}\sqrt{F_{yw}}} \qquad \begin{array}{l} \text{for } C_v \ge 0.8 \\ \text{(inelastic buckling)} \end{array} \qquad (11.11.8)$$

$$C_v = \frac{240,000}{(h/t_w)^2 F_{yw}} \qquad \begin{array}{l} \text{for } C_v < 0.8 \\ \text{(elastic buckling)} \end{array} \qquad (11.11.9)$$

---

*For SI units, with F_{yw} in MPa,

1. When $\dfrac{1100}{\sqrt{F_{yw}}} \le \dfrac{h}{t_w} \le \dfrac{1380}{\sqrt{F_{yw}}}$,

$$C_v = \frac{1100}{\dfrac{h}{t_w}\sqrt{F_{yw}}} \qquad (11.11.2)$$

2. When $\dfrac{h}{t_w} > \dfrac{1380}{\sqrt{F_{yw}}}$,

$$C_v = \frac{578,000}{(h/t_w)^2 F_{yw}} \qquad (11.11.3)$$

Placement Criteria Including Tension-Field Action—LRFD

When the factored shear V_u exceeds ϕV_n, with $\phi = 0.90$ and V_n given by Eq. 11.11.5 with $k_v = 5$, stiffeners are required. When h/t_w exceeds 260, stiffeners are always required. The use of intermediate stiffeners reduces the a/h ratio and increases V_n. Equation 11.11.5 logically applies for situations *with* and *without* intermediate stiffeners when the objective is to prevent buckling resulting from shear.

Under LRFD-Appendix G3 both buckling strength and post-buckling strength are recognized. The post-buckling behavior, known as *tension-field action,* is similar to truss action as shown in Figs. 11.9.2 and 11.9.3, and the total nominal strength V_n is given by Eq. 11.9.27,

$$V_n = 0.6F_{yw}A_w\left(C_v + \frac{1 - C_v}{1.15\sqrt{1 + (a/h)^2}}\right) \qquad [11.9.27]$$

which is LRFD-Formula (A-G3-2). Since C_v is a function of h/t_w, V_n is a function of both h/t_w and a/h, making evaluation difficult without a design aid. The design shear ϕV_n is tabulated in LRFD "NUMERICAL VALUES" TABLE 10 for $F_y = 36$ and 50 ksi.

While theoretically the only upper limits on h/t_w are those of LRFD-Appendix G1 to prevent vertical buckling of the flange, practical considerations relating to fabrication, handling, and erection [11.14] give rise to the traditional ASD restriction

$$\frac{a}{h} \le \left(\frac{260}{h/t_w}\right)^2 \le 3.0 \qquad (11.11.10)$$

LRFD-Appendix G3 does not allow use of tension-field action when Eq. 11.11.10 is exceeded; it also requires using $k_v = 5$ when the above a/h limit is exceeded. The *Proposed Criteria for Load and Resistance Factor Design of Steel Building Structures* [11.15], the source for the AISC LRFD plate girder provisions, indicates Eq. 11.11.10 as a limit. The LRFD "NUMERICAL VALUES" TABLE 10 also uses Eq. 11.11.10 as a limit. The authors recommend not exceeding the a/h limit of Eq. 11.11.10 even if the Specification may permit it.

Placement Criteria Including Tension-Field Action—ASD

As discussed in Sec. 11.9, the basis for strength to include tension-field action is the same for ASD and LRFD. For Allowable Stress Design Eq. 11.9.33 is used for intermediate stiffener placement when tension-field action is included,

$$F_v = \frac{F_y}{2.89}\left(C_v + \frac{1 - C_v}{1.15\sqrt{1 + (a/h)^2}}\right) \qquad [11.9.33]$$

which is ASD-Formula (G3-1). In addition, Eq. 11.11.10 is the a/h limit of ASD-F5.

End Panels and Interior Panels Having an Opening in an Adjacent Panel

Figure 11.9.5 shows that at the junction of intermediate stiffener and flange, equilibrium requires an axial tension to develop in the flange of the *adjacent* panel. When no such flange is available, as in an end panel, the tension-field cannot adequately develop. Both LRFD and ASD, therefore, consider that only buckling strength (*no tension-field action*) is available in that end panel. Thus, for end panels (that is, panels having *no adjacent panel* and interior panels having a significant hole in an adjacent panel):

1. *Load and Resistance Factor Design* (LRFD-Appendix G3)

$$V_n = C_v(0.6F_{yw})A_w \qquad [11.11.5]$$

where C_v is given by Eqs. 11.11.2 and 11.11.3.

2. *Allowable Stress Design* (ASD-F4)

$$F_v = \frac{F_y C_v}{2.89} \le 0.40F_y \qquad [11.11.7]$$

where C_v is given by Eqs. 11.11.8 and 11.11.9.

Stiffness Requirement

Intermediate stiffeners must be sufficiently rigid to keep the web *at the stiffener* from deflecting out-of-plane when buckling of the web occurs. The stiffener must have a rigidity EI_{st} that is related to the web plate rigidity $Et_w^2a/[12(1 - \mu^2)]$.

1. *Load and Resistance Factor Design* (LRFD-Appendix F2.3)

$$I_{st} \ge jat_w^3 \qquad (11.11.11)$$

where I_{st} = moment of inertia of the cross-sectional area of a transverse stiffener taken about the center of the web thickness when the stiffener consists of a pair of plates, and about the face of the stiffener in contact with the web when single plate stiffeners are used

$$j = \frac{2.5}{(a/h)^2} - 2 \ge 0.5 \qquad (11.11.12)$$

2. *Allowable Stress Design* (ASD-G4)

$$I_{st} \ge \left(\frac{h}{50}\right)^4 \qquad (11.11.13)$$

Various theoretical relationships have been developed for the ratio γ_0 of the stiffener rigidity to one panel of web plate rigidity, which may be expressed

$$\gamma_0 = \frac{EI_{st}}{Da} = \frac{EI_{st}[12(1 - \mu^2)]}{Et_w^3a} \qquad (11.11.14)$$

where I_{st} = optimum stiffener moment of inertia
$D = Et_w^3/[12(1 - \mu^2)]$ = flexural rigidity per unit length of web plate.

Equation 11.11.11 used in LRFD is essentially the following proposed by Bleich [6.9, p. 417]

$$\gamma_0 = 4\left(\frac{7}{(a/h)^2} - 5\right) \tag{11.11.15}$$

According to Eq. 11.11.14 with $\mu = 0.3$,

$$\text{Required } I_{st} = \frac{\gamma_0 t_w^3 a}{12(1 - \mu^2)} = \frac{\gamma_0 t_w^3 a}{10.92} \tag{11.11.16}$$

Substitution of Eq. 11.11.15 into Eq. 11.11.16 gives

$$I_{st} \geq \left(\frac{2.56}{(a/h)^2} - 1.83\right) a t_w^3 \tag{11.11.17}$$

which was simplified by Vincent [11.16] for use in AASHTO Load Factor Design for Bridges to become Eq. 11.11.11 used in AISC Load and Resistance Factor Design.

In order to rationalize the ASD requirement of Eq. 11.11.13, multiply and divide Eq. 11.11.1 by h^2 giving

$$I_{st} \geq \frac{h^4}{(h/t_w)^2 (a/t_w)} [2.5 - 2(a/h)^2] \tag{11.11.18}$$

In order to compare with the ASD requirement of Eq. 11.11.13, let $h/t = 200$ and a/t_w be 170; consequently, $a/h = 0.85$. Substitution into Eq. 11.11.18 gives

$$I_{st} \leq \left(\frac{h}{53}\right)^4 \tag{11.11.19}$$

The Allowable Stress Design requirement seems in the proper order of magnitude; clearly any rational requirement must include the stiffener spacing a, as does the LRFD requirement.

Strength Requirement

Intermediate stiffeners carry a compression load only after buckling of the web has occurred. As the post-buckling truss-like "tension-field action" increases, the stiffener force increases. The maximum force P_s in the stiffener, reached simultaneously with reaching the nominal shear strength V_n, is given by Eq. 11.9.22. The stiffener area required will be the force P_s divided by the yield stress F_{yst} of the stiffener steel, as follows:

1. *Load and Resistance Factor Design* (LRFD-Appendix G4).

$$\text{Required } A_{st} = \frac{F_{yw}}{F_{yst}}\left(0.15 D A_w (1 - C_v)\frac{V_u}{\phi V_n} - 18 t_w^2\right) \tag{11.9.29}$$

2. *Allowable Stress Design* (ASD-G4).

$$\text{Required } A_{st} = \frac{1 - C_v}{2}\left(\frac{a}{h} - \frac{(a/h)^2}{\sqrt{1 + (a/h)^2}}\right)\frac{F_{yw}}{F_{yst}}DA_w \qquad [11.9.35]$$

where $\quad C_v$ = values from Eqs. 11.11.2 and 11.11.3 for LRFD

$\qquad\qquad$ = values from Eqs. 11.11.8 and 11.11.9 for ASD

$\qquad\quad D$ = factor to account for eccentric loading on stiffeners

$\qquad\qquad$ = 1.0 for stiffeners in pairs on each side of web

$\qquad\qquad$ = 1.8 for single angle stiffeners

$\qquad\qquad$ = 2.4 for single plate stiffeners

$\qquad F_{yst}$ = specified yield stress for stiffener steel

$\qquad F_{yw}$ = yield stress of web steel

Equation 11.9.35 for required A_{st} is based on the assumption that the panel is loaded to its strength; when full strength of a panel is not required the area A_{st} required may be reduced proportionally.

Connection to Web

When tension-field action is utilized, the force P_s = required A_{st} times F_{yst} must be transferred between the web and the stiffener. Basler [11.3] recommended the force P_s be considered transferred over one-third the girder height. On that basis, Basler [11.3] recommended that intermediate stiffeners be designed to provide a shear *flow* (kips/in.) strength f_{nv} given by

$$f_{nv} = 0.045h\sqrt{\frac{F_{yw}^3}{E}} \qquad (11.11.20)$$

The Specifications [1.5,1.16] require:

1. *Load and Resistance Factor Design.* There is no LRFD shear flow requirement, though one could use Eq. 11.11.20, considering f_{nv} as ϕR_{nw} for fillet welds (see Chap. 5). In general when intermittent fillet welds are used, the minimum segment length and minimum clear between segments will likely be sufficient (LRFD-J2.2).

2. *Allowable Stress Design* (ASD-G4). Taking the shear strength flow requirement f_{nv} of Eq. 11.11.20, and dividing by the factor 1.65 to put the equation into the service load range so that the shear flow becomes the service load value f_{vs}, gives

$$f_{vs} = h\sqrt{\left(\frac{F_{yw}, \text{ksi}}{340}\right)^3} \qquad (11.11.21)*$$

which is ASD-Formula (G4-3).

---

*For SI units, with F_y in MPa, h in mm, and f_{vs} in kN/m,

$$f_{vs} = h\sqrt{\left(\frac{F_{yw}}{650}\right)^3} \qquad (11.11.21)$$

For both LRFD and ASD, h is in inches, and f_{nv} (for LRFD) or f_{vs} (for ASD) are kips/in. As with the area requirement, when panels adjacent to the stiffener are not loaded to their full strength, the shear flow used as the connection requirement may be reduced in proportion that the strength required is below that provided.

Connection to Flanges

Intermediate stiffeners are provided to assist the web; to stiffen and create nodal lines during buckling of the web and to accept compression forces transmitted directly from the web. At the compression flange, welding of the stiffener across the flange as shown in Fig. 11.11.1 provides stability to the stiffener and holds it perpendicular to the web; in addition, such welding provides restraint against torsional buckling (Fig. 11.3.1b) of the compression flange.

On the tension flange, the effects of stress concentration increase the fatigue or brittle fracture possibilities, i.e., welding in no way helps the tension flange. Since the work of Basler [11.17] has shown that welding of stiffeners to the tension flange is unnecessary for proper functioning of stiffeners, LRFD-Appendix F2.3 and ASD-G4 permit stopping stiffeners "short of the tension flange provided bearing is not needed to transmit a concentrated load or reaction." The weld by which the stiffener is attached to the web "shall be terminated not closer than 4 times the web thickness nor more than 6 times the web thickness from the near toe of the web-to-flange weld."

For situations where the stiffener serves as the attachment for lateral bracing, the welding to the compression flange should be designed to transmit 1 percent (rule-of-thumb) of the compressive force in the flange. For important lateral bracing design in situations involving long unsupported lengths, the strength of lateral bracing connections should be designed using the principles of Sec. 9.13.

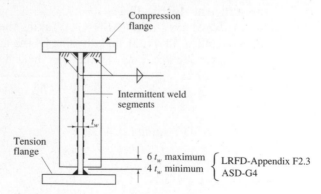

Figure 11.11.1   Intermediate stiffener connection to flange.

11.12 BEARING STIFFENER DESIGN

Concentrated loads, such as at unframed end reactions, must be carried by stiffeners placed in pairs. Whenever concentrated loads, such as end reactions or columns supported by plate girders, exceed the local web yielding, web crippling, or sidesway web buckling strengths, bearing stiffeners must be provided. Local web yielding and web crippling were discussed in Sec. 7.8 since they are also of concern on rolled beams. Local web yielding (formerly called web crippling) is provided for in LRFD and ASD-K1.3; web crippling (formerly called web buckling) is in LRFD and ASD-K1.4. Sidesway web buckling generally is of concern only on narrow flange plate girders and is in LRFD and ASD-K1.5. These three phenomena all are related to the strength of a thin web in the vicinity of concentrated loads.

Bearing stiffeners, unlike intermediate stiffeners, should be close fitting at the bearing end, and when the concentrated load is compression against the flange, the stiffener may either be connected to, or bear against, the flange transmitting the concentrated load. When the concentrated load is a tension pull on the flange, the stiffener must be attached to the flange being pulled. For plate girders the usual situation is compression against the flange. In general, compression load transmitting bearing stiffeners should extend "approximately to the edge of the flange plates . . ." according to the 1978 AISC Specification (ASD-1.10.5). There is no such requirement in current Specifications, but the authors believe it to be good practice.

Column Stability Criterion

Bearing stiffeners transmitting compression loads are designed as columns under the provisions of LRFD-K1.8 and K1.9 and ASD-K1.8. The column consists of the stiffeners, plus a portion of the web tributary to them, as defined in LRFD-K1.9 and ASD-K1.8, and shown in Fig. 11.12.1.

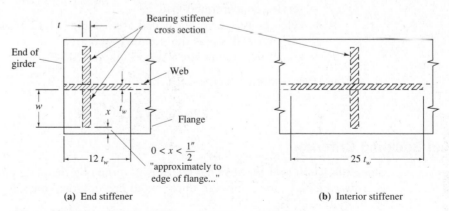

(a) End stiffener   **(b)** Interior stiffener

Figure 11.12.1 Bearing stiffener effective cross-sections.

The effective length KL of the "column" is less than the depth h of the web plate because of the restraint provided by the flanges. The effective length KL, according to LRFD-K1.9 and ASD-K1.8, is to be taken "equal to $0.75h$".

The slenderness ratio is computed,

$$\frac{KL}{r} = 0.75\frac{h}{r} \qquad (11.12.1)$$

where h = web plate depth

r = radius of gyration of the shaded portion shown in Fig. 11.12.1 taken about the mid-thickness of the web

The effective area A_e required is then computed using the column strength P_n in accordance with LRFD-E2, or the allowable stress F_a for a column in accordance with ASD-E2.

Load and Resistance Factor Design.   The strength requirement is

$$\phi_c P_n \geq P_u \qquad (11.12.2)$$

where ϕ_c = resistance factor = 0.85

$P_n = F_{cr}A_e$

P_u = factored concentrated compression load

F_{cr} = column buckling stress according to LRFD-E2

A_e = column area; i.e., the shaded area of Fig. 11.12.1 which includes the stiffener plates plus the tributary web area.

Thus, the required effective area A_e is

$$\text{Required } A_e = \frac{P_u}{\phi_c F_{cr}} \qquad (11.12.3)$$

Allowable Stress Design.   The allowable stress requirement is

$$f_a \leq F_a \qquad (11.12.4)$$

where $f_a = P/A_e$

P = service concentrated load

A_e = column area; i.e., the shaded area of Fig. 11.12.1 which includes the stiffener plates plus the tributary web area

F_a = allowable column stress according to ASD-E2

Thus, the required effective area A_e is

$$\text{Required } A_e = \frac{P}{F_a} \qquad (11.12.5)$$

Local Buckling Criterion

Since the width w of the stiffener plates is governed by the plate girder flange width (see Fig. 11.12.1), the minimum thickness to prevent local buckling is

$$\text{Min } t = \frac{w}{95/\sqrt{F_y}} \qquad (11.12.6)$$

as governed by $\lambda \le \lambda_r$ for a uniformly stressed unstiffened compression element according to LRFD-B5 or to satisfy the "noncompact" limit of ASD-B5. When the thickness exceeds the limit of Eq. 11.12.6, the "slender" compression element will have reduced efficiency (i.e., $Q < 1$ as discussed in Sec. 6.18) and must be treated in accordance with LRFD-Appendix B5.3a or ASD-Appendix B5.2. Because a bearing stiffener is an important element of a plate girder, the authors recommend satisfying Eq. 11.12.6 so that $Q = 1$.

The wording of 1993 LRFD-K1.9(2) mandates satisfying Eq. 11.12.6; thus, designing for $Q < 1$ is no longer an option.

Bearing Criterion

In order to bring bearing stiffener plates tight against the flanges, one corner of each stiffener plate must be cut off so as to clear the flange-to-web fillet weld. The remaining area of direct bearing is less than the gross area of the stiffener plates. The strength in bearing under LRFD-J8.1 or the service load stress in bearing under ASD-J8.1 must be satisfactory.

Load and Resistance Factor Design. The bearing requirement of LRFD-J8 is

$$\phi R_n \ge P_u \tag{11.12.7}$$

where $\phi = 0.75$
R_n = nominal bearing strength = $1.8F_y A_{pb}$ (11.12.8)
P_u = factored concentrated load
A_{pb} = contact area of stiffener bearing against the flange

Allowable Stress Design. The bearing requirement of ASD-J8.1 is

$$f_p \le 0.90F_y \tag{11.12.9}$$

where $f_p = P/A_{pb}$
P = service concentrated load
A_{pb} = contact area of stiffener bearing against the flange

11.13 LONGITUDINAL WEB STIFFENERS

Longitudinal stiffeners, as shown in Fig. 11.13.1, can increase the bending and shear strengths of a plate girder. In general, they are not as effective as transverse stiffeners; however, they are frequently desired on highway bridge girders for esthetic reasons. Studies of longitudinal stiffener effectiveness, as related to stiffener size and location, have been made by Cooper [11.18, 11.19] and others at Lehigh University. These studies and others are summarized in the *SSRC Guide* [6.8, pp. 211–223] and by Bleich [6.9, pp. 418–423]. The ASCE-AASHTO Task Committee [11.20] provides a full review of the theory and design of longitudinally stiffened plate girders.

The principal use of longitudinal stiffeners is in highway bridge design, where transverse stiffeners are used on both sides of a steel girder except on the exterior side of the exterior girder for more pleasing appearance. Rarely are longitudinal stiffeners used on *both* sides of a web as shown in Fig. 11.13.1.

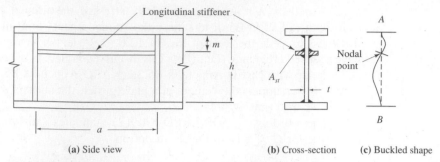

(a) Side view   **(b)** Cross-section   **(c)** Buckled shape

Figure 11.13.1   Effect of longitudinal stiffener on plate girder web stability.

The primary function of longitudinal stiffeners is to control lateral web deflections [11.20], and hence the bend-buckling strength as discussed in Sec. 11.6.

As discussed in Sec. 11.6, the elastic buckling strength of the web plate in bending (Fig. 11.6.1) may be written

$$F_{cr} = \frac{\pi^2 E k}{12(1 - \mu^2)(h/t)^2} \tag{11.13.1}$$

If the plate is stiffened by a longitudinal stiffener, as shown in Fig. 11.13.1, the value of k will be significantly greater than for the unstiffened case. The stiffener used should be stiff enough so that when buckling occurs, a nodal line will be formed along the line of the stiffener.

Under bending alone, the value of the buckling coefficient k has been found to be as high as 142.6 for the case where the flanges are assumed to provide full restraint to rotation at points A and B of Fig. 11.13.1c and $m = h/5$. For the case where the flanges provide no moment restraint at A and B (simply supported) the stiffener located at $m = h/5$ is also the optimum location. Such stiffener placement in the compression zone serves the purpose of maintaining the full effectiveness of the web in resisting bending stress, which is really the stiffener's principal function.

For webs subjected to shear alone, the longitudinal stiffener should be located at mid-height. For combined shear and bending the stiffener should be located so that $h/5 < m < h/2$; because of its principal function, however, it should preferably be closer to $h/5$.

For design there are two requirements: (1) a moment of inertia to insure adequate stiffness to create a nodal line along the stiffener, and (2) an area adequate to carry axial compression force while acting integrally with the web.

The design requirement for stiffness can be expressed as a function of the rigidity of the web, using the same approach as discussed for transverse stiffeners. Substituting the web height h for the transverse stiffener spacing a in Eq. 11.11.16 gives

$$\text{Required } I_{st} = \frac{\gamma_0 t_w^3 h}{10.92} \tag{11.13.2}$$

The AISC Specifications [1.5, 1.16] give no information regarding longitudinal stiffeners. For highway bridges, AASHTO-10.48.6.3 [1.3] gives the following expression for Load Factor Design:

$$\text{Required } I_{st} = t_w^3 h \left[2.4 \left(\frac{a}{h} \right)^2 - 0.13 \right] \tag{11.13.3}$$

The moment of inertia I_{st} is to be that of the stiffener plate(s) combined with a centrally located web strip not more than $18t_w$ in width.

The location of the longitudinal stiffener shall be at $m = h/5$, and the local buckling requirement (AASHTO-10.48.6) is

$$\text{Min } t = \frac{w}{82/\sqrt{F_y}, \text{ ksi}} \tag{11.13.4}$$

In addition, the radius of gyration r of the stiffener combined with a centrally located web strip not more than $18t_w$ in width shall be at least

$$\text{Min } r \geq \frac{a}{727/\sqrt{F_y}, \text{ ksi}} \tag{11.13.5}$$

Note the AASHTO coefficients have been converted from their stated values to accommodate using F_y in ksi as used throughout this text instead of psi as in AASHTO.

11.14 PROPORTIONING THE SECTION

The cross-section of a girder must be selected such that it adequately performs its functions and requires minimum cost. The function requirements may be summarized as:

1. Strength to carry bending moment (adequate section modulus S_x).
2. Vertical stiffness to satisfy any deflection limitations (adequate moment of inertia I_x).
3. Lateral stiffness to prevent lateral-torsional buckling of compression flange (adequate lateral bracing or low L_b/r_T).
4. Strength to carry shear (adequate web area).
5. Stiffness to improve buckling or post-buckling strength of the web (related to h/t and a/h ratios).

To satisfy these function requirements at minimum cost, it will be assumed in what follows that minimum cost is equivalent to minimum weight.

Flange-Area Formula

For simplicity in design it is convenient to replace the real system of Fig. 11.14.1a with a substitute system, Fig. 11.14.1b, which allows the moment to be replaced by a couple

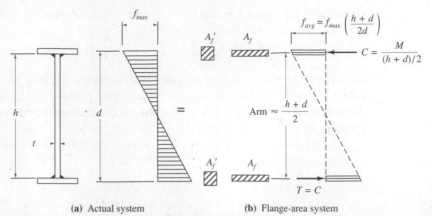

(a) Actual system   **(b)** Flange-area system

Figure 11.14.1   Flange-area formula development.

with the forces of the couple acting at the flange centroids. The forces can then be treated as axial load situations. If the distance between flange forces is approximately $(h + d)/2$, the forces of the couple are

$$C = T = \frac{M}{(h + d)/2} \qquad (11.14.1)$$

The effective area on which these forces act is equal to the flange plate area A_f plus additional area A_f' to represent the effectiveness of the web in resisting moment.

The average stress on the total effective area is

$$f_{avg} = \frac{\text{Force}}{\text{Area}} = \frac{M}{(h + d)/2}\left(\frac{1}{A_f + A_f'}\right) \qquad (11.14.2)$$

The area A_f' must be taken such that the bending moment carried by the web is the same for both the real and substitute systems:

$$M_{\text{real system}} = f_{max}\left(\frac{h}{d}\right)\frac{th^2}{6} \qquad (11.14.3)$$

$$M_{\text{substitute system}} = f_{max}\left(\frac{h + d}{2d}\right)A_f'\left(\frac{h + d}{2}\right) \qquad (11.14.4)$$

Equating Eqs. 11.14.3 and 11.14.4 gives

$$A_f' = \frac{h}{d}\left(\frac{th^2}{6}\right)\left(\frac{2d}{h + d}\right)\left(\frac{2}{h + d}\right) = \frac{th}{6}\left(\frac{2h}{h + d}\right)^2 \qquad (11.14.5)$$

which if $A_w = th$, and the squared term is neglected, becomes

$$A_f' = \frac{A_w}{6} \qquad (11.14.6)$$

Next, solving Eq. 11.14.2 for A_f gives

$$A_f = \frac{M}{[(h + d)/2]f_{avg}} - A_f' \qquad (11.14.7)$$

which, using Eq. 11.14.5 and $f_{avg} = f_{max}(h + d)/2d$, gives

$$A_f = \left[\frac{M}{f_{max}h}\left(\frac{d}{h}\right) - \frac{A_w}{6}\right]\left(\frac{2h}{h + d}\right)^2 \tag{11.14.8}$$

Letting the squared term equal unity overestimates slightly the value of A_f, while letting $d/h = 1$ underestimates the value. For preliminary design purposes these simplifications are justified to give a simple expression for the required area of one flange plate,

$$A_f = \frac{M}{fh} - \frac{A_w}{6} \tag{11.14.9}$$

In the use of Eq. 11.14.9, if f is taken as the average stress on the flange, the d/h term will be nearly accounted for. When checking a section, of course, the correct moment of inertia must be obtained and the maximum strength for LRFD or the stress for ASD computed.

Optimum Girder Depth

The variation in girder cross-sectional area is to be examined as a function of web depth to determine the depth which will give minimum area. Extended treatment of this subject has been given by Shedd [11.1], Bresler, Lin, and Scalzi [11.21] and Blodgett [11.22]. Schilling [11.23], Azad [11.24], Fleischer [11.25], and Anderson and Chong [11.26] have provided extended treatment of optimum proportioning, including the hybrid girder having a web of lower yield strength than the flanges.

The average gross area A_g of the girder for the entire span may be expressed

$$A_g = 2C_1A_f + C_2ht \tag{11.14.10}$$

where C_1 = factor to account for reducing flange size at regions of lower than
maximum moment
C_2 = factor to account for reducing web thickness at regions of reduced shear

Substituting Eq. 11.14.9 into Eq. 11.14.10 gives

$$A_g = 2C_1\left(\frac{M}{fh} - \frac{ht}{6}\right) + C_2ht \tag{11.14.11}$$

To find the minimum average gross area,

$$\frac{\partial A_g}{\partial h} = 0 \tag{11.14.12}$$

(a) *Case 1.* No depth restriction; desire large h/t. Assume β_w = constant = h/t; $t = h/\beta_w$. Equation 11.14.11 becomes

$$A_g = 2C_1\left(\frac{M}{fh} - \frac{h^2}{6\beta_w}\right) + C_2\frac{h^2}{\beta_2} \tag{11.14.13}$$

$$\frac{\partial A_g}{\partial h} = 0 = \frac{-2C_1Mf}{f^2h^2} - \frac{4C_1h}{6\beta_w} + \frac{2C_2h}{\beta_w} \tag{11.14.14}$$

$$0 = -6C_1M\beta_w - 2C_1h^3f + 6C_2h^3f \tag{11.14.15}$$

from which

$$h = \sqrt[3]{\frac{3MC_1\beta_w}{f(3C_2 - C_1)}} \qquad (11.14.16)$$

and if one neglects the section reduction in regions of lower stress, $C_1 = C_2 = 1$, Eq. 11.14.16 becomes

$$h = \sqrt[3]{\frac{3M\beta_w}{2f}} \qquad (11.14.17)$$

where $M = M_u/\phi$ =factored service load moment divided by $\phi = 0.90$ for LRFD, or
   = service load moment moment (unfactored) for ASD
   f = average stress on flange using F_{cr} as extreme fiber value when nominal moment strength is achieved according to LRFD-Appendix G2 for LRFD, or
   = average stress on flange using F_b as extreme fiber allowable value according to ASD-G2 for ASD

Using Eq. 11.14.13 with $C_1 = C_2 = 1$ and substituting for M/f from Eq. 11.14.17 gives

$$A_g = \frac{4h^2}{3\beta_w} - \frac{h^2}{3\beta_w} + \frac{h^2}{\beta_w} \doteq \frac{2h^2}{\beta_w} \qquad (11.14.18)$$

from which the girder weight per foot can be estimated using the fact that steel weight is 3.4 lb/sq in./linear ft (0.00784 kg/mm²/linear metre).

$$\text{lb/ft} = 3.4A_g = \frac{6.8h^2}{\beta_w} = 8.9\sqrt[3]{\frac{M^2}{f^2\beta_w}} \qquad (11.14.19)*$$

using inch units for the variables. Stiffeners will generally increase this value by 5 to 10 percent. M and f are defined for LRFD and ASD following Eq. 11.14.17.

(b) *Case 2.* Minimum web thickness; $t =$ const. Differentiating Eq. 11.14.11, $\partial A_g/\partial h = 0$, gives

$$\frac{-2C_1 Mf}{f^2 h^2} - \frac{C_1 t}{3} + C_2 t = 0 \qquad (11.14.20)$$

$$h = \sqrt{\frac{6C_1 M}{ft(3C_2 - C_1)}} \qquad (11.14.21)$$

---

*For SI units, the mass per metre is

$$\text{kg/m} = 1.72\sqrt[3]{\frac{M^2}{f^2\beta_w}} \qquad (11.14.19)$$

$$\text{kg/m} = 0.0181\sqrt{\frac{Mt}{f}} \qquad (11.14.23)$$

using mm units for the variables.

If $C_1 = C_2 = 1$,

$$h = \sqrt{\frac{3M}{ft}} \qquad (11.14.22)$$

where M and f are defined for LRFD and ASD following Eq. 11.14.17. The weight per foot can be obtained using A_g from Eq. 11.14.11,

$$\text{lb/ft} = 3.4A_g = 4.53ht = 7.85\sqrt{\frac{Mt}{f}} \qquad (11.14.23)*$$

using inch units for the variables. M and f are defined, as before, for LRFD and ASD following Eq. 11.14.17. Again, an estimate for the weight of stiffeners should be added to the equation value.

(c) *Case 3*. Heavy shear which governs web area; $A_w = \text{constant} = \text{web area}, ht$. Equation 11.14.11 becomes

$$A_g = 2C_1\left(\frac{M}{fh} - \frac{A_w}{6}\right) + C_2A_w \qquad (11.14.24)$$

from which it is apparent that minimum A_g results from maximum depth h. This case usually does not govern.

If the same kind of steel is used throughout, the value of C_1 may vary from 0.7 to 0.9 when used with the maximum positive moment: 0.85 to 0.90 is the usual range. The value of C_2 is not as likely to vary except on continuous structures where it might be 1.05 when used with maximum positive moment, or 0.95 when used with maximum negative moment. Because of the complexity of evaluating C_1 and C_2 for continuous structures, it might be well to take them as unity.

Flange Plate Changes in Size

It is usually economical to reduce the size of flange plates in the region of low moment. While no specific rules can be made to help the designer to determine when it is desirable to change flange plate size, certain simple relationships are possible if only one change in flange size is desired.

(a) *Case 1*. Linear variation in moment—two flange plate sizes. Consider the situation of Fig. 11.14.2a, and assuming both plates are fully utilized for bending moment, the flange-area formula can be used for each plate.

$$A_f = \frac{M}{hf} - \frac{A_w}{6} \qquad (11.14.25)$$

$$A_{f1} = \frac{M(x/L)}{hf} - \frac{A_w}{6} \qquad (11.14.26)$$

The total flange volume in the length L is

$$\begin{aligned}
\text{Vol} &= A_f(L - x) + A_{f1}x \\
&= \frac{M(L - x)}{hf} - \frac{A_w}{6}(L - x) + \frac{M}{hf}\left(\frac{x^2}{L}\right) - \frac{A_w}{6}x \\
&= \frac{M}{hf}\left(\frac{L^2 - xL + x^2}{L}\right) - \frac{A_w L}{6} \qquad (11.14.27)
\end{aligned}$$

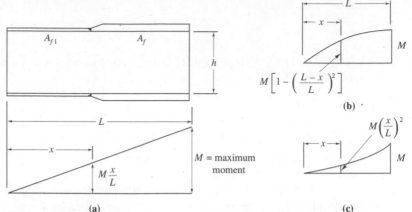

Figure 11.14.2   Common moment variations for determining changes in flange plate size.

For minimum volume,

$$\frac{\partial(\text{Vol})}{\partial x} = 0 = 2x - L; \qquad x = \frac{L}{2}$$

which means

$$\frac{A_{f1}}{A_f} = \frac{\dfrac{M}{2hf} - \dfrac{A_w}{6}}{\dfrac{M}{hf} - \dfrac{A_w}{6}} \approx \frac{1}{2} \qquad\qquad (11.14.28)$$

(b) *Case 2.* Parabolic variation as for uniformly loaded simple beam (see Fig. 11.14.2b). The total volume in the length L is

$$\text{Vol} = \frac{M}{hf}\left(\frac{L^3 - L^2 x + 2Lx^2 - x^3}{L^2}\right) - \frac{A_w L}{6}$$

$$\frac{\partial(\text{Vol})}{\partial x} = 0 = x^2 - \frac{4}{3}Lx + \frac{L^2}{3}; \qquad x = \frac{L}{3}$$

and

$$\frac{A_{f1}}{A_f} \approx \frac{5}{9} \qquad\qquad (11.14.29)$$

(c) *Case 3.* Parabolic variation as for uniformily loaded cantilever (see Fig. 11.14.2c). The total volume in the length L is

$$\text{Vol} = \frac{M}{hf}\left(\frac{L^2 - L^2 x + x^3}{L^2}\right) - \frac{A_w L}{6}$$

$$\frac{\partial(\text{Vol})}{\partial x} = 0 = 3x^2 - L^2; \qquad x = \frac{L}{\sqrt{3}}$$

and

$$\frac{A_{f1}}{A_f} \approx \frac{1}{3} \qquad\qquad (11.14.30)$$

The foregoing developments can provide a guide for change of plate sizes. Since making a change involves a groove welded butt joint of the flange plates, enough material must be saved to more than offset the welding cost.

As a rule of thumb, unless 200 to 300 lb of material are saved in a flange plate per added splice the added cost of the butt splice (considering a plate about 2 ft wide and 2 in. thick) is not justified.

Flange Plate Proportions

According to most of the theory developed earlier in this chapter, the flange plate can be any width and thickness as long as it contributes to the girder the properties necessary to satisfy the functional requirements. However, nearly all testing has used flange plate dimensions which were considered reasonable.

In order to assist the engineer who is unsure of what constitutes such reasonable dimensions, the following guidelines are suggested.

1. Typically the ratio of girder flange width to girder depth, b_f/d, varies from about 0.3 for shallow girders to about 0.2 for deep girders.
2. Plate widths should be in 2-in. increments.
3. Plate thickness increments should be as follows:

   $$\frac{1}{16} \text{ in.} \qquad t \le \frac{9}{16} \text{ in.}$$
   $$\frac{1}{8} \text{ in.} \qquad \frac{5}{8} = t \le 1\frac{1}{2} \text{ in.}$$
   $$\frac{1}{4} \text{ in.} \qquad t > 1\frac{1}{2} \text{ in.}$$

4. Where lateral stability is of concern, the flange plate width-to-thickness ratios $b_f/2t_f$ should be kept at about the λ_p value of LRFD-B5, or at about one-half to two-thirds of the $b_f/2t_f$ "noncompact" limit in ASD-B5, *in the maximum moment regions*. This will permit reducing the thickness of flange plates in regions of low moment. For such cases, flange plate area reduction should be made by reducing the thickness rather than the width.
5. For laterally stable girders, the flange plate area reduction in regions of lower moment may be accomplished by reducing the thickness, reducing the width, or reducing both thickness and width. A slight advantage in fatigue strength accrues by reducing the width rather than the thickness [11.22]. The transition slope should not exceed 1 in $2\frac{1}{2}$ for either width or thickness, and is usually 1 in 4 to 1 in 12 for the transition in width [11.22].

11.15 PLATE GIRDER DESIGN EXAMPLE—LRFD

Partially design a two-span continuous welded plate girder to support uniform load w of 0.8 kips/ft dead load and 3.2 kips/ft live load, plus two fixed position concentrated loads W of 15 kips dead load and 60 kips live load in each span as shown in

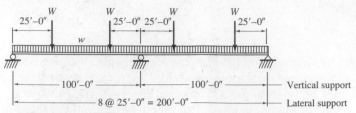

Figure 11.15.1   Girder loading and support for design example.

Fig. 11.15.1. Lateral support is provided at each support and every 25 ft between supports. The girder is to have constant depth web plate for the two spans. Use A36 steel in the positive moment zone and A572 Grade 50 for the negative moment zone. Use Load and Resistance Factor Design.

Additional specifications and general comments:

1. The live load is to be applied in its correct manner; that is, applied as necessary to obtain the maximum and minimum moments and shears at every location along the girder. This will mean two live load cases for the factored bending moment envelope as shown in Fig. 11.15.2. For the shear envelope, the maximum values at the exterior and interior supports are obtained from the two bending moment loading cases. For all other shear envelope values, the live load was placed using partial loading in the manner dictated by the shear influence lines.[*]
2. Assume $\frac{5}{16}$ in. is minimum practical web plate thickness.
3. Assume no depth restriction; also, that any deep thin-web girder that is selected can be feasibly fabricated, transported to the construction site, and erected without excessive difficulty or cost.

Solution

(a) *Factored Loads.* Using the basic factored gravity load combination of LRFD-A4.1,

$$\text{Uniform live load } w_u = 1.6(3.2) = 5.12 \text{ kips/ft}$$
$$\text{Uniform dead load } w_u = 1.2(0.8) = 0.96 \text{ kips/ft}$$
$$\text{Concentrated live load } W_u = 1.6(60) = 96 \text{ kips}$$
$$\text{Concentrated dead load } W_u = 1.2(15) = 18 \text{ kips}$$

(b) *Structural Analysis.* Any method of statically indeterminate structural analysis can be used to obtain the elastic moment and shear envelopes under factored loads. The results are presented in Fig. 11.15.2.

(c) *Estimate of Weight.* The maximum factored moments due to the superimposed loads (that is, without girder weight) of Fig. 11.15.1 for the positive and negative moment regions are

$$+M_u = +7640 \text{ ft-kips}$$
$$-M_u = -10,800 \text{ ft-kips}$$

---

[*] See Chu-Kia Wang and Charles G. Salmon, *Introductory Structural Analysis*. Englewood Cliffs, NJ: Prentice-Hall, Inc., 1984, pp. 258–275.

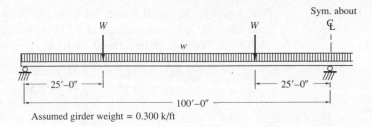

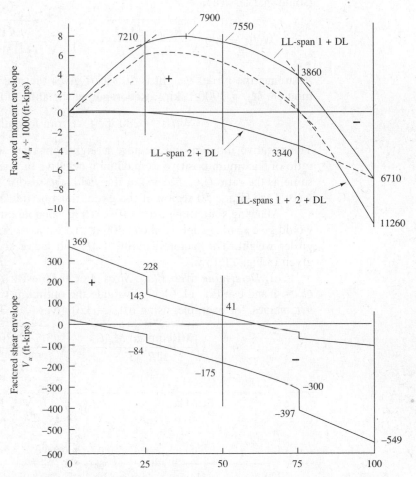

Figure 11.15.2   Factored moment and factored shear envelopes for two-span continuous beam of illustrative example.

Since there is no depth limitation, the depth based on maximum h/t_w may be desired. The LRFD-Appendix G1 or ASD-G1 limits are obtained, referring to Table 11.3.1,

$$\text{Max } h/t_w = 322 \quad (333 \text{ when } a/h \leq 1.5) \qquad \text{A36 steel}$$

$$\text{Max } h/t_w = 243 \quad (283 \text{ when } a/h \leq 1.5) \qquad \text{A572 Grade 50}$$

There will be strength reduction from the bend-buckling limit state when h/t_w exceeds $970/\sqrt{F_{cr}}$ according to LRFD-Appendix G2, or $760/\sqrt{F_b}$ according to ASD-G2. For this design those limits are

$$h/t_w > 162 \qquad \text{A36 when } F_{cr} = F_y$$
$$h/t_w > 137 \qquad \text{A572 Grade 50 when } F_{cr} = F_y$$

For a weight estimate, try $\beta_w = h/t_w = 300$ and use Eq. 11.14.19, $M =$ Required $M_n = M_u/\phi_b = 7640/0.90 = 8490$ ft-kips, and $f = R_{PG}F_{cr} \approx 34$ ksi (somewhat less than F_y):

$$\text{Wt/ft} = 8.9\sqrt[3]{\frac{(\text{Required } M_n)^2}{(R_{PG}F_{cr})^2 \beta_w}} = 8.9\sqrt[3]{\frac{[8490(12)]^2}{(34)^2 300}} = 276 \text{ lb/ft}$$

Assuming the girder weight is 300 lb/ft gives the maximum factored positive moment $M_u = 7900$ ft-kips. Recomputing the above formula gives

$$\text{Wt/ft} \approx 280 \text{ lb/ft} \quad \text{(from formula)}$$

The negative moment will require a slightly heavier section even though the ratio of maximum positive to maximum negative moment is approximately the same as the ratio (i.e., 50/36) of the yield stresses for the materials used. The web of the Grade 50 region of the girder must be thicker than the A36 region.

Allowing something extra (10% is a reasonable estimate) for the stiffeners would give a value slightly above 300 lb/ft. Use $w = 300$ lb/ft as the estimated girder weight. The factored moment M_u and factored shear V_u envelopes are given in Fig. 11.15.2.

(d) *Determine Web Plate Sizes.* For $+M_u$ with A36 steel assume $C_1 = C_2 = 1$ and use Eq. 11.14.17. Evaluate the optimum value for h using various h/t_w values, for example, using $h/t_w = 320$ gives

$$h = \sqrt[3]{\frac{3(\text{Required } M_n)\beta_w}{2R_{PG}F_{cr}}} = \sqrt[3]{\frac{3(7900/0.90)(12)320}{2(34)}} = 114 \text{ in.}$$

$\dfrac{h}{t_w} = \beta_w$	Formula h (in.)	h (in.)	t (in.)	A_w (sq in.)	$\dfrac{V_u}{A_w}$ (ksi)	Actual $\dfrac{h}{t}$
320	114.1	114	3/8	42.8	8.6	304
333	115.7	114	3/8	42.8	8.6	304
Max for 3/8 in.		125.0 using $h/t_w = 333$				
		124	3/8	46.5	7.9	331
Max for 3/8 in.		120.8 using $h/t_w = 322$				
		120	3/8	45.0	8.2	320
Max for 5/16 in.		104.2 using $h/t_w = 333$				
		104	5/16	32.5	11.4	333
Max for 5/16 in.		100.6 using $h/t_w = 322$				
		100	5/16	31.3	11.8	320

Based on the positive moment requirement, the economical depth appears at first to be 114 in. from the formula. However, if the lightest weight girder is desired, the web area must be used efficiently. Once the $\frac{3}{8}$ in. thickness is obtained from the 114 in. depth, the maximum depth for a $\frac{3}{8}$-in. web should be determined, which gives an h of 120 to 124 in. This is considerably deeper than the formula value. Also, there is considerable reserve shear strength; V_u/A_w will typically be satisfactory as high as 12 to 16 ksi. Note that $\phi V_n/A_w$ can be about 16 ksi for $h/t_w = 320$ according to LRFD TABLE 10-36. With the design factored shear stress only about 8 to 9 ksi with the $\frac{3}{8}$-in. web, it is probable that a lighter girder will result using a thinner (say $\frac{5}{16}$ in.) web. The $\frac{5}{16}$-in. web would indicate a girder shallower than 114 in., say around 100 in.

Evaluate optimum h for the negative moment region initially using Eq. 11.14.17 with $M = M_u/\phi = 11260/0.90$ and $h/t_w = 240$.

$\dfrac{h}{t_w} = \beta_w$	Formula h (in.)	h (in.)	t (in.)	A_w (sq in.)	$\dfrac{V_u}{A_w}$ (ksi)	Actual $\dfrac{h}{t}$
240	104.0	104	7/16	45.5	12.1	238
283	109.9	108	7/16	47.3	11.6	247
Max for 7/16 in.		123.7 using $h/t_w = 283$				
		122	7/16	53.4	10.3	279
Max for 7/16 in.		106.2 using $h/t_w = 243$				
		106	7/16	46.4	11.8	242
Max for 3/8 in.		106.1 using $h/t_w = 283$				
		106	3/8	39.8	13.8	283
Max for 3/8 in.		91.0 using $h/t_w = 243$				
		90	3/8	33.8	16.3	240

On the basis of these preliminary computations, a depth of 100 in. is chosen. The formula indicates somewhat deeper; however, the low shear indicates that the web can be thinner. Since intermediate stiffeners are to be used, the reader may note that when h/t_w ratios are larger the stiffener spacing requirement over most of the girder length will be controlled by Eq. 11.11.10 and will not be dependent on the magnitude of the shear force. Thus the deeper the girder, the longer must be the stiffener plates, with little advantage to offset the extra weight. Generally, multiples of 2 in. should be used for the web plate depth.

Combined shear and moment strength (LRFD-Appendix G5 and ASD-G5) must be considered in regions where high shear and high moment simultaneously occur. In this design, such a location is adjacent to the interior support. The factored shear V_u should be kept somewhat below the design shear strength ϕV_u so that the factored moment M_u can equal ϕM_n.

Allowing for some flexibility in the design, use $h = 100$ in. which will mean that the web slenderness h/t_w is near but not at the upper limit. Try

$$\frac{5}{16} \times 100 \qquad (h/t_w = 320) \text{ for } +M, \qquad F_y = 36 \text{ ksi}$$
$$\frac{3}{8} \times 100 \qquad (h/t_w = 267) \text{ for } -M, \qquad F_y = 50 \text{ ksi}$$

(e) *Select Flange Plates for Negative Moment.* $M_u = 11,260$ ft-kips, web plate $= \frac{3}{8} \times 100$ ($A_w = 37.5$ sq in.). The lateral-torsional buckling limit state must be examined with regard to the 25-ft distance between lateral supports. The slenderness parameter λ involves the radius of gyration r_T, as discussed following Eq. 11.4.15, and as traditionally controlling in Allowable Stress Design for plate girders. If a typical ratio $b_f/d \approx 0.25$ is assumed, then $b_f = 24$ in. Using the radius of gyration of a rectangle about its mid-depth,

$$r = b_f/\sqrt{12} = 0.289 b_f$$

which gives $r \approx 6.9$ in., say 7 in.

$$\text{Estimated } \lambda = \frac{L_b}{r_T} = \frac{25(12)}{7} = 43$$

Check λ_p for the lateral-torsional buckling limit state,

$$\lambda_p = \frac{300}{\sqrt{F_{yf}}} = \frac{300}{\sqrt{50}} = 42.4$$

Since λ slightly exceeds λ_p, the moment strength might be reduced; however, the moment gradient is favorable. Referring to the 25-ft laterally unbraced segment adjacent to the interior support in Fig. 11.15.2, when maximum M_u occurs the moment gradient is closely a linear one varying from 11,260 ft-kips to zero. Thus, $C_b = 1.67$ for that segment when maximum moment is acting. It is highly probable that the lateral-torsional buckling limit state does not affect the strength of this girder.

The high slenderness ratio h/t_w for the web will, however, reduce the moment strength. To obtain an estimated reduction from the bend-buckling limit state, use Eq. 11.4.3 (LRFD-Appendix G2) with estimated h/t_w of 267 and $a_r = A_w/A_f \approx 1.5$,

$$R_{PG} = 1 - \frac{a_r}{1200 + 300a_r}\left(\frac{h}{t_w} - \frac{970}{\sqrt{F_{cr}}}\right) \leq 1.0 \qquad [11.4.3]$$

$$= 1 - \frac{1.5}{1200 + 300(1.5)}\left(267 - \frac{970}{\sqrt{\approx 48}}\right) = 0.88$$

which makes $R_{PG} F_{cr} \approx 43$ ksi.

Using the flange-area formula, Eq. 11.14.9, gives the requirement for one flange as

$$A_f = \frac{M_u/\phi_b}{R_{PG} F_{cr} h} - \frac{A_w}{6} = \frac{11,260(12)/0.90}{43(101)} - \frac{37.5}{6} = 28.3 \text{ sq in.}$$

Some possible choices considering 43 ksi to be a low estimate:

$$1\frac{1}{4} \times 22, \qquad A_f = 27.5 \text{ sq in.}, \qquad b_f/2t_f = 8.8$$
$$1\frac{1}{8} \times 24, \qquad A_f = 27.0 \text{ sq in.}, \qquad b_f/2t_f = 10.7$$
$$1\frac{1}{8} \times 26, \qquad A_f = 29.25 \text{ sq in.}, \qquad b_f/2t_f = 11.6$$

Try plates—$1\frac{1}{4} \times 22$: $A_f = 27.5$ sq in.

In this selection, the width-to-thickness ratio $\lambda = b_f/2t_f$ should be kept near the λ_p value (9.2 for $F_y = 50$ ksi from Table 11.4.1). Some strength reduction may be tolerated from the flange local buckling limit state, but preferably not.

Check F_{cr} for lateral-torsional buckling (LTB) limit state (LRFD-Appendix G2), Eqs. 11.4.4 through 11.4.9:

$$r_T = \sqrt{\frac{\frac{1}{12}(22)^3(1.25)}{27.5 + 37.5/6}} = 5.73 \text{ in.}$$

$$\lambda_p = \frac{300}{\sqrt{F_{yf}}} = 42.4$$

$$\lambda_{\text{LTB}} = \frac{L_b}{r_T} = \frac{25(12)}{5.73} = 52.4 > (\lambda_p = 42.4)$$

Since $\lambda > \lambda_p$, F_{cr} may be reduced below F_y. However, in this case at the negative moment zone, $C_b \approx 1.67$ from Table 9.6.3 assuming moment varies linearly over the 25-ft segment; thus $C_b F_{cr}$ is quite likely above F_y (the upper limit). Check Eq. 11.4.7, first computing λ_r for the lateral-torsional buckling limit state,

$$F_{cr} = C_b F_{yf}\left[1 - \frac{1}{2}\left(\frac{\lambda - \lambda_p}{\lambda_r - \lambda_p}\right)\right] \leq F_{yf} \qquad [11.4.7]$$

$$F_{cr} = 1.67(50)\left[1 - \frac{1}{2}\left(\frac{52.4 - 42.4}{107 - 42.4}\right)\right] = 1.67(46.2) > F_{yf}$$

Thus, F_{cr} based on lateral-torsional buckling is $F_{yf} = 50$ ksi.

Compute F_{cr} for the flange local buckling (FLB) limit state (LRFD-Appendix G2) when $\lambda_{\text{FLB}} > \lambda_p$. This was satisfied in this design when the flange plates were selected; otherwise this limit state must be treated according to Eqs. 11.4.10 through 11.4.13.

Next, evaluate the strength reduction resulting from the bend-buckling limit state when $h/t_w > 970/\sqrt{F_{cr}}$, using Eqs. 11.4.2 and 11.4.3,

$$a_r = \frac{A_w}{A_f} = \frac{37.5}{27.5} = 1.36$$

$$R_{\text{PG}} = 1 - \frac{a_r}{1200 + 300a_r}\left(\frac{h}{t_w} - \frac{970}{\sqrt{F_{cr}}}\right) \leq 1 \qquad [11.4.3]$$

In evaluating Eq. 11.4.3, the F_{cr} is the lesser value from the lateral-torsional buckling (LTB) and the flange local buckling (FLB) limit states. In this design, $F_{cr} = F_y = 50$ ksi. Thus, Eq. 11.4.3 is evaluated using $a_r = 1.36$, $h/t_w = 267$, and $F_{cr} = 50$ ksi, to obtain $R_{\text{PG}} = 0.89$.

The moment of inertia of the cross-section must be evaluated,

$$
\begin{array}{lll}
1\frac{1}{4} \times 22: & 27.5(2)(101.25/2)^2 = & 140,960 \\
\frac{3}{8} \times 100: & (0.375)(100)^3/12 = & \underline{31,250} \\
& I = & 172,210 \text{ in.}^4
\end{array}
$$

$$S_x = \frac{I}{(h/2 + t_f)} = \frac{172,210}{50 + 1.25} = 3360 \text{ in.}$$

Then, using Eq. 11.4.2, the nominal strength M_n can be evaluated,

$$M_n = F_{cr} S_x R_{PG} = 50(3360)(0.89)/12 = 12,460 \text{ ft-kips}$$

$$\phi M_n = 0.90(12,460) = 11,200 \text{ ft-kips} \approx (M_u = 11,260 \text{ ft-kips})$$

Accept plates $1\frac{1}{4} \times 22$.

(f) *Select Flange Plates for Positive Moment.* $M_u = 7900$ ft-kips, web plate $= \frac{5}{16} \times 100$ ($A_w = 31.25$ sq in.). Estimate the flange width to be about 22 in., the same as for negative moment region, which makes $L_b/r_T \approx 52$ as computed in part (e) above. In this positive moment region $C_b \approx 1.0$ conservatively, which means F_{cr} for the lateral-torsional buckling limit state might be a little below F_y.

Again estimate $R_{PG} \approx 0.90$ for the effect of the bend-buckling limit state,

$$R_{PG} F_{cr} \approx 0.90(36) = 32 \text{ ksi}$$

Using the flange-area formula, Eq. 11.14.9, gives the requirement for one flange as

$$A_f = \frac{M_u/\phi_b}{R_{PG} F_{cr} h} - \frac{A_w}{6} = \frac{7900(12)/0.90}{32(101)} - \frac{31.25}{6} = 27.4 \text{ sq in.}$$

Some possible choices:

$1\frac{1}{4} \times 22$,	$A_f = 27.5$ sq in.,	$b_f/2t_f = 8.8$
$1\frac{1}{8} \times 24$,	$A_f = 27.0$ sq in.,	$b_f/2t_f = 10.7$
$1\frac{1}{8} \times 26$,	$A_f = 29.25$ sq in.,	$b_f/2t_f = 11.6 > 10.8$ NG

Try plates—$1\frac{1}{8} \times 24$: $A_f = 27.0$ sq in.

The choices here are identical to those listed for the negative moment zone. Here, however, the $1\frac{1}{8} \times 24$ may provide enough strength because $\lambda = b_f/2t_f = 10.7$ does not exceed $\lambda_p = 10.8$ for A36 steel. That lighter plate will be investigated in this example; however, the practical choice would be the same $1\frac{1}{4} \times 22$ used in the negative moment zone.

Check F_{cr} for lateral-torsional buckling (LTB) limit state (LRFD-Appendix G2), Eqs. 11.4.4 through 11.4.9:

$$r_T = \sqrt{\frac{\frac{1}{12}(24)^3(1.125)}{27.0 + 31.25/6}} = 6.34 \text{ in.}$$

$$\lambda_p = \frac{300}{\sqrt{36}} = 50.0$$

$$\lambda_{LTB} = \frac{L_b}{r_T} = \frac{25(12)}{6.34} = 47.3 < (\lambda_p = 50.0)$$

Since $\lambda < \lambda_p$, $F_{cr} = F_y = 36$ ksi.

Next, evaluate the strength reduction resulting from the bend-buckling limit state when $h/t_w > 970/\sqrt{F_{cr}}$, using Eqs. 11.4.2 and 11.4.3 with $h/t_w = 320$,

$$a_r = \frac{A_w}{A_f} = \frac{31.25}{27.0} = 1.16$$

$$R_{PG} = 1 - \frac{1.16}{1200 + 300(1.16)}\left(320 - \frac{970}{\sqrt{36}}\right) = 0.88$$

The moment of inertia of the cross-section must be evaluated,

$1\frac{1}{8} \times 24$: $27.0(2)(101.125/2)^2 = 138,050$
$\frac{5}{16} \times 100$: $(0.3125)(100)^3/12 = \underline{26,040}$
$I = \overline{164,090}$ in.^{4}

$$S_x = \frac{I}{(h/2 + t_f)} = \frac{164,090}{50 + 1.125} = 3210 \text{ in.}$$

Then, using Eq. 11.4.2, the nominal strength M_n can be evaluated,

$$M_n = F_{cr}S_xR_{PG} = 36(3210)(0.88)/12 = 8474 \text{ ft-kips}$$

$$\phi_b M_n = 0.90(8474) = 7630 \text{ ft-kips} < [M_u = 7900 \text{ ft-kips}] \quad \text{NG}$$

This understrength is not acceptable. Increase flange plates to $1\frac{1}{4} \times 24$, with $A_f = 30.0$ sq in., $I_x = 179,800$ in.^{4}, $r_y = 5.62$ in., $r_T = 6.40$ in., and $S_x = 3509$ in.^{3} The $b_f/2t_f = 9.6$ does not exceed $\lambda_p = 10.8$; thus, local flange buckling does not control. For this section,

$$R_{PG} = 1 - \frac{a_r}{1200 + 300a_r}\left(\frac{h}{t_w} - \frac{970}{\sqrt{F_{cr}}}\right) \le 1.0$$

$$= 1 - \frac{1.04}{1200 + 300(1.04)}\left(320 - \frac{970}{\sqrt{36}}\right) = 0.89 \qquad [11.4.3]$$

The nominal strength M_n is

$$M_n = F_{cr}S_xR_{PG} = 36(3509)(0.89)/12 = 9370 \text{ ft-kips}$$

$$\phi_b M_n = 0.90(9370) = 8430 \text{ ft-kips} > [M_u = 7900 \text{ ft-kips}] \quad \text{OK}$$

Use flange plates $1\frac{1}{4} \times 24$.

(g) *Intermediate Stiffeners—Placement in Positive Moment Zone.* Web plate $= \frac{5}{16} \times 100$, $A_w = 31.25$ sq in., $F_y = 36$ ksi.

Exterior end, $V_u = 369$ kips (Fig. 11.15.2). In end panels post-buckling strength (i.e., tension-field action) is *not permitted to be used*. Equation 11.11.5 represents the nominal shear strength V_n of such an end panel,

$$V_n = C_v(0.6F_{yw})A_w$$

$$\text{Required } C_v = \frac{V_u/\phi_v}{0.6F_{yw}A_w} = \frac{369/0.90}{0.6(36)31.25} = 0.607$$

[11.11.5]

The coefficient C_v is given by Eq. 11.8.10 for $C_v < 0.8$,

$$C_v = \frac{44{,}000k_v}{(h/t_w)^2 F_{yw}} = \frac{44{,}000k_v}{(320)^2 36} = \frac{k_v}{83.8}$$

Required $k_v = 0.607(83.8) = 50.9$

$$k_v = 50.9 = 5 + \frac{5}{(a/h)^2}$$

$$\text{Max } \frac{a}{h} = 0.33$$

$$\text{Max } a = 0.33(100) = 33 \text{ in.}$$

Use 2'-9".

Frequently, this maximum a/h can be obtained by using LRFD "NUMERICAL VALUES" TABLE 9, which contains $\phi_v V_n/A_w$ values as a function of h/t_w and a/h. In this case,

$$\text{Required } \frac{\phi_v V_n}{A_w} = \frac{V_u}{A_w} = \frac{369}{31.25} = 11.8 \text{ ksi}$$

Then enter TABLE 9-36 with $h/t_w = 320$, look for 11.8 ksi, and determine a/h. In this case, values as high as 11.8 ksi indicate only that a/h is *less than* 0.5. Thus, a detailed calculation as above would be necessary.

Panel 2: Assume the shear envelope is linear (a conservative assumption) over the 25 ft to the bearing stiffener under the concentrated load. Thus, at 2.75 ft from the support,

$$V_{u2} = 360 - \frac{369 - 228}{25}(2.75) = 353 \text{ kips}$$

Equation 11.9.27 which includes tension-field action is applicable in all panels of this girder, except the exterior panel treated above. Use LRFD "NUMERICAL VALUES" TABLE 10-36, entering with $h/t_w = 320$ and

$$\text{Required } \frac{\phi_v V_n}{A_w} = \frac{V_u}{A_w} = \frac{353}{31.25} = 11.3 \text{ ksi}$$

Find no values tabulated for 11.3 ksi and $h/t_w = 320$. This means that the upper limit according to Eq. 11.11.10 controls,

$$\frac{a}{h} \le \left(\frac{260}{h/t_w}\right)^2 = \left(\frac{260}{320}\right)^2 = 0.66 \le 3.0$$

$$a = 0.66(100) = 66 \text{ in. } (5.5 \text{ ft})$$

Since lateral support occurs every 25 ft and a bearing stiffener is required at concentrated loads, the stiffener spacing is usually arranged to fit these limitations. Considering the first stiffener at 2'-9", 22'-3" remains. Four spaces at the maximum of 5.5 ft will be less than 22'-3". Thus, five spaces must be used. It is preferable to use 3-in. multiples for stiffener spacing.

Use stiffener spaces as follows starting from the exterior support: 2 @ 2'-6",
4 @ 5'-0".

For the region from 25 to 50 ft from the end, the a/h limitation of 0.66 still
governs.
Use 5 spaces @5'-0".

For the region from 50 to 75 ft, the maximum shear stress is still below the
value in panel 2, so that a/h maximum still governs.
Use 5 spaces @ 5'-0".

(h) *Intermediate Stiffeners—Placement in Negative Moment Zone.* Web
plate $= \frac{3}{8} \times 100$, $A_w = 37.5$ sq in., $F_y = 50$ ksi.
Interior end:

$$V_u = 549 \text{ kips}$$

Use Eqs. 11.9.27 and 11.11.10 (LFRD-Appendix G3) for this interior panel
(there is an adjacent panel in the other span):

$$\text{Required } \frac{\phi_v V_n}{A_w} = \frac{V_u}{A_w} = \frac{549}{37.5} = 14.6 \text{ ksi}, \qquad \frac{h}{t_w} = \frac{100}{0.375} = 267$$

Using LRFD "NUMERICAL VALUES" TABLE 10-50, again find the maximum a/h is controlled by the arbitrary limit (see also Fig. 11.9.9).

$$\text{Max } \frac{a}{h} = \left(\frac{260}{h/t_w}\right)^2 = \left(\frac{260}{267}\right)^2 = 0.95$$

This panel adjacent to the interior support on a continuous span has high
shear *and bending moment* at the same location. Thus, LRFD-Appendix G5 is
likely to control stiffener spacing. Thus, Eq. 11.10.14 must be investigated. The
design strength $\phi_b M_n$ was computed in part (e) to be 11,200 ft-kips at the
maximum negative moment location; thus, the moment strength is fully utilized.
Since more than 75% of the moment strength is used, the full shear strength may
not be used, as shown in Fig. 11.10.2. Actually this is a restriction *related to the
web;* however, for practical purposes in LRFD, the overall moment strength is
used in the interaction formula, Eq. 11.10.14. In ASD, the *stress on the web* is
used in the moment-related term of the interaction equation, Eq. 11.10.21.

The flexure-shear strength interaction limitation can be satisfied either by
increasing the moment strength (i.e., reducing the percent utilization) or by
increasing the shear strength. The practical procedure at this stage of design is
to place the intermediate stiffeners at sufficiently close intervals to make the
precent utilization of the shear strength low enough to satisfy the interaction
equation.

Solving Eq. 11.10.14 for the *required* $V_u/\phi_v V_n$ gives

$$\frac{M_u}{\phi M_n} + 0.625\left(\frac{V_u}{\phi V_n}\right) \le 1.375 \qquad\qquad [11.10.14]$$

$$\text{Required } \frac{V_u}{\phi_v V_n} = \frac{1.375 - \dfrac{M_u}{\phi_b M_n}}{0.625} = 2.2 - 1.6(1.0) = 0.6$$

The design strength $\phi_v V_n$ that must be provided in this panel is

$$\text{Required } \phi_v V_n = V_u/0.6 = 549/0.6 = 915 \text{ kips}$$

Using LRFD "NUMERICAL VALUES" TABLE 10-50 with $h/t_w = 267$ and

$$\text{Required } \phi_v V_n/A_w = 915/37.5 = 24.4 \text{ ksi}$$

Find maximum $a/h < 0.5$ (i.e., off table)

The actual formulas must be used to solve by trial for a/h; find $a/h = 0.38$. Calculations are as follows,

$$k_v = 5 + \frac{5}{(a/h)^2} = 5 + \frac{5}{(0.38)^2} = 39.6$$

Assuming $C_v < 0.8$ by referring to Fig. 11.9.9; then using Eq. 11.8.10 for C_v,

$$C_v = \frac{44,000k_v}{(h/t_w)^2 F_{yw}} = \frac{44,000(39.6)}{(267)^2 50} = 0.489$$

$$V_n = 0.6F_{yw}A_w\left(C_v + \frac{1 - C_v}{1.15\sqrt{1 + (a/h)^2}}\right) \qquad [11.9.27]$$

$$= 0.6(50)(37.5)\left(0.489 + \frac{1 - 0.489}{1.15\sqrt{1 + (0.38)^2}}\right) = 1020 \text{ kips}$$

$$\phi_v V_n = 0.90(1020) = 916 \text{ kips} > (\text{Required } \phi_v V_n = 915 \text{ kips}) \quad \text{OK}$$

Thus, the maximum distance to the first intermediate stiffener in the negative moment region is

$$\text{Max } a = 0.38(100) = 38 \text{ in.}$$

Use 3'-2".

Panel 2: Assume the shear envelope is linear (a consecutive assumption) over the 25 ft to the bearing stiffener under the concentrated load. Thus, at 3.17 ft from the interior support,

$$V_{u2} = 549 - \frac{549 - 397}{25}(3.17) = 530 \text{ kips}$$

Use LRFD "NUMERICAL VALUES" TABLE 10-50, entering with $h/t_w = 267$ and

$$\frac{\text{Required } \phi_v V_n}{A_w} = \frac{V_u}{A_w} = \frac{530}{37.5} = 14.1 \text{ ksi}$$

Find no values tabulated for 14.1 ksi and $h/t_w = 267$. However, as in the first panel, strength in combined bending and flexure is likely to control. From the moment envelope, Fig. 11.15.2, it is reasonable to use a linear variation from 11,260 ft-kips to zero over 25 ft to the concentrated load. At 3.17 ft from the support the factored moment is 9830 ft-kips, which makes

$$\frac{M_u}{\phi_b M_n} = \frac{9830}{11,200} = 0.88 > 0.75$$

The interaction requirement will also control this panel. The maximum $V_u/\phi_v V_n$ is

$$\text{Required } \frac{V_u}{\phi_v V_n} = \frac{1.375 - \dfrac{M_u}{\phi_b M_n}}{0.625} = 2.2 - 1.6(0.88) = 0.79$$

The design strength $\phi_v V_n$ that must be provided in this panel is

$$\text{Required } \phi_v V_n = V_u/0.79 = 530/0.79 = 670 \text{ kips}$$

Using LRFD "NUMERICAL VALUES" TABLE 10-50 with $h/t_w = 267$ and

$$\text{Required } \phi_v V_n/A_w = 670/37.5 = 17.9 \text{ ksi}$$
$$\text{Find maximum } a/h > [(260/267)^2 = 0.95]$$

Thus, the arbitrary upper limit, Eq. 11.11.10 controls.

$$\text{Max } a = 0.95(100) = 95 \text{ in. } (7'\text{-}11'')$$

With the first stiffener at 3'-2" from the center of support, the remaining distance to the bearing stiffener at the concentrated load will require 3 spaces @ ≈7'-3". (Note: As treated below, the use of the maximum stiffener spacing of 5'-6" on the $\frac{5}{16}$ web panel to the right of the concentrated load 25 ft from the interior support will require changing the aforementioned stiffener spacing.) The arrangement of intermediate stiffeners is shown on the design sketch, Fig. 11.15.6.

(i) *Location of Flange and Web Splices.* The location of splices is partially dependent on the type of splices used; for a bolted field splice joining the A36 with the A572 sections, both flanges and web usually would be spliced at the same location; for a field welded splice it might be preferable to splice the flanges at a location offset from the web splice by as much as 10 ft. Such an offset reduces stress concentration and probably assists in getting proper alignment at the splice. For this example assume splices for flanges and web are to occur at the same location.

From previous calculations for stiffener spacing it is apparent that splicing could be done at, say, 26 ft from the interior support, just beyond the bearing stiffener. Perhaps, however, it can be made closer to the interior support.

Try 5'-6" from the concentrated load (19'-16" from interior support) which would be the maximum spacing based on the maximum $a/h = 0.66$ for the $\frac{5}{16}$-in. web. See part (g).

Examine a panel of $\frac{5}{16}$-in. web immediately to the right of the concentrated load that is 25 ft from the interior support. The maximum moment of about 3500 ft-kips occurs only when no live load is on span 1; thus, the shear will be low. The most critical case will be the loading for maximum negative moment (LL spans 1 and 2) in combination with the high shear from the shear envelope. For that situation, the factored forces at 5'-6" to the right of the concentrated load are

$$M_u \approx 2500 \text{ ft-kips;} \qquad V_u = 430 \text{ kips}$$
$$\frac{M_u}{\phi_b M_n} = \frac{2500}{8430 \text{ (from part f)}} < 0.75$$

The combined shear and flexure interaction relationship does not control. For shear strength, use LRFD "NUMERICAL VALUES" TABLE 10-36 with $h/t_w = 320$ and

$$\frac{\text{Required } \phi_v V_n}{A_w} = \frac{V_u}{A_w} = \frac{430}{31.25} = 13.8 \text{ ksi}$$

Find $a/h > 0.7$; therefore, maximum a/h based on Eq. 11.11.10 controls as in part (g). Maximum $a/h = 0.66$, giving $a = 66$ in. Thus, splice the web at 20 ft from the interior support. The arrangement indicated below uses the minimum number of stiffeners based on LRFD strength and other limitations. A desirable 4-in. offset will exist between the splice and the nearest stiffener.

Use stiffener spaces as follows starting from the interior support: 1 @ 3'-2", 3 @ 5'-6", and 1 @ 5'-4".

A summary showing stiffener spacing appears in Fig. 11.15.6.

(j) *Size of Intermediate Stiffeners.* Frequently, A36 stiffeners are suitable, with higher yield stress material offering little if any saving. Try A36 steel for all stiffeners.

Panel 2 from exterior support: This is the first panel in which tension-field action is presumed to occur.

Having reduced the spacing of the first stiffener to 2'-6" from 2'-9", the maximum shear force in the panel will be somewhat higher than used for the panel 2 computation in part (g). Thus,

$$V_u = 355 \text{ kips}$$
$$V_u/A_w = \text{Required } \phi_v V_n/A_w = 355/31.25 = 11.4 \text{ ksi}$$
$$a/h = 30/100 = 0.30$$
$$h/t_w = 100/0.3125 = 320$$

Entering LRFD "NUMERICAL VALUES" TABLE 10-36 with the above values, find for $a/h = 0.5$ (the smallest value in the table),

$$\phi_v V_n/A_w \approx 16.4 \text{ ksi}$$
$$A_{st}/A_w \approx 0.049$$

This stiffener area ratio requirement from the table is the same as using Eq. 11.9.28 (LRFD-Appendix G4). The requirement may be reduced in proportion that the shear strength is underutilized,

$$\frac{V_u/A_w}{\phi_v V_n/A_w} = \frac{11.4}{16.4} = 0.70$$

The stiffener area A_{st} required is

$$\text{Required } A_{st} = 0.049(0.70)A_w$$
$$= 0.049(0.70)31.25 = 1.07 \text{ sq in.}$$

This area requirement assumes stiffeners are to be used in pairs. Furthermore, local buckling of this unstiffened element must be precluded; i.e., $\lambda \leq \lambda_r$, ac-

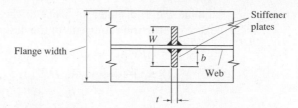

Figure 11.15.3   Cross-section of intermediate stiffener plates.

cording to LRFD-B5. The width b and thickness t, as shown in Fig. 11.15.3, must satisfy

$$\left(\lambda = \frac{b}{t}\right) \leq \left(\lambda_r = \frac{95}{\sqrt{F_{yst}}} = \frac{95}{\sqrt{36}} = 15.8\right)$$

The requirement for stiffness, Eq. 11.11.11 gives

$$I_{st} \geq jat_w^3 \qquad\qquad [11.11.11]$$

where

$$j = \frac{2.5}{(a/h)^2} - 2 = \frac{2.5}{(30/100)^2} - 2 = 25.8$$

$$I_{st} \geq 25.8(30)(0.3125)^3 = 23.5 \text{ in.}^4$$

To find minimum acceptable stiffener width (Fig. 11.15.3),

$$\text{Required } r^2 = \frac{I_{st}}{A_{st}} = \frac{23.5}{1.07} = 22.0 \text{ in.}^2$$

$$\text{Provided } r^2 = \frac{tW^3}{12tW} = \frac{W^2}{12}$$

$$\text{Required } W = \sqrt{12(22.0)} = 16.2 \text{ in.}$$

This would indicate 8-in. wide plates, and to satisfy $b/t \leq 15.8$ the thickness would be about 0.5 in. These would be inordinately large and A_{st} would be about 8 sq in. Try pairs of plates $\frac{3}{8} \times 5$, giving $b/t = 13.3$ which is less than λ_r.

Check the moment of inertia I_{st},

$$I_{st} = \frac{tW^3}{12} = \frac{0.375(10.3125)^3}{12} = 34.3 \text{ in.}^4$$

Thus, the provided I_{st} exceeds the required 23.5 in.[4] If the b/t limit of 15.8 can be accepted as 16 (i.e., 1% high), plate $\frac{5}{16} \times 5$ will be acceptable; the moment of inertia I_{st} for those plates is 28.6 in.[4] The authors prefer the $\frac{5}{16} \times 5$ plates.

Since the stiffeners for the panel are about the minimum size based on local buckling provisions, the reduced area required for the more interior panels on the $\frac{5}{16}$-in. web is of little consequence.

Use 2PLs—$\frac{5}{16} \times 5$ for all intermediate stiffeners attached to the $\frac{5}{16}$-in. web.

Examine the end panel adjacent to the interior support. Because of the shear-flexure interaction requirement the design shear strength $\phi_v V_n$ required is:

$$\text{Required } \phi_v V_n = 915 \text{ kips [from part (h)]}$$
$$\text{Required } \phi_v V_n/A_w = 915/37.5 = 24.4 \text{ ksi}$$
$$a/h = 38/100 = 0.38$$
$$h/t_w = 100/0.375 = 267$$

As in the calculation to determine the stiffener spacing, at $a/h = 0.38$ the panel uses 100% of the shear strength. Entering LRFD "NUMERICAL VALUES" TABLE 10-50 with the above values, find Required $\phi_v V_n/A_w = 24.4$ ksi is not within the table; however, extrapolation indicates a very small (or zero) value. Check by calculating Required A_{st} from Eq. 11.9.29,

$$\text{Required } A_{st} = \frac{F_{yw}}{F_{yst}}\left(0.15 D A_w(1 - C_v)\frac{V_u}{\phi_v V_n} - 18t_w^2\right) \qquad [11.9.29]$$

$$= \frac{50}{36}\left[0.15(1.0)37.5(1 - 0.489)\frac{549}{915} - 18(0.375)^2\right]$$

$$= \text{negative}$$

There is no strength requirement for the stiffeners. TABLE 10-50 gives values of A_{st}/A_w in percent for $F_y = 50$ ksi (both web and stiffeners). Thus stiffness controls rather than strength. Use A36 stiffeners with limit $\lambda_r = 15.8$. Try $\frac{5}{16} \times 5$ plates in pairs.
Check required I_{st},

$$I_{st} \geq jat_w^3 = 10.3(45)(0.375)^3 = 24.4 \text{ in.}^4$$

where

$$j = \frac{2.5}{(a/h)^2} - 2 = \frac{2.5}{(45/100)^2} - 2 = 10.3$$

$$I_{st} = \frac{tW^3}{12} = \frac{0.3125(10.375)^3}{12} = 29.0 \text{ in.}^4 > 24.4 \text{ in.}^4 \quad \text{OK}$$

The $\frac{5}{16} \times 5$ plates are satisfactory for use on the $\frac{3}{8}$-in. web of Grade 50 in the negative moment zone.
Use 2 PLs—$\frac{5}{16} \times 5$ of A36 steel for all intermediate stiffeners connected to the $\frac{3}{8}$ in. web.

(k) *Connections of Intermediate Stiffeners to Web.* For the $\frac{5}{16}$-in. A36 web, using Eq. 11.11.20,

$$f_{nv} = 0.045h\sqrt{\frac{F_{yw}^3}{E}} \qquad [11.11.20]$$

Even though there is no specific force requirement for connecting intermediate stiffeners except the tension-field force P_s used for the A_{st} requirement, the use of Eq. 11.11.20 as recommended by Basler [11.3] seems appropriate.

$$f_{nv} = 0.045(100)\sqrt{\frac{(36)^3}{29,000}} = 5.7 \text{ kips/in.}$$

The stiffener spacing at the first panel at the exterior support was finally made 2′-6″ instead of the 2′-9″ permitted. Thus the f_{nv} value could logically be reduced in proportion that the shear is underutilized. Using Eq. 11.8.10 for $C_v < 0.8$ with $a/h = 30/100 = 0.30$, $h/t_w = 320$, and $k = 60.6$ from Eq. 11.8.8 gives

$$C_v = \frac{44,000(60.6)}{(320)^2(36)} = 0.72$$

$$V_n = C_v(0.6F_{yw})A_w = 0.72(0.6)(36)(31.25) = 486 \text{ kips}$$

$$\frac{V_u}{\phi V_n} = \frac{369}{0.90(486)} = 0.84$$

Required $f_{nv} = 5.7(0.84) = 4.8$ kips/in.

Min weld size $a = \frac{3}{16}$ in. (LRFD-Table J2.5)

Determine maximum effective weld size (LRFD-J2.4 using shear strength on throat of fillet with E70 electrodes) (see also text, Sec. 5.14), Eq. 5.14.9,

$$a_{\text{max eff}} = 0.707\frac{F_u t_1}{F_{\text{EXX}}} = 0.707\frac{58(5/16)}{70} = 0.183 \text{ in.}$$

Try $\frac{3}{16}$-in. weld of E70 electrodes. The nominal weld strength R_{nw} is

$$R_{nw} = a(0.707)(0.6F_{\text{EXX}})$$
$$= 0.183(0.707)(0.6)70 = 5.44 \text{ kips/in.}$$

Since four lines of fillet welds (Fig. 11.15.3) are to provide 4.8 kips/in., then $4.8/4 = 1.2$ kips/in. are required along each line.

$$\% \text{ of continuous } \tfrac{3}{16}\text{-in. weld required} = \frac{1.2}{5.44}(100) = 22\%$$

For intermittent welding, minimum segment is 1.5 in. (or 4 times the weld size, if longer), according to LRFD-J2.2b; the nominal strength of this segment is

$$L_w R_{nw} = 1.5(5.44) = 8.2 \text{ kips}$$

$$\text{Max pitch} = \frac{8.2}{1.2} = 6.8 \text{ in.}$$

Use $\frac{3}{16}$ in.—$1\frac{1}{2}$ in. segments @ $6\frac{1}{2}″$ pitch, E70 electrodes, for connecting $\frac{5}{16} \times 5$ plates to $\frac{5}{16}$ in. web.

For this $\frac{3}{8}$-in. A572 Grade 50 web:

$$\text{Required } f_{nv} = 0.045(100)\sqrt{\frac{(50)^3}{29,000}} = 9.3 \text{ kips/in.}$$

where the yield stress of the web is used in f_{nv}.

At the interior support, the actual shear strength equals approximately the required strength; thus no reduction in f_{nv} is appropriate.

$$\text{Min weld size } a = \tfrac{3}{16} \text{ in.}$$

$$a_{\text{max eff}} = 0.707\frac{F_u t_1}{F_{\text{EXX}}} = 0.707\frac{58(3/8)}{70} = 0.183 \text{ in.}$$

based on the A36 stiffeners. Use $\frac{3}{16}$-in. fillet weld with E70 electrodes,

$$R_n = 0.183(0.707)(0.6)70 = 5.44 \text{ kips/in.}$$

For $1\frac{1}{2}$-in. segments, the maximum pitch is

$$\text{Max pitch} = \frac{1.5(5.44)}{9.3/4} = 3.5 \text{ in.}$$

which means nearly continuous $\frac{3}{16}$-in. weld.

Use $\frac{3}{16}$ *in. continuous weld, E70 electrodes, for connecting* $\frac{5}{16} \times 5$ *plates*

(A36) *to* $\frac{3}{8}$ *in. web* ($F_y = 50$ ksi).

(1) *Flange to Web Connection—A36 steel region.* The welding of the flanges to the web must provide for the factored horizontal shear flow $V_u Q/I_x$ at the joint. Though this requirement does not explicitly occur in the LRFD Specification, it is certainly implied in a provision such as LRFD-E4, 1st Par. for built-up members where connection is required to be " . . . adequate to provide for the transfer of the required forces." The factored load shear flow is

$$\text{Shear flow} = \frac{V_u Q}{I_x} \text{ kips/in.}$$

where V_u = factored shear at section
  Q = statical moment of flange area about x-axis
  I_x = moment of inertia of section about x-axis

Welding along both sides of the web provides a shear flow strength, which, if it exceeds the $V_u Q/I_x$ requirement, may be reduced by the use of intermittent welding. Normally flange to web welding is made continuous, primarily because automatic fabricating procedures usually make it more economical to do so. However, for design purposes the minimum percent of continuous weld acceptable in each panel between stiffeners will be prescribed. If it is more economical for the fabricator to use more weld, it is permissible to do so. The following calculations conservatively assume shielded metal arc welding (SMAW) is to be used.

$$\text{Min weld size } a = \tfrac{5}{16} \text{ in.}   \text{(LRFD-Table J2.4)}$$

$$a_{\text{max eff}} = 0.707\frac{F_u t_1}{F_{\text{EXX}}} = 0.707\frac{58(5/16)}{70} = 0.183 \text{ in.}$$

for E70 electrodes. The shear flow $V_u Q/I_x$ is

$$Q = A_f\left(\frac{h}{2} + \frac{t_f}{2}\right) = 24(1.125)(50.65) = 1370 \text{ in.}^3$$

$$\frac{V_u Q}{I_x} = \frac{369(1370)}{164,000} = 3.1 \text{ kips/in.}$$

Equating the strength of two fillets to $V_u Q/I_x$ gives

$$2a(0.707)(0.6)70 = 3.1$$

$$\text{Required } a = 0.05 \text{ in.} < 0.183 \text{ in.}$$

Use $\frac{5}{16}$ in. weld, E70 electrodes   (effective   size = 0.183   in.).   The   design strength ϕR_{nw} for continuous weld on both sides of the web is

$$\phi R_{nw} = 0.75(2)(0.707)(0.183)(0.6)70 = 8.2 \text{ kips/in.}$$

$$\text{Min \% continuous weld} = \frac{3.1}{8.2}(100) = 38\%$$

The maximum spacing of weld segments, according to LRFD-E4, 1st par., is

$$\text{Spacing} \leq \left(\frac{127t}{\sqrt{F_y}} = \frac{127(5/16)}{\sqrt{36}} = 6.6 \text{ in.} \right) \leq 12 \text{ in.}$$

(m) *Flange to Web Connection—A572 steel region.*

Min weld size $a = \frac{5}{16}$ in.   (LRFD-Table J2.4)

$$a_{\text{max eff}} = \frac{F_u t_1}{2(0.707)F_{\text{EXX}}} = 0.707 \frac{65(3/8)}{70} = 0.246 \text{ in.}$$

The maximum shear flow strength requirement is

$$\frac{V_u Q}{I_x} = \frac{549(1390)}{172,000} = 4.4 \text{ kips/in.}$$

where

$$Q = A_f(h/2 + t_f/2) = 22(1.25)(50.625) = 1390 \text{ in.}^3$$

The design strength ϕR_{nw} for continuous $\frac{5}{16}$-in. weld on both sides of the web is

$$\phi R_{nw} = 0.75(2)(0.707)(0.246)(0.6)70 = 11.0 \text{ kips/in.}$$

$$\text{Min \% continuous weld} = \frac{4.4}{11.0}(100) = 41\%$$

The maximum spacing of weld segments, according to LRFD-E4 is

$$\text{Spacing} \leq \left(\frac{127t}{\sqrt{F_y}} = \frac{127(3/8)}{\sqrt{50}} = 6.7 \text{ in.} \right) \leq 12 \text{ in.}$$

Use $\frac{5}{16}$ in. weld, E70 electrodes. The minimum percentage of continuous weld required for each panel along the $\frac{3}{8}$-in. A572 Grade 50 web is summarized in Fig. 11.15.6. LRFD-E4 requires at least 22% of continuous weld (that is, $1\frac{1}{2}$-in. segments @ 6.7 in. maximum pitch)

(n) *Design of Bearing Stiffeners—Interior Support.* As discussed in Sec. 11.12, bearing stiffeners generally are required on plate girders at all concentrated loads and reactions. Specifically, LRFD-K1.8 requires a pair of transverse stiffeners extending the full depth of the web "at unframed ends of beams and girders not otherwise restrained against rotation about their longitudinal axes." On rolled beams it may be possible to satisfy the requirements of LRFD-K1.3 to prevent local web yielding, LRFD-K1.4 to prevent web crippling, and LRFD-K1.5 to prevent sidesway web buckling; however, satisfying LRFD-K1.3, K1.4, and K1.5 on a plate girder would be rare. LRFD-K1.9 provides the specific requirements for bearing stiffeners.

The interior factored reaction is

$$\text{Reaction} = 2(549) = 1098 \text{ kips}$$

Bearing stiffeners should extend approximately to the edge of the flange plate even though not required to do so by either LRFD or ASD. The width b should then be

$$b = \frac{22 - 0.375}{2} = 10.8 \text{ in.}   \text{(say 10 in.)}$$

Column strength criterion according to LRFD-K1.9, 3rd par. (Fig. 11.15.4 for cross-section):

$$r \approx 0.25(20.375) = 5.1 \text{ in.}$$

$$\frac{KL}{r} \approx \frac{0.75(100)}{5.1} = 14.7$$

Using LRFD "NUMERICAL VALUES" TABLE 3-36, find

$$\phi_c F_{cr} = 30.2 \text{ ksi}   \text{(for A36 steel)}$$

$$\text{Required } A_{st} = \frac{1098}{30.2} = 36.4 \text{ sq in.}$$

$$\text{Required } t = \frac{36.4 - 9.37(0.375)}{2(10)} = 1.64 \text{ in.}$$

Local buckling criterion [LRFD-K1.9(2)]:

$$\left(\lambda = \frac{b}{t} \right) \leq \left(\lambda_r = \frac{95}{\sqrt{F_{yst}}} = \frac{95}{\sqrt{36}} = 15.8 \right)$$

$$\text{Required } t = \frac{10}{15.8} = 0.63 \text{ in.}$$

Note that the proportioning requirements of LRFD-K1.9 that refer to the "flange" are primarily relating to a rigid beam-to-column connection of the type treated in Sec. 13.6; the term "flange" in this context does *not* mean the plate girder flange.

Bearing criterion (LRFD-J8.1) using the contact area A_{pb}:

$$\text{Reqd } A_{pb} = \frac{1098}{\phi(1.8)F_y} = \frac{1098}{0.75(1.8)36} = 22.6 \text{ sq in.}$$

$$A_{pb} = 2(10 - 0.5)t$$
$$\nearrow$$
$$\text{est. to allow for fillet weld}$$

$$\text{Required } t = \frac{22.6}{2(9.5)} = 1.19 \text{ in.}$$

Column strength criterion governs. So as to properly distribute this reaction, try 4PLs—$\frac{7}{8} \times 10$ to serve as bearing stiffeners. (Alternative, 2PLs—$1\frac{3}{4} \times 10$.)

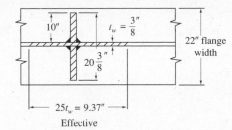

Figure 11.15.4 Cross-section (trial) of bearing stiffener at interior support.

Recheck r for columns strength:

$$r = \sqrt{\frac{\frac{1}{12}(1.75)(20.375)^3}{20(1.75) + 9.375(0.375)}} = 5.7 \text{ in.} > \text{estimated } 5.1 \text{ in.}$$

By inspection, column strength is adequate.

Use 4PLs—$\frac{7}{8} \times 10$ for $R = 1098$ kips (A36 steel), as shown in Fig. 11.15.5.

(o) _Connection for Bearing Stiffeners to Web._

$$\text{Required } f_{nv} = \frac{1098}{8(100)} = 1.37 \text{ kips/in.}$$

for eight lines of fillet weld (Fig. 11.15.5). Use $\frac{5}{16}$ in., with maximum effective size = 0.246 and E70 electrodes, as determined in part (m). The design strength ϕR_{nw} for one line of continuous weld is

$$\phi R_{nw} = 0.75(0.707)(0.246)(0.6)70 = 5.5 \text{ kips/in.}$$

$$\text{Min \% continuous weld} = \frac{1.37}{5.5}(100) = 25\%$$

(p) _Bearing Stiffeners at Concentrated Loads 25 ft from Supports._ At these locations, check whether or not the intermediate stiffeners will be adequate to serve as bearing stiffeners. The factored loads W_u are

$$W_u = 1.2(15) + 1.6(60) = 114 \text{ kips}$$

$$r = \sqrt{\frac{\frac{1}{12}(5/16)(10.3125)^3}{10(0.3125) + 25(0.3125)(0.3125)}} = 2.27 \text{ in.}$$

$$\frac{KL}{r} = \frac{0.75(100)}{2.27} = 33$$

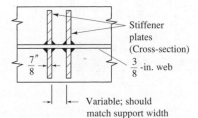

Figure 11.15.5 Cross-section (final) of bearing stiffener at interior support.

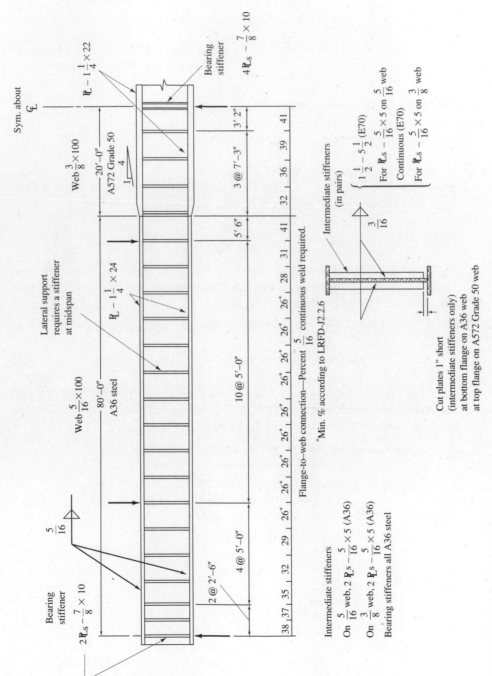

Figure 11.15.6   Design sketch.

Using LRFD "NUMERICAL VALUES" TABLE 3-36, find

$$\phi_c F_{cr} = 28.9 \text{ ksi (for A36 steel)}$$
$$A_{st} = 10(5/16) + 25(5/16)(5/16) = 5.57 \text{ sq in.}$$
$$\phi_c P_n = 28.9(5.57) = 161 \text{ kips} > (W_u = 114 \text{ kips}) \quad \text{OK}$$

The $\frac{5}{16} \times 5$ intermediate stiffeners are satisfactory for bearing stiffeners under the W_u concentrated loads.

(q) *Design Sketch.* Every design must have all decisions summarized on a design sketch, such as Fig. 11.15.6. The design of the bearing stiffeners for the exterior support has been omitted because the procedure has been adequately illustrated in parts (o) and (p). The girder weight, multiplying a plate cross-sectional area by 3.4 to get lb/ft, is (for one span):

A572 plates:	$1\frac{1}{4} \times 22 \times 20'$	= 93.5(40)	= 3,740
	$\frac{3}{8} \times 100 \times 20'$	= 127.5(20)	= 2,550
A36 plates:	$1\frac{1}{4} \times 24 \times 80'$	= 102.1(160)	= 16,328
	$\frac{5}{16} \times 100 \times 80'$	= 106.2(80)	= 8,500
Stiffeners:	$\frac{5}{16} \times 5 \times 8.33 \times 38$	= 5.31(8.33)38	= 1,682
	$\frac{7}{8} \times 10 \times 8.33 \times 4$	= 29.75(8.33)4	= 991
		Total	= 33,791 lb

Average weight = 338 lb/ft. ■■

SELECTED REFERENCES

11.1. Thomas C. Shedd, *Structural Design in Steel*. New York: John Wiley & Sons, Inc., 1934, Chap. 3.

11.2. Edwin H. Gaylord, Jr. and Charles N. Gaylord, *Design of Steel Structures*. New York: McGraw-Hill Book Company, Inc., 1957, Chap. 8.

11.3. Konrad Basler, "Strength of Plate Girders in Shear," *Transactions,* ASCE, **128,** Part II (1963), 683–719. (Also as Paper No. 2967, *Journal of the Structural Division,* ASCE, October 1961.)

11.4. I. Lyse and H. J. Godfrey. "Investigation of Web Buckling in Steel Beams," *Transactions,* ASCE, **100** (1935), 675–706.

11.5. Konrad Basler. "Strength of Plate Girders in Bending." *Transactions,* ASCE, **128,** Part II (1963), 655–686. (Also as Paper No. 2913, *Journal of the Structural Division,* ASCE, August 1961.)

11.6. Ronald W. Frost and Charles G. Schilling. "Behavior of Hybrid Beams Subjected to Static loads," *Journal of the Structural Division,* ASCE, **90,** ST3 (June 1964), 55–88.

11.7. Charles G. Schilling. "Web Crippling Tests on Hybrid Girders," *Journal of the Structural Division,* ASCE, **93,** ST1 (February 1967), 59–70.

11.8. Phillip S. Carskaddan. "Shear Buckling of Unstiffened Hybrid Beams," *Journal of the Structural Division,* ASCE, **94,** ST8 (August 1968), 1965–1990.

11.9. A. Anthony Toprac and Murugesam Natarajan, "Fatigue Strength of Hybrid Plate Girders," *Journal of the Structural Division,* ASCE, **97,** ST4 (April 1971), 1203–1225.

11.10. C. G. Schilling, Chairman. "Design of Hybrid Steel beams," Report of Subcommittee 1 on Hybrid Beams and Girders, Joint ASCE-AASHO Committee on Flexural Members, *Journal of the Structural Division,* ASCE, **94,** ST6 (June 1968), 1397–1426.

11.11. Peter B. Cooper, Theodore V. Galambos, and Mayasandra K. Ravindra. "LRFD Criteria for Plate Girders," *Journal of the Structural Division,* ASCE, **104,** ST9 (September 1978), 1389–1407.

11.12. John L. Dawe and Geoffery L. Kulak. "Local Buckling of W Shape Columns and Beams," *Journal of Structural Engineering,* **110,** 6 (June 1984), 1292–1304.

11.13. Cynthia J. Zahn. "Plate Girder Design Using LRFD," *Engineering Journal,* AISC, **24,** 1 (First Quarter 1987), 11–20.

11.14. Konrad Basler. "New Provisions for Plate Girder Design," *Proceedings, AISC National Engineering Conference.* Chicago, IL: American Institute of Steel Construction, 1961, 65–74.

11.15. AISI. "Proposed Criteria for Load and Resistance Factor Design of Steel Building Structures," *Bulletin No. 27.* Washington, DC: American Iron and Steel Institute, January 1978.

11.16. George S. Vincent. "Tentative Criteria for Load Factor Design of Steel Bridges," *Bulletin No. 15.* Washington, DC: American Iron and Steel Institute, March 1969.

11.17. Konrad Basler. "Strength of Plate Girders Under Combined Bending and Shear," *Journal of the Structural Division,* ASCE, **87,** ST7 (October 1961), 181–197. See also *Transactions,* ASCE, **128** (1963), Part II, 720–735.

11.18. Peter B. Cooper. "Strength of Longitudinally Stiffened Plate Girders," *Journal of the Structural Division,* ASCE, **93,** ST2 (April 1967), 419–451.

11.19. M. A. D'Apice, D. J. Fielding, and P. B. Cooper. "Static Tests on Longitudinally Stiffened Plate Girders," *Welding Research Council Bulletin No. 117,* October 1966. Also, *Strength of Plate Girders with Longitudinal Stiffeners, Bulletin No. 16,* American Iron and Steel Institute, April 1969, Paper No. III. (Includes historical survey and bibliography on longitudinally stiffened plates.)

11.20. Task Committee on Longitudinally Stiffened Plate Girders of the ASCE-AASHTO Committee on Flexural Members of the Committee on Metals of the Structural Division. "Theory and Design of Longitudinally Stiffened Plate Girders," *Journal of the Structural Division,* ASCE, **104,** ST4 (April 1978), 697–716.

11.21. Boris Bresler, T. Y. Lin, and John Scalzi. *Design of Steel Structures,* 2nd ed. New York: John Wiley & Sons, Inc., 1968, pp. 497–554.

11.22. Omer W. Blodgett. *Design of Welded Structures.* Cleveland, Ohio: James F. Lincoln Arc Welding Foundation, 1966.

11.23. Charles G. Schilling. "Optimum Proportions for I-shaped Beams," *Journal of the Structural Division,* ASCE, **100,** ST12 (December 1974), 2385–2401.

11.24. Abul K. Azad. "Continuous Steel I-Girders: Optimum Proportioning," *Journal of the Structural Division,* ASCE, **106,** ST7 (July 1980), 1543–1555.

11.25. Walter H. Fleischer. "Design and Optimization of Plate Girders and Weld-fabricated Beams for Building Construction," *Engineering Journal,* AISC, **22,** 1 (First Quarter 1985), 1–10.

11.26. Katherine E. Anderson and Ken P. Chong. "Least Cost Computer-aided Design of Steel Girders," *Engineering Journal,* AISC, **23,** 4 (Fourth Quarter 1986), 151–156.

PROBLEMS

All problems are to be done according to the AISC Load and Resistance Factor Design or Allowable Stress Design, as indicated by the instructor. All given loads are service loads unless otherwise indicated. A design sketch showing all decisions is required in all design problems.

Analysis

11.1. (This problem solution is used for Probs. 11.2 to 11.6.) A plate girder supported as shown must carry a dead load of 2 kips/ft and live load of 8 kips/ft, not including the girder weight. In addition, a concentrated dead load of 100 kips must be carried at the end of the cantilever. Compute and draw to scale for later use the moment and shear envelopes for this girder.

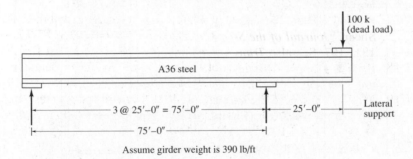

Assume girder weight is 390 lb/ft

Problem 11.1

11.2. (Use results of Prob. 11.1.) For the given plate girder designed for the conditions of Prob. 11.1, compute and draw to scale the moment capacity diagram, ϕM_n for LRFD (or allowable moment $F_b S_x$ for ASD) vs location along span. Neglect any reduction that might result from combined shear and moment interaction according to LRFD-Appendix G5 or ASD-G5. Compare capacity with moment envelope requirements from Prob. 11.1.

11.3. (Use results of Prob. 11.1.) For the given plate girder, compute and draw to scale the shear capacity diagram, ϕV_n for LRFD (or allowable shear $F_v A_w$ for ASD) vs location along span *based on location of intermediate stiffeners.*

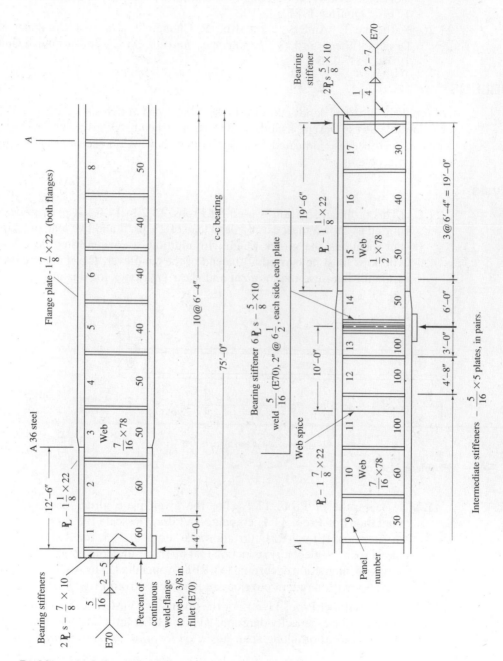

Problems 11.2 through 11.6

Neglect any consideration of combined shear and bending moment under LRFD-Appendix G5 or ASD-G5. Compare shear capacity with shear envelope requirements from Prob. 11.1.

11.4. (Use results of Prob. 11.1.) Investigate combined shear and moment strength by checking interaction relationship at all critical locations.

11.5. (Use results of Prob. 11.1.) Compute and draw to scale the shear capacity diagram, $\phi_v V_n$ for LRFD (or allowable shear $F_v A_w$ for ASD) vs location along span, for the girder *based on the flange-to-web connection*.

11.6. (Use results of Prob. 11.1.) Check the adequacy of each of the bearing stiffeners, including connection to the web.

11.7. Given the data for the 50-ft simply supported span, having lateral support at the ends and at the concentrated loads located 18 ft from each end;
  a. Investigate the acceptability of the 84-in. spacing for stiffener panel 4.
  b. Investigate combined shear and bending at its most critical location in the girder.
  c. Investigate the adequacy of the size of intermediate stiffeners.
  d. Investigate the adequacy of the size of the bearing stiffener ($2\text{PLs}—\frac{1}{2} \times 8$) at the support.
  e. Specify the flange-to-web connection.
  f. Specify the connection for intermediate stiffeners.
  g. Specify the connection for the support bearing stiffener.

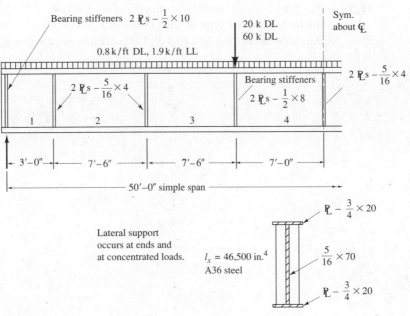

Problem 11.7

11.8. Using the given information concerning the portion of the plate girder, determine how close stiffener B must be to stiffener A in order for the design to be acceptable according to Load and Resistance Factor Design.

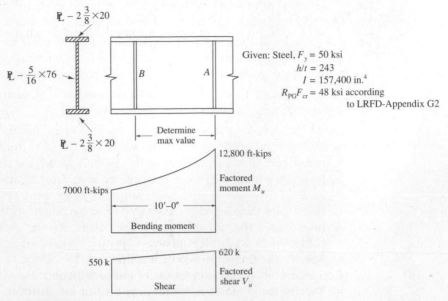

Given: Steel, $F_y = 50$ ksi
$h/t = 243$
$I = 157{,}400$ in.^{4}
$R_{PG}F_{cr} = 48$ ksi according
to LRFD-Appendix G2

Problem 11.8

11.9. Using the given information concerning the portion of the plate girder, determine how close stiffener B must be to stiffener A in order for the design to be acceptable according to Load and Resistance Factor Design.

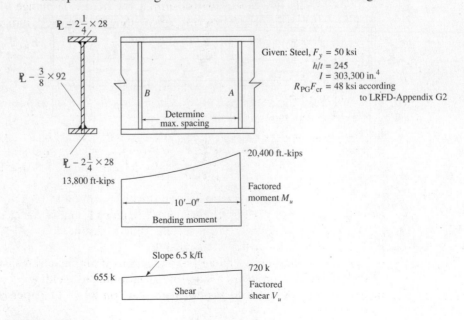

Given: Steel, $F_y = 50$ ksi
$h/t = 245$
$I = 303{,}300$ in.^{4}
$R_{PG}F_{cr} = 48$ ksi according
to LRFD-Appendix G2

Problem 11.9

11.10. Given the plate girder interior panel as shown, of A36 material.

    a. Disregarding combined bending and shear, determine the shear strength V_n (kips) of the given panel. What percent of the capacity represents elastic buckling strength and what percent comes from "tension-field action"?

    b. If $M_u/\phi M_n = 0.92$ at the extreme fiber of the web, what is the design shear strength ϕV_n (kips)?

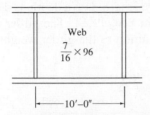

Web
$\frac{7}{16} \times 96$

$\longmapsto$ —10′-0″—$\longmapsto$

Problem 11.10

11.11. The cross-section near the end of a simply supported girder consists of flange plates, $1\frac{1}{2} \times 18$, and a web plate, $\frac{3}{8} \times 90$ ($I = 135{,}780$ in.^{4}). The girder is of A36 steel and has lateral support of the compression flange every 14 ft.

    a. Determine the service moment capacity of this section if the loading is 70% live load and 30% dead load.

    b. Determine the maximum distance from the reaction to the first intermediate stiffener if the end reaction is 120 kips dead load and 200 kips live load.

    c. Using E70 electrodes, determine the necessary flange to web welding in the end panel. If intermittent welding can be used, indicate the length of weld segment and the pitch distance.

Design

The following requirements apply to all design problems:

    a. Consider live load to be applied as necessary to give maximum range of internal forces (moments and shears).

    b. Moment and shear envelopes are to be drawn to be scale.

    c. For *continuous girders,* use $F_y = 50$ ksi in the negative moment zone and A36 in the positive moment zone. For statically determinate girders use the steel specified in the problems.

    d. Consider $\frac{5}{16}$ in. as the minimum web thickness available.

    e. Follow the proportioning guidelines of Chapter 11, unless you clearly indicate that you are doing otherwise for a specific stated reason.

    f. Use A36 steel for stiffeners if possible.

    g. Specify intermittent welding for connections if any material saving results, even though a cost analysis may later dictate continuous welding.

    h. Submit design sketch to approximate scale on $8\frac{1}{2} \times 11$ paper showing all final decisions.

    i. Compute the total average weight per foot of the girder, including all stiffeners.

j. For the *two-span continuous girder* designs, assume one splice is required in each span. Any extra butt-welded flange splice (two flanges) should be treated as adding 10 lb/ft to the average weight. (This may approximate the added cost effect of such splices.) Web splices in excess of one required per span should be considered as adding 6 lb/ft to the average girder weight.

11.12. For the case indicated by the instructor, design a simply supported I-shaped plate girder cross-section and specify the location and size of intermediate stiffeners. Omit design of connections and bearing stiffeners. The cross-section is to be constant over the entire length. Use a depth as dictated by the optimum depth equations discussed in Sec. 11.14. The girder must carry a uniform dead load of 0.5 kip/ft in addition to the girder weight.

Case	Span (ft)	Live loading (kips/ft)	Yield stress (ksi)	Lateral support
1	60	3	36	Continuous
2	70	3	50	Every 17.5 ft
3	80	3	60	Every 20 ft
4	60	5	36	Every 20 ft
5	70	5	50	Every 17.5 ft
6	80	5	60	Every 20 ft
7	90	3	36	Every 15 ft

11.13. For the case assigned by the instructor, completely design a *two-span continuous* girder, including bearing stiffeners and all connections. In addition to the uniform loading, two concentrated loads W (60% dead load, 40% live load) are located at the distance a from each end in each span. Lateral support is provided at all vertical supports, at concentrated loads, and at the intervals indicated over the interior of the spans.

Case	w_D Dead load (kip/ft)	w_L Live load (kip/ft)	Span length (ft)	W (kips)	a (ft)	Lateral support
1	1.25	2.25	140	60	35	Every 35 ft
2	1.25	2.25	125	60	25	Every 25 ft
3	1.25	2.25	130	70	30	Every 25 ft
(for Case 3, only *one* concentrated load at 30 ft from exterior supports)						
4	1.00	2.50	100	100	20	Every 30 ft
5	0.80	3.00	100	80	20	Every 30 ft
6	0.80	3.00	125	0	—	Every 25 ft
7	0.80	4.00	125	0	—	Every 25 ft
8	1.25	2.25	200	0	—	Continuous
9	1.25	2.25	200	0	—	Every 40 ft

Review of Theory

11.14. Explain the physical significance of the following C_v values for shear in plate girders. Describe the limit state based on shear strength in each of the following categories: (a) $C_v \leq 0.8$; (b) $0.8 < C_v \leq 1.0$; (c) $C_v > 1.0$

11.15. What is the specific behavior that is prevented when h/t_w is kept below $970/\sqrt{F_{cr}}$ for LRFD or $760/\sqrt{F_b}$ for ASD?

11.16. Show by a diagram of forces in equilibrium why stiffener spacing requirements are different for end panels and panels containing large holes than they are for interior panels of a plate girder.

11.17. Consider the LRFD-Appendix G1 or ASD-G1 limitation on the clear distance between flanges.
 a. Show the stress condition on the web plate that gives rise to the limitation equation.
 b. State explicitly what the number 16.5 in the equation means.
 c. If h/t_w equals the limit value, show the effective girder cross-section that might be used to compute moment capacity.

11.18. On a plate girder with $F_y = 50$ ksi, if the web $h/t_w = 185$, what will happen to the girder before the nominal moment strength is reached? Particularly describe what happens in a panel between intermediate stiffeners where the moment is high and the shear is low. Be specific and use a sketch.

Chapter 12

Combined Bending and Axial Load

12.1 INTRODUCTION

Nearly all members in a structure are subjected to both bending moment and axial load—either tension or compression. When the magnitude of one or the other is relatively small, its effect is usually neglected and the member is designed as a beam, as an axially loaded column, or as a tension member. For many situation neither effect can properly be neglected and the behavior under the combined loading must be considered in design. The member subjected to axial compression and bending is referred to as a *beam-column,* and is the major element treated in this chapter. The general subject of strength and stability considerations and design procedures for beam-columns has been extensively treated by Massonnet [12.1] and Austin [12.2], and a recent summary is given in the *SSRC Guide* [6.8]. The history of steel beam-column design has been reviewed by Sputo [12.56].

Since bending is involved, all of the factors considered in Chapters 7 and 9 apply here also; particulary, the stability related factors, such as lateral-torsional buckling and local buckling of compression elements. When bending is combined with axial tension, the chance of instability is reduced and yielding usually governs the design. For bending in combination with axial compression, the possibility of instability is increased with all of the considerations of Chapter 6 applying. Furthermore, when axial compression is present, a secondary bending moment arises equal to the axial compression force times the deflection.

A number of categories of combined bending and axial load along with the likely mode of failure may be summarized as follows:

1. Axial tension and bending; failure usually by yielding.
2. Axial compression and bending about one axis; failure by instability in the plane of bending, without twisting. (Transversely loaded beam-columns that are stable with regard to lateral-torsional buckling are an example of this category.)
3. Axial compression and bending about the strong axis: failure by lateral-torsional buckling.

4. Axial compression and biaxial bending—torsionally stiff sections; failure by instability in one of the principal directions. (W shapes are usually in this category.)
5. Axial compression and biaxial bending—thin-walled open sections; failure by combined twisting and bending on these torsionally weak sections.
6. Axial compression, biaxial bending, and torsion; failure by combined twisting and bending when plane of bending does not contain the shear center.

Because of the number of failure modes, no simple design procedure is likely to account for such varied behavior. Design procedures generally are in one of three categories: (1) limitation on combined stress; (2) semi-empirical interaction formulas, based on working stress procedures, and (3) semi-empirical interaction procedures based on strength. Limitations on combined stress ordinarily cannot provide a proper criterion unless instability is prevented, or large safety factors are used. Interaction equations come closer to describing the true behavior since they account for the stability situations commonly encountered. The AISC Specification formulas for beam-columns are of the interaction type in both Load and Resistance Factor Design and Allowable Stress Design.

12.2 DIFFERENTIAL EQUATION FOR AXIAL COMPRESSION AND BENDING

In order to understand the behavior, the basic situation of case 2, Sec. 12.1, will be treated. Failure by instability in the plane of bending is assumed. Consider the general case of Fig. 12.2.1, where the lateral loading $w(z)$ in combination with any end moments, M_1 and M_2, constitute the primary bending moment M_i which is a function of z. The primary moment causes the member to deflect y giving rise to a secondary moment Py. Stating the moment M_z at the location z of Fig. 12.2.1, gives

$$M_z = M_i + Py = -EI\frac{d^2y}{dz^2} \tag{12.2.1}$$

for sections with constant EI, and dividing by EI gives

$$\frac{d^2y}{dz^2} + \frac{P}{EI}y = -\frac{M_i}{EI} \tag{12.2.2}$$

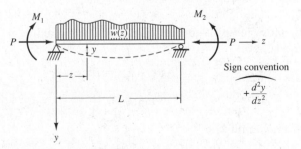

Figure 12.2.1   General loading of beam-column.

For design purposes, the general expression for moment M_z is of greater importance than the deflection y. Differentiating Eq. 12.2.2 twice gives

$$\frac{d^4y}{dz^4} + \frac{P}{EI}\frac{d^2y}{dz^2} = -\frac{1}{EI}\frac{d^2M_i}{dz^2} \tag{12.2.3}$$

From Eq. 12.2.1,

$$\frac{d^2y}{dz^2} = -\frac{M_z}{EI} \quad \text{and} \quad \frac{d^4y}{dz^4} = -\frac{1}{EI}\frac{d^2M_z}{dz^2}$$

Substitution in Eq. 12.2.3 gives

$$-\frac{1}{EI}\frac{d^2M_z}{dz^2} + \frac{P}{EI}\left(\frac{-M_z}{EI}\right) = -\frac{1}{EI}\frac{d^2M_i}{dz^2}$$

or, simplifying and letting $k^2 = P/EI$,

$$\frac{d^2M_z}{dz^2} + k^2M_z = \frac{d^2M_i}{dz^2} \tag{12.2.4}$$

which is of the same form as the deflection differential equation, Eq. 12.2.2.

The homogeneous solution for Eq. 12.2.4 is

$$M_z = A\sin kz + B\cos kz$$

as first discussed in Sec. 6.2. To this must be added the particular solution that will satisfy the right-hand side of the differential equation. Since $M_i = f(z)$, where $f(z)$ is usually a polynomial in z, the particular solution will be of the same form; thus the complete solution may be written

$$M_z = A\sin kz + B\cos kz + f_1(z) \tag{12.2.5}$$

where $f_1(z)$ = value of M_z satisfying Eq. 12.2.4. When M_z is a continuous function, the maximum value of M_z may be found by differentiation:

$$\frac{dM_z}{dz} = 0 = Ak\cos kz - Bk\sin kz + \frac{df_1(z)}{dz} \tag{12.2.6}$$

For most ordinary loading cases, such as concentrated loads, uniform loads, end moments, or combinations thereof, it may be shown that

$$\frac{df_1(z)}{dz} = 0$$

in which case a general expression for maximum M_z can be established; from Eq. 12.2.6,

$$Ak\cos kz = Bk\sin kz$$

$$\tan kz = \frac{A}{B} \tag{12.2.7}$$

At maximum M_z,

$$\sin kz = \frac{A}{\sqrt{A^2 + B^2}}, \qquad \cos kz = \frac{B}{\sqrt{A^2 + B^2}} \tag{12.2.8}$$

and substitution of Eqs. 12.2.8 into Eq. 12.2.5 gives

$$M_{z\,max} = \frac{A^2}{\sqrt{A^2 + B^2}} + \frac{B^2}{\sqrt{A^2 + B^2}} + f_1(z)$$
$$= \sqrt{A^2 + B^2} + f_1(z) \tag{12.2.9}$$

It is noted that whenever $df_1(z)/dz \neq 0$, Eq. 12.2.6 must be solved for kz and the result substituted into Eq. 12.2.5.

Case 1—Unequal End Moments Without Transverse Loading

Referring to Fig. 12.2.2, the primary moment M_i may be expressed

$$M_i = M_1 + \frac{M_2 - M_1}{L}z \tag{12.2.10}$$

Since

$$\frac{d^2M_1}{dz^2} = 0$$

Eq. 12.2.4 becomes a homogeneous equation, in which case $f_1(z) = 0$ for Eq. 12.2.5. The maximum moment, Eq. 12.2.9, is

$$M_{x\,max} = \sqrt{A^2 + B^2} \tag{12.2.11}$$

The constants A and B are evaluated by applying the boundary conditions to Eq. 12.2.5. The general equation is

$$M_z = A \sin kz + B \cos kz$$

and the conditions are

(1) at $z = 0$, $\qquad\qquad M_z = M_1$
$$\therefore B = M_1$$

(2) at $z = L$, $\qquad\qquad M_z = M_2$
$$M_2 = A \sin kL + M_1 \cos kL$$
$$\therefore A = \frac{M_2 - M_1 \cos kL}{\sin kL}$$

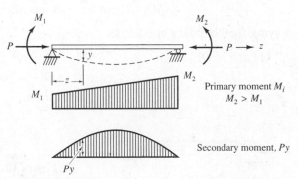

Figure 12.2.2 Case 1—end moments without transverse loading.

so that

$$M_z = \left(\frac{M_2 - M_1 \cos kL}{\sin kL}\right)\sin kz + M_1 \cos kz \qquad (12.2.12)$$

and

$$M_{z\,max} = \sqrt{\left(\frac{M_2 - M_1 \cos kL}{\sin kL}\right)^2 + M_1^2}$$

$$= M_2 \sqrt{\frac{1 - 2(M_1/M_2)\cos kL + (M_1/M_2)^2}{\sin^2 kL}} \qquad (12.2.13)$$

Case 2—Transverse Uniform Loading

Referring to Fig. 12.2.3, the primary moment M_i may be expressed as

$$M_i = \frac{w}{2} z(L - z) \qquad (12.2.14)$$

Since

$$\frac{d^2 M_i}{dz^2} = -w$$

$f_i(z) \neq 0$; the particular solution for the differential equation is required. Let $f_1(z) = C_1 + C_2 z$; i.e., any polynomial. Substitute the particular solution into Eq. 12.2.4,

$$\frac{d^2[f_1(z)]}{dz^2} = 0$$

$$0 + k^2(C_1 + C_2 z) = -w$$

Thus

$$C_1 = -w/k^2$$
$$C_2 = 0$$

Equation 12.2.5 then becomes

$$M_z = A \sin kz + B \cos kz - w/k^2 \qquad (12.2.15)$$

Applying the boundary conditions,

(1) at $z = 0$, $\qquad\qquad M_z = 0$

$$0 = B - w/k^2$$

$$\therefore B = w/k^2$$

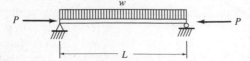

Figure 12.2.3 Case 2—transverse uniform loading.

(2) at $z = L$, $\qquad\qquad M_z = 0$

$$0 = A \sin kL + \frac{w}{k^2} \cos kL - \frac{w}{k^2}$$

$$\therefore A = \frac{w}{k^2}\left(\frac{1 - \cos kL}{\sin kL}\right)$$

Since $df_1(z)/dz = 0$, Eq. 12.2.9 gives maximum moment,

$$M_{z\,\text{max}} = \frac{w}{k^2}\sqrt{\left(\frac{1 - \cos kL}{\sin kL}\right)^2 + 1} - \frac{w}{k^2}$$

$$= \frac{w}{k^2}\left(\sec\frac{kL}{2} - 1\right)$$

$$= \frac{wL^2}{8}\underbrace{\left(\frac{8}{(kL)^2}\right)\left(\sec\frac{kL}{2} - 1\right)}_{\substack{\text{magnification factor due}\\\text{to axial compression}}} \tag{12.2.16}$$

Case 3—Equal End Moments Without Transverse Loading (Secant Formula)

Consider that $M_1 = M_2 = M$ in Fig. 12.2.2, in which case Eq. 12.2.13 becomes

$$M_{z\,\text{max}} = M\sqrt{\frac{2(1 - \cos kL)}{\sin^2 kL}} \tag{12.2.17}$$

$$= M\sqrt{\frac{2(1 - \cos kL)}{1 - \cos^2 kL}} = M\left(\frac{1}{\cos kL/2}\right)$$

$$= M \sec\frac{kL}{2} \tag{12.2.18}$$

Loading a beam-column with constant moment over its length causes a constant compression along one entire flange and constitutes the most severe loading on such a member. In view of this it would always be conservative to multiply the maximum primary moment due to *any* loading by sec $kL/2$, which is excessively conservative in most cases.

12.3 MOMENT MAGNIFICATION—SIMPLIFIED TREATMENT FOR MEMBERS IN SINGLE CURVATURE WITHOUT END TRANSLATION

As an alternate to the differential equation approach, a simple approximate procedure is satisfactory for many common situations.

Assume a beam-column is subject to lateral loading $w(z)$ that causes a deflection δ_0 at midspan, as shown in Fig. 12.3.1. The secondary bending moment may be assumed to vary as a sine curve, which is nearly correct for members with no end restraint whose primary bending moment and deflection are maximum at midspan.

The portion of the midspan deflection y_1, due to the secondary bending moment, equals the moment of the M/EI diagram between the support and midspan (shaded

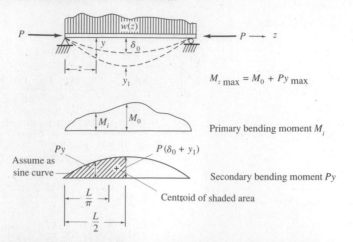

Figure 12.3.1   Primary and secondary bending moment.

portion of Fig. 12.3.1) taken about the support, according to the moment-area principle:

$$y_1 = \frac{P}{EI}(y_1 + \delta_0)\left(\frac{L}{2}\right)\frac{2}{\pi}\left(\frac{L}{\pi}\right) = (y_1 + \delta_0)\frac{PL^2}{\pi^2 EI} \tag{12.3.1}$$

or

$$y_1 = (y_1 + \delta_0)\frac{P}{P_e} \tag{12.3.2}$$

where $P_e = \pi^2 EI/L^2$. Solving for y_1,

$$y_1 = \delta_0\left[\frac{P/P_e}{1 - P/P_e}\right] = \delta_0\left(\frac{\alpha}{1 - \alpha}\right) \tag{12.3.3}$$

where $\alpha = P/P_e$. Since y_{max} is the sum of δ_0 and y_1,

$$y_{max} = \delta_0 + y_1 = \delta_0 + \delta_0\left(\frac{\alpha}{1 - \alpha}\right) = \frac{\delta_0}{1 - \alpha} \tag{12.3.4}$$

The maximum bending moment including the axial effect becomes

$$M_{z\,max} = M_0 + Py_{max} \tag{12.3.5}$$

Substituting the expression for y_{max} in Eq. 12.3.5 and setting $P = \alpha P_e = \alpha\pi^2 EI/L^2$, Eq. 12.3.5 may be written as the primary moment M_0 multiplied by a magnification factor B_1; thus

$$M_{z\,max} = M_0 B_1 \tag{12.3.6}$$

where

$$B_1 = \frac{C_m}{1 - \alpha} \tag{12.3.7}$$

and

$$C_m = 1 + \left(\frac{\pi^2 EI \,\delta_0}{M_0 L^2} - 1\right)\alpha \tag{12.3.8}$$

which can be expressed as

$$C_m = 1 + \psi\alpha \qquad (12.3.9)$$

For usual cases of single curvature, the magnification factor B_1 to be applied to the primary bending moment is equal to $C_m/(1 - \alpha)$. Rigorous differential equation solutions for Cases 1 to 7 shown in Table 12.3.1 are available in Timoshenko and Gere [6.67, Chap. 1]. Calculation of approximations for ψ of Eq. 12.3.9 have been made by Yura, whose calculations are presented by Iwankiw [12.3].

The approximate values for C_m for positive moment shown in Table 12.3.1 are computed using Eq. 12.3.8 and they are in general agreement with theoretical results

TABLE 12.3.1 SUGGESTED VALUES FOR C_m FOR SITUATIONS WITH NO JOINT TRANSLATION[a]

	Case	C_m (positive) moment	C_m (negative moment)	Primary Bending Moment
1		$1 + 0.2\alpha$[†]	—	
2		1.0	—	
3		$1 - 0.2\alpha$	—	
4		$1 - 0.3\alpha$	$1 - 0.4\alpha$	
5		$1 - 0.4\alpha$	$1 - 0.4\alpha$	
6		$1 - 0.4\alpha$	$1 - 0.3\alpha$	
7		$1 - 0.6\alpha$	$1 - 0.2\alpha$	
8		Eq. (12.3.8)	not available	

[a] Adapted from LRFD Commentary-Table C-C1.1 [1.17].

[†] $\alpha = \dfrac{P}{P_e/FS}$ for ASD ; $\alpha = \dfrac{P_u}{P_e} = \dfrac{P_u}{\pi^2 E/(KL/r)^2}$ for LRFD

even though Eq. 12.3.8 is derived using a sine curve deflection. The negative moment values have been included by AISC beginning with the 1980 AISC Manual. It is noted that the magnification of negative moment is necessary in order to maintain zero slope at the fixed supports when the beam deflects in the positive moment region. From a practical point of view, it is doubtful that zero slope is maintained; thus, the negative moment magnification is probably overestimated.

The correct value for C_m will be close to 1.0 for all cases because α rarely will exceed about 0.3. For this reason LRFD-C1 and ASD-H1 provide that C_m can be determined by "rational analysis" or 1.0 can be used for members with unrestrained ends and 0.85 is acceptable for members with restrained ends. The authors consider that C_m from Table 12.3.1 will satisfy the requirement of "rational analysis."

EXAMPLE 12.3.1 _____

Compare the differential equation magnification factor for the loading of Fig. 12.2.3, Eq. 12.2.16, with the approximate value, Eq. 12.3.7.

Solution.   For the differential equation,

$$B_1 = \text{magnification factor} = \frac{2}{(kL/2)^2}\left(\sec\frac{kL}{2} - 1\right) \tag{a}$$

where

$$\frac{kL}{2} = \frac{L}{2}\sqrt{\frac{P}{EI}} = \frac{\pi}{2}\sqrt{\alpha}$$

For the approximate solution,

$$B_1 = \frac{C_m}{1 - \alpha}$$

$$\delta_0 = \frac{5wL^4}{384EI}; \qquad M_0 = \frac{wL^2}{8}$$

$$\frac{\delta_0}{M_0} = \frac{5L^2}{48EI}$$

$$C_m = 1 + \left(\frac{\pi^2 EI}{L^2}\frac{5L^2}{48EI} - 1\right)\alpha = 1 + 0.028\alpha$$

$$B_1 = \frac{1 + 0.028\alpha}{1 - \alpha} \tag{b}$$

α	sec $kL/2$	Eq. (a)	Eq. (b)
0.1	1.137	1.114	1.114
0.2	1.310	1.257	1.257
0.3	1.533	1.441	1.441
0.4	1.832	1.686	1.685
0.5	2.252	2.030	2.028
0.6	2.884	2.546	2.542
0.7	3.941	3.405	3.399
0.8	6.058	5.125	5.112
0.9	12.419	10.284	10.253

Obviously, there is no significant difference between the differential equation solution and the approximate solution for this case. ■ ■

12.4 MOMENT MAGNIFICATION—MEMBERS SUBJECT TO END MOMENTS ONLY; NO JOINT TRANSLATION

For the situation shown in Fig. 12.2.2, which has no transverse loading, the theoretical maximum moment is given by Eq. 12.2.13,

$$M_{z\,max} = M_2 \sqrt{\frac{(M_1/M_2)^2 - 2(M_1/M_2)\cos kL + 1}{\sin^2 kL}} \qquad [12.2.13]$$

For this situation the maximum moment may be either (1) the larger end moment M_2 at the braced location (Fig. 12.4.1a), or (2) the magnified moment given by Eq. 12.2.13 which occurs at various locations out along the span (Fig. 12.4.1b), depending on the ratio M_1/M_2 and the value of α, since $kL = \pi \sqrt{\alpha}$. In order to make an analysis, one needs to know whether the maximum moment occurs at a location away from the support, and if so, the correct *distance*. To eliminate the need for such information, the concept of equivalent uniform moment (Fig. 12.4.1c) is used. Thus when investigating a member at a location away from the supported point, use of the equivalent moment assumes $M_{z\,max}$ to be at midspan.

To establish the equivalent moment M_E, let the solution for uniform moment, Eq. 12.2.17 with $M = M_E$, be equated with Eq. 12.2.13. One obtains

$$M_E = M_2 \sqrt{\frac{(M_1/M_2)^2 - 2(M_1/M_2)\cos kL + 1}{2(1 - \cos kL)}} \qquad (12.4.1)$$

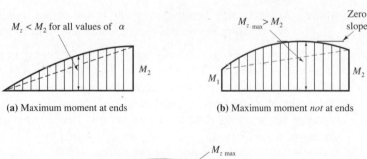

(a) Maximum moment at ends

(b) Maximum moment *not* at ends

(c) Equivalent uniform moment with maximum magnified moment at midspan

Figure 12.4.1 Primary plus secondary bending moment for members acted upon by end moments only.

By the procedure used in Example 12.3.1 it may be shown that for uniform moment, the magnification factor is obtained from Eq. 12.2.18:

$$B_1 = \sec \frac{kL}{2} = \frac{1}{1 - \alpha} \tag{12.4.2}$$

and using the equivalent uniform moment M_E to replace M_1 and M_2, the full maximum moment may be expressed as

$$M_{z \, max} = M_E\left(\frac{1}{1 - \alpha}\right) \tag{12.4.3}$$

which when compared with Eq. 12.3.6 may be written

$$M_{z \, max} = C_m M_2\left(\frac{1}{1 - \alpha}\right) \tag{12.4.4}$$

where $C_m = M_E/M_2$

$$= \sqrt{\frac{(M_1/M_2)^2 - 2(M_1/M_2)\cos kL + 1}{2(1 - \cos kL)}} \tag{12.4.5}$$

Equation 12.4.5 does not consider lateral-torsional buckling, or fully cover the double-curvature cases where M_1/M_2 lies between -0.5 and -1.0. The actual failure of members bent in double curvature with bending moment ratios -0.5 to -1.0 is generally one of "unwinding" through from double to single curvature in a sudden type of buckling, as discussed by Ketter [12.4], among others.

Austin [12.2] has discussed proposals of Massonnet [12.1] and Horne [12.5] to approximate C_m. For many years the AISC Specifications [1.5, 1.16] have used the following simple approximation,

$$C_m = 0.6 + 0.4\frac{M_1}{M_2} \tag{12.4.6}$$

where M_1 = smaller bending moment at one end of a member
$\quad\quad\quad M_2$ = larger bending moment at one end of a member

Note that M_1 and M_2 use bending moment signs rather than rotational moment signs. Prior to the 1986 LRFD Specification, Eq. 12.4.6 was not permitted to be less than 0.4; a very conservative procedure. However, since C_m as used in current design interaction relationships by LRFD and ASD is *not* directly related to lateral-torsional buckling in the manner of the Massonnet and Horne equations, there is no reason for the lower limit. A study of C_m has been presented by Chen and Zhou [12.6], and the inelastic amplification factor has been discussed by Sohal and Syed [12.55].

The 1993 LRFD Specification [1.16] and 1989 ASD Specification [1.5] do not have the 0.4 lower limit on C_m.

A comparison of Eqs. 12.4.5 and 12.4.6 is shown in Fig. 12.4.2. Note that for a given value of $\alpha = P/P_e$, the curve terminates when the moment M_2 at the end of the member exceeds the magnified moment within the span. The most important situations are those where the magnified moment within the span exceeds the moment at the end. The straight line used by AISC lies near the upper limit for C_m at any given bending moment ratio, and thus seems to be a reasonable approximation.

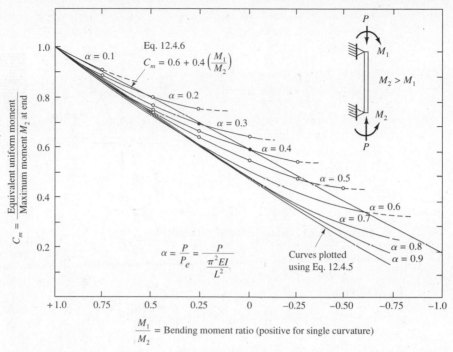

Figure 12.4.2 Comparison of theoretical C_m with AISC straight line approximation for members acted upon by end moments only, without joint translation.

12.5 MOMENT MAGNIFICATION—MEMBERS WITH SIDESWAY POSSIBLE

The unbraced frame, i.e., the frame where joint translation may occur when instability arises due to slenderness of the compression elements, does not lend itself to the simple but relatively accurate treatment presented in the last two sections. More complete treatment of braced and unbraced elastic frames is in Chapter 14 and in the *SSRC Guide* [6.8].

A simple approximation of C_m for this case may be obtained by starting with Eq. 12.3.6 which applies for the single curvature case,

$$M_{\max} = M_0 B = M_0 \left(\frac{C_m}{1 - \alpha} \right) \tag{12.5.1}$$

where B = magnification factor.

Next consider the situation of Fig. 12.5.1. Whatever the degree of restraint at the top and bottom of the two-story member, the deflection curve, and therefore the secondary bending moment (P times deflection), may reasonably be assumed to be a sine curve, in which case, the development used when no sidesway occurs (Fig. 12.3.1) is also valid here. Since $2L$ from Fig. 12.5.1 equals L for Fig. 12.3.1, Eq. 12.3.7 for

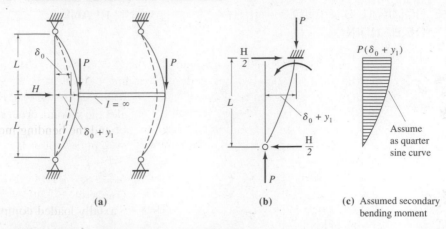

(a)   (b)   (c) Assumed secondary
bending moment

Figure 12.5.1   Beam-column with sidesway.

C_m becomes

$$C_m = 1 + \left(\frac{\pi^2 EI \delta_0}{4L^2 M_0} - 1\right)\alpha \qquad (12.5.2)$$

The larger effective length ($2L$ instead of L) is also used in the computation of α. Next, referring to Fig. 12.5.1,

$$\delta_0 = \frac{(H/2)L^3}{3EI} \qquad (12.5.3)$$

$$M_0 = \frac{HL}{2} \qquad (12.5.4)$$

Substitution of Eqs. 12.5.3 and 12.5.4 into Eq. 12.5.2 gives

$$C_m = 1 + \left[\frac{\pi^2 EI}{4L^2}\left(\frac{HL^3}{6EI}\right)\left(\frac{2}{HL}\right) - 1\right]\alpha$$

or

$$C_m = 1 + \left(\frac{\pi^2}{12} - 1\right)\alpha = 1 - 0.18\alpha \qquad (12.5.5)$$

as suggested in the ASD Commentary-H1 [1.6]. However, ASD-H1 states that C_m "shall be taken as" 0.85 for frames subject to joint translation, a generally *unconservative* value. On the other hand, LRFD-C1 uses $C_m = 1.0$ for the sway case.

A direct comparison of the effect of these different values in design is not possible because Allowable Stress Design uses the magnifier B to increase the first-order moment caused by the total load acting, whereas Load and Resistance Factor Design uses a braced frame magnifier B_1 to account for the second-order effects of the gravity (nonsway) portion of the load, and a sway magnifier B_2 to account for the second-order effects of the lateral (sway-inducing) portion of the load.

12.6   NOMINAL STRENGTH—INSTABILITY IN THE PLANE OF BENDING

The basic strength of a beam-column where lateral-torsional buckling and local buckling are adequately prevented, and bending is about one axis, will be achieved when instability occurs *in the plane of bending* (*without twisting*) (case 2 of Sec. 12.1). The differential equation solution, which includes the second-order P times deflection y term, shows that the axial compression effect and the bending moment effect cannot be determined separately and combined by superposition. It is a nonlinear relationship.

Furthermore, residual stresses cause some fibers to reach the yield stress before others, even when the elastically computed stresses due to applied load are the same at those fibers. This is similar to the effect on axially loaded compression members discussed in Secs. 6.5 and 6.6.

An analysis to determine the strength interaction between axial compression P and bending moment M for a beam-column is complicated. First, the M–θ–P (moment–curvature–axial compression) relationship must be developed. This can be done by assuming the yield penetration depth γh (Fig. 12.6.1) at various values. For each γh there will be a complete range of M–θ–P, or M/M_p–θ/θ_y–P/P_y in nondimensional form, where M_p is the plastic moment, θ_y is the curvature when the extreme fiber reaches stress F_y, and $P_y = A_g F_y$. From these M/M_p–θ/θ_y–P/P_y curves, a series of M/M_p–θ/θ_y curves can be obtained; one for each value of P/P_y.

Once M/M_p–θ/θ_y curves have been obtained, a specific combination of P_u/P_y and slenderness ratio KL/r is selected; the moment M/M_p is then applied incrementally to as high a value (M_u/M_p) such that the deflection is still stable. This combination of P_u/P_y, M_u/M_p, and KL/r represents one point on a strength interaction relationship, such as shown in Fig. 12.6.2. Various aspects of this procedure have been explained by Ketter, Kaminsky, and Beedle [12.7], Galambos and Ketter [12.8], and Ketter [12.4].

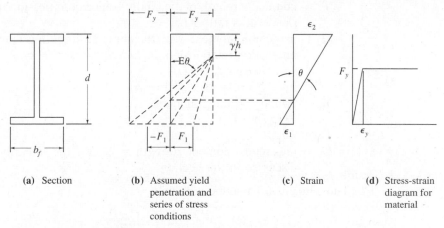

   (a) Section   **(b)** Assumed yield   **(c)** Strain   **(d)** Stress-strain
                                   penetration and   diagram for
                                   series of stress   material
                                   conditions

Figure 12.6.1   Member under axial compression and bending, for $\epsilon_1 \leq \epsilon_y$.

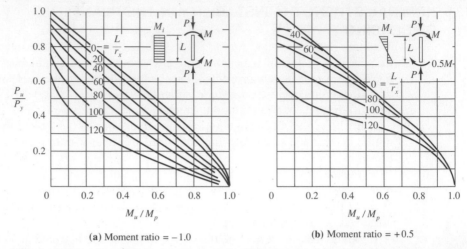

Figure 12.6.2   Strength interaction curves (W8×31, F_y = 33 ksi, linear residual stress = F_r = 0.3F_y, strong axis bending) for braced frame members. *Note*: For F_y > 33 ksi, use adjusted L/r = (actual L/r)$\sqrt{F_y/33}$. (Adapted from Ketter [12.4])

Similar procedures to obtain interaction relationships have been presented by Hauck and Lee [12.9, 12.10], Lee and Anand [12.11], and Rossow, Barney, and Lee [12.12].

An interaction equation representing this behavior is the following,

$$\frac{P_u}{P_n} + \frac{M_u}{M_n} \le 1 \tag{12.6.1}$$

where P_u = maximum axial compression load when nominal strength is reached based on the interaction relationship

  P_n = nominal strength of an axially loaded compression member based on slenderness ratio KL/r

  M_u = maximum moment when the nominal strength is reached based on the interaction relationship, *including the second-order effect* (*the so-called P–Δ effect*)

  = Eq. 12.3.6 = $M_0 C_m/(1 - \alpha)$

  M_0 = primary bending moment

  α = $P_u L^2/\pi^2 EI$

  C_m = Eq. 12.3.9

  M_n = maximum moment strength = M_p for the laterally stable situation discussed in this section

Thus, Eq. 12.6.1 may be written

$$\frac{P_u}{P_n} + \frac{M_u}{M_n}\left(\frac{C_m}{1 - \alpha}\right) \le 1 \tag{12.6.2}$$

In studying the interaction curves of Fig. 12.6.2, the reader is reminded that the strength P_n when $M_u = 0$ is based on the slenderness ratio KL/r_x. As stated at the beginning of the section, the member was assumed to fail by instability in the plane of bending.

Other studies of instability in the plane of bending have included the effects of transverse loading [12.4, 12.13–12.17].

12.7 NOMINAL STRENGTH—FAILURE BY COMBINED BENDING AND TORSION

The ordinary beam-column unbraced over a finite length involves consideration of instability transverse or oblique to the plane of bending, involving torsional effects. This subject is an extension of lateral-torsional buckling in beams (Chapter 9) and involves both elastic and inelastic considerations. Interaction curves for a number of elastic buckling situations have been developed, including (1) I-section columns with eccentric end loads in the plane of the web [12.18]; (2) I-columns with unequal end moments but without restraint to rotation about principal axes at the ends of the member [12.19]; and (3) I-columns with unequal end moments—hinged at ends for rotation about strong axis, but elastically restrained for rotation about the weak axis at the ends [12.20]. An excellent summary of the topic is presented by Massonnet [12.1] which includes discussion of plastic effects.

A number of studies are available containing both analytical and experimental treatments of inelastic lateral-torsional buckling [12.21–12.26].

Tests by Massonnet [12.1] and computer studies [12.21, 12.22] including inelastic effects indicate that interaction diagrams similar to those in Fig. 12.6.2 will result when including lateral-torsional buckling. The main difference is that P_u/P_y will be lower when based on KL/r_y (weak axis) rather than on KL/r_x (strong axis). Of course, M_n may also be less than M_p because of lateral-torsional buckling.

Torsional-Flexural Buckling of Thin-Walled Open Sections

A singly symmetric section subject to flexure and axial compression will have a torsional moment acting even without being in the slightly buckled position. This is because the shear center and center of gravity do not coincide (see Chap. 8 on torsion). The differential equation development has been given by Pekoz and Winter [12.27], Pekoz and Celebi [12.28], and the topic has been well summarized by Yu [12.29]. The AISI Specification [1.11] provides detailed rules for design to include the possibility of torsional-flexural buckling.

12.8 NOMINAL STRENGTH—INTERACTION EQUATIONS

Strength interaction equations relating axial compression P_u to bending moment M_u have long been recognized as the practical procedure for design.

Case 1—No Instability

For the braced location where instability cannot occur (i.e., $Kl/r = 0$), the uppermost curves of Fig. 12.6.2 apply and may be approximated by

$$\frac{P_u}{P_y} + \frac{M_u}{1.18M_p} = 1.0 \qquad (12.8.1)$$

and $M_u/M_p \leq 1.0$. In the above equation, $P_y = A_g F_y$ and $M_p = $ maximum moment strength of the member in the absence of axial load (equals the plastic moment for all cases where premature local buckling is prevented). The comparison with the theoretical result in Fig. 12.8.1 shows Eq. 12.8.1 to be a good approximation.

Case 2—Instability in the Plane of Bending

The curves of Fig. 12.6.2 for various combinations of moments and values of L/r_x may be approximated by

$$\frac{P_u}{P_n} + \frac{M_E}{M_p(1 - P_u/P_e)} = 1.0 \qquad (12.8.2)$$

where $P_n = $ compression nominal strength under axial load based on slenderness ratio for the *axis of bending*

$M_E = $ Eq. 12.4.1 (or its alternate, $C_m M_2$, where $C_m = $ Eq. 12.4.6)

$P_e = \pi^2 EI/L^2$

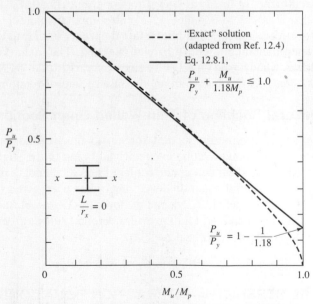

Figure 12.8.1 "Exact" nominal strength interaction relationship for typical wide-flange sections (including residual stress) compared with interaction equation— Case 1, no instability.

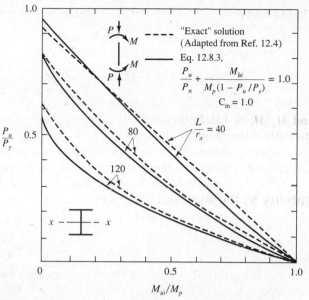

Figure 12.8.2 "Exact" nominal strength interaction curves for typical wide-flange sections (including residual stress) compared with interaction equation—Case 2, instability in the plane of bending ("exact" solution from Fig. 7 of Ref. 12.4). *Note*: For actual use, use adjusted $L/r = $ (actual L/r) $\sqrt{F_y/33}$.

Massonnet [12.1] has shown that Eq. 12.8.2 is a good approximation by comparing it with the curves of Galambos and Ketter [12.8]. A comparison in Fig. 12.8.2 of Eq. 12.8.2 with some curves from Fig. 12.6.2a shows the correlation.

For primary bending moment from transverse loading, Lu and Kamalvand [12.13] have shown that when M_E is replaced by $C_m M_{ui}$, using C_m as given by Eq. 12.3.9, Eq. 12.8.2 (actually Eq. 12.8.3 with $C_m = 1.0$) is also a good representation for the "exact" solutions. Thus, in general, the interaction equation may be written

$$\frac{P_u}{P_n} + \frac{C_m M_{ui}}{M_p(1 - P_u/P_e)} = 1.0 \qquad (12.8.3)$$

for all cases of instability *in the plane of bending*.

Case 3—Instability by Lateral-Torsional Buckling

Massonnet [12.1] has shown that, with only slight error, the form of Eq. 12.8.2 may also be used for this case. P_n for this case may be governed by the slenderness ratio for the axis normal to the axis of bending. Further, since lateral-torsional buckling as a beam may occur for a moment less than M_p, use M_n instead of M_p in Eq. 12.8.3. Using the various definitions of C_m from Secs. 12.3 through 12.5, the general notation

$C_m M_{ui}$ should be used. Thus for instability under Cases 2 or 3, the following interaction equation may be considered to apply:

$$\frac{P_u}{P_n} + \frac{M_{ui} C_m}{M_n(1 - P_u/P_e)} = 1.0 \tag{12.8.4}$$

where P_u = applied axial compression load

M_{ui} = applied primary bending moment

$P_n = A_g F_{cr}$ = nominal strength for axially loaded compression member (Eqs. 6.7.7 or 6.7.8 for F_{cr} would be used in LRFD)

M_n = nominal moment strength in the absence of axial load computed by methods of Chapter 9. For adequately braced members of low slenderness ratio where local buckling is precluded, $M_n = M_p$.

C_m = factors discussed in Secs. 12.3–12.5

$P_e = \dfrac{\pi^2 EI}{L^2}$

Other proposals for interaction equations have been given by Cheong-Siat-Moy and Downs [12.30], Duan and Chen [12.31], Sohal, Duan, and Chen [12.32], and Cal, Liu, and Chen [12.33].

12.9 BIAXIAL BENDING

The nominal strength of members under axial compression and biaxial bending has been studied by Birnstiel and Michalos [12.34], Culver [12.35, 12.36], Harstead, Birnstiel, and Leu [12.37], Syal and Sharma [12.38], Santathadaporn and Chen [12.39], and Chen and Atsuta [12.38]. Even with a number of simplifying assumptions, the analysis is complex. Some tests have been performed [12.41] which, though limited, have shown agreement with computer studies. Tebedge and Chen [12.43] have given interaction surfaces in the form of tables for design. The status of work on biaxial bending of compression members is summarized by Chen and Santathadaporn [12.42] and more recently by the *SSRC Guide* [6.8].

Simple plastic theory becomes inadequate when moments exist about two principal axes. When only one moment exists, plastic behavior (constant moment with increasing rotation θ) is exhibited no matter what the value of axial compression. The effective plastic moment reduces as axial compression increases, but plastic behavior does occur.

Upon application of an additional moment about the other principal axis, one might consider an interaction surface relating P, M_x, and M_y. Even for ideal elastic-plastic material, however, present plastic analysis theorems neglect the influence of deformation on equilibrium. For zero length compression members, the concept of an interaction surface (see Fig. 12.9.1) may be thought of as a first step to obtaining the strength under biaxial bending.

While few designers concern themselves greatly about the sequence of load application, nevertheless loading sequence affects strength. This is also true for uniax-

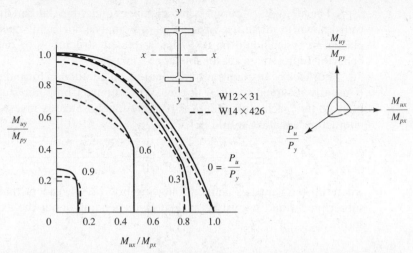

Figure 12.9.1 Contours on strength interaction surface for short members where instability does not occur. (Adapted from Chen and Santathadaporn [12.42])

ial bending and compression, but it has less effect on that case than for the biaxial loading.

Figure 12.9.2 illustrates several loading sequences to reach point A, a particular value of P_u, M_{ux}, and M_{uy}. Point A may be reached by the following paths: (1) Apply P_u first, then M_{uy}, then M_{ux} (path $0-1-2-A$); (2); apply P_u first, then apply M_{ux} and M_{uy} proportionally (path $0-1-A$); (3) apply P_u, M_{ux} and M_{uy} by increasing magnitude in constant proportion (path $O-A$); (4) apply M_{ux} and M_{uy} in constant proportion, then apply P_u (path $O-3-A$).

Other combinations are possible and in general the loading may become applied via any path through space to get from O to A on Fig. 12.9.2. A given section will exhibit a different strength for each path of loading. Nearly all investigators to date (1995) have used proportional loading (path $O-A$).

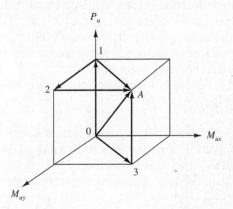

Figure 12.9.2 Paths of loading for biaxial bending combined with axial force. (Adapted from Chen and Santathadaporn [12.42])

The strength of compression members under biaxial bending is not sufficiently well known to make use of it for plastic analysis of rigid space frames; therefore, plastic analysis should be restricted to planar structures, or ones for which planar behavior represents a reasonable approximation.

For lack of any contrary information, an interaction formula, such as Eq. 12.8.4, is usually assumed to apply for biaxial bending. Computer studies and some tests indicate that such a procedure is realistic for those cases investigated. Thus the full interaction equation would be

$$\frac{P_u}{P_n} + \frac{M_{ux}C_{mx}}{M_{nx}(1 - P_u/P_{ex})} + \frac{M_{uy}C_{my}}{M_{ny}(1 - P_u/P_{ey})} \leq 1 \qquad (12.9.1)$$

where all terms are as defined following Eq. 12.8.4, except that now the quantities subscripted x and y must be evaluated for bending about the axis indicated by subscript.

12.10 LOAD AND RESISTANCE FACTOR DESIGN CRITERIA

Long tradition in Allowable Stress Design of beam-columns has used two equations for design; one for yielding or maximum strength and one for stability. That would correspond to using Eq. 12.8.1 for yielding and Eq. 12.8.4 for stability. The early LRFD recommendations [12.44] used those equations.

According to Yura [12.45], Task Group 10 developing LRFD design rules for beam-columns had the following objectives:

1. The rules should apply to a wide range of variables, such as strong and weak axis bending, effect of residual stress, braced (nonsway) and unbraced (sway possible) structures, inelastic behavior, all column slenderness ratios, leaned column systems, and second-order effects.
2. The second-order effect should be kept separate and identifiable in the criterion so that the designer could perform a second-order analysis if desired.
3. Elastic analysis, including elastic second-order analysis if desired, should be the basis for determining internal forces, since practical inelastic analysis techniques are not available for office use.
4. The strength interaction equation using elastic second-order analysis should not be greater than 5% unconservative when compared with "exact" inelastic second-order analysis theoretical solutions.
5. Mathematically identical problems should give the same results using the strength interaction criterion.

There was an attempt to eliminate use of the effective length factor K for compression strength P_n in the criterion; however, this was found not feasible. It was, however, found possible to avoid using separate "yielding" and "stability" equations; instead the single criterion (which has two applicable regions) applies for both. A thorough review of the LRFD provisions for beam-columns has been provided by Chen and Lui [12.46].

Load and Resistance Factor Design in LRFD-H1 prescribes the following for compression and bending moment:

1. For $\dfrac{P_u}{\phi_c P_n} \geq 0.2$

$$\frac{P_u}{\phi_c P_n} + \frac{8}{9}\left(\frac{M_{ux}}{\phi_b M_{nx}} + \frac{M_{uy}}{\phi_b M_{ny}}\right) \leq 1.0 \qquad (12.10.1)$$

2. For $\dfrac{P_u}{\phi_c P_n} < 0.2$

$$\frac{P_u}{2\phi_c P_n} + \left(\frac{M_{ux}}{\phi_b M_{nx}} + \frac{M_{uy}}{\phi_b M_{ny}}\right) \leq 1.0 \qquad (12.10.2)$$

where P_u = factored axial compression load
P_n = nominal compressive strength considering the member as loaded by axial compression only in accordance with LRFD-E2 (See Chapter 6)
ϕ_c = strength reduction factor (resistance factor) for compression members = 0.85
M_{ux} = factored bending moment acting about the x-axis, *including second-order effects*
= $B_{1x}M_{ntx} + B_{2x}M_{\ell tx}$, if moment magnification is used in lieu of computing the elastic second-order moments (The term $B_{2x}M_{\ell tx}$ applies only for the unbraced frame member; i.e., where sway is possible. This case is treated in Sec. 12.11.) Provisions for computing B_1 and B_2 are in LRFD-C1.
M_{nx} = nominal moment strength for bending about the x-axis, in accordance with LRFD-F1 (See Chapter 9)
ϕ_b = strength reduction factor (resistance factor) for flexural members = 0.90
M_{uy} = same as M_{ux} except referred to the y-axis
M_{ny} = same as M_{nx} except referred to the y-axis

Moment Magnifier—Braced Frame

The moment magnifier B_1 for members having *no* joint translation was treated in Secs. 12.3 and 12.4. For this *nonsway* case, LRFD-C1 gives the magnifier as (Eq. 12.3.7 where $\alpha = P_u/P_{e1}$; subscript 1 refers to braced case)

$$B_1 = \frac{C_m}{1 - P_u/P_{e1}} \geq 1.0 \qquad (12.10.3)$$

where C_m = factors discussed in Secs. 12.3 and 12.4, to be taken as follows:

1. For *braced frame* members having transverse loading between supports, C_m is an integral part of the moment magnifier B_1, whose value may be determined by rational analysis,

$$C_m = 1 + \psi\alpha = 1 + \psi\frac{P_u}{P_{e1}} \text{ (LRFD Com-C1)} \qquad (12.10.4)$$

$$= \text{Values from Table 12.3.1}$$

$$C_m = 1.0 \text{ (LRFD-C1)} \qquad (12.10.5)$$
(ends of member unrestrained)

$$C_m = 0.85 \text{ (LRFD-C1)} \qquad (12.10.6)$$
(ends of member restrained)

2. For *braced frame* members without transverse loading between supports but having end moments M_1 (smaller one) and M_2 (larger one), C_m converts the linearly varying primary bending moment into an equivalent uniform moment $M_E = C_m M_2$,

$$C_m = 0.6 - 0.4M_1/M_2 \qquad (12.10.7)$$

The moments M_1 and M_2 are *rotational* moments, rather than bending moments as used in Sec. 12.4. Therefore, the ratio is negative $(-)$ for single curvature and positive $(+)$ for double curvature. Since M_E is the primary moment, C_m is really not part of the magnification factor.

P_u = factored axial compression load
P_{e1} = Euler load, using KL/r (or λ_c) for the *axis of bending* and with $K \leq 1.0$ (for *braced* frame)

$$P_{e1} = \frac{\pi^2 EA_g}{(KL/r)^2} = \frac{A_g F_y}{\lambda_c^2} \qquad (12.10.8)$$

The braced frame (no translation) beam-column total factored moment is

$$M_u = B_1 M_{nt} \qquad (12.10.9)$$

where M_{nt} = the primary factored moment (for the *no t*ranslation case; hence the subscript *nt*)

EXAMPLE 12.10.1 _____

Investigate the acceptability of a W16×67 used as a beam-column in a braced frame under the loading shown in Fig. 12.10.1. The steel is A572 Grade 60.

Solution

(a) Factored loads.

$$P_u = 1.2P_D + 1.6P_L = 1.2(87.5) + 1.6(262.5) = 525 \text{ kips}$$
$$M_{nt} = 1.2M_D + 1.6M_L = 1.2(15) + 1.6(45) = 90 \text{ ft-kips}$$
$$M_u = B_1 M_{nt}$$

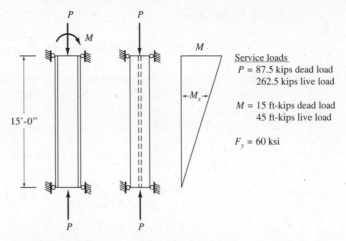

Figure 12.10.1 Beam-column for Examples 12.10.1 and 12.14.1.

Note that the factored primary moment is referred to using the subscript (*no t*ranslation).

(b) Column effect. Calculate λ_c using Eq. 6.7.3,

$$\text{Largest } \frac{KL}{r} = \frac{KL}{r_y} = \frac{1.0(15)12}{2.46} = 73.2$$

$$\lambda_c = \frac{KL}{r}\sqrt{\frac{F_y}{\pi^2 E}} = 73.2\sqrt{\frac{60}{\pi^2 29{,}000}} = 1.06$$

Find $\dfrac{\phi_c F_{cr}}{F_y} = 0.531$ (LRFD-"NUMERICAL VALUES" TABLE 4)

$$\phi_c F_{cr} = 0.531(60) = 31.9 \text{ ksi}$$
$$\phi_c P_n = \phi_c F_{cr} A_g = 31.9(19.7) = 628 \text{ kips}$$

Check $P_u/(\phi_c P_n) \geq 0.2$,

$$\left(\frac{P_u}{\phi_c P_n} = \frac{525}{628} = 0.836 \right) > 0.2; \quad \underline{\textit{Use} \text{ Eq. 12.10.1}}$$

Note that for the web $\lambda = h/t_w = 35.9$, which exceeds the λ_r limit of $253/\sqrt{F_y} = 32.7$ given in LRFD-B5. However, Q will be less than 1.0 only when h/t_w exceeds $253/\sqrt{F_{cr}} = 41.3$. (See Sec. 6.18 and LRFD-Appendix B5.3b.)

(c) Beam effect. The laterally unbraced length L_b is 15 ft.

$$M_p = F_y Z_x = 60(130)/12 = 650 \text{ ft-kips}$$
$$M_r = (F_y - F_r)S_x = (60 - 10)(117)/12 = 488 \text{ ft-kips}$$
$$L_p = \frac{300}{\sqrt{F_y}} r_y = \frac{300(2.46)}{\sqrt{60}\,(12)} = 7.9 \text{ ft}$$

Using $X_1 = 2350$ ksi and $X_2 = 4690 \times 10^{-6}$ in.^{4}/kips^2,

$$L_r = \frac{r_y X_1}{(F_y - F_r)} \sqrt{1 + \sqrt{1 + X_2(F_y - F_r)^2}} = 20.6 \text{ ft}$$

Since $L_p < L_b < L_r$, M_n lies between M_p and M_r for $C_b = 1.0$. In this case, however, $C_b = 1.67$ using Eq. 9.6.11 or Table 9.6.3; thus, it is probable that M_n will equal M_p. Check whether or not the W16×67 is "compact" for $F_y = 60$ ksi according to LRFD-B5,

$$\left(\frac{b_f}{2t_f} = \frac{10.235}{2(0.665)} = 7.7 \right) < (\lambda_p = 8.4 \quad \text{Table 9.6.1}) \quad \text{OK}$$

$$\left(\frac{P_u}{\phi_b P_y} = \frac{525}{0.9(60)19.7} = 0.494 \right) > 0.125$$

$$\lambda_p = \frac{191}{\sqrt{F_y}} \left(2.33 - \frac{P_u}{\phi_b P_y} \right) \geq \frac{253}{\sqrt{F_y}}$$

$$\lambda_p = \left[\frac{191}{\sqrt{60}} (2.33 - 0.494) = 45.3 \right] > \left[\frac{253}{\sqrt{60}} = 32.7 \right]$$

$$\left[\lambda = \frac{h}{t_w} = 35.9 \quad \begin{array}{c} \text{From } LRFD \\ \textit{Manual} \end{array} \right] < [\lambda_p = 45.3] \quad \text{OK}$$

The section is "compact". Next use Eq. 9.6.4 to compute M_n,

$$M_n = C_b \left[M_p - (M_p - M_r) \left(\frac{L_b - L_p}{L_r - L_p} \right) \right] \leq M_p \qquad [9.6.4]$$

$$= 1.67 \left[650 - (650 - 488) \left(\frac{15 - 7.9}{20.6 - 7.9} \right) \right] = 934 \text{ ft-kips}$$

Since M_n cannot exceed M_p, $M_n = M_p$ for this analysis.

(d) Moment magnification. The slenderness ratio KL/r involved in moment magnification must relate to the *axis of bending,* in this case the x-axis,

$$\text{Axis of bending } \frac{KL}{r} = \frac{KL}{r_x} = \frac{1.0(15)12}{6.96} = 25.9$$

$$C_m = 0.6 - 0.4(M_1/M_2) = 0.6 - 0.4(0/90) = 0.6$$

$$P_{e1} = \frac{\pi^2 EA_g}{(KL/r)^2} = \frac{\pi^2(29,000)19.7}{(25.9)^2} = 8430 \text{ kips}$$

$$B_1 = \frac{C_m}{1 - P_u/P_{e1}} = \frac{0.6}{1 - 525/8430} = 0.6(1.07) = 0.64$$

Since B_1 is computed to be less than 1.0, use 1.0. In this case, the moment varying from 90 ft-kips to zero over the 15 ft length is the same as if $C_m M_2 = 0.6(90) = 54$ ft-kips were constant over that length. The 54 ft-kips is then magnified to $54(1.07) = 58$ ft-kips, *but it is still less than the actual value* 90 ft-kips *at the end of the member.* The use of $B_1 = 1.0$ is the same as saying the magnified value out in the span is less than the value M_2 at the end of the member.

(e) Check LRFD Formula (H1-1a), Eq. 12.10.1, omitting the bending term for the y-axis,

$$\frac{P_u}{\phi_c P_n} + \frac{8}{9}\left(\frac{M_{ux}}{\phi_b M_{nx}}\right) \le 1.0 \qquad [12.10.1]$$

$$0.836 + \frac{8}{9}\left(\frac{90}{0.90(650)}\right) = 0.836 + 0.137 = 0.973 < 1 \quad \text{OK}$$

The W16×67 section *is* acceptable according to LRFD. ∎∎

12.11 UNBRACED FRAME—LOAD AND RESISTANCE FACTOR DESIGN

As shown in Figs. 6.9.2b and c and 6.9.3, an unbraced frame must rely on the flexural interaction of its beams and columns to limit horizontal displacement. Under lateral loads, a "braced" frame will resist the lateral force by such components as diagonal bracing or shear walls so that lateral distortion will be of small magnitude. Thus secondary bending moments $P\Delta$ from sidesway (the P–Δ effect) may ordinarily be neglected. However, for "unbraced" frames, the relatively larger sidesway deflection Δ due to lateral load will give rise to secondary moments $P\Delta$ (P is the gravity load) that must be provided for in design. Thus, an unbraced frame requires an analysis to accomplish the following tasks:

1. Provide strength under factored loads to resist gravity load, neglecting any sidesway effect except in cases of unbalanced loading or unsymmetric structural configuration where a "significant restraining force" (LRFD Commentary C1, par. 8) would be necessary to prevent sway. Out-of-plumbness can typically be expected to have greater effect than sway under gravity load.
2. Provide strength under factored loads to resist lateral load (i.e., factored wind or earthquake load). The moments under lateral load include the primary moments from first-order elastic analysis plus secondary moments due to P–Δ effect.
3. Provide stiffness such that the relative horizontal deflection (sway) between adjacent floors, and for the entire frame, is within specified limits (usually, say, equal to the height L_s divided by 400 or 500 under service loads).

Referring to Fig. 12.11.1, first-order equilibrium requires

$$M_{\ell t1} + M_{\ell t2} = H_u L_s \qquad (12.11.1)$$

The first-order sway deflection Δ_{1u} causes the total gravity load ΣP_u to be acting at the eccentricity Δ_{1u}. The lateral load moment $H_u L_s$ is thus increased by the amount $\Sigma P_u \Delta_{1u}$. Since the total moment now acting is $H_u L_s + \Sigma P_u \Delta_{1u}$, the relative lateral deflection (sway) will increase to Δ_{2u} when the structure reaches equilibrium in the final displaced position, as shown in Fig. 12.11.1b.

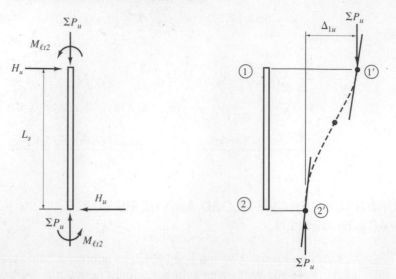

(a) First-order analysis: For equilibrium, $M_{\ell t1} + M_{\ell t2} = H_u L_s$

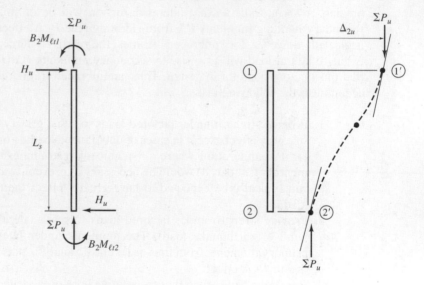

(b) Second-order analysis: For equilibrium, $B_2(M_{\ell t1} + M_{\ell t2}) = H_u L_s + \Sigma P_u \Delta_{2u}$

Figure 12.11.1   Summation of forces acting on all columns in one story of a multistory building frame.

The final (including $P-\Delta$ effect) moment equilibrium may be expressed

$$B_2(M_{\ell t1} + M_{\ell t2}) = H_u L_s + \Sigma P_u \, \Delta_{2u} \qquad (12.11.2)$$

in which B_2 is the magnification factor, and $M_{\ell t1}$ and $M_{\ell t2}$ are primary moments (called $M_{\ell t}$ in LRFD-C1, with the subscript ℓt referring to lateral translation) resulting from a sway analysis under factored loads.

Comparing Eq. 12.11.2 with Eq. 12.11.1 gives

$$B_2 = \frac{H_u L_s + \Sigma P_u \Delta_{2u}}{H_u L_s} \qquad (12.11.3)$$

Referring to Fig. 12.11.1, and using a proportionality factor η, let

$$\Delta_{u1} = \eta H_u \qquad (12.11.4)$$

which is valid for first-order analysis. The equivalent magnified lateral load in Fig. 12.11.1b, that is, the total moment divided by L_s, may be taken as

$$\text{Equivalent lateral load} = H_u + \frac{\Sigma P_u \Delta_{2u}}{L_s} \qquad (12.11.5)$$

Thus,

$$\Delta_{2u} = \eta (\text{Equivalent lateral load})$$

$$= \eta \left(H_u + \frac{\Sigma P_u \Delta_{2u}}{L_s} \right) \qquad (12.11.6)$$

Substituting Δ_{1u} for ηH_u gives

$$\Delta_{2u} = \Delta_{1u} + \frac{\Delta_{1u} \Sigma P_u \Delta_{2u}}{H_u L_s} \qquad (12.11.7)$$

from which solving for Δ_{2u} gives

$$\Delta_{2u} = \frac{\Delta_{1u}}{1 - \Sigma P_u \left(\dfrac{\Delta_{1u}}{H_u L_s} \right)} \qquad (12.11.8)$$

Substituting Eq. 12.11.8 into Eq. 12.11.3 gives

$$B_2 = \frac{1}{1 - \Sigma P_u \left(\dfrac{\Delta_{1u}}{H_u L_s} \right)} \qquad (12.11.9)$$

Note that H_u in Eq. 12.11.9 is the *total* lateral force (shear) acting in the story, called ΣH in LRFD-C1. Even though the deflection Δ_{1u} and force H_u are for factored loads in the derivation, the ratio Δ_{1u}/H_u for factored loads or Δ_1/H for service loads will be the same, since the first-order analysis is to be *elastic*.

There are two ways in which the magnified end moments $B_2 M_{\ell t 1}$ and $B_2 M_{\ell t 2}$ may be obtained:

1. *Moment Magnifier Method.* LRFD-C1 indicates that the total factored moment M_u for use in LRFD Formulas (H1-1a) or (H1-1b), Eqs. 12.10.1 or 12.10.2, is obtained as follows:

$$M_u = B_1 M_{nt} + B_2 M_{\ell t} \qquad (12.11.10)$$

where *two* first-order elastic analyses are required; (1) a gravity-only analysis assuming no sway to obtain the M_{nt} values and B_1; and (2) a lateral-load-only sway analysis to obtain the $M_{\ell t}$ values and B_2. The nonsway magnifier B_1 has beeen treated in Secs. 12.3, 12.4, and 12.10. Duan, Sohal, and Chen [12.47] have presented a discussion of the B_1 factor and suggested alternative formulation.

The sway magnifier B_2 is given by LRFD-C1 as

$$B_2 = \frac{1}{1 - \dfrac{\Sigma P_u}{\Sigma P_{e2}}} \tag{12.11.11}$$

or

$$B_2 = \frac{1}{1 - \Sigma P_u \left(\dfrac{\Delta_{oh}}{\Sigma HL} \right)} \tag{12.11.12}$$

where ΣP_u = factored axial compression load for all columns in a story subject to sway

Δ_{oh} = translation deflection (sway deflection) of the story under consideration; under factored load when factored horizontal loads H_u are used, or under service load when service horizontal loads H are used

ΣH = sum of all story horizontal forces producing Δ_{oh}

L = story height

P_{e2} = Same as Eq. 12.10.8, except that K in the plane of bending will be based on the unbraced frame action and will be ≥ 1.0. Figure 6.9.4b for *sidesway not prevented* may be used.

2. *Second-Order Analysis.* An alternate to computing the magnified primary moments using Eq. 12.11.10 is to directly use a second-order analysis under factored loads. A second-order analysis is one in which the equations of equilibrium are based on the deformed structure instead of the original undeformed geometry as in first-order analysis. The reader is referred to the methods of MacGregor and Hage [12.48], LeMessurier, McNamara, and Scrivener [12.49], LeMessurier [12.50, 12.51], Wood, Beaulieu, and Adams [12.52, 12.53], and Scholz [12.54] for suggestions on second-order analysis.

Equation 12.11.11 is of the form derived in Secs. 12.3 and 12.5, and Eq. 12.11.12 was derived as Eq. 12.11.9, replacing Δ_{1u} by the symbol Δ_{oh}, recognizing that H_u is the total horizontal force acting in the story by using the Σ sign. Further, noting that the quantity $\Delta_{oh}/\Sigma H$ may be either for factored loads or for service loads, the subscript u is not used.

The story stiffness concept of assuming each story behaves independently of the other stories and that the $P-\Delta$ effect is equivalent to the effect of a lateral force $\Sigma P_u \Delta/L$ has been discussed by Chen and Lui [12.46] and MacGregor and Hage [12.48], among others.

EXAMPLE 12.11.1 _____

Investigate the acceptability of W14×145 columns in a single-bay multistory unbraced frame, part of which is shown in Fig. 12.11.2. The axial compression P is the

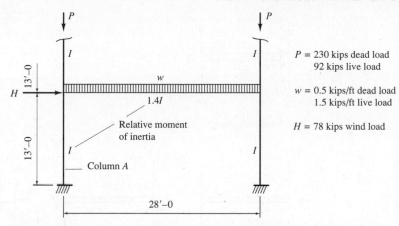

Figure 12.11.2 Lowest portion of single-bay multistory unbraced frame for Example 12.11.1.

total load acting on the bottom story columns and the wind load H is the total horizontal service wind load acting on the entire multi-story frame. The steel is A36. Use Load and Resistance Factor Design.

Solution

(a) Compute factored loads. In accordance with LRFD-A4.1, at least two factored load combinations are necessary when wind or earthquake act on the structure,

1. Gravity Load [LRFD-A4.1, Formula (A4-2)]

$$P_u = 1.2(230) + 1.6(92) = 423 \text{ kips}$$
$$w_u = 1.2(0.5) + 1.6(1.5) = 3.0 \text{ kips/ft on each floor}$$

2. Gravity Load + Wind [LRFD-A4.1, Formula (A4-4)]

$$P_u = 1.2(230) + 0.5(92) = 322 \text{ kips}$$
$$H_u = 1.3(78) = 101.4 \text{ kips}$$
$$w_u = 1.2(0.5) + 0.5(1.5) = 1.35 \text{ kips/ft on all floors except roof}$$

(b) First-order elastic structural analysis. Unless a second-order analysis is to be performed, the factored moments used in the investigation must be obtained using moment magnification, Eq. 12.11.10. Assume the factored gravity load (loading 1) in part (a) gives the factored moments in Fig. 12.11.3a.

The gravity plus wind (loading 2) in part (a) is divided into two separate first-order elastic analyses under factored loads; a nonsway analysis under gravity loads only, and a sway analysis under lateral load only. Assume the results on the column to be investigated are as shown in Fig. 12.11.3c and d; the sway analysis also causes an additional factored compressive load of 15 kips on the leeward column. Loading 2 will be initially assumed to govern the beam-column strength.

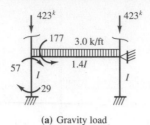

(a) Gravity load

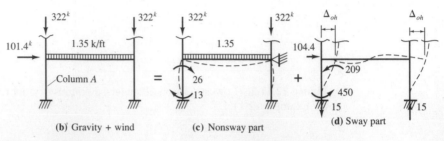

(b) Gravity + wind   **(c)** Nonsway part   **(d)** Sway part

Figure 12.11.3   Forces from first-order elastic analyses under factored loads— Example 12.11.1.

(c) Column strength. Verify that local buckling will not occur; i.e., that $Q = 1.0$:

$$\left[\lambda = \frac{b_f}{2t_f} = 7.1\right] < \left[\lambda_r = \frac{95}{\sqrt{F_y}} = \frac{95}{\sqrt{36}} = 15.8\right] \quad \text{OK}$$

$$\left[\lambda = \frac{h}{t_w} = 16.8\right] < \left[\lambda_r = \frac{253}{\sqrt{F_y}} = \frac{253}{\sqrt{36}} = 42.2\right] \quad \text{OK}$$

The effective length factor K_x in the plane of the frame is determined using restraint factors G with Fig. 6.9.4.

$$G_{\text{top}} = \frac{\Sigma I/L, \text{ columns}}{\Sigma I/L, \text{ beams}} = \frac{2(I/13)}{1.4I/28} = 3.08$$

$G_{\text{bottom}} = 1.0$ (The practical recommendation of LRFD Commentary-Fig. C-C2.2 for nominally fixed base.)

Find $K_x = 1.57$ (From Fig. 6.9.4)

Since in the y-direction the column is considered pinned at the top and bottom, $K_y = 1.0$; thus,

$$\frac{K_x L_x}{r_x} = \frac{1.57(13)12}{6.33} = 38.7 \qquad \frac{K_y L_y}{r_y} = \frac{1.0(13)12}{3.98} = 39.2$$

$\phi_c F_{cr} = 28.2$ ksi (LRFD "NUMERICAL VALUES" TABLE 5)

$\phi_c P_n = \phi_c F_{cr} A_g = 28.2(42.7) = 1204$ Kips

Check $P_u/(\phi_c P_n) \ge 0.2$,

$$\left(\frac{P_u}{\phi_c P_n} = \frac{322 + 15 + 1.35(14)}{1204} = 0.30\right) > 0.2; \quad \text{Use Eq. 12.10.1}$$

(d) Beam effect. The laterally unbraced length L_b is 13 ft.

$$M_p = F_y Z_x = 36(260)/12 = 780 \text{ ft-kips}$$

$$L_p = \frac{300}{\sqrt{F_y}} r_y = \frac{300(3.98)}{\sqrt{36}\,(12)} = 16.6 \text{ ft}$$

Since $L_b < L_p$, M_n equals M_p. Check whether or not the W14×145 is "compact" for $F_y = 36$ ksi according to LRFD-B5,

$$\left(\lambda = \frac{b_f}{2t_f} = \frac{15.500}{2(1.090)} = 7.1 \right) < (\lambda_p = 10.8 \quad \text{Table 9.6.1}) \quad \text{OK}$$

$$\left(\frac{P_u}{\phi_b P_y} = \frac{525}{0.9(60)19.7} = 0.494 \right) > 0.125$$

$$\lambda_p = \frac{191}{\sqrt{F_y}} \left(2.33 - \frac{P_u}{\phi_b P_y} \right) \geq \frac{253}{\sqrt{F_y}}$$

$$\lambda_p = \left[\frac{191}{\sqrt{60}}(2.33 - 0.494) = 45.3 \right] > \left[\frac{253}{\sqrt{60}} = 32.7 \right]$$

$$\left[\lambda = \frac{h}{t_w} = 35.9 \, \substack{\text{From } LRFD \\ \text{Manual}} \right] < [\lambda_p = 45.3] \quad \text{OK}$$

Note that when $h/t_w < \lambda_r$ for axial compression alone, as in part c, then $\lambda < \lambda_p$ for beam action is automatically satisfied. For most rolled I-shaped sections this is the case. The properties section of the *LRFD Manual* identifies sections that are compact for beam action, regardless of the magnitude of axial compression, by giving no value (i.e., —) for F_y'''. When a value of F_y''' is given lower than the value being used, then the effect of axial compression on web local buckling must be investigated as above, as well as investigate flange local buckling.

The section is "compact"; therefore, $M_n = M_p = 780$ ft-kips: $\phi M_n = 0.90(780) = 702$ ft-kips.

(e) Moment magnification—nonsway magnifier B_1 for structure of Fig. 12.11.3c. The slenderness ratio ratio KL/r involved in moment magnification must relate to the *axis of bending*, in this case the x-axis. Thus $K_x \leq 1.0$ for the nonsway part. Since the structure is actually an *unbraced* one, the authors recommend using 1.0 rather than any value less than 1.0 (say, from Fig. 6.9.4a).

$$\text{Axis of bending} \quad \frac{KL}{r} = \frac{KL}{r_x} = \frac{1.0(13)12}{6.33} = 24.6$$

$$C_m = 0.6 - 0.4(M_1/M_2) = 0.6 - 0.4(13/26) = 0.4$$

$$P_{e1} = \frac{\pi^2 E A_g}{(KL/r)^2} = \frac{\pi^2(29,000)42.7}{(24.6)^2} = 20,200 \text{ kips}$$

$$P_u = 332 + 1.35(14) = 341 \text{ kips}$$

$$B_1 = \frac{C_m}{1 - P_u/P_{e1}} = \frac{0.4}{1 - 341/20,200} = 0.4(1.02) = 0.41$$

When B_1 is computed to be less than 1.0, the magnified moment between the ends of the column in Fig. 12.11.3c is less than the moment at the end of the member (26 ft-kips). Use $B_1 M_{nt} = 26$ ft-kips.

(f) Moment magnification—sway magnifier B_2 for structure of Fig. 12.11.3d. The total factored compression load ΣP_u to be carried by all columns of the frame within one story (in this case two columns) is

$$\Sigma P_u = 2(322) + 1.35(28) = 682 \text{ kips}$$

The Euler load P_{e2} for the column being investigated must be computed using KL/r for the axis of bending, and the K value must be for the *unbraced* frame, that is, $K \geq 1.0$. In this case, $K_x = 1.57$ and $K_x L_x/r_x = 38.7$ as determined in part (c),

$$P_{e2} = \frac{\pi^2 E A_g}{(KL/r)^2} = \frac{\pi^2 (29,000)42.7}{(38.7)^2} = 8160 \text{ kips}$$

and the $\Sigma P_{e2} = 2(8160) = 16,320$ kips because both columns resisting sway are identical. Thus, the sway magnifier B_2 is

$$B_2 = \cfrac{1}{1 - \cfrac{\Sigma P_u}{\Sigma P_{e2}}} = \cfrac{1}{1 - \cfrac{682}{16,320}} = 1.04$$

The maximum magnified moment M_u, Eq. 12.11.10, for column A is

$$M_u = B_1 M_{nt} + B_2 M_{\ell t}$$
$$= 1.0(26) + 1.04(450) = 496 \text{ ft-kips}$$

Note that LRFD-C1 requires the *maximum* M_{nt} and *maximum* $M_{\ell t}$ to be used in computing M_u. Logically, the values of M_{nt} and $M_{\ell t}$ simultaneously should be at the same end of the member. Even that procedure is not entirely correct because the maximum magnified moment in the nonsway case may occur out in the span (and will when B_1 exceeds 1.0) while the magnified sway moment occurs at the end of the member.

(g) Check LRFD Formula (H1-1a), Eq. 12.10.1, omitting the bending term for the y-axis,

$$\frac{P_u}{\phi_c P_n} + \frac{8}{9}\left(\frac{M_{ux}}{\phi_b M_{nx}}\right) \leq 1.0 \qquad [12.10.1]$$

$$0.30 + \frac{8}{9}\left(\frac{496}{702}\right) = 0.30 + 0.63 = 0.93 < 1 \quad \text{OK}$$

The W14×145 section *is* acceptable according to LRFD. ■■

12.12 DESIGN PROCEDURES— LOAD AND RESISTANCE FACTOR DESIGN

To aid in selection of a beam-column section, it is usually advantageous to convert, in at least an approximate way, the resulting bending moment M_u into an equivalent axial compression load P_{uEQ}. Occasionally, conversion of the axial compression into equivalent moment will speed the selection process.

The interaction equation for $P_u/\phi_c P_n \geq 0.2$, Eq. 12.10.1, is

$$\frac{P_u}{\phi_c P_n} + \frac{8}{9}\left(\frac{M_{ux}}{\phi_b M_{nx}} + \frac{M_{uy}}{\phi_b M_{ny}}\right) \leq 1.0 \qquad [12.10.1]$$

Multiplying by $\phi_c P_n$ gives

$$P_u + \frac{8}{9}\left(\frac{\phi_c P_n}{\phi_b M_{nx}}\right)M_{ux} + \frac{8}{9}\left(\frac{\phi_c P_n}{\phi_b M_{ny}}\right)M_{uy} \leq (\phi_c P_n = P_{uEQ}) \qquad (12.12.1)$$

Then, let $M_{nx} = F_y Z_x$ and $M_{ny} = F_y Z_y$ for preliminary design. M_{nx} may be less than the plastic moment strength as controlled by the limit states of lateral-torsional buckling, local flange buckling, or local web buckling; however, commonly M_{nx} will equal the plastic moment strength. $M_n = M_p$ will usually be a good assumption at the start of a design. Next, factoring out the cross-sectional properties A_g/Z_x and A_g/Z_y, Eq. 12.12.1 becomes

$$P_{uEQ} = P_u + \frac{8}{9}\left(\frac{\phi_c F_{cr}}{\phi_b F_y}\right)\frac{A_g}{Z_x}M_{ux} + \frac{8}{9}\left(\frac{\phi_c F_{cr}}{\phi_b F_y}\right)\frac{A_g}{Z_y}M_{uy} \qquad (12.12.2)$$

For a *braced* frame under uniaxial bending and compression where $M_{ux} = M_{ntx}[C_{mx}/(1 - P_u/P_{ex})]$, Eq. 12.12.2 (neglecting the third term for bending about the y-axis) becomes

$$P_{uEQ} = P_u + \frac{8}{9}\left(\frac{\phi_c F_{cr}}{\phi_b F_y}\right)\frac{A_g}{Z_x}\left(\frac{C_{mx}}{1 - \dfrac{P_u}{P_{ex}}}\right)M_{ntx} \qquad (12.12.3)$$

Next, the magnification term, where $\pi^2 E = 286{,}000$, becomes

$$\frac{1}{1 - \dfrac{P_u}{P_{ex}}} = \frac{P_{ex}}{P_{ex} - P_u} = \frac{286{,}000 I_x}{286{,}000 I_x - P_u(K_x L_x)^2} \qquad (12.12.4)$$

Thus, the equivalent factored column load P_{uEQ} may be expressed using Eq. 12.12.2, corresponding to LRFD-Formula (H1-1a) for uniaxial bending *when* $M_n = M_p$,

$$P_{uEQ} = P_u + M_{nt}\beta_{az}\left(\frac{8\phi_c F_{cr}}{9\phi_b F_y}\right)\left(\frac{\beta_m}{\beta_m - P_u(KL)^2}\right) \qquad (12.12.5)$$

where M_{nt} = factored moment using first-order analysis on a *braced* frame
β_{az} = ratio A_g/Z, using Z_x or Z_y as appropriate for the axis of bending
β_m = $286{,}000 I$, using I_x or I_y as appropriate for the axis of bending
KL = $K_x L_x$ or $K_y L_y$ as appropriate for the axis of bending

Equation 12.12.5 is a reasonable way to approach the design of a beam-column. Estimate β_{az} from Table 12.12.1. Taking $\beta_{az} = 0.2$/in. will be a reasonable general approach when the depth of section has not been predetermined. The first bracketed term (stress ratio) will be a reduction term and may be estimated as 0.7 to begin the

TABLE 12.12.1   APPROXIMATE VALUES FOR RATIOS β_{az} AND β_b TO USE IN LFRD DESIGN OF BEAM-COLUMNS

Sections	$\beta_{az} = \dfrac{A_g}{Z_x}$			$\beta_b = \dfrac{Z_x}{Z_y}$
	Light weight (1/in.)	Medium weight (1/in.)	Heavy weight (1/in.)	
W8	0.33	0.30	0.29	2.2
W10	0.22	0.24	0.28	2.1
W12	0.16	0.20	0.20	2.2
W14	0.13	0.16	0.20	2.1
W16	0.15	0.16	0.17	5.0
W18	0.12	0.14	0.16	5.0
W21	0.10	0.12	0.14	6.0
W24	0.09	0.11	0.12	7.0
W27	0.08	0.09	0.10	7.0
W30	0.08	0.09	0.09	7.5
W33	0.07	0.08	0.08	8.0
W36	0.07	0.07	0.08	8.5

process. The last bracketed term is the magnification term and might be neglected for the first approximation.

Yura [12.45] has suggested using the following as an initial approximation for biaxial bending in both the *braced* frame and *unbraced* frame case,

$$P_{uEQ} = P_u + M_{ux}\left(\frac{2}{d}\right) + M_{uy}\left(\frac{7.5}{b_f}\right) \tag{12.12.6}$$

where d = nominal depth of the section

        b_f = nominal width of the flange

       M_{ux} = factored moment about the x-axis ($B_1 M_{nt}$ for the *braced* frame or $B_1 M_{nt} + B_2 M_{\ell t}$ for the *unbraced* frame)

       M_{uy} = factored moment about the y-axis

Note that $Z_x \approx 2A_f(d/2)$ and $A_g \approx 2A_f$; therefore $(\beta_{az})_x = A_g/Z_x \approx 2/d$ as in Eq. 12.12.6. Further, taking $Z_y \approx A_f b_f/4$ gives $(\beta_{az})_y \approx 8/b_f$, close to the $7.5/b_f$ recommended by Yura. As a first approximation, Yura is essentially suggesting to neglect the reduction effect of the stress ratio term and the increasing effect of the magnification term; a logical beginning.

When bending moment is the predominant internal force, the practical approach to design may be to compute an estimated equivalent bending moment M_{uEQ}.

Multiply Eq. 12.10.1 by $\phi_b M_{nx}$ giving

$$P_u \frac{\phi_b M_{nx}}{\phi_c P_n} + \frac{8}{9} M_{ux} + \frac{8}{9} M_{uy} \frac{\phi_b M_{nx}}{\phi_b M_{ny}} \le \phi_b M_{nx} \tag{12.12.7}$$

Letting $M_{nx} = F_y Z_x$, $M_{ny} = F_y Z_y$, and $P_n = F_{cr} A_g$,

$$P_u \frac{\phi_b}{\phi_c} \frac{F_y}{F_{cr}} \frac{Z_x}{A_g} + \frac{8}{9} M_{ux} + \frac{8}{9} M_{uy} \frac{F_y}{F_y} \frac{Z_x}{Z_y} \le \phi_b M_{nx} \tag{12.12.8}$$

Next, defining $\beta_{az} = A_g/Z_x$ and $\beta_b = Z_x/Z_y$, and calling the right hand side M_{uEQ}, gives

$$M_{uEQ} = \frac{8}{9}M_{ux} + \frac{8}{9}M_{uy}\beta_b + P_u\left(\frac{1}{\beta_{az}}\right)\left(\frac{\phi_b F_y}{\phi_c F_{cr}}\right) \tag{12.12.9}$$

which for *uniaxial bending* becomes

$$M_{uEQ} = \frac{8}{9}M_u + P_u\left(\frac{1}{\beta_{az}}\right)\left(\frac{\phi_b F_y}{\phi_c F_{cr}}\right) \tag{12.12.10}$$

For *biaxial bending*, approximating $(\beta_{az})_x = A_g/Z_x \approx 2/d$ and $(\beta_{az})_y = A_g/Z_y \approx 8/b_f$ makes $\beta_b = Z_x/Z_y \approx 4d/b_f$. Further, assuming the stress ratio term is 1.4 and $1/\beta_{az}$ is $d/2$, Eq. 12.12.9 then becomes,

$$M_{uEQ} = \frac{8}{9}\left[M_{ux} + \left(\frac{4d}{b_f}\right)M_{uy}\right] + P_u(0.7d) \tag{12.12.11}$$

where M_{ux} and M_{uy} are the factored moments about the x- and y-axes, respectively (that is, $B_1 M_{nt}$ for the braced column, and $B_1 M_{nt} + B_2 M_{\ell t}$ for the unbraced frame column).

The above equations for design assume that axial compression is such that the interaction equation, Eq. 12.10.1, for $P_u/\phi_c P_n \geq 0.2$ applies. For low axial compression where Eq. 12.10.2 is to be used, it is likely to be more practical to use the M_{uEQ} approach keeping in mind to use one-half of the P_u term and eliminate the 8/9 fraction.

12.13 EXAMPLES—LRFD

Several examples are included to illustrate application of the interaction formulas using principles and procedures for columns from Chapter 6, and beams from Chapters 7 and 9.

EXAMPLE 12.13.1 ───

Select the lightest W14 section to carry an axial compression load of 120 kips dead load and 30 kips live load, in combination with a bending moment of 100 ft-kips dead load, 100 ft-kips live load, and 300 ft-kips wind load. The member is part of a *braced* system, with transverse support provided in each principal direction at the top and bottom of a 14-ft length. Conservatively assume the moment causes single curvature and varies as shown in Fig. 12.13.1. Use A36 steel and Load and Resistance Factor Design.

Solution.

(a) Compute factored loads. Assume a first-order elastic analysis was performed to obtain the given forces. Using LRFD-A4.1, Formulas (A4-2) and (A4-4),

1. Gravity Load

$$P_u = 1.2(120) + 1.6(30) = 192 \text{ kips}$$
$$M_{nt} = 1.2(100) + 1.6(100) = 280 \text{ ft-kips}$$

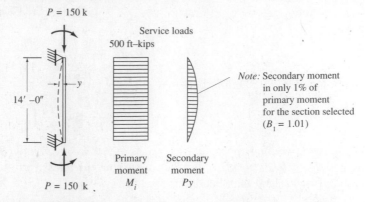

Figure 12.13.1   Examples 12.13.1 and 12.15.1.

2. Gravity Load + Wind

$$P_u = 1.2(120) + 0.5(30) = 159 \text{ kips}$$
$$M_{nt} = 1.2(100) + 0.5(100) + 1.3(300) = 560 \text{ ft-kips}$$

It appears that loading Case 2 will be the more severe loading; thus, use that case to select the section. Then any other loading cases can be checked.

(b) Estimate the P_{uEQ} for the given conditions using either Yura's equation, Eq. 12.12.6, or the more detailed Eq. 12.12.5. For Eq. 12.12.6,

$$P_{uEQ} = P_u + M_x\left(\frac{2}{d}\right) = 159 + 560(12)\left(\frac{2}{14}\right) = 1120 \text{ kips}$$

or, using Eq. 12.12.5 and estimating β_{az} as 0.16 from Table 12.12.1 for medium weight W14 sections, and using 0.7 for the "stress ratio" term,

$$\begin{aligned} P_{uEQ} &= P_u + M_{nt}\beta_{az}(0.7) \\ &= 159 + 560(12)(0.16)0.7 = 912 \text{ kips} \end{aligned}$$

The difference in the estimates is the 0.7 reduction for the "stress ratio" term. Conservatively assume the effective length factor $K = 1.0$. If adjacent member stiffnesses are known, the alignment chart, Fig. 6.9.4a, could be used to determine $K < 1.0$ for this braced frame.

Use *LRFD Manual* [1.17] load tables "COLUMNS" giving $\phi_c P_n$. Enter with $P_u = 912$ kips and $KL = 14.0$ ft,

Find: W14×120, $\phi_c P_n = 971$ kips
          W14×132, $\phi_c P_n = 1070$ kips

Check the W14×120.

(c) Column effect. Both the flange and the web satisfy $\lambda < \lambda_r$ to preclude local buckling prior to achieving column strength based on KL/r.

$$\text{Largest } \frac{KL}{r} = \frac{KL}{r_y} = \frac{1.0(14)12}{3.74} = 44.9$$

Find $\phi_c F_{cr} = 27.5$ ksi (LRFD "NUMERICAL VALUES" TABLE 3-36)

$$\phi_c P_n = \phi_c F_{cr} A_g = 27.5(35.3) = 971 \text{ kips}$$

Check $P_u/(\phi_c P_n) \geq 0.2$,

$$\left(\frac{P_u}{\phi_c P_n} = \frac{159}{971} = 0.164 \right) < 0.2; \quad \text{Use Eq. 12.10.2}$$

(d) Beam effect. The laterally unbraced length L_b is 14 ft.

$$M_p = F_y Z_x = 36(212)/12 = 636 \text{ ft-kips}$$

$$L_p = \frac{300}{\sqrt{F_y}} r_y = \frac{300(3.74)}{\sqrt{36}(12)} = 15.6 \text{ ft}$$

Since $L_b < L_p$, $M_n = M_p$ regardless of C_b. In this case, $C_b = 1.0$. Check the "compact" section requirement; i.e., is $\lambda \leq \lambda_p$ for both flange and web? The *LRFD Manual* W shapes properties indicates *no value* for F_y''', identifying the section as "compact" for all yield stresses up through 65 ksi. The detailed check is illustrated in Example 12.11.1, part (d). Thus,

$$\phi_b M_n = \phi_b M_p = 0.90(636) = 572 \text{ ft-kips}$$

Noting the $M_{nt}/\phi_b M_n = 560/572 = 0.98$ without the axial compression term or the magnification, it is clear this section is not adequate.
Try W14×132 section.

(e) Recheck column and beam strength for W14×132.

$$\text{Largest } \frac{KL}{r} = \frac{KL}{r_y} = \frac{1.0(14)12}{3.76} = 44.7; \quad \phi_c F_{cr} = 27.5 \text{ ksi}$$

$$\phi_c P_n = \phi_c F_{cr} A_g = 27.5(38.8) = 1070 \text{ kips}$$

The ratio $P_u/\phi_c P_n = 159/1070 = 0.149$, which is still less than 0.2. For this section, which is also "compact",

$$L_p = 15.7 \text{ ft} > (L_b = 14 \text{ ft})$$

From *LRFD Manual* "LOAD FACTOR DESIGN SELECTION TABLE,"

$$\phi_b M_n = \phi_b M_p = 632 \text{ ft-kips}$$

For this section, the ratio $M_{nt}/\phi_b M_n = 560/632 = 0.89$, making it likely this section is adequate.

(f) Moment magnification for W14×132. The slenderness ratio KL/r involved in moment magnification must relate to the *axis of bending,* in this case the x-axis,

$$\text{Axis of bending } \frac{KL}{r} = \frac{KL}{r_x} = \frac{1.0(14)12}{6.28} = 26.8$$

$$C_m = 1.0 \quad \text{(constant moment)}$$

$$P_{e1} = \frac{\pi^2 E A_g}{(KL/r)^2} = \frac{\pi^2 (29,000)38.8}{(26.8)^2} = 15,500 \text{ kips}$$

$$B_1 = \frac{C_m}{1 - P_u/P_{e1}} = \frac{1.0}{1 - 159/15,500} = 1.01$$

(g) Check LRFD Formula (H1-1b), Eq. 12.10.2, omitting the bending term for the y-axis,

$$M_{ux} = M_{nt}B_1 = 560(1.01) = 566 \text{ ft-kips}$$

$$\frac{P_u}{2\phi_c P_n} + \left(\frac{M_{ux}}{\phi_b M_{nx}}\right) \leq 1.0 \qquad\qquad [12.10.2]$$

$$\frac{0.149}{2} + \left(\frac{566}{632}\right) = 0.074 + 0.896 = 0.97 < 1 \quad \text{OK}$$

The other loading case from part (a) clearly does not control. *Use* W14×132, A36 steel. ■■

EXAMPLE 12.13.2

Select the lightest W section to carry axial compression of 24 kips dead load and 96 kips live load applied at an eccentricity $e = 5$ in. as shown in Fig. 12.13.2. The member is a part of a *braced* frame, and is conservatively assumed loaded in single curvature with constant e. Use A36 steel and Load and Resistance Factor Design.

Solution

(a) Compute factored loads. Assume a first-order elastic analysis was performed to obtain the given forces. using LRFD-A4.1, Formula (A4-2),

$$P_u = 1.2(24) + 1.6(96) = 182 \text{ kips}$$
$$M_{nt} = P_u(e) = 182(5/12) = 76.0 \text{ ft-kips}$$

(b) Estimate the P_{uEQ} for the given conditions using either Yura's equation, Eq. 12.12.6, or the more detailed Eq. 12.12.5. For Eq. 12.12.6,

$$P_{uEQ} = P_u + M_x\left(\frac{2}{d}\right) = 182 + 76.0(12)\left(\frac{2}{d}\right)$$

$$\begin{array}{lll}
\text{W10,} & P_{uEQ} \approx 182 + 1824/10 = 365 \text{ kips} \\
\text{W12,} & P_{uEQ} \approx 182 + 1824/12 = 334 \text{ kips} \\
\text{W14,} & P_{uEQ} \approx 182 + 1824/14 = 313 \text{ kips}
\end{array}$$

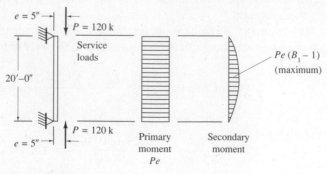

Figure 12.13.2   Example 12.13.2.

or, using Eq. 12.12.5 and estimating β_{az} as 0.24 for W10, 0.20 for W12, and 0.16 for W14, from Table 12.12.1 for medium weight sections, and using 0.7 for the "stress ratio" term,

$$P_{uEQ} = P_u + M_{nt}\beta_{az}(0.7)$$

W10, $\quad P_{uEQ} \approx 182 + 76.0(12)(0.24)0.7 = 336$ kips

W12, $\quad P_{uEQ} \approx 182 + 76.0(12)(0.20)0.7 = 310$ kips

W14, $\quad P_{uEQ} \approx 182 + 76.0(12)(0.16)0.7 = 285$ kips

Since the use of 0.7 for the "stress ratio" factor is likely to cause the latter calculation to be close but possibly too low, and Yura's estimate is likely to be too high, use both for obtaining the trial sections; however, assume the lower values of P_{uEQ} are likely to be closer.

Using the *LRFD Manual* "COLUMNS" load tables, select for $KL = 20$ ft,

W10×60 $\quad \phi_c P_n = 340$ kips

W12×58 $\quad \phi_c P_n = 321$ kips

W14×61 $\quad \phi_c P_n = 331$ kips

(c) Check W12×58 column action. Note that all sections in the *LRFD Manual* "COLUMNS" load tables for A36 and $F_y = 50$ ksi steels, where a value of L_p is given satisfy $\lambda \leq \lambda_r$ for local buckling in axial compression, and are "compact" for beam action; that is, $\lambda \leq \lambda_p$.

$$\text{Largest } \frac{KL}{r} = \frac{KL}{r_y} = \frac{1.0(20)12}{2.51} = 95.6; \qquad \phi_c F_{cr} = 18.9 \text{ ksi}$$

$$\phi_c P_n = \phi_c F_{cr} A_g = 18.9(17.0) = 321 \text{ kips}$$

The ratio $P_u/\phi_c P_n = 182/321 = 0.567$, which exceeds 0.2; therefore, LRFD-Formula (H1-1a), Eq. 12.11.1, applies.

(d) Beam action.

$$(L_p = 10.5 \text{ ft}) < (L_b = 20 \text{ ft}) < (L_r = 38.4 \text{ ft})$$

Therefore, $\phi_b M_n$ must be linearly interpolated between $\phi_b M_p$ (233 ft-kips from *LRFD Manual* p. 4-19) and $\phi_b M_r$ (152 ft-kips) according to Eq. 9.6.4 multiplied by ϕ_b,

$$M_n = C_b\left[M_p - (M_p - M_r)\left(\frac{L_b - L_p}{L_r - L_p}\right)\right] \leq M_p \qquad [9.6.4]$$

$$\phi_b M_n = 1.0\left[233 - (233 - 152)\left(\frac{20 - 10.5}{38.4 - 10.5}\right)\right] = 205 \text{ ft-kips}$$

(e) Moment magnification for W12×58. The slenderness ratio KL/r involved in moment magnification must relate to the *axis of bending,* in this case the *x*-axis,

$$\text{Axis of bending } \frac{KL}{r} = \frac{KL}{r_x} = \frac{1.0(20)12}{5.28} = 45.5$$

$$C_m = 1.0 \quad \text{(constant moment)}$$

$$P_{e1} = \frac{\pi^2 E A_g}{(KL/r)^2} = \frac{\pi^2 (29,000)17.0}{(45.5)^2} = 2350 \text{ kips}$$

$$B_1 = \frac{C_m}{1 - P_u/P_{e1}} = \frac{1.0}{1 - 182/2350} = 1.08$$

(f) Check LRFD Formula (H1-1a), Eq. 12.10.1, omitting the bending term for the y-axis,

$$M_{ux} = M_{nt} B_1 = 76.0(1.08) = 82.1 \text{ ft-kips}$$

$$\frac{P_u}{\phi_c P_n} + \frac{8}{9}\left(\frac{M_{ux}}{\phi_b M_{nx}}\right) \le 1.0 \qquad\qquad [12.10.1]$$

$$\frac{182}{321} + \frac{8}{9}\left(\frac{82.1}{205}\right) = 0.567 + 0.356 = 0.92 < 1 \quad \text{OK}$$

Use W12×58, A36 steel. Though the criterion indicates exces strength, the result for the W12×53 is 1.03 which exceeds the LRFD limit but within about 3% might be acceptable. A check of W10 and W14 sections shows no lighter section meets the acceptability criterion. ■■

EXAMPLE 12.13.3 ————————————————————————

Design a beam-column W section for the service loading conditions shown in Fig. 12.13.3. The compression load P is 30 kips dead load and 70 kips live load. The bracket load W is 2 kips dead load and 18 kips live load, as might be caused by a crane, and the horizontal load H is 5 kips live load, as might be the horizontal effect of a crane. The member is part of a *braced* system, has support in the weak direction at mid-height, but only at the top and bottom for the strong direction. Use A36 steel and Load and Resistance Factor Design.

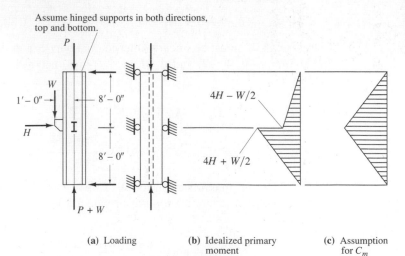

(a) Loading   (b) Idealized primary   (c) Assumption
                          moment   for C_m

Figure 12.13.3   Examples 12.13.3 and 12.15.2.

Solution.   The particular features of this example are (a) the bracing is not at the same locations for both principal directions; and (b) the lateral transverse loading causes the primary bending moment.

(a) Compute factored loads. Assume a first-order elastic analysis was performed to obtain the given forces. Using LRFD-A4.1, Formula (A4-2),

$$P_u = 1.2(30) + 1.6(70) = 148 \text{ kips at top}$$
$$W_u = 1.2(2) + 1.6(18) = 31.2 \text{ kips}$$
$$P_u = 148 + 31.2 = 179 \text{ kips at bottom} \quad (controls)$$
$$H_u = 1.6(5) = 8 \text{ kips}$$

The horizontal reaction at the bottom of the column is $(H/2 + W/16)$, making the maximum moment,

$$\text{Maximum moment} = (H/2 + W/16)8$$
$$M_{nt} = (8/2 + 31.2/16)8 = 47.6 \text{ ft-kips}$$

(b) Establish effective lengths. The member must be viewed as a column without bending moment, then as a beam without column load, as in Fig. 12.13.4,

For column action:

$$K_x L_x = 16 \text{ ft} \qquad K_y L_y = 8 \text{ ft}$$

Required $r_x/r_y \geq 2.0$ if $KL = 8$ ft is valid for entering
LRFD Manual "COLUMNS" load tables

For beam action:

$$\text{Laterally unbraced length } L_b = 8 \text{ ft}$$
$$KL \text{ for moment magnification} = K_x L_x = 16 \text{ ft}$$

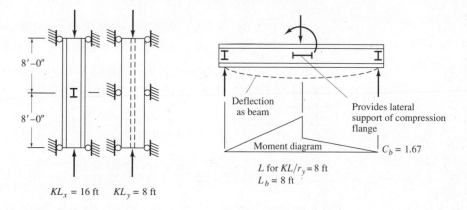

(a) Column action   (b) Beam action

Figure 12.13.4   Separate beam action and column action from Examples 12.13.3 and 12.15.2.

(c) Estimate the P_{uEQ} for the given conditions using either Yura's equation, Eq. 12.12.6, or the more detailed Eq. 12.12.5. Using Eq. 12.12.5 and estimating β_{az} as 0.24 for W10, 0.20 for W12, and 0.16 for W14, from Table 12.12.1 for medium weight sections, and using 0.7 for the "stress ratio" term,

$$P_{uEQ} = P_u + M_{nt}\beta_{az}(0.7)$$

$$
\begin{aligned}
\text{W10,} \quad & P_{uEQ} \approx 179 + 47.6(12)(0.24)0.7 = 275 \text{ kips} \\
\text{W12,} \quad & P_{uEQ} \approx 179 + 47.6(12)(0.20)0.7 = 259 \text{ kips} \\
\text{W14,} \quad & P_{uEQ} \approx 179 + 47.6(12)(0.16)0.7 = 243 \text{ kips}
\end{aligned}
$$

Using the *LRFD Manual* "COLUMNS" load tables, select for $KL = 8$ft,

$$
\begin{aligned}
\text{W10} \times 39 \quad & \phi_c P_n = 311 \text{ kips} \quad & r_x/r_y = 2.16 \\
\text{W12} \times 40 \quad & \phi_c P_n = 269 \text{ kips} \quad & r_x/r_y = 2.66 \\
\text{W14} \times 43 \quad & \phi_c P_n = 337 \text{ kips} \quad & r_x/r_y = 3.08
\end{aligned}
$$

All three sections are "compact" since L_p values are given in these tables. Weak axis does control column action since all have $r_x/r_y > K_x L_x/K_y r_y = 2.0$. Note the W12×40 and the W14×43 are the lightest tabulated in the column load tables; however, the next lighter sections, W12×35 or W14×38, may work.

(d) Check W10×39 column action.

$$\frac{K_x L_x}{r_x} = \frac{1.0(16)12}{4.27} = 45.0 \qquad \frac{K_y L_y}{r_y} = \frac{1.0(8)12}{1.98} = 48.5$$

Find $\phi_c F_{cr} = 27.0$ ksi   (based on $KL/r = 48.5$)

$$\phi_c P_n = \phi_c F_{cr} A_g = 27.0(11.5) \doteq 311 \text{ kips}$$

The ratio $P_u/\phi_c P_n = 179/311 = 0.576$, which exceeds 0.2; therefore, LRFD-Formula (H1-1a), Eq. 12.11.1, applies.

(e) Beam action.

$$(L_p = 8.3 \text{ ft}) > (L_b = 8.0 \text{ ft})$$

Therefore,

$$M_n = M_p = F_y Z_x = 36(46.8)/12 = 140 \text{ ft-kips}$$
$$\phi_b M_n = 0.90(140) \doteq 126 \text{ ft-kips}$$

(f) Moment magnification for W10×39. The slenderness ratio KL/r involved in moment magnification must relate to the *axis of bending*, in this case the x-axis,

$$\text{Axis of bending } \frac{KL}{r} = \frac{K_x L_x}{r_x} = 45.0$$

$$P_{e1} = \frac{\pi^2 E A_g}{(KL/r)^2} = \frac{\pi^2(29,000)11.5}{(45.0)^2} = 1630 \text{ kips}$$

For transverse loading, C_m may be evaluated using Table 12.3.1, Case 3,

$$C_m \approx 1 - 0.2\frac{P_u}{P_{e1}} = 1 - 0.2\left(\frac{179}{1630}\right) = 0.978$$

$$B_1 = \frac{C_m}{1 - P_u/P_{e1}} = \frac{0.978}{1 - 179/1630} = 1.10$$

(g) Check LRFD Formula (H1-1a), Eq. 12.10.1, omitting the bending term for the y-axis,

$$M_{ux} = M_{nt}B_1 = 47.6(1.10) = 52.4 \text{ ft-kips}$$

$$\frac{P_u}{\phi_c P_n} + \frac{8}{9}\left(\frac{M_{ux}}{\phi_b M_{nx}}\right) = 0.576 + \frac{8}{9}\left(\frac{52.4}{126}\right) = 0.94 < 1.0 \quad \text{OK}$$

Use W10×39, A36 steel. Checks of lighter sections give 1.03 for W12×35, 0.91 for W14×38, and 1.02 for W14×34. All of these sections satisfy $\lambda < \lambda_p$ for "compact" section. The lightest section that satisfies is the W14×38; however, the extra 4 in. of depth may not be desirable. ■■

EXAMPLE 12.13.4 _____

Investigate the adequacy of the W8×24 section in Fig. 12.13.5, of A572 Grade 50 steel loaded as a beam-column. The axial compression is 15 kips dead load and 60 kips live load, and the uniformly distributed superimposed lateral load of 0.1 kip/ft dead load and 0.4 kip/ft live load causes bending about the weak axis. Use Load and Resistance Factor Design.

Solution

(a) Compute factored loads. Assume a first-order elastic analysis was performed to obtain the given forces. Using LRFD-A4.1, Formula (A4-2),

$$w_u = 1.2(0.124) + 1.6(0.4) = 0.79 \text{ kips/ft}$$
$$\text{(adding in 0.024 kip/ft beam weight)}$$
$$M_{nt} = w_u L^2/8 = 0.79(10)^2/8 = 9.86 \text{ ft-kips}$$
$$P_u = 1.2(15) + 1.6(60) = 114 \text{ kips}$$

(d) Column action. In this example buckling as a column occurs in the plane of bending, whereas in the three previous examples column buckling occurred about the axis which was not the axis of bending. Using *LRFD Manual* "NUMERICAL VALUES" TABLE 3-50, or "COLUMNS" load tables,

$$\text{Largest } \frac{KL}{r} = \frac{KL}{r_y} = \frac{1.0(10)12}{1.61} = 74.5; \quad \phi_c F_{cr} = 28.3 \text{ ksi}$$

$$\phi_c P_n = \phi_c F_{cr} A_g = 28.3(7.08) = 200 \text{ kips}$$

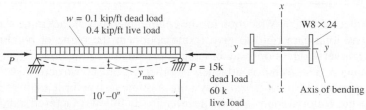

Figure 12.13.5 Examples 12.13.4 and 12.15.3.

The ratio $P_u/\phi_c P_n = 114/200 = 0.57$, which exceeds 0.2; therefore, LRFD-Formula (H1-1a), Eq. 12.11.1, applies.

(c) Beam action. Because bending occurs in the weak direction, lateral-torsional buckling is not a possible controlling limit state; nor is web local buckling. The section is "compact" according to LRFD-B5 if flange local buckling is precluded; that is, when $\lambda \leq \lambda_p$ for the flange,

$$\left(\lambda = \frac{b_f}{2t_f} = \frac{6.495}{2(0.400)} = 8.1\right) < \left(\lambda_p = \frac{65}{\sqrt{F_y}} = 9.2\right) \quad \text{OK}$$

Therefore,

$$M_n = M_p = F_y Z_y = 50(8.57)/12 = 35.7 \text{ ft-kips,}$$

$$\phi_b M_n = 0.90(35.7) = 32.1 \text{ ft-kips}$$

This value *cannot* be obtained from *LRFD Manual* "LOAD FACTOR DESIGN SELECTION TABLE" because those tables are only for x-axis bending.

(d) Moment magnification. The slenderness ratio KL/r involved in moment magnification must relate to the *axis of bending,* in this case the y-axis,

$$\text{Axis of bending } \frac{KL}{r} = \frac{K_y L_y}{r_y} = 74.5$$

$$P_{e1} = \frac{\pi^2 E A_g}{(KL/r)^2} = \frac{\pi^2(29,000)7.08}{(74.5)^2} = 365 \text{ kips}$$

For transverse loading, C_m may be evaluated using Table 12.3.1, Case 1, which indicates

$$C_m = 1.0$$

$$B_1 = \frac{C_m}{1 - P_u/P_{e1}} = \frac{1.0}{1 - 114/365} = 1.45$$

(e) Check LRFD Formula (H1-1a), Eq. 12.10.1, omitting the bending term for the x-axis,

$$M_{uy} = M_{nt} B_1 = 9.86(1.45) = 14.3 \text{ ft-kips}$$

$$\frac{P_u}{\phi_c P_n} + \frac{8}{9}\left(\frac{M_{uy}}{\phi_b M_{ny}}\right) = 0.57 + \frac{8}{9}\left(\frac{14.3}{32.1}\right) = 0.97 < 1.0 \quad \text{OK}$$

The W8×24, A572 Grade 50, *is acceptable.* ∎∎

EXAMPLE 12.13.5 —————————————————————————

Select a suitable W section for the column member of the frame shown in Fig. 12.13.6. The joints are rigid to give frame action in the plane of bending; however, in the transverse direction sway bracing is provided and the attachments are assumed to be hinged. The frame is attached to other braced construction such that this frame is considered part of a *braced* system. Assume a structural analysis has been performed using factored loads and the results are as follows: factored axial compression $P_u =$

Assume girder monent of inertia three times that of column

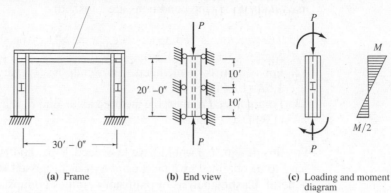

(a) Frame   (b) End view   (c) Loading and moment
                                              diagram

Figure 12.13.6 Examples 12.13.5, 12.13.6, and 12.15.4.

92 kips; and factored moments at the top of column $M_{nt2} = 168$ ft-kips and at the
bottom $M_{nt1} = 84$ ft-kips. Use A36 steel and Load and Resistance Factor Design.

Solution

(a) Establish effective lengths. The member must be viewed as a column
without bending moment, then as a beam without column load. For the plane of
bending, the effective length factor K_x may be estimated as ≈ 0.7 from Fig.
6.9.5, or may be obtained from Alignment Chart, Fig. 6.9.4a, as

$$G_A = 1.0 \text{ (fixed)}, \qquad G_B = \frac{I/20}{3I/30} = 0.5, \qquad K_x = 0.73$$

For the transverse direction, the member is assumed hinged at the top and
bottom; thus, $K_y = 1.0$. For beam action, $L_b = 10$ ft and for moment
magnification, $K_x L_x = 0.73(20) = 14.6$ ft.

(b) Estimate the P_{uEQ} for the given conditions using either Yura's equation,
Eq. 12.12.6, or the more detailed Eq. 12.12.5. Using Eq. 12.12.5 and estimating
β_{az} as 0.24 for W10, 0.20 for W12, and 0.16 for W14, from Table 12.12.1 for
medium weight sections, and using 0.7 for the "stress ratio" term,

$$P_{uEQ} = P_u + M_{nt}\beta_{az}(0.7)$$

W10, $P_{uEQ} \approx 92 + 168(12)(0.24)0.7 = 431$ kips
W12, $P_{uEQ} \approx 92 + 168(12)(0.20)0.7 = 374$ kips
W14, $P_{uEQ} \approx 92 + 168(12)(0.16)0.7 = 318$ kips

Using the *LRFD Manual* "COLUMNS" load tables, select for $KL = 10$ ft,

W10×54 $\phi_c P_n = 431$ kips $r_x/r_y = 1.71$
W12×53 $\phi_c P_n = 422$ kips $r_x/r_y = 2.11$
W14×53 $\phi_c P_n = 351$ kips $r_x/r_y = 3.06$

For the tables to give the correct $\phi_c P_n$, r_x/r_y for the section must exceed
$K_x L_x/K_y L_y = 0.7(20)/10 = 1.40$; the above sections all qualify. Try
W14×48.

(c) Can plastic analysis be used for this design? Plastic analysis may be used if the following conditions are satisfied:

1. The section is "compact"; i.e., $\lambda \le \lambda_p$ for the flange and the web, and $L_b \le L_{pd}$ (LRFD-F1.2d).
2. For compression members, the slenderness parameter λ_c may not exceed $1.5K$ (LRFD-E1.2).
3. For unbraced frames, the factored axial load P_u may not exceed $0.75\phi_c A_g F_y$ (LRFD-C2.2).

For this design, L_{pd} would have to exceed 10 ft, and the factored forces would have to be obtained by a plastic analysis as discussed in Chapters 10 and 15. In general, plastic analysis of multistory frames is impractical.

(d) Investigate equivalent beam approach to selection. Since column action represents only about 25 to 30% of the total load as indicated by P_{uEQ} in part (b) above, an alternative approach using M_{uEQ}, given by Eq. 12.12.10, may be preferred.

$$M_{uEQ} = \frac{8}{9}M_u + P_u\left(\frac{1}{\beta_{az}}\right)\left(\frac{\phi_b F_y}{\phi_c F_{cr}}\right) \qquad [12.12.10]$$

Estimating β_{az} as 0.16 and $\phi_c F_{cr}/\phi_b F_y$ as 0.7, as used in part (a), the M_{uEQ} for a W14 would be

$$W14, \qquad M_{uEQ} = \frac{8}{9}(168) + \frac{92}{[12(0.16)0.7]} = 218 \text{ ft-kips}$$

Then, assuming that M_n can equal M_p, the *LRFD Manual* "LOAD FACTOR DESIGN SELECTION TABLE" can be used, or the charts "BEAM DESIGN MOMENTS" can be used when L_b is larger. In this example, the lightest section to carry the given loading will likely be deeper than W14. From Table 12.12.1 estimate β_{az} as 0.14 (a median value for W16, W18, and W21),

$$M_{uEQ} = \frac{8}{9}(168) + \frac{92}{[12(0.14)0.7]} = 228 \text{ ft-kips}$$

$$\text{Required } Z_x = \frac{M_{uEQ}}{\phi_b F_y} = \frac{228(12)}{0.90(36)} = 84 \text{ in.}^3$$

Because of the steep moment gradient, $C_b \ne 1.0$; therefore the moment strength will be C_b times M_{uEQ}. This can approximately be accounted for by entering the beam charts with M_{uEQ}/C_b.

Over the upper 10 ft laterally unbraced length, $M_1/M_2 = -42/168 = -0.25$. Table 9.6.3 gives $C_b = 1.43$. Thus, the required $M_{uEQ} \approx 228/1.43 = 159$ ft-kips.

From *LRFD Manual* "LOAD FACTOR DESIGN SELECTION TABLE," look for $Z = 84$ in.^{3} and $L_p = 10$ ft. None of the deeper sections have L_p anywhere near 10 ft. Lightest section indicated is W21×44 with $L_p = 5.3$ ft.

The reduced strength with $L_b = 10$ ft is likely recovered when $C_b = 1.43$. Try the charts, "BEAM DESIGN MOMENTS" with $M_{uEQ} = 159$ ft-kips and $L_b = 10$ ft. The indicated possibilities from shallowest to deepest are:

$$W10 \times 49, \quad W12 \times 45, \quad W14 \times 43, \quad W18 \times 40, \quad \text{and} \quad W21 \times 44$$

From the equivalent beam approach, the W10, W12, and W14 sections are one section lighter than obtained by using P_{uEQ} in the "COLUMNS" load tables. Sections deeper than W14 do not appear in the "COLUMNS" load tables because such deeper sections are not commonly used as columns.

Assume here the deeper W21 is desired because about 70% of the action is beam action.

(e) Check W21×44 section for column action.

$$\frac{K_x L_x}{r_x} = \frac{0.7(20)12}{8.06} = 20.8 \qquad \frac{K_y L_y}{r_y} = \frac{1.0(10)12}{1.26} = 95.2$$

Find $\phi_c F_{cr} = 19.0$ ksi   (based on $KL/r = 95.2$)

$$\phi_c P_n = \phi_c F_{cr} A_g = 19.0(13.0) = 247 \text{ kips}$$

The ratio $P_u/\phi_c P_n = 92/247 = 0.372$, which exceeds 0.2; therefore, LRFD-Formula (H1-1a), Eq. 12.11.1, applies.

(f) Beam action. Since L_p is given in *LRFD Manual*, W21×44 is "compact."

$$(L_p = 5.3 \text{ ft}) < (L_b = 10 \text{ ft}) < (L_r = 15.4 \text{ ft})$$

Therefore, M_n must be linearly interpolated between M_p and M_r according to Eq. 9.6.4. From *LRFD Manual* "LOAD FACTOR DESIGN SELECTION TABLE" find

$$\phi_b M_p = 258 \text{ ft-kips;} \qquad \phi_b M_r = 159 \text{ ft-kips}$$

$$M_n = C_b \left[M_p - (M_p - M_r)\left(\frac{L_b - L_p}{L_r - L_p}\right) \right] \le M_p \qquad [9.6.4]$$

Using ϕM_p and ϕM_r will give ϕM_n,

$$\phi M_n = 1.43 \left[258 - (258 - 159)\left(\frac{10 - 5.3}{15.4 - 5.3}\right) \right] = 336 \text{ ft-kips}$$

Since M_n cannot exceed M_p, $M_n = M_p$; $\phi_b M_n = \phi_b M_p = 258$ ft-kips.

(g) Moment magnification for W21×44. The slenderness ratio KL/r involved in moment magnification must relate to the *axis of bending*, in this case the x-axis,

$$\text{Axis of bending } \frac{KL}{r} = \frac{K_x L_x}{r_x} = 20.8$$

$$P_{e1} = \frac{\pi^2 E A_g}{(KL/r)^2} = \frac{\pi^2 (29,000) 13.0}{(20.8)^2} = 8600 \text{ kips}$$

For end moment loading, C_m is LRFD Formula (H1-4), Eq. 12.4.6,

$$C_m = 0.6 - 0.4\frac{M_1}{M_2} = 0.6 - 0.4\left(\frac{84}{168}\right) = 0.4$$

$$B_1 = \frac{C_m}{1 - P_u/P_{e1}} = \frac{0.4}{1 - 92/8600} = 0.4(1.01) < 1.0$$

Thus, the magnified moment out away from the support does not exceed the primary moment at the support; thus, $B_1 = 1.0$.

(h) Check LRFD Formula (H1-1a), Eq. 12.10.1, omitting the bending term for the y-axis,

$$M_{ux} = M_{nt}B_1 = 168(1.0) = 168 \text{ ft-kips}$$

$$\frac{P_u}{\phi_c P_n} + \frac{8}{9}\left(\frac{M_{ux}}{\phi_b M_{nx}}\right) = 0.372 + \frac{8}{9}\left(\frac{168}{258}\right) = 0.95 < 1.0 \quad \text{OK}$$

Use W21×44, A36 steel. Checks of other sections give 0.98 for W16×45, 0.96 for W18×46, and 0.97 for W14×48. ■■

EXAMPLE 12.13.6 —————————————————————————————

Repeat Example 12.13.5 except treat the frame as *unbraced* in the plane of bending. Though the total factored loads acting in this example are the same as in Example 12.13.5, they would have been determined by an entirely different procedure. Two elastic analyses under factored loads are required: a nonsway analysis for gravity loads, and a sway analysis for lateral loads. Assume the factored moments M_{nt} for the nonsway analysis are 30 ft-kips and 15 ft-kips at the top and bottom of the column, respectively. The factored moments $M_{\ell t}$ from the sway analysis are 138 ft-kips both at the top and the bottom. All moments cause double curvature in the column.

Solution

(a) Effective lengths for column and beam action. A significant difference between this case and the preceding one is that the effective length factor K_x for the plane of bending exceeds one. Using the Alignment Chart, Fig. 6.9.4b for the *unbraced* frame (sidesway not prevented),

$$G_A = 1.0 \text{ (fixed)}, \qquad G_B = \frac{I/20}{3I/30} = 0.5, \qquad K_x \approx 1.24$$

This value of K may be adjusted for inelastic buckling according to the discussion in Sec. 6.9.

(b) Estimate the P_{uEQ} for the given conditions using either Yura's equation, Eq. 12.12.6, or the more detailed Eq. 12.12.5. Using Eq. 12.12.5 and estimating β_{az} as 0.24 for W10, 0.20 for W12, and 0.16 for W14, from Table 12.12.1 for medium weight sections, and using 0.7 for the "stress ratio" term.

$$P_{uEQ} = P_u + M_u\beta_{az}(0.7)$$

This time the equation includes M_u instead of M_{nt} used for the braced frame. It is relatively easier to estimate B_1 for the braced frame (and for the nonsway part of the unbraced frame analysis) than it is to estimate B_2 for the sway analysis. Often B_1 is 1.0 or not much larger; however, B_2 is commonly in the range 1.2 to 1.5 and it may be larger. Practical designs should have B_2 not exceeding about 1.5.

In addition to estimating β_{az} and the "stress ratio" term of Eq. 12.12.5, B_2 must be estimated. Using B_2 estimated at 1.3 and B_1 at 1.0,

$$M_u = M_{nt}B_1 + M_{\ell t}B_2$$

Estimated $M_u \approx 30(1.0) + 138(1.3) = 209$ ft-kips

Next, note that the *LRFD Manual* "COLUMNS" load tables give the correct $\phi_c P_n$ for $K_y L_y = 10$ ft only when r_x/r_y exceeds

$$K_x L_x/K_y L_y = 1.24(20)/10 = 2.48$$

Selection of W10 or W12 sections may likely have their column strength controlled by column buckling in the plane of bending, while deeper sections will be controlled by weak-axis buckling.

The estimated equivalent column loads P_{uEQ} are

W10, $P_{uEQ} \approx 92 + 209(12)(0.24)0.7 = 513$ kips
W12, $P_{uEQ} \approx 92 + 209(12)(0.20)0.7 = 443$ kips
W14, $P_{uEQ} \approx 92 + 209(12)(0.16)0.7 = 373$ kips

Using the *LRFD Manual* "COLUMNS" load tables, select for $KL = 10$ ft, or $KL = K_x L_x/(r_x/r_y) = 1.24(20)/(r_x/r_y) = 24.8/(r_x/r_y)$ when $r_x/r_y < 2.48$,

W10,   for $KL = 24.8/1.71 = 14.5$ ft,
    W10×68 $\phi_c P_n \approx 485$ kips $r_x/r_y = 1.71$

W12,   for $KL = 24.8/2.10 = 11.8$ ft,
    W12×58 $\phi_c P_n \approx 440$ kips $r_x/r_y = 2.10$

W14 and deeper,   for $KL = 10$ ft,
    W14×53 $\phi_c P_n \approx 389$ kips $r_x/r_y = 3.07$

Alternatively, using the M_{uEQ} approach with Eq. 12.12.10, using β_{az} of 0.14 for deeper sections (as in Example 12.13.5),

$$M_{uEQ} = \frac{8}{9}(209) + \frac{92}{[12(0.14)0.7]} = 264 \text{ ft-kips}$$

$$\text{Required } Z_x = \frac{M_{uEQ}}{\phi_b F_y} = \frac{264(12)}{0.90(36)} = 98 \text{ in.}^3$$

From *LRFD Manual* "LOAD FACTOR DESIGN SELECTION TABLE", find W18×50 with $Z_x = 101$ in.³ and W21×44 with slightly lower $Z_x = 95.4$ in.³ Considering the required 98 to be approximate, check the W21×44.

(c) Check W21×44 section for column action. Here $K_x = 1.24$ instead of 0.7 for Example 12.13.6,

$$\frac{K_x L_x}{r_x} = \frac{1.24(20)12}{8.06} = 36.9 \qquad \frac{K_y L_y}{r_y} = 95.2 \quad \text{(Example 12.13.5)}$$

Since the y-axis slenderness still is the larger, the value of $\phi_c P_n = 247$ kips obtained in Example 12.13.5 still prevails. The ratio $P_u/\phi_c P_n$ still equals $92/247 = 0.372$, which exceeds 0.2; therefore, LRFD-Formula (H1-1a), Eq. 12.11.1, applies.

(d) Beam action on W21×44. These calculations are identical to Example 12.13.5, part (f). $\phi_b M_n = 258$ ft-kips.

(e) Nonsway part—magnification factor B_1. This calculation is identical to Example 12.13.5, part (g). In this example it would be preferable to use $K_x = 1.0$ for P_{e1}. However, $B_1 = 1.0$ whether $K_x = 0.7$ or $K_x = 1.0$.

(f) Sway part—magnification factor B_2. The slenderness ratio in the plane of the frame increases from the nonsway part because here $K_x = 1.24$; thus, using $K_x L_x/r_x = 36.9$ gives

$$P_{e2} = \frac{\pi^2 E A_g}{(KL/r)^2} = \frac{\pi^2 (29,000) 13.0}{(36.9)^2} = 2730 \text{ kips}$$

Since in this example two equal sized columns having equal loading are participating in the sway resistance,

$$\Sigma P_{e2} = 2(2730) = 5460 \text{ kips}$$

$$\Sigma P_u = 2(92) = 184 \text{ kips}$$

Then Eq. 12.11.11 for the sway magnifier B_2 gives

$$B_2 = \frac{1}{1 - \dfrac{\Sigma P_u}{\Sigma P_{e2}}} = \frac{1}{1 - \dfrac{184}{5460}} = 1.035$$

(g) Check LRFD Formula (H1-1a), Eq. 12.10.1, omitting the bending term for the y-axis,

$$M_{ux} = M_{nt} B_1 + M_{\ell t} B_2 = 30(1.0) + 138(1.035) = 173 \text{ ft-kips}$$

$$\frac{P_u}{\phi_c P_n} + \frac{8}{9}\left(\frac{M_{ux}}{\phi_b M_{nx}}\right) = 0.372 + \frac{8}{9}\left(\frac{173}{258}\right) = 0.97 < 1.0 \quad \text{OK}$$

Use W21×44, A36 steel. Checks of other sections give 0.98 for W18×46, 1.01 for W16×45, and 1.00 for W14×48. ∎∎

EXAMPLE 12.13.7 ──

Select the lightest W12 section to carry an axial compression in addition to biaxial bending, loaded as shown in Fig. 12.13.7. Assume a first-order structural analysis has been performed using factored loads. The results give $P_u = 375$ kips, $M_{ntx} = 38$ ft-kips about the x-axis at the top of the column, and $M_{nty} = 14$ ft-kips about the y-axis at the top of the column. Use A572 Grade 50 steel and Load and Resistance Factor Design.

Solution

(a) Effective lengths for column and beam action.

Column action:
$$K_x L_x = 1.0(15) = 15 \text{ ft}$$
$$K_y L_y = 1.0(7.5) = 7.5 \text{ ft}$$

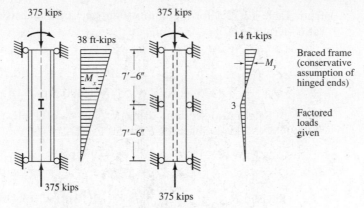

Figure 12.13.7   Example 12.13.7.

Beam action: L_b = 7.5 ft for x-axis bending

C_{bx} = 1.25 (Eq. 9.6.11)(LRFD-F1.3)

(from Table 9.6.3 for

$M_1/M_2 = -19/38 = -0.5$)

L_b = not applicable for y-axis bending

C_{by} = not applicable for y-axis bending

Moment magnification: KL = 15 ft for P_{ex}

C_{mx} = 0.6

KL = 7.5 ft for P_{ey}

C_{my} = 0.6 − 0.4(3/14) = 0.51

Note that lateral-torsional buckling is not a possible limit state for weak axis bending of W shapes; therefore, L_b and C_b are not applicable for the y-axis.

(b) Estimate the P_{uEQ} for the given conditions using either Yura's equation, Eq. 12.12.6, or the more detailed Eq. 12.12.5. Using Eq. 12.12.5 for uniaxial bending,

$$P_{uEQ} = P_u + M_{nt}\beta_{az}\left(\frac{8\phi_c F_{cr}}{9\phi_b F_y}\right)\left(\frac{\beta_m}{\beta_m - P_u(KL)^2}\right) \quad [12.12.5]$$

There will be two bending moment terms for biaxial bending. Neglecting the magnification term (last bracket term), Eq. 12.12.5 becomes for biaxial bending,

$$P_{uEQ} = P_u + M_{ntx}(\beta_{az})_x\left(\frac{8\phi_c F_{cr}}{9\phi_b F_y}\right) + M_{nty}(\beta_{az})_y\left(\frac{8\phi_c F_{cr}}{9\phi_b F_y}\right)$$

The designer may directly use the above equation by estimating both $(\beta_{az})_x$ and $(\beta_{az})_y$, or perhaps more practical use the same β_{az} for both terms and multiply the y-axis term by $\beta_b = Z_x/Z_y$. The β_b may also be estimated from Table 12.12.1. Using the latter approach, estimate β_{az} as 0.20 and β_b as 2.2 for W12 from Table 12.12.1 for medium weight sections, and using 0.7 for the "stress ratio" term,

$$P_{uEQ} = P_u + M_{ntx}\beta_{az}(0.7) + M_{nty}\beta_{az}\beta_b(0.7)$$

$$P_{uEQ} = 375 + 38(12)(0.20)0.7 + 14(12)(0.20)(2.2)0.7$$

$$= 375 + 63.8 + 51.7 = 490 \text{ kips}$$

(b) Select trial section. Using $K_y L_y = 7.5$ ft, select from *LRFD Manual* "COLUMNS" load tables for $F_y = 50$ ksi,

$$W12\times45, \qquad \phi_c P_n = 479 \text{ kips}; \qquad r_x/r_y = 2.10$$

$$\beta_{az} = A_g/Z_x = 0.20; \quad \beta_b = Z_x/Z_y = 3.4$$

Revised $P_{uEQ} = 375 + 63.8 + 14(12)(0.20)(3.4)0.7 = 519$ kips

It appears the next heavier W12 will be needed; W12×50, having $\phi_c P_n = 535$ kips. Try the W12×50.

(c) Column action. Since W12×50, A572 Grade 50, is in the "COLUMNS" load tables, $\lambda \leq \lambda_r$ for axial compression.

$$\frac{K_x L_x}{r_x} = \frac{15(12)}{5.18} = 34.7 \qquad \frac{K_y L_y}{r_y} = \frac{7.5(12)}{1.96} = 45.9$$

Find $\phi_c F_{cr} = 36.4$ ksi   (based on $KL/r = 45.9$)

$$\phi_c P_n = \phi_c F_{cr} A_g = 36.4(14.7) = 535 \text{ kips}$$

The ratio $P_u/\phi_c P_n = 375/535 = 0.70$, which exceeds 0.2; therefore, LRFD-Formula (H1-1a), Eq. 12.11.1, applies.

(d) Beam action—x-axis. From *LRFD Manual* "LOAD FACTOR DESIGN SELECTION TABLE," find

$$L_p = 6.9 \text{ ft}$$

$$L_r = 21.7 \text{ ft}$$

$$\phi_b M_{px} = 271 \text{ ft-kips}$$

$$\phi_b M_{rx} = 194 \text{ ft-kips}$$

Since $L_p < L_b < L_r$, $\phi_b M_{nx}$ must be obtained by linear interpolation using Eq. 9.6.4 where the interpolation result is multiplied by C_b. Using Eq. 9.6.4 with $C_b = 1.25$,

$$M_n = C_b\left[M_p - (M_p - M_r)\left(\frac{L_b - L_p}{L_r - L_p}\right)\right] \leq M_p \qquad [9.6.4]$$

$$\phi_b M_n = 1.25\left[271 - (271 - 194)\left(\frac{7.5 - 6.9}{21.7 - 6.9}\right)\right] = 335 \text{ ft-kips}$$

The result for $\phi_b M_n$ cannot exceed $\phi_b M_p = 271$ ft-kips. Thus, $\phi_b M_n = 271$ ft-kips. Check $8\phi_c F_{cr}/(9\phi_b F_y) = 8(36.4)/[9(45)] = 0.72$, which exceeds the 0.7 assumed in estimating P_{uEQ}. One could revise the P_{uEQ} at this stage before proceeding,

Revised $P_{uEQ} = 375 + 38(12)(0.20)(0.72) + 14(12)(0.20)(3.4)(0.72)$

$$= 375 + 66 + 82 = 523 \text{ kips} < (\phi_c P_n = 536 \text{ kips}) \quad \text{OK}$$

(e) Beam action—y-axis. Since lateral-torsional buckling and web local buckling cannot be applicable limit states, and $\lambda < \lambda_p$ for the flange precludes flange local buckling,

$$M_{ny} = M_{py} = F_y Z_y = 50(21.4)/12 = 89.2 \text{ ft-kips}$$
$$\phi_b M_{ny} = 0.90(89.2) = 80.3 \text{ ft-kips}$$

Note that M_{py} exceeds $1.5 S_y F_y$, which is not permitted according to LRFD-F1.1.

All standard I-shaped sections have M_p slightly greater than the $1.5 M_y$ limit, with respect to y-axis bending. The authors do not believe that limit was intended to apply for weak-axis bending of standard W, S, or M sections, but rather was intended for unsymmetrical sections, such as T sections. If the limit $1.5 M_y$ were intended to apply for y-axis bending of W, S, and M sections, the limit controls on all of them; there would be no need to tabulate Z_y in the *LRFD Manual*.

(f) Moment magnification—x-axis. The slenderness ratio KL/r involved in moment magnification must relate to the *axis of bending,* in this case the x-axis,

$$\text{Axis of bending } \frac{KL}{r} = \frac{K_x L_x}{r_x} = 34.7$$

$$P_{e1} = \frac{\pi^2 E A_g}{(KL/r)^2} = \frac{\pi^2(29,000)14.7}{(34.7)^2} = 3480 \text{ kips}$$

For end moment loading, C_m is LRFD Formula (H1-4), Eq. 12.4.6,

$$C_m = 0.6 - 0.4\frac{M_1}{M_2} = 0.6 - 0.4\left(\frac{0}{38}\right) = 0.6$$

$$B_1 = \frac{C_m}{1 - P_u/P_{e1}} = \frac{0.6}{1 - 375/3480} = 0.6(1.12) < 1.0$$

Thus, the magnified moment out away from the support does not exceed the primary moment at the support; thus, $B_{1x} = 1.0$.

(g) Moment magnification—y-axis. The slenderness ratio KL/r involved in moment magnification must relate to the *axis of bending,* in this case the y-axis,

$$\text{Axis of bending } \frac{Kl}{r} = \frac{K_y L_y}{r_y} = 45.9$$

$$P_{e1} = \frac{\pi^2 E A_g}{(KL/r)^2} = \frac{\pi^2(29,000)14.7}{(45.9)^2} = 2000 \text{ kips}$$

For end moment loading, C_m is LRFD Formula (H1-4), Eq. 12.4.6,

$$C_m = 0.6 - 0.4\frac{M_1}{M_2} = 0.6 - 0.4\left(\frac{3}{14}\right) = 0.51$$

$$B_1 = \frac{C_m}{1 - P_u/P_{e1}} = \frac{0.51}{1 - 375/2000} = 0.51(1.23) < 1.0$$

Thus, the magnified moment out away from the support does not exceed the primary moment at the support; thus, $B_{1y} = 1.0$.

(h) Check LRFD Formula (H1-1a), Eq. 12.10.1,

$$M_{ux} = M_{ntx}B_{1x} = 38(1.0) = 38 \text{ ft-kips}$$
$$M_{uy} = M_{nty}B_{1y} = 14(1.0) = 14 \text{ ft-kips}$$

$$\frac{P_u}{\phi_c P_n} + \frac{8}{9}\left(\frac{M_{ux}}{\phi_b M_{nx}}\right) + \frac{8}{9}\left(\frac{M_{uy}}{\phi_b M_{ny}}\right) \le 1.0 \qquad [12.10.1]$$

$$\frac{375}{535} + \frac{8}{9}\left(\frac{38}{271}\right) + \frac{8}{9}\left(\frac{14}{80.3}\right) = 0.700 + 0.125 + 0.155 = 0.98 < 1 \qquad \text{OK}$$

Use W12×50, A572 Grade 50 steel. ■■

EXAMPLE 12.13.8 ──

Select an economical structural tee for use as the top chord in a roof truss, as shown in Fig. 12.13.8. The transverse concentrated load W is 1 kip dead load (neglect the weight of the structural tee) and 4 kips live load. The axial compression P is 17 kips dead load and 68 kips live load. Assume the primary moment M on the member occurs as if the ends were fixed; that is, $WL/8$ for both positive and negative moment zones. The chord is designed to be continuous over 12-ft spans, with transverse lateral support at 6-ft intervals. Use $F_y = 50$ ksi and Load and Resistance Factor Design.

Solution

(a) Compute factored loads. Assume a first-order elastic analysis was performed to obtain the given forces. Using LRFD-A4.1, Formula (A4-2),

$$W_u = 1.2(1.0) + 1.6(4.0) = 7.6 \text{ kips}$$
$$P_u = 1.2(17) + 1.6(68) = 129 \text{ kips}$$
$$M_{nt} = \pm W_u L/8 = 7.6(12)/8 = 11.4 \text{ ft-kips}$$

(b) Effective lengths for column and beam action. For members of a truss, SSRC recommends using $K = 1.0$ (see Sec. 6.9).

Column action: $K_x L_x = 1.0(12) = 12$ ft
                   $K_y L_y = 1.0(6) = 6$ ft

Beam action: $L_b = 6$ ft for x-axis bending
                       (applies only when $r_x > r_y$)
                   $C_b = 2.27$ max (Table 9.6.3)
                       (applies only when $r_x > r_y$)

Moment magnification: $K_x L_x = 12$ ft for P_{ex}
                   $C_m = 0.85$ for restrained ends

Note that the strength criterion must be checked at midspan where the flange is in compression, and at the supports where the stem is in compression. Also, the laterally unbraced length L_b and moment gradient term C_b have meaning only when $r_x > r_y$; when bending occurs in the weak direction, lateral-torsional buckling is not a possible limit state.

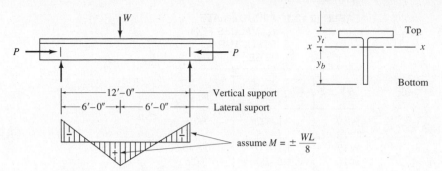

Figure 12.13.8   Example 12.13.8.

(c) Estimate the P_{uEQ} using Eq. 12.12.5. For structural tees, as discussed in Sec. 9.12, the denominator M_{nx} of the bending term in the LRFD-H1 beam-column criterion cannot exceed (LRFD-F1.2c):

$$M_n = 1.5M_y \text{ when the stem is in compression}$$
$$M_n = 1.0M_y \text{ when the stem is in tension}$$

This means the β_{az} term of Eq. 12.12.5 becomes $A_g/S_{x,stem}$ for the usual case where the strength M_n is controlled by stress at the extreme fiber of the stem.

Estimating the "stress ratio" term as 0.7, estimating β_{az} as 1.6 for WT7, 1.4 for WT8, and 1.2 for WT9 from Table 12.13.1, and neglecting the magnification term,

$$P_{uEQ} = P_u + M_{nt}\beta_{az}\left(\frac{8\phi_c F_{cr}}{9\phi_b F_y}\right)\left(\frac{\beta_m}{\beta_m - P(KL)^2}\right) \qquad [12.12.5]$$

$$P_{uEQ} = P_u + M_{nt}\beta_{az}(0.7)$$

WT7, $P_{uEQ} \approx 129 + 11.4(12)(1.6)0.7 = 282$ kips
WT8, $P_{uEQ} \approx 129 + 11.4(12)(1.4)0.7 = 263$ kips
WT9, $P_{uEQ} \approx 129 + 11.4(12)(1.2)0.7 = 244$ kips

Using the *LRFD Manual* "COLUMNS Structural tees cut from W shapes" load tables, select for $K_y L_y = 6$ ft or $K_x L_x = 12$ ft,

	For $K_x L_x = 12$ ft	For $K_y L_y = 6$ ft
WT7×37	$\phi_c P_n = 293$ kips	$\phi_c P_n = >405$ kips
WT8×28.5	$\phi_c P_n = 262$ kips	$\phi_c P_n = 268$ kips
WT9×30	$\phi_c P_n = 253$ kips	$\phi_c P_n = \approx 240$ kips

Additional selection data may be obtained by noting the proportion of P_{uEQ} that represents the bending effect. For the WT7 above, the bending proportion is 54% (153/282) of the total effect, thus,

$$\frac{8}{9}\left(\frac{M_u}{\phi_b M_n}\right) \approx 0.54$$

TABLE 12.13.1   APPROXIMATE β_{az} VALUES FOR STRUCTURAL TEES

WT	$\dfrac{A}{S_{x,\,stem}}$	$\dfrac{A}{S_{x,\,flange}}$
WT9	1–1.2	0.3
WT8	1–1.5	0.3
WT7	1.2–2.0	0.3–0.5
WT6	1.5–2.0	0.5
WT5	2.0–2.5	0.6
WT4	2.5–3.5	0.7

which, noting that $M_n = F_y S_{x,\,stem}$ maximum, gives an approximate required $S_{x,\,stem}$,

$$\text{Required } S_{x,\,stem} \approx \frac{8M_u}{9\phi_b F_y(0.54)} = \frac{8(11.4)12}{9(0.90)(50)0.54} = 5.0 \text{ in.}^3$$

For the tentative selections, WT7×37 has $S_{x,\,stem} = 6.25$ in.^{3}, WT8×28.5 has $S_{x,\,stem} = 7.77$ in.^{3}, and WT9×30 has $S_{x,\,stem} = 9.3$ in.^{3}. Further checking the WT8×28.5, $A_g/S_{x,\,stem} = 8.38/7.77 = 1.1$; revised $P_{uEQ} \approx 234$ kips. From "COLUMNS" load tables it still appears the WT8×28.5 may be the appropriate choice.

Try WT8×28.5.

(d) Column action. Check WT8×28.5,

$$\frac{K_x L_x}{r_x} = \frac{12(12)}{2.41} = 59.8 \qquad \frac{K_y L_y}{r_y} = \frac{6(12)}{1.60} = 45.0$$

Since the stem of a structural tee is a thin compression element, local buckling may be a controlling limit state; check $\lambda \geq \lambda_r$ from LRFD-B5,

$$\left(\lambda = \frac{d}{t_w} = \frac{8.215}{0.430} = 19.1 \right) > \left(\lambda_r = \frac{127}{\sqrt{F_y}} = 18.0 \right) \quad \text{NG}$$

The efficiency of the stem is reduced; i.e., $Q_s < 1.0$. Using LRFD-Appendix B5.3a for $\lambda_r < \lambda < 176/\sqrt{F_y}$,

$$Q_s = 1.908 - 0.00715(d/t_w)\sqrt{F_y}$$

$$Q_s = 1.908 - 0.00715(19.1)\sqrt{50} = 0.942$$

Compute $\lambda_c \sqrt{Q}$ according to LRFD-E2,

$$\lambda_c \sqrt{Q_s} = \frac{KL/r}{\pi} \sqrt{\frac{F_y Q_s}{E}} = \frac{59.8}{\pi} \sqrt{\frac{50(0.942)}{29,000}} = 0.766$$

Use $\lambda_c \sqrt{Q_c}$ to enter *LRFD Manual* "NUMERICAL VALUES" TABLE 4,

$$\frac{\phi_c F_{cr}}{Q_s F_y} = 0.665; \qquad \phi_c F_{cr} = 0.665(0.942)50 = 31.3 \text{ ksi*}$$

$$\phi_c P_n = \phi_c F_{cr} A_g = 31.3(8.38) = 262 \text{ kips}$$

The ratio $P_u/\phi_c P_n = 129/262 = 0.492$, which exceeds 0.2; therefore, LRFD-Formula (H1-1a), Eq. 12.11.1, applies.

(e) Beam action—flange in compression at midspan. Check the ratio $r_x/r_y = 2.41/1.60 > 1.0$; thus, the lateral-torsional buckling limit state must be considered. The maximum nominal moment strength M_n at midspan according to LRFD-F1.2c occurs when either the flange extreme fiber in compression reaches F_y (or a lesser value in the unlikely event that local buckling makes Q_s less than unity), or the stem extreme fiber reaches F_y in tension. The latter is the usual case,

$$M_n = M_y = F_y S_{x,stem} = 50(7.77)/12 = 32.3 \text{ ft-kips}$$

It will be highly unlikely that lateral-torsional buckling will control on structural tees. However, LRFD-F1.2c gives an equation for M_{cr}. Using $C_b = 2.27$ in LRFD Formula (F1-15) as illustrated in Example 9.12.2, $L_b = 141$ ft when $M_{cr} = M_y = 32.3$ ft-kips.

(f) Beam action—web in compression at supports. Since $\lambda > \lambda_r$ as determined in part (d), $Q_s = 0.942$. For flexural members, according to LRFD-Appendix B5.3a, the maximum bending stress "shall not exceed $\phi_b F_y Q_s$." Thus,

$$M_n = M_y = F_y Q_s S_{x,stem} = 50(0.942)(7.77)/12 = 30.5 \text{ ft-kips}$$

In this example, the factored forces are the same at the supports and at midspan; therefore, the strength criterion at the supports ($M_n = 30.5$ ft-kips) will control.

(g) Moment magnification. The slenderness ratio KL/r involved in moment magnification must relate to the *axis of bending,* in this case $K_x L_x/r_x = 59.8$ from part (d); thus.

$$P_{e1} = \frac{\pi^2 E A_g}{(KL/r)^2} = \frac{\pi^2(29,000)8.38}{(59.8)^2} = 671 \text{ kips}$$

For restrained ends with concentrated load at midspan, one may take $C_m = 0.85$ in accordance with LRFD-C1. Alternatively, a more accurate expression from

---

*If flexural-torsional buckling is considered in addition to flexural buckling about the x- and y-axes, as apparently is required by LRFD-Appendix E3 for tee-sections, that lengthy procedure will give $\phi_c F_{cr} = 33.2$ ksi using a torsional effective length $K_z L$ of 12 ft. Thus, flexural-torsional buckling does not control (since 33.2 ksi exceeds the flexural buckling value of 31.3 ksi). In the authors' view, this complexity for usual design seems unnecessary. Torsional restraint exists at the ends and also from the lateral support at mid-span.

LRFD-Commentary TABLE C-C1.1 (Table 12.3.1, Case 7, negative moment) is

$$C_m = 1 - 0.2 \frac{P_u}{P_{e1}} = 1.0 - 0.2 \left(\frac{129}{671}\right) = 0.96$$

$$B_1 = \frac{C_m}{1 - P_u/P_{e1}} = \frac{0.96}{1 - 129/671} = 1.19$$

If $C_m = 0.85$, $B_1 = 1.05$.

(h) Check LRFD Formula (H1-1a), Eq. 12.10.1, omitting the term for bending about the y-axis,

$$M_{ux} = M_{ntx} B_{1x} = 11.4(1.19) = 13.6 \text{ ft-kips}$$
$$\phi_b M_{nx} = \phi_b M_{y} = 0.90(30.5) = 27.4 \text{ ft-kips}$$

$$\frac{P_u}{\phi_c P_n} + \frac{8}{9} \left(\frac{M_{ux}}{\phi_b M_{nx}}\right) \le 1.0 \qquad \text{[12.10.1]}$$

$$\frac{129}{262} + \frac{8}{9} \left(\frac{13.6}{27.4}\right) = 0.492 + 0.441 = 0.93 < 1 \quad \text{OK}$$

Use WT8×28.5, A572 Grade 50 steel. ■■

12.14 ALLOWABLE STRESS DESIGN CRITERIA

For Allowable Stress Design, the strength interaction equations, Eqs. 12.8.1 through 12.8.4, may be converted to unit stresses and a factor of safety (FS) applied to bring them into the service load range.

Stability Interaction Criterion

The nominal strength interaction equation, Eq. 12.8.4, including lateral-torsional buckling is

$$\frac{P_u}{P_n} + \frac{M_{ui}}{M_n} \frac{C_m}{(1 - P_u/P_e)} = 1 \qquad \text{[12.8.4]}$$

where P_u and M_{ui} are the axial force and primary bending moment, respectively, that occur *when failure is imminent*. When both the numerator and denominator are divided by a factor (FS) to bring all terms into the service load range,

$$\frac{P_u/(A_g\text{FS})}{P_n/(A_g\text{FS})} + \frac{M_{ui}/[S(\text{FS})]}{M_n/[S(\text{FS})]} \frac{C_m}{\left[1 - \dfrac{P_u/(A_g\text{FS})}{P_e/(A_g\text{FS})}\right]} = 1.0 \qquad (12.14.1)$$

or

$$\frac{P/A_g}{P_n/(A_g\text{FS})} + \frac{M_i/S}{M_n/[S(\text{FS})]} \frac{C_m}{\left[1 - \dfrac{P/A_g}{P_e/(A_g\text{FS})}\right]} = 1.0 \qquad (12.14.2)$$

which gives as a design requirement,

$$\frac{f_a}{F_a} + \frac{f_b}{F_b} \frac{C_m}{(1 - f_a/F'_e)} \le 1.0 \tag{12.14.3}$$

for *uniaxial* bending and compression.

By analogy, for bending about both x- and y-axes, Eq. 12.14.3 would become

$$\frac{f_a}{F_a} + \frac{f_{bx}C_{mx}}{F_{bx}(1 - f_a/F'_{ex})} + \frac{f_{by}C_{my}}{F_{by}(1 - f_a/F'_{ey})} \le 1.0 \tag{12.14.4}$$

which is the stability interaction equation, ASD Formula (H1-1),

where $f_a = P/A_g$ = axial compression stress at service load

f_{bx}, f_{by} = flexural stresses at service load based on primary bending moment about the x- and y-axes, respectively

F_a = allowable compression stress considering the member as loaded by axial compression only

F_{bx}, F_{by} = allowable flexural stresses for the x- and y-axes, respectively, considering the member loaded in bending only. According to the definition of C_b in ASD-F1.3, when the stability equation, Eq. 12.14.4, is used for *braced frames* $C_b = 1.0$, but when Eq. 12.14.4 is used for *unbraced frames*,
$C_b = 1.75 + 1.05(M_1/M_2) + 0.3(M_1/M_2)^2 \le 2.3$

C_m = factors discussed in Secs. 12.3 through 12.5, to be taken as follows:

1. For *braced frame* members having transverse loading between supports, C_m is an integral part of the moment magnifier B_1, and whose value may be determined by rational analysis,

$$C_m = 1 + \psi\alpha \quad \text{(ASD Commentary-H1)} \tag{12.14.5}$$
$$= \text{Values from Table 12.3.1}$$
$$C_m = 1.0 \quad \text{(ASD-H1)} \tag{12.14.6}$$
(ends of member unrestrained)
$$C_m = 0.85 \quad \text{(ASD-H1)} \tag{12.14.7}$$
(ends of member restrained)

2. For *braced frame* members without transverse loading between supports but having end moments M_1 (smaller one) and M_2 (larger one), C_m converts the linearly varying primary bending moment into an equivalent uniform moment $M_E = C_m M_2$,

$$C_m = 0.6 - 0.4M_1/M_2 \tag{12.14.8}$$

The moments M_1 and M_2 are *rotational* moments, rather than bending moments as used in Sec. 12.4. Therefore, the ratio is negative $(-)$ for single curvature and positive $(+)$ for double curvature. Since M_E is the primary moment, C_m is really not part of the magnification factor.

3. For *unbraced frame* members,

$$C_m = 0.85 \quad \text{(ASD-H1)} \tag{12.14.9}$$

or from ASD-Commentary H1,

$$C_m = 1 - 0.18 f_a/F'_e \tag{12.14.10}$$

as discussed in Sec. 12.5.

In the application of Eq. 12.14.4 [ASD Formula (H1-1)], the term F'_e refers to the *effective pin-end length* in the *plane of bending*:

$$F'_e = \frac{P_e}{A_g \text{FS}} = \frac{\pi^2 EI}{A_g L^2 \text{FS}}$$

$$= \frac{12\pi^2 E}{23(KL/r)^2} = \frac{149{,}000}{(KL/r)^2} \text{ ksi} \tag{12.14.11}*$$

where K = effective length factor (see Sec. 6.9)
   L = actual unbraced length in the plane of bending
   r = radius of gyration for the axis of bending

It is noted that a nominal safety factor of $23/12 = 1.92$ is used; this is the maximum factor used for long axially loaded members (see Sec. 6.11) and is therefore conservative in the magnification term.

Yielding Interaction Criterion

At support locations in *braced frames* and for low slenderness situations in *unbraced frames*, yielding (plastic strength under the action of P and M) under combined axial compression and bending may govern. The strength interaction equation, Eq. 12.8.1, forms the basis for the Allowable Stress Design criterion,

$$\frac{P_u}{P_y} + \frac{M_u}{1.18 M_p} = 1.0 \tag{12.8.1}$$

where P_u and M_u are the axial force and primary bending moment, respectively, that occur when maximum strength is achieved. When the numerator and denominator are divided by a factor (FS) to bring all terms into the service load range,

$$\frac{P_u/(A_g \text{FS})}{P_y/(A_g \text{FS})} + \frac{M_u/[S(\text{FS})]}{1.18 M_p/[S(\text{FS})]} = 1.0 \tag{12.14.12}$$

or

$$\frac{P/A_g}{P_y/(A_g \text{FS})} + \frac{M/S}{1.18 M_p/[S(\text{FS})]} = 1.0 \tag{12.14.13}$$

---

* For SI units, $\qquad F'_e = \dfrac{1027{,}000}{(KL/r)^2} \text{MPa}$ $\qquad\qquad$ (12.14.11)

which gives as a design requirement,

$$\frac{f_a}{0.60F_y} + \frac{f_b}{1.18M_p/[S(\text{FS})]} = 1.0 \tag{12.14.14}$$

The ASD Specification has used an expression more conservative than Eq. 12.14.14 by using 1.0 instead of 1.18, and instead of $M_p/[S(\text{FS})]$, which for I-shaped sections would correspond to $0.66F_y$ for x-axis bending and $0.75F_y$ for y-axis bending, ASD-H1 has used F_b. For instance, the allowable stress F_{bx} might require reduction below $0.66F_y$ or even below $0.60F_y$ because of the laterally unbraced length adjacent to a braced location.

Thus ASD-H1 uses for the yield criterion,

$$\frac{f_a}{0.60F_y} + \frac{f_b}{F_b} \leq 1.0 \tag{12.14.15}$$

for *uniaxial* bending and compression.

For biaxial bending, the general equation is

$$\frac{f_a}{0.60F_y} + \frac{f_{bx}}{F_{bx}} + \frac{f_{by}}{F_{by}} \leq 1.0 \tag{12.14.16}$$

which is the yield interaction equation, ASD Formula (H1-2), where the quantities are as defined following Eq. 12.14.4, except that in evaluating F_{bx} and F_{by} the moment gradient term C_b is used in exactly the way it is used for bending alone, rather than in the special way it is used for the stability interaction criterion, Eq. 12.14.4.

Simplified Interaction Criterion for Small Axial Compression

When f_a/F_a does not exceed 0.15, ASD-H1 *permits* use of the following instead of the two formulas, ASD Formulas (H1-1) and (H1-2):

$$\frac{f_a}{F_a} + \frac{f_{bx}}{F_{bx}} + \frac{f_{by}}{F_{by}} \leq 1.0 \tag{12.14.17}$$

which is ASD Formula (H1-3).

As is usual with design specifications, current requirements reflect some historical developments. Until 1961, the AISC Specification did not consider the secondary bending moment due to deflection. When axial compression is relatively small, neglect of the secondary effect makes little difference. This simplified alternative of neglecting the secondary effect is not included in LRFD.

Redistribution of Moments to Approximate Plastic Behavior

Regarding bending moment adjustments to account for plastic-behavior moment redistribution, as discussed in Chapter 10, ASD-F1.1 allows the 10% reduction in negative moment on a beam to be used in proportioning the column if the *beam or girder* has $L_b \leq L_c$ for lateral-torsional buckling, satisfies "compact section" requirements, and the compressive stress f_a on the column does not exceed $0.15F_a$.

EXAMPLE 12.14.1

Investigate the acceptability of a W16×67 used as a beam-column under the loading shown in Fig. 12.10.1. The total service loads are $P = 350$ kips and $M = 60$ ft-kips, and $F_y = 60$ ksi. Use Allowable Stress Design.

Solution

(a) Column effect. Check local buckling slenderness ratio limits for axial compression.

$$\left[\frac{b_f}{2t_f} = 7.7 \begin{array}{c} \text{From } LRFD \\ Manual \end{array}\right] < \left[\frac{95}{\sqrt{F_y}} = \frac{95}{\sqrt{60}} = 12.3\right] \quad \text{OK}$$

$$\left[\frac{h}{t_w} = 35.9 \begin{array}{c} \text{From } LRFD \\ Manual \end{array}\right] > \left[\frac{253}{\sqrt{F_y}} = \frac{253}{\sqrt{60}} = 32.7\right] \quad \text{NG}$$

There is a potential for $Q < 1.0$. Assume $Q = 1.0$ and determine allowable stress F_a,

$$\frac{K_x L_x}{r_x} = \frac{15(12)}{2.46} = 73$$

$$\frac{K_x L_x/r_x}{C_c} = \frac{73}{97.7} = 0.747$$

From ASD "NUMERICAL VALUES" TABLE 3, find $C_a = 0.381$, giving

$$F_a = C_a F_y = 0.381(60) = 22.8 \text{ ksi}$$

From ASD-Appendix B5.2b, check b_e with trial $f = 22.8$ ksi,

$$b_e = \frac{253t}{\sqrt{f}}\left[1 - \frac{44.3}{(b/f)\sqrt{f}}\right] \le b$$

$b = $ width of stiffened element $= d - 2t_f = 16.33 - 2(0.665) = 15.0$ in.

$$b_e = \frac{253(0.395)}{\sqrt{22.8}}\left[1 - \frac{44.3}{(15.0/0.395)\sqrt{22.8}}\right] = 15.8 \text{ in.} > [b = 15 \text{ in.}]$$

Therefore, $b_e = b$; there is no web local buckling effect and $Q = 1.0$.

$$f_a = \frac{P}{A_g} = \frac{350}{19.7} = 17.8 \text{ ksi}$$

$$\frac{f_a}{F_a} = \frac{17.8}{22.8} = 0.78 > 0.15; \quad \text{use ASD Formulas (H1-1) and (H1-2)}$$

(b) Beam effect. Since local buckling limits on $b_f/2t_f$ and h/t_w are satisfied for axial compression (part a), those limits to achieve $F_b = 0.6F_y$ are automatically satisfied. Determine L_c from Eqs. 9.7.1 and 9.7.2,

$$L_c = \frac{76b_f}{\sqrt{F_y}} = \frac{76(10.235)}{\sqrt{60}\,(12)} = 8.4 \text{ ft} \quad (controls) \qquad [9.7.1]$$

or

$$L_c = \frac{20,000}{(d/A_f)F_y} = \frac{20,000}{2.40(60)12} = 11.6 \text{ ft} \qquad [9.7.2]$$

Thus, $L_c = 8.4$ ft. Since L_b exceeds L_c, the allowable stress F_b will be less than $0.66F_y$. Check L_u using Eqs. 9.7.25 (one of which is Eq. 9.7.2 above); the other is

$$L_u = r_T \sqrt{\frac{102,000}{F_y}} = 2.75 \sqrt{\frac{102,000}{60} \frac{1}{12}} = 9.4 \text{ ft}$$

Thus, $L_u = 11.6$ ft (the larger of 11.6 and 9.4 ft). Since L_b exceeds L_u the allowable stress will be reduced below $0.60F_y$. Use ASD Formulas (F1-6) and (F1-8),

$$F_b(\text{F1-8}) = \frac{12,000}{L_b d/A_f} = \frac{12,000}{15(12)(2.40)} = 27.8 \text{ ksi} < 0.60F_y$$

$$\frac{L_b}{r_T} = \frac{15(12)}{2.75} = 65.5$$

$$F_b(\text{F1-6}) = 40.0 - \frac{(L_b/r_T)^2}{425} = 40.0 - \frac{(65.5)^2}{425} = 29.9 \text{ ksi}$$

Since $29.9 > 27.8$, F_b (F1-6) controls. In both formulas C_b is taken as 1.0 (i.e., not used), because C_m in this situation converts the moment diagram into an equivalent uniform moment, for which C_b would be 1.0.

$$C_m = 0.6 - 0.4(M_1/M_2) = 0.60$$

$$f_b = \frac{60(12)}{117} = 6.15 \text{ ksi}$$

$$\frac{C_m f_b}{F_b} = \frac{0.6(6.15)}{29.9} = 0.12$$

(c) Moment magnification. Using *ASD Manual* "NUMERICAL VALUES" TABLE 8,

$$\frac{KL}{r_x} = \frac{15(12)}{6.96} = 25.9; \qquad F'_e = 223 \text{ ksi}$$

where the x-axis is the axis of bending. The magnification factor is

$$\frac{1}{1 - f_a/F'_e} = \frac{1.0}{1 - 17.8/223} = \frac{1.0}{1 - 0.0798} = 1.09$$

(d) Check of ASD Formulas:
For stability, Formula (H1-1),

$$\frac{f_a}{F_a} + \frac{C_m f_b}{F_b} \left(\frac{1.0}{1 - f_a/F'_e} \right) = 0.78 + 0.12(1.09) = 0.91 < 1.0$$

For yielding, Formula (H1-2), at the braced point,

$$\frac{f_a}{0.60F_y} + \frac{f_b}{F_b} = \frac{17.8}{36} + \frac{6.15}{36} = 0.66 < 1.0$$

For the above equation which does not involve C_m, F_b (F1-8) should use $C_b = 1.75$ for this problem. In which case $F_b > 0.60F_y$; use $0.60F_y = 36$ ksi. The W16×67 is acceptable for the given loading. ■■

12.15 DESIGN PROCEDURES—ALLOWABLE STRESS DESIGN

To aid in selection of a beam-column section, it is usually advantageous to convert, in an approximate way, the resulting bending moment into an equivalent axial compression load and then to make use of column tables, as was done in Sec. 12.12 for LRFD. Occasionally, conversion of the axial load into equivalent moment will be helpful.

The stability interaction equation, Eq. 12.14.3, may be written

$$\frac{P}{A_g F_a} + \frac{M}{F_b S}\left(\frac{C_m}{1 - f_a/F_e'}\right) = 1.0 \qquad (12.15.1)$$

Multiplying by $A_g F_a$ gives

$$P + M\left(\frac{A_g}{S}\right)\left(\frac{F_a}{F_b}\right)\left(\frac{C_m}{1 - f_a/F_e'}\right) = F_a A_g = P_{EQ} \qquad (12.15.2)$$

Next, examine the magnification term, which may be changed in form, using Eq. 12.14.11 for F_e',

$$\frac{1}{1 - f_a/F_e'} = \frac{F_e'}{F_e' - f_a} = \frac{149{,}000 r^2}{(KL)^2\left(\dfrac{149{,}000 r^2}{(KL)^2} - \dfrac{P}{A_g}\right)} \qquad (12.15.3)$$

$$= \frac{149{,}000 A_g r^2}{149{,}000 A_g r^2 - P(KL)^2} \qquad (12.15.4)$$

Thus the equivalent column load P_{EQ} may be expressed using Eq. 12.15.2, corresponding to ASD Formula (H1-1) for uniaxial bending.

$$P_{EQ} = P + M\beta_{as}\left(\frac{F_a}{F_b}\right)\left(\frac{C_m a}{a - P(KL)^2}\right) \qquad (12.15.5)$$

where β_{as} = bending factor = A_g/S
    $a = 149{,}000 A_g r^2$ for axis of bending

Note is made that the allowable stress ratio (F_a/F_b) reduces P_{EQ} while the term $[C_m a/(a - P(KL)^2)]$ usually increases P_{EQ}.

When the yield criterion controls, the equivalent column load P_{EQ} may be expressed,

$$P_{EQ} = P + M\beta_{as}\left(\frac{0.60 F_y}{F_b}\right) \qquad (12.15.6)$$

corresponding to ASD Formula (H1-2) for uniaxial bending.

When $f_a/F_a < 0.15$, P_{EQ} may be computed similarly from Eq. 12.14.17 as

$$P_{EQ} = P + M\beta_{as}\left(\frac{F_a}{F_b}\right) \qquad (12.15.7)$$

corresponding to ASD Formula (H1-3) for uniaxial bending.

Several examples follow which demonstrate application of the interaction formulas, using principles established in Chapters 6, 7, and 9 and the approach outlined above.

EXAMPLE 12.15.1 _____

Select the lightest W14 section to carry service loads consisting of an axial compression P of 150 kips in combination with a moment M of 500 ft-kips. The member is part of a *braced* system (see Fig. 12.13.1), with support provided in each direction at top and bottom of a 14-ft length. Conservatively assume the moment causes single curvature and varies as shown in Fig. 12.13.1. Use A36 steel and Allowable Stress Design.

Solution. Conservatively assume the effective length factor $K = 1.0$. If adjacent member stiffness is known, the alignment chart which appears as Fig. 9.6.4a in Chapter 6 may be used to determine $K < 1.0$ for this braced frame.

Since at the start one may have no idea whether $f_a/F_a < 0.15$, use the simplest P_{EQ} expression, Eq. 12.15.7,

$$P_{EQ} = P + M\beta_{as}\left(\frac{F_a}{F_b}\right)$$

Referring to *ASD Manual* "COLUMNS" load tables: for W14 sections, find the average bending factor with respect to the strong axis, $\beta_{as} \approx 0.19$. Neglect temporarily F_a/F_b; i.e., assume $F_a \approx F_b$.

$$P_{EQ} \approx 150 + 500(12)(0.19) = 150 + 1140 = 1290 \text{ kips}$$

At this point, since the column part represents less than 15% (150/1290) of the total, assume that ASD Formula (H1-3) will apply.

Obtain trial sections from *ASD Manual* using $P = 1290$ kips and $L = 14$ ft, and try one section lighter than indicated.

$$\text{W14}\times211, \qquad \beta_{as} = 0.183, \qquad r_y = 4.07 \text{ in.}, \qquad L_c = 16.7 \text{ ft}$$

$$P = 1183 \text{ kips}$$

Check: Use *ASD Manual* "NUMERICAL VALUES" TABLE 3.

$$\frac{K_y L_y}{r_y} = \frac{14(12)}{4.08} = 41.3; \qquad \frac{KL/r}{C_c} = \frac{41.3}{126.1} = 0.328$$

Find $C_a = 0.530;$ $\qquad F_a = C_a F_y = 0.530(36) = 19.1 \text{ ksi}$

$L_b = 14 \text{ ft} < L_c$ (Sec. 9.7); $\qquad F_b = 0.66F_y = 24 \text{ ksi}$

$$\frac{F_a}{F_b} = \frac{19.1}{24} = 0.80$$

Revise using closer approximations for the variables,

$$P_{EQ} = 150 + 500(12)(0.183)(0.80) = 1028 \text{ kips}$$

Select for the revised P_{EQ},

$$\text{W14}\times193, \qquad \beta_{as} = 0.183, \qquad r_y = 4.05 \text{ in.}, \qquad L_c = 16.6 \text{ ft}$$

$$P = 1083 \text{ kips}$$

Little change in properties from preliminary check. Width/thickness limits to preclude flange and web local buckling are satisfied. Beam-column criterion check:

$$\frac{KL}{r_y} = \frac{14(12)}{4.05} = 41.5; \qquad F_a = 19.1 \text{ ksi}$$

$$L_b = 14 \text{ ft} < L_c; \qquad F_b = 24 \text{ ksi}$$

$$f_a = \frac{P}{A_g} = \frac{150}{56.8} = 2.6 \text{ ksi}$$

$$\frac{f_a}{F_a} = \frac{2.6}{19.1} = 0.14 < 0.15; \qquad \text{Use ASD Formula (H1-3)}$$

$$f_b = \frac{M}{S_x} = \frac{500(12)}{310} = 19.4 \text{ ksi}$$

$$\frac{f_a}{F_a} + \frac{f_b}{F_b} = 0.14 + 0.81 = 0.95 < 1.0 \qquad \text{OK}$$

Use W14×193. Note that a W14×132 was suitable for this loading using LRFD in Example 12.13.1. ■■

EXAMPLE 12.15.2 _____

Investigate using Allowable Stress Design the W10×39 section designed as a beam-column according to Load and Resistance Factor Design in Example 12.13.3 (Figs. 12.13.3 and 12.13.4). A36 steel.

Solution

(a) Establish effective lengths. The member must be viewed separately as a column, then as a beam, as in Fig. 12.13.4.
For column action:

$$K_x L_x = 16 \text{ ft}; \qquad K_y L_y = 8 \text{ ft}$$

Required $r_x/r_y \geq 2$ if $KL = 8$ ft is valid for entering *ASD Manual*, "COLUMNS" load tables

For beam action:

$$L_b = \text{Laterally unbraced length} = 8 \text{ ft}$$

$$L \text{ for moment magnification} = 16 \text{ ft}$$

(b) Check W10×39:
Column action:

$$f_a = \frac{P}{A_g} = \frac{120}{11.5} = 10.4 \text{ ksi}$$

$$\frac{KL}{r_y} = \frac{8(12)}{1.98} = 48.5, \qquad \frac{r_x}{r_y} = 2.16 > 2.0$$

$$F_a = 18.5 \text{ ksi}$$

$$\frac{f_a}{F_a} = \frac{10.4}{18.5} = 0.56 > 0.15 \quad \text{Use ASD Formula (H1-1)}$$

Beam action:

$$f_b = \frac{M}{S_x} = \frac{360}{42.1} = 8.55 \text{ ksi}$$

$$L_c = 8.4 \text{ ft} > (L_b = 8 \text{ ft})$$

$$F_b = 0.66F_y = 24 \text{ ksi}$$

$$\frac{f_b}{F_b} = \frac{8.55}{24} = 0.36$$

Note: Since an L_c value is given in the *ASD Manual* tables, the local buckling limitations of ASD-B5 are automatically satisfied for A36 steel. When no tables are applicable, the following checks must be made in order to use $0.66F_y$:

$$\frac{L_b}{b_f} \leq \left[\text{smaller of } \frac{76}{\sqrt{F_y}} \text{ or } \frac{20{,}000}{b_f(d/A_f)F_y} \right] \quad \begin{array}{l} \text{for lateral-torsional buckling} \\ \text{prevention (ASD-F1.1)} \end{array}$$

$$\frac{b_f}{2t_f} \leq \frac{65}{\sqrt{F_y}} \quad \text{for flange local buckling prevention (ASD-B5)}$$

$$\frac{d}{t_w} \leq \frac{640}{\sqrt{F_y}}\left(1 - 3.74\frac{f_a}{F_y}\right) \geq \frac{257}{\sqrt{F_y}} \quad \begin{array}{l} \text{for web local buckling prevention} \\ \text{(ASD-B5)} \end{array}$$

This has been discussed in Secs. 6.17 and 9.5.

Beam-column moment magnification:

$$\frac{KL}{r_x} = \frac{16(12)}{4.27} = 45 \quad \text{(for axis of bending)}$$

$$F'_e = 73.7 \text{ ksi } (\textit{ASD Manual} \text{ "NUMERICAL VALUES" TABLE 8)}$$

$$\frac{f_a}{F'_e} = \frac{10.4}{73.7} = 0.14, \qquad 1 - \frac{f_a}{F'_e} = 0.86$$

$$C_m \approx 1 - 0.2\frac{f_a}{F'_e} = 0.97 \quad \text{(Fig. 12.3.1, Case 3)}$$

This C_m expression assumes moment variation of Fig. 12.13.3c may approximate that of Fig. 12.13.3b.

$$\text{Magnification factor} = \frac{C_m}{1 - f_a/F'_e} = \frac{0.97}{0.86} = 1.13$$

ASD Formula (H1-1):

$$\frac{f_a}{F_a} + \frac{f_b}{F_b}\frac{C_m}{(1 - f_a/F'_e)} = 0.56 + 0.36(1.13) = 0.97 < 1.0 \quad \text{OK}$$

Use W10×39. ■■

EXAMPLE 12.15.3

Investigate the adequacy according to Allowable Stress Design of the W8×24 section shown in Fig. 12.13.5. The member is loaded to cause bending about its weak axis, and the steel is A572 Grade 50.

Solution

(a) ASD Specification check; since maximum moment occurs away from the supports, the stability criterion will govern. Check ASD Formula (H1-1).

Note that for this problem, buckling as a column occurs in the plane of bending, whereas in the previous two examples column buckling occurred about the axis which was not the axis of bending.

(b) Column action. Since W8×24 appears in the *ASD Manual* "COLUMNS" load tables, the limits to preclude local buckling are automatically satisfied.

$$\frac{KL}{r_y} = \frac{1.0(10)(12)}{1.61} = 74.5; \qquad F_a = 20.1 \text{ ksi}$$

$$f_a = \frac{P}{A_g} = \frac{70}{7.08} = 9.9 \text{ ksi}$$

$$\frac{f_a}{F_a} = \frac{9.9}{20.1} = 0.49$$

(c) Beam action. Since bending is about the weak axis, lateral-torsional buckling cannot occur. Assuming the section is "compact" with regard to flange local buckling, it would be able to develop its plastic moment strength. Checking ASD-B5,

$$\left(\frac{b_f}{2t_f} = \frac{6.495}{2(0.400)} = 8.1\right) < \left(\frac{65}{\sqrt{50}} = 9.2\right) \quad \text{OK}$$

$$F_b = 0.75F_y = 37.5 \text{ ksi}$$

$$M = \tfrac{1}{8}(0.4)(10)^2 = 5 \text{ ft-kips}$$

$$f_b = \frac{M}{S_y} = \frac{5(12)}{5.63} = 10.7 \text{ ksi}$$

$$\frac{f_b}{F_b} = \frac{10.7}{37.5} = 0.29$$

(d) Moment magnification. Using *ASD Manual* "NUMERICAL VALUES" TABLE 8,

$$F'_e = 26.9 \text{ ksi}, \quad \text{using } \frac{KL}{r_y} = 74.5$$

$$\frac{f_a}{F'_e} = \frac{9.9}{26.9} = 0.37; \qquad 1 - \frac{f_a}{F'_e} = 0.63$$

$$C_m = 1.0 \quad \text{(from Table 12.3.1, Case 2)}$$

$$\text{Magnification factor} = \frac{C_m}{1 - f_a/F'_e} = \frac{1}{0.63} = 1.58$$

(e) Interaction criterion, ASD Formula (H1-1).

$$\frac{f_a}{F_a} + \frac{f_b}{F_b}\left(\frac{C_m}{1 - f_a/F_e'}\right) = 0.49 + 0.29(1.58) = 0.95 < 1 \quad \text{OK}$$

W8×24 is acceptable.   ∎∎

EXAMPLE 12.15.4

Select a suitable W section for the column member of the frame shown in Fig. 12.13.6, using service loads $P = 60$ kips and $M = 112$ ft-kips. The joints are rigid to give frame action in the plane of bending, but in the transverse direction sway bracing is provided and the attachments may be considered hinged. Use A36 steel and Allowable Stress Design. Solve for the following two cases:

1. *Braced* frame in the plane of bending.
2. *Unbraced* frame in the plane of bending.

Solution

(a) Frame *braced* in the plane of bending. Estimate equivalent column load P_{EQ}:

$$K_x \approx 0.7,$$

or can be obtained from Alignment Chart, Fig. 6.9.4a, as

$$G_A = 1.0 \text{ (fixed)}, \qquad G_B = \frac{I/20}{3I/30} = 0.5, \qquad K_x = 0.73$$

$$P(K_x L)^2 = 60(0.7)^2(240)^2 = 1.69 \times 10^6$$

Using Eq. 12.15.6 with the stress ratio term taken as unity,

$$P_{EQ} \approx P + \beta_{as} M$$

W10,	$P_{EQ} = 60 + 0.26(112)(12) = 60 + 350 = 410$ kips
W12,	$P_{EQ} = 60 + 0.22(112)(12) = 60 + 296 = 356$ kips
W14,	$P_{EQ} = 60 + 0.20(112)(12) = 60 + 269 = 329$ kips

(b) Braced frame—select section. Using *ASD Manual* "COLUMNS" load tables with $K_y L_y = 10$ ft, preliminary trial sections are

W10×68	$P = 373$ kips @ $L = 10$ ft	$r_x/r_y = 1.71$
W12×65	$P = 367$ kips	$r_x/r_y = 1.75$
W14×61	$P = 330$ kips	$r_x/r_y = 2.44$

For values in the tables to give correct column allowable loads, r_x/r_y of section must exceed $K_x L_x/K_y L_y = 0.7(20)/10 = 1.40$.

Because of the moment gradient indicating double curvature, with $C_m = 0.6 - 0.4(56/112) = 0.4$, it is likely that the yield criterion at the braced point will control. Thus ASD Formula (H1-2), where both denominator terms are likely to be $0.60F_y$, may be used to obtain required area for a more accurate choice of section. For a W14,

$$\text{Required } A_g = \frac{P_{EQ}}{0.60F_y} = \frac{329}{22} = 14.9 \text{ sq in.}$$

Try W14×53, W16×50, W18×46, or W21×44. The sections deeper than W14 do not appear in column load tables but may be satisfactory since P_{EQ} decreases as the depth increases. (For W21, $P_{EQ} \approx 60 + 0.16(112)12 = 275$ kips. Required $A_g = 12.5$ sq in.) Of course, changing depth will affect the relative moments of interia. For the *braced* frame such changes will have less effect than on the *unbraced* frame.

Try W21×44.

(c) Braced frame—check yield equation, ASD Formula (H1-2). Check W21×44: At the braced point,

$$f_a = \frac{60}{13.0} = 4.6 \text{ ksi}; \qquad f_b = \frac{112(12)}{81.6} = 16.5 \text{ ksi}$$

$$C_b = 1.75 + 1.05(-28/112) + 0.3(28/112)^2 = 1.51$$

$$(L_c = 6.6 \text{ ft}) < (L_b = 10 \text{ ft}) \quad \text{cannot use } F_b = 0.66F_y!$$

$$(L_u = 7.0 \text{ ft}) < (L_b = 10 \text{ ft})$$

For the yield equation, ASD Formula (H1-2), C_b is used in computing F_b.

$$F_b(\text{F1-8}) = \frac{12,000(1.51)}{10(12)7.06} = 21.4 \text{ ksi} \quad \text{(controls)}$$

$$L_b/r_T = 10(12)/1.57 = 76.4$$

$$F_b(\text{F1-6}) = 24.0 - \frac{(76.4)^2}{1181(1.51)} = 20.7 \text{ ksi}$$

$$\frac{f_a}{0.60F_y} + \frac{f_b}{F_b} = \frac{4.6}{22} + \frac{16.5}{21.4} = 0.98 < 1.0 \quad \text{OK}$$

(d) Braced frame—check stability equation, ASD Formula (H1-1). Check W21×44 for stability out in the span,

$$\frac{K_y L_y}{r_y} = \frac{1.0(10)(12)}{1.26} = 95.2; \qquad F_a = 13.6 \text{ ksi}$$

For the *braced* frame, C_b is *not used* for computing F_b in the formula where C_m is used.

$$F_b(\text{F1-6}) = 24.0 - \frac{(76.4)^2}{1181} = 19.1 \text{ ksi}$$

$$\frac{K_x L_x}{r_x} = \frac{0.7(20)(12)}{8.06} = 20.8; \qquad F_e' = 345 \text{ ksi}$$

$$\frac{f_a}{F_e'} = \frac{4.6}{345} = 0.013; \qquad 1 - \frac{f_a}{F_e'} = 0.987$$

$$\frac{f_a}{F_a} + \frac{f_b}{F_b}\left(\frac{C_m}{1 - f_a/F_e'}\right) = \frac{4.6}{13.6} + \frac{16.5}{19.1}\left(\frac{0.4}{0.987}\right)$$

$$= 0.34 + 0.86(0.41) = 0.69 < 1$$

Use W21×44.

(e) Unbraced frame—estimate equivalent column load P_{EQ}. The particular feature of the unbraced frame is that K_x for the axis of bending exceeds one. Using the alignment chart from Chapter 6 (Fig. 6.9.4b) for *unbraced* frame (sidesway not prevented) for $G_A = 1.0$ (fixed at bottom) and $G_B = 0.5$ [see part (a)], find $K_x \approx 1.24$.

Equivalent column load estimates are the same as for the braced frame case, but in using column tables, correct allowable loads are obtained for $K_y L_y = 10$ ft only when r_x/r_y, exceeds

$$K_x L_x/K_y L_y = 1.24(20)/10 = 2.48$$

Selection of W10 or W12 sections, if the stability criterion governs, will be controlled by column buckling in the plane of bending, while deeper sections will be controlled by weak-axis column buckling.

From *ASD Manual* "COLUMNS" load tables,

W10, for $L = 1.24(20)/1.71 = 14.5$ ft, $P_{EQ} \approx 410$ kips, find W10×77 with
$\quad P \approx 380$ kips.
W12, for $L = 1.24(20)/1.75 = 14.2$ ft, $P_{EQ} \approx 356$ kips, find W12×65 with
$\quad P \approx 340$ kips.
W14, for $L = 1.24(20)/2.44 = 10.2$ ft, $P_{EQ} \approx 329$ kips, find W14×61 with
$\quad P \approx 330$ kips.

These preliminary sections are likely to be too heavy because F_a/F_b is usually a significant reduction effect, while magnification is normally not more than 10 to 20%. If the yielding equation controls, the section and procedure are as for the braced frame where W21×44 was acceptable.

Try W21×44.

(f) Unbraced frame—check stability equation, ASD Formula (H1-1). For W21×44,

$K_y L_y/r_y = 95.2 \quad$ as from part (d)
$K_x L_x/r_x = 1.24(240)/8.06 = 36.9$

$\quad F_a = 13.6$ ksi; weak axis governs
$\quad F_b = 21.4$ ksi; (for $C_b = 1.51$) calculation shown in part (c)

Note that for the *unbraced* frame, the C_b ordinarily used for beams in braced systems is prescribed by ASD to be used in the stability equation.

$$F'_e = 110 \text{ ksi}; \qquad \text{based on } KL/r = 36.9$$
$$C_m = 0.85 \quad \text{for unbraced frame}$$

$$\text{Magnification factor} = \frac{C_m}{1 - f_a/F'_e} = \frac{0.85}{1 - 4.6/110} = 0.89$$

$$\frac{f_a}{F_a} + \frac{f_b}{F_b}\left(\frac{C_m}{1 - f_a/F'_e}\right) = 0.34 + 0.77(0.89) = 1.03 > 1$$

For this unbraced frame the stability requirement governs. Thus if a small overstress is acceptable for the unbraced frame, the same W21×44 could be

used whether the frame is braced or unbraced. Without exceeding the safety criterion, the W16×50 would be the choice (W18×46 gives 1.01 for ASD Formula H1-1).
Use W16×50. ■■

SELECTED REFERENCES

12.1. Charles Massonnet. "Stability Considerations in the Design of Steel Columns," *Journal of the Structural Division*, ASCE, **85,** ST7(September 1959), 75–111.

12.2. Walter J. Austin. "Strength and Design of Metal Beam-Columns," *Journal of the Structural Division*, ASCE, **87,** ST4(April 1961), 1–32.

12.3. Nestor R. Iwankiw. "Note on Beam-Column Moment Amplification Factor," *Engineering Journal,* AISC, **21,** 1 (First Quarter 1984), 21–23; Disc. by Le-Wu Lu, **22,** 1 (First Quarter 1985), 47–48; Disc. by Joseph A. Yura, **22,** 1 (First Quarter 1985), 48.

12.4. Robert L. Ketter. "Further Studies of the Strength of Beam-Columns," *Journal of the Structural Division,* ASCE, **87,** ST6 (August 1961), 135–152. Also *Transactions,* ASCE, **127** (1962), Part II, 244–266.

12.5. M. R. Horne. "The Stanchion Problem in Frame Structures Designed According to Ultimate Carrying Capacity," *Proc. Inst. Civil Engrs.,* **5,** 1, Part III (April 1956), 105–146.

12.6. Wai-Fah Chen and Suiping Zhou. "C_m Factor in Load and Resistance Factor Design," *Journal of Structural Engineering,* ASCE, **113,** 8 (August 1987), 1738–1754.

12.7. Robert L. Ketter, Edmund L. Kaminsky, and Lynn S. Beedle. "Plastic Deformation of Wide-Flange Beam-Columns," *Transactions,* ASCE, **120** (1955), 1028–1069.

12.8. Theodore V. Galambos and Robert L. Ketter. "Columns Under Combined Bending and Thrust," *Journal of the Engineering Mechanics Division*, ASCE, **85,** EM2 (April 1959), 1–30. Also *Transactions,* ASCE, **126** (1961), Part I, 1–25.

12.9. George F. Hauck and Seng-Lip Lee. "Stability of Elasto-Plastic Wide-Flange Columns," *Journal of the Structural Division,* ASCE, **89,** ST6 (December 1963), 297–324.

12.10. S. L. Lee and G. F. Hauck. "Buckling of Steel Columns Under Arbitrary End Loads," *Journal of the Structural Division*, ASCE, **90,** ST2 (April 1964), 179–200.

12.11. S. L. Lee and S. C. Anand. "Buckling of Eccentrically Loaded Steel Columns," *Journal of the Structural Division,* ASCE, **92,** ST2 (April 1966), 351–370.

12.12. Edwin C. Rossow, George B. Barney, and Seng-Lip Lee. "Eccentrically Loaded Steel Columns with Initial Curvature," *Journal of the Structural Division,* ASCE, **93,** ST2 (April 1967), 339–358.

12.13. Le-Wu Lu and Hassan Kamalvand. "Ultimate Strength of Laterally Loaded Columns," *Journal of the Structural Division*, ASCE, **94,** ST6 (June 1968), 1505–1524.

12.14. W. F. Chen. "Further Studies of Inelastic Beam-Column Problem," *Journal of the Structural Division,* ASCE, **97,** ST2 (February 1971), 529–544.

12.15. Gordon W. English and Peter F. Adams. "Experiments on Laterally Loaded Steel Beam-Columns," *Journal of the Structural Division,* ASCE, **99,** ST7 (July 1973), 1457–1470.

12.16. Francois Cheong-Siat-Moy. "Methods of Analysis of Laterally Loaded Columns," *Journal of the Structural Division,* ASCE, **100,** ST5 (May 1974), 953–970.

12.17. Francois Cheong-Siat-Moy. "General Analysis of Laterally Loaded Beam-Columns," *Journal of the Structural Division,* ASCE, **100,** ST6 (June 1974), 1263–1278.

12.18. B. G. Johnston. "Lateral Buckling of I-Section Columns with Eccentric End Loads in the Plane of the Web," *Transactions,* ASME, **62** (1941), A–176.

12.19. Mario G. Salvadori. "Lateral Buckling of I-Beams," *Transactions,* ASCE, **120** (1955), 1165–1182.

12.20. M. Salvadori. "Lateral Buckling of Eccentrically Loaded I-Columns," *Transactions,* ASCE, **121** (1956), 1163–1178.

12.21. Constancio Miranda and Morris Ojalvo. "Inelastic Lateral-Torsional Buckling of Beam Columns," *Journal of the Engineering Mechanics Division,* ASCE, **91,** EM6 (December 1965), 21–37.

12.22. Yushi Fukumoto and T. V. Galambos. "Inelastic Lateral-Torsional Buckling of Beam Columns," *Journal of the Structural Division,* ASCE, **92,** ST2 (April 1966), 41–61.

12.23. T. V. Galambos, P. F. Adams, and Y. Fukumoto. "Further Studies on the Lateral-Torsional Buckling of Steel Beam-Columns," *Bulletin No. 115,* Welding Research Council, July 1966.

12.24. Ralph C. Van Kuren and T. V. Galambos. "Beam Column Experiments," *Journal of the Structural Division,* ASCE, **90,** ST2 (April 1964), 223–256.

12.25. Mark A. Bradford and Nicholas S. Trahair. "Inelastic Buckling Tests on Beam-Columns," *Journal of Structural Engineering,* ASCE, **112,** 3 (March 1986), 538–549.

12.26. Mark A. Bradford, Peter E. Cuk, Marian A. Gizejowski, and Nicholas Trahair. "Inelastic Lateral Buckling of Beam-Columns," *Journal of Structural Engineering,* ASCE, **113,** 11 (November 1987), 2259–2277.

12.27. T. B. Pekoz and G. Winter. "Torsional-Flexural Buckling of Thin-Walled Sections Under Eccentric Load," *Journal of the Structural Division,* ASCE, **95,** ST5 (May 1969), 941–963.

12.28. Teoman B. Pekoz and N. Celebi. "Torsional Flexural Buckling of Thin-Walled Sections Under Eccentric Load," *Cornell Engineering Research Bulletin 69-1.* Ithaca, NY: Cornell University, 1969.

12.29. Wei-Wen Yu. *Cold-Formed Steel Design,* 2nd ed. New York: John Wiley & Sons, Inc., 1991, Chap. 6.

12.30. Francois Cheong-Siat-Moy and Tom Downs. "New Interaction Equation for Steel Beam-Columns," *Journal of the Structural Division,* ASCE, **106,** ST5 (May 1980), 1047–1061.

12.31. Lian Duan and Wai-Fah Chen. "Design Interaction Equation for Steel Beam-Columns," *Journal of Structural Engineering,* ASCE, **115,** 5 (May 1989), 1225–1243. Disc. **117,** 8 (August 1991), 2553–2561.

12.32. Iqbal S. Sohal, Lian Duan, and Wai-Fah Chen. "Design Interaction Equations for Steel Members," *Journal of Structural Engineering,* ASCE, **115,** 7 (July 1989), 1650–1665. Disc., **117,** 7 (July 1991), 2189–2196.

12.33. C. S. Cal, X. L. Liu, and W. F. Chen. "Further Verifications of Beam-Column Strength Equations," *Journal of Structural Engineering,* ASCE, **117,** 2 (February 1991), 501–513.

12.34. Charles Birnstiel and James Michalos. "Ultimate Load of H-Columns Under Biaxial Bending," *Journal of the Structural Division,* ASCE, **89,** ST2 (April 1963), 161–197.

12.35. Charles G. Culver. "Exact Solution of Biaxial Bending Equations," *Journal of the Structural Division,* ASCE, **92,** ST2 (April 1966), 63–83.

12.36. Charles G. Culver, "Initial Imperfections in Biaxial Bending," *Journal of the Structural Division,* ASCE, **92,** ST3 (June 1966), 119–135.

12.37. Gunnar A. Harstead, Charles Birnstiel, and Keh-Chun Leu. "Inelastic Behavior of H-Columns Under Biaxial Bending," *Journal of the Structural Division,* ASCE, **94,** ST10 (October 1968), 2371–2398.

12.38. Ishwar C. Syal and Satya S. Sharma. "Biaxially Loaded Beam-Column Analysis," *Journal of the Structural Division,* ASCE, **97,** ST9 (September 1971), 2245–2259.

12.39. Sakda Santathadaporn and Wai F. Chen. "Analysis of Biaxially Loaded Steel H-Columns," *Journal of the Structural Division,* ASCE, **99,** ST3 (March 1973), 491–509.

12.40. Wai F. Chen and Toshio Atsuta. "Ultimate Strength of Biaxially Loaded Steel H-Columns," *Journal of the Structural Division,* ASCE, **99,** ST3 (March 1973), 469–489.

12.41. Charles Birnstiel. "Experiments on H-Columns Under Biaxial Bending," *Journal of the Structural Division,* ASCE, **94,** ST10 (October 1968), 2429–2449.

12.42. Wai F. Chen and Sakda Santathadaporn. "Review of Column Behavior Under Biaxial Loading," *Journal of the Structural Division,* ASCE, **94,** ST12 (December 1968), 2999–3021.

12.43. Negussie Tebedge and Wai F. Chen. "Design Criteria for H-Columns Under Biaxial Loading," *Journal of the Structural Division,* ASCE, **100,** ST3 (March 1974), 579–598.

12.44. Reidar Bjorhovde, Theodore V. Galambos, and Mayasandra K. Ravindra. "LRFD Criteria for Steel Beam-Columns," *Journal of the Structural Division,* ASCE, **104,** ST9 (September 1978), 1371–1387.

12.45. Joseph A. Yura. "Combined Bending and Axial Load," Notes distributed by AISC at 1988 National Steel Construction Conference, Miami Beach, FL to assist classroom teaching of LRFD. Chicago, IL: American Institute of Steel Construction, 1988.

12.46. W. F. Chen and E. M. Lui. "Columns with End Restraint and Bending in Load and Resistance Factor Design," *Engineering Journal*, AISC, **22,** 3 (Third Quarter 1985), 105–132. Errata: **25,** 2 (Second Quarter 1988), 59.

12.47. Lian Duan, Iqbal S. Sohal, and Wai-Fah Chen. "On Beam-Column Moment Amplification Factor," *Engineering Journal*, AISC, **26,** 4 (Fourth Quarter 1989), 130–135. Disc. **27,** 4 (Fourth Quarter 1990), 168–172.

12.48. James G. MacGregor and Sven E. Hage. "Stability Analysis and Design of Concrete Frames," *Journal of the Structural Division*, ASCE, **103,** ST10 (October 1977), 1953–1970.

12.49. William J. LeMessurier, Robert J. McNamara, and J. C. Scrivener. "Approximate Analytical Model for Multistory Frames," *Engineering Journal*, AISC, **11,** 4 (Fourth Quarter 1974), 92–98.

12.50. Wm. J. LeMessurier. "A Practical Method of Second Order Analysis: Part 1—Pin Jointed Systems," *Engineering Journal*, AISC, **13,** 4 (Fourth Quarter 1976), 89–96.

12.51. Wm. J. LeMessurier. "A Practical Method of Second Order Analysis: Part 2—Rigid Frames," *Engineering Journal*, AISC, **14,** 2 (Second Quarter 1977), 49–67.

12.52. Brian R. Wood, Denis Beaulieu, and Peter F. Adams. "Column Design by P Delta Method," *Journal of the Structural Division*, ASCE, **102,** ST2 (February 1976), 411–427.

12.53. Brian R. Wood, Denis Beaulieu, and Peter F. Adams. "Further Aspects of Design by *P*-Delta Method," *Journal of the Structural Division*, ASCE, **102,** ST3 (March 1976), 487–500.

12.54. H. Scholz. "*P*-Delta Effect in Elastic Analysis of Sway Frames," *Journal of Structural Engineering*, ASCE, **113,** 3 (March 1987), 534–545.

12.55. I. S. Sohal and N. A. Syed. "Inelastic Amplification Factor for Design of Steel Beam-Columns," *Journal of Structural Engineering*, ASCE, **118**, 7 (July 1992), 1822–1839.

12.56. Thomas Sputo. "History of Steel Beam-Column Design," *Journal of Structural Engineering,* ASCE, **119,** 2 (February 1993), 547–557.

PROBLEMS

All problems (except Probs. 12.27–12.31) are to be done according to AISC Load and Resistance Factor Design or Allowable Stress Design, as indicated by the instructor. The requirement of W section is intended to include W, S, and M sections. All given loads are service loads unless otherwise indicated. Assume lateral support consists of translational restraint but not moment (rotational) restraint, unless otherwise indicated. Assume all standard sections are equally readily available in the indicated grade of steel (even though actually they are not). A figure showing span and loading is required, and after making a design selection a final check of the strength interaction equation (involving factored loads P_u and M_u, and design strengths ϕP_n and ϕM_n for LRFD) or stress interaction equations (involving service load stresses f_a and f_b, and allowable stresses F_a and F_b for ASD) is required.

Problems 12.1 through 12.26 relate to design considerations. Problems 12.27 through 12.31 relate to other theoretical considerations.

12.1. Investigate the adequacy of the section assuming the loading is 30% dead load and 70% live load. No translation of joints can occur, and external lateral support is provided at the ends only.

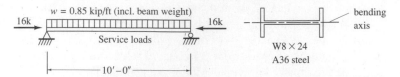

Problem 12.1

12.2. Investigate the given section with regard to safety if primary bending is in the weak direction and the loading w is 0.02 kip/ft dead load and 0.08 kip/ft live load and P is 2 kips dead load and 18 kips live load. A572 Grade 50 steel.

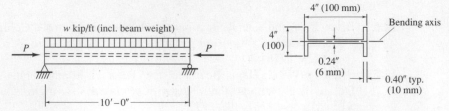

Problem 12.2

12.3. Determine the maximum service load W (kips) at the mid-height of the beam-column shown. Assume the member is hinged with respect to bending in both x and y directions at the top and bottom. Additionally, lateral support occurs in the weak direction at mid-height.

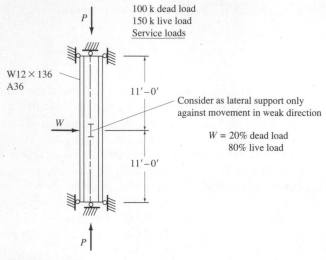

Problem 12.3

12.4. Investigate the adequacy of the given section. No joint translation can occur and external lateral support is provided at the ends only.

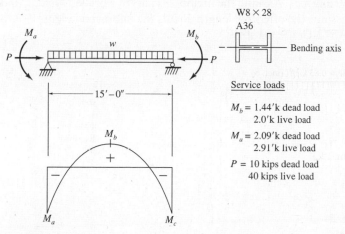

W8 × 28
A36

Bending axis

Service loads

$M_b = 1.44'$k dead load
2.0'k live load

$M_a = 2.09'$k dead load
2.91'k live load

$P = 10$ kips dead load
40 kips live load

Problem 12.4

12.5. Determine the safe service load W permitted for this braced frame beam-column.

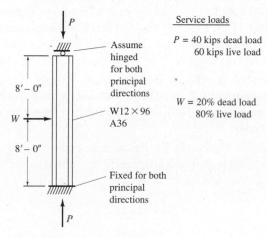

Assume hinged for both principal directions

W12 × 96
A36

Fixed for both principal directions

Service loads

$P = 40$ kips dead load
60 kips live load

$W = 20\%$ dead load
80\% live load

Problem 12.5

12.6. Determine the service axial load P which the W12×45 may be permitted to carry. Lateral support is provided at ends and at midspan. Compare for A36 and A572 Grade 50 steels.

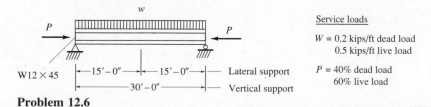

W12 × 45

15'–0"   15'–0"
30'–0"

Lateral support
Vertical support

Service loads

$W = 0.2$ kips/ft dead load
0.5 kips/ft live load

$P = 40\%$ dead load
60\% live load

Problem 12.6

12.7. Select the lightest W14 section to carry a service load P as shown in the accompanying figure, with an eccentricity $e = 12$ in. with respect to the strong axis. Assume the member is part of a braced system, and conservatively assume the effective length equals the unbraced height. Use (a) A36 steel; (b) A572 Grade 60 steel.

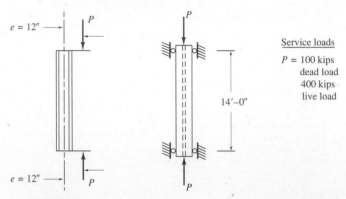

Service loads

P = 100 kips
     dead load
     400 kips
     live load

Problem 12.7

12.8. Select the lightest W14 section to carry an axial compression P of 100 kips dead load and 140 kips live load along with a bending moment M of 125 ft-kips dead load and 325 ft-kips live load, which, for conservative simplicity, is assumed to be constant along the 15-ft equivalent pin-end length of the member in the braced structure. Use (a) A36 steel; (b) A572 Grade 50.

12.9. If the service load P is 30 kips dead load and 95 kips live load and w is 0.8 kips/ft dead load and 1.2 kips/ft live load for the beam-column span and support conditions of Prob. 12.6, select the lightest W section. Use (a) A36 steel; (b) A572 Grade 60.

12.10. Select the lightest W14 section for the beam-column of the accompanying figure. Assume lateral buckling is adequately prevented such that $L_b < L_p$ for LRFD or $L_b < L_c$ for ASD. Use A36 steel.

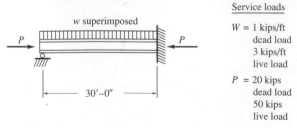

Service loads

W = 1 kips/ft
     dead load
     3 kips/ft
     live load

P = 20 kips
     dead load
     50 kips
     live load

Problem 12.10

12.11. For the member of a braced system in the accompanying figure, select the lightest W section. Use (a) A36 steel and (b) A572 Grade 60 steel.

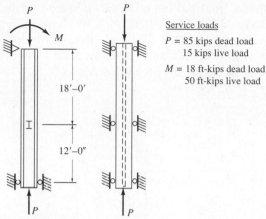

Service loads

P = 85 kips dead load
     15 kips live load

M = 18 ft-kips dead load
     50 ft-kips live load

Problem 12.11

12.12. Redesign the section for Example 12.13.5 for A572 Grade 60 steel.

12.13. A frame braced against sidesway has a beam-column loaded as in the accompanying figure resulting from an elastic analysis. The horizontal member has lateral support at its ends and every 9 ft. Select the lightest W section acceptable using A36 steel.

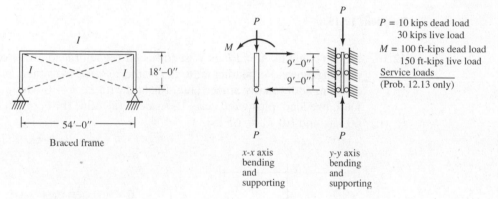

P = 10 kips dead load
     30 kips live load

M = 100 ft-kips dead load
      150 ft-kips live load

Service loads
(Prob. 12.13 only)

Braced frame

x-x axis
bending
and
supporting

y-y axis
bending
and
supporting

Problems 12.13 and 12.14

12.14. Redesign the column member of Prob. 12.13 as part of an unbraced frame, but disregard the service loads of that problem. Assume two factored load first-order analyses have been performed: (1) the gravity load nonsway analysis giving P_u = 65 kips and M_{nt} = 100 ft-kips; and (2) a sway analysis giving $M_{\ell t}$ = 275 ft-kips and P_u = 10 kips. Assume both columns must carry the same load.

12.15. Design the column A as a W section for the unbraced frame member of the accompanying figure. The frame is unbraced in the plane of the frame and analysis has determined K_x = 1.4; in the perpendicular direction the structure is braced such that $K_y = 1.0$. First-order factored load analyses under dead load, live load, and wind ($1.2\,D + 0.5L + 1.3W$) have been performed; the *nonsway* gravity analysis ($1.2D + 0.5L$), and the *sway* wind ($1.3W$) analysis giving the values as shown. Assume all columns in the story are the same and the load on each is identical. Use steels: (a) A36; (b) $F_y = 50$ ksi; (c) $F_y = 60$ ksi; and (d) $F_y = 65$ ksi.

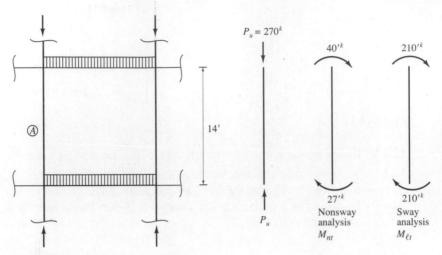

Problem 12.15

12.16. Design columns A and B as W sections for the unbraced frame of the accompanying figure. Perpendicular to the frame, assume the system braced and the columns additionally braced at mid-height with $K_y = 1.0$. Design for the dead load, live load, plus wind case. Use steels: (a) A36; (b) $F_y = 50$ ksi; (c) $F_y = 60$ ksi; and (d) $F_y = 65$ ksi.

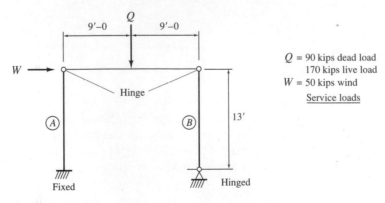

Q = 90 kips dead load
        170 kips live load
W = 50 kips wind
        Service loads

Problem 12.16

12.17. Design columns A and B as W sections for the unbraced frame shown in the accompanying figure. The system is braced in the direction perpendicular to the frame such that $K_y = 1.0$. Design for the dead load, live load, plus wind load case using the uniform live load acting on *both* spans. Use steels: (a) A36; (b) $F_y = 50$ ksi; (c) $F_y = 60$ ksi; and (d) $F_y = 65$ ksi.

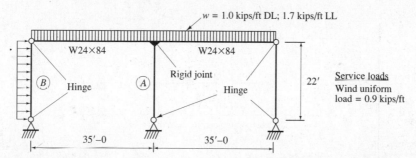

$w = 1.0$ kips/ft DL; 1.7 kips/ft LL

W24×84   W24×84

Rigid joint

B.   Hinge   *A*   Hinge

22′   Service loads
Wind uniform
load = 0.9 kips/ft

35′–0   35′–0

Problem 12.17

12.18. Design the lightest W12 section for column A of the unbraced frame in the accompanying figure. Use the dead load plus live load plus wind loading case. Preliminary design has selected W27×94 for all adjacent beams, and it has been decided column A will be approximately the same size as those above and below it. Assume the forces given are the result of two first-order analyses; a nonsway analysis giving factored moments M_{nt}, and a sway analysis giving factored moments $M_{\ell t}$. In the weak direction, assume the system braced such that $K_y = 1.0$ and $L_y = 13$ ft. Use steels: (a) A36; (b) $F_y = 50$ ksi; (c) $F_y = 60$ ksi; and (d) $F_y = 65$ ksi.

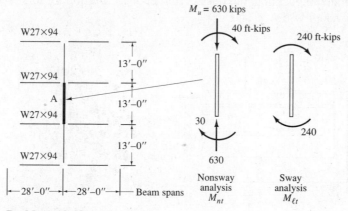

W27×94

W27×94

A

W27×94

W27×94

13′–0″

13′–0″

13′–0″

28′–0″   28′–0″   Beam spans

$M_u = 630$ kips

40 ft-kips   240 ft-kips

30   240

630   240

Nonsway   Sway
analysis   analysis
M_{nt} $M_{\ell t}$

Problem 12.18

12.19. For the vierendeel truss (rigid frame) shown, investigate the adequacy of members *A* and *B*. The uniform loading is 90% live load and 10% dead load including the weight of the steel section. Assume all other forces are 10% dead load and 90% live load. The steel is A36. Assume simple cross-bracing between given frame and an adjacent parallel one.

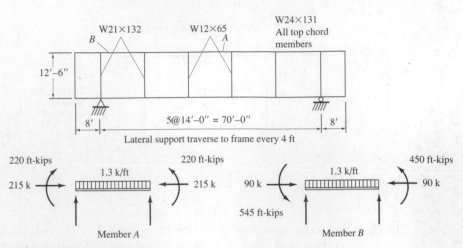

Problems 12.19 and 12.20

12.20. Redesign member *A* for A572 Grade 60 steel.

12.21. Design the lightest W section to carry two eccentric 20-kip loads causing primary bending plus an axial compression of 275 kips. Assume all loads are 20% dead load and 80% live load. Assume torsional fixity at the vertical supports; assume lateral bracing at midspan is adequate to prevent lateral-torsional buckling, but does permit enough rotation for torsional moments to develop. Use A36 steel.

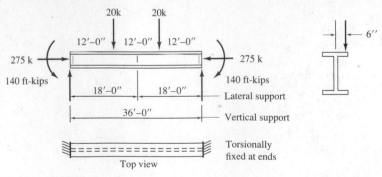

Problem 12.21

12.22. Investigate the given structural tee for the loading shown in the accompanying figure. The uniform loading is delivered through construction which prevents lateral buckling of the member. Assume continuity over the end supports which resulted in given end moments. Assume all loads are 25% dead load and 75% live load. Use A36 steel.

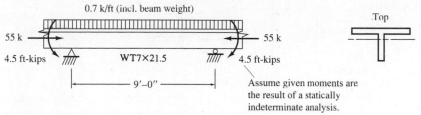

Problem 12.22

12.23. Select the lightest WT7 structural tee for the loading shown in the accompanying figure. Assume vertical and lateral supports at ends only. Use A572 Grade 50 steel.

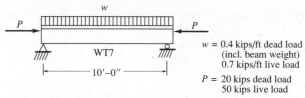

Problem 12.23

12.24. Select an economical structural tee to serve as a continuous compression chord member of a truss to carry service loads as shown. For design purposes assume fixed ends on the member. Use the more economical of A36 or A572 Grade 50 steel, assuming the Grade 50 costs 12% more per pound than A36.

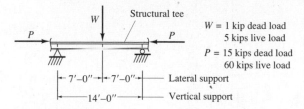

Problem 12.24

12.25. A column in a building has the factored load reactions at the top from beams framing into it, as shown in the accompanying figure. Assume the framing beams contribute moments at the top of the column, but the bottom of the column is hinged (no moments). The beams framing into the web are assumed to rest on seats, where reactions are assumed to be 2 in. from the center of the web. The reaction from the other beam is assumed to be acting at the face of the flange. Use A36 steel and select lightest W section.

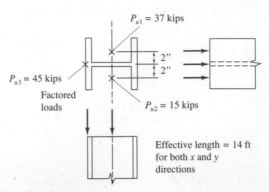

Problem 12.25

12.26. For the factored loading shown, select lightest sections for the following conditions:
a. W14 using A36 steel
b. W in any depth using $F_y = 50$ ksi
c. W14 using $F_y = 60$ ksi
d. W14 using $F_y = 70$ ksi

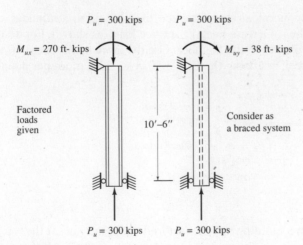

Problem 12.26

Problems Relating to Theoretical Considerations

12.27. **through 12.31.** For the given loading and support conditions, develop the differential equation for the moment M_z in the plane of bending, and determine the maximum value for M_z.

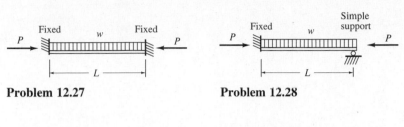

Problem 12.27

Problem 12.28

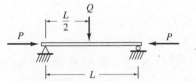

Problem 12.29

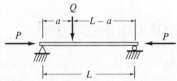

Problem 12.30

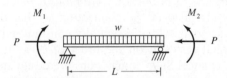

Problem 12.31

Chapter 13

Connections

13.1 TYPES OF CONNECTIONS

Steel construction is categorized by LRFD-A2.2 and ASD-A2.2 into various "Types" depending on the amount of restraint developed by the connections. Three types are identified:

Fully Restrained (also called *Rigid Frame,* or *Continuous Frame*). This situation occurs when full continuity is provided at the connection so that original angles between intersecting members are maintained essentially constant during loading of the structure; i.e., with rotational restraint on the order of 90% or more of that necessary to prevent any angle change. Such connections are designated "Type FR" (for fully restrained) by LRFD-A2.2 and have for many years been known in Allowable Stress Design as "Type 1" (ASD-A2.2).

Simple Framing (also called *Unrestrained* or *Free-Ended*). This situation occurs where rotational restraint at the ends of members is as little as practicable. For beams, simple framing is intended to provide only shear transfer at the ends. Simple framing is usually assumed to exist when the original angle between intersecting members may change approximately 80% or more of the amount it theoretically would change if frictionless hinged connections could be used. When a simply supported beam is designed, simple framing connections must be used. When plastic analysis is used, continuity is inherently assumed; therefore, in that case it is inappropriate to use simple framing. Two or more planar systems designed using plastic analysis may, however, be linked together by simple framing connections combined with a bracing system (such as cross-bracing). Structures using simple framing connections have long been called "Type 2" construction in Allowable Stress Design (ASD-A2.2) and are called "Type PR" (for partially restrained) in LRFD-A2.2. The designation PR for these connections is in recognition of the fact that some restraint is always present.

Welded connections for rigid frame construction, showing beam-to column connections with column web stiffeners. Rural Mutual Insurance Building, Madison, Wis. (Photo by C. G. Salmon)

Load and Resistance Factor Design requires that when Type PR constrution is intended to be "simple framing," three specific requirements (LRFD-A2.2) apply, paraphrased as follows:

a. The simply supported beam reactions under factored loads must be adequately carried by such connections.

b. The structure and its connections must be adequate to resist factored lateral loads.

c. Connections must have sufficient inelastic rotation capacity so that angle changes inherent in the simple framing assumption can occur under the combination of factored gravity and lateral loading *without overloading the end fastening system.*

Semi-Rigid Framing. Semi-rigid framing occurs when rotational restraint is approximately between 20% and 90% of that necessary to prevent relative angle change. This means that with semi-rigid framing the moment transmitted across the joint is neither zero (or a small amount) as in simple framing, nor is it the full continuity moment as assumed in elastic rigid-frame analysis. Semi-rigid framing is specifically designated as "Type 3" under ASD-A2.2. In Load and Resistance Factor Design, semi-rigid framing is included in "Type PR" where its use "depends on the evidence of predictable proportion of full end restraint." In ASD, the design of semi-rigid connections requires a "dependable and known moment capacity intermediate in degree between the rigidity of Type 1 and the flexibility of Type 2."

Semi-rigid connections are not used in structures when plastic analysis is used in design, and are not commonly used in Allowable Stress Design because of the difficulty in obtaining the moment–rotation relationship for a given connection. However, with greater availability of high-strength steels required in designing this type of connection, the authors believe use of semi-rigid connections will increase.

Beam Line

In order to better understand the practical distinction between the AISC framing types, the beam line developed by Batho and Rowan [13.1] and used by Sourochnikoff [13.5] is a useful graphical device.

As shown in Fig. 13.1.1, consider a beam AB loaded in any manner and subject to end moments M_a and M_b, and with end slopes θ_a and θ_b. The moments necessary

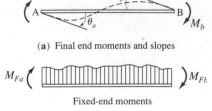

(**a**) Final end moments and slopes

Fixed-end moments

Figure 13.1.1 Moments and rotations for slope-deflection equations (shown with positive signs).

to have $\theta_a = \theta_b = 0$ are designated M_{Fa} and M_{Fb}, the fixed-end moments. Writing the slope deflection equations,*

$$\left. \begin{array}{l} M_a = M_{Fa} + \dfrac{4EI}{L}\theta_a + \dfrac{2EI}{L}\theta_b \\[2mm] M_b = M_{Fb} + \dfrac{2EI}{L}\theta_a + \dfrac{4EI}{L}\theta_b \end{array} \right\} \qquad (13.1.1)$$

Solving Eqs. 13.1.1 for θ_a and θ_b gives

$$\left. \begin{array}{l} \dfrac{6EI}{L}\theta_a = 2(M_a - M_{Fa}) - (M_b - M_{Fb}) \\[2mm] \dfrac{6EI}{L}\theta_b = -(M_a - M_{Fa}) + 2(M_b - M_{Fb}) \end{array} \right\} \qquad (13.1.2)$$

Subtracting the second equation from the first gives

$$\frac{6EI}{L}(\theta_a - \theta_b) = 3(M_a - M_b) - 3(M_{Fa} - M_{Fb}) \qquad (13.1.3)$$

If symmetrical loading is considered, then

$$M_b = -M_a, \qquad \theta_b = -\theta_a, \qquad M_{Fb} = -M_{Fa} \qquad (13.1.4)$$

in which case Eq. 13.1.3 becomes

$$\frac{2EI}{L}\theta_a = M_a - M_{Fa}$$

or

$$M_a = M_{Fa} + \frac{2EI}{L}\theta_a \qquad (13.1.5)$$

which may be called the *beam-line equation*. When $\theta_a = 0$ (a full fixity condition), $M_a = M_{Fa}$; and for a hinged end where $M_a = 0$, the slope becomes $\theta_a = -M_{Fa}/(2EI/L)$.

Figure 13.1.2 shows a diagram of the beam-line equation and also the moment–rotation behavior of typical connections of ASD Types 1, 2, and 3; and LRFD Types FR and PR. The typical rigid connection would have to carry an end moment M_1, about 90% or more of M_{Fa}; hence its degree of restraint may be said to be 90%. The simple connection (Type 2) may have to resist only 20% or less of the moment M_{Fa}, as indicated by the moment M_2, while the semi-rigid connection would be expected to resist some intermediate value M_3, at perhaps 50% of the fixed-end moment M_{Fa}.

When the moment–rotation characteristics of a particular connection are available, the strength can be designed so that the resulting end rotation θ is compatible with that caused by the loads. Discussions of semi-rigid connections and moment–rotation characteristics of various connection arrangements are given by Hechtman

---

*For instance, see Chu-Kia Wang and Charles G. Salmon, *Introductory Structural Analysis*. Englewood Cliffs, NJ: Prenctice-Hall, Inc., 1984 (Chap. 9).

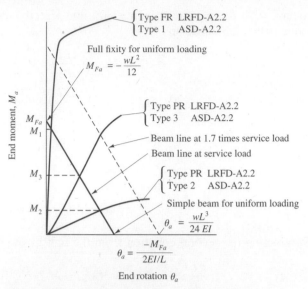

Figure 13.1.2 Moment–rotation characteristics of AISC connection types.

and Johnston [13.6], Schenker, Salmon, and Johnston [13.7], Lindsay, Ioannides, and Goverdhan [13.8], Brown [13.9], Ammerman and Leon [13.10], Leon, Ammerman, Lin, and McCauley [13.11], Nethercot, Davison, and Kirby [13.12], Chen and Kishi [13.13], Azizinamini and Radziminski [13.14], Leon and Ammerman [13.15], Bjorhovde, Colson, Brozzetti [13.16], Zandonini and Zanon [13.17], and Leon [13.18].

13.2 SIMPLE SHEAR CONNECTIONS

Simple shear connections (commonly known as *framed beam connections*) are used to connect beams to other beams or to column flanges or webs when simple support of the beam has been assumed. Design of such connections has become somewhat standardized using *LRFD Manual* [1.18] Table 9–2, "All-Bolted Double Angle Connections," Table 9–3, "Combination Bolted/Welded Double-Angle Connections," Table 9–4, "All-Welded Double Angle Connections," and Table 9–5, "Bolted/Welded Shear End-Plate Connections." For ASD, the *ASD Manual* [1.7] contains tables "FRAMED BEAM CONNECTIONS" and *Manual of Steel Construction, Volume II Connections* [13.3] contains Chapter 3 on Simple Shear Connections. Seated beam connections, treated in Sec. 13.3 are also considered simple shear connections.

Typical bolted and welded framed connections are shown in Fig. 13.2.1. It is intended in such connections that the angles be as flexible as possible. The connection to the column (2 rows of 5 fasteners shown in Fig. 13.2.1a) is usually made in the field, while the connection to the beam web (one row of 5 fasteners shown in Fig. 13.2.1b) is usually made in the shop. Generally on plans, shop fastener holes are shown as in

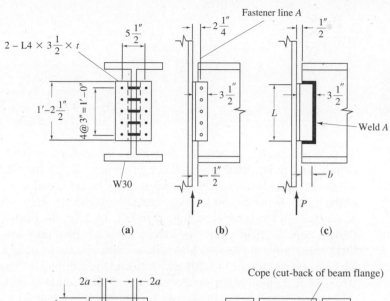

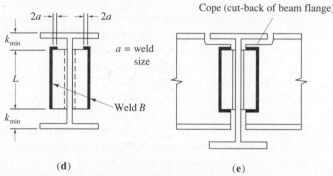

Figure 13.2.1 Simple shear double-angle connections.

Fig. 13.2.1b, while field fastener holes are shown as solid black dots. There have been many studies [13.19–13.28] of simple shear double-angle connections.

In today's fabrication practice, the shop connection is usually welded, while the field connection may be either bolted or welded; thus, any combination in Fig. 13.2.1 of (a) with (b) or (c); or (d) with (b) or (c) may be used.

The *single-plate framing connection* is a modification where a single plate (instead of the pair of angles) is bolted flat against the beam web and then is welded perpendicular to the beam web or column flange or web to which it is attached. The design of single plate framing connections has been studied by Richard, Gillett, Kriegh, and Lewis [13.29], Young and Disque [13.30]. Richard, Kriegh, and Hormby [13.31], Hormby, Richard, and Kriegh [13.32], and Astaneh, Call, and McMullin [13.33].

Another type of simple shear connection is the *tee framing connection* as studied by Astaneh and Nader [13.35, 13.36], where the tee flange attaches to the supporting column (or beam) and the tee web laps against the loaded beam to transmit its shear.

Another single-plate framing connection, studied by Kennedy [13.34], uses the plate in the vertical position welded flat against the end of the beam with the connection to the beam or column made with bolts.

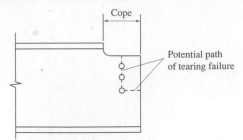

Figure 13.2.2   Tearing failure at coped ends of framed beam connection.

When angles, sometimes known as *clip angles,* are used to attach a beam to a column, there is a clearance setback of about $\frac{1}{2}$ in. so that if the beam is too long, within acceptable tolerances, the angles may be relocated without cutting off a piece of the beam. When beams intersect and are attached to other beams so that the flanges of both are at the same elevation, as in Fig. 13.2.1e, the beams framing in have their flanges coped, or cut away. The loss of section is primarily loss of flange that carries little shear anyway, so that normally a cope results in little loss of shear strength. Birkemoe and Gilmor [13.20] have shown that a coped web subject to high bearing stress in a high-strength bolted beam end connection may fail in a tearing mode (known as "block shear") along a line through the holes, as shown in Fig. 13.2.2. Additional study of block shear in such situations has been made by Yura, Birkemoe, and Ricles [13.21] and Ricles and Yura [13.23].

"Block shear" was first discussed in regard to tension members in Sec. 3.6. Block shear may be critical in simple shear connections when there is a short connection using few bolts that do not extend uniformly over the entire depth of the web. LRFD-J4.3 requires consideration of the block shear limit state "where the top flange is coped and in similar situations, such as tension members and gusset plates." ASD-J4 contains a similar requirement.

In addition to block shear, copes of the flanges of beams may affect local web buckling, as reported by Cheng and Yura [13.24], and lateral-torsional buckling, as reported by Gupta [13.25], Cheng, Yura, and Johnson [13.26], and Cheng and Yura [13.27].

The number of high-strength bolts is based on the direct shear, neglecting any eccentricity of loading, while the weld length and size include the effect of eccentric loading. The fasteners, bolts or welds, are designed in accordance with procedures of Chapters 4 and 5, respectively.

The thickness of the framing angles or plate is usually controlled by the "block shear" strength. In addition, angles should be thick enough such that bearing does not control. The angles are *expected to bend* so that the assumed rotation of the supported beam at its ends can occur.

Flexural Deformation and Strength of Connection Angles

Referring to Fig. 13.2.3, the tensile force T per inch acts at the top of framing angles of length L as shown in Fig. 13.2.1 when an end moment acts. This end moment arises from the reaction P acting at an eccentricity e measured, as in Fig. 13.2.1b or c, from the point of action of P to the centroid of the fastener line A or to the centroid of weld A.

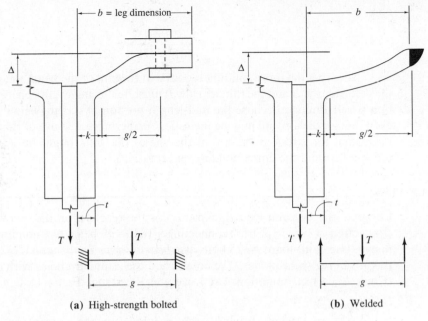

(a) High-strength bolted   **(b)** Welded

Figure 13.2.3   Behavior at tension edge of framing angles.

The concentrated load T acting on the connection angles may be considered as acting on a fixed-end beam for the bolted connection and on a simply supported beam for a welded connection. The true situation for each case is partially end restrained. The higher restraint for the bolted connection arises from the clamping action between the pieces caused by initial tension in the bolts, while the welding causes little clamping action.

A simple approximation of the deflection Δ at the tension end of the framing angles will therefore be that of a fixed-end beam (high-strength bolted connection) having concentrated load T at mid-length of a span g,

$$\Delta = \frac{Tg^3}{192EI} \qquad (13.2.1)$$

and for the concentrated load T acting at mid-length of a simply supported span g (welded connection),

$$\Delta = \frac{Tg^3}{48EI} \qquad (13.2.2)$$

Noting that the maximum force T will occur when the top of the angle yields; i.e., Max $T = 2F_y t$ per unit length at the top, and that $I = t^3/12$ per unit length, Eqs. 13.2.1 and 13.2.2 become

$$\Delta = \frac{2F_y tg^3}{192E(t^3/12)} = \frac{F_y g^3}{8Et^2} \qquad (13.2.3)$$

and

$$\Delta = \frac{2F_y tg^3}{48E(t^3/12)} = \frac{F_y g^3}{2Et^2} \qquad (13.2.4)$$

Note that the more rotation required at the end of the beam the greater must be the deformation Δ and the thinner (low t) must be the angles. In general, this deformation is self-limiting because the mid-length portion of the angles will remain elastic even though the top end may be inelastic. Furthermore, the use of thick angles would mean that the stress at the top of the angles due to T might be less than F_y, thus reducing Δ (and the corresponding end rotation).

EXAMPLE 13.2.1 _____

Compute the factored load P_u capacity for the reaction on the 5 row framed beam connection of Fig. 13.2.1 for connecting a W30×99 beam to a column having a $\frac{3}{4}$-in. flange. Use $\frac{3}{4}$-in.-diam A325 bolts as a *bearing-type connection* (A325-X) having no threads in the shear planes. The connection uses standard holes with clean mill scale (Class A) surface condition. Use Load and Resistance Factor Design and A36 steel.

Solution

(a) Design strengths ϕR_n per bolt for the connection to the web of W30×99, $t_w = 0.520$ in. Using Eq. 4.7.3, the design strength in double shear ($m = 2$) is

$$\phi R_n = \phi(0.50F_u^b)mA_b$$
$$= 0.75(0.50)(120)(2)0.4418 = 39.8 \text{ kips/bolt}$$

Since the bolt spacing exceeds $3d_b = 2.25$ in., the design strength in bearing on the 0.520-in. web, from Eq. 4.7.9, is

$$\phi R_n = \phi(2.4F_u)d_b t$$
$$= 0.75(2.4)(58)(0.75)0.520 = 40.7 \text{ kips/bolt}$$

(b) Total connection factored load P_u capacity based on the web connection to the W30×99 beam. Common practice has been to neglect the eccentricity e with respect to the fastener line. Thus, the maximum factored reaction P_u would equal the design strength per bolt times the number n of bolts,

$$\text{Maximum } P_u = n\phi R_n = 5(39.8) = 199 \text{ kips}$$

If the eccentricity with respect to the fastener line were considered,

$$e = 2\tfrac{1}{4} \text{ in.}$$

assuming the reaction to be along fastener line A. For this eccentric shear loading, ultimate strength analysis or elastic vector analysis, as discussed in Sec. 4.12, may be used. Using ultimate strength analysis as represented by the *LRFD Manual* Table 8-18, "Coefficients C for Eccentrically Loaded Bolt Groups," p. 8-40, for vertical spacing of fasteners $s = 3$ in., the number of fasteners $n = 5$, and eccentricity $e_x = 2.25$ in.,

Find coefficient $C = 4.27$

$$P_u = C\phi R_n = 4.27(39.8) = 170 \text{ kips}$$

This is 15% less than when eccentricity is neglected.

(c) Design strengths ϕR_n per bolt for the connection to the flange of the $\frac{3}{4}$-in. column. The design strength in single shear $(m = 1)$ is

$$\phi R_n \text{ (single shear)} = \phi(0.50 F_u^b) m A_b$$
$$\phi R_n = 0.75(0.50)(120)(1)0.4418 = 19.9 \text{ kips/bolt}$$

The design strength in bearing on the angles will be less than bearing on the $\frac{3}{4}$-in. flange. The minimum thickness for the angles so that bearing will not govern is obtained by setting the strength in bearing equal to the strength (19.9 kips/bolt) in shear.

$$\phi R_n \text{ (bearing)} = \phi(2.4 F_u)dt = 19.9 \text{ kips}$$
$$\text{Min } t = \frac{19.9}{\phi(2.4)F_u d} = \frac{19.9}{0.75(2.4)(58)0.75} = 0.25 \text{ in.}$$

The design strength in tension will be dependent on the applied shear load; however, the upper limit on that strength is

$$\phi R_n \text{ (tension)} = \phi(0.75 F_u^b) A_b$$
$$= 0.75(90.0)0.4418 = 29.8 \text{ kips}$$

(d) Total connection factored load P_u capacity based on the connection to the $\frac{3}{4}$-in. flange. Neglecting eccentricity, the ten fasteners in single shear give

$$P_u = 10(19.9) = 199 \text{ kips}$$

If the combined shear and tension effect is considered, and the elastic vector method is used, the factored tension T_u at the most heavily loaded bolt is

$$T_u = \frac{Mc}{I} A = \frac{P_u ec}{I} A = \frac{P_u(2.25)6}{4(3)^2 + 4(6)^2} = 0.075 P_u$$

and the direct shear per bolt is

$$V_u = \frac{P}{A} A = \frac{P_u}{10} = 0.10 P_u$$

From Table 4.14.1 (LRFD-Table J3.5) the nominal load $F_{ut}' A_b$ in tension permitted in the presence of shear is, for bearing-type A325-X connections using $\frac{3}{4}$-in. diam bolts $(A_b = 0.4418 \text{ sq in.})$,

$$\phi F_{ut}' A_b = 0.75[117 - 1.5 f_{uv}] A_b \le [\phi(90)A_b = 0.75(90)A_b]$$
$$= 87.8(0.4418) - 1.13 V_u \le 67.5(0.4418)$$
$$\text{Max } T_u = 38.8 - 1.13(0.10 P_u) \le 29.8 \text{ kips}$$
$$0.075 P_u = 38.8 - 1.13(0.10 P_u)$$
$$P_u = 207 \text{ kips}$$

Check that $0.075P_u$ does not exceed 29.8 kips; $0.075P_u = 0.075(207) = 15.5$ kips, which means the upper limit on $\phi F'_{ut} A_b$ is not exceeded. In addition to that check, the strength based on combined tension and shear cannot exceed the strength based on shear alone. Referring to Fig. 4.14.3, f_{uv} cannot exceed 45 ksi; however, $f_{uv} = 0.10P_u/A_b = 0.10(207)/0.4418 = 46.9$ ksi from the above computation. Thus, P_u cannot exceed 199 kips based on shear alone.

(e) Summary of factored load capacity results.

Connection to beam web:	$P_u = 199$ kips (neglect e)
	$P_u = 170$ kips (consider e)
Connection to column flange:	$P_u = 199$ kips (neglect e)
	$P_u = 199$ kips (consider e)

For the illustration of edge and end distance requirements, the value of 199 kips neglecting eccentricity will be used.

(f) End and edge distances. When bearing strength based on $2.4F_u$ is used according to LRFD-J3.10, as shown in parts (a) and (c), a minimum end distance of $1.5d_b$ is acceptable. In this case,

$$\text{Minimum end distance} \geq [1.5d_b = 1.5(0.75) = 1.13 \text{ in.}]$$

When a lesser distance is desired, Eq. 4.7.16 can be used, where P is the factored load per bolt. Equation 4.7.16 should also be used to check the end distance on the beam web,

$$\text{End distance } L_e \geq \left[\frac{P}{\phi F_u t} = \frac{199/5}{0.75(58)0.52} = 1.76 \text{ in.} \right]$$

The end distance requirement is conservative (i.e., large) because the full load on the uppermost bolt is assumed to be directed toward the nearest edge; actually only the horizontal component need be used. The 1.75 in. provided is certainly adequate.

(g) Shear on the net section through the angles. For shear as well as tension, the net area A_{nv} according to LRFD-B2, is based on the nominal dimension of the hole plus $\frac{1}{16}$-in. Thus, for standard holes,

$$A_{nv} = t\left[14.5 - 5\left(\frac{13}{16} + \frac{1}{16} \right) \right] = 10.13t \text{ per angle}$$

Shear rupture design strength (LRFD-J4.1) is

$$\phi R_n = \phi(0.6F_u)A_{nv}$$

$$\text{Required } t = \frac{\text{Reaction}}{\phi(0.6F_u)A_{nv}} = \frac{199/2}{0.75(0.6)(58)10.13} = 0.376 \text{ in.}$$

Select $\frac{3}{8}$-in. angles.

(h) Examine the degree of simple support provided by $\frac{3}{8}$-in. angles. Using Eq. 13.2.3, the elastic deflection is

$$\Delta = \frac{F_y g^3}{8Et^2} = \frac{36[2(2.5 - 13/16)]^3}{8(29,000)(0.375)^2} = 0.04 \text{ in.}$$

The longest beam span uniformly loaded for a W30×99 having a factored reaction of 199 kips is

$$\phi_b M_p = 0.90 Z_x F_y = 0.90(312)36/12 = 842 \text{ ft-kips}$$

$$M_u = \frac{W_u L}{8} = \frac{199(2)L}{8}$$

$$L = \frac{8(842)}{199(2)} = 16.9 \text{ ft}$$

Assuming the service moment is about 0.5 of M_p, the elastically computed end slope is

$$\theta = \frac{WL^2}{24EI} = \frac{ML}{3EI} = \frac{(842/2)(12)(16.9)12}{3(29,000)3990} = 0.0030 \text{ radian}$$

Assuming rotation about the bottom of the angle, the deformation required at the top of the angles to accommodate the service load rotations at the ends of the beam would be

$$\Delta = (d/2)\theta = 14.5(0.0030) = 0.04 \text{ in.} \approx 0.04$$

Based on these approximate calculations, the angles must yield at service load to accommodate rotation at the ends of the beam. Even if such rotation occurs, the full simply supported beam moment is not likely to occur; there will be some end moment. ■■

The conclusion is that simple shear connections should use the thinnest angles consistent with the bearing and shear fracture limit states, as well as practical limitations. Bertwell [13.37] has provided some additional discussion on the behavior of simple shear connections.

EXAMPLE 13.2.2 _____

Investigate the 5 row simple shear connection of Example 13.2.1 (Fig. 13.2.1) as a *slip-critical connection* (A325-SC) assuming 75% live load and 25% dead load.

Solution

(a) Strength in shear and bearing. A slip-critical connection has the same strength requirements for shear and bearing as a bearing-type connection. From Example 13.2.1,

$$\phi R_n = 39.8 \text{ kips/bolt} \quad \text{(double shear)}$$

$$\phi P_n = 40.7 \text{ kips/bolt} \quad \text{(bearing on 0.52-in.web)}$$

Using $\frac{3}{8}$-in.-thick angles means bearing on the angles will not control, nor will the shear fracture limit state.

(b) Capacity based on strength limit states. The maximum factored load P_u that can be carried was determined in Example 13.2.1 to be 199 kips (neglecting eccentricity). Based on this the service load capacity is

$$1.2(0.25P) + 1.6(0.75P) = 199 \text{ kips}$$

$$P = 133 \text{ kips}$$

(c) Check the slip serviceability limit state. The service load capacity per bolt is, from Table 4.9.1 for Class A surface condition and standard holes,

$$R = F_v m A_b = 17(2)0.4418 = 15.0 \text{ kips/bolt}$$

When eccentricity is neglected,

$$P = R(\text{number of bolts}) = 15.0(5) = 75 \text{ kips}$$

Thus, comparing the results from parts (b) and (c), the serviceability limit state controls. This will usually be the case for slip-critical connections. When a surface condition is used having a high slip-resistance coefficient and/or threads may exist in the shear planes, it will be possible for the strength limit state to control.

(d) Consider eccentricity for the slip-critical connection. The combination of shear and tension on the attachment to the column flange will be more critical than eccentric shear at fastener line A (Fig. 13.2.1). The service load components at the most heavily loaded fasteners are

$$T = \frac{Mc}{I} A = \frac{P(2.25)6}{4(3)^2 + 4(6)^2} = 0.075P \quad \text{(tension)}$$

and the direct shear per bolt is

$$V = \frac{P}{A} A = \frac{P}{10} = 0.10P \quad \text{(direct shear)}$$

The service load capacity in shear for a bolt simultaneously subject to tension is, according to LRFD-J3.9a,

$$F_v = 17\left(1 - \frac{T}{T_b}\right) = 17\left(1 - \frac{0.075P}{28}\right)$$

For fasteners fully loaded in shear, $f_v = 0.10P/A$:

$$\frac{0.10P}{0.4418} = 17\left[1 - \frac{0.075P}{28}\right]$$

$$P = 63 \text{ kips} < 75 \text{ kips} \quad \text{(neglect } e\text{)}$$

(e) Summary.

$$P = 136 \text{ kips} \quad \text{(based on strength limit states)}$$
$$P = 75 \text{ kips} \quad \text{(based on serviceability limit state of slip neglecting eccentricity)}$$
$$P = 63 \text{ kips} \quad \text{(based on serviceability limit state of slip considering eccentricity)}$$

The *LRFD Manual* Tables 9–2 implicity accept the neglect of eccentricity for double-angle simple shear connections. To verify this, one must use the option of computing the factored load capacity of slip-critical connections under LRFD-J3.9b which directs the designer to LRFD-Appendix J3.8b. Though it is illogical to calculate for a service load design criterion (slip resistance) using factored loads, the 1993 LRFD Specification has added this option. The design slip resistance ϕR_{str} per bolt at factored service loads is given as

$$\phi R_{str} = \phi 1.13 \mu T_i m \qquad [4.9.1]$$

where the terms are defined in Sec. 4.9. For this example, $\phi = 1.0$ for standard holes; $\mu = 0.33$ for Class A surface condition; $T_i =$ initial tension $= 28$ kips for $\frac{3}{4}$-in. diam bolts; and $m = 2$ slip (shear) planes. Thus,

$$\phi R_{str} = 1.0(1.13)(0.33)(28)2 = 20.9 \text{ kips/bolt}$$

This compares with $R = 15$ kips per bolt at service load. The factored load P_u for the connection, neglecting eccentricity, is

$$P_u = \phi R_{str}(\text{number of bolts}) = 20.9(5) = 104.5 \text{ kips}$$

LRFD Manual, Table 9–2, pp. 9-36, 37, give 104 kips for SC, Class A, standard holes. Thus, eccentricity was neglected! ■■

EXAMPLE 13.2.3 _____

Design a double-angle simple shear connection for a W10×68 beam having a factored load reaction P_{u1} of 70 kips and a W24×104 beam having a factored load reaction P_{u2} of 210 kips. These two beams are to frame into opposite sides of a plate girder having a $\frac{3}{8}$-in. web as shown in Fig. 13.2.4. The connection is to be of $\frac{3}{4}$-in.-diam A325 bolts in a bearing-type connection with threads excluded from the shear planes. Use LRFD and A572 Grade 50 steel.

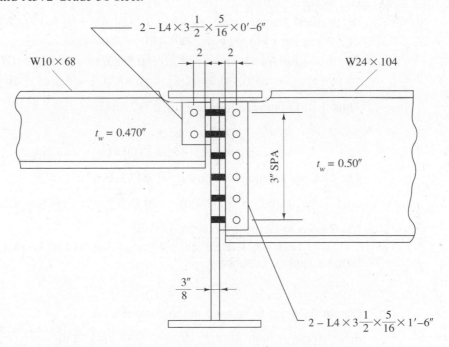

Figure 13.2.4 Framed beam connection having unequal reactions.

Solution

(a) Compute design bolt values relating to connections to webs of W10×68 and W24×104. Initially assume that $s \geq 3d_b$, $L_e \geq 1.5d_b$, and two or more bolts in a line of force, so that $R_n = 2.4F_u d_b t$ for bearing,

ϕR_n (bearing) $= \phi(2.4F_u)d_b t$

$\phi R_n = 0.75(2.4)(65)(0.75)0.470 = 41.2$ kips/bolt (W10)

$\phi R_n = 0.75(2.4)(65)(0.75)0.500 = 43.9$ kips/bolt (W24)

ϕR_n (double shear) $= \phi(0.50F_u^b)mA_b$

$\phi R_n = 0.75(60.0)(2)0.4418 = 39.8$ kips/bolt

$$\text{Number of bolts} = \frac{70}{39.8} = 1.8, \quad \text{say 2 (W10)}$$

$$\text{Number of bolts} = \frac{210}{39.8} = 5.3, \quad \text{say 6 (W24)}$$

Use 3-in. bolt spacing with 1.5-in. end distance at cope (top of angles).

(b) Check block shear on W10 according to LRFD-J4.3. This is discussed in Sec. 3.6. To see which formula applies, compute

$$(F_u A_{nt} \geq 0.6F_u A_{nv})$$

Obtain the areas to be used in the computation,

A_{gv} = shear yielding area = $4.5(0.470) = 2.11$ sq in.

A_{nv} = shear fracture area = $[4.5 - 1.5(13/16 + 1/16)]0.470 = 1.50$ sq in.

A_{gt} = tension yield area = $2.0(0.470) = 0.94$ sq in.

A_{nt} = tension fracture area = $[2.0 - 0.5(13/16 + 1/16)]0.470 = 0.73$ sq in.

$$[F_u A_{nt} = 65(0.73) = 47.5] < [0.6F_u A_{nv} = 0.6(65)1.50 = 58.5]$$

Thus, LRFD Formula (J4-3b), Eq. 3.6.2, applies,

$$T_n = 0.6F_u A_{nv} + F_y A_{gt} \qquad [3.6.2]$$
$$T_n = 58.5 + 50(2.0)0.470 = 106 \text{ kips}$$

The factored reaction capacity $P_u = \phi T_n$, where $\phi = 0.75$,

$$[P_u = \phi T_n = 0.75(106) = 79 \text{ kips}] > 70 \text{ kips required} \quad \text{OK}$$

Use 2 bolts to connect the W10 section.

(c) Check block shear on W24 according to LRFD-J4.3. To see which formula applies, compute

$$(F_u A_{nt} \geq 0.6F_u A_{nv})$$

Obtain the areas to be used in the computation,

A_{gv} = shear yielding area = $16.5(0.50) = 8.25$ sq in.

A_{nv} = shear fracture area = $[16.5 - 5.5(13/16 + 1/16)]0.50 = 5.84$ sq in.

A_{gt} = tension yield area = 2.0(0.50) = 1.00 sq in.

A_{nt} = tension fracture area = $[2.0 - 0.5(13/16 + 1/16)]0.50 = 0.78$ sq in.

$$[F_u A_{nt} = 65(0.78) = 50.7] < [0.6F_u A_{nv} = 0.6(65)5.84 = 228]$$

Thus LRFD Formula (J4-3b), Eq. 3.6.2, applies,

$$T_n = 0.6F_u A_{nv} + F_y A_{gt} \qquad\qquad [3.6.2]$$
$$T_n = 228 + 50(1.00) = 278 \text{ kips}$$

The factored reaction capacity $P_u = \phi T_n$, where $\phi = 0.75$,

$$[P_u = \phi T_n = 0.75(278) = 209 \text{ kips}] \approx 210 \text{ kips required} \text{ OK}$$

Use 6 bolts to connect the W24 section.

(d) Connections to plate girder web. For this connection, the two bolts common to both sides will be governed by double shear or bearing on the $\frac{3}{8}$-in. plate, while the remainder are governed by single shear or bearing on the $\frac{3}{8}$-in. plate.

$$\phi R_n \text{ (bearing)} = \phi(2.4F_u)dt$$
$$\phi R_n = 0.75(2.4)(65)(0.75)0.375 = 32.9 \text{ kips/bolt}$$
$$\phi R_n \text{ (double shear)} = 39.8 \text{ kips/bolt}   [\text{from (a) above}]$$
$$\phi R_n \text{ (single shear)} = 39.8/2 = 19.9 \text{ kips/bolt}$$

For the bolts common to both sides, bearing governs. The four bolts in common carry 70/4 = 17.5 kips from the W10×68. The remainder is available for the W24×104 reaction; i.e., 32.9 − 17.5 = 15.4 kips.

If all bolts were to carry equal load,

$$\text{Number of bolts} = \frac{210}{19.9} = 10.6, \quad \text{say } 12$$

When 12 bolts are used, the average load per bolt will be 210/12 = 17.5 kips. Even though this exceeds the 15.4 kips available on the top 4 bolts, the lower 8 bolts are not fully loaded. Accept this arrangement.

(e) Angles thickness. Bearing will not control angles thickness unless end distance $L_e < 1.5d_b$ or bolt spacing $s < 3d_b$, according to LRFD-J3.10 because $2.4F_u$ was used to compute bearing strength. For the $\frac{3}{4}$-in. diam bolts, $1.5d_b = 1.13$ in. and $3d_b = 2.26$ in. In this design, the end distance $L_e = 1.25$ in. and the minimum spacing is 2.5 in.

The thickness of angles might be controlled by their shear rupture strength in accordance with LRFD-J4.1,

$$\phi(0.6F_u)A_{nv} \geq P_u$$

For the angles connected to the W10,

$$0.75(0.6)(65)[6.0 - 2(\tfrac{13}{16} + \tfrac{1}{16})]2t \geq 70$$
$$t \geq \frac{70}{0.75(0.6)65[6.0 - 1.75]2} = 0.28 \text{ in.}$$

For the angles connected to the W24,

$$0.75(0.6)(65)[18.0 - 6(\tfrac{13}{16} + \tfrac{1}{16})]2t \geq 210$$

$$t \geq \frac{210}{0.75(0.6)65[18.0 - 5.25]2} = 0.28 \text{ in.}$$

Occasionally, the angle strength might be controlled by gross shear yield on the angles under LRFD-J5.3; in this case, the thickness required is 0.22 in., less than the 0.28 in. required above. The bearing value ϕR_n for $\tfrac{5}{16}$-in. angles is 27.4 kips/bolt, which is more than adequate to carry the factored load of 17.5 kips/bolt from the beams (a coincidence that the same factored load per bolt is contributed by both the W10 and the W24); therefore, $\tfrac{5}{16}$-in. angles are satisfactory as connecting angles.

Use 2—L4×3$\tfrac{1}{2}$×$\tfrac{5}{16}$×0′−6″ for W10×68.

Use 2—L4×3$\tfrac{1}{2}$×$\tfrac{5}{16}$×1′−6″ for W24×104.

The length of angle should not exceed the dimension T, which is $7\tfrac{5}{8}$ in. for the W10×68. The girder flange thickness is such that a cope is required on the beams that encroaches on the T dimension. ■■

Weld Capacity in Eccentric Shear on Angle Connections

Since no initial tension is involved with welded connections, the eccentricity of loading, even though small, is considered. The principles of Chapter 5 (See. 5.18) are used with the welds treated as lines.

EXAMPLE 13.2.4 _____

Compute the factored load P_u capacity for weld A on the angle connection shown in Fig. 13.2.1. The beam is a W30×99 and the weld is $\tfrac{1}{4}$ in. with E70 electrodes. The angles are 4×3$\tfrac{1}{2}$×$\tfrac{5}{16}$×1′-2$\tfrac{1}{2}$″ in length. Use A36 steel and Load and Resistance Factor Design.

Solution   Analysis of this eccentric shear situation may be done using strength analysis as presented in Sec. 5.17 or the elastic (vector) method presented in Sec. 5.18.

(a) Elastic (vector) method. Using I_p from Table 5.18.1 and referring to Fig. 13.2.1c,

$$I_p = \frac{8(3)^3 + 6(3)(14.5)^2 + (14.5)^3}{12} - \frac{(3)^4}{2(3) + 14.5} = 583.5 \text{ in.}^3$$

Using the moment of inertia computed with a 1-in. effective throat, the force per unit length at critical locations can be computed.

$$R_v = \frac{P_u}{2(20.5)} = 0.0244P_u \quad \text{(direct shear component)} \downarrow$$

$$\bar{x} = \frac{(3)^2}{2(3) + 14.5} = 0.44 \text{ in.}$$

The x and y components of force due to torsional moment are

$$R_y = \frac{P_u(3.50 - 0.44)(3.50 - 0.44 - 0.50)}{2(583.5)} = 0.00671P_u \downarrow$$

$$R_x = \frac{P_u(3.50 - 0.44)(7.25)}{2(583.5)} = 0.0190P_u \rightarrow$$

$$R_u = P_u\sqrt{(0.0244 + 0.0067)^2 + (0.0190)^2} = 0.0364P_u$$

The design strength ϕR_{nw} per inch of weld is

$$\phi R_{nw} = \phi(0.707a)(0.60F_{EXX})$$
$$= 0.75(0.707)(\tfrac{1}{4})(0.60)70 = 5.57 \text{ kips/in.}$$

Check base metal shear fracture for the beam web and the angles,

$$(\phi R_{nw})_{\text{base metal}} = 0.75(0.60F_u)t = 0.45F_u t$$
$$(\phi R_{nw})_{\text{angle}} = 0.45(58)0.3125 = 8.16 \text{ kips/in.}$$
$$(\phi R_{nw})_{\text{web}} = 0.45(58)(0.520/2) = 6.79 \text{ kips/in.}$$

Weld strength controls; $\phi R_{nw} = 5.57$ kips/in.

$$P_u = \frac{5.57}{0.0364} = 153 \text{ kips}$$

(b) Strength analysis. Use *LRFD Manual* [1.18], Table 8–42, "Coefficients C for Eccentrically Loaded Weld Groups" with $\theta = 0°$. For $\tfrac{1}{4}$-in. weld using E70 electrodes,

$$a = (e - xL)/L = (3.5 - 0.44)/14.5 = 0.211$$
$$k = kL/L \quad = 3.0/14.5 = 0.207$$

$a =$	$k = 0.2$	0.207	0.3	
0.2	1.98		2.33	
0.211	1.958	1.982	2.306	$C = 1.982$
0.25	1.88		2.22	

Table value $= \phi P_n$

$$\phi P_n = CC_1DL = 1.982(1.0)(4)14.5 = 115 \text{ kips}$$

where C_1 = coefficient for electrode = (Electrode used)/70
D = number of $\tfrac{1}{16}$s of an inch in weld size
L = length of vertical weld, in.

Since there are two angles, the factored load reaction capacity is

$$P_u = 2(115) = 230 \text{ kips}$$

As expected, the strength analysis gives the higher value.   ■■

Tests of welded angle connections by Johnston and Green [13.19] and Johnston and Diets [13.38] have demonstrated that performance of web angles agrees generally with assumptions.

Weld Capacity in Tension and Shear on Angle Connections

This is the field-welded connection shown in Fig. 13.2.1d. There is no agreement regarding the strength analysis for this situation. Blodgett [13.2] considers the strength as an eccentric shear situation in the plane of the welds. With the eccentric load as in Fig. 13.2.5b, the angles bear against themselves for a distance of $L/6$ from the top, and the torsional stress over the remaining $\frac{5}{6}$ of the length L is resisted by the weld. Neglecting the effects of the returns at the top, the horizontal component R_x can be obtained from moment equilibrium. Equilibrium in the plane of the load P and weld leg B requires

$$\underbrace{\frac{1}{2}R_x\left(\frac{5}{6}L\right)}_{\text{force}}\underbrace{\frac{2}{3}L}_{\text{arm}} = \frac{P}{2}e_2 \tag{13.2.5}$$

$$R_x = \frac{9Pe_2}{5L^2} \text{ force/unit length} \tag{13.2.6}$$

The direct shear component is

$$R_v = \frac{P}{2L} \text{ force/unit length} \tag{13.2.7}$$

$$\text{Actual } R = \sqrt{\left(\frac{P}{2L}\right)^2 + \left(\frac{9}{5}\frac{Pe_2}{L^2}\right)^2}$$

$$R = \frac{P}{2L^2}\sqrt{L^2 + 12.9e_2^2} \text{ force/unit length} \tag{13.2.8}$$

Equation 13.2.8 neglects eccentricity e_1, which tends to cause tension at the top of the weld lines. The authors believe it is more appropriate to consider the flexural stress distribution of Fig. 13.2.5c to be a more appropriate approach. The flexural tension component R_x at the top of the weld B is

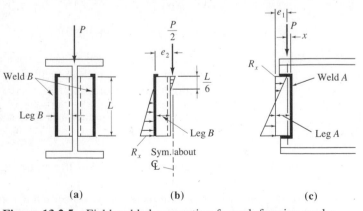

Figure 13.2.5　Field-welded connection for web framing angles.

$$R_x = \frac{Mc}{I} = \frac{Pe_1(L/2)}{2L^3/12} = \frac{3Pe_1}{L^2} \tag{13.2.9}$$

when the returns at the tops of the angles are neglected. The direct shear component R_v is

$$R_v = \frac{P}{2L} \text{ force/unit length} \tag{13.2.10}$$

$$\text{Actual } R = \sqrt{\left(\frac{P}{2L}\right)^2 + \left(\frac{3Pe_1}{L^2}\right)^2}$$

$$R = \frac{P}{2L^2}\sqrt{L^2 + 36e_1^2} \text{ force/unit length} \tag{13.2.11}$$

Or, if returns are considered (distance b of Fig. 13.2.6) the expression becomes complicated. The *LRFD Manual* p. 9-89 indicate the returns to be twice the weld size. The returns have the greatest effect when the angle length L is short. It may be reasonable to consider the returns to be $L/12$ (2 times $\frac{1}{4}$ in. weld for $L = 6$ in.).

Using, from Table 5.18.1 (Case 4), $S = I/\bar{y}$ referred to the tension fiber at the top of the configuration,

$$S = 2\left(\frac{4bd + d^2}{6}\right) \tag{13.2.12}$$

which for $d = L$ and $b = L/12$ becomes

$$S = \frac{4L^2}{9} \tag{13.2.13}$$

The flexural component, as shown in Fig. 13.2.6, is

$$R_x = \frac{M}{S} = \frac{Pe_1}{S} = \frac{Pe_1}{4L^2/9} = \frac{9Pe_1}{4L^2} \tag{13.2.14}$$

Since little of the shear is carried by the returns, they are neglected for the direct shear component, giving

$$R_v = \frac{P}{2L} \tag{13.2.15}$$

$$\text{Actual } R = \sqrt{\left(\frac{P}{2L}\right)^2 + \left(\frac{9Pe_1}{4L^2}\right)^2}$$

$$R = \frac{P}{2L^2}\sqrt{L^2 + 20.25e_1^2} \text{ kips/in.} \tag{13.2.16}$$

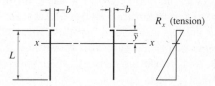

Figure 13.2.6 Weld configuration for web angles and beam seats.

EXAMPLE 13.2.5

Determine the factored load capacity P_u of weld B on Fig. 13.2.5 if $\frac{5}{16}$-in. weld is used and $L = 20$ in. E70 electrodes are used in shielded metal arc welding (SMAW). $4{\times}3{\times}\frac{3}{8}$ angles are used. Assume base material is thick enough to preclude shear fracture as the controlling limit state; i.e., the fillet weld strength controls. Use Load and Resistance Factor Design.

Solution

(a) Best procedure, Eq. 13.2.16

$$\phi R_{nw} = 0.75(0.707)(\tfrac{5}{16})(0.60)70 = 6.96 \text{ kips/in.}$$

$$\text{Actual } R_u = \frac{P_u}{2L^2}\sqrt{L^2 + 20.25e_1^2}$$

$$e_1 = 3.00 - \bar{x} = 3.00 - 0.25 = 2.75 \text{ in.}$$

$$\bar{x} = \frac{2(2.5)(1.25)}{2(2.5) + 20} = 0.25 \text{ in.}$$

$$\text{Actual } R_u = \frac{P_u}{2(20)^2}\sqrt{(20)^2 + 20.25(2.75)^2} = 0.0294P_u$$

$$P_u = \frac{6.96}{0.0294} = 237 \text{ kips}$$

(b) Neglecting returns entirely, Eq. 13.2.11,

$$\text{Actual } R_u = \frac{P_u}{2(20)^2}\sqrt{(20)^2 + 36(2.75)^2} = 0.0324P_u$$

$$P_u = \frac{6.96}{0.0324} = 215 \text{ kips}$$

(c) Using Ref. 13.2 equation, Eq. 13.2.8,

$$\text{Actual } R_u = \frac{P_u}{2(20)^2}\sqrt{(20)^2 + 12.9e_2^2}$$

$$e_2 = 4\text{-in. leg}$$

$$\text{Actual } R_u = 0.0308P_u$$

$$P_u = \frac{6.96}{0.0308} = 226 \text{ kips}$$

The authors believe method (a) to be appropriate, $P_u = 237$ kips. ■■

Note that when the elastic (vector) method is used, the same formulas apply for Allowable Stress Design and for Load and Resistance Factor Design. In ASD, P is the service load connection capacity, R is the service load resultant force per unit length at the most highly stressed weld segment, and the allowable resistance is R_w. In LRFD, P_u and R_u represent factored loads, and the design strength ϕR_{nw} is used.

13.3 SEATED BEAM CONNECTIONS—UNSTIFFENED

As an alternative to simple shear connections using web angles, or other attachments to the beam web, a beam may be supported on a seat, either unstiffened or stiffened. In this section the unstiffened seat is treated, as shown in Fig. 13.3.1. The unstiffened seat (an angle) is shown in Fig. 13.3.1 and is designed to carry the entire reaction. This type of connection must always, however, be used with a top clip angle, whose intended function is to provide lateral support of the compression flange.

As with the case of the simple shear angle connection, the seated connection is intended to transfer only the vertical reaction and should not give significant restraining moment on the end of the beam; thus the seat and the top angle should be relatively flexible. The behavior of welded seat angle connections has been studied by Lyse and Schreiner [13.39], Roeder [13.40] and Roeder and Dailey [13.41].

The thickness of seat angle is determined by the flexural strength at a critical section of the angle, as shown in Fig. 13.3.2. If a bolted connection is used without attachment to the beam (Fig. 13.3.2a), the critical section should probably be taken as the net section through the upper bolt line. When the beam is attached to the seat (as it should be) as in Fig. 13.3.2b, the rotation of the beam at the end creates a force that tends to restrain the pull away from the column. The critical section for flexure will then be at or near the base of the fillet on the outstanding leg. Similarly for the welded seat, the weld completely along the end holds the angle tight against the column, in which case the critical section is as shown in Fig. 13.3.2c, whether or not the beam is attached to the seat. As a practical matter, rarely will the beam be left unattached from the seat, so the design procedures of this section use a critical section as in Figs. 13.3.2b and c, taken at $\frac{3}{8}$ in. from the face of the angle.

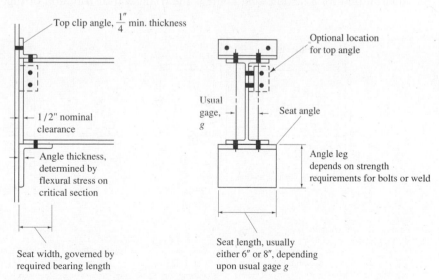

Figure 13.3.1   Seated beam connections—unstiffened.

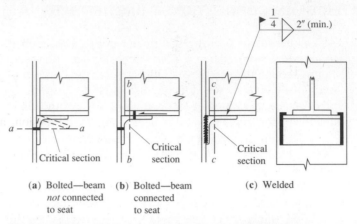

Figure 13.3.2   Critical section for flexure on seats.

The bending moments on the critical section of the angle and on the connection to the column flange are determined by taking the beam reaction times the distances to the critical sections. The beam reaction occurs at the centroid of the bearing stress distribution, as shown in Fig. 13.3.3. While the LRFD and ASD Specifications do not state how the computation of this bending moment is to be made, a conservative approach is to assume the reaction at the center of the full contact width (Fig. 13.3.3a). This will lead to excessively thick angles in most cases. The less conservative approach of assuming the reaction at the center of the *required* bearing length N measured from the end of the beam (Fig. 13.3.3b) has been used by Blodgett [13.2] and has been the approach used for AISC Manual tables. Another rational distribution for a flexible seat angle is the triangular distribution of Fig. 13.3.3c, and

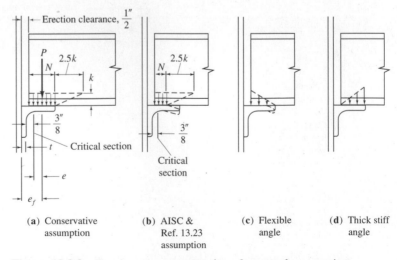

Figure 13.3.3   Bearing stress assumptions for seated connections.

if the angle is very stiff the reaction may become heavier on the outer edge, as in Fig. 13.3.3d.

The design of unstiffened seats involves the following steps:

1. Determine the seat width.
2. Determine the moment arms e and e_f.
3. Determine the length and thickness of the angle.
4. Determine the supporting angle leg dimension, and the weld size; or the number and placement of bolts.

The design of seated beam connections and the background for the LRFD load tables have been reviewed by Garrett and Brockenbrough [13.42] and by Brockenbrough [13.43].

The seat width is determined from the bearing length N required based on (a) the local web yielding limit state, as given by LRFD-K1.3 or ASD-K1.3, or (b) the web crippling limit state, as given by LRFD-K1.4 or ASD-K1.4. Local web yielding usually controls the bearing length.

Local Web Yielding—LRFD [LRFD Formula (K1-3)],

$$N = \frac{P_u}{\phi F_{yw} t_w} - 2.5k \tag{13.3.1}$$

Local Web Yielding—ASD [ASD Formula (K1–3)],

$$N = \frac{P}{0.66 F_{yw} t_w} - 2.5k \geq k \tag{13.3.2}$$

where
P_u = factored load reaction for LRFD
P = service load reaction for ASD
ϕ = resistance factor = 1.0
t_w = web thickness of supported beam
F_{yw} = yield stress of web of supported beam
k = distance from outer face of flange to web toe of fillet

The required bearing length N may not be taken less than k. Generally the seat width should not be less than 3 in. with *LRFD Manual* Table 9–6, "All-Bolted Unstiffened Seated Connections," indicating a standard 4-in. seat width.

Web Crippling—LRFD [LRFD Formulas (K1-5a) and (K1-56)],

For $N/d \geq 0.2$,

$$P_n = 68 t_w^2 \left[1 + 3 \left(\frac{N}{d} \right) \left(\frac{t_w}{t_f} \right)^{1.5} \right] \sqrt{\frac{F_{yw} t_f}{t_w}} \tag{13.3.3}$$

For $N/d > 0.2$,

$$P_n = 68 t_w^2 \left[1 + \left(\frac{4N}{d} - 0.2 \right) \left(\frac{t_w}{t_f} \right)^{1.5} \right] \sqrt{\frac{F_{yw} t_f}{t_w}} \tag{13.3.4}$$

Web Crippling—ASD [ASD Formula (K1-5)],

$$P = 34\, t_w^2 \left[1 + 3\left(\frac{N}{d}\right)\left(\frac{t_w}{t_f}\right)^{1.5} \right] \sqrt{\frac{F_{yw} t_f}{t_w}} \tag{13.3.5}$$

where ϕ = resistance factor = 0.75
  P_n = nominal reaction strength
  P = service load reaction.
  t_w = beam web thickness
  t_f = beam flange thickness
  d = beam overall depth

The moment arms e and e_f are obtained as follows, referring to Fig. 13.3.3a,

$$e_f = \text{erection clearance} + \frac{N}{2} \tag{13.3.6}$$

$$e = e_f - t - \frac{3}{8} \tag{13.3.7}$$

The bending moment on the critical section of the angle is

$$M_u = P_u e \quad \text{(for LRFD)}$$
$$M = Pe \quad \text{(for ASD)}$$

The thickness t of the angle is obtained letting $M_u = \phi_b M_n$ for LRFD or $M =$ allowable moment for ASD. The strength is that of a solid rectangular section bent about its weak axis.

Angle Thickness—LRFD (LRFD-F1.1),

$$\phi_b M_n = \phi_b M_p = \phi_b Z F_y = \phi_b \frac{L t^2}{4} F_y \tag{13.3.8}$$

$$t^2 = \frac{4 M_u}{\phi_b F_y L} = \frac{4 P_u e}{\phi_b F_y L} \tag{13.3.9}$$

Angle Thickness—ASD (ASD-F2.2),

$$M = S(0.75 F_y) = \frac{L t^2}{6}(0.75 F_y) \tag{13.3.10}$$

$$t^2 = \frac{6M}{0.75 F_y L} = \frac{8 Pe}{F_y L} \tag{13.3.11}$$

where P_u = factored reaction to be carried using LRFD
  P = service load reaction using ASD
  ϕ_b = resistance factor = 0.90
  e = eccentricity of load to the critical section on angle, such as Eq. 13.3.7
  L = length of seat angle (i.e., width of rectangular section being bent)
  F_y = yield stress of seat angle steel

This length of the seat angle is generally taken as either 6 in. or 8 in. for a beam gage g of $3\frac{1}{2}$ in. and $5\frac{1}{2}$ in., respectively.

The number of bolts, which are in combined shear and tension, is determined in accordance with the principles of Sec. 4.15.

The weld size and length are obtained using the principles of Sec. 5.19 with Eq. 13.2.16 applicable to this case; direct shear and bending about the x–x axis for the configuration of Fig. 13.2.5 with the returns $b \approx L/12$.

EXAMPLE 13.3.1 _____

Design the seat angle to support a W12×40 beam on a 25-ft span, assuming the beam has adequate lateral support. Use A36 steel and Load and Resistance Factor Design.

Solution   In many cases it will be wise practice to design the seat for the maximum reaction when the beam is fully loaded in flexure.

(a) Determine the seat width, length, and thickness. The flexural strength is

$$\phi_b M_n = \phi_b M_p = \phi_b Z_x F_y = 0.90(57.5)36/12 = 155 \text{ ft-kips}$$

$$P_u = \frac{w_u L}{2} = \frac{8\phi_b M_n}{2L} = \frac{8(155)}{2(25)} = 24.8 \text{ kips}$$

Bearing length N required based on local web yielding, Eq. 13.3.1, is

$$N = \frac{P_u}{\phi F_{yw} t_w} - 2.5k = \frac{24.8}{1.0(36)(0.295)} - 2.5(1.25) = \text{negative}$$

A minimum bearing length must be used. Following *LRFD Manual,* Table 9–6, "All-Bolted Unstiffened Seated Connections," suggestion, use 4-in. seat (i.e., angle leg). According to LRFD-K1.3, $N \geq k$. With a 4-in. angle leg, N will exceed k (i.e., 1.25 in.). Check web crippling with actual N using a clearance of $\frac{3}{4}$-in. to allow possible mill underrun.

$$N = 4 - 0.75 = 3.25 \text{ in.}$$

$$\text{Compute } \frac{N}{d} = \frac{3.25}{11.94} = 0.27 > 0.2, \text{ use Eq. 13.3.4}$$

$$\phi P_n = \phi 68 t_w^2 \left[1 + \left(\frac{4N}{d} - 0.2 \right) \left(\frac{t_w}{t_f} \right)^{1.5} \right] \sqrt{\frac{F_{yw} t_f}{t_w}} \qquad [13.3.4]$$

$$\phi P_n = 0.75(68)(0.295)^2 \left[1 + \left(\frac{4(3.25)}{11.94} - 0.2 \right) \left(\frac{0.295}{0.515} \right)^{1.5} \right] \sqrt{\frac{36(0.515)}{0.295}}$$

$$= 48.7 \text{ kips}$$

$$[\phi P_n = 48.7 \text{ kips}] > [P_u = 24.8 \text{ kips}] \qquad \text{OK}$$

As expected, web crippling does not control. Following the usual practice $N = k$ is used for determining angle thickness,

$$e_f = \frac{1.25}{2} + \frac{3}{4} = 1.375 \text{ in.}$$

Trying $t = \frac{1}{2}$ in.,

$$e = e_f - t - \tfrac{3}{8} = 1.375 - 0.50 - 0.375 = 0.50 \text{ in.}$$

Since the usual gage $g = 5\tfrac{1}{2}$ in.* for W12×40, use angle length of 8 in. The angle thickness required is then, by Eq. 13.3.9,

$$t^2 = \frac{4P_u e}{\phi_b F_y L} = \frac{4(24.8)0.50}{0.90(36)8} = 0.19 \text{ in.}; \qquad t = 0.44 \text{ in.}$$

Use seat angle, $\tfrac{1}{2}$ in. thick and 8 in. long.

(b) Determine bolted connection to column, using $\tfrac{3}{4}$-in.-diam A325 bolts in a bearing-type connection with no threads in the shear plane.

$\phi R_n = 19.9$ kips   (single shear) (*LRFD Manual*, Table 8–11)

$\phi R_n = 39.1$ kips   (bearing on $\tfrac{1}{2}$-in.-thick angle) (*LRFD Manual*, Table 8–13)

$\phi R_n = 29.8$ kips   (tension) (*LRFD Manual*, Table 8–15)

Using Eq. 4.12.29, obtain a rough estimate of the number of fasteners per vertical line for two vertical rows of fasteners at a 3-in. pitch.

$$n \approx \sqrt{\frac{6M}{Rp}} = \sqrt{\frac{6(24.8)1.38}{19.9(3)2}} = 1.3$$

Try 2 bolts (i.e., $n = 1$) as shown in Fig. 13.3.4. The direct shear component is

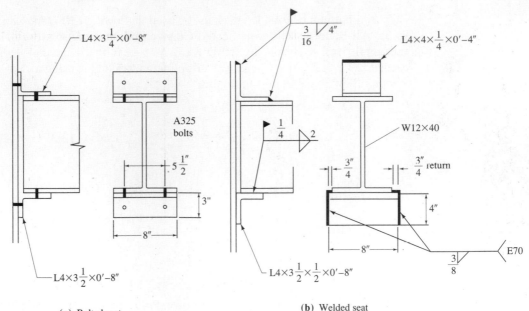

(a) Bolted seat

(b) Welded seat

Figure 13.3.4   Designs for Example 13.3.1.

---

*The distance g is not given in the current AISC Manuals [1.7, 1.18]; see p. 1–17 of the 7th Edition of the AISC Manual.

$$R_v = \frac{P_u}{n} = \frac{24.8}{2} = 12.4 \text{ kips} < 19.9 \text{ kips} \qquad \text{OK}$$

Since the bolts lie on the center of gravity, no moment of inertia can be computed using Σy^2. However, since initial tension exists, the initial compression, according to Eq. 4.15.1, is

$$f_{bi} = \frac{\Sigma T_b}{bd} = \frac{2(28)}{8(3)} = 2.33 \text{ ksi}$$

The change in stress due to moment, Eq. 4.15.2, is

$$f_{tb} = \frac{6M}{bd^2} = \frac{6Pe}{bd^2} = \frac{6(24.8/1.5)1.38}{8(3)^2} = 1.9 \text{ ksi}$$

Since $1.9 < 2.33$, the initial precompression is not eliminated and the connection can be considered safe. The factored load reaction $P_u = 24.8$ kips was divided by an average overload factor of 1.5 to approximate service load. Under full factored load the initial tension would be overcome at the top of the angle; however, compression would exist at the toes of the angles and the tension would be carried in the bolts. The initial tension approach seems justified for this case. *Use* 2 bolts, with seat angle, L4×3×$\frac{1}{2}$ × 0′-8″.

(c) Determine welded connection to column, using Eq. 13.2.16, with E70 electrodes with shielded metal arc welding:

Max weld size = $\frac{1}{2} - \frac{1}{16} = \frac{7}{16}$ in.

Min weld size = LRFD-Table J2.4 based on thickest material being joined.

Try L = 4-in. supported leg:

$$R_u = \frac{P_u}{2L^2}\sqrt{L^2 + 20.25e_1^2} \qquad [13.2.16]$$

where $e_1 = e_f$

$$R_u = \frac{24.8}{2(4)^2}\sqrt{(4)^2 + 20.25(1.38)^2} = 5.72 \text{ kips/in.}$$

$$\phi R_{nw} = \phi(0.707a)(0.60F_{EXX}) = 0.75(0.707a)42.0 = 22.3a$$

Base metal shear fracture does not control on $\frac{1}{2}$-in. angles.

$$\text{Weld size } a = \frac{5.72}{22.3} = 0.26 \text{ in.,} \quad \text{say } \tfrac{5}{16} \text{ in.}$$

A more conservative approach is to measure e_f to the center of the contact bearing width of the seat (Fig. 13.3.3a). This traditional AISC method for tables giving weld capacity for seats gives

$$e_f = \frac{N}{2} + \frac{3}{4} = \frac{3.5 - 0.75}{2} + 0.75 = 2.13 \text{ in.}$$

which upon substitution into Eq. 13.2.16 with $\phi R_{nw} = 22.3a = 22.3(0.3125) = 6.96$ kips/in. (for $\frac{5}{16}$ in. weld), gives a factored load capacity of

$$P_u = \frac{\phi R_{nw}(2L^2)}{\sqrt{L^2 + 20.25e_f^2}} = \frac{6.96(2)(4)^2}{\sqrt{(4)^2 + 20.25(2.13)^2}} = 21.5 \text{ kips}$$

The weld size required would then be

$$\text{Required } a = \tfrac{5}{16}\left(\frac{24.8}{21.5}\right) = 0.36 \text{ in., say } \tfrac{3}{8} \text{ in.}$$

Use L4×3$\frac{1}{2}$×$\frac{1}{2}$ × 0'-8" with $\frac{3}{8}$ in. weld. Since the W12×40 flange width is 8 in., the beam must be "blocked" or "cut" to have a reduced flange width over the seat so that the necessary welding can be done, or if the column flange permits, the seat may be longer than 8 in. The final designs are shown in Fig. 13.3.4. ■■

13.4 STIFFENED SEAT CONNECTIONS

When reactions become heavier than desirable for unstiffened seats, stiffeners may be used with the seat angle in bolted construction, or a T-shaped stiffened seat may be used in welded construction. The unstiffened seat may become excessively thick when the factored beam reaction exceeds about 60 kips. There are no AISC restrictions, however, to the maximum load that may be carried by unstiffened seats.

The stiffened seat as discussed herein is not intended to be part of a moment resisting connection, but rather it is only to support vertical loads. Here the stiffened seat is treated as "simple framing" under LRFD Type PR or ASD Type 2. Behavior of welded brackets has been studied by Jensen [13.44].

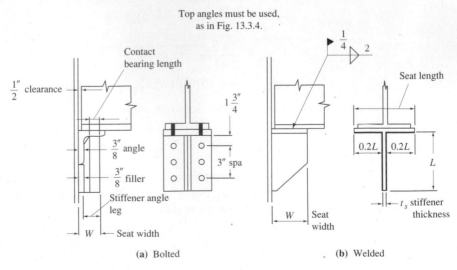

Figure 13.4.1   Stiffened seat—beam web in line with stiffener.

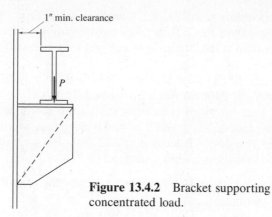

1″ min. clearance

P

Figure 13.4.2 Bracket supporting concentrated load.

There are two basic types of loading used on stiffened seats; the common one where the reaction is carried with the beam web directly in line with the stiffener, as shown in Fig. 13.4.1; the other is with a beam oriented so the plane of the web is at 90 degrees to the plane of the stiffener, as in Fig. 13.4.2. Furthermore, a difference in behavior arises depending on the angle at which the stiffener is cut, as shown in Fig. 13.4.3. If the angle θ is approximately 90 degrees, the stiffener behaves similarly to an unstiffened element under uniform compression, and local buckling may be prevented by satisfying LRFD or ASD-B5. When the supporting plate is cut to create a triangular bracket plate a different behavior results, and this case is discussed in Sec. 13.5.

The steps in the design of stiffened seats are as follows:

1. Determine the seat width.
2. Determine the eccentricity e_s of load.
3. Determine the stiffener thickness t_s.
4. Determine the angle sizes and arrangement of bolts; or the weld size and length.

The seat width is based on the required bearing length N (1) to prevent local web yielding, according to LRFD-K1.3 or ASD-K1.3, Eqs. 13.3.1 or 13.3.2; and (2) to

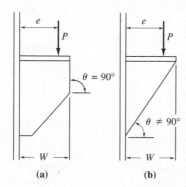

e P

$\theta = 90°$

W

(a)

e P

$\theta \neq 90°$

W

(b)

Figure 13.4.3 Two cases of inclination angle at free endge of stiffener.

prevent web crippling, according to LRFD-K1.4 or ASD-K1.4, Eqs. 13.3.3 through 13.3.5. Because of the rigidity of the stiffener, the most highly stressed portion is at the edge of the seat rather than at the interior side as it was for the unstiffened seat (see Fig. 13.4.4).

For stiffened seats, the *LRFD Manual,* Table 9–8, "All-Bolted Stiffened Seated Connections," and Table 9–9, "Bolted/Welded Stiffened Seated Connections," contain maximum factored load reactions (i.e., design strength) for bolted connection stiffener angle outstanding legs of $3\frac{1}{2}$, 4, and 5 in. and welded seat widths of 4 to 9 in.

Assuming the beam reaction P is located at $N/2$ from the edge of the seat the stiffener thickness t_s should satisfy several criteria:

1. Stiffener thickness t_s should be equal to or greater than the thickness t_w of the supported beam web,

$$t_s \geq t_w \tag{13.4.1}$$

2. Local buckling of the stiffener must be prevented in accordance with LRFD-B5 or ASD-B5, also given in LRFD-K1.9(2),

$$t_s \geq \frac{W}{95/\sqrt{F_y}} \tag{13.4.2}$$

where $W =$ width of stiffener.

3. Bearing on the contact area of stiffener must satisfy LRFD-J8.1 or ASD-J8. For *angle stiffeners*, it is assumed that $\frac{1}{2}$ in. of the angle is cut off in order to get close bearing under the seat angle:

a. *Load and Resistance Factor Design* ($\phi = 0.75$):

$$t_s \geq \frac{P_u}{\phi(1.8F_y)2(W - 0.5)} \quad \text{(for angle stiffeners)} \tag{13.4.3}$$

b. *Allowable Stress Design:*

$$t_s \geq \frac{P}{0.90F_y(2)(W - 0.5)} \quad \text{(for angle stiffeners)} \tag{13.4.4}$$

A structural tee might be used instead of two angles; in which case the 2 in

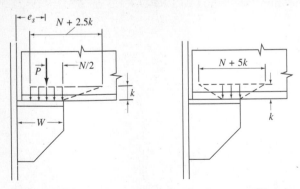

Figure 13.4.4   Bearing load distribution on stiffened seats.

the denominator of Eqs. 13.4.3 and 13.4.4 would *not* be used. Equations 13.4.3 and 13.4.4 assume no eccentricity of load with respect to center of bearing contact length assumed.

4. For *eccentric loading* on stiffener, bearing strength according to LRFD-J8 or ASD-J8 must be satisfied. In this situation, a single welded plate stiffener is generally used. The highest bearing stress at the outer edge of the stiffener may not exceed $\phi(1.8F_y)$ under factored load for LRFD, or $0.90F_y$ under service load for ASD. Using combined stress (see Fig. 13.4.4),

$$f_b = \frac{P}{A} + \frac{M}{S} = \frac{P}{Wt_s} + \frac{P(e_s - W/2)}{t_s W^2/6} = \frac{P}{t_s W^2}(6e_s - 2W) \qquad (13.4.5)$$

a. *Load and Resistance Factor Design* ($\phi = 0.75$):

$$t_s \geq \frac{P_u(6e_s - 2W)}{\phi(1.8F_y)W^2} \quad \text{(for welded stiffener)} \qquad (13.4.6)$$

b. *Allowable Stress Design*:

$$t_s \geq \frac{P(6e_s - 2W)}{0.90F_y W^2} \quad \text{(for welded stiffener)} \qquad (13.4.7)$$

5. Plate thickness must be adequate to develop the fillet welds used to attach it, according to LRFD-J2.4 or ASD-J2.4. The concept for maximum effective weld size was developed in Sec. 5.14, giving Eq. 5.14.9 relating weld size to base material thickness.

a. *Load and Resistance Factor Design* (LRFD-J2.4):

$$a_{\text{max eff}} = \frac{0.60F_u t_1}{2(0.707)0.60F_{\text{EXX}}} = 0.707 \frac{F_u t_1}{F_{\text{EXX}}} \qquad [5.14.9]$$

where t_1 = thickness of base material (= t_s here)
          F_u = tensile strength of base material
          F_{EXX} = tensile strength of electrode material
          (70 ksi for E70 electrodes)

Assuming two lines of fillet weld of size *a* using E70 electrodes, the stiffener thickness t_s required such that the stiffener plate will not be overloaded in shear is, from Eq. 5.14.9,

$$t_s \geq 1.71a \quad \text{(for A36 steel)} \qquad (13.4.8)$$
$$t_s \geq 1.52a \quad \text{(for } F_y = 50 \text{ ksi)} \qquad (13.4.9)$$

b. *Allowable Stress Design* (ASD-J2.4): Though not developed in Chapter 5, the following is obtained by equating the allowable capacity for two lines of fillet weld to the allowable shear capacity of stiffener plate of thickness t_s:

$$a_{\text{max eff}} = \frac{0.40F_y t_s}{2(0.707)0.30F_{\text{EXX}}} = 0.943 \frac{F_y t_s}{F_{\text{EXX}}} \qquad (13.4.10)$$

where $\qquad F_y$ = yield stress of base material

$\qquad\qquad F_{EXX}$ = tensile strength of electrode material

$\qquad\qquad\qquad$ (70 ksi for E70 electrodes)

Assuming two lines of fillet weld of size a using E70 electrodes, the stiffener thickness t_s required such that the stiffener plate will not be overloaded in shear is, from Eq. 13.4.10,

$$t_s \geq 2.06a \quad \text{(for A36 steel)} \qquad (13.4.11)$$

$$t_s \geq 1.48a \quad \text{(for } F_y = 50 \text{ ksi)} \qquad (13.4.12)$$

Once the stiffener dimensions have been established, the connection must be designed to transmit the reaction at the moment arm e_s. For the bolted connection, *LRFD Manual,* Table 9–8, "All-Bolted Stiffened Seated Connections," uses only direct shear in determining fastener group capacities. One may reason, as in Example 13.3.1, that as long as initial compression between the pieces in contact is not reduced to zero due to flexure, the moment component need not be considered.

For the welded connection suggested by the *LRFD Manual,* p. 9-138, as shown in Fig. 13.4.1b, the weld configuration is subject to direct shear and flexure using the combined stress at the top of the weld as the critical one. Thus, the configuration is identical to that used for web framing angles (see Fig. 13.2.1d) except the return is longer. Using $d = L$ and $b = 0.2L$ in the S values for Case 4 from Table 5.18.1 gives

$$\bar{y} = \frac{L^2}{2(L + b)} = \frac{L^2}{2(1.2L)} = \frac{L}{2.4}$$

$$S_x = \frac{2(4bL + L^2)}{6} = \frac{4(0.2L)L + L^2}{3} = 0.6L^2$$

Then,

$$R_x = \frac{M}{S_x} = \frac{Pe_s}{0.6L^2} \quad \text{force/unit length} \rightarrow$$

$$R_v = \frac{P}{2(L + 0.2L)} = \frac{P}{2.4L} \quad \text{force/unit length} \downarrow$$

$$R = \sqrt{\left(\frac{Pe_s}{0.6L^2}\right)^2 + \left(\frac{P}{2.4L}\right)^2}$$

$$R = \frac{P}{2.4L^2}\sqrt{16e_s^2 + L^2} \quad \text{force/unit length} \qquad (13.4.13)$$

Equation 13.4.13 is used for obtaining loads for welded stiffened beam seats in *LRFD Manual,* Table 9–9, "Bolted/Welded Stiffened Seated Connections," when e_s is taken as $0.8W$. For LRFD, R is the design strength ϕR_{nw} and P is the factored load reaction P_u. For ASD, R is the allowable weld capacity R_w and P is the service load reaction.

EXAMPLE 13.4.1 _____

Design a welded stiffened seat to support a W30×99 beam having a factored load reaction $P_u = 160$ kips. Use A572 Grade 50 steel and Load and Resistance Factor Design.

Solution The bearing length N required is obtained from (a) local web yielding and (b) web crippling criteria. From local web yielding (LRFD-K1.3),

$$N = \frac{P_u}{\phi F_{yw} t_w} - 2.5k = \frac{160}{1.0(50)(0.520)} - 2.5\left(1\frac{7}{16}\right) = 2.6 \text{ in.}$$

From the web crippling criterion (LRFD-K1.4), assuming $N/d \leq 0.2$ use Eq. 13.3.3,

$$\phi P_n = 0.75(68)(0.520)^2\left[1 + 3\left(\frac{N}{29.65}\right)\left(\frac{0.520}{0.670}\right)^{1.5}\right]\sqrt{\frac{50(0.670)}{0.520}}$$

Using $N = 3.0$ in. gives $\phi P_n = 134$ kips, which is not enough. Solving by trial using $\phi P_n = 160$ kips gives $N = 6.5$ in.

With $N = 6.5$ in., $N/d = 6.5/29.65 = 0.22 > 0.2$. This means Eq. 13.3.4 must be used to check web crippling. Thus,

$$\phi P_n = \phi 68 t_w^2\left[1 + \left(\frac{4N}{d} - 0.2\right)\left(\frac{t_w}{t_f}\right)^{1.5}\right]\sqrt{\frac{F_{yw} t_f}{t_w}} \qquad [13.3.4]$$

$$= 0.75(68)(0.520)^2\left[1 + \left(\frac{4(6.5)}{29.65} - 0.2\right)\left(\frac{0.520}{0.670}\right)^{1.5}\right]\sqrt{\frac{50(0.670)}{0.520}}$$

$$= 162 \text{ kips}$$

$$[\phi P_n = 162 \text{ kips}] > [P_u = 160 \text{ kips}] \qquad \text{OK}$$

$$\text{Required } W = 6.5 + 0.5(\text{setback}) = 7.0 \text{ in.} \quad \textit{Use } 7 \text{ in.}$$

For the seat plate thickness, use a thickness comparable to the flange of the W30×99 supported beam; *use* $\frac{5}{8}$ in. Minimum weld size for welding on $\frac{5}{8}$-in. seat and 0.67-in. flange is $\frac{1}{4}$ in.

The stiffener thickness is next to be established:

$$t_s \geq t_w = 0.520 \text{ in.} \qquad [13.4.1]$$

$$t_s \geq \frac{W}{95/\sqrt{F_y}} = \frac{7}{13.4} = 0.52 \text{ in.} \qquad [13.4.2]$$

$$e_s = W - \frac{N}{2} = 7.0 - \frac{6.5}{2} = 3.8 \text{ in.}$$

$$t_s \geq \frac{P_u(6e_s - 2W)}{\phi(1.8F_y)W^2} = \frac{160(22.5 - 14)}{0.75(90)(7)^2} = 0.41 \text{ in.} \qquad [13.4.6]$$

The use of a $\frac{5}{8}$-in. stiffener plate would mean a maximum effective weld size of

$$t_s \geq 1.52a \qquad [13.4.9]$$

$$a_{\text{max eff}} = \frac{t_s}{1.52} = \frac{0.625}{1.52} = 0.41 \text{ in.}$$

Thus, weld size is not of concern since a weld smaller than 0.41 in. would be preferred. Generally, the maximum weld that can be placed in one pass would be used; in this case, $\frac{5}{16}$ in.

For estimating the length L of weld required, assume that e_s in Eq. 13.4.13 is approximately $L/4$; i.e., that $e_s = 0.8W = 0.8(7) = 5.6$ in. is roughly $L/4$.

$$R = \frac{P}{2.4L^2}\sqrt{16\left(\frac{L^2}{16}\right) + L^2} = 0.59\frac{P}{L}$$

For LRFD, R becomes ϕR_{nw} and using $\frac{5}{16}$-in. E70 weld,

$$\phi R_{nw} = \phi(0.707a)(0.60F_{\text{EXX}})$$
$$= 0.75(0.707)\left(\tfrac{5}{16}\right)(0.60)70 = 6.96 \text{ kips/in.}$$

$$\text{Required } L \approx \frac{0.59(160)}{6.96} = 13.6 \text{ in.}$$

For $L = 14$ in., 5.6 in. is $0.4L$; which when used as e_s/L in Eq. 13.4.13 gives required $L \approx 18$ in. The answer lies between 14 and 18 in. Try $L = 16$ in. with $\frac{5}{16}$-in. weld,

$$R_u = \frac{160}{2.4(16)^2}\sqrt{16(5.6)^2 + (16)^2} = 7.2 \text{ kips/in.} > 6.96 \text{ kips/in.} \qquad \text{NG}$$

The length $L = 16$ in. is not adequate; one could use 17 in. but 18 in. may be preferred.

Use $\frac{5}{16}$-in. weld with $L = 18$ in. Use stiffener plate, $\frac{5}{8}\times7 \times 1'\text{-}6''$; and seat plate $\frac{5}{8}\times7 \times 1' - 0''$. The seat plate width equals the flange width (10.45 in.) plus enough to easily make the welds (approx. 4 times the weld size is often used). The final design is shown in Fig. 13.4.5.

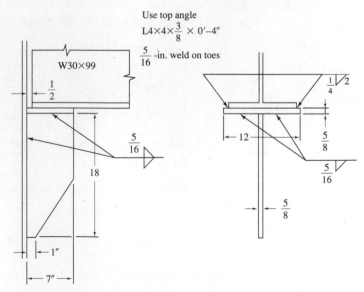

Figure 13.4.5   Design for Example 13.4.1.

13.5 TRIANGULAR BRACKET PLATES

When the stiffener for a bracket is cut into a triangular shape, as in Fig. 13.4.3b, the plate behaves in a different manner than when the free edge is parallel to the direction of applied load in the region where the greatest stress occurs, as in Fig. 13.4.5. The triangular bracket plate arrangement and notation are shown in Fig. 13.5.1.

The behavior of triangular bracket plates has been studied analytically by Salmon [13.45] and experimentally by Salmon, Buettner, and O'Sheridan [13.46] and design suggestions have been proposed by Beedle et al. [13.47]. For small stiffened plates to support beam reactions there is little danger of buckling or failure of the stiffener if cut into a triangular shape.

Most Exact Analysis and Design Recommendations

For many years, design of such brackets was either empirical without benefit of theory or tests, or when in doubt, angle or plate stiffeners were used along the diagonal edge. The recommendations presented here are based on certain assumptions: (1) the top plate is solidly attached to the support; (2) the load P is distributed (though not necessarily uniformly) and has its centroid at approximately $0.6b$ from the support; and (3) the ratio b/a, loaded edge to supported edge, lies between 0.50 and 2.0.

The original theoretical analysis was concerned with elastic buckling; however, the experimental work showed that triangular bracket plates have considerable post-buckling strength. Yielding along the free edge frequently occurs prior to buckling, at which point redistribution of stresses occurs. A considerable margin of safety against collapse was observed indicating the ultimate capacity may be expected to be at least 1.6 times the buckling load.

The maximum stress was found to occur at the free edge; however because of the complex nature of the stress distribution, the stress on the free edge is not obtainable by any simple process. Because of this difficulty, a ratio z was established between the average stress, P/bt, on the loaded edge to the maximum stress f_{max} on the free edge. The original theoretical expression [13.45] for z was revised as a result of the tests

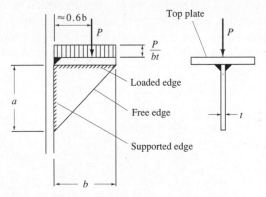

Figure 13.5.1 Triangular bracket plate.

[13.46], which conformed closely to what one could realistically expect in practice. The relationship is given [13.46] as

$$z = \frac{P/bt}{f_{max}} = 1.39 - 2.2\left(\frac{b}{a}\right) + 1.27\left(\frac{b}{a}\right)^2 - 0.25\left(\frac{b}{a}\right)^3 \qquad (13.5.1)$$

which for practical purposes may be obtained from Fig. 13.5.2.

The nominal strength P_n when the free edge reaches the yield stress is

$$P_n = F_y z b t \qquad (13.5.2)$$

For the plate buckling limit state, the width/thickness ratio b/t must be restricted in accordance with a relationship of the type of Eq. 6.16.4,

$$\frac{b}{t} \leq \frac{constant}{\sqrt{F_y}} \qquad (13.5.3)$$

Figure 13.5.3 gives the variation in $(b/t)\sqrt{F_y}$ with b/a for the theoretical studies [13.45] (fixed and simply supported), the welded bracket tests result [13.46], and the authors' suggested design curve. The design requirement may be expressed (with F_y in ksi) as

$$\text{For } 0.5 \geq \frac{b}{a} \leq 1.0; \qquad \frac{b}{t} \leq \frac{250}{\sqrt{F_y}} \qquad (13.5.4a)*$$

$$\text{For } 1.0 \leq \frac{b}{a} \leq 2.0; \qquad \frac{b}{t} \leq \frac{250(b/a)}{\sqrt{F_y}} \qquad (13.5.4b)*$$

Satisfying the above limits means that yielding along the *diagonal free edge* will occur prior to buckling, and with conservatism compared to the welded tests, as shown by Fig. 13.5.3.

It is noted that the b/t limits suggested here are higher than those of Ref. 13.47 (p. 552), which were based solely on the theoretical studies [13.45]. Reference 13.47 suggests a coefficient of 180 instead of 250 in Eq. 13.5.4 and reaches a maximum of 300 instead of 500 as indicated by Eq. 13.5.4b for $b/a = 2.0$. The reason for the higher values is found in the test results which showed the principal stress along the diagonal free edge to be lower relative to the stress on the loaded edge than had been established by the theoretical study. In other words, the z value, Eq. 13.5.1, as determined by tests is substantially smaller than assumed for the design suggestion of Ref 13.46.

---

* For SI units, with F_y in MPa,

$$\text{For } 0.5 \leq \frac{b}{a} \leq 1.0; \qquad \frac{b}{t} \leq \frac{656}{\sqrt{F_y}} \qquad (13.5.4a)$$

$$\text{For } 1.0 \leq \frac{b}{a} \leq 2.0; \qquad \frac{b}{t} \leq \frac{656(b/a)}{\sqrt{F_y}} \qquad (13.5.4b)$$

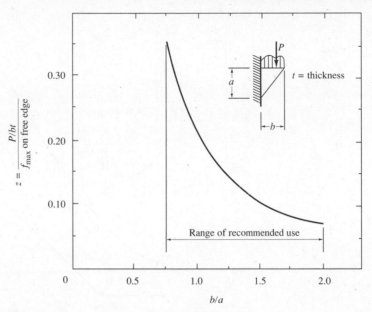

Figure 13.5.2   Coefficient used to obtain maximum stress on free edge.

Plastic Strength of Bracket Plates

Reference 13.47 suggests that to develop the full plastic strength of brackets the b/t ratios should be restricted to about $\frac{1}{3}$ of those limitations for achieving first yield on the free edge. The test results [13.46] indicated that ultimate strengths of at least 1.6 times buckling strengths could be achieved due to post-buckling strength. To be certain of developing the *plastic capacity* of the bracket, it may be realistic to use half of the limitations of Eqs. 13.5.4a and b.

To establish the plastic strength of a bracket plate used in rigid-frame structures, one may follow the approach of Beedle et. al. [13.47] as shown in Fig. 13.5.4. This method assumes that plastic strength develops on the critical section. Taking force equilibrium parallel to the free edge and moment equilibrium about point O gives the Beedle et al. [13.47] equation for the ultimate load,

$$P_n = F_y t \sin^2\alpha(\sqrt{4e^2 + b^2} - 2e) \qquad (13.5.5)$$

In addition to the triangular plate being adequate, the top plate must carry the nominal load $P_n \cot \alpha$.

EXAMPLE 13.5.1 _____

Determine the thickness required for a triangular bracket plate 25 in. by 20 in. to carry a factored load of 60 kips. Assume the load is located 15 in. from the face of support as shown in Fig. 13.5.5, and that A36 material is used. Use Load and Resistance Factor Design.

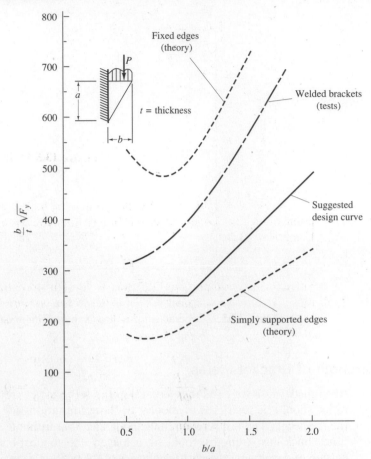

Figure 13.5.3 Critical b/t values so that yield stress is reached along diagonal free edge without buckling.

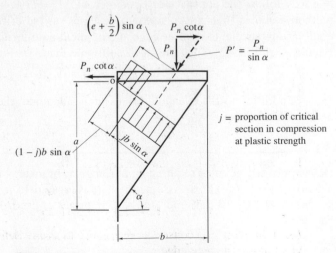

Figure 13.5.4 Plastic strength analysis.

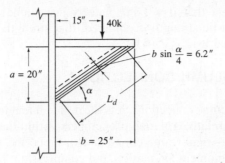

Figure 13.5.5   Bracket for Example 13.5.1.

Solution
(a) Use the more exact method. Since the load is approximately at the 0.6 point along the loaded edge, the bracket fits the assumption of this method. Using Eq. 13.5.2,

$$P_n = F_y z b t \qquad\qquad [13.5.2]$$

Since this is a compression situation, the design strength ϕP_n should be equated to P_u. It may be reasonable to use the ϕ for compression; $\phi_c = 0.85$. From Fig. 13.5.2, $b/a = 25/30 = 1.25$, find $z = 0.135$. Then the strength requirement gives

$$P_u = \phi_c P_n = 0.85 F_y z b t = 60 \text{ kips}$$

$$t \geq \frac{P_u}{\phi F_y z b} = \frac{60}{0.85(36)(0.135)25} = 0.58 \text{ in.}$$

The stability requirement, Eq. 13.5.4b, gives

$$t \geq \frac{b\sqrt{F_y}}{250(b/a)} = \frac{25\sqrt{36}}{250(1.25)} = 0.48 \text{ in.}$$

Use $\frac{5}{8}$-in. plate.

(b) Plastic strength method. Using Eq. 13.5.5 for the strength requirement,

$$t \geq \frac{P_u}{\phi F_y \sin^2\alpha \left[\sqrt{4e^2 + b^2} - 2e\right]}$$

Using $e = 15 - 25/2 = 2.5$ in.,

$$t \geq \frac{60}{0.85(36)(0.39)[\sqrt{4(2.5)^2 + (25)^2} - 2(2.5)]} = 0.25 \text{ in.}$$

For stability, using *one-half* of Eq. 13.5.4b

$$t \geq \frac{b\sqrt{F_y}}{125(b/a)} = \frac{25\sqrt{36}}{125(1.25)} = 0.96 \text{ in.}$$

Use 1-in. plate as a conservative practice to assure deformation well beyond first yield along the free edge.

The authors note that if a 1-in. plate is just stable enough to inhibit buckling until the plastic strength is obtained, that strength would be 4 times (1.0/0.25) the factored load P_u. ■■

13.6 CONTINUOUS BEAM-TO-COLUMN CONNECTIONS

In continuous beam-to-column connections, it is the design intent to have full transfer of moment and little or no relative rotation of members within the joint (i.e., LRFD Type FR or ASD Type 1—rigid-frame connections). Since the flanges of a beam carry most of the bending moment via tension and compression flange forces acting at a moment arm approximately equal to the beam depth, it is the transfer of these essentially axial forces for which provision must be made. Since the shear is carried primarily by the web of a beam, full continuity requires that it be transferred directly from the web.

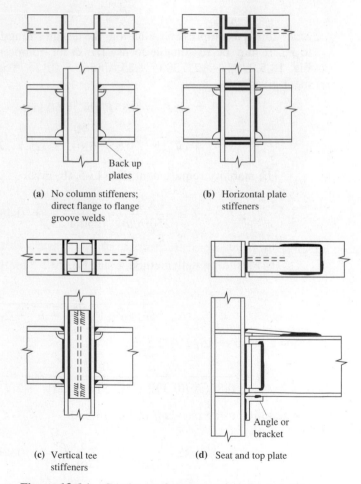

(a) No column stiffeners; direct flange to flange groove welds

(b) Horizontal plate stiffeners

(c) Vertical tee stiffeners

(d) Seat and top plate

Figure 13.6.1   Continuous beam-to-column connections: welded attachment to column flange.

Columns being rigidly framed by beams may have attachments to both flanges, as in Figs. 13.6.1a, b, and c, or only to one flange, as in Fig. 13.6.1d, and in Fig. 13.6.2. Alternatively, the rigid attachment of beams may be to the web, from either or both sides, as in Fig. 13.6.3. When the rigid system has rigid attachments *either* to the flanges or the web (but not to both) the system is said to be a two-way, or planar, rigid frame. When the rigid frame system consists of continuous connections to both flange (or flanges) and web (either or both sides), the system becomes a three-dimensional system, or space frame.

The variety of arrangements for a continuous beam-to-column connection is so great as to preclude any complete listing or illustration; however, those shown in Figs. 13.6.1, 13.6.2, and 13.6.3 are believed to be common in current (1995) design use. Most connections are partly shop welded and then completed in the field by either welding or fastening with high-strength bolts.

The principal design concern is with transmission of concentrated loads caused by flange forces on beams to the adjacent columns. The web of a column may be unable to accept the compression force from a beam flange without additional stiffening; the flange of a column may exhibit excessive deformation caused by a tension force from a beam flange.

Rigid framing is used to greatest advantage in (a) structures designed using plastic analysis, either under Load and Resistance Factor Design or Plastic Design, or (b) structures designed using elastic analysis where the sections are "compact" and the

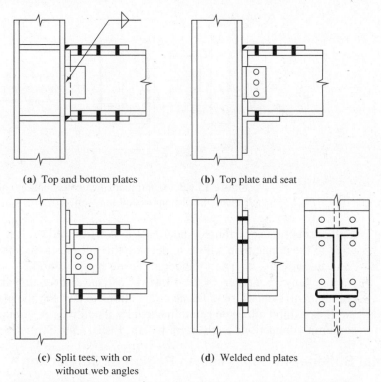

(a) Top and bottom plates (b) Top plate and seat

(c) Split tees, with or without web angles (d) Welded end plates

Figure 13.6.2   Continuous beam-to-column connections: bolted attachments.

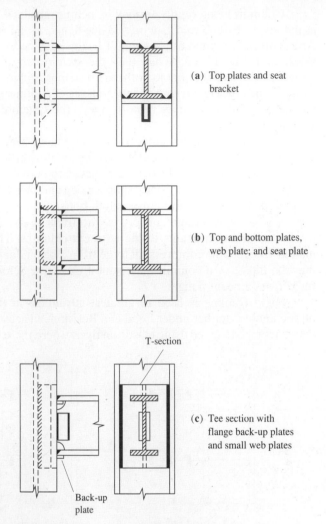

(a) Top plates and seat bracket

(b) Top and bottom plates, web plate; and seat plate

T-section

(c) Tee section with flange back-up plates and small web plates

Back-up plate

Figure 13.6.3 Continuous beam-to-column connections: welded attactment to column web.

10% reduction is permitted under LRFD-A5.1 or ASD-C1. In these cases, the objective of the connection will be to develop the full plastic moment strength at the joint, and in addition, be able to undergo plastic hinge rotation.

Many researchers [7.28, 13.48-13.59] have adequately demonstrated the ability of beam-to-column rigid frame connections to develop the plastic hinge moment, as well as exhibit adequate rotation capacity (ductility). A summary of bolted beam-to-column connections is given by Kulak, Fisher, and Struik [3.1].

Horizontal Stiffener in Compression Region of Connection

When the forces in beam flanges are transmitted to column flanges as compression or tension forces, horizontal stiffeners, as in Figs. 13.6.1b and d and Fig. 13.6.2a, may be required. Such stiffeners prevent local flange bending (LRFD-K1.2 or ASD-K1.2)

from the tension force, local web yielding (LRFD-K1.3 or ASD-K1.3), web crippling (LRFD-K1.4 or ASD-K1.4), and compression buckling (LRFD-K1.6 or ASD-K1.6) caused by the compression force. Local web yielding and web crippling were treated in Sec. 7.8.

Local Web Yielding.   Both LRFD and ASD treat this phenomenon, formerly called "web crippling," in an identical manner based on the work of Graham, Sherbourne, Khabbaz, and Jensen [7.28].

Consider a beam compression flange bearing against a column as in Fig. 13.6.4a. When the maximum local yielding strength of the column web is reached, the load will have been distributed along the base of the fillet (k from the face of the flange) on a 1:2.5 slope. Using Eq. 7.8.2 for an interior concentrated load according to LRFD and Eq. 7.8.6 for ASD and taking the bearing length N as the beam flange thickness t_{fb} and the reaction R or ϕR_n as the flange force P_{bf} gives

Local Web Yielding—Load and Resistance Factor Design

$$P_{bf} = \phi(5k + t_{fb})F_{yw}t_{wc} \qquad (13.6.1)$$

where P_{bf} is the design strength of the column web to resist a beam flange transmitted force under *factored* loads, ϕ = resistance factor = 1.0, F_{yw} is the yield stress of the column web, and t_{wc} is the column web thickness. The required P_{bf} could be as high as $F_{yb}A_f$ when the beam is expected to develop its plastic moment strength.

Local Web Yielding—Allowable Stress Design

$$P_{bf} = (5k + t_{fb})(0.66F_{yw})t_{wc} \qquad (13.6.2)$$

where P_{bf} is the service load flange force that must be transmitted.

Web Crippling.   In accordance with Eq. 7.8.8, using t_{bf} for the bearing length N,

Web Crippling—Load and Resistance Factor Design

$$P_{bf} = \phi 135 t_{wc}^2 \left[1 + 3 \left(\frac{t_{fb}}{d} \right) \left(\frac{t_{wc}}{t_{fc}} \right)^{1.5} \right] \sqrt{\frac{F_{yw}t_{fc}}{t_{wc}}} \qquad (13.6.3)$$

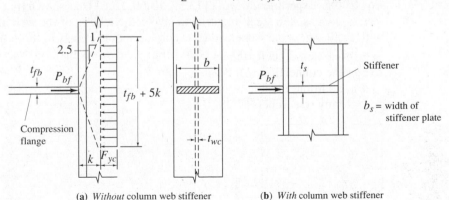

(a) *Without* column web stiffener   (b) *With* column web stiffener

Figure 13.6.4   Strength of column web in compression region of connection—AISC approach.

where ϕ = resistance factor = 0.75
P_{bf} = factored force from beam flange
t_{wc} = column web thickness
t_{fb} = beam flange thickness
t_{fc} = column flange thickness
d = column overall depth

Web Crippling—Allowable Stress Design

$$P_{bf} = 67.5t_{wc}^2 \left[1 + 3\left(\frac{t_{fb}}{d}\right)\left(\frac{t_{wc}}{t_{fc}}\right)^{1.5} \right] \sqrt{\frac{F_{yw}t_{fc}}{t_{wc}}} \tag{13.6.4}$$

where P_{bf} = service load capacity.

Compression Buckling of the Web.   Overall buckling of the column web must also be prevented as a controlling limit state. This limit state is of concern when concentrated loads from beam flanges are applied to *both* column flanges. When only one column flange is subjected to a concentrated load, the overall web buckling limit state need not be checked.

Equation 6.14.28 applies to elastic buckling of a plate,

$$F_{cr} = \frac{k\pi^2 E}{12(1 - \mu^2)(b/t)^2} \tag{6.14.28}$$

Chen and Newlin [13.60] and Chen and Oppenheim [13.61] have treated web buckling in a column as analogous to a plate subject to equal and opposite concentrated loads as shown in Fig. 13.6.5. The elastic buckling load P_{cr} for that situation using a simply supported plate having a large ratio a/h, is given by Timoshenko and Gere [6.67, pp. 387–389] as

$$P_{cr} = \frac{4\pi^2 E t^3}{12(1 - \mu^2)h} = \frac{33,400t^3}{h} \tag{13.6.5}$$

If the rotational restraint provided by the column flanges were the fully fixed condition, the buckling strength would be theoretically twice as great as that given by Eq. 13.6.5. Experimental work [13.52, 13.60, 13.61] has shown that when lower yield stress (such as A36 steel) material is involved, yielding of the web along the junction with the flange occurs at a load level corresponding closely to the simple support case; i.e., Eq. 13.6.5. When 100 ksi (690 MPa) yield stress material was used in tests, the

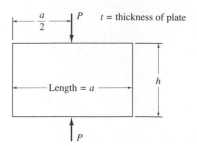

Figure 13.6.5 Plate subject to equal and opposite concentric concentrated loads.

higher yield stress material provided a high degree of rotational restraint along the junction of flange to web, giving a buckling strength about twice that obtained when A36 steel was used. Thus, Chen and Newlin [13.60] suggested that the increase in effective degree of fixity at the loaded edge may be accounted for in practical design by making the strength proportional to the square root of the yield stress. Their stability criterion is

$$P_{cr} = \frac{33,400t^3}{h} \sqrt{\frac{F_{yw}}{36}} = \frac{5570t^3 \sqrt{F_{yw}}}{h} \qquad (13.6.6)$$

The coefficient 5570 of this semirational expression was adjusted downward to 4100, representing a lower bound for all test results. Then for the beam-to-column connection t is the column web thickness t_{wc} and the concentrated load P_{cr} is the nominal reaction strength P_n, giving the LRFD-K1.6 expression

Compression Buckling of Column Web—LRFD

$$P_n = \frac{4100t_{wc}^3 \sqrt{F_{yw}}}{h} \qquad (13.6.7)$$

where P_n = nominal strength provided by column web, kips
  h = web depth clear of fillets = $d - 2k$, in.
  t_{wc} = column web thickness, in.
  F_{yw} = column web yield stress, ksi

Substitution of the required nominal strength P_{bf}/ϕ for P_n, and using $\phi = 0.90$ according to LRFD-K1.6, gives the factored load P_{bf} that can be carried without overall column web buckling when no stiffeners are used,

$$P_{bf} = \frac{\phi 4100t_{wc}^3 \sqrt{F_{yw}}}{h} \qquad (13.6.8)$$

where $\phi = 0.90$
  P_{bf} = maximum factored concentrated load from beam flanges

Note that Eq. 13.6.8 applies *only* when P_{bf} acts at *both* flanges.

Compression Buckling of Column Web—ASD

$$\frac{h}{t_{wc}} \le \frac{4100t_{wc}^2 \sqrt{F_{yw}}}{P_{bf}} \qquad (13.6.9)$$

where P_{bf} = factored concentrated load from beam flanges. Note that the LRFD-K1.6 uses $\phi = 0.90$, whereas ASD-K1.6 does not, seemingly giving ASD a slightly less conservative result. However, in ASD the "factor" to be used is 5/3 according to ASD-K1.2 in the definition of P_{bf}. Thus, the end result will be comparable using either LRFD or ASD.

When any or all of the three concentrated load related limit states (i.e., local web yielding, web crippling, or compression buckling of the web) indicate inadequacy of the column web to transmit the beam flange compression force P_{bf}, stiffeners in pairs (total area A_{st}) must be provided on the column to resist the force P_{bf}.

Compression Stiffener Rules—LRFD The following are LRFD requirements for stiffener design:

1. When the web crippling (LRFD-K1.4) or compression buckling of the web (LRFD-K1.6) limit states indicate the need for stiffeners, they are to be designed as axially loaded compression members (LRFD-K1.9). Alternatively, doubler plates may be used in accordance with LRFD-K1.10.

2. When compression buckling of the web (LRFD-K1.6) controls, the stiffeners (if used) must extend the entire depth of the column.

3. When local web yielding (LRFD-K1.3) controls, and the concentrated load P_{bf} is applied at only one column flange, the stiffeners need not extend more than one-half the depth of the column web.

4. When local web yielding (LRFD-K1.3) controls, the area A_{st} of stiffeners (in pairs) required is the excess of the factored force P_{bf} over the design resistance ϕR_n, divided by the stiffener design yield stress ϕF_{yst}. Thus,

$$A_{st} \geq \frac{P_{bf} - \phi F_{yw}(t_{fb} + 5k)t_{wc}}{\phi F_{yst}} \qquad (13.6.10)$$

The resistance factor ϕ for local web yielding is prescribed as 1.0; thus simplifying Eq. 13.6.10 and making it applicable for both LRFD and ASD design. P_{bf} is the factored load; in ASD the service load is factored by 5/3.

5. Proportioning of stiffeners. The following proportioning requirement appears in ASD-K1.8 and in LRFD-K1.9(1). The stiffener width b_{st} plus $\frac{1}{2}$ the column web thickness t_{wc} may not be less than $\frac{1}{3}$ of the beam flange width b_{fb} or moment plate connection width delivering the force P_{bf},

$$b_{st} + \frac{t_{wc}}{2} \geq \frac{b_{fb}}{3} \qquad (13.6.11)$$

which makes

$$b_{st} \geq \frac{b_{fb}}{3} - \frac{t_{wc}}{2} \qquad (13.6.12)$$

6. The local buckling limits of LRFD-B5 or ASD-B5, restated in LRFD-K1.9(2), for unstiffened compression elements must be satisfied. Since these limits relate to $b_f/2t_f$ for the beam transmitting the force to the column, the limit will be satisfied when the stiffener thickness t_s is not less than one-half the beam flange thickness t_{fb}. Thus,

$$t_s \geq \frac{t_{fb}}{2} \qquad (13.6.13)$$

7. The weld joining stiffeners to the column web should be sized to carry the force in the stiffener caused by unbalanced moments on opposite sides of the column.

Horizontal Stiffener in Tension Region of Connection

At the beam tension flange attachment to a column, the pull on the column flange, as shown in Fig. 13.6.6, may cause sufficient deformation as to impair the strength of the column. A yield line analysis was performed by Graham et al. [7.28] on the portion of the column flange of width q and length p, as in Fig. 13.6.6. Placing a line load on the system, the nominal strength P_n was approximated as

$$P_n = 7t_{fc}^2 F_{yf} + t_{fb}mF_{yw} \tag{13.6.14}$$

where the first term represents the bending resistance of the column flange as two plate elements (one on each side of the web) and the second term is the portion of the load that goes directly into the column web. Using conservatism comparable to the local web yielding in compression criterion, Eq. 13.6.14 was multiplied by 0.8; then solving for t_{fc} gives

$$t_{fc} = \sqrt{\frac{P_n}{7F_{yf}}\left(1.25 - \frac{t_{fb}mF_{yw}}{P_n}\right)} \tag{13.6.15}$$

From tests [7.28] the minimum value of $t_{fb}mF_{yw}/P_n$ was determined to be 0.15. Thus, using 0.15 for the second term in the bracket of Eq. 13.6.15 gives the conservative expression used by both LRFD and ASD for the minimum column flange thickness t_{fc} to avoid needing a column stiffener to assist in carrying the tension force from a beam flange,

$$t_{fc} \geq 0.4\sqrt{\frac{P_n}{F_{yf}}} \tag{13.6.16}$$

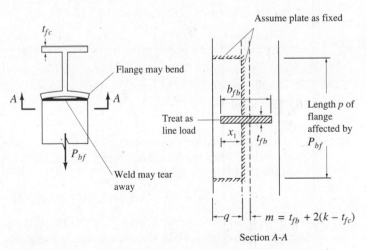

Figure 13.6.6 Strength of column flange in tension region of connection.

Local Flange Bending from Tension Force—LRFD   In order for more general applicability to any situation, Eq. 13.6.16 is solved for the nominal strength P_n in LRFD-K1.2,

$$P_n = 6.25t_{fc}^2 F_{yf} \qquad (13.6.17)$$

Since the factored flange force P_{bf} cannot exceed the design strength ϕP_n, either Eq. 13.6.17 can be used to determine the available strength or Eq. 13.6.16 can be used to compute the minimum flange thickness needed to avoid stiffeners. Thus, in LRFD, Eq. 13.6.16 for minimum column flange thickness t_{fc} would be

$$t_{fc} \geq 0.4\sqrt{\frac{P_{bf}}{\phi F_{yf}}} \qquad (13.6.18)$$

where P_{bf} = factored concentrated tension load
ϕ = 0.90 (LRFD-K1.2)
F_{yf} = yield stress of the column flange being bent

Local Flange Bending from Tension Force—ASD   In ASD-K1.2, Eq. 13.6.18 with the ϕ included in P_{bf},

$$t_{fc} \geq 0.4\sqrt{\frac{P_{bf}}{F_{yf}}} \qquad (13.6.19)$$

where P_{bf} = 5/3 times the *service* concentrated tension load; the 5/3 approximates factored load divided by ϕ = 0.90

Since, generally, the compression-related stiffener requirements are more likely to control, the reader should particularly note those requirements. The same proportioning requirements for compression stiffeners should be used for tension stiffeners. Usually the same size stiffeners would be used for both compression and tension (if any tension stiffener is needed). Examples of the design of beam-to-column moment connections have been presented by Miller [13.62].

EXAMPLE 13.6.1 _____

Design the connection for the rigid framing of two W16×40 beams to the flanges of a W12×65 column using A572 Grade 50 steel, as shown in Fig. 13.6.7. Use A36 steel for stiffeners if needed. Use LRFD.

Solution

(a) Compression region. Design for the maximum flange force transmitted by the beam. Thus the maximum factored force will be assumed to be the design strength of the beam flange,

$$P_{bf} = \phi A_f F_y = 0.90(6.995)(0.505)50 = 159 \text{ kips}$$

since the beam is "compact" with F_y = 50 ksi. If the beam were "noncompact" this approach might still be reasonable unless the actual factored moment is used to establish the connection force.

To avoid stiffeners, the column web-related strength must satisfy LRFD-K1.3 (local web yielding), LRFD-K1.4 (web crippling), and LRFD-K1.6 (com-

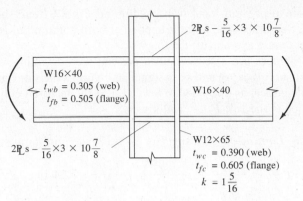

Figure 13.6.7 Connection with horizontal stiffeners for Example 13.6.1.

pression buckling of the web). Solving Eqs. 13.6.1, 13.6.3, and 13.6.8 for the maximum factored load P_{bf} that can be carried without stiffeners,

$$P_{bf} = \phi(5k + t_{fb})F_{yw}t_{wc} \qquad\qquad [13.6.1]$$
$$= 1.0[5(1.3125) + 0.505](50)0.390 = 138 \text{ kips}$$

$$P_{bf} = \phi\, 135t_{wc}^2\left[1 + 3\left(\frac{t_{fb}}{d}\right)\left(\frac{t_{wc}}{t_{fc}}\right)^{1.5}\right]\sqrt{\frac{F_{yw}t_{fc}}{t_{wc}}} \qquad\qquad [13.6.3]$$

$$P_{bf} = 0.75(135)(0.390)^2\left[1 + 3\left(\frac{0.505}{12.12}\right)\left(\frac{0.390}{0.605}\right)^{1.5}\right]\sqrt{\frac{50(0.605)}{0.390}} = 144 \text{ kips}$$

$$P_{bf} = \frac{\phi\, 4100t_{wc}^3\sqrt{F_{yw}}}{h} \qquad\qquad [13.6.8]$$

Note that in Eq. 13.6.8, h is the "clear distance between flanges less the fillet or corner radius for rolled shapes; distance between adjacent lines of fasteners or the clear distance between flanges when welds are used for built-up shapes." For rolled shapes h/t_w is a property given in the AISC Manuals. Using $h/t_w = 24.9$ from *LRFD Manual* p. 1-38 in Eq. 13.6.8 gives

$$P_{bf} = \frac{0.90(4100)(0.390)^2\sqrt{50}}{24.9} = 159 \text{ kips}$$

Since the factored load capacities P_{bf} without stiffeners for local web yielding ($P_{bf} = 138$ kips) and for web crippling ($P_{bf} = 144$ kips) are less than the 159 kips required, compression stiffeners are required under LRFD-K1.3 and LRFD-K1.4. The extra strength must be obtained from stiffeners.

Local web yielding (Eq. 13.6.10) indicates the need for compression stiffeners. Taking the lowest ($P_{bf} = 138$ kips) of the three strengths,

$$\text{Required } A_{st} = \frac{P_u - 138}{\phi F_{yst}} = \frac{159 - 138}{0.85(36)} = 0.69 \text{ sq in.}$$

Stiffeners to prevent local web yielding need not extend the full depth of the column web unless the force is applied at both column flanges; in this case, the stiffeners must extend full depth.

Stiffeners are also needed to prevent web crippling, and such stiffeners must extend full depth and be designed as columns the same as bearing stiffeners in a plate girder, as discussed in Sec. 11.12. The factored column load that must be carried is the full $P_{bf} = 159$ kips.

Size Limitations.   Using Eq. 13.6.12, the minimum stiffener width b_{st} is

$$\text{Min } b_{st} \geq \frac{b_{fb}}{3} - \frac{t_{wc}}{2} = \frac{6.995}{3} - \frac{0.390}{2} = 2.1 \text{ in.,} \quad \text{say 3 in.}$$

The minimum stiffener thickness t_s according to Eq. 13.6.13 is

$$\text{Min } t_s = \frac{t_{fb}}{2} = \frac{0.505}{2} = 0.252 \text{ in.,} \quad \text{say } \tfrac{5}{16} \text{ in.}$$

Two plates—$\tfrac{5}{16} \times 3$ give more than the minimum area $A_{st} = 0.69$ sq in. required.

Local Buckling Limits.   When elastic analysis was used,

$$\left(\lambda = \frac{b_{st}}{t_s} \right) \leq \left(\lambda_r = \frac{95}{\sqrt{F_y}} = \frac{95}{\sqrt{36}} = 15.8 \right)$$

$$\text{Min } t_s = \frac{3}{15.8} = 0.19 \text{ in.}$$

or, when plastic analysis was used,

$$\left(\lambda = \frac{b_{st}}{t_s} \right) \leq \left(\lambda_p = \frac{65}{\sqrt{F_y}} = \frac{65}{\sqrt{36}} = 10.8 \right)$$

$$\text{Min } t_s = \frac{3}{10.8} = 0.28 \text{ in.} \quad \text{say, } \textit{Use } \tfrac{5}{16} \text{ in.}$$

Column Strength of Stiffeners. The column consists of the stiffeners acting in combination with a length of web equal to $25t_{wc} = 25(0.390) = 9.75$ in. (see LRFD-K1.9 and text Sec. 11.8),

$$r = \sqrt{\frac{\frac{1}{12}(0.3125)(3 + 3 + 0.39)^3}{2(3)0.3125 + 9.75(0.390)}} = 1.1 \text{ in.}$$

Using an effective length KL equal to 0.75 of the column depth; i.e., approximately 9 in.,

$$\frac{KL}{r} = \frac{9}{1.1} \approx 8; \quad \text{Thus, } \phi_c F_{cr} = 30.5 \text{ ksi}$$

$$A_g = 2(3)0.3125 + 9.75(0.390) = 5.68 \text{ sq in.}$$

$$\phi_c P_n = A_g(\phi_c F_{cr}) = 5.68(30.5) = 173 \text{ kips} > 159 \text{ kips} \quad \text{OK}$$

(b) Tension region. To prevent excessive bending of a column flange from the action of a beam flange tension force, the minimum column flange thickness required by LRFD-K1.2 is given by Eq. 13.6.18,

$$\text{Required } t_{fc} \geq 0.4\sqrt{\frac{P_{bf}}{\phi F_{yf}}} = 0.4\sqrt{\frac{159}{0.90(50)}} = 0.75 \text{ in.}$$

Since the flange thickness of the W12×65 ($t_f = 0.605$ in.) is less than the required $t_{fc} = 0.75$ in., tension stiffeners are required. The local buckling limits do not apply to tension elements; however, for practical purposes the same stiffeners used as compression stiffeners should be used here.

Use 2 PLs—$\frac{5}{16}$×3 × $10\frac{7}{8}$, A36, for both compression and tension sides.

(c) Connection of plates to column. The forces to be used for the design of welding are shown in Fig. 13.6.8. When beams frame from both sides and contribute equal flange forces P_{bf}, the welds on the ends of the plates must carry the portion of the force P_{bf} which is not taken directly by the web.

For this example, Fig. 13.6.8a applies. The weld should be designed to carry the maximum stiffener force $A_{st}F_{yst}$; actually, when the stiffeners act in combination with the web, each part will try to carry load in proportion to the stiffness. Thus, the stiffeners will likely carry more than just the excess that the web alone could not carry. Maximum effective weld size along ends of stiffener plates in tension, using SMAW with E70 electrodes, is

$$a_{\max \text{ eff}} = \frac{0.90F_{yst}t_s}{2(0.707)0.75(0.60F_{\text{EXX}})} = 1.414\frac{F_{yst}t_s}{F_{\text{EXX}}} = 1.414\frac{36t_s}{70} = 0.73t_s$$

Using $\frac{5}{16}$-in. stiffener plates,

$$a_{\max \text{ eff}} = 0.73t_s = 0.73(0.3125) = 0.23 \text{ in}$$
$$\text{Min weld size } a = \tfrac{1}{4} \text{ in.}$$

The maximum factored force to be carried by the stiffeners is

$$\text{Force} = \phi_c A_{st}F_{yst} = 0.85(2)(3)(0.3125)36 = 57 \text{ kips}$$

For fillet welds top and bottom of plates, the strength required per inch is

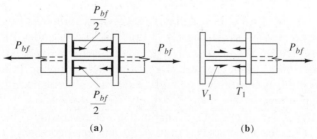

Figure 13.6.8 Weld requirements for horizontal plates.

$$\text{Required } \phi R_{nw} = \frac{57}{2(6)} = 4.8 \text{ kips/in.}$$

From Table 5.14.2, $\frac{1}{4}$-in. weld provides $\phi R_{nw} = 5.57 \, (0.23/0.25) = 5.1$ kips/in. Use $\frac{1}{4}$-in. fillet weld, top and bottom, on both tension and compression plates where they bear against the column flanges. Along the column web, fillet weld is required only on one side of the plate.

When a beam frames in from one side only, as in Fig. 13.6.8b, the weld forces T_1 are designed as in the symmetrical case. Additionally, however, shear forces V_1 must be developed in accordance with LRFD-K1.7 to take the proportion of the unbalanced flange force which comes to the stiffener plates; in this case $V_1 = T_1$. ■■

Vertical Plate and Tee Stiffeners

Sometimes it may be desirable to use vertical plates, or structural tee sections as shown in Fig. 13.6.1c. Particularly, they may be useful in a four-way system where beams are attached to the tee sections. Research [7.28] indicates that a vertical stiffener at the toe of the flange is only one-half as effective as is the web of the column.

Thus, assuming two vertical stiffeners (one at each flange toe) each having one-half the web strength indicated by Eq. 13.6.1, gives the following for the maximum factored load P_{bf} that can be transmitted based on local web yielding:

$$P_{bf} = \phi F_{yw}(t_{fb} + 5k)t_{wc} + 2\phi\left(\frac{F_{yst}}{2}\right)(t_{fb} + 5k)t_s \qquad (13.6.20)$$

which upon solving for the required stiffener thickness t_s gives

$$t_s \geq \frac{P_{bf}}{\phi(t_{fb} + 5k)F_{yst}} - t_{wc}\left(\frac{F_{yw}}{F_{yst}}\right) \qquad (13.6.21)$$

where P_{bf} = factored concentrated load from beam flange
t_{fb} = flange thickness of beam transmitting force
t_{wc} = column web thickness
k = distance from outer face of column flange to web toe of fillet
F_{yst} = yield stress of stiffener
F_{yw} = yield stress of column web

If the vertical stiffener is a plate, then compression buckling of the column web must be prevented by satisfying Eq. 13.6.8 or 13.6.9, under LRFD-K1.6 or ASD-K1.6, respectively. When structural tees are used, the web attachment precludes compression buckling of the web. Regarding web crippling (LRFD-K1.4 or ASD-K1.4), it is unlikely this limit state will control when vertical stiffeners are used at the column flange tips.

For the design of the tee stiffener and connection when a beam is attached to it, as in Fig. 13.6.3c, some special considerations are necessary. The beam flange, if equal in width to the tee, acts on the tee as shown in Fig. 13.6.9a. For analysis, one might consider uniform loading on a two-span beam as shown in Fig. 13.6.9b, wherein

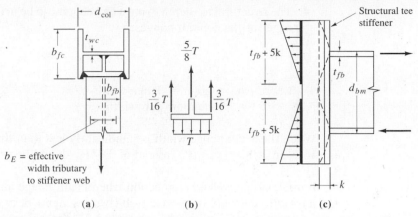

Figure 13.6.9 Structural tee stiffener.

the load would be transmitted $\frac{5}{8}$ into the tee web and $\frac{3}{16}$ each to the column flanges. Blodgett [13.2] suggests when the beam flange extends full width of the tee, one may assume the effective flange width b_E (Fig. 13.6.9a) tributary to the tee web to be $\frac{3}{4}$ of the beam flange width.

One may summarize the design requirements when the *beam flange width is approximately the same as the tee flange width*:

1. The tee web thickness t_w must satisfy Eq. 13.6.1:

$$t_w \geq \frac{0.75 P_{bf}}{(t_{fb} + 5k) F_{yst}} \qquad (13.6.22)$$

   where $0.75 P_{bf}$ = the portion of the factored load transmitted by the beam flange tributary to the tee stiffener

   k = distance to root of fillet; tee section

   t_{fb} = flange thickness of beam

2. The structural tee flange thickness t_s must be able to carry the beam tensile flange force without excessive deformation; hence Eq. 13.6.18 (or 13.6.19) should be satisfied. This will be conservative since the equation was derived for the free-edge condition at the flange toes, whereas here there is a welded connection. Using $0.75 P_{bf}$ as in 1,

$$t_s \geq 0.4 \sqrt{\frac{0.75 P_{bf}}{\phi F_{yst}}} = 0.35 \sqrt{\frac{P_{bf}}{\phi F_{yst}}} \qquad (13.6.23)$$

3. The structural tee flange width b_s must extend fully between the column flanges:

$$b_s = d_{col} - 2t_{fc} \qquad (13.6.24)$$

   where d_{col} = column overall depth

   t_{fc} = flange thickness of column

4. The structural tee depth d_s must be adequate to be nearly flush with the outer edges of the column flanges:

$$d_s = \frac{b_{fc} - t_{wc}}{2} \qquad (13.6.25)$$

where b_{fc} = column flange width
t_{wc} = column web width

When the beam flange width is significantly less than the tee flange width (say, more than an inch or two), P_{bf} instead of $0.75P_{bf}$ should be used in Eqs. 13.6.22 and 13.6.23.

In making the welded connection when beam flange and tee flange are nearly equal in width, the weld on the tee web (two segments of two fillets) is to resist the moment assuming $0.75P_{bf}$ must be carried [13.2]. At the flange tips of the tee, Blodgett suggests [13.2, p. 5.7–13] to design for $\frac{1}{3}$ of the beam flange force (somewhat greater than the $\frac{3}{16}$ of Fig. 13.6.9b) to be carried.

EXAMPLE 13.6.2

Design a vertical tee stiffener connection to frame a W14×61 beam into the web of a W12×65 column. Use A572 Grade 50 steel. Use the type of connection shown in Fig. 13.6.9, and Load and Resistance Factor Design.

Solution.   Since the beam flange width (9.995 in.) is approximately the same as the clear distance between column flanges (12.12–1.21 = 10.91 in.), Eqs. 13.6.22 and 13.6.23 may be applied.

(a) Determine the stiffener web thickness t_w required to prevent local web yielding (LRFD-K1.2). The maximum *factored* beam flange force P_{bf} that can be carried is

$$P_{bf} = \phi A_f F_y = 0.90(9.995)(0.645)50 = 290 \text{ kips}$$

Using Eq. 13.6.22, and estimating $k = 1$ in. for the tee section,

$$\text{Required } t_w = \frac{0.75P_{bf}}{(t_{fb} + 5k)F_{yst}} = \frac{0.75(290)}{[0.645 + 5(1.0)]50} = 0.77 \text{ in.}$$

(b) Determine stiffener flange thickness to prevent distortion under tension. Using Eq. 13.6.23,

$$\text{Required } t_s = 0.35 \sqrt{\frac{P_{bf}}{\phi F_{yst}}} = 0.35 \sqrt{\frac{290}{0.90(50)}} = 0.89 \text{ in.}$$

(c) Select a structural tee section. The maximum flange width permitted is

$$\text{Max } b_s = d_{col} - 2t_{fc} = 12.12 - 2(0.605) = 10.91 \text{ in.}$$
$$\text{Max depth} = 0.5(12.00 - 0.390) = 5.81 \text{ in.}$$

Try W12×96 cut into tees:

$$t_s = 0.900 \text{ in.} \quad \text{(flange thickness)}$$
$$t_w = 0.550 \text{ in. with } k = 1\tfrac{5}{8} \text{ in.}$$

Rechecking Eq. 13.6.22,

$$\text{Required } t_w = \frac{0.75(290)}{[0.645 + 5(1.625)]50} = 0.50 \text{ in.} < \text{provided } t_w \quad \text{OK}$$

Try W12×96 cut as shown in Fig. 13.6.10a.

(d) Welding on stiffener web. Referring to Fig. 13.6.9c, the length of tee stiffener required is

$$\text{Length} = d_{bm} + 5k = 13.89 + 5(1.625) = 22.02 \text{ in.}$$

Try length of tee = 2'-0".

Length of weld, upper, and lower ends:

$$t_{fb} + 5k = 0.645 + 5(1.625) = 8.77 \text{ in.}$$

Try 9-in. weld at each end, connecting web of tee to web of column.

Assume the moment contributed by the center $\frac{3}{4}$ of the flange is tributary to the tee web,

$$\phi M_n = 0.75\phi F_y Z_x = [0.75(0.90)(50)102]\tfrac{1}{12} = 287 \text{ ft-kips}$$

where Z_x is the plastic modulus for the W14×61 beam, which is "compact" for $F_y = 50$ ksi. The strength of the welding is conservatively treated with the elastic "vector" analysis using the section modulus S of the two 9-in. weld segments treated as lines,

$$S = 2\left(\frac{1}{12}\right)\left[\frac{(24)^3 - (24 - 18)^3}{12}\right] = 189 \text{ in.}^2$$

The factored load R_u at the top of the weld is

$$R_u = \frac{\phi M_n}{S} = \frac{287(12)}{189} = 18.2 \text{ kips/in.}$$

The weld size a required is

$$a = \frac{R_u}{\phi(0.707)(0.60F_{EXX})} = \frac{18.2}{0.75(0.707)(0.60)70} = 0.82 \text{ in.}$$

Check flexural strength of stiffener web as a rectangular section of width $t_w = 0.55$ in. and depth 24 in.,

$$\phi M_p = \phi F_y Z = 0.90(50)\left(\frac{0.55(24)^2}{4}\right)\frac{1}{12} = 297 \text{ ft-kips} > 287 \quad \text{OK}$$

Though the elastic "vector" analysis is a conservative treatment for welds, nevertheless the weld force to be transmitted to base material should not exceed the base material strength. At the top of the weld lines, the maximum effective weld size is

$$\begin{array}{c} \text{Tension strength} \\ \text{of stiffener web} \end{array} = \text{Strength of two fillets}$$

$$t_w \phi_t F_y = 2[\phi(0.707a)0.60F_{EXX}]$$

$$0.55(0.90)50 = 2(0.75)(0.707a)(0.60)70$$

$$24.75 = 44.54a; \quad a_{\text{max eff}} = 0.56 \text{ in.}$$

The use of $\frac{13}{16}$-in. welds on the 0.55-in. web is not acceptable; therefore, increase the section to W12×120 from which to cut the tee. The web is still not quite thick enough; however, this seems acceptable because of the uncertainty in the proportion of the beam moment transmitted through the tee web and that transmitted between flanges of the tee and the column, as well as the conservatism in the weld resistance calculation.

Use W12×120 as shown in Fig. 13.6.10b.

(e) Welding on stiffener flanges. Assume conservatively that the tee connection to the column flange (Figs. 13.6.9a and b) may have to carry as much as $\frac{1}{3}$ of the beam flange force. The concentrated forces from the beam flange may be considered as distributed along the column flange over the distance $t_{fb} + 5t_s$, as shown in Fig. 13.6.11,

Assuming the factored force T_u attributable to the groove weld along one flange is

$$T_u = \phi F_y t_{fb}\left(\frac{b}{3}\right) = 0.90(50)(0.645)\left(\frac{9.995}{3}\right) = 97 \text{ kips}$$

$$t_{fb} + 5t_s = 0.645 + 5(1.105) = 6.2 \text{ in.}$$

Factored weld force $R_u = \dfrac{97}{6.2} = 15.6$ kips/in.

Using a partial joint penetration U-groove weld with E70 electrode material, the design strength ϕR_{nw} is

$$\phi R_{nw} = 0.80(0.60 F_{EXX}) = 0.80(0.60)70 = 33 \text{ kips/in.}$$

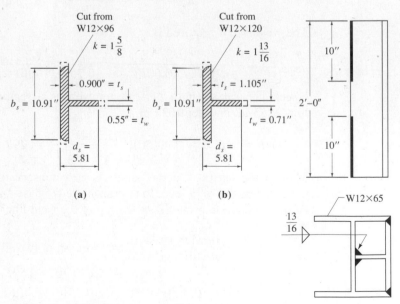

Figure 13.6.10 Example 13.6.2. Tee selection and welding to web.

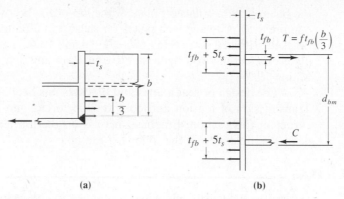

Figure 13.6.11 Forces carried by weld along tee stiffener flanges.

for tension normal to effective area (LRFD-Table J2.5). The required effective throat dimension is

$$\text{Required effective throat} = \frac{15.6}{33.6} = 0.46 \text{ in.}$$

According to LRFD-Table J2.1, the effective throat for a partial joint penetration U-groove is the "depth of chamfer." The minimum effective throat (LRFD-Table J2.3) is $\frac{5}{16}$ in. for the stiffener flange of 1.105 in.

Use $\frac{11}{16}$-in. single-U-groove (AWS-2.10, BC-P6 prequalified joint; also *LRFD Manual* p. 8-151) weld along the edges of the flange of the tee.

   (f) Effect of beam shear force. Ordinarily the length of weld used is so large that the additional capacity required to carry end shear is negligible. A W14×61 might be expected to carry something on the order of 70 kips factored load shear, in which case it is resisted by

$$\text{Total length of weld} = \underset{\text{web}}{4(10)} + \underset{\text{flanges}}{2(24)} = 88 \text{ in.}$$

$$\text{Required extra weld strength} = \frac{70}{88} = 0.80 \text{ kips/in.}$$

$$\text{Added size for web fillets} = \frac{0.80}{0.75(0.707)42} = 0.04 \text{ in.}$$

$$\text{Added size for flange groove welds} = \frac{0.80}{33.6} = 0.03 \text{ in.}$$

   The shear component and the flexure component actually act at 90 degrees to one another so that adding the requirements algebraically would be overly safe. The authors consider the weld design adequate without an increase for direct shear. ■■

Top Tension Plates

When the beam is connected to the column flange and the column is stiffened by vertical or horizontal plate stiffeners, or when the beam is connected to the column web through a vertical tee stiffener, a simple means of transmitting the moment from the beam is with a tension plate at the top of the beam combined with (1) a bottom compression plate and web plates for shear; (2) a bottom seat angle; or (3) a bottom bracket (stiffened seat). See Figs. 13.6.1d and 13.6.2a and b. The behavior of such top plate connections has been studied by Pray and Jensen [13.63] and Brandes and Mains [13.64].

The design of seat angles and brackets has been treated in Secs. 13.3 and 13.4 Transmission of tension and compression forces into the column has been treated earlier in this section. Emphasis here is on the top plate, a tension member whose design is illustrated in the following example.

EXAMPLE 13.6.3

Design the top tension plate and its connection by welding or by A325 high-strength bolts for transferring the full end moment from a W14×61 to a column. Use A572 Grade 50 steel. Assume the connection to be of the type shown in Figs. 13.6.1d or 13.6.2a.

Solution

(a) Design plate as a tension member.

Since the W14×61 is a "compact section" for Grade 50 steel, the factored flange force T_u is approximately

$$T_u = \frac{\phi M_n}{d_{bm}} = \frac{\phi F_y Z_x}{d_{bm}} = \frac{0.90(50)102}{\approx 14} = 328 \text{ kips}$$

When there are no holes, the yielding limit state for tension members will control over the fracture on net section limit state. Thus, using Eq. 3.9.2, the required gross plate are A_g is

$$A_g = \frac{T_u}{\phi_t F_y} = \frac{328}{0.90(50)} = 7.3 \text{ sq in.}$$

The plate width must be *less* than the W14×61 flange width of 10.0 in. *Use* PL—$\frac{7}{8}$×9, A_g = 7.9 sq in., for welded construction.

(b) Determine welding for plate to beam flange. Try $\frac{3}{8}$-in. fillet weld with E70 electrodes using the shielded metal arc process,

$$\phi R_{nw} = \phi(0.707a)(0.60 F_{EXX}) = 0.75(0.707a)(0.60)70 = 22.3a$$
$$= 22.3(0.375) = 8.35 \text{ kips/in.}$$

The required length L_w of weld is

$$L_w = \frac{T_u}{\phi R_{nw}} = \frac{328}{8.35} = 39 \text{ in.}$$

To reduce the length of weld, use $\frac{1}{2}$-in. weld. This will require $L_w = 29.4$ in. *Use* $\frac{1}{2}$-in. weld, 9 in. on end and 11 in. on each side for a total of 31 in. The design is summarized in Fig. 13.6.12a.

The plate is connected to the column flange by a full penetration single bevel groove weld, using a backup bar. When a seat is used it will serve as

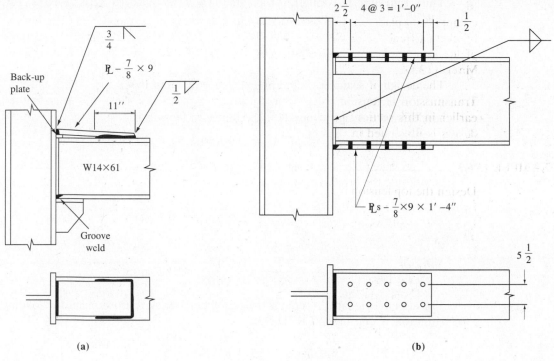

Figure 13.6.12 Example 13.6.3. Moment connection using top plate.

backup to make the flange groove weld. Otherwise, a backup plate is needed below the compression flange. Frequently it may be satisfactory to use fillet welds along the top and bottom of the plate to make the connection to the column flange.

(c) Design bolted connection for plate to beam flange. Use $\frac{7}{8}$-in.-diam A325 bolts in a bearing-type connection with threads excluded from the shear plane (A325-X).

$$\phi R_n = \phi(0.50F_u^b)mA_b = 0.75(0.50)1(0.6013) = 27.1 \text{ kips}$$
(single shear)

$$\phi R_n = \phi(2.4F_ud_bt) = 0.75(2.4)(65)(0.875)0.645 = 66.0 \text{ kips}$$
(bearing on 0.645-in. beam flange)

The use of $2.4F_u$ assumes $s \geq 3d_b$ and $L_e \geq 1.5d_b$.

When bolts are used, the holes in the W14×61 flange may indicate a reduction in strength according to LRFD-B10, as discussed in Sec. 7.9. No reduction is to be made when LRFD Formula (B10-1) is satisfied,

$$0.75F_uA_{fn} \geq 0.9F_yA_{fg}$$

where A_{fn} = net flange area
        A_{fg} = gross flange area

$$0.75(65)[9.995(0.645) - 2(1)0.645] \geq 0.9(50)[9.995(0.645)]$$
$$[0.75(65)5.16 = 252 \text{ kips}] < [0.9(50)6.45 = 290 \text{ kips}]$$

Thus, reduction in strength is indicated. The effective area A_{fe} is given by Eq. 7.9.2,

$$A_{fe} = \frac{5}{6}\frac{F_u}{F_y}A_{fn} \qquad\qquad [7.9.2]$$

$$= \frac{5}{6}\left(\frac{65}{50}\right)5.16 = 5.59 \text{ sq in.}$$

The effective design strength of the flange in tension is

$$\phi T_n = \phi F_y A_{fe} = 0.90(50)5.59 = 252 \text{ kips}$$

For the plate as a tension member, the effect of holes must be deducted in computing the effective net area A_e,

$$A_g = \frac{T_u}{\phi_t F_y} = \frac{252}{0.90(50)} = 5.6 \text{ sq in.}$$

or

$$A_e = \frac{T_u}{\phi_t F_u} = \frac{252}{0.75(65)} = 5.2 \text{ sq in.}$$

The effective net area A_e equals the actual net area A_n in this concentric loading situation, according to LRFD-B3.

Try PL—1×9. Check A_n and LRFD-J5.2(b),

$$A_n = [9.00 - 2(1)]0.875 = 6.1 \text{ sq in.} > [A_e = 5.2 \text{ sq in.}] \quad \text{OK}$$
$$0.85A_g = 0.85(9.0)0.875 = 6.7 \text{ sq in.} > A_n \quad \text{OK}$$

$$\text{No. of bolts required} = \frac{252}{27.1} = 9.3; \quad \text{Say 10}$$

Use 10—$\frac{7}{8}$-in.-diam A325 bolts, as shown in Fig. 13.6.12b. ■■

Split-Beam Tee Connections—Prying Action

For bolted moment connections, the split-beam tee as shown in Fig. 13.6.2c is not often any longer used. However, the design of a connection involving the transfer of a tensile force through a thick-plate bolted connection, such as the tee connected to the flange, will illustrate treatment of "prying action." Prying action was first mentioned in Sec. 4.13.

Consider the deformation of a tee section, as in Fig. 13.6.13, where as the pull on the web deforms the flange and deflects it outward, the edges of the flange tips bear against the connected piece giving rise to the force Q, known as the "prying force." Inclusion of this force is required by LRFD-J3.6 wherein it states, "The applied load shall be the sum of the factored loads and any tension resulting from prying action produced by deformation of the connected parts." A similar statement appears in ASD-J3.4.

When bolted connections are subject to distortion such as in Fig. 13.6.13, the treatment of Sec. 4.13 for bolts in tension is not valid. When a thick-flanged tee distorts, the flange tips tend to dig in, giving rise to the force Q.

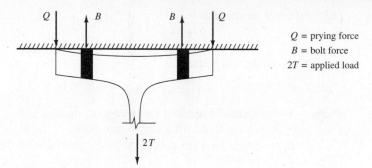

Q = prying force
B = bolt force
$2T$ = applied load

Figure 13.6.13 Prying action.

Kulak, Fisher, and Struik [3.1, pp. 274–282] have reviewed the various theories relating to prying action. They have recommended the procedure used in the *LRFD Manual* [1.18]. The model, formulation, and design procedure is well explained by Thornton [13.65, 13.67] and Astaneh [13.66], particularly with regard to Allowable Stress Design.

The analytical model used in the *LRFD Manual* [1.18] procedure is shown in Fig. 13.6.14. Moment equilibrium requires

$$M_1 + M_2 - T_b = 0 \qquad (13.6.26)$$

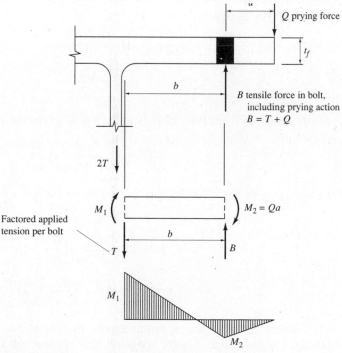

Q prying force

t_f

B tensile force in bolt, including prying action
$B = T + Q$

$2T$

M_1 $M_2 = Qa$

Factored applied tension per bolt

T b B

M_1

M_2

Figure 13.6.14 Analytical model for prying action.

The cantilever moment is

$$M_2 = Qa \tag{13.6.27}$$

and force equilibrium requires

$$T + Q - B = 0 \tag{13.6.28}$$

where T = externally applied factored load on one bolt
Q = prying force tributary to one bolt
B = tensile load in the bolt (to be compared with the design strength ϕR_n of the bolt)

Eqs. 13.6.26 through 13.6.28 are the independent equations of equilibrium that must be satisfied.

Next, referring to "width" as the dimension along the length of the tee section tributary to one bolt, let

$$\delta = \frac{\text{net width at bolt line}}{\text{gross width at critical section near web face}} \tag{13.6.29}$$

M_2 may be multiplied and divided by δM_1 without any effect, and the ratio $M_2/\delta M_1$ may be called α; thus

$$M_2 = M_2\left(\frac{\delta M_1}{\delta M_1}\right) = \left(\frac{M_2}{\delta M_1}\right)\delta M_1 = \alpha \delta M_1 \tag{13.6.30}$$

Substituting Eq. 13.6.30 for M_2 into Eqs. 13.6.26 and 13.6.27 gives the following two equations for M_1,

$$M_1 = \frac{Tb}{1 + \alpha\delta} \tag{13.6.31}$$

$$M_1 = \frac{Qa}{\alpha\delta} \tag{13.6.32}$$

Eliminating M_1 from Eqs. 13.6.31 and 13.6.32 gives the prying force Q as,

$$Q = T\left(\frac{\alpha\delta}{1 + \alpha\delta}\right)\left(\frac{b}{a}\right) \tag{13.6.33}$$

Substitution of Eq. 13.6.33 into the force equilibrium equation, Eq. 13.6.28, gives

$$T + Q - B = 0$$

$$T + T\left(\frac{\alpha\delta}{1 + \alpha\delta}\right)\left(\frac{b}{a}\right) = B$$

$$B = T\left[1 + \left(\frac{\alpha\delta}{1 + \alpha\delta}\right)\left(\frac{b}{a}\right)\right] \tag{13.6.34}$$

There are two design requirements that must be satisfied; (1) the moment strength of the flange must be adequate, and (2) the bolt strength in tension must be adequate.

Moment Strength of the Tee Flange   The LRFD requirement would be

$$\phi M_n \geq M_1 \tag{13.6.35}$$

$$\phi M_n = \phi M_p = \phi Z F_y = \phi \frac{wt_f^2}{4} F_y \tag{13.6.36}$$

where w is the length of flange, parallel to stem, tributary to one bolt, and ϕ is the resistance factor $\phi_b = 0.90$ for flexure. Putting Eqs. 13.6.36 and 13.6.31 into Eq. 13.6.35 gives the design requirement for the flange thickness t_f,

$$t_f \geq \sqrt{\frac{4Tb}{\phi_b w F_y (1 + \alpha\delta)}} \tag{13.6.37}$$

Tension Strength of Bolts   The LRFD requirement would be

$$\phi R_n \geq B \tag{13.6.38}$$

where B is the factored load force on one bolt, given by Eq. 13.6.34.

The design strength of bolts in tension is given by Eq. 4.7.4,

$$\phi R_n = \phi F_u^b (0.75 A_b) = 0.75 F_u^b (0.75 A_b) \tag{4.7.4}$$

or for A325 bolts having $F_u^b = 120$ ksi and A490 bolts having $F_u^b = 150$ ksi,

$$\phi R_n = 67.5 A_b \quad \text{(A325)} \tag{13.6.39}$$

$$\phi R_n = 84.4 A_b \quad \text{(A490)} \tag{13.6.40}$$

Thornton [13.65] has shown there is *no unique solution* for Eqs. 13.6.37 and 13.6.38. Thus, a trial procedure to obtain one satisfactory solution is necessary. One may note that when $\alpha = 0$ there will be no prying action and single curvature bending; and when $\alpha = 1.0$ there will be maximum prying action and double curvature bending.

The aforementioned model and development has been calibrated by tests [3.1] indicating the b dimension in Fig. 13.6.14 should be larger and the a dimension smaller by an amount equal to one-half the bolt diameter d_b. Thus, a' and b' are used in place of a and b in *all* of the foregoing equations,

$$a' = a + \frac{d_b}{2} \tag{13.6.41}$$

$$b' = b - \frac{d_b}{2} \tag{13.6.42}$$

Load and Resistance Factor Design   The flange thickness t_f required from modified (using b' for b) Eq. 13.6.36 is

$$t_f \geq \sqrt{\frac{4Tb'}{\phi_b w F_y (1 + \alpha\delta)}} \tag{13.6.43}$$

and using $\phi_b = 0.90$ for flexure, Eq. 13.6.43 becomes

$$t_f \geq \sqrt{\frac{4.44 Tb'}{w F_y (1 + \alpha\delta)}} \tag{13.6.44}$$

where T = factored externally applied tension per bolt

$b' = b - d/2$

d_b = bolt diameter

w = tributary width of resisting section (i.e., length of tee section) tributary to one bolt

$\alpha = (0 \leq \alpha \leq 1.0) = M_2/\delta M_1$

δ = ratio of net area at bolt line (where M_2 acts) to gross area where M_1 acts = $(w - d_b)/w$

The bolt design strength required according to Eq. 13.6.38 is

$$\phi R_n \geq T\left[1 + \frac{\alpha\delta}{1 + \alpha\delta}\left(\frac{b'}{a'}\right)\right] \tag{13.6.45}$$

where $a' = a + d_b/2$.

Allowable Stress Design   Equations similar to those for LRFD are used in ASD. The reader is referred to Thornton [13.65] and Astaneh [13.66] for details. The trial procedure is illustrated for LRFD in the following example.

EXAMPLE 13.6.4 ━━

Design a split-beam tee connection, such as in Fig. 13.6.2c, to enable a plastic hinge to develop in a W14×61 beam framing to the flange of a W14×159 column. Use A572 Grade 50 steel with $\frac{3}{4}$-in. diam A325 bolts in a bearing-type connection (A325-N). Use Load and Resistance Factor Design.

Solution

(a) Compute the factored tensile force to be carried. The W14×61 is compact (i.e., $\lambda \leq \lambda_p$ for local flange buckling and local web buckling); in addition, assume that lateral-torsional buckling is precluded (i.e., $L_b \leq L_p$).

Since there are flange holes in the W14×61 beam for this bolted connection, the nominal moment strength M_n of the beam may be less than M_p. Check in accordance with LRFD-B10. As discussed in Sec. 7.9, in order to use full gross properties, Eq. 7.9.1 must be satisfied,

$$0.75F_u A_{fn} \geq 0.9 F_y A_{fg} \tag{[7.9.1]}$$

$A_{fn} = [9.995 - 2(0.75 + 0.125)]0.645 = 5.32$ sq in.

$A_{fg} = (9.995)0.645 = 6.45$ sq in.

$[0.75(65)5.32 = 259$ kips$] < [0.9(50)6.45 = 290$ kips$]$

This means a reduced effective flange area A_{fe} must be used, as follows:

$$A_{fe} = \frac{5}{6}\frac{F_u}{F_y}A_{fn} = \frac{5}{6}\left(\frac{65}{50}\right)5.32 = 5.76 \text{ sq in.}$$

Only the tension flange needs to be reduced; however, it will be practical to use A_{fe} as the area of each flange. Because of the shift in neutral axis if the deduction is from only one flange, the difference in deducting from both flanges compared to deducting from only one flange is small.

Compute the reduced plastic modulus Z_x,

$$Z_x = 2A_{fe}\left(\frac{d - t_f}{2}\right) + t_w\left(\frac{d}{2} - t_f\right)^2$$

$$Z_x = 2(5.76)\left(\frac{13.89 - 0.645}{2}\right) + 0.375\left(\frac{13.89}{2} - 0.645\right)^2 = 91.2 \text{ in.}^3$$

$$M_u = \phi_b M_n = \phi_b Z_x F_y = [0.90(91.2)50]\tfrac{1}{12} = 342 \text{ ft-kips}$$

If all bending moment is carried by the tees, the force of the internal couple is

$$\text{Force} = \frac{M_u}{d_{bm}} = \frac{342(12)}{13.89} = 295 \text{ kips}$$

(b) Check whether or not the tensile force can be accommodated by the bolts in tension.

$$\phi R_n = \phi F_u^b(0.75A_b) = 0.75F_u^b(0.75A_b) \qquad [4.7.4]$$
$$\phi R_n = 67.5A_b = 67.5(0.4418) = 29.8 \text{ kips}$$

Only 8 bolts will fit, as shown in Fig. 13.6.15; therefore the maximum factored tensile force that may be carried is

$$T_u = 8(29.8) = 238 \text{ kips} < 295 \text{ kips} \quad \text{NG}$$

When this difficulty arises, one may use a stub beam or a tee stub attached to the bottom of the main beam (Fig. 13.6.15) to increase the moment arm of the couple. Actually, when designing for the support moment, one might have used a beam size required for the midspan moment and then used the stub beam to gain the increased capacity required at the support. The necessary moment arm is

$$\text{Required arm} = \frac{342(12)}{238} = 17.2 \text{ in.}$$

$$\text{Extra depth required} = 17.2 - 13.89 = 3.3 \text{ in.}$$

Try as stub beam a WT5×24.5, $t_w = 0.340$ in., $t_f = 0.560$ in., $b_f = 10.000$ in., $d = 4.990$ in., whose dimensions are comparable to the main W14×61 beam.

$$\text{Force of couple} = \frac{342(12)}{13.89 + 4.990} = 217 \text{ kips}$$

Using 8 bolts in tension,

$$\left(R_u = \frac{217}{8} = 27.2 \text{ kips}\right) < (\phi R_n = 29.8 \text{ kips}) \quad \text{OK}$$

(c) Check shear strength on web (section $c-c$ of Fig. 13.6.15) of WT5×24.5. Applying LRFD-F2.2, the length L of tee required is

$$\text{Required } L = \frac{\text{Force}}{\phi(0.6F_{yw})t_w} = \frac{217}{0.90(0.6)(50)0.340} = 23.6 \text{ in.}$$

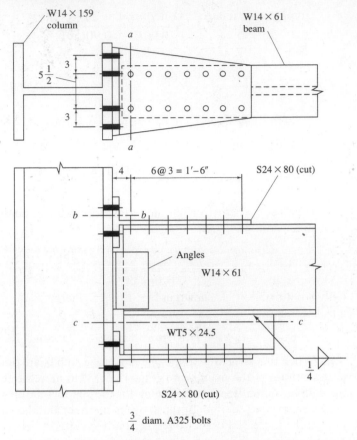

Figure 13.6.15   Example 13.6.4. Split-beam tee connection with tee stub.

(d) Determine bolts required to transmit tension and compression forces at the top and bottom of beam.

$$\phi R_n = 0.4418(0.75)48 = 15.9 \text{ kips} \quad \text{(controls)}$$
$$\text{(single shear } LRFD \text{ Manual, Table 8–11)}$$

$$\phi R_n = 0.75(2.4F_u d_b t) = 0.75(2.4)(65)(0.75)t = 87.8t$$
$$\text{(bearing assuming } s \geq 3d_b \text{ and } L_e \geq 1.5d_b)$$

$$\text{Number of bolts} = \frac{217}{15.9} = 13.6, \quad \text{use } 14$$

The minimum length of WT5×24.5 required using 7 bolts per line at 3-in. spacing is 21 in.: _Use_ WT5×24.5 stub tee, 2′-0″ long, welded to the bottom of W14×61, as shown in Fig. 13.6.15.

(e) Determine thickness required to transmit tension on section *a–a* (Fig. 13.6.15):

$$\text{Required } A_g = \frac{T_u}{0.90F_y} = \frac{217}{0.90(50)} = 4.8 \text{ sq in.}$$

and

$$\text{Required } A_n = \frac{T_u}{0.75F_u} = \frac{217}{0.75(65)} = 4.5 \text{ sq in.}$$

Using the length of section $a-a$ as 13 in. (column flange width = 15.565 in.), and deducting two holes, gives

$$t \geq \frac{4.5}{13 - 2(0.875)} = 0.40 \text{ in.}$$

$$t \geq \frac{15.9}{87.8} = 0.18 \text{ in.} \quad \text{(bearing does not control)}$$

(f) Determine the flange thickness for the tee section attached to the column. Equation 13.6.44 has been developed to provide flexural strength on section $b-b$ of Fig. 13.6.15. Estimate the adjusted distance b'. Estimate the usual gage g as about 4 in.; thus,

$$b' = b - \frac{d_b}{2} = \frac{g}{2} - \frac{t_w}{2} - \frac{d_b}{2} = \frac{4}{2} - \frac{t_w}{2} - \frac{3}{8} \approx \text{say } 1.25 \text{ in.}$$

Assuming *no prying action* $Q = 0$; thus, $\alpha = 0$. Then, from Eq. 13.6.44, assuming a length w of 14 in. at the critical section (section $b-b$) of the tee,

$$t_f \geq \sqrt{\frac{4.44Tb'}{wF_y(1 + \alpha\delta)}} = \sqrt{\frac{4.44(217/2)1.25}{14(50)(1 + 0)}} = 0.91 \text{ in.}$$

As an alternative, Thornton [13.65] recommends computing β, which from Eq. 13.6.34 (again using a' and b' instead of a and b) is a function of $\alpha\delta$,

$$\beta = \left(\frac{B}{T} - 1\right)\frac{a'}{b'} = \left(\frac{29.8}{25.9} - 1\right)(\approx 1.25) = 0.2$$

Then use α as follows:

if $\beta \geq 1$, use $\alpha = 1$ (meaning large prying force)

if $\beta < 1$, use $\alpha = $ lesser of $\dfrac{1}{\delta}\left(\dfrac{\beta}{1 - \beta}\right)$ and 1.0

In this example, δ may be estimated using Eq. 13.6.29 as

$$\delta = \frac{3 - (0.75 + 1/16)}{3} = 0.73$$

$$\frac{1}{\delta}\left(\frac{\beta}{1 - \beta}\right) = \frac{1}{0.73}\left(\frac{0.2}{1 - 0.2}\right) = 0.34 < 1; \quad \text{use } \alpha = 0.34$$

In which case, Thornton's procedure [13.65] would give $\alpha\delta = 0.34(0.73) = 0.25$ for use in Eq. 13.6.44 for t_f,

$$t_f \geq \sqrt{\frac{4.44Tb'}{wF_y(1 + \alpha\delta)}} = \sqrt{\frac{4.44(217/2)1.25}{14(50)(1 + 0.25)}} = 0.81 \text{ in.}$$

Try a tee cut from S24×80, $t_f = 0.870$ in., $t_w = 0.500$ in.

(g) Check the prying force using Eq. 13.6.33 with a' and b' instead of a and b,

$$a' = a + \frac{d_b}{2} = \frac{b_f - g}{2} + \frac{d_b}{2} = \frac{7.000 - 4}{2} + \frac{3}{8} = 1.875 \text{ in.}$$

$$b' = b - \frac{d_b}{2} = \frac{g}{2} - \frac{t_w}{2} - \frac{d_b}{2} = \frac{4}{2} - \frac{0.500}{2} - \frac{3}{8} = 1.375 \text{ in.}$$

Kulak, Fisher, and Struik [3.1] recommended that $a \leq 1.25b$. In this example, $a = 1.5$ in. and $b = 1.75$ in.; thus, $a < 1.25b$. Taking the length of the tee section as 14 in. (see top view of Fig. 13.6.15) with four holes deducted,

$$\delta = \frac{14 - 4(0.75 + 1/16)}{14} = 0.77$$

Using the same trial value $\alpha = 0.34$ as previously gives $\alpha\delta = 0.26$. Then find Q,

$$Q = T\left(\frac{\alpha\delta}{1 + \alpha\delta}\right)\left(\frac{b'}{a'}\right) = T\left(\frac{0.26}{1 + 0.26}\right)\frac{1.375}{1.875} = 0.15T$$

Then, compare $(T + Q)$ per bolt to ϕR_n,

$$T + Q = 1.15T$$
$$1.15R_u = 1.15(25.9) = 29.8 \text{ kips} = \phi R_n \quad \text{OK}$$

Since the factored load R_u per bolt, increased by the prying force, exactly equals the design tension strength ϕR_n of a bolt, the bolts are satisfactory. Thornton [13.65] has shown that when the actual t_f exceeds the required t_f, the actual Q will be less than the Q computed above. More detail on this procedure is provided in Ref. 13.65 and in the *LRFD Manual*.

(h) Recheck the thickness t_f required.

$$t_f \geq \sqrt{\frac{4.44Tb'}{wF_y(1 + \alpha\delta)}} = \sqrt{\frac{4.44(217/2)1.375}{14(50)(1 + 0.26)}} = 0.85 \text{ in.}$$

The flange thickness t_f provided is 0.870 in. The design is satisfactory.
Use tees cut from S24×80 to carry tensile and compressive forces.

Not discussed in this example is the development of the shear strength of the W14×61. A pair of angles may be attached to the beam web for the purpose of providing whatever shear is required. The final design is shown in Fig. 13.6.15. ■■

End-Plate Connections

The practical alternative to the split-beam tee connection is the end-plate moment connection, as shown in Fig. 13.6.2d. Having much simpler fabrication details, a single plate welded on the end of a beam has become relatively common.

An excellent summary of the history of the use of end-plate connections is given by Griffiths [13.68]. The behavior of end-plate moment connections has been studied by Beedle and Christopher [13.49], Onderdonk, Lathrop, and Coel [13.48], Douty and McGuire [13.50], Agerskov [13.69, 13.70], Krishnamurthy [13.71], Krishnamurthy, Huang, Jeffrey, and Avery [13.72], Mann and Morris [13.73], Murray [13.74], Hendrick and Murray [13.75], Yee and Melchers [13.76], Murray and Kukreti [13.77], and Curtis and Murray [13.78], among others.

A conservative approach to end-plate connection design is to use the prying action concept discussed previously in this section. The region near the tension flange of the beam is designed similarly to the split-beam tee connection. This fastener group is designed for shear and tension, including any effect of prying action.

Krishnamurthy [13.71] has recommended a modified split-beam tee method based on analytical study and correlation with available tests. This method involves equations containing statistical coefficients and is suitable only with adequate design aids. Murray et al. [13.74, 13.77] has proposed procedures for 4-bolt and 8-bolt stiffened end plates.

EXAMPLE 13.6.5 _____

Design an end-plate connection for a W14×53 beam attachment to a W14×176 column, both of A36 steel. Design for maximum factored beam moment and 60 kips factored shear. Use A325 bolts in a bearing-type connection (A325-X). Use Load and Resistance Factor Design.

Solution

(a) Determine the number of bolts required to carry the maximum factored tensile force T_u from the bending moment. The design moment strength $\phi_b M_n$ of the beam is

$$\phi_b M_n = \phi_b M_p = \phi_b Z_x F_y = 0.90(87.1)(36)\tfrac{1}{12} = 235 \text{ ft-kips}$$

$$\text{Max } T_u = \frac{\phi_b M_n}{d - t_{fb}} = \frac{235(12)}{13.92 - 0.660} = 213 \text{ kips}$$

For $\tfrac{7}{8}$-in.-diam bolts (A325-X), the design strength in tension is

$$\phi R_n = \phi(0.75 A_b)F_u^b = 0.75(90.0)A_b = 0.75(90)0.6013 = 40.6 \text{ kips}$$

$$\text{Number of bolts} = \frac{T_u}{\phi R_n} = \frac{213}{40.6} = 5.2$$

For symmetrical placement above and below the tension flange, either 4 or 8 bolts would be needed. Four 1-in.-diam bolts will carry 4(53.0) = 212 kips; accept this as satisfactory.

(b) Establish the plan dimensions of the end plate. For determining the distance s (Fig. 13.6.16) the fillet weld size (for E70 electrodes) and the bolt installation clearance are needed.

$$\text{Required } \phi R_{nw} = \frac{T_u}{L_w} = \frac{T_u}{2b_f - t_w} = \frac{213}{2(8.060) - 0.370} = 13.5 \text{ kips/in.}$$

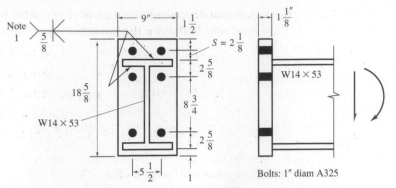

Note 1:   Weld on compression flange same as on tension flange

Figure 13.6.16   End-plate connection for Example 13.6.5.

Because the maximum effective fillet weld size along the 0.660 in. flange is 0.48 in., 13.5 kips/in. cannot be developed (see Table 5.14.2, which indicates $\frac{5}{8}$-in. weld is needed to obtain 13.9 kips/in.). Thus, use full penetration groove weld. The minimum assembling clearance for 1-in.-diam bolts is given by *LRFD Manual,* Table 8–4 "Entering and Tightening Clearances," p. 8-13, as $1\frac{7}{16}$ in. (typically about $\frac{1}{2}$ in. more than bolt diameter).

$$\text{Distance } s = 1\tfrac{7}{16} + \tfrac{5}{8} = 2.06 \text{ in.}$$

Try a plate 9 in. wide and about 18 in. long.

(c) Estimate the plate thickness t_p required. A conservative procedure is to use Eq. 13.6.44 with the prying action force Q set equal to zero. The distance b' will be

$$b' = s - \tfrac{1}{2}d_b = 2.06 - \tfrac{1}{2} = 1.56 \text{ in.}$$

$$t_p \geq \sqrt{\frac{4.44Tb'}{wF_y(1 + \alpha\delta)}} = \sqrt{\frac{4.44(213/2)1.56}{9(36)(1 + 0)}} = 1.51 \text{ in.}$$

The designer could probably use the procedure in Example 13.6.4 relating to the prying action force Q to justify use of a thinner plate.

An alternative is to use Krishnamurthy's method [13.71] where the bending moment on the plate is empirically adjusted such that the prying action force does not directly enter the calculation. In this procedure the effective span b' is taken as

$$b' = s - \tfrac{1}{4}d_b - \text{weld size} = 2.06 - \tfrac{1}{4} - \tfrac{5}{8} = 1.19 \text{ in.}$$

The resisting moment $M_e = Tb'$ is given by

$$M_e = \frac{\alpha_m T_u b'}{4} \tag{a}$$

where $\quad \alpha_m = C_a C_b \left(\frac{A_f}{A_w}\right)^{1/3} \left(\frac{b'}{d_b}\right)^{1/4} \tag{b}$

C_a = constant dependent on the yield stress of the beam and end plate; assuming both have the same F_y:

$$F_y = 36 \text{ ksi}; \quad C_a = 1.36 \text{ for A325 bolts}$$
$$C_a = 1.38 \text{ for A490 bolts}$$
$$F_y = 50 \text{ ksi}; \quad C_a = 1.31 \text{ for A325 bolts}$$
$$C_a = 1.33 \text{ for A490 bolts}$$

(Other values in Ref. 13.71 and the *LRFD Manual* p. 10-25)

$$C_b = \sqrt{b_f/b_s} \tag{c}$$

b_f = beam tension flange width
b_s = width of end plate
b' = effective arm (computed above)
d_b = bolt diameter
A_f = area of beam tension flange
A_w = web area, clear of flanges = $(d - 2t_f)t_w$
d = overall depth of W section

For this example,

$$C_b = \sqrt{b_f/b_s} = \sqrt{8.06/9} = 0.95$$
$$C_a = 1.36$$

$$\alpha_m = C_a C_b \left(\frac{A_f}{A_w}\right)^{1/3} \left(\frac{b'}{d_b}\right)^{1/4}$$

$$= 1.36(0.95)\left(\frac{8.060(0.660)}{0.370[13.92 - 2(0.660)]}\right)^{1/3}\left(\frac{1.19}{1.00}\right)^{1/4} = 1.42$$

$$M_e = \frac{\alpha_m T_u b'}{4} = \frac{1.42(213)1.19}{4} = 90.0 \text{ in.-kips}$$

Use M_e in place of Tb' in Eq. 13.6.44 with $Q = 0$,

$$t_p \geq \sqrt{\frac{4.44Tb'}{wF_y(1 + \alpha\delta)}} = \sqrt{\frac{4.44(90.0)}{9(36)(1 + 0)}} = 1.11 \text{ in.}$$

Thus, a $1\frac{1}{8}$-in. plate is acceptable according to the Krishnamurthy method.

(d) Check combined shear and tension on bolts. Referring to Sec. 4.14, compute the stress f_{uv} resulting from factored shear.

$$f_{uv} = \frac{V_u}{A_b} = \frac{60}{6(0.7854)} = 12.7 \text{ ksi}$$

The factored load tensile stress limit F'_{ut} from LRFD-Table J3.5 (Table 4.14.1) for A325-X is

$$\phi F'_{ut} = \phi(117 - 1.5f_{uv}) = 0.75[117 - 1.5(12.7)] = 73.5 \text{ ksi}$$

This exceeds the upper limit on $\phi F'_{ut} = 0.75(90) = 68$ ksi. Thus, $\phi F'_{ut} = 68$ ksi.

$$f_{ut} = \frac{T_u}{A} = \frac{213}{4(0.7854)} = 67.8 \text{ ksi} < \phi F'_{ut} \quad \text{OK}$$

The above calculation was made in part (a). If the factored shear f_{uv} had reduced F'_{ut} below 68 ksi, two more bolts could be used in the compression region to reduce the factored shear stress so that the tension bolts would be acceptable.

(e) *Length of end-plate.* At the end near the compression flange of the beam, it is sometimes desirable to extend the plate outside the beam flange an amount equal to the end-plate thickness t_p. This will increase the length of the critical section used in computing the strength based on local web yielding. The strength in local web yielding under a concentrated load is given by LRFD-K1.3; however, a concentrated load passing through a thick end-plate distributes over a longer critical length. Hendrick and Murray [13.75] have recommended the following expression for the *factored* compression reaction strength P_{bf},

$$P_{bf} = F_{yc} t_{wc} (t_{fb} + 6k + 2t_p + 2a) \tag{d}$$

where P_{bf} = factored compression force from beam flange
  t_{wc} = column web thickness
  t_{fb} = beam flange thickness
  k = distance from face of column flange to root of fillet
  t_p = thickness of end-plate
  a = fillet weld leg dimension for beam flange to end-plate weld

In this example,

$$P_{bf} = 36(0.830)[0.660 + 6(2.0) + 2(1.125) + 2(0.625)] = 483 \text{ kips}$$

This exceeds the factored applied compression force of 213 kips and is therefore satisfactory without stiffeners.

The overall web buckling of the column should also be checked according to LRFD-K1.6, as well as web crippling according to LRFD-K1.4. These checks and the design of any stiffeners required have been previously shown at the beginning of this section in Example 13.6.1.

Use PL—$1\frac{1}{8} \times 9 \times 1'$-$6\frac{5}{8}''$, with 6–1-in.-diam bolts in a bearing-type connection (A325-X), as shown in Fig. 13.6.16.

Murray [13.79] has provided an overall treatment of extended end-plate moment connections that is endorsed by AISC. ■■

Beam to Column-Web Direct Connection

Instead of using vertical tee stiffeners (Fig. 13.6.1c) or complicated details such as shown in Fig. 13.6.3, occasionally the designer would prefer to attach a beam directly to the column web, using either a welded or bolted connection. The column web must then resist the moment effect by plate action.

Abolitz and Warner [13.80], Stockwell [13.81], and Kapp [13.82] have presented yield line analyses to determine the strength of a column web when directly atached by a moment connection or a direct tension connection. Rentschler, Chen, and Driscoll [13.58, 13.83], Hoptay and Ainso [13.84], and Hopper, Batson, and

Ainso [13.85] have conducted tests of such connections. Attachment of beams to box columns presents a similar situation. Practical design data is presented in Ref. 13.81; other design recommendations appear in Ref.13.85. A detailed discussion of yield line analysis and design of these connections is outside the scope of this text.

13.7 CONTINUOUS BEAM-TO-BEAM CONNECTIONS

When beams frame transversely to other beams or girders, they may be attached to either or both sides of the girder web using simple shear (framed beam) connections (Sec. 13.2) or using simple seats in combination with a framed beam connection (Sec. 13.3). When full continuity of the beam is desired, the connection must develop a higher degree of rigidity than provided by simple shear connections. For beam-to-beam connections, the principal objective is to provide a means of allowing the tensile force developed in one beam flange to be carried across to the adjacent beam framing opposite the girder web. These connections may be divided into two categories; (1) those with intersecting tension flanges not rigidly attached to one another, as in Fig. 13.7.1; and (2) those with rigidly attached intersecting tension flanges, as in Fig. 13.7.2.

When the intersecting tension flanges are not rigidly attached (Fig. 13.7.1) the connection design is essentially a tension member design at the tension flange along with a shear connection.

On the other hand, the designer should be cautious when designing intersecting beams with tension flanges attached, since this case becomes one of biaxial instead of uniaxial stress. With biaxial stress the possibility of brittle fracture increases (see Sec. 2.9). In addition, a biaxial stress yield criterion must be used, such as Eq. 2.6.2, to establish safety:

$$\sigma_y^2 = \sigma_1^2 + \sigma_2^2 - \sigma_1\sigma_2 \qquad [2.6.2]$$

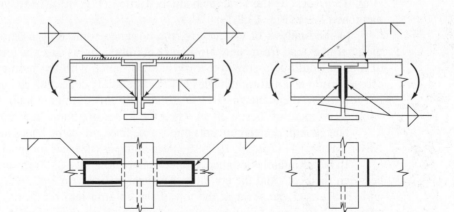

Figure 13.7.1 Intersecting beam connections: tension flanges *not attached* to each other.

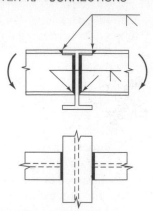

Figure 13.7.2   Intersecting beam connection: tension flanges *attached* to each other.

where σ_1, σ_2 are the principal stresses acting. When the beams frame at 90 degrees to one another the axial stresses in the flanges are principal stresses.

The designer should make certain the σ_y of Eq. 2.6.2 does not exceed $0.60F_y$ for Allowable Stress Design, nor the tension–compression yield stress F_y under factored loads for Load and Resistance Factor Design.

13.8 RIGID-FRAME KNEES

In the design of rigid frames according to either LRFD or ASD, the safe transmission of load at the junction of beam and column is of great importance. When members join with their webs lying in the plane of the frame, the junction is frequently referred to as a knee joint. Typical knee joints are (1) the square knee, with or without a diagonal or other stiffener (Fig. 13.8.1a and b); (2) the square knee with a bracket (Fig. 13.8.1c); (3) the straight haunched knee (Fig. 13.8.1d); and (4) the curved haunched knee (Fig. 13.8.1e).

In the analysis of a structure, the designer commonly assumes the span of a member to extend from center-to-center of joints; in this case the knees. The moment of inertia would be assumed to vary in accordance with the moment of inertia of a cross-section taken at right angles to lines extending center-to-center of knees. Internal forces are then determined using statically indeterminate analysis, including the variable moment of inertia effect when the knees are haunched or curved.

The general design concepts applicable for rigid frame knees are summarized in ASCE Manual 41 [7.2, pp. 167–186] which forms the basis for much of what follows.

To be adequately designed, a knee connection must (1) transfer the end moment between the beam and the column, (2) transfer the beam end shear into the column, and (3) transfer the shear at the top of the column into the beam. Furthermore, in performing the three functions relating to strength, the knee must deform in a manner consistent with the analysis by which moments and shears were determined.

If a plastic hinge associated with the failure mechanism is expected to form at or near the knee, adequate rotation capacity must be built into the connection. Square knees have the greatest rotation capacity but are also the most flexible (i.e., deform

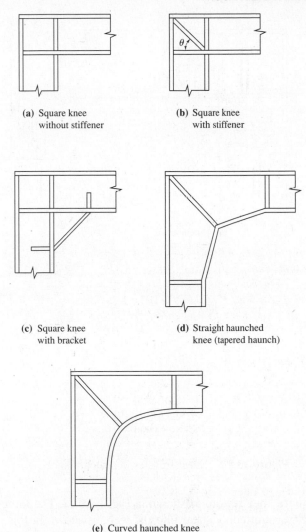

(a) Square knee
without stiffener

(b) Square knee
with stiffener

(c) Square knee
with bracket

(d) Straight haunched
knee (tapered haunch)

(e) Curved haunched knee

Figure 13.8.1 Rigid-frame knees.

elastically the most under service load conditions.) Curved knees are the stiffest but have the least rotation capacity. Since straight tapered knees provide reasonable stiffness along with adequate rotation capacity, in addition to the fact that they are cheaper than curved haunches to fabricate, the straight haunched knees are commonly used.

Shear Transfer in Square Knees

In the design of a rigid frame having square knees, two rolled sections may come together at right angles as shown in Fig. 13.8.1a. A frame analysis, either elastic or plastic, will have established what moments and shears act on the boundaries of the

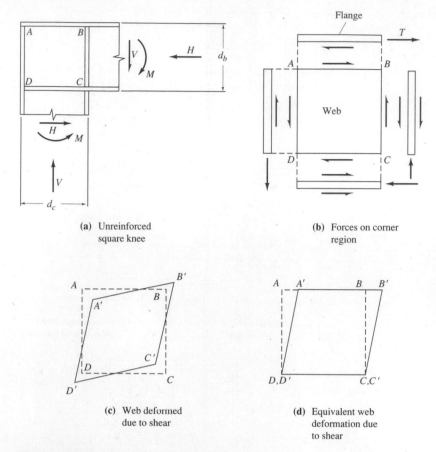

Figure 13.8.2   Shear transfer in a square knee.

square knee region, as shown in Fig. 13.8.2a. The forces carried by the flanges must be transmitted by shear into the web, as shown in Fig. 13.8.2b.

Assuming all bending moment to be carried by the flanges, and approximating the distance between flange centroids as $0.95d_b$, the flange force is

$$T = \frac{M}{0.95d_b} \tag{13.8.1}$$

where d_b is the overall depth of the beam. The nominal strength in shear of the web across AB is

$$V_n = V_{ab} = \tau_y t_w d_c \tag{13.8.2}$$

where d_c is the overall depth of the column.

Using Load and Resistance Factor Design philosophy, the factored moment M_u gives rise to the factored tension force T_u. The design strength ϕV_n may then be equated to T_u to give the required web thickness t_w,

$$\phi V_n = T_u \tag{13.8.3}$$

$$\phi \tau_y t_w d_c = \frac{M_u}{0.95 d_b} \tag{13.8.4}$$

In accordance with LRFD-K1.7, $\tau_y = 0.60 F_y$ and $\phi = 0.90$ for the yielding limit state. Equation 13.8.4 when solved for the required web thickness t_w gives

$$\text{Required } t_w = \frac{1.95 M_u}{F_y d_b d_c} = \frac{1.95 M_u}{F_y A_{bc}} \tag{13.8.5}$$

where $A_{bc} = d_b d_c$ = the planar area within the knee. Equations 13.8.4 and 13.8.5 assume that interaction between axial load P_u and column shear V_u does not control, i.e., that $P_u \leq 0.4 P_y$ for the column.

This same approach must be used when beams frame on opposite faces of a column producing a high shear on the column web, as shown in Fig. 13.8.3. The total shear to be transferred across AB is

$$\text{Total factored shear} = C_{u1} + T_{u2} - V_u \tag{13.8.6}$$

where C_{u1} = compression force produced by M_{u1}
T_{u2} = tensile force produced by M_{u2}
V_u = factored shear in the column

The total factored shear given by Eq. 13.8.6 would replace T_u in the development of Eq. 13.8.5. Thus, for two beams of equal depth framing on opposite faces of a column,

$$\text{Required } t_w = \frac{1.95(M_{u1} + M_{u2})}{F_y A_{bc}} - \frac{1.85 V_u}{F_y d_c} \tag{13.8.7}$$

To illustrate Allowable Stress Design philosophy, the maximum allowable shear

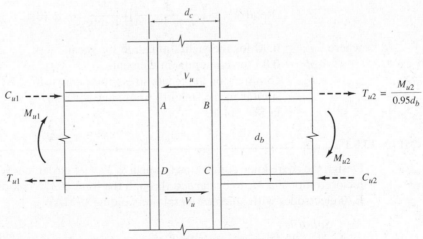

Figure 13.8.3 General case of shear transfer within a rigid joint.

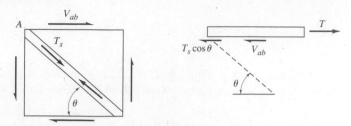

Figure 13.8.4   Diagonal stiffener effect.

stress $0.40F_y$ replaces $\phi(0.60F_y)$ and service loads M_1, M_2, and V replace factored loads in the equilibrium requirement. Thus, Eq. 13.8.7 would become for ASD

$$\text{Required } t_w = \frac{2.63(M_1 + M_2)}{F_y A_{bc}} - \frac{2.5V}{F_y d_c} \tag{13.8.8}$$

In a rigid frame knee, the required web thickness will typically exceed that provided by a W section; thus, the reinforcement will be required. A doubler plate is sometimes used to thicken the web region; a generally impractical solution because of the difficulty making the attachment to the column web. Usually a pair of diagonal stiffeners is the best solution.

When diagonal stiffeners are used, the horizontal component $C_s \cos\theta$ of the stiffener force C_s participates with the web. Equilibrium (Fig. 13.8.4) then requires

$$T = V_{ab} + C_s \cos\theta \tag{13.8.9}$$

or

$$\frac{M_u}{0.95d_b} = \phi_v(0.60F_y)t_w d_c + A_{st}\phi_c F_{cr} \cos\theta \tag{13.8.10}$$

Solving for required A_{st} gives

$$\text{Reqd } A_{st} = \left(\frac{1}{\phi_c F_{cr} \cos\theta}\right)\left[\frac{M_u}{0.95d_b} - \phi_v(0.60F_y)t_w d_c\right] \tag{13.8.11}$$

where $\phi_v = 0.90$ for the yield limit state for shear
$\phi_c = 0.85$ for compression elements
$F_{cr} =$ compression limit state stress computed using LRFD-E2 in the manner similar to the stability limit state for bearing stiffeners, as discussed in Sec. 11.12.

EXAMPLE 13.8.1 _____

Design the squre knee connection to join a W27×94 girder to a W14×74 column. The factored moment M_u to be carried through the joint is 376 ft-kips. Use A36 steel and E70 electrodes with shielded metal arc welding (SMAW).

Solution

(a) Determine whether or not a diagonal stiffener is required. Using Eq. 13.8.4, and referring to Fig. 13.8.5,

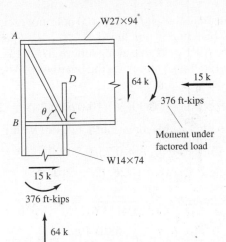

Figure 13.8.5 Square knee of Example 13.8.1.

$$\text{Required } t_w = \frac{1.95M_u}{F_y A_{bc}} = \frac{1.95(376)12}{36(26.92)14.17} = 0.64 \text{ in.}$$

$$\text{Actual } t_w = 0.490 \text{ in. for W27}\times 94 < 0.64 \text{ in.}$$

A diagonal stiffener is required.

(b) Determine stiffener size.

$$\tan \theta = \frac{26.92}{14.17} = 1.900; \qquad \cos \theta = 0.466$$

Using Eq. 13.8.11 with the assumption $F_{cr} \approx F_y$,

$$\text{Reqd } A_{st} = \left(\frac{1}{0.85(36)0.466}\right)\left(\frac{376(12)}{0.95(26.92)} - 0.90(0.6)(36)(0.49)14.17\right)$$

$$\text{Required } A_{st} = 2.9 \text{ sq in.}$$

(c) Evaluate the strength of the stiffeners acting as a column. The slenderness ratio KL/r for a compression member consisting of two plates 3 in. wide attached to the web is

$$\text{Overall depth} = 2(3.0) + t_w = 6.49 \text{ in.}$$

$$r = \sqrt{\frac{t_{st}(\text{depth})^3}{12 t_{st}(\text{depth})}} = \text{depth} \sqrt{\frac{1}{12}} = 6.49 \sqrt{\frac{1}{12}} = 1.9 \text{ in.}$$

$$\frac{KL}{r} = \frac{d_c/\cos \theta}{r} = \frac{14.17/0.466}{1.9} = 16; \qquad \phi F_{cr} = 30 \text{ ksi}$$

Note that the upper bound for $\phi_c F_{cr} = 0.85(36) = 30.6$ ksi. As a practical matter, except for unusually long stiffeners, F_{cr} may be approximated as F_y for design of diagonal stiffeners. Thus, the required A_{st} as computed in part (b) is essentially correct.

Use 2 PLs—$\frac{1}{2}\times3$. Note that $\lambda = 3.0/0.5 = 6.0$ does not exceed λ_r of 15.8 from LRFD-B5; thus, local buckling is precluded.

(d) Determine the fillet weld size along length AB in Fig. 13.8.5. The weld must transmit the factored flange force into the beam web. The maximum design flange force that can develop is $\phi_t F_y A_f$,

$$\text{Flange force } \phi_t F_y A_f = 0.90(36)(10.070)0.785 = 256 \text{ kips}$$

The design strength of fillet welds along both sides of web is

$$\phi R_{nw} = 0.75t_e(0.60F_{EXX}) = 0.75(0.707)2a(42) = 44.6a$$

$$\text{Available length for weld} = 26.92 - 2(0.745) = 25.4 \text{ in.}$$

$$\text{Required } a = \frac{256}{25.4(44.6)} = 0.23 \text{ in.}$$

$$\text{Check: } a_{\text{max eff}} = 0.707\frac{F_u t_{wb}}{F_{EXX}} = 0.707\frac{58(0.490)}{70} = 0.29 \text{ in.} > 0.23 \text{ in.} \quad \text{OK}$$

$$\text{Min } a = \tfrac{5}{16}\text{-in. from Table 5.11.1} \quad (\text{for } t_{fc} = 0.785 \text{ in.})$$

Use $\frac{1}{4}$-in. E70 fillet weld along length AB (both sides of girder web).

(e) Determine fillet weld size along length BC. The connection of the column web to the beam flange must carry the force resulting from flexure and axial load combined with the shear. At the most highly stressed location there will be tension and shear acting simultaneously on the weld. A conservative approach will be to compute the resultant of the shear and tension components.

$$\text{Tensile component} = \phi_t F_y t_w = 0.90(36)0.450 = 14.6 \text{ kips/in.}$$

$$\text{Shear component} = \frac{V_u}{d_c - 2t_f} = \frac{15}{14.17 - 2(0.785)} = 1.2 \text{ kips/in.}$$

$$\text{Resultant loading} = \sqrt{(14.6)^2 + (1.2)^2} = 14.6 \text{ kips/in.}$$

$$\text{Required } a = \frac{14.6}{2(0.707)(0.75)42} = 0.33 \text{ in.}$$

Since the tensile component is dominant, maximum effective weld size is

$$t_{wc}\phi_t F_y = 2\phi(0.707)a(0.60F_{EXX})$$

$$a_{\text{max eff}} = \frac{0.450(0.90)36}{2(0.75)(0.707)(0.60)70} = 0.33 \text{ in.} \quad \text{OK}$$

Use $\frac{3}{8}$-in. E70 fillet weld along length BC (both sides of column web).

(f) Determine the weld required along stiffeners. The weld must develop the stiffener strength,

$$\phi C_s = \phi F_{yst} A_{st} = 0.90(36)(2)(3.0)0.50 = 97 \text{ kips}$$

$$\text{Required } a = \frac{97/(14.17/\cos\theta)}{4(0.707)(0.75)42} = 0.04 \text{ in.}$$

$$\text{Min } a = \tfrac{3}{16}\text{-in. (Table 5.11.1 for } \tfrac{1}{2}\text{-in. stiffener)}$$

Use $\frac{3}{16}$-in. E70 fillet weld along diagonal AC (both sides of girder web).

(g) Determine the extent of stiffener required from point C vertically into girder. The design strength based on local web yielding (LRFD-K1.3) of the W27×94 web to carry the design compression force P_{bf} from the inside column flange at C is, using Eq. 13.6.1,

$$P_{bf} = \phi(5k + t_{fb})F_{yw}t_{wc} \qquad\qquad [13.6.1]$$
$$= 1.0[5(1.4375) + 0.785](36)0.490 = 141 \text{ kips}$$

where t_{fb} is the thickness of the W14×74 flange transmitting the force, and t_{wc} is the web thickness of the W27×94 resisting the force.

Since the flange concentrated load from the column is 256 kips, as computed in part (d), and the diagonal stiffeners are already performing another task, vertical stiffeners must be used along CD. When stiffeners are needed to prevent local flange yielding (LRFD-K1.3) or local flange bending (LRFD-K1.2), they must extend "at least one-half the depth of the web". Perhaps the designer should also investigate the strength based on web crippling (LRFD-K1.4); however, with the stiffeners along CD and the diagonal stiffeners along AC (which were investigated as a column) web crippling is adequately prevented. When the local web yielding limit state does not require stiffeners, the web crippling limit state should be investigated.

(h) Establish plate size for stiffeners along CD. The required area of stiffeners,

$$\text{Required area} = \frac{(256 - 141)/2}{\phi_c F_y} = \frac{57.5}{0.85(36)} = 1.9 \text{ sq in./plate}$$

$$\text{Width available} = \frac{b_{fb} - t_{wb}}{2} = \frac{9.990 - 0.490}{2} = 4.8 \text{ in.}$$

$$\text{Minimum } t_s = \frac{1.9}{4.0} = 0.48 \text{ in.} \quad \text{assuming 4-in. plates}$$

Use $\frac{1}{2}$-in. plates. To preclude the local buckling limit state, λ must not exceed λ_r according to LRFD-B5. Here $\lambda = 8.0$, which does not exceed 15.8 (i.e., λ_r) and is acceptable. When the strength M_n required within the knee is the plastic moment strength M_p, it will be preferable to keep $\lambda \leq \lambda_p$. Since $\lambda_p = 65/\sqrt{F_y} = 10.8$, in this case $\lambda < \lambda_p$.
Use 2PLs—$\frac{1}{2}$×4 × 1'-0", tapered from full width at C to zero width at D (Fig. 13.8.5). ■■

Straight Haunch and Curved Haunch Knees

Straight haunch knees (also called *tapered haunches*), as shown in Fig. 13.8.1d, and curved haunch knees, as shown in Fig. 13.8.1e, may extend over a significant portion of a span; in which case they are not really connections but rather an intergral part of a variable moment of inertia frame. Detailed treatment of tapered and curved haunches is available in Blodgett [13.2] and ASCE Manual 41 [7.2]. For analysis of rigid frames, the reader is referred to *Single Span Rigid Frames in Steel* [13.86].

13.9 COLUMN BASE PLATES

Column base plates distribute the concentrated loads acting in the elements of columns to the supporting medium, commonly a concrete pedestal or footing. These heavy loads must be distributed to prevent crusing of the concrete support. Another concern is the connection, or anchorage of the base plate and column to the concrete foundation. In frame analysis, evaluation of degree of fixity may be of interest.

Base Plates Under Axial Load

The design of base plates involves several considerations:

1. The area of the base plate will depend on the bearing strength of the concrete or grout under the steel plate.
2. The thickness of the plate will be controlled by the bending strength of the plate. When the plate dimensions B and N, shown in Fig. 13.9.1a, are significantly larger than the profile dimensions b and d of the steel section, the traditional approach is to design the plate as having cantilever spans m and n uniformly loaded.
3. For plates not extending much beyond the profile limits of the steel section, an alternate approach is required. This situation arises when the column load is relatively small. In this case the lightly loaded plate may be treated as uniformly loaded over an H-shaped area A_H adjacent to the inside perimeter of the column, as shown in Fig. 13.9.1b.

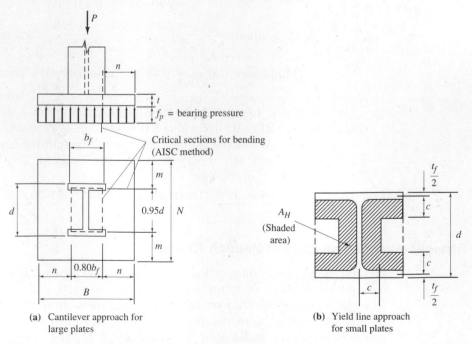

(a) Cantilever approach for large plates

(b) Yield line approach for small plates

Figure 13.9.1 Column base plates.

The LRFD Specification does not give a specific design procedure for base plates. The *LRFD Manual* section "COLUMN BASE PLATES," pp. 11-54 to 11-64, contains the procedure presented in 1990 by Thornton [13.87], which includes design of both heavily and lightly loaded plates. Lightly loaded base plates have traditionally given designers difficulty because the cantilever distances n and m of Fig. 13.9.1a were small, giving little or no calculated thickness to resist bending. Stockwell [13.88] suggested a yield-line solution, and Murray [13.89] significantly extended the yield-line approach. Ahmed and Kreps [13.90] demonstrated the inconsistencies in the early design suggestions. DeWolf [13.91] and Fling [13.92] have also treated to subject. Thornton [13.87, 13.93] summarized and synthesized these studies to obtain the currently accepted procedure.

Establish the Plan Dimensions $B \times N$ This will be controlled by bearing on the concrete in accordance with LRFD-J9. The design bearing strength $\phi_c P_p$ must at least equal the factored column load P_u,

$$\phi_c P_p \geq P_u \tag{13.9.1}$$

There are two categories for nominal strength P_p in bearing:

1. Bearing on the full area A_1 of a concrete support,

$$P_p = 0.85 f_c' A_1 \tag{13.9.2}$$

2. Bearing on area A_1 which is *less* than full area A_2 of a concrete support,

$$P_p = 0.85 f_c' A_1 \sqrt{A_2/A_1} \leq 0.85 f_c' (2A_1) \tag{13.9.3}$$

where $\phi_c = 0.60$ for bearing on concrete
 A_1 = area ($B \times N$ in Fig. 13.9.1) of steel plate concentrically bearing on a concrete support
 A_2 = maximum area of the portion of the supporting surface that is geometrically similar to and concentric with the loaded area
 f_c' = specified 28-day compressive strength for concrete
 $f_p = P_u/BN$

Plate Thickness According to Cantilever Method Assuming the bearing pressure is uniform over the entire bearing plate, the factored cantilever moment is

$$M_u = f_p \left(\frac{Nn^2}{2} \right) \quad \text{(on section parallel to column web)} \tag{13.9.4a}$$

$$M_u = f_p \left(\frac{Bm^2}{2} \right) \quad \text{(on section parallel to column flanges)} \tag{13.9.4b}$$

The yielding limit state for bending on the plate requires

$$\phi_b M_n \geq M_u \tag{13.9.5}$$

where

$$\phi_b M_n = \phi_b M_p = \phi_b Z F_y = \phi_b (N t_p^2 / 4) F_y \tag{13.9.6a}$$

or

$$\phi_b M_n = \phi_b M_p = \phi_b (B t_p^2 / 4) F_y \qquad (13.9.6b)$$

Equating $\phi_b M_p$ with M_u, Eq. 13.9.4a with Eq. 13.9.6a, and Eq. 13.9.4b with Eq. 13.9.6b, gives the following two equations

$$\phi_b \left(\frac{N t_p^2}{4} \right) F_y = f_p \left(\frac{N n^2}{2} \right) \qquad (13.9.7a)$$

and

$$\phi_b \left(\frac{B t_p^2}{4} \right) F_y = f_p \left(\frac{B m^2}{2} \right) \qquad (13.9.7b)$$

Solving for t_p in Eqs. 13.9.7 gives

$$t_p = n \sqrt{\frac{2 f_p}{\phi_b F_y}} \quad \text{or} \quad t_p = m \sqrt{\frac{2 f_p}{\phi_b F_y}} \qquad (13.9.8)$$

or, with $\phi = 0.90$, Eqs. 13.9.8 become

$$t_p = 1.5 n \sqrt{\frac{f_p}{F_y}} \quad \text{or} \quad t_p = 1.5 m \sqrt{\frac{f_p}{F_y}} \qquad (13.9.9)$$

The larger of n or m would be used to obtain the required thickness.

Plate Thickness According to Yield Line Approach   In this semi-empirical approach, the effective cantilever length $\lambda n'$ replaces m or n in Eq. 13.9.9. Thornton's formulation [13.93] of the yield-line analysis indicates,

$$n' = \tfrac{1}{4} \sqrt{d b_f} \qquad (13.9.10)$$

where d = overall depth of the steel W section
b_f = flange width of W section

The factor λ relates to the load P_0 carried by the area $d b_f$ actually loaded. Thornton [13.87] gives

$$\lambda = \frac{2 \sqrt{X}}{1 + \sqrt{1 - X}} \qquad (13.9.11)$$

where

$$X = \frac{4 P_0}{(d + b_f)^2 F_p} \le 1.0 \qquad (13.9.12)$$

where F_p = limiting bearing pressure $0.85 f_c'$ from LRFD-J9
$P_0 = P_u (d b_f / B N)$

Substituting for P_0 in Eq. 13.9.12 and noting that $P_p = F_p(BN)$ gives

$$X = \frac{4}{(d + b_f)^2 F_p} \left(P_u \frac{d b_f}{B N} \right) = \frac{4 d b_f}{(d + b_f)^2} \frac{P_u}{P_p} \le 1.0 \qquad (13.9.13)$$

Thornton [13.87] has developed λ as the link between the Murray [13.89]–Stockwell [13.86] method and his method [13.93]. Essentially, $\lambda = 1$ is the divide between lightly loaded ($\lambda < 1$) and heavily loaded situations. From Eq. 13.9.11, $\lambda = 1$ when $X = 0.64$.

For design purposes, *LRFD Manual* (p. 11-59) had placed the ϕ_c with P_p in the denominator of Eq. 13.9.13. Thus,

$$X = \frac{4db_f}{(d + b_f)^2} \frac{P_u}{\phi_c P_p} \le 1.0 \qquad (13.9.14)$$

The introduction for ϕ_c is unrelated to the derivation of X, and within the accuracy of the method makes no difference; however, in the use of the equation, the link to Eq. 13.9.1 is a practical one.

Design Equation The required thickness for an axially loaded base plate is, referring to Eqs. 13.9.9,

$$t_p = 1.5(\text{largest of } n, m, \text{ or } \lambda n') \sqrt{\frac{f_p}{F_y}} \qquad (13.9.15)$$

where $n = (B - 0.8b_f)/2$
  $m = (N - 0.95d)/2$
  $n' = \frac{1}{4}\sqrt{db_f}$

  $\lambda = \dfrac{2\sqrt{X}}{1 + \sqrt{1 - X}} \le 1$; conservatively $\lambda = 1$

  $X = \dfrac{4db_f}{(d + b_f)^2} \dfrac{P_u}{\phi_c P_p} \le 1.0$

Inherently, λ can exceed 1.0; however, when it does, the plate is heavily loaded and the simple cantilever method is appropriate.

Some practical aspects of column base selection are given by Ricker [13.94]. DeWolf and Ricker [13.95] have presented an overall treatment of column base plate design which is endorsed by AISC.

EXAMPLE 13.9.1 _____

Design a base plate for a W14×145 column of A36 steel to carry factored axial loads of 400 kips dead load, 275 kips live load, and 100 kips wind. Assume a concrete pedestal will be under the base plate and the pedestal will have a dimension 6 in. larger than the base plate in each direction. The steel is A36 and the concrete has $f'_c = 3$ ksi. Use Load and Resistance Factor Design.

Solution
  (a) Compute the factored load P_u.

$$P_u = 1.2P_D + 1.6P_L = 1.2(400) + 1.6(275) = 920 \text{ kips}$$
$$P_u = 1.2P_D + 0.5P_L + 1.3P_W$$
$$= 1.2(400) + 0.5(275) + 1.3(100) = 748 \text{ kips}$$

The factored load P_u to be carried is 920 kips.

(b) Determine the maximum base plate area A_1 required. Assuming the pedestal is exactly the dimensions of the base plate, the required area using Eq. 13.9.2 would be

$$\text{Required } A_1 = \frac{P_u}{\phi(0.85)f_c'} = \frac{920}{0.60(0.85)3} = 601 \text{ sq in.}$$

The plate should be approximately square, with slight differences in dimensions B and N to give nearly equal values for m and n.

$$0.80b_f = 0.80(15.50) = 12.40 \text{ in.}$$
$$0.95d = 0.95(14.78) = 14.04 \text{ in.}$$

Try $B = 24$ in. and $N = 25$ in. This would make the concrete pedestal 30 in. by 31 in. with an area $A_2 = 930$ sq in. Using Eq. 13.9.3,

$$\sqrt{A_2/A_1} = \sqrt{930/600} = 1.24 \leq 2.0$$

$$\text{Required } A_1 = \frac{\text{Preliminary required } A_1}{1.24} = \frac{601}{1.24} = 485 \text{ sq in.}$$

Use $B = 21$ in. and $N = 23$ in. giving $A_1 = 483$ sq. in. Adding the 6 in. to each dimension will make the pedestal 27 in. by 29 in. and $A_2 = 783$ sq in. Using Eq. 13.9.3 again gives required $A_1 = 472$ sq in. Keeping bearing plate dimensions in whole inches, this calculation converges quickly.

(c) Determine the plate thickness. Using Eq. 13.9.15,

$$t_p = 1.5(\text{largest of } n, m, \text{ or } \lambda n')\sqrt{\frac{f_p}{F_y}} \qquad [13.9.15]$$

$$n = 0.5(B - 0.8b_f) = 0.5(21 - 12.40) = 4.30 \text{ in.}$$
$$m = 0.5(N - 0.95d) = 0.5(23 - 14.04) = 4.48 \text{ in.}$$

Using the selected $B = 21$ in. and $N = 23$ in.,

$$\phi_c P_p = 0.60(0.85 f_c')A_1\sqrt{A_2/A_1}$$
$$= 0.60(0.85)(3)(483)\sqrt{783/483}$$
$$= 1.53(483)1.27 = 941 \text{ kips}$$

$$X = \frac{4db_f}{(d + b_f)^2}\frac{P_u}{\phi_c P_p} = \frac{4(14.78)15.50}{(14.78 + 15.50)^2}\frac{920}{941} = 0.97 > 0.64$$

When $X > 0.64$, $\lambda = 1$. Then

$$n' = \tfrac{1}{4}\sqrt{db_f} = \tfrac{1}{4}\sqrt{14.78(15.50)} = 3.78 \text{ in.}$$

The largest of n, m, and n' is 4.48 in.

$$t_p = 1.5(\text{largest of } n, m, \text{ or } \lambda n')\sqrt{\frac{f_p}{F_y}}$$

$$= 1.5(4.48)\sqrt{\frac{P_u}{BNF_y}} = 1.5(4.48)\sqrt{\frac{920}{21(23)36}} = 1.54 \text{ in.}$$

Thus, the cantilever method requires the greater thickness (1.54 in.). *Use* plate—$1\frac{1}{2}\times 21 \times 1'$-$11''$. Only rarely will the yield line approach formulas give the greater thickness. ■ ■

Column Bases to Resist Moment

Column bases frequently must resist moment in addition to axial compression. The situation has some similarities to the behavior of bolted connections discussed in Chapter 4, and in many respects is analogous to the situation of reinforcing bars in concrete construction. The axial force causes a precompression between the base plate and the contact surface (frequently a concrete wall or footing). When the moment is applied, the precompression on the tension side in flexure is reduced, often to zero, leaving only the anchor bolt to provide the tensile force resistance. On the compression side, the contact area remains in compression. The anchorage will have an ability to undergo rotational deformation, depending primarily on the length of anchor bolt available to deform elastically. Also, the behavior is influenced by whether or not the anchor bolts are given an initial pretension (similar to the installation of high-strength bolts as discussed in Chapter 4). The moment-rotation characteristics of column anchorages are treated in detail by Salmon, Schenker, and Johnston [13.96].

A number of elaborate methods are available for designing moment-resisting bases, with variations depending on the magnitude of the eccentricity of loading and the specific details of the anchorage. Some simple details are shown in Fig. 13.9.2.

When the eccentricity of loading, $e = M/P$, is small such that it does not exceed $\frac{1}{6}$ of the plate dimension N in the direction of bending (i.e., within the middle third of the plate dimension), the ordinary combined stress formula applies. Thus for small e,

$$f_p = \frac{P}{A} \pm \frac{M}{S} \qquad (13.9.16)$$

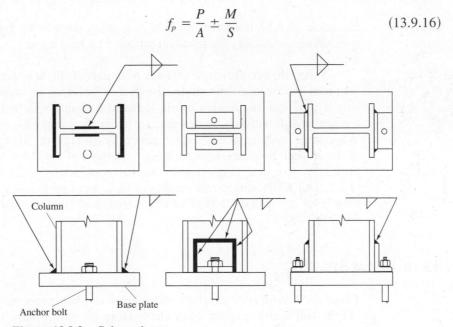

Figure 13.9.2 Column bases.

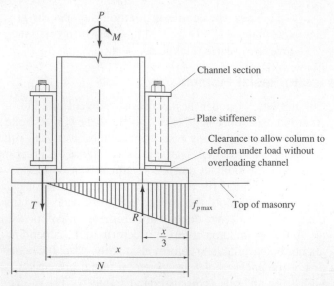

Figure 13.9.3   Moment-resisting column anchorage.

where $M = Pe$.

$$S = Ar^2/(N/2) = AN/6$$
$$r^2 = N^2/12$$
$$f_p = \frac{P}{A} \pm \frac{6Pe}{AN} = \frac{P}{A}\left[1 \pm \frac{6e}{N}\right] \qquad (13.9.17)$$

Equation 13.9.17 is correct for $e \leq N/6$ when there is no bolt pretension, and is considered satisfactory for practical purposes at least up to $e = N/2$ without serious error.

When the eccentricity e exceeds $N/6$, part of the base plate at the tension face becomes inactive and the stress distribution becomes as shown in Fig. 13.9.3. A simple practical assumption to use for such a situation is that the resultant of the triangular distribution is located directly under the compression flange of the column. When large moment is designed for, generally the attachment arrangement to the base of the column becomes more complex than those of Fig. 13.9.2. For a more detailed presentation on designing column bases for moment, the reader is referred to Blodgett [13.2, Sec. 3.3]. Experimental studies of base plates subject to axial load and moment have been reported by DeWolf and Sarisley [13.97] and Thambiratnam and Paramasivan [13.98].

13.10 BEAM SPLICES

There are several reasons why a rolled beam or plate girder may be spliced, such as: (1) the full length may not be available from the mill; (2) the fabricator may find it economical to splice even when the full length could be obtained; (3) the designer may desire to use splice points as an aid to cambering; and (4) the designer may desire a change in section to fit the variation in strength required along the span.

On each side of a joint, one must provide for the transverse shear and bending moment. For welded plate girders, and frequently for rolled beams, the splice may be accomplished by a full-penetration groove weld. Splices of material made entirely in the shop are nearly always groove welded. Field splices are becoming more frequently all welded, though usually using lapping of pieces and fillet-welded connections, instead of groove welding where dimensional control is a critical factor. Since full-penetration groove-welded splices are as strong as the base material, they require no further comment here.

Many field splices for beams and welded plate girders use laping splice plates and high-strength bolts as connectors. In this section the four-plate beam splice is treated.

Splices are designed for the moment M and the shear V occurring at the spliced point, or they are designed for some arbitrary or specification-prescribed higher values. LRFD-J7 and ASD-J7 require groove-welded splices to "develop the full strength of the smaller spliced section." "Other types of splices in cross sections of plate girders and beams shall develop the strength required by the forces at the point of splice." Earlier AISC Specifications required splices to develop not less than 50% of the effective strength of the material spliced.

For beam splices, each element of the splice is designed to do the work the material underlying the splice plates could do if uncut. Plates on the flange do the work of the flange, while plates over the web do the work of the web. One typical arrangement of a four-plate splice is shown in Fig. 13.10.1.

When determining the forces for which to design, one should recognize that the splice actually covers a finite distance along the span (perhaps 1 or 2 ft). Along this distance the moment and shear vary. In accordance with the principles used for designing bolted connections, the forces designed for should be those acting on the center of gravity of the bolt group. From Fig. 13.10.2 it may be noted that theroetically the moment at the centroid of the connectors on one side of the splice is different from that at the centroid on the other. Some designers, therefore, proportion the connection on each side of the splice for a moment, $M_1 = M + Ve$. Such a procedure rarely seems justified. Design of such splices has been treated by Kulak and Green [13.99]. In those unusual situations when a splice is located where both shear and bending moment are high, such a procedure might seem desirable.

Most splices are located in regions where either the shear or the bending moment are low so that often the design is for a specification-required minimum strength. As discussed in Sec. 10.6 for plastic analysis, one must be wary of designing a splice for

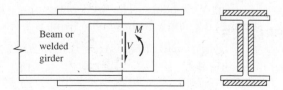

Figure 13.10.1 Four-plate beam splice.

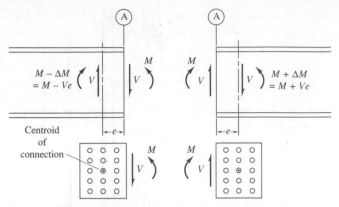

Figure 13.10.2   Forces acting on web splice plates.

a low moment just because the spice is near an inflection point. *If moments for the structure were computed using theory of statically indeterminate structures without a hinge within the span, one should not later design a splice which has low stiffness to act as a hinge.*

For most situations the authors recommend designing for the actual or specification-required forces *at the splice* and neglecting any eccentricity effect.

EXAMPLE 13.10.1

Design a rolled beam splice (four-plate type of Fig. 13.10.1) for a W24×84 beam of A36 steel to be located where the factored moment M_u is 405 ft-kips and the factored shear V_u is 140 kips. Use A36 steel and $\frac{3}{4}$-in.-diam A325 bolts in a bearing-type connection (A325-X). Use Load and Resistance Factor Design.

Solution

(a) Check the maximum design strengths ϕM_n and ϕV_n of the W24×84 beam.

$$\phi_b M_n = \phi_b M_p = 0.90 Z_x F_y = [0.90(224)(36)]\tfrac{1}{12} = 605 \text{ ft-kips}$$
$$\phi_v V_n = \phi_v(0.60 F_y)A_w = 0.90(0.60)(36)(24.1)0.47 = 220 \text{ kips}$$

The factored loads are 67% and 64% of the design moment and design shear strengths, respectively. Even though LRFD does not require any minimum proportion of the strength to be developed by a splice, it is prudent to design splices for a significant (say, at least 50% of strength; the authors prefer at least 2/3 of the strength) proportion of the member strength.

(b) Web plates. The web plates must carry all the shear. The design strength ϕR_n for bolts in double shear is

$$\phi R_n = \phi(0.50 F_u^b)mA_b = 0.75(0.50)(120)(2)0.4418 = 39.8 \text{ kips}$$

$$\text{Number of bolts} = \frac{140}{39.8} = 3.5, \text{ say } 4$$

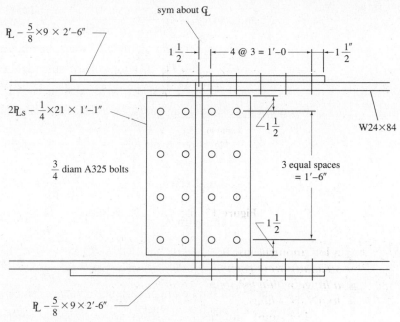

Figure 13.10.3 Design for Example 13.10.1.

Since 4 bolts are needed for shear alone, and the web plates will also carry of portion of the bending moment the beam web would carry, try 2 lines of 4 bolts each as shown in Fig. 13.10.3. There will be 3 spaces vertically of approximately 6 in. when the plate depth is made 21 in. (which is the available flat dimension T of the beam web).

The web plate thickness to prevent shear rupture along a net section is, according to LRFD-J4.1,

$$\phi(0.6F_u)A_{nv} = V_u$$

$$\text{Required } A_{nv} = \frac{V_u}{\phi(0.6F_u)} = \frac{140}{0.75(0.6)58} = 5.4 \text{ sq in.}$$

Using 4 bolts, the thickness required for each of two plates based on the shear rupture limit state will be

$$\text{Required } t = \frac{A_{nv}}{2[21 - 4(13/16)]} = \frac{5.4}{2(17.8)} = 0.15 \text{ in.}$$

A thickness of $\frac{1}{4}$ in. should be considered the practical minimum.
Use 2 PLs—$\frac{1}{4} \times 21$.

(c) Flange plates. The splice plates are designed as tension members. The plate width is made 9 in.; this is slightly less than the beam flange width of 9.02 in. Both the fracture and yielding limit states must be investigated in accordance with LRFD-J5.2,

$$\text{Flange force} = \frac{M_u}{\text{Arm}} = \frac{M_u}{d + \text{est. } t} = \frac{405(12)}{24.10 + 0.5} = 198 \text{ kips}$$

$$\phi T_n = \phi A_g F_y \quad \text{yielding of the splice plate } (\phi = 0.90)$$

$$\phi T_n = \phi A_n F_u \quad \text{fracture of the splice plate } (\phi = 0.75)$$

$$A_n \leq 0.85 A_g$$

In addition, "block shear" must be considered, as shown in Example 4.8.3.

$$\text{Required } A_g = \frac{T_u}{\phi F_y} = \frac{198}{0.90(36)} = 6.11 \text{ sq in.}$$

$$\text{Required } A_n = \frac{T_u}{\phi F_u} = \frac{198}{0.75(58)} = 4.55 \text{ sq in.}$$

$$\text{Minimum } A_g = \frac{\text{Required } A_n}{0.85} = \frac{4.55}{0.85} = 5.35 \text{ sq in.}$$

$$\text{Required } t = \frac{4.55}{9 - 2(3/4 + 1/8)} = 0.63 \text{ in.}$$

Block shear will not be of concern with plates as thick as $\frac{5}{8}$ in.; particularly here where 5 bolts per line are used.

Use 2PLs—$\frac{5}{8} \times 9$.

   (d) Flange bolts. The bolts are in single shear,

$$\phi R_n = 19.9 \text{ kips} \quad \text{(single shear, } LRFD \text{ } Manual, \text{ Table 8–11)}$$

$$\phi R_n = \phi(2.4 d_b t F_u) \quad \text{(bearing: LRFD-J3.10)}$$

$$= 0.75(2.4)(3/4)(0.625)58 = 48.9 \text{ kips}$$

$$\text{Number of bolts} = \frac{T_u}{\phi R_n} = \frac{198}{19.9} = 10 \text{ bolts}$$

Use 2 rows of 5 bolts, each side of splice.

   (e) Web bolts. Compute the moment carried by the web plates when F_y is reached at the center of the tension flange plates,

$$\phi_b M_n = \phi_b \left(\frac{t d^2}{6} \right) F_y \left(\frac{10.5}{12.3} \right) = 0.90 \left(\frac{2(0.25)(21)^2}{6} \right) \frac{36(0.85)}{12} = 84.3 \text{ ft-kips}$$

A conservative approach is to determine the force on the web bolts nearest the flange using the elastic vector method,

$$\Sigma x^2 + \Sigma y^2 = 8(1.5)^2 + 4(9)^2 + 4(3)^2 = 378 \text{ in.}^2$$

$$R_{ux} = \frac{My}{\Sigma x^2 + \Sigma y^2} = \frac{84.3(12)9}{378} = 24.1 \text{ kips} \rightarrow$$

$$R_{uy} = \frac{Mx}{\Sigma x^2 + \Sigma y^2} = \frac{84.3(12)1.5}{378} = 4.0 \text{ kips} \downarrow$$

$$R_{uv} = \frac{P}{\Sigma N} = \frac{140}{8} = 17.5 \text{ kips} \downarrow$$

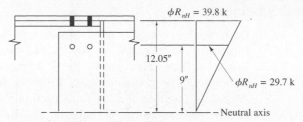

Figure 13.10.4 Check for adequacy of bolts at top of web plates, Example 13.10.1.

$$R_u = \sqrt{[R_{uy} + R_{uv}]^2 + R_{ux}^2}$$
$$R_u = \sqrt{(4.0 + 17.5)^2 + (24.1)^2} = 32.3 \text{ kips} < 36.8 \text{ kips} \quad \text{OK}$$

where, for the W24×84 web (0.470 in.),

$$\phi R_n = 0.75(2.4)(0.75)(0.470)58 = 36.8 \text{ kips.}$$

One further check should be made to see whether the design bolt strength used for the flange splice plate bolts is compatible with the design bolt strength used at the outer web bolts. If a linear deformation relationship as would be caused by bending moment is assumed (Fig. 13.10.4),

Flange horizontal force = 2(19.9) = 39.8 kips   (two bolts)

$$\text{Max } R_{nH} = 39.8\left(\frac{9}{12.05}\right) = 29.7 \text{ kips} > (R_{ux} = 24.1 \text{ kips}) \quad \text{OK}$$

Use 2 vertical rows of 4 bolts each, each side of joint.

The final design is shown in Fig. 13.10.3. End and edge distances should also be checked in accordance with procedures discussed in Chapter 4. ■■

SELECTED REFERENCES

13.1. C. Batho and H. C. Rowan. "Investigations of Beam and Stanchion Connections," 2nd Report, Steel Structures Reseach Committee, Dept. of Scientific and Industrial Research of Great Britain, His Majesty's Stationery Office, London, 1934.

13.2. Omer W. Blodgett. *Design of Welded Structures.* Cleveland, Ohio: James F. Lincoln Arc Welding Foundation, 1966.

13.3. AISC. *Manual of Steel Construction, Volume II Connections*, ASD 9th edition/LRFD 1st edition. Chicago: American Institute of Steel Construction, 1992.

13.4. William A. Thornton. "Load and Resistance Factor Design of Connections," *Proceedings, Solutions in Steel,* National Engineering Conference. Chicago, IL: American Institute of Steel Construction, June 12–14, 1986 (pp. 33-1 to 33-22).

Semi-Rigid Connections

13.5.   Basil Sourochnikoff. "Wind Stresses in Semi-Rigid Connections of Steel Framework," *Transactions,* ASCE, **115** (1950), 382–402.

13.6.   Robert A. Hechtman and Bruce G. Johnston. *Riveted Semi–Rigid Beam-to-Column Building Connections.* Progress Report Number 1. Chicago, IL: American Institute of Steel Construction, November 1947 (118 pp.).

13.7.   Leo Schenker, Charles G. Salmon, and Bruce G. Johnston. *Structural Steel Conections,* Armed Forces Special Weapons Project, Report No. 352, Engineering Research Institute, University of Michigan, June 1954.

13.8.   Stanley D. Lindsey, Socrates A. Ioannides, and Arvind Goverdhan. "LRFD Analysis and Design of Beams with Partially Restrained Connections," *Engineering Journal,* AISC, **19,** 4 (Fourth Quarter 1985), 157–162.

13.9.   Jack H. Brown. "Moments on Beam-Columns with Flexible Connections in Braced Frames," *Engineering Journal,* AISC, **20,** 4 (Fourth Quarter 1986), 157–165.

13.10. Douglas J. Ammerman and Roberto T. Leon. "Behavior of Semi-rigid Composite Connections," *Engineering Journal,* AISC, **21,** 2 (Second Quarter 1987), 53–62.

13.11. Roberto T. Leon, Douglas J. Ammerman, Jihshya Lin, and Robert D. McCauley. "Semi-rigid Composite Steel Frames," *Engineering Journal,* AISC, **21,** 4 (Fourth Quarter 1987), 147–155.

13.12. David A. Nethercot, J. Buick Davison, and Patrick A. Kirby. "Connection Flexibility and Beam Design in Non-sway Frames," *Engineering Journal*, AISC, **22,** 3 (Third Quarter 1988), 99–108.

13.13. Wai-Fah Chen and N. Kishi. "Semirigid Steel Beam-to-Column Connections: Data Base and Modeling," *Journal of Structural Engineering*, ASCE, **115,** 1 (January 1989), 105–119.

13.14. Atorod Azizinamini and James B. Radziminski. "Static and Cyclic Performance of Semirigid Steel Beam-to-Column Connections," *Journal of Structural Engineering,* ASCE, **115,** 12 (December 1989), 2979–2999.

13.15. Roberto T. Leon and Douglas J. Ammerman. "Semi-Rigid Composite Connections for Gravity Loads," *Engineering Journal,* AISC, **27,** 1 (First Quarter 1990), 1–11.

13.16. Reidar Bjorhovde, Andre Colson, and Jacques Brozzetti. "Classification System for Beam-to-Column Connections," *Journal of Structural Engineering,* ASCE, **116,** 11 (November 1990), 3059–3076.

13.17. Riccardo Zandonini and Paolo Zanon. "Beam Design in PR Braced Steel Frames," *Engineering Journal,* AISC, **25,** 3 (Third Quarter 1991), 85–97.

13.18. Roberto Leon. "Composite Semi-Rigid Connections," *Modern Steel Construction,* AISC, October 1992, 18–23.

Simple Shear Connections—Double Angle

13.19. Bruce Johnston and Lloyd F. Green. "Flexible Welded Angle Connections," *Welding Journal*, **19,** 10 (October 1940), 402s–408s.

13.20. Peter C. Birkemoe and Michael I. Gilmor. "Behavior of Bearing Critical

Double-Angle Beam Connections," *Engineering Journal*, AISC, **15,** 4 (4th Quarter 1978), 109–115.

13.21. Joseph A. Yura, Peter C. Birkemoe, and James M. Ricles. "Beam Web Shear Connections: An Experimental Study," *Journal of the Structural Division*, ASCE, **108,** ST2 (February 1982), 311–325.

13.22. AISC. "Predesigned Bolted Framing Angle Connections," *Engineering Journal*, AISC, **19,** 1 (First Quarter 1982), 1–11.

13.23. James M. Ricles and Joseph A. Yura. Strength of Double-Row Bolted-Web Connections, *Journal of Structural Engineering*, ASCE, **109,** 1 (January 1983), 126–142. (Also Ref. 3.12).

13.24. Jung-June R. Cheng and Joseph A. Yura. "Local Web Buckling of Coped Beams," *Journal of Structural Engineering*, ASCE, **112,** 10 (October 1986), 2314–2331.

13.25. Ajaya K. Gupta. "Buckling of Coped Steel Beams," *Journal of Structural Engineering*, ASCE, **110,** 9 (September 1984), 1977–1987; Disc. by Jung-June Cheng and Joseph A. Yura, **112,** 1 (January 1986), 201–204.

13.26. Jung-June R. Cheng, Joseph A. Yura, and C. Philip Johnson. "Lateral Buckling Coped Steel Beams," *Journal of Structural Engineering*, ASCE, **114,** 1 (January 1988), 1–15.

13.27. Jung-June R. Cheng and Joseph A. Yura. "Lateral Buckling Tests on Coped Steel Beams," *Journal of Structural Engineering*, ASCE, **114,** 1 (January 1988), 16–30.

13.28. Abolhassan Astaneh, Marwan N. Nader, and Lincoln Malik. "Cyclic Behavior of Double Angle Connections," *Journal of Structural Engineering*, ASCE, **115,** 5 (May 1989), 1101–1118.

Simple Shear Connections—Single Plate

13.29. Ralph M. Richard, Paul E. Gillett, James D. Kriegh, and Brett A. Lewis. "The Analysis and Design of Single Plate Framing Connections," *Engineering Journal*, AISC, **17,** 2 (Second Quarter 1980), 38–51.

13.30. Ned W. Young and Robert O. Disque. "Design Aids for Single Plate Framing Connections," *Engineering Journal*, AISC, **18,** 4 (Fourth Quarter 1981), 129–148.; Disc. by John D. Griffiths, **19,** 3 (Third Quarter 1982), 179.

13.31. Ralph M. Richard, James D. Kriegh, and David E. Hormby. "Design of Single Plate Framing Connections with A307 Bolts," *Engineering Journal*, AISC, **19,** 4 (Fourth Quarter 1982), 209–213; Disc. by Edward P. Becker, **22,** 1 (First Quarter 1985), 50–51.

13.32. David E. Hormby, Ralph M. Richard, and James D. Kriegh. "Single-Plate Framing Connections with Grade-50 Steel and Composite Construction," *Engineering Journal*, AISC, **21,** 3 (Third Quarter 1984), 125-138.

13.33. Abolhassan Astaneh, Steven M. Call, and Kurt M. McMullin. "Design of Single Plate Shear Connections," *Engineering Journal*, AISC, **26,** 1 (1st Quarter 1989), 21–32. Disc. by Ralph M. Richard, **27,** 3 (Third Quarter 1990), 121–126.

13.34. D. J. L. Kennedy. "Moment–Rotation Characteristics of Shear Connections," *Engineering Journal*, AISC, **6,** 4, (October 1969), 105–115.

Simple Shear Connections—Tee Framing

13.35. Abolhassan Astaneh and Marwan N. Nader. "Design of Tee Framing Shear Connections," *Engineering Journal*, AISC, **26,** 1 (First Quarter 1989), 9–20.

13.36. Abolhassan Astaneh and Marwan N. Nader. "Experimental Studies and Design of Steel Tee Shear Connections," *Journal of Structural Engineering,* ASCE, **116,** 10 (October 1990), 2882–2902.

Seated Beam Connections

13.37. W. Bertwell. Discussion of "Design of Bolts or Rivets Subject to Combined Shear and Tension," by Alfred Zweig, *Engineering Journal*, AISC, **3,** 4 (October 1966), 165–167.

13.38. Bruce G. Johnston and Gordon R. Diets. "Tests of Miscellaneous Welded Building Connections," *Welding Journal*, **21,** 1 (January 1942), 5s–27s.

13.39. Inge Lyse and Norman G. Schreiner. "An Investigation of Welded Seat Angle Connections," *Welding Journal*, **14,** 2 (February 1935), Research Supplement, 1–15.

13.40. Charles W. Roeder. "Results of Experiments on Seated-beam Connections," *Proceedings,* National Engineering Conference & Conference of Operating Personnel. Chicago, IL: American Institute of Steel Construction, April 29–May 2, 1987 (pp. 43-1 to 43-12).

13.41. Charles W. Roeder and Ronald H. Dailey. "The Results of Experiments on Seated Beam Connections," *Engineering Journal,* AISC, **26,** 3 (Third Quarter 1989), 90–95.

13.42. J. H. Garrett, Jr. and R. L. Brockenbrough. "Design Loads for Seated-beam Connections in LRFD," *Engineering Journal,* AISC, **23,** 2 (Second Quarter 1986) 84–88.

13.43. Roger L. Brockenbrough. "Design Loads for Seated-beam Connections," *Proceedings,* National Engineering Conference & Conference of Operating Personnel. Chicago, IL: American Institute of Steel Construction, April 29–May 2, 1987 (pp. 12-1 to 12-16).

13.44. Cyril D. Jensen. "Welded Structural Brackets," *Welding Journal,* **15,** 10 (October 1936), Research Supplement, 9–15.

Triangular Bracket Plates

13.45. Charles G. Salmon. "Analysis of Triangular Bracket-Type Plates," *Journal of the Engineering Mechanics Division,* ASCE, **88,** EM6 (December 1962), 41–87.

13.46. Charles G. Salmon, Donald R. Buettner, and Thomas C. O'Sheridan. "Laboratory Investigation of Unstiffened Triangular Bracket Plates," *Journal of the Structural Division,* ASCE, **90,** ST2 (April 1964) 257–278.

13.47. Lynn S. Beedle et al. *Structural Steel Design.* New York: The Ronald Press Company, 1964 (pp. 550–555).

Continuous Beam-to-Column Connections

13.48. A. B. Onderdonk, R. P. Lathrop, and Joseph Coel. "End Plate Connections in Plastically Designed Structures," *Engineering Journal,* AISC, **1,** 1 (January 1964), 24–27.

13.49. Lynn S. Beedle and Richard Christopher. "Tests of Steel Moment Connections," *Engineering Journal,* AISC, **1,** 4 (October 1964), 116–125.

13.50. Richard T. Douty and William McGuire. "High Strength Bolted Moment Connections," *Journal of the Structural Division,* ASCE, **91,** ST2 (April 1965). 101–128.

13.51. J. S. Huang, W. F. Chen, and L. S. Beedle. "Behavior and Design of Steel Beam-to-Column Moment Connections," *WRC Bulletin 188,* Welding Research Council, New York, October 1973, 1–23.

13.52. J. E. Regec, J. S. Huang, and W. F. Chen. "Test of a Fully Welded Beam-to-Column Connection," *WRC Bulletin 188,* Welding Research Council, New York, October 1973, 24–35.

13.53. John Parfitt, Jr. and Wai F. Chen. "Test of Welded Steel Beam-to-Column Connections," *Journal of the Structural Division*, ASCE, **102,** ST1 (January 1976), 189–202.

13.54. Bruce D. Macdonald. "Moment Connections Weakened by Laminations," *Journal the Structural Division,* ASCE, **105,** ST8 (August 1979), 1605–1619.

13.55. Paul Grundy, Ian R. Thomas, and Ian D. Bennetts. "Beam-to-Column Moment Connections," *Journal of the Structural Division,* ASCE, **106,** ST1 (January 1980), 313–330. Disc. by N. Krishnamurthy, **106,** ST12 (December 1980), 2573–2574.

13.56. W. F. Chen and K. V. Patel. "Static Behavior of Beam-to-Column Moment Connections," *Journal of the Structural Division,* ASCE, **107,** ST9 (September 1981), 1815–1838.

13.57. Jelle Witteveen, Jan W. B. Stark, Frans S. K. Bijlaard, and Piet Zoetemeijer. "Welded and Bolted Beam-to-Column Connections," *Journal of the Structural Division*, ASCE, **108,** ST2 (February 1982), 433–455.

13.58. G. P. Rentschler, W. F. Chen, and G. C. Driscoll. "Beam-to-Column Web Connection Details," *Journal of the Structural Division*, ASCE, **108,** ST2 (February 1982), 393–409.

13.59. Kirit V. Patel and Wai F. Chen. "Nonlinear Analysis of Steel Moment Connections," *Journal of Structural Engineering*, ASCE, **110**, 8 (August 1984), 1861–1874.

13.60. Wai F. Chen and David E. Newlin. "Column Web Strength in Beam-to-Column Connections," *Journal of the Structural Division*, ASCE, **99,** ST9 (September 1973), 1978–1984.

13.61. Wai F. Chen and Irving J. Oppenheim. "Web Buckling Strength of Beam-to-Column Connections," *Journal of the Structural Division*, ASCE, **100,** ST1 (January 1974), 279–285.

13.62. Eugene W. Miller. "Load and Resistance Factor Design of Moment Connections," *Proceedings, Solutions in Steel*, National Engineering Conference.

Chicago, IL: American Institute of Steel Construction, June 12–14, 1986 (pp. 29-1 to 29-22).

13.63. R. Ford Pray and Cyril Jensen. "Welded Top Plate Beam-Column Connections," *Welding Journal*, **35,** 7 (July 1956), 338s-347s.

13.64. J. L. Brandes and R. M. Mains. "Report of Tests of Welded Top-Plate and Seat Building Connections," *Welding Journal*, **24,** 3 (March 1944), 146s–165s.

13.65. William A. Thornton. "Prying Action—A General Treatment," *Engineering Journal*, AISC, **22,** 2 (Second Quarter 1985), 67–75.

13.66. Abolhassan Astaneh. "Procedure for Design and Analysis of Hanger-type Connections," *Engineering Journal*, AISC, **22,** 2 (Second Quarter 1985), 63–66.

13.67. W. A. Thornton. "Strength and Serviceability of Hanger Connections," *Engineering Journal*, AISC, **29,** 4 (Fourth Quarter 1992), 145–149.

End-Plate Moment Connections

13.68. John D. Griffiths. "End Plate Moment Connections–Their Use and Misuse," *Engineering Journal*, AISC, **21,** 1 (First Quarter 1984), 32–34.

13.69. Henning Agerskov. "High-Strength Bolted Connections Subject to Prying," *Journal of the Structural Division*, ASCE, **102,** ST1 (January 1976), 161–175.

13.70. Henning Agerskov. "Analysis of Bolted Connections Subject to Prying," *Journal of the Structural Division*, ASCE, **103,** ST11 (November 1977), 2145–2163.

13.71. N. Krishnamurthy. "A Fresh Look at Bolted End-Plate Behavior and Design," *Engineering Journal*, AISC, **15,** 2 (2nd Quarter 1978), 39–49. Disc. by H. Agerskov, W. McGuire, J. D. Griffiths and J. M. Wooten, N. W. Rimmer, and author, **16,** 2 (2nd Quarter 1979), 54–64.

13.72. Natarajau Krishnamurthy, Horng-Te Huang, Paul K. Jeffrey, and Louie K. Avery. "Analytical $M-\theta$ Curves for End-Plate Connections," *Journal of the Structural Division*, ASCE, **105,** ST1 (January 1979), 133–145.

13.73. Allan P. Mann and Linden J. Morris. "Limit Design of Extended End-Plate Connections," *Journal of the Structural Division*, ASCE, **105,** ST3 (March 1979), 511–526.

13.74. Thomas M. Murray. "Beam-to-Column End-Plate Connections—Column Design Considerations," presented at AISC National Engineering Conference, March 28–30, 1984, Tampa, FL (10 pp.).

13.75. Alan Hendrick and Thomas M. Murray. "Column Web Compression Strength at End-Plate Connections," *Engineering Journal*, AISC, **21,** 3 (Third Quarter 1984), 161–169.

13.76. Yoke Leong Yee and Robert E. Melchers. "Moment–Rotation Curves for Bolted Connections," *Journal of Structural Engineering*, ASCE, **112,** 3 (March 1986), 615–635.

13.77. Thomas M. Murray and Anant R. Kukreti. "Design of 8-bolt Stiffened Moment End Plates," *Engineering Journal*, AISC, **25,** 2 (Second Quarter 1988) 45–53.

13.78. Larry E. Curtis and Thomas M. Murray. "Column Flange Strength at Moment End-plate Connections," *Engineering Journal*, AISC, **26,** 2 (Second Quarter 1989), 41–50.

13.79. Thomas M. Murray. *Extended End-Plate Moment Connections*, Steel Design Guide Series 4 (Publ. No. D804). Chicago: American Institute of Steel Construction, Inc., 1990.

Beam-to-Column Web Moment Connections

13.80. A. Leon Abolitz and Marvin E. Warner. "Bending Under Seated Connections," *Engineering Journal*, AISC, **2,** 1 (January 1965), 1–5.

13.81. Frank W. Stockwell, Jr. "Yield Line Analysis of Column Webs with Welded Beam Connections," *Engineering Journal*, AISC, **11,** 1 (First Quarter 1974), 12–17.

13.82. Richard H. Kapp. "Yield Line Analysis of a Web Connection in Direct Tension," *Engineering Journal*, AISC, **11,** 1 (Second Quarter 1974), 38–41.

13.83. Glenn P. Rentschler, Wai F. Chen, and George C. Driscoll. "Tests of Beam-to-Column Web Moment Connections," *Journal of the Structural Division*, ASCE, **106,** ST5 (May 1980), 1005–1022.

13.84. Joseph M. Hoptay and Heino Ainso. "An Experimantal Look at Bracket-Loaded Webs," *Engineering Journal*, AISC, **18,** 1 (First Quarter 1981), 1–7.

13.85. Bruce E. Hopper, Gordon B. Batson, and Heino Ainso. "Bracket Loaded Webs with Low Slenderness Ratios," *Engineering Journal*, AISC, **22,** 1 (First Quarter 1985), 11–18.

Rigid Frame Knees

13.86. John D. Griffiths. *Single Span Rigid Frames in Steel*. New York: American Institute of Steel Construction, Inc., 1948.

Column Base Plates

13.87. W. A. Thornton. "Design of Base Plates for Wide Flange Columns—A Concatenation of Methods," *Engineering Journal*, AISC, **27,** 4 (Fourth Quarter 1990), 173–174.

13.88. Frank W. Stockwell, Jr. "Preliminary Base Plate Selection," *Engineering Journal*, AISC, **12,** 3 (Third Quarter 1975), 92–99.

13.89. Thomas M. Murray. "Design of Lightly Loaded Steel Column Base Plates," *Engineering Journal*, AISC, **20,** 4 (Fourth Quarter 1983), 143–152.

13.90. Salahuddin Ahmed and Robert R. Kreps. "Inconsistencies in Column Base Plate Design in the New AISC ASD Manual," *Engineering Journal*, AISC, **27,** 3 (Third Quarter 1990). 106–107.

13.91. John T. DeWolf. "Axially Loaded Column Base Plates," *Journal of the Structural Division*, ASCE, **104,** ST5 (May 1978), 781–794.

13.92. Russell S. Fling. "Design of Steel Bearing Plates," *Engineering Journal*, AISC, **7,** 2 (April 1970), 37–40.

13.93. W. A. Thornton. "Design of Small Base Plates for Wide Flange Columns," *Engineering Journal*, AISC, **27,** 3 (Third Quarter 1990), 108–110.

13.94. David T. Ricker. "Some Practical Aspects of Column Base Selection," *Engineering Journal*, AISC, **26,** 3 (Third Quarter 1989), 81–89.

13.95. John T. DeWolf and David T. Ricker. *Column Base Plates,* Steel Design Guide Series 1 (Publ. No. D801). Chicago: American Institute of Steel Construction, Inc., 1990.

13.96. Charles G. Salmon, Leo Schenker, and Bruce G. Johnston. "Moment–Rotation Characteristics of Column Anchorages," *Transactions*, ASCE, **122** (1957), 132–154.

13.97. John T. DeWolf and Edward F. Sarisley. "Column Base Plates with Axial Loads and Moments," *Journal of the Structural Division*, ASCE, **106,** ST11 (November 1980), 2176–2184.

13.98. David P. Thambiratnam and P. Paramasivam. "Base Plates under Axial Loads and Moments," *Journal of Structural Engineering*, ASCE, **112,** 5 (May 1986), 1166–1181.

Splices

13.99. Geoffrey L. Kulak and Deborah L. Green. "Design of Connectors in Web–Flange Beam or Girder Splices," *Engineering Journal*, AISC, **27,** 2 (Second Quarter 1990), 41–48.

PROBLEMS

All problems are to be done according to the AISC Load and Resistance Factor Design or Allowable Stress Design, as indicated by the instructor. The requirement of W section is intended to include W, S, and M sections. All given loads are service loads unless otherwise indicated. If needed, assume service loads are 2/3 of factored loads unless otherwise indicated. Assume lateral support consists of translational restraint but not moment (rotational) restraint, unless otherwise indicated. Assume all standard sections are equally readily available in the indicated grade of steel (even though actually they are not). Assume all contact surfaces are Class A and all holes are standard holes unless otherwise indicated. A design sketch to scale is required for all design problems.

13.1. Compute the maximum factored load (20% dead load and 80% gravity live load) reaction P_u that can be developed for the simple shear (framed beam) connection shown in the accompanying figure. The fasteners are $\frac{7}{8}$-in.-diam A325 bolts in bearing-type connections with threads excluded from the shear planes (A325-X).

a. Neglect eccentricity as is customary on simple shear connections; compare detailed calculation results with *LRFD* or *ASD Manual,* as appropriate.

b. Consider eccentricity of loading; use the ultimate strength approach of Sec. 4.12 for eccentric shear, and use one of the combined shear and tension approaches of Sec. 4.15 for the connection to the column flange.

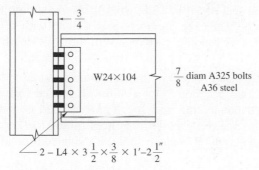

Problems 13.1 and 13.2

13.2. Repeat the analysis requested in Prob. 13.1 except the bolts are in slip-critical connections (A325-SC).

13.3. Design a double angle simple shear (framed beam) connection for a W16×50 beam having a reaction of 20 kips dead load and 30 kips live load to connect to the column flange of a W8×67. Determine the angles to be used, and show the number and placement of fasteners. Use A36 steel. Use an *economical size* A325 bolt in bearing-type connections, with threads excluded from the shear planes (A325-X).

13.4. Redesign the double angle simple shear connection of Prob. 13.3 as a slip-critical connection (A325-SC).

13.5. Design a double angle welded simple shear (framed beam) connection for a W33×118 beam connecting to a column flange which is $\frac{3}{4}$ in. thick. The beam reaction is 25 kips dead load and 100 kips live load. E70 electrodes are to be used with shielded metal arc welding (SMAW), and the base steel is A572 Grade 50. Design using basic principles; then compare with applicable *LRFD* or *ASD Manual* tables.

13.6. Compute the maximum factored load (25% dead load and 75% gravity live load) reaction P_u the W14×30 beam may transmit to the seat angle B. What should be the size for angle A and what is its function? Use $\frac{7}{8}$-in.-diam A325 bolts in a bearing-type connection with threads excluded from the shear planes (A325-X).

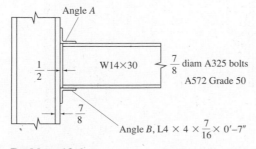

Problem 13.6

13.7. For the unstiffened seat of Prob. 13.6, what is the maximum factored load reaction P_u available when the connection is welded with E70 electrodes using shielded metal arc welding (SMAW). Will the 7-in. length of seat angle be satisfactory?

13.8. Assume a W16×57 beam is to have the maximum factored load (20% dead load and 75% live load) reaction P_u for a uniformly loaded 15-ft span. Design an unstiffened seat connection and specify its welded connection. Show detailed calculations and compare with any applicable *LRFD* or *ASD Manual* tables. Use A36 steel, and E70 electrodes with shielded metal arc welding (SMAW).

13.9. Redesign the unstiffened seat connection of Prob. 13.8 using $\frac{3}{4}$-in.-diam A325 bolts in a *bearing-type* connection (A325-X).

13.10. Redesign the unstiffened seat connection of Prob. 13.8 using $\frac{3}{4}$ in.-diam A325 bolts in a *slip-critical* connection (A325-SC).

13.11. Design an unstiffened seat angle to support a W10×22 beam on the web of a W12×65 column. The beam reaction is 6 kips dead load and 16 kips live load. There is no beam on the opposite side of the web of the column. Use $\frac{3}{4}$-in.-diam A325 bolts in a bearing-type connection with no threads in the shear planes (A325-X) and A36 steel. Show detailed calculations for the design; then compare with applicable *LRFD* or *ASD Manual* tables. Specify a size for the top clip angle.

13.12. Design a welded stiffened seat similar to that of Fig. 13.4.4 attached to a W14×74 column flange. The seat must support a W21×83 having a reaction of 30 kips dead load and 100 kips live load. Use E70 electrodes with shielded metal arc welding (SMAW) and A572 Grade 60 steel.

13.13. Design a welded stiffened seat to support a W18×65 beam having a reaction of 20 kips dead load and 70 kips live load. The beam lies at right angles to the stiffened seat as in Fig. 13.4.2. Use E70 electrodes, shielded metal arc welding (SMAW), and A572 Grade 60 steel.

13.14. Specify the weld size and triangular bracket plate thickness for the situation of the accompanying figure. Use $P = 8$ kips dead load and 25 kips live load, $e = 6$ in., and $L_w = 10$ in. Use E70 electrodes with shielded metal arc welding (SMAW), and A36 steel for base material.

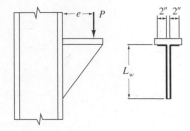

Problems 13.14 and 13.15

13.15. Repeat Prob. 13.14 using $P = 15$ kips dead load and 30 kips live load, $e = 3$ in., and $L_w = 12$ in.

13.16. Design a stiffened seat (bracket plate) consisting of a structural tee that is bolted to the 1-in. flange of a column as in the accompanying figure. The seat plate is welded to the web of the tee. Use $\frac{7}{8}$-in.-diam A325 bolts in a bearing-type connection with threads excluded from the shear planes (A325-X) and A572 Grade 50 steel.

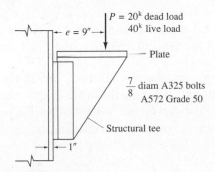

Problem 13.16

13.17. Investigate the adequacy of the connection shown in the accompanying figure. If the number of connectors is not adequate, determine the correct number. Then specify the dimensions of the plates, including their thickness. The load P is 5 kips dead load and 15 kips live load. Use $\frac{7}{8}$-in.-diam A325 bolts in a slip-critical connection (A325-SC).

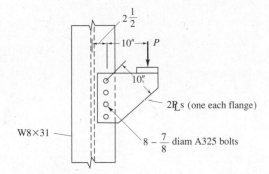

Problem 13.17

13.18. Design a continuous beam-to-column connection for W21×55 beams to connect both flanges of a W14×84 column. Use the type of connection shown in Fig. 13.6.1b if stiffeners are required. Assume the nominal moment strength M_n at the connection is the plastic moment strength M_p of the beams and that elastic analysis has been used to obtain the factored load (20% dead load and 80% live load) internal forces. Use E70 electrodes with shielded metal arc welding and A36 steel for the elements being welded. The factored shear V_u in the beams adjacent to the column is 50 kips.

13.19. Design for the conditions of Prob. 13.18 but consider the W21×55 to frame in only on one side. Use the plate stiffeners as in Fig. 13.6.1b if necessary.

13.20. Design for the conditions of Prob. 13.18 except the beam frames in on only one side and the type of connection shown in Fig. 13.6.1d is to be used.

13.21. The split-tee beam connection shown in the accompanying figure is subject to factored moment M_u of 135 ft-kips and a factored end reaction P_u of 50 kips.

   a. Determine whether or not the tees are adequate, and if not, increase their size.

   b. Determine the number of $\frac{7}{8}$-in.-diam A325 bolts (A325-X) to connect the tees to the column flanges, the tees to the beam flanges, the clip angles to the column flange, and the clip angles to the beam web.

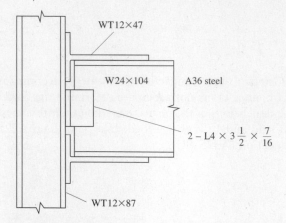

Problem 13.21

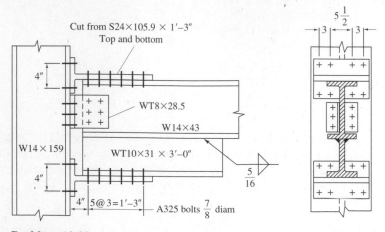

Problem 13.22

13.22. For the split-tee beam connection with the stub beam shown, determine the design moment strength ϕM_n and design beam reaction strength ϕP_n at the face of the column. Use $\frac{7}{8}$-in.-diam A325 bolts in a bearing-type connection with no threads in the shear planes (A325-X), and A36 steel.

13.23. Design a split-tee connection as shown in Fig. 13.6.2c to connect a W16×50 beam to a W14×132 column. Assume the nominal moment strength M_n to be developed is the plastic moment strength M_p of the beam. The factored shear V_u to be carried is 35 kips. Use $\frac{3}{4}$-in.-diam A325 bolts, or if it seems advantageous use A490 bolts, in a bearing-type connection with threads excluded from the shear planes (A325-X or A490-X).

13.24. Redesign the connection of Prob. 13.23 using an end-plate connection as shown in Fig. 13.6.2d.

13.25. Redesign the connection of Prob. 13.21 using an end-plate connection as shown in Fig. 13.6.2d. The factored moment M_u to be carried is 135 ft-kips, and the factored shear V_u is 50 kips.

13.26. Design a vertical tee stiffener (as in Figs. 13.6.3 and 13.6.9) connection to frame a W16×67 beam into the web of a W12×96 column. The beam must develop its plastic moment strength M_p and the factored shear V_u is 52 kips. Use A 572 Grade 60 steel, and shielded metal arc welding (SMAW) with E80 electrodes.

13.27. Redesign the connection of Prob. 13.23 using top plate and seat connection similar to that of Fig. 13.6.12a. Use shielded metal arc welding (SWAW) with E70 electrodes.

13.28. Redesign the connection of Prob. 13.23 using a top and bottom plate connection similar to that of Fig. 13.6.12b. Use shielded metal arc welding (SMAW) with E70 electrodes.

13.29. For the continuous beam connection of the accompanying figure, specify the plate size, weld size using E70 electrodes, and length of weld to develop full plastic moment strength M_p of the W10×39 beam. Consider only plate A and not other aspects of the connection. Use A36 steel and shielded metal arc welding (SMAW) with E70 electrodes.

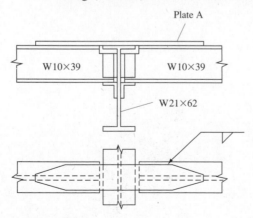

Problem 13.29

13.30. Design a square knee connection (Fig. 13.8.1a or b) to join a W24×84 girder to a W30×108 column. Assume the plastic moment strength M_p must develop in the girder. Use A36 steel and shielded metal arc welding with E70 electrodes.

13.31. Design the column base plate for a W14×211 of A572 Grade 50 steel subject to factored axial load P_u of 1100 kips. The plate is to be supported on a 6-ft square concrete footing having $f'_c = 3000$ psi.

13.32. Design the column base plate for a W12×65 column to carry a factored axial load P_u of 370 kips. The base plate is supported on a 5-ft square concrete footing having $f'_c = 3500$ psi.

13.33. Design a four-plate beam splice for a W14×53 beam of A36 steel. Design for 80% of the moment strength and 60% of the shear strength of the section. Use $\frac{3}{4}$-in.-diam A325 bolts in a bearing-type connection with threads excluded from the shear planes (A325-X).

13.34. Redo Prob. 13.33 for W24×84 beam.

13.35. Redo Prob. 13.33 for W21×55 beam.

13.36. Design an eight-plate splice (plates inside and outside of the flanges and both sides of the web) for a welded girder consisting of $1\frac{7}{8}$×22 flanges and $\frac{7}{16}$×78 web. Use factored moment M_u of 6800 ft-kips and factored shear V_u of 510 kips. Use A36 steel and $\frac{7}{8}$-in.-diam A325 bolts in a slip-critical connection (A325-SC).

Chapter 14

Frames—Braced and Unbraced

14.1 GENERAL

As discussed in Sec. 6.9, the effective length of the column members in frames is dependent on whether the frame is *braced* or *unbraced*. For the *braced* frame the effective length KL is equal to or less than the actual length. For the *unbraced* frame, the effective length KL always is greater than the actual length.

In order to understand frame behavior, consider in Fig. 14.1.1 the forces that arise in a column member of a frame as a result of lateral deflection due to a force such as wind. The moments M_Δ and shears Q_Δ are those portions of the moments and shears required to balance the moment $P\Delta$. In addition, there will be moments and shears caused by gravity loads at the particular floor level. Equilibrium in Fig. 14.1.1a requires

$$P\Delta = Q_\Delta h + 2M_\Delta \qquad (14.1.1)$$

The lateral deflection Δ is commonly called *drift* when it results from wind loading in multistory frames, as shown in Fig. 14.1.2. Drift consists of two parts; that

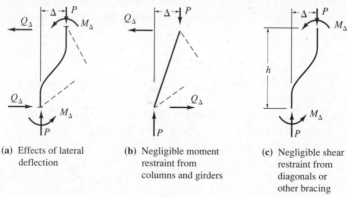

(a) Effects of lateral deflection

(b) Negligible moment restraint from columns and girders

(c) Negligible shear restraint from diagonals or other bracing

Figure 14.1.1   Secondary bending moment due to $P\Delta$ in frames.

Welded unbraced rigid frame (Vierendeel truss).
(Courtesy Bethlehem Steel Corporation)

resulting directly from horizontal load, and that arising from vertical load times the drift.

A frame will deflect under lateral loading such as wind regardless of the pattern of its component members. However, the manner in which equilibrium is maintained against the moment $P\Delta$ differs depending on the restraint conditions. If the building were a vertical pin-jointed truss under lateral loading, there would be no continuity at the joints to allow the moment M_Δ to develop. In which case, as in Fig. 14.1.1b,

$$Q_\Delta = \frac{P\Delta}{h} \tag{14.1.2}$$

Diagonal and horizontal members (web members of the truss) would have to carry the entire shear Q_Δ.

On the other hand, if the members are rigidly joined together but without diagonal members, there would be little shear resistance. Neglecting the shear resistance entirely would result in

$$M_\Delta = \frac{P\Delta}{2} \tag{14.1.3}$$

as in Fig. 14.1.1c. In this case, the girders and columns would have to accommodate the moment M_Δ.

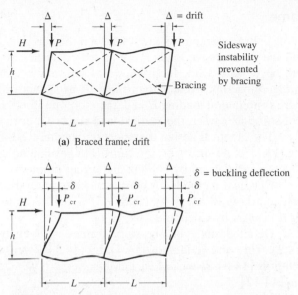

(a) Braced frame; drift

(b) Unbraced frame; drift and sidesway buckling

Figure 14.1.2 Comparison of braced and unbraced frames.

Braced Frame

A braced frame has relatively small moment resistance from columns and girders to counterbalance $P\Delta$, in comparison with the actual restraint from diagonals or other bracing. In other words, Eq. 14.1.2 is assumed to represent the braced frame in simplified design procedures. As is shown later in this chapter, there is both flexural resistance and shear resistance developed in the braced and the unbraced frame. It is the relative magnitudes of these resistances that make the difference between the braced and unbraced frame.

Basically, a braced frame is more appropriately defined as one in which *sidesway buckling is prevented* by bracing elements of the structure other than the structural frame itself. As will be seen in the next section, the theoretical elastic stability analysis of a braced frame assumes no relative joint displacements, which obviously could occur only with infinitely stiff bracing. However, it is practical for design and reasonably correct to assume negligible moment resistance as implied by Eq. 14.1.2, and to assume that for stability purposes the sidesway mode is thereby also prevented.

The term "sidesway" is used to refer to stable elastic lateral movement of a frame, usually due to lateral loads, such as wind. Sidesway buckling is the sudden lateral movement, such as δ of Fig. 14.1.2b, caused by axial loads reaching a certain critical value.

In conclusion, the braced frame accommodates the $P\Delta$ moments by developing shears Q_Δ in the bracing system.

Unbraced Frame

In the unbraced frame, as Fig. 14.1.2b, if the horizontal load H is maintained constant and the compressive loads P are increased sufficiently to cause failure, such failure will occur with a side lurch known as sidesway buckling. The lateral deflection will *suddenly* become greater than the drift as shown in Fig. 14.1.2b. For cases where there is no lateral loading H and, therefore, no initial deflection, the sudden sidesway will still occur when the vertical load reaches a critical value.

The practical design treatment of the unbraced frame assumes that, referring to Fig. 14.1.1c, no shears Q_Δ are capable of developing and Eq. 14.1.3 applies. Any $P\Delta$ effects are balanced by column and girder moments in the unbraced frame.

Classical methods for elastic buckling analysis of frames are widely available, particularly in Bleich [6.9], the *SSRC Guide* [6.8], and recently in *Stability Design of Steel Frames* by Chen and Lui [14.1].

Detailed study of one-story frames is given by Goldberg [14.2], Galambos [14.3], Yura and Galambos [14.4], Lu [14.5], Zweig and Kahn [14.6], Lind [14.7], Schilling [14.8], Zweig [14.9], LeMessurier [12.50, 12.51], Scholz [14.10, 14.11], and Liu [14.12].

Multistory unbraced frames have been treated by Levi, Driscoll, and Lu [14.13], Korn and Galambos [14.14], Switzky and Wang [14.15], Liapunov [14.16], Cheong-Siat-Moy [14.17, 14.18], Cheong-Siat-Moy, Ozer, and Lu [14.19], Lu, Ozer, Daniels, Okten, and Morino [14.20], Haris [14.21], and Gaiotti and Smith [14.22]. Schultz [14.45] has reviewed methods of approximating lateral stiffness of stories in elastic frames. The space frame has been treated by Razzaq and Naim [14.23].

As a practical matter, unbraced multistory frames analysis is done using computer methods, at the very least using a first-order elastic analysis, wherein the original undeformed geometry is used. Computer capability is making possible sophisticated analyses. The link between research and practice has been discussed by White and Hajjar [14.24].

Matrix formulation using stiffness and/or flexibility coefficients is the standard procedure for frame analysis. Wang [14.25] and Halldorsson and Wang [14.26] illustrate the matrix formulation. Frame analysis methods are generally outside the scope of this text. Stiffness and flexibility coefficients are developed in Sec. 14.2 to provide introductory material; stiffness coefficients in the well-known slope-deflection method are used in explaining frame behavior in the remainder of this chapter. Elastic computer-aided analysis of multistory frames is generally available, for example, by Mehringer, Pierson, and Orbison [14.27].

Inelastic Buckling

Since some fibers of a cross-section usually yield prior to frame buckling occurring, inelastic buckling would usually govern the actual strength of a frame. Many studies of inelastic buckling have been conducted, including for braced frames the work of Ojalvo and Levi [14.28], and Levi, Driscoll, and Lu [14.29]. For unbraced frames, the reader is referred to Merchant [14.30], Yura and Galambos [14.4], Lu [14.5], Levi, Driscoll, and Lu [14.13], Korn and Galambos [14.14], Springfield and Adams [14.31], Daniels and Lu [14.32], Liapunov [14.16], Cheong-Siat-Moy [14.17–14.19,

14.33], Lu, Ozer, Daniels, Okten, and Morino [14.20], and Haris [14.21]. Inelastic buckling studies have been extended to the hybrid frame by Arnold, Adams, and Lu [14.34], and to space frames by McVinnie and Gaylord [14.35].

Recently, second-order inelastic analysis has been recommended by King, White, and Chen [14.46], among others.

14.2 ELASTIC BUCKLING OF FRAMES

The distinction between the braced and unbraced frame has been made in Sec. 14.1. In addition, there are two kinds of loading that may contribute to instability. For the *braced* frame of Fig. 14.2.1a, there are no primary bending moments; the only loading is axial compression. The critical load for such a situation is usually defined as a buckling load. No bending of members occurs until the buckling load is reached.

For Fig. 14.2.1b, bending moments exist when the structure is entirely stable. Due to of the rigid joints, moments are transmitted into column members. In addition to the primary bending moments in the column members, the axial compression induces secondary moments equal to P times the deflection. This was discussed in Chapter 12. Under certain combinations of axial compression and moment, the lateral deflection of the column increases without achieving equilibrium; this is usually referred to as instability, or instability in the presence of primary bending moment.

When primary bending moments are present, enough plastic hinges may develop prior to achieving frame instability so that a mechanism forms, in which case for the braced frame the nominal strength is the plastic strength (Fig. 14.2.1c).

The strength of unbraced frames, as shown in Fig. 14.2.2, also may be separated into three categories; buckling in the absence of primary moment (Fig. 14.2.2a), instability in the presence of primary moment (Fig. 14.2.2b), and plastic strength (Fig. 14.2.2c). For the unbraced frame, achieving plastic strength frequently (though not always) means achieving a mechanism associated with overall geometric instability.

The remainder of this section treats the case of elastic buckling in the absence of primary bending moment, with the purpose of having the reader understand the difference in behavior of braced and unbraced frames.

To investigate the elastic stability of a rigid frame, it is first necessary to establish

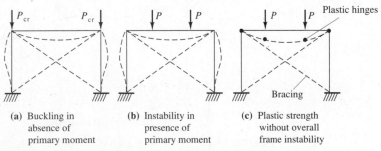

(a) Buckling in absence of primary moment

(b) Instability in presence of primary moment

(c) Plastic strength without overall frame instability

Figure 14.2.1 Strength of braced frames.

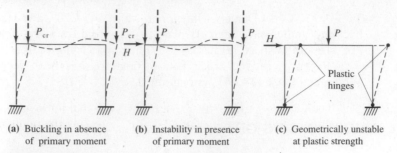

(a) Buckling in absence
of primary moment

(b) Instability in presence
of primary moment

(c) Geometrically unstable
at plastic strength

Figure 14.2.2 Strength of unbraced frames.

the relationships between end moments and end slopes for the individual frame members and then apply the compatibility of deformations requirement for rigid joints.

General Flexibility and Stiffness Coefficients for Beam-Columns

The reader is assumed to have some familiarity with the slope-deflection equations used in frame analysis when axial effect is not considered. For a prismatic section without axial load and without transverse load, as in Fig 14.2.3a,

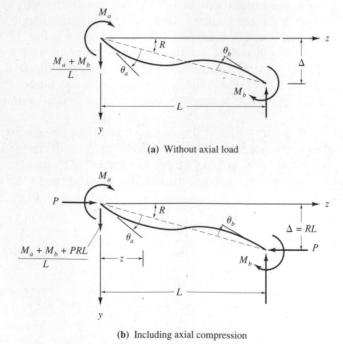

(a) Without axial load

(b) Including axial compression

Figure 14.2.3 Definition of terms and sign convention for slope-deflection equations.

$$M_a = \theta_a\left(\frac{4EI}{L}\right) + \theta_b\left(\frac{2EI}{L}\right) \qquad (14.2.1a)$$

$$M_b = \theta_a\left(\frac{2EI}{L}\right) + \theta_b\left(\frac{4EI}{L}\right) \qquad (14.2.1b)$$

The following treatment, though similar to that for beam-columns in Sec. 12.2, is a more general approach that begins by expressing the moment at any section z of Fig. 14.2.3b:

$$-EI\frac{d^2y}{dz^2} = M_z = M_a + Py - \left(\frac{M_a + M_b + PRL}{L}\right)z \qquad (14.2.2)$$

$$\frac{d^2y}{dz^2} + \frac{P}{EI}y = -\frac{M_a}{EI} + \frac{M_a + M_b}{EIL}z + \frac{P}{EI}Rz \qquad (14.2.3)$$

Letting $k^2 = P/EI$, the solution for Eq. 14.2.3 is

$$y = A\sin kz + B\cos kz - \frac{M_a}{P} + \frac{M_a + M_b}{PL}z + Rz \qquad (14.2.4)$$

Applying the boundary conditions of zero deflection at $z = 0$ and $y = RL$ at $z = L$ gives

$$B = \frac{M_a}{P}$$

and

$$A = \frac{1}{P\sin kL}(M_a\cos kL + M_b)$$

Then

$$y = \frac{M_a}{P}\left[\frac{\sin k(L-z)}{\sin kL} - \frac{(L-z)}{L}\right] - \frac{M_b}{P}\left(\frac{\sin kz}{\sin kL} - \frac{z}{L}\right) + Rz \qquad (14.2.5)$$

Differentiating once to obtain the slope,

$$\frac{dy}{dz} = \frac{M_a}{P}\left[\frac{-k\cos k(L-z)}{\sin kL} + \frac{1}{L}\right] - \frac{M_b}{P}\left(\frac{k\cos kz}{\sin kL} - \frac{1}{L}\right) + R \qquad (14.2.6)$$

when $z = 0$, $\dfrac{dy}{dz} = \theta_a + R$ and when $z = L$, $\dfrac{dy}{dz} = \theta_b + R$.

Letting $\phi^2/L^2 = k^2 = P/EI$, θ_a and θ_b after some manipulation of terms may be expressed* as

---

*The use of the symbol ϕ as the stability parameter throughout this chapter should not be confused with the resistance factor (strength reduction factor) ϕ used in Load and Resistance Factor Design.

$$\theta_a = \frac{M_a L}{EI}\left(\frac{\sin\phi - \phi\cos\phi}{\phi^2\sin\phi}\right) + \frac{M_b L}{EI}\left(\frac{\sin\phi - \phi}{\phi^2\sin\phi}\right) \qquad (14.2.7a)$$

$$\theta_b = \frac{M_a L}{EI}\left(\frac{\sin\phi - \phi}{\phi^2\sin\phi}\right) + \frac{M_b L}{EI}\left(\frac{\sin\phi - \phi\cos\phi}{\phi^2\sin\phi}\right) \qquad (14.2.7b)$$

where the ϕ functions (i.e., expressions within parentheses) are known as *flexibility coefficients*, $f_{ii}, f_{ij}, f_{ji},$ and f_{jj}. To obtain the beam-column counterparts to Eq. 14.2.1, solve Eqs. 14.2.7 (i.e., invert the matrix of coefficients) to obtain

$$M_a = \theta_a\frac{EI}{L}\left(\frac{\phi\sin\phi - \phi^2\cos\phi}{2 - 2\cos\phi - \phi\sin\phi}\right) + \theta_b\frac{EI}{L}\left(\frac{\phi^2 - \phi\sin\phi}{2 - 2\cos\phi - \phi\sin\phi}\right) \qquad (14.2.8a)$$

$$M_b = \theta_a\frac{EI}{L}\left(\frac{\phi^2 - \phi\sin\phi}{2 - 2\cos\phi - \phi\sin\phi}\right) + \theta_b\frac{EI}{L}\left(\frac{\phi\sin\phi - \phi^2\cos\phi}{2 - 2\cos\phi - \phi\sin\phi}\right) \qquad (14.2.8b)$$

where the ϕ functions are known as *stiffness coefficients*. The θ values are the end slopes measured with reference to the axis of the member.

Note that since $\phi^2 = PL^2/EI$, $\phi = 0$ means no axial compression and Eq. 14.2.8 should become Eq. 14.2.1. To verify the coefficient 4 when $\phi = 0$ in Eq. 14.2.8a, the numerator and denominator of the bracketed term must be differentiated four times in accordance with L'Hospital's Rule and then apply the $\phi = 0$ limit.

To simplify the use of Eqs. 14.2.8 for the slope-deflection solution of frame buckling problems, let the stiffness coefficients be referred to as $S_{ii}, S_{ij}, S_{ji},$ and S_{jj}. Because they are symmetrical $S_{ji} = S_{ij}$ and $S_{jj} = S_{ii}$. Thus Eqs. 14.2.8 become

$$M_a = \theta_a\frac{EI}{L}S_{ii} + \theta_b\frac{EI}{L}S_{ij} \qquad (14.2.9a)$$

$$M_b = \theta_a\frac{EI}{L}S_{ij} + \theta_b\frac{EI}{L}S_{ii} \qquad (14.2.9b)$$

Braced Frame—Slope-Deflection Method

The analysis of the rigid frame of Fig. 14.2.4 is presented using the slope-deflection method. Using clockwise rotations and rotational end moments as positive, the slope-deflection equations are as follows, using Eqs. 14.2.9:

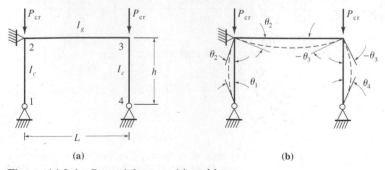

Figure 14.2.4 Braced frame—hinged base.

$$M_{12} = \theta_1 \frac{EI_c}{h} S_{ii} + \theta_2 \frac{EI_c}{h} S_{ij} \qquad (14.2.10)$$

$$M_{21} = \theta_1 \frac{EI_c}{h} S_{ij} + \theta_2 \frac{EI_c}{h} S_{ii} \qquad (14.2.11)$$

$$M_{23} = \theta_2 \frac{EI_g}{L} S_{ii} + \theta_3 \frac{EI_g}{L} S_{ij} = \frac{2EI_g}{L} \theta_2 \qquad (14.2.12)$$

Because no axial compression acts on member 2–3, $S_{ii} = 4$ and $S_{ij} = 2$ for Eq. 14.2.12. Use of symmetry reduces the number of moment equations from six (two for each frame member) to three, and $\theta_3 = -\theta_2$. The equilibrium equations for the joints are

$$M_{12} = 0 \qquad (14.2.13)$$

$$M_{21} + M_{23} = 0 \qquad (14.2.14)$$

Substitution of Eqs. 14.2.10 through 14.2.12 into Eqs. 14.2.13 and 14.2.14 gives

$$\left. \begin{array}{l} \theta_1 \dfrac{EI_c}{h} S_{ii} + \theta_2 \dfrac{EI_c}{h} S_{ij} \qquad\qquad = 0 \\[3mm] \theta_1 \dfrac{EI_c}{h} S_{ij} + \theta_2 \left(\dfrac{EI_c}{h} S_{ii} + \dfrac{2EI_g}{L} \right) = 0 \end{array} \right\} \qquad (14.2.15)$$

Since the θ values cannot be zero when buckling occurs, the determinant of the coefficients of the θs must be zero. Thus the determinant, which is the stability equation, is

$$\left(\frac{EI_c}{h} \right)^2 \left(S_{ii}^2 + \frac{2I_g h}{I_c L} S_{ii} - S_{ij}^2 \right) = 0 \qquad (14.2.16)$$

Since EI_c/h cannot be zero, the other term must be zero. Thus

$$S_{ii} - \frac{S_{ij}^2}{S_{ii}} = -\frac{2I_g h}{I_c L} \qquad (14.2.17)$$

which in terms of ϕ becomes

$$\frac{\phi^2 \sin \phi}{\sin \phi - \phi \cos \phi} = -\frac{2 I_g h}{I_c L} \qquad (14.2.18)$$

EXAMPLE 14.2.1 _____

Determine the buckling load P_{cr} and effective length KL for a braced rigid frame, as in Fig. 14.2.4, which has $I_g = 2100$ in.^{4} (W24×76); $I_c = 796$ in.^{4} (W14×74); $L = 36$ ft; and $h = 14$ ft.

Solution The stability equation to be satisfied is Eq. 14.2.18, which inverted is

$$\frac{\sin \phi - \phi \cos \phi}{\phi^2 \sin \phi} = -\frac{I_c L}{2I_g h} = \frac{-796(36)}{2(2100)14} = -0.487$$

The smallest value of ϕ satisfying the buckling equation is the critical value; i.e.,

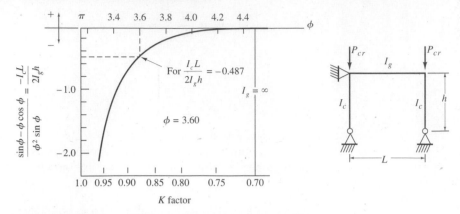

Figure 14.2.5 Braced frame—hinged base, Example 14.2.1.

the one that governs. As I_g approaches zero, an isolated pinned column is indicated, with $\phi^2 = \pi^2$. As girder stiffness increases, ϕ should become greater than π. For an infinitely stiff girder, according to Fig. 6.9.1 which indicates $K = 0.7$, one would expect $\phi^2 = (\pi/0.7)^2 - (4.49)^2$.

The solution for $\phi = 3.60$ obtained by trial using values greater than π but less than 4.49 is shown in Fig. 14.2.5.

Comparing with the isolated pinned column,

$$\frac{\phi^2 EI}{h^2} = \frac{\pi^2 EI}{(Kh)^2} \qquad (14.2.19)$$

it may be noted that the effective length factor K may be expressed

$$K = \frac{\pi}{\phi} \qquad (14.2.20)$$

which for this problem means $K = \pi/3.60 = 0.87$. In other words, to account for frame buckling the column member could be designed using $0.87h$ as the pinned length of the isolated member. The K factor axis is also shown on Fig. 14.2.5 so that for various frame properties one could obtain the K factor directly. Note that a large increase in girder stiffness achieves only a small reduction in K.

The elastic buckling load is

$$P_{cr} = \frac{(3.60)^2 EI_c}{h^2} = \frac{(3.60)^2(29,000)796}{(14)^2(12)^2} = 10,600 \text{ kips} \quad \blacksquare\blacksquare$$

Unbraced Frame—Slope-Deflection Method

Next, the same frame that was treated as braced against sidesway at joint 2 is now to be analyzed as unbraced; i.e., with the horizontal support at joint 2 removed, as in Fig. 14.2.6. Even for the unbraced frame it is conceivable to consider a symmetrical

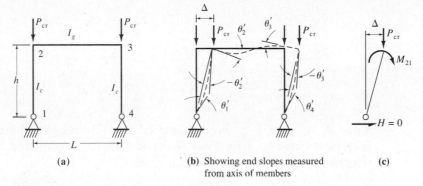

(a)

(b) Showing end slopes measured from axis of members

(c)

Figure 14.2.6 Unbraced frame—hinged base. ($\theta' = \theta$ from Eqs. 14.2.9 and Fig. 14.2.3.)

buckling mode (Fig. 14.2.1a) exactly as when it was braced. It will be demonstrated that the sidesway buckling mode (Fig. 14.2.6b) will occur at a smaller load than the value obtained for the symmetrical case.

Equations 14.2.9 are applied with the added factor that the axes of some members are tilted due to sidesway, so that letting θ represent *total* rotation Δ/h must be subtracted from it to get the end slope θ' measured with respect to the member axis.

$$M_{12} = \left(\theta_1 - \frac{\Delta}{h} \right) \frac{EI_c}{h} S_{ii} + \left(\theta_2 - \frac{\Delta}{h} \right) \frac{EI_c}{h} S_{ij} \qquad (14.2.21)$$

$$M_{21} = \left(\theta_1 - \frac{\Delta}{h} \right) \frac{EI_c}{h} S_{ij} + \left(\theta_2 - \frac{\Delta}{h} \right) \frac{EI_c}{h} S_{ii} \qquad (14.2.22)$$

$$M_{23} = \theta_2 \frac{EI_g}{L} S_{ii} + \theta_3 \frac{EI_g}{L} S_{ij} \qquad (14.2.23)$$

With no axial compression considered on member 2–3, $S_{ii} = 4$ and $S_{ij} = 2$ for Eq. 14.2.23. This time if the structure is symmetrical, the sidesway buckling gives an antisymmetrical deflected curve; thus $\theta_3 = \theta_2$ and only three end moment equations are needed instead of six. The equilibrium equations are

$$M_{12} = 0 \qquad (14.2.24)$$

$$M_{21} + M_{23} = 0 \qquad (14.2.25)$$

which are the same as for the braced case. In addition, the sum of the base shears must be zero since no external horizontal force is acting. Because of antisymmetry $H_1 = -H_4$ so that only one column member needs to be considered. Referring to Fig. 14.2.6c,

$$H = \frac{M_{21} + P_{cr}\Delta}{h} = 0 \qquad (14.2.26)$$

Equations 14.2.24, 14.2.25, and 14.2.26 then become

$$\left.\begin{array}{l} \theta_1 \dfrac{EI_c}{h} S_{ii} + \theta_2 \dfrac{EI_c}{h} S_{ij} \qquad\qquad + \Delta \dfrac{EI_c}{h^2}(-S_{ii} - S_{ij}) \qquad = 0 \\[3mm] \theta_1 \dfrac{EI_c}{h} S_{ij} + \theta_2 \left(\dfrac{EI_c}{h} S_{ii} + \dfrac{6EI_g}{L}\right) + \Delta \dfrac{EI_c}{h^2}(-S_{ii} - S_{ij}) \qquad = 0 \\[3mm] \theta_1 \dfrac{EI_c}{h} S_{ij} + \theta_2 \dfrac{EI_c}{h} S_{ii} \qquad\qquad + \Delta \dfrac{EI_c}{h^2}(\phi^2 - S_{ii} - S_{ij}) = 0 \end{array}\right\} \quad (14.2.27)$$

Since θ_1, θ_2, and Δ cannot be zero (if buckling occurs), the determinate of the coefficients must be zero. Thus algebraic elimination of θ_1 and computation of the remaining four element determinant gives

$$\frac{EI_c}{h^2}(S_{ii}^2 - S_{ij}^2)\left[\frac{6EI_g}{L}\phi^2 \frac{S_{ii}}{(S_{ii}^2 - S_{ij}^2)} + \frac{\phi^2 EI_c}{h} - \frac{6EI_g}{L}\right] = 0 \qquad (14.2.28)$$

Since $S_{ii} \neq S_{ij}$ the bracketed term must be zero. Substitution of the S_{ii} and S_{ij} expressions in terms of ϕ from Eqs. 14.2.8 gives the stability equation,

$$\frac{1}{\phi^2} - \frac{\sin\phi - \phi\cos\phi}{\phi^2 \sin\phi} = \frac{I_c L}{6I_g h} \qquad (14.2.29)$$

or

$$\phi \tan \phi = \frac{6I_g h}{I_c L} \qquad (14.2.30)$$

EXAMPLE 14.2.2 _____

Determine the buckling load P_{cr} and the effective length KL for the unbraced frame consisting of the same members and span as in Example 14.2.1.

Solution

$$\phi \tan \phi = \frac{6I_g h}{I_c L} = \frac{6(2100)(14)}{796(36)} = +6.156$$

which is solved by trial to obtain $\phi = 1.354$, the smallest value satisfying the equation (see Fig. 14.2.7).

$$K = \frac{\pi}{\phi} = \frac{\pi}{1.354} = 2.32$$

Thus if the frame is unbraced it may be designed as an isolated member of effective length $2.32h$, whereas if it were braced the effective length would be $0.87h$. The range of K values may be studied for this frame in Fig. 14.2.7. With hinged bases, even with an infinitely rigid beam the effective length factor K is never less than 2 for an unbraced frame.

The buckling load is

$$P_{cr} = \frac{(1.354)^2 EI_c}{h^2} = \frac{(1.354)^2 29,000(796)}{(14)^2(12)^2} = 1500 \text{ kips}$$

or about $\frac{1}{7}$ of what it was for the same frame when braced. ■■

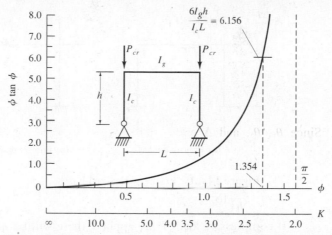

Figure 14.2.7   Unbraced frame—hinged base, Example 14.2.2.

14.3 GENERAL PROCEDURE FOR EFFECTIVE LENGTH

For ordinary design, it is impractical to analyze an entire frame to determine its buckling strength and its effective length (equivalent pinned-end length). Thus it is desirable to have some general way of obtaining the K factor without the full analysis.

The usual procedure for obtaining the effective length factor K is to use the alignment charts of Julian and Lawrence [6.54]. These appear as Fig. 6.9.4 and their use is discussed in Sec. 6.9. The full mathematical development, including particularly the assumptions used, appears in Sec. 14.3 of the first and second editions of this book. Hajjar and White [14.47] have provided an excellent general treatment of effective length as related to unbraced frames. The practical application material from this section appears in Sec. 6.9. For unsymmetrical frames, Chu and Chow [14.36] have presented modifications to permit use of the alignment charts.

14.4 STABILITY OF FRAMES UNDER PRIMARY BENDING MOMENTS

Referring to Figs. 14.2.1 and 14.2.2, the distinction should be noted between (1) buckling in the absence of primary bending moments; i.e., *no* moments exist until buckling occurs; and (2) magnification of primary bending moments (which exist even without the presence of compressive loads) due to axial compression P times deflection Δ. The buckling of frames, which involved solving for the compressive loads P that make a determinant equal to zero, was treated in Sec. 14.2.

This section considers frame behavior when primary moments also exist. Consider a simple rectangular fixed base frame subject to primary moments as in Fig. 14.4.1a. If simple bending theory is used, and the stiffness of the girder is *not* considered reduced due to axial compression, any ordinary procedure of statically indeterminate analysis method will give the moments.

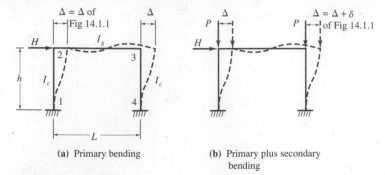

(a) Primary bending

(b) Primary plus secondary bending

Figure 14.4.1   Frame with fixed bases.

If compressive loads P are applied, there will be additional moments $P\Delta$, as in Fig. 14.4.1. The total effect may be treated as a magnification factor times the primary bending moments. This has been discussed in detail in Chapter 12 for members with no translation of joints (braced frames such as in Fig. 14.2.1b), and has been approximately presented for the unbraced frame beam-column in Sec. 12.5.

The following treatment is intended to provide an understanding of the mathematics and compare the actual magnification factor with the simple expression, Eq. 12.5.1, as used in LRFD-C1 and as suggested by ASD Commentary-H1.

Primary Bending of Fixed-Base Frame

To determine magnification of primary moments, the primary moments must first be determined. While design practice could use any method of elastic frame analysis, the slope-deflection method is used here so that the reader may compare what follows with the modifications required to include the $P\Delta$ effect.

For the frame of Fig. 14.4.1a, use Eqs. 14.2.1 noting that θ is the angle measured from the axis of the member, and that for θ to represent the total rotation, Δ/h must be subtracted from it. The end moment equations are

$$\left.\begin{aligned}
M_{12} &= \left(\theta_1 - \frac{\Delta}{h}\right)\frac{4EI_c}{h} + \left(\theta_2 - \frac{\Delta}{h}\right)\frac{2EI_c}{h} \\
M_{21} &= \left(\theta_1 - \frac{\Delta}{h}\right)\frac{2EI_c}{h} + \left(\theta_2 - \frac{\Delta}{h}\right)\frac{4EI_c}{h} \\
M_{23} &= \theta_2\frac{4EL_g}{L} + \theta_3\frac{2EI_g}{L}
\end{aligned}\right\} \qquad (14.4.1)$$

Because of symmetry in the structure,

$$\left.\begin{aligned}
M_{43} &= M_{12} \\
M_{34} &= M_{21} \\
\theta_2 &= \theta_3
\end{aligned}\right\} \qquad (14.4.2)$$

Also, because of the fixed base, $\theta_1 = 0$.

For equilibrium, it is required that

$$M_{21} + M_{23} = 0 \tag{14.4.3}$$

and also the shears on the columns must equal H:

$$\frac{M_{12} + M_{21}}{h} + \frac{M_{43} + M_{34}}{h} + H = 0 \tag{14.4.4}$$

Substitution of Eqs. 14.4.1 into Eqs. 14.4.3 and 14.4.4 gives

$$\left.\begin{array}{l} \theta_2\left(\dfrac{4EI_c}{h} + \dfrac{6EI_g}{L}\right) - \dfrac{\Delta}{h}\left(\dfrac{6EI_c}{h}\right) = 0 \\[3mm] \theta_2\left(\dfrac{12EI_c}{h}\right) \qquad\quad - \dfrac{\Delta}{h}\left(\dfrac{24EI_c}{h}\right) = -Hh \end{array}\right\} \tag{14.4.5}$$

Solving for θ_2 and Δ/h gives

$$\theta_2 = \frac{Hh^2}{4EI_c}\left(\frac{I_c L/I_g h}{I_c L/I_g h + 6}\right) \tag{14.4.6}$$

and

$$\frac{\Delta}{h} = \frac{Hh^2}{12EI_c}\left(\frac{2I_c L/I_g h + 3}{I_c L/I_g h + 6}\right) \tag{14.4.7}$$

Substituting into the M_{12} and M_{21}, Eqs. 14.4.1, gives

$$M_{12} = \frac{-Hh}{2}\left(\frac{I_c L/I_g h + 3}{I_c L/I_g h + 6}\right) \tag{14.4.8}$$

and

$$M_{21} = \frac{-Hh}{2}\left(\frac{3}{I_c L/I_g h + 6}\right) \tag{14.4.9}$$

M_{12} and M_{21} are the primary moments which are magnified when the axial loads P are applied.

Magnification Factor for Fixed-Base Frame

The frame of Fig. 14.4.1b is investigated next to determine the magnified moments M_{12} and M_{21}. Use the slope-deflection equations, Eqs. 14.2.9, that include effect of axial compression on stiffness. Again the rotation angle Δ/h must be subtracted from the full angle to obtain the value measured from the member axis:

$$\left.\begin{array}{l} M_{12} = \left(\theta_1 - \dfrac{\Delta}{h}\right)\dfrac{EI_c}{h}S_{ii} + \left(\theta_2 - \dfrac{\Delta}{h}\right)\dfrac{EI_c}{h}S_{ij} \\[3mm] M_{21} = \left(\theta_1 - \dfrac{\Delta}{h}\right)\dfrac{EI_c}{h}S_{ij} + \left(\theta_2 - \dfrac{\Delta}{h}\right)\dfrac{EI_c}{h}S_{ii} \\[3mm] M_{23} = \theta_2\dfrac{4EI_g}{L} + \theta_3\dfrac{2EI_g}{L} \end{array}\right\} \tag{14.4.10}$$

whereas for primary moment determination, the compression effect on the girder stiffness is not considered. The same symmetry and fixed-base condition ($\theta_1 = 0$) as in Eqs. 14.4.2 apply.

The equilibrium conditions are

$$M_{21} + M_{23} = 0 \tag{14.4.11}$$

the same as when the axial force P was not present; and

$$\frac{M_{12} + M_{21}}{h} + \frac{M_{43} + M_{34}}{h} + \frac{2P\Delta}{h} + H = 0 \tag{14.4.12}$$

where the term $2P\Delta/h$ is added to the previous condition of Eq. 14.4.4. Let $P = \phi^2 EI_c/h^2$.

Substitution of Eqs. 14.4.10 into Eqs. 14.4.11 and 14.4.12, and recalling $M_{43} = M_{12}$ and $M_{34} = M_{21}$, gives

$$\left.\begin{array}{l} \theta_2\left(S_{ii} + \dfrac{6I_g h}{I_c L}\right) - \dfrac{\Delta}{h}(S_{ii} + S_{ij}) \qquad\qquad = 0 \\[3mm] \theta_2[2(S_{ii} + S_{ij})] - \dfrac{\Delta}{h}[4(S_{ii} + S_{ij}) - 2\phi^2] = \dfrac{-Hh^2}{EI_c} \end{array}\right\} \tag{14.4.13}$$

Equations 14.4.13 compare with Eqs. 14.4.5 when Eqs. 14.4.5 are divided by EI_c/h.

Solving Eqs. 14.4.13 for θ_2 and Δ/h gives

$$\frac{\Delta}{h} = \left(\frac{S_{ii} + 6I_g h/I_c L}{S_{ii} + S_{ij}}\right)\theta_2 \tag{14.4.14}$$

and

$$\theta_2 = \frac{Hh^2}{2EI_c}\left(\frac{(S_{ii} + S_{ij})(I_c L/I_g h)}{(S_{ii}^2 - S_{ij}^2 - \phi^2 S_{ii})(I_c L/I_g h) + 6\left[2(S_{ii} + S_{ij}) - \phi^2\right]}\right) \tag{14.4.15}$$

Note that Eq. 14.4.15 becomes Eq. 14.4.6 when $P = 0$. When $P = 0$, $\phi^2 = Ph^2/EI_c = 0$, $S_{ii} = 4$, and $S_{ij} = 2$.

Substituting Eqs. 14.4.14 and 14.4.15 into Eqs. 14.4.10 gives

$$M_{12} = \theta_2\frac{EI_c}{h}(S_{ij} - S_{ii} - 6I_g h/I_c L)$$

$$= \frac{-Hh}{2}\left\{\frac{(S_{ii} + S_{ij})(S_{ii} - S_{ij} + 6I_g h/I_c L)(I_c L/I_g h)}{(S_{ii}^2 - S_{ij}^2 - \phi^2 S_{ii})I_c L/I_g h + 6[2(S_{ii} + S_{ij}) - \phi^2]}\right\}$$

$$M_{12} = \frac{-Hh}{2}\left\{\frac{(S_{ii}^2 - S_{ij}^2)I_c L/I_g h + 6(S_{ii} + S_{ij})}{(S_{ii}^2 - S_{ij}^2 - \phi^2 S_{ii})I_c L/I_g h + 6[2(S_{ii} + S_{ij}) - \phi^2]}\right\} \tag{14.4.16}$$

and

$$M_{21} = \frac{-Hh}{2}\left\{\frac{6(S_{ii} + S_{ij})}{(S_{ii}^2 - S_{ij}^2 - \phi^2 S_{ii})I_c L/I_g h + 6[2(S_{ii} + S_{ij}) - \phi^2]}\right\} \tag{14.4.17}$$

which become the same as Eqs. 14.4.8 and 14.4.9 when $\phi = 0$; i.e., no axial compression.

The magnification factor, the *sway magnifier* B_2 in Load and Resistance Factor terminology, is the ratio of the moment including the $P\Delta$ effect to the primary moment without $P\Delta$. Thus, dividing Eq. 14.4.16 by Eq. 14.4.8 gives the magnifier B_2 for the moment at the bottom of the column,

$$(B_2)_{\text{bottom}} = \frac{\text{Eq. 14.4.16}}{\text{Eq. 14.4.8}} \qquad (14.4.18)$$

and the magnifier B_2 for the moment at the top of the column,

$$(B_2)_{\text{top}} = \frac{\text{Eq. 14.4.17}}{\text{Eq. 14.4.9}} \qquad (14.4.19)$$

Note is made that the magnifier B_2 is applicable only in rectangular frames.

In actual practice it would be easier to do the second-order analysis to obtain the moments as obtained in this section than to compute correctly the magnification factor. The purpose here is to show the relative correctness of the sway magnification factor B_2 as given by LRFD-C1,

$$B_2 = \frac{1}{1 - \dfrac{\Sigma P_u}{\Sigma P_{e2}}} = \frac{1}{1 - \alpha} \qquad (14.4.20)$$

where $\alpha = \Sigma P_u / \Sigma P_{e2}$
ΣP_u = sum of all factored loads in the columns resisting sway
ΣP_{e2} = sum of the Euler buckling loads in the columns resisting sway
$P_{e2} = \pi^2 EI/(KL)^2$
K = effective length factor in the plane of bending ≥ 1.0

Note that columns resisting sway in a story, such as the two columns in Fig. 14.4.1, cannot act independently; thus, α must include the effects on all columns resisting sway in the story. In ASD-Commentary H1, the following sway magnifier is suggested,

$$B_2 = \frac{1 - 0.18\alpha}{1 - \alpha} \qquad (14.4.21)$$

The term P_{e2} is the buckling load occurring in the absence of primary moment. Following the procedure of Sec. 14.3, P_{e2} may be determined for the frame of Fig. 14.4.1 by letting $H = 0$; then setting the determinant of the coefficients in Eqs. 14.4.13 equal to zero. Substitution of the S_{ii} and S_{ij} in terms of the stability parameter ϕ into that determinant will finally give the stability equation,

$$\phi \sin \phi \cos \phi \left(\frac{\cot \phi}{\phi} - \frac{I_c L}{6 I_g h} \right) = 1 \qquad (14.4.22)$$

which for example, if $I_c L / I_g h = 1$, one obtains $\phi_{\text{cr}} = 0.865\pi$. For the fixed-base frame, the buckling load would be

$$P_{e2} = \frac{(0.865\pi)^2 \, EI_c}{h^2}$$

and

$$\alpha = \frac{P}{P_{e2}} = \left(\frac{\phi}{0.865\pi}\right)^2, \quad \text{for} \quad \frac{I_c L}{I_g h} = 1.0$$

The effective length factor K is

$$K = \frac{\pi}{\phi_{cr}} = \frac{1.0}{0.865} = 1.16$$

Table 14.4.1 provides a comparison of the theoretical magnification factor, Eq. 14.4.19 for the top of the column, with the LRFD Eq. 14.4.20 and ASD Eq. 14.4.21.

TABLE 14.4.1 COMPARISON OF THEORETICAL MAGNIFICATION FACTOR B_2 WITH $B_2 = 1/(1 - \alpha)$ for LRFD and $B_2 = (1 - 0.18\alpha)/(1 - \alpha)$ for ASD.[†]

	$\frac{I_c L}{I_g h} = 0.5$		$\frac{I_c L}{I_g h} = 1.0$		$\frac{I_c L}{I_g h} = 5.0$		LRFD	ASD
α	ϕ	Exact*	ϕ	Exact*	ϕ	Exact*	$\dfrac{1}{1-\alpha}$	$\dfrac{1-0.18\alpha}{1-\alpha}$
0.10	0.92	1.102	0.86	1.107	0.66	1.114	1.111	1.091
0.20	1.30	1.230	1.22	1.242	0.94	1.257	1.250	1.205
0.30	1.59	1.394	1.49	1.415	1.15	1.441	1.429	1.351
0.40	1.84	1.612	1.72	1.645	1.32	1.687	1.667	1.547
0.50	2.05	1.918	1.92	1.968	1.48	2.031	2.000	1.820
0.60	2.25	2.377	2.10	2.453	1.62	2.548	2.500	2.230
0.70	2.43	3.143	2.27	3.261	1.75	3.410	3.333	2.913

[†] $\alpha = \Sigma P_u / \Sigma P_{e2}$
*B_2 from Eq. 14.4.19.

Comparison of the values for B_2 in Table 14.4.1 shows that LRFD Eq. 14.4.20 compares favorably with the theoretical values; the ASD Commentary Eq. 14.4.21 is somewhat lower over the entire range of α. In ASD the procedure is to magnify the *entire* moment (both the sway and nonsway parts) by the sway magnifier. This is inherently conservative; thus, the slightly lower magnifier is reasonable.

Actual "unbraced frames" always have attachments which are designed as non-structural elements; however, such items do actually contribute some bracing to the structure. In other words, a real building can never be as flexible as the skeleton elastic frame. In addition, such items as exterior walls, partitions, and stairways, all tend to add to the overall stiffness.

For further study of the instability of frames in the presence of primary moment, the reader is referred to the work of Bleich [6.9], Lu [14.37], McGuire [3.11], Galambos [14.38], and the *SSRC Guide* [6.8].

14.5 BRACING REQUIREMENTS—BRACED FRAME

One of the decisions faced by the designer is the determination of whether a frame is braced or unbraced. Most efficient use of material in compression members is obtained when the frame is braced so that sidesway buckling or instability cannot occur. Some design guidelines are provided by Galambos [13.39]. As to what constitutes bracing, LRFD-C2.1 indicates that a frame is braced when lateral stability is provided by "diagonal bracing, shear walls or equivalent means." ASD-C2.1 indicates a braced frame is one having "adequate attachment to diagonal bracing, to shear walls, to an adjacent structure having adequate lateral stability, or to floor slabs or roof decks secured horizontally by walls or bracing systems parallel to the plane of the frame."

The *amount* of stiffness required to prevent sidesway buckling is not indicated by the AISC Specifications [1.5, 1.16]. It has been suggested [14.39] that "in many instances nonstructural building elements, curtain walls for example, can provide the necessary stiffness against sidesway buckling."

The following presentation from Galambos [14.38] may provide assistance in making the necessary engineering judgment regarding the strength required to create a braced frame. This discussion complements that in Sec. 9.13, where the emphasis was on beam and column bracing.

Stiffness Required from Bracing

It is the objective to use bracing to convert Fig. 14.2.2a into Fig. 14.2.1a. A simple and conservative procedure is to idealize the braced frame, as in Fig. 14.5.1. The following assumptions are made:

1. Columns do not participate in resisting sidesway.
2. Columns are hinged at ends.
3. Bracing acts independently as a spring at the top of columns.

Equilibrium using summation of moments about point 1 of Fig. 14.5.1 gives

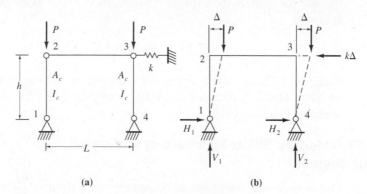

Figure 14.5.1 Idealized bracing arrangements.

$$PΔ + P(Δ + L) - kΔh - V_2L = 0$$

$$V_2 = P + 2P\left(\frac{Δ}{L}\right) - kΔ\frac{h}{L} \qquad (14.5.1)$$

Similarly, summation of moments about point 4 gives

$$V_1 = P - 2P\left(\frac{Δ}{L}\right) + kΔ\frac{h}{L} \qquad (14.5.2)$$

Moments about the hinges at 2 and 3 for free-body diagrams of columns 1-2 and 4-3, respectively, give

$$H_1 = V_1\frac{Δ}{h} \qquad (14.5.3)$$

$$H_2 = V_2\frac{Δ}{h} \qquad (14.5.4)$$

Summation of horizontal forces gives

$$H_1 + H_2 - kΔ = 0 \qquad (14.5.5)$$

Substitution of Eqs. 14.5.1 and 14.5.2 into Eqs. 14.5.3 and 14.5.4, and then into Eq. 14.5.5, gives

$$\frac{Δ}{h}(2P - kh) = 0 \qquad (14.5.6)$$

where Eq. 14.5.6 is the approximate buckling equation. For a simple rectangular frame, as in Fig. 14.2.4, the exact buckling equation is given by Eq. 14.2.18.

From Eq. 14.5.6, substituting the nominal column strength P_n for P, gives for *two* columns

$$k = \frac{2P_n}{h} \qquad (14.5.7)$$

For a frame of several bays, the buckling load P (equal to nominal strength P_n) must include the sum of the loads carried on all columns resisting sway; thus Eq. 14.5.7 becomes

$$\text{Required } k = \frac{\Sigma P_n}{h} \qquad (14.5.8)$$

Equation 14.5.8 gives the stiffness k required for all bracing for the columns involved in ΣP_n. The ΣP_n includes the loads (factored loads $\Sigma P_u/\phi_c$) to be carried on all columns resisting sway.

Bracing Provided by Stiffer Members in an Overall Unbraced System

Overall sidesway buckling can occur only when the total lateral or sidesway resistance to horizontal movement is overcome. When the loading on individual members is less than their strength, the reserve strength can be utilized to provide bracing for other members. Yura [6.55] has presented the generally accepted design approach. More recently, Cheong-Siat-Moy [14.40], de Buen [14.41], Geschwindner [14.42], and Aristizabal-Ochoa [14.43] have treated the concept.

Examine the unbraced frame of Fig. 14.5.2a. Assuming the columns hinged at the junction with the beam, $K = 2.0$ for this cantilever situation. Members A, B, and C were proportioned for the axial loads 100, 300, and 400 kips, respectively. When sidesway occurs as in Fig. 14.5.2b, moments $P\Delta$ are produced at the base and the total load is the 800 kips.

If the system of Fig. 14.5.2 were a *braced* one, the practical effective length

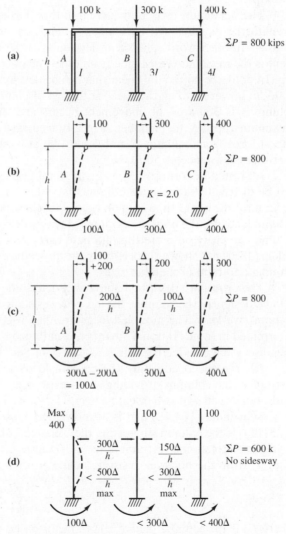

Figure 14.5.2 Stiffer members bracing less stiff member in frame.

factor K would be 1.0 instead of 2.0, and the strengths of the members would be four times as great; 400, 1200, and 1600 kips, respectively, at columns A, B, and C.

Suppose the loads applied to columns B and C are only 200 and 300 kips instead of the maximum strengths of 300 and 400 kips. As in Fig. 14.5.2c, columns B and C will not sidesway buckle until the moments developed at the bottom reach 300Δ and 400Δ, respectively. Those columns can therefore sustain horizontal forces at their tops sufficient to cause sidesway buckling. Column C and column B can each develop $100\Delta/h$ as a shear. These resistances are additive and act as horizontal restraint at the top of column A. Thus column A can carry the 100 kips original load, plus 200 kips as a result of the extra bracing from columns B and C. The total frame load is still 800 kips.

The maximum increase in strength that horizontal bracing can provide is the strength the member will have when its ends cannot translate with respect to one another. In other words, the shear developed at the top to prevent horizontal movement is the maximum resistance from other members that is usable. For instance, in Fig. 14.5.2d, the horizontal shear that can be developed based on columns B and C is $500\Delta/h$; however, when the shear at the top of column A is $300\Delta/h$ and the load on column A is 300 kips, no movement occurs and the member is fully braced. The maximum capacity for member A in a braced system is 400 kips and the resisting lateral force from columns B and C must be at least $400\Delta/h$, which is less than the maximum resistance of $500\Delta/h$.

As stated by Yura [6.55], ". . . the total gravity loads which produce sidesway can be distributed among the columns in a story in any manner. Sidesway will not occur until the total frame load on a story reaches the sum of the *potential* individual column loads for the *unbraced* frame." There still is the limitation that an individual column can carry no more than it could carry in a braced frame; i.e., with $K = 1$. Salem [14.6] has shown theoretically this procedure is valid regardless of the type of framing and ratio of member sizes.

The procedure described above will determine the load for which each column is to be designed; the columns having sway resistance carry more than their tributary compressive load. The modified loads are then used with effective length factors K determined in some "rational" manner, usually from the alignment chart, Fig. 6.9.4b (see discussion of effective length factors K in Sec. 6.9).

Alternative to adjusting the column loads to reflect the differences in sway resistance, the columns providing sway resistance could be designed for their actual loads, but use an *adjusted* effective length factor K. This approach has been presented by Geschwindner [14.42] and is acknowledged as acceptable by the *LRFD Manual* (p. 3-11). If the column supporting the "leaner" column (i.e., a column idealized as having zero rotational restraint at its ends) must carry its actual load P_u along with a load Q_u from the nonsway-resisting columns, then the Euler buckling load will be

$$P_u + Q_u = \frac{\pi^2 EI}{K_0^2 L^2} \tag{14.5.9}$$

where K_0 is the actual effective length factor based on actual frame members.

If one wishes the design of the column to be based on the actual column load P_u, then an adjusted effective length factor K_n must be used. The Euler equation becomes

$$P_u = \frac{\pi^2 EI}{K_n^2 L^2} \tag{14.5.10}$$

Solving Eq. 14.5.10 for $\pi^2 EI/L^2$ and substituting into Eq. 14.5.9 gives

$$K_n = K_0 \sqrt{\frac{P_u + Q_u}{P_u}} \tag{14.5.11}$$

where P_u is the actual factored gravity load to be carried by the frame sway-resisting column, and Q_u is the total factored gravity on all the "leaner" columns whose sway-resistance must be provided by the column carrying P_u.

Example 15.3.4 employing this concept is given in Chapter 15 on the design of rigid frames.

Diagonal Bracing

When cross-bracing is used, it is generally assumed that it can only act in tension; i.e., the diagonal which would be in compression buckles slightly and becomes inactive. Under a horizontal force F the diagonal brace in Fig. 14.5.3 must carry the force

$$\text{Brace force} = \frac{F}{\cos \alpha} \tag{14.5.12}$$

and the elongation of the brace, $\Delta \cos \alpha$, is

$$\text{Elongation} = \frac{(\text{brace force})(\text{brace length})}{(\text{area of brace})E} \tag{14.5.13}$$

or

$$\Delta \cos \alpha = \frac{(F/\cos \alpha)\sqrt{h^2 + L^2}}{A_b E} \tag{14.5.14}$$

Solving for F, using $\cos \alpha = L/\sqrt{h^2 + L^2}$, gives

$$F = \frac{A_b E L^2}{(h^2 + L^2)^{3/2}} \Delta \tag{14.5.15}$$

Since $F = k\Delta$ according to Fig. 14.5.1b,

$$k = \frac{A_b E L^2}{(h^2 + L^2)^{3/2}} \tag{14.5.16}$$

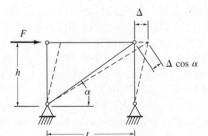

Figure 14.5.3 Deformation of the bracing member.

Substitution of $k = \Sigma P_n/h$ into Eq. 14.5.16 and solving for A_b gives

$$A_b = \frac{\left[1 + \left(\dfrac{L}{h}\right)^2\right]^{3/2} \Sigma P_n}{\left(\dfrac{L}{h}\right)^2 E} \tag{14.5.17}$$

EXAMPLE 14.5.1 _____

Determine the area for the diagonal brace shown in Fig. 14.5.3 required to convert the unbraced frame of Example 14.2.2, Fig. 14.2.6a, into a braced frame.

Solution   For a frame with $I_g = 2100$ in.^{4}, $I_c = 796$ in.^{4}, $L = 36$ ft, and $h = 14$ ft, the *braced* frame buckling load from Example 14.2.1 is

$$P_{cr} = \frac{(3.60)^2 EI_c}{h^2}$$

Since there are two columns, $\Sigma P_n = 2P_{cr}$. Substituting ΣP_n into Eq. 14.5.17 gives the required area A_b of the bracing as

$$A_b = \frac{\left[1 + \left(\dfrac{36}{14}\right)^2\right]^{3/2} 2(3.60)^2 E(796)}{\left(\dfrac{36}{14}\right)^2 E(14)^2(144)}$$

$$= \frac{2(7.6)^{3/2}(12.95)(796)}{6.6(196)(144)} = 2.3 \text{ sq in.}$$

The brace area required is about 10% of the column cross-sectional area (21.8 in. for W14×74). It is noted that the example unbraced frame was very flexible with a higher than usual effective length factor K of 2.32; thus a somewhat heavier than usual brace was required.   ■■

14.6 OVERALL STABILITY WHEN PLASTIC HINGES FORM

General

This chapter has focused primarily on basic concepts of frame stability, using the single-story frame as an example. LRFD-C1 states "Second order ($P\Delta$) effects shall be considered in the design of frames." For one- and two-story frames, the second-order $P\Delta$ effect rarely affects the design, however. When plastic analysis is used as permitted under LRFD-A5.1, consideration of second-order effects will be impractical and will offset any advantage of using plastic analysis. Traditionally, in one- and perhaps two-story lightly loaded frames, the $P\Delta$ effect has been neglected (seemingly without detrimental effects on the structure). In fact, for one- and two-story frames, plastic design in ASD-N3* states "the maximum strength is permitted to be determined by a routine plastic analysis procedure and ignore the frame instability effect ($P\Delta$)." Load and Resistance Factor Design does not provide such a waiver. Under these conditions, plastic design may be the method of choice for such frames.

For multistory *braced* frames, the design of the bracing system must include the $P\Delta$ effect. For multistory *unbraced* frames, the $P\Delta$ effect must be included directly in the calculations for maximum strength.

Braced Frames

Braced frames are usually designed to cause any plastic hinges associated with the failure mechanism to form in the girders. For one- and two-story frames designed using plastic design, the $P\Delta$ effect may be ignored according to ASD-N.3. However, under LRFD-C1 such effects must be included. The columns are then designed as beam-columns in accordance with concepts treated in Chapter 12. Design procedures for one- and two-story braced frames are presented in Chapter 15.

Multistory braced frames may be designed in accordance with *Plastic Design of Braced Multistory Steel Frames* [14.44].

Unbraced Frames

The plastic design provisions of ASD-N3.2 state "The strength of an unbraced multistory frame shall be determined by a rational analysis which includes the effect of frame instability and column axial deformation." Such a frame must be stable under factored gravity loads, as well as factored gravity plus factored horizontal loads (such as wind or earthquake). Further, for either plastic design (ASD-N3.2) or plastic analysis under LRFD-C2.2 the factored axial force in columns may not exceed $0.75A_g F_y$.

The treatment of plastic analysis for multistory unbraced frames is beyond the scope of this text; there is currently (1995) no generally recognized acceptable procedure for such an analysis. The reader is referred to work of Cheong-Siat-Moy, Ozer, and Lu [14.19], Springfield and Adams [14.31], Liapunov [14.16], Daniels and Lu [14.32], and LeMessurier [12.50, 12.51] for techniques of evaluating the strength of multistory unbraced frames.

SELECTED REFERENCES

14.1. W. F. Chen and E. M. Lui. *Stability Design of Steel Frames.* Boca Raton, FL:CRC Press, 1991.

14.2. John E. Goldberg. "Buckling of One-Story Frames and Buildings," *Journal of the Structural Division,* ASCE, **86**, ST10 (October 1960), 53–85. Also *Transactions,* **126** (1961), Part II, 482–515.

---

* The notation ASD is also used here for plastic design since that topic occurs in Chapter N of the 1989 ASD Specification; all other chapters actually relate to Allowable Stress Design (ASD).

14.3. Theodore V. Galambos. "Influence of Partial Base Fixity on Frame Stability," *Journal of the Structural Division*, ASCE, **86,** ST5 (May 1960), 85–108. Also *Transactions,* **126** (1961), Part II, 929–969.

14.4. Joseph A. Yura and Theodore V. Galambos. "Strength of Single-Story Steel Frames," *Journal of the Structural Division*, ASCE, **91,** ST5 (October 1965), 81–101.

14.5. Le-Wu Lu. "Inelastic Buckling of Steel Frames," *Journal of the Structural Division*, ASCE, **91,** ST6 (December 1965), 185–214.

14.6. Alfred Zweig and Albert Kahn. "Buckling Analysis of One-Story Frames," *Journal of the Structural Division*, ASCE, **94,** ST9 (September 1968), 2107–2134. Disc. by Adel Helmy Salem, **95,** ST5 (May 1969), 1017–1029.

14.7. Niels C. Lind. "Simple Illustration of Frame Instability," *Journal of the Structural Division*, ASCE, **103,** ST1 (January 1977), 1–8.

14.8. C. G. Schilling. "Buckling of One-Story Frames," *Engineering Journal*, AISC, **20,** 2 (Second Quarter 1983), 49–57. Disc. by Dan S. Correnti, Alfred Zweig, John Springfield, and author, **21,** 4 (Fourth Quarter 1984), 207–215.

14.9. Alfred Zweig. "Force Method for Frame Buckling Analysis," *Journal of Structural Engineering*, ASCE, **110,** 8 (August 1984), 1893–1912.

14.10. H. Scholz. "*P*-Delta Effect in Elastic Analysis of Sway Frames," *Journal of Structural Engineering*, ASCE, **113,** 3 (March 1987), 534–545. Errata, **113,** 12 (December 1987), 2525. Also REF 12.54.

14.11. H. Scholz. "*P*-Delta Effect under Repeated Loading," *Journal of Structural Engineering*, ASCE, **116,** 8 (August 1990), 2070–2082.

14.12. Eric M. Lui. "A Practical *P*-Delta Analysis Method for Type FR and PR Frames," *Engineering Journal*, AISC, **25,** 3 (Third Quarter 1988), 85–98.

14.13. Victor Levi, George C. Driscoll, Jr., and Le-Wu Lu. "Analysis of Restrained Columns Permitted to Sway," *Journal of the Structural Division*, ASCE, **93,** ST1 (February 1967), 87–107.

14.14. Alfred Korn and Theodore V. Galambos. "Behavior of Elastic-Plastic Frames," *Journal of the Structural Division*, ASCE, **94,** ST5 (May 1968), 1119–1142.

14.15. Harold Switzky and Ping Chun Wang. "Design and Analysis of Frames for Stability," *Journal of the Structural Division*, ASCE, **95,** ST4 (April 1969), 695–713.

14.16. Sviatoslav Liapunov. "Ultimate Strength of Multistory Steel Rigid Frames," *Journal of the Structural Division*, ASCE, **100,** ST8 (August 1974), 1643–1655.

14.17. Francois Cheong-Siat-Moy. "Inelastic Sway Buckling of Multistory Frames," *Journal of the Structural Division*, ASCE, **102,** ST1 (January 1976), 66–75.

14.18. Francois Cheong-Siat-Moy. "Multistory Frame Design Using Story Stiffness Concept," *Journal of the Structural Division*, ASCE, **102,** ST6 (June 1976), 1197–1212.

14.19. Francois Cheong-Siat-Moy, Erkan Ozer, and Le-Wu Lu. "Strength of Steel Frames under Gravity Loads," *Journal of the Structural Division*, ASCE, **103,** ST6 (June 1977), 1223–1235.

14.20. Le-Wu Lu, Erkan Ozer, J. Hartley Daniels, Omer S. Okten, and Shosuke Morino. "Strength and Drift Characteristics of Steel Frames," *Journal of the Structural Division,* ASCE, **103,** ST11 (November 1977), 2225–2241.

14.21. Ali A. K. Haris. "Approximate Stiffness Analysis of High-Rise Buildings," *Journal of the Structural Division,* ASCE, **104,** ST4 (April 1978), 681–696.

14.22. Regina Gaiotti and Bryan Stafford Smith. "*P*-Delta Analysis of Building Structures," *Journal of Structural Engineering,* ASCE, **115,** 4 (April 1989), 755–770.

14.23. Zia Razzaq and Moossa M. Naim. "Elastic Instability of Unbraced Space Frames," *Journal of the Structural Division,* ASCE, **106,** ST7 (July 1980), 1389–1400.

14.24. Donald W. White and Jerome F. Hajjar. "Application of Second-Order Elastic Analysis in LRFD: Research to Practice," *Engineering Journal,* AISC, **28,** 1 (Fourth Quarter 1991), 133–148.

14.25. Chu-Kia Wang. "Stability of Rigid Frames with Nonuniform Members," *Journal of the Structural Division,* ASCE, **93,** ST1 (February 1967), 275–294.

14.26. Ottar P. Halldorsson and Chu-Kia Wang. "Stability Analysis of Frameworks by Matrix Methods," *Journal of the Structural Division,* ASCE, **94,** ST7 (July 1968), 1745–1760.

14.27. Vincent Mehringer, George Pierson, and James G. Orbison. "Computer-Aided Analysis and Design of Steel Frames," *Engineering Journal,* AISC, **22,** 3 (Third Quarter 1985), 143–149.

14.28. M. Ojalvo and V. Levi. "Columns in Planar Continuous Structures," *Journal of the Structural Division,* ASCE, **89,** ST1 (February 1963), 1–23.

14.29. Victor Levi, George C. Driscoll, Jr., and Le-Wu Lu. "Structural Subassemblages Prevented from Sway," *Journal of the Structural Division,* ASCE, **91,** ST5 (October 1965), 103–127.

14.30. W. Merchant. "The Failure Load of Rigid Jointed Frameworks as Influenced by Stability," *Structural Engineer,* **32** (July 1954), 185–190.

14.31. John Springfield and Peter F. Adams. "Aspects of Column Design in Tall Steel Buildings," *Journal of the Structural Division,* ASCE, **98,** ST5 (May 1972), 1069–1083.

14.32. J. Hartley Daniels and Le-Wu Lu. "Plastic Subassemblage Analysis for Unbraced Frames," *Journal of the Structural Division,* ASCE, **98,** ST8 (August 1972), 1769–1788.

14.33. Francois Cheong-Siat-Moy. "Consideration of Secondary Effects in Frame Design," *Journal of the Structural Division,* ASCE, **103,** ST10 (October 1977), 2005–2019.

14.34. Peter Arnold, Peter F. Adams, and Le-Wu Lu. "Strength and Behavior of an Inelastic Hybrid Frame," *Journal of the Structural Division,* ASCE, **94,** ST1 (January 1968), 243–266.

14.35. William W. McVinnie and Edwin H. Gaylord. "Inelastic Buckling of Unbraced Space Frames," *Journal of the Structural Division,* ASCE, **94,** ST8 (August 1968), 1863–1885.

14.36. Kuang-Han Chu and Hsueh-Lien Chow. "Effective Column Length in Unsymmetrical Frames," *Publications,* International Association for Bridge and Structural Engineering, **29–I,** 1969, 1–15.

14.37. Le-Wu Lu. "Stability of Frames Under Primary Bending Moments," *Journal of the Structural Division,* ASCE, **89,** ST3 (June 1963), 35–62.

14.38. Theodore V. Galambos. *Structural Members and Frames.* Englewood Cliffs, NJ: Prentice-Hall, Inc., 1968 (pp. 176–189).

14.39. Theodore V. Galambos. "Lateral Support for Tier Building Frames," *Engineering Journal,* AISC, **1,** 1 (January 1964), 16–19. Disc. by Ira Hooper, **1,** 4 (October 1964), 141.

14.40. Francois Cheong-Siat-Moy. "Column Design in Gravity-Loaded Frames," *Journal of Structural Engineering,* ASCE, **117,** 2 (May 1991), 1448–1461.

14.41. Oscar de Buen. "Column Design in Steel Frames Under Gravity Loads," *Journal of Structural Engineering,* ASCE, **118,** 10 (October 1992), 2928–2935.

14.42. Louis F. Geschwindner. "A Practical Approach to the 'Leaning' Column," *Engineering Journal,* AISC, **32,** 2 (Second Quarter 1995), 63–72.

14.43. J. Dario Aristizabal-Ochoa. "Slenderness K Factor for Leaning Columns," *Journal of Structural Engineering,* ASCE, **120,** 10 (October 1994), 2977–2991.

14.44. AISC. *Plastic Design of Braced Multi-Story Frames.* New York: American Institute of Steel Construction, 1968.

14.45. Arturo E. Schultz. "Approximating Lateral Stiffness of Stories in Elastic Frames," *Journal of Structural Engineering,* ASCE, **118,** 1 (January 1992), 243–263.

14.46. W. S. King, D. W. White, and W. F. Chen. "Second-Order Inelastic Analysis Methods for Steel-Frame Design," *Journal of Structural Engineering,* ASCE, **118,** 2 (February 1992), 408–428.

14.47. Jerome F. Hajjar and Donald W. White. "The Accuracy of Column Stability Calculations in Unbraced Frames and the Influence of Columns with Effective Length Factors Less Then One," *Engineering Journal,* AISC, **31,** 3 (3rd Quarter 1994), 81–97.

Chapter 15

Design of Rigid Frames

15.1 INTRODUCTION

Concepts and procedures developed in previous chapters are combined in this chapter by the use of illustrative examples. Plastic analysis is extended from the continuous beam treatment in Chapter 10 to one-story frames. The reader is assumed to be familiar with elastic statically indeterminate analysis methods.

There are four approaches that can be used for designing one-story frames: (1) factored load elastic analysis according to Load and Resistance Factor Design [1.16]; (2) factored load plastic analysis according to LRFD [1.16] provided special conditions of LRFD-A5.1 are satisfied; (3) plastic analysis and design according the Chapter N of 1989 ASD Specification [1.5]; and (4) service load elastic analysis according to Allowable Stress Design [1.5].

Plastic design [1.5, Chapter N] is likely the method of choice for one-story frames consisting of rolled W sections. When sections used have their limit states controlled by stability (lateral-torsional buckling, local flange buckling, or local web buckling), factored load elastic analysis under LRFD will be the preferred method. Plastic analysis under LRFD *requires* consideration of the $P\Delta$ effect, even in one-story frames, whereas plastic design [1.5, Chapter N] specifically permits neglecting such effect.

15.2 PLASTIC ANALYSIS OF ONE-STORY FRAMES

As discussed in Chapter 10 for continuous beams, the plastic strength of a structure may be obtained either by using the equilibrium method, or the energy method. In braced frames, where joints cannot displace (i.e., no sidesway), the plastic strength may be obtained exactly as for continuous beams. For unbraced frames the sidesway mechanism creates a somewhat more complicated analysis than that for continuous beams. The following examples are intended to illustrate principles of plastic analysis for one-story unbraced frames.

A more extended treatment of plastic analysis is available in *Plastic Design in Steel* [15.1], in books by Beedle [15.2] and Massonnet and Save [15.3] devoted entirely to plastic analysis and design, and papers by Beedle [15.4] and Estes [15.5].

Welded multistory rigid frame. (Courtesy Bethlehem Steel Corporation)

Equilibrium Method

As discussed in Sec. 10.2, equilibrium must be satisfied at every stage of loading from a small load until the collapse mechanism has been achieved. When a sufficient number of plastic hinges have been developed to allow instantaneous hinge rotations without developing increased resistance, a mechanism is said to have occurred.

EXAMPLE 15.2.1

Determine the plastic strength for the frame of Fig. 15.2.1, using the equilibrium method.

Solution   The elastic moment diagram shape is shown in Fig. 15.2.1b. For ease in solution it is often desirable to show the diagram with a horizontal baseline as shown in Fig. 15.2.1d. Assuming overall frame stability is adequate, the collapse mechanism is that shown in Fig. 15.2.1c.

The simple beam moment for the girder is

$$M_s = \frac{W_n L}{4} \tag{a}$$

which when superimposed on the moments at the ends of the girder ($H \times h$) gives for equilibrium when the mechanism occurs,

$$M_p = M_s - M_p = \frac{W_n L}{4} - M_p \tag{b}$$

$$W_n = \frac{8M_p}{L} \tag{c}$$

From the concepts of plastic analysis it might have been expected that only two plastic hinges should be required for a collapse mechanism, since the structure is statically inderterminate to the first degree. In this case both corner hinges form simultaneously, since the horizontal base reactions are equal and opposite. Depending on the ratio of h to L either the midspan plastic hinge, or the two

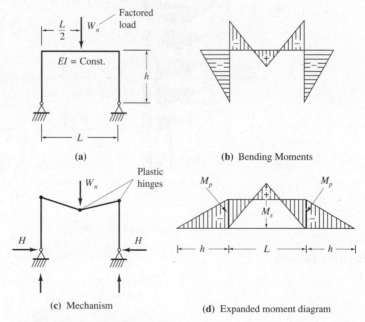

(a)

(b) Bending Moments

(c) Mechanism

(d) Expanded moment diagram

Figure 15.2.1   Example 15.2.1

corner plastic hinges form. If, however, h/L is large, the corner plastic hinges occur first and the structure has reached its collapse condition. If the beam and columns are of different cross-section the structure can be designed so the positive moment plastic hinge occurs first. This should be the objective. The occurrence first of the corner plastic hinges creates overall frame instability prior to utilizing the flexural strength of the girder; such a result should be avoided. ■■

EXAMPLE 15.2.2 _____

Determine the plastic strength of the same frame as in Example 15.2.1 except in addition apply a horizontal load $0.5W_n$ at the top of the column (see Fig. 15.2.2a). Use the equilibrium method.

Solution Again, two plastic hinges should provide the collapse mechanism. This time, however, additional mechanisms are possible, as shown in Fig. 15.2.2. It is possible that (a) plastic hinges form at points 2 and 4 (Fig. 15.2.2b); (b) plastic hinges form at points 3 and 4 (Fig. 15.2.2c); and also (c) plastic hinges form at points 2, 3, and 4. The last situation would occur only if two of the hinges formed simultaneously.

(a) Plastic hinges at points 2 and 4,

$$M_s = \frac{W_n L}{4} \tag{a}$$

Equilibrium requires the positive moment M_p at point 2 $(0.5W_n h - Hh)$ to equal the negative moment M_p at point 4 (Hh):

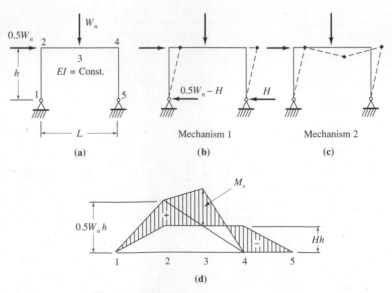

Figure 15.2.2   Example 15.2.2.

$$M_p = 0.5W_n h - M_p \tag{b}$$

$$M_p = \frac{W_n h}{4}; \qquad W_n = \frac{4M_p}{h} \tag{c}$$

Further, in order that Eq. (c) is valid, the resulting moment at point 3 cannot exceed M_p:

$$M_3 = M_s + \tfrac{1}{2}(0.5W_n h) - M_p \tag{d}$$

$$= \frac{W_n L}{4} + \frac{W_n h}{4} - \frac{W_n h}{4} = \frac{W_n L}{4} \le M_p \tag{e}$$

Equation (e) requires $W_n L/4 < W_n h/4$, which means if $L < h$ plastic hinges form at points 2 and 4. If $L > h$ the plastic hinges will form at points 3 and 4.

(b) Plastic hinges at points 3 and 4. For this, Eq. (d) is to be used letting $M_3 = M_p$,

$$M_p = M_s + \tfrac{1}{2}(0.5W_n h) - M_p \tag{f}$$

$$M_p = \frac{W_n}{8}(L + h) \tag{g}$$

$$W_n = \frac{8M_p}{L + h} \tag{h}$$

The above analysis has assumed the same moment of inertia (constant M_p) for both girder and column. If the girder and column are different, a combined mechanism with plastic hinges at points 2, 3, and 4 could be achieved. ■■

EXAMPLE 15.2.3 _____

Determine the plastic strength W_n for the gabled frame of Fig. 15.2.3 using the equilibrium method. All elements of the frame are identical, having a plastic moment strength M_p of 200 ft-kips.

Solution   Equilibrium requires a compatibility between the moment diagrams of the component loadings as shown in Fig. 15.2.3b. The moment due to H may be thought of as causing negative bending, while the other two components are causing positive bending. The moment diagrams for the components are shown on a horizontal baseline in Fig. 15.2.3c.

The maximum so-called positive moments, M_{s1} and M_{s2} are

$$M_{s1} = 7.5W_n$$
$$M_{s2} = 0.5W_n h = 7.5W_n$$

Several possible collapse mechanisms must be considered.

(a) Consider a sidesway mechanism (Fig. 15.2.3d) with plastic hinges at points 2 and 6. Equilibrium requires at point 2:

$$M_2 = M_p = M_{s2} - Hh \tag{a}$$

and at point 6:

$$M_6 = M_p = Hh = 15H \tag{b}$$

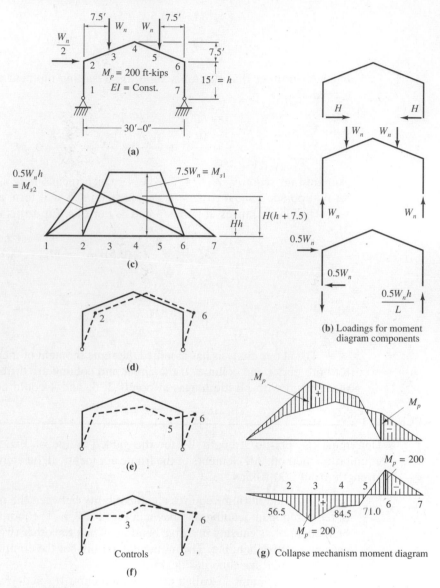

Figure 15.2.3   Gabled frame analysis—Example 15.2.3.

Thus

$$M_p = M_{s2} - M_p = 7.5W_n - M_p$$
$$M_p = 3.75W_n \tag{c}$$

(b) Consider a combination mechanism with plastic hinges at points 5 and 6 (Fig. 15.2.3e). At point 5,

$$M_5 = M_p = M_{s1} + \tfrac{1}{4}M_{s2} - H(h + 3.75)$$

$$M_p = 7.5W_n + \frac{7.5}{4}W_n - \frac{M_p}{15}(15 + 3.75)$$

$$M_p = 4.17W_n$$

Actually, the mechanism consisting of plastic hinges at points 5 and 6 could have been eliminated as a possibility by comparing the total positive moment at points 3 and 4 (Fig. 15.2.3c) which will indicate that point 3 will achieve a plastic hinge before point 5 can attain it.

(c) Consider the mechanism with plastic hinges at points 3 and 6:

$$M_3 = M_p = M_{s1} + \tfrac{3}{4}M_{s2} - H(h + 3.75)$$

$$M_p = 7.5W_n + \frac{3}{4}(7.5W_n) - \frac{M_p}{15}(15 + 3.75)$$

$$M_p = 5.83W_n \quad \text{Governs}$$

The largest M_p, or the smallest W_n, for a mechanism to occur indicates the governing one. A check may be made by determining the collapse mechanism moment diagram assuming $M_p = 200$ ft-kips. Thus

$$W_n = \frac{200}{5.83} = 34.2 \text{ kips}$$

$$H = \frac{M_p}{15} = \frac{200}{15} = 13.33 \text{ kips}$$

$$M_{s1} = 7.5W_n = 7.5(34.2) = 256.5 \text{ ft-kips}$$

$$M_{s2} = 256.5 \text{ ft-kips}$$

At the critical locations, the collapse mechanism moment is

$$M_2 = 256.5 - 200 = 56.5 \text{ ft-kips} < M_p \quad \text{OK}$$
$$M_3 = 256.5 + \tfrac{3}{4}(256.5) - 13.33(18.75) = 200 \text{ ft-kips} = M_p$$
$$M_4 = 256.5 + \tfrac{1}{2}(256.5) - 13.33(22.5) = 84.5 \text{ ft-kips} < M_p \quad \text{OK}$$
$$M_5 = 256.5 + \tfrac{1}{4}(256.5) - 13.33(18.75) = 71.0 \text{ ft-kips} < M_p \quad \text{OK}$$
$$M_6 = 13.33(15) = 200 \text{ ft-kips} = M_p$$

The final collapse mechanism moment diagram is shown in Fig. 15.2.3g, both as it would be from graphical superposition and as it would be if the net moment is plotted on a horizontal baseline. ■■

EXAMPLE 15.2.4

Determine the plastic moment required for the frame of Fig. 15.2.4. Consider (a) uniform gravity loading w_n alone; and (b) uniform gravity loading in combination with uniform lateral loading.

Solution

(a) Gravity loading only. The simple beam moment is

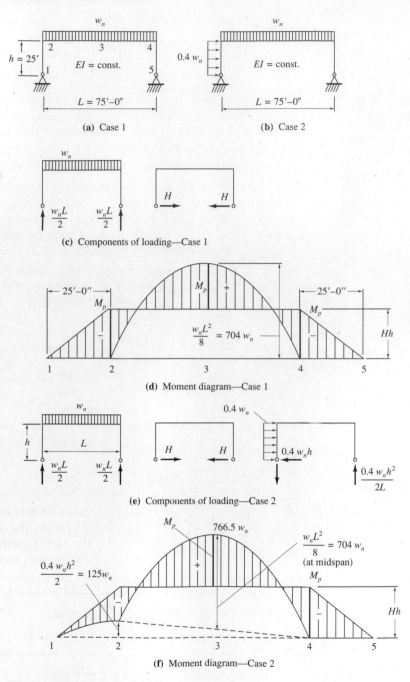

(a) Case 1

(b) Case 2

(c) Components of loading—Case 1

(d) Moment diagram—Case 1

(e) Components of loading—Case 2

(f) Moment diagram—Case 2

Figure 15.2.4 Example 15.2.4.

$$M_s = \frac{w_n L^2}{8} = \frac{w_n (75)^2}{8} = 704 w_n \qquad \text{(a)}$$

For a properly designed frame, the plastic hinges will occur at the ends and midspan of the girder:

$$M_p = M_s - M_p$$

$$M_p = \frac{1}{2} M_s = \frac{w_n L^2}{16} = 352 w_n \qquad \text{(b)}$$

and the superposition of the bending moments due to the loading components is shown in Fig. 15.2.4d.

(b) Combined gravity and lateral loading. There is an additional component of loading as seen in Fig. 15.2.4e which contributes an unsymmetrical effect of the same bending moment sign as caused by the gravity uniform loading. It may be observed from Fig. 15.2.4f that the maximum moments will occur slightly to the left of point 3 and at point 4. One could mathematically solve for the exact location and magnitude of the plastic moment. However, for design purposes one may use a graphical construction and divide the maximum value of the parabolic curve by two to establish the horizontal line Hh. In other words, equalize the positive and negative moments. The moment at point 2 is obviously less than that at point 4 so that no plastic hinge will form at point 2.

By scaling, the maximum ordinate of the parabola near point 3 is $768 w_n$. In which case,

$$M_p = \frac{768 w_n}{2} = 384 w_n \qquad \text{(c)}$$

The uniform lateral loading causes about $9\frac{1}{2}\%$ reduction in gravity carrying capacity, or alternatively it would require a section having about $9\frac{1}{2}\%$ greater moment strength. ∎∎

Energy Method

Just as discussed for beams in Chapter 10, the energy method may also be used for the plastic analysis of frames. For certain frame arrangements, the energy method will be found easier. The following examples show how the energy method may be applied to the same structures previously analyzed by the equilibrium method.

EXAMPLE 15.2.5 _____

Repeat Example 15.2.1 except use the energy method.

Solution The mechanism is shown in Fig. 15.2.5. The plastic hinge locations are assumed and from the geometry the angles θ are established. The external work done by the load equals the internal strain energy of the plastic moments moving through their angles of rotation:

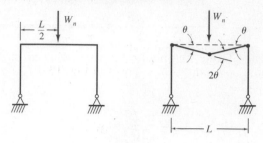

Figure 15.2.5 Example 15.2.5.

External work = Internal work

$$W_n \frac{\theta L}{2} = M_p(\theta + 2\theta + \theta)$$

$$W_n = \frac{8M_p}{L}$$

exactly as obtained in Example 15.2.1. ■■

EXAMPLE 15.2.6 _____

Repeat Example 15.2.2 using the energy method.

Solution The possible mechanisms are shown in Fig. 15.2.6.
  (a) Mechanism 1

$$0.5W_n \theta h = M_p(\theta + \theta)$$

$$W_n = \frac{4M_p}{h}$$

(b) Mechanism 2

$$0.5W_n \theta h + W_n \theta \frac{L}{2} = M_p(2\theta + 2\theta)$$

$$W_n = \frac{8M_p}{L + h}$$

The result are identical to those of Example 15.2.2.

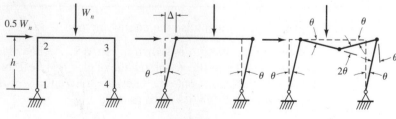

Figure 15.2.6 Example 15.2.6. ■■

EXAMPLE 15.2.7 _____

Repeat Example 15.2.3 using the energy method.

Solution   The possible mechanisms are shown in Fig. 15.2.7.
(a) Mechanism 1

$$0.5W_n(15\theta) = M_p(2\theta)$$
$$M_p = 3.75W_n$$

(b) Mechanism 2. In order to treat this more complex mechanism, the concept of *instantaneous center* is used. When plastic hinges form at points 5 and 6 three rigid bodies remain which rotate as the structure moves. Segment 1–2–3–4–5 rotates about point 1; segment 6–7 rotates about point 7; segment 5–6 rotates and translates an amount which is controlled by the movement of points 5 and 6 on the adjacent rigid segments. If the body is rigid, point 5′ is perpendicular to line 1–5, and point 6′ is perpendicular to line 6–7. Thus, the points 5 and 6 may be thought of as rotating about point 0, the intersection of line 1–5 and line 6–7; i.e., the instantaneous center.

The first step in the energy method using the instantaneous center is to determine its location; since point 5 is 22.5 ft horizontally and 18.75 ft vertically from point 1, the vertical distance to point 0 from point 7 is

$$\frac{x}{30} = \frac{18.75}{22.5}, \qquad x = 25 \text{ ft}$$

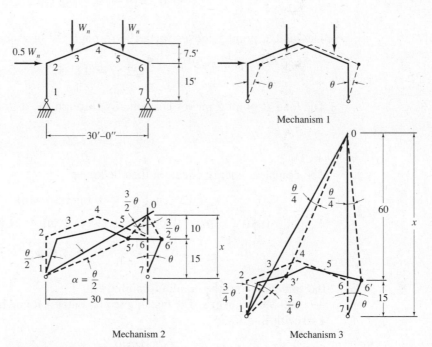

Figure 15.2.7   Example 15.2.7.

Next, a reference angle θ is established arbitrarily as shown in Fig. 15.2.7. By proportion, the angle of rotation with respect to point 0 is $3\theta/2$. The rigid-body segment 5–6 rotates through this angle $3\theta/2$. By inverse proportion as distance 0–5 is to 1–5, the rotation of rigid-body 1–2–3–4–5 about point 1 is

$$\frac{\frac{1}{4}}{\frac{3}{4}} = \frac{\alpha}{3\theta/2}, \qquad \alpha = \frac{\theta}{2}$$

The relative plastic hinge rotation at point 5 is

$$\frac{\theta}{2} + \frac{3\theta}{2} = 2\theta$$

The relative plastic hinge rotation at point 6 is

$$\theta + \frac{3\theta}{2} = 2.5\theta$$

To compute external work done by the applied loads, the vertical distance moved due to rotation of points 3 and 5 and the horizontal distance moved at point 2 are required.

The vertical displacement of point 3 equals the angle of rotation times the horizontal projection from points 2 to 3. The load at point 3 moves vertically through a distance

$$\frac{\theta}{2}(7.5) = 3.75\theta$$

The load at point 5 moves vertically through the distance

$$\frac{\theta}{2}(22.5) = 11.25\theta$$

The load at point 2 moves horizontally through the distance

$$\frac{\theta}{2}(15) = 7.5\theta$$

The complete energy equation then becomes

$$\text{External work} = \text{Internal work}$$

$$0.5W_n(7.5\theta) + W_n(3.75\theta) + W_n(11.25\theta) = M_p(2\theta + 2.5\theta)$$

$$M_p = \frac{18.75}{4.5}W_n = 4.17W_n$$

the same as from the equilibrium method.

(c) Mechanism 3. The instantaneous center is found by intersecting the line 1–3 with line 6–7:

$$\frac{x}{30} = \frac{18.75}{7.5}, \qquad x = 75 \text{ ft}$$

If θ is defined by Fig. 15.2.7, then the angle of rotation with respect to 0 is $\theta/4$, since the distance 0–6 is four times the distance 6–7. Since the distance 0–3 is three times the distance 3–1, the angle 3–1–3′ is $3\theta/4$ (3 times the rotation angle about 0).

The external work done by the various loads is

$$\text{Load at 2,} \qquad 0.5W_n\left(\frac{3\theta}{4}\right)(15) = \frac{22.5}{4}W_n\theta$$

$$\text{Load at 3,} \qquad W_n\left(\frac{3\theta}{4}\right)(7.5) = \frac{22.5}{4}W_n\theta$$

$$\text{Load at 5,} \qquad W_n\left(\frac{\theta}{4}\right)(7.5) = \frac{7.5}{4}W_n\theta$$

The internal strain energy is

$$\text{Moment at 3,} \qquad M_p\left(\frac{3\theta}{4} + \frac{\theta}{4}\right) = M_p\theta$$

$$\text{Moment at 6,} \qquad M_p\left(\theta + \frac{\theta}{4}\right) = M_p\frac{5\theta}{4}$$

$$\text{External work} = \text{Internal work}$$

$$W_n\theta\left(\frac{22.5}{4} + \frac{22.5}{4} + \frac{7.5}{4}\right) = M_p\theta\left(1 + \frac{5}{4}\right)$$

$$W_n\left(\frac{52.5}{4}\right) = M_p\left(\frac{9}{4}\right)$$

$$M_p = \frac{52.5}{9}W_n = 5.83W_n \quad \text{Governs}$$

The result is identical with Example 15.2.3. ■■

15.3 LOAD AND RESISTANCE FACTOR DESIGN— ONE-STORY FRAMES

There are two approaches that may be used under AISC Load and Resistance Factor Design: (1) elastic analysis using factored loads: and (2) plastic analysis using factored loads. These two approaches are compared in several examples.

Plastic analysis is permitted under LRFD-A5.1 for steels having a yield stress F_y not exceeding 65 ksi. The following other conditions must be satisfied:

1. In regions where plastic hinge rotation is necessary to reach the plastic strength collapse mechanism, local flange buckling must be precluded as a limit state by limiting $b_f/2t_f \le \lambda_p$ (LRFD-B5.2), where for flanges of I-shaped sections

$$\frac{b_f}{2t_f} \le \left(\lambda_p = \frac{65}{\sqrt{F_{yf}}}\right) \tag{15.3.1}$$

2. The web local buckling limit state must be precluded in beams and beam-columns by limiting $h/t_w \leq \lambda_p$ (LRFD-B5.2), where for webs in flexural and axial compression,

   a. For $P_u/(\phi_b P_y) \leq 0.125$:

   $$\lambda_p = \frac{640}{\sqrt{F_y}}\left(1 - \frac{2.75 P_u}{\phi_b P_y}\right)$$ (15.3.2)

   b. For $P_u/(\phi_b P_y) > 0.125$:

   $$\lambda_p = \frac{191}{\sqrt{F_y}}\left(2.33 - \frac{P_u}{\phi_b P_y}\right) \geq \frac{253}{\sqrt{F_y}}$$ (15.3.3)

3. The factored axial compression force in members of frames is restricted.

   a. In *braced* frames (LRFD-C2.1),

   $$P_u \leq 0.85 \phi_c A_g F_y$$ (15.3.4)

   b. In *unbraced* frames (LRFD-C2.2),

   $$P_u \leq 0.75 \phi_c A_g F_y$$ (15.3.5)

   where P_u is the axial force in columns "caused by factored gravity plus factored horizontal loads."

4. The column slenderness parameter λ_c (as defined in LRFD-E2) may not exceed 1.5 times the effective length factor K (LRFD-E1.2).

5. At plastic hinge locations associated with the plastic strength collapse mechanism, except in the region of the last plastic hinge to form, the lateral-torsional buckling limit state must be precluded by limiting $L_b \leq L_{pd}$ (LRFD-F1.2d), where for I-shaped sections, having compression flange equal to or larger than the tension flange, and bent about the major axis,

   $$L_{pd} = \frac{3600 + 2200(M_1/M_p)}{F_y} r_y$$ (15.3.6)

   where M_1/M_p is positive when moments cause reverse curvature, and negative for single curvature.

6. In the region of the last plastic hinge to form, and in regions not adjacent to a plastic hinge, the ordinary flexural strength requirements of LRFD-F1.2 apply, as discussed in Chapters 7, 9, 10, and 12.

7. For beam-columns, the factored second-order moment M_u for use in the interaction equations of LRFD-H1 must satisfy the $P\Delta$ effects (LRFD-C1) and frame stability (LRFD-C2). It is interesting to note that LRFD-A5.1 in listing all the sections applicable to plastic analysis included LRFD-C2 *but did not include the $P\Delta$ requirement of LRFD-C1*. However, the referenced section H1 includes the requirement of LRFD-C2. As previously stated, long practice with plastic design has specifically *neglected* the $P\Delta$ effect in one-story frames.

8. For flexural composite members (see Chapter 16), the nominal moment strength M_n must be computed using plastic stress distributions in accor-

dance with LRFD-I3.2. In accordance with LRFD-I3.2, h/t_w cannot exceed $640/\sqrt{F_{yf}}$.

EXAMPLE 15.3.1

Design a rectangular frame of 75-ft span and 25-ft height to carry 0.2 kip/ft dead load, 0.8 kip/ft snow load, 0.1 kip/ft gravity directional wind load, and horizontal uniform wind of 0.44 kip/ft. Lateral bracing from purlins occurs every 6 ft on the girder, and bracing is provided by wall girts on the columns every 5 ft. Use A36 steel and *plastic analysis* under Load and Resistance Factor Design.

Solution The frame is shown in Fig. 15.3.1.

(a) Factored load combinations according to LRFD-A4.1

Case 1: Gravity load; LRFD Formula (A4-3)

$$w_u = 1.2D + 1.6S = 1.2(0.2) + 1.6(0.8) = 1.52 \text{ kips/ft}$$

Case 2: Gravity + wind; LRFD Formula (A4-3)

$$w_u = 1.2D + 1.6S + 0.8W$$
$$= 1.2(0.2) + 1.6(0.8) + 0.8(0.1) = 1.60 \text{ kips/ft}$$
$$w_{uh} = 0.8W = 0.8(0.44) = 0.352 \text{ kip/ft} \quad \text{(horizontal)}$$

Case 3: Gravity + wind; LRFD Formula (A4-4)

$$w_u = 1.2D + 0.5S + 1.3W$$
$$= 1.2(0.2) + 0.5(0.8) + 1.3(0.1) = 0.77 \text{ kip/ft}$$
$$w_{uh} = 1.3W = 1.3(0.44) = 0.572 \text{ kip/ft} \quad \text{(horizontal)}$$

(b) Gravity loading only—Case 1. Using the plastic analysis illustrated in Example 15.2.4,

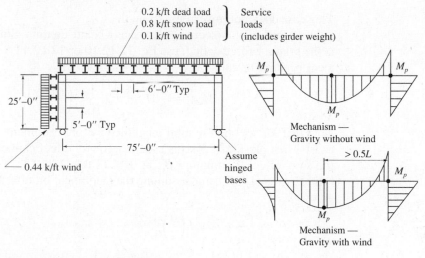

Figure 15.3.1 Example 15.3.1.

$$\text{Required } M_p = \frac{w_n L^2}{16} = \frac{w_u L^2}{\phi_b 16} = \frac{1.52(75)^2}{0.90(16)} = 594 \text{ ft-kips}$$

$$\text{Required } Z_x = \frac{\text{Required } M_p}{F_y} = \frac{594(12)}{36} = 198 \text{ in.}^3$$

Try W24×76, $Z_x = 200$ in.^{3}, $M_p = 36(200)/12 = 600$ ft-kips, or a deeper heavier section if deflection is a critical factor. Check the flange and web local buckling limits of LRFD-B5; i.e., Eqs. 15.3.1 and 15.3.2,

$$\left(\frac{b_f}{2t_f} = \frac{8.990}{2(0.680)}\right) \le \left(\lambda_p = \frac{65}{\sqrt{F_{yf}}} = 10.8\right) \quad \text{OK}$$

$$\left[\frac{h}{t_w} = 49.0\right] \le \left[\lambda_p = \frac{640}{\sqrt{F_y}}\left(1 - \frac{2.75(61.5)}{0.90(22.4)36}\right) = 82\right] \quad \text{OK}$$

(c) Gravity plus wind—Case 2. Though the ratio of lateral load to gravity load here is different than that used in Example 15.2.4, the procedure will be the same. Following that procedure for this case where $w_{uh}/w_u = 0.352/1.60 = 0.22$ gives

$$\text{Required } \phi_b M_p \approx 370 w_u = 370(1.60) = 592 \text{ ft-kips}$$

$$\text{Required } M_p \approx 592/\phi_b = 592/0.90 = 658 \text{ ft-kips}$$

$$\text{Required } Z_x = 658(12)/36 = 219 \text{ in.}^3$$

Increase section at least to W24×84; $\phi_b M_p = 605$ ft-kips. For this section $\lambda < \lambda_p$ for flange and web local buckling.

(d) Gravity plus wind—Case 3. Following the same procedure as in Example 15.2.4 with $w_{uh}/w_u = 0.572/0.77 = 0.74$ gives

$$\text{Required } \phi_b M_p \approx 410 w_u = 410(0.77) = 316 \text{ ft-kips}$$

$$\text{Required } M_p \approx 316/\phi_b = 316/0.90 = 351 \text{ ft-kips}$$

This case does not govern for strength.

(e) Preliminary section selected as a beam-column. Since bending moment is the primary force acting, use Eq. 12.12.10 or 12.12.11 to estimate the equivalent bending moment M_{uEQ},

$$M_{uEQ} = \frac{8}{9}M_u + P_u\left(\frac{1}{\beta_{az}}\right)\left(\frac{\phi_b F_y}{\phi_c F_{cr}}\right) \qquad [12.12.10]$$

Note that M_u will be the total magnified factored moment, $B_2(M_{nt} + M_{\ell t})$, because when plastic analysis is used the gravity effect cannot be separated from the sway effect. Estimating $\beta_{az} = 0.16$ (Table 12.12.1), estimating the stress ratio term as $(1/0.7)$, and assuming the magnification offsets the 8/9 reduction, gives

$$P_u = \frac{w_u L}{2} + w_u h\left(\frac{h}{2}\right)\frac{1}{L}$$

$$= \frac{1.6(75)}{2} + 0.352(25)\left(\frac{25}{2}\right)\frac{1}{75} = 60 + 1.5 = 61.5 \text{ kips}$$

$$M_{uEQ} = 592 + 61.5\left(\frac{1}{0.16}\right)\left(\frac{1}{0.7}\right)\left(\frac{1}{12}\right) = 592 + 46 = 638 \text{ ft-kips}$$

Note that W24×84 is not adequate. Try W27×84; $\phi_b M_p = 659$ ft-kips.

(f) Check lateral-torsional buckling for W27×84. In order to have used plastic analysis L_b must not exceed L_{pd} (Eq. 15.3.4). The minimum L_{pd} will be when $M_1/M_p = -1$, giving $L_{pd} = 3.2$ ft. This means that the column lateral bracing at 5 ft and the beam lateral bracing at 6 ft are both adequate regardless of the moment gradient. Thus, $L_b < L_{pd}$ and the lateral-torsional buckling limit state does not control.

(g) Check the W27×84 as a beam-column according to LRFD-H2. The effective length factor K_x for buckling in the plane of the frame must be determined by "structural analysis" according to LRFD-C2.2 and may not be less than unity. Using the alignment chart, Fig. 6.9.4,

$$G_A \approx 10 \quad \text{(estimated for hinge at bottom of column)}$$

$$G_B = \frac{I/L \text{ for column}}{I/L \text{ for girder}} = \frac{I/25}{I/75} = 3.0 \quad \text{(top of column)}$$

From Fig. 6.9.4b, unbraced frame, find $K_x = 2.25$.

$$\text{Axis of bending } \frac{KL}{r} = \frac{K_x L_x}{r_x} = \frac{2.25(25)12}{10.7} = 63$$

At this point the G factor may be adjusted as explained in Sec. 6.9 to account for inelastic behavior. For A36 steel, Table 6.9.1 may be used to obtain the adjustment factor β_s. For $KL/r = 63$, find $\beta_s = 0.47$ for a column having $P_u/A_g = \phi_c F_{cr}$. In this case the axial compression $P_u/A_g = 61.5/24.8 = 2.5$ ksi; considerably below $\phi_c F_{cr}$. Thus, 2.5 ksi should be used as F_{cr} to obtain $\beta_s = 1.0$; no correction for inelastic behavior.

$$P_{e2} = \frac{\pi^2 E A_g}{(KL/r)^2} = \frac{\pi^2(29,000)24.8}{(63)^2} = 1780 \text{ kips}$$

Use the sway magnifier B_2 on the entire moment M_u as a conservative treatment of the $P\Delta$ effect,

$$B_2 = \frac{1}{1 - \dfrac{\Sigma P_u}{\Sigma P_{e2}}} = \frac{1}{1 - \dfrac{1.60(75)}{2(1780)}} = 1.03$$

Examine the minor axis slenderness ratio,

$$\frac{K_y L_y}{r_y} = \frac{1.0(5)12}{2.07} = 29$$

Since $K_x L_x/r_x > K_y L_y/r_y$, the major axis controls $\phi_c F_{cr}$,

$$K_x L_x/r_x = 63; \qquad \phi_c F_{cr} = 24.8 \text{ ksi}$$

Check the beam-column interaction equation (LRFD-H1.2), Eqs. 12.10.1 or 12.10.2,

$$\frac{P_u}{\phi_c P_n} = \frac{P_u}{\phi_c F_{cr} A_g} = \frac{61.5}{24.8(24.8)} = 0.101$$

Use LRFD Formula (H1-1b), Eq. 12.10.2, for uniaxial bending,

$$\frac{P_u}{2\phi_c P_n} + \left(\frac{M_{ux}}{\phi_b M_{nx}}\right) \leq 1.0 \qquad [12.10.2]$$

$$M_{nt} + M_{\ell t} = 370(1.60) = 592 \text{ ft-kips}$$

$$M_{ux} = B_2(M_{nt} + M_{\ell t}) = 1.03(592) = 610 \text{ ft-kips}$$

Note that in plastic analysis the combined gravity plus lateral loading is used; thus, M_{nt} and $M_{\ell t}$ are not separately computed. Magnifying the combined moment using the sway magnifier is conservative. Thus, Eq. 12.10.2 becomes

$$\frac{0.101}{2} + \frac{610}{659} = 0.051 + 0.926 = 0.97 \quad \text{OK}$$

If the shallower W24×94 is desired, $\phi_b M_p = 686$ ft-kips. The $K_x L_x/r_x$ becomes 68.4, $P_e = 1695$ kips, $B_2 = 1.037$, and $\phi_c F_{cr} = 23.9$ ksi. The strength criterion, Eq. 12.10.2, becomes

$$\frac{61.5}{2(23.9)27.7} + \frac{1.037(592)}{686} = 0.046 + 0.895 = 0.94 < 1 \quad \text{OK}$$

Recheck Eq. 15.3.2 for h/t_w limit. For $P_u/(\phi_b P_y) \leq 61.5/[0.90(36)(27.7)] = 0.070 < 0.125$ limit,

$$\lambda_p = \frac{640}{\sqrt{F_y}}\left(1 - \frac{2.75 P_u}{\phi_b P_y}\right) = \frac{640}{\sqrt{36}}(1 - 2.75(0.070)) = \frac{518}{\sqrt{36}} = 86$$

The actual h/t_w for the W24×94 is 41.9; well below the limit.
Use W24×94. ∎∎

It is clear that the $P\Delta$ effect is not significant (less than 4%) in this frame and the traditional neglect in plastic design (ASD-N) of the $P\Delta$ effect in one-story frames probably results in error no greater than the strength prediction variability of the interaction equation. To complete the design of the frame, the knee joint must be investigated, providing stiffeners if necessary, in a manner similar to Example 13.8.1.

EXAMPLE 15.3.2 _____

Redesign the frame of Example 15.3.1 (Fig. 15.3.1) if the column cannot be deeper than 14.5 in. actual depth. Use A36 steel and Load and Resistance Factor Design.

Solution   For this solution the column and the beam will have different plastic moment strengths. The factored loads appear in Example 15.3.1, part (a).

(a) Select section for plastic strength using loading Case 1 (gravity loading). There are many possible solutions using various sizes of column. Try a W14×82 section having $\phi_b M_p = 375$ ft-kips.

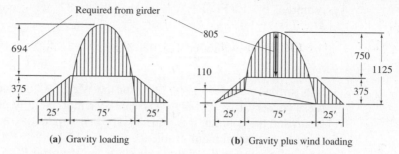

Figure 15.3.2 Factored load collapse moment diagrams for plastic analysis on Example 15.3.2, assuming W14×82 columns ($\phi_b M_p = 375$ ft-kips).

Assuming gravity loading governs, the design plastic moment $\phi_b M_p$ required at midspan ($+M$) for the girder is (see Fig. 15.3.2)

$$\phi_b M_p = \frac{w_u L^2}{8} - \phi_b M_p(\text{column}) = \frac{1.52(75)^2}{8} - 375 = 694 \text{ ft-kips}$$

A W27×94 having $\phi_b M_p = 751$ ft-kips might be selected to satisfy strength. For practicality of making the corner connection, the two members should have approximately the same flange width as do these two members. To fully utilize the $\phi_b M_p$ of 751 ft-kips, perhaps the column can be reduced to W14×74 having $\phi_b M_p$ of 340 ft-kips. The design plastic moment strength required for the girder will become

$$\text{Required } \phi_b M_p = 1069 - 340 = 729 \text{ ft-kips}$$

Try W27×94 girder, $\phi_b M_p = 751$ ft-kips.

(b) Check requirement for strength for the Case 2 loading. Refer to Fig. 15.3.2 for the plastic analysis using different column and girder sections.

$$M_s \quad (\text{at knee}) = \tfrac{1}{2} w_{uh} h^2 = 0.5(0.352)(25)^2 = 110 \text{ ft-kips}$$
$$M_s \quad (\text{at girder midspan}) = \tfrac{1}{8} w_u L^2 = \tfrac{1}{8}(1.60)(75)^2 = 1125 \text{ ft-kips}$$

The required design plastic moment from the girder, assuming the plastic hinge to occur approximately at midspan, is

$$\phi_b M_p = 1125 + 0.5(110) - \phi_b M_p(\text{column})$$
$$= 1125 + 55 - 340 = 840 \text{ ft-kips}$$

This exceeds the $\phi_b M_p$ of 751 provided by the W27×94; revise to W14×82 column ($\phi_b M_p = 375$ ft-kips) and W27×102 beam ($\phi_b M_p = 824$ ft-kips). The required design plastic moment strength from the girder would be

$$\text{Required } \phi_b M_p = 1125 + 55 - 375 = 805 \text{ ft-kips}$$

(c) Check flange local buckling, web local buckling, and lateral-torsional buckling. One may determine from the *LRFD Manual* [1.18] that both the W14×82 and W27×102 are compact shapes. Check L_{pd}, Eq. 15.3.4, as required by LRFD-F1.1. For the W14×82 with constant moment (i.e, $M_1/M_p = -1$),

$$L_{pd} = \frac{3600 + 2200(M_1/M_p)}{F_y} r_y = \frac{1400(2.48)}{36(12)} = 8.0 \text{ ft}$$

In a similar manner, $L_{pd} = 7.0$ ft for the W27×102 having $r_y = 2.15$ in. Since these conservative L_{pd} values exceed the actual laterally unbraced length on the beam and the column, it is unnecessary to compute the increase in L_{pd} resulting from favorable moment gradient.

(d) Compute the column strength and the ratio $P_u/\phi_c P_n$. For this unbraced frame use the alignment chart, Fig. 6.9.4, to evaluate the effective length factor in the plane of the frame,

$$G_A \approx 10 \quad \text{(estimated for hinge at bottom of column)}$$

$$G_B = \frac{I/L \text{ for column}}{I/L \text{ for girder}} = \frac{822/25}{3620/75} = 0.73 \quad \text{(top of column)}$$

From Fig. 6.9.4b, unbraced frame, find $K_x = 1.8$.

$$\text{Axis of bending} \quad \frac{KL}{r} = \frac{K_x L_x}{r_x} = \frac{1.8(25)12}{6.05} = 89.3$$

Since the axial load on the column is low and the slenderness is high, there will be no adjustment in K_x for inelastic effect as discussed in Sec. 6.9.

$$\frac{K_y L_y}{r_y} = \frac{1.0(5)12}{2.48} = 24$$

Since $K_x L_x/r_x > K_y L_y/r_y$, the major axis controls $\phi_c F_{cr}$,

$$K_x L_x/r_x = 89.3; \qquad \phi_c F_{cr} = 20.1 \text{ ksi}$$

The axial compression effect of the beam-column interaction equation is

$$\frac{P_u}{\phi_c P_n} = \frac{P_u}{\phi_c F_{cr} A_g} = \frac{61.5}{20.1(24.1)} = 0.127$$

Since this is less than 0.2, LRFD Formula (H1-lb), Eq. 12.10.2, applies.

(e) Compute the sway magnifier B_2. Again as in Example 15.3.1, use the sway magnifier on the total moment obtained in the plastic analysis of the frame. In evaluating P_{e2}, use the axis of bending KL/r of 89.3,

$$P_{e2} = \frac{\pi^2 E A_g}{(KL/r)^2} = \frac{\pi^2(29,000)24.1}{(89.3)^2} = 867 \text{ kips}$$

Use the sway magnifier B_2 on the entire moment M_u as a conservative treatment of the $P\Delta$ effect,

$$B_2 = \frac{1}{1 - \dfrac{\Sigma P_u}{\Sigma P_{e2}}} = \frac{1}{1 - \dfrac{1.60(75)}{2(867)}} = 1.074$$

(f) Check the beam-column interaction criterion. Use Eq. 12.10.2,

$$\frac{P_u}{2\phi_c P_n} + \left(\frac{M_{ux}}{\phi_b M_{nx}}\right) \leq 1.0 \qquad [12.10.2]$$

$$M_{ux} = B_2(\phi_b M_p \text{ at column top}) = 1.074(375) = 403 \text{ ft-kips}$$

Note that in plastic analysis the combined gravity plus lateral loading is used; thus, M_{nt} and $M_{\ell t}$ are not separately computed. In the plastic analysis, it was assumed the full plastic moment strength of the column was developed when the $\phi_b M_p$ requirement for the girder was computed. Since the actual $\phi_b M_p$ of 824 ft-kips for the girder exceeds the requirement of 805 ft-kips, one could compute the actual column $\phi_b M_p$ that is utilized and find M_u utilized is 355 ft-kips. M_{ux} would then become 1.074(355) = 381 ft-kips. This exceeds the design strength of the W14×82. Increase the column to W14×90 having $\phi_b M_p = 424$ ft-kips.

Recomputing for W14×90, $r_x = 6.14$ in., $K_x L_x / r_x = 87.9$, $\phi_c F_{cr} = 20.4$ ksi, $A_g = 26.5$ sq in., $P_{e2} = 980$ kips, and $B_2 = 1.065$. Thus, Eq. 12.10.2 becomes

$$\frac{61.5}{2(20.4)26.5} + \frac{1.065(355)}{425} = 0.057 + 0.890 = 0.95 < 1 \quad \text{OK}$$

The design is considered adequate since the hinged-base assumption certainly understimates the moment restraint and lateral stiffness of the frame.

(g) Check shear in the knee region using stiffeners as necessary, as illustrated in Example 13.8.1.

Use W27×102 for the girder and W14×90 for the column. ■■

EXAMPLE 15.3.3 _____

Redesign the rectangular frame of Example 15.3.1 using Load and Resistance Factor Design *without using plastic analysis.* Use A36 steel.

Solution

(a) Factored loads. The factored load combinations were computed in Example 15.3.1, part (a). The gravity plus wind (Case 2) was determined to control. Design here only for that case. The factored loads are

Case 2: Gravity + wind; LRFD Formula (A4-3)

$$w_u = 1.2D + 1.6S + 0.8W$$
$$= 1.2(0.2) + 1.6(0.8) + 0.8(0.1) = 1.60 \text{ kips/ft}$$
$$w_{uh} = 0.8W = 0.8(0.44) = 0.352 \text{ kip/ft} \quad \text{(horizontal)}$$

(b) Structural analysis. Elastic analysis of the structure under factored loads is needed; *assume the EI for the column and girder are equal,* as was assumed in Example 15.3.1. Statically indeterminate analysis is used to obtain the factored moments in Fig. 15.3.3. The details of statically indeterminate

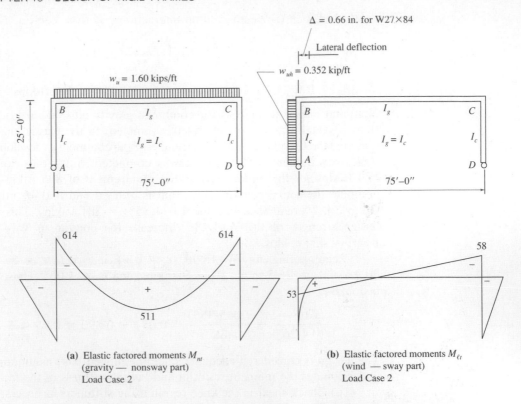

Figure 15.3.3   Design under LRFD without using plastic analysis; Example 15.3.3.

analysis are outside the scope of this text; many standard references are available.* According to LRFD-H1.2, two analyses are needed; one for the gravity part (1.60 kips/ft), and one for the sway part (horizontal 0.352 kip/ft). The alternative of using a second-order analysis to obtain M_u is not used here.

(c) Select a preliminary trial section. Neglecting the magnification effect, the maximum factored moment $(M_{nt} + M_{\ell t})$ is

$$M_{nt} + M_{\ell t} = 614 + 58 = 672 \text{ ft-kips}$$

and the factored axial compression P_u is

$$P_u = 1.60(75)/2 + (53 + 58)/75 = 61.5 \text{ kips}$$

Since bending moment is the primary force acting, use Eq. 12.12.10 or 12.12.11 to estimate the equivalent bending moment $M_{u\text{EQ}}$,

---

*See for example, Chu-Kia Wang and Charles G. Salmon, *Introductory Structural Analysis*. Englewood Cliffs, NJ: Prentice-Hall, Inc., 1984; C. K. Wang, *Intermediate Structural Analysis*. New York: McGraw-Hill Book Company, 1983; Chu-Kia Wang, *Structural Analysis on Microcomputers*. New York: Macmillan Publishing Company, 1986.

$$M_{uEQ} = \frac{8}{9}M_u + P_u\left(\frac{1}{\beta_{az}}\right)\left(\frac{\phi_b F_y}{\phi_c F_{cr}}\right) \qquad [12.12.10]$$

If the actual magnification factors B_1 on the gravity moments and B_2 on the sway moments are assumed to be 1.0 to begin the process, then $M_u = M_{nt} + M_{\ell t}$.

Estimating $\beta_{az} = 0.16$ (Table 12.12.1), estimating the stress ratio term as $(1/0.7)$, and assuming the magnification offsets the 8/9 reduction, gives

$$M_{uEQ} = 672 + 61.5\left(\frac{1}{0.16}\right)\left(\frac{1}{0.7}\right)\left(\frac{1}{12}\right) = 672 + 46 = 718 \text{ ft-kips}$$

Since lateral bracing is at relatively close spacing, the section may best be selected assuming maximum moment strength; i.e., $\phi_b M_n = \phi_b M_p$. Thus using the *LRFD Manual* "LOAD FACTOR DESIGN SELECTION TABLE" and entering with required $\phi_b M_p = 718$ ft-kips, find W27×94 section as the lightest having the required strength,

$$\text{Try W27×94;} \qquad \phi_b M_p = 751 \text{ ft-kips}$$

(c) Check column action. The effective length factor K_x in the plane of the unbraced frame is determined for $I_g = I_c$ exactly as in Example 15.3.1. Thus, as before

$$K_x = 2.25 \quad \text{(unbraced)} \qquad K_y = 1.0 \quad \text{(braced)}$$

$$\text{Axis of bending } \frac{KL}{r} = \frac{K_x L_x}{r_x} = \frac{2.25(25)12}{10.9} = 61.9$$

$$\frac{K_y L_y}{r_y} = \frac{1.0(5)12}{2.12} = 28.3$$

Since $K_x L_x/r_x > K_y L_y/r_y$, the major axis controls $\phi_c F_{cr}$,

$$K_x L_x/r_x = 61.9; \qquad \phi_c F_{cr} = 25.0 \text{ ksi}$$

The axial compression effect of the beam-column interaction equation is

$$\frac{P_u}{\phi_c P_n} = \frac{P_u}{\phi_c F_{cr} A_g} = \frac{61.5}{25.0(27.7)} = 0.088$$

Since this is less than 0.2, LRFD Formula (H1-1b), Eq. 12.10.2, applies.

(d) Compute nonsway magnifier B_1. Although $K \leq 1.0$ according to LRFD-H1.2a, the authors believe that $K = 1.0$ should be used. When $K = 1.0$, $K_x L_x/r_x = 27.5$ for the axis of bending.

$$P_{e1} = \frac{\pi^2 EA_g}{(KL/r)^2} = \frac{\pi^2(29,000)27.7}{(27.5)^2} = 10,500 \text{ kips}$$

$$B_1 = \frac{C_m}{1 - P_u/P_{e1}} = \frac{0.6}{1 - 61.5/10,500} < 1$$

Since $B_1 < 1.0$ it means the maximum moment at the end of the member is to be used; i.e., $B_1 = 1.0$.

(e) Compute the sway magnifier B_2. Using $K_x L_x / r_x$ for the axis of bending of 61.9 when $K_x = 2.25$,

$$P_{e2} = \frac{\pi^2 E A_g}{(KL/r)^2} = \frac{\pi^2 (29,000) 27.7}{(61.9)^2} = 2070 \text{ kips}$$

$$B_2 = \frac{1}{1 - \dfrac{\Sigma P_u}{\Sigma P_{e2}}} = \frac{1}{1 - \dfrac{1.60(75)}{2(2070)}} = 1.03$$

(f) Check the beam-column interaction equation, Eq. 12.10.2 [LRFD Formula (H1-1b)],

$$\frac{P_u}{2\phi_c P_n} + \left(\frac{M_{ux}}{\phi_b M_{nx}} \right) \leq 1.0 \qquad [12.10.2]$$

$$M_{ux} = B_1 M_{nt} + B_2 M_{\ell t}$$
$$= 1.0(614) + 1.03(58) = 674 \text{ ft-kips}$$

The interaction criterion becomes

$$\frac{0.088}{2} + \frac{674}{751} = 0.044 + 0.897 = 0.94 \quad \text{say OK}$$

(g) Check the local flange buckling, local web buckling, and lateral-torsional buckling limit states. The W27×94 is a compact section since $\lambda < \lambda_p$ of LRFD-B5. For the lateral-torsional buckling limit state, the L_b of 5 ft does not exceed L_p,

$$L_p = \frac{300}{\sqrt{F_y}} r_y = \frac{300(2.12)}{\sqrt{36} \, (12)} = 8.8 \text{ ft}$$

Thus, $\phi_b M_n = \phi_b M_p$ as assumed. Had the laterally unbraced length been larger, the possibility of M_n less than M_p would have been considered earlier in the design process. Note that the girder with $L_b = 6$ ft also has $M_n = M_p$. *Use* W27×94 (revised below).

(h) Reduction in end moment according to LRFD-A5.1, 3rd par. When the conditions of LRFD-A5.1, 2nd par. for using plastic analysis are satisfied *but plastic analysis is not used,* the beam-column (and the beam) may be proportioned for "nine-tenths of the negative moments produced by gravity loading at points of support, provided that the maximum positive moment is increased by one-tenth of the average negative moments."

In this design, the conditions for using plastic analysis *are* met. Thus, instead of M_{nt} of 614 ft-kips, $0.90(614) = 553$ ft-kips could have been used. The positive moment would then have increased to $511 + 0.10(614 + 614)/2 = 572$ ft-kips, which exceeds the reduced negative moment value 553 ft-kips. In this case, equalizing the positive and negative moments would be appropriate to give $M_{nt} = 563$ ft-kips.

If $M_{nt} = 563$ ft-kips were used in the design, the section could be reduced to W27×84. For that section, $B_2 = 1.036$, $M_u = 622$ ft-kips, $P_u/\phi_c P_{cr} = 0.100$, and $\phi_b M_p = 659$ ft-kips. The beam-column criterion, Eq. 12.10.2 [LRFD Formula (H1-1b)], gives

$$\frac{0.100}{2} + \frac{622}{659} = 0.050 + 0.944 = 0.994 \quad \text{OK}$$

(i) Examine alternate calculation of magnification factor B_2 using Eq. 12.11.12 [LRFD Formula (C1-4)]. The lateral load deflection Δ_{oh} at the top of the frame is computed as 0.66 in. for the loading in Fig. 15.3.3b using W27×84 section for both beam and column. The total factored load shear ΣH in the story is $0.352(25) = 8.8$ kips, and the total factored gravity load $\Sigma P_u = 1.60(75) = 120$ kips. Equation 12.11.12 then gives

$$B_2 = \frac{1}{1 - \Sigma P_u \left(\dfrac{\Delta_{oh}}{\Sigma HL}\right)} = \frac{1}{1 - \dfrac{120(0.66)}{8.8(25)12}} = 1.03$$

which is slightly less than computed using ΣP_{e2}. Check B_2 for loading Case 3 which has larger lateral load with lower gravity load. The factored lateral load deflection $\Delta_{oh} = 1.07$ in., the total factored load shear $\Sigma H = 0.572(25) = 14.3$ kips, and the total factored gravity load $\Sigma P_u = 0.77(75) = 57.8$ kips. Equation 12.11.12 gives

$$B_2 = \frac{1}{1 - \Sigma P_u \left(\dfrac{\Delta_{oh}}{\Sigma HL}\right)} = \frac{1}{1 - \dfrac{57.8(1.07)}{14.3(25)12}} = 1.02$$

As expected, Case 3 does not control.
   (j) Conclusion.
Use W27×84. The knee joint still must be checked for shear, and reinforced with stiffeners if necessary, as illustrated in Example 13.8.1. ■■

EXAMPLE 15.3.4 _____

Select W sections for the column of the frame of Fig. 15.3.4a, containing rigid joints at the tops of columns A and D. The tops of columns B and C are to be considered pinned. The bases of columns A and D are restrained such that $G = (\Sigma I/L)_{\text{col}}/(\Sigma I/L)_{\text{beam}} = 3.5$. The structure is braced in the direction perpendicular to the given frame, with each column braced at the top and bottom. Use A36 steel. (This example illustrates the unbraced frame concept discussed in the latter part of Sec. 14.5.) Use Load and Resistance Factor Design.

Solution
   (a) Compute factored loads.

$$P_{u1} = 1.2(10) + 1.6(40) = 76 \text{ kips}$$
$$P_{u2} = 1.2(15) + 1.6(70) = 130 \text{ kips}$$

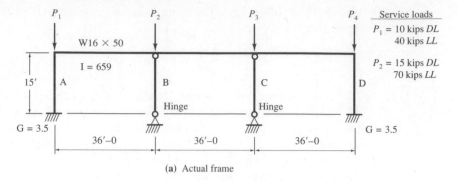

(a) Actual frame

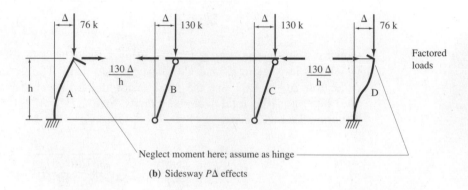

(b) Sidesway $P\Delta$ effects

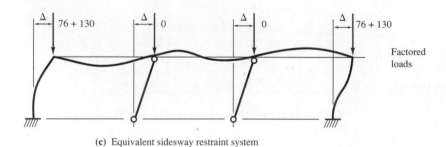

(c) Equivalent sidesway restraint system

Figure 15.3.4   Frame of Example 15.3.4.

(b) Design of columns B and C. These columns are designed as pinned-end columns within a *braced* system; thus, $K = 1.0$. The bracing is provided by the rigid framing between the girder and column at points A and D.

$$K_y L_y = 15 \text{ ft}; \qquad P_u = 130 \text{ kips}$$

Select from the *LRFD Manual* "COLUMNS" design axial strength $\phi_c P_n$ tables,

Find W8×28; $[\phi_c P_n = 132 \text{ kips}] > P_u$ OK

Check the section:

$$\frac{K_y L_y}{r_y} = \frac{1.0(15)12}{1.62} = 111; \qquad \phi_c F_{cr} = 16.0 \text{ ksi}$$

$$[\phi_c P_n = \phi_c F_{cr} A_g = 16.0(8.25) = 132 \text{ kips}] > [P_u = 130 \text{ kips}] \quad \text{OK}$$

Use W8×28 *for interior columns B and C.*

(c) Design exterior columns *A* and *D* using Yura's method [6.55] as discussed in Sec. 14.5. These columns must brace the entire assemblage. In order to resist sidesway the interior columns *B* and *C* develop combined $P\Delta$ moments of $2(130)\Delta = 260\Delta$, as shown in Fig. 15.3.4b. To stabilize the system requires development of a horizontal bracing force $260\Delta/h$ at the top of the frame. Since *two* identical exterior columns are intended to provide the bracing, each exterior column must contribute $130\Delta/h$. Thus on columns *A* and *D* the total sidesway moment is

$$\text{Total sidesway moment} = \left(\frac{130\Delta}{h}\right)h + 76\Delta = 206\Delta$$

The above neglects end moments that may be developed by the partial restraint at the column bases. In other words, the extra bracing force required is the same as if the member itself were carrying 206 kips. Thus, when the exterior columns are designed to carry $P_u = 206$ kips those columns will brace the entire system.

The K_x factor to use for columns *A* and *D* in the frame will depend on the stiffness of those columns; assume to start that $K_x = 2.0$. Then $K_x L_x = 2(15) = 30$ ft; $K_y L_y = 1.0(15) = 15$ ft. Use the *LRFD Manual* tables "COLUMNS" where $K_y L_y$ is tabulated vs $\phi_c P_n$. Since the $P_u = 206$ kips must be carried only for the frame action, $K_x L_x$ *governs* for design to carry 206 kips. Since the "COLUMNS" tables are based on $K_y L_y$, one must enter the tables with an equivalent $K_y L_y = K_x L_x/(r_x/r_y)$. Using those tables, find

$$\text{Equivalent } K_y L_y \text{ for W10} = \frac{30}{2.16} = 13.9 \text{ ft}; \qquad \text{Gives W10×33}$$

$$\text{Equivalent } K_y L_y \text{ for W12} = \frac{30}{2.66} = 11.3 \text{ ft}; \qquad \text{Gives W12×40}$$

For the W12 section, the tables "COLUMNS" give only W12×40 which carries much more load than required; try a lighter W12×30 for which there is no tabulated load. Check that $b_f/2t_f \leq 95/\sqrt{F_y}$ and $h/t_w \leq 253/\sqrt{F_y}$ are satisfied.

For the *y-y* axis (minor-axis buckling), the actual P_u of 76 kips must be carried.

$$\frac{K_y L_y}{r_y} = \frac{1.0(15)12}{1.52} = 118; \qquad \phi_c F_{cr} = 14.6 \text{ ksi}$$

$$\phi_c P_n = \phi_c F_{cr} A_g = 14.6(8.79) = 128 \text{ kips} > 76 \text{ kips} \quad \text{OK}$$

For the *x-x* axis, the end restraint must be evaluated:

$$G_{\text{top}} = \frac{\Sigma I/L \text{ for column}}{\Sigma I/L \text{ for girder}} = \frac{238/15}{659/36} = 0.87$$

The alignment chart, Fig. 6.9.4b, is based on far ends of girders restrained. When the far end is not restrained, as assumed here, the G factor should be increased; multiplied by two when the far end is hinged. Thus,

$$G_{top} = 0.87(2) = 1.74$$
$$G_{bottom} = 3.5 \quad (given)$$

From the alignment chart, Fig. 6.9.4b, find $K_x = 1.7$.

$$\frac{K_x L_x}{r_x} = \frac{1.7(15)12}{5.21} = 59; \qquad \phi_c F_{cr} = 25.5 \text{ ksi}$$

$$\phi_c P_n = \phi_c F_{cr} A_g = 25.5(8.79) = 224 \text{ kips} > [P_u = 206 \text{ kips}] \quad OK$$

Check LRFD-C2.2, Eq. 15.3.5,

$$[0.75\phi_c A_g F_y = 0.75(0.85)(8.79)36 = 202 \text{ kips}] \approx [P_u = 206 \text{ kips}]$$

This does not quite satisfy LRFD-C2.2; however, the 206 kips is not the true column load but rather a pseudo load such that the columns A and D can have adequate lateral stiffness when interacting with the beam to provide the sway resistance for the entire assemblage. Consider the design acceptable!
Use W12×30 for exterior column.

(d) Repeat the design of columns A and D using Geschwindner's method [14.42]. In this method, the actual factored load $P_u = 76$ kips is used on columns A and D, but the effective length factor K_0 becomes K_n given by Eq. 14.5.11. In part (b) above, the effective length factor K (now called K_0) was found to be 1.7. That value is now adjusted, as follows:

$$K_n = K_0 \sqrt{\frac{P_u + Q_u}{P_u}} \qquad\qquad [14.5.11]$$

$$= 1.7 \sqrt{\frac{76 + 130}{76}} = 2.80$$

Try the W12×30 obtained in part (b).

$$\frac{K_x L_x}{r_x} = \frac{2.80(15)12}{5.21} = 97; \qquad \phi_c F_{cr} = 18.7 \text{ ksi}$$

$$[\phi_c P_n = \phi_c F_{cr} A_g = 18.7(8.79) = 164 \text{ kips}] > [P_u = 76 \text{ kips}]$$

Because of the low slenderness ratio (i.e., 59) of the original column, the answer using the increased K_n is less conservative than using the increased load for design.
Try W12×26. Check local buckling limits,

$$\left(\frac{b_f}{2t_f} = 8.5\right) < \left(\frac{95}{\sqrt{F_y}} = 15.8\right) \quad OK$$

$$\left(\frac{h}{t_w} = 47.2\right) > \left(\frac{253}{\sqrt{F_y}} = 42.2\right) \quad NG$$

There is a potential web local buckling problem; Q may be less than 1.0. Initially assume $Q = 1.0$,

$$G_{\text{top}} = 2\frac{204/15}{659/36} = 2(0.74) = 1.48$$

$$G_{\text{bottom}} = 3.5 \text{ (given)}$$

From alignment chart, Fig. 6.9.4b, find $K_x = 1.65$, and $K_n = 2.72$.

$$\frac{K_x L_x}{r_x} = \frac{2.72(15)12}{5.17} = 95$$

$$\frac{K_y L_y}{r_y} = \frac{1.0(15)12}{1.51} = 119 \quad \text{Controls!}$$

$$\phi_c F_{\text{cr}} = 14.5 \text{ ksi}$$

Check h/t_w according to LRFD-B5.3b using $f = 14.5$ ksi,

$$\left(\frac{h}{t_w} = 47.2\right) > \left(\frac{253}{\sqrt{f}} = \frac{253}{\sqrt{14.5}} = 62.5\right) \quad \text{OK}$$

Thus, $Q = 1.0$.

$$[\phi_c P_n = \phi_c F_{\text{cr}} A_g = 14.5(7.65) = 111 \text{ kips}] > [P_u = 76 \text{ kips}] \quad \text{OK}$$

Check LRFD-C2.2, Eq. 15.3.5,

$$[0.75\phi_c A_g F_y = 0.75(0.85)(7.65)36 = 176 \text{ kips}] > [P_u = 76 \text{ kips}] \quad \text{OK}$$

This does satisfy LRFD-C2.2; and here the check is against the actual P_u of 76 kips. The next lightest W12 will not work.

Use W12×26 for exterior column. ■■

When lateral load acts on a system such as the frame of Fig. 15.3.4, the same procedure is used as illustrated in Example 15.3.3. The factored gravity load elastic analysis is made for the nonsway situation (assuming the frame and loading are relatively symmetrical and cause insignificant sway). Then the factored lateral load elastic analysis is made without the gravity load. When the sway magnifier B_2 is computed, the total P_u participating in the $P\Delta$ effect is used in ΣP_u, including the loads on columns B and C. However, ΣP_e includes only the contributions of those columns actually resisting sway; i.e., columns A and D in this example.

15.4 MULTISTORY FRAMES

The design of multistory frames, either braced or unbraced, is outside the scope of this text. *Braced* multistory frames have been designed using Allowable Stress Design (Working Stress Method) since the earliest design of steel structures. Plastic Design of multistory braced frames was accepted by AISC with the adoption of the 1969 AISC Specification. The theoretical and experimental background, developed primarily at Lehigh University, was first presented to the engineering profession in 1965

[15.6]. Additional background material is available in the work of Driscoll and Beedle [15.7], Lu [15.8], Williams and Galambos [15.9], Goldberg [15.10], and Yura and Lu [15.11]. The design procedure is well-described in *Plastic Design of Braced Multi-story Steel Frames* [14.37].

Regarding *unbraced* multistory frames, procedures are continuing to be developed. Load and Resistance Factor Design [1.16] has no restriction on the number of stories, and the $P\Delta$ effect *must* be considered. Plastic Design in Chapter N of the 1989 ASD Specification [1.5] has no restriction on the number of stories; however, for one- and two-story frames, the frame stability effect ($P\Delta$) may be ignored (ASD-N3). Chapter N does require that "The strength of an unbraced multistory frame shall be determined by a rational analysis which includes the effect of frame instability . . . ." There presently (1995) seems to be no practical accepted procedure for using Plastic Design on multistory unbraced frames.

The *SSRC Guide* [6.8, Chap. 16] contains excellent treatment of methods to evaluate frame stability for multistory frames. Background on unbraced frame stability is available in the Lehigh Lecture Notes [15.6], Driscoll and Beedle [15.7], and Daniels [15.12]. Other studies on the stability of unbraced frames are given by Cheong-Siat-Moy [15.13, 15.15, 14.18], and Cheong-Siat-Moy and Lu [15.14].

SELECTED REFERENCES

15.1. AISC. *Plastic Design in Steel.* New York: American Institute of Steel Construction, Inc., 1959.

15.2. Lynn S. Beedle. *Plastic Design of Steel Frames.* New York: John Wiley & Sons, Inc., 1958.

15.3. C. E. Massonnet and M. A. Save. *Plastic Analysis and Design, Vol. I, Beams and Frames.* Waltham, MA: Blaisdell Publishing Company, 1965.

15.4. Lynn S. Beedle. "Plastic Strength of Steel Frames," *Proceedings,* ASCE, **81,** Paper No. 764 (August 1955). Also *Transactions,* **122** (1957), 1139–1168.

15.5. Edward R. Estes, Jr. "Design of Multi-span Rigid Frames by Plastic Analysis," *Proceedings of National Engineering Conference,* AISC, 1955, 27–39.

15.6. George C. Driscoll, Jr., et al. *Plastic Design of Multi-Story Frames,* Lecture Notes, Fritz Engineering Laboratory Report No. 273.20, Lehigh University, Bethlehem, PA, 1965.

15.7. George C. Driscoll and Lynn S. Beedle. "Research in Plastic Design of Multi-story Frames," *Engineering Journal,* AISC, **1,** 3 (July 1964), 92–100.

15.8. Le-Wu Lu. "Design of Braced Multi-Story Frames by the Plastic Method," *Engineering Journal,* AISC, **4,** 1 (January 1967), 1–9.

15.9. James B. Williams and Theodore V. Galambos. "Economic Study of a Braced Multi-Story Steel Frame" *Engineering Journal,* AISC **5,** 1 (January 1968), 2–11.

15.10. John E. Goldberg. "Lateral Buckling of Braced Multistory Frames," *Journal of the Structural Division,* ASCE, **94,** ST12 (December 1968), 2963–2983.

15.11. Joseph A. Yura and Le-Wu Lu. "Ultimate Load Tests of Braced Multistory Frames," *Journal of the Structural Division,* ASCE, **95,** ST10 (October 1969), 2243–2263.

15.12. J. Hartley Daniels. "A Plastic Method for Unbraced Frame Design," *Engineering Journal,* AISC, **3,** 4 (October 1966), 141–149.

15.13. F. Cheong-Siat-Moy. "Stiffness Design of Unbraced Steel Frames," *Engineering Journal,* AISC, **13,** 1 (First Quarter 1976), 8–10.

15.14. F. Cheong-Siat-Moy and Le-Wu Lu. "Stiffness and Strength Design of Multistory Frames," *Publications*, International Association for Bridge and Structural Engineering, **36**-II, 1976, 31–47.

15.15. Francois Cheong-Siat-Moy. "Frame Design Without Using Effective Column Length," *Journal of the Structural Division,* ASCE, **104,** ST1 (January 1978), 23–33.

PROBLEMS

All problems are to be done in accordance with Load and Resistance Factor Design. Use the plastic analysis option under LRFD-A5.1, if applicable, only when specifically assigned by the instructor; otherwise use factored load elastic analysis. All given loads are service loads.

15.1. Design using W shapes an unbraced rectangular frame of 60-ft span as shown in the accompanying figure, to carry uniformly distributed gravity dead and live loads of 1.5 and 2.5 kips/ft, respectively, not including the weight of the girder. Use a lateral wind load of 0.8 kip/ft acting uniformly over the 18-ft height. Lateral bracing occurs every 5 ft on the girder and 4.5 ft on the columns. Include, if assigned by the instructor, the design of the knee region at the top of the column as discussed in Sec. 13.8. Use A36 steel.

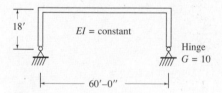

Problem 15.1

15.2. Design for the conditions of Prob. 15.1 except assume the moment of inertia for the beam to be three times that of the column.

15.3. Design using W shapes the unbraced rectangular frame of the accompanying figure to carry uniformly distributed gravity dead and live loads of 1.4 and 2.2 kips/ft, respectively. Use a lateral wind load of 0.8 kip/ft acting uniformly over the 14-ft height. Assume the structure is braced perpendicular to the plane of the frame such that columns are braced at the top and bottom with respect to the minor axis, and that $K_y = 1.0$. Assume all joints at the tops of columns are rigid; i.e., fully moment resisting. Use A36 steel.

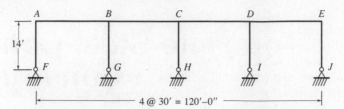

Problem 15.3

15.4. Design as in Prob. 15.3 except use simple connections (assume as pinned) at the tops of columns *BG*, *CH*, and *DI*, with rigid moment resisting connections only at *A* and *E*.

Chapter 16

Composite Steel-Concrete Construction

16.1 HISTORICAL BACKGROUND

Steel framing supporting cast-in-place reinforced concrete slab construction was historically designed on the assumption that the concrete slab acts independently of the steel in resisting loads. No consideration was given to the composite effect of the steel and concrete acting together. This neglect was justified on the basis that the bond between the concrete floor or deck and the top of the steel beam could not be depended upon. However, with the advent of welding, it became practical to provide mechanical shear connectors to resist the horizontal shear which develops during bending.

Steel beams encased in concrete were widely used from the early 1900s until the development of lightweight materials for fire protection in the past 40 years. Some such beams were designed compositely and some were not. In the early 1930s bridge construction began to use composite sections. Not until the early 1960s was it economical to use composite construction for buildings. However, current practice (1995) utilizes composite action in nearly all situations where concrete and steel are in contact, both on bridges and buildings.

Composite construction, as treated in this chapter, consists of a solid cast-in-place concrete slab placed upon and interconnected to a steel rolled W section or welded I-shaped girder, as shown in Fig. 16.1.1. The concrete slab is also commonly

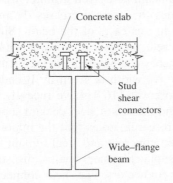

Concrete slab

Stud shear connectors

Wide–flange beam

Figure 16.1.1 Conventional composite steel-concrete beam.

Shear stud connectors on flanges of bridge girders to be embedded in the concrete slab in order to make steel section and concrete slab act as a unit (i.e., compositely). (Photo by C. G. Salmon)

cast upon cold-formed steel deck (Fig. 16.1.2), which itself is supported on a steel I-shaped section. The corrugations (ribs) may be either parallel to or perpendicular to the supporting beam. When the ribs are parallel to the beam, the behavior is essentially that of a variable thickness slab supported directly on the steel beam. When the ribs are perpendicular to the steel W section, special treatment is required. The many varieties of composite steel–concrete construction are discussed in the State-of-the Art Report [16.1]. In this chapter, treatment is restricted to the solid slab supported by a steel I-shaped section.

The composite beam is one having a wide flange (concrete slab), typically spanning 8 to 15 ft between parallel beams. Ordinary beam theory, where the stress is assumed constant across the width of a beam at a given distance from the neutral axis, *does not apply*. Plate theory indicates the stress decreases the more distant a point is from the stiff part (steel section in this case) of the beam. Similarly to the treatment of T-sections in reinforced concrete, an *equivalent width* is used in place of the actual width, so that ordinary beam theory can be used. An excellent summary of the factors involved in obtaining an effective width is given by Brendel [16.2] and Heins and Fan [16.3]. Vallenilla and Bjorhovde [16.4] have recently reviewed the effective width in the context of LRFD and the use of steel deck to support the slab.

Viest [16.5], in his 1960 review of research, notes that the important factor in composite action is that the bond between concrete and steel remain unbroken. As designers began to place slabs on top of supporting steel beams, investigators began to study the behavior of mechanical shear connectors. The shear connectors provided

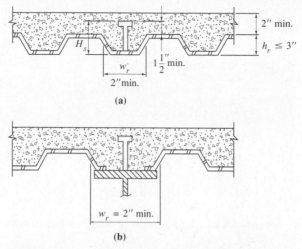

Figure 16.1.2 Composite section using formed steel deck. Steel beam supporting deck and slab may be parallel to ribs of formed deck (as in b.) or perpendicular to the ribs. (LRFD-I3.5 or ASD-I5)(Adapted from LRFD Commentary [1.17])

the interaction necessary for the concrete slab and steel beam to act as a unit; i.e., no slip between the concrete and steel beam parallel to the beam. For the earlier encased beams there had been sufficient contact area between concrete and steel so that friction provided the necessary interaction between the two materials.

The State-of-the-Art Report of 1974 [16.1] provides an overall survey of the subject of composite construction, including bibliography. Hansell, Galambos, Ravindra, and Viest [16.6] have provided the background for Load and Resistance Factor Design. Iyengar and Iqbal [16.7] have provided a modern review of composite construction in building design, and Lorenz and Stockwell [16.8] and Lorenz [16.9] have provided treatment of basic design concepts for Load and Resistance Factor Design.

A thorough treatment of steel-concrete composite construction in the context of *Eurocode 4* has been developed by IABSE [16.42].

16.2 COMPOSITE ACTION

Composite action is developed when two load-carrying structural members such as a concrete floor system and the supporting steel beam (Fig. 16.2.1a) are integrally connected and deflect as a single unit as in Fig. 16.2.1b. The extent to which composite action is developed depends on the provisions made to insure a single linear strain from the top of the concrete slab to the bottom of the steel section.

In developing the concept of composite behavior, consider first the noncomposite beam of Fig. 16.2.1a, wherein if friction between the slab and beam is neglected, the beam and slab each carry separately a part of the load. This is further

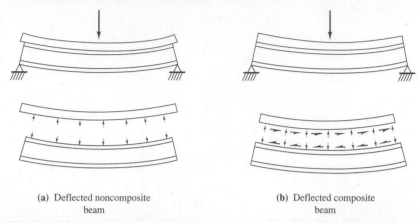

(a) Deflected noncomposite
beam

(b) Deflected composite
beam

Figure 16.2.1   Comparison of deflected beams with and without composite
action.

shown in Fig. 16.2.2a. When the slab deforms under vertical load, its lower surface is
in tension and elongates; while the upper surface of the beam is in compression and
shortens. Thus a discontinuity will occur at the plane of contact. Since friction is
neglected, only vertical internal forces act between the slab and beam.

When a system acts compositely (Fig. 16.2.1b and 16.2.2c) no relative slip
occurs between the slab and beam. Horizontal forces (shears) are developed that act

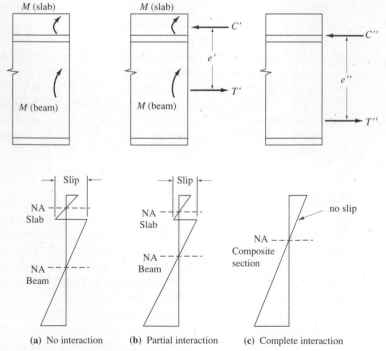

(a) No interaction   **(b)** Partial interaction   **(c)** Complete interaction

Figure 16.2.2   Strain variation in composite beams.

on the lower surface of the slab to compress and shorten it, while simultaneously they act on the upper surface of the beam to elongate it.

By an examination of the strain distribution that occurs when there is no interaction between the concrete slab and the steel beam (Fig. 16.2.2a), it is seen that the total resisting moment is equal to

$$\Sigma M = M_{\text{slab}} + M_{\text{beam}} \qquad (16.2.1)$$

It is noted that for this case there are two neutral axes; one at the center of gravity of the slab and the other at the center of gravity of the beam. The horizontal slip resulting from the bottom of the slab in tension and the top of the beam in compression is also indicated.

Consider next the case where only partial interaction is present, Fig. 16.2.2b. The neutral axis of the slab is closer to the beam and that of the beam closer to the slab. Due to the partial interaction, the horizontal slip has now decreased. The result of the partial interaction is the partial development of the maximum compressive and tensile forces C' and T', in the concrete slab and steel beam, respectively. The resisting moment of the section would then be increased by the amount $T'e'$ or $C'e'$.

When complete interaction (known as full composite action) between the slab and the beam is developed, no slip occurs and the resulting strain diagram is shown in Fig. 16.2.2c. Under this condition, a single neutral axis occurs which lies below that of the slab and above that of the beam. In addition, the compressive and tensile forces C'' and T'', respectively, are larger than the C' and T' existing with partial interaction. The resisting moment of the fully developed composite section then becomes

$$\Sigma M = T''e'' \quad \text{or} \quad C''e'' \qquad (16.2.2)$$

16.3 ADVANTAGES AND DISADVANTAGES

The basic advantages resulting from composite design are

1. Reduction in the weight of steel
2. Shallower steel beams
3. Increased floor stiffness
4. Increased span length for a given number

A weight savings in steel of 20 to 30% is often possible by taking full advantage of a composite system. Such a weight reduction in the supporting steel beams usually permits the use of a shallower as well as a lighter member. This advantage may reduce the height of a multistoried building significantly so as to provide savings in other building materials such as outside walls and stairways. The overall economy of using composite construction when considering total building cost appears to be increasingly favorable [16.10, 16.11].

The stiffness of a composite floor is substantially greater than that of a concrete floor with its supporting beams acting independently. Normally the concrete slab acts as a one-way plate spanning between the supporting beams. In composite design, an additional use is made of the slab by its action in a direction parallel to and in combination with the supporting steel beams. The net effect is to greatly increase the moment of inertia of the floor system in the direction of the steel beams. The increased stiffness considerably reduces the live load deflections and, if shoring is provided during construction, also reduces dead load deflections. Assuming full composite action, the nominal strength of the section greatly exceeds the sum of the strengths of the slab and the beam considered separately, providing high overload capacity.

While there are no major disadvantages, some limitations should be recognized. In continuous construction, the negative moment region will have a different stiffness because the concrete slab in tension is expected to be cracked and not participating. In general, it is considered acceptable to assume the moment of inertia to be constant through both positive and negative moment regions, using the positive moment composite section moment of inertia (LRFD-I1). Tension in the concrete is neglected.

Long-term deflection caused by concrete creep and shrinkage could be important when the composite section resists a substantial part of the dead load, or when the live load is of long duration. This is discussed in Sec. 16.12.

16.4 EFFECTIVE WIDTH

The concept of *effective width* is useful in design when strength must be determined for an element subject to nonuniform distribution of stress. Referring to Fig. 16.4.1, the concrete slab of a composite section is considered to be infinitely wide. The

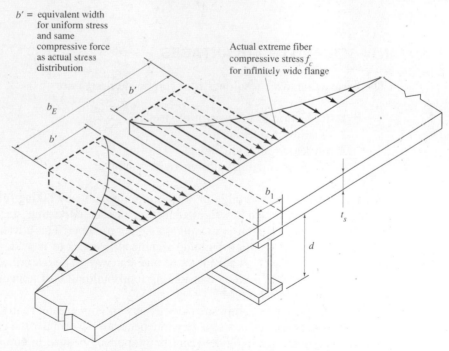

Figure 16.4.1   Actual and equivalent stress distribution over flange width.

intensity of extreme fiber stress f_c is a maximum over the steel beam and decreases nonlinearly as the distance from the supporting beam increases.

The effective width b_E of a flange for a composite member may be expressed

$$b_E = b_f + 2b' \qquad (16.4.1)$$

where $2b'$ times the maximum stress f_c equals the area under the curves for f_c. Various investigators, including Timoshenko and Goodier [16.12] and von Kármán [16.13], have derived expressions for the effective width of homogeneous beams having wide flanges; and Johnson and Lewis [16.14] have shown such expressions are valid for beams in which the flange and web are of different materials.

The analysis for effective width involves theory of elasticity applied to plates, using an infinitely long continuous beam on equidistant supports, with an infinitely wide flange having a small thickness compared to the beam depth. The total compression force carried by the equivalent system must be the same as that carried by the real system.

The practical simplifications for design purposes are given by LRFD-I3.1 and ASD-I1; the *same* for service load calculations as for nominal strength calculations when failure is imminent.

1. For an *interior girder,* referring to Fig. 16.4.2,

$$b_E \leq \frac{L}{4} \qquad (16.4.2a)$$

$$b_E \leq b_0 \quad \text{(for equal beam spacing)} \qquad (16.4.2b)$$

2. For an *exterior girder,*

$$b_E \leq \frac{L}{8} + \left(\begin{array}{l} \text{distance from beam center} \\ \text{to edge of slab} \end{array} \right) \qquad (16.4.3a)$$

$$b_E \leq \tfrac{1}{2}b_0 + \left(\begin{array}{l} \text{distance from beam center} \\ \text{to edge of slab} \end{array} \right) \qquad (16.4.3b)$$

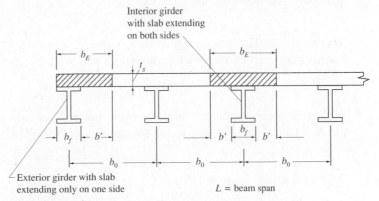

Figure 16.4.2 Dimensions governing effective width b_E on composite steel-concrete beams.

The American Concrete Institute (ACI) Code [16.15] has long used the following effective flange widths for *T*-sections:

1. For an *interior girder*, referring to Fig. 16.4.2,

$$b_E \le \frac{L}{4} \tag{16.4.4a}$$

$$b_E \le b_0 \quad \text{(for equal beam spacing)} \tag{16.4.4b}$$

$$b_E \le b_f + 16t_s \tag{16.4.4c}$$

2. For an *exterior girder*,

$$b_E \le b_f + \frac{L}{12} \tag{16.4.5a}$$

$$b_E \le b_f + 6t_s \tag{16.4.5b}$$

$$b_E \le b_f + 0.5(\text{clear distance to next beam}) \tag{16.4.5c}$$

These 1989 ACI Code effective widths are identical to AISC effective widths used prior to the 1986 LRFD Specification. The new AISC rules are simpler; eliminating the beam flange width b_f and the slab thickness t_s as variables.

16.5 COMPUTATION OF ELASTIC SECTION PROPERTIES

The elastic section properites of a composite section can be computed by the transformed section method. In contrast to reinforced concrete, where the reinforcing bar steel is transformed into an equivalent concrete area, the concrete slab in the composite section is transformed into equivalent steel. As a result, the concrete area is reduced by using a slab width equal to b_E/n, where n is the modulus of elasticity ratio E_s/E_c. E_s is the modulus of elasticity of steel, taken as 29,000 ksi, and E_c in psi is given by the ACI Code [16.15], as follows:

$$E_c = 33(w^{1.5})\sqrt{f'_c}, \text{ psi} \tag{16.5.1}*$$

where w is the density of concrete in pcf and f'_c is in psi. Since the AISC Specifications [1.5, 1.16] use stress in ksi for all formulas, LRFD-I2.2 converts Eq. 16.5.1 approximately to the following for E_c in ksi:

$$E_c = w^{1.5}\sqrt{f'_c}, \text{ ksi} \tag{16.5.2}*$$

Note the $\sqrt{1000}$ is 31.6; thus, Eq. 16.5.2 gives E_c about 4% lower than the ACI Code. For normal-weight concrete, weighing approximately 145 pcf, Eq. 16.5.2 gives E_c in ksi as

$$E_c = 1750\sqrt{f'_c}, \text{ ksi} \tag{16.5.3}*$$

Within the accuracy that the modulus of elasticity of concrete may be predicted, either the ACI Code [16.15] value or the suggested value of LRFD-I2.2 is acceptable. The

---

*For SI units, giving E_c in MPa,

$$E_c = w^{1.5}(0.043)\sqrt{f'_c} \tag{16.5.1}$$

$$E_c = w^{1.5}(0.041)\sqrt{f'_c} \tag{16.5.2}$$

$$E_c = 4600\sqrt{f'_c} \tag{16.5.3}$$

where w is in kg/m³ and f'_c is in MPa.

TABLE 16.5.1   PRACTICAL VALUES FOR MODULAR RATIO n

f'_c (psi)	Modular ratio $n = E_s/E_c$	f'_c (MPa)
3000	9	21
3500	$8\frac{1}{2}$	24
4000	8	28
4500	$7\frac{1}{2}$	31
5000	7	35
6000	$6\frac{1}{2}$	42

modulus of elasticity ratio n is commonly taken to the nearest whole number. Table 16.5.1 indicates practical values usually used in computing elastic section properties.

Effective Elastic Section Modulus

A complete beam may be considered as a steel member to which has been added a cover plate on the top flange. This "cover plate" being concrete is considered to be effective only when the top flange is in compression. In continuous beams, the concrete slab is usually ignored in regions of negative moment. If the neutral axis falls within the concrete slab, present practice is to consider only that portion of the concrete slab which is in compression.

LRFD-I3.2 and ASD-I2 permit reinforcement parallel to the steel beam and lying within the effective slab width to be included in computing properties of composite sections. These reinforcing bars usually make little difference to the composite section modulus in the positive moment region and their effect is frequently neglected.

EXAMPLE 16.5.1 _____

Compute the elastic section properties of the composite section shown in Fig. 16.5.1 assuming $f'_c = 3000$ psi and $n = 9$. Use the effective flange width according to LRFD-I3.1 and ASD-I1.

Solution   First, determine effective width (LRFD-I3.1 or ASD-I1).

$$b_E = L/4 = 0.25(30)12 = 90 \text{ in.} \quad controls$$
$$b_E = b_0 = 8(12) = 96 \text{ in.}$$

The width of equivalent steel is $b_E/n = 10.0$ in. The computation of the moment of inertia I_x about the center of gravity of the W21×62 is shown, as follows:

Element	Transformed Area A (sq in.)	Moment Arm from Centroid y (in.)	Ay (in.³)	Ay^2 (in.⁴)	I_0 (in.⁴)
Slab	40.0	+12.495	+500	6245	53
W21×62	18.3	0	0	0	1330
Cover plate	7.0	−10.995	−77	846	1
	65.3		+423	7091	1384

$$I_x = I_0 + Ay^2 = 1384 + 7091 = 8475 \text{ in.}^4$$

$$\bar{y} = \frac{+423}{65.3} = +6.48 \text{ in.}$$

$$I_{tr} = I_x - A\bar{y}^2 = 8475 - 65.3(6.48)^2 = 5737 \text{ in.}^4$$

$$y_t = 10.50 - 6.48 + 4.0 = 8.02 \text{ in.}$$

$$y_b = 10.50 + 6.48 + 1.0 = 17.98 \text{ in.}$$

The symbol I_{tr} is used for the *fully composite uncracked transformed section* moment of inertia. The elastic section modulus S_{conc} referred to the top fiber of the concrete slab is

$$S_{conc} = I_{tr}/y_t = 5737/8.02 = 715 \text{ in.}^3$$

The elastic section modulus S_{tr} referred to the extreme fiber at the tension flange of the steel section (in this case the cover plate) is

$$S_{tr} = I_{tr}/y_b = 5737/17.98 = 319 \text{ in.}^3$$

The addition of a cover plate at the tension flange brings the neutral axis down and permits more economical use of the composite section. However, the cost

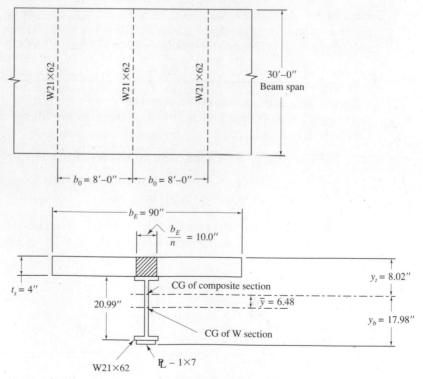

Figure 16.5.1   Composite section for Example 16.5.1.

of welding a cover plate to the rolled section usually exceeds any material saving; thus, a cover plate is rarely used.   ■■

16.6 SERVICE LOAD STRESSES WITH AND WITHOUT SHORING

The actual stresses that result due to a given loading on a composite member are dependent upon the manner of construction.

The simplest construction occurs when the steel beams are placed first and used to support the concrete slab formwork. In this case the steel beam acting noncompositely (i.e., by itself) supports the weight of the forms, the wet concrete, and its own weight. Once forms are removed and concrete has cured, the section will act compositely to resist all dead and live loads placed after the curing of concrete. Such construction is said to be *without temporary shoring* (i.e., unshored).

Alternatively, to reduce the service load stresses, the steel beams may be supported on temporary shoring; in which case, the steel beam, forms, and wet concrete, are carried by the shores. After curing of the concrete, the shores are removed and the section acts compositely to resist all loads. This system is called *shored* construction.

The following example illustrates the difference in service load stresses under the two system of construction.

EXAMPLE 16.6.1 ───

For the steel W21×62 with the 1 by 7-in. plate of Fig. 16.5.1, determine the service load stresses considering that (a) construction is without temporary shoring, and (b) construction uses temporary shores. The dead- and live-load moment M_L to be superimposed on the system after the concrete has cured is 560 ft-kips.

Solution   The composite section properties as computed in Example 16.5.2 are

$$S_{top} = 715 \text{ in.}^3 \quad (\text{top of concrete})$$

$$S_{bottom} = S_{tr} = 319 \text{ in.}^3 \quad (\text{bottom of steel})$$

The noncomposite properties for the steel section alone (see Fig. 16.6.1) are computed as follows:

$$\bar{y} = \frac{7.0(10.995)}{7.0 + 18.3} = 3.04 \text{ in.}$$

$$y_b = 10.495 - 3.04 + 1.00 = 8.45 \text{ in.}$$

$$I_s = I_0 \text{ (W21×62)} + A_p y^2 - A\bar{y}^2$$

$$= 1330 + 7.0(10.995)^2 - 25.3(3.04)^2$$

$$= 1330 + 846 - 234 = 1942 \text{ in.}^4$$

$$S_{st} = \frac{1942}{13.55} = 143 \text{ in.}^3 \quad (\text{top})$$

$$S_{sb} = \frac{1942}{8.45} = 230 \text{ in.}^3 \quad (\text{bottom})$$

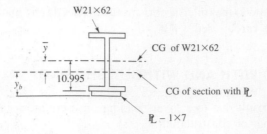

Figure 16.6.1   Steel section for Example 16.6.1.

(a) *Without Temporary Shores*. Weight due to the concrete slab and steel beam,

$$w \text{ (concrete slab), } (4/12)(8)0.15 = 0.40$$
$$w \text{ (steel beam)} \qquad\qquad = \underline{0.06}$$
$$\qquad\qquad\qquad\qquad\qquad\quad 0.46 \text{ kips/ft}$$

$$M_D \text{ (DL on noncomposite)} = \tfrac{1}{8}(0.46)(30)^2 = 51.8 \text{ ft-kips}$$

$$f_{\text{top}} = \frac{M_D}{S_{st}\text{(steel section)}} = \frac{51.8(12)}{143} = 4.3 \text{ ksi}$$

$$f_{\text{bottom}} = \frac{M_D}{S_{sb}\text{(steel section)}} = \frac{51.8(12)}{230} = 2.7 \text{ ksi}$$

The additional stresses after the concrete has cured are

$$f_{\text{top}} = \frac{M_L}{nS_{\text{top}}\text{(composite)}} = \frac{560(12)}{9(715)} = 1.04 \text{ ksi (concrete stress)}$$

where the stress in the concrete is $1/n$ times the stress on equivalent steel (transformed section).

$$f_{\text{bottom}} = \frac{M_L}{S_{\text{tr}}} = \frac{560(12)}{319} = 21.1 \text{ ksi}$$

The total maximum tensile stress in the steel is

$$f = f\text{(noncomposite)} + f\text{(composite)} = 2.7 + 21.1 = 23.8 \text{ ksi}$$

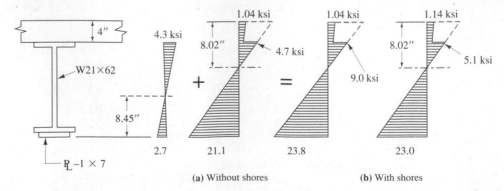

(a) Without shores   (b) With shores

Figure 16.6.2   Service load stresses for Example 16.6.1.

(b) *With Temporary Shores.* Under this condition all loads are resisted by the composite section.

$$f_{\text{top}} = \frac{M_D + M_L}{S_{\text{top}}(\text{composite})} = \frac{(560 + 51.8)12}{715(9)} = 1.14 \text{ ksi} \quad \text{on concrete}$$

$$f_{\text{bottom}} = \frac{M_D + M_L}{S_{\text{tr}}} = \frac{(560 + 51.8)12}{319} = 23.0 \text{ ksi}$$

Stress distributions for both with and without shores are given in Fig. 16.6.2. Since the dead load was small in this example, use of shores gave insignificant reduction in service load stress. Where thicker slabs are used, the dead load stresses may become as high as 30%, in which case using or not using shores will make a significant difference. ■■

16.7 NOMINAL MOMENT STRENGTH OF FULLY COMPOSITE SECTIONS

The nominal strength M_n of a composite section having its slab in compression (positive moment) depends on the yield stress F_y and section properties (including slenderness $\lambda = h/t_w$ for the web) for the steel beam, the concrete slab strength f_c', and the strength of shear connectors providing the interface shear transfer between slab and beam.

The nominal strength (commonly called *ultimate strength*) concepts were first applied to design practice as recommended by the ASCE-ACI Joint Committee on Composite Construction [16.16], and further modified by Slutter and Driscoll [16.17]. Ultimate strength was reviewed in the State-of-the-Art Report [16.1], and treated in the context of Load and Resistance Factor Design by Hansell et al. [16.6].

Traditionally, since the Joint Committee Report [16.16] the design of composite beams has been based on nominal moment strength even though Allowable Stress Design was used. Load and Resistance Factor Design is particularly adapted to using composite flexural members since the concepts of strength are easier to understand without trying to convert them into a service load based Allowable Stress Design.

The nominal moment strength M_n when the slab is in compression (positive moment) is divided into two categories according to LRFD-I3.2, depending on web slenderness, as follows:

1. For $h/t_w \leq (\lambda_p = 640/\sqrt{F_{yf}})$:

M_n = nominal moment strength based on plastic stress distribution on the composite section

$\phi_b = 0.85$

2. For $h/t_w > (\lambda_p = 640/\sqrt{F_{yf}})$:

M_n = nominal moment strength based on superposition of elastic stresses (shown in Sec. 16.6), considering the effects of shoring

$\phi_b = 0.90$

where F_{yf} is the yield stress of the flange; this will give a conservative (low) limit for λ_p because the web of a hybrid girder will have a lower yield stress than the flange.

Since the elastic properties and effects of shoring have been treated in Sec. 16.6, this section focuses on strength based on plastic stress distribution.

The nominal strength M_n based on plastic stress distribution may be divided into two general categories: (1) the plastic neutral axis (PNA) occurs in the slab; and (2) the plastic neutral axis occurs in the steel section. When the PNA occurs in the steel section, the nominal strength M_n calculation will differ depending on whether the PNA is in the flange or the web.

The concrete is assumed to develop only compression forces. Although concrete is able to sustain a limited amount of tension, the tensile strength is negligible at the strains occurring when nominal strength is reached.

Case 1—Plastic Neutral Axis (PNA) in the Slab

Referring to Fig. 16.7.1b and assuming the Whitney rectangular stress distribution* (uniform stress of $0.85 f_c'$ acting over a depth a), the compressive force C is

$$C = 0.85 f_c' a b_E \qquad (16.7.1)$$

The tensile force T is the yield stress on the beam times its area:

$$T = A_s F_y \qquad (16.7.2)$$

Equating the compressive force C to the tensile force T gives

$$a = \frac{A_s F_y}{0.85 f_c' b_E} \qquad (16.7.3)$$

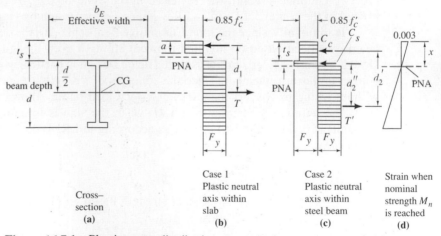

Figure 16.7.1 Plastic stress distribution at nominal moment strength M_n. (PNA = plastic neutral axis)

---

*For the development of the concept of replacing the true distribution of compressive stress by a rectangular stress distribution, see for example, Chu-Kia Wang and Charles G. Salmon, *Reinforced Concrete Design*, 5th ed. (HarperCollins, New York, 1992, Chap. 3).

According to the ACI-accepted [16.15, Sec. 10.2.7] rectangular stress distribution, the neutral axis distance x, as shown in Fig. 16.7.1d, equals $a/0.85$ for $f'_c \leq$ 4000 psi. The nominal moment strength M_n, from Fig. 16.7.1b, becomes

$$M_n = Cd_1 \quad \text{or} \quad Td_1 \tag{16.7.4}$$

When the slab is capable of developing a compressive force at least equal to the full yield strength of the steel beam, the PNA will be in the slab, the common situation for fully composite sections. Expressing the nominal strength in terms of the steel force gives

$$M_n = A_s F_y \left(\frac{d}{2} + t_s - \frac{a}{2} \right) \tag{16.7.5}$$

The usual procedure for computing nominal strength is to assume the depth a for the rectangular stress distribution will not exceed t_s; i.e., use Eq. 16.7.3. If a is verified to not exceed t_s, Eq. 16.7.5 can be used to obtain nominal strength M_n.

In the past, Case 1 has been referred to as "slab adequate"; meaning that the slab is capable of developing in compression the full nominal strength of the steel beam in tension.

Case 2—Plastic Neutral Axis (PNA) in the Steel Beam

If the depth a of the stress block as determined in Eq. 16.7.3 exceeds the slab thickness, the stress distribution will be as shown in Fig. 16.7.1c. The compressive force C_c in the slab is

$$C_c = 0.85 f'_c b_E t_s \tag{16.7.6}$$

The compressive force in the steel beam resulting from the portion of the beam above the neutral axis is shown in Fig. 16.7.1c as C_s.

The tensile force T' which is now less than $A_s F_y$ must equal the sum of the compressive forces:

$$T' = C_c + C_s \tag{16.7.7}$$

Also,

$$T' = A_s F_y - C_s \tag{16.7.8}$$

Equating Eqs. 16.7.7 and 16.7.8, C_s becomes

$$C_s = \frac{A_s F_y - C_c}{2}$$

or

$$C_s = \frac{A_s F_y - 0.85 f'_c b_E t_s}{2} \tag{16.7.9}$$

Considering the compressive forces C_c and C_s, the nominal moment strength M_n for Case 2 is

$$M_n = C_c d'_2 + C_s d''_2 \tag{16.7.10}$$

where the moment arms d'_2 and d''_2 are as shown in Fig. 16.7.1c.

When the Case 2 situation occurs, the steel beam must be capable of accommodating plastic strain in both tension and compression to achieve the nominal strength condition. The lower the PNA occurs in the steel section the more local buckling may influence the behavior. As indicated earlier in this section, in order to use the plastic stress distribution at all, LRFD-I3.2 requires the web $\lambda \leq \lambda_p$.

When the flange of the steel section adjacent to the slab is in compression, there might be concern regarding flange local buckling. The combination of concrete bearing against the compression flange and the shear connectors used to attach the slab and steel beam together eliminates flange local buckling as well as lateral-torsional buckling as controlling limit states. LRFD-I3.2 addresses only the issue of web local buckling; it is silent regarding flange local buckling. ASD-I2 specifically states that when shear connectors are used ". . . the steel section is exempt from compact flange criteria and there is no limit on unsupported length of compression flange."

EXAMPLE 16.7.1

Determine the nominal moment strength M_n of the composite section shown in Fig. 16.7.2. Use A36 steel, $f'_c = 3000$ psi and $n = 9$.

Solution   Assume the plastic neutral axis (PNA) is within the slab; i.e., that $a \leq t_s$ (Case 1),

$$a = \frac{A_s F_y}{0.85 f'_c b_E} = \frac{10.6(36)}{0.85(3)60} = 2.49 \text{ in.} < t_s \quad \text{OK}$$

$$C = 0.85 f'_c a b_E = 0.85(3)(2.49)(60) = 382 \text{ kips}$$

$$T = A_s F_y = 10.6(36) = 382 \text{ kips} \quad \text{(checks)}$$

$$\text{Arm } d_1 = \frac{d}{2} + t - \frac{a}{2} = 7.925 + 4.0 - 1.245 = 10.68 \text{ in.}$$

The nominal moment strength M_n is then

$$M_n = C d_1 = T d_1$$
$$= 382(10.68)\tfrac{1}{12} = 340 \text{ ft-kips} \quad \blacksquare\blacksquare$$

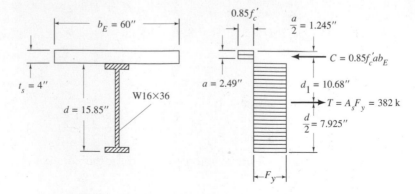

Figure 16.7.2   Example 16.7.1.

EXAMPLE 16.7.2

Determine the nominal moment strength M_n of the composite section shown in Fig. 16.7.3. Use A36 steel with $f'_c = 3000$ psi, and $n = 9$.

Solution   Referring to Fig. 16.7.3, assume the plastic neutral axis (PNA) is within the flange (i.e., Case 1),

$$a = \frac{A_s F_y}{0.85 f'_c b_E} = \frac{47.0(36)}{0.85(3)(72)} = 9.24 \text{ in.} > t_s = 7 \text{ in.}   \text{NG}$$

Since the concrete slab is only 7 in. thick, the slab cannot develop enough strength to balance the tension force $A_s F_y$ capable of developing in the steel section; thus the PNA will be within the steel section; thus, Case 2 applies. Using Eq. 16.7.6,

$$C_c = 0.85 f'_c b_E t_s = 0.85(3)72(7) = 1285 \text{ kips}$$

Using Eq. 16.7.9,

$$C_s = \frac{A_s F_y - 0.85 f'_c b_E t_s}{2} = \frac{47.0(36) - 1285}{2} = 205 \text{ kips}$$

Assuming only the flange of the W36×160 ($b_f = 12.00$ in.) is in compression, the portion of the flange d_f to the neutral axis is

$$d_f = \frac{205}{36(12.00)} = 0.475 \text{ in.} < [t_f = 1.020 \text{ in.}]$$

Thus, the PNA is within the flange. The location of the centroid of the tension portion of the steel beam from the bottom is

$$\bar{y} = \frac{47.0(18) - 0.475(12)35.76}{47.0 - 0.475(12)} = 15.55 \text{ in.}$$

Referring to Fig. 16.7.3, the nominal composite moment strength M_n from Eq. 16.7.10 is

$$M_n = C_c d'_2 + C_s d''_2$$
$$= [1285(23.95) + 205(20.21)]/12 = 2910 \text{ ft-kips}   \blacksquare\blacksquare$$

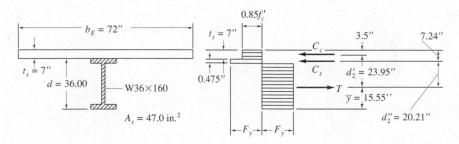

Figure 16.7.3   Example 16.7.2.

The nominal strength M_n has inherently assumed that shear connectors will provide sufficient shear transfer at the slab-to-flange interface to develop however much of the slab compressive strength that is required to balance the tension force developed in the steel beam. Shear connectors are treated in Sec. 16.8.

The nominal strength M_n is *independent* of whether or not the system is shored during construction. Even though service load stresses are different, as illustrated in Sec. 16.6, the nominal strength is the same, shored or unshored.

16.8 SHEAR CONNECTORS

The horizontal shear that develops between the concrete slab and the steel beam during loading must be resisted so that the slip shown in Fig. 16.2.2 will be restrained. A fully composite section will have no slip at the concrete-steel interface. Although some bond may develop between the steel and the concrete, it is not sufficiently predictable to provide the required interface shear strength. Neither can friction between the concrete slab and the steel beam develop such strength.

Instead, mechanical shear connectors are *required* (LRFD-I5.2 and ASD-I1), except for the totally concrete-encased steel beam. Some mechanical shear connectors are shown in Fig. 16.8.1. The only connectors specifically provided for in the AISC Specifications [1.5, 1.16] are stud connectors (LRFD-I5.3 and ASD-I4) and

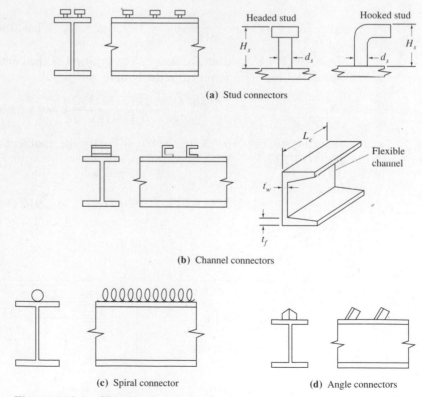

(a) Stud connectors

(b) Channel connectors

(c) Spiral connector

(d) Angle connectors

Figure 16.8.1   Shear connectors.

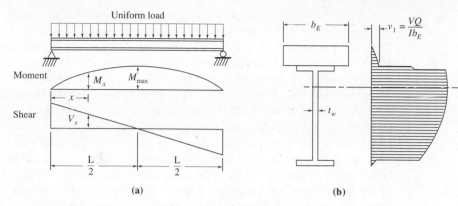

Figure 16.8.2 Shear variation for uniform loading and distribution of shear stress over the depth of a steel-concrete composite section.

channel connectors (LRFD-I5.4 and ASD-I4). Currently (1995), nearly all shear connectors are headed studs.

Ideally, to obtain a *fully composite section* the shear connectors should be stiff enough to provide the complete interaction (i.e., no slip at the interface) shown in Fig. 16.2.2c. This, however, would require that the connectors be infinitely rigid. Also, by referring to the shear diagram for a uniformly loaded beam as shown in Fig. 16.8.2, it would be inferred, theoretically at least, that more shear connectors are required near the ends of the span where the shear is high, than near midspan where the shear is low.

Consider the shear stress distribution of Fig. 16.8.2b wherein the stress v_1 must be developed by the connection between the slab and beam. Under elastic conditions the shear stress at any point in the cross-section will vary from a maximum at the support to zero at midspan. Next, examine the equilibrium of an elemental slice of the beam, as in Fig. 16.8.3. The shear force per unit distance along the span is $dC/dx = v_1 b_E = V(\int y\, dA)/I$. (The $\int y\, dA$ is commonly given the symbol Q in elastic beam theory; this should not be confused with the nominal connector strength Q_n used below.) Thus, if a given connector has an allowable service load capacity q (kips), the maximum spacing p to provide the required strength is

$$p = \frac{q}{V(\int y\, dA)/I} \qquad (16.8.1)$$

where $\int y\, dA$ is the statical moment of the transformed compressive concrete area (the

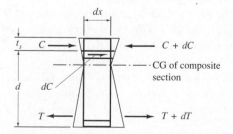

Figure 16.8.3 Force required from shear connectors at service load level.

slab) taken about the neutral axis of the composite section. Equation 16.8.1 is based on *elastic* beam theory and a fully composite section.

Until recent years, Eq. 16.8.1 was used to space shear connectors. AASHTO-10.38.5.1.1 [1.3] requires using Eq. 16.8.1 to design for fatigue, a service load limit state related to the *range* of force applied, in this case the range of shear V_r resulting from live load (and impact). AASHTO also requires a strength limit state check.

According to the strength limit state, the shear connectors at nominal moment strength share equally in transmitting the shear at the interface between concrete slab and steel beam. This means, referring to Fig. 16.8.2a, that shear connectors are required to transfer the compressive force developed in the slab at midspan to the steel beam in the distance $L/2$, since no compressive force can exist in the slab at the end of the span where zero moment exists. The nominal shear transfer strength cannot exceed the maximum force the concrete can develop, namely

$$C_{max} = 0.85 f'_c b_E t_s \qquad (16.8.2)$$

where b_E is the effective slab width and t_s is the slab thickness. When the maximum force T_{max} that can develop in the steel is less than C_{max}, the maximum shear transfer strength will be

$$T_{max} = A_s F_y \qquad (16.8.3)$$

where A_s is the cross-sectional area of the steel section.

Thus, when the nominal strength Q_n of one shear connector is known, the total number N of shear connectors required between points of maximum and zero bending moment is

$$N = \frac{C_{max}}{Q_n} \quad \text{or} \quad \frac{T_{max}}{Q_n}, \quad \text{whichever is } smaller \qquad (16.8.4)$$

Thus, the strength is achieved when the total number N of shear connectors is placed between the maximum moment and zero moment locations. Uniform spacing will be the simplest procedure, because the number of connectors rather than the spacings affects the strength.

The determination of the connector capacity analytically is complex, since the shear connector deforms under load and the concrete which surrounds it is also a deformable material. Moreover, the amount of deformation a shear connector undergoes is dependent upon factors such as its own shape and size, its location along the beam, the location of the maximum moment, and the manner in which it is attached to the top flange of the steel beam. In addition, any particular shear connector may yield sufficiently to cause slip between the beam and the slab. In the latter case the adjacent shear connectors pick up the additional shear.

As a result of the complex behavior of shear connectors, their capacities are not based solely on a theoretical analysis. In order to develop a rational approach, a number of research programs, summarized by Viest [16.1, 16.5], were undertaken to develop the strengths of the various types of shear connectors.

Investigators determined that shear connectors will not fail when the average load per connector is below that causing 0.003 in. (0.076 mm) residual slip between concrete and steel. The amount of slip is also a function of the strength of the concrete that surrounds the shear connector. Relating connector capacity to a specified slip may be realistic for bridge design where fatigue strength is important, but it is overly conservative with respect to failure loads. So-called "ultimate" capacities used prior to 1965 [16.17] were based on slip limitation, giving values about one-third of the strengths obtained when failure of a connector is the criterion.

When flexural strength of the composite section is the basis for design, the connectors must be adequate to satisfy equilibrium of the concrete slab between the points of maxmum and zero moment, as discussed in the development of Eqs. 16.8.2, 16.8.3, and 16.8.4. Slip is not a criterion for this equilibrium requirement. As stated by Slutter and Driscoll [16.17], "the magnitude of slip will not reduce the ultimate moment provided that (1) the equilibrium condition is satisfied, and (2) the magnitude of slip is no greater than the lowest value of slip at which an individual connector might fail." More recent studies by Ollgaard, Slutter, and Fisher [16.19] and McGarraugh and Baldwin [16.20] included the effect of lightweight concrete on stud connector capacity.

Two currently accepted expressions for the nominal strength Q_n of shear connectors are as follows:

1. *Headed steel stud connectors welded to flange* (Fig. 16.8.1a). Load and Resistance Factor Design (LRFD-I5.3) gives essentially the expression developed at Lehigh [16.19],

$$Q_n = 0.5 A_{sc} \sqrt{f_c' E_c} \le A_{sc} F_u^b \qquad (16.8.5)*$$

where Q_n = nominal strength of one stud, kips
  A_{sc} = cross-sectional area of stud = $\pi d_s^2 / 4$, sq in.
  F_u^b = minimum specified tensile strength of stud, ksi
  f_c' = 28-day compressive strength of concrete, ksi
  E_c = modulus of elasticity of concrete, ksi
    = $(w^{1.5}) \sqrt{f_c'}$, according to LRFD-I5.3, using f_c' in ksi.
    For normal-weight concrete having density w = 145 pcf,
    $E_c = 1746 \sqrt{f_c'}$. Note that the ACI Code [16.15] gives slightly different values using $E_c = w^{1.5} 33 \sqrt{f_c'}$, with f_c' in psi instead of ksi.

2. *Channel connectors* (Fig. 16.8.1b). LRFD-I5.4 gives for the nominal connector strength Q_n,

$$Q_n = 0.3(t_f + 0.5 t_w) L_c \sqrt{f_c' E_c} \qquad (16.8.6)*$$

---

*For SI units, $Q_n = 0.0005 A_{sc} \sqrt{f_c' E_c}$ (for stud)   (16.8.5)

*For SI units, $Q_n = 0.0003(t_f + 0.5 t_w) L_c \sqrt{f_c' E_c}$ (for channel)   (16.8.6)

with A_{sc}, t_f, t_u, and L_c, mm; f_c', MPa; and Q_n, kN. For these, $E_c = w^{1.5}(0.041) \sqrt{f_c'}$ with w in kg/m³; normal-weight concrete (145 pcf) is 2320 kg/m³.

where Q_n = nominal strength of one channel, kips
t_f = channel flange thickness (Fig. 16.8.1), in.
t_w = channel web thickness, in.
L_c = length of channel, in.
f_c' = 28-day compressive strength of concrete, ksi
E_c = modulus of elasticity of concrete (defined following Eq. 16.8.5), ksi

Connector Design—Load and Resistance Factor Design

The nominal strength Q_n of the connectors is directly used in Load and Resistance Factor Design. LRFD-I5.2 requires ". . . the entire horizontal shear at the interface between the steel beam and the concrete slab shall be assumed to be transferred by shear connectors." For fully composite sections, the nominal horizontal shear strength V_{nh} to be provided by connectors is the smaller of Eqs. 16.8.2 and 16.8.3.

The section may also be designed as *partially composite,* where the forces utilized of the internal couple are less than either the nominal compression strength available from the concrete, or the nominal tension strength available from the steel section. In partially composite sections, the strength ΣQ_n of the shear connectors determines the magnitude of the forces of the internal couple and nominal moment strength M_n, and correspondingly the required nominal horizontal shear strength V_{nh}. Lorenz and Stockwell [16.7] have discussed stresses in partial composite beams. Bradford and Gilbert [16.44] have provided recent work on partial interaction under sustained loads.

For the positive moment situations (i.e., compression in the concrete slab), the shear strength V_{nh} required is, therefore, the *smallest* of the following:

a. $\qquad$ V_{nh} required $= 0.85 f_c' b_E t_s$ $\qquad$ [16.8.2]

b. $\qquad$ V_{nh} required $= A_s F_y$ $\qquad$ [16.8.3]

c. $\qquad$ V_{nh} required $= \Sigma Q_n$ provided

When c. applies, the number of connectors controls the nominal strength M_n of the section.

As stated earlier, and as specifically stated in LRFD-I5.6, the strength V_{nh} must be provided " . . . each side of the point of maximum moment . . ." to the points of zero moment. Further, LRFD-I5.6 states that the connectors *shall* be distributed uniformly between the point of maximum moment and the point of zero moment. This is "unless otherwise specified", whatever that may mean. As long as adequate strength is provided, the spacing of the connectors is not important.

The nominal strengths Q_n for stud and channel connectors from LRFD-I5.3 and I5.4 are given by Eqs. 16.8.5 and 16.8.6; values for common stud diameters and some channels are given in Table 16.8.1.

When a formed steel deck is used (see Fig. 16.1.2) with shear studs embedded in the supported concrete slab, reduction factors must be applied to Q_n in accordance with LRFD-I5.3.

In the case of continuous beams (also see Sec. 16.13), the longitudinal reinforcing bar steel within the effective width of the concrete slab is permitted (LRFD-I1,

TABLE 16.8.1 NOMINAL STRENGTH Q_n (kips) FOR STUD AND
CHANNEL SHEAR CONNECTORS USED WITH
NORMAL-WEIGHT CONCRETE[†]

	Concrete strength f'_c (ksi)		
Connector	3.0	3.5	4.0
1/2″ diam × 2″ headed stud	9.4	10.5	11.6
5/8″ diam × 2-1/2″ headed stud	14.6	16.4	18.1
3/4″ diam × 3″ headed stud	21.0	23.6	26.1
7/8″ diam × 3-1/2″ headed stud	28.6	32.1	35.5
Channel C3×4.1	$10.2L_c^*$	$11.5L_c$	$12.7L_c$
Channel C4×5.4	$11.1L_c$	$12.4L_c$	$13.8L_c$
Channel C5×6.7	$11.9L_c$	$13.3L_c$	$14.7L_c$

[†] LRFD-Formula (I5-1), Eq. 16.8.5, used for studs; LRFD-Formula (15–2),
Eq. 16.8.6, used for channels. Studs, A108 Type 2, $F_u^b = 60$ ksi.
$*L_c$ = Length of channel, in.

par. 5) to be assumed to act compositely with the steel beam in the areas of negative
moment. The total nominal horizontal strength V_{nh} needed from shear connectors
between the interior support and each adjacent point of inflection (zero moment)
equals the tension force available from the reinforcement (since the tension in the
concrete is neglected),

$$T_{slab} = A_r F_{yr} \qquad (16.8.7)$$

where A_r = total area of adequately developed longitudinal reinforcing steel
within the effective width b_E of the concrete slab
F_{yr} = minimum specified yield stress of the reinforcing steel

EXAMPLE 16.8.1

Determine the number of $\frac{3}{4}$-in.-diam × 3-in. shear stud connectors required to
develop the fully composite section of Fig. 16.8.4. Assume the applied loading is
uniform and the beam is simply supported. Use $F_y = 36$ ksi, $f'_c = 3$ ksi, and Load and
Resistance Factor Design.

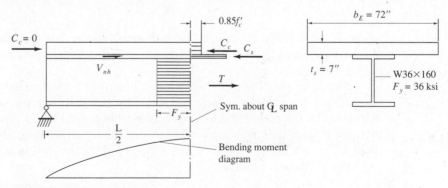

Figure 16.8.4   Example 16.8.1.

Solution   Using Eqs. 16.8.2 and 16.8.3,

$$V_{nh} = C_{max} = 0.85 f'_c b_E t_s = 0.85(3.0)(72)7 = 1285 \text{ kips}$$

or

$$V_{nh} = T_{max} = A_s F_y = 47.0(36) = 1692 \text{ kips}$$

As found from the analysis in Example 16.7.2, the neutral axis is located within the steel section; thus, $C_{max} < T_{max}$. The force in the concrete to be carried by shear connectors is 1285 kips.

The nominal strength Q_n per connector, from Eq. 16.8.5 or Table 16.8.1, is 21.0 kips. The number N of shear connectors required for each half span is

$$N = \frac{1285}{21.0} = 61$$

Use $61 - \frac{3}{4}$*-in.-diam* $\times$ *3-in. studs per half span.*   ■■

Connector Design—Allowable Stress Design

Allowable Stress Design of connectors uses the same strength concepts described above. ASD-I4 assumes the service load horizontal shear forces V_h are one-half the values occurring when nominal strength is reached; i.e., Eqs. 16.8.5 and 16.8.6 divided by 2. The allowable connector loads q (kips) are approximately the nominal strength Q_n values divided by approximately 2. Thus, the end result should be about the same by either LRFD or ASD. Actually, the q values of ASD are slightly more than one-half the Q_n values; the ratio is about 0.54.

Connector Design—Elastic Concept for Fatigue Strength

The 1992 AASHTO Specifications [1.3] requirements for fatigue are based largely on the work of Slutter and Fisher [16.21]. For fatigue, the *range* of service load shear rather than strength under overload is the major concern. Fatigue strength may be expressed

$$\log N = A + BS_r \tag{16.8.8}$$

where S_r is the range of service load horizontal shear; N is the number of cycles to failure; and A and B are empirical constants. The equation used for design is shown in Fig. 16.8.5.

Since the magnitude of shear force transmitted by individual connectors when service loads act agrees well with prediction by elastic theory, the horizontal shear is calculated by the elastic relation VQ/I. Fatigue is critical under repeated application of service load; thus it is reasonable to determine variation in shear using elastic theory. The spacing of the connectors will vary along the span in accordance with V.

For *cyclical load*, Eq. 16.8.1 gives

$$\frac{(V_{max} - V_{min})Q}{I} = \frac{\text{Allowable range } \Sigma Z_r}{p} \tag{16.8.9}$$

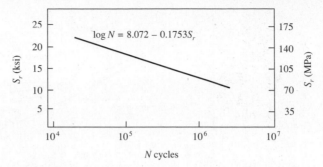

Figure 16.8.5 Fatigue strength of stud shear connectors. (From Ref. 16.21)

where p is the connector spacing. AASHTO-10.38.5.1.1 [1.3] gives Eq. 16.8.9 as

$$\left[S_r = \frac{V_r Q}{I} \right] \le \frac{\Sigma Z_r}{p} \tag{16.8.10}$$

where $V_r = V_{max} - V_{min}$ due to live load only
Z_r = allowable range of load per connector
     = αd_s^2 for welded studs, lb/stud connector
S_r = range of horizontal shear, kips/in.
d_s = stud diameter, in.
α = 13,000 for 100,000 cycles
  10,600 for 500,000 cycles
   7850 for 2 million cycles
   5500 for over 2 million cycles

EXAMPLE 16.8.2

Redesign the shear conncetors for the beam of Example 16.8.1 (Fig 16.8.4) using the service load stress fatigue requirement of AASHTO with $\frac{3}{4}$-in.-diam × 3-in. stud connectors. Design for 500,000 cycles of loading of live load. Whether or not the beam is shored, only the live load is the cylical load. Use uniform live load of 3.5 kips/ft, a spacing of 6 ft for beams, a beam span of 45 ft, $F_y = 36$ ksi, and $f'_c = 3$ ksi.

Solution

(a) Loads and shears. For the fatigue requirement in AASHTO-10.38.5.1.1 only the *range* of service live load is needed. At the support with full span loaded,

$$V = \tfrac{1}{2}wL = 0.5(3.5)45 = 78.8 \text{ kips}$$

Using partial span loading of live load,

$$\text{Max } V(\text{at } \tfrac{1}{4} \text{ point}) = 3.5(45)(0.75)(0.375) = 44.3 \text{ kips}$$
$$\text{Max } V(\text{at midspan}) = \tfrac{1}{8}wL = \tfrac{1}{8}(3.5)45 = 19.7 \text{ kips}$$

The envelope showing the *range* of live load shear is given in Fig. 16.8.6.

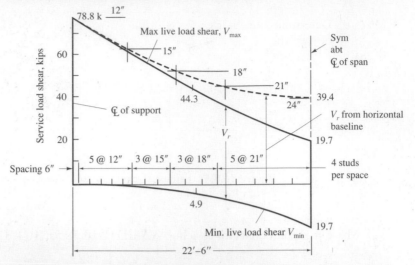

Figure 16.8.6   Shear range diagram and stud spacing according to elastic fatigue theory used by AASHTO-Example 16.8.2.

Inclusion of dead load shear would change both V_{max} and V_{min} by the same amount at any section along the beam; however, $(V_{max} - V_{min})$, that is, the range V_r would not be affected.

(b) Compute elastic composite section properties $(n = 9)$ (see Fig. 16.8.4).

Effective slab width $b_E = b_0 = 72$ in.

Element	Effective area, A (sq in.)	Arm from CG of steel beam, y (in.)	Ay (sq in.)	Ay^2 (in.³)	I_0 (in.⁴)
Slab, 72(7)/9	56.0	21.5	1204	25,900	230
W36×160	47.0	—	—	—	9760
	103.0		1204	25,900	9990

$$I_x = Ay^2 + I_0 = 25{,}900 + 9900 = 35{,}900 \text{ in.}^4$$

$$\bar{y} = \frac{1204}{103.0} = 11.68 \text{ in.}$$

$$I_{tr} = 35{,}900 - 103.0(11.68)^2 = 21{,}800 \text{ in.}^4$$

$$y_t = 18.0 + 7.0 - 11.68 = 13.32 \text{ in.}$$

$$y_b = 18.0 + 11.68 = 29.68 \text{ in.}$$

$$S_t = \frac{21{,}800}{13.32} = 1640 \text{ in.}^3 \quad \text{(concrete at top)}$$

$$S_b = S_{tr} = \frac{21{,}800}{29.68} = 735 \text{ in.}^3 \quad \text{(steel at bottom)}$$

Determine the static moment of the effective concrete area about the centroid of the composite section,

$$Q = 56.0(y_t - 3.5) = 56.0(9.82) = 550 \text{ in.}^3$$

(c) Determine the allowable load for $\frac{3}{4}$-in.-diam $\times$ 3-in. stud connectors. AASHTO-10.38.5.1.1 gives an allowable service load range Z_r per connector based on fatigue for 500,000 cycles of loading as

$$Z_r = 10.6d_s^2$$
$$= 10.6(0.75)^2 = 5.96 \text{ kips}$$

The AASHTO allowable values are based on a slip limitation.

(d) Determine spacing of connectors. Use 4 studs across the beam flange width at each location:

$$\Sigma Z_r \text{ for 4 studs} = 4(5.96) = 23.8 \text{ kips}$$

Using Eq. 16.8.1,

$$p = \frac{\Sigma Z_r}{V_r Q/I} = \frac{\Sigma Z_r I}{V_r Q}$$

where $I/Q = 21{,}800/550 = 39.6$ in.

$$p = \frac{23.8(39.6)}{V_r} = \frac{943}{V_r (\text{kips})}$$

The values are computed in the table below and the spacing is determined graphically on the shear diagram of Fig. 16.8.6.

p (in.)	V_r (kips)	p (in.)	V_r (kips)
12	79	21	45
15	63	24	39
18	52		

The fatigue service load criterion requires 8% more connectors (66 vs 61 per half span) than the procedure based on strength.   ■■

16.9 HYBRID COMPOSITE GIRDERS

The general discussion of the hybrid plate girder appears in Sec. 11.7. The hybrid girder is one that has either the tension flange or both flanges of the steel section made with a higher strength grade of steel than used for the web. There are particular economic advantages to the hybrid girder in composite construction where the concrete slab provides a large compression capacity. The neutral axis will lie near the compression face of the composite section, causing the higher stressed tension flange to become the element controlling the strength. Use of a stronger grade of steel for the tension flange will avoid the use of a large plate.

The design and behavior of hybrid girders has been presented in detail in a Joint ASCE-AASHO Committee Report [11.10]. The theoretical aspects of the bending behavior of composite members have been given by Schilling [16.22]. The primary behavioral feature of practical concern is the yielding of the web prior to reaching maximum strength in the flanges. As discussed in Sec. 11.7, the flange is designed to have extra strength to make up for the reduced moment strength of the web. This reduced moment strength available from the web reduces the overall nominal strength M_n of the girder. The correction for this is made by multiplying the nominal strength computed considering all other factors by a reduction factor R_e, given by Eq. 11.7.2 (from LRFD-Appendix G2).

When extending the hybrid concept to composite steel-concrete members, the behavior is essentially the same as when the hybrid girder is noncomposite [11.10, 16.22]. The additional complicating factor is that the neutral axis of the composite section is not at mid-depth and therefore requires an evaluation of an unsymmetrical hybrid member. Since a greater percentage of the depth of the web is located below the neutral axis (on the tension side) the early yielding of the lower strength web means a greater reduction in overall nominal strength M_n for a composite hybrid beam than for a noncomposite symmetrical hybrid beam.

To account for a variable distance to the tension flange from the neutral axis, AASHTO-10.53.1.4 [1.3] uses the following equation for the strength reduction factor R_e as recommended by the Joint ASCE-AASHO Committee [11.10],

$$R_e = \left[1 - \frac{\beta\psi(1-\alpha)^2(3-\psi+\alpha\psi)}{6 + \beta\psi(3-\psi)} \right] \qquad (16.9.1)$$

where ψ is the ratio of the neutral axis distance from the outer edge of the tension flange to the depth of the steel section. The other variables are defined following Eq. 11.7.3. The neutral axis location is that based on the elastic transformed composite section. Remember that the maximum strength of a plate girder is governed by first yielding of the flanges, as discussed in Chapter 11; thus, it is assumed that the web is "slender". Equation 16.9.1 is used *only* for unsymmetrical sections, which would usually be the situation for a composite girder. This equation is *not* to be used if the top flange has a higher yield strength or larger area than the bottom flange.

16.10 COMPOSITE FLEXURAL MEMBERS CONTAINING FORMED STEEL DECK

Composite flexural members may be made using formed steel deck, as shown in Fig. 16.1.2. The formed metal deck may be placed perpendicular to or parallel with the supporting beam. Furthermore, the beam may actually be an open web joist. Typically, the deck plate varies in thickness from 22 ga. (0.0336 in., 0.853 mm) to 12 ga. (0.1084 in., 2.75 mm). The deck rib height typically is $1\frac{1}{2}$, 2, and 3 in. for spans of, say, 8, 10, and 15 ft. As shown in Fig. 16.1.2, the thickness of the concrete slab above the top of the ribs must be at least 2 in. (LRFD-I3.5 a and ASD-I5.1) and the embedment of the stud connectors into the concrete above the top of the ribs must be at least $1\frac{1}{2}$ in.

When the steel deck ribs are perpendicular to the steel beam, the stud strength Q_n may have to be reduced from that given by Eq. 16.8.5 by a reduction factor in LRFD-I5.3. Easterling, Gibbings, and Murray [16.43] provide a recent study of strength of shear studs in steel deck on composite beams.

Full treatment of formed steel deck supported slab composite beams is outside the scope of this chapter. The reader is referred to Grant, Fisher, and Slutter [16.23], and particularly with regard to LRFD design, to Vinnakota, Foley, and Vinnakota [16.24]. Composite open-web joists have been treated by Tide and Galambos [16.25] and Rongoe [16.26]. Two-way acting composite slabs with steel deck have been treated by Porter [16.29], and design *Specifications* and *Commentary* [16.27, 16.28] are available from ASCE. The special considerations regarding the design of "stub-girders" are treated by Buckner, Deville, and McKee [16.30].

16.11 DESIGN PROCEDURE—LOAD AND RESISTANCE FACTOR DESIGN

The design of composite beams involves providing sufficient plastic strength ϕM_p of the composite section to equal the factored moment. Using rolled W shapes, local buckling ordinarily is not a controlling limit state although h/t_w should be checked when the PNA is in the web, and because the compression flange is attached to the concrete slab lateral-torsional buckling is precluded as a controlling limit state. Thus, it is required that

$$\phi_b M_p \geq M_u \tag{16.11.1}$$

where $\phi_b = 0.85$ for a composite beam.

In general, the design should be started by assuming the plastic neutral axis (PNA) is within the slab (Case 1—Fig. 16.7.1b). Thus, using Eq. 16.7.5, the required area A_s for the steel section is

$$\text{Required } A_s = \frac{M_u}{\phi_b F_y \left(\dfrac{d}{2} + t_s - \dfrac{a}{2} \right)} \tag{16.11.2}$$

Typically $a/2$ can be estimated as 1 in. for preliminary design.

In addition to the strength requirement under full dead and live load, LRFD-I3.4 requires that when temporary shores are not used during construction, the steel section alone must have adequate strength "to support all loads applied prior to the concrete attaining 75% of its specified strength f_c'." For this condition, local buckling of the beam elements and lateral-torsional buckling must be considered.

16.12 DESIGN PROCEDURE—ALLOWABLE STRESS DESIGN

Allowable Stress Design is based on the strength of a composite beam being independent of whether or not the beam is shored during construction.

1. Thus, the approach of ASD-I2.2 is to design for dead and live load acting on the composite section, giving the required composite section modulus

S_{tr}, referred to the tension extreme fiber, as

$$\text{Required } S_{tr} = \frac{M_D + M_L}{0.66\, F_y} \qquad (16.12.1)$$

where M_D = service load moment caused by loads applied *prior* to the time the concrete has achieved 75% of its required strength (non-composite loading)

M_L = service load moment caused by loads applied *after* the concrete has achieved 75% of its required strength.

The allowable stress $0.66F_y$ is based on the condition that the section satisfies the "compact section" requirement of ASD-B5 for the *web*. The flange local buckling and lateral-torsional buckling limit states are not applicable in positive moment regions (ASD-I2). Equation 16.12.1 applies *whether or not shores are used.*

2. When temporary shores are not actually used, the service load stress on the steel section must be checked and verified not to exceed $0.90F_y$. Even though the strength is satisfactory based on Eq. 16.12.1, it is deemed desirable to preclude yielding at service load. Thus, ASD-I2, 5th par., requires

$$f_b = \frac{M_D}{S_s} + \frac{M_L}{S_{tr}} \le 0.90F_y \qquad (16.12.2)$$

where S_s = section modulus of the steel section alone referred to the tension extreme fiber

S_{tr} = effective section modulus of the transformed composite section referred to the tension flange, computed when the concrete strength is 75% of its required strength.

3. When temporary shores are not used, the steel beam alone must support all loads applied prior to the concrete achieving 75% of its specified strength. Under this condition, flange local buckling and lateral-torsional buckling must be considered. The requirement is

$$\text{Required } S_s = \frac{M_D + M_{const}}{F_b} \qquad (16.12.3)$$

where F_b may be $0.66F_y$ or a lesser value if flange local buckling or lateral-torsional buckling control. When the tension flange in the positive moment region is larger than the compression flange, the compression flange may control. Any construction load moment M_{const} must be considered at this loading stage. Thus, S_s of Eq. 16.12.3 refers to the more critical flange on the steel section.

4. Partial composite action. When fewer shear connectors are used than necessary to develop a fully composite section, an effective section modulus S_{eff} may be used, determined as follows according to ASD-I2,

$$S_{eff} = S_s + \sqrt{\frac{V'_h}{V_h}}(S_{tr} - S_s) \qquad (16.12.4)$$

where $V_h = V_{nh}$ from Eqs. 16.8.2 or 16.8.3 (use smaller) divided by the factor of safety of 2 (ASD-I4)

$V'_h = N$ (i.e., number of shear connectors used) times the allowable capacity q per connector

S_s = elastic section modulus of steel section alone, referred to its tension flange

S_{tr} = elastic section modulus of the transformed composite section referred to its tension flange

16.13 LRFD EXAMPLES—SIMPLY SUPPORTED BEAMS

EXAMPLE 16.13.1 _____

Design an interior composite beam for the floor whose plan is shown in Fig. 16.13.1 assuming the beam is to be constructed without temporary shoring. Use $F_y = 36$ ksi, $f'_c = 3$ ksi ($n = 9$), a 4-in. slab, and Load and Resistance Factor Design.

Solution
  (a) Compute factored loads.
Loads carried on steel beam:

$$\text{concrete slab, } \tfrac{4}{12}(0.15)8 = 0.40 \text{ kip/ft}$$
$$\text{beam weight (estimated)} = \underline{0.04} \text{ kip/ft}$$
$$0.44 \text{ kip/ft} \times 1.2 = 0.53 \text{ kip/ft}$$

Load carried by composite action:

$$\text{live load, } 0.15(8) = 1.20 \text{ kips/ft} \times 1.6 = 1.92 \text{ kips/ft}$$

  (b) Compute service load and factored load moments.

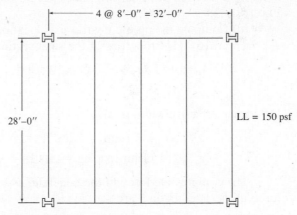

4 @ 8'-0" = 32'-0"

28'-0"

LL = 150 psf

Figure 16.13.1 Beam framing plan for Examples 16.13.1 and 16.14.1.

$$M_D = \tfrac{1}{8}(0.44)(28)^2 = 43 \text{ ft-kips} \quad \text{(service load)}$$
$$M_L = \tfrac{1}{8}(1.20)(28)^2 = 118 \text{ ft-kips} \quad \text{(service load)}$$
$$M_u = \tfrac{1}{8}(0.53 + 1.92)(28)^2 = 240 \text{ ft-kips} \quad \text{(factored)}$$

(c) Select the section. Use Eq. 16.11.2 assuming the PNA (plastic neutral axis) is within the slab. Estimate $a \approx 1.0$ for preliminary selection.

$$\text{Required } A_s = \frac{M_u}{\phi_b F_y \left(\dfrac{d}{2} + t_s - \dfrac{a}{2} \right)} \qquad [16.11.2]$$

From Eq. 16.11.2, the design strength $\phi_b M_n$ provided can be computed as A_s times the denominator. For a given value of $(t_s - a/2)$, $\phi_b M_n$ can be tabulated for a steel W section for any given yield stress; such tabulated information is given in the *LRFD Manual*. Thus, for the 4-in. slab and estimated a of 1 in.,

$$t_s - \frac{a}{2} = 4 - 0.50 = 3.5 \text{ in.}$$

$$\text{Required } A_s = \frac{240(12)}{0.85(36)(7 + 3.5)} = 9.0 \text{ sq in.} \quad \text{(for W14)}$$

$$\text{Required } A_s = \frac{240(12)}{0.85(36)(8 + 3.5)} = 8.2 \text{ sq in.} \quad \text{(for W16)}$$

Using *LRFD Manual*, "COMPOSITE BEAM SELECTION TABLE," entering with $Y2 = t_s - a/2 = 3.5$ in. and required $\phi M_n = 240$ ft-kips, find

W16×31 $\phi M_n = 266$ ft-kips; $A_s = 9.12$ sq in.

W14×34 $\phi M_n = 267$ ft-kips; $A_s = 10.0$ sq in.

The tabulated values selected are for the PNA within the slab (that is, $Y1 = $ distance from PNA to top of steel beam $= 0$ in.). When these tables are available, their use will be faster and more accurate than putting estimated d into Eq. 16.11.2.

(d) Compute the plastic neutral axis location and check strength.
Try W16×31: Properties of the steel section alone are:

$$A_s = 9.12 \text{ sq in.} \qquad I_x = 375 \text{ in.}^4 \qquad b_f = 5.525 \text{ in.}$$
$$d = 15.88 \text{ in.}$$

Determine effective width of slab:

$$b_E = \tfrac{1}{4} \text{ of span} = 0.25(28)12 = 84 \text{ in.} \quad \textit{controls}$$
$$\text{or} \quad b_E = \text{beam spacing} = 8(12) = 96 \text{ in.}$$

The compressive force in the concrete, assuming $a < t_s$, and the tension force in the steel section, are

$$C = 0.85 f_c' b_E a = 0.85(3)84a = 214a$$
$$T = A_s F_y = 9.12(36) = 328 \text{ kips}$$

Statics requires

$$C = T$$

$$a = 1.53 \text{ in.} < t_s \quad \text{OK as assumed}$$

The nominal moment strength M_n is

$$M_n = T\left(\frac{d}{2} + t_s - \frac{a}{2}\right)$$

$$M_n = 328\left(\frac{15.88}{2} + 4.0 - \frac{1.53}{2}\right)\frac{1}{12} = 306 \text{ ft-kips}$$

$$\phi_b M_n = 0.85(306) = 260 \text{ ft-kips} > (M_u = 239 \text{ ft-kips}) \quad \text{OK}$$

Note that M_u has been revised to include the correct beam weight.

(e) Check the strength of the steel section to support construction loads (LRFD-I3.4). This check is required when shores are not used. Assume adequate lateral support is provided during construction such that $L_b \le L_p$ and the section is compact for local buckling; therefore $\phi_b M_n = \phi_b M_p$, and $\phi_b = 0.90$ for the steel section acting noncompositely. There are no LRFD-prescribed construction loads. It is prudent to consider that some of the wet concrete load should be treated as live load, say 50% of it (accomplished by using an average overload factor of 1.4). Further, other construction live load on the order of 20 to 25 psf should be included (20 psf used here).

$$
\begin{array}{lll}
\text{Slab} & = 0.40(1.4) & = 0.56 \quad \text{kip/ft} \\
\text{Construction} & = 0.02(8)1.6 & = 0.26 \quad \text{kip/ft} \\
\text{Steel section} & = 0.031(1.2) & = \underline{0.04} \quad \text{kip/ft} \\
& & w_u = \overline{0.86} \quad \text{kip/ft}
\end{array}
$$

$$M_u = \tfrac{1}{8}(0.86)(28)^2 = 84 \text{ ft-kips}$$

$$\phi_b M_p \text{ for W16×31} = 146 \text{ ft-kips} > 84 \text{ ft-kips} \quad \text{OK}$$

(f) Design shear connectors. The compressive force in the slab must be carried by shear connectors,

$$C = 0.85 f'_c b_E a = 0.85(3)84a = 214a = 214(1.53) = 328 \text{ kips}$$

Since $a < t_s$, V_{nh} will be based on the 328 kips, which does equal $T_{max} = A_s F_y$. Using $\tfrac{3}{4}$-in.-diam × 3-in. headed studs, $Q_n = 21.0 \text{ kips/stud}$ from Table 16.8.1. The number N of connectors required to carry 328 kips is

$$N = \frac{V_{nh}}{Q_n} = \frac{328}{21.0} = 15.6, \quad \text{say } 16$$

which is the number of connectors required for the region between maximum moment and the support (zero moment location). Thus, 32 studs are needed for the entire span. Using a uniform spacing with two studs at each location, the spacing p required would be

$$p = \frac{L}{N} = \frac{28(12)}{16} = 21 \text{ in.}$$

Maximum $p = 8t_s = 8(4) = 32$ in.   (LRFD-I5.6)

Minimum $p = 6(\text{diam}) = 6(0.75) = 4.5$ in.   (LRFD-I5.6)

Use W16×31 section of A36 steel, along with 32—$\frac{3}{4}$-in.-diam × 3-in. headed stud connectors over the entire span, spaced at 21 in. The connectors are to be placed in pairs starting at the support. ■■

EXAMPLE 16.13.2

Design an interior composite beam to span 30 ft with a beam spacing of 8 ft, using the minimum number of $\frac{3}{4}$-in.-diam × 3-in. stud shear connectors. The slab is 5 in. thick. The beam is to be constructed without shores. The beam must support a ceiling of 7 psf, partitions and other dead load of 25 psf, and live load of 150 psf. Use A572 Grade 50 steel and $f'_c = 3$ ksi ($n = 9$) concrete. Use Load and Resistance Factor Design.

Solution

(a) Compute factored loads and bending moments. The dead load and moment that must be carried by the steel beam alone during construction are

$$5\text{-in. slab, } \tfrac{5}{12}(8)0.15 \quad = 0.50 \text{ kips/ft}$$
$$\text{Steel beam (assumed)} = \underline{0.03}$$
$$\phantom{\text{Steel beam (assumed)} = } 0.53 \text{ kips/ft}$$

$$M_D = \tfrac{1}{8}(0.53)(30)^2 = 60 \text{ ft-kips}$$
$$M_{u1} = 1.2(60) = 72 \text{ ft-kips}$$

The partition and ceiling dead loads, and the live load that must be carried by the composite section are

$$\text{Live load } 0.15(8) \quad = 1.2 \text{ kips/ft}$$
$$\text{Partitions } 0.025(8) = 0.2$$
$$\text{Ceiling } 0.007(8) \quad = \underline{0.05}$$
$$\phantom{\text{Ceiling } 0.007(8) \quad = } 1.45 \text{ kips/ft}$$

$$M_L = \tfrac{1}{8}(1.45)(30)^2 = 163 \text{ ft-kips}$$
$$w_{u2} = 1.2(0.25) + 1.6(1.2) = 2.22 \text{ kips/ft}$$
$$M_{u2} = \tfrac{1}{8}(2.22)(30)^2 = 250 \text{ ft-kips}$$
$$M_u = M_{u1} + M_{u2} = 72 + 250 = 322 \text{ ft-kips}$$

(b) Select the section. One could use Eq. 16.11.2 assuming the PNA (plastic neutral axis) is within the slab and solve for required A_s as illustrated in Example 16.13.1 (part c). Alternatively, it will be simpler to use *LRFD Manual,* "COMPOSITE BEAM SELECTION TABLE" for $F_y = 50$ ksi; Eq. 16.11.2 is tabulated for the steel W shapes for various values of $Y2$. Estimate $a \approx 1.0$ for preliminary selection as in Example 16.14.1. For the 5-in. slab,

$$Y2 = t_s - \frac{a}{2} = 5 - 0.50 = 4.5 \text{ in.}$$

Required $\phi M_n = 322$ ft-kips

Find: W16×26 $\phi M_n = 336$ ft-kips
                             W14×30 $\phi M_n = 358$ ft-kips

The tabulated values selected are for the PNA within the slab (that is, $Y1 =$ distance from PNA to top of steel beam $= 0$ in.).

(c) Investigate the W16×26 further. For fully composite action, compute the plastic neutral axis location and check strength.

Try W16×26: Properties of the steel section alone are:

$$A_s = 7.68 \text{ sq in.} \qquad I_x = 301 \text{ in.}^4 \qquad b_f = 5.500 \text{ in.}$$
$$d = 15.69 \text{ in.}$$

Determine effective width of slab:

$$b_E = \tfrac{1}{4} \text{ of span} = 0.25(30)12 = 90 \text{ in.} \quad \textit{controls}$$
$$\text{or} \quad b_E = \text{beam spacing} = 8(12) = 96 \text{ in.}$$

The compressive force in the concrete, assuming $a < t_s$, and the tension force in the steel section are

$$C = 0.85 f'_c b_E a = 0.85(3)90a = 229.5a$$
$$T = A_s F_y = 7.68(50) = 384 \text{ kips}$$

Statics requires

$$C = T$$
$$a = 1.67 \text{ in.} < t_s \quad \text{OK as assumed}$$

The nominal moment strength M_n is

$$M_n = T\left(\frac{d}{2} + t_s - \frac{a}{2}\right)$$

$$M_n = 384\left(\frac{15.69}{2} + 5.0 - \frac{1.67}{2}\right)\frac{1}{12} = 384 \text{ ft-kips}$$

$$[\phi_b M_n = 0.85(384) = 327 \text{ ft-kips}] > [M_u = 322 \text{ ft-kips}] \quad \text{OK}$$

The W16×26 section is adequate as a *fully composite section*. However, when a minimum number of shear connectors is desired and only partial composite action is used, the steel section usually must be heavier. Try W16×31 section.

(d) Minimum number of shear connectors required. The maximum spacing p along the span is

$$\text{Maximum } p = 8t_s = 8(5) = 40 \text{ in.} \quad \text{(LRFD-I5.6)}$$

$$N = \frac{L}{p} = \frac{30(12)}{40} = 9 \text{ spaces}$$

The connectors would be in pairs which would mean 20 connectors for the 30-ft span, with 5 pairs (10 connectors) located between midspan and the end of the beam. When $\tfrac{3}{4}$-in.-diam studs are used, 10 connectors provide nominal strength ΣQ_n,

$$\Sigma Q_n = 10(21.0) = 210 \text{ kips}$$

Since the force in the slab based on connector strength is less than the maximum steel force,

$$T_{max} = A_s F_y = 9.12(50) = 456 \text{ kips}$$

the plastic neutral axis (PNA) is within the steel section.

(e) Locate plastic neutral axis (PNA) and compute nominal strength. Check if PNA occurs within the flange,

$$\Sigma Q_n = 210 \text{ kips}$$

$$\text{Max force in flange} = t_f b_f F_y = 0.440(5.525)50 = 121.6 \text{ kips}$$

$$T_{max} - 121.6 = 334.5 \text{ kips} > \Sigma Q_n$$

Thus, PNA is in the web. For equilibrium of internal forces, referring to Fig. 16.13.2, compute the compression force in the web,

$$\Sigma Q_n + C_f + C_w = T_{max} - C_f - C_w$$
$$210 + 121.6 + C_w = 334.5 - C_w$$
$$2C_w = 2.9 \text{ kips}$$

$$\begin{array}{c}\text{Depth to PNA from} \\ \text{inside of flange}\end{array} = \frac{C_w}{F_y t_w} = \frac{1.45}{50(0.275)} = 0.11 \text{ in.}$$

$$\text{PNA from top of slab} = t_s + t_f + 0.11$$
$$= 5 + 0.44 + 0.11 = 5.55 \text{ in.}$$

Locate the centroid y_1 of the portion of the steel section in tension measured from the bottom of the steel section,

	Area, A	Arm, y	Ay
W section	9.12	7.94	72.41
Flange	−2.43	15.66	−38.07
Web	−0.03	15.39	−0.45
	6.66 sq in.		33.90 in.³

$$y_1 = \frac{33.90}{6.66} = 5.09 \text{ in.}$$

(f) Compute the nominal moment strength M_n. Since the ΣQ_n representing the strength of the shear connectors used is less than the force in the concrete when there is fully composite action, force ΣQ_n is taken equivalent to $C_c = 0.85 f'_c b_E a$, the concrete force represented by the rectangular stress distribution in the concrete. That means

$$a = \frac{\Sigma Q_n}{0.85 f'_c b_E} = \frac{210}{0.85(3)90} = 0.91 \text{ in.}$$

Referring to Fig. 16.13.2, taking internal moments about the point of action of T_s gives

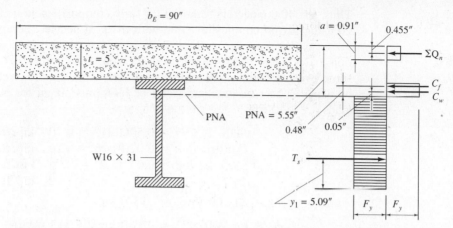

Figure 16.13.2 Example 16.13.2, showing stress distribution to obtain plastic neutral axis.

ΣQ_n:

$$M_{n1} = \Sigma Q_n(d - 5.09 + t_s - a/2)$$
$$= 210(15.88 - 5.09 + 5 - 0.91/2)\tfrac{1}{12}$$
$$= 210(15.33)\tfrac{1}{12} = 268.3 \text{ ft-kips}$$

C_f:

$$M_{n2} = C_f(d - 5.09 - t_f/2)$$
$$= 121.6(15.88 - 5.09 - 0.440/2)\tfrac{1}{12}$$
$$= 121.6(10.57)\tfrac{1}{12} = 108.3 \text{ ft-kips}$$

C_w:

$$M_{n3} = C_w(d - 5.09 - t_f - 0.11/2)\tfrac{1}{12}$$
$$= 1.45(15.88 - 5.09 - 0.440 - 0.055)\tfrac{1}{12}$$
$$= 1.45(10.30)\tfrac{1}{12} = 1.2 \text{ ft-kips}$$

$$M_n = M_{n1} = M_{n2} + M_{n3}$$
$$= 268.3 + 108.3 + 1.2 = 377.8 \text{ ft-kips}$$

$$\phi_b M_n = 0.85(377.8) = 321 \text{ ft-kips}$$

After correcting the dead load for the W16×31 section, the factored moment M_u becomes 321 ft-kips. Thus, $\phi_b M_n = M_u$ and the design is acceptable.

The designer should compare the economics of the W16×26 using connectors to develop a fully composite section with W16×31 using the minimum 20 connectors needed for this span length. To obain a fully composite section the force to be carried by shear connectors would have been $T_{max} = A_s F_y = 384$ kips for the W16×26 section. The number of $\tfrac{3}{4}$-in.-diam studs needed would be

$$N = \frac{384}{21.0} = 18.3, \quad \text{say 20 for half the span}$$

Thus, the 40 connectors required for fully composite action can be reduced to 20 using partial composite action with the next heavier section.

(g) Check the strength of the W16×31 steel section to support construction loads (LRFD-I3.4). Refer to discussion in Example 16.14.1, part (e). Assume construction live load consists of 50% of the wet concrete (accomplished by using an average overload factor of 1.4), plus 20 psf for other construction loads.

$$
\begin{array}{llll}
\text{Slab} & = 0.50(1.4) & = 0.70 & \text{kip/ft} \\
\text{Construction} & = 0.02(8)1.6 & = 0.26 & \text{kip/ft} \\
\text{Steel section} & = 0.031(1.2) & = \underline{0.04} & \text{kip/ft} \\
& & w_u = \overline{1.00} & \text{kip/ft}
\end{array}
$$

$$
M_u = \tfrac{1}{8}(1.00)(30)^2 = 113 \text{ ft-kips}
$$

$$
\phi_b M_p \text{ for W16×31} = 203 \text{ ft-kips} > 113 \text{ ft-kips} \quad \text{OK}
$$

Use W16×31 section ($F_y = 50$ ksi), with 20—$\tfrac{3}{4}$-in.-diam connectors over the entire span, spaced at 40 in.   ■■

16.14 ASD EXAMPLE—SIMPLY SUPPORTED BEAM

EXAMPLE 16.14.1 _____

Redesign the composite beam of Example 16.13.1 (see Fig. 16.13.1) using Allowable Stress Design. The materials are $F_y = 36$ ksi, $f'_c = 3$ ksi ($n = 9$), and a 4-in. slab.

Solution

(a) Service load bending moments. From Example 16.13.1,

$$
M_D = 43 \text{ ft-kips}
$$
$$
M_L = 118 \text{ ft-kips}
$$

(b) Select steel section as if shores were to be used (ASD-I2.2). For $M_D + M_L$ the allowable stress is $0.66F_y$ (assuming a compact web; i.e., $d/t_w \le 640/\sqrt{F_y}$) on the composite section.

$$
\text{Required } S_{\text{tr}} = \frac{(43 + 118)12}{24} = 80.5 \text{ in.}^3
$$

For M_D acting on the steel section alone, the allowable stress is at least $0.60F_y$ if adequate lateral support is provided,

$$
\text{Required } S_s = \frac{M_D}{0.60F_y} = \frac{43(12)}{22} = 23.4 \text{ in.}^3
$$

Enter ASD *Manual,* "COMPOSITE BEAM SELECTION TABLE" to select a section. Entry requires

$$
Y2 = \text{top of steel beam to center of slab} = \frac{t_s(\text{i.e., slab thickness})}{2} = \frac{4}{2} = 2 \text{ in.}
$$

$$A_{\text{ctr}} = \text{transformed effective concrete area} = \frac{b_E}{n} t_s = \frac{84(4)}{9} = 37.3 \text{ in.}^3$$

Looking for S_{tr} of 80.5 in.^{3}, select from p. 2–278 a W16×36 section ($A_s = 10.6$ sq in.; $I_x = 448$ in.^{4}; and $S_x = 56.5$ in.^{3})

(c) Compute composite elastic section properties. Referring to Fig. 16.14.1, determine effective width b_E (ASD-I1),

$$b_E = \tfrac{1}{4} \text{ of span} = 0.25(28)12 = 84 \text{ in.} \quad \textit{controls}$$

$$\text{or} \quad b_E = \text{beam spacing} = 8(12) = 96 \text{ in.}$$

The width of equivalent steel in $b_E/n = 84/9 = 9.33$ in. The moment of inertia and elastic section modulus values are computed as follows:

Element	Transformed area A (sq in.)	Moment arm from centroid y (in.)	Ay (in.^{3})	Ay^2 (in.^{4})	I_0 (in.^{4})
Slab	37.33	9.93	370.7	3681	49.8
W16×36	10.6	0	0	0	448
	47.9		370.7	3681	497.8

$$I_x = I_0 + Ay^2 = 498 + 3681 = 4179 \text{ in.}^4$$

$$\overline{y} = \frac{370.7}{47.9} = 7.73 \text{ in. above centroid of W16×36}$$

$$I_{\text{tr}} = I_x - A\overline{y}^2 = 4179 - 47.9(7.73)^2 = 1312 \text{ in.}^4$$

$$y_t = 7.93 - 7.73 + 4.0 = 4.20 \text{ in. to top of slab}$$

$$y_b = 7.93 + 7.73 = 15.66 \text{ in. to bottom of steel}$$

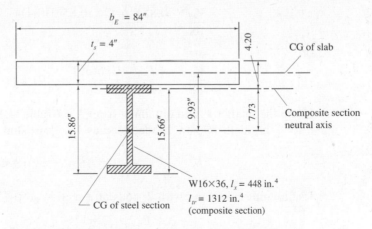

Figure 16.14.1 Beam cross-section for Example 16.14.1.

$$S_{\text{conc}} = I_{\text{tr}}/y_t = 1312/4.20 = 313 \text{ in.}^3$$
$$S_{\text{tr}} = I_{\text{tr}}/y_b = 1312/15.66 = 84 \text{ in.}^3$$

(d) Check the stresses. At the top of the concrete slab,

$$f_c = \frac{(43 + 118)12}{9(313)} = 0.69 \text{ ksi}$$

which does not exceed allowable $f_c = 0.45 f'_c = 1.35$ ksi and, therefore, is acceptable.

At the bottom of the steel beam; $F_b = 0.66F_y = 24$ ksi,

$$f_b = \frac{(43 + 118)12}{84} = 23.0 \text{ ksi} < F_b \quad \text{OK}$$

(e) Check the stress at service load when the beam is *not* shored during construction (ASD-I2), second to last paragraph).

$$f_{b1} = \frac{M_D}{S_s} = \frac{43(12)}{56.5} = 9.1 \text{ ksi} \quad \text{(noncomposite)}$$

$$f_{b2} = \frac{M_L}{S_{\text{tr}}} = \frac{118(12)}{84} = 16.9 \text{ ksi} \quad \text{(composite)}$$

$$f_b = f_{b1} + f_{b2} = 9.1 + 16.9 = 26.0 \text{ ksi}$$

The allowable stress is $0.90F_y = 32.4$ ksi; thus, this is acceptable. This check is to be certain there will be at least a small margin against having permanent deformation at full service load. The check in part (d) relates to having adequate strength.
Use W16×36, A36 steel.

(f) Design shear connectors. As discussed in Sec. 16.8, ASD-I4 uses the nominal strength V_{nh} divided by 2 as a factor of safety (FS) to obtain a service load requirement,

$$V_h = \frac{V_{nh}}{\text{FS}} = \frac{0.85 f'_c b_E t_s}{\text{FS}} = \frac{0.85(3)(84)4}{2} = 428 \text{ kips}$$

or

$$V_h = \frac{V_{nh}}{\text{FS}} = \frac{A_s F_y}{\text{FS}} = \frac{10.6(36)}{2} = 191 \text{ kips}$$

Using the smaller V_h and obtaining from ASD-Table I4.1 the allowable capacity q for $\frac{3}{4}$-in.-diam × 3-in. headed stud as 11.5 kips/stud,

$$N = \frac{V_h}{q} = \frac{191}{11.5} = 16.6, \quad \text{say 18 studs}$$

Use a uniform spacing with 2 studs at a section across the beam width:

$$p = \frac{L/2}{N/2} = \frac{28(12)}{24} = 14.1 \text{ in.}$$

Use a 14-in. spacing for the pairs of stud connectors, starting at the support. _Use 18—$\frac{3}{4}$-in.-diam $\times$ 3-in. headed stud connectors on each side of the centerline at midspan._ ■■

16.15 DEFLECTIONS

The deflection of a composite beam will depend on whether it is shored or unshored during construction. Creep and shrinkage of the concrete in the slab also affect the result. Calculation of deflection requires obtaining the elastic cracked transformed section moment of inertia I_{tr} for the composite beam, and if unshored, also the elastic moment of inertia of the steel section alone.

When the steel beam is shored from below during the hardening of the concrete slab, the composite section will carry both the dead and live load. However, if the steel beam is _not_ shored, the steel beam alone must carry the dead load.

When the construction is _without_ shoring, the total deflection will be the sum of the dead load deflection of the steel beam alone and the live load deflection of the composite section.

When shoring provides the support during the hardening of the concrete slab, the composite section resists the entire load. Account should be taken to reflect the fact that concrete is subject to creep under long time load and that shrinkage will occur. This inelastic behavior may be approximated by multiplying the modulus of elasticity ratio n by a time-dependent factor such as two; thus reducing the effective width b_E/n. The result is a reduced moment of inertia I_{tr} to be used for computing the sustained load (dead load) deflection. The live load deflection would be computed using the elastic cracked transformed section moment of inertia.

Because the concrete slab in building construction is normally not too thick (say $t_s \leq 6$ in.) creep deflection is often not considered. The AISC Specifications [1.5, 1.16] give no indication of any concern with creep of a concrete slab in composite construction. However, as discussed in Sec. 7.6, LRFD-L3.1 states "Deformations in structural members and structural systems due to service loads shall not impair the serviceability of the structure." and ASD-L3.1 states "Beams and girders supporting floors and roofs shall be proportioned with due regard to the deflection produced by the design loads."

The ACI-ASCE Joint Committee [16.16] recommends using $E_c/2$ as the sustained concrete modulus of elasticity instead of E_c when computing sustained load creep deflection. AASHTO-10.38.1.4 [1.3] uses $E_c/3$ instead of E_c. Such arbitrary procedures can at best give an estimate of creep effects, probably no better than $\pm 30\%$. The steel section, exhibiting no creep, and representing the principal carrying element, ensures that creep problems will usually be minimal.

More accurate procedures for computing deflections to account for creep and shrinkage on composite steel-concrete beams are available in a paper by Roll [16.31], and particularly in _Deformation of Concrete Structures_ by Branson [16.32]. Lamport and Porter [16.45] have treated deflection prediction for concrete slabs reinforced with steel decking.

EXAMPLE 16.15.1 _____

Compute the service dead and live load deflections for the composite beam consisting of W16×36 with 4-in. slab designed in Example 16.14.1 (see Fig. 16.14.1).

Solution   Regardless of whether the selection of the steel section has been done by Load and Resistance Factor Design or by Allowable Stress Design, the deflections must be computed for *service* loads acting on the elastic section.

(a) Compute the dead load deflection. From Example 16.13.1, part a, the service dead load is 0.44 kip/ft, and all must be carried by the steel beam alone when the beam is unshored.

$$\Delta_{DL} = \frac{5wL^4}{384E_s I_s} = \frac{5(0.44)(28)^4(12)^3}{384(29,000)448} = 0.47 \text{ in.}, \quad \text{say } \tfrac{1}{2} \text{ in.}$$

The beam can be cambered or the slab can be thickened toward midspan so that this amount of deflection is compensated for during construction.

(b) Compute the live load deflection. From Example 16.13.1, part a, the service live load is 1.2 kips/ft. This load must be carried by the composite section; thus, the elastic composite moment of inertia I_{tr} is required. This was computed in Example 16.14.1 to be $I_{tr} = 1312$ in.^{4} Often some of the dead load, such as partition loads and other items placed after the concrete slab has cured, acts on the composite section.

$$\Delta_{LL} = \frac{5wL^4}{384E_s I_{tr}} = \frac{5(1.2)(28)^4(12)^3}{384(29,000)1312} = 0.44 \text{ in.}, \quad \text{say } \tfrac{1}{2} \text{ in.}$$

As discussed in Sec. 7.6, it has been traditional to consider that live load deflection exceeding $L/360$ may cause cracking of plaster. On the other hand, the ACI Code [16.15] restricts the live load plus creep and shrinkage deflection to a maximum of $L/480$. Thus, in the absence of any specific LRFD limitation, a limit of approximately $L/400$ will likely give satisfactory serviceability for the floor system. In this case,

$$\Delta_{limit} = \frac{L}{400} = \frac{28(12)}{400} = 0.84 \text{ in.} > \Delta_{LL} \quad \text{OK}$$

One may conclude that deflection will not cause conern. Note that $L/400$ is *not* an AISC limit under either LRFD or ASD. It is the designer's responsibility to establish any limit.

16.16 CONTINUOUS BEAMS

Traditionally on continuous beams the positive moment region has been designed as a composite section and the negative moment region where the concrete slab is in tension as a noncomposite section. However, some composite action has been known to exist in the negative moment region. Continuous composite beams have been studied by Barnard and Johnson [16.33], Johnson, Van Dalen, and Kemp [16.34],

Daniels and Fisher [16.35], Hamada and Longworth [16.36, 16.37], and Kubo and Galambos [16.38]. Kubo and Galambos extended the treatment to plate girders (that is, beams having h/t_w exceeding $970/\sqrt{F_y}$).

According to LRFD-I3.2 and ASD-I2.2, 3rd par., the steel reinforcement that extends parallel to the steel beam and lies within the concrete slab effective width b_E *may be used* as part of the effective composite section. This is true for both positive and negative moment regions. The inclusion of such steel reinforcement has little effect in positive moment regions but can help in negative moment regions. In the negative moment region the concrete ordinarily is all in tension and is therefore not considered effective (LRFD-I1, ASD-I2.2, 3rd par., and AASHTO-10.38.1.6).

When the steel reinforcing bars in the concrete slab are utilized to contribute to composite action, the force developed by such bars must be transferred by shear connectors. The nominal strength developed would be

$$T_n(\text{for } -M \text{ region}) = A_{sr}F_{yr} \tag{16.16.1}$$

$$C_n(\text{for } +M \text{ region}) = A_s'F_{yr} \tag{16.16.2}$$

where A_{sr} = total area of longitudinal reinforcing steel at the interior support located within the effective flange width b_E

A_s' = total area of longitudinal compression steel acting with the concrete slab at the location of maximum positive moment and lying within the effective width b_E

F_{yr} = specified minimum yield stress of the longitudinal reinforcing steel

Thus, the nominal strength V_{nh} for which shear connectors must be provided in the *negative moment zone* is

$$V_{nh} = A_{sr}F_{yr} \tag{16.16.3}$$

In the *positive moment zone,* when compression steel is included in the computation of composite section properties (plastic neutral axis for LRFD or elastic section modulus for ASD), the nominal strength V_{nh2} from the compression steel is

$$V_{nh2} = A_s'F_{yr} \tag{16.16.4}$$

The total horizontal shear force between the point of zero moment and the point of maximum moment is the smallest of $(0.85f_c'A_c + V_{nh2})$, A_sF_y, and ΣQ_n. LRFD has no specific mention of the compression reinforcement in the positive moment zone; thus, inclusion of V_{nh2} is optional.

As discussed in Sec. 16.7, the usual limit state for composite sections in the positive moment zones is crushing of the concrete at the top of the slab. This assumes no shear connector failure, no longitudinal splitting because of inadequate reinforcing bar development, and no shear failure in the slab. In the negative moment region, the usual limit state is flange local buckling [16.37].

Regarding the lateral-torsional buckling limit state, the usual provisions for noncomposite steel sections apply to the negative moment regions of continuous composite beams. The limits on λ from LRFD-B5 or ASD-B5 for the flange and web local buckling limit states must be applied in the negative moment zone.

EXAMPLE 16.16.1

Compute the plastic neutral axis (PNA) location and the nominal strength M_n for the section of Fig. 16.16.1 subject to negative bending moment. The W12×26 steel section is of A36 steel and the reinforcement in the slab has $F_{yr} = 60$ ksi.

Solution

(a) Determine the plastic neutral axis location. The concrete slab will be in tension; therefore, none of the concrete is assumed to be effective. The reinforcing bars contribute the nominal tension strength T_{sr},

$$T_{sr} = A_{sr}F_{yr} = 10(0.31)60 = 186 \text{ kips}$$

The maximum nominal compression force in the W12 section is

$$C_{max} = A_s F_y = 7.65(36) = 275.4 \text{ kips}$$

Since C_{max} exceeds T_{sr}, the PNA is within the steel W12 section. In which case the force equilibrium requirement may be expressed,

$$T_{sr} + T_s = C_{max} - T_s$$
$$2T_s = C_{max} - T_{sr} = 275.4 - 186 = 89.4$$
$$T_s = 44.7 \text{ kips}$$

Assuming that the PNA is within the flange of the W12,

$$\text{From top of flange to PNA} = \frac{T_s}{F_y b_f} = \frac{44.7}{36(6.49)} = 0.19 \text{ in.}$$

The assumption that PNA is within the flange is confirmed since $0.19 < (t_f = 0.38 \text{ in.})$. Thus, the distance PNA from top of slab is

$$\text{PNA} = t_s + 0.19 = 4.0 + 0.19 = 4.19 \text{ in.}$$

(b) Compute the nominal moment strength M_n. Locate the center of grav-

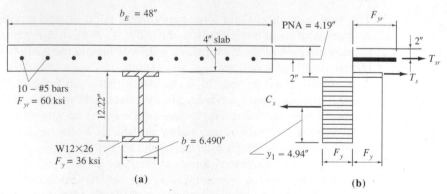

Figure 16.16.1 Composite section for negative bending of Example 16.16.1, including plastic stress distribution according to LRFD-I1.

ity y_1 of the compression force C_s in the steel section, measured from the bottom of the steel section,

	Area, A	Arm, y	Ay
W12 section	7.65	6.11	46.74
Flange	−1.24	12.12	−15.05
	6.41 sq in.		31.69 in.^{3}

$$y_1 = \frac{31.69}{6.41} = 4.94 \text{ in.}$$

Referring to Fig. 16.16.1, taking internal moments about the point of action of C_s gives

T_{sr}:
$$\begin{aligned} M_{n1} &= T_{sr}\,(d - 4.94 + t_s - 2.00) \\ &= 186(12.22 - 4.94 + 4 - 2.00)\tfrac{1}{12} \\ &= 186(10.28)\tfrac{1}{12} = 159.3 \text{ ft-kips} \end{aligned}$$

T_s:
$$\begin{aligned} M_{n2} &= T_s\,(d - 4.94 - 0.19/2) \\ &= 44.7(12.22 - 4.94 - 0.19/2)\tfrac{1}{12} \\ &= 44.7(7.18)\tfrac{1}{12} = 26.7 \text{ ft-kips} \end{aligned}$$

$$\begin{aligned} M_n &= M_{n1} + M_{n2} \\ &= 159.3 + 26.7 = 186 \text{ ft-kips} \end{aligned}$$

$$\phi_b M_n = 0.85(186) = 158 \text{ ft-kips}$$

Note that for composite action in the negative moment region, shear connectors must be used throughout the entire region. The required ΣQ_n equals the force T_{sr} in the reinforcement. When partial composite action is used, ΣQ_n will be less than T_{sr}. In such a case, the PNA location and the nominal moment strength are computed using ΣQ_n instead of T_{sr}. ■■

16.17 COMPOSITE COLUMNS

A composite column is defined by LRFD-I1 as "A steel column fabricated from rolled or built-up steel shapes and encased in structural concrete or fabricated from steel pipe or tubing and filled with structural concrete." An example of the former is shown in Fig. 16.17.1, where a steel W section is encased in concrete; the concrete must contain longitudinal reinforcing bars and these must be surrounded by lateral ties in the manner of a reinforced concrete column.

The steel section must comprise at least 4% of the total cross-sectional area, otherwise the column must be designed as an ordinary reinforced concrete column. Research by Furlong [16.39, 16.40] and others was reviewed by Task Group 20 of the Structural Stability Research Council, chaired by Furlong [16.41]. This SSRC Task Group Report forms the basis for design of composite columns under LRFD-I2; ASD Chapter I does not contain provisions for composite columns.

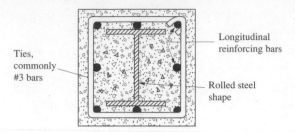

Figure 16.17.1   Composite column section; rolled steel shape encased in concrete.

Limitations

In order to qualify as a composite column, the limitations of LRFD-I2.1 must be satisfied:

1.
$$A_s \geq 0.04A_g \tag{16.17.1}$$

2. For a concrete encasement:
   a. Longitudinal reinforcing bars must be used; load carrying bars must be continuous at framed levels (wherever a beam or slab frames to the column); other longitudinal bars used only to restrain concrete may be interrupted at framed levels.
   b. Lateral ties must be used; spacing of ties may not exceed $\frac{2}{3}$ of least lateral column dimension.
   c. Area of lateral ties and longitudinal reinforcement each must be at least 0.007 sq in./in. of bar spacing.
   d. Clear cover of at least 1.5 in. is required.

3. Concrete strength f'_c:
   a. Normal-weight concrete: 3 ksi $\leq f'_c \leq$ 8 ksi
   b. Structural lightweight concrete: $f'_c \geq$ 4 ksi

4. Maximum yield stress of steel used in strength computations is 55 ksi for either structural steel or reinforcing bars.

5. Minimum wall thickness t for concrete-filled pipe or tubing:
   a. For each face width b in rectangular sections:

$$t \geq b\sqrt{\frac{F_y}{3E}} \tag{16.17.2}$$

   b. For outside diameter D in circular sections:

$$t \geq D\sqrt{\frac{F_y}{8E}} \tag{16.17.3}$$

Nominal Strength

The nominal strength P_n of a composite column is computed using the regular column strength provisions of LRFD-E2, as discussed in Chapter 6, except the yield stress F_y becomes the modified yield stress F_{my}, the modulus of elasticity E becomes the modified modulus E_m, and the radius of gyration r becomes the modified radius r_m. The modified expressions are:

For *concrete-filled pipe or tube:*

$$F_{my} = F_y = F_{yr}\left(\frac{A_r}{A_s}\right) + 0.85 f_c'\left(\frac{A_c}{A_s}\right) \qquad (16.17.4)$$

with F_y and $F_{yr} \leq 55$ ksi

$$E_m = 29,000 + 0.4E_c\left(\frac{A_c}{A_s}\right) \qquad (16.17.5)$$

$$r_m = r_s \qquad (16.17.6)$$

For *concrete-encased structural steel:*

$$F_{my} = F_y + 0.7F_{yr}\left(\frac{A_r}{A_s}\right) + 0.6 f_c'\left(\frac{A_c}{A_s}\right) \qquad (16.17.7)$$

with F_y and $F_{yr} \leq 55$ ksi

$$E_m = 29,000 + 0.2E_c\left(\frac{A_c}{A_s}\right) \qquad (16.17.8)$$

$$r_m = r_s \geq 0.3d_{\text{buck}} \qquad (16.17.9)$$

where A_c = area of concrete
A_r = area of longitudinal reinforcing bars
A_s = gross area of steel shape, pipe, or tube
E_c = modulus of elasticity of concrete, ksi
  = $(w^{1.5})\sqrt{f_c'}$, where w is the density of concrete in pcf (i.e., 145 pcf for normal-weight concrete) and f_c' is in ksi
F_y = specified minimum yield stress of steel shape, pipe, or tube
F_{yr} = specified minimum yield stress of longitudinal reinforcing bars
f_c' = specified 28-day compressive strength of concrete
r_s = radius of gyration of steel shape, pipe, or tube
d_{buck} = overall dimension of the composite section in the plane of buckling

The *LRFD Manual* contains tables, "COMPOSITE COLUMNS" giving axial design strength $\phi_c P_n$. Note that $\phi_c = 0.85$ for both steel columns and composite columns.

Composite beam-column design has been treated by Uang, Wattar, and Leet [16.46].

SELECTED REFERENCES

16.1. Ivan M. Viest. Chairman, "Composite Steel-Concrete Construction," Report of the Subcommittee on the State-of-the-Art Survey of the Task Committee on Composite Construction of the Committee on Metals of the Structural Division, *Journal of the Structural Division*, ASCE, **100,** ST5 (May 1974), 1085–1139.

16.2.   Gottfried Brendel. "Strength of the Compression Slab of T-Beams Subject to Simple Bending," *ACI Journal, Proceedings,* **61,** January 1964, 57–76.

16.3.   Conrad P. Heins and Horn Ming Fan. "Effective Composite Beam Width at Ultimate Load," *Journal of the Structural Division,* ASCE, **102,** ST11 (November 1978), 2163–2179.

16.4.   Cesar R. Vallenilla and Reidar Bjorhovde. "Efective Width Criteria for Composite Beams," *Engineering Journal,* AISC, **22,** 4 (Fourth Quarter 1985), 169–175.

16.5.   Ivan M. Viest. "Review of Research on Composite Steel-Concrete Construction," *Journal of the Structural Division,* ASCE, **86,** ST6 (June 1960), 1–21; Also *Transactions,* ASCE, **126** (1961), Part II, 1101–1123.

16.6.   William C. Hansell, Theodore V. Galambos, Mayasandra K. Ravindra, and Ivan M. Viest. "Composite Beam Criteria in LRFD," *Journal of the Structural Division,* ASCE, **104,** ST9 (September 1978), 1409–1426; Disc. **106,** ST2 (February 1980), 571–572.

16.7.   Srinivasa H. Iyengar and Mohammad Iqbal. "Composite Construction," *Building Structural Design Handbook,* Chapter 23. New York: John Wiley & Sons, 1987.

16.8.   Robert F. Lorenz and Frank W. Stockwell, Jr. "Concrete Slab Stresses in Partial Composite Beams and Girders," *Engineering Journal,* AISC, **21,** 3 (Third Quarter 1984), 185–188.

16.9.   Robert F. Lorenz. "Understanding Composite Beam Design Methods Using LRFD," *Engineering Journal,* AISC, **25,** 1 (Second Quarter 1988), 35–38.

16.10. Robert F. Lorenz. "Some Economic Considerations for Composite Floor Beams," *Engineering Journal,* AISC, **20,** 2 (Second Quarter 1983), 78–81.

16.11. Mark C. Zahn. "The Economies of LRFD in Composite Floor Beams," *Engineering Journal,* AISC, **24,** 2 (Second Quarter 1987), 87–92.

16.12. S. Timoshenko and J. Goodier. *Theory of Elasticity.* New York: McGraw-Hill Book Company, Inc., 1959, Chap. 6.

16.13. Theodore von Kármán. "Die Mittragende Breitte," *August-Föppel-Festschrift,* 1924. (See also Collected Works of Theodore von Kármán, Volume II, p. 176.)

16.14. John E. Johnson and Albert D. M. Lewis. "Structural Behavior in a Gypsum Roof-Deck System," *Journal of the Structural Division,* ASCE, **92,** ST2 (April 1966), 283–296.

16.15. ACI Committee 318. *Building Code Requirements for Reinforced Concrete.* Detroit. MI: American Concrete Institute, 1989.

16.16. Joint ASCE-ACI Committee on Composite Construction. "Tentative Recommendations for the Design and Construction of Composite Beams and Girders for Buildings," "*Journal of the Structural Division,*" ASCE, **86,** ST12 (December 1960), 73–92.

16.17. Roger G. Slutter and George C. Driscoll. "Flexural Strength of Steel-Concrete Composite Beams," *Journal of the Structural Division,* ASCE, **91,** ST2 (April 1965), 71–99.

16.18. Peter Ansourian and Jack William Roderick. "Analysis of Composite Beams,"

Journal of the Structural Division, ASCE, **104,** ST10 (October 1978), 1631–1645.

16.19. Jorgen G. Ollgaard, Roger G. Slutter, and John W. Fisher. "Shear Strength of Stud Connectors in Lightweight and Normal-Weight Concrete," *Engineering Journal,* AISC, **8,** 2 (April 1971), 55–64.

16.20. Jay B. McGarraugh and J. W. Baldwin, Jr. "Lightweight Concrete-on-Steel Composite Beams," *Engineering Journal,* AISC, **8,** 3 (July 1971), 90–98.

16.21. Roger G. Slutter and John W. Fisher. "Fatigue Strength of Shear Connectors," *Highway Research Record No. 147,* Highway Research Board, 1966, pp. 65–88.

16.22. Charles G. Schilling. "Bending Behavior of Composite Hybrid Beams," *Journal of the Structural Division,* ASCE, **94,** ST8 (August 1968), 1945–1964.

16.23. John A. Grant, Jr., John W. Fisher, and Roger G. Slutter. "Composite Beams with Formed Steel Deck," *Engineering Journal,* AISC, **14,** 2 (First Quarter 1977), 24–43.

16.24. Sriramulu Vinnakota, Christopher M. Foley, and Murthy R. Vinnakota. "Design of Partially or Fully Composite Beams, with Ribbed Metal Deck, Using LRFD Specifications," *Engineering Journal,* AISC, **25,** 2 (Second Quarter 1988), 60–78.

16.25. R. H. R. Tide and T. V. Galambos. "Composite Open-Web Steel Joists," *Engineering Journal,* AISC, **7,** 1 (January 1970), 27–36.

16.26. James Rongoe. "A Composite Girder System for Joist Supported Slabs," *Engineering Journal,* AISC, **21,** 3 (Second Quarter 1984), 155–160.

16.27. ASCE. *Specifications for the Design and Construction of Composite Slabs.* New York: Technical Council on Codes and Standards, American Society of Civil Engineers, 1984.

16.28. ASCE. *Commentary on Specifications for the Design and Construction of Composite Slabs.* New York: Technical Council on Codes and Standards, American Society of Civil Engineers, 1984.

16.29. Max L. Porter. "Analysis of Two-Way Acting Composite Slabs," *Journal of Structural Engineering,* ASCE, **111,** 1 (January 1985), 1–17.

16.30. C. Dale Buckner, Danny J. Deville, and Dean C. McKee. "Shear Strength of Slabs in Stub Girders," *Journal of the Structural Division.* ASCE, **107,** ST2 (February 1981), 273–280.

16.31. Frederic Roll. "Effects of Differential Shrinkage and Creep on a Composite Steel-Concrete Structure," *Designing for Effects of Creep, Shrinkage, Temperature in Concrete Structures* (SP-27). Detroit, MI: American Concrete Institute, 1971 (pp. 187–214).

16.32. Dan E. Branson. *Deformation of Concrete Structures.* New York: McGraw-Hill Book Company, Inc., 1977.

16.33. P. R. Barnard and R. P. Johnson. "Plastic Behavior of Continuous Composite Beams," *Proceedings,* Institution of Civil Engineers, October 1965.

16.34. R. P. Johnson, K. Van Dalen, and A. R. Kemp, "Ultimate Strength of Continuous Composite Beams," *Proceedings of the Conference on Structural Steelwork,* British Constructional Steelwork Association, November 1967.

16.35. J. H. Daniels and J. W. Fisher. "Static Behavior of Continuous Composite Beams," *Fritz Engineering Laboratory Report* No. 324.2, Lehigh University, Bethlehem, PA, March 1967.

16.36. Sumio Hamada and Jack Longworth. "Buckling of Composite Beams in Negative Bending," *Journal of the Structural Division,* ASCE, **100,** ST11 (November 1974), 2205–2222.

16.37. Sumio Hamada and Jack Longworth. "Ultimate Strength of Continuous Composite Beams," *Journal of the Structural Division,* ASCE, **102,** ST7 (July 1976), 1463–1478.

16.38. Masahiro Kubo and Theodore V. Galambos. "Plastic Collapse Load of Continuous Composite Plate Girders," *Engineering Journal,* AISC, **25,** 4 (Fourth Quarter 1988), 145–155.

16.39. Richard W. Furlong. "Strength of Steel Encased Concrete Beam Columns," *Journal of the Structural Divison,* ASCE, **94,** ST1 (January 1968), 267–281.

16.40. Richard W. Furlong. "AISC Column Design Logic Makes Sense for Composite Columns, Too," *Engineering Journal,* AISC, **13,** 1 (First Quarter 1976), 1–7.

16.41. Task Group 20, Structural Stability Research Council. "A Specification for the Design of Steel-Concrete Composite Columns," *Engineering Journal,* AISC, **16,** 4 (Fourth Quarter 1979), 101–115.

16.42. IABSE. *Composite Steel-Concrete Construction and Eurocode 4,* Short Course Notes, International Association for Bridge and Structural Engineering, Brussels, 1990, 191 pp.

16.43. W. Samuel Easterling, David R. Gibbings and Thomas M. Murray. "Strength of Shear Studs in Steel Deck on Composite Beams and Joists," *Engineering Journal,* AISC, **30,** 2 (2nd Quarter 1993), 44–55.

16.44. Mark Andrew Bradford and R. Ian Gilbert. "Composite Beams with Partial Interaction Under Sustained Loads," *Journal of Structural Engineering,* ASCE, **118,** 7 (July 1992), 1871–1883.

16.45. William B. Lamport and Max L. Porter. "Deflection Predictions for Concrete Slabs Reinforced with Steel Decking," *ACI Structural Journal,* **87,** September-October 1990, 546–570.

16.46. Chia-Ming Uang, Samer W. Wattar, and Kenneth M. Leet. "Proposed Revision of the Equivalent Axial Load Method for LRFD Steel and Composite Beam-Column Design," *Engineering Journal,* AISC, **27,** 4 (4th Quarter 1990), 150–157.

PROBLEMS

All problems are to be done according to AISC Load and Resistance Factor Design or Allowable Stress Design, as indicated by the instructor. All given loads are service loads unless otherwise indicated.

16.1. For the case (or cases) assigned by the instructor, compute the location of the transformed composite section neutral axis and moment of inertia I_{tr}. Refer to the accompanying figure.

Case	Steel section	F_y (ksi)	Slab t_s (in.)	b_E (in.)	f'_c (ksi)	
1	W12×26	36	4	72	3	($n = 9$)
2	W12×26	60	4	72	3	($n = 9$)
3	W16×31	36	4	84	3.5	($n = 8.5$)
4	W21×44	50	4	84	4	($n = 8$)
5	W24×55	50	4.5	90	4	($n - 8$)
6	W18×50	50	5	72	4	($n = 8$)
7	W18×50	50	4	72	3	($n = 9$)
8	W24×76	50	4.5	72	4	($n = 8$)
9	W24×94	50	4.5	72	4	($n = 8$)
10	W21×62	50	5	96	4	($n = 8$)
11	W21×147	50	4.5	96	4	($n = 8$)

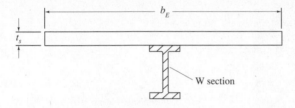

Problems 16.1, 16.2, and 16.3

16.2. For the case (or cases) listed for Prob. 16.1 and assigned by the instructor, compute the location of plastic neutral axis (PNA) measured from the top of the slab, as well as the nominal strength M_n. Assume the sections are *fully composite*. Refer to the accompanying figure.

16.3. For the case (or cases) listed for Prob. 16.1 and assigned by the instructor, select an economical size of headed stud shear connector from Table 16.8.1, determine the total number needed to develop a *fully composite* section for the beam, and specify the spacing. Assume the simply supported beam span equals $4b_E$.

16.4. For the case (or cases) assigned by the instructor, select a W section to design a *fully composite* section for span *BD* of the accompanying figure. Assume for simplicity that the slab is the only dead load to be considered. Also, select an economical size of headed stud shear connector from Table 16.8.1, determine the total number needed for the beam, and specify the spacing, to develop a fully composite beam. No shoring is to be used; therefore, assume that during construction wet concrete of 75 psf is live load and that an additional construction live load of 25 psf may act. The final composite beam may not have live load deflection exceeding $L/360$.

Case	Live Load (psf)	F_y (ksi)	Slab t_s (in.)	Span (ft)	Beam Spacing (ft)	f'_c (ksi)	
1	100	36	4	36	8	3	$(n = 9)$
2	100	60	4	36	8	3	$(n = 9)$
3	80	36	4	36	7	3	$(n = 9)$
4	80	50	4	36	7	4	$(n = 8)$
5	80	50	4	40	7	4	$(n = 8)$
6	80	50	4	40	7	4	$(n = 8)$
7	125	36	4.5	28	8	3	$(n = 9)$
8	125	50	4.5	40	8	3	$(n = 9)$
9	125	50	4.5	42	8	4	$(n = 8)$
10	125	50	4.5	45	8	4	$(n = 8)$
11	125	50	5	45	9	4	$(n = 8)$
12	125	50	5.5	48	9	4	$(n = 8)$

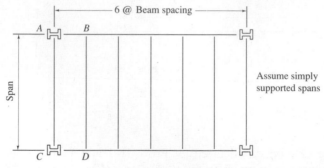

Problems 16.4 and 16.5. Framing plan.

16.5. For the case (or cases) given in Prob. 16.4 assigned by the instructor, select a W section to design a *partially composite* section for span *BD* of the accompanying figure, using the minimum number of $\frac{3}{4}$-in.-diam headed stud shear connectors. Specify number of studs and the spacing. In addition to the slab dead load, using a ceiling load of 7 psf and partition load of 25 psf. No shoring is to be used; therefore, assume that during construction wet concrete of 75 psf is live load and that an additional construction live load of 25 psf may act. The dead load deflection before composite action is effective may not exceed $\frac{7}{8}$ in. and composite beam deflection resulting from superimposed dead load (i.e., ceiling and partitions) and live load may not exceed $L/360$.

16.6. Design a composite encased W shape column to resist a factored axial compression load P_u of 900 kips. The effective length $KL = 12$ ft, $F_y = 50$ ksi for structural steel, and $f'_c = 4.5$ ksi for concrete.

Appendix

TABLE A1   APPROXIMATE RADIUS OF GYRATION

$r_x = 0.29h$ $\quad r_y = 0.29b$	$r_x = 0.42h$ $\quad r_y = 0.42b$	$r_x = 0.31h$ $\quad r_y = 0.48b$
$r_x = 0.40h$ $\quad h = $ mean h	$r_y = $ same as for 2 L	$r_x = 0.37h$ $\quad r_y = 0.28b$
$r_x = 0.25h$	$r_x = 0.42h$ $\quad r_y = $ same as for 2 L	$r_x = 0.31h$
$r = \sqrt{\dfrac{H^2 + h^2}{16}}$ $\quad r = 0.35H_m$	$r_x = 0.39h$ $\quad r_y = 0.21b$	$r_x = 0.31h$
$r_x = 0.31h$ $\quad r_y = 0.31h$ $\quad r_z = 0.197h$	$r_x = 0.45h$ $\quad r_y = 0.235b$	$r_x = 0.40h$ $\quad r_y = 0.21b$
$r_x = 0.29h$ $\quad r_y = 0.32b$ $\quad r_z = 0.18\dfrac{h+b}{2}$	$r_x = 0.36h$ $\quad r_y = 0.45b$	$r_x = 0.38h$ $\quad r_y = 0.22b$
$r_x = 0.31h$ $\quad r_y = 0.215b = b(0.21 + 0.02s)$	$r_x = 0.36h$ $\quad r_y = 0.60b$	$r_x = 0.39h$
$r_x = 0.32h$ $\quad r_y = 0.21b = b(0.19 + 0.02s)$	$r_x = 0.36h$ $\quad r_y = 0.53b$	$r_x = 0.35h$
$r_x = 0.29h$ $\quad r_y = 0.24b = b(0.23 + 0.02s)$	$r_x = 0.39h$ $\quad r_y = 0.55b$	$r_x = 0.435h$ $\quad r_y = 0.25b$
$r_x = 0.30h$ $\quad r_y = 0.17b$	$r_x = 0.42h$ $\quad r_y = 0.32b$	$r_x = 0.42h$
$r_x = 0.25h$ $\quad r_y = 0.21b$	$r_x = 0.44h$ $\quad r_y = 0.28b$	$r_x = 0.42h$
$r_x = 0.21h$ $\quad r_y = 0.21b$ $\quad r_z = 0.19h$	$r_x = 0.50h$ $\quad r_y = 0.28b$	$r_x = 0.285h$ $\quad r_y = 0.37b$
$r_x = 0.38h$ $\quad r_y = 0.19b$	$r_x = 0.39h$ $\quad r_y = 0.21b$	$r_x = 0.42h$ $\quad r_y = 0.23b$

* J.A.L. Waddell. *Bridge Engineering*, Vol. 1. New York: John Wiley & Sons, Inc., 1916, p. 504. Used by permission.

TABLE A2 TORSIONAL PROPERTIES

O = shear center J = torsion constant, C_w = warping constant
G = centroid I_p = polar moment of inertia about shear center

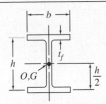

$$J = \tfrac{1}{3}(2bt_f^3 + ht_w^3)$$

$$C_w = \frac{I_f h^2}{2} = \frac{t_f b^3 h^2}{24} = \frac{h^2 I_y}{4}$$

$$I_p = I_x + I_y$$

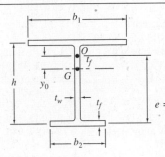

$$J = \tfrac{1}{3}(b_1 t_f^3 + b_2 t_f^3 + ht_w^3)$$

$$C_w = \frac{t_f h^2}{12}\left(\frac{b_1^3 b_2^3}{b_1^3 + b_2^3}\right)$$

$$e = h\frac{b_1^3}{b_1^3 + b_2^3}$$

$$I_p = I_y + I_x + Ay_0^2$$

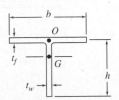

$$J = \tfrac{1}{3}(bt_f^3 + ht_w^3)$$

$$C_w = \frac{1}{36}\left(\frac{b^3 t_f^3}{4} + h^3 t_w^3\right)$$

$\approx$ zero for small t

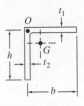

$$J = \tfrac{1}{3}(bt_1^3 + ht_2^3)$$

$$C_w = \frac{1}{36}(b^3 t_1^3 + h^3 t_2^3)$$

$\approx$ zero for small t

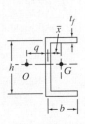

$$J = \tfrac{1}{3}(2bt_f^3 + ht_w^3)$$

$$C_w = \frac{t_f b^3 h^2}{12}\left(\frac{3bt_f + 2ht_w}{6bt_f + ht_w}\right) = \frac{h^2}{4}(I_y + A\bar{x}^2 - q\bar{x}A)$$

$$q = \frac{th^2 b^2}{4I_x}$$

Index

Q_n	= nominal strength of a shear connector (Sec. 16.8)
Q_s	= shape factor for unstiffened compression element (Sec. 6.18)
r	= radius of gyration, $\sqrt{I/A_g}$; radial distance from centroid to point of stress (Sec. 5.18)
r_i	= distance from instantaneous center to a weld element (Fig. 5.17.2)
r_0	= distance from instantaneous center to vertical weld line
r_{max}	= distance to weld element farthest from instantaneous center
r_T	= radius of gyration of a section comprising the compression flange plus one-third of the compression web area, taken about an axis in the plane of the web; used in ASD, Eq. 9.7.14
r_x, r_y, r_z	= radius of gyration about x-, y-, or z-axes, respectively
R	= service load per bolt; service load resistance (allowable load) per bolt (Chap. 4); service load per inch on welds
R_e	= moment strength reduction factor for hybrid girder (Secs. 11.7 and 16.9)
R_i	= resistance of a bolt at any deformation Δ (Chap. 4); strength of a fillet weld segment per unit length (Sec. 5.17)
$R_{i,\text{ult}}$	= ultimate shear load on an element, Eq. 5.17.3
R_n	= nominal strength of one fastener in tension, shear, or bearing; nominal reaction strength (Sec. 7.8)
R_{nt}	= nominal strength of bolt in tension
R_{nv}	= nominal strength of bolt in shear
R_{nw}	= nominal strength of weld per inch of length
R_{PG}	= reduction factor for "bend-buckling" of the web, Eq. 11.4.3
R_s	= direct shear component of bolt resistance
R_u	= factored load per bolt; factored load per unit length of weld; factored reaction (Sec. 7.8)
R_{ult}	= ultimate shear resistance in a bolt, $\tau_u A_b$
R_{us}	= factored direct shear on bolt subject to eccentric load
R_{ut}	= factored tension load on bolt
R_{uv}	= factored direct shear component on bolt
R_{ux}, R_{uy}	= factored shear on bolt, in x- or y-direction, respectively
R_v	= direct shear component of bolt resistance; shear component of eccentric force on fillet welds; direct shear component of weld resistance/per unit length
R_w	= allowable load per inch on welds
R_x, R_y	= x- or y-direction component of bolt resistance; x- or y-component of torsional moment force on fillet welds
s	= stagger of bolt holes measured in the line of force (Chap. 3); distance from free edge along a thin wall section (Chap. 8); band width for tension-field force T (Sec. 11.9)
S, S_x, S_y	= elastic section modulus, $I/\bar{y}$ (Table 5.18.1), with respect to x- or y-axes (I_x/c_y or I_y/c_x), according to subscript
S_s	= elastic section modulus of steel section alone, referred to its tension flange
S_{tr}	= elastic section modulus of composite section, I_{tr}/y_b
S_{xc}, S_{xt}	= section modulus S_x referred to the compression flange, S_{xc}, or the tension flange, S_{xt}
t	= thickness; thickness of material against which bolt bears
t_e	= effective throat dimension of a weld (Sec. 5.12)
t_f, t_{fb}, t_{fc}	= flange thickness; for beam, t_{fb}; for column, t_{fc}
t_s	= thickness of stiffener; slab thickness
t_w, t_{wb}, t_{wc}	= web thickness; for beam, t_{wb}; for column, t_{wc}
T	= tensile force; service load tensile force; torsional moment or torsional service load moment (Chap. 8); base metal thickness (Table 5.11.1)
T_b	= initial force in bolt resulting from installation
T_n	= nominal strength of a tension member
T_u	= factored tension load; factored torsional moment (Chap. 8)
u	= displacement in the x-direction
u_f	= lateral deflection of flange
U	= reduction factor to account for shear lag (Sec. 3.9)

v	= shear stress; displacement in the y-direction
v_s	= St. Venant torsion shear stress (Chap. 8)
v_w	= warping torsion shear stress (Chap. 8)
V	= shear; service load shear force on a bolt
V_f	= warping torsion shear force in flange
V_h	= V_{nh} from Eqs. 16.8.2 or 16.8.3 (use smaller) divided by the factor of safety of 2 (ASD-I4)
V_h'	= N (i.e., number of shear connectors used) times the allowable capacity Q per connector
V_n	= nominal shear strength
V_n'	= nominal shear strength in the presence of bending moment
V_{nh}	= nominal horizontal shear strength across interface between slab and steel section in a composite beam
V_r	= range of service load shear force, Eq. 16.8.9
V_u	= factored shear force
V_x, V_y	= shear in the x- and y-directions, respectively
w	= uniform loading; service uniformly distributed load on beam; displacement in z-direction (Fig. 6.14.2); width of stiffener plate (Chap. 11); density of concrete, Eq. 16.5.1
w_D, w_L	= service uniform dead and live load, respectively
w_n	= w_u/ϕ_b = required nominal uniform load causing collapse mechanism (Chap. 10)
w_u	= factored uniform load
w_{uh}	= factored uniform horizontal load
W	= total service load on a span; concentrated load on beam; width of stiffener (Chap. 11); seat width (Chap. 13)
W_n	= W_u/ϕ_b = required nominal concentrated load causing collapse (Chap. 10)
W_u	= factored concentrated load
x_0, y_0	= shear center distances from centroid measured along the x- and y-axes, respectively
X_1	= constant (units = ksi), Eq. 9.6.7
X_2	= constant (units = in.^{4}/kips^2), Eq. 9.6.8
y	= deflection at a location z along axis of member
$\bar{y}$	= center of gravity (CG) of composite section measured from CG of gravity of steel W section
y_0	= $(V_n'/V_n)h$
y_1	= total deflection (including second-order deflection) of beam-column
y_b	= distance to bottom of steel section from CG of composite section
y_c, y_t	= distances from CG of the section to the compression and tension extreme fibers, respectively
y_t	= distance to top of slab from CG of composite section
Z, Z_x, Z_y	= plastic modulus, $\int y\,dA$, with respect to the axes indicated by subscript
α	= constant $GJ/(2EC_w)$, Eq. 9.4.7; ratio of web yield stress to flange yield stress, F_{yw}/F_{yf}, (Sec. 11.7); P_u/P_e or $\Sigma P_u/\Sigma P_e$ for LRFD, and P/P_e for ASD (Chap. 12)(Sec. 6.8)
β	= flexure analogy modification factor (Chap. 8); A_w/A_f, ratio of web cross-sectional area to cross-sectional area of the *compression* flange (Sec. 11.7)
β_{as}	= A_g/S for ASD, using S_x or S_y as appropriate for the axis of bending (Chap. 12)
β_{az}	= A_g/Z for LRFD, using Z_x or Z_y as appropriate for the axis of bending (Chap. 12)
β_b	= Z_x/Z_y (see Table 12.12.1 for approximate values)
β_m	= 286,000 I, using I_x or I_y as appropriate for the axis of bending
β_s	= E_t/E, Eq. 6.9.2 (Table 6.9.1)
β_w	= selected ratio h/t_w for design (Sec. 11.14)
γ	= general term for overload factor; strain angle; angle between the plane of bending and the xz plane (Sec. 7.10)
γ_i	= overload factors (LRFD-A4.1)
δ	= deflection; virtual displacement; sidesway buckling deflection
δ_0	= first-order deflection of beam-column
ϵ	= strain, in./in. or mm/mm